Design Examples and Design Problems (DP)

Modern
Control Systems

ELEVENTH EDITION

Richard C. Dorf
University of California, Davis

Robert H. Bishop
The University of Texas at Austin

PEARSON

Prentice
Hall

Pearson Education International

Vice President and Editorial Director, ECS: *Marcia J. Horton*
Acquistions Editor: *Michael McDonald*
Senior Managing Editor: *Scott Disanno*
Senior Production Editor: *Irwin Zucker*
Art Editor: *Greg Dulles*
Manufacturing Manager: *Alexis Heydt-Long*
Manufacturing Buyer: *Lisa McDowell*
Senior Marketing Manager: *Tim Galligan*

© 2008 Pearson Education, Inc.
Pearson Prentice Hall
Pearson Education, Inc.
Upper Saddle River, NJ 07458

MATLAB is a registered trademark of The Math Works, Inc., 24 Prime Park Way, Natick, MA 01760-1520

LabVIEW is a registered trademark of National Instruments, Inc., 11500 North Mopac Expwy, Austin, TX 78759-3504.

The author and publisher of this book have used their best efforts in preparing this book. These efforts include the development, research, and testing of the theories and programs to determine their effectiveness. The author and publisher make no warranty of any kind, expressed or implied, with regard to these programs or the documentation contained in this book. The author and publisher shall not be liable in any event for incidental or consequential damages in connection with, or arising out of, the furnishing, performance, or use of these programs.

Printed in the United States of America

10 9 8 7 6 5 4 3

ISBN 13: 978-0-13-245192-5

ISBN 10: 0-13-245192-1

Pearson Education LTD., London
Pearson Education Australia PTY, Limited
Pearson Education Singapore, Pte. Ltd
Pearson Education North Asia Ltd
Pearson Education Canada, Inc.
Pearson Educación de Mexico, S.A. de C.V.
Pearson Education -- Japan
Pearson Education Malaysia, Pte. Ltd
Pearson Education, Upper Saddle River, New Jersey

Of the greater teachers—
when they are gone,
their students will say:
we did it ourselves.

Dedicated to

Lynda Ferrera Bishop

and

Joy MacDonald Dorf

In grateful appreciation

Contents

CHAPTER **3**   *State Variable Models   144*

CHAPTER **4**   *Feedback Control System Characteristics   212*

CHAPTER **5**   *The Performance of Feedback Control Systems   277*

CHAPTER **6** *The Stability of Linear Feedback Systems* **355**

CHAPTER **7** *The Root Locus Method* **407**

CHAPTER **8**   *Frequency Response Methods   493*

CHAPTER **9**   *Stability in the Frequency Domain   567*

CHAPTER 10 *The Design of Feedback Control Systems* 667

CHAPTER 11 *The Design of State Variable Feedback Systems* 756

Preface

MODERN CONTROL SYSTEMS—THE BOOK

The Mars Exploration Rover (MER-A), also known as *Spirit,* was launched on a Delta II rocket, in June 2003 to Mars, the Red Planet. *Spirit* entered the Martian atmosphere seven months later in January, 2004. When the spacecraft entered the Martian atmosphere it was traveling 19,300 kilometers per hour. For about four minutes in the upper atmosphere, the spacecraft aeroshell decelerated the vehicle to a velocity of 1,600 kilometers per hour. Then a parachute was deployed to slow the spacecraft to about 300 kilometers per hour. At an altitude of about 100 meters, retrorockets slowed the descent and airbags were inflated to cushion the shock of landing. The *Spirit* struck the Martian ground at around 50 km/hr and bounced and rolled until it stopped near the target point in the Gusev Crater. The target landing site was chosen because it looks like a crater lakebed. The *Spirit* mobile rover has reached interesting places in the Gusev Crater to perform in-situ tests to help scientists answer many of the lingering questions about the history of our neighbor planet. In fact, *Spirit* discovered evidence of an ancient volcanic explosion near the landing site in Gusev Crater. The successful entry, descent, and landing of *Spirit* is an astonishing illustration of the power of control systems. Given the large distances to Mars, it is not possible for a spacecraft to fly through the atmosphere while under ground control—the entry, descent, and landing must be controlled autonomously on-board the spacecraft. Designing systems capable of performing planetary entry is one of the great challenges facing control system engineers.

The precursor NASA Mars mission, known as the Mars Pathfinder, also journeyed to the Red Planet and landed on July 4, 1997. The Pathfinder mission, one of the first of the NASA Discovery-class missions, was the first mission to land on Mars since the successful Viking spacecraft in the 1970s. Pathfinder deployed the first-ever autonomous rover vehicle, known as the *Sojourner,* to explore the landing site area. The mobile *Sojourner* had a mass of 10.5 kilograms and traveled a total of 100 meters (never straying more than 12 meters or so from the lander) in its 30-day mission. By comparison, the *Spirit* rover has a mass of 180 kilograms and is designed to roam about 40 meters per day. *Spirit* has spent four years exploring Mars and has driven over 7 kilometers. The fast pace of development of more capable planetary rovers is evident. Plans for the Mars Science Laboratory planetary rover (scheduled for launch in 2009) call for a 1000-kilogram rover with a mission duration of 500 days and the capability to traverse 30 kilometers over the mission lifetime.

Control engineers play a critical role in the success of the planetary exploration program. The role of autonomous vehicle spacecraft control systems will continue to increase as flight computer hardware and operating systems improve. Pathfinder used a commercially produced, multitasking computer operating system hosted in a 32-bit radiation-hardened workstation with 1-gigabyte storage, programmable in C.

This was quite an advancement over the Apollo computers, which had a fixed (read-only) memory of 36,864 words (one word was 16 bits) together with an erasable memory of 2,048 words. The Apollo "programming language" was a pseudocode notation encoded and stored as a list of data words "interpreted" and translated into a sequence of subroutine links.[1] The MER computer in the Spirit rover utilizes a 32-bit Rad 6000 microprocessor operating at a speed of 20 million instructions per second. This is a radiation-hardened version of the PowerPC chip used in many Macintosh computers. The on-board memory includes 128 megabytes of random access memory, 256 megabytes of flash memory, and smaller amounts of other nonvolatile memory to protect against power-off cycles so that data will not be unintentionally erased. The total memory and power of the MER computers is approximately the equivalent memory of a typical powerful laptop. As with all space mission computers, the *Spirit* computer contains special memory to tolerate the extreme radiation environment from space. Interesting real-world problems, such as planetary mobile rovers like *Spirit* and *Sojourner*, are used as illustrative examples throughout the book. For example, a mobile rover design problem is discussed in the Design Example in Section 4.8.

Control engineering is an exciting and a challenging field. By its very nature, control engineering is a multidisciplinary subject, and it has taken its place as a core course in the engineering curriculum. It is reasonable to expect different approaches to mastering and practicing the art of control engineering. Since the subject has a strong mathematical foundation, we might approach it from a strictly theoretical point of view, emphasizing theorems and proofs. On the other hand, since the ultimate objective is to implement controllers in real systems, we might take an ad hoc approach relying only on intuition and hands-on experience when designing feedback control systems. Our approach is to present a control engineering methodology that, while based on mathematical fundamentals, stresses physical system modeling and practical control system designs with realistic system specifications.

We believe that the most important and productive approach to learning is for each of us to rediscover and re-create anew the answers and methods of the past. Thus, the ideal is to present the student with a series of problems and questions and point to some of the answers that have been obtained over the past decades. The traditional method—to confront the student not with the problem but with the finished solution—is to deprive the student of all excitement, to shut off the creative impulse, to reduce the adventure of humankind to a dusty heap of theorems. The issue, then, is to present some of the unanswered and important problems that we continue to confront, for it may be asserted that what we have truly learned and understood, we discovered ourselves.

The purpose of this book is to present the structure of feedback control theory and to provide a sequence of exciting discoveries as we proceed through the text and problems. If this book is able to assist the student in discovering feedback control system theory and practice, it will have succeeded.

---

[1]For further reading on the Apollo guidance, navigation, and control system, see R. H. Battin, *An Introduction to the Mathematics and Methods of Astrodynamics,* AIAA Education Series, J. S. Pzemieniecki/Series Editor-in-Chief, 1987.

THE AUDIENCE

This text is designed for an introductory undergraduate course in control systems for engineering students. There is very little demarcation between aerospace, chemical, electrical, industrial, and mechanical engineering in control system practice; therefore, this text is written without any conscious bias toward one discipline. Thus, it is hoped that this book will be equally useful for all engineering disciplines and, perhaps, will assist in illustrating the utility of control engineering. The numerous problems and examples represent all fields, and the examples of the sociological, biological, ecological, and economic control systems are intended to provide the reader with an awareness of the general applicability of control theory to many facets of life. We believe that exposing students of one discipline to examples and problems from other disciplines will provide them with the ability to see beyond their own field of study. Many students pursue careers in engineering fields other than their own. For example, many electrical and mechanical engineers find themselves in the aerospace industry working alongside aerospace engineers. We hope this introduction to control engineering will give students a broader understanding of control system design and analysis.

In its first ten editions, *Modern Control Systems* has been used in senior-level courses for engineering students at more than 400 colleges and universities. It also has been used in courses for engineering graduate students with no previous background in control engineering.

THE ELEVENTH EDITION

A companion website is available to students and faculty using the eleventh edition. The website contains practice exercises, all the m-files in the book, Laplace and z-transform tables, written materials on matrix algebra, complex numbers, and symbols, units, and conversion factors. An icon will appear in the book margin whenever there is additional related material on the website. Also, since the website provides a mechanism for continuously updating and adding control-related materials of interest to students and professors, it is advisable to visit the website regularly during the semester or quarter when taking the course. The MCS website address is http://www.prenhall.com/dorf.

With the eleventh edition, we continue to evolve the design emphasis that historically has characterized *Modern Control Systems*. Using the real-world engineering problems associated with designing a controller for a disk drive read system, we present the *Sequential Design Example* (identified by an arrow icon in the text), which is considered sequentially in each chapter using the methods and concepts in that chapter. Disk drives are used in computers of all sizes and they represent an important application of control engineering. Various aspects of the design of controllers for the disk drive read system are considered in each chapter. For example, in Chapter 1 we identify the control goals, identify the variables to be controlled, write the control specifications, and establish the preliminary system configuration for the disk drive. Then, in Chapter 2, we obtain models of the process, sensors, and actuators. In the remaining chapters, we continue the design process, stressing the main points of the chapters.

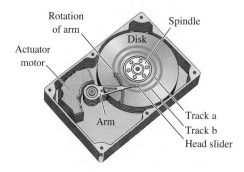

In the same spirit as the *Sequential Design Example,* we present a design problem that we call the *Continuous Design Problem* (identified by a triple arrow icon in the text) to give students the opportunity to build upon a design problem from chapter to chapter. High-precision machinery places stringent demands on table slide systems. In the *Continuous Design Problem,* students apply the techniques and tools presented in each chapter to the development of a design solution that meets the specified requirements.

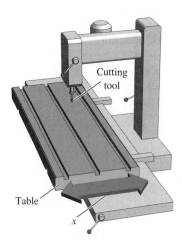

The computer-aided design and analysis component of the book continues to evolve and improve. The end-of-chapter computer problem set is identified by the graphical icon in the text. Also, many of the solutions to various components of the *Sequential Design Example* utilize m-files with corresponding scripts included in the figures.

PEDAGOGY

The book is organized around the concepts of control system theory as they have been developed in the frequency and time domains. An attempt has been made to make the selection of topics, as well as the systems discussed in the examples and problems, modern in the best sense. Therefore, this book includes discussions on robust control systems and system sensitivity, state variable models, controllability and observability, computer control systems, internal model control, robust PID controllers, and computer-aided design and analysis, to name a few. However, the classical topics of control theory that have proved to be so very useful in practice have been retained and expanded.

Building Basic Principles: From Classical to Modern. Our goal is to present a clear exposition of the basic principles of frequency- and time-domain design techniques. The classical methods of control engineering are thoroughly covered: Laplace transforms and transfer functions; root locus design; Routh–Hurwitz stability analysis; frequency response methods, including Bode, Nyquist, and Nichols; steady-state error for standard test signals; second-order system approximations; and phase and gain margin and bandwidth. In addition, coverage of the state variable method is significant. Fundamental notions of controllability and observability for state variable models are discussed. Full state feedback design with Ackermann's formula for pole placement is presented, along with a discussion on the limitations of state variable feedback. Observers are introduced as a means to provide state estimates when the complete state is not measured.

Upon this strong foundation of basic principles, the book provides many opportunities to explore topics beyond the traditional. Advances in robust control theory are introduced in Chapter 12. The implementation of digital computer control systems is discussed in Chapter 13. Each chapter (but the first) introduces the student to the notion of computer-aided design and analysis. The book concludes with an extensive references section, divided by chapter, to guide the student to further sources of information on control engineering.

Progressive Development of Problem-Solving Skills. Reading the chapters, attending lectures and taking notes, and working through the illustrated examples are all part of the learning process. But the real test comes at the end of the chapter with the problems. The book takes the issue of problem solving seriously. In each chapter, there are five problem types:

- ❏ Exercises
- ❏ Problems
- ❏ Advanced Problems
- ❏ Design Problems
- ❏ Computer Problems

For example, the problem set for The Root Locus Method, Chapter 7 (see page 407) includes 27 exercises, 39 problems, 13 advanced problems, 13 design problems, and 9 computer-based problems. The exercises permit the students to readily utilize the concepts and methods introduced in each chapter by solving relatively straightforward exercises before attempting the more complex problems. Answers to one-third of the exercises are provided. The problems require an extension of the concepts of the chapter to new situations. The advanced problems represent problems of increasing complexity. The design problems emphasize the design task; the computer-based problems give the student practice with problem solving using computers. In total, the book contains more than 800 problems. Also, the MCS website contains practice exercises that are instantly graded, so they provide quick feedback for students. The abundance of problems of increasing complexity gives students confidence in their problem-solving ability as they work their way from the exercises to the design and computer-based problems. A complete instructor manual, available for all adopters of the text for course use, contains complete solutions to all end-of-chapter problems.

A set of m-files, the *Modern Control Systems Toolbox,* has been developed by the authors to supplement the text. The m-files contain the scripts from each computer-based example in the text. You may retrieve the m-files from Prentice Hall at http://www.prenhall.com/dorf.

Design Emphasis without Compromising Basic Principles.   The all-important topic of design of real-world, complex control systems is a major theme throughout the text. Emphasis on design for real-world applications addresses interest in design by ABET and industry. The design process consists of seven main building blocks which we arrange into three groups:

1. Establishment of goals and variables to be controlled, and definition of specifications (metrics) against which to measure performance
2. System definition and modeling
3. Control system design and integrated system simulation and analysis

In each chapter of this book, we highlight the connection between the design process and the main topics of that chapter. The objective is to demonstrate different aspects of the design process through illustrative examples. Various aspects of the control system design process are illustrated in detail in the following examples:

❑   insulin delivery control system (Section 1.8, page 27)

❑   fluid flow modeling (Section 2.8, page 83)

❑   space station orientation modeling (Section 3.8, page 176)

❑   blood pressure control during anesthesia (Section 4.8, page 237)

❑   attitude control of an airplane (Section 5.9, page 319)

❑   robot-controlled motorcycle (Section 6.5, page 375)

❑   automobile velocity control (Section 7.7, page 452)

❑   control of one leg of a six-legged robot (Section 8.6, page 526)

❑   hot ingot robot control (Section 9.8, page 610)

❑   milling machine control system (Section 10.12, page 714)

❑   diesel electric locomotive control (Section 11.9, page 798)

❑   digital audio tape controller (Section 12.8, page 861)

❑   fly-by-wire aircraft control surface (Section 13.10, page 928)

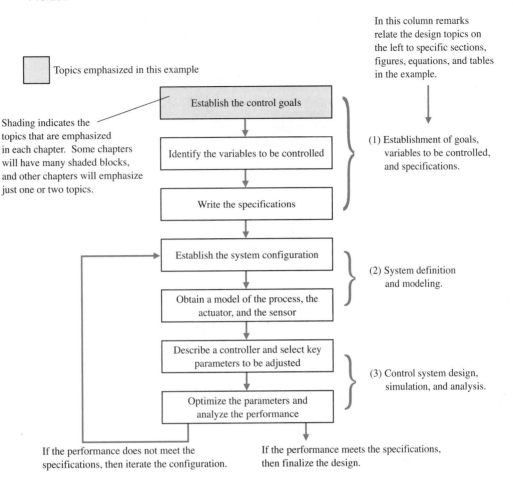

In this column remarks relate the design topics on the left to specific sections, figures, equations, and tables in the example.

Topics emphasized in this example

Establish the control goals

Shading indicates the topics that are emphasized in each chapter. Some chapters will have many shaded blocks, and other chapters will emphasize just one or two topics.

Identify the variables to be controlled

(1) Establishment of goals, variables to be controlled, and specifications.

Write the specifications

Establish the system configuration

(2) System definition and modeling.

Obtain a model of the process, the actuator, and the sensor

Describe a controller and select key parameters to be adjusted

(3) Control system design, simulation, and analysis.

Optimize the parameters and analyze the performance

If the performance does not meet the specifications, then iterate the configuration.

If the performance meets the specifications, then finalize the design.

Each chapter includes a section to assist students in utilizing computer-aided design and analysis concepts and rework many of the design examples. In Chapter 5, the Sequential Design Example: Disk Drive Read System is analyzed using computer-based methods. An m-file script that can be used to analyze the design is presented in Figure 5.47, p. 335. In general, each script is annotated with comment boxes that highlight important aspects of the script. The accompanying output of the script (generally a graph) also contains comment boxes pointing out significant elements. The scripts can also be utilized with modifications as the foundation for solving other related problems.

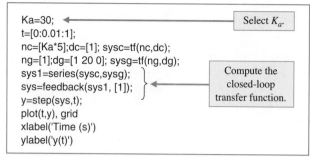

(a)

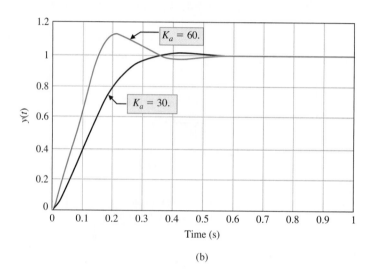

(b)

Learning Enhancement.   Each chapter begins with a chapter preview describing the topics the student can expect to encounter. The chapters conclude with an end-of-chapter summary, as well as terms and concepts. These sections reinforce the important concepts introduced in the chapter and serve as a reference for later use.

A second color is used to add emphasis when needed and to make the graphs and figures easier to interpret. Design Problem 4.4, page 217, asks the student to determine the value of K of the controller so that the response, denoted by $Y(s)$, to a step change in the position, denoted by $R(s)$, is satisfactory and the effect of the disturbance, denoted by $T_d(s)$, is minimized. The associated Figure DP4.4, p. 272, assists the student with (a) visualizing the problem and (b) taking the next step to develop the transfer function model and to complete the design.

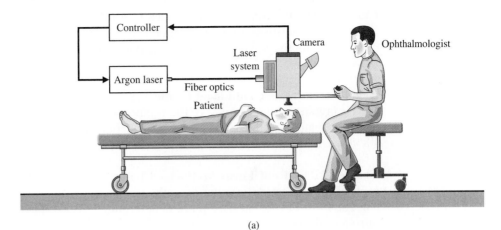

(a)

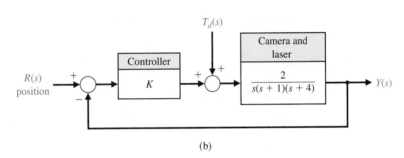

(b)

THE ORGANIZATION

Chapter 1 Introduction to Control Systems. Chapter 1 provides an introduction to the basic history of control theory and practice. The purpose of this chapter is to describe the general approach to designing and building a control system.

Chapter 2 Mathematical Models of Systems. Mathematical models of physical systems in input–output or transfer function form are developed in Chapter 2. A wide range of systems (including mechanical, electrical, and fluid) are considered.

Chapter 3 State Variable Models. Mathematical models of systems in state variable form are developed in Chapter 3. Using matrix methods, the transient response of control systems and the performance of these systems are examined.

Chapter 4 Feedback Control System Characteristics. The characteristics of feedback control systems are described in Chapter 4. The advantages of feedback are discussed, and the concept of the system error signal is introduced.

Chapter 5 The Performance of Feedback Control Systems.   In Chapter 5, the performance of control systems is examined. The performance of a control system is correlated with the s-plane location of the poles and zeros of the transfer function of the system.

Chapter 6 The Stability of Linear Feedback Systems.   The stability of feedback systems is investigated in Chapter 6. The relationship of system stability to the characteristic equation of the system transfer function is studied. The Routh–Hurwitz stability criterion is introduced.

Chapter 7 The Root Locus Method.   Chapter 7 deals with the motion of the roots of the characteristic equation in the s-plane as one or two parameters are varied. The locus of roots in the s-plane is determined by a graphical method. We also introduce the popular PID controller.

Chapter 8 Frequency Response Methods.   In Chapter 8, a steady-state sinusoid input signal is utilized to examine the steady-state response of the system as the frequency of the sinusoid is varied. The development of the frequency response plot, called the Bode plot, is considered.

Chapter 9 Stability in the Frequency Domain.   System stability utilizing frequency response methods is investigated in Chapter 9. Relative stability and the Nyquist criterion are discussed.

Chapter 10 The Design of Feedback Control Systems.   Several approaches to designing and compensating a control system are described and developed in Chapter 10. Various candidates for service as compensators are presented and it is shown how they help to achieve improved performance.

Chapter 11 The Design of State Variable Feedback Systems.   The main topic of Chapter 11 is the design of control systems using state variable models. Full-state feedback design and observer design methods based on pole placement are discussed. Tests for controllability and observability are presented, and the concept of an internal model design is discussed.

Chapter 12 Robust Control Systems.   Chapter 12 deals with the design of highly accurate control systems in the presence of significant uncertainty. Five methods for robust design are discussed, including root locus, frequency response, ITAE methods for robust PID controllers, internal models, and pseudo-quantitative feedback.

Chapter 13 Digital Control Systems.   Methods for describing and analyzing the performance of computer control systems are described in Chapter 13. The stability and performance of sampled-data systems are discussed.

Appendixes.   The appendixes are as follows:

    A MATLAB Basics
    B MathScript Basics

ACKNOWLEDGMENTS

We wish to express our sincere appreciation to the following individuals who have assisted us with the development of this eleventh edition, as well as all previous editions: Mahmoud A. Abdallah, Central Sate University (OH); John N. Chiasson, University of Pittsburgh; Samy El-Sawah, California State Polytechnic University, Pomona; Peter J. Gorder, Kansas State University; Duane Hanselman, University of Maine; Ashok Iyer, University of Nevada, Las Vegas; Leslie R. Koval, University of Missouri-Rolla; L. G. Kraft, University of New Hampshire; Thomas Kurfess, Georgia Institute of Technology; Julio C. Mandojana, Mankato State University; Jure Medanic, University of Illinois at Urbana-Champaign; Eduardo A. Misawa, Oklahoma State University; Medhat M. Morcos, Kansas State University; Mark Nagurka, Marquette University; Carla Schwartz, The MathWorks, Inc.; D. Subbaram Naidu, Idaho State University; Ron Perez, University of Wisconsin-Milwaukee; Murat Tanyel, Dordt College; Hal Tharp, University of Arizona; John Valasek, Texas A & M University; Paul P. Wang, Duke University; and Ravi Warrier, GMI Engineering and Management Institute.

OPEN LINES OF COMMUNICATION

The authors would like to establish a line of communication with the users of *Modern Control Systems*. We encourage all readers to send comments and suggestions for this and future editions. By doing this, we can keep you informed of any general-interest news regarding the textbook and pass along interesting comments of other users.

Keep in touch!

Richard C. Dorf   dorf@ece.ucdavis.edu
Robert H. Bishop   rhbishop@mail.utexas.edu

About the Authors

Richard C. Dorf is a Professor of Electrical and Computer Engineering at the University of California, Davis. Known as an instructor who is highly concerned with the discipline of electrical engineering and its application to social and economic needs, Professor Dorf has written and edited several successful engineering textbooks and handbooks, including the best selling *Engineering Handbook*, second edition and the third edition of the *Electrical Engineering Handbook*. Professor Dorf is also co-author of *Technology Ventures*, a leading textbook on technology entrepreneurship. Professor Dorf is a Fellow of the IEEE and a Fellow of the ASEE. He is active in the fields of control system design and robotics. Dr. Dorf holds a patent for the PIDA controller.

Robert H. Bishop is the Chairman of the Department of Aerospace Engineering and Engineering Mechanics at The University of Texas at Austin. He holds the Joe J. King Professorship and in 2002 was inducted into the UT Academy of Distinguished Teachers. A talented educator, Professor Bishop has been recognized for his contributions in the classroom with the coveted Lockheed Martin Tactical Aircraft Systems Award for Excellence in Engineering Teaching. He received the John Leland Atwood Award from the American Society of Engineering Educators and the American Institute of Aeronautics and Astronautics, which is periodically given to "a leader who has made lasting and significant contributions to aerospace engineering education." Professor Bishop is a Fellow of AIAA and is active in the IEEE and ASEE. He is a distinguished researcher with an interest in guidance, navigation, and control of aerospace vehicles.

Introduction to Control Systems

PREVIEW

In this chapter, we discuss open- and closed-loop feedback control systems. A control system consists of interconnected components to achieve a desired purpose. We examine examples of control systems through the course of history. These early systems incorporated many of the same ideas of feedback that are employed in modern manufacturing processes, alternative energy, complex hybrid automobiles, and sophisticated robots. A design process is presented that encompasses the establishment of goals and variables to be controlled, definition of specifications, system definition, modeling, and analysis. The iterative nature of design allows us to handle the design gap effectively while accomplishing necessary trade-offs in complexity, performance, and cost. Finally, we introduce the Sequential Design Example: Disk Drive Read System. This example will be considered sequentially in each chapter of this book. It represents a very important and practical control system design problem while simultaneously serving as a useful learning tool.

DESIRED OUTCOMES

Upon completion of Chapter 1, students should:

❑   Possess a basic understanding of control system engineering and be able to offer some illustrative examples and their relationship to key contemporary issues.

❑   Be able to recount a brief history of control systems and their role in society.

❑   Be capable of discussing the future of controls in the context of their evolutionary pathways.

❑   Recognize the elements of control system design and possess an appreciation of controls in the context of engineering design.

1.1 INTRODUCTION

Engineering is concerned with understanding and controlling the materials and forces of nature for the benefit of humankind. Control system engineers are concerned with understanding and controlling segments of their environment, often called **systems**, to provide useful economic products for society. The twin goals of understanding and controlling are complementary because effective systems control requires that the systems be understood and modeled. Furthermore, control engineering must often consider the control of poorly understood systems such as chemical process systems. The present challenge to control engineers is the modeling and control of modern, complex, interrelated systems such as traffic control systems, chemical processes, and robotic systems. Simultaneously, the fortunate engineer has the opportunity to control many useful and interesting industrial automation systems. Perhaps the most characteristic quality of control engineering is the opportunity to control machines and industrial and economic processes for the benefit of society.

Control engineering is based on the foundations of feedback theory and linear system analysis, and it integrates the concepts of network theory and communication theory. Therefore control engineering is not limited to any engineering discipline but is equally applicable to aeronautical, chemical, mechanical, environmental, civil, and electrical engineering. For example, a control system often includes electrical, mechanical, and chemical components. Furthermore, as the understanding of the dynamics of business, social, and political systems increases, the ability to control these systems will also increase.

A **control system** is an interconnection of components forming a system configuration that will provide a desired system response. The basis for analysis of a system is the foundation provided by linear system theory, which assumes a cause–effect relationship for the components of a system. Therefore a component or **process** to be controlled can be represented by a block, as shown in Figure 1.1. The input–output relationship represents the cause-and-effect relationship of the process, which in turn represents a processing of the input signal to provide an output signal variable, often with a power amplification. An **open-loop control system** uses a controller and an actuator to obtain the desired response, as shown in Figure 1.2. An open-loop system is a system without feedback.

> **An open-loop control system utilizes an actuating device to control the process directly without using feedback.**

FIGURE 1.1
Process to be controlled.

Input ⟶ Process ⟶ Output

FIGURE 1.2
Open-loop control system (without feedback).

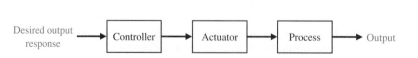

Desired output response ⟶ Controller ⟶ Actuator ⟶ Process ⟶ Output

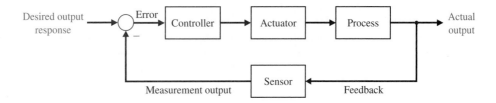

FIGURE 1.3
Closed-loop
feedback control
system (with
feedback).

In contrast to an open-loop control system, a closed-loop control system utilizes an additional measure of the actual output to compare the actual output with the desired output response. The measure of the output is called the **feedback signal**. A simple **closed-loop feedback control system** is shown in Figure 1.3. A feedback control system is a control system that tends to maintain a prescribed relationship of one system variable to another by comparing functions of these variables and using the difference as a means of control. With an accurate sensor, the measured output is a good approximation of the actual output of the system.

A feedback control system often uses a function of a prescribed relationship between the output and reference input to control the process. Often the difference between the output of the process under control and the reference input is amplified and used to control the process so that the difference is continually reduced. In general, the difference between the desired output and the actual output is equal to the error, which is then adjusted by the controller. The output of the controller causes the actuator to modulate the process in order to reduce the error. The sequence is such, for instance, that if a ship is heading incorrectly to the right, the rudder is actuated to direct the ship to the left. The system shown in Figure 1.3 is a **negative feedback** control system, because the output is subtracted from the input and the difference is used as the input signal to the controller. The feedback concept has been the foundation for control system analysis and design.

> **A closed-loop control system uses a measurement of the output and feedback of this signal to compare it with the desired output (reference or command).**

Due to the increasing complexity of the system under control and the interest in achieving optimum performance, the importance of control system engineering has grown in the past decade. Furthermore, as the systems become more complex, the interrelationship of many controlled variables must be considered in the control scheme. A block diagram depicting a **multivariable control system** is shown in Figure 1.4.

A common example of an open-loop control system is a microwave oven set to operate for a fixed time. An example of a closed-loop control system is a person steering an automobile (assuming his or her eyes are open) by looking at the auto's location on the road and making the appropriate adjustments.

The introduction of feedback enables us to control a desired output and can improve accuracy, but it requires attention to the issue of stability of response.

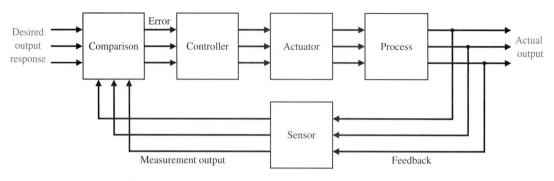

FIGURE 1.4   Multivariable control system.

1.2 BRIEF HISTORY OF AUTOMATIC CONTROL

The use of feedback to control a system has a fascinating history. The first applications of feedback control appeared in the development of float regulator mechanisms in Greece in the period 300 to 1 B.C. [1, 2, 3]. The water clock of Ktesibios used a float regulator (refer to Problem 1.11). An oil lamp devised by Philon in approximately 250 B.C. used a float regulator in an oil lamp for maintaining a constant level of fuel oil. Heron of Alexandria, who lived in the first century A.D., published a book entitled *Pneumatica*, which outlined several forms of water-level mechanisms using float regulators [1].

The first feedback system to be invented in modern Europe was the temperature regulator of Cornelis Drebbel (1572–1633) of Holland [1]. Dennis Papin (1647–1712) invented the first pressure regulator for steam boilers in 1681. Papin's pressure regulator was a form of safety regulator similar to a pressure-cooker valve.

The first automatic feedback controller used in an industrial process is generally agreed to be James Watt's **flyball governor**, developed in 1769 for controlling the speed of a steam engine [1, 2]. The all-mechanical device, shown in Figure 1.5,

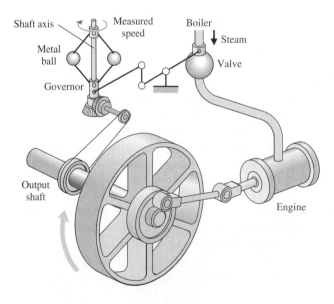

FIGURE 1.5
Watt's flyball
governor.

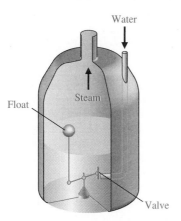

FIGURE 1.6
Water-level float
regulator.

measured the speed of the output shaft and utilized the movement of the flyball to control the steam valve and therefore the amount of steam entering the engine. As depicted in Figure 1.5, the governor shaft axis is connected via mechanical linkages and beveled gears to the output shaft of the steam engine. As the steam engine output shaft speed increases, the ball weights rise and move away from the shaft axis and through mechanical linkages the steam valve closes and the engine slows down.

The first historical feedback system, claimed by Russia, is the water-level float regulator said to have been invented by I. Polzunov in 1765 [4]. The level regulator system is shown in Figure 1.6. The float detects the water level and controls the valve that covers the water inlet in the boiler.

The next century was characterized by the development of automatic control systems through intuition and invention. Efforts to increase the accuracy of the control system led to slower attenuation of the transient oscillations and even to unstable systems. It then became imperative to develop a theory of automatic control. In 1868, J.C. Maxwell formulated a mathematical theory related to control theory using a differential equation model of a governor [5]. Maxwell's study was concerned with the effect various system parameters had on the system performance. During the same period, I. A. Vyshnegradskii formulated a mathematical theory of regulators [6].

Prior to World War II, control theory and practice developed differently in the United States and western Europe than in Russia and eastern Europe. The main impetus for the use of feedback in the United States was the development of the telephone system and electronic feedback amplifiers by Bode, Nyquist, and Black at Bell Telephone Laboratories [7–10, 12].

Harold S. Black graduated from Worcester Polytechnic Institute in 1921 and joined Bell Laboratories of American Telegraph and Telephone (AT&T). In 1921, the major task confronting Bell Laboratories was the improvement of the telephone system and the design of improved signal amplifiers. Black was assigned the task of linearizing, stabilizing, and improving the amplifiers that were used in tandem to carry conversations over distances of several thousand miles.

Black reports [8]:

Then came the morning of Tuesday, August 2, 1927, when the concept of the negative feedback amplifier came to me in a flash while I was crossing the Hudson River on the Lackawanna Ferry, on my way to work. For more than 50 years I have pondered how and why the idea came, and I can't say any more today than I could that morning. All I know is that after several years of hard work on the problem, I suddenly realized that if I fed the amplifier output back to the input, in reverse phase, and kept the device from oscillating (singing, as we called it then), I would have exactly what I wanted: a means of canceling out the distortion in the output. I opened my morning newspaper and on a page of *The New York Times* I sketched a simple canonical diagram of a negative feedback amplifier plus the equations for the amplification with feedback. I signed the sketch, and 20 minutes later, when I reached the laboratory at 463 West Street, it was witnessed, understood, and signed by the late Earl C. Blessing.

I envisioned this circuit as leading to extremely linear amplifiers (40 to 50 dB of negative feedback), but an important question is: How did I know I could avoid self-oscillations over very wide frequency bands when many people doubted such circuits would be stable? My confidence stemmed from work that I had done two years earlier on certain novel oscillator circuits and three years earlier in designing the terminal circuits, including the filters, and developing the mathematics for a carrier telephone system for short toll circuits.

The frequency domain was used primarily to describe the operation of the feedback amplifiers in terms of bandwidth and other frequency variables. In contrast, the eminent mathematicians and applied mechanicians in the former Soviet Union inspired and dominated the field of control theory. Therefore, the Russian theory tended to utilize a time-domain formulation using differential equations.

The control of an industrial process (manufacturing, production, and so on) by automatic rather than manual means is often called **automation**. Automation is prevalent in the chemical, electric power, paper, automobile, and steel industries, among others. The concept of automation is central to our industrial society. Automatic machines are used to increase the production of a plant per worker in order to offset rising wages and inflationary costs. Thus industries are concerned with the productivity per worker of their plants. **Productivity** is defined as the ratio of physical output to physical input [26]. In this case, we are referring to labor productivity, which is real output per hour of work.

The transformation of the U.S. labor force in the country's brief history follows the progressive mechanization of work that attended the evolution of the agrarian republic into an industrial world power. In 1820, more than 70 percent of the labor force worked on the farm. By 1900, less than 40 percent were engaged in agriculture. Today, less than 5 percent works in agriculture [15].

In 1925, some 588,000 people—about 1.3 percent of the nation's labor force—were needed to mine 520 million tons of bituminous coal and lignite, almost all of it from underground. By 1980, production was up to 774 million tons, but the work force had been reduced to 208,000. Furthermore, only 136,000 of that number were employed in underground mining operations. The highly mechanized and highly productive surface mines, with just 72,000 workers, produced 482 million tons, or 62 percent of the total [27].

A large impetus to the theory and practice of automatic control occurred during World War II when it became necessary to design and construct automatic airplane piloting, gun-positioning systems, radar antenna control systems, and other military systems based on the feedback control approach. The complexity and expected performance of these military systems necessitated an extension of the available control techniques and fostered interest in control systems and the development of new insights and methods. Prior to 1940, for most cases, the design of control systems was an art involving a trial-and-error approach. During the 1940s, mathematical and analytical methods increased in number and utility, and control engineering became an engineering discipline in its own right [10–12].

Another example of the discovery of an engineering solution to a control system problem was the creation of a gun director by David B. Parkinson of Bell Telephone Laboratories. In the spring of 1940, Parkinson was a 29-year-old engineer intent on improving the automatic level recorder, an instrument that used strip-chart paper to plot the record of a voltage. A critical component was a small potentiometer used to control the pen of the recorder through an actuator.

Parkinson had a dream about an antiaircraft gun that was successfully felling airplanes. Parkinson described the situation [13]:

> After three or four shots one of the men in the crew smiled at me and beckoned me to come closer to the gun. When I drew near he pointed to the exposed end of the left trunnion. Mounted there was the control potentiometer of my level recorder!

The next morning Parkinson realized the significance of his dream:

> If my potentiometer could control the pen on the recorder, something similar could, with suitable engineering, control an antiaircraft gun.

After considerable effort, an engineering model was delivered for testing to the U.S. Army on December 1, 1941. Production models were available by early 1943, and eventually 3000 gun controllers were delivered. Input to the controller was provided by radar, and the gun was aimed by taking the data of the airplane's present position and calculating the target's future position.

Frequency-domain techniques continued to dominate the field of control following World War II with the increased use of the Laplace transform and the complex frequency plane. During the 1950s, the emphasis in control engineering theory was on the development and use of the s-plane methods and, particularly, the root locus approach. Furthermore, during the 1980s, the use of digital computers for control components became routine. The technology of these new control elements to perform accurate and rapid calculations was formerly unavailable to control engineers. There are now over 400,000 digital process control computers installed in the United States [14, 27]. These computers are employed especially for process control systems in which many variables are measured and controlled simultaneously by the computer.

With the advent of Sputnik and the space age, another new impetus was imparted to control engineering. It became necessary to design complex, highly accurate control systems for missiles and space probes. Furthermore, the necessity to minimize the weight of satellites and to control them very accurately has spawned the important field of optimal control. Due to these requirements, the time-domain methods developed by Liapunov, Minorsky, and others have been met with great interest in the last two decades. Recent theories of optimal control developed by L. S. Pontryagin in the former Soviet Union and R. Bellman in the United States, as well

Table 1.1   Selected Historical Developments of Control Systems

1769	James Watt's steam engine and governor developed. The Watt steam engine is often used to mark the beginning of the Industrial Revolution in Great Britain. During the Industrial Revolution, great strides were made in the development of mechanization, a technology preceding automation.
1800	Eli Whitney's concept of interchangeable parts manufacturing demonstrated in the production of muskets. Whitney's development is often considered to be the beginning of mass production.
1868	J. C. Maxwell formulates a mathematical model for a governor control of a steam engine.
1913	Henry Ford's mechanized assembly machine introduced for automobile production.
1927	H. S. Black conceives of the negative feedback amplifier and H. W. Bode analyzes feedback amplifiers.
1932	H. Nyquist develops a method for analyzing the stability of systems.
1941	Creation of first antiaircraft gun with active control.
1952	Numerical control (NC) developed at Massachusetts Institute of Technology for control of machine-tool axes.
1954	George Devol develops "programmed article transfer," considered to be the first industrial robot design.
1957	Sputnik launches the space age leading, in time, to miniaturization of computers and advances in automatic control theory.
1960	First Unimate robot introduced, based on Devol's designs. Unimate installed in 1961 for tending die-casting machines.
1970	State-variable models and optimal control developed.
1980	Robust control system design widely studied.
1983	Introduction of the personal computer (and control design software soon thereafter) brought the tools of design to the engineer's desktop.
1990	Export-oriented manufacturing companies emphasize automation.
1994	Feedback control widely used in automobiles. Reliable, robust systems demanded in manufacturing.
1997	First ever autonomous rover vehicle, known as Sojourner, explores the Martian surface.
1998–2003	Advances in micro- and nanotechnology. First intelligent micromachines are developed and functioning nanomachines are created.

as recent studies of robust systems, have contributed to the interest in time-domain methods. It now is clear that control engineering must consider both the time-domain and the frequency-domain approaches simultaneously in the analysis and design of control systems.

A selected history of control system development is summarized in Table 1.1.

1.3 EXAMPLES OF CONTROL SYSTEMS

Control engineering is concerned with the analysis and design of goal-oriented systems. Therefore the mechanization of goal-oriented policies has grown into a hierarchy of goal-oriented control systems. Modern control theory is concerned with systems that have self-organizing, adaptive, robust, learning, and optimum qualities.

Feedback control is a fundamental fact of modern industry and society. Driving an automobile is a pleasant task when the auto responds rapidly to the driver's commands. Many cars have power steering and brakes, which utilize hydraulic amplifiers for amplification of the force to the brakes or the steering wheel. A simple block diagram of an automobile steering control system is shown in Figure 1.7(a). The desired course is compared with a measurement of the actual course in order to generate a measure of the error, as shown in Figure 1.7(b). This measurement is obtained by visual and tactile (body movement) feedback, as provided by the feel of the steering wheel by the hand (sensor). This feedback system is a familiar version of the steering control system in an ocean liner or the flight controls in a large airplane. A typical direction-of-travel response is shown in Figure 1.7(c).

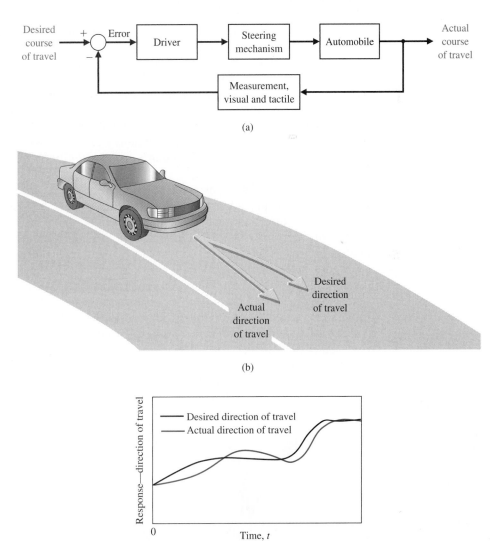

FIGURE 1.7
(a) Automobile steering control system. (b) The driver uses the difference between the actual and the desired direction of travel to generate a controlled adjustment of the steering wheel. (c) Typical direction-of-travel response.

A basic, manually controlled closed-loop system for regulating the level of fluid in a tank is shown in Figure 1.8. The input is a reference level of fluid that the operator is instructed to maintain. (This reference is memorized by the operator.) The power amplifier is the operator, and the sensor is visual. The operator compares the actual level with the desired level and opens or closes the valve (actuator), adjusting the fluid flow out, to maintain the desired level.

Other familiar control systems have the same basic elements as the system shown in Figure 1.3. A refrigerator has a temperature setting or desired temperature, a thermostat to measure the actual temperature and the error, and a compressor motor for power amplification. Other examples in the home are the oven, furnace, and water heater. In industry, there are many examples, including speed controls; process temperature and pressure controls; and position, thickness, composition, and quality controls [14, 17, 18].

In its modern usage, automation can be defined as a technology that uses programmed commands to operate a given process, combined with feedback of information to determine that the commands have been properly executed. Automation is often used for processes that were previously operated by humans. When automated, the process can operate without human assistance or interference. In fact, most automated systems are capable of performing their functions with greater accuracy and precision, and in less time, than humans are able to do. A semiautomated process is one that incorporates both humans and robots. For instance, many automobile assembly line operations require cooperation between a human operator and an intelligent robot.

Feedback control systems are used extensively in industrial applications. Thousands of industrial and laboratory robots are currently in use. Manipulators can pick up objects weighing hundreds of pounds and position them with an accuracy of one-tenth of an inch or better [28]. Automatic handling equipment for home, school, and industry is particularly useful for hazardous, repetitive, dull, or simple tasks. Machines that automatically load and unload, cut, weld, or cast are used by industry to obtain accuracy, safety, economy, and productivity [14, 27, 28, 41]. The use of computers integrated with machines that perform tasks like a human worker has been foreseen by several authors. In his famous 1923 play, entitled *R.U.R.* [48], Karel Capek called artificial workers *robots*, deriving the word from the Czech noun *robota*, meaning "work."

FIGURE 1.8
A manual control system for regulating the level of fluid in a tank by adjusting the output valve. The operator views the level of fluid through a port in the side of the tank.

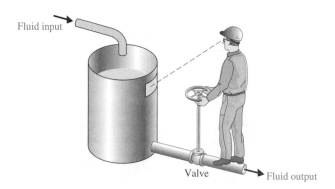

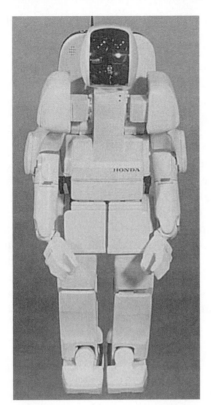

FIGURE 1.9
The Honda P3
humanoid robot. P3
walks, climbs stairs,
and turns corners.
Photo courtesy of
American Honda
Motor, Inc.

A **robot** is a computer-controlled machine and involves technology closely associated with automation. Industrial robotics can be defined as a particular field of automation in which the automated machine (that is, the robot) is designed to substitute for human labor [18, 27, 33]. Thus robots possess certain humanlike characteristics. Today, the most common humanlike characteristic is a mechanical manipulator that is patterned somewhat after the human arm and wrist. Some devices even have anthropomorphic mechanisms, including what we might recognize as mechanical arms, wrists, and hands [14, 27, 28]. An example of an anthropomorphic robot is shown in Figure 1.9. We recognize that the automatic machine is well suited to some tasks, as noted in Table 1.2, and that other tasks are best carried out by humans.

Another very important application of control technology is in the control of the modern automobile [19, 20]. Control systems for suspension, steering, and engine

Table 1.2 Task Difficulty: Human Versus Automatic Machine

Tasks Difficult for a Machine	Tasks Difficult for a Human
Inspect seedlings in a nursery.	Inspect a system in a hot, toxic
Drive a vehicle through rugged terrain.	environment.
Identify the most expensive jewels on	Repetitively assemble a clock.
a tray of jewels.	Land an airliner at night, in bad weather.

control have been introduced. Many new autos have a four-wheel-steering system, as well as an antiskid control system.

A three-axis control system for inspecting individual semiconductor wafers is shown in Figure 1.10. This system uses a specific motor to drive each axis to the desired position in the *x-y-z*-axis, respectively. The goal is to achieve smooth, accurate movement in each axis. This control system is an important one for the semiconductor manufacturing industry.

There has been considerable discussion recently concerning the gap between practice and theory in control engineering. However, it is natural that theory precedes the applications in many fields of control engineering. Nonetheless, it is interesting to note that in the electric power industry, the largest industry in the United States, the gap is relatively insignificant. The electric power industry is primarily interested in energy conversion, control, and distribution. It is critical that computer control be increasingly applied to the power industry in order to improve the efficient use of energy resources. Also, the control of power **plants** for minimum waste emission has become increasingly important. The modern, large-capacity plants, which exceed several hundred megawatts, require automatic control systems that account for the interrelationship of the process variables and optimum power production. It is common to have 90 or more manipulated variables under

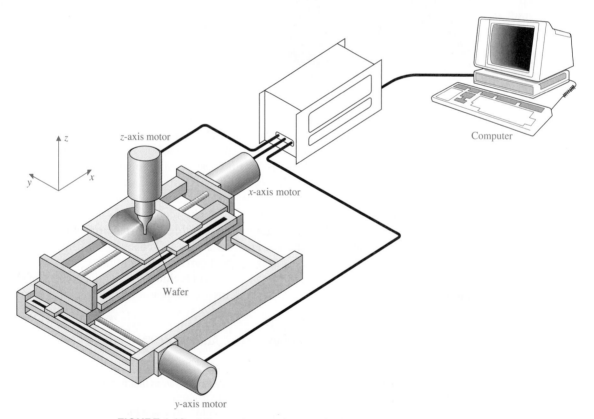

FIGURE 1.10   A three-axis control system for inspecting individual semiconductor wafers with a highly sensitive camera.

coordinated control. A simplified model showing several of the important control variables of a large boiler–generator system is shown in Figure 1.11. This is an example of the importance of measuring many variables, such as pressure and oxygen, to provide information to the computer for control calculations.

The electric power industry has used the modern aspects of control engineering for significant and interesting applications. It appears that in the process industry, the factor that maintains the applications gap is the lack of instrumentation to measure all the important process variables, including the quality and composition of the product. As these instruments become available, the applications of modern control theory to industrial systems should increase measurably.

Another important industry, the metallurgical industry, has had considerable success in automatically controlling its processes. In fact, in many cases, the control theory is being fully implemented. For example, a hot-strip steel mill, which involves a $100-million investment, is controlled for temperature, strip width, thickness, and quality.

Rapidly rising energy costs coupled with threats of energy curtailment are resulting in new efforts for efficient automatic energy management. Computer controls are used to control energy use in industry and to stabilize and connect loads evenly to gain fuel economy.

There has been considerable interest recently in applying the feedback control concepts to automatic warehousing and inventory control. Furthermore, automatic control of agricultural systems (farms) is receiving increased interest. Automatically controlled silos and tractors have been developed and tested. Automatic control of

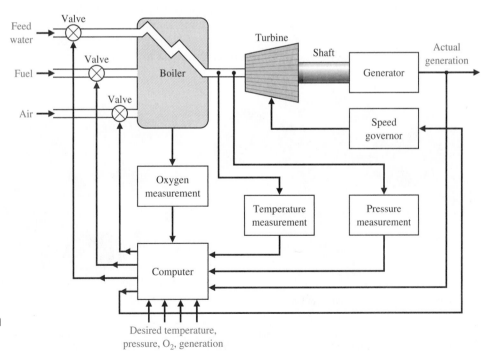

FIGURE 1.11
Coordinated control system for a boiler–generator.

wind turbine generators, solar heating and cooling, and automobile engine performance are important modern examples [20, 21].

Also, there have been many applications of control system theory to biomedical experimentation, diagnosis, prosthetics, and biological control systems [22, 23, 51]. The control systems under consideration range from the cellular level to the central nervous system and include temperature regulation and neurological, respiratory, and cardiovascular control. Most physiological control systems are closed-loop systems. However, we find not one controller but rather control loop within control loop, forming a hierarchy of systems. The modeling of the structure of biological processes confronts the analyst with a high-order model and a complex structure. Prosthetic devices that aid the 46 million handicapped individuals in the United States are designed to provide automatically controlled aids to the disabled [22, 27, 42]. The robotic hand shown in Figure 1.12 belongs to Obrero, a humanoid robot developed at MIT that is capable of sensitive manipulation. The Obrero robot is responsive to the properties of the object it holds and does not rely on vision as the main sensor. The hand has position and force control of the fingers employing very sensitive tactile sensors and series elastic actuators in its joints.

Finally, it has become interesting and valuable to attempt to model the feedback processes prevalent in the social, economic, and political spheres. This approach is undeveloped at present but appears to have a reasonable future. Society, of course, is composed of many feedback systems and regulatory bodies, such as the Federal Reserve Board, which are controllers exerting the forces on society necessary to maintain a desired output. A simple lumped model of the national income feedback control system is shown in Figure 1.13. This type of model helps the analyst to understand the effects of government control—granted its existence—and the dynamic effects of government spending. Of course, many other loops not shown also exist, since, theoretically, government spending cannot exceed the tax collected without generating a deficit, which is itself a control loop containing the Internal Revenue Service and the Congress. In a socialist country, the loop due to consumers is de-emphasized and

(a) Computer-aided drawing (Courtesy of Eduardo Torres-Jara).   (b) The Obrero robotic hand (Photo by Iuliu Vasilescu).

FIGURE 1.12   The Obrero robot is responsive to the properties of the object it holds and does not rely on vision as the main sensor but as a complement. Obrero is part of the Humanoid Robotics Group at the MIT Computer Science and Artificial Intelligence Laboratory.

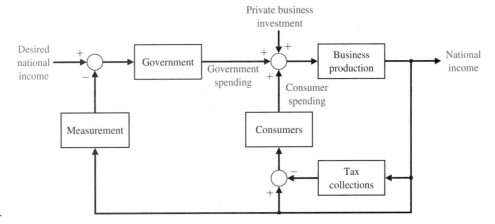

FIGURE 1.13
A feedback control system model of the national income.

government control is emphasized. In that case, the measurement block must be accurate and must respond rapidly; both are very difficult characteristics to realize from a bureaucratic system. This type of political or social feedback model, while usually nonrigorous, does impart information and understanding.

The ongoing area of research and development of unmanned aerial vehicles (UAVs) is full of potential for the application of control systems. An example of a UAV is shown in Figure 1.14. UAVs are unmanned but are usually controlled by ground operators. Typically they do not operate autonomously and their inability to provide the level of safety of a manned plane keeps them from flying freely in the commercial airspace. One significant challenge is to develop control systems that will avoid in-air collisions. Ultimately, the goal is to employ the UAV autonomously in such applications as aerial photography to assist in disaster mitigation, survey work to assist in construction projects, crop monitoring, and continuous weather monitoring. In a military setting, UAVs can perform intelligence, surveillance, and reconnaissance missions [83]. Smart unmanned aircraft will require significant deployment of advanced control systems throughout the airframe.

FIGURE 1.14
An unmanned aerial vehicle. (Used with permission. Credit: DARPA.)

1.4 ENGINEERING DESIGN

Engineering design is the central task of the engineer. It is a complex process in which both creativity and analysis play major roles.

> **Design is the process of conceiving or inventing the forms, parts, and details of a system to achieve a specified purpose.**

Design activity can be thought of as planning for the emergence of a particular product or system. Design is an innovative act whereby the engineer creatively uses knowledge and materials to specify the shape, function, and material content of a system. The design steps are (1) to determine a need arising from the values of various groups, covering the spectrum from public policy makers to the consumer; (2) to specify in detail what the solution to that need must be and to embody these values; (3) to develop and evaluate various alternative solutions to meet these specifications; and (4) to decide which one is to be designed in detail and fabricated.

An important factor in realistic design is the limitation of time. Design takes place under imposed schedules, and we eventually settle for a design that may be less than ideal but considered "good enough." In many cases, time is the *only* competitive advantage.

A major challenge for the designer is writing the specifications for the technical product. **Specifications** are statements that explicitly state what the device or product is to be and do. The design of technical systems aims to provide appropriate design specifications and rests on four characteristics: complexity, trade-offs, design gaps, and risk.

Complexity of design results from the wide range of tools, issues, and knowledge to be used in the process. The large number of factors to be considered illustrates the complexity of the design specification activity, not only in assigning these factors their relative importance in a particular design, but also in giving them substance either in numerical or written form, or both.

The concept of **trade-off** involves the need to resolve conflicting design goals, all of which are desirable. The design process requires an efficient compromise between desirable but conflicting criteria.

In making a technical device, we generally find that the final product does not appear as originally visualized. For example, our image of the problem we are solving does not appear in written description and ultimately in the specifications. Such **design gaps** are intrinsic in the progression from an abstract idea to its realization.

This inability to be absolutely sure about predictions of the performance of a technological object leads to major uncertainties about the actual effects of the designed devices and products. These uncertainties are embodied in the idea of unintended consequences or **risk**. The result is that designing a system is a risk-taking activity.

Complexity, trade-off, gaps, and risk are inherent in designing new systems and devices. Although they can be minimized by considering all the effects of a given design, they are always present in the design process.

Within engineering design, there is a fundamental difference between the two major types of thinking that must take place: engineering analysis and synthesis. Attention is focused on models of the physical systems that are analyzed to provide insight and that indicate directions for improvement. On the other hand, **synthesis** is the process by which these new physical configurations are created.

Design is a process that may proceed in many directions before the desired one is found. It is a deliberate process by which a designer creates something new in response to a recognized need while recognizing realistic constraints. The design process is inherently iterative—we must start somewhere! Successful engineers learn to simplify complex systems appropriately for design and analysis purposes. A gap between the complex physical system and the design model is inevitable. Design gaps are intrinsic in the progression from the initial concept to the final product. We know intuitively that it is easier to improve an initial concept incrementally than to try to create a final design at the start. In other words, engineering design is not a linear process. It is an iterative, nonlinear, creative process.

The main approach to the most effective engineering design is parameter analysis and optimization. Parameter analysis is based on (1) identification of the key parameters, (2) generation of the system configuration, and (3) evaluation of how well the configuration meets the needs. These three steps form an iterative loop. Once the key parameters are identified and the configuration synthesized, the designer can **optimize** the parameters. Typically, the designer strives to identify a limited set of parameters to be adjusted.

1.5 CONTROL SYSTEM DESIGN

The design of control systems is a specific example of engineering design. The goal of control engineering design is to obtain the configuration, specifications, and identification of the key parameters of a proposed system to meet an actual need.

The control system design process is illustrated in Figure 1.15. The design process consists of seven main building blocks, which we arrange into three groups:

1. Establishment of goals and variables to be controlled, and definition of specifications (metrics) against which to measure performance
2. System definition and modeling
3. Control system design and integrated system simulation and analysis

In each chapter of this book, we will highlight the connection between the design process illustrated in Figure 1.15 and the main topics of that chapter. The objective is to demonstrate different aspects of the design process through illustrative examples. We have established the following connections between the chapters in this book and the design process block diagram:

1. Establishment of goals, control variables, and specifications: Chapters 1, 3, 4, and 13.
2. System definition and modeling: Chapters 2–4, and 11–13.
3. Control system design, simulation, and analysis: Chapters 4–13.

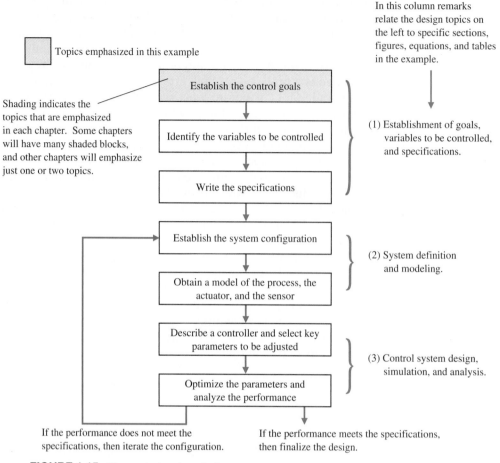

In this column remarks relate the design topics on the left to specific sections, figures, equations, and tables in the example.

Topics emphasized in this example

Shading indicates the topics that are emphasized in each chapter. Some chapters will have many shaded blocks, and other chapters will emphasize just one or two topics.

Establish the control goals

Identify the variables to be controlled

Write the specifications

(1) Establishment of goals, variables to be controlled, and specifications.

Establish the system configuration

Obtain a model of the process, the actuator, and the sensor

(2) System definition and modeling.

Describe a controller and select key parameters to be adjusted

Optimize the parameters and analyze the performance

(3) Control system design, simulation, and analysis.

If the performance does not meet the specifications, then iterate the configuration.

If the performance meets the specifications, then finalize the design.

FIGURE 1.15   The control system design process.

The first step in the design process consists of establishing the system goals. For example, we may state that our goal is to control the velocity of a motor accurately. The second step is to identify the variables that we desire to control (for example, the velocity of the motor). The third step is to write the specifications in terms of the accuracy we must attain. This required accuracy of control will then lead to the identification of a sensor to measure the controlled variable. The performance specifications will describe how the closed-loop system should perform and will include (1) good regulation against disturbances, (2) desirable responses to commands, (3) realistic actuator signals, (4) low sensitivities, and (5) robustness.

As designers, we proceed to the first attempt to configure a system that will result in the desired control performance. This system configuration will normally consist of a sensor, the process under control, an actuator, and a controller, as shown in Figure 1.3. The next step consists of identifying a candidate for the actuator. This will, of course, depend on the process, but the actuation chosen must be capable of effectively adjusting the performance of the process. For example, if we wish to control the speed of a rotating flywheel, we will select a motor as the actuator. The sensor, in this case, must be capable of accurately measuring the speed. We then obtain a model for each of these elements.

Students studying controls are often given the models, frequently represented in transfer function or state variable form, with the understanding that they represent the underlying physical systems, but without further explanation. An obvious question is, where did the transfer function or state variable model come from? Within the context of a course in control systems, there is a need to address key questions surrounding modeling. To that end, in the early chapters, we will provide insight into key modeling concerns and answer fundamental questions: How is the transfer function obtained? What basic assumptions are implied in the model development? How general are the transfer functions? However, mathematical modeling of physical systems is a subject in and of itself. We cannot hope to cover the mathematical modeling in its entirety, but interested students are encouraged to seek outside references (see for example [85–89]).

The next step is the selection of a controller, which often consists of a summing amplifier that will compare the desired response and the actual response and then forward this error-measurement signal to an amplifier.

The final step in the design process is the adjustment of the parameters of the system to achieve the desired performance. If we can achieve the desired performance by adjusting the parameters, we will finalize the design and proceed to document the results. If not, we will need to establish an improved system configuration and perhaps select an enhanced actuator and sensor. Then we will repeat the design steps until we are able to meet the specifications, or until we decide the specifications are too demanding and should be relaxed.

The design process has been dramatically affected by the advent of powerful and inexpensive computers and effective control design and analysis software. For example, the Boeing 777, which incorporates the most advanced flight avionics of any U.S. commercial aircraft, was almost entirely computer-designed [62, 63]. Verification of final designs in high-fidelity computer simulations is essential. In many applications, the certification of the control system in realistic simulations represents a significant cost in terms of money and time. The Boeing 777 test pilots flew about 2400 flights in high-fidelity simulations before the first aircraft was even built.

Another notable example of computer-aided design and analysis is the McDonnell Douglas Delta Clipper experimental vehicle DC-X, which was designed, built, and flown in 24 months. Computer-aided design tools and automated code-generation contributed to an estimated 80 percent cost savings and 30 percent time savings [64].

In summary, the controller design problem is as follows: Given a model of the system to be controlled (including its sensors and actuators) and a set of design goals, find a suitable controller, or determine that none exists. As with most of engineering design, the design of a feedback control system is an iterative and nonlinear process. A successful designer must consider the underlying physics of the plant under control, the control design strategy, the controller design architecture (that is, what type of controller will be employed), and effective controller tuning strategies. In addition, once the design is completed, the controller is often implemented in hardware, and hence issues of interfacing with hardware can appear. When taken together, these different phases of control system design make the task of designing and implementing a control system quite challenging [82].

1.6 MECHATRONIC SYSTEMS

A natural stage in the evolutionary process of modern engineering design is encompassed in the area known as **mechatronics** [70]. The term mechatronics was coined in Japan in the 1970s [71–73]. Mechatronics is the synergistic integration of mechanical, electrical, and computer systems and has evolved over the past 30 years, leading to a new breed of intelligent products. Feedback control is an integral aspect of modern mechatronic systems. One can understand the extent that mechatronics reaches into various disciplines by considering the components that make up mechatronics [74–77]. The key elements of mechatronics are (1) physical systems modeling, (2) sensors and actuators, (3) signals and systems, (4) computers and logic systems, and (5) software and data acquisition. Feedback control encompasses aspects of all five key elements of mechatronics, but is associated primarily with the element of signals and systems, as illustrated in Figure 1.16.

Advances in computer hardware and software technology coupled with the desire to increase the performance-to-cost ratio has revolutionized engineering design. New products are being developed at the intersection of traditional disciplines of engineering, computer science, and the natural sciences. Advancements in traditional disciplines are fueling the growth of mechatronics systems by providing "enabling technologies." A critical enabling technology was the microprocessor which has had a profound effect on the design of consumer products. We should

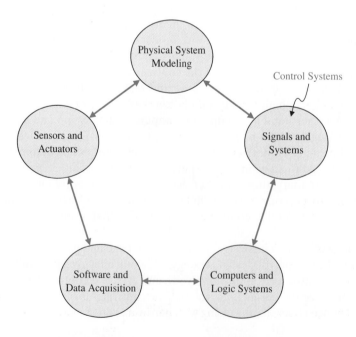

FIGURE 1.16
The key elements of mechatronics [70].

expect continued advancements in cost-effective microprocessors and microcontrollers, novel sensors and actuators enabled by advancements in applications of microelectromechanical systems (MEMS), advanced control methodologies and real-time programming methods, networking and wireless technologies, and mature computer-aided engineering (CAE) technologies for advanced system modeling, virtual prototyping, and testing. The continued rapid development in these areas will only accelerate the pace of smart (that is, actively controlled) products.

An exciting area of future mechatronic system development in which control systems will play a significant role is the area of alternative energy production and consumption. Hybrid fuel automobiles and efficient wind power generation are two examples of systems that can benefit from mechatronic design methods. In fact, the mechatronic design philosophy can be effectively illustrated by the example of the evolution of the modern automobile [70]. Before the 1960s, the radio was the only significant electronic device in an automobile. Today, many automobiles have 30–60 microcontrollers, up to 100 electric motors, about 200 pounds of wiring, a multitude of sensors, and thousands of lines of software code. A modern automobile can no longer be classified as a strictly mechanical machine—it has been transformed into a comprehensive mechatronic system.

EXAMPLE 1.1   **Hybrid fuel vehicles**

Recent research and development has led to the next-generation **hybrid fuel automobile**, depicted in Figure 1.17. The hybrid fuel vehicle utilizes a conventional internal combustion engine in combination with a battery (or other energy storage device such as a fuel cell or flywheel) and an electric motor to provide a propulsion system capable of doubling the fuel economy over conventional automobiles. Although these hybrid vehicles will never be zero-emission vehicles (since they have internal combustion engines), they can reduce the level of harmful emissions by one-third to one-half, and with future improvements, these emissions may reduce even further. As stated earlier, the modern automobile requires many advanced control systems to

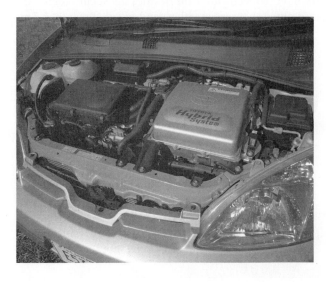

FIGURE 1.17
The hybrid fuel automobile can be viewed as a mechatronic system. (Used with permission of DOE/NREL. Credit: Warren Gretz.)

operate. The control systems must regulate the performance of the engine, including fuel–air mixtures, valve timing, transmissions, wheel traction control, antilock brakes, and electronically controlled suspensions, among many other functions. On the hybrid fuel vehicle, there are additional control functions that must be satisfied. Especially necessary is the control of power between the internal combustion engine and the electric motor, determining power storage needs and implementing the battery charging, and preparing the vehicle for low-emission start-ups. The overall effectiveness of the hybrid fuel vehicle depends on the combination of power units that are selected (e.g., battery versus fuel cell for power storage). Ultimately, however, the control strategy that integrates the various electrical and mechanical components into a viable transportation system strongly influences the acceptability of the hybrid fuel vehicle concept in the marketplace. ∎

The second example of a mechatronic system is the advanced wind power generation system.

EXAMPLE 1.2   **Wind power**

Many nations in the world today are faced with unstable energy supplies, often leading to rising fuel prices and energy shortages. Additionally, the negative effects of fossil fuel utilization on the quality of our air are well documented. Many nations have an imbalance in the supply and demand of energy, consuming more than they produce. To address this imbalance, many engineers are considering developing advanced systems to access other sources of energy, such as wind energy. In fact, wind energy is one of the fastest-growing forms of energy generation in the United States and in other locations around the world. A wind farm now in use in western Texas is illustrated in Figure 1.18.

In 2006, the installed global wind energy capacity was over 59,000 MW. In the United States, there was enough energy derived from wind to power over 2.5 million homes, according to the American Wind Energy Association. For the past 35 years, researchers have concentrated on developing technologies that work well in high wind areas (defined to be areas with a wind speed of at least 6.7 m/s at a height of 10 m).

FIGURE 1.18
Efficient wind power generation in west Texas. (Used with permission of DOE/NREL. Credit: Lower Colorado River Authority.)

Most of the easily accessible high wind sites in the United States are now utilized, and improved technology must be developed to make lower wind areas more cost effective. New developments are required in materials and aerodynamics so that longer turbine rotors can operate efficiently in the lower winds, and in a related problem, the towers that support the turbine must be made taller without increasing the overall costs. In addition, advanced controls will be required to achieve the level of efficiency required in the wind generation drive train. ∎

EXAMPLE 1.3   Embedded computers

Many contemporary control systems are **embedded control** systems [90]. Embedded control systems employ on-board special-purpose digital computers as integral components of the feedback loop. Fig. 1.19 illustrates a student-built rover constructed around the Compact RIO by National Instruments, Inc. that serves as the on-board embedded computer. In the rover design, the sensors include an optical encoder for measuring engine speed, a rate gyro and accelerometer to measure turns, and a Global Positioning System (GPS) unit to obtain position and velocity estimates of the vehicle. The actuators include two linear actuators to turn the front wheels and to brake and accelerate. The communications device permits the rover to stay in contact with the ground station.

Advances in sensors, actuators, and communication devices are leading to a new class of embedded control systems that are networked using wireless technology, thereby enabling spatially-distributed control. Embedded control system designers must be able to understand and work with various network protocols, diverse operating systems and programming languages. While the theory of systems and controls serves as the foundation for the modern control system design, the design process is

FIGURE 1.19   A rover using an embedded computer in the feedback loop. *(Photo by R.H. Bishop.)*

rapidly expanding into a multi-disciplinary enterprise encompassing multiple engineering areas, as well as information technology and computer science. ∎

Advances in alternate energy products, such as the hybrid automobile and the generation of efficient wind power generators, provide vivid examples of mechatronics development. There are numerous other examples of intelligent systems poised to enter our everyday life, including autonomous rovers, smart home appliances (e.g., dishwashers, vacuum cleaners, and microwave ovens), wireless network-enabled devices, "human-friendly machines" [81] that perform robot-assisted surgery, and implantable sensors and actuators.

1.7 THE FUTURE EVOLUTION OF CONTROL SYSTEMS

The continuing goal of control systems is to provide extensive flexibility and a high level of autonomy. Two system concepts are approaching this goal by different evolutionary pathways, as illustrated in Figure 1.20. Today's industrial robot is perceived as quite autonomous—once it is programmed, further intervention is not normally required. Because of sensory limitations, these robotic systems have limited flexibility in adapting to work environment changes; improving perception is the motivation of computer vision research. The control system is very adaptable, but it relies on human supervision. Advanced robotic systems are striving for task adaptability through enhanced sensory feedback. Research areas concentrating on artificial intelligence, sensor integration, computer vision, and off-line CAD/CAM programming will make systems more universal and economical. Control systems are moving toward autonomous operation as an enhancement to human control. Research in supervisory control, human–machine interface methods, and computer database management are intended to reduce operator burden and improve operator efficiency. Many research activities are common to robotics

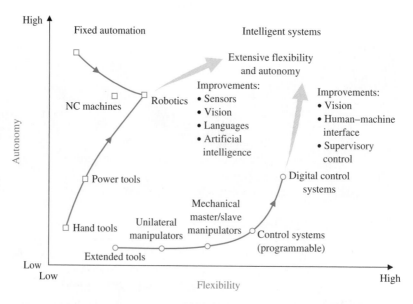

FIGURE 1.20
Future evolution of control systems and robotics.

and control systems and are aimed at reducing implementation cost and expanding the realm of application. These include improved communication methods and advanced programming languages.

The easing of human labor by technology, a process that began in prehistory, is entering a new stage. The acceleration in the pace of technological innovation inaugurated by the Industrial Revolution has until recently resulted mainly in the displacement of human muscle power from the tasks of production. The current revolution in computer technology is causing an equally momentous social change, the expansion of information gathering and information processing as computers extend the reach of the human brain [16].

Control systems are used to achieve (1) increased productivity and (2) improved performance of a device or system. Automation is used to improve productivity and obtain high-quality products. Automation is the automatic operation or control of a process, device, or system. We use automatic control of machines and processes to produce a product reliably and with high precision [28]. With the demand for flexible, custom production, a need for flexible automation and robotics is growing [17, 25].

The theory, practice, and application of automatic control is a large, exciting, and extremely useful engineering discipline. One can readily understand the motivation for a study of modern control systems.

1.8 DESIGN EXAMPLES

In this section we present illustrative design examples. This is a pattern that we will follow in all subsequent chapters. Each chapter will contain a number of interesting examples in a special section entitled Design Examples meant to highlight the main topics of the chapter. At least one example among those presented in the Design Example section will be a more detailed problem and solution that demonstrates one of more of the steps in the design process shown in Figure 1.15. In the first example presented here, a rotating disk speed control illustrates the concept of open-loop and closed-loop feedback control. The second example is an insulin delivery control system in which we determine the design goals, the variables to control, and a preliminary closed-loop system configuration.

EXAMPLE 1.4   Rotating disk speed control

Many modern devices employ a rotating disk held at a constant speed. For example, a CD player requires a constant speed of rotation in spite of motor wear and variation and other component changes. Our goal is to design a system for rotating disk speed control that will ensure that the actual speed of rotation is within a specified percentage of the desired speed [43, 46]. We will consider a system without feedback and a system with feedback.

To obtain disk rotation, we will select a DC motor as the actuator because it provides a speed proportional to the applied motor voltage. For the input voltage to the motor, we will select an amplifier that can provide the required power.

The open-loop system (without feedback) is shown in Figure 1.21(a). This system uses a battery source to provide a voltage that is proportional to the desired speed. This

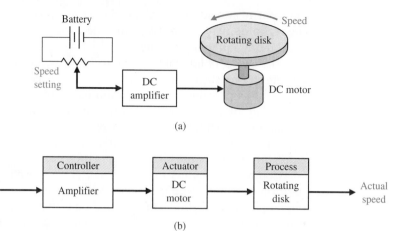

(a)

FIGURE 1.21
(a) Open-loop
(without feedback)
control of the speed
of a rotating disk.
(b) Block diagram
model.

Desired speed (voltage) → | Controller — Amplifier | → | Actuator — DC motor | → | Process — Rotating disk | → Actual speed

(b)

voltage is amplified and applied to the motor. The block diagram of the open-loop system identifying the controller, actuator, and process is shown in Figure 1.21(b).

To obtain a feedback system, we need to select a sensor. One useful sensor is a tachometer that provides an output voltage proportional to the speed of its shaft. Thus the closed-loop feedback system takes the form shown in Fig. 1.22(a). The block diagram model of the feedback system is shown in Fig. 1.22(b). The error voltage is generated by the difference between the input voltage and the tachometer voltage.

We expect the feedback system of Figure 1.22 to be superior to the open-loop system of Figure 1.21 because the feedback system will respond to errors and act to

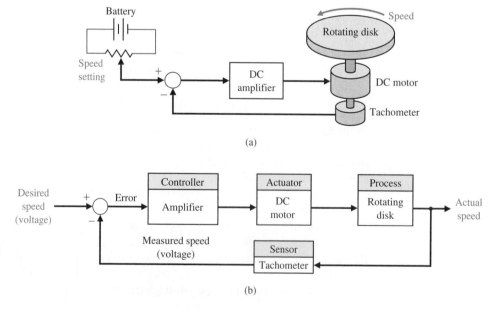

(a)

FIGURE 1.22
(a) Closed-loop
control of the speed
of a rotating disk.
(b) Block diagram
model.

(b)

reduce them. With precision components, we could expect to reduce the error of the feedback system to one-hundredth of the error of the open-loop system. ■

EXAMPLE 1.5   Insulin delivery control system

Control systems have been utilized in the biomedical field to create implanted automatic drug-delivery systems to patients [29–31]. Automatic systems can be used to regulate blood pressure, blood sugar level, and heart rate. A common application of control engineering is in the field of open-loop system drug delivery, in which mathematical models of the dose–effect relationship of the drugs are used. A drug-delivery system implanted in the body uses an open-loop system, since miniaturized glucose sensors are not yet available. The best solutions rely on individually programmable, pocket-sized insulin pumps that can deliver insulin according to a preset time history. More complicated systems will use closed-loop control for the measured blood glucose levels.

The blood glucose and insulin concentrations for a healthy person are shown in Figure 1.23. The system must provide the insulin from a reservoir implanted within the diabetic person. Therefore, the control goal is:

Control Goal
>   Design a system to regulate the blood sugar concentration of a diabetic by controlled dispensing of insulin.

Referring to Figure 1.23, the next step in the design process is to define the variable to be controlled. Associated with the control goal we can define the variable to be controlled to be:

Variable to Be Controlled
>   Blood glucose concentration

In subsequent chapters, we will have the tools to quantitatively describe the control design specifications using a variety of steady-state performance specifications and transient response specifications, both in the time-domain and in the frequency domain. At this point, the control design specifications will be qualitative and imprecise. In that regard, for the problem at hand, we can state the design specification as:

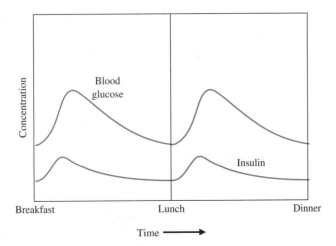

FIGURE 1.23
The blood glucose and insulin levels for a healthy person.

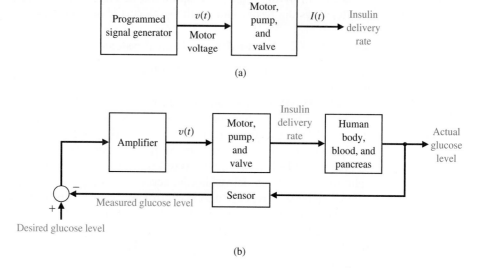

FIGURE 1.24
(a) Open-loop
(without feedback)
control and
(b) closed-loop
control of blood
glucose.

Control Design Specifications

    Provide a blood glucose level for the diabetic that closely approximates (tracks) the glucose level of a healthy person.

Given the design goals, variables to be controlled, and control design specifications, we can now propose a preliminary system configuration. An open-loop system would use a preprogrammed signal generator and miniature motor pump to regulate the insulin delivery rate as shown in Figure 1.24(a). The feedback control system would use a sensor to measure the actual glucose level and compare that level with the desired level, thus turning the motor pump on when it is required, as shown in Figure 1.24(b). ■

1.9 SEQUENTIAL DESIGN EXAMPLE: DISK DRIVE READ SYSTEM

This design example, identified by the arrow icon, will be considered sequentially in each chapter. We will use the design process of Figure 1.15 in each chapter to identify the steps that we are accomplishing. For example, in Chapter 1 we (1) identify the control goal, (2) identify the variables to control, (3) write the initial specifications for the variables, and (4) establish the preliminary system configuration.

    Information can be readily and efficiently stored on magnetic disks. Disk drives are used in notebook computers and larger computers of all sizes and are essentially all standardized as defined by ANSI standards [54, 69]. Worldwide sales of disk drives are greater than 250 million units [55, 68]. In the past, disk drive designers have concentrated on increasing data density and data access times. Beginning in the early 1990s, disk drive densities increased at rates of over 60 percent per year and very recently, these rates exceed 100 percent per year. Figure 1.25 shows the disk drive density trends. Designers are now considering employing disk drives to perform tasks historically delegated to central processing units (CPUs), thereby leading to improvements in the computing environment [69]. Three areas of "intelligence" under investigation

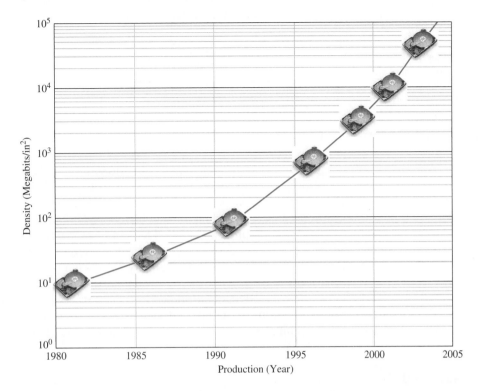

include off-line error recovery, disk drive failure warnings, and storing data across multiple disk drives. Consider the basic diagram of a disk drive shown in Fig. 1.26. The goal of the disk drive reader device is to position the reader head to read the data stored on a track on the disk. The variable to accurately control is the position of the reader head (mounted on a slider device). The disk rotates at a speed between 1800 and 7200 rpm, and the head "flies" above the disk at a distance of less than 100 nm. The initial specification for the position accuracy is 1 μm. Furthermore, we plan to be able to move the head from track a to track b within 50 ms, if possible. Thus, we establish an initial system configuration as shown in Figure 1.27. This proposed closed-loop system uses a motor to actuate (move) the arm to the desired location on the disk. We will consider the design of the disk drive further in Chapter 2.

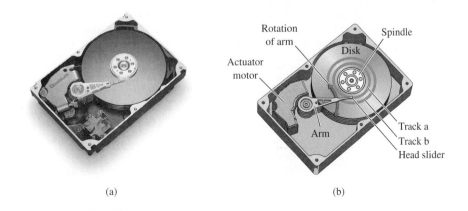

FIGURE 1.26
(a) A disk drive
© 1999 Quantum
Corporation. All
rights reserved.
(b) Diagram of a
disk drive.

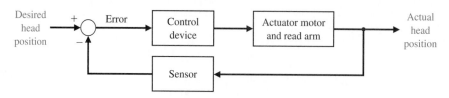

FIGURE 1.27
Closed-loop control system for disk drive.

1.10 SUMMARY

In this chapter, we discussed open- and closed-loop feedback control systems. Examples of control systems through the course of history were presented to motivate and connect the subject to the past. In terms of contemporary issues, key areas of application were discussed, including humanoid robots, unmanned aerial vehicles, wind energy, hybrid automobiles, and embedded control. The central role of controls in mechatronics was discussed. Mechatronics is the synergistic integration of mechanical, electrical, and computer systems. Finally, the design process was presented in a structured form and included the following steps: the establishment of goals and variables to be controlled, definition of specifications, system definition, modeling, and analysis. The iterative nature of design allows us to handle the design gap effectively while accomplishing necessary trade-offs in complexity, performance, and cost.

EXERCISES

Exercises are straightforward applications of the concepts of the chapter.

The following systems can be described by a block diagram showing the cause–effect relationship and the feedback (if present). Identify the function of each block and the desired input variable, output variable, and measured variable. Use Figure 1.3 as a model where appropriate.

E1.1   A precise optical signal source can control the output power level to within 1 percent [32]. A laser is controlled by an input current to yield the power output. A microprocessor controls the input current to the laser. The microprocessor compares the desired power level with a measured signal proportional to the laser power output obtained from a sensor. Complete the block diagram representing this closed-loop control system shown in Figure E1.1, identifying

the output, input, and measured variables and the control device.

E1.2   An automobile driver uses a control system to maintain the speed of the car at a prescribed level. Sketch a block diagram to illustrate this feedback system.

E1.3   Fly-fishing is a sport that challenges the person to cast a small feathery fly using a light rod and line. The goal is to place the fly accurately and lightly on the distant surface of the stream [65]. Describe the fly-casting process and a model of this process.

E1.4   An autofocus camera will adjust the distance of the lens from the film by using a beam of infrared or ultrasound to determine the distance to the subject [45]. Sketch a block diagram of this open-loop control system, and briefly explain its operation.

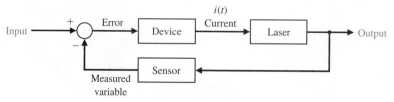

FIGURE E1.1   Partial block diagram of an optical source.

E1.5 Because a sailboat can't sail directly into the wind, and traveling straight downwind is usually slow, the shortest sailing distance is rarely a straight line. Thus sailboats tack upwind—the familiar zigzag course—and jibe downwind. A tactician's decision of when to tack and where to go can determine the outcome of a race.

Describe the process of tacking a sailboat as the wind shifts direction. Sketch a block diagram depicting this process.

E1.6 Modern automated highways are being implemented around the world. Consider two highway lanes merging into a single lane. Describe a feedback control system carried on the automobile trailing the lead automobile that ensures that the vehicles merge with a prescribed gap between the two vehicles.

E1.7 Describe the block diagram of the speed control system of a motorcycle with a human driver.

E1.8 Describe the process of human biofeedback used to regulate factors such as pain or body temperature. Biofeedback is a technique whereby a human can, with some success, consciously regulate pulse, reaction to pain, and body temperature.

E1.9 Future advanced commercial aircraft will be E-enabled. This will allow the aircraft to take advantage of continuing improvements in computer power and network growth. Aircraft can continuously communicate their location, speed, and critical health parameters to ground controllers, and gather and transmit local meteorological data. Sketch a block diagram showing how the meteorological data from multiple aircraft can be transmitted to the ground, combined using ground-based powerful networked computers to create an accurate weather situational awareness, and then transmitted back to the aircraft for optimal routing.

E1.10 Unmanned aerial vehicles (UAVs) are being developed to operate in the air autonomously for long

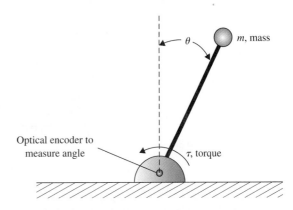

FIGURE E1.11 Inverted pendulum control.

periods of time (see Section 1.3). By autonomous, we mean that there is no interaction with human ground controllers. Sketch a block diagram of an autonomous UAV that is tasked for crop monitoring using aerial photography. The UAV must photograph and transmit the entire land area by flying a pre-specified trajectory as accurately as possible.

E1.11 Consider the inverted pendulum shown in Figure E1.11. Sketch the block diagram of a feedback control system using Figure 1.3 as the model. Identify the process, sensor, actuator, and controller. The objective is keep the pendulum in the upright position, that is to keep $\theta = 0$, in the presence of disturbances.

E1.12 Describe the block diagram of a person playing a video game. Suppose that the input device is a joystick and the game is being played on a desktop computer. Use Figure 1.3 as a model of the block diagram.

PROBLEMS

Problems require extending the concepts of this chapter to new situations.

The following systems may be described by a block diagram showing the cause–effect relationship and the feedback (if present). Each block should describe its function. Use Figure 1.3 as a model where appropriate.

P1.1 Many luxury automobiles have thermostatically controlled air-conditioning systems for the comfort of the passengers. Sketch a block diagram of an air-conditioning system where the driver sets the desired interior temperature on a dashboard panel. Identify the function of each element of the thermostatically controlled cooling system.

P1.2 In the past, control systems used a human operator as part of a closed-loop control system. Sketch the

block diagram of the valve control system shown in Figure P1.2.

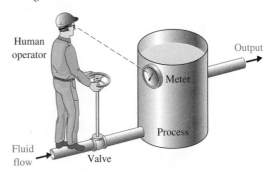

FIGURE P1.2 Fluid-flow control.

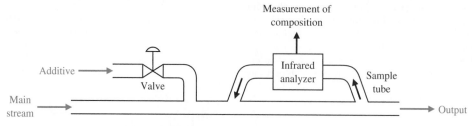

FIGURE P1.3   Chemical composition control.

P1.3   In a chemical process control system, it is valuable to control the chemical composition of the product. To do so, a measurement of the composition can be obtained by using an infrared stream analyzer, as shown in Figure P1.3. The valve on the additive stream may be controlled. Complete the control feedback loop, and sketch a block diagram describing the operation of the control loop.

P1.4   The accurate control of a nuclear reactor is important for power system generators. Assuming the number of neutrons present is proportional to the power level, an ionization chamber is used to measure the power level. The current i_o is proportional to the power level. The position of the graphite control rods moderates the power level. Complete the control system of the nuclear reactor shown in Figure P1.4 and sketch the block diagram describing the operation of the feedback control loop.

P1.5   A light-seeking control system, used to track the sun, is shown in Figure P1.5. The output shaft, driven by the motor through a worm reduction gear, has a bracket attached on which are mounted two photocells. Complete the closed-loop system so that the system follows the light source.

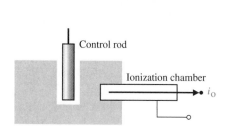

FIGURE P1.4   Nuclear reactor control.

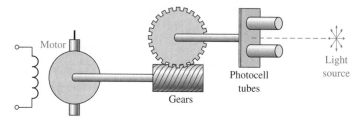

FIGURE P1.5   A photocell is mounted in each tube. The light reaching each cell is the same in both only when the light source is exactly in the middle as shown.

P1.6   Feedback systems do not always involve negative feedback. Economic inflation, which is evidenced by continually rising prices, is a **positive feedback** system. A positive feedback control system, as shown in Figure P1.6, adds the feedback signal to the input signal, and the resulting signal is used as the input to the process. A simple model of the price–wage inflationary spiral is shown in Figure P1.6. Add additional feedback loops, such as legislative control or control of the tax rate, to stabilize the system. It is assumed that an increase in workers' salaries, after some time delay, results in an increase in prices. Under what conditions could prices be stabilized by falsifying or delaying the availability of cost-of-living data? How would a national wage and price economic guideline program affect the feedback system?

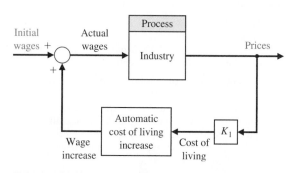

FIGURE P1.6   Positive feedback.

P1.7   The story is told about the sergeant who stopped at the jewelry store every morning at nine o'clock and compared and reset his watch with the chronometer in the window. Finally, one day the sergeant went into the store and complimented the owner on the accuracy of the chronometer.

"Is it set according to time signals from Arlington?" asked the sergeant.

"No," said the owner, "I set it by the five o'clock cannon fired from the fort each afternoon. Tell me, Sergeant, why do you stop every day and check your watch?"

The sergeant replied, "I'm the gunner at the fort!"

Is the feedback prevalent in this case positive or negative? The jeweler's chronometer loses two minutes each 24-hour period and the sergeant's watch loses three minutes during each eight hours. What is the net time error of the cannon at the fort after 12 days?

P1.8   The student–teacher learning process is inherently a feedback process intended to reduce the system error

P1.10   The role of air traffic control systems is increasing as airplane traffic increases at busy airports. Engineers are developing air traffic control systems and collision avoidance systems using the Global Positioning System (GPS) navigation satellites [34, 61]. GPS allows each aircraft to know its position in the airspace landing corridor very precisely. Sketch a block diagram depicting how an air traffic controller might use GPS for aircraft collision avoidance.

P1.11   Automatic control of water level using a float level was used in the Middle East for a water clock [1, 11]. The water clock (Figure P1.11) was used from sometime before Christ until the 17th century. Discuss the operation of the water clock, and establish how the float provides a feedback control that maintains the accuracy of the clock. Sketch a block diagram of the feedback system.

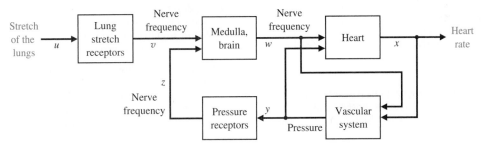

FIGURE P1.9   Heart-rate control.

to a minimum. With the aid of Figure 1.3, construct a feedback model of the learning process and identify each block of the system.

P1.9   Models of physiological control systems are valuable aids to the medical profession. A model of the heart-rate control system is shown in Figure P1.9 [23, 24, 51]. This model includes the processing of the nerve signals by the brain. The heart-rate control system is, in fact, a multivariable system, and the variables x, y, w, v, z, and u are vector variables. In other words, the variable x represents many heart variables $x_1, x_2, \ldots, x_n$. Examine the model of the heart-rate control system and add or delete blocks, if necessary. Determine a control system model of one of the following physiological control systems:

1. Respiratory control system
2. Adrenaline control system
3. Human arm control system
4. Eye control system
5. Pancreas and the blood-sugar-level control system
6. Circulatory system

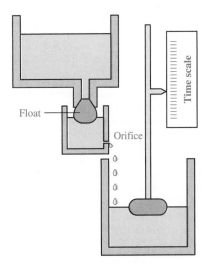

FIGURE P1.11   Water clock. (From Newton, Gould, and Kaiser, *Analytical Design of Linear Feedback Controls*. Wiley, New York, 1957, with permission.)

P1.12   An automatic turning gear for windmills was invented by Meikle in about 1750 [1, 11]. The fantail gear shown in Figure P1.12 automatically turns the windmill into the wind. The fantail windmill at right angle to the mainsail is used to turn the turret. The gear ratio is of the order of 3000 to 1. Discuss the operation of the windmill, and establish the feedback operation that maintains the main sails into the wind.

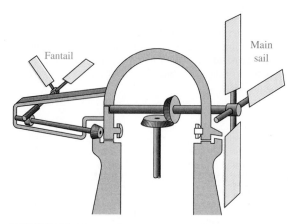

FIGURE P1.12   Automatic turning gear for windmills. (From Newton, Gould, and Kaiser, *Analytical Design of Linear Feedback Controls*. Wiley, New York, 1957, with permission.)

P1.13   A common example of a two-input control system is a home shower with separate valves for hot and cold water. The objective is to obtain (1) a desired temperature of the shower water and (2) a desired flow of water. Sketch a block diagram of the closed-loop control system.

P1.14   Adam Smith (1723–1790) discussed the issue of free competition between the participants of an economy in his book *Wealth of Nations*. It may be said that Smith employed social feedback mechanisms to explain his theories [44]. Smith suggests that (1) the available workers as a whole compare the various possible employments and enter that one offering the greatest rewards, and (2) in any employment the rewards diminish as the number of competing workers rises. Let r = total of rewards averaged over all trades, c = total of rewards in a particular trade, and q = influx of workers into the specific trade. Sketch a feedback system to represent this system.

P1.15   Small computers are used in automobiles to control emissions and obtain improved gas mileage. A computer-controlled fuel injection system that automatically adjusts the fuel–air mixture ratio could improve gas mileage and reduce unwanted polluting emissions significantly. Sketch a block diagram for such a system for an automobile.

P1.16   All humans have experienced a fever associated with an illness. A fever is related to the changing of the control input in the body's thermostat. This thermostat, within the brain, normally regulates temperature near 98°F in spite of external temperatures ranging from 0° to 100°F or more. For a fever, the input, or desired, temperature is increased. Even to many scientists, it often comes as a surprise to learn that fever does not indicate something wrong with body temperature control but rather well-contrived regulation at an elevated level of desired input. Sketch a block diagram of the temperature control system and explain how aspirin will lower a fever.

P1.17   Baseball players use feedback to judge a fly ball and to hit a pitch [35]. Describe a method used by a batter to judge the location of a pitch so that he can have the bat in the proper position to hit the ball.

P1.18   A cutaway view of a commonly used pressure regulator is shown in Figure P1.18. The desired pressure is set by turning a calibrated screw. This compresses the spring and sets up a force that opposes the upward motion of the diaphragm. The bottom side of the diaphragm is exposed to the water pressure that

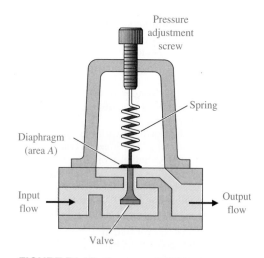

FIGURE P1.18   Pressure regulator.

is to be controlled. Thus the motion of the diaphragm is an indication of the pressure difference between the desired and the actual pressures. It acts like a comparator. The valve is connected to the diaphragm and moves according to the pressure difference until it reaches a position in which the difference is zero. Sketch a block diagram showing the control system with the output pressure as the regulated variable.

P1.19 Ichiro Masaki of General Motors has patented a system that automatically adjusts a car's speed to keep a safe distance from vehicles in front. Using a video camera, the system detects and stores a reference image of the car in front. It then compares this image with a stream of incoming live images as the two cars move down the highway and calculates the distance. Masaki suggests that the system could control steering as well as speed, allowing drivers to lock on to the car ahead and get a "computerized tow." Sketch a block diagram for the control system.

P1.20 A high-performance race car with an adjustable wing (airfoil) is shown in Figure P1.20. Develop a block diagram describing the ability of the airfoil to keep a constant road adhesion between the car's tires and the race track surface. Why is it important to maintain good road adhesion?

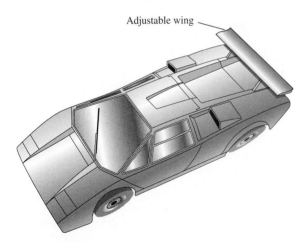

FIGURE P1.20 A high-performance race car with an adjustable wing.

P1.21 The potential of employing two or more helicopters for transporting payloads that are too heavy for a single helicopter is a well-addressed issue in the civil and military rotorcraft design arenas [38]. Overall requirements can be satisfied more efficiently with a smaller aircraft by using multilift for

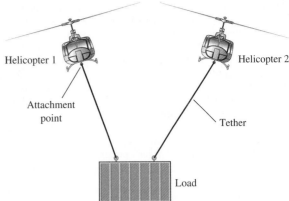

FIGURE P1.21 Two helicopters used to lift and move a large load.

infrequent peak demands. Hence the principal motivation for using multilift can be attributed to the promise of obtaining increased productivity without having to manufacture larger and more expensive helicopters. A specific case of a multilift arrangement where two helicopters jointly transport payloads has been named **twin lift**. Figure P1.21 shows a typical "two-point pendant" twin lift configuration in the lateral/vertical plane.

Develop the block diagram describing the pilots' action, the position of each helicopter, and the position of the load.

P1.22. Engineers want to design a control system that will allow a building or other structure to react to the force of an earthquake much as a human would. The structure would yield to the force, but only so much, before developing strength to push back [50]. Develop a block diagram of a control system to reduce the effect of an earthquake force.

P1.23 Engineers at the Science University of Tokyo are developing a robot with a humanlike face [56]. The robot can display facial expressions, so that it can work cooperatively with human workers. Sketch a block diagram for a facial expression control system of your own design.

P1.24 An innovation for an intermittent automobile windshield wiper is the concept of adjusting its wiping cycle according to the intensity of the rain [60]. Sketch a block diagram of the wiper control system.

P1.25 In the past 40 years, over 20,000 metric tons of hardware have been placed in Earth's orbit. During the same time span, over 15,000 metric tons of hardware returned to Earth. The objects remaining in

Earth's orbit range in size from large operational spacecraft to tiny flecks of paint. There are about 150,000 objects in Earth's orbit 1 cm or larger in size. About 10,000 of the space objects are currently tracked from groundstations on the Earth. Space traffic control [67] is becoming an important issue, especially for commercial satellite companies that plan to "fly" their satellites through orbit altitudes where other satellites are operating, and through areas where high concentrations of space debris may exist. Sketch a block diagram of a space traffic control system that commercial companies might use to keep their satellites safe from collisions while operating in space.

P1.26 NASA is developing a compact rover designed to transmit data from the surface of an asteroid back to Earth, as illustrated in Figure P1.26. The rover will use a camera to take panoramic shots of the asteroid surface. The rover can position itself so that the camera can be pointed straight down at the surface or straight up at the sky. Sketch a block diagram illustrating how the microrover can be positioned to point the camera in the desired direction. Assume that the pointing commands are relayed from the Earth to the microrover and that the position of the camera is measured and relayed back to Earth.

P1.27 A direct methanol fuel cell is an electrochemical device that converts a methanol water solution to electricity [84]. Like rechargeable batteries, fuel cells directly convert chemicals to energy; they are very

FIGURE P1.26   Microrover designed to explore an asteroid. (Photo courtesy of NASA.)

often compared to batteries, specifically rechargeable batteries. However, one significant difference between rechargeable batteries and direct methanol fuel cells is that, by adding more methanol water solution, the fuel cells recharge instantly. Sketch a block diagram of the direct methanol fuel cell recharging system that uses feedback (refer to Figure 1.3) to continuously monitor and recharge the fuel cell.

ADVANCED PROBLEMS

Advanced problems represent problems of increasing complexity.

AP1.1 The development of robotic microsurgery devices will have major implications on delicate eye and brain surgical procedures. The microsurgery devices employ feedback control to reduce the effects of the surgeon's muscle tremors. Precision movements by an articulated robotic arm can greatly help a surgeon by providing a carefully controlled hand. One such device is shown in Figure AP1.1. The microsurgical devices have been evaluated in clinical procedures and are now being commercialized. Sketch a block diagram of the surgical process with a microsurgical device in the loop being operated by a surgeon. Assume that the position of the end-effector on the microsurgical device can be measured and is available for feedback.

AP1.2 Advanced wind energy systems are being installed in many locations throughout the world as a way for nations to deal with rising fuel prices and energy shortages, and to reduce the negative effects of fossil fuel utilization on the quality of the air (refer to Example 1.2 in Section 1.6). The modern windmill can be viewed as a mechatronic system. Consider Figure 1.16, which illustrates the key elements of mechatronic systems. Using Figure 1.16 as a guide, think about how an advanced wind energy system would be designed as a mechatronic system. List the various components of the wind energy system and associate each component with one of the five elements of a mechatronic system: physical system modeling, signals and systems, computers and logic systems, software and data acquisition, and sensors and actuators.

FIGURE AP1.1 Microsurgery robotic manipulator. (Photo courtesy of NASA.)

AP1.3 Many modern luxury automobiles have an autopark option. This feature will parallel park an automobile without driver intervention. Figure AP1.3 illustrates the parallel parking scenario. Using Figure 1.3 as a model, sketch a block diagram of the automated parallel parking feedback control system. In your own words, describe the control problem and the challenges facing the designers of the control system.

AP1.4 Adaptive optics has applications to a wide variety of key control problems, including imaging of the human retina and large-scale, ground-based astronomical observations [91]. In both cases, the approach is to use a wavefront sensor to measure distortions in the incoming light and to actively control and compensate to the errors induced by the distortions. Consider the case of an extremely large ground-based optical telescope, possibly an optical telescope up to 100 meters in diameter. The telescope components include deformable mirrors actuated by micro-electro-mechanical (MEMS) devices and sensors to measure the distortion of the incoming light as it passes through the turbulent and uncertain atmosphere of Earth.

There is at least one major technological barrier to constructing a 100-m optical telescope. The numerical computations associated with the control and compensation of the extremely large optical telescope can be on the order of 10^{10} calculations each 1.5 ms. To date, this computational power is unachievable. If we assume that the computational capability will ultimately be available, then one can consider the design of a feedback control system that uses the available computational power. We can consider many control issues associated with the large-scale optical telescope. Some of the controls problems that might be considered include controlling the pointing of the main dish, controlling the individual deformable mirrors, and attenuating the deformation of the dish due to changes in outside temperature.

Employing Figure 1.3 as a model for the block diagram, describe a closed-loop feedback control system to control one of the deformable mirrors to compensate for the distortions in the incoming light. Figure AP1.4 shows a diagram of the telescope with a single deformable mirror. Suppose that the mirror has an associated MEMS actuator that can be used to vary the orientation. Also, assume that the wavefront sensor and associated algorithms provide the desired configuration of the deformable mirror to the feedback control system.

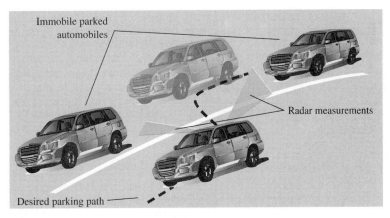

FIGURE AP1.3 Automated parallel parking of an automobile.

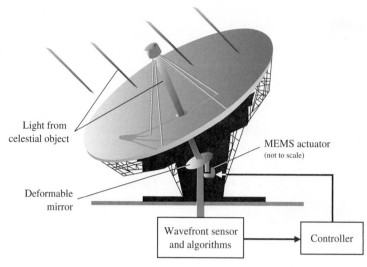

FIGURE AP1.4   Extremely large optical telescope with deformable mirrors for atmosphere compensation.

DESIGN PROBLEMS

Design problems emphasize the design task. Continuous design problems (CDP) build upon a design problem from chapter to chapter.

CDP1.1   Increasingly stringent requirements of modern, high-precision machinery are placing increasing demands on slide systems [57]. The typical goal is to accurately control the desired path of the table shown in Figure CDP1.1. Sketch a block diagram model of a

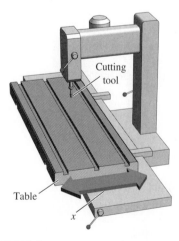

FIGURE CDP1.1   Machine tool with table.

feedback system to achieve the desired goal. The table can move in the x direction as shown.

DP1.1   The road and vehicle noise that invade an automobile's cabin hastens occupant fatigue [66]. Design the block diagram of an "antinoise" feedback system that will reduce the effect of unwanted noises. Indicate the device within each block.

DP1.2   Many cars are fitted with cruise control that, at the press of a button, automatically maintains a set speed. In this way, the driver can cruise at a speed limit or economic speed without continually checking the speedometer. Design a feedback-control in block diagram form for a cruise control system.

DP1.3   As part of the automation of a dairy farm, the automation of cow milking is under study [37]. Design a milking machine that can milk cows four or five times a day at the cow's demand. Sketch a block diagram and indicate the devices in each block.

DP1.4   A large, braced robot arm for welding large structures is shown in Figure DP1.4. Sketch the block diagram of a closed-loop feedback control system for accurately controlling the location of the weld tip.

DP1.5   Vehicle traction control, which includes antiskid braking and antispin acceleration, can enhance vehicle performance and handling. The objective of this control is to maximize tire traction by preventing locked brakes as well as tire spinning during acceleration.

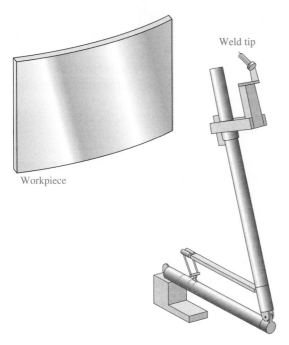

Weld tip

Workpiece

FIGURE DP1.4 Robot welder.

Wheel slip, the difference between the vehicle speed and the wheel speed, is chosen as the controlled variable because of its strong influence on the tractive force between the tire and the road [19]. The adhesion coefficient between the wheel and the road reaches a maximum at a low slip. Develop a block diagram model of one wheel of a traction control system.

DP1.6 The Hubble space telescope was repaired and modified in space on several occasions [47, 49, 52]. One challenging problem with controlling the Hubble is damping the jitter that vibrates the spacecraft each time it passes into or out of the Earth's shadow. The worst vibration has a period of about 20 seconds, or a frequency of 0.05 hertz. Design a feedback system that will reduce the vibrations of the Hubble space telescope.

DP1.7 A challenging application of control design is the use of nanorobots in medicine. Nanorobots will require onboard computing capability, and very tiny sensors and actuators. Fortunately, advances in bio-molecular computing, bio-sensors, and actuators are promising to enable medical nanorobots to emerge within the next decade [92]. Many interesting medical applications will benefit from nanorobotics. For example, one use might be to use the robotic devices to precisely deliver anti-HIV drugs or to combat cancer by targeted delivering of chemotherapy.

At the present time, we cannot construct practical nanorobots, but we can consider the control design process that would enable the eventual development and installation of these tiny devices in the medical field. Consider the problem of designing a nanorobot to deliver a cancer drug to a specific location within the human body. The target site might be the location of a tumor, for example. Using the control design process illustrated in Figure 1.15, suggest one or more control goals that might guide the design process. Recommend the variables that should be controlled and provide a list of reasonable specifications for those variables.

FIGURE DP1.7 An artist illustration of a nanorobot interacting with human blood cells.

TERMS AND CONCEPTS

Automation The control of a process by automatic means.

Closed-loop feedback control system A system that uses a measurement of the output and compares it with the desired output to control the process.

Complexity of design The intricate pattern of interwoven parts and knowledge required.

Control system An interconnection of components forming a system configuration that will provide a desired response.

Design The process of conceiving or inventing the forms, parts, and details of a system to achieve a specified purpose.

Design gap   A gap between the complex physical system and the design model intrinsic to the progression from the initial concept to the final product.

Engineering design   The process of designing a technical system.

Feedback signal   A measure of the output of the system used for feedback to control the system.

Flyball governor   A mechanical device for controlling the speed of a steam engine.

Hybrid fuel automobile   An automobile that uses a conventional internal combustion engine in combination with an energy storage device to provide a propulsion system.

Mechatronics   The synergistic integration of mechanical, electrical, and computer systems.

Multivariable control system   A system with more than one input variable or more than one output variable.

Negative feedback   An output signal fed back so that it subtracts from the input signal.

Open-loop control system   A system that uses a device to control the process without using feedback. Thus the output has no effect upon the signal to the process.

Optimization   The adjustment of the parameters to achieve the most favorable or advantageous design.

Plant   *See* Process.

Positive feedback   An output signal fed back so that it adds to the input signal.

Process   The device, plant, or system under control.

Productivity   The ratio of physical output to physical input of an industrial process.

Risk   Uncertainties embodied in the unintended consequences of a design.

Robot   Programmable computers integrated with a manipulator. A reprogrammable, multifunctional manipulator used for a variety of tasks.

Specifications   Statements that explicitly state what the device or product is to be and to do. A set of prescribed performance criteria.

Synthesis   The process by which new physical configurations are created. The combining of separate elements or devices to form a coherent whole.

System   An interconnection of elements and devices for a desired purpose.

Trade-off   The result of making a judgment about how to compromise between conflicting criteria.

Mathematical Models of Systems

PREVIEW

Mathematical models of physical systems are key elements in the design and analysis of control systems. The dynamic behavior is generally described by ordinary differential equations. We will consider a wide range of systems, including mechanical, hydraulic, and electrical. Since most physical systems are nonlinear, we will discuss linearization approximations, which allow us to use Laplace transform methods. We will then proceed to obtain the input–output relationship for components and subsystems in the form of transfer functions. The transfer function blocks can be organized into block diagrams or signal-flow graphs to graphically depict the interconnections. Block diagrams (and signal-flow graphs) are very convenient and natural tools for designing and analyzing complicated control systems. We conclude the chapter by developing transfer function models for the various components of the Sequential Design Example: Disk Drive Read System.

DESIRED OUTCOMES

Upon completion of Chapter 2, students should:

❑   Recognize that differential equations can describe the dynamic behavior of physical systems.

❑   Be able to utilize linearization approximations through the use of Taylor series expansions.

❑   Understand the application of Laplace transforms and their role in obtaining transfer functions.

❑   Be aware of block diagrams (and signal-flow graphs) and their role in analyzing control systems.

❑   Understand the important role of modeling in the control system design process.

2.1 INTRODUCTION

To understand and control complex systems, one must obtain quantitative **mathematical models** of these systems. It is necessary therefore to analyze the relationships between the system variables and to obtain a mathematical model. Because the systems under consideration are dynamic in nature, the descriptive equations are usually **differential equations**. Furthermore, if these equations can be **linearized**, then the **Laplace transform** can be used to simplify the method of solution. In practice, the complexity of systems and our ignorance of all the relevant factors necessitate the introduction of **assumptions** concerning the system operation. Therefore we will often find it useful to consider the physical system, express any necessary assumptions, and linearize the system. Then, by using the physical laws describing the linear equivalent system, we can obtain a set of linear differential equations. Finally, using mathematical tools, such as the Laplace transform, we obtain a solution describing the operation of the system. In summary, the approach to dynamic system modeling can be listed as follows:

1. Define the system and its components.

2. Formulate the mathematical model and fundamental necessary assumptions based on basic principles.

3. Obtain the differential equations representing the mathematical model.

4. Solve the equations for the desired output variables.

5. Examine the solutions and the assumptions.

6. If necessary, reanalyze or redesign the system.

2.2 DIFFERENTIAL EQUATIONS OF PHYSICAL SYSTEMS

The differential equations describing the dynamic performance of a physical system are obtained by utilizing the physical laws of the process [1–3]. This approach applies equally well to mechanical [1], electrical [3], fluid, and thermodynamic systems [4]. Consider the torsional spring–mass system in Figure 2.1 with applied torque $T_a(t)$. Assume the torsional spring element is massless. Suppose we want to measure the torque $T_s(t)$ transmitted to the mass m. Since the spring is massless, the sum of the torques acting on the spring itself must be zero, or

$$T_a(t) - T_s(t) = 0,$$

which implies that $T_s(t) = T_a(t)$. We see immediately that the external torque $T_a(t)$ applied at the end of the spring is transmitted *through* the torsional spring. Because of this, we refer to the torque as a **through-variable**. In a similar manner, the angular rate difference associated with the torsional spring element is

$$\omega(t) = \omega_s(t) - \omega_a(t).$$

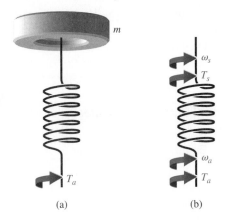

FIGURE 2.1
(a) Torsional spring–mass system. (b) Spring element.

(a)   (b)

Thus, the angular rate difference is measured across the torsional spring element and is referred to as an **across-variable**. These same types of arguments can be made for most common physical variables (such as force, current, volume, flow rate, etc.). A more complete discussion on through- and across-variables can be found in [30, 33]. A summary of the through- and across-variables of dynamic systems is given in Table 2.1 [5]. Information concerning the International System (SI) of units associated with the various variables discussed in this section can be found at the MCS website.[†] For example, variables that measure temperature are degrees Kelvin in SI units, and variables that measure length are meters. Important conversions between SI and English units are also given at the MCS website. A summary of the describing equations for lumped,

Table 2.1   Summary of Through- and Across-Variables for Physical Systems

System	Variable Through Element	Integrated Through-Variable	Variable Across Element	Integrated Across-Variable
Electrical	Current, i	Charge, q	Voltage difference, v_{21}	Flux linkage, λ_{21}
Mechanical translational	Force, F	Translational momentum, P	Velocity difference, v_{21}	Displacement difference, y_{21}
Mechanical rotational	Torque, T	Angular momentum, h	Angular velocity difference, ω_{21}	Angular displacement difference, θ_{21}
Fluid	Fluid volumetric rate of flow, Q	Volume, V	Pressure difference, P_{21}	Pressure momentum, γ_{21}
Thermal	Heat flow rate, q	Heat energy, H	Temperature difference, $\mathcal{T}_{21}$	

[†]The companion website is found at www.prenhall.com/dorf.

linear, dynamic elements is given in Table 2.2 [5]. The equations in Table 2.2 are idealized descriptions and only approximate the actual conditions (for example, when a linear, lumped approximation is used for a distributed element).

Table 2.2   Summary of Governing Differential Equations for Ideal Elements

Type of Element	Physical Element	Governing Equation	Energy E or Power $\mathcal{P}$	Symbol
Inductive storage	Electrical inductance	$v_{21} = L\dfrac{di}{dt}$	$E = \dfrac{1}{2}Li^2$	$v_2 \circ$—L—$\circ v_1$, i
	Translational spring	$v_{21} = \dfrac{1}{k}\dfrac{dF}{dt}$	$E = \dfrac{1}{2}\dfrac{F^2}{k}$	$v_2 \circ$—k—$\circ\, F$, v_1
	Rotational spring	$\omega_{21} = \dfrac{1}{k}\dfrac{dT}{dt}$	$E = \dfrac{1}{2}\dfrac{T^2}{k}$	$\omega_2 \circ$—k—$\circ\, T$, ω_1
	Fluid inertia	$P_{21} = I\dfrac{dQ}{dt}$	$E = \dfrac{1}{2}IQ^2$	$P_2 \circ$—I—$\circ\, P_1$, Q
Capacitive storage	Electrical capacitance	$i = C\dfrac{dv_{21}}{dt}$	$E = \dfrac{1}{2}Cv_{21}{}^2$	$v_2 \circ$—C—$\circ v_1$, i
	Translational mass	$F = M\dfrac{dv_2}{dt}$	$E = \dfrac{1}{2}Mv_2{}^2$	F—M—$v_1 =$ constant, v_2
	Rotational mass	$T = J\dfrac{d\omega_2}{dt}$	$E = \dfrac{1}{2}J\omega_2{}^2$	T—J—$\omega_1 =$ constant, ω_2
	Fluid capacitance	$Q = C_f\dfrac{dP_{21}}{dt}$	$E = \dfrac{1}{2}C_f P_{21}{}^2$	Q—C_f—P_1, P_2
	Thermal capacitance	$q = C_t\dfrac{d\mathcal{T}_2}{dt}$	$E = C_t\mathcal{T}_2$	q—C_t—$\mathcal{T}_1 =$ constant, $\mathcal{T}_2$
Energy dissipators	Electrical resistance	$i = \dfrac{1}{R}v_{21}$	$\mathcal{P} = \dfrac{1}{R}v_{21}{}^2$	$v_2 \circ$—R—$\circ v_1$, i
	Translational damper	$F = bv_{21}$	$\mathcal{P} = bv_{21}{}^2$	F—b—v_1, v_2
	Rotational damper	$T = b\omega_{21}$	$\mathcal{P} = b\omega_{21}{}^2$	T—b—ω_1, ω_2
	Fluid resistance	$Q = \dfrac{1}{R_f}P_{21}$	$\mathcal{P} = \dfrac{1}{R_f}P_{21}{}^2$	$P_2 \circ$—R_f—$\circ P_1$, Q
	Thermal resistance	$q = \dfrac{1}{R_t}\mathcal{T}_{21}$	$\mathcal{P} = \dfrac{1}{R_t}\mathcal{T}_{21}$	$\mathcal{T}_2 \circ$—R_t—$\circ \mathcal{T}_1$, q

Nomenclature

- *Through-variable:* F = force, T = torque, i = current, Q = fluid volumetric flow rate, q = heat flow rate.
- *Across-variable:* v = translational velocity, ω = angular velocity, v = voltage, P = pressure, $\mathcal{T}$ = temperature.
- *Inductive storage:* L = inductance, $1/k$ = reciprocal translational or rotational stiffness, I = fluid inertance.
- *Capacitive storage:* C = capacitance, M = mass, J = moment of inertia, C_f = fluid capacitance, C_t = thermal capacitance.
- *Energy dissipators:* R = resistance, b = viscous friction, R_f = fluid resistance, R_t = thermal resistance.

The symbol v is used for both voltage in electrical circuits and velocity in translational mechanical systems and is distinguished within the context of each differential equation. For mechanical systems, one uses Newton's laws; for electrical systems, Kirchhoff's voltage laws. For example, the simple spring-mass-damper mechanical system shown in Figure 2.2(a) is described by Newton's second law of motion. (This system could represent, for example, an automobile shock absorber.) The free-body diagram of the mass M is shown in Figure 2.2(b). In this spring-mass-damper example, we model the wall friction as a **viscous damper**, that is, the friction force is linearly proportional to the velocity of the mass. In reality the friction force may behave in a more complicated fashion. For example, the wall friction may behave as a **Coulomb damper**. Coulomb friction, also known as dry friction, is a nonlinear function of the mass velocity and possesses a discontinuity around zero velocity. For a well-lubricated, sliding surface, the viscous friction is appropriate and will be used here and in subsequent spring-mass-damper examples. Summing the forces acting on M and utilizing Newton's second law yields

$$M\frac{d^2y(t)}{dt^2} + b\frac{dy(t)}{dt} + ky(t) = r(t), \tag{2.1}$$

where k is the spring constant of the ideal spring and b is the friction constant. Equation (2.1) is a second-order linear constant-coefficient differential equation.

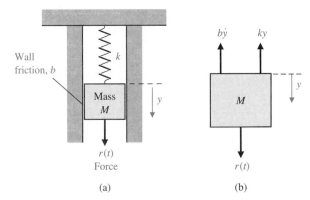

FIGURE 2.2
(a) Spring-mass-damper system.
(b) Free-body diagram.

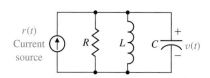

FIGURE 2.3
RLC circuit.

Alternatively, one may describe the electrical RLC circuit of Figure 2.3 by utilizing Kirchhoff's current law. Then we obtain the following integrodifferential equation:

$$\frac{v(t)}{R} + C\frac{dv(t)}{dt} + \frac{1}{L}\int_0^t v(t)\, dt = r(t). \tag{2.2}$$

The solution of the differential equation describing the process may be obtained by classical methods such as the use of integrating factors and the method of undetermined coefficients [1]. For example, when the mass is initially displaced a distance $y(0) = y_0$ and released, the dynamic response of an **underdamped system** is represented by an equation of the form

$$y(t) = K_1 e^{-\alpha_1 t} \sin(\beta_1 t + \theta_1). \tag{2.3}$$

A similar solution is obtained for the voltage of the RLC circuit when the circuit is subjected to a constant current $r(t) = I$. Then the voltage is

$$v(t) = K_2 e^{-\alpha_2 t} \cos(\beta_2 t + \theta_2). \tag{2.4}$$

A voltage curve typical of an underdamped RLC circuit is shown in Figure 2.4.

To reveal further the close similarity between the differential equations for the mechanical and electrical systems, we shall rewrite Equation (2.1) in terms of velocity:

$$v(t) = \frac{dy(t)}{dt}.$$

Then we have

$$M\frac{dv(t)}{dt} + bv(t) + k\int_0^t v(t)\, dt = r(t). \tag{2.5}$$

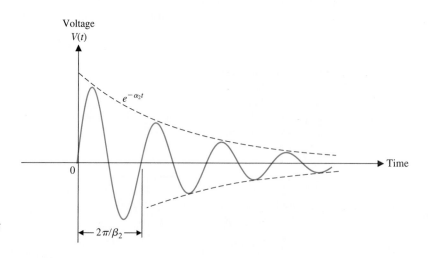

FIGURE 2.4
Typical voltage
response for
underdamped RLC
circuit.

One immediately notes the equivalence of Equations (2.5) and (2.2) where velocity $v(t)$ and voltage $v(t)$ are equivalent variables, usually called **analogous variables**, and the systems are analogous systems. Therefore the solution for velocity is similar to Equation (2.4), and the response for an underdamped system is shown in Figure 2.4. The concept of analogous systems is a very useful and powerful technique for system modeling. The voltage–velocity analogy, often called the force–current analogy, is a natural one because it relates the analogous through- and across-variables of the electrical and mechanical systems. Another analogy that relates the velocity and current variables is often used and is called the force–voltage analogy [22, 24].

Analogous systems with similar solutions exist for electrical, mechanical, thermal, and fluid systems. The existence of analogous systems and solutions provides the analyst with the ability to extend the solution of one system to all analogous systems with the same describing differential equations. Therefore what one learns about the analysis and design of electrical systems is immediately extended to an understanding of fluid, thermal, and mechanical systems.

2.3 LINEAR APPROXIMATIONS OF PHYSICAL SYSTEMS

A great majority of physical systems are linear within some range of the variables. In general, systems ultimately become nonlinear as the variables are increased without limit. For example, the spring-mass-damper system of Figure 2.2 is linear and described by Equation (2.1) as long as the mass is subjected to small deflections $y(t)$. However, if $y(t)$ were continually increased, eventually the spring would be overextended and break. Therefore the question of linearity and the range of applicability must be considered for each system.

A system is defined as linear in terms of the system excitation and response. In the case of the electrical network, the excitation is the input current $r(t)$ and the response is the voltage $v(t)$. In general, a **necessary condition** for a linear system can be determined in terms of an excitation $x(t)$ and a response $y(t)$. When the system at rest is subjected to an excitation $x_1(t)$, it provides a response $y_1(t)$. Furthermore, when the system is subjected to an excitation $x_2(t)$, it provides a corresponding response $y_2(t)$. For a linear system, it is necessary that the excitation $x_1(t) + x_2(t)$ result in a response $y_1(t) + y_2(t)$. This is usually called the **principle of superposition**.

Furthermore, the magnitude scale factor must be preserved in a **linear system**. Again, consider a system with an input $x(t)$ that results in an output $y(t)$. Then the response of a linear system to a constant multiple β of an input x must be equal to the response to the input multiplied by the same constant so that the output is equal to βy. This is called the property of **homogeneity**

> **A linear system satisfies the properties of superposition and homogeneity.**

A system characterized by the relation $y = x^2$ is not linear, because the superposition property is not satisfied. A system represented by the relation $y = mx + b$ is not linear, because it does not satisfy the homogeneity property. However, this second system may be considered linear about an operating point x_0, y_0 for small changes Δx and Δy. When $x = x_0 + \Delta x$ and $y = y_0 + \Delta y$, we have

$$y = mx + b$$

or

$$y_0 + \Delta y = mx_0 + m \, \Delta x + b.$$

Therefore, $\Delta y = m \, \Delta x$, which satisfies the necessary conditions.

The linearity of many mechanical and electrical elements can be assumed over a reasonably large range of the variables [7]. This is not usually the case for thermal and fluid elements, which are more frequently nonlinear in character. Fortunately, however, one can often linearize nonlinear elements assuming small-signal conditions. This is the normal approach used to obtain a linear equivalent circuit for electronic circuits and transistors. Consider a general element with an excitation (through-) variable $x(t)$ and a response (across-) variable $y(t)$. Several examples of dynamic system variables are given in Table 2.1. The relationship of the two variables is written as

$$y(t) = g(x(t)), \tag{2.6}$$

where $g(x(t))$ indicates $y(t)$ is a function of $x(t)$. The normal operating point is designated by x_0. Because the curve (function) is continuous over the range of interest, a **Taylor series** expansion about the operating point may be utilized [7]. Then we have

$$y = g(x) = g(x_0) + \left.\frac{dg}{dx}\right|_{x=x_0} \frac{(x - x_0)}{1!} + \left.\frac{d^2g}{dx^2}\right|_{x=x_0} \frac{(x - x_0)^2}{2!} + \cdots. \tag{2.7}$$

The slope at the operating point,

$$\left.\frac{dg}{dx}\right|_{x=x_0},$$

is a good approximation to the curve over a small range of $(x - x_0)$, the deviation from the operating point. Then, as a reasonable approximation, Equation (2.7) becomes

$$y = g(x_0) + \left.\frac{dg}{dx}\right|_{x=x_0} (x - x_0) = y_0 + m(x - x_0), \tag{2.8}$$

where m is the slope at the operating point. Finally, Equation (2.8) can be rewritten as the linear equation

$$(y - y_0) = m(x - x_0)$$

or

$$\Delta y = m \, \Delta x. \tag{2.9}$$

Consider the case of a mass, M, sitting on a nonlinear spring, as shown in Figure 2.5(a). The normal operating point is the equilibrium position that occurs when the spring force balances the gravitational force Mg, where g is the gravitational constant. Thus, we obtain $f_0 = Mg$, as shown. For the nonlinear spring with $f = y^2$, the equilibrium position is $y_0 = (Mg)^{1/2}$. The linear model for small deviation is

$$\Delta f = m \, \Delta y,$$

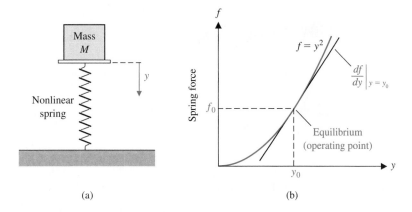

FIGURE 2.5
(a) A mass sitting on a nonlinear spring. (b) The spring force versus y.

where

$$m = \left.\frac{df}{dy}\right|_{y_0},$$

as shown in Figure 2.5(b). Thus, $m = 2y_0$. A **linear approximation** is as accurate as the assumption of small signals is applicable to the specific problem.

If the dependent variable y depends upon several excitation variables, $x_1, x_2, \ldots, x_n$, then the functional relationship is written as

$$y = g(x_1, x_2, \ldots, x_n). \tag{2.10}$$

The Taylor series expansion about the operating point $x_{1_0}, x_{2_0}, \ldots, x_{n_0}$ is useful for a linear approximation to the nonlinear function. When the higher-order terms are neglected, the linear approximation is written as

$$y = g(x_{1_0}, x_{2_0}, \ldots, x_{n_0}) + \left.\frac{\partial g}{\partial x_1}\right|_{x=x_0}(x_1 - x_{1_0}) + \left.\frac{\partial g}{\partial x_2}\right|_{x=x_0}(x_2 - x_{2_0}) \quad (2.11)$$

$$+ \cdots + \left.\frac{\partial g}{\partial x_n}\right|_{x=x_0}(x_n - x_{n_0}),$$

where x_0 is the operating point. Example 2.1 will clearly illustrate the utility of this method.

EXAMPLE 2.1   **Pendulum oscillator model**

Consider the pendulum oscillator shown in Figure 2.6(a). The torque on the mass is

$$T = MgL \sin \theta, \tag{2.12}$$

where g is the gravity constant. The equilibrium condition for the mass is $\theta_0 = 0°$. The nonlinear relation between T and θ is shown graphically in Figure 2.6(b). The first derivative evaluated at equilibrium provides the linear approximation, which is

$$T - T_0 \cong MgL\left.\frac{\partial \sin \theta}{\partial \theta}\right|_{\theta=\theta_0}(\theta - \theta_0),$$

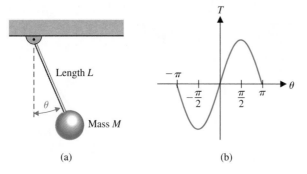

FIGURE 2.6
Pendulum
oscillator.

(a)   (b)

where $T_0 = 0$. Then, we have

$$T = MgL(\cos 0°)(\theta - 0°)$$

$$= MgL\theta. \tag{2.13}$$

This approximation is reasonably accurate for $-\pi/4 \leq \theta \leq \pi/4$. For example, the response of the linear model for the swing through $\pm 30°$ is within 5% of the actual nonlinear pendulum response. ∎

2.4 THE LAPLACE TRANSFORM

The ability to obtain linear approximations of physical systems allows the analyst to consider the use of the **Laplace transformation**. The Laplace transform method substitutes relatively easily solved algebraic equations for the more difficult differential equations [1, 3]. The time-response solution is obtained by the following operations:

   1. Obtain the linearized differential equations.
   2. Obtain the Laplace transformation of the differential equations.
   3. Solve the resulting algebraic equation for the transform of the variable of interest.

The Laplace transform exists for linear differential equations for which the transformation integral converges. Therefore, for $f(t)$ to be transformable, it is sufficient that

$$\int_{0^-}^{\infty} |f(t)|e^{-\sigma_1 t}\, dt < \infty,$$

for some real, positive σ_1 [1]. The 0^- indicates that the integral should include any discontinuity, such as a delta function at $t = 0$. If the magnitude of $f(t)$ is $|f(t)| < Me^{\alpha t}$ for all positive t, the integral will converge for $\sigma_1 > \alpha$. The region of convergence is therefore given by $\infty > \sigma_1 > \alpha$, and σ_1 is known as the abscissa of absolute convergence. Signals that are physically realizable always have a Laplace transform. The **Laplace transformation** for a function of time, $f(t)$, is

$$F(s) = \int_{0^-}^{\infty} f(t)e^{-st}\, dt = \mathscr{L}\{f(t)\}. \tag{2.14}$$

The **inverse Laplace transform** is written as

$$f(t) = \frac{1}{2\pi j} \int_{\sigma - j\infty}^{\sigma + j\infty} F(s)e^{+st}\, ds. \tag{2.15}$$

The transformation integrals have been employed to derive tables of Laplace transforms that are used for the great majority of problems. A table of important Laplace transform pairs is given in Table 2.3, and a more complete list of Laplace transform pairs can be found at the MCS website.

Table 2.3   Important Laplace Transform Pairs

$f(t)$	$F(s)$
Step function, $u(t)$	$\dfrac{1}{s}$
e^{-at}	$\dfrac{1}{s+a}$
$\sin \omega t$	$\dfrac{\omega}{s^2 + \omega^2}$
$\cos \omega t$	$\dfrac{s}{s^2 + \omega^2}$
t^n	$\dfrac{n!}{s^{n+1}}$
$f^{(k)}(t) = \dfrac{d^k f(t)}{dt^k}$	$s^k F(s) - s^{k-1}f(0^-) - s^{k-2}f'(0^-)$ $- \cdots - f^{(k-1)}(0^-)$
$\displaystyle\int_{-\infty}^{t} f(t)\,dt$	$\dfrac{F(s)}{s} + \dfrac{1}{s}\displaystyle\int_{-\infty}^{0} f(t)\,dt$
Impulse function $\delta(t)$	1
$e^{-at}\sin \omega t$	$\dfrac{\omega}{(s+a)^2 + \omega^2}$
$e^{-at}\cos \omega t$	$\dfrac{s+a}{(s+a)^2 + \omega^2}$
$\dfrac{1}{\omega}[(\alpha - a)^2 + \omega^2]^{1/2} e^{-at}\sin(\omega t + \phi),$ $\phi = \tan^{-1}\dfrac{\omega}{\alpha - a}$	$\dfrac{s + \alpha}{(s+a)^2 + \omega^2}$
$\dfrac{\omega_n}{\sqrt{1 - \zeta^2}}e^{-\zeta \omega_n t}\sin \omega_n \sqrt{1 - \zeta^2}\,t,\ \zeta < 1$	$\dfrac{\omega_n^2}{s^2 + 2\zeta\omega_n s + \omega_n^2}$
$\dfrac{1}{a^2 + \omega^2} + \dfrac{1}{\omega\sqrt{a^2 + \omega^2}}e^{-at}\sin(\omega t - \phi),$ $\phi = \tan^{-1}\dfrac{\omega}{-a}$	$\dfrac{1}{s[(s+a)^2 + \omega^2]}$
$1 - \dfrac{1}{\sqrt{1 - \zeta^2}}e^{-\zeta \omega_n t}\sin\left(\omega_n \sqrt{1 - \zeta^2}\,t + \phi\right),$ $\phi = \cos^{-1}\zeta,\ \zeta < 1$	$\dfrac{\omega_n^2}{s(s^2 + 2\zeta\omega_n s + \omega_n^2)}$
$\dfrac{\alpha}{a^2 + \omega^2} + \dfrac{1}{\omega}\left[\dfrac{(\alpha - a)^2 + \omega^2}{a^2 + \omega^2}\right]^{1/2}e^{-at}\sin(\omega t + \phi).$ $\phi = \tan^{-1}\dfrac{\omega}{\alpha - a} - \tan^{-1}\dfrac{\omega}{-a}$	$\dfrac{s + \alpha}{s[(s+a)^2 + \omega^2]}$

Alternatively, the Laplace variable s can be considered to be the differential operator so that

$$s \equiv \frac{d}{dt}. \tag{2.16}$$

Then we also have the integral operator

$$\frac{1}{s} \equiv \int_{0^-}^{t} dt. \tag{2.17}$$

The inverse Laplace transformation is usually obtained by using the Heaviside partial fraction expansion. This approach is particularly useful for systems analysis and design because the effect of each characteristic root or eigenvalue can be clearly observed.

To illustrate the usefulness of the Laplace transformation and the steps involved in the system analysis, reconsider the spring-mass-damper system described by Equation (2.1), which is

$$M\frac{d^2y}{dt^2} + b\frac{dy}{dt} + ky = r(t). \tag{2.18}$$

We wish to obtain the response, y, as a function of time. The Laplace transform of Equation (2.18) is

$$M\left(s^2Y(s) - sy(0^-) - \frac{dy}{dt}(0^-)\right) + b(sY(s) - y(0^-)) + kY(s) = R(s). \tag{2.19}$$

When

$$r(t) = 0, \quad \text{and} \quad y(0^-) = y_0, \quad \text{and} \quad \left.\frac{dy}{dt}\right|_{t=0^-} = 0,$$

we have

$$Ms^2Y(s) - Msy_0 + bsY(s) - by_0 + kY(s) = 0. \tag{2.20}$$

Solving for $Y(s)$, we obtain

$$Y(s) = \frac{(Ms + b)y_0}{Ms^2 + bs + k} = \frac{p(s)}{q(s)}. \tag{2.21}$$

The denominator polynomial $q(s)$, when set equal to zero, is called the **characteristic equation** because the roots of this equation determine the character of the time response. The roots of this characteristic equation are also called the **poles** of the system. The roots of the numerator polynomial $p(s)$ are called the **zeros** of the system; for example, $s = -b/M$ is a zero of Equation (2.21). Poles and zeros are critical frequencies. At the poles, the function $Y(s)$ becomes infinite, whereas at the zeros, the function becomes zero. The complex frequency **s-plane** plot of the poles and zeros graphically portrays the character of the natural transient response of the system.

For a specific case, consider the system when $k/M = 2$ and $b/M = 3$. Then Equation (2.21) becomes

$$Y(s) = \frac{(s + 3)y_0}{(s + 1)(s + 2)}. \tag{2.22}$$

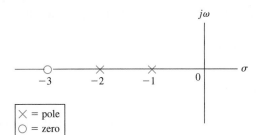

FIGURE 2.7
An s-plane pole and
zero plot.

$\times$ = pole
$\bigcirc$ = zero

The poles and zeros of $Y(s)$ are shown on the s-plane in Figure 2.7.

Expanding Equation (2.22) in a partial fraction expansion, we obtain

$$Y(s) = \frac{k_1}{s + 1} + \frac{k_2}{s + 2}, \tag{2.23}$$

where k_1 and k_2 are the coefficients of the expansion. The coefficients k_i are called
residues and are evaluated by multiplying through by the denominator factor of
Equation (2.22) corresponding to k_i and setting s equal to the root. Evaluating k_1
when $y_0 = 1$, we have

$$k_1 = \left. \frac{(s - s_1)p(s)}{q(s)} \right|_{s=s_1} \tag{2.24}$$

$$= \left. \frac{(s + 1)(s + 3)}{(s + 1)(s + 2)} \right|_{s_1=-1} = 2$$

and $k_2 = -1$. Alternatively, the residues of $Y(s)$ at the respective poles may be evaluated graphically on the s-plane plot, since Equation (2.24) may be written as

$$k_1 = \left. \frac{s + 3}{s + 2} \right|_{s=s_1=-1} \tag{2.25}$$

$$= \left. \frac{s_1 + 3}{s_1 + 2} \right|_{s_1=-1} = 2.$$

The graphical representation of Equation (2.25) is shown in Figure 2.8. The graphical method of evaluating the residues is particularly valuable when the order of the
characteristic equation is high and several poles are complex conjugate pairs.

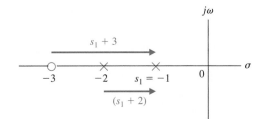

FIGURE 2.8
Graphical
evaluation of the
residues.

The inverse Laplace transform of Equation (2.22) is then

$$y(t) = \mathcal{L}^{-1}\left\{\frac{2}{s+1}\right\} + \mathcal{L}^{-1}\left\{\frac{-1}{s+2}\right\}. \tag{2.26}$$

Using Table 2.3, we find that

$$y(t) = 2e^{-t} - 1e^{-2t}. \tag{2.27}$$

Finally, it is usually desired to determine the **steady-state** or **final value** of the response of $y(t)$. For example, the final or steady-state rest position of the spring-mass-damper system may be calculated. The **final value theorem** states that

$$\boxed{\lim_{t \to \infty} y(t) = \lim_{s \to 0} sY(s),} \tag{2.28}$$

where a simple pole of $Y(s)$ at the origin is permitted, but poles on the imaginary axis and in the right half-plane and repeated poles at the origin are excluded. Therefore, for the specific case of the spring-mass-damper, we find that

$$\lim_{t \to \infty} y(t) = \lim_{s \to 0} sY(s) = 0. \tag{2.29}$$

Hence the final position for the mass is the normal equilibrium position $y = 0$.

To illustrate clearly the salient points of the Laplace transform method, let us reconsider the spring-mass-damper system for the underdamped case. The equation for $Y(s)$ may be written as

$$Y(s) = \frac{(s + b/M)y_0}{s^2 + (b/M)s + k/M} = \frac{(s + 2\zeta\omega_n)y_0}{s^2 + 2\zeta\omega_n s + \omega_n^2}, \tag{2.30}$$

where ζ is the dimensionless **damping ratio**, and ω_n is the **natural frequency** of the system. The roots of the characteristic equation are

$$s_1, s_2 = -\zeta\omega_n \pm \omega_n\sqrt{\zeta^2 - 1}, \tag{2.31}$$

where, in this case, $\omega_n = \sqrt{k/M}$ and $\zeta = b/(2\sqrt{kM})$. When $\zeta > 1$, the roots are real; when $\zeta < 1$, the roots are complex and conjugates. When $\zeta = 1$, the roots are repeated and real, and the condition is called **critical damping**.

When $\zeta < 1$, the response is underdamped, and

$$s_{1,2} = -\zeta\omega_n \pm j\omega_n\sqrt{1 - \zeta^2}. \tag{2.32}$$

The s-plane plot of the poles and zeros of $Y(s)$ is shown in Figure 2.9, where $\theta = \cos^{-1}\zeta$. As ζ varies with ω_n constant, the complex conjugate roots follow a circular

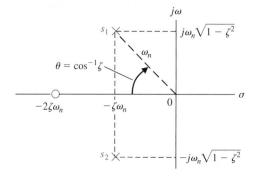

FIGURE 2.9
An s-plane plot of the poles and zeros of Y(s).

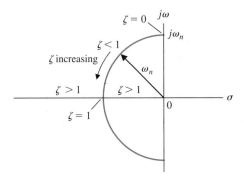

FIGURE 2.10
The locus of roots as ζ varies with ω_n constant.

locus, as shown in Figure 2.10. The transient response is increasingly oscillatory as the roots approach the imaginary axis when ζ approaches zero.

The inverse Laplace transform can be evaluated using the graphical residue evaluation. The partial fraction expansion of Equation (2.30) is

$$Y(s) = \frac{k_1}{s - s_1} + \frac{k_2}{s - s_2}. \tag{2.33}$$

Since s_2 is the complex conjugate of s_1, the residue k_2 is the complex conjugate of k_1 so that we obtain

$$Y(s) = \frac{k_1}{s - s_1} + \frac{k_1^*}{s - s_1^*},$$

where the asterisk indicates the conjugate relation. The residue k_1 is evaluated from Figure 2.11 as

$$k_1 = \frac{y_0(s_1 + 2\zeta\omega_n)}{s_1 - s_1^*} = \frac{y_0 M_1 e^{j\theta}}{M_2 e^{j\pi/2}}, \tag{2.34}$$

where M_1 is the magnitude of $s_1 + 2\zeta\omega_n$, and M_2 is the magnitude of $s_1 - s_1^*$. (A review of complex numbers can be found on the MCS website.) In this case, we obtain

$$k_1 = \frac{y_0(\omega_n e^{j\theta})}{2\omega_n\sqrt{1 - \zeta^2}e^{j\pi/2}} = \frac{y_0}{2\sqrt{1 - \zeta^2}e^{j(\pi/2-\theta)}}, \tag{2.35}$$

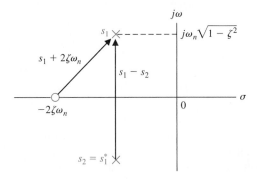

FIGURE 2.11
Evaluation of the residue k_1.

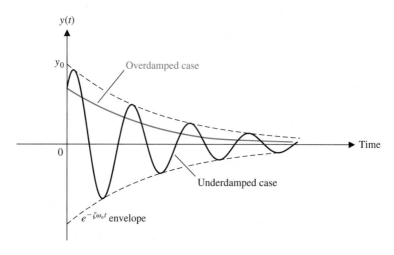

FIGURE 2.12
Response of the
spring-mass-
damper system.

where $\theta = \cos^{-1} \zeta$. Therefore,

$$k_2 = \frac{y_0}{2\sqrt{1 - \zeta^2}} e^{j(\pi/2 - \theta)}. \tag{2.36}$$

Finally, letting $\beta = \sqrt{1 - \zeta^2}$, we find that

$$y(t) = k_1 e^{s_1 t} + k_2 e^{s_2 t}$$

$$= \frac{y_0}{2\sqrt{1 - \zeta^2}} (e^{j(\theta - \pi/2)} e^{-\zeta \omega_n t} e^{j\omega_n \beta t} + e^{j(\pi/2 - \theta)} e^{-\zeta \omega_n t} e^{-j\omega_n \beta t})$$

$$= \frac{y_0}{\sqrt{1 - \zeta^2}} e^{-\zeta \omega_n t} \sin(\omega_n \sqrt{1 - \zeta^2} t + \theta). \tag{2.37}$$

The solution, Equation (2.37), can also be obtained using item 11 of Table 2.3. The transient responses of the overdamped ($\zeta > 1$) and underdamped ($\zeta < 1$) cases are shown in Figure 2.12. The transient response that occurs when $\zeta < 1$ exhibits an oscillation in which the amplitude decreases with time, and it is called a **damped oscillation**.

The relationship between the s-plane location of the poles and zeros and the form of the transient response can be interpreted from the s-plane pole–zero plots. For example, as seen in Equation (2.37), adjusting the value of $\zeta \omega_n$ varies the $e^{-\zeta \omega_n t}$ envelope, hence the response $y(t)$ shown in Figure 2.12. The larger the value of $\zeta \omega_n$, the faster the damping of the response, $y(t)$. In Figure 2.9, we see that the location of the complex pole s_1 is given by $s_1 = -\zeta \omega_n + j\omega_n \sqrt{1 - \zeta^2}$. So, making $\zeta \omega_n$ larger moves the pole further to the left in the s-plane. Thus, the connection between the location of the pole in the s-plane and the step response is apparent—moving the pole s_1 farther in the left half-plane leads to a faster damping of the transient step response. Of course, most control systems will have more than one complex pair of poles, so the transient response will be the result of the contributions of all the poles. In fact, the magnitude of the response of each pole, represented by the residue, can be visualized by examining the graphical residues on the s-plane. We will discuss the connection between the pole and zero locations and the transient and steady-state response more in subsequent chapters. We will find that the Laplace transformation and the s-plane approach are very useful techniques for system analysis and design where emphasis is placed on the transient and steady-state performance. In fact, because the study of control systems is concerned primarily with the transient and steady-state performance of dynamic systems, we have real cause to appreciate the value of the Laplace transform techniques.

2.5 THE TRANSFER FUNCTION OF LINEAR SYSTEMS

The **transfer function** of a linear system is defined as the ratio of the Laplace transform of the output variable to the Laplace transform of the input variable, with all initial conditions assumed to be zero. The transfer function of a system (or element) represents the relationship describing the dynamics of the system under consideration.

A transfer function may be defined only for a linear, stationary (constant parameter) system. A nonstationary system, often called a time-varying system, has one or more time-varying parameters, and the Laplace transformation may not be utilized. Furthermore, a transfer function is an input–output description of the behavior of a system. Thus, the transfer function description does not include any information concerning the internal structure of the system and its behavior.

The transfer function of the spring-mass-damper system is obtained from the original Equation (2.19), rewritten with zero initial conditions as follows:

$$Ms^2Y(s) + bsY(s) + kY(s) = R(s). \tag{2.38}$$

Then the transfer function is

$$\frac{\text{Output}}{\text{Input}} = G(s) = \frac{Y(s)}{R(s)} = \frac{1}{Ms^2 + bs + k}. \tag{2.39}$$

The transfer function of the RC network shown in Figure 2.13 is obtained by writing the Kirchhoff voltage equation, yielding

$$V_1(s) = \left(R + \frac{1}{Cs} \right)I(s), \tag{2.40}$$

expressed in terms of transform variables. We shall frequently refer to variables and their transforms interchangeably. The transform variable will be distinguishable by the use of an uppercase letter or the argument (s).

The output voltage is

$$V_2(s) = I(s)\left(\frac{1}{Cs} \right). \tag{2.41}$$

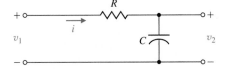

FIGURE 2.13
An RC network.

Therefore, solving Equation (2.40) for $I(s)$ and substituting in Equation (2.41), we have

$$V_2(s) = \frac{(1/Cs)V_1(s)}{R + 1/Cs}.$$

Then the transfer function is obtained as the ratio $V_2(s)/V_1(s)$, which is

$$G(s) = \frac{V_2(s)}{V_1(s)} = \frac{1}{RCs + 1} = \frac{1}{\tau s + 1} = \frac{1/\tau}{s + 1/\tau}, \qquad (2.42)$$

where $\tau = RC$, the **time constant** of the network. The single pole of $G(s)$ is $s = -1/\tau$. Equation (2.42) could be immediately obtained if one observes that the circuit is a voltage divider, where

$$\frac{V_2(s)}{V_1(s)} = \frac{Z_2(s)}{Z_1(s) + Z_2(s)}, \qquad (2.43)$$

and $Z_1(s) = R$, $Z_2 = 1/Cs$.

A multiloop electrical circuit or an analogous multiple-mass mechanical system results in a set of simultaneous equations in the Laplace variable. It is usually more convenient to solve the simultaneous equations by using matrices and determinants [1, 3, 16]. An introduction to matrices and determinants can be found on the MCS website.

Let us consider the long-term behavior of a system and determine the response to certain inputs that remain after the transients fade away. Consider the dynamic system represented by the differential equation

$$\frac{d^n y}{dt^n} + q_{n-1}\frac{d^{n-1}y}{dt^{n-1}} + \cdots + q_0 y = p_{n-1}\frac{d^{n-1}r}{dt^{n-1}} + p_{n-2}\frac{d^{n-2}r}{dt^{n-2}} + \cdots + p_0 r, \quad (2.44)$$

where $y(t)$ is the response, and $r(t)$ is the input or forcing function. If the initial conditions are all zero, then the transfer function is the coefficient of $R(s)$ in

$$Y(s) = G(s)R(s) = \frac{p(s)}{q(s)}R(s) = \frac{p_{n-1}s^{n-1} + p_{n-2}s^{n-2} + \cdots + p_0}{s^n + q_{n-1}s^{n-1} + \cdots + q_0}R(s). \qquad (2.45)$$

The output response consists of a natural response (determined by the initial conditions) plus a forced response determined by the input. We now have

$$Y(s) = \frac{m(s)}{q(s)} + \frac{p(s)}{q(s)}R(s),$$

where $q(s) = 0$ is the characteristic equation. If the input has the rational form

$$R(s) = \frac{n(s)}{d(s)},$$

then

$$Y(s) = \frac{m(s)}{q(s)} + \frac{p(s)}{q(s)}\frac{n(s)}{d(s)} = Y_1(s) + Y_2(s) + Y_3(s), \qquad (2.46)$$

where $Y_1(s)$ is the partial fraction expansion of the natural response, $Y_2(s)$ is the partial fraction expansion of the terms involving factors of $q(s)$, and $Y_3(s)$ is the partial fraction expansion of terms involving factors of $d(s)$.

Taking the inverse Laplace transform yields

$$y(t) = y_1(t) + y_2(t) + y_3(t).$$

The transient response consists of $y_1(t) + y_2(t)$, and the steady-state response is $y_3(t)$.

EXAMPLE 2.2   Solution of a differential equation

Consider a system represented by the differential equation

$$\frac{d^2y}{dt^2} + 4\frac{dy}{dt} + 3y = 2r(t),$$

where the initial conditions are $y(0) = 1, \dfrac{dy}{dt}(0) = 0$, and $r(t) = 1, t \geq 0$.

The Laplace transform yields

$$[s^2Y(s) - sy(0)] + 4[sY(s) - y(0)] + 3Y(s) = 2R(s).$$

Since $R(s) = 1/s$ and $y(0) = 1$, we obtain

$$Y(s) = \frac{s+4}{s^2+4s+3} + \frac{2}{s(s^2+4s+3)},$$

where $q(s) = s^2 + 4s + 3 = (s+1)(s+3) = 0$ is the characteristic equation, and $d(s) = s$. Then the partial fraction expansion yields

$$Y(s) = \left[\frac{3/2}{s+1} + \frac{-1/2}{s+3}\right] + \left[\frac{-1}{s+1} + \frac{1/3}{s+3}\right] + \frac{2/3}{s} = Y_1(s) + Y_2(s) + Y_3(s).$$

Hence, the response is

$$y(t) = \left[\frac{3}{2}e^{-t} - \frac{1}{2}e^{-3t}\right] + \left[-1e^{-t} + \frac{1}{3}e^{-3t}\right] + \frac{2}{3},$$

and the steady-state response is

$$\lim_{t\to\infty} y(t) = \frac{2}{3}. \quad \blacksquare$$

EXAMPLE 2.3   Transfer function of an op-amp circuit

The operational amplifier (op-amp) belongs to an important class of analog integrated circuits commonly used as building blocks in the implementation of control systems and in many other important applications. Op-amps are active elements (that is, they have external power sources) with a high gain when operating in their linear regions. A model of an ideal op-amp is shown in Figure 2.14.

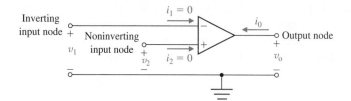

FIGURE 2.14
The ideal op-amp.

The operating conditions for the ideal op-amp are (1) $i_1 = 0$ and $i_2 = 0$, thus implying that the input impedance is infinite, and (2) $v_2 - v_1 = 0$ (or $v_1 = v_2$). The input–output relationship for an ideal op-amp is

$$v_0 = K(v_2 - v_1) = -K(v_1 - v_2),$$

where the gain K approaches infinity. In our analysis, we will assume that the linear op-amps are operating with high gain and under idealized conditions.

Consider the inverting amplifier shown in Figure 2.15. Under ideal conditions, we have $i_1 = 0$, so that writing the node equation at v_1 yields

$$\frac{v_1 - v_{in}}{R_1} + \frac{v_1 - v_0}{R_2} = 0.$$

Since $v_2 = v_1$ (under ideal conditions) and $v_2 = 0$ (see Figure 2.15 and compare it with Figure 2.14), it follows that $v_1 = 0$. Therefore,

$$-\frac{v_{in}}{R_1} - \frac{v_0}{R_2} = 0,$$

and rearranging terms, we obtain

$$\frac{v_0}{v_{in}} = -\frac{R_2}{R_1}.$$

We see that when $R_2 = R_1$, the ideal op-amp circuit inverts the sign of the input, that is, $v_0 = -v_{in}$ when $R_2 = R_1$. ∎

EXAMPLE 2.4   Transfer function of a system

Consider the mechanical system shown in Figure 2.16 and its electrical circuit analog shown in Figure 2.17. The electrical circuit analog is a force–current analog as outlined in Table 2.1. The velocities $v_1(t)$ and $v_2(t)$ of the mechanical system are directly

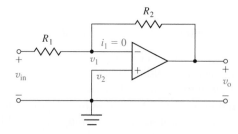

FIGURE 2.15
An inverting amplifier operating with ideal conditions.

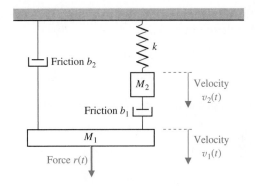

FIGURE 2.16
Two-mass
mechanical system.

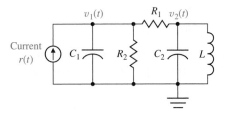

FIGURE 2.17
Two-node electric
circuit analog
$C_1 = M_1$, $C_2 = M_2$,
$L = 1/k$, $R_1 = 1/b_1$,
$R_2 = 1/b_2$.

analogous to the node voltages $v_1(t)$ and $v_2(t)$ of the electrical circuit. The simultaneous equations, assuming that the initial conditions are zero, are

$$M_1 s V_1(s) + (b_1 + b_2)V_1(s) - b_1 V_2(s) = R(s), \qquad (2.47)$$

and

$$M_2 s V_2(s) + b_1(V_2(s) - V_1(s)) + k\frac{V_2(s)}{s} = 0. \qquad (2.48)$$

These equations are obtained using the force equations for the mechanical system of Figure 2.16. Rearranging Equations (2.47) and (2.48), we obtain

$$(M_1 s + (b_1 + b_2))V_1(s) + (-b_1)V_2(s) = R(s),$$

$$(-b_1)V_1(s) + \left(M_2 s + b_1 + \frac{k}{s}\right)V_2(s) = 0,$$

or, in matrix form,

$$\begin{bmatrix} M_1 s + b_1 + b_2 & -b_1 \\ -b_1 & M_2 s + b_1 + \dfrac{k}{s} \end{bmatrix} \begin{bmatrix} V_1(s) \\ V_2(s) \end{bmatrix} = \begin{bmatrix} R(s) \\ 0 \end{bmatrix}. \qquad (2.49)$$

Assuming that the velocity of M_1 is the output variable, we solve for $V_1(s)$ by matrix inversion or Cramer's rule to obtain [1, 3]

$$V_1(s) = \frac{(M_2s + b_1 + k/s)R(s)}{(M_1s + b_1 + b_2)(M_2s + b_1 + k/s) - b_1{}^2}. \tag{2.50}$$

Then the transfer function of the mechanical (or electrical) system is

$$G(s) = \frac{V_1(s)}{R(s)} = \frac{(M_2s + b_1 + k/s)}{(M_1s + b_1 + b_2)(M_2s + b_1 + k/s) - b_1{}^2}$$

$$= \frac{(M_2s^2 + b_1s + k)}{(M_1s + b_1 + b_2)(M_2s^2 + b_1s + k) - b_1{}^2s}. \tag{2.51}$$

If the transfer function in terms of the position $x_1(t)$ is desired, then we have

$$\frac{X_1(s)}{R(s)} = \frac{V_1(s)}{sR(s)} = \frac{G(s)}{s}. \tag{2.52} \;\blacksquare$$

As an example, let us obtain the transfer function of an important electrical control component, the **DC motor** [8]. A DC motor is used to move loads and is called an **actuator**.

> **An actuator is a device that provides the motive power to the process.**

EXAMPLE 2.5   **Transfer function of the DC motor**

The DC motor is a power actuator device that delivers energy to a load, as shown in Figure 2.18(a); a sketch of a DC motor is shown in Figure 2.18(b). The DC motor converts direct current (DC) electrical energy into rotational mechanical energy. A major fraction of the torque generated in the rotor (armature) of the motor is available to drive an external load. Because of features such as high torque, speed controllability over a wide range, portability, well-behaved speed–torque charac-teristics, and adaptability to various types of control methods, DC motors are widely used in numerous control applications, including robotic manipulators, tape trans-port mechanisms, disk drives, machine tools, and servovalve actuators.

The transfer function of the DC motor will be developed for a linear approxi-mation to an actual motor, and second-order effects, such as hysteresis and the volt-age drop across the brushes, will be neglected. The input voltage may be applied to the field or armature terminals. The air-gap flux ϕ of the motor is proportional to the field current, provided the field is unsaturated, so that

$$\phi = K_f i_f. \tag{2.53}$$

The torque developed by the motor is assumed to be related linearly to ϕ and the armature current as follows:

$$T_m = K_1 \phi i_a(t) = K_1 K_f i_f(t) i_a(t). \tag{2.54}$$

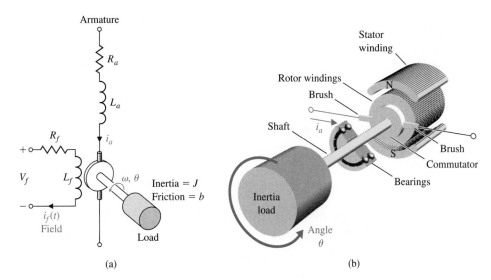

FIGURE 2.18
A DC motor
(a) electrical
diagram and
(b) sketch.

(a)   (b)

It is clear from Equation (2.54) that, to have a linear system, one current must be maintained constant while the other current becomes the input current. First, we shall consider the **field current controlled motor**, which provides a substantial power amplification. Then we have, in Laplace transform notation,

$$T_m(s) = (K_1 K_f I_a)I_f(s) = K_m I_f(s), \tag{2.55}$$

where $i_a = I_a$ is a constant armature current, and K_m is defined as the motor constant. The field current is related to the field voltage as

$$V_f(s) = (R_f + L_f s)I_f(s). \tag{2.56}$$

The motor torque $T_m(s)$ is equal to the torque delivered to the load. This relation may be expressed as

$$T_m(s) = T_L(s) + T_d(s), \tag{2.57}$$

where $T_L(s)$ is the load torque and $T_d(s)$ is the disturbance torque, which is often negligible. However, the disturbance torque often must be considered in systems subjected to external forces such as antenna wind-gust forces. The load torque for rotating inertia, as shown in Figure 2.18, is written as

$$T_L(s) = Js^2\theta(s) + bs\theta(s). \tag{2.58}$$

Rearranging Equations (2.55)–(2.57), we have

$$T_L(s) = T_m(s) - T_d(s), \tag{2.59}$$

$$T_m(s) = K_m I_f(s), \tag{2.60}$$

$$I_f(s) = \frac{V_f(s)}{R_f + L_f s}. \tag{2.61}$$

FIGURE 2.19
Block diagram
model of field-
controlled DC
motor.

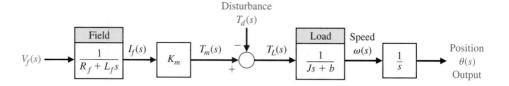

Therefore, the transfer function of the motor–load combination, with $T_d(s) = 0$, is

$$\frac{\theta(s)}{V_f(s)} = \frac{K_m}{s(Js + b)(L_f s + R_f)} = \frac{K_m/(JL_f)}{s(s + b/J)(s + R_f/L_f)}. \qquad (2.62)$$

The block diagram model of the field-controlled DC motor is shown in Figure 2.19. Alternatively, the transfer function may be written in terms of the time constants of the motor as

$$\frac{\theta(s)}{V_f(s)} = G(s) = \frac{K_m/(bR_f)}{s(\tau_f s + 1)(\tau_L s + 1)}, \qquad (2.63)$$

where $\tau_f = L_f/R_f$ and $\tau_L = J/b$. Typically, one finds that $\tau_L > \tau_f$ and often the field time constant may be neglected.

The **armature-controlled DC motor** uses the armature current i_a as the control variable. The stator field can be established by a field coil and current or a permanent magnet. When a constant field current is established in a field coil, the motor torque is

$$T_m(s) = (K_1 K_f I_f)I_a(s) = K_m I_a(s). \qquad (2.64)$$

When a permanent magnet is used, we have

$$T_m(s) = K_m I_a(s),$$

where K_m is a function of the permeability of the magnetic material.

The armature current is related to the input voltage applied to the armature by

$$V_a(s) = (R_a + L_a s)I_a(s) + V_b(s), \qquad (2.65)$$

where $V_b(s)$ is the back electromotive-force voltage proportional to the motor speed. Therefore, we have

$$V_b(s) = K_b \omega(s), \qquad (2.66)$$

where $\omega(s) = s\theta(s)$ is the transform of the angular speed and the armature current is

$$I_a(s) = \frac{V_a(s) - K_b \omega(s)}{R_a + L_a s}. \qquad (2.67)$$

Equations (2.58) and (2.59) represent the load torque, so that

$$T_L(s) = Js^2\theta(s) + bs\theta(s) = T_m(s) - T_d(s). \qquad (2.68)$$

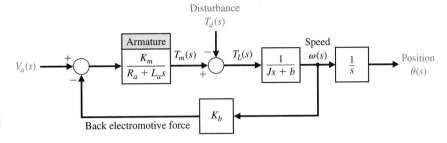

FIGURE 2.20
Armature-controlled
DC motor.

The relations for the armature-controlled DC motor are shown schematically in Figure 2.20. Using Equations (2.64), (2.67), and (2.68) or the block diagram, and letting $T_d(s) = 0$, we solve to obtain the transfer function

$$G(s) = \frac{\theta(s)}{V_a(s)} = \frac{K_m}{s[(R_a + L_a s)(Js + b) + K_b K_m]}$$

$$= \frac{K_m}{s(s^2 + 2\zeta\omega_n s + \omega_n^2)}. \tag{2.69}$$

However, for many DC motors, the time constant of the armature, $\tau_a = L_a/R_a$, is negligible; therefore,

$$G(s) = \frac{\theta(s)}{V_a(s)} = \frac{K_m}{s[R_a(Js + b) + K_b K_m]} = \frac{K_m/(R_a b + K_b K_m)}{s(\tau_1 s + 1)}, \tag{2.70}$$

where the equivalent time constant $\tau_1 = R_a J/(R_a b + K_b K_m)$.

Note that K_m is equal to K_b. This equality may be shown by considering the steady-state motor operation and the power balance when the rotor resistance is neglected. The power input to the rotor is $(K_b\omega)i_a$, and the power delivered to the shaft is $T\omega$. In the steady-state condition, the power input is equal to the power delivered to the shaft so that $(K_b\omega)i_a = T\omega$; since $T = K_m i_a$ (Equation 2.64), we find that $K_b = K_m$.

Electric motors are used for moving loads when a rapid response is not required and for relatively low power requirements. Typical constants for a fractional horsepower motor are provided in Table 2.4. Actuators that operate as a result of hydraulic pressure are used for large loads. Figure 2.21 shows the usual ranges of use for electromechanical drives as contrasted to electrohydraulic drives. Typical applications are also shown on the figure. ■

Table 2.4   Typical Constants for a Fractional Horsepower DC Motor

Motor constant K_m	$50 \times 10^{-3}\,\text{N}\cdot\text{m/A}$
Rotor inertia J_m	$1 \times 10^{-3}\,\text{N}\cdot\text{m}\cdot\text{s}^2/\text{rad}$
Field time constant τ_f	1 ms
Rotor time constant τ	100 ms
Maximum output power	$\frac{1}{4}$ hp, 187 W

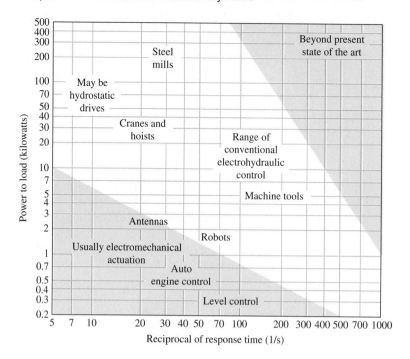

FIGURE 2.21
Range of control response time and power to load for electromechanical and electrohydraulic devices.

EXAMPLE 2.6 **Transfer function of a hydraulic actuator**

A useful actuator for the linear positioning of a mass is the hydraulic actuator shown in Table 2.5, item 9 [9, 10]. The hydraulic actuator is capable of providing a large power amplification. It will be assumed that the hydraulic fluid is available from a constant pressure source and that the compressibility of the fluid is negligible. A downward input displacement x moves the control valve; thus, fluid passes into the upper part of the cylinder, and the piston is forced downward. A small, low-power displacement of $x(t)$ causes a larger, high-power displacement, $y(t)$. The volumetric fluid flow rate Q is related to the input displacement $x(t)$ and the differential pressure across the piston as $Q = g(x, P)$. Using the Taylor series linearization as in Equation (2.11), we have

$$Q = \left(\frac{\partial g}{\partial x}\right)_{x_0, P_0} x + \left(\frac{\partial g}{\partial P}\right)_{x_0, P_0} P = k_x x - k_P P, \qquad (2.71)$$

where $g = g(x, P)$ and (x_0, P_0) is the operating point. The force developed by the actuator piston is equal to the area of the piston, A, multiplied by the pressure, P. This force is applied to the mass, so we have

$$AP = M\frac{d^2y}{dt^2} + b\frac{dy}{dt}. \qquad (2.72)$$

Thus, substituting Equation (2.71) into Equation (2.72), we obtain

$$\frac{A}{k_P}(k_x x - Q) = M\frac{d^2 y}{dt^2} + b\frac{dy}{dt}.$$ (2.73)

Furthermore, the volumetric fluid flow is related to the piston movement as

$$Q = A\frac{dy}{dt}.$$ (2.74)

Then, substituting Equation (2.74) into Equation (2.73) and rearranging, we have

$$\frac{Ak_x}{k_P}x = M\frac{d^2 y}{dt^2} + \left(b + \frac{A^2}{k_P}\right)\frac{dy}{dt}.$$ (2.75)

Therefore, using the Laplace transformation, we have the transfer function

$$\frac{Y(s)}{X(s)} = \frac{K}{s(Ms + B)},$$ (2.76)

where

$$K = \frac{Ak_x}{k_P} \quad \text{and} \quad B = b + \frac{A^2}{k_P}.$$

Note that the transfer function of the hydraulic actuator is similar to that of the electric motor. For an actuator operating at high pressure levels and requiring a rapid response of the load, we must account for the effect of the compressibility of the fluid [4, 5].

    Symbols, units, and conversion factors associated with many of the variables in Table 2.5 are located at the MCS website. The symbols and units for each variable can be found in tables with corresponding conversions between SI and English units. ∎

    The transfer function concept and approach is very important because it provides the analyst and designer with a useful mathematical model of the system elements. We shall find the transfer function to be a continually valuable aid in the attempt to model dynamic systems. The approach is particularly useful because the *s*-plane poles and zeros of the transfer function represent the transient response of the system. The transfer functions of several dynamic elements are given in Table 2.5.

    In many situations in engineering, the transmission of rotary motion from one shaft to another is a fundamental requirement. For example, the output power of an automobile engine is transferred to the driving wheels by means of the gearbox and differential. The gearbox allows the driver to select different gear ratios depending on the traffic situation, whereas the differential has a fixed ratio. The speed of the engine in this case is not constant, since it is under the control of the driver. Another example is a set of gears that transfer the power at the shaft of an electric motor to the shaft of a rotating antenna. Examples of mechanical converters are gears, chain drives, and belt drives. A commonly used electric converter is the electric transformer. An example of a device that converts rotational motion to linear motion is the rack-and-pinion gear shown in Table 2.5, item 17.

Table 2.5   Transfer Functions of Dynamic Elements and Networks

Element or System	G(s)

1. Integrating circuit, filter

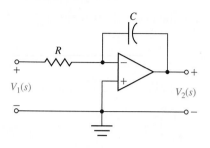

$$\frac{V_2(s)}{V_1(s)} = -\frac{1}{RCs}$$

2. Differentiating circuit

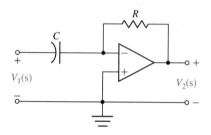

$$\frac{V_2(s)}{V_1(s)} = -RCs$$

3. Differentiating circuit

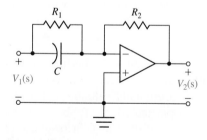

$$\frac{V_2(s)}{V_1(s)} = -\frac{R_2(R_1Cs + 1)}{R_1}$$

4. Integrating filter

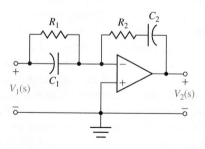

$$\frac{V_2(s)}{V_1(s)} = -\frac{(R_1C_1s + 1)(R_2C_2s + 1)}{R_1C_2s}$$

(continued)

Table 2.5 *Continued*

Element or System	G(s)

5. DC motor, field-controlled, rotational actuator

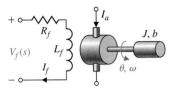

$$\frac{\theta(s)}{V_f(s)} = \frac{K_m}{s(Js + b)(L_f s + R_f)}$$

6. DC motor, armature-controlled, rotational actuator

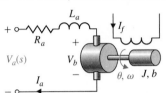

$$\frac{\theta(s)}{V_a(s)} = \frac{K_m}{s[(R_a + L_a s)(Js + b) + K_b K_m]}$$

7. AC motor, two-phase control field, rotational actuator

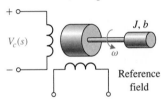

$$\frac{\theta(s)}{V_c(s)} = \frac{K_m}{s(\tau s + 1)}$$

$$\tau = J/(b - m)$$

$m =$ slope of linearized torque-speed curve (normally negative)

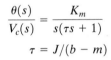

$$\frac{V_0(s)}{V_c(s)} = \frac{K/(R_c R_q)}{(s\tau_c + 1)(s\tau_q + 1)}$$

$$\tau_c = L_c/R_c, \qquad \tau_q = L_q/R_q$$

for the unloaded case, $i_d \approx 0$, $\tau_c \approx \tau_q$,
$$0.05 \text{ s} < \tau_c < 0.5 \text{ s}$$

$$V_q, V_{34} = V_d$$

9. Hydraulic actuator

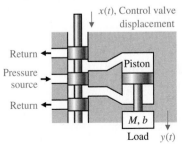

$$\frac{Y(s)}{X(s)} = \frac{K}{s(Ms + B)}$$

$$K = \frac{Ak_x}{k_p}, \qquad B = \left(b + \frac{A^2}{k_p}\right)$$

$$k_x = \frac{\partial g}{\partial x}\bigg|_{x_0}, \qquad k_p = \frac{\partial g}{\partial P}\bigg|_{P_0},$$

$$g = g(x, P) = \text{flow}$$

$$A = \text{area of piston} \qquad \textit{(continued)}$$

Table 2.5 *Continued*

Element or System	G(s)

10. Gear train, rotational transformer

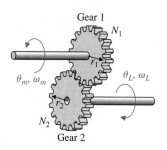

$$\text{Gear ratio} = n = \frac{N_1}{N_2}$$

$$N_2\theta_L = N_1\theta_m, \qquad \theta_L = n\theta_m$$

$$\omega_L = n\omega_m$$

11. Potentiometer, voltage control

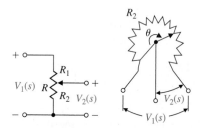

$$\frac{V_2(s)}{V_1(s)} = \frac{R_2}{R} = \frac{R_2}{R_1 + R_2}$$

$$\frac{R_2}{R} = \frac{\theta}{\theta_{\max}}$$

12. Potentiometer, error detector bridge

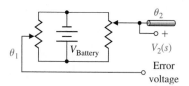

$$V_2(s) = k_s(\theta_1(s) - \theta_2(s))$$

$$V_2(s) = k_s\theta_{\text{error}}(s)$$

$$k_s = \frac{V_{\text{Battery}}}{\theta_{\max}}$$

13. Tachometer, velocity sensor

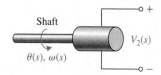

$$V_2(s) = K_t\omega(s) = K_ts\theta(s)$$

$$K_t = \text{constant}$$

14. DC amplifier

$$\frac{V_2(s)}{V_1(s)} = \frac{k_a}{s\tau + 1}$$

$R_o = $ output resistance

$C_o = $ output capacitance

$\tau = R_oC_o, \tau \ll 1\text{s}$

and is often negligible for
controller amplifier

(continued)

Table 2.5 *Continued*

Element or System	$G(s)$
15. Accelerometer, acceleration sensor	

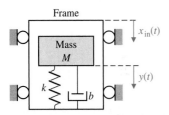

$$x_o(t) = y(t) - x_{in}(t),$$

$$\frac{X_o(s)}{X_{in}(s)} = \frac{-s^2}{s^2 + (b/M)s + k/M}$$

For low-frequency oscillations, where

$$\omega < \omega_n,$$

$$\frac{X_o(j\omega)}{X_{in}(j\omega)} \simeq \frac{\omega^2}{k/M}$$

16. Thermal heating system

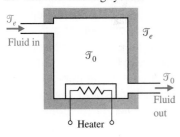

$$\frac{\mathcal{T}(s)}{q(s)} = \frac{1}{C_t s + (QS + 1/R_t)}, \text{ where}$$

$\mathcal{T} = \mathcal{T}_o - \mathcal{T}_e$ = temperature difference due to thermal process

C_t = thermal capacitance

Q = fluid flow rate = constant

S = specific heat of water

R_t = thermal resistance of insulation

$q(s)$ = transform of rate of heat flow of heating element

17. Rack and pinion

$x = r\theta$

converts radial motion

to linear motion

2.6 BLOCK DIAGRAM MODELS

The dynamic systems that comprise automatic control systems are represented mathematically by a set of simultaneous differential equations. As we have noted in the previous sections, the Laplace transformation reduces the problem to the solution of a set of linear algebraic equations. Since control systems are concerned with the control of specific variables, the controlled variables must relate to the controlling variables. This relationship is typically represented by the transfer function of the subsystem relating

FIGURE 2.22
Block diagram of a
DC motor.

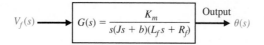

$$V_f(s) \longrightarrow \boxed{G(s) = \dfrac{K_m}{s(Js + b)(L_f s + R_f)}} \begin{array}{c}\text{Output}\\ \longrightarrow\end{array} \theta(s)$$

FIGURE 2.23
General block
representation of
two-input, two-
output system.

the input and output variables. Therefore, one can correctly assume that the transfer function is an important relation for control engineering.

The importance of this cause-and-effect relationship is evidenced by the facility to represent the relationship of system variables by diagrammatic means. The **block diagram** representation of the system relationships is prevalent in control system engineering. Block diagrams consist of unidirectional, operational blocks that represent the transfer function of the variables of interest. A block diagram of a field-controlled DC motor and load is shown in Figure 2.22. The relationship between the displacement $\theta(s)$ and the input voltage $V_f(s)$ is clearly portrayed by the block diagram.

To represent a system with several variables under control, an interconnection of blocks is utilized. For example, the system shown in Figure 2.23 has two input variables and two output variables [6]. Using transfer function relations, we can write the simultaneous equations for the output variables as

$$Y_1(s) = G_{11}(s)R_1(s) + G_{12}(s)R_2(s), \tag{2.77}$$

and

$$Y_2(s) = G_{21}(s)R_1(s) + G_{22}(s)R_2(s), \tag{2.78}$$

where $G_{ij}(s)$ is the transfer function relating the ith output variable to the jth input variable. The block diagram representing this set of equations is shown in Figure 2.24. In general, for J inputs and I outputs, we write the simultaneous equation in matrix form as

$$\begin{bmatrix} Y_1(s) \\ Y_2(s) \\ \vdots \\ Y_I(s) \end{bmatrix} = \begin{bmatrix} G_{11}(s) & \cdots & G_{1J}(s) \\ G_{21}(s) & \cdots & G_{2J}(s) \\ \vdots & & \vdots \\ G_{I1}(s) & \cdots & G_{IJ}(s) \end{bmatrix} \begin{bmatrix} R_1(s) \\ R_2(s) \\ \vdots \\ R_J(s) \end{bmatrix} \tag{2.79}$$

or simply

$$\mathbf{Y} = \mathbf{GR}. \tag{2.80}$$

FIGURE 2.24
Block diagram of
interconnected
system.

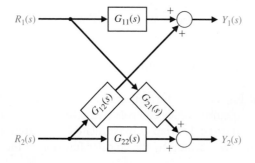

Here the **Y** and **R** matrices are column matrices containing the I output and the J input variables, respectively, and **G** is an I by J transfer function matrix. The matrix representation of the interrelationship of many variables is particularly valuable for complex multivariable control systems. An introduction to matrix algebra is provided on the MCS website for those unfamiliar with matrix algebra or who would find a review helpful [22].

The block diagram representation of a given system often can be reduced to a simplified block diagram with fewer blocks than the original diagram. Since the transfer functions represent linear systems, the multiplication is commutative. Thus, in Table 2.6, item 1, we have

$$X_3(s) = G_2(s)X_2(s) = G_1(s)G_2(s)X_1(s).$$

Table 2.6   Block Diagram Transformations

Transformation	Original Diagram	Equivalent Diagram
1. Combining blocks in cascade	$X_1 \rightarrow \boxed{G_1(s)} \xrightarrow{X_2} \boxed{G_2(s)} \xrightarrow{X_3}$	$X_1 \rightarrow \boxed{G_1 G_2} \xrightarrow{X_3}$ or $X_1 \rightarrow \boxed{G_2 G_1} \xrightarrow{X_3}$
2. Moving a summing point behind a block	$X_1 \xrightarrow{+} \bigcirc \pm \rightarrow \boxed{G} \xrightarrow{X_3}$, feedback X_2	$X_1 \rightarrow \boxed{G} \rightarrow \bigcirc \xrightarrow{+} X_3$, $\pm$, $X_2 \rightarrow \boxed{G}$
3. Moving a pickoff point ahead of a block	$X_1 \rightarrow \boxed{G} \xrightarrow{X_2}$, pickoff X_2	$X_1 \rightarrow \boxed{G} \xrightarrow{X_2}$, $X_2 \leftarrow \boxed{G}$
4. Moving a pickoff point behind a block	$X_1 \rightarrow \boxed{G} \xrightarrow{X_2}$, pickoff X_1	$X_1 \rightarrow \boxed{G} \xrightarrow{X_2}$, $X_1 \leftarrow \boxed{\dfrac{1}{G}}$
5. Moving a summing point ahead of a block	$X_1 \rightarrow \boxed{G} \xrightarrow{+} \bigcirc \xrightarrow{X_3}$, $\pm$, X_2	$X_1 \xrightarrow{+} \bigcirc \rightarrow \boxed{G} \xrightarrow{X_3}$, $\pm$, $\boxed{\dfrac{1}{G}} \xleftarrow{} X_2$
6. Eliminating a feedback loop	$X_1 \xrightarrow{+} \bigcirc \pm \rightarrow \boxed{G} \xrightarrow{X_2}$, $\boxed{H} \leftarrow$	$X_1 \rightarrow \boxed{\dfrac{G}{1 \mp GH}} \xrightarrow{X_2}$

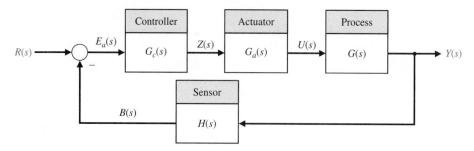

FIGURE 2.25
Negative feedback
control system.

When two blocks are connected in cascade, as in Table 2.6, item 1, we assume that

$$X_3(s) = G_2(s)G_1(s)X_1(s)$$

holds true. This assumes that when the first block is connected to the second block, the effect of loading of the first block is negligible. Loading and interaction between interconnected components or systems may occur. If the loading of interconnected devices does occur, the engineer must account for this change in the transfer function and use the corrected transfer function in subsequent calculations.

Block diagram transformations and reduction techniques are derived by considering the algebra of the diagram variables. For example, consider the block diagram shown in Figure 2.25. This negative feedback control system is described by the equation for the actuating signal, which is

$$E_a(s) = R(s) - B(s) = R(s) - H(s)Y(s). \tag{2.81}$$

Because the output is related to the actuating signal by $G(s)$, we have

$$Y(s) = G(s)U(s) = G(s)G_a(s)Z(s) = G(s)G_a(s)G_c(s)E_a(s); \tag{2.82}$$

thus,

$$Y(s) = G(s)G_a(s)G_c(s)[R(s) - H(s)Y(s)]. \tag{2.83}$$

Combining the $Y(s)$ terms, we obtain

$$Y(s)[1 + G(s)G_a(s)G_c(s)H(s)] = G(s)G_a(s)G_c(s)R(s). \tag{2.84}$$

Therefore, the transfer function relating the output $Y(s)$ to the input $R(s)$ is

$$\boxed{\frac{Y(s)}{R(s)} = \frac{G(s)G_a(s)G_c(s)}{1 + G(s)G_a(s)G_c(s)H(s)}.} \tag{2.85}$$

This **closed-loop transfer function** is particularly important because it represents many of the existing practical control systems.

The reduction of the block diagram shown in Figure 2.25 to a single block representation is one example of several useful techniques. These diagram transformations are given in Table 2.6. All the transformations in Table 2.6 can be derived by simple algebraic manipulation of the equations representing the blocks. System analysis by the method of block diagram reduction affords a better understanding of the contribution of each component element than possible by the manipulation of

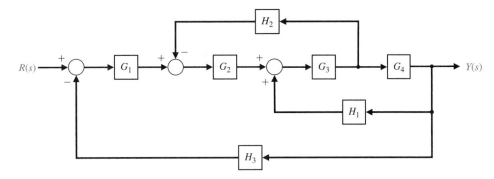

FIGURE 2.26
Multiple-loop
feedback control
system.

equations. The utility of the block diagram transformations will be illustrated by an example using block diagram reduction.

EXAMPLE 2.7   **Block diagram reduction**

The block diagram of a multiple-loop feedback control system is shown in Figure 2.26. It is interesting to note that the feedback signal $H_1(s)Y(s)$ is a positive feedback signal, and the loop $G_3(s)G_4(s)H_1(s)$ is a **positive feedback loop**. The block diagram reduction procedure is based on the use of Table 2.6, transformation 6, which eliminates feedback loops. Therefore the other transformations are used to transform the diagram to a form ready for eliminating feedback loops. First, to eliminate the loop $G_3G_4H_1$, we move H_2 behind block G_4 by using transformation 4, and obtain Figure 2.27(a). Eliminating the loop $G_3G_4H_1$ by using transformation 6, we obtain Figure 2.27(b). Then, eliminating the inner loop containing H_2/G_4, we obtain Figure 2.27(c). Finally, by reducing the loop containing H_3, we obtain the closed-loop system transfer function as shown in Figure 2.27(d). It is worthwhile to examine the form of the numerator and denominator of this closed-loop transfer function. We note that the numerator is composed of the cascade transfer function of the feed-forward elements connecting the input $R(s)$ and the output $Y(s)$. The denominator is composed of 1 minus the sum of each loop transfer function. The loop $G_3G_4H_1$ has a plus sign in the sum to be subtracted because it is a positive feedback loop, whereas the loops $G_1G_2G_3G_4H_3$ and $G_2G_3H_2$ are negative feedback loops. To illustrate this point, the denominator can be rewritten as

$$q(s) = 1 - (+G_3G_4H_1 - G_2G_3H_2 - G_1G_2G_3G_4H_3). \qquad (2.86)$$

This form of the numerator and denominator is quite close to the general form for multiple-loop feedback systems, as we shall find in the following section. ∎

The block diagram representation of feedback control systems is a valuable and widely used approach. The block diagram provides the analyst with a graphical representation of the interrelationships of controlled and input variables. Furthermore, the designer can readily visualize the possibilities for adding blocks to the existing system block diagram to alter and improve the system performance. The transition from the block diagram method to a method utilizing a line path representation instead of a block representation is readily accomplished and is presented in the following section.

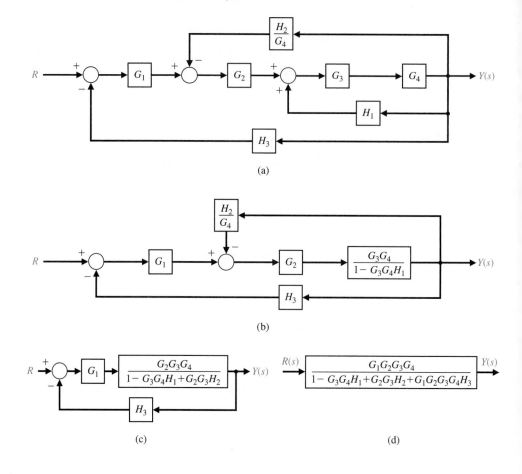

(a)

(b)

(c)

(d)

FIGURE 2.27
Block diagram
reduction of the
system of Figure
2.26.

2.7 SIGNAL-FLOW GRAPH MODELS

Block diagrams are adequate for the representation of the interrelationships of controlled and input variables. However, for a system with reasonably complex interrelationships, the block diagram reduction procedure is cumbersome and often quite difficult to complete. An alternative method for determining the relationship between system variables has been developed by Mason and is based on a representation of the system by line segments [4, 25]. The advantage of the line path method, called the signal-flow graph method, is the availability of a flow graph gain formula, which provides the relation between system variables without requiring any reduction procedure or manipulation of the flow graph.

The transition from a block diagram representation to a directed line segment representation is easy to accomplish by reconsidering the systems of the previous section. A **signal-flow graph** is a diagram consisting of nodes that are connected by several directed branches and is a graphical representation of a set of linear relations. Signal-flow graphs are particularly useful for feedback control systems because feedback theory is primarily concerned with the flow and processing of signals in systems. The basic element of a signal-flow graph is a unidirectional path segment called a **branch**, which relates the dependency of an input and an output variable in

FIGURE 2.28
Signal-flow graph
of the DC motor.

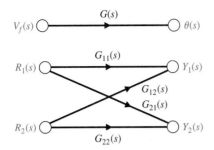

FIGURE 2.29
Signal-flow graph
of interconnected
system.

a manner equivalent to a block of a block diagram. Therefore, the branch relating the output $\theta(s)$ of a DC motor to the field voltage $V_f(s)$ is similar to the block diagram of Figure 2.22 and is shown in Figure 2.28. The input and output points or junctions are called **nodes**. Similarly, the signal-flow graph representing Equations (2.77) and (2.78), as well as Figure 2.24, is shown in Figure 2.29. The relation between each variable is written next to the directional arrow. All branches leaving a node will pass the nodal signal to the output node of each branch (unidirectionally). The summation of all signals entering a node is equal to the node variable. A **path** is a branch or a continuous sequence of branches that can be traversed from one signal (node) to another signal (node). A **loop** is a closed path that originates and terminates on the same node, with no node being met twice along the path. Two loops are said to be **nontouching** if they do not have a common node. Two touching loops share one or more common nodes. Therefore, considering Figure 2.29 again, we obtain

$$Y_1(s) = G_{11}(s)R_1(s) + G_{12}(s)R_2(s), \qquad (2.87)$$

and

$$Y_2(s) = G_{21}(s)R_1(s) + G_{22}(s)R_2(s). \qquad (2.88)$$

The flow graph is simply a pictorial method of writing a system of algebraic equations that indicates the interdependencies of the variables. As another example, consider the following set of simultaneous algebraic equations:

$$a_{11}x_1 + a_{12}x_2 + r_1 = x_1 \qquad (2.89)$$

$$a_{21}x_1 + a_{22}x_2 + r_2 = x_2. \qquad (2.90)$$

The two input variables are r_1 and r_2, and the output variables are x_1 and x_2. A signal-flow graph representing Equations (2.89) and (2.90) is shown in Figure 2.30. Equations (2.89) and (2.90) may be rewritten as

$$x_1(1 - a_{11}) + x_2(-a_{12}) = r_1, \qquad (2.91)$$

and

$$x_1(-a_{21}) + x_2(1 - a_{22}) = r_2. \qquad (2.92)$$

The simultaneous solution of Equations (2.91) and (2.92) using Cramer's rule results in the solutions

$$x_1 = \frac{(1 - a_{22})r_1 + a_{12}r_2}{(1 - a_{11})(1 - a_{22}) - a_{12}a_{21}} = \frac{1 - a_{22}}{\Delta}r_1 + \frac{a_{12}}{\Delta}r_2, \qquad (2.93)$$

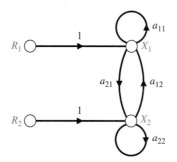

FIGURE 2.30
Signal-flow graph
of two algebraic
equations.

and

$$x_2 = \frac{(1 - a_{11})r_2 + a_{21}r_1}{(1 - a_{11})(1 - a_{22}) - a_{12}a_{21}} = \frac{1 - a_{11}}{\Delta}r_2 + \frac{a_{21}}{\Delta}r_1. \qquad (2.94)$$

The denominator of the solution is the determinant Δ of the set of equations and is rewritten as

$$\Delta = (1 - a_{11})(1 - a_{22}) - a_{12}a_{21} = 1 - a_{11} - a_{22} + a_{11}a_{22} - a_{12}a_{21}. \qquad (2.95)$$

In this case, the denominator is equal to 1 minus each self-loop a_{11}, a_{22}, and $a_{12}a_{21}$, plus the product of the two nontouching loops a_{11} and a_{22}. The loops a_{22} and $a_{21}a_{12}$ are touching, as are a_{11} and $a_{21}a_{12}$.

The numerator for x_1 with the input r_1 is 1 times $1 - a_{22}$, which is the value of Δ excluding terms that touch the path 1 from r_1 to x_1. Therefore the numerator from r_2 to x_1 is simply a_{12} because the path through a_{12} touches all the loops. The numerator for x_2 is symmetrical to that of x_1.

In general, the linear dependence T_{ij} between the independent variable x_i (often called the input variable) and a dependent variable x_j is given by Mason's signal-flow gain formula [11, 12],

$$T_{ij} = \frac{\sum_k P_{ijk} \Delta_{ijk}}{\Delta}, \qquad (2.96)$$

P_{ijk} = gain of kth path from variable x_i to variable x_j,

Δ = determinant of the graph,

Δ_{ijk} = cofactor of the path P_{ijk},

and the summation is taken over all possible k paths from x_i to x_j. The path gain or transmittance P_{ijk} is defined as the product of the gains of the branches of the path, traversed in the direction of the arrows with no node encountered more than once. The cofactor Δ_{ijk} is the determinant with the loops touching the kth path removed. The determinant Δ is

$$\Delta = 1 - \sum_{n=1}^{N} L_n + \sum_{\substack{n, m \\ \text{nontouching}}} L_n L_m - \sum_{\substack{n, m, p \\ \text{nontouching}}} L_n L_m L_p + \cdots, \qquad (2.97)$$

where L_q equals the value of the qth loop transmittance. Therefore the rule for evaluating Δ in terms of loops $L_1, L_2, L_3, \ldots, L_N$ is

$\Delta = 1 - $ (sum of all different loop gains)

$\quad$ + (sum of the gain products of all combinations of two nontouching loops)

$\quad$ − (sum of the gain products of all combinations of three nontouching loops)

$\quad$ + $\cdots$.

The gain formula is often used to relate the output variable $Y(s)$ to the input variable $R(s)$ and is given in somewhat simplified form as

$$T = \frac{\Sigma_k P_k \Delta_k}{\Delta},\qquad (2.98)$$

where $T(s) = Y(s)/R(s)$.

Several examples will illustrate the utility and ease of this method. Although the gain Equation (2.96) appears to be formidable, one must remember that it represents a summation process, not a complicated solution process.

EXAMPLE 2.8   Transfer function of an interacting system

A two-path signal-flow graph is shown in Figure 2.31(a) and the corresponding block diagram is shown in Figure 2.31(b). An example of a control system with multiple signal paths is a multilegged robot. The paths connecting the input $R(s)$ and output $Y(s)$ are

$$P_1 = G_1 G_2 G_3 G_4 \text{ (path 1)} \quad \text{and} \quad P_2 = G_5 G_6 G_7 G_8 \text{ (path 2)}.$$

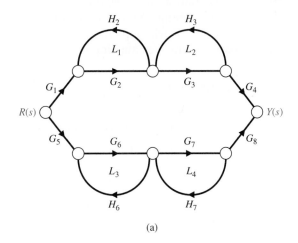

(a)

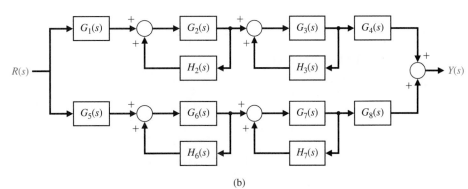

(b)

FIGURE 2.31
Two-path
interacting system.
(a) Signal-flow
graph. (b) Block
diagram.

There are four self-loops:

$$L_1 = G_2H_2, \qquad L_2 = H_3G_3, \qquad L_3 = G_6H_6, \quad \text{and} \quad L_4 = G_7H_7.$$

Loops L_1 and L_2 do not touch L_3 and L_4. Therefore, the determinant is

$$\Delta = 1 - (L_1 + L_2 + L_3 + L_4) + (L_1L_3 + L_1L_4 + L_2L_3 + L_2L_4). \qquad (2.99)$$

The cofactor of the determinant along path 1 is evaluated by removing the loops that touch path 1 from Δ. Hence, we have

$$L_1 = L_2 = 0 \quad \text{and} \quad \Delta_1 = 1 - (L_3 + L_4).$$

Similarly, the cofactor for path 2 is

$$\Delta_2 = 1 - (L_1 + L_2).$$

Therefore, the transfer function of the system is

$$\frac{Y(s)}{R(s)} = T(s) = \frac{P_1\Delta_1 + P_2\Delta_2}{\Delta} \qquad (2.100)$$

$$= \frac{G_1G_2G_3G_4(1 - L_3 - L_4) + G_5G_6G_7G_8(1 - L_1 - L_2)}{1 - L_1 - L_2 - L_3 - L_4 + L_1L_3 + L_1L_4 + L_2L_3 + L_2L_4}.$$

A similar analysis can be accomplished using block diagram reduction techniques. The block diagram shown in Figure 2.31(b) has four inner feedback loops within the overall block diagram. The block diagram reduction is simplified by first reducing the four inner feedback loops and then placing the resulting systems in series. Along the top path, the transfer function is

$$Y_1(s) = G_1(s)\left[\frac{G_2(s)}{1 - G_2(s)H_2(s)}\right]\left[\frac{G_3(s)}{1 - G_3(s)H_3(s)}\right]G_4(s)R(s)$$

$$= \left[\frac{G_1(s)G_2(s)G_3(s)G_4(s)}{(1 - G_2(s)H_2(s))(1 - G_3(s)H_3(s))}\right]R(s).$$

Similarly across the bottom path, the transfer function is

$$Y_2(s) = G_5(s)\left[\frac{G_6(s)}{1 - G_6(s)H_6(s)}\right]\left[\frac{G_7(s)}{1 - G_7(s)H_7(s)}\right]G_8(s)R(s)$$

$$= \left[\frac{G_5(s)G_6(s)G_7(s)G_8(s)}{(1 - G_6(s)H_6(s))(1 - G_7(s)H_7(s))}\right]R(s).$$

The total transfer function is then given by

$$Y(s) = Y_1(s) + Y_2(s) = \left[\frac{G_1(s)G_2(s)G_3(s)G_4(s)}{(1 - G_2(s)H_2(s))(1 - G_3(s)H_3(s))}\right.$$

$$\left. + \frac{G_5(s)G_6(s)G_7(s)G_8(s)}{(1 - G_6(s)H_6(s))(1 - G_7(s)H_7(s))}\right]R(s). \quad \blacksquare$$

EXAMPLE 2.9   **Armature-controlled motor**

The block diagram of the armature-controlled DC motor is shown in Figure 2.20. This diagram was obtained from Equations (2.64)–(2.68). The signal-flow diagram can be obtained either from Equations (2.64)–(2.68) or from the block diagram and is shown in Figure 2.32. Using Mason's signal-flow gain formula, let us obtain the transfer function for $\theta(s)/V_a(s)$ with $T_d(s) = 0$. The forward path is $P_1(s)$, which touches the one loop, $L_1(s)$, where

$$P_1(s) = \frac{1}{s}G_1(s)G_2(s) \quad \text{and} \quad L_1(s) = -K_bG_1(s)G_2(s).$$

Therefore, the transfer function is

$$T(s) = \frac{P_1(s)}{1 - L_1(s)} = \frac{(1/s)G_1(s)G_2(s)}{1 + K_bG_1(s)G_2(s)} = \frac{K_m}{s[(R_a + L_as)(Js + b) + K_bK_m]},$$

which is exactly the same as that derived earlier (Equation 2.69). ∎

The signal-flow graph gain formula provides a reasonably straightforward approach for the evaluation of complicated systems. To compare the method with block diagram reduction, which is really not much more difficult, let us reconsider the complex system of Example 2.7.

EXAMPLE 2.10   **Transfer function of a multiple-loop system**

A multiple-loop feedback system is shown in Figure 2.26 in block diagram form. There is no need to redraw the diagram in signal-flow graph form, and so we shall proceed as usual by using Mason's signal-flow gain formula, Equation (2.98). There is one forward path $P_1 = G_1G_2G_3G_4$. The feedback loops are

$$L_1 = -G_2G_3H_2, \qquad L_2 = G_3G_4H_1, \quad \text{and} \quad L_3 = -G_1G_2G_3G_4H_3. \qquad (2.101)$$

All the loops have common nodes and therefore are all touching. Furthermore, the path P_1 touches all the loops, so $\Delta_1 = 1$. Thus, the closed-loop transfer function is

$$T(s) = \frac{Y(s)}{R(s)} = \frac{P_1\Delta_1}{1 - L_1 - L_2 - L_3}$$

$$= \frac{G_1G_2G_3G_4}{1 + G_2G_3H_2 - G_3G_4H_1 + G_1G_2G_3G_4H_3}. \qquad (2.102) \ \blacksquare$$

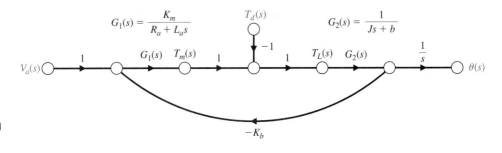

FIGURE 2.32
The signal-flow graph of the armature-controlled DC motor.

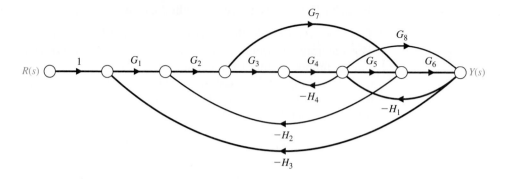

FIGURE 2.33
Multiple-loop
system.

EXAMPLE 2.11   **Transfer function of a complex system**

Finally, we shall consider a reasonably complex system that would be difficult to re-duce by block diagram techniques. A system with several feedback loops and feed-forward paths is shown in Figure 2.33. The forward paths are

$$P_1 = G_1G_2G_3G_4G_5G_6, \qquad P_2 = G_1G_2G_7G_6, \quad \text{and} \quad P_3 = G_1G_2G_3G_4G_8.$$

The feedback loops are

$$L_1 = -G_2G_3G_4G_5H_2, \qquad L_2 = -G_5G_6H_1, \qquad L_3 = -G_8H_1, \qquad L_4 = -G_7H_2G_2$$
$$L_5 = -G_4H_4 \qquad L_6 = -G_1G_2G_3G_4G_5G_6H_3, \qquad L_7 = -G_1G_2G_7G_6H_3, \quad \text{and}$$
$$L_8 = -G_1G_2G_3G_4G_8H_3.$$

Loop L_5 does not touch loop L_4 or loop L_7, and loop L_3 does not touch loop L_4; but all other loops touch. Therefore, the determinant is

$$\Delta = 1 - (L_1 + L_2 + L_3 + L_4 + L_5 + L_6 + L_7 + L_8) + (L_5L_7 + L_5L_4 + L_3L_4). \tag{2.103}$$

The cofactors are

$$\Delta_1 = \Delta_3 = 1 \quad \text{and} \quad \Delta_2 = 1 - L_5 = 1 + G_4H_4.$$

Finally, the transfer function is

$$T(s) = \frac{Y(s)}{R(s)} = \frac{P_1 + P_2\Delta_2 + P_3}{\Delta}. \tag{2.104} \blacksquare$$

Signal-flow graphs and Mason's signal-flow gain formula may be used prof-itably for the analysis of feedback control systems, electronic amplifier circuits, sta-tistical systems, and mechanical systems, among many other examples.

2.8 DESIGN EXAMPLES

In this section we present five illustrative design examples. In the first example, we present a detailed look at modeling of the fluid level in a reservoir. The modeling is presented in a very detailed manner to emphasize the effort required to obtain a linear model in the form of a transfer function. The design process depicted in Figure 1.15 is highlighted in this example. The remaining four examples include an electric traction motor model development, a look at a mechanical accelerometer aboard a rocket sled, an overview of a laboratory robot and the associated hardware specifications, and the design of a low-pass filter.

EXAMPLE 2.12   Fluid flow modeling

A fluid flow system is shown in Figure 2.34. The reservoir (or tank) contains water that evacuates through an output port. Water is fed to the reservoir through a pipe controlled by an input valve. The variables of interest are the fluid velocity V (m/s), fluid height in the reservoir H (m), and pressure p (N/m^2). The pressure is defined as the force per unit area exerted by the fluid on a surface immersed (and at rest with respect to) the fluid. Fluid pressure acts normal to the surface. For further reading on fluid flow modeling, see [34–36].

The elements of the control system design process emphasized in this example are shown in Figure 2.35. The strategy is to establish the system configuration and then obtain the appropriate mathematical models describing the fluid flow reservoir from an input–output perspective.

The general equations of motion and energy describing fluid flow are quite complicated. The governing equations are coupled nonlinear partial differential equations. We must make some selective assumptions that reduce the complexity of the mathematical model. Although the control engineer is not required to be a fluid dynamicist, and a deep understanding of fluid dynamics is not necessarily acquired during the control system design process, it makes good engineering sense to gain at least a rudimentary understanding of the important simplifying assumptions. For a more complete discussion of fluid motion, see [37–39].

To obtain a realistic, yet tractable, mathematical model for the fluid flow reservoir, we first make several key assumptions. We assume that the water in the tank is incompressible and that the flow is inviscid, irrotational and steady. An incompressible fluid has a constant density ρ (kg/m^3). In fact, all fluids are compressible to some extent. The compressibility factor, k, is a measure of the compressibility of a fluid. A smaller value of k indicates less compressibility. Air (which is a compressible fluid) has a compressibility factor of $k_{air} = 0.98$ m^2/N, while water has a compressibility factor of $k_{H_2O} = 4.9 \times 10^{-10}$ m^2/N $= 50 \times 10^{-6}$ atm^{-1}. In other words, a given volume of

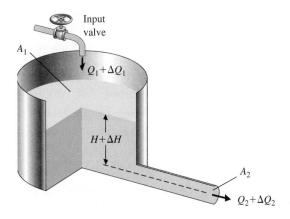

FIGURE 2.34
The fluid flow
reservoir
configuration.

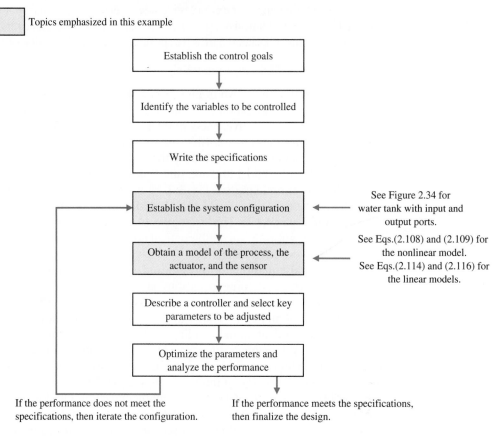

FIGURE 2.35   Elements of the control system design process emphasized in the fluid flow reservoir example.

water decreases by 50 one-millionths of the original volume for each atmosphere (atm) increase in pressure. Thus the assumption that the water is incompressible is valid for our application.

Consider a fluid in motion. Suppose that initially the flow velocities are different for adjacent layers of fluid. Then an exchange of molecules between the two layers tends to equalize the velocities in the layers. This is internal friction, and the exchange of momentum is known as viscosity. Solids are more viscous than fluids, and fluids are more viscous than gases. A measure of viscosity is the coefficient of viscosity μ (N s/m^2). A larger coefficient of viscosity implies higher viscosity. The coefficient of viscosity (under standard conditions, 20°C) for air is

$$\mu_{\text{air}} = 0.178 \times 10^{-4} \text{ N s/m}^2,$$

and for water we have

$$\mu_{\text{H}_2\text{O}} = 1.054 \times 10^{-3} \text{ N s/m}^2.$$

So water is about 60 times more viscous than air. Viscosity depends primarily on temperature, not pressure. For comparison, water at 0°C is about 2 times more viscous than water at 20°C. With fluids of low viscosity, such as air and water, the effects of friction are important only in the boundary layer, a thin layer adjacent to the wall of the reservoir and output pipe. We can neglect viscosity in our model development. We say our fluid is inviscid.

If each fluid element at each point in the flow has no net angular velocity about that point, the flow is termed irrotational. Imagine a small paddle wheel immersed in the fluid (say in the output port). If the paddle wheel translates without rotating, the flow is irrotational. We will assume the water in the tank is irrotational. For an inviscid fluid, an initially irrotational flow remains irrotational.

The water flow in the tank and output port can be either steady or unsteady. The flow is steady if the velocity at each point is constant in time. This does not necessarily imply that the velocity is the same at every point but rather that at any given point the velocity does not change with time. Steady-state conditions can be achieved at low fluid speeds. We will assume steady flow conditions. If the output port area is too large, then the flow through the reservoir may not be slow enough to establish the steady-state condition that we are assuming exists and our model will not accurately predict the fluid flow motion.

To obtain a mathematical model of the flow within the reservoir, we employ basic principles of science and engineering, such as the principle of conservation of mass. The mass of water in the tank at any given time is

$$m = \rho A_1 H, \tag{2.105}$$

where A_1 is the area of the tank, ρ is the water density, and H is the height of the water in the reservoir. The constants for the reservoir system are given in Table 2.7.

In the following formulas, a subscript 1 denotes quantities at the input, and a subscript 2 refers to quantities at the output. Taking the time derivative of m in Equation (2.105) yields

$$\dot{m} = \rho A_1 \dot{H},$$

where we have used the fact that our fluid is incompressible (that is, $\dot{\rho} = 0$) and that the area of the tank, A_1, does not change with time. The change in mass in the reservoir is equal to the mass that enters the tank minus the mass that leaves the tank, or

$$\dot{m} = \rho A_1 \dot{H} = Q_1 - \rho A_2 v_2, \tag{2.106}$$

where Q_1 is the steady-state input mass flow rate, v_2 is the exit velocity, and A_2 is the output port area. The exit velocity, v_2, is a function of the water height. From Bernoulli's equation [39] we have

$$\frac{1}{2}\rho v_1^2 + P_1 + \rho g H = \frac{1}{2}\rho v_2^2 + P_2,$$

Table 2.7   Water Tank Physical Constants

ρ (kg/m³)	g (m/s²)	A_1 (m²)	A_2 (m²)	H^* (m)	Q^* (kg/s)
1000	9.8	$\pi/4$	$\pi/400$	1	34.77

where v_1 is the water velocity at the mouth of the reservoir, and P_1 and P_2 are the atmospheric pressures at the input and output, respectively. But P_1 and P_2 are equal to atmospheric pressure, and A_2 is sufficiently small ($A_2 = A_1/100$), so the water flows out slowly and the velocity v_1 is negligible. Thus Bernoulli's equation reduces to

$$v_2 = \sqrt{2gH}. \tag{2.107}$$

Substituting Equation (2.107) into Equation (2.106) and solving for $\dot{H}$ yields

$$\dot{H} = -\left[\frac{A_2}{A_1}\sqrt{2g}\right]\sqrt{H} + \frac{1}{\rho A_1}Q_1. \tag{2.108}$$

Using Equation (2.107), we obtain the exit mass flow rate

$$Q_2 = \rho A_2 v_2 = (\rho\sqrt{2g}A_2)\sqrt{H}. \tag{2.109}$$

To keep the equations manageable, define

$$k_1 := -\frac{A_2\sqrt{2g}}{A_1},$$

$$k_2 := \frac{1}{\rho A_1},$$

$$k_3 := \rho\sqrt{2g}A_2.$$

Then, it follows that

$$\dot{H} = k_1\sqrt{H} + k_2 Q_1,$$
$$Q_2 = k_3\sqrt{H}. \tag{2.110}$$

Equation (2.110) represents our model of the water tank system, where the input is Q_1 and the output is Q_2. Equation (2.110) is a nonlinear, first-order, ordinary differential equation model. The nonlinearity comes from the $H^{1/2}$ term. The model in Equation (2.110) has the functional form

$$\dot{H} = f(H, Q_1),$$
$$Q_2 = h(H, Q_1),$$

where

$$f(H, Q_1) = k_1\sqrt{H} + k_2 Q_1 \quad \text{and} \quad h(H, Q_1) = k_3\sqrt{H}.$$

A set of linearized equations describing the height of the water in the reservoir is obtained using Taylor series expansions about an equilibrium flow condition. When the tank system is in equilibrium, we have $\dot{H} = 0$. We can define Q^* and H^* as the equilibrium input mass flow rate and water level, respectively. The relationship between Q^* and H^* is given by

$$Q^* = -\frac{k_1}{k_2}\sqrt{H^*} = \rho\sqrt{2g}A_2\sqrt{H^*}. \tag{2.111}$$

This condition occurs when just enough water enters the tank in A_1 to make up for the amount leaving through A_2. We can write the water level and input mass flow rate as

$$H = H^* + \Delta H, \tag{2.112}$$
$$Q_1 = Q^* + \Delta Q_1,$$

where ΔH and ΔQ_1 are small deviations from the equilibrium (steady-state) values. The Taylor series expansion about the equilibrium conditions is given by

$$\dot{H} = f(H, Q_1) = f(H^*, Q^*) + \left.\frac{\partial f}{\partial H}\right|_{\substack{H=H^* \\ Q_1=Q^*}} (H - H^*) \tag{2.113}$$

$$+ \left.\frac{\partial f}{\partial Q_1}\right|_{\substack{H=H^* \\ Q_1=Q^*}} (Q_1 - Q^*) + \cdots,$$

where

$$\left.\frac{\partial f}{\partial H}\right|_{\substack{H=H^* \\ Q_1=Q^*}} = \left.\frac{\partial(k_1\sqrt{H} + k_2 Q_1)}{\partial H}\right|_{\substack{H=H^* \\ Q_1=Q^*}} = \frac{1}{2}\frac{k_1}{\sqrt{H^*}},$$

and

$$\left.\frac{\partial f}{\partial Q_1}\right|_{\substack{H=H^* \\ Q_1=Q^*}} = \left.\frac{\partial(k_1\sqrt{H} + k_2 Q_1)}{\partial Q_1}\right|_{\substack{H=H^* \\ Q_1=Q^*}} = k_2.$$

Using Equation (2.111), we have

$$\sqrt{H^*} = \frac{Q^*}{\rho\sqrt{2g}\,A_2},$$

so that

$$\left.\frac{\partial f}{\partial H}\right|_{\substack{H=H^* \\ Q_1=Q^*}} = -\frac{A_2{}^2}{A_1}\frac{g\rho}{Q^*}.$$

It follows from Equation (2.112) that

$$\dot{H} = \Delta\dot{H},$$

since H^* is constant. Also, the term $f(H^*, Q^*)$ is identically zero, by definition of the equilibrium condition. Neglecting the higher order terms in the Taylor series expansion yields

$$\Delta\dot{H} = -\frac{A_2{}^2}{A_1}\frac{g\rho}{Q^*}\Delta H + \frac{1}{\rho A_1}\Delta Q_1. \tag{2.114}$$

Equation (2.114) is a linear model describing the deviation in water level ΔH from the steady-state due to a deviation from the nominal input mass flow rate ΔQ_1.

Similarly, for the output variable Q_2 we have

$$Q_2 = Q_2^* + \Delta Q_2 = h(H, Q_1) \tag{2.115}$$

$$\approx h(H^*, Q^*) + \left.\frac{\partial h}{\partial H}\right|_{\substack{H=H^* \\ Q_1=Q^*}}\Delta H + \left.\frac{\partial h}{\partial Q_1}\right|_{\substack{H=H^* \\ Q_1=Q^*}}\Delta Q_1,$$

where ΔQ_2 is a small deviation in the output mass flow rate and

$$\left.\frac{\partial h}{\partial H}\right|_{\substack{H=H^* \\ Q_1=Q^*}} = \frac{g\rho^2 A_2{}^2}{Q^*},$$

and

$$\left.\frac{\partial h}{\partial Q_1}\right|_{\substack{H=H^* \\ Q_1=Q^*}} = 0.$$

Therefore, the linearized equation for the output variable Q_2 is

$$\Delta Q_2 = \frac{g\rho^2 A_2^{\;2}}{Q^*}\Delta H. \tag{2.116}$$

For control system design and analysis, it is convenient to obtain the input–output relationship in the form of a transfer function. The tool to accomplish this is the Laplace transform, discussed in Section 2.4. Taking the time-derivative of Equation (2.116) and substituting into Equation (2.114) yields the input–output relationship

$$\Delta\dot{Q}_2 + \frac{A_2^{\;2}}{A_1}\frac{g\rho}{Q^*}\Delta Q_2 = \frac{A_2^{\;2}g\rho}{A_1 Q^*}\Delta Q_1.$$

If we define

$$\Omega := \frac{A_2^{\;2}}{A_1}\frac{g\rho}{Q^*}, \tag{2.117}$$

then we have

$$\Delta\dot{Q}_2 + \Omega\Delta Q_2 = \Omega\Delta Q_1. \tag{2.118}$$

Taking the Laplace transform (with zero initial conditions) yields the transfer function

$$\Delta Q_2(s)/\Delta Q_1(s) = \frac{\Omega}{s+\Omega}. \tag{2.119}$$

Equation (2.119) describes the relationship between the change in the output mass flow rate $\Delta Q_2(s)$ due to a change in the input mass flow rate $\Delta Q_1(s)$. We can also obtain a transfer function relationship between the change in the input mass flow rate and the change in the water level in the tank, $\Delta H(s)$. Taking the Laplace transform (with zero initial conditions) of Eq. (2.114) yields

$$\Delta H(s)/\Delta Q_1(s) = \frac{k_2}{s+\Omega}. \tag{2.120}$$

Given the linear time-invariant model of the water tank system in Equation (2.118), we can obtain solutions for step and sinusoidal inputs. Remember that our input $\Delta Q_1(s)$ is actually a change in the input mass flow rate from the steady-state value Q^*.

Consider the step input

$$\Delta Q_1(s) = q_o/s,$$

where q_o is the magnitude of the step input, and the initial condition is $\Delta Q_2(0) = 0$. Then we can use the transfer function form given in Eq. (2.119) to obtain

$$\Delta Q_2(s) = \frac{q_o\Omega}{s(s+\Omega)}.$$

The partial fraction expansion yields

$$\Delta Q_2(s) = \frac{-q_o}{s + \Omega} + \frac{q_o}{s}.$$

Taking the inverse Laplace transform yields

$$\Delta Q_2(t) = -q_o e^{-\Omega t} + q_o.$$

Note that $\Omega > 0$ (see Equation (2.117), so the term $e^{-\Omega t}$ approaches zero as t approaches ∞. Therefore, the steady-state output due to the step input of magnitude q_o is

$$\Delta Q_{2_{ss}} = q_o.$$

We see that in the steady state, the deviation of the output mass flow rate from the equilibrium value is equal to the deviation of the input mass flow rate from the equilibrium value. By examining the variable Ω in Equation (2.117), we find that the larger the output port opening A_2, the faster the system reaches steady state. In other words, as Ω gets larger, the exponential term $e^{-\Omega t}$ vanishes more quickly, and steady state is reached faster.

Similarly for the water level we have

$$\Delta H(s) = \frac{-q_o k_2}{\Omega}\left(\frac{1}{s + \Omega} - \frac{1}{s}\right).$$

Taking the inverse Laplace transform yields

$$\Delta H(t) = \frac{-q_o k_2}{\Omega}(e^{-\Omega t} - 1).$$

The steady-state change in water level due to the step input of magnitude q_o is

$$\Delta H_{ss} = \frac{q_o k_2}{\Omega}.$$

Consider the sinusoidal input

$$\Delta Q_1(t) = q_o \sin \omega t,$$

which has Laplace transform

$$\Delta Q_1(s) = \frac{q_o \omega}{s^2 + \omega^2}.$$

Suppose the system has zero initial conditions, that is, $\Delta Q_2(0) = 0$. Then from Equation (2.119) we have

$$\Delta Q_2(s) = \frac{q_o \omega \Omega}{(s + \Omega)(s^2 + \omega^2)}.$$

Expanding in a partial fraction expansion and taking the inverse Laplace transform yields

$$\Delta Q_2(t) = q_o \Omega \omega\left(\frac{e^{-\Omega t}}{\Omega^2 + \omega^2} + \frac{\sin(\omega t - \phi)}{\omega(\Omega^2 + \omega^2)^{1/2}}\right),$$

where $\phi = \tan^{-1}(\omega/\Omega)$. So, as $t \to \infty$, we have

$$\Delta Q_2(t) \quad \to \quad \frac{q_o\Omega}{\sqrt{\Omega^2 + \omega^2}} \sin(\omega t - \phi).$$

The maximum change in output flow rate is

$$|\Delta Q_2(t)|_{max} = \frac{q_o\Omega}{\sqrt{\Omega^2 + \omega^2}}. \tag{2.121}$$

The above analytic analysis of the linear system model to step and sinusoidal inputs is a valuable way to gain insight into the system response to test signals. Analytic analysis is limited, however, in the sense that a more complete representation can be obtained with carefully constructed numerical investigations using computer simulations of both the linear and nonlinear mathematical models. A computer simulation uses a model and the actual conditions of the system being modeled, as well as actual input commands to which the system will be subjected.

Various levels of simulation fidelity (that is, accuracy) are available to the control engineer. In the early stages of the design process, highly interactive design software packages are effective. At this stage, computer speed is not as important as the time it takes to obtain an initial valid solution and to iterate and fine tune that solution. Good graphics output capability is crucial. The analysis simulations are generally low fidelity in the sense that many of the simplifications (such as linearization) made in the design process are retained in the simulation.

As the design matures usually it is necessary to conduct numerical experiments in a more realistic simulation environment. At this point in the design process, the computer processing speed becomes more important, since long simulation times necessarily reduce the number of computer experiments that can be obtained and correspondingly raise costs. Usually these high-fidelity simulations are programmed in FORTRAN, C, C++, Matlab, LabVIEW or similar languages.

Assuming that a model and the simulation are reliably accurate, computer simulation has the following advantages [14]:

1. System performance can be observed under all conceivable conditions.

2. Results of field-system performance can be extrapolated with a simulation model for prediction purposes.

3. Decisions concerning future systems presently in a conceptual stage can be examined.

4. Trials of systems under test can be accomplished in a much-reduced period of time.

5. Simulation results can be obtained at lower cost than real experimentation.

6. Study of hypothetical situations can be achieved even when the hypothetical situation would be unrealizable at present.

7. Computer modeling and simulation is often the only feasible or safe technique to analyze and evaluate a system.

The nonlinear model describing the water level flow rate is as follows (using the constants given in Table 2.7):

$$\dot{H} = -0.0443\sqrt{H} + 1.2732 \times 10^{-3} Q_1, \tag{2.122}$$

$$Q_2 = 34.77\sqrt{H}.$$

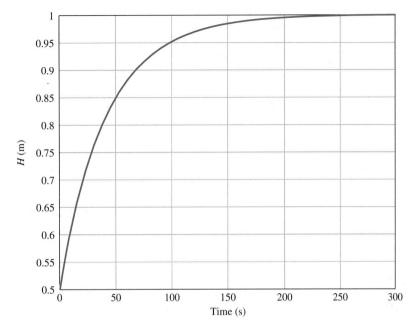

FIGURE 2.36
The tank water level time history obtained by integrating the nonlinear equations of motion in Equation (2.122) with $H(0) = 0.5$ m and $Q_1(t) = Q^* = 34.77$ kg/s.

With $H(0) = 0.5$ m and $Q_1(t) = 34.77$ kg/s, we can numerically integrate the non-linear model given by Equation (2.122) to obtain the time history of $H(t)$ and $Q_2(t)$. The response of the system is shown in Figure 2.36. As expected from Equation (2.111), the system steady-state water level is $H^* = 1$ m when $Q^* = 34.77$ kg/m³.

It takes about 250 seconds to reach steady-state. Suppose that the system is at steady state and we want to evaluate the response to a step change in the input mass flow rate. Consider

$$\Delta Q_1(t) = 1 \text{ kg/s}.$$

Then we can use the transfer function model to obtain the unit step response. The step response is shown in Figure 2.37 for both the linear and nonlinear models. Using the linear model, we find that the steady-state change in water level is $\Delta H = 5.75$ cm. Using the nonlinear model, we find that the steady-state change in water level is $\Delta H = 5.835$ cm. So we see a small difference in the results obtained from the linear model and the more accurate nonlinear model.

As the final step, we consider the system response to a sinusoidal change in the input flow rate. Let

$$\Delta Q_1(s) = \frac{q_o \omega}{s^2 + \omega^2},$$

where $\omega = 0.05$ rad/s and $q_o = 1$. The total water input flow rate is

$$Q_1(t) = Q^* + \Delta Q_1(t),$$

where $Q^* = 34.77$ kg/s. The output flow rate is shown in Figure 2.38.

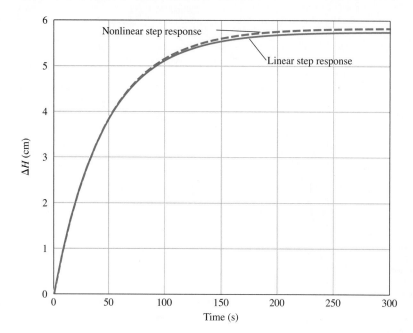

FIGURE 2.37
The response showing the linear versus nonlinear response to a step input.

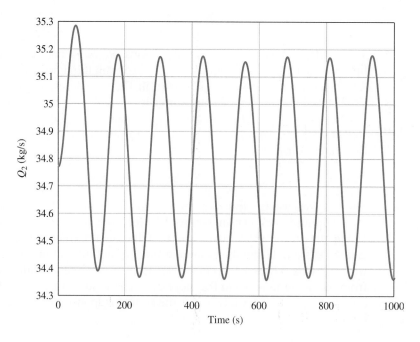

FIGURE 2.38
The output flow rate response to a sinusoidal variation in the input flow.

The response of the water level is shown in Figure 2.39. The water level is sinusoidal, with an average value of $H_{av} = H^* = 1$ meter. As shown in Equation (2.121), the output flow rate is sinusoidal in the steady-state, with

$$|\Delta Q_2(t)|_{max} = \frac{q_o \Omega}{\sqrt{\Omega^2 + \omega^2}} = 0.4 \text{ kg/s.}$$

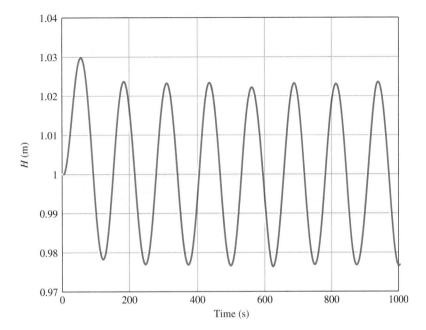

FIGURE 2.39
The water level response to a sinusoidal variation in the input flow.

Thus in the steady-state (see Figure 2.38) we expect that the output flow rate will oscillate at a frequency of $\omega = 0.05$ rad/s, with a maximum value of

$$Q_{2_{max}} = Q^* + |\Delta Q_2(t)|_{max} = 35.18 \text{ kg/s.} \quad \blacksquare$$

EXAMPLE 2.13   **Electric traction motor control**

A majority of modern trains and local transit vehicles utilize electric traction motors. The electric motor drive for a railway vehicle is shown in block diagram form in Figure 2.40(a), incorporating the necessary control of the velocity of the vehicle. The goal of the design is to obtain a system model and the closed-loop transfer function of the system, $\omega(s)/\omega_d(s)$, select appropriate resistors R_1, R_2, R_3, and R_4, and then predict the system response.

The first step is to describe the transfer function of each block. We propose the use of a tachometer to generate a voltage proportional to velocity and to connect that voltage, v_t, to one input of a difference amplifier, as shown in Figure 2.40(b). The power amplifier is nonlinear and can be approximately represented by $v_2 = 2e^{3v_1} = g(v_1)$, an exponential function with a normal operating point, $v_{10} = 1.5$ V. Using the technique in Section 2.3, we then obtain a linear model:

$$\Delta v_2 = \left.\frac{dg(v_1)}{dv_1}\right|_{v_{10}} \Delta v_1 = 2[3 \exp(3v_{10})] \Delta v_1 = 2(270) \Delta v_1 = 540 \Delta v_1. \quad (2.123)$$

Then, discarding the delta notation and using the Laplace transform, we find that

$$V_2(s) = 540V_1(s).$$

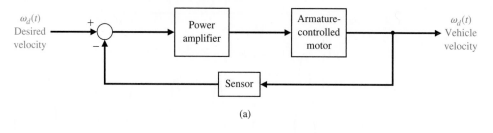

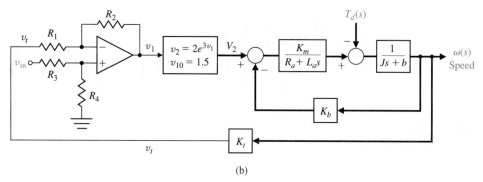

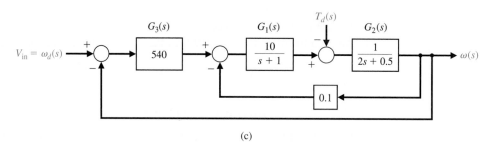

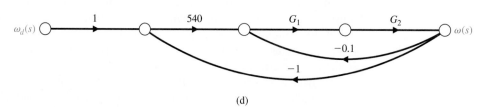

FIGURE 2.40
Speed control of an electric traction motor.

(a)

(b)

(c)

(d)

Also, for the differential amplifier, we have

$$v_1 = \frac{1 + R_2/R_1}{1 + R_3/R_4} v_{in} - \frac{R_2}{R_1} v_t. \tag{2.124}$$

We wish to obtain an input control that sets $\omega_d(t) = v_{in}$, where the units of ω_d are rad/s and the units of v_{in} are volts. Then, when $v_{in} = 10$ V, the steady-state speed is $\omega = 10$ rad/s. We note that $v_t = K_t \omega_d$ in steady state, and we expect, in balance, the steady-state output to be

$$v_1 = \frac{1 + R_2/R_1}{1 + R_3/R_4} v_{in} - \frac{R_2}{R_1} K_t v_{in}. \tag{2.125}$$

Table 2.8   Parameters of a Large DC Motor

$K_m = 10$	$J = 2$
$R_a = 1$	$b = 0.5$
$L_a = 1$	$K_b = 0.1$

When the system is in balance, $v_1 = 0$, and when $K_t = 0.1$, we have

$$\frac{1 + R_2/R_1}{1 + R_3/R_4} = \frac{R_2}{R_1} K_t = 1.$$

This relation can be achieved when

$$R_2/R_1 = 10 \quad \text{and} \quad R_3/R_4 = 10.$$

The parameters of the motor and load are given in Table 2.8. The overall system is shown in Figure 2.40(b). Reducing the block diagram in Figure 2.40(c) or the signal-flow graph in Figure 2.40(d) yields the transfer function

$$\frac{\omega(s)}{\omega_d(s)} = \frac{540G_1(s)G_2(s)}{1 + 0.1G_1G_2 + 540G_1G_2} = \frac{540G_1G_2}{1 + 540.1G_1G_2}$$

$$= \frac{5400}{(s + 1)(2s + 0.5) + 5401} = \frac{5400}{2s^2 + 2.5s + 5401.5}$$

$$= \frac{2700}{s^2 + 1.25s + 2700.75}. \tag{2.126}$$

Since the characteristic equation is second order, we note that $\omega_n = 52$ and $\zeta = 0.012$, and we expect the response of the system to be highly oscillatory (underdamped). ∎

EXAMPLE 2.14   **Mechanical accelerometer**

A mechanical accelerometer is used to measure the acceleration of a levitated test sled, as shown in Figure 2.41. The test sled is magnetically levitated above a guide rail a small distance δ. The accelerometer provides a measurement of the acceleration $a(t)$ of the sled, since the position y of the mass M, with respect to the accelerometer case, is proportional to the acceleration of the case (and the sled). The goal is to design an accelerometer with an appropriate dynamic responsiveness. We wish to design an accelerometer with an acceptable time for the desired measurement characteristic, $y(t) = qa(t)$, to be attained (q is a constant).

The sum of the forces acting on the mass is

$$-b\frac{dy}{dt} - ky = M\frac{d^2}{dt^2}(y + x)$$

or

$$M\frac{d^2y}{dt^2} + b\frac{dy}{dt} + ky = -M\frac{d^2x}{dt^2}. \tag{2.127}$$

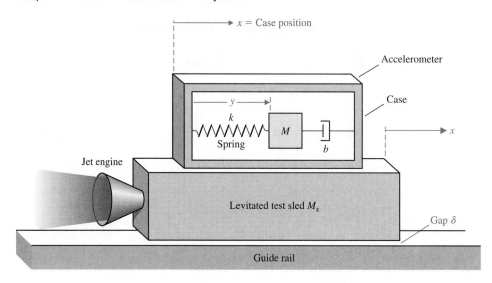

FIGURE 2.41
An accelerometer mounted on a jet-engine test sled.

Since

$$M_s \frac{d^2x}{dt^2} = F(t),$$

is the engine force, we have

$$M\ddot{y} + b\dot{y} + ky = -\frac{M}{M_s}F(t),$$

or

$$\ddot{y} + \frac{b}{M}\dot{y} + \frac{k}{M}y = -\frac{F(t)}{M_s}. \tag{2.128}$$

We select the coefficients where $b/M = 3$, $k/M = 2$, $F(t)/M_s = Q(t)$, and we consider the initial conditions $y(0) = -1$ and $\dot{y}(0) = 2$. We then obtain the Laplace transform equation, when the force, and thus $Q(t)$, is a step function, as follows:

$$(s^2 Y(s) - sy(0) - \dot{y}(0)) + 3(sY(s) - y(0)) + 2Y(s) = -Q(s). \tag{2.129}$$

Since $Q(s) = P/s$, where P is the magnitude of the step function, we obtain

$$(s^2 Y(s) + s - 2) + 3(sY(s) + 1) + 2Y(s) = -\frac{P}{s},$$

or

$$(s^2 + 3s + 2)Y(s) = \frac{-(s^2 + s + P)}{s}. \tag{2.130}$$

Thus the output transform is

$$Y(s) = \frac{-(s^2 + s + P)}{s(s^2 + 3s + 2)} = \frac{-(s^2 + s + P)}{s(s + 1)(s + 2)}. \tag{2.131}$$

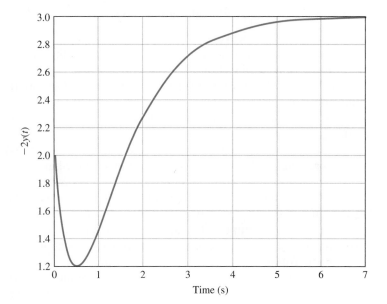

FIGURE 2.42
Accelerometer
response.

Expanding in partial fraction form yields

$$Y(s) = \frac{k_1}{s} + \frac{k_2}{s + 1} + \frac{k_3}{s + 2}. \tag{2.132}$$

We then have

$$k_1 = \frac{-(s^2 + s + P)}{(s + 1)(s + 2)}\bigg|_{s=0} = -\frac{P}{2}. \tag{2.133}$$

Similarly, $k_2 = +P$ and $k_3 = \dfrac{-P - 2}{2}$. Thus,

$$Y(s) = \frac{-P}{2s} + \frac{P}{s + 1} + \frac{-P - 2}{2(s + 2)}. \tag{2.134}$$

Therefore, the output measurement is

$$y(t) = \frac{1}{2}[-P + 2Pe^{-t} - (P + 2)e^{-2t}], \quad t \ge 0.$$

A plot of $y(t)$ is shown in Figure 2.42 for $P = 3$. We can see that $y(t)$ is proportional to the magnitude of the force after 5 seconds. Thus in steady state, after 5 seconds, the response $y(t)$ is proportional to the acceleration, as desired. If this period is excessively long, we must increase the spring constant, k, and the friction, b, while reducing the mass, M. If we are able to select the components so that $b/M = 12$ and $k/M = 32$, the accelerometer will attain the proportional response in 1 second. (It is left to the reader to show this.) ∎

EXAMPLE 2.15   **Design of a laboratory robot**

In this example, we endeavor to show the physical design of a laboratory device and demonstrate its complex design. We will also exhibit the many components commonly used in a control system.

A robot for laboratory use is shown in Figure 2.43. A laboratory robot's work volume must allow the robot to reach the entire bench area and access existing analytical instruments. There must also be sufficient area for a stockroom of supplies for unattended operation.

The laboratory robot can be involved in three types of tasks during an analytical experiment. The first is sample introduction, wherein the robot is trained to accept a number of different sample trays, racks, and containers and to introduce them into the system. The second set of tasks involves the robot transporting the samples between individual dedicated automated stations for chemical preparation and instrumental analysis. Samples must be scheduled and moved between these stations as necessary to complete the analysis. In the third set of tasks for the robot, flexible automation provides new capability to the analytical laboratory. The robot must be programmed to emulate the human operator or work with various devices. All of these types of operations are required for an effective laboratory robot.

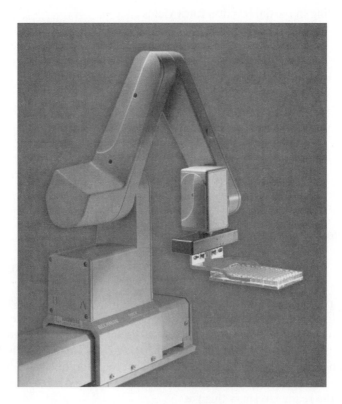

FIGURE 2.43
Laboratory robot used for sample preparation. The robot manipulates small objects, such as test tubes, and probes in and out of tight places at relatively high speeds [15]. (Photo courtesy of Beckman Coulter, Inc.)

Table 2.9   ORCA Robot Arm Hardware Specifications

Arm	Articulated, Rail-Mounted	Teach Pendant	Joy Stick with Emergency Stop
Degrees of freedom	6	Cycle time	4 s (move 1 inch up, 12 inch across, 1 inch down, and back)
Reach	±54 cm	Maximum speed	75 cm/s
Height	78 cm	Dwell time	50 ms typical (for moves within a motion)
Rail	1 and 2 m	Payload	0.5 kg continuous, 2.5 kg transient (with restrictions)
Weight	8.0 kg	Vertical deflection	<1.5 mm at continuous payload
Precision	±0.25 mm	Cross-sectional work envelope	1 m²
Finger travel (gripper)	40 mm		
Gripper rotation	±77 revolutions		

The ORCA laboratory robot is an anthropomorphic arm, mounted on a rail, designed as the optimum configuration for the analytical laboratory [15]. The rail can be located at the front or back of a workbench, or placed in the middle of a table when access to both sides of the rail is required. Simple software commands permit moving the arm from one side of the rail to the other while maintaining the wrist position (to transfer open containers) or locking the wrist angle (to transfer objects in virtually any orientation). The rectilinear geometry, in contrast to the cylindrical geometry used by many robots, permits more accessories to be placed within the robot workspace and provides an excellent match to the laboratory bench. Movement of all joints is coordinated through software, which simplifies the use of the robot by representing the robot positions and movements in the more familiar Cartesian coordinate space.

The physical and performance specifications of the ORCA system are shown in Table 2.9. The design for the ORCA laboratory robot progressed to the selection of the component parts required to obtain the total system. The exploded view of the robot is shown in Figure 2.44. This device uses six DC motors, gears, belt drives, and a rail and carriage. The specifications are challenging and require the designer to model the system components and their interconnections accurately. ∎

EXAMPLE 2.16   **Design of a low-pass filter**

Our goal is to design a first-order low-pass filter that passes signals at a frequency below 106.1 Hz and attenuates signals with a frequency above 106.1 Hz. In addition, the DC gain should be $1/2$.

A ladder network with one energy storage element, as shown in Figure 2.45(a), will act as a first-order low-pass network. Note that the DC gain will be equal to $1/2$ (open-circuit the capacitor). The current and voltage equations are

$$I_1 = (V_1 - V_2)G,$$

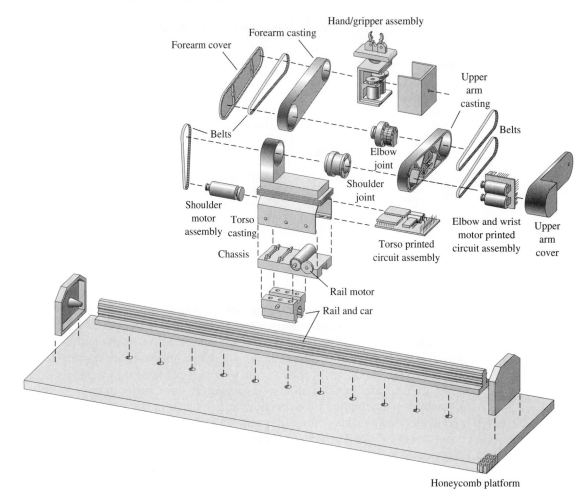

FIGURE 2.44 Exploded view of the ORCA robot showing the components [15]. (Courtesy of Beckman Coulter, Inc.)

$$I_2 = (V_2 - V_3)G,$$

$$V_2 = (I_1 - I_2)R,$$

$$V_3 = I_2Z,$$

where $G = 1/R$, $Z(s) = 1/Cs$, and $I_1(s) = I_1$ (we omit the (s)). The signal-flow graph constructed for the four equations is shown in Figure 2.45(b), and the corresponding block diagram is shown in Figure 2.45(c). The three loops are $L_1 = -GR = -1$, $L_2 = -GR = -1$, and $L_3 = -GZ$. All loops touch the forward path. Loops L_1 and L_3 are nontouching. Therefore, the transfer function is

$$T(s) = \frac{V_3}{V_1} = \frac{P_1}{1 - (L_1 + L_2 + L_3) + L_1L_3} = \frac{GZ}{3 + 2GZ}$$

$$= \frac{1}{3RCs + 2} = \frac{1/(3RC)}{s + 2/(3RC)}.$$

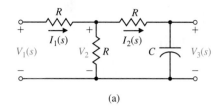

(a)

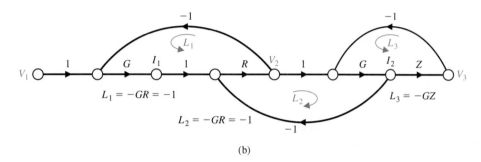

(b)

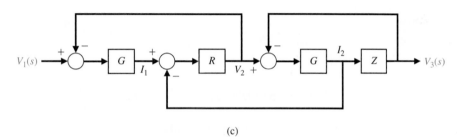

FIGURE 2.45
(a) Ladder network,
(b) its signal-flow
graph, and (c) its
block diagram.

(c)

If one prefers to utilize block diagram reduction techniques, one can start at the output with

$$V_3(s) = ZI_2(s).$$

But the block diagram shows that

$$I_2(s) = G(V_2(s) - V_3(s)).$$

Therefore,

$$V_3(s) = ZGV_2(s) - ZGV_3(s)$$

so

$$V_2(s) = \frac{1 + ZG}{ZG} V_3(s).$$

We will use this relationship between $V_3(s)$ and $V_2(s)$ in the subsequent development. Continuing with the block diagram reduction, we have

$$V_3(s) = -ZGV_3(s) + ZGR(I_1(s) - I_2(s)),$$

but from the block diagram, we see that

$$I_1 = G(V_1(s) - V_2(s)), \qquad I_2 = \frac{V_3(s)}{Z}.$$

Therefore,

$$V_3(s) = -ZGV_3(s) + ZG^2R(V_1(s) - V_2(s)) - GRV_3(s).$$

Substituting for $V_2(s)$ yields

$$V_3(s) = \frac{(GR)(GZ)}{1 + 2GR + GZ + (GR)(GZ)}V_1(s).$$

But we know that $GR = 1$; hence, we obtain

$$V_3(s) = \frac{GZ}{3 + 2GZ}V_1(s).$$

Note that the DC gain is $\frac{1}{2}$, as expected. The pole is desired at $p = 2\pi(106.1) = 666.7 = 2000/3$. Therefore, we require $RC = 0.001$. Select $R = 1\,\mathrm{k\Omega}$ and $C = 1\,\mu\mathrm{F}$. Hence, we achieve the filter

$$T(s) = \frac{333.3}{(s + 666.7)}. \quad \blacksquare$$

2.9 THE SIMULATION OF SYSTEMS USING CONTROL DESIGN SOFTWARE

Application of the many classical and modern control system design and analysis tools is based on mathematical models. Most popular control design software packages can be used with systems given in the form of transfer function descriptions. In this book, we will focus on m-file scripts containing commands and functions to analyze and design control systems. Various commercial control system packages are available for student use. The m-files described here are compatible with the MATLAB[†] Control System Toolbox and the LabVIEW MathScript.[‡]

We begin this section by analyzing a typical spring-mass-damper mathematical model of a mechanical system. Using an m-file script, we will develop an interactive analysis capability to analyze the effects of natural frequency and damping on the unforced response of the mass displacement. This analysis will use the fact that we have an analytic solution that describes the unforced time response of the mass displacement.

Later, we will discuss transfer functions and block diagrams. In particular, we are interested in manipulating polynomials, computing poles and zeros of transfer functions, computing closed-loop transfer functions, computing block diagram reductions, and computing the response of a system to a unit step input. The section concludes with the electric traction motor control design of Example 2.13.

[†]See Appendix A for an introduction to MATLAB.
[‡]See Appendix B for an introduction to LabVIEW MathScipt.

The functions covered in this section are roots, poly, conv, polyval, tf, pzmap, pole, zero, series, parallel, feedback, minreal, and step.

Spring-Mass-Damper System. A spring-mass-damper mechanical system is shown in Figure 2.2. The motion of the mass, denoted by $y(t)$, is described by the differential equation

$$M\ddot{y}(t) + b\dot{y}(t) + ky(t) = r(t).$$

The unforced dynamic response $y(t)$ of the spring-mass-damper mechanical system is

$$y(t) = \frac{y(0)}{\sqrt{1 - \zeta^2}} e^{-\zeta \omega_n t} \sin\left(\omega_n \sqrt{1 - \zeta^2}\, t + \theta\right),$$

where $\omega_n = \sqrt{k/M}$, $\zeta = b/(2\sqrt{kM})$, and $\theta = \cos^{-1}\zeta$. The initial displacement is $y(0)$. The transient system response is **underdamped** when $\zeta < 1$, **overdamped** when $\zeta > 1$, and **critically damped** when $\zeta = 1$. We can visualize the unforced time response of the mass displacement following an initial displacement of $y(0)$. Consider the underdamped case:

$$\Box \; y(0) = 0.15 \text{ m}, \quad \omega_n = \sqrt{2}\frac{\text{rad}}{\text{sec}}, \quad \zeta = \frac{1}{2\sqrt{2}} \quad \left(\frac{k}{M} = 2, \frac{b}{M} = 1\right)$$

The commands to generate the plot of the unforced response are shown in Figure 2.46. In the setup, the variables $y(0)$, ω_n, t, and ζ are input at the command level. Then the script unforced.m is executed to generate the desired plots. This creates an interactive analysis capability to analyze the effects of natural frequency and damping on the unforced response of the mass displacement. One can investigate the effects of the natural frequency and the damping on the time response by simply entering new

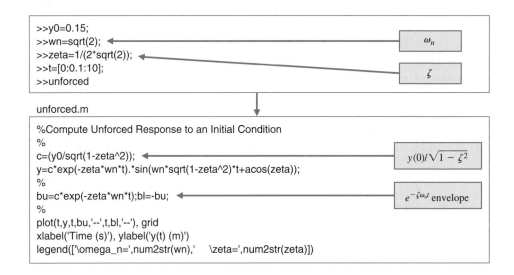

FIGURE 2.46
Script to analyze the spring-mass-damper.

```
>>y0=0.15;
>>wn=sqrt(2);
>>zeta=1/(2*sqrt(2));
>>t=[0:0.1:10];
>>unforced
```

ω_n

ζ

unforced.m

```
%Compute Unforced Response to an Initial Condition
%
c=(y0/sqrt(1-zeta^2));
y=c*exp(-zeta*wn*t).*sin(wn*sqrt(1-zeta^2)*t+acos(zeta));
%
bu=c*exp(-zeta*wn*t);bl=-bu;
%
plot(t,y,t,bu,'--',t,bl,'--'), grid
xlabel('Time (s)'), ylabel('y(t) (m)')
legend(['\omega_n=',num2str(wn),'   \zeta=',num2str(zeta)])
```

$y(0)/\sqrt{1 - \zeta^2}$

$e^{-\zeta \omega_n t}$ envelope

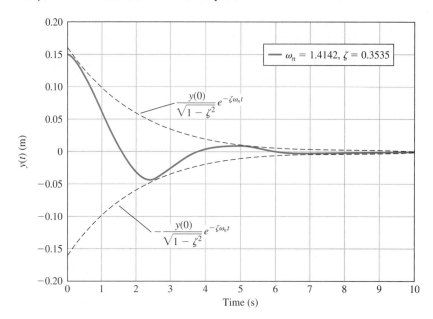

FIGURE 2.47
Spring-mass-damper unforced response.

values of ω_n and ζ at the command prompt and running the script unforced.m again. The time-response plot is shown in Figure 2.47. Notice that the script automatically labels the plot with the values of the damping coefficient and natural frequency. This avoids confusion when making many interactive simulations. Using scripts is an important aspect of developing an effective interactive design and analysis capability.

For the spring-mass-damper problem, the unforced solution to the differential equation was readily available. In general, when simulating closed-loop feedback control systems subject to a variety of inputs and initial conditions, it is difficult to obtain the solution analytically. In these cases, we can compute the solutions numerically and to display the solution graphically.

Most systems considered in this book can be described by transfer functions. Since the transfer function is a ratio of polynomials, we begin by investigating how to manipulate polynomials, remembering that working with transfer functions means that both a numerator polynomial and a denominator polynomial must be specified.

Polynomials are represented by row vectors containing the polynomial coefficients in order of descending degree. For example, the polynomial

$$p(s) = s^3 + 3s^2 + 4$$

is entered as shown in Figure 2.48. Notice that even though the coefficient of the s term is zero, it is included in the input definition of $p(s)$.

If **p** is a row vector containing the coefficients of $p(s)$ in descending degree, then roots(**p**) is a column vector containing the roots of the polynomial. Conversely, if **r** is a column vector containing the roots of the polynomial, then poly(**r**) is a row vector with the polynomial coefficients in descending degree. We can compute the roots of the polynomial $p(s) = s^3 + 3s^2 + 4$ with the roots function as shown in Figure 2.48. In this figure, we show how to reassemble the polynomial with the poly function.

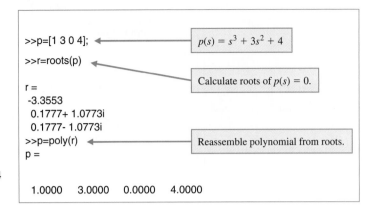

FIGURE 2.48
Entering the polynomial
$p(s) = s^3 + 3s^2 + 4$
and calculating its roots.

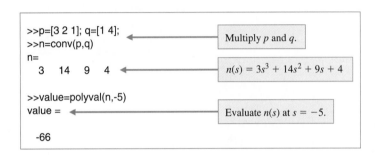

FIGURE 2.49
Using **conv** and **polyval** to multiply and evaluate the polynomials
$(3s^2 + 2s + 1)$
$(s + 4)$.

Multiplication of polynomials is accomplished with the conv function. Suppose we want to expand the polynomial

$$n(s) = (3s^2 + 2s + 1)(s + 4).$$

The associated commands using the conv function are shown in Figure 2.49. Thus, the expanded polynomial is

$$n(s) = 3s^3 + 14s^2 + 9s + 4.$$

The function polyval is used to evaluate the value of a polynomial at the given value of the variable. The polynomial $n(s)$ has the value $n(-5) = -66$, as shown in Figure 2.49.

Linear, time-invariant system models can be treated as *objects*, allowing one to manipulate the system models as single entities. In the case of transfer functions, one creates the system models using the tf function; for state variable models one employs the ss function (see Chapter 3). The use of tf is illustrated in Figure 2.50(a). For example, if one has the two system models

$$G_1(s) = \frac{10}{s^2 + 2s + 5} \quad \text{and} \quad G_2(s) = \frac{1}{s + 1},$$

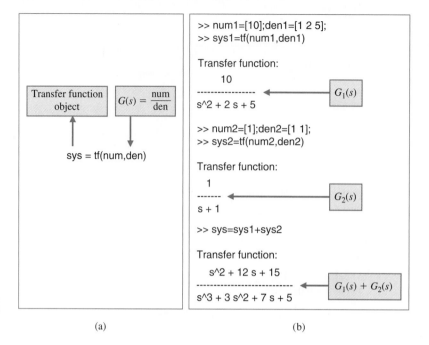

FIGURE 2.50
(a) The **tf** function.
(b) Using the **tf** function to create transfer function objects and adding them using the "+" operator.

one can add them using the "+" operator to obtain

$$G(s) = G_1(s) + G_2(s) = \frac{s^2 + 12s + 15}{s^3 + 3s^2 + 7s + 5}.$$

The corresponding commands are shown in Figure 2.50(b) where sys1 represents $G_1(s)$ and sys2 represents $G_2(s)$. Computing the poles and zeros associated with a transfer function is accomplished by operating on the system model object with the pole and zero functions, respectively, as illustrated in Figure 2.51.

In the next example, we will obtain a plot of the pole–zero locations in the complex plane. This will be accomplished using the pzmap function, shown in Figure 2.52. On the pole–zero map, zeros are denoted by an "o" and poles are denoted by an "×". If the pzmap function is invoked without left-hand arguments, the plot is generated automatically.

EXAMPLE 2.17   **Transfer functions**

Consider the transfer functions

$$G(s) = \frac{6s^2 + 1}{s^3 + 3s^2 + 3s + 1} \quad \text{and} \quad H(s) = \frac{(s + 1)(s + 2)}{(s + 2i)(s - 2i)(s + 3)}.$$

Using an m-file script, we can compute the poles and zeros of $G(s)$, the characteristic equation of $H(s)$, and divide $G(s)$ by $H(s)$. We can also obtain a plot of the pole–zero map of $G(s)/H(s)$ in the complex plane.

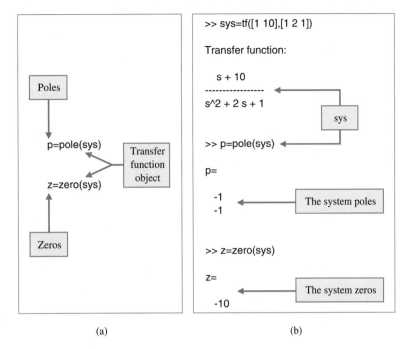

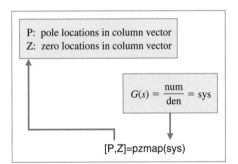

FIGURE 2.51
(a) The **pole** and **zero** functions.
(b) Using the pole and zero functions to compute the **pole** and **zero** locations of a linear system.

(a)   (b)

FIGURE 2.52
The **pzmap** function.

The pole–zero map of the transfer function $G(s)/H(s)$ is shown in Figure 2.53, and the associated commands are shown in Figure 2.54. The pole–zero map shows clearly the five zero locations, but it appears that there are only two poles. This cannot be the case, since we know that for physical systems the number of poles must be greater than or equal to the number of zeros. Using the roots function, we can ascertain that there are in fact four poles at $s = -1$. Hence, multiple poles or multiple zeros at the same location cannot be discerned on the pole–zero map. ∎

Block Diagram Models.   Suppose we have developed mathematical models in the form of transfer functions for a process, represented by $G(s)$, and a controller, represented by $G_c(s)$, and possibly many other system components such as sensors and actuators. Our objective is to interconnect these components to form a control system.

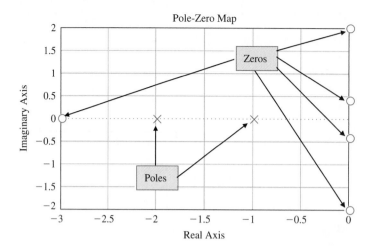

FIGURE 2.53
Pole–zero map for
$G(s)/H(s)$.

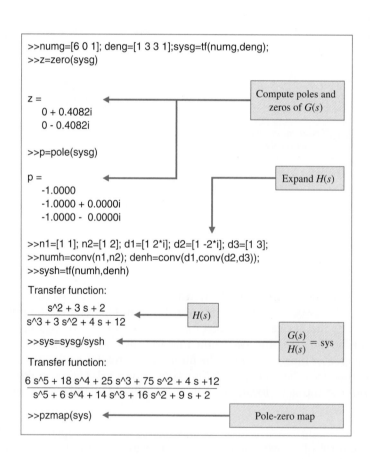

FIGURE 2.54
Transfer function
example for $G(s)$
and $H(s)$.

FIGURE 2.55
Open-loop control
system (without
feedback).

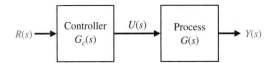

A simple open-loop control system can be obtained by interconnecting a process and a controller in series as illustrated in Figure 2.55. We can compute the transfer function from $R(s)$ to $Y(s)$, as follows.

EXAMPLE 2.18   **Series connection**

Let the process represented by the transfer function $G(s)$ be

$$G(s) = \frac{1}{500s^2},$$

and let the controller represented by the transfer function $G_c(s)$ be

$$G_c(s) = \frac{s + 1}{s + 2}.$$

We can use the **series** function to cascade two transfer functions $G_1(s)$ and $G_2(s)$, as shown in Figure 2.56.

The transfer function $G_c(s)G(s)$ is computed using the **series** function as shown in Figure 2.57. The resulting transfer function is

$$G_c(s)G(s) = \frac{s + 1}{500s^3 + 1000s^2} = \text{sys},$$

where **sys** is the transfer function name in the m-file script. ■

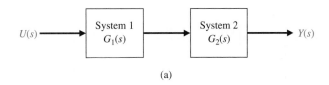

(a)

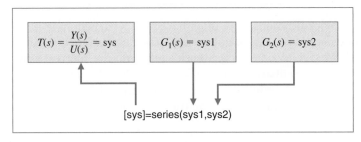

FIGURE 2.56
(a) Block diagram.
(b) The **series**
function.

(b)

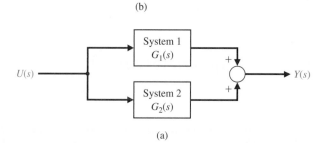

FIGURE 2.57
Application of the
series function.

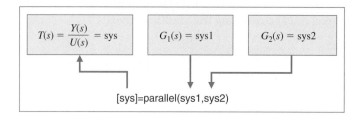

FIGURE 2.58
(a) Block diagram.
(b) The **parallel**
function.

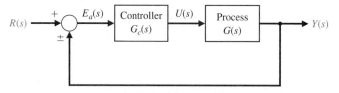

FIGURE 2.59   A
basic control
system with unity
feedback.

Block diagrams quite often have transfer functions in parallel. In such cases, the function parallel can be quite useful. The parallel function is described in Figure 2.58.

We can introduce a **feedback signal** into the control system by closing the loop with **unity feedback**, as shown in Figure 2.59. The signal $E_a(s)$ is an **error signal**; the signal $R(s)$ is a **reference input**. In this control system, the controller is in the forward path, and the closed-loop transfer function is

$$T(s) = \frac{G_c(s)G(s)}{1 \mp G_c(s)G(s)}.$$

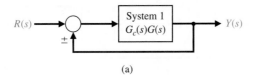

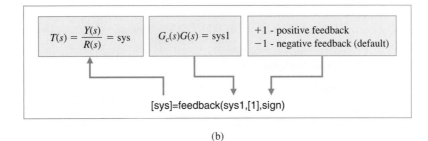

FIGURE 2.60
(a) Block diagram.
(b) The **feedback**
function with unity
feedback.

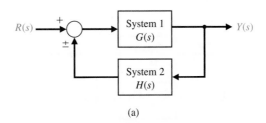

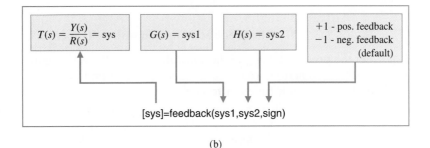

FIGURE 2.61
(a) Block diagram.
(b) The **feedback**
function.

We can utilize the feedback function to aid in the block diagram reduction process to compute closed-loop transfer functions for single- and multiple-loop control systems.

It is often the case that the closed-loop control system has unity feedback, as illustrated in Figure 2.59. We can use the feedback function to compute the closed-loop transfer function by setting $H(s) = 1$. The use of the feedback function for unity feedback is depicted in Figure 2.60.

The feedback function is shown in Figure 2.61 with the associated system configuration, which includes $H(s)$ in the feedback path. If the input "sign" is omitted, then negative feedback is assumed.

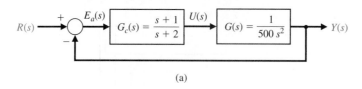

(a)

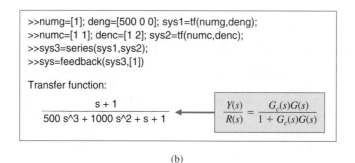

FIGURE 2.62
(a) Block diagram.
(b) Application of
the **feedback**
function.

(b)

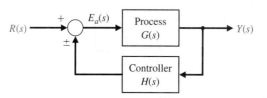

FIGURE 2.63
A basic control
system with the
controller in the
feedback loop.

EXAMPLE 2.19   **The feedback function with unity feedback**

Let the process, $G(s)$, and the controller, $G_c(s)$, be as in Figure 2.62(a). To apply the feedback function, we first use the series function to compute $G_c(s)G(s)$, followed by the feedback function to close the loop. The command sequence is shown in Figure 2.62(b). The closed-loop transfer function, as shown in Figure 2.62(b), is

$$T(s) = \frac{G_c(s)G(s)}{1 + G_c(s)G(s)} = \frac{s + 1}{500s^3 + 1000s^2 + s + 1} = \text{sys.} \blacksquare$$

Another basic feedback control configuration is shown in Figure 2.63. In this case, the controller is located in the feedback path. The closed-loop transfer function is

$$T(s) = \frac{G(s)}{1 \mp G(s)H(s)}.$$

EXAMPLE 2.20   **The feedback function**

Let the process, $G(s)$, and the controller, $H(s)$, be as in Figure 2.64(a). To compute the closed-loop transfer function with the controller in the feedback loop, we use

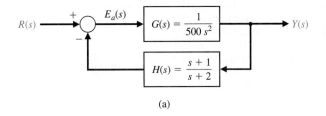

(a)

```
>>numg=[1]; deng=[500 0 0]; sys1=tf(numg,deng);
>>numh=[1 1]; denh=[1 2]; sys2=tf(numh,denh);
>>sys=feedback(sys1,sys2);
>>sys

Transfer function:
```

FIGURE 2.64
Application of the
feedback function:
(a) block diagram,
(b) m-file script.

$$\frac{s+2}{500\ s\mathord{\wedge}3 + 1000\ s\mathord{\wedge}2 + s + 1} \qquad \boxed{\dfrac{Y(s)}{R(s)} = \dfrac{G(s)}{1 + G(s)H(s)}}$$

(b)

the feedback function. The command sequence is shown in Figure 2.64(b). The closed-loop transfer function is

$$T(s) = \frac{s+2}{500s^3 + 1000s^2 + s + 1} = \text{sys.} \ \blacksquare$$

The functions series, parallel, and feedback can be used as aids in block diagram manipulations for multiple-loop block diagrams.

EXAMPLE 2.21   **Multiloop reduction**

A multiloop feedback system is shown in Figure 2.26. Our objective is to compute the closed-loop transfer function

$$T(s) = \frac{Y(s)}{R(s)}$$

when

$$G_1(s) = \frac{1}{s+10}, \qquad G_2(s) = \frac{1}{s+1},$$

$$G_3(s) = \frac{s^2 + 1}{s^2 + 4s + 4}, \qquad G_4(s) = \frac{s+1}{s+6},$$

and

$$H_1(s) = \frac{s+1}{s+2}, \qquad H_2(s) = 2, \quad \text{and} \quad H_3(s) = 1.$$

```
>>ng1=[1]; dg1=[1 10]; sysg1=tf(ng1,dg1);
>>ng2=[1]; dg2=[1 1]; sysg2=tf(ng2,dg2);
>>ng3=[1 0 1]; dg3=[1 4 4]; sysg3=tf(ng3,dg3);
>>ng4=[1 1]; dg4=[1 6]; sysg4=tf(ng4,dg4);          Step 1
>>nh1=[1 1]; dh1=[1 2]; sysh1=tf(nh1,dh1);
>>nh2=[2]; dh2=[1]; sysh2=tf(nh2,dh2);
>>nh3=[1]; dh3=[1]; sysh3=tf(nh3,dh3);
>>sys1=sysh2/sysg4;                                 Step 2
>>sys2=series(sysg3,sysg4);
>>sys3=feedback(sys2,sysh1,+1);                     Step 3
>>sys4=series(sysg2,sys3);
>>sys5=feedback(sys4,sys1);                         Step 4
>>sys6=series(sysg1,sys5);
>>sys=feedback(sys6,sysh3);                         Step 5
```

Transfer function:

$$\frac{s^5 + 4\,s^4 + 6\,s^3 + 6\,s^2 + 5\,s + 2}{12\,s^6 + 205\,s^5 + 1066\,s^4 + 2517\,s^3 + 3128\,s^2 + 2196\,s + 712}$$

FIGURE 2.65
Multiple-loop block
reduction.

For this example, a five-step procedure is followed:

❏ Step 1. Input the system transfer functions.

❏ Step 2. Move H_2 behind G_4.

❏ Step 3. Eliminate the $G_3G_4H_1$ loop.

❏ Step 4. Eliminate the loop containing H_2.

❏ Step 5. Eliminate the remaining loop and calculate $T(s)$.

The five steps are utilized in Figure 2.65, and the corresponding block diagram reduction is shown in Figure 2.27. The result of executing the commands is

$$\text{sys} = \frac{s^5 + 4s^4 + 6s^3 + 6s^2 + 5s + 2}{12s^6 + 205s^5 + 1066s^4 + 2517s^3 + 3128s^2 + 2196s + 712}.$$

We must be careful in calling this the closed-loop transfer function. The transfer function is defined as the input–output relationship after pole–zero cancellations. If we compute the poles and zeros of $T(s)$, we find that the numerator and denominator polynomials have $(s + 1)$ as a common factor. This must be canceled before we can claim we have the closed-loop transfer function. To assist us in the pole–zero cancellation, we will use the minreal function. The minreal function, shown in Figure 2.66, removes common pole–zero factors of a transfer function. The final step in the block reduction process is to cancel out the common factors, as shown in Figure 2.67. After the application of the minreal function, we find that the order of the denominator polynomial has been reduced from six to five, implying one pole–zero cancellation. ■

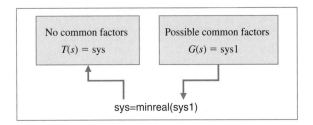

FIGURE 2.66 The **minreal** function.

```
>>num=[1 4 6 6 5 2]; den=[12 205 1066 2517 3128 2196 712];
>>sys1=tf(num,den);
>>sys=minreal(sys1);   <---- Cancel common factors.

Transfer function:
   0.08333 s^4 + 0.25 s^3 + 0.25 s^2 + 0.25 s + 0.1667
-------------------------------------------------------------
s^5 + 16.08 s^4 + 72.75 s^3 + 137 s^2 + 123.7 s + 59.33
```

FIGURE 2.67 Application of the **minreal** function.

EXAMPLE 2.22 Electric traction motor control

Finally, let us reconsider the electric traction motor system from Example 2.13. The block diagram is shown in Figure 2.40(c). The objective is to compute the closed-loop transfer function and investigate the response of $\omega(s)$ to a commanded $\omega_d(s)$. The first step, as shown in Figure 2.68, is to compute the closed-loop transfer function $\omega(s)/\omega_d(s) = T(s)$. The closed-loop characteristic equation is second order with $\omega_n = 52$ and $\zeta = 0.012$. Since the damping is low, we expect the response to be highly oscillatory. We can investigate the response $\omega(t)$ to a reference input, $\omega_d(t)$, by utilizing the step function. The step function, shown in Figure 2.69, calculates the unit step response of a linear system. The step function is very important, since control system performance specifications are often given in terms of the unit step response.

```
>>num1=[10]; den1=[1 1]; sys1=tf(num1,den1);
>>num2=[1]; den2=[2 0.5]; sys2=tf(num2,den2);
>>num3=[540]; den3=[1]; sys3=tf(num3,den3);
>>num4=[0.1]; den4=[1]; sys4=tf(num4,den4);
>>sys5=series(sys1,sys2);
>>sys6=feedback(sys5,sys4);      <---- Eliminate inner loop
>>sys7=series(sys3,sys6);
>>sys=feedback(sys7,[1])         <---- Compute closed-loop transfer function

Transfer function:
       5400
-----------------------       <---- ω(s)/ω_d(s)
2 s^2 + 2.5 s + 5402
```

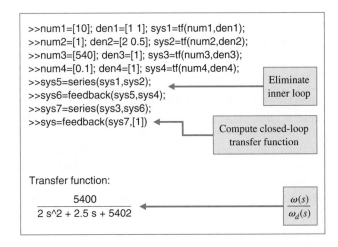

FIGURE 2.68 Electric traction motor block reduction.

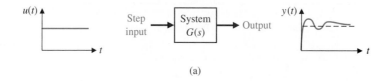

(a)

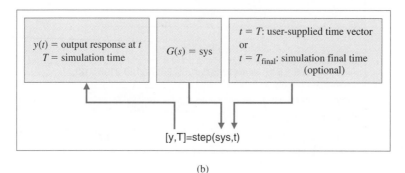

FIGURE 2.69
The **step** function.

(b)

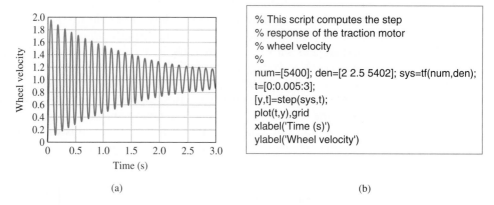

```
% This script computes the step
% response of the traction motor
% wheel velocity
%
num=[5400]; den=[2 2.5 5402]; sys=tf(num,den);
t=[0:0.005:3];
[y,t]=step(sys,t);
plot(t,y),grid
xlabel('Time (s)')
ylabel('Wheel velocity')
```

(a)   (b)

FIGURE 2.70   (a) Traction motor wheel velocity step response. (b) m-file script.

If the only objective is to plot the output, $y(t)$, we can use the **step** function without left-hand arguments and obtain the plot automatically with axis labels. If we need $y(t)$ for any purpose other than plotting, we must use the **step** function with left-hand arguments, followed by the plot function to plot $y(t)$. We define t as a row vector containing the times at which we wish the value of the output variable $y(t)$. We can also select $t = t_{final}$, which results in a step response from $t = 0$ to $t = t_{final}$ and the number of intermediate points are selected automatically.

The step response of the electric traction motor is shown in Figure 2.70. As expected, the wheel velocity response, given by $y(t)$, is highly oscillatory. Note that the output is $y(t) \equiv \omega(t)$. ∎

2.10 SEQUENTIAL DESIGN EXAMPLE: DISK DRIVE READ SYSTEM

  In Section 1.9, we developed an initial goal for the disk drive system: to position the reader head accurately at the desired track and to move from one track to another within 10 ms, if possible. We need to identify the plant, the sensor, and the controller. We will obtain a model of the plant $G(s)$ and the sensor. The disk drive reader uses a permanent magnet DC motor to rotate the reader arm (see Figure 1.26). The DC motor is called a voice coil motor in the disk drive industry. The read head is mounted on a slider device, which is connected to the arm as shown in Figure 2.71. A flexure (spring metal) is used to enable the head to float above the disk at a gap of less than 100 nm. The thin-film head reads the magnetic flux and provides a signal to an amplifier. The error signal of Figure 2.72(a) is provided by reading the error from a prerecorded index track. Assuming an accurate read head, the sensor has a transfer function $H(s) = 1$, as shown in Figure 2.72(b). The model of the permanent magnet DC motor and a linear amplifier is shown in Figure 2.72(b). As a good approximation, we use the model of the armature-controlled DC motor as shown earlier in

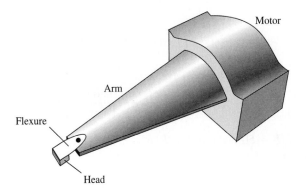

FIGURE 2.71
Head mount for reader, showing flexure.

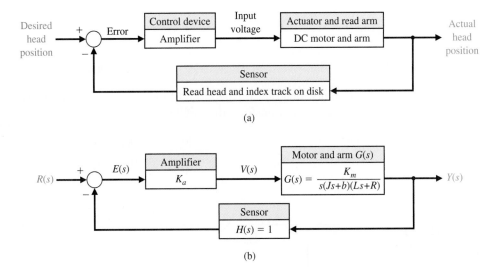

FIGURE 2.72
Block diagram model of disk drive read system.

Table 2.10   Typical Parameters for Disk Drive Reader

Parameter	Symbol	Typical Value
Inertia of arm and read head	J	$1\,\mathrm{N\,m\,s^2/rad}$
Friction	b	$20\,\mathrm{N\,m\,s/rad}$
Amplifier	K_a	10–1000
Armature resistance	R	$1\,\Omega$
Motor constant	K_m	$5\,\mathrm{N\,m/A}$
Armature inductance	L	$1\,\mathrm{mH}$

Figure 2.20 with $K_b = 0$. The model shown in Figure 2.72(b) assumes that the flexure is entirely rigid and does not significantly flex. In Chapter 4, we will consider the model when the flexure cannot be assumed to be completely rigid.

Typical parameters for the disk drive system are given in Table 2.10. Thus, we have

$$G(s) = \frac{K_m}{s(Js + b)(Ls + R)}$$

$$= \frac{5000}{s(s + 20)(s + 1000)}. \tag{2.135}$$

We can also write

$$G(s) = \frac{K_m/(bR)}{s(\tau_L s + 1)(\tau s + 1)}, \tag{2.136}$$

where $\tau_L = J/b = 50$ ms and $\tau = L/R = 1$ ms. Since $\tau \ll \tau_L$, we often neglect τ. Then, we would have

$$G(s) \approx \frac{K_m/(bR)}{s(\tau_L s + 1)} = \frac{0.25}{s(0.05s + 1)},$$

or

$$G(s) = \frac{5}{s(s + 20)}.$$

The block diagram of the closed-loop system is shown in Figure 2.73. Using the block diagram transformation of Table 2.6, we have

$$\frac{Y(s)}{R(s)} = \frac{K_a G(s)}{1 + K_a G(s)}. \tag{2.137}$$

FIGURE 2.73
Block diagram of closed-loop system.

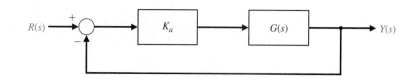

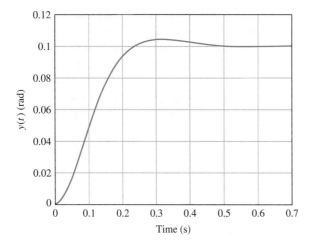

FIGURE 2.74
The system response of the system shown in Figure 2.73 for $R(s) = \dfrac{0.1}{s}$

Using the approximate second-order model for $G(s)$, we obtain

$$\frac{Y(s)}{R(s)} = \frac{5K_a}{s^2 + 20s + 5K_a}.$$

When $K_a = 40$, we have

$$Y(s) = \frac{200}{s^2 + 20s + 200} R(s).$$

We obtain the step response for $R(s) = \dfrac{0.1}{s}$ rad, as shown in Figure 2.74.

2.11 SUMMARY

In this chapter, we have been concerned with quantitative mathematical models of control components and systems. The differential equations describing the dynamic performance of physical systems were utilized to construct a mathematical model. The physical systems under consideration included mechanical, electrical, fluid, and thermodynamic systems. A linear approximation using a Taylor series expansion about the operating point was utilized to obtain a small-signal linear approximation for nonlinear control components. Then, with the approximation of a linear system, one may utilize the Laplace transformation and its related input–output relationship given by the transfer function. The transfer function approach to linear systems allows the analyst to determine the response of the system to various input signals in terms of the location of the poles and zeros of the transfer function. Using transfer function notations, block diagram models of systems of interconnected components were developed. The block relationships were obtained. Additionally, an alternative use of transfer function models in signal-flow graph form was investigated. Mason's signal-flow gain formula was investigated and was found to be useful for obtaining the relationship between system variables in a complex feedback system. The advantage of the signal-flow graph method was the availability of Mason's signal-flow gain formula, which provides the relationship between system variables without requiring any reduction or manipulation of the flow graph. Thus, in Chapter 2, we have obtained a useful mathematical model for feedback control systems by developing the concept of a transfer function of a linear system and the relationship among system variables using block diagram and signal-flow graph models. We considered the utility of the computer simulation of linear and nonlinear systems to determine the response of a system for several conditions of the system parameters and the environment. Finally, we continued the development of the Disk Drive Read System by obtaining a model in transfer function form of the motor and arm.

EXERCISES

Exercises are straightforward applications of the concepts of the chapter.

E2.1 A unity, negative feedback system has a nonlinear function $y = f(e) = e^2$, as shown in Figure E2.1. For an input r in the range of 0 to 4, calculate and plot the open-loop and closed-loop output versus input and show that the feedback system results in a more linear relationship.

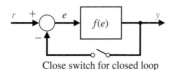

Close switch for closed loop

FIGURE E2.1   Open and closed loop.

E2.2 A thermistor has a response to temperature represented by

$$R = R_o e^{-0.1T},$$

where $R_o = 10{,}000 \ \Omega$, R = resistance,   and T = temperature in degrees Celsius. Find the linear model for the thermistor operating at $T = 20°C$ and for a small range of variation of temperature.

Answer: $\Delta R = -135\Delta T$

E2.3 The force versus displacement for a spring is shown in Figure E2.3 for the spring-mass-damper system of Figure 2.1. Graphically find the spring constant for the equilibrium point of $y = 0.5$ cm and a range of operation of ± 1.5 cm.

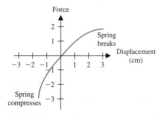

FIGURE E2.3   Spring behavior.

E2.4 A laser printer uses a laser beam to print copy rapidly for a computer. The laser is positioned by a control input $r(t)$, so that we have

$$Y(s) = \frac{4(s + 50)}{s^2 + 30s + 200} R(s).$$

The input $r(t)$ represents the desired position of the laser beam.

(a) If $r(t)$ is a unit step input, find the output $y(t)$.
(b) What is the final value of $y(t)$?

Answer: (a) $y(t) = 1 + 0.6e^{-20t} - 1.6e^{-10t}$, (b) $y_{ss} = 1$

E2.5 A noninverting amplifier uses an op-amp as shown in Figure E2.5. Assume an ideal op-amp model and determine v_o/v_{in}.

Answer: $\dfrac{v_o}{v_{in}} = 1 + \dfrac{R_2}{R_1}$

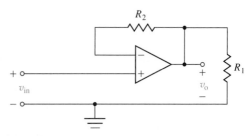

FIGURE E2.5   A noninverting amplifier using an op-amp.

E2.6 A nonlinear device is represented by the function

$$y = f(x) = x^{1/2},$$

where the operating point for the input x is $x_o = 1/2$. Determine the linear approximation in the form of Equation (2.9).

Answer: $\Delta y = \Delta x/\sqrt{2}$

E2.7 A lamp's intensity stays constant when monitored by an optotransistor-controlled feedback loop. When the voltage drops, the lamp's output also drops, and optotransistor Q_1 draws less current. As a result, a power transistor conducts more heavily and charges a capacitor more rapidly [25]. The capacitor voltage controls the lamp voltage directly. A block diagram of the system is shown in Figure E2.7. Find the closed-loop transfer function, $I(s)/R(s)$ where $I(s)$ is the lamp intensity, and $R(s)$ is the command or desired level of light.

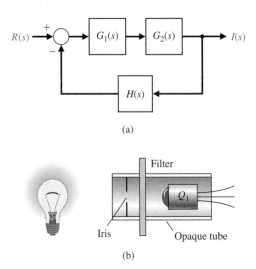

(a)

(b)

FIGURE E2.7 Lamp controller.

E2.8 A control engineer, N. Minorsky, designed an innovative ship steering system in the 1930s for the U.S. Navy. The system is represented by the block diagram shown in Figure E2.8, where $Y(s)$ is the ship's course, $R(s)$ is the desired course, and $A(s)$ is the rudder angle [17]. Find the transfer function $Y(s)/R(s)$.

Answer: $\dfrac{Y(s)}{R(s)} =$

$$\frac{KG_1(s)G_2(s)/s}{1 + G_1(s)H_3(s) + G_1(s)G_2(s)[H_1(s) + H_2(s)] + KG_1(s)G_2(s)/s}$$

E2.9 A four-wheel antilock automobile braking system uses electronic feedback to control automatically the brake force on each wheel [16]. A block diagram model of a brake control system is shown in Figure E2.9, where $F_f(s)$ and $F_R(s)$ are the braking force of the front and rear wheels, respectively, and $R(s)$ is the desired automobile response on an icy road. Find $F_f(s)/R(s)$.

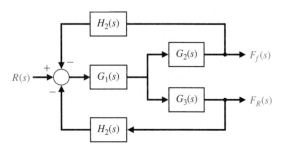

FIGURE E2.9 Brake control system.

E2.10 One of the most potentially beneficial applications of an automotive control system is the active control of the suspension system. One feedback control system uses a shock absorber consisting of a cylinder filled with a compressible fluid that provides both spring and damping forces [18]. The cylinder has a plunger activated by a gear motor, a displacement-measuring sensor, and a piston. Spring force is generated by piston displacement, which compresses the fluid. During piston displacement, the pressure imbalance across the piston is used to control damping. The plunger varies the internal volume of the cylinder. This feedback system is shown in Figure E2.10. Develop a linear model for this device using a block diagram model.

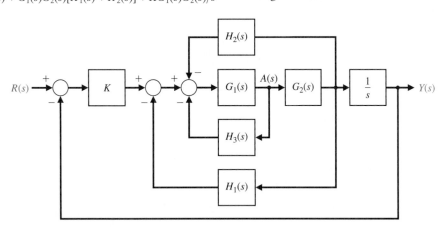

FIGURE E2.8 Ship steering system.

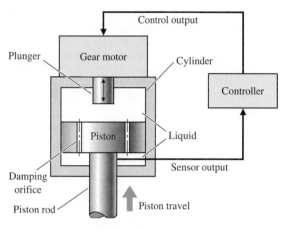

FIGURE E2.10   Shock absorber.

E2.11 A spring exhibits a force-versus-displacement characteristic as shown in Figure E2.11. For small deviations from the operating point x_o, find the spring constant when x_o is (a) -1.4; (b) 0; (c) 3.5.

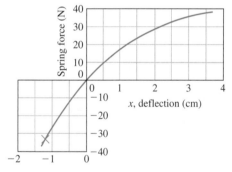

FIGURE E2.11   Spring characteristic.

E2.12 Off-road vehicles experience many disturbance inputs as they traverse over rough roads. An active suspension system can be controlled by a sensor that looks "ahead" at the road conditions. An example of a simple suspension system that can accommodate the bumps is shown in Figure E2.12. Find the appropriate gain K_1 so that the vehicle does not bounce when the desired deflection is $R(s) = 0$ and the disturbance is $T_d(s)$.

Answer: $K_1 K_2 = 1$

E2.13 Find the transfer function

$$\frac{Y_1(s)}{R_2(s)}$$

for the multivariable system in Figure E2.13.

E2.14 Obtain the differential equations for the circuit in Figure E2.14 in terms of i_1 and i_2.

E2.15 The position control system for a spacecraft platform is governed by the following equations:

$$\frac{d^2p}{dt^2} + 2\frac{dp}{dt} + 4p = \theta$$

$$v_1 = r - p$$

$$\frac{d\theta}{dt} = 0.6v_2$$

$$v_2 = 7v_1.$$

The variables involved are as follows:

$r(t)$ = desired platform position

$p(t)$ = actual platform position

$v_1(t)$ = amplifier input voltage

$v_2(t)$ = amplifier output voltage

$\theta(t)$ = motor shaft position

Sketch a signal-flow diagram or a block diagram of the system, identifying the component parts and their transmittances; then determine the system transfer function $P(s)/R(s)$.

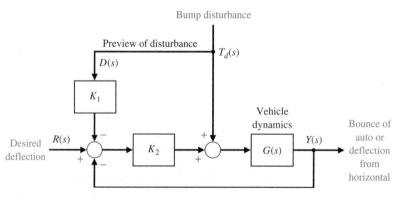

FIGURE E2.12   Active suspension system.

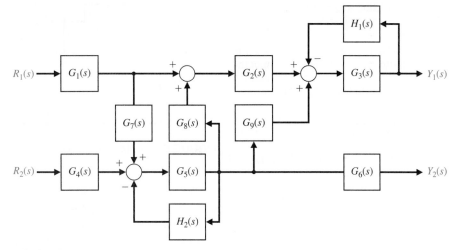

FIGURE E2.13   Multivariable system.

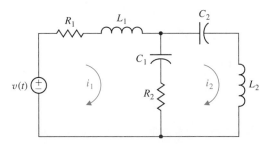

FIGURE E2.14   Electric circuit.

E2.16   A spring used in an auto shock absorber develops a force f represented by the relation

$$f = kx^4,$$

where x is the displacement of the spring. Determine a linear model for the spring when $x_o = 1$.

E2.17   The output y and input x of a device are related by

$$y = x + 1.4x^3.$$

(a) Find the values of the output for steady-state operation at the two operating points $x_o = 1$ and $x_o = 2$.
(b) Obtain a linearized model for both operating points and compare them.

E2.18   The transfer function of a system is

$$\frac{Y(s)}{R(s)} = \frac{10(s + 2)}{s^2 + 8s + 15}.$$

Determine $y(t)$ when $r(t)$ is a unit step input.

Answer: $y(t) = 1.33 + 1.67e^{-3t} - 3e^{-5t}, t \geq 0$

E2.19   Determine the transfer function $V_0(s)/V(s)$ of the operational amplifier circuit shown in Figure E2.19. Assume

an ideal operational amplifier. Determine the transfer function when $R_1 = R_2 = 100 \text{ k}\Omega$, $C_1 = 10 \text{ }\mu\text{F}$, and $C_2 = 5 \text{ }\mu\text{F}$.

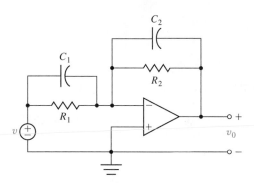

FIGURE E2.19   Op-amp circuit.

E2.20   A high-precision positioning slide is shown in Figure E2.20. Determine the transfer function $X_p(s)/X_{in}(s)$

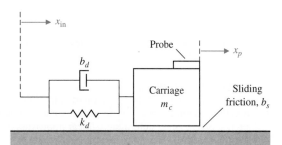

FIGURE E2.20   Precision slide.

when the drive shaft friction is $b_d = 0.7$, the drive shaft spring constant is $k_d = 2, m_c = 1$, and the sliding friction is $b_s = 0.8$.

E2.21   The rotational velocity ω of the satellite shown in Figure E2.21 is adjusted by changing the length of the beam L. The transfer function between $\omega(s)$ and the incremental change in beam length $\Delta L(s)$ is

$$\frac{\omega(s)}{\Delta L(s)} = \frac{2.5(s + 2)}{(s + 5)(s + 1)^2}.$$

The beam length change is $\Delta L(s) = 1/(4s)$. Determine the response of the velocity $\omega(t)$.

Answer: $\omega(t) = \dfrac{1}{4} + \dfrac{3}{128}e^{-5t} - \dfrac{35}{128}e^{-t} - \dfrac{5}{32}te^{-t}$

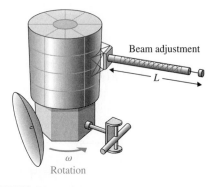

Beam adjustment

L

ω

Rotation

FIGURE E2.21   Satellite with adjustable rotational velocity.

E2.22   Determine the closed-loop transfer function $T(s) = Y(s)/R(s)$ for the system of Figure E2.22.

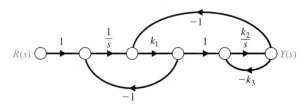

$R(s)$ 1 $\dfrac{1}{s}$ k_1 1 -1 $\dfrac{k_2}{s}$ $Y(s)$ $-k_3$ -1

FIGURE E2.22   Control system with three feedback loops.

E2.23   The block diagram of a system is shown in Figure E2.23. Determine the transfer function $T(s) = Y(s)/R(s)$.

E2.24   An amplifier may have a region of deadband as shown in Figure E2.24. Use an approximation that uses a cubic equation $y = ax^3$ in the approximately linear region. Select a and determine a linear approximation for the amplifier when the operating point is $x = 0.6$.

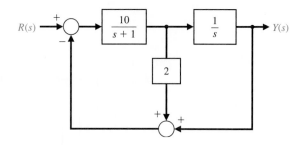

$R(s)$ $\dfrac{10}{s + 1}$ $\dfrac{1}{s}$ $Y(s)$ 2

FIGURE E2.23   Multiloop feedback system.

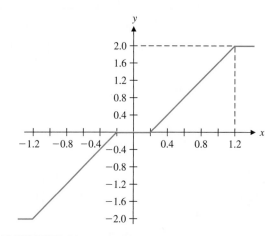

FIGURE E2.24   An amplifier with a deadband region.

E2.25   Determine the transfer function $X_2(s)/F(s)$ for the system shown in Figure E2.25. Both masses slide on a frictionless surface, and $k = 1$ N/m.

Answer: $\dfrac{X_2(s)}{F(s)} = \dfrac{1}{s^2(s^2 + 2)}$

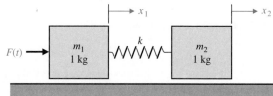

x_1 x_2

$F(t)$ m_1 1 kg k m_2 1 kg

FIGURE E2.25   Two connected masses on a frictionless surface.

E2.26   Find the transfer function $Y(s)/T_d(s)$ for the system shown in Figure E2.26.

Answer: $\dfrac{Y(s)}{T_d(s)} = \dfrac{G_2(s)}{1 + G_1(s)G_2(s)H(s)}$

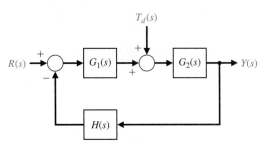

FIGURE E2.26 System with disturbance.

E2.27 Determine the transfer function $V_o(s)/V(s)$ for the op-amp circuit shown in Figure E2.27 [1]. Let $R_1 = 167$ kΩ, $R_2 = 240$ kΩ, $R_3 = 1$ kΩ, $R_4 = 100$ kΩ, and $C = 1$ μF. Assume an ideal op-amp.

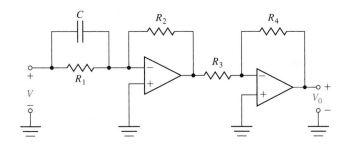

FIGURE E2.27
Op-amp circuit.

E2.28 A system is shown in Fig. E2.28(a).

(a) Determine $G(s)$ and $H(s)$ of the block diagram shown in Figure E2.28(b) that are equivalent to those of the block diagram of Figure E2.28(a).

(b) Determine $Y(s)/R(s)$ for Figure E2.28(b).

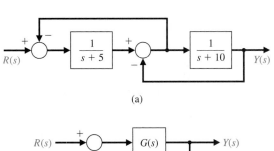

(a)

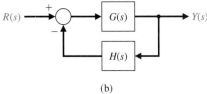

(b)

FIGURE E2.28 Block diagram equivalence.

E2.29 A system is shown in Figure E2.29.

(a) Find the closed-loop transfer function $Y(s)/R(s)$ when $G(s) = \dfrac{10}{s^2 + 2s + 10}$.

(b) Determine $Y(s)$ when the input $R(s)$ is a unit step.

(c) Compute $y(t)$.

E2.30 Determine the partial fraction expansion for $V(s)$ and compute the inverse Laplace transform. The transfer function $V(s)$ is given by:

$$V(s) = \frac{400}{s^2 + 8s + 400}.$$

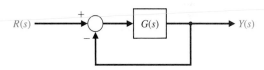

FIGURE E2.29 Unity feedback control system.

PROBLEMS

Problems require an extension of the concepts of the chapter to new situations.

P2.1   An electric circuit is shown in Figure P2.1. Obtain a set of simultaneous integrodifferential equations representing the network.

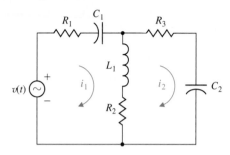

FIGURE P2.1   Electric circuit.

P2.2   A dynamic vibration absorber is shown in Figure P2.2. This system is representative of many situations involving the vibration of machines containing unbalanced components. The parameters M_2 and k_{12} may be chosen so that the main mass M_1 does not vibrate in the steady state when $F(t) = a \sin(\omega_0 t)$. Obtain the differential equations describing the system.

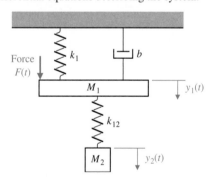

FIGURE P2.2   Vibration absorber.

P2.3   A coupled spring–mass system is shown in Figure P2.3. The masses and springs are assumed to be equal. Obtain the differential equations describing the system.

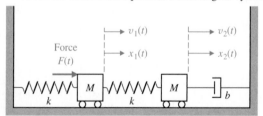

FIGURE P2.3   Two-mass system.

P2.4   A nonlinear amplifier can be described by the following characteristic:

$$v_0(t) = \begin{cases} v_{in}^2 & v_{in} \geq 0 \\ -v_{in}^2 & v_{in} < 0 \end{cases}.$$

The amplifier will be operated over a range of ± 0.5 volts around the operating point for v_{in}. Describe the amplifier by a linear approximation (a) when the operating point is $v_{in} = 0$ and (b) when the operating point is $v_{in} = 1$ volt. Obtain a sketch of the nonlinear function and the approximation for each case.

P2.5   Fluid flowing through an orifice can be represented by the nonlinear equation

$$Q = K(P_1 - P_2)^{1/2},$$

where the variables are shown in Figure P2.5 and K is a constant [2]. (a) Determine a linear approximation for the fluid-flow equation. (b) What happens to the approximation obtained in part (a) if the operating point is $P_1 - P_2 = 0$?

FIGURE P2.5   Flow through an orifice.

P2.6   Using the Laplace transformation, obtain the current $I_2(s)$ of Problem 2.1. Assume that all the initial currents are zero, the initial voltage across capacitor C_1 is zero, $v(t)$ is zero, and the initial voltage across C_2 is 10 volts.

P2.7   Obtain the transfer function of the differentiating circuit shown in Figure P2.7.

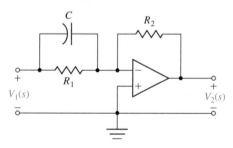

FIGURE P2.7   A differentiating circuit.

P2.8   A bridged-T network is often used in AC control systems as a filter network [8]. The circuit of one

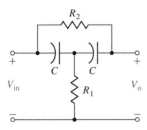

FIGURE P2.8 Bridged-T network.

bridged-T network is shown in Figure P2.8. Show that the transfer function of the network is

$$\frac{V_o(s)}{V_{in}(s)} = \frac{1 + 2R_1Cs + R_1R_2C^2s^2}{1 + (2R_1 + R_2)Cs + R_1R_2C^2s^2}.$$

Sketch the pole–zero diagram when $R_1 = 0.5$, $R_2 = 1$, and $C = 0.5$.

P2.9 Determine the transfer function $X_1(s)/F(s)$ for the coupled spring–mass system of Problem 2.3. Sketch the s-plane pole–zero diagram for low damping when $M = 1$, $b/k = 1$, and

$$\zeta = \frac{1}{2}\frac{b}{\sqrt{kM}} = 0.1.$$

P2.10 Determine the transfer function $Y_1(s)/F(s)$ for the vibration absorber system of Problem 2.2. Determine the necessary parameters M_2 and k_{12} so that the mass M_1 does not vibrate in the steady state when $F(t) = a \sin(\omega_0 t)$.

P2.11 For electromechanical systems that require large power amplification, rotary amplifiers are often used [8, 19]. An amplidyne is a power amplifying rotary amplifier. An amplidyne and a servomotor are shown in Figure P2.11. Obtain the transfer function $\theta(s)/V_c(s)$, and draw the block diagram of the system. Assume $v_d = k_2i_q$ and $v_q = k_1i_c$.

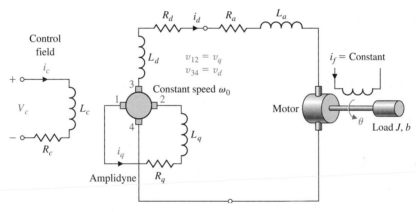

FIGURE P2.11 Amplidyne and armature-controlled motor.

P2.12 For the open-loop control system described by the block diagram shown in Figure P2.12, determine the value of K such that $y(t) \to 10$ as $t \to \infty$ when $r(t)$ is a unit step input. Assume zero initial conditions.

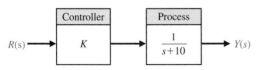

FIGURE P2.12 Open-loop control system.

P2.13 An electromechanical open-loop control system is shown in Figure P2.13. The generator, driven at a constant speed, provides the field voltage for the motor. The motor has an inertia J_m and bearing friction b_m. Obtain

the transfer function $\theta_L(s)/V_f(s)$ and draw a block diagram of the system. The generator voltage v_g can be assumed to be proportional to the field current i_f.

P2.14 A rotating load is connected to a field-controlled DC electric motor through a gear system. The motor is assumed to be linear. A test results in the output load reaching a speed of 1 rad/s within 0.5 s when a constant 80 V is applied to the motor terminals. The output steady-state speed is 2.4 rad/s. Determine the transfer function $\theta(s)/V_f(s)$ of the motor, in rad/V. The inductance of the field may be assumed to be negligible (see Figure 2.18). Also, note that the application of 80 V to the motor terminals is a step input of 80 V in magnitude.

P2.15 Consider the spring-mass system depicted in Figure P2.15. Determine a differential equation to describe the motion of the mass m. Obtain the system response $x(t)$ with the initial conditions $x(0) = x_0$ and $\dot{x}(0) = 0$.

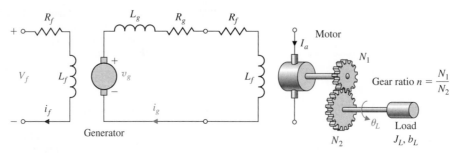

FIGURE P2.13   Motor and generator.

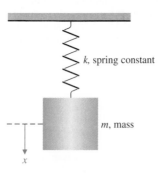

FIGURE P2.15   Suspended spring–mass system.

P2.16   Obtain a signal-flow graph to represent the following set of algebraic equations where x_1 and x_2 are to be considered the dependent variables and 6 and 11 are the inputs:

$$x_1 + 1.5x_2 = 6, \qquad 2x_1 + 4x_2 = 11.$$

Determine the value of each dependent variable by using the gain formula. After solving for x_1 by Mason's signal-flow gain formula, verify the solution by using Cramer's rule.

P2.17   A mechanical system is shown in Figure P2.17, which is subjected to a known displacement $x_3(t)$ with respect to the reference. (a) Determine the two independent equations of motion. (b) Obtain the equations of motion in terms of the Laplace transform, assuming that the initial conditions are zero. (c) Sketch a signal-flow graph representing the system of equations. (d) Obtain the relationship $T_{13}(s)$ between $X_1(s)$ and $X_3(s)$ by using Mason's signal-flow gain formula. Compare the work necessary to obtain $T_{13}(s)$ by matrix methods to that using Mason's signal-flow gain formula.

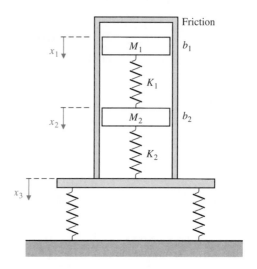

FIGURE P2.17   Mechanical system.

P2.18   An LC ladder network is shown in Figure P2.18. One may write the equations describing the network as follows:

$$I_1 = (V_1 - V_a)Y_1, \qquad V_a = (I_1 - I_a)Z_2,$$
$$I_a = (V_a - V_2)Y_3, \qquad V_2 = I_aZ_4.$$

Construct a flow graph from the equations and determine the transfer function $V_2(s)/V_1(s)$.

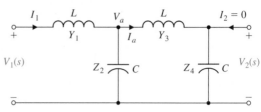

FIGURE P2.18 LC ladder network.

P2.19 A voltage follower (buffer amplifier) is shown in Figure P2.19. Show that $T = v_0/v_{in} = 1$. Assume an ideal op-amp.

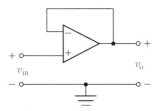

FIGURE P2.19 A buffer amplifier.

P2.20 The source follower amplifier provides lower output impedance and essentially unity gain. The circuit diagram is shown in Figure P2.20(a), and the small-signal model is shown in Figure P2.20(b). This circuit uses an FET and provides a gain of approximately unity. Assume that $R_2 \gg R_1$ for biasing purposes and that

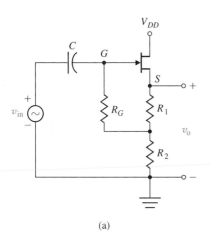

(a)

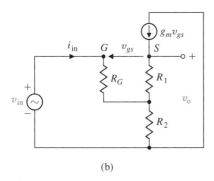

(b)

FIGURE P2.20 The source follower or common drain amplifier using an FET.

$R_g \gg R_2$. (a) Solve for the amplifier gain. (b) Solve for the gain when $g_m = 2000 \ \mu\Omega$ and $R_s = 10 \ k\Omega$ where $R_s = R_1 + R_2$. (c) Sketch a block diagram that represents the circuit equations.

P2.21 A hydraulic servomechanism with mechanical feedback is shown in Figure P2.21 [19]. The power piston has an area equal to A. When the valve is moved a small amount Δz, the oil will flow through to the cylinder at a rate $p \cdot \Delta z$, where p is the port coefficient. The input oil pressure is assumed to be constant. From the geometry, we find that $\Delta z = k \dfrac{l_1 - l_2}{l_1}(x - y) - \dfrac{l_2}{l_1}y$. (a) Determine the closed-loop signal-flow graph or block diagram for this mechanical system. (b) Obtain the closed-loop transfer function $Y(s)/X(s)$.

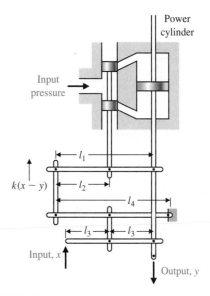

FIGURE P2.21 Hydraulic servomechanism.

P2.22 Figure P2.22 shows two pendulums suspended from frictionless pivots and connected at their midpoints by a spring [1]. Assume that each pendulum can be represented by a mass M at the end of a massless bar of length L. Also assume that the displacement is small and linear approximations can be used for $\sin \theta$ and $\cos \theta$. The spring located in the middle of the bars is unstretched when $\theta_1 = \theta_2$. The input force is represented by $f(t)$, which influences the left-hand bar only. (a) Obtain the equations of motion, and sketch a block diagram for them. (b) Determine the transfer function $T(s) = \theta_1(s)/F(s)$. (c) Sketch the location of the poles and zeros of $T(s)$ on the s-plane.

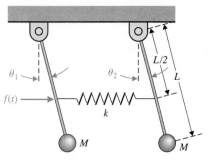

FIGURE P2.22 The bars are each of length L and the spring is located at $L/2$.

P2.23 The small-signal circuit equivalent to a common-emitter transistor amplifier is shown in Figure P2.23.

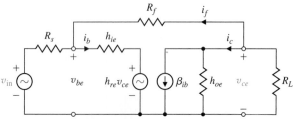

FIGURE P2.23 *CE* amplifier.

The transistor amplifier includes a feedback resistor R_f. Determine the input–output ratio v_{ce}/v_{in}.

P2.24 A two-transistor series voltage feedback amplifier is shown in Figure P2.24(a). This AC equivalent circuit neglects the bias resistors and the shunt capacitors. A block diagram representing the circuit is shown in Figure P2.24(b). This block diagram neglects the effect of h_{re}, which is usually an accurate approximation, and assumes that $R_2 + R_L \gg R_1$. (a) Determine the voltage gain v_o/v_{in}. (b) Determine the current gain i_{c2}/i_{b1}. (c) Determine the input impedance v_{in}/i_{b1}.

P2.25 H. S. Black is noted for developing a negative feedback amplifier in 1927. Often overlooked is the fact that three years earlier he had invented a circuit design technique known as feedforward correction [20]. Recent experiments have shown that this technique offers the potential for yielding excellent amplifier stabilization. Black's amplifier is shown in Figure P2.25(a) in the form recorded in 1924. The block diagram is shown in Figure P2.25(b). Determine the transfer function between the output $Y(s)$ and the input $R(s)$ and between the output and the disturbance $T_d(s)$. $G(s)$ is used to denote the amplifier represented by μ in Figure P2.25(a).

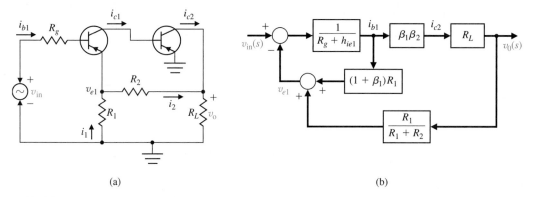

(a)   (b)

FIGURE P2.24 Feedback amplifier.

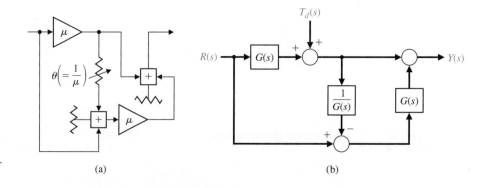

FIGURE P2.25 H. S. Black's amplifier.

(a)   (b)

P2.26 A robot includes significant flexibility in the arm members with a heavy load in the gripper [6, 21]. A two-mass model of the robot is shown in Figure. P2.26. Find the transfer function $Y(s)/F(s)$.

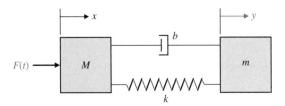

FIGURE P2.26 The spring-mass-damper model of a robot arm.

P2.27 Magnetic levitation trains provide a high-speed, very low friction alternative to steel wheels on steel rails. The train floats on an air gap as shown in Figure P2.27 [27]. The levitation force F_L is controlled by the coil current i in the levitation coils and may be approximated by

$$F_L = k \frac{i^2}{z^2},$$

where z is the air gap. This force is opposed by the downward force $F = mg$. Determine the linearized relationship between the air gap z and the controlling current near the equilibrium condition.

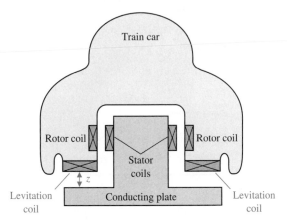

FIGURE P2.27 Cutaway view of train.

P2.28 A multiple-loop model of an urban ecological system might include the following variables: number of people in the city (P), modernization (M), migration into the city (C), sanitation facilities (S), number of diseases (D), bacteria/area (B), and amount of garbage/area (G), where the symbol for the variable is given in parentheses. The following causal loops are hypothesized:

1. $P \rightarrow G \rightarrow B \rightarrow D \rightarrow P$
2. $P \rightarrow M \rightarrow C \rightarrow P$
3. $P \rightarrow M \rightarrow S \rightarrow D \rightarrow P$
4. $P \rightarrow M \rightarrow S \rightarrow B \rightarrow D \rightarrow P$

Sketch a signal-flow graph for these causal relationships, using appropriate gain symbols. Indicate whether you believe each gain transmission is positive or negative. For example, the causal link S to B is negative because improved sanitation facilities lead to reduced bacteria/area. Which of the four loops are positive feedback loops and which are negative feedback loops?

P2.29 We desire to balance a rolling ball on a tilting beam as shown in Figure P2.29. We will assume the motor input current i controls the torque with negligible friction. Assume the beam may be balanced near the horizontal ($\phi = 0$); therefore, we have a small deviation of ϕ. Find the transfer function $X(s)/I(s)$, and draw a block diagram illustrating the transfer function showing $\phi(s)$, $X(s)$, and $I(s)$.

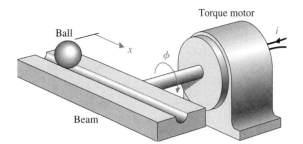

FIGURE P2.29 Tilting beam and ball.

P2.30 The measurement or sensor element in a feedback system is important to the accuracy of the system [6]. The dynamic response of the sensor is important. Most sensor elements possess a transfer function

$$H(s) = \frac{k}{\tau s + 1}$$

Suppose that a position-sensing photo detector has $\tau = 4 \, \mu s$ and $0.999 < k < 1.001$. Obtain the step response of the system, and find the k resulting in the fastest response—that is, the fastest time to reach 98% of the final value.

P2.31 Consider the cable reel control system given in Figure P2.31. Find the value of K such that for a desired velocity of 50 m/s, the percent overshoot is less than 9%.

P2.32 An interacting control system with two inputs and two outputs is shown in Figure P2.32. Solve for $Y_1(s)/R_1(s)$ and $Y_2(s)/R_1(s)$ when $R_2 = 0$.

P2.33 A system consists of two electric motors that are coupled by a continuous flexible belt. The belt also

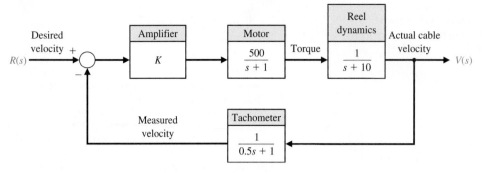

FIGURE P2.31   Cable reel control system.

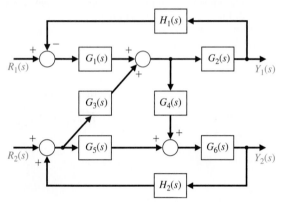

FIGURE P2.32   Interacting System.

passes over a swinging arm that is instrumented to allow measurement of the belt speed and tension. The basic control problem is to regulate the belt speed and tension by varying the motor torques.

An example of a practical system similar to that shown occurs in textile fiber manufacturing processes when yarn is wound from one spool to another at high speed. Between the two spools, the yarn is processed in a way that may require the yarn speed and tension to be controlled within defined limits. A model of the system is shown in Figure P2.33. Find $Y_2(s)/R_1(s)$. Determine a relationship for the system that will make Y_2 independent of R_1.

P2.34   Find the transfer function for $Y(s)/R(s)$ for the idle-speed control system for a fuel-injected engine as shown in Figure P2.34.

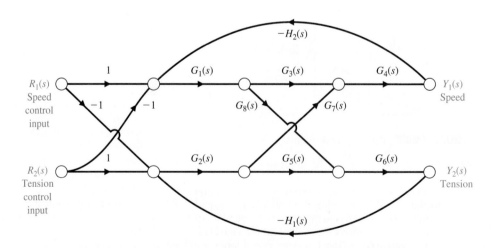

FIGURE P2.33
A model of the coupled motor drives.

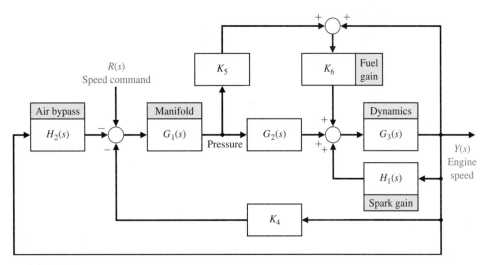

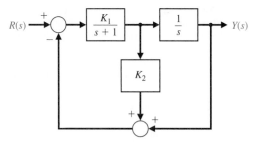

FIGURE P2.34   Idle speed control system.

P2.35   The suspension system for one wheel of an old-fashioned pickup truck is illustrated in Figure P2.35. The mass of the vehicle is m_1 and the mass of the wheel is m_2. The suspension spring has a spring constant k_1 and the tire has a spring constant k_2. The damping constant of the shock absorber is b. Obtain the transfer function $Y_1(s)/X(s)$, which represents the vehicle response to bumps in the road.

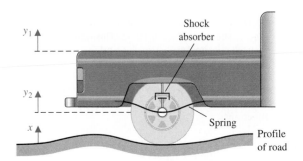

FIGURE P2.35   Pickup truck suspension.

P2.36   A feedback control system has the structure shown in Figure P2.36. Determine the closed-loop transfer function $Y(s)/R(s)$ (a) by block diagram manipulation and (b) by using a signal-flow graph and Mason's signal-flow gain formula. (c) Select the gains K_1 and K_2 so that the closed-loop response to a step input is critically damped with two equal roots at $s = -10$. (d) Plot the critically damped response for a unit step input. What is the time required for the step response to reach 90% of its final value?

FIGURE P2.36   Multiloop feedback system.

P2.37   A system is represented by Figure P2.37. (a) Determine the partial fraction expansion and $y(t)$ for a ramp input, $r(t) = t, t \geq 0$. (b) Obtain a plot of $y(t)$ for part (a), and find $y(t)$ for $t = 1.5$ s. (c) Determine the impulse response of the system $y(t)$ for $t \geq 0$. (d) Obtain a plot of $y(t)$ for part (c) and find $y(t)$ for $t = 1.5$ s.

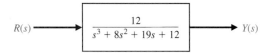

FIGURE P2.37   A third-order system.

P2.38   A two-mass system is shown in Figure P2.38 with an input force $u(t)$. When $m_1 = m_2 = 1$ and $K_1 = K_2 = 1$, find the set of differential equations describing the system.

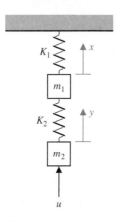

FIGURE P2.38   Two-mass system.

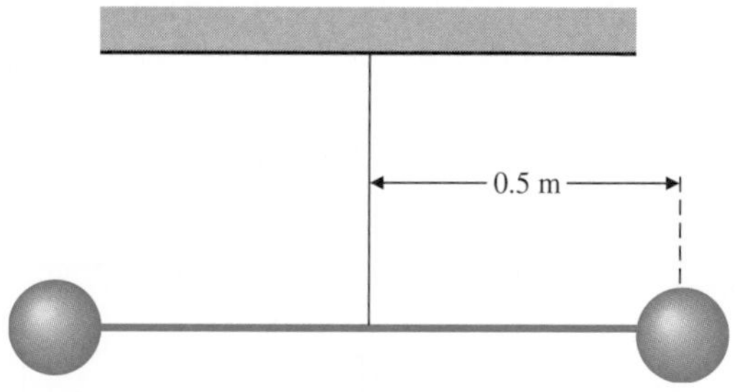

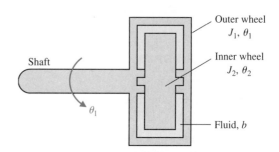

FIGURE P2.39   Winding oscillator.

P2.39   A winding oscillator consists of two steel spheres on each end of a long slender rod, as shown in Figure P2.39. The rod is hung on a thin wire that can be twisted many revolutions without breaking. The device will be wound up 4000 degrees. How long will it take until the motion decays to a swing of only 10 degrees? Assume that the thin wire has a rotational spring constant of 2×10^{-4} N m/rad and that the viscous friction coefficient for the sphere in air is 2×10^{-4} N m s/rad. The sphere has a mass of 1 kg.

P2.40   For the circuit of Figure P2.40, determine the transform of the output voltage $V_0(s)$. Assume that the circuit is in steady state when $t < 0$. Assume that the switch moves instantaneously from contact 1 to contact 2 at $t = 0$.

P2.41   A damping device is used to reduce the undesired vibrations of machines. A viscous fluid, such as a heavy oil, is placed between the wheels, as shown in Figure P2.41. When vibration becomes excessive, the relative motion of the two wheels creates damping. When the device is rotating without vibration, there is no relative motion and no damping occurs. Find $\theta_1(s)$ and $\theta_2(s)$. Assume that the shaft has a spring constant K and that b is the damping constant of the fluid. The load torque is T.

FIGURE P2.41   Cutaway view of damping device.

P2.42   The lateral control of a rocket with a gimbaled engine is shown in Figure P2.42. The lateral deviation from the desired trajectory is h and the forward rocket speed is V. The control torque of the engine is T_c and the disturbance torque is T_d. Derive the describing equations of a linear model of the system, and draw the block diagram with the appropriate transfer functions.

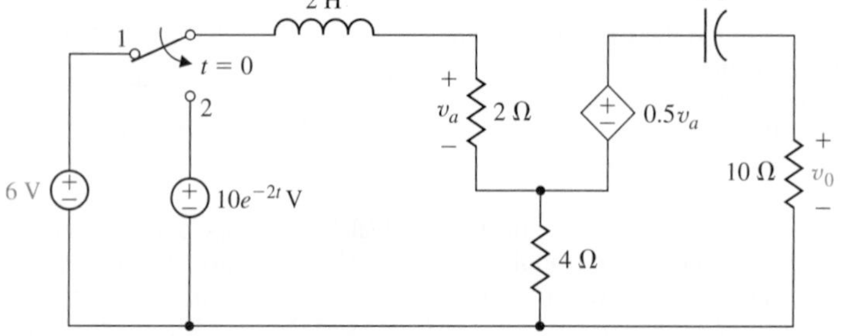

FIGURE P2.40
Model of an
electronic circuit.

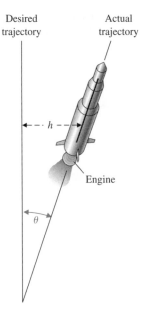

Desired trajectory   Actual trajectory

Engine

FIGURE P2.42   Rocket with gimbaled engine.

P2.43   In many applications, such as reading product codes in supermarkets and in printing and manufacturing, an optical scanner is utilized to read codes, as shown in Figure P2.43. As the mirror rotates, a friction force is developed that is proportional to its angular speed. The friction constant is equal to 0.06 N s/rad, and the moment of inertia is equal to 0.1 kg m². The output variable is the velocity $\omega(t)$. (a) Obtain the differential equation for the motor. (b) Find the response of the system when the input motor torque is a unit step and the initial velocity at $t = 0$ is equal to 0.7.

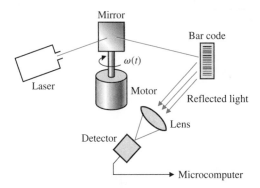

FIGURE P2.43   Optical scanner.

P2.44   An ideal set of gears is shown in Table 2.5, item 10. Neglect the inertia and friction of the gears and assume that the work done by one gear is equal to that

of the other. Derive the relationships given in item 10 of Table 2.5. Also, determine the relationship between the torques T_m and T_L.

P2.45   An ideal set of gears is connected to a solid cylinder load as shown in Figure P2.45. The inertia of the motor shaft and gear G_2 is J_m. Determine (a) the inertia of the load J_L and (b) the torque T at the motor shaft. Assume the friction at the load is b_L and the friction at the motor shaft is b_m. Also assume the density of the load disk is ρ and the gear ratio is n. Hint: The torque at the motorshaft is given by $T = T_1 + T_m$.

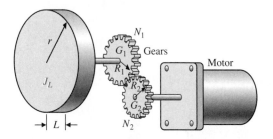

FIGURE P2.45   Motor, gears, and load.

P2.46   To exploit the strength advantage of robot manipulators and the intellectual advantage of humans, a class of manipulators called **extenders** has been examined [23]. The extender is defined as an active manipulator worn by a human to augment the human's strength. The human provides an input $U(s)$, as shown in Figure P2.46. The endpoint of the extender is $P(s)$. Determine the output $P(s)$ for both $U(s)$ and $F(s)$ in the form

$$P(s) = T_1(s)U(s) + T_2(s)F(s).$$

P2.47   A load added to a truck results in a force F on the support spring, and the tire flexes as shown in Figure

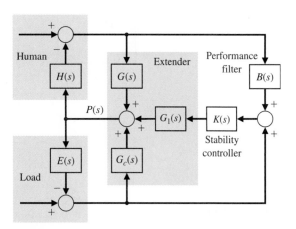

FIGURE P2.46   Model of extender.

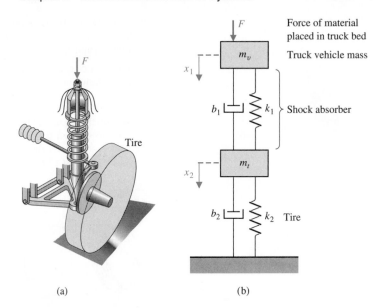

FIGURE P2.47
Truck support
model.   (a)   (b)

P2.47(a). The model for the tire movement is shown in Figure P2.47(b). Determine the transfer function $X_1(s)/F(s)$.

P2.48   The water level $h(t)$ in a tank is controlled by an open-loop system, as shown in Figure P2.48. A DC motor controlled by an armature current i_a turns a shaft, opening a valve. The inductance of the DC motor is negligible, that is, $L_a = 0$. Also, the rotational friction of the motor shaft and valve is negligible, that is, $b = 0$. The height of the water in the

tank is

$$h(t) = \int [1.6\theta(t) - h(t)]\, dt,$$

the motor constant is $K_m = 10$, and the inertia of the motor shaft and valve is $J = 6 \times 10^{-3}\ \text{kg m}^2$. Determine (a) the differential equation for $h(t)$ and $v(t)$ and (b) the transfer function $H(s)/V(s)$.

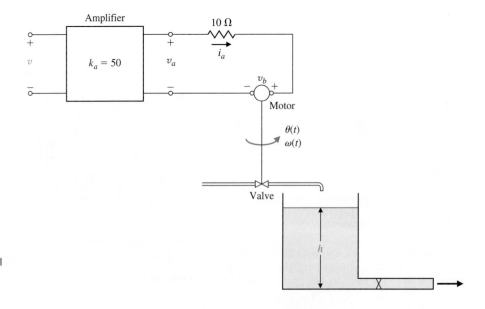

FIGURE P2.48
Open-loop control
system for the
water level of a
tank.

P2.49 The circuit shown in Figure P2.49 is called a lead-lag filter.

(a) Find the transfer function $V_2(s)/V_1(s)$. Assume an ideal op-amp.
(b) Determine $V_2(s)/V_1(s)$ when $R_1 = 100\,k\Omega$, $R_2 = 200\,k\Omega$, $C_1 = 1\,\mu F$, and $C_2 = 0.1\,\mu F$.
(c) Determine the partial fraction expansion for $V_2(s)/V_1(s)$.

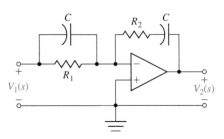

FIGURE P2.49 Lead-lag filter.

P2.50 A closed-loop control system is shown in Figure P2.50.

(a) Determine the transfer function

$$T(s) = Y(s)/R(s).$$

(b) Determine the poles and zeros of $T(s)$.
(c) Use a unit step input, $R(s) = 1/s$, and obtain the partial fraction expansion for $Y(s)$ and the value of the residues.
(d) Plot $y(t)$ and discuss the effect of the real and complex poles of $T(s)$. Do the complex poles or the real poles dominate the response?

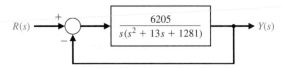

FIGURE P2.50 Unity feedback control system.

P2.51 A closed-loop control system is shown in Figure P2.51.

(a) Determine the transfer function $T(s) = Y(s)/R(s)$.
(b) Determine the poles and zeros of $T(s)$.
(c) Use a unit step input, $R(s) = 1/s$, and obtain the partial fraction expansion for $Y(s)$ and the value of the residues.
(d) Plot $y(t)$ and discuss the effect of the real and complex poles of $T(s)$. Do the complex poles or the real poles dominate the response?
(e) Predict the final value of $y(t)$ for the unit step input.

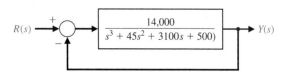

FIGURE P2.51 Third-order feedback system.

ADVANCED PROBLEMS

AP2.1 An armature-controlled DC motor is driving a load. The input voltage is 5 V. The speed at $t = 2$ seconds is 30 rad/s, and the steady speed is 70 rad/s when $t \to \infty$. Determine the transfer function $\omega(s)/V(s)$.

AP2.2 A system has a block diagram as shown in Figure AP2.2. Determine the transfer function

$$T(s) = \frac{Y_2(s)}{R_1(s)}.$$

It is desired to decouple $Y_2(s)$ from $R_1(s)$ by obtaining $T(s) = 0$. Select $G_5(s)$ in terms of the other $G_i(s)$ to achieve decoupling.

AP2.3 Consider the feedback control system in Figure AP2.3. Define the tracking error as

$$E(s) = R(s) - Y(s).$$

(a) Determine a suitable $H(s)$ such that the tracking error is zero for any input $R(s)$ in the absence of a disturbance input (that is, when $T_d(s) = 0$). (b) Using $H(s)$ determined in part (a), determine the response $Y(s)$ for a disturbance $T_d(s)$ when the input $R(s) = 0$. (c) Is it possible to obtain $Y(s) = 0$ for an arbitrary disturbance $T_d(s)$ when $G_d(s) \neq 0$? Explain your answer.

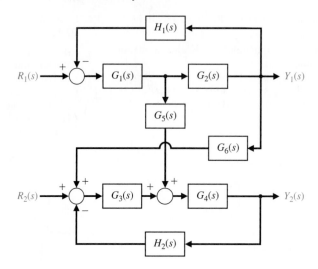

FIGURE AP2.2
Interacting control
system.

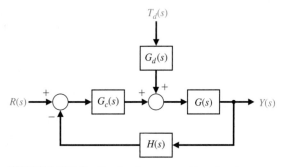

FIGURE AP2.3   Feedback system with a disturbance
input.

AP2.4   Consider a thermal heating system given by

$$\frac{\mathcal{T}(s)}{q(s)} = \frac{1}{C_t s + (QS + 1/R_t)},$$

where the output $\mathcal{T}(s)$ is the temperature difference
due to the thermal process, the input $q(s)$ is the rate of
heat flow of the heating element. The system parameters are $C_t, Q, S,$ and R_t. The thermal heating system is
illustrated in Table 2.5. (a) Determine the response of
the system to a unit step $q(s) = 1/s$. (b) As $t \to \infty$,
what value does the step response determined in part
(a) approach? This is known as the steady-state response. (c) Describe how you would select the system
parameters $C_t, Q, S,$ and R_t to increase the speed of response of the system to a step input.

AP2.5   For the three-cart system illustrated in Figure
AP2.5, obtain the equations of motion. The system has
three inputs $u_1, u_2,$ and u_3 and three outputs $x_1, x_2,$
and x_3. Obtain three second-order ordinary differential equations with constant coefficients. If possible,
write the equations of motion in matrix form.

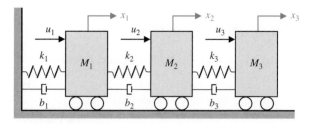

FIGURE AP2.5   Three-cart system with Three inputs and
three outputs.

DESIGN PROBLEMS

CDP2.1 We want to accurately position a table for a machine as shown in Figure CDP2.1. A traction-drive motor with a capstan roller possesses several desirable characteristics compared to the more popular ball screw. The traction drive exhibits low friction and no backlash. However, it is susceptible to disturbances. Develop a model of the traction drive shown in Figure CDP2.1(a) for the parameters given in Table CDP2.1. The drive uses a DC armature-controlled motor with a capstan roller attached to the shaft. The drive bar moves the linear slide-table. The slide uses an air bearing, so its friction is negligible. We are considering the open-loop model, Figure CDP2.1(b), and its transfer function in this problem. Feedback will be introduced later.

Table CDP2.1 Typical Parameters for the Armature-Controlled DC Motor and the Capstan and Slide

M_s	Mass of slide	5.693 kg
M_b	Mass of drive bar	6.96 kg
J_m	Inertia of roller, shaft, motor and tachometer	$10.91 \cdot 10^{-3}$ kg m^2
r	Roller radius	$31.75 \cdot 10^{-3}$ m
b_m	Motor damping	0.268 N ms/rad
K_m	Torque constant	0.8379 N m/amp
K_b	Back emf constant	0.838 V s/rad
R_m	Motor resistance	1.36 Ω
L_m	Motor inductance	3.6 mH

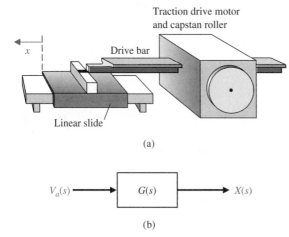

Traction drive motor
and capstan roller

Drive bar

x

Linear slide

(a)

$V_a(s) \longrightarrow \boxed{G(s)} \longrightarrow X(s)$

(b)

FIGURE CDP2.1 (a) Traction drive, capstan roller, and linear slide. (b) The block diagram model.

DP2.1 A control system is shown in Figure DP2.1. The transfer functions $G_2(s)$ and $H_2(s)$ are fixed. Determine the transfer functions $G_1(s)$ and $H_1(s)$ so that

the closed-loop transfer function $Y(s)/R(s)$ is exactly equal to 1.

DP2.2 The television beam circuit of a television is represented by the model in Figure DP2.2. Select the unknown conductance G so that the voltage v is 24 V. Each conductance is given in siemens (S).

DP2.3 An input $r(t) = t, t \geq 0$, is applied to a black box with a transfer function $G(s)$. The resulting output response, when the initial conditions are zero, is

$$y(t) = e^{-t} - \frac{1}{4}e^{-2t} - \frac{3}{4} + \frac{1}{2}t, t \geq 0.$$

Determine $G(s)$ for this system.

DP2.4 An operational amplifier circuit that can serve as a filter circuit is shown in Figure DP2.4. (a) Determine the transfer function of the circuit, assuming an ideal op-amp. Find $v_0(t)$ when the input is $v_1(t) = At$, $t \geq 0$.

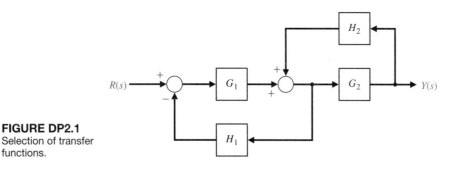

FIGURE DP2.1
Selection of transfer functions.

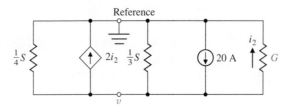

FIGURE DP2.2
Television beam
circuit.

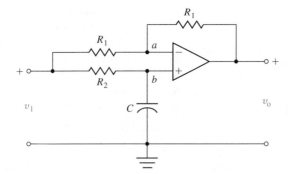

FIGURE DP2.4
Operational
amplifier circuit.

 COMPUTER PROBLEMS

CP2.1 Consider the two polynomials

$$p(s) = s^2 + 5s + 4$$

and

$$q(s) = s + 10.$$

Compute the following

(a) $p(s)q(s)$

(b) poles and zeros of $G(s) = \dfrac{q(s)}{p(s)}$

(c) $p(-1)$

CP2.2 Consider the feedback system depicted in Figure CP2.2.

(a) Compute the closed-loop transfer function using the series and feedback functions.

(b) Obtain the closed-loop system unit step response with the step function, and verify that final value of the output is 2/5.

CP2.3 Consider the differential equation

$$\ddot{y} + 4\dot{y} + 4y = u,$$

where $y(0) = \dot{y}(0) = 0$ and $u(t)$ is a unit step. Determine the solution $y(t)$ analytically and verify by co-plotting the analytic solution and the step response obtained with the step function.

CP2.4 Consider the mechanical system depicted in Figure CP2.4. The input is given by $f(t)$, and the output

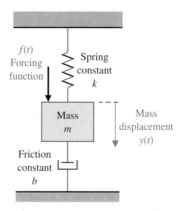

FIGURE CP2.4 A mechanical spring-mass-damper system.

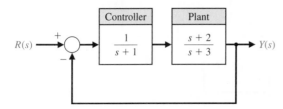

FIGURE CP2.2 A negative feedback control system.

is $y(t)$. Determine the transfer function from $f(t)$ to $y(t)$ and, using an m-file, plot the system response to a unit step input. Let $m = 10$, $k = 1$, and $b = 0.5$. Show that the peak amplitude of the output is about 1.8.

CP2.5 A satellite single-axis attitude control system can be represented by the block diagram in Figure CP2.5. The variables k, a, and b are controller parameters, and J is the spacecraft moment of inertia. Suppose the nominal moment of inertia is $J = 10.8\text{E}8\,(\text{slug ft}^2)$, and the controller parameters are $k = 10.8\text{E}8$, $a = 1$, and $b = 8$.

(a) Develop an m-file script to compute the closed-loop transfer function $T(s) = \theta(s)/\theta_d(s)$.
(b) Compute and plot the step response to a 10° step input.
(c) The exact moment of inertia is generally unknown and may change slowly with time. Compare the step response performance of the spacecraft when J is reduced by 20% and 50%. Use the controller parameters $k = 10.8\text{E}8$, $a = 1$, and $b = 8$ and a 10° step input. Discuss your results.

CP2.6 Consider the block diagram in Figure CP2.6.

(a) Use an m-file to reduce the block diagram in Figure CP2.6, and compute the closed-loop transfer function.
(b) Generate a pole–zero map of the closed-loop transfer function in graphical form using the pzmap function.

(c) Determine explicitly the poles and zeros of the closed-loop transfer function using the pole and zero functions and correlate the results with the pole–zero map in part (b).

CP2.7 For the simple pendulum shown in Figure CP2.7, the nonlinear equation of motion is given by

$$\ddot{\theta}(t) + \frac{g}{L}\sin\theta = 0,$$

where $L = 0.5$ m, $m = 1$ kg, and $g = 9.8$ m/s². When the nonlinear equation is linearized about the equilibrium point $\theta = 0$, we obtain the linear time-invariant model,

$$\ddot{\theta} + \frac{g}{L}\theta = 0.$$

Create an m-file to plot both the nonlinear and the linear response of the simple pendulum when the initial angle of the pendulum is $\theta(0) = 30°$ and explain any differences.

CP2.8 A system has a transfer function

$$\frac{X(s)}{R(s)} = \frac{(15/z)(s + z)}{s^2 + 3s + 15}.$$

Plot the response of the system when $R(s)$ is a unit step for the parameter $z = 3, 6$, and 12.

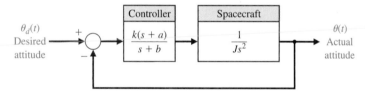

FIGURE CP2.5 A spacecraft single-axis attitude control block diagram.

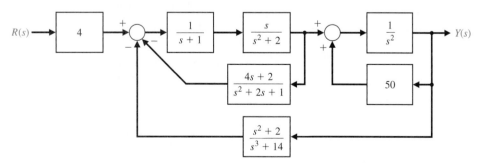

FIGURE CP2.6 A multiple-loop feedback control system block diagram.

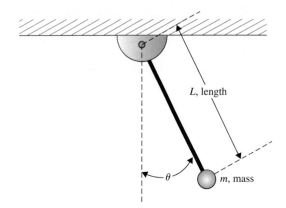

FIGURE CP2.7
Simple pendulum.

CP2.9 Consider the feedback control system in Figure CP2.9, where

$$G(s) = \frac{s + 1}{s + 2} \quad \text{and} \quad H(s) = \frac{1}{s + 1}.$$

(a) Using an m-file, determine the closed-loop transfer function.

(b) Obtain the pole–zero map using the pzmap function. Where are the closed-loop system poles and zeros?

(c) Are there any pole–zero cancellations? If so, use the minreal function to cancel common poles and zeros in the closed-loop transfer function.

(d) Why is it important to cancel common poles and zeros in the transfer function?

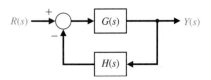

FIGURE CP2.9   Control system with nonunity feedback.

TERMS AND CONCEPTS

Actuator   The device that causes the process to provide the output. The device that provides the motive power to the process.

Assumptions   Statements that reflect situations and conditions that are taken for granted and without proof. In control systems, assumptions are often employed to simplify the physical dynamical models of systems under consideration to make the control design problem more tractable.

Block diagrams   Unidirectional, operational blocks that represent the transfer functions of the elements of the system.

Characteristic equation   The relation formed by equating to zero the denominator of a transfer function.

Closed-loop transfer function   A ratio of the output signal to the input signal for an interconnection of systems when all the feedback or feedforward loops have been closed or otherwise accounted for. Generally obtained by block diagram or signal-flow graph reduction.

Critical damping   The case where damping is on the boundary between underdamped and overdamped.

Damped oscillation   An oscillation in which the amplitude decreases with time.

Damping ratio   A measure of damping. A dimensionless number for the second-order characteristic equation.

DC motor   An electric actuator that uses an input voltage as a control variable.

Differential equation An equation including differentials of a function.

Error signal The difference between the desired output $R(s)$ and the actual output $Y(s)$; therefore $E(s) = R(s) - Y(s)$.

Final value The value that the output achieves after all the transient constituents of the response have faded. Also referred to as the steady-state value.

Final value theorem The theorem that states that $\lim_{t \to \infty} y(t) = \lim_{s \to 0} sY(s)$, where $Y(s)$ is the Laplace transform of $y(t)$.

Homogeneity The property of a linear system in which the system response, $y(t)$, to an input $u(t)$ leads to the response $\beta y(t)$ when the input is $\beta u(t)$.

Laplace transform A transformation of a function $f(t)$ from the time domain into the complex frequency domain yielding $F(s)$.

Linear approximation An approximate model that results in a linear relationship between the output and the input of the device.

Linear system A system that satisfies the properties of superposition and homogeneity.

Linearized Made linear or placed in a linear form. Taylor series approximations are commonly employed to obtain linear models of physical systems.

Mason loop rule A rule that enables the user to obtain a transfer function by tracing paths and loops within a system.

Mathematical models Descriptions of the behavior of a system using mathematics.

Natural frequency The frequency of natural oscillation that would occur for two complex poles if the damping were equal to zero.

Necessary condition A condition or statement that must be satisfied to achieve a desired effect or result. For example, for a linear system it is necessary that the input $u_1(t) + u_2(t)$ results in the response $y_1(t) + y_2(t)$, where the input $u_1(t)$ results in the response $y_1(t)$ and the input $u_2(t)$ results in the response $y_2(t)$.

Overdamped The case where the damping ratio is $\zeta > 1$.

Poles The roots of the denominator polynomial (i.e., the roots of the characteristic equation) of the transfer function.

Principle of superposition The law that states that if two inputs are scaled and summed and routed through a linear, time-invariant system, then the output will be identical to the sum of outputs due to the individual scaled inputs when routed through the same system.

Reference input The input to a control system often representing the desired output, denoted by $R(s)$.

Residues The constants k_i associated with the partial fraction expansion of the output $Y(s)$, when the output is written in a residue-pole format.

Signal-flow graph A diagram that consists of nodes connected by several directed branches and that is a graphical representation of a set of linear relations.

Simulation A model of a system that is used to investigate the behavior of a system by utilizing actual input signals.

Steady state The value that the output achieves after all the transient constituents of the response have faded. Also referred to as the final value.

s-plane The complex plane where, given the complex number $s = s + jw$, the x-axis (or horizontal axis) is the s-axis, and the y-axis (or vertical axis) is the jw-axis.

Taylor series A power series defined by $g(x) = \sum_{m=0}^{\infty} \frac{g^{(m)}(x_0)}{m!}(x - x_0)^m$. For $m < \infty$, the series is an approximation which is used to linearize functions and system models.

Time constant The time interval necessary for a system to change from one state to another by a specified percentage. For a first order system, the time constant is the time it takes the output to manifest a 63.2% change due to a step input.

Transfer function The ratio of the Laplace transform of the output variable to the Laplace transform of the input variable.

Underdamped The case where the damping ratio is $\zeta < 1$.

Unity feedback A feedback control system wherein the gain of the feedback loop is one.

Zeros The roots of the numerator polynomial of the transfer function.

State Variable Models

P R E V I E W

In this chapter, we consider system modeling using time-domain methods. As before, we will consider physical systems described by an nth-order ordinary differential equation. Utilizing a (nonunique) set of variables, known as state variables, we can obtain a set of first-order differential equations. We group these first-order equations using a compact matrix notation in a model known as the state variable model. The time-domain state variable model lends itself readily to computer solution and analysis. The relationship between signal-flow graph models and state variable models will be investigated. Several interesting physical systems, including a space station and a printer belt drive, are presented and analyzed. The chapter concludes with the development of a state variable model for the Sequential Design Example: Disk Drive Read System.

DESIRED OUTCOMES

Upon completion of Chapter 3, students should:

❏   Understand the concept of state variables, state differential equations, and output equations.

❏   Recognize that state variable models can describe the dynamic behavior of physical systems and can be represented by block diagrams and signal flow graphs.

❏   Know how to obtain the transfer function model from a state variable model, and vice versa.

❏   Be aware of solution methods for state variable models and the role of the state transition matrix in obtaining the time responses.

❏   Understand the important role of state variable modeling in control system design.

3.1 INTRODUCTION

In the preceding chapter, we developed and studied several useful approaches to the analysis and design of feedback systems. The Laplace transform was used to transform the differential equations representing the system to an algebraic equation expressed in terms of the complex variable s. Using this algebraic equation, we were able to obtain a transfer function representation of the input–output relationship.

The ready availability of digital computers makes it practical to consider the time-domain formulation of the equations representing control systems. The time-domain techniques can be used for nonlinear, time-varying, and multivariable systems.

> **A time-varying control system is a system in which one or more of the parameters of the system may vary as a function of time.**

For example, the mass of a missile varies as a function of time as the fuel is expended during flight. A multivariable system, as discussed in Section 2.6, is a system with several input and output signals.

The solution of a time-domain formulation of a control system problem is facilitated by the availability and ease of use of digital computers. Therefore we are interested in reconsidering the time-domain description of dynamic systems as they are represented by the system differential equation. The **time domain** is the mathematical domain that incorporates the response and description of a system in terms of time, t.

The time-domain representation of control systems is an essential basis for modern control theory and system optimization. In Chapter 11, we will have an opportunity to design an optimum control system by utilizing time-domain methods. In this chapter, we develop the time-domain representation of control systems and illustrate several methods for the solution of the system time response.

3.2 THE STATE VARIABLES OF A DYNAMIC SYSTEM

The time-domain analysis and design of control systems uses the concept of the state of a system [1–3, 5].

> **The state of a system is a set of variables whose values, together with the input signals and the equations describing the dynamics, will provide the future state and output of the system.**

For a dynamic system, the state of a system is described in terms of a set of **state variables** $[x_1(t), x_2(t), \ldots, x_n(t)]$. The state variables are those variables that determine the future behavior of a system when the present state of the system and the excitation signals are known. Consider the system shown in Figure 3.1, where $y_1(t)$

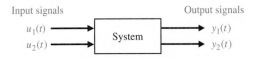

FIGURE 3.1
System block
diagram.

and $y_2(t)$ are the output signals and $u_1(t)$ and $u_2(t)$ are the input signals. A set of state variables $(x_1, x_2, \ldots, x_n)$ for the system shown in the figure is a set such that knowledge of the initial values of the state variables $[x_1(t_0), x_2(t_0), \ldots, x_n(t_0)]$ at the initial time t_0, and of the input signals $u_1(t)$ and $u_2(t)$ for $t \geq t_0$, suffices to determine the future values of the outputs and state variables [2].

> **The state variables describe the present configuration of a system and can be used to determine the future response, given the excitation inputs and the equations describing the dynamics.**

The general form of a dynamic system is shown in Figure 3.2. A simple example of a state variable is the state of an on–off light switch. The switch can be in either the on or the off position, and thus the state of the switch can assume one of two possible values. Thus, if we know the present state (position) of the switch at t_0 and if an input is applied, we are able to determine the future value of the state of the element.

The concept of a set of state variables that represent a dynamic system can be illustrated in terms of the spring-mass-damper system shown in Figure 3.3. The number of state variables chosen to represent this system should be as small as possible in order to avoid redundant state variables. A set of state variables sufficient to describe this system includes the position and the velocity of the mass. Therefore, we will define a set of state variables as (x_1, x_2), where

$$x_1(t) = y(t) \quad \text{and} \quad x_2(t) = \frac{dy(t)}{dt}.$$

The differential equation describes the behavior of the system and is usually written as

$$M\frac{d^2y}{dt^2} + b\frac{dy}{dt} + ky = u(t). \tag{3.1}$$

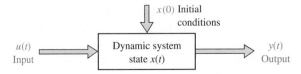

FIGURE 3.2
Dynamic system.

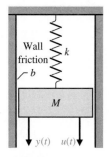

FIGURE 3.3
A spring-mass-
damper system.

To write Equation (3.1) in terms of the state variables, we substitute the state variables as already defined and obtain

$$M\frac{dx_2}{dt} + bx_2 + kx_1 = u(t). \tag{3.2}$$

Therefore, we can write the equations that describe the behavior of the spring-mass-damper system as the set of two first-order differential equations

$$\frac{dx_1}{dt} = x_2 \tag{3.3}$$

and

$$\frac{dx_2}{dt} = \frac{-b}{M}x_2 - \frac{k}{M}x_1 + \frac{1}{M}u. \tag{3.4}$$

This set of differential equations describes the behavior of the state of the system in terms of the rate of change of each state variable.

As another example of the state variable characterization of a system, consider the *RLC* circuit shown in Figure 3.4. The state of this system can be described by a set of state variables (x_1, x_2), where x_1 is the capacitor voltage $v_c(t)$ and x_2 is the inductor current $i_L(t)$. This choice of state variables is intuitively satisfactory because the stored energy of the network can be described in terms of these variables as

$$\mathscr{E} = \frac{1}{2}Li_L^2 + \frac{1}{2}Cv_c^2. \tag{3.5}$$

Therefore $x_1(t_0)$ and $x_2(t_0)$ provide the total initial energy of the network and the state of the system at $t = t_0$. For a passive *RLC* network, the number of state variables required is equal to the number of independent energy-storage elements. Utilizing Kirchhoff's current law at the junction, we obtain a first-order differential equation by describing the rate of change of capacitor voltage as

$$i_c = C\frac{dv_c}{dt} = +u(t) - i_L. \tag{3.6}$$

Kirchhoff's voltage law for the right-hand loop provides the equation describing the rate of change of inductor current as

$$L\frac{di_L}{dt} = -Ri_L + v_c. \tag{3.7}$$

The output of this system is represented by the linear algebraic equation

$$v_o = Ri_L(t).$$

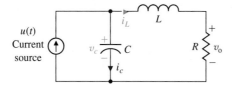

FIGURE 3.4
An *RLC* circuit.

We can rewrite Equations (3.6) and (3.7) as a set of two first-order differential equations in terms of the state variables x_1 and x_2 as follows:

$$\frac{dx_1}{dt} = -\frac{1}{C}x_2 + \frac{1}{C}u(t), \tag{3.8}$$

and

$$\frac{dx_2}{dt} = +\frac{1}{L}x_1 - \frac{R}{L}x_2. \tag{3.9}$$

The output signal is then

$$y_1(t) = v_o(t) = Rx_2. \tag{3.10}$$

Utilizing Equations (3.8) and (3.9) and the initial conditions of the network represented by $[x_1(t_0), x_2(t_0)]$, we can determine the system's future behavior and its output.

The state variables that describe a system are not a unique set, and several alternative sets of state variables can be chosen. For example, for a second-order system, such as the spring-mass-damper or RLC circuit, the state variables may be any two independent linear combinations of $x_1(t)$ and $x_2(t)$. For the RLC circuit, we might choose the set of state variables as the two voltages, $v_c(t)$ and $v_L(t)$, where v_L is the voltage drop across the inductor. Then the new state variables, x_1^* and x_2^*, are related to the old state variables, x_1 and x_2, as

$$x_1^* = v_c = x_1, \tag{3.11}$$

and

$$x_2^* = v_L = v_c - Ri_L = x_1 - Rx_2. \tag{3.12}$$

Equation (3.12) represents the relation between the inductor voltage and the former state variables v_c and i_L. In a typical system, there are several choices of a set of state variables that specify the energy stored in a system and therefore adequately describe the dynamics of the system. It is usual to choose a set of state variables that can be readily measured.

An alternative approach to developing a model of a device is the use of the bond graph. Bond graphs can be used for electrical, mechanical, hydraulic, and thermal devices or systems as well as for combinations of various types of elements. Bond graphs produce a set of equations in the state variable form [7].

The state variables of a system characterize the dynamic behavior of a system. The engineer's interest is primarily in physical systems, where the variables are voltages, currents, velocities, positions, pressures, temperatures, and similar physical variables. However, the concept of system state is not limited to the analysis of physical systems and is particularly useful in analyzing biological, social, and economic systems. For these systems, the concept of state is extended beyond the concept of the current configuration of a physical system to the broader viewpoint of variables that will be capable of describing the future behavior of the system.

3.3 THE STATE DIFFERENTIAL EQUATION

The response of a system is described by the set of first-order differential equations written in terms of the state variables $(x_1, x_2, \ldots, x_n)$ and the inputs $(u_1, u_2, \ldots, u_m)$. These first-order differential equations can be written in general form as

$$
\dot{x}_1 = a_{11}x_1 + a_{12}x_2 + \cdots + a_{1n}x_n + b_{11}u_1 + \cdots + b_{1m}u_m,
$$

$$
\dot{x}_2 = a_{21}x_1 + a_{22}x_2 + \cdots + a_{2n}x_n + b_{21}u_1 + \cdots + b_{2m}u_m,
$$

$$
\vdots
$$

$$
\dot{x}_n = a_{n1}x_1 + a_{n2}x_2 + \cdots + a_{nn}x_n + b_{n1}u_1 + \cdots + b_{nm}u_m, \tag{3.13}
$$

where $\dot{x} = dx/dt$. Thus, this set of simultaneous differential equations can be written in matrix form as follows [2, 5]:

$$
\frac{d}{dt}\begin{bmatrix} x_1 \\ x_2 \\ \vdots \\ x_n \end{bmatrix} = \begin{bmatrix} a_{11} & a_{12} \cdots & a_{1n} \\ a_{21} & a_{22} \cdots & a_{2n} \\ \vdots & \cdots & \vdots \\ a_{n1} & a_{n2} \cdots & a_{nn} \end{bmatrix}\begin{bmatrix} x_1 \\ x_2 \\ \vdots \\ x_n \end{bmatrix} + \begin{bmatrix} b_{11} \cdots b_{1m} \\ \vdots \quad \vdots \\ b_{n1} \cdots b_{nm} \end{bmatrix}\begin{bmatrix} u_1 \\ \vdots \\ u_m \end{bmatrix}. \tag{3.14}
$$

The column matrix consisting of the state variables is called the **state vector** and is written as

$$
\mathbf{x} = \begin{bmatrix} x_1 \\ x_2 \\ \vdots \\ x_n \end{bmatrix}, \tag{3.15}
$$

where the boldface indicates a vector. The vector of input signals is defined as **u**. Then the system can be represented by the compact notation of the **state differential equation** as

$$
\boxed{\dot{\mathbf{x}} = \mathbf{A}\mathbf{x} + \mathbf{B}\mathbf{u}.} \tag{3.16}
$$

The differential equation (3.16) is also commonly called the state equation.

The matrix $\mathbf{A}$ is an $n \times n$ square matrix, and $\mathbf{B}$ is an $n \times m$ matrix.[†] The state differential equation relates the rate of change of the state of the system to the state of the system and the input signals. In general, the outputs of a linear system can be related to the state variables and the input signals by the **output equation**

$$
\boxed{\mathbf{y} = \mathbf{C}\mathbf{x} + \mathbf{D}\mathbf{u},} \tag{3.17}
$$

[†]Boldfaced lowercase letters denote vector quantities and boldfaced uppercase letters denote matrices. For an introduction to matrices and elementary matrix operations, refer to the MCS website and references [1] and [2].

where **y** is the set of output signals expressed in column vector form. The **state-space representation** (or state-variable representation) comprises the state differential equation and the output equation.

We use Equations (3.8) and (3.9) to obtain the state variable differential equation for the *RLC* of Figure 3.4 as

$$\dot{\mathbf{x}} = \begin{bmatrix} 0 & \dfrac{-1}{C} \\[2mm] \dfrac{1}{L} & \dfrac{-R}{L} \end{bmatrix} \mathbf{x} + \begin{bmatrix} \dfrac{1}{C} \\[2mm] 0 \end{bmatrix} u(t) \tag{3.18}$$

and the output as

$$y = \begin{bmatrix} 0 & R \end{bmatrix} \mathbf{x}. \tag{3.19}$$

When $R = 3$, $L = 1$, and $C = 1/2$, we have

$$\dot{\mathbf{x}} = \begin{bmatrix} 0 & -2 \\ 1 & -3 \end{bmatrix} \mathbf{x} + \begin{bmatrix} 2 \\ 0 \end{bmatrix} u$$

and

$$y = \begin{bmatrix} 0 & 3 \end{bmatrix} \mathbf{x}.$$

The solution of the state differential equation (Equation 3.16) can be obtained in a manner similar to the method for solving a first-order differential equation. Consider the first-order differential equation

$$\dot{x} = ax + bu, \tag{3.20}$$

where $x(t)$ and $u(t)$ are scalar functions of time. We expect an exponential solution of the form e^{at}. Taking the Laplace transform of Equation (3.20), we have

$$sX(s) - x(0) = aX(s) + bU(s);$$

therefore,

$$X(s) = \frac{x(0)}{s - a} + \frac{b}{s - a} U(s). \tag{3.21}$$

The inverse Laplace transform of Equation (3.21) can be shown to be

$$x(t) = e^{at} x(0) + \int_0^t e^{+a(t-\tau)} bu(\tau)\, d\tau. \tag{3.22}$$

We expect the solution of the general state differential equation to be similar to Equation (3.22) and to be of exponential form. The **matrix exponential function** is defined as

$$e^{\mathbf{A}t} = \exp(\mathbf{A}t) = \mathbf{I} + \mathbf{A}t + \frac{\mathbf{A}^2 t^2}{2!} + \cdots + \frac{\mathbf{A}^k t^k}{k!} + \cdots, \tag{3.23}$$

which converges for all finite t and any $\mathbf{A}$ [2]. Then the solution of the state differential equation is found to be

$$\mathbf{x}(t) = \exp(\mathbf{A}t)\mathbf{x}(0) + \int_0^t \exp[\mathbf{A}(t - \tau)]\mathbf{B}\mathbf{u}(\tau)\, d\tau. \tag{3.24}$$

Equation (3.24) may be verified by taking the Laplace transform of Equation (3.16) and rearranging to obtain

$$\mathbf{X}(s) = [s\mathbf{I} - \mathbf{A}]^{-1}\mathbf{x}(0) + [s\mathbf{I} - \mathbf{A}]^{-1}\mathbf{B}\mathbf{U}(s), \tag{3.25}$$

where we note that $[s\mathbf{I} - \mathbf{A}]^{-1} = \boldsymbol{\Phi}(s)$ is the Laplace transform of $\boldsymbol{\Phi}(t) = \exp(\mathbf{A}t)$. Taking the inverse Laplace transform of Equation (3.25) and noting that the second term on the right-hand side involves the product $\boldsymbol{\Phi}(s)\mathbf{B}\mathbf{U}(s)$, we obtain Equation (3.24). The matrix exponential function describes the unforced response of the system and is called the **fundamental** or **state transition matrix** $\boldsymbol{\Phi}(t)$. Thus, Equation (3.24) can be written as

$$\mathbf{x}(t) = \boldsymbol{\Phi}(t)\mathbf{x}(0) + \int_0^t \boldsymbol{\Phi}(t - \tau)\mathbf{B}\mathbf{u}(\tau)\, d\tau. \tag{3.26}$$

The solution to the unforced system (that is, when $\mathbf{u} = 0$) is simply

$$\begin{bmatrix} x_1(t) \\ x_2(t) \\ \vdots \\ x_n(t) \end{bmatrix} = \begin{bmatrix} \phi_{11}(t) & \cdots & \phi_{1n}(t) \\ \phi_{21}(t) & \cdots & \phi_{2n}(t) \\ \vdots & & \vdots \\ \phi_{n1}(t) & \cdots & \phi_{nn}(t) \end{bmatrix} \begin{bmatrix} x_1(0) \\ x_2(0) \\ \vdots \\ x_n(0) \end{bmatrix}. \tag{3.27}$$

We note therefore that to determine the state transition matrix, all initial conditions are set to 0 except for one state variable, and the output of each state variable is evaluated. That is, the term $\phi_{ij}(t)$ is the response of the ith state variable due to an initial condition on the jth state variable when there are zero initial conditions on all the other variables. We shall use this relationship between the initial conditions and the state variables to evaluate the coefficients of the transition matrix in a later section. However, first we shall develop several suitable signal-flow state models of systems and investigate the stability of the systems by utilizing these flow graphs.

EXAMPLE 3.1   Two rolling carts

Consider the system shown in Figure 3.5. The variables of interest are noted on the figure and defined as: M_1, M_2 = mass of carts, p, q = position of carts, u = external force acting on system, k_1, k_2 = spring constants, and b_1, b_2 = damping coefficients. The free-body diagram of mass M_1 is shown in Figure 3.6(b), where $\dot{p}, \dot{q}$ = velocity of M_1 and M_2, respectively. We assume that the cars have negligible rolling friction. We consider any existing rolling friction to be lumped into the damping coefficients, b_1 and b_2.

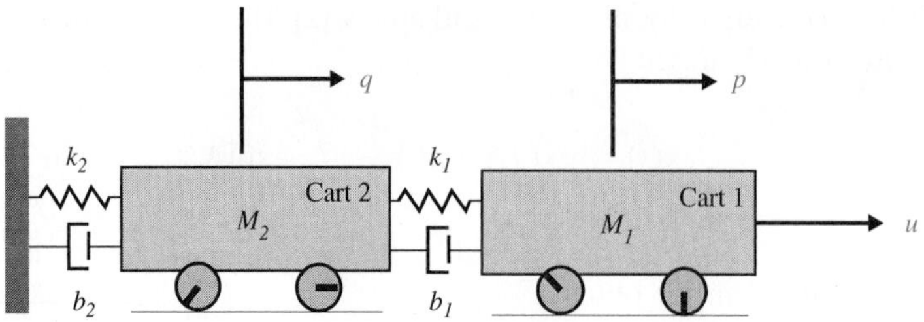

FIGURE 3.5
Two rolling carts attached with springs and dampers.

Now, given the free-body diagram with forces and directions appropriately applied, we use Newton's second law (sum of the forces equals mass of the object multiplied by its acceleration) to obtain the equations of motion—one equation for each mass. For mass M_1 we have

$$M_1\ddot{p} = u + f_s + f_d = u - k_1(p - q) - b_1(\dot{p} - \dot{q}),$$

or

$$M_1\ddot{p} + b_1\dot{p} + k_1p = u + k_1q + b_1\dot{q}, \tag{3.28}$$

where

$$\ddot{p}, \ddot{q} = \text{acceleration of } M_1 \text{ and } M_2, \text{ respectively.}$$

Similarly, for mass M_2 we have

$$M_2\ddot{q} = k_1(p - q) + b_1(\dot{p} - \dot{q}) - k_2q - b_2\dot{q},$$

or

$$M_2\ddot{q} + (k_1 + k_2)q + (b_1 + b_2)\dot{q} = k_1p + b_1\dot{p}. \tag{3.29}$$

We now have a model given by the two second-order ordinary differential equations in Equations (3.28) and (3.29). We can start developing a state-space model by defining

$$x_1 = p,$$
$$x_2 = q.$$

We could have alternatively defined $x_1 = q$ and $x_2 = p$. The state-space model is not unique. Denoting the derivatives of x_1 and x_2 as x_3 and x_4, respectively, it follows that

$$x_3 = \dot{x}_1 = \dot{p}, \tag{3.30}$$

$$x_4 = \dot{x}_2 = \dot{q}. \tag{3.31}$$

Taking the derivative of x_3 and x_4 yields, respectively,

$$\dot{x}_3 = \ddot{p} = -\frac{b_1}{M_1}\dot{p} - \frac{k_1}{M_1}p + \frac{1}{M_1}u + \frac{k_1}{M_1}q + \frac{b_1}{M_1}\dot{q}, \tag{3.32}$$

$$\dot{x}_4 = \ddot{q} = -\frac{k_1 + k_2}{M_2}q - \frac{b_1 + b_2}{M_2}\dot{q} + \frac{k_1}{M_2}p + \frac{b_1}{M_2}\dot{p}, \tag{3.33}$$

where we use the relationship for $\dot{p}$ given in Equation (3.28) and the relationship for $\ddot{q}$ given in Equation (3.29). But $\dot{p} = x_3$ and $\dot{q} = x_4$, so Equation (3.32) can be written as

$$\dot{x}_3 = -\frac{k_1}{M_1}x_1 + \frac{k_1}{M_1}x_2 - \frac{b_1}{M_1}x_3 + \frac{b_1}{M_1}x_4 + \frac{1}{M_1}u \tag{3.34}$$

and Equation (3.33) as

$$\dot{x}_4 = \frac{k_1}{M_2}x_1 - \frac{k_1 + k_2}{M_2}x_2 + \frac{b_1}{M_2}x_3 - \frac{b_1 + b_2}{M_2}x_4. \tag{3.35}$$

In matrix form, Equations (3.30), (3.31), (3.34), and (3.35) can be written as

$$\dot{\mathbf{x}} = \mathbf{A}\mathbf{x} + \mathbf{B}u$$

where

$$\mathbf{x} = \begin{pmatrix} x_1 \\ x_2 \\ x_3 \\ x_4 \end{pmatrix} = \begin{pmatrix} p \\ q \\ \dot{p} \\ \dot{q} \end{pmatrix},$$

$$\mathbf{A} = \begin{bmatrix} 0 & 0 & 1 & 0 \\ 0 & 0 & 0 & 1 \\ -\dfrac{k_1}{M_1} & \dfrac{k_1}{M_1} & -\dfrac{b_1}{M_1} & \dfrac{b_1}{M_1} \\ \dfrac{k_1}{M_2} & -\dfrac{k_1 + k_2}{M_2} & \dfrac{b_1}{M_2} & -\dfrac{b_1 + b_2}{M_2} \end{bmatrix}, \quad \text{and} \quad \mathbf{B} = \begin{bmatrix} 0 \\ 0 \\ \dfrac{1}{M_1} \\ 0 \end{bmatrix},$$

and u is the external force acting on the system (see Figure 3.6). If we choose p as the output, then

$$y = \begin{bmatrix} 1 & 0 & 0 & 0 \end{bmatrix}\mathbf{x} = \mathbf{C}\mathbf{x}.$$

Suppose that the two rolling carts have the following parameter values: $k_1 = 150$ N/m; $k_2 = 700$ N/m; $b_1 = 15$ N s/m; $b_2 = 30$ N s/m; $M_1 = 5$ kg; and $M_2 = 20$ kg. The

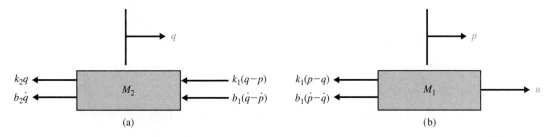

(a)   (b)

FIGURE 3.6   Free-body diagrams of the two rolling carts. (a) Cart 2; (b) Cart 1.

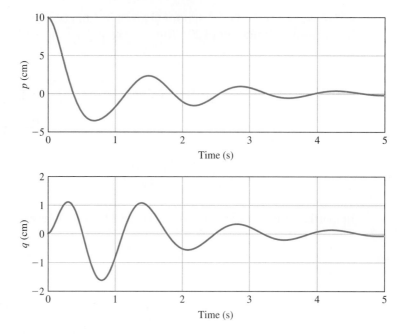

FIGURE 3.7
Initial condition
response of the two
cart system.

response of the two rolling cart system is shown in Figure 3.7 when the initial conditions are $p(0) = 10$ cm, $q(0) = 0$, and $\dot{p}(0) = \dot{q}(0) = 0$ and there is no input driving
force, that is, $u(t) = 0$. ■

3.4 SIGNAL-FLOW GRAPH AND BLOCK DIAGRAM MODELS

The state of a system describes that system's dynamic behavior where the dynamics
of the system are represented by a set of first-order differential equations. Alternatively, the dynamics of the system can be represented by a state differential equation
as in Equation (3.16). In either case, it is useful to develop a graphical model of the
system and use this model to relate the state variable concept to the familiar transfer
function representation. The graphical model can be represented via signal-flow
graphs or block diagrams.

As we have learned in previous chapters, a system can be meaningfully described by an input–output relationship, the transfer function $G(s)$. For example, if
we are interested in the relation between the output voltage and the input voltage of
the network of Figure 3.4, we can obtain the transfer function

$$G(s) = \frac{V_0(s)}{U(s)}.$$

The transfer function for the RLC network of Figure 3.4 is of the form

$$G(s) = \frac{V_0(s)}{U(s)} = \frac{\alpha}{s^2 + \beta s + \gamma}, \qquad (3.36)$$

where α, β, and γ are functions of the circuit parameters R, L, and C, respectively. The values of α, β, and γ can be determined from the differential equations that describe the circuit. For the RLC circuit (see Equations 3.8 and 3.9), we have

$$\dot{x}_1 = -\frac{1}{C}x_2 + \frac{1}{C}u(t), \tag{3.37}$$

$$\dot{x}_2 = \frac{1}{L}x_1 - \frac{R}{L}x_2, \tag{3.38}$$

and

$$v_o = Rx_2. \tag{3.39}$$

The flow graph representing these simultaneous equations is shown in Figure 3.8(a), where $1/s$ indicates an integration. The corresponding block diagram model is shown in Figure 3.8(b). The transfer function is found to be

$$\frac{V_o(s)}{U(s)} = \frac{+R/(LCs^2)}{1 + R/(Ls) + 1/(LCs^2)} = \frac{+R/(LC)}{s^2 + (R/L)s + 1/(LC)}. \tag{3.40}$$

Unfortunately many electric circuits, electromechanical systems, and other control systems are not as simple as the RLC circuit of Figure 3.4, and it is often a difficult task to determine a set of first-order differential equations describing the system. Therefore, it is often simpler to derive the transfer function of the system by the techniques of Chapter 2 and then derive the state model from the transfer function.

The signal-flow graph state model and the block diagram model can be readily derived from the transfer function of a system. However, as we noted in Section 3.3,

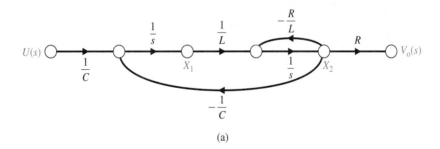

(a)

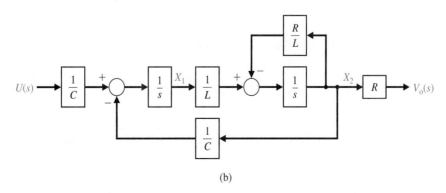

(b)

FIGURE 3.8
RLC network.
(a) Signal-flow
graph. (b) Block
diagram.

there is more than one alternative set of state variables, and therefore there is more than one possible form for the signal-flow graph and block diagram models. There are several key **canonical forms** of the state-variable representation, such as the phase variable canonical form, that we will investigate in this chapter. In general, we can represent a transfer function as

$$G(s) = \frac{Y(s)}{U(s)} = \frac{b_m s^m + b_{m-1} s^{m-1} + \cdots + b_1 s + b_0}{s^n + a_{n-1} s^{n-1} + \cdots + a_1 s + a_0} \tag{3.41}$$

where $n \geq m$, and all the a and b coefficients are real numbers. If we multiply the numerator and denominator by s^{-n}, we obtain

$$G(s) = \frac{b_m s^{-(n-m)} + b_{m-1} s^{-(n-m+1)} + \cdots + b_1 s^{-(n-1)} + b_0 s^{-n}}{1 + a_{n-1} s^{-1} + \cdots + a_1 s^{-(n-1)} + a_0 s^{-n}}. \tag{3.42}$$

Our familiarity with Mason's signal-flow gain formula allows us to recognize the familiar feedback factors in the denominator and the forward-path factors in the numerator. Mason's signal-flow gain formula was discussed in Section 2.7 and is written as

$$G(s) = \frac{Y(s)}{U(s)} = \frac{\sum_k P_k \Delta_k}{\Delta}. \tag{3.43}$$

When all the feedback loops are touching and all the forward paths touch the feedback loops, Equation (3.43) reduces to

$$G(s) = \frac{\sum_k P_k}{1 - \sum_{q=1}^{N} L_q} = \frac{\text{Sum of the forward-path factors}}{1 - \text{sum of the feedback loop factors}}. \tag{3.44}$$

There are several flow graphs that could represent the transfer function. Two flow graph configurations based on Mason's signal-flow gain formula are of particular interest, and we will consider these in greater detail. In the next section, we will consider two additional configurations: the physical state variable model and the diagonal (or Jordan canonical) form model.

To illustrate the derivation of the signal-flow graph state model, let us initially consider the fourth-order transfer function

$$G(s) = \frac{Y(s)}{U(s)} = \frac{b_0}{s^4 + a_3 s^3 + a_2 s^2 + a_1 s + a_0}$$

$$= \frac{b_0 s^{-4}}{1 + a_3 s^{-1} + a_2 s^{-2} + a_1 s^{-3} + a_0 s^{-4}}. \tag{3.45}$$

First we note that the system is fourth order, and hence we identify four state variables (x_1, x_2, x_3, x_4). Recalling Mason's signal-flow gain formula, we note that the denominator can be considered to be 1 minus the sum of the loop gains. Furthermore, the numerator of the transfer function is equal to the forward-path factor of the flow graph. The flow graph must include a minimum number of integrators equal to the order of the system. Therefore, we use four integrators to represent this system. The necessary flow graph nodes and the four integrators are shown in Figure 3.9. Considering the simplest series interconnection of integrators, we can

FIGURE 3.9
Flow graph nodes
and integrators for
fourth-order
system.

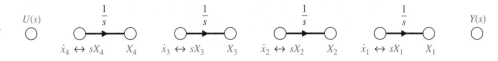

represent the transfer function by the flow graph of Figure 3.10. Examining this figure, we note that all the loops are touching and that the transfer function of this flow graph is indeed Equation (3.45). The reader can readily verify this by noting that the forward-path factor of the flow graph is b_0/s^4 and the denominator is equal to 1 minus the sum of the loop gains.

We can also consider the block diagram model of Equation (3.45). Rearranging the terms in Equation (3.45) and taking the inverse Laplace transform yields the differential equation model

$$\frac{d^4(y/b_0)}{dt^4} + a_3\frac{d^3(y/b_0)}{dt^3} + a_2\frac{d^2(y/b_0)}{dt^2} + a_1\frac{d(y/b_0)}{dt} + a_0(y/b_0) = u.$$

Define the four state variables as follows:

$$x_1 = y/b_0$$
$$x_2 = \dot{x}_1 = \dot{y}/b_0$$
$$x_3 = \dot{x}_2 = \ddot{y}/b_0$$
$$x_4 = \dot{x}_3 = \dddot{y}/b_0.$$

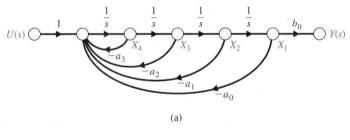

(a)

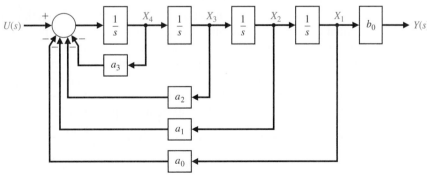

(b)

FIGURE 3.10
Model for $G(s)$ of
Equation (3.45).
(a) Signal-flow
graph. (b) Block
diagram.

Then it follows that the fourth-order differential equation can be written equivalently as four first-order differential equations, namely,

$$\dot{x}_1 = x_2,$$
$$\dot{x}_2 = x_3,$$
$$\dot{x}_3 = x_4,$$

and

$$\dot{x}_4 = -a_0 x_1 - a_1 x_2 - a_2 x_3 - a_3 x_4 + u;$$

and the corresponding output equation is

$$y = b_0 x_1.$$

The block diagram model can be readily obtained from the four first-order differential equations as illustrated in Figure 3.10(b).

Now consider the fourth-order transfer function when the numerator is a polynomial in s, so that we have

$$
\begin{aligned}
G(s) &= \frac{b_3 s^3 + b_2 s^2 + b_1 s + b_0}{s^4 + a_3 s^3 + a_2 s^2 + a_1 s + a_0} \\
&= \frac{b_3 s^{-1} + b_2 s^{-2} + b_1 s^{-3} + b_0 s^{-4}}{1 + a_3 s^{-1} + a_2 s^{-2} + a_1 s^{-3} + a_0 s^{-4}}.
\end{aligned}
\tag{3.46}
$$

The numerator terms represent forward-path factors in Mason's signal-flow gain formula. The forward paths will touch all the loops, and a suitable signal-flow graph realization of Equation (3.46) is shown in Figure 3.11(a). The forward-path factors are b_3/s, b_2/s^2, b_1/s^3, and b_0/s^4 as required to provide the numerator of the transfer function. Recall that Mason's signal-flow gain formula indicates that the numerator of the transfer function is simply the sum of the forward-path factors. This general form of a signal-flow graph can represent the general transfer function of Equation (3.46) by utilizing n feedback loops involving the a_n coefficients and m forward-path factors involving the b_m coefficients. The general form of the flow graph state model and the block diagram model shown in Figure 3.11 is called the **phase variable canonical form**.

The state variables are identified in Figure 3.11 as the output of each energy storage element, that is, the output of each integrator. To obtain the set of first-order differential equations representing the state model of Equation (3.46), we will introduce a new set of flow graph nodes immediately preceding each integrator of Figure 3.11(a) [5, 6]. The nodes are placed before each integrator, and therefore they represent the derivative of the output of each integrator. The signal-flow graph, including the added nodes, is shown in Figure 3.12. Using the flow graph of this figure, we are able to obtain the following set of first-order differential equations describing the state of the model:

$$\dot{x}_1 = x_2, \quad \dot{x}_2 = x_3, \quad \dot{x}_3 = x_4,$$
$$\dot{x}_4 = -a_0 x_1 - a_1 x_2 - a_2 x_3 - a_3 x_4 + u. \tag{3.47}$$

In this equation, $x_1, x_2, \ldots x_n$ are the n **phase variables**.

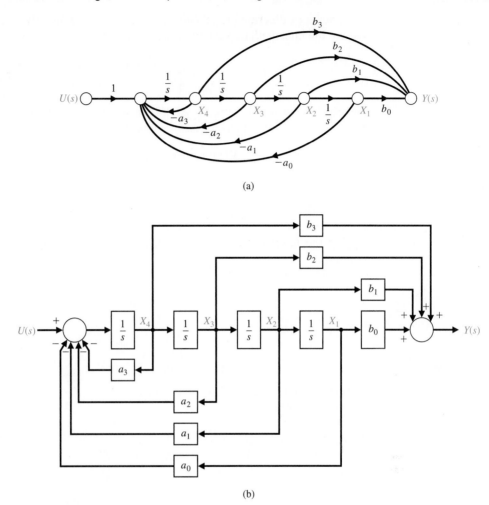

(a)

(b)

FIGURE 3.11
Model for $G(s)$ of Equation (3.46) in the phase variable format. (a) Signal-flow graph. (b) Block diagram.

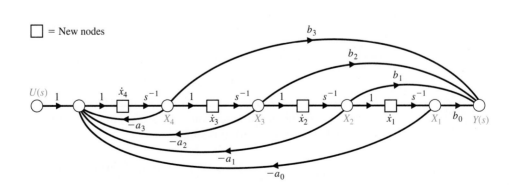

FIGURE 3.12
Flow graph of Figure 3.11 with nodes inserted.

The block diagram model can also be constructed directly from Equation (3.46). Define the intermediate variable $Z(s)$ and rewrite Equation (3.46) as

$$G(s) = \frac{Y(s)}{U(s)} = \frac{b_3 s^3 + b_2 s^2 + b_1 s + b_0}{s^4 + a_3 s^3 + a_2 s^2 + a_1 s + a_0} \frac{Z(s)}{Z(s)}.$$

Notice that, by multiplying by $Z(s)/Z(s)$, we do not change the transfer function, $G(s)$. Equating the numerator and denominator polynomials yields

$$Y(s) = [b_3 s^3 + b_2 s^2 + b_1 s + b_0] Z(s)$$

and

$$U(s) = [s^4 + a_3 s^3 + a_2 s^2 + a_1 s + a_0] Z(s).$$

Taking the inverse Laplace transform of both equations yields the differential equations

$$y = b_3 \frac{d^3 z}{dt^3} + b_2 \frac{d^2 z}{dt^2} + b_1 \frac{dz}{dt} + b_0 z$$

and

$$u = \frac{d^4 z}{dt^4} + a_3 \frac{d^3 z}{dt^3} + a_2 \frac{d^2 z}{dt^2} + a_1 \frac{dz}{dt} + a_0 z.$$

Define the four state variables as follows:

$$x_1 = z$$
$$x_2 = \dot{x}_1 = \dot{z}$$
$$x_3 = \dot{x}_2 = \ddot{z}$$
$$x_4 = \dot{x}_3 = \dddot{z}.$$

Then the differential equation can be written equivalently as

$$\dot{x}_1 = x_2,$$
$$\dot{x}_2 = x_3,$$
$$\dot{x}_3 = x_4,$$

and

$$\dot{x}_4 = -a_0 x_1 - a_1 x_2 - a_2 x_3 - a_3 x_4 + u,$$

and the corresponding output equation is

$$y = b_0 x_1 + b_1 x_2 + b_2 x_3 + b_3 x_4.$$

The block diagram model can be readily obtained from the four first-order differential equations and the output equation as illustrated in Figure 3.11(b).

Furthermore, the output is simply

$$y(t) = b_0 x_1 + b_1 x_2 + b_2 x_3 + b_3 x_4. \tag{3.48}$$

In matrix form, we can represent the system in Equation (3.46) as

$$\dot{\mathbf{x}} = \mathbf{A}\mathbf{x} + \mathbf{B}u, \tag{3.49}$$

or

$$\frac{d}{dt}\begin{bmatrix} x_1 \\ x_2 \\ x_3 \\ x_4 \end{bmatrix} = \begin{bmatrix} 0 & 1 & 0 & 0 \\ 0 & 0 & 1 & 0 \\ 0 & 0 & 0 & 1 \\ -a_0 & -a_1 & -a_2 & -a_3 \end{bmatrix}\begin{bmatrix} x_1 \\ x_2 \\ x_3 \\ x_4 \end{bmatrix} + \begin{bmatrix} 0 \\ 0 \\ 0 \\ 1 \end{bmatrix}u(t). \tag{3.50}$$

The output is then

$$y(t) = \mathbf{C}\mathbf{x} = [b_0 \quad b_1 \quad b_2 \quad b_3]\begin{bmatrix} x_1 \\ x_2 \\ x_3 \\ x_4 \end{bmatrix}. \tag{3.51}$$

The graphical structures of Figure 3.11 are not unique representations of Equation (3.46); another equally useful structure can be obtained. A flow graph that represents Equation (3.46) equally well is shown in Figure 3.13(a). In this case, the forward-path factors are obtained by feeding forward the signal $U(s)$. We will call this model the **input feedforward canonical form**.

Then the output signal $y(t)$ is equal to the first state variable $x_1(t)$. This flow graph structure has the forward-path factors b_0/s^4, b_1/s^3, b_2/s^2, b_3/s, and all the forward paths touch the feedback loops. Therefore, the resulting transfer function is indeed equal to Equation (3.46).

Associated with the input feedforward format, we have the set of first-order differential equations

$$\dot{x}_1 = -a_3x_1 + x_2 + b_3u, \qquad \dot{x}_2 = -a_2x_1 + x_3 + b_2u,$$
$$\dot{x}_3 = -a_1x_1 + x_4 + b_1u, \quad \text{and} \quad \dot{x}_4 = -a_0x_1 + b_0u. \tag{3.52}$$

Thus, in matrix form, we have

$$\frac{d\mathbf{x}}{dt} = \begin{bmatrix} -a_3 & 1 & 0 & 0 \\ -a_2 & 0 & 1 & 0 \\ -a_1 & 0 & 0 & 1 \\ -a_0 & 0 & 0 & 0 \end{bmatrix}\mathbf{x} + \begin{bmatrix} b_3 \\ b_2 \\ b_1 \\ b_0 \end{bmatrix}u(t) \tag{3.53}$$

and

$$y(t) = [1 \quad 0 \quad 0 \quad 0]\mathbf{x} + [0]u(t).$$

Although the input feedforward canonical form of Figure 3.13 represents the same transfer function as the phase variable canonical form of Figure 3.11, the state variables of each graph are not equal. Furthermore we recognize that the initial conditions of the system can be represented by the initial conditions of the integrators, $x_1(0), x_2(0), \ldots, x_n(0)$. Let us consider a control system and determine the state differential equation by utilizing the two forms of flow graph state models.

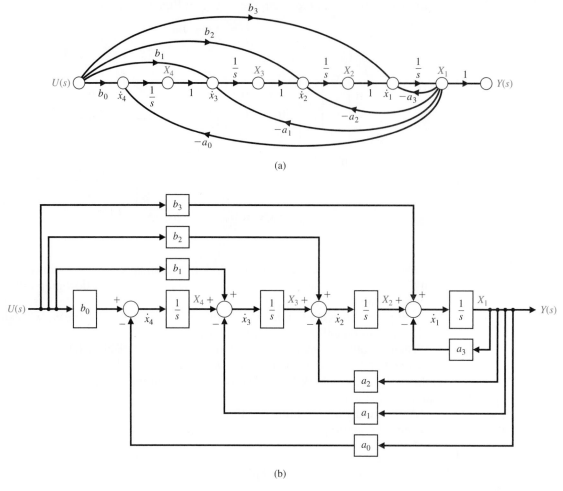

(a)

(b)

FIGURE 3.13   (a) Alternative flow graph state model for Equation (3.46). This model is called the input feedforward canonical form. (b) Block diagram of the input feedforward canonical form.

EXAMPLE 3.2   **Two state variable models**

A single-loop control system is shown in Figure 3.14. The closed-loop transfer function of the system is

$$T(s) = \frac{Y(s)}{U(s)} = \frac{2s^2 + 8s + 6}{s^3 + 8s^2 + 16s + 6}.$$

FIGURE 3.14
Single-loop control
system.

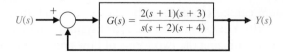

Multiplying the numerator and denominator by s^{-3}, we have

$$T(s) = \frac{Y(s)}{U(s)} = \frac{2s^{-1} + 8s^{-2} + 6s^{-3}}{1 + 8s^{-1} + 16s^{-2} + 6s^{-3}}. \tag{3.54}$$

The first model is the phase variable state model using the feedforward of the state variables to provide the output signal. The signal-flow graph and block diagram are shown in Figures 3.15(a) and (b), respectively. The state differential equation is

$$\dot{\mathbf{x}} = \begin{bmatrix} 0 & 1 & 0 \\ 0 & 0 & 1 \\ -6 & -16 & -8 \end{bmatrix} \mathbf{x} + \begin{bmatrix} 0 \\ 0 \\ 1 \end{bmatrix} u(t), \tag{3.55}$$

and the output is

$$y(t) = \begin{bmatrix} 6 & 8 & 2 \end{bmatrix} \begin{bmatrix} x_1 \\ x_2 \\ x_3 \end{bmatrix}. \tag{3.56}$$

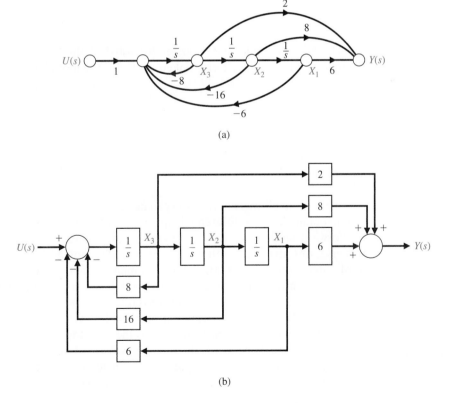

(a)

(b)

FIGURE 3.15
(a) Phase variable flow graph state model for $T(s)$.
(b) Block diagram for the phase variable canonical form.

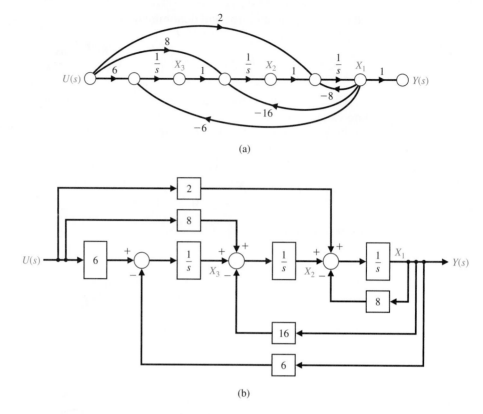

(a)

FIGURE 3.16
(a) Alternative flow graph state model for $T(s)$ using the input feedforward canonical form.
(b) Block diagram model.

(b)

The second model uses the feedforward of the input variable, as shown in Figure 3.16. The vector differential equation for the input feedforward model is

$$\dot{\mathbf{x}} = \begin{bmatrix} -8 & 1 & 0 \\ -16 & 0 & 1 \\ -6 & 0 & 0 \end{bmatrix} \mathbf{x} + \begin{bmatrix} 2 \\ 8 \\ 6 \end{bmatrix} u(t), \qquad (3.57)$$

and the output is $y(t) = x_1(t)$. ∎

We note that it was not necessary to factor the numerator or denominator polynomial to obtain the state differential equations for the phase variable model or the input feedforward model. Avoiding the factoring of polynomials permits us to avoid the tedious effort involved. Both models require three integrators because the system is third order. However, it is important to emphasize that the state variables of the state model of Figure 3.15 are not identical to the state variables of the state model of Figure 3.16. Of course, one set of state variables is related to the other set of state variables by an appropriate linear transformation of variables. A linear matrix transformation is represented by $\mathbf{z} = \mathbf{Mx}$, which transforms the $\mathbf{x}$-vector into the $\mathbf{z}$-vector by means of the $\mathbf{M}$ matrix (see Appendix E on the MCS website). Finally, we note that the transfer function of Equation (3.41) represents a single-output linear constant coefficient system; thus, the transfer function can represent an nth-order differential equation

$$\frac{d^n y}{dt^n} + a_{n-1}\frac{d^{n-1} y}{dt^{n-1}} + \cdots + a_0 y(t) = \frac{d^m u}{dt^m} + b_{m-1}\frac{d^{m-1} u}{dt^{m-1}} + \cdots + b_0 u(t). \quad (3.58)$$

Accordingly, we can obtain the n first-order equations for the nth-order differential equation by utilizing the phase variable model or the input feedforward model of this section.

3.5   ALTERNATIVE SIGNAL-FLOW GRAPH AND BLOCK DIAGRAM MODELS

Often the control system designer studies an actual control system block diagram that represents physical devices and variables. An example of a model of a DC motor with shaft velocity as the output is shown in Figure 3.17 [9]. We wish to select the **physical variables** as the state variables. Thus, we select: $x_1 = y(t)$, the velocity output; $x_2 = i(t)$, the field current; and the third state variable, x_3, is selected to be $x_3 = \frac{1}{4}r(t) - \frac{1}{20}u(t)$, where $u(t)$ is the field voltage. We may draw the models for these physical variables, as shown in Figure 3.18. Note that the state variables x_1, x_2, and x_3 are identified on the models. We will denote this format as the physical state variable model. This model is particularly useful when we can measure the physical state variables. Note that the model of each block is separately determined. For example, note that the transfer

FIGURE 3.17
A block diagram model of an open-loop DC motor control with velocity as the output.

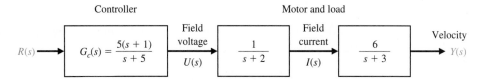

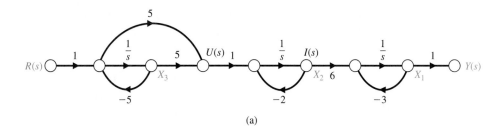

(a)

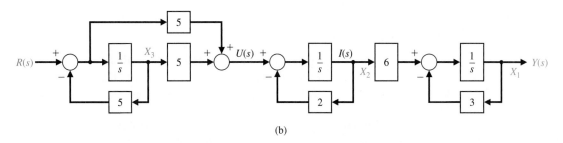

(b)

FIGURE 3.18   (a) The physical state variable signal-flow graph for the block diagram of Figure 3.17. (b) Physical state block diagram.

function for the controller is

$$\frac{U(s)}{R(s)} = G_c(s) = \frac{5(s + 1)}{s + 5} = \frac{5 + 5s^{-1}}{1 + 5s^{-1}},$$

and the flow graph between $R(s)$ and $U(s)$ represents $G_c(s)$.

The state variable differential equation is directly obtained from Figure 3.18 as

$$\dot{\mathbf{x}} = \begin{bmatrix} -3 & 6 & 0 \\ 0 & -2 & -20 \\ 0 & 0 & -5 \end{bmatrix} \mathbf{x} + \begin{bmatrix} 0 \\ 5 \\ 1 \end{bmatrix} r(t) \tag{3.59}$$

and

$$y = \begin{bmatrix} 1 & 0 & 0 \end{bmatrix} \mathbf{x}. \tag{3.60}$$

A second form of the model we need to consider is the decoupled response modes. The overall input–output transfer function of the block diagram system shown in Figure 3.17 is

$$\frac{Y(s)}{R(s)} = T(s) = \frac{30(s + 1)}{(s + 5)(s + 2)(s + 3)} = \frac{q(s)}{(s - s_1)(s - s_2)(s - s_3)},$$

and the transient response has three modes dictated by s_1, s_2, and s_3. These modes are indicated by the partial fraction expansion as

$$\frac{Y(s)}{R(s)} = T(s) = \frac{k_1}{s + 5} + \frac{k_2}{s + 2} + \frac{k_3}{s + 3}. \tag{3.61}$$

Using the procedure described in Chapter 2, we find that $k_1 = -20$, $k_2 = -10$, and $k_3 = 30$. The decoupled state variable model representing Equation (3.61) is shown in Figure 3.19. The state variable matrix differential equation is

$$\dot{\mathbf{x}} = \begin{bmatrix} -5 & 0 & 0 \\ 0 & -2 & 0 \\ 0 & 0 & -3 \end{bmatrix} \mathbf{x} + \begin{bmatrix} 1 \\ 1 \\ 1 \end{bmatrix} r(t)$$

and

$$y(t) = \begin{bmatrix} -20 & -10 & 30 \end{bmatrix} \mathbf{x}. \tag{3.62}$$

Note that we chose x_1 as the state variable associated with $s_1 = -5$, x_2 associated with $s_2 = -2$, and x_3 associated with $s_3 = -3$, as indicated in Figure 3.19. This choice of state variables is arbitrary; for example, x_1 could be chosen as associated with the factor $s + 2$.

The decoupled form of the state differential matrix equation displays the distinct model poles $-s_1, -s_2, \ldots, -s_n$, and this format is often called the **diagonal canonical form**. A system can always be written in diagonal form if it possesses distinct poles; otherwise, it can only be written in a block diagonal form, known as the **Jordan canonical form** [29].

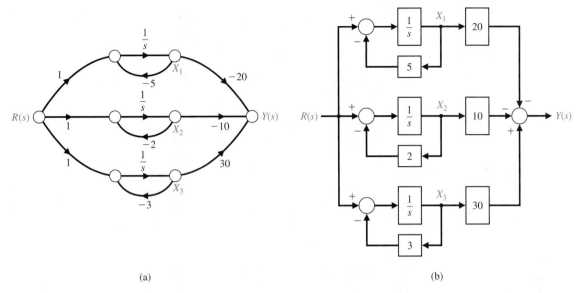

(a)   (b)

FIGURE 3.19   (a) The decoupled state variable flow graph model for the system shown in block diagram form in Figure 3.17. (b) The decoupled state variable block diagram model.

EXAMPLE 3.3   **Spread of an epidemic disease**

The spread of an epidemic disease can be described by a set of differential equations. The population under study is made up of three groups, x_1, x_2, and x_3, such that the group x_1 is susceptible to the epidemic disease, group x_2 is infected with the disease, and group x_3 has been removed from the initial population. The removal of x_3 will be due to immunization, death, or isolation from x_1. The feedback system can be represented by the following equations:

$$\frac{dx_1}{dt} = -\alpha x_1 - \beta x_2 + u_1(t),$$

$$\frac{dx_2}{dt} = \beta x_1 - \gamma x_2 + u_2(t),$$

$$\frac{dx_3}{dt} = \alpha x_1 + \gamma x_2.$$

The rate at which new susceptibles are added to the population is equal to $u_1(t)$, and the rate at which new infectives are added to the population is equal to $u_2(t)$. For a closed population, we have $u_1(t) = u_2(t) = 0$. It is interesting to note that these equations could equally well represent the spread of information or a new idea through a population.

The physical state variables for this system are x_1, x_2, and x_3. The model that represents this set of differential equations is shown in Figure 3.20. The vector differential equation is equal to

$$\frac{d}{dt}\begin{bmatrix} x_1 \\ x_2 \\ x_3 \end{bmatrix} = \begin{bmatrix} -\alpha & -\beta & 0 \\ \beta & -\gamma & 0 \\ \alpha & \gamma & 0 \end{bmatrix}\begin{bmatrix} x_1 \\ x_2 \\ x_3 \end{bmatrix} + \begin{bmatrix} 1 & 0 \\ 0 & 1 \\ 0 & 0 \end{bmatrix}\begin{bmatrix} u_1(t) \\ u_2(t) \end{bmatrix}. \qquad (3.63)$$

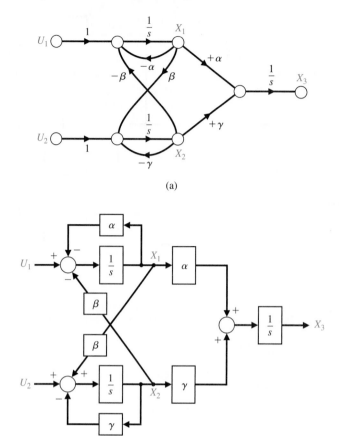

FIGURE 3.20
Model for the
spread of an
epidemic disease.
(a) Signal-flow
graph. (b) Block
diagram model.

(b)

By examining Equation (3.63) and the models depicted in Figure 3.20, we find that the state variable x_3 is dependent on x_1 and x_2 and does not affect the variables x_1 and x_2.

Let us consider a closed population, so that $u_1(t) = u_2(t) = 0$. The equilibrium point in the state space for this system is obtained by setting $d\mathbf{x}/dt = \mathbf{0}$. The equilibrium point in the state space is the point at which the system settles in the equilibrium, or rest, condition. Examining Equation (3.63), we find that the equilibrium point for this system is $x_1 = x_2 = 0$. Thus, to determine whether the epidemic disease is eliminated from the population, we must obtain the characteristic equation of the system. From the signal-flow graph shown in Figure 3.20, we obtain the flow graph determinant

$$\Delta(s) = 1 - (-\alpha s^{-1} - \gamma s^{-1} - \beta^2 s^{-2}) + (\alpha \gamma s^{-2}), \qquad (3.64)$$

where there are three loops, two of which are nontouching. Thus, the characteristic equation is

$$q(s) = s^2 \Delta(s) = s^2 + (\alpha + \gamma)s + (\alpha \gamma + \beta^2) = 0. \qquad (3.65)$$

The roots of this characteristic equation will lie in the left-hand s-plane when $\alpha + \gamma > 0$ and $\alpha \gamma + \beta^2 > 0$. When roots are in the left-hand plane, we expect the unforced response to decay to zero as $t \rightarrow \infty$. ∎

EXAMPLE 3.4   **Inverted pendulum control**

The problem of balancing a broomstick on a person's hand is illustrated in Figure 3.21. The only equilibrium condition is $\theta(t) = 0$ and $d\theta/dt = 0$. The problem of balancing a broomstick on one's hand is not unlike the problem of controlling the attitude of a missile during the initial stages of launch. This problem is the classic and intriguing problem of the inverted pendulum mounted on a cart, as shown in Figure 3.22. The cart must be moved so that mass m is always in an upright position. The state variables must be expressed in terms of the angular rotation $\theta(t)$ and the position of the cart $y(t)$. The differential equations describing the motion of the system can be obtained by writing the sum of the forces in the horizontal direction and the sum of the moments about the pivot point [2, 3, 10, 28]. We will assume that $M \gg m$ and the angle of rotation θ is small so that the equations are linear. The sum of the forces in the horizontal direction is

$$M\ddot{y} + ml\ddot{\theta} - u(t) = 0, \tag{3.66}$$

where $u(t)$ equals the force on the cart, and l is the distance from the mass m to the pivot point. The sum of the torques about the pivot point is

$$ml\ddot{y} + ml^2\ddot{\theta} - mlg\theta = 0. \tag{3.67}$$

The state variables for the two second-order equations are chosen as $(x_1, x_2, x_3, x_4) = (y, \dot{y}, \theta, \dot{\theta})$. Then Equations (3.66) and (3.67) are written in terms of the state variables as

$$M\dot{x}_2 + ml\dot{x}_4 - u(t) = 0 \tag{3.68}$$

FIGURE 3.21
An inverted pendulum balanced on a person's hand by moving the hand to reduce $\theta(t)$. Assume, for ease, that the pendulum rotates in the x–y plane.

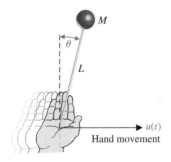

FIGURE 3.22
A cart and an inverted pendulum. The pendulum is constrained to pivot in the vertical plane.

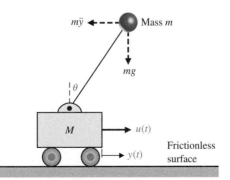

and

$$\dot{x}_2 + l\dot{x}_4 - gx_3 = 0. \tag{3.69}$$

To obtain the necessary first-order differential equations, we solve for $l\dot{x}_4$ in Equation (3.69) and substitute into Equation (3.68) to obtain

$$M\dot{x}_2 + mgx_3 = u(t), \tag{3.70}$$

since $M \gg m$. Substituting $\dot{x}_2$ from Equation (3.68) into Equation (3.69), we have

$$Ml\dot{x}_4 - Mgx_3 + u(t) = 0. \tag{3.71}$$

Therefore, the four first-order differential equations can be written as

$$\dot{x}_1 = x_2, \qquad \dot{x}_2 = -\frac{mg}{M}x_3 + \frac{1}{M}u(t),$$

$$\dot{x}_3 = x_4, \quad \text{and} \quad \dot{x}_4 = \frac{g}{l}x_3 - \frac{1}{Ml}u(t). \tag{3.72}$$

Thus, the system matrices are

$$\mathbf{A} = \begin{bmatrix} 0 & 1 & 0 & 0 \\ 0 & 0 & -mg/M & 0 \\ 0 & 0 & 0 & 1 \\ 0 & 0 & g/l & 0 \end{bmatrix}, \quad \mathbf{B} = \begin{bmatrix} 0 \\ 1/M \\ 0 \\ -1/(Ml) \end{bmatrix} \tag{3.73} \blacksquare$$

3.6 THE TRANSFER FUNCTION FROM THE STATE EQUATION

Given a transfer function $G(s)$, we can obtain the state variable equations using the signal-flow graph model. Now we turn to the matter of determining the transfer function $G(s)$ of a single-input, single-output (SISO) system. Recalling Equations (3.16) and (3.17), we have

$$\dot{\mathbf{x}} = \mathbf{Ax} + \mathbf{B}u \tag{3.74}$$

and

$$y = \mathbf{Cx} + \mathbf{D}u \tag{3.75}$$

where y is the single output and u is the single input. The Laplace transforms of Equations (3.74) and (3.75) are

$$s\mathbf{X}(s) = \mathbf{AX}(s) + \mathbf{B}U(s) \tag{3.76}$$

and

$$Y(s) = \mathbf{CX}(s) + \mathbf{D}U(s) \tag{3.77}$$

where $\mathbf{B}$ is an $n \times 1$ matrix, since u is a single input. Note that we do not include initial conditions, since we seek the transfer function. Rearranging Equation (3.76), we obtain

$$(s\mathbf{I} - \mathbf{A})\mathbf{X}(s) = \mathbf{B}U(s).$$

Since $[s\mathbf{I} - \mathbf{A}]^{-1} = \Phi(s)$, we have

$$\mathbf{X}(s) = \Phi(s)\mathbf{B}U(s).$$

Substituting $\mathbf{X}(s)$ into Equation (3.77), we obtain

$$Y(s) = [\mathbf{C}\Phi(s)\mathbf{B} + \mathbf{D}]U(s). \tag{3.78}$$

Therefore, the transfer function $G(s) = Y(s)/U(s)$ is

$$\boxed{G(s) = \mathbf{C}\Phi(s)\mathbf{B} + \mathbf{D}} \tag{3.79}$$

EXAMPLE 3.5   **Transfer function of an *RLC* circuit**

Let us determine the transfer function $G(s) = Y(s)/U(s)$ for the *RLC* circuit of Figure 3.4 as described by the differential equations (see Equations 3.18 and 3.19):

$$\dot{\mathbf{x}} = \begin{bmatrix} 0 & \dfrac{-1}{C} \\ \dfrac{1}{L} & \dfrac{-R}{L} \end{bmatrix}\mathbf{x} + \begin{bmatrix} \dfrac{1}{C} \\ 0 \end{bmatrix}u$$

$$y = \begin{bmatrix} 0 & R \end{bmatrix}\mathbf{x}.$$

Then we have

$$[s\mathbf{I} - \mathbf{A}] = \begin{bmatrix} s & \dfrac{1}{C} \\ \dfrac{-1}{L} & s + \dfrac{R}{L} \end{bmatrix}.$$

Therefore, we obtain

$$\Phi(s) = [s\mathbf{I} - \mathbf{A}]^{-1} = \frac{1}{\Delta(s)}\begin{bmatrix} \left(s + \dfrac{R}{L}\right) & \dfrac{-1}{C} \\ \dfrac{1}{L} & s \end{bmatrix},$$

where

$$\Delta(s) = s^2 + \frac{R}{L}s + \frac{1}{LC}.$$

Then the transfer function is

$$G(s) = \begin{bmatrix} 0 & R \end{bmatrix}\begin{bmatrix} \dfrac{s + \dfrac{R}{L}}{\Delta(s)} & \dfrac{-1}{C\Delta(s)} \\ \dfrac{1}{L\Delta(s)} & \dfrac{s}{\Delta(s)} \end{bmatrix}\begin{bmatrix} \dfrac{1}{C} \\ 0 \end{bmatrix}$$

$$= \frac{R/(LC)}{\Delta(s)} = \frac{R/(LC)}{s^2 + \dfrac{R}{L}s + \dfrac{1}{LC}},$$

which agrees with the result Equation (3.40) obtained from the flow graph model using Mason's signal-flow gain formula. ∎

3.7 THE TIME RESPONSE AND THE STATE TRANSITION MATRIX

It is often desirable to obtain the time response of the state variables of a control system and thus examine the performance of the system. The transient response of a system can be readily obtained by evaluating the solution to the state vector differential equation. In Section 3.3, we found that the solution for the state differential equation (3.26) was

$$\mathbf{x}(t) = \mathbf{\Phi}(t)\mathbf{x}(0) + \int_0^t \mathbf{\Phi}(t - \tau)\mathbf{B}\mathbf{u}(\tau)\, d\tau. \tag{3.80}$$

Clearly, if the initial conditions $\mathbf{x}(0)$, the input $\mathbf{u}(\tau)$, and the state transition matrix $\mathbf{\Phi}(t)$ are known, the time response of $\mathbf{x}(t)$ can be numerically evaluated. Thus the problem focuses on the evaluation of $\mathbf{\Phi}(t)$, the state transition matrix that represents the response of the system. Fortunately, the state transition matrix can be readily evaluated by using the signal-flow graph techniques with which we are already familiar.

Before proceeding to the evaluation of the state transition matrix using signal-flow graphs, we should note that several other methods exist for evaluating the transition matrix, such as the evaluation of the exponential series

$$\mathbf{\Phi}(t) = \exp(\mathbf{A}t) = \sum_{k=0}^{\infty} \frac{\mathbf{A}^k t^k}{k!} \tag{3.81}$$

in a truncated form [2, 8]. Several efficient methods exist for the evaluation of $\mathbf{\Phi}(t)$ by means of a computer algorithm [21].

In Equation (3.25), we found that $\mathbf{\Phi}(s) = [s\mathbf{I} - \mathbf{A}]^{-1}$. Therefore, if $\mathbf{\Phi}(s)$ is obtained by completing the matrix inversion, we can obtain $\mathbf{\Phi}(t)$ by noting that $\mathbf{\Phi}(t) = \mathcal{L}^{-1}\{\mathbf{\Phi}(s)\}$. The matrix inversion process is generally unwieldy for higher-order systems.

The usefulness of the signal-flow graph state model for obtaining the state transition matrix becomes clear upon consideration of the Laplace transformation version of Equation (3.80) when the input is zero. Taking the Laplace transformation of Equation (3.80) when $\mathbf{u}(\tau) = 0$, we have

$$\mathbf{X}(s) = \mathbf{\Phi}(s)\mathbf{x}(0). \tag{3.82}$$

Therefore, we can evaluate the Laplace transform of the transition matrix from the signal-flow graph by determining the relation between a state variable $X_i(s)$ and the state initial conditions $[x_1(0), x_2(0), \ldots, x_n(0)]$. Then the state transition matrix is simply the inverse transform of $\Phi(s)$; that is,

$$\Phi(t) = \mathcal{L}^{-1}\{\Phi(s)\}. \tag{3.83}$$

The relationship between a state variable $X_i(s)$ and the initial conditions $\mathbf{x}(0)$ is obtained by using Mason's signal-flow gain formula. Thus, for a second-order system, we would have

$$X_1(s) = \phi_{11}(s)x_1(0) + \phi_{12}(s)x_2(0),$$
$$X_2(s) = \phi_{21}(s)x_1(0) + \phi_{22}(s)x_2(0), \tag{3.84}$$

and the relation between $X_2(s)$ as an output and $x_1(0)$ as an input can be evaluated by Mason's signal-flow gain formula. All the elements of the state transition matrix, $\phi_{ij}(s)$, can be obtained by evaluating the individual relationships between $X_i(s)$ and $x_j(0)$ from the state model flow graph. An example will illustrate this approach to determining the transition matrix.

EXAMPLE 3.6   **Evaluation of the state transition matrix**

We will consider the *RLC* network of Figure 3.4. We seek to evaluate $\Phi(s)$ by (1) determining the matrix inversion $\Phi(s) = [s\mathbf{I} - \mathbf{A}]^{-1}$ and (2) using the signal-flow diagram and Mason's signal-flow gain formula.

First, we determine $\Phi(s)$ by evaluating $\Phi(s) = [s\mathbf{I} - \mathbf{A}]^{-1}$. We note from Equation (3.18) that

$$\mathbf{A} = \begin{bmatrix} 0 & -2 \\ 1 & -3 \end{bmatrix}.$$

Then

$$[s\mathbf{I} - \mathbf{A}] = \begin{bmatrix} s & 2 \\ -1 & s + 3 \end{bmatrix}. \tag{3.85}$$

The inverse matrix is

$$\Phi(s) = [s\mathbf{I} - \mathbf{A}]^{-1} = \frac{1}{\Delta(s)}\begin{bmatrix} s + 3 & -2 \\ 1 & s \end{bmatrix}, \tag{3.86}$$

where $\Delta(s) = s(s + 3) + 2 = s^2 + 3s + 2 = (s + 1)(s + 2)$.

The signal-flow graph state model of the *RLC* network of Figure 3.4 is shown in Figure 3.8. This *RLC* network, which was discussed in Sections 3.3 and 3.4, can be represented by the state variables $x_1 = v_c$ and $x_2 = i_L$. The initial conditions, $x_1(0)$ and $x_2(0)$, represent the initial capacitor voltage and inductor current, respectively. The flow graph, including the initial conditions of each state variable, is shown in Figure 3.23. The initial conditions appear as the initial value of the state variable at the output of each integrator.

To obtain $\Phi(s)$, we set $U(s) = 0$. When $R = 3$, $L = 1$, and $C = 1/2$, we obtain the signal-flow graph shown in Figure 3.24, where the output and input nodes are deleted because they are not involved in the evaluation of $\Phi(s)$. Then, using Mason's

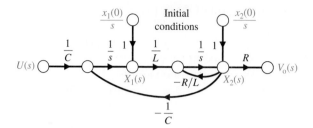

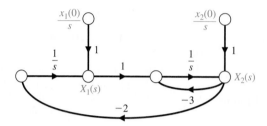

FIGURE 3.23
Flow graph of the
RLC network.

FIGURE 3.24
Flow graph of the
RLC network with
U(s) = 0.

signal-flow gain formula, we obtain $X_1(s)$ in terms of $x_1(0)$ as

$$X_1(s) = \frac{1 \cdot \Delta_1(s) \cdot [x_1(0)/s]}{\Delta(s)}, \tag{3.87}$$

where $\Delta(s)$ is the graph determinant, and $\Delta_1(s)$ is the path cofactor. The graph determinant is

$$\Delta(s) = 1 + 3s^{-1} + 2s^{-2}.$$

The path cofactor is $\Delta_1 = 1 + 3s^{-1}$ because the path between $x_1(0)$ and $X_1(s)$ does not touch the loop with the factor $-3s^{-1}$. Therefore, the first element of the transition matrix is

$$\phi_{11}(s) = \frac{(1 + 3s^{-1})(1/s)}{1 + 3s^{-1} + 2s^{-2}} = \frac{s + 3}{s^2 + 3s + 2}. \tag{3.88}$$

The element $\phi_{12}(s)$ is obtained by evaluating the relationship between $X_1(s)$ and $x_2(0)$ as

$$X_1(s) = \frac{(-2s^{-1})(x_2(0)/s)}{1 + 3s^{-1} + 2s^{-2}}.$$

Therefore, we obtain

$$\phi_{12}(s) = \frac{-2}{s^2 + 3s + 2}. \tag{3.89}$$

Similarly, for $\phi_{21}(s)$ we have

$$\phi_{21}(s) = \frac{(s^{-1})(1/s)}{1 + 3s^{-1} + 2s^{-2}} = \frac{1}{s^2 + 3s + 2}. \tag{3.90}$$

Finally, for $\phi_{22}(s)$, we obtain

$$\phi_{22}(s) = \frac{1(1/s)}{1 + 3s^{-1} + 2s^{-2}} = \frac{s}{s^2 + 3s + 2}. \tag{3.91}$$

Therefore, the state transition matrix in Laplace transformation form is

$$\Phi(s) = \begin{bmatrix} (s + 3)/(s^2 + 3s + 2) & -2/(s^2 + 3s + 2) \\ 1/(s^2 + 3s + 2) & s/(s^2 + 3s + 2) \end{bmatrix}. \tag{3.92}$$

The factors of the characteristic equation are $(s + 1)$ and $(s + 2)$, so that

$$(s + 1)(s + 2) = s^2 + 3s + 2.$$

Then the state transition matrix is

$$\Phi(t) = \mathcal{L}^{-1}\{\Phi(s)\} = \begin{bmatrix} (2e^{-t} - e^{-2t}) & (-2e^{-t} + 2e^{-2t}) \\ (e^{-t} - e^{-2t}) & (-e^{-t} + 2e^{-2t}) \end{bmatrix}. \tag{3.93}$$

The evaluation of the time response of the RLC network to various initial conditions and input signals can now be evaluated by using Equation (3.80). For example, when $x_1(0) = x_2(0) = 1$ and $u(t) = 0$, we have

$$\begin{bmatrix} x_1(t) \\ x_2(t) \end{bmatrix} = \Phi(t) \begin{bmatrix} 1 \\ 1 \end{bmatrix} = \begin{bmatrix} e^{-2t} \\ e^{-2t} \end{bmatrix}. \tag{3.94}$$

The response of the system for these initial conditions is shown in Figure 3.25. The trajectory of the state vector $[x_1(t), x_2(t)]$ on the (x_1, x_2)-plane is shown in Figure 3.26. The evaluation of the time response is facilitated by the determination of the state transition matrix. Although this approach is limited to linear systems, it is a powerful method and utilizes the familiar signal-flow graph to evaluate the transition matrix. ∎

FIGURE 3.25
Time response
of the state
variables of the
RLC network for
$x_1(0) = x_2(0) = 1$.

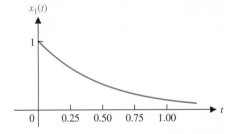

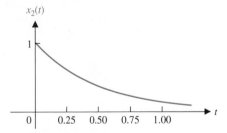

FIGURE 3.26
Trajectory of the
state vector in the
(x_1, x_2)-plane.

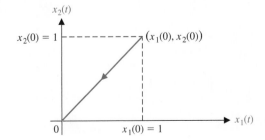

3.8 DESIGN EXAMPLES

In this section we present two illustrative design examples. In the first example, we present a detailed look at modeling a large space vehicle (such as a space station) using a state variable model. The state variable model is then used to take a look at the stability of the orientation of the spacecraft in a low earth orbit. The design process depicted in Figure 1.15 is highlighted in this example. The second example is a printer belt drive modeling exercise. The relationship between the state variable model and the block diagram discussed in Chapter 2 is illustrated and, using block diagram reduction methods, the transfer function equivalent of the state variable model is obtained.

EXAMPLE 3.7   **Modeling the orientation of a space station**

The International Space Station, shown in Figure 3.27, is a good example of a multi-purpose spacecraft that can operate in many different configurations. An important step in the control system design process is to develop a mathematical model of the spacecraft motion. In general, this model describes the translation and attitude motion of the spacecraft under the influence of external forces and torques, and controller and actuator forces and torques. The resulting spacecraft dynamic model is a set of highly coupled, nonlinear ordinary differential equations. Our objective is to simplify the model while retaining important system characteristics. This is not a trivial task, but an important, and often neglected component of control engineering. In this example, the rotational motion is considered. The translational motion, while critically important to orbit maintenance, can be decoupled from the rotational motion.

Many spacecraft (such as the International Space Station) will maintain an earth-pointing attitude. This means that cameras and other scientific instruments pointing down will be able to sense the earth, as depicted in Figure 3.27. Conversely,

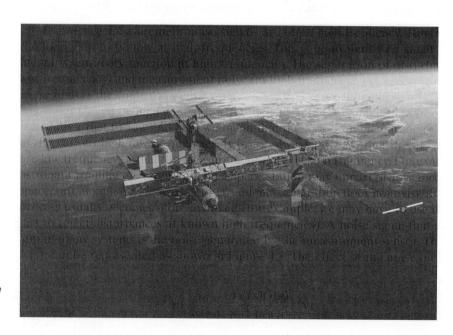

FIGURE 3.27
An artist's conception of the International Space Station. (Courtesy of NASA.)

scientific instruments pointing up will see deep space, as desired. To achieve earth-pointing attitude, the spacecraft needs an attitude hold control system capable of applying the necessary torques. The torques are the inputs to the system, in this case, the space station. The attitude is the output of the system. The International Space Station employs control moment gyros and reaction control jets as actuators to control the attitude. The control moment gyros are momentum exchangers and are preferable to reaction control jets because they do not expend fuel. They are actuators that consist of a constant-rate flywheel mounted on a set of gimbals. The flywheel orientation is varied by rotating the gimbals, resulting in a change in direction of the flywheel angular momentum. In accord with the basic principle of conservation of angular momentum, changes in control moment gyro momentum must be transferred to the space station, thereby producing a reaction torque. The reaction torque can be employed to control the space station attitude. However, there is a maximum limit of control that can be provided by the control moment gyro. When that maximum is attained, the device is said to have reached saturation. So, while control moment gyros do not expend fuel, they can provide only a limited amount of control. In practice, it is possible to control the attitude of the space station while simultaneously desaturating the control moment gyros.

Several methods for desaturating the control moment gyros are available, but using existing natural environmental torques is the preferred method because it minimizes the use of the reaction control jets. A clever idea is to use gravity gradient torques (which occur naturally and come free of charge) to continuously desaturate the momentum exchange devices. Due to the variation of the earth's gravitational field over the International Space Station, the total moment generated by the gravitational forces about the spacecraft's center of mass is nonzero. This nonzero moment is called the **gravity gradient torque**. A change in attitude changes the gravity gradient torque acting on the vehicle. Thus, combining attitude control and momentum management becomes a matter of compromise.

The elements of the design process emphasized in this example are illustrated in Figure 3.28. We can begin the modeling process by defining the attitude of the space station using the three angles, θ_2 (the pitch angle), θ_3 (the yaw angle), and θ_1 (the roll angle). These three angles represent the attitude of the space station relative to the desired earth-pointing attitude. When $\theta_1 = \theta_2 = \theta_3 = 0$, the space station is oriented in the desired direction. The goal is to keep the space station oriented in the desired attitude while minimizing the amount of momentum exchange required by the control momentum gyros (keeping in mind that we want to avoid saturation). The control goal can be stated as

Control Goal

Minimize the roll, yaw, and pitch angles in the presence of persistent external disturbances while simultaneously minimizing the control moment gyro momentum.

The time rate of change of the angular momentum of a body about its center of mass is equal to the sum of the external torques acting on that body. Thus the attitude dynamics of a spacecraft are driven by externally acting torques. The main external torque acting on the space station is due to gravity. Since we treat the earth as a point mass, the gravity gradient torque [30] acting on the spacecraft is given by

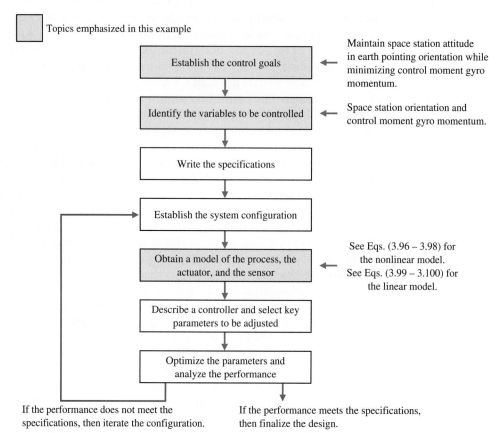

FIGURE 3.28
Elements of the control system design process emphasized in the spacecraft control example.

$$\mathbf{T}_g = 3n^2\mathbf{c} \times \mathbf{Ic}, \tag{3.95}$$

where n is the orbital angular velocity ($n = 0.0011$ rad/s for the space station), and $\mathbf{c}$ is

$$\mathbf{c} = \begin{bmatrix} -\sin\theta_2\cos\theta_3 \\ \sin\theta_1\cos\theta_2 + \cos\theta_1\sin\theta_2\sin\theta_3 \\ \cos\theta_1\cos\theta_2 - \sin\theta_1\sin\theta_2\sin\theta_3 \end{bmatrix}.$$

The notation '$\times$' denotes vector cross-product. Matrix $\mathbf{I}$ is the spacecraft inertia matrix and is a function of the space station configuration. It also follows from Equation (3.95) that the gravity gradient torques are a function of the attitude θ_1, θ_2, and θ_3. We want to maintain a prescribed attitude (that is earth-pointing $\theta_1 = \theta_2 = \theta_3 = 0$), but sometimes we must deviate from that attitude so that we can generate gravity gradient torques to assist in the control moment gyro momentum management. Therein lies the conflict; as engineers we often are required to develop control systems to manage conflicting goals.

Now we examine the effect of the aerodynamic torque acting on the space station. Even at the high altitude of the space station, the aerodynamic torque does affect the attitude motion. The aerodynamic torque acting on the space station is generated by the atmospheric drag force that acts through the center of pressure. In general, the center of pressure and the center of mass do not coincide, so aerodynamic torques develop. In low earth orbit, the aerodynamic torque is a sinusoidal function that tends to oscillate around a small bias. The oscillation in the torque is primarily a result of the earth's diurnal atmospheric bulge. Due to heating, the atmosphere closest to the sun extends further into space than the atmosphere on the side of the earth away from the sun. As the space station travels around the earth (once every 90 minutes or so), it moves through varying air densities, thus causing a cyclic aerodynamic torque. Also, the space station solar panels rotate as they track the sun. This results in another cyclic component of aerodynamic torque. The aerodynamic torque is generally much smaller than the gravity gradient torque. Therefore, for design purposes we can ignore the atmospheric drag torque and view it as a disturbance torque. We would like the controller to minimize the effects of the aerodynamic disturbance on the spacecraft attitude.

Torques caused by the gravitation of other planetary bodies, magnetic fields, solar radiation and wind, and other less significant phenomena are much smaller than the earth's gravity-induced torque and aerodynamic torque. We ignore these torques in the dynamic model and view them as disturbances.

Finally, we need to discuss the control moment gyros themselves. First, we will lump all the control moment gyros together and view them as a single source of torque. We represent the total control moment gyro momentum with the variable **h**. We need to know and understand the dynamics in the design phase to manage the angular momentum. But since the time constants associated with these dynamics are much shorter than for attitude dynamics, we can ignore the dynamics and assume that the control moment gyros can produce precisely and without a time delay the torque demanded by the control system.

Based on the above discussion, a simplified nonlinear model that we can use as the basis for the control design is

$$\dot{\Theta} = \mathbf{R}\Omega + \mathbf{n}, \tag{3.96}$$

$$\mathbf{I}\dot{\Omega} = -\Omega \times \mathbf{I}\Omega + 3n^2\mathbf{c} \times \mathbf{Ic} - \mathbf{u}, \tag{3.97}$$

$$\dot{\mathbf{h}} = -\Omega \times \mathbf{h} + \mathbf{u}, \tag{3.98}$$

where

$$\mathbf{R}(\Theta) = \frac{1}{\cos\theta_3}\begin{bmatrix} \cos\theta_3 & -\cos\theta_1\sin\theta_3 & \sin\theta_1\sin\theta_3 \\ 0 & \cos\theta_1 & -\sin\theta_1 \\ 0 & \sin\theta_1\cos\theta_3 & \cos\theta_1\cos\theta_3 \end{bmatrix}$$

$$\mathbf{n} = \begin{bmatrix} 0 \\ n \\ 0 \end{bmatrix}, \quad \Omega = \begin{bmatrix} \omega_1 \\ \omega_2 \\ \omega_3 \end{bmatrix}, \quad \Theta = \begin{bmatrix} \theta_1 \\ \theta_2 \\ \theta_3 \end{bmatrix}, \quad \mathbf{u} = \begin{bmatrix} u_1 \\ u_2 \\ u_3 \end{bmatrix},$$

where **u** is the control moment gyro input torque, Ω in the angular velocity, **I** is the moment of inertia matrix, and **n** is the orbital angular velocity. Two good references that describe the fundamentals of spacecraft dynamic modeling are [31] and [32]. There have been many papers dealing with space station control and momentum management. One of the first to present the nonlinear model in Equations (3.96–3.98) is Wie et al. [33]. Other related information about the model and the control problem in general appears in [34–38]. Articles related to advanced control topics on the space station can be found in [39–45]. Researchers are developing nonlinear control laws based on the nonlinear model in Equations (3.96)–(3.98). Several good articles on this topic appear in [46–55].

Equation (3.96) represents the kinematics—the relationship between the Euler angles, denoted by Θ, and the angular velocity vector, Ω. Equation (3.97) represents the space station attitude dynamics. The terms on the right side represent the sum of the external torques acting on the spacecraft. The first torque is due to inertia cross-coupling. The second term represents the gravity gradient torque, and the last term is the torque applied to the spacecraft from the actuators. The disturbance torques (due to such factors as the atmosphere) are not included in the model used in the design. Equation (3.98) represents the control moment gyro total momentum.

The conventional approach to spacecraft momentum management design is to develop a linear model, representing the spacecraft attitude and control moment gyro momentum by linearizing the nonlinear model. This linearization is accomplished by a standard Taylor series approximation. Linear control design methods can then be readily applied. For linearization purposes we assume that the spacecraft has zero products of inertia (that is, the inertia matrix is diagonal) and the aerodynamic disturbances are negligible. The equilibrium state that we linearize about is

$$\Theta = \mathbf{0},$$

$$\Omega = \begin{bmatrix} 0 \\ -n \\ 0 \end{bmatrix}$$

$$\mathbf{h} = \mathbf{0},$$

and where we assume that

$$\mathbf{I} = \begin{bmatrix} I_1 & 0 & 0 \\ 0 & I_2 & 0 \\ 0 & 0 & I_3 \end{bmatrix}.$$

In reality, the inertia matrix, $\mathbf{I}$, is not a diagonal matrix. Neglecting the off-diagonal terms is consistent with the linearization approximations and is a common assumption. Applying the Taylor series approximations yields the linear model, which as it turns out decouples the pitch axis from the roll/yaw axis.

The linearized equations for the pitch axis are

$$\begin{bmatrix} \dot{\theta}_2 \\ \dot{\omega}_2 \\ \dot{h}_2 \end{bmatrix} = \begin{bmatrix} 0 & 1 & 0 \\ 3n^2\Delta_2 & 0 & 0 \\ 0 & 0 & 0 \end{bmatrix} \begin{bmatrix} \theta_2 \\ \omega_2 \\ h_2 \end{bmatrix} + \begin{bmatrix} 0 \\ -\frac{1}{I_2} \\ 1 \end{bmatrix} u_2, \qquad (3.99)$$

where

$$\Delta_2 := \frac{I_3 - I_1}{I_2}.$$

The subscript 2 refers to the pitch axis terms, the subscript 1 is for the roll axis terms, and 3 is for the yaw axis terms. The linearized equations for the roll/yaw axes are

$$
\begin{bmatrix} \dot\theta_1 \\ \dot\theta_3 \\ \dot\omega_1 \\ \dot\omega_3 \\ \dot h_1 \\ \dot h_3 \end{bmatrix}
=
\begin{bmatrix}
0 & n & 1 & 0 & 0 & 0 \\
-n & 0 & 0 & 1 & 0 & 0 \\
-3n^2\Delta_1 & 0 & 0 & -n\Delta_1 & 0 & 0 \\
0 & 0 & -n\Delta_3 & 0 & 0 & 0 \\
0 & 0 & 0 & 0 & 0 & n \\
0 & 0 & 0 & 0 & -n & 0
\end{bmatrix}
\begin{bmatrix} \theta_1 \\ \theta_3 \\ \omega_1 \\ \omega_3 \\ h_1 \\ h_3 \end{bmatrix}
$$

$$
+
\begin{bmatrix}
0 & 0 \\
0 & 0 \\
-\frac{1}{I_1} & 0 \\
0 & -\frac{1}{I_3} \\
1 & 0 \\
0 & 1
\end{bmatrix}
\begin{bmatrix} u_1 \\ u_3 \end{bmatrix},
\tag{3.100}
$$

where

$$
\Delta_1 := \frac{I_2 - I_3}{I_1} \quad \text{and} \quad \Delta_3 := \frac{I_1 - I_2}{I_3}.
$$

Consider the analysis of the pitch axis. Define the state-vector as

$$
\mathbf{x}(t) := \begin{pmatrix} \theta_2(t) \\ \omega_2(t) \\ h_2(t) \end{pmatrix},
$$

and the output as

$$
y(t) = \theta_2(t) = [1 \quad 0 \quad 0]\mathbf{x}(t).
$$

Here we are considering the spacecraft attitude, $\theta_2(t)$, as the output of interest. We can just as easily consider both the angular velocity, ω_2, and the control moment gyro momentum, h_2, as outputs. The state variable model is

$$
\dot{\mathbf{x}} = \mathbf{A}\mathbf{x} + \mathbf{B}u, \tag{3.101}
$$
$$
y = \mathbf{C}\mathbf{x} + \mathbf{D}u,
$$

where

$$
\mathbf{A} = \begin{bmatrix} 0 & 1 & 0 \\ 3n^2\,\Delta_2 & 0 & 0 \\ 0 & 0 & 0 \end{bmatrix}, \quad
\mathbf{B} = \begin{bmatrix} 0 \\ -\frac{1}{I_2} \\ 1 \end{bmatrix},
$$

$$
\mathbf{C} = [1 \quad 0 \quad 0], \quad \mathbf{D} = [0],
$$

and where u is the control moment gyro torque in the pitch axis. The solution to the state differential equation, given in Equation (3.101), is

$$
\mathbf{x}(t) = \mathbf{\Phi}(t)\mathbf{x}(0) + \int_0^t \mathbf{\Phi}(t - \tau)\mathbf{B}u(\tau)\,d\tau,
$$

where

$$\Phi(t) = \exp(\mathbf{A}t) = \mathcal{L}^{-1}\{(s\mathbf{I} - \mathbf{A})^{-1}\}$$

$$= \begin{bmatrix} \dfrac{1}{2}(e^{\sqrt{3n^2\Delta_2}\,t} + e^{-\sqrt{3n^2\Delta_2}\,t}) & \dfrac{1}{2\sqrt{3n^2\Delta_2}}(e^{\sqrt{3n^2\Delta_2}\,t} - e^{-\sqrt{3n^2\Delta_2}\,t}) & 0 \\[3mm] \dfrac{\sqrt{3n^2\Delta_2}}{2}(e^{\sqrt{3n^2\Delta_2}\,t} - e^{-\sqrt{3n^2\Delta_2}\,t}) & \dfrac{1}{2}(e^{\sqrt{3n^2\Delta_2}\,t} + e^{-\sqrt{3n^2\Delta_2}\,t}) & 0 \\[3mm] 0 & 0 & 1 \end{bmatrix}.$$

We can see that if $\Delta_2 > 0$, then some elements of the state transition matrix will have terms of the form e^{at}, where $a > 0$. As we shall see (in Chapter 6) this indicates that our system is unstable. Also, if we are interested in the output, $y(t) = \theta_2(t)$, we have

$$y(t) = \mathbf{C}\mathbf{x}(t).$$

With $\mathbf{x}(t)$ given by

$$\mathbf{x}(t) = \Phi(t)\mathbf{x}(0) + \int_0^t \Phi(t - \tau)\mathbf{B}u(\tau)d\tau,$$

it follows that

$$y(t) = \mathbf{C}\Phi(t)\mathbf{x}(0) + \int_0^t \mathbf{C}\Phi(t - \tau)\mathbf{B}u(\tau)d\tau.$$

The transfer function relating the output $Y(s)$ to the input $U(s)$ is

$$G(s) = \frac{Y(s)}{U(s)} = \mathbf{C}(s\mathbf{I} - \mathbf{A})^{-1}\mathbf{B} = -\frac{1}{I_2(s^2 - 3n^2\Delta_2)}.$$

The characteristic equation is

$$s^2 - 3n^2\Delta_2 = \left(s + \sqrt{3n^2\Delta_2}\right)\left(s - \sqrt{3n^2\Delta_2}\right) = 0.$$

If $\Delta_2 > 0$ (that is, if $I_3 > I_1$), then we have two real poles—one in the left half-plane and the other in the right half-plane. For spacecraft with $I_3 > I_1$, we can say that an earth-pointing attitude is an unstable orientation. This means that active control is necessary.

Conversely, when $\Delta_2 < 0$ (that is, when $I_1 > I_3$), the characteristic equation has two imaginary roots at

$$s = \pm j\sqrt{3n^2|\Delta_2|}.$$

This type of spacecraft is marginally stable. In the absence of any control moment gyro torques, the spacecraft will oscillate around the earth-pointing orientation for any small initial deviation from the desired attitude. ∎

EXAMPLE 3.8   **Printer belt drive modeling**

A commonly used low-cost printer for a computer uses a belt drive to move the print-ing device laterally across the printed page [11]. The printing device may be a laser printer, a print ball, or thermal printhead. An example of a belt drive printer with a DC motor actuator is shown in Figure 3.29. In this model, a light sensor is used to measure the position of the printing device, and the belt tension adjusts the spring flexibility of the belt. The goal of the design is to determine the effect of the belt spring constant k and select appropriate parameters for the motor, the belt pulley, and the controller. To achieve the analysis, we will determine a model of the belt-drive system and select many of its parameters. Using this model, we will obtain the signal-flow graph model and select the state variables. We then will determine an appropriate transfer function for the system and select its other parameters, except for the spring constant. Finally, we will examine the effect of varying the spring constant within a realistic range.

We propose the model of the belt-drive system shown in Figure 3.30. This model assumes that the spring constant of the belt is k, the radius of the pulley is r, the angular rotation of the motor shaft is θ, and the angular rotation of the right-hand pulley is θ_p. The mass of the printing device is m, and its position is $y(t)$. A light sensor is used to measure y, and the output of the sensor is a voltage v_1, where $v_1 = k_1 y$. The controller

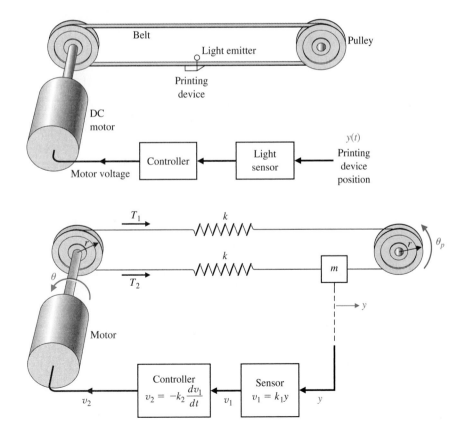

FIGURE 3.29
Printer belt-drive
system.

FIGURE 3.30
Printer belt-drive
model.

provides an output voltage v_2, where v_2 is a function of v_1. The voltage v_2 is connected to the field of the motor. Let us assume that we can use the linear relationship

$$v_2 = -\left[k_2\frac{dv_1}{dt} + k_3v_1\right],$$

and elect to use $k_2 = 0.1$ and $k_3 = 0$ (velocity feedback).

The inertia of the motor and pulley is $J = J_{motor} + J_{pulley}$. We plan to use a moderate-DC motor. Selecting a typical 1/8-hp DC motor, we find that $J = 0.01$ kg m^2, the field inductance is negligible, the field resistance is $R = 2\,\Omega$, the motor constant is $K_m = 2$ Nm/A, and the motor and pulley friction is $b = 0.25$ Nms/rad. The radius of the pulley is $r = 0.15$ m. The system parameters are summarized in Table 3.1.

We now proceed to write the equations of the motion for the system; note that $y = r\theta_p$. Then the tension from equilibrium T_1 is

$$T_1 = k(r\theta - r\theta_p) = k(r\theta - y).$$

The tension from equilibrium T_2 is

$$T_2 = k(y - r\theta).$$

The net tension at the mass m is

$$T_1 - T_2 = m\frac{d^2y}{dt^2} \tag{3.102}$$

and

$$T_1 - T_2 = k(r\theta - y) - k(y - r\theta) = 2k(r\theta - y) = 2kx_1, \tag{3.103}$$

where the first state variable is $x_1 = r\theta - y$. Let the second state variable be $x_2 = dy/dt$, and use Equations (3.102) and (3.103) to obtain

$$\frac{dx_2}{dt} = \frac{2k}{m}x_1. \tag{3.104}$$

The first derivative of x_1 is

$$\frac{dx_1}{dt} = r\frac{d\theta}{dt} - \frac{dy}{dt} = rx_3 - x_2 \tag{3.105}$$

Table 3.1   Parameters of Printing Device

Mass	$m = 0.2$ kg
Light sensor	$k_1 = 1$ V/m
Radius	$r = 0.15$ m
Motor	
Inductance	$L \approx 0$
Friction	$b = 0.25$ Nms/rad
Resistance	$R = 2\,\Omega$
Constant	$K_m = 2$ Nm/A
Inertia	$J = J_{motor} + J_{pulley}$: $J = 0.01$ kg m^2

when we select the third state variable as $x_3 = d\theta/dt$. We now require a differential equation describing the motor rotation. When $L = 0$, we have the field current $i = v_2/R$ and the motor torque $T_m = K_m i$. Therefore,

$$T_m = \frac{K_m}{R} v_2,$$

and the motor torque provides the torque to drive the belts plus the disturbance or undesired load torque, so that

$$T_m = T + T_d.$$

The torque T drives the shaft to the pulley, so that

$$T = J\frac{d^2\theta}{dt^2} + b\frac{d\theta}{dt} + r(T_1 - T_2).$$

Therefore,

$$\frac{dx_3}{dt} = \frac{d^2\theta}{dt^2}.$$

Hence,

$$\frac{dx_3}{dt} = \frac{T_m - T_d}{J} - \frac{b}{J}x_3 - \frac{2kr}{J}x_1,$$

where

$$T_m = \frac{K_m}{R} v_2, \quad \text{and} \quad v_2 = -k_1 k_2 \frac{dy}{dt} = -k_1 k_2 x_2.$$

Thus, we obtain

$$\frac{dx_3}{dt} = \frac{-K_m k_1 k_2}{JR}x_2 - \frac{b}{J}x_3 - \frac{2kr}{J}x_1 - \frac{T_d}{J}. \qquad (3.106)$$

Equations (3.104)–(3.106) are the three first-order differential equations required to describe this system. The matrix differential equation is

$$\dot{\mathbf{x}} = \begin{bmatrix} 0 & -1 & r \\ \dfrac{2k}{m} & 0 & 0 \\ \dfrac{-2kr}{J} & \dfrac{-K_m k_1 k_2}{JR} & \dfrac{-b}{J} \end{bmatrix} \mathbf{x} + \begin{bmatrix} 0 \\ 0 \\ \dfrac{-1}{J} \end{bmatrix} T_d. \qquad (3.107)$$

The signal-flow graph and block diagram models representing the matrix differential equation are shown in Figure 3.31, where we include the identification of the node for the torque disturbance torque T_d.

We can use the flow graph to determine the transfer function $X_1(s)/T_d(s)$. The goal is to reduce the effect of the disturbance T_d, and the transfer function will show us how to accomplish this goal. Using Mason's signal-flow gain formula, we obtain

$$\frac{X_1(s)}{T_d(s)} = \frac{-\dfrac{r}{J}s^{-2}}{1 - (L_1 + L_2 + L_3 + L_4) + L_1L_2},$$

where

$$L_1 = \frac{-b}{J}s^{-1}, \quad L_2 = \frac{-2k}{m}s^{-2}, \quad L_3 = \frac{-2kr^2s^{-2}}{J}, \quad \text{and} \quad L_4 = \frac{-2kK_mk_1k_2rs^{-3}}{mJR}.$$

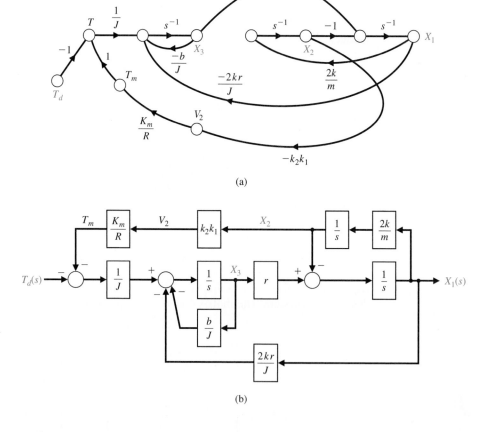

(a)

(b)

FIGURE 3.31
Printer belt drive.
(a) Signal-flow
graph. (b) Block
diagram model.

We therefore have

$$\frac{X_1(s)}{T_d(s)} = \frac{-\left(\dfrac{r}{J}\right)s}{s^3 + \left(\dfrac{b}{J}\right)s^2 + \left(\dfrac{2k}{m} + \dfrac{2kr^2}{J}\right)s + \left(\dfrac{2kb}{Jm} + \dfrac{2kK_mk_1k_2r}{JmR}\right)}.$$

We can also determine the closed-loop transfer function using block diagram reduction methods, as illustrated in Figure 3.32. Remember, there is no unique path to follow in reducing the block diagram; however, there is only one correct solution in the end. The original block diagram is shown in Figure 3.31(b). The result of the first step is shown in 3.32(a), where the upper feedback loop has been reduced to a single transfer function.

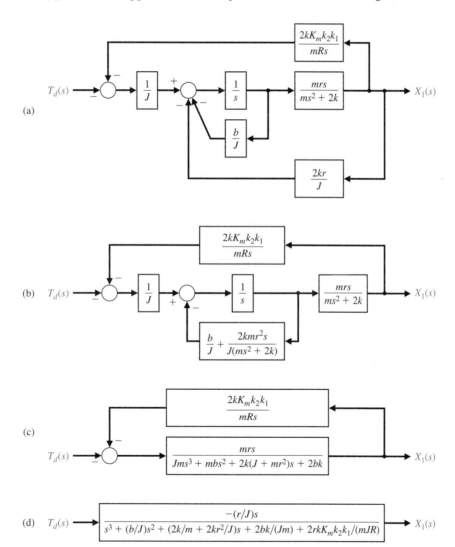

FIGURE 3.32
Printer belt drive block diagram reduction.

The second step illustrated in Figure 3.32(b) then reduces the two lower feedback loops to a single transfer function. In the third step shown in Figure 3.32(c), the lower feedback loop is closed and then the remaining transfer functions in series in the lower loop are combined. The final step closed-loop transfer function is shown in Figure 3.32(d).

Substituting the parameter values summarized in Table 3.1, we obtain

$$\frac{X_1(s)}{T_d(s)} = \frac{-15s}{s^3 + 25s^2 + 14.5ks + 1000k(0.25 + 0.15k_2)}. \tag{3.108}$$

We wish to select the spring constant k and the gain k_2 so that the state variable x_1 will quickly decline to a low value when a disturbance occurs. For test purposes, consider a step disturbance $T_d(s) = a/s$. Recalling that $x_1 = r\theta - y$, we thus seek a small magnitude for x_1 so that y is nearly equal to the desired $r\theta$. If we have a perfectly stiff belt with $k \to \infty$, then $y = r\theta$ exactly. With a step disturbance, $T_d(s) = a/s$, we have

$$X_1(s) = \frac{-15a}{s^3 + 25s^2 + 14.5ks + 1000k(0.25 + 0.15k_2)}. \tag{3.109}$$

The final value theorem gives

$$\lim_{t \to \infty} x_1(t) = \lim_{s \to 0} sX_1(s) = 0, \tag{3.110}$$

and thus the steady-state value of $x_1(t)$ is zero. We need to use a realistic value for k in the range $1 \le k \le 40$. For an average value of $k = 20$ and $k_2 = 0.1$, we have

$$X_1(s) = \frac{-15a}{s^3 + 25s^2 + 290s + 5300}$$

$$= \frac{-15a}{(s + 22.56)(s^2 + 2.44s + 234.93)}. \tag{3.111}$$

The characteristic equation has one real root and two complex roots. The partial fraction expansion yields

$$\frac{X_1(s)}{a} = \frac{A}{s + 22.56} + \frac{Bs + C}{(s + 1.22)^2 + (15.28)^2}, \tag{3.112}$$

where we find $A = -0.0218$, $B = 0.0218$, and $C = -0.4381$. Clearly with these small residues, the response to the unit disturbance is relatively small. Because A and B are small compared to C, we may approximate $X_1(s)$ as

$$\frac{X_1(s)}{a} \cong \frac{-0.4381}{(s + 1.22)^2 + (15.28)^2}.$$

Using Table 2.3, we obtain

$$\frac{x_1(t)}{a} \cong -0.0287e^{-1.22t} \sin 15.28t. \tag{3.113}$$

The actual response of x_1 is shown in Figure 3.33. This system will reduce the effect of the unwanted disturbance to a relatively small magnitude. Thus we have achieved our design objective. ■

3.9 ANALYSIS OF STATE VARIABLE MODELS USING CONTROL DESIGN SOFTWARE

The time-domain method utilizes a state-space representation of the system model, given by

$$\dot{\mathbf{x}} = \mathbf{A}\mathbf{x} + \mathbf{B}u \quad \text{and} \quad y = \mathbf{C}\mathbf{x} + \mathbf{D}u. \tag{3.114}$$

The vector $\mathbf{x}$ is the state of the system, $\mathbf{A}$ is the constant $n \times n$ system matrix, $\mathbf{B}$ is the constant $n \times m$ input matrix, $\mathbf{C}$ is the constant $p \times n$ output matrix, and $\mathbf{D}$ is a constant $p \times m$ matrix. The number of inputs, m, and the number of outputs, p, are taken to be one, since we are considering only single-input, single-output (SISO) problems. Therefore y and u are not bold (matrix) variables.

The main elements of the state-space representation in Equation (3.114) are the state vector $\mathbf{x}$ and the constant matrices $(\mathbf{A}, \mathbf{B}, \mathbf{C}, \mathbf{D})$. Two new functions covered in this section are ss and lsim. We also consider the use of the expm function to calculate the state transition matrix.

Given a transfer function, we can obtain an equivalent state-space representation and vice versa. The function tf can be used to convert a state-space representation to a transfer function representation; the function ss can be used to convert a transfer function representation to a state-space representation. These functions are shown in Figure 3.34, where sys_tf represents a transfer function model and sys_ss is a state-space representation.

For instance, consider the third-order system

$$T(s) = \frac{Y(s)}{R(s)} = \frac{2s^2 + 8s + 6}{s^3 + 8s^2 + 16s + 6}. \tag{3.115}$$

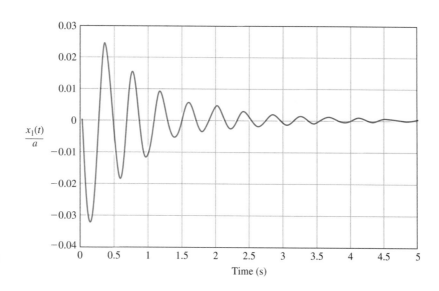

FIGURE 3.33
Response of $x_1(t)$ to a step disturbance: peak value $= -0.0325$.

We can obtain a state-space representation using the **ss** function, as shown in Figure 3.35. A state-space representation of Equation (3.115) is given by Equation (3.114), where

$$A = \begin{bmatrix} -8 & -4 & -1.5 \\ 4 & 0 & 0 \\ 0 & 1 & 0 \end{bmatrix}, \quad B = \begin{bmatrix} 2 \\ 0 \\ 0 \end{bmatrix},$$

$$C = [1 \quad 1 \quad 0.75], \quad \text{and} \quad D = [0].$$

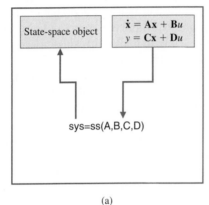

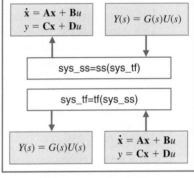

FIGURE 3.34
(a) The **ss** function.
(b) Linear system model conversion.

(a)   (b)

convert.m

```
% Convert G(s) = (2s^2+8s+6)/(s^3+8s^2+16s+6)
% to a state-space representation
%
num=[2 8 6]; den=[1 8 16 6]; sys_tf=tf(num,den);
sys_ss=ss(sys_tf);
```

```
>>convert
a =
            x1      x2      x3
    x1      -8      -4      -1.5
    x2       4       0       0
    x3       0       1       0

b =
            u1
    x1       2
    x2       0
    x3       0

c =
            x1      x2      x3
    y1       1       1       0.75

d =
            u1
    y1       0
```

(a)   (b)

FIGURE 3.35   Conversion of Equation (3.115) to a state-space representation. (a) m-file script. (b) Output printout.

The state-space representation of the transfer function in Equation (3.115) is depicted in Figure 3.36.

The state variable representation is not unique. For example, another equally valid state variable representation is given by

$$
\mathbf{A} = \begin{bmatrix} -8 & -2 & -0.75 \\ 8 & 0 & 0 \\ 0 & 1 & 0 \end{bmatrix}, \mathbf{B} = \begin{bmatrix} 0.125 \\ 0 \\ 0 \end{bmatrix}, \mathbf{C} = [16 \quad 8 \quad 6], \mathbf{D} = [0].
$$

It is possible that when using the ss function, the state variable representation provided by your control design software will be different from the above two examples depending on the specific software and version.

The time response of the system in Equation (3.114) is given by the solution to the vector integral equation

$$
\mathbf{x}(t) = \exp(\mathbf{A}t)\mathbf{x}(0) + \int_0^t \exp[\mathbf{A}(t - \tau)]\mathbf{B}u(\tau)\, d\tau. \tag{3.116}
$$

The matrix exponential function in Equation (3.116) is the state transition matrix, $\Phi(t)$, where (Equation 3.23)

$$
\Phi(t) = \exp(\mathbf{A}t).
$$

We can use the function expm to compute the transition matrix for a given time, as illustrated in Figure 3.37. The expm(A) function computes the matrix exponential. In contrast, the exp(A) function calculates $e^{a_{ij}}$ for each of the elements $a_{ij} \in \mathbf{A}$.

For example, let us consider the RLC network of Figure 3.4 described by the state-space representation of Equation (3.18) with

$$
\mathbf{A} = \begin{bmatrix} 0 & -2 \\ 1 & -3 \end{bmatrix}, \quad \mathbf{B} = \begin{bmatrix} 2 \\ 0 \end{bmatrix}, \quad \mathbf{C} = [1 \quad 0], \quad \text{and} \quad \mathbf{D} = 0.
$$

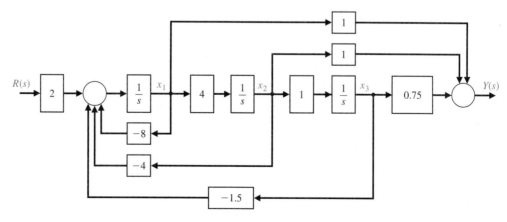

FIGURE 3.36   Block diagram with x_1 defined as the leftmost state variable.

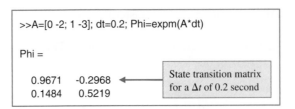

FIGURE 3.37
Computing the state transition matrix for a given time, $\Delta t = dt$.

The initial conditions are $x_1(0) = x_2(0) = 1$ and the input $u(t) = 0$. At $t = 0.2$, the state transition matrix is as given in Figure 3.37. The state at $t = 0.2$ is predicted by the state transition methods to be

$$\begin{bmatrix} x_1 \\ x_2 \end{bmatrix}_{t=0.2} = \begin{bmatrix} 0.9671 & -0.2968 \\ 0.1484 & 0.5219 \end{bmatrix} \begin{bmatrix} x_1 \\ x_2 \end{bmatrix}_{t=0} = \begin{bmatrix} 0.6703 \\ 0.6703 \end{bmatrix}.$$

The time response of the system of Equation (3.115) can also be obtained by using the lsim function. The lsim function can accept as input nonzero initial conditions as well as an input function, as shown in Figure 3.38. Using the lsim function, we can calculate the response for the RLC network as shown in Figure 3.39.

The state at $t = 0.2$ is predicted with the lsim function to be $x_1(0.2) = x_2(0.2) = 0.6703$. If we can compare the results obtained by the lsim function and by multiplying the initial condition state vector by the state transition matrix, we find identical results.

3.10 SEQUENTIAL DESIGN EXAMPLE: DISK DRIVE READ SYSTEM

Advanced disks have as many as 5000 tracks per cm. These tracks are typically 1 μm wide. Thus, there are stringent requirements on the accuracy of the reader head position and of the movement from one track to another. In this chapter, we

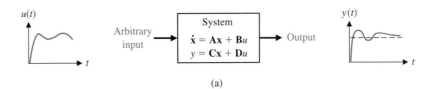

(a)

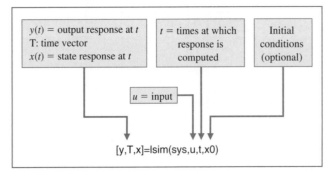

FIGURE 3.38
The **lsim** function for calculating the output and state response.

(b)

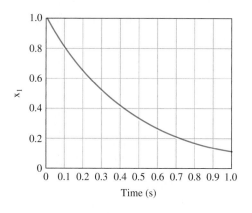
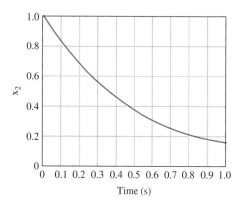

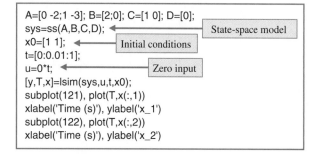

FIGURE 3.39
Computing the time response for nonzero initial conditions and zero input using **lsim**.

will develop a state variable model of the disk drive system that will include the effect of the flexure mount.

Consider again the head mount shown in Figure 2.71. Since we want a lightweight arm and flexure for rapid movement, we must consider the effect of the flexure, which is a very thin mount made of spring steel. Again, we wish to accurately control the position of the head $y(t)$ as shown in Figure 3.40(a). We will attempt to derive a model for the system shown in Figure 3.40(a). Here we identify the motor mass as M_1 and the head mount mass as M_2. The flexure spring is represented by the spring constant k. The force $u(t)$ to drive the mass M_1 is generated by the DC motor. If the spring is absolutely rigid (nonspringy), then we obtain the simplified model shown in Figure 3.40(b). Typical parameters for the two-mass system are given in Table 3.2.

Let us obtain the transfer function model of the simplified system of Figure 3.40(b). Note that $M = M_1 + M_2 = 20.5$ g $= 0.0205$ kg. Then we have

$$M\frac{d^2y}{dt^2} + b_1\frac{dy}{dt} = u(t). \qquad (3.117)$$

Therefore, the transfer function model is

$$\frac{Y(s)}{U(s)} = \frac{1}{s(Ms + b_1)}.$$

Table 3.2   Typical Parameters of the Two-Mass Model

Parameter	Symbol	Value
Motor mass	M_1	20 g = 0.02 kg
Flexure spring	k	$10 \leq k \leq \infty$
Head mounting mass	M_2	0.5 g = 0.0005 kg
Head position	$x_2(t)$	variable in mm
Friction at mass 1	b_1	410×10^{-3} N/(m/s)
Field resistance	R	$1\ \Omega$
Field inductance	L	1 mH
Motor constant	K_m	0.1025 N m/A
Friction at mass 2	b_2	4.1×10^{-3} N/(m/s)

FIGURE 3.40
(a) Model of the
two-mass system
with a spring
flexure.
(b) Simplified model
with a rigid spring.

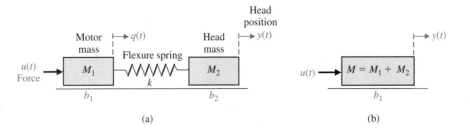

(a)   (b)

For the parameters of Table 3.2, we obtain

$$\frac{Y(s)}{U(s)} = \frac{1}{s(0.0205s + 0.410)} = \frac{48.78}{s(s + 20)}.$$

The transfer function model of the head reader, including the effect of the motor coil, is shown in Figure 3.41. When $R = 1\ \Omega$, $L = 1$ mH, and $K_m = 0.1025$, we obtain

$$G(s) = \frac{Y(s)}{V(s)} = \frac{5000}{s(s + 20)(s + 1000)}, \tag{3.118}$$

which is exactly the same model we obtained in Chapter 2.

Now let us obtain the state variable model of the two-mass system shown in Figure 3.40(a). Write the differential equations as

$$\text{Mass } M_1: M_1 \frac{d^2q}{dt^2} + b_1 \frac{dq}{dt} + k(q - y) = u(t)$$

FIGURE 3.41
Transfer function
model of head
reader device with
a rigid spring.

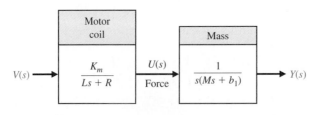

$$\text{Mass } M_2\text{: } M_2\frac{d^2y}{dt^2} + b_2\frac{dy}{dt} + k(y - q) = 0.$$

To develop the state variable model, we choose the state variables as $x_1 = q$ and $x_2 = y$. Then we have

$$x_3 = \frac{dq}{dt} \quad \text{and} \quad x_4 = \frac{dy}{dt}.$$

Then, in matrix form,

$$\dot{\mathbf{x}} = \mathbf{Ax} + \mathbf{B}u,$$

and we have

$$\mathbf{x} = \begin{bmatrix} q \\ y \\ \dot{q} \\ \dot{y} \end{bmatrix},$$

$$\mathbf{B} = \begin{bmatrix} 0 \\ 0 \\ 1/M_1 \\ 0 \end{bmatrix},$$

and

$$\mathbf{A} = \begin{bmatrix} 0 & 0 & 1 & 0 \\ 0 & 0 & 0 & 1 \\ -k/M_1 & k/M_1 & -b_1/M_1 & 0 \\ k/M_2 & -k/M_2 & 0 & -b_2/M_2 \end{bmatrix}. \tag{3.119}$$

Note that the output is $\dot{y}(t) = x_4$. Also, for $L = 0$ or negligible inductance, then $u(t) = K_m v(t)$. For the typical parameters and for $k = 10$, we have

$$\mathbf{B} = \begin{bmatrix} 0 \\ 0 \\ 50 \\ 0 \end{bmatrix}$$

and

$$\mathbf{A} = \begin{bmatrix} 0 & 0 & 1 & 0 \\ 0 & 0 & 0 & 1 \\ -500 & +500 & -20.5 & 0 \\ +20000 & -20000 & 0 & -8.2 \end{bmatrix}.$$

The response of $\dot{y}$ for $u(t) = 1, t > 0$ is shown in Figure 3.42. This response is quite oscillatory, and it is clear that we want a very rigid flexure with $k > 100$.

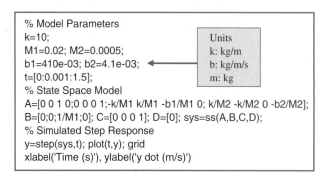

```
% Model Parameters
k=10;
M1=0.02; M2=0.0005;                    Units
b1=410e-03; b2=4.1e-03;  ◄────         k: kg/m
t=[0:0.001:1.5];                       b: kg/m/s
% State Space Model                    m: kg
A=[0 0 1 0;0 0 0 1;-k/M1 k/M1 -b1/M1 0; k/M2 -k/M2 0 -b2/M2];
B=[0;0;1/M1;0]; C=[0 0 0 1]; D=[0]; sys=ss(A,B,C,D);
% Simulated Step Response
y=step(sys,t); plot(t,y); grid
xlabel('Time (s)'), ylabel('y dot (m/s)')
```

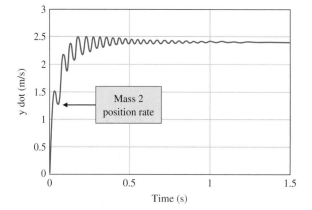

FIGURE 3.42
Response of y for a step input for the two-mass model with $k = 10$.

3.11 SUMMARY

In this chapter, we have considered the description and analysis of systems in the time domain. The concept of the state of a system and the definition of the state variables of a system were discussed. The selection of a set of state variables of a system was examined, and the nonuniqueness of the state variables was noted. The state differential equation and the solution for $\mathbf{x}(t)$ were discussed. Alternative signal-flow graph and block diagram model structures were considered for representing the transfer function (or differential equation) of a system. Using Mason's signal-flow gain formula, we noted the ease of obtaining the flow graph model. The state differential equation representing the flow graph and block diagram models was also examined. The time response of a linear system and its associated transition matrix was discussed, and the utility of Mason's signal-flow gain formula for obtaining the transition matrix was illustrated. A detailed analysis of a space station model development was presented for a realistic scenario where the attitude control is accomplished in conjunction with minimizing the actuator control. The relationship between modeling with state variable forms and control system design was established. The use of control design software to convert a transfer function to state variable form and calculate the state transition matrix was discussed and illustrated. The chapter concluded with the development of a state variable model for the Sequential Design Example: Disk Drive Read System.

EXERCISES

E3.1 For the circuit shown in Figure E3.1 identify a set of state variables.

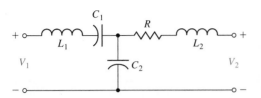

FIGURE E3.1 *RLC* circuit.

E3.2 A robot-arm drive system for one joint can be represented by the differential equation [8]

$$\frac{dv(t)}{dt} = -k_1 v(t) - k_2 y(t) + k_3 i(t),$$

where $v(t)$ = velocity, $y(t)$ = position, and $i(t)$ is the control-motor current. Put the equations in state variable form and set up the matrix form for $k_1 = k_2 = 1$.

E3.3 A system can be represented by the state vector differential equation of Equation (3.16), where

$$\mathbf{A} = \begin{bmatrix} 0 & 1 \\ -1 & -1 \end{bmatrix}.$$

Find the characteristic roots of the system.

Answer: $s = -1/2 \pm j\sqrt{3}/2$

E3.4 Obtain a state variable matrix for a system with a differential equation

$$\frac{d^3 y}{dt^3} + 4\frac{d^2 y}{dt^2} + 6\frac{dy}{dt} + 8y = 20u(t).$$

E3.5 A system is represented by a block diagram as shown in Figure E3.5. Write the state equations in the form of Equations (3.16) and (3.17).

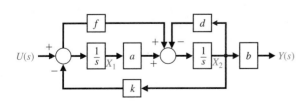

FIGURE E3.5 Block diagram.

E3.6 A system is represented by Equation (3.16), where

$$\mathbf{A} = \begin{bmatrix} 0 & 1 \\ 0 & 0 \end{bmatrix}.$$

(a) Find the matrix $\Phi(t)$. (b) For the initial conditions $x_1(0) = x_2(0) = 1$, find $\mathbf{x}(t)$.

Answer: (b) $x_1 = 1 + t, x_2 = 1, t \geq 0$

E3.7 Consider the spring and mass shown in Figure 3.3 where $M = 1$ kg, $k = 100$ N/m, and $b = 20$ N s/m. (a) Find the state vector differential equation. (b) Find the roots of the characteristic equation for this system.

Answer: (a) $\dot{\mathbf{x}} = \begin{bmatrix} 0 & 1 \\ -100 & -20 \end{bmatrix} \mathbf{x} + \begin{bmatrix} 0 \\ 1 \end{bmatrix} u$

(b) $s = -10, -10$

E3.8 The manual, low-altitude hovering task above a moving landing deck of a small ship is very demanding, particularly in adverse weather and sea conditions. The hovering condition is represented by the matrix

$$\mathbf{A} = \begin{bmatrix} 0 & 1 & 0 \\ 0 & 0 & 1 \\ 0 & -5 & -2 \end{bmatrix}.$$

Find the roots of the characteristic equation.

E3.9 A multi-loop block diagram is shown in Figure E3.9. The state variables are denoted by x_1 and x_2. (a) Determine a state variable representation of the closed-loop system where the output is denoted by $y(t)$ and the input is $r(t)$. (b) Determine the characteristic equation.

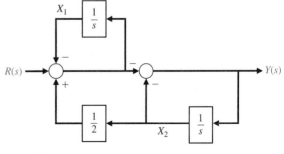

FIGURE E3.9 Multi-loop feedback control system.

E3.10 A hovering vehicle control system is represented by two state variables, and [15]

$$A = \begin{bmatrix} 0 & 6 \\ -1 & -5 \end{bmatrix}.$$

(a) Find the roots of the characteristic equation.
(b) Find the state transition matrix $\Phi(t)$.

Answer: (a) $s = -3, -2$

(b) $\Phi(t) = \begin{bmatrix} 3e^{-2t} - 2e^{-3t} & -6e^{-3t} + 6e^{-2t} \\ e^{-3t} - e^{-2t} & 3e^{-3t} - 2e^{-2t} \end{bmatrix}$

E3.11 Determine a state variable representation for the system described by the transfer function

$$T(s) = \frac{Y(s)}{R(s)} = \frac{4(s + 3)}{(s + 2)(s + 6)}.$$

E3.12 Use a state variable model to describe the circuit of Figure E3.12. Obtain the response to an input unit step when the initial current is zero and the initial capacitor voltage is zero.

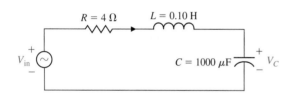

FIGURE E3.12   *RLC* series circuit.

E3.13 A system is described by the two differential equations

$$\frac{dy}{dt} + y - 2u + aw = 0,$$

and

$$\frac{dw}{dt} - by + 4u = 0,$$

where w and y are functions of time, and u is an input $u(t)$. (a) Select a set of state variables. (b) Write the matrix differential equation and specify the elements of the matrices. (c) Find the characteristic roots of the system in terms of the parameters a and b.

Answer: (c) $s = -1/2 \pm \sqrt{1 - 4ab}/2$

E3.14 Develop the state-space representation of a radioactive material of mass M to which additional radioactive material is added at the rate $r(t) = Ku(t)$, where K is a constant. Identify the state variables.

E3.15 Consider the case of the two masses connected as shown in Figure E3.15. The sliding friction of each mass has the constant b. Determine a state variable matrix differential equation.

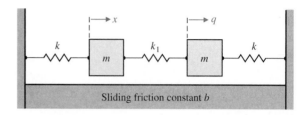

FIGURE E3.15   Two-mass system.

E3.16 Two cars with negligible rolling friction are connected as shown in Figure E3.16. An input force is $u(t)$. The output is the position of cart 2, that is, $y(t) = q(t)$. Determine a state space representation of the system.

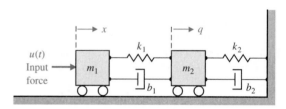

FIGURE E3.16   Two cars with negligible rolling friction.

E3.17 Determine a state variable differential matrix equation for the circuit shown in Figure E3.17:

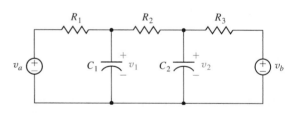

FIGURE E3.17   *RC* circuit.

E3.18 Consider a system represented by the following differential equations:

$$Ri_1 + L_1 \frac{di_1}{dt} + v = v_a$$

$$L_2 \frac{di_2}{dt} + v = v_b$$

$$i_1 + i_2 = C \frac{dv}{dt}$$

where R, L_1, L_2 and C are given constants, and v_a and v_b are inputs. Let the state variables be defined as $x_1 = i_1$, $x_2 = i_2$, and $x_3 = v$. Obtain a state variable representation of the system where the output is x_3.

E3.19 A single-input, single-output system has the matrix equations

$$\dot{\mathbf{x}} = \begin{bmatrix} 0 & 1 \\ -3 & -4 \end{bmatrix}\mathbf{x} + \begin{bmatrix} 0 \\ 1 \end{bmatrix}u$$

and

$$y = [10 \quad 0]\mathbf{x}.$$

Determine the transfer function $G(s) = Y(s)/U(s)$.

Answer : $G(s) = \dfrac{10}{s^2 + 4s + 3}$

E3.20 For the simple pendulum shown in Figure E3.20, the nonlinear equations of motion are given by

$$\ddot{\theta} + \frac{g}{L}\sin\theta + \frac{k}{m}\dot{\theta} = 0,$$

where g is gravity, L is the length of the pendulum, m is the mass attached at the end of the pendulum (we assume the rod is massless), and k is the coefficient of friction at the pivot point.

(a) Linearize the equations of motion about the equilibrium condition $\theta = 0°$.

(b) Obtain a state variable representation of the system. The system output is the angle θ.

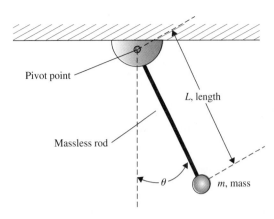

FIGURE E3.20 Simple pendulum.

E3.21 A single-input, single-output system is described by

$$\dot{\mathbf{x}}(t) = \begin{bmatrix} 0 & 1 \\ -1 & -2 \end{bmatrix}\mathbf{x}(t) + \begin{bmatrix} 1 \\ 0 \end{bmatrix}u(t)$$

$$y(t) = [0 \quad 1]\mathbf{x}(t)$$

Obtain the transfer function $G(s) = Y(s)/U(s)$ and determine the response of the system to a unit step input.

PROBLEMS

P3.1 An RLC circuit is shown in Figure P3.1. (a) Identify a suitable set of state variables. (b) Obtain the set of first-order differential equations in terms of the state variables. (c) Write the state differential equation.

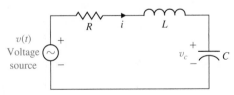

FIGURE P3.1 RLC circuit.

P3.2 A **balanced** bridge network is shown in Figure P3.2.

(a) Show that the **A** and **B** matrices for this circuit are

$$\mathbf{A} = \begin{bmatrix} -2/((R_1 + R_2)C) & 0 \\ 0 & -2R_1R_2/((R_1 + R_2)L) \end{bmatrix},$$

$$\mathbf{B} = 1/(R_1 + R_2)\begin{bmatrix} 1/C & 1/C \\ R_2/L & -R_2/L \end{bmatrix}.$$

(b) Sketch the block diagram. The state variables are $(x_1, x_2) = (v_c, i_L)$.

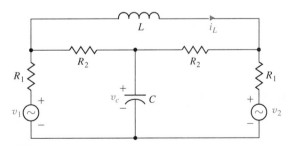

FIGURE P3.2 Balanced bridge network.

P3.3 An RLC network is shown in Figure P3.3. Define the state variables as $x_1 = i_L$ and $x_2 = v_c$. Obtain the state differential equation.

Partial answer :

$$\mathbf{A} = \begin{bmatrix} 0 & 1/L \\ -1/C & -1/(RC) \end{bmatrix}.$$

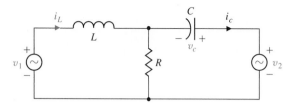

FIGURE P3.3   *RLC* circuit.

P3.4   The transfer function of a system is

$$T(s) = \frac{Y(s)}{R(s)} = \frac{s^2 + 2s + 5}{s^3 + 2s^2 + 3s + 10}.$$

Sketch the block diagram and determine the state variable matrix differential equation for the following formats: (a) phase variables; (b) input feedforward.

P3.5   A closed-loop control system is shown in Figure P3.5. (a) Determine the closed-loop transfer function $T(s) = Y(s)/R(s)$. (b) Sketch a block diagram model for the system, and determine the matrix differential equation.

P3.6   Determine the state variable matrix equation for the circuit shown in Figure P3.6. Let $x_1 = v_1$, $x_2 = v_2$, and $x_3 = i$.

P3.7   An automatic depth-control system for a robot submarine is shown in Figure P3.7. The depth is measured by a pressure transducer. The gain of the stern plane actuator is $K = 1$ when the vertical velocity is 25 m/s. The submarine has the approximate transfer function

$$G(s) = \frac{(s + 1)^2}{s^2 + 1},$$

and the feedback transducer is $H(s) = 2s + 1$. (a) Determine a state-space representation for the system. (b) Determine whether the system is stable.

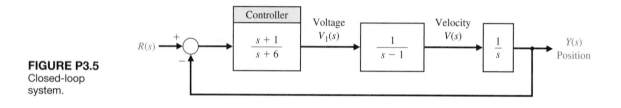

FIGURE P3.5
Closed-loop
system.

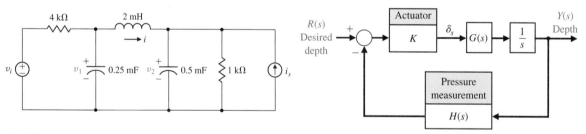

FIGURE P3.6   *RLC* circuit.

FIGURE P3.7   Submarine depth control.

P3.8   The soft landing of a lunar module descending on the moon can be modeled as shown in Figure P3.8. Define the state variables as $x_1 = y$, $x_2 = dy/dt$, $x_3 = m$ and the control as $u = dm/dt$. Assume that g is the gravity constant on the moon. Find the state-space equations for this system. Is this a linear model?

P3.9   A speed control system using fluid flow components is to be designed. The system is a pure fluid control system because it does not have any moving

mechanical parts. The fluid may be a gas or a liquid. A system is desired that maintains the speed within 0.5% of the desired speed by using a tuning fork reference and a valve actuator. Fluid control systems are insensitive and reliable over a wide range of temperature, electromagnetic and nuclear radiation, acceleration, and vibration. The amplification within the system is achieved by using a fluid jet deflection amplifier. The system can be designed for a 500-kW

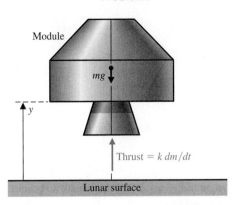

Module

mg

y

Thrust $= k \, dm/dt$

Lunar surface

FIGURE P3.8   Lunar module landing control.

steam turbine with a speed of 12,000 rpm. The block diagram of the system is shown in Figure P3.9. In dimensionless units, we have $b = 0.1, J = 1$, and $K_1 = 0.5$. (a) Determine the closed-loop transfer function

$$T(s) = \frac{\omega(s)}{R(s)}.$$

(b) Determine a state space representation. (c) Determine the characteristic equation obtained from the **A** matrix.

P3.10   Many control systems must operate in two dimensions, for example, the x- and the y-axes. A two-axis control system is shown in Figure P3.10, where a set of state variables is identified. The gain of each axis is K_1 and K_2, respectively. (a) Obtain the state differential equation. (b) Find the characteristic equation from the **A** matrix. (c) Determine the state transition matrix for $K_1 = 1$ and $K_2 = 2$.

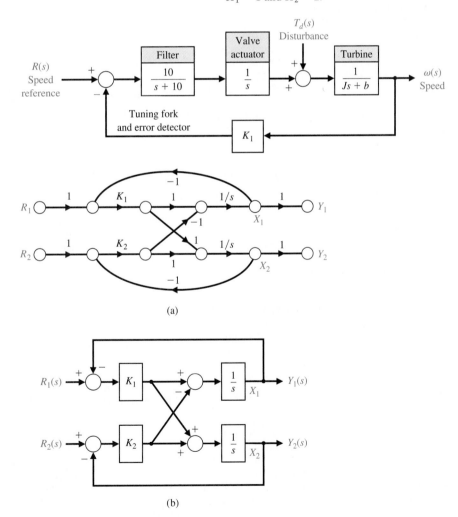

FIGURE P3.9
Steam turbine
control.

(a)

FIGURE P3.10
Two-axis system.
(a) Signal-flow
graph. (b) Block
diagram model.

(b)

P3.11   A system is described by Equation (3.16) with

$$\mathbf{A} = \begin{bmatrix} 1 & -2 \\ 2 & -3 \end{bmatrix}$$

with $u(t) = 0$, and $x_1(0) = x_2(0) = 10$. Determine $x_1(t)$ and $x_2(t)$.

P3.12   A system is described by its transfer function

$$\frac{Y(s)}{R(s)} = T(s) = \frac{8(s + 5)}{s^3 + 12s^2 + 44s + 48}.$$

(a) Determine the phase variable canonical form.
(b) Determine the diagonal canonical form of the state variable matrix equation.
(c) Determine $\Phi(t)$, the state transition matrix.

P3.13   Consider again the *RLC* circuit of Problem P3.1 when $R = 2.5$, $L = 1/4$, and $C = 1/6$. (a) Determine whether the system is stable by finding the characteristic equation with the aid of the **A** matrix. (b) Determine the transition matrix of the network. (c) When the initial inductor current is 0.1 amp, $v_c(0) = 0$, and $v(t) = 0$, determine the response of the system. (d) Repeat part (c) when the initial conditions are zero and $v(t) = E$, for $t > 0$, where E is a constant.

P3.14   Determine a state space representation for a system with the transfer function

$$\frac{Y(s)}{R(s)} = T(s) = \frac{s + 10}{s^4 + 12s^3 + 23s^2 + 34s + 40}.$$

P3.15   Obtain a block diagram and a state variable representation of this system.

$$\frac{Y(s)}{R(s)} = T(s) = \frac{5(s + 4)}{s^3 + 10s^2 + 31s + 20}.$$

P3.16   A system for dispensing radioactive fluid into capsules is shown in Figure P3.16(a). The horizontal axis moving the tray of capsules is actuated by a linear motor. The *x*-axis control is shown in Figure P3.16(b). Assume $K = 500$. Obtain (a) a state variable representation and (b) the unit step-response of the system. (c) Determine the characteristic roots of the system.

P3.17   The dynamics of a controlled submarine are significantly different from those of an aircraft, missile, or surface ship. This difference results primarily from the moment in the vertical plane due to the buoyancy effect. Therefore, it is interesting to consider the control of the depth of a submarine. The equations describing the dynamics of a submarine can be obtained by using Newton's laws and the angles defined in Figure P3.17. To simplify the equations, we will assume that θ is a small angle and the velocity v is constant and equal to 25 ft/s. The state variables of the submarine, considering

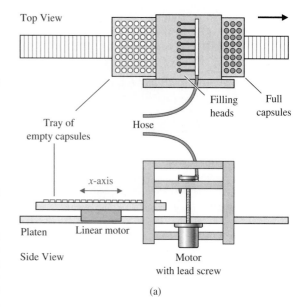

Top View

Filling heads
Full capsules

Tray of empty capsules
Hose

x-axis

Platen   Linear motor

Side View

Motor with lead screw

(a)

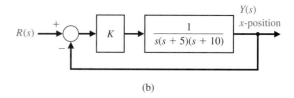

$Y(s)$
x-position

$R(s)$ → K → $\dfrac{1}{s(s + 5)(s + 10)}$ →

(b)

FIGURE P3.16   Automatic fluid dispenser.

only vertical control, are $x_1 = \theta$, $x_2 = d\theta/dt$, and $x_3 = \alpha$, where α is the angle of attack. Thus the state vector differential equation for this system, when the submarine has an Albacore type hull, is

$$\dot{\mathbf{x}} = \begin{bmatrix} 0 & 1 & 0 \\ -0.0071 & -0.111 & 0.12 \\ 0 & 0.07 & -0.3 \end{bmatrix} \mathbf{x} + \begin{bmatrix} 0 \\ -0.095 \\ +0.072 \end{bmatrix} u(t),$$

where $u(t) = \delta_s(t)$, the deflection of the stern plane. (a) Determine whether the system is stable. (b) Determine the response of the system to a stern plane step command of 0.285° with the initial conditions equal to zero.

P3.18   A system is described by the state variable equations

$$\dot{\mathbf{x}} = \begin{bmatrix} 1 & 1 & -1 \\ 4 & 3 & 0 \\ -2 & 1 & 10 \end{bmatrix} \mathbf{x} + \begin{bmatrix} 0 \\ 0 \\ 4 \end{bmatrix} u,$$

$$y = [-1 \quad 2 \quad 0]\mathbf{x}.$$

Determine $G(s) = Y(s)/U(s)$.

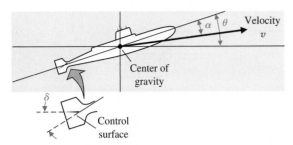

FIGURE P3.17   Submarine depth control.

P3.19   Consider the control of the robot shown in Figure P3.19. The motor turning at the elbow moves the wrist through the forearm, which has some flexibility as shown [16]. The spring has a spring constant k and friction-damping constant b. Let the state variables be $x_1 = \phi_1 - \phi_2$ and $x_2 = \omega_1/\omega_0$, where

$$\omega_0^2 = \frac{k(J_1 + J_2)}{J_1 J_2}.$$

Write the state variable equation in matrix form when $x_3 = \omega_2/\omega_0$.

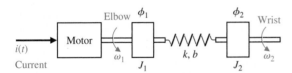

FIGURE P3.19   An industrial robot. (Courtesy of GCA Corporation.)

P3.20   Consider the system described by

$$\dot{\mathbf{x}}(t) = \begin{bmatrix} 0 & 1 \\ -2 & -3 \end{bmatrix} \mathbf{x}(t),$$

where $\mathbf{x}(t) = [x_1(t) \quad x_2(t)]^T$. (a) Compute the state transition matrix $\Phi(t, 0)$. (b) Using the state transition matrix from (a) and for the initial conditions $x_1(0) = 1$ and $x_2(0) = -1$, find the solution $\mathbf{x}(t)$ for $t \geq 0$.

P3.21   A nuclear reactor that has been operating in equilibrium at a high thermal-neutron flux level is suddenly shut down. At shutdown, the density X of xenon 135 and the density I of iodine 135 are 7×10^{16} and 3×10^{15} atoms per unit volume, respectively. The half-lives of I_{135} and Xe_{135} nucleides are 6.7 and 9.2 hours, respectively. The decay equations are [17, 23]

$$\dot{I} = -\frac{0.693}{6.7} I, \qquad \dot{X} = -\frac{0.693}{9.2} X - I.$$

Determine the concentrations of I_{135} and Xe_{135} as functions of time following shutdown by determining (a) the transition matrix and the system response. (b) Verify that the response of the system is that shown in Figure P3.21.

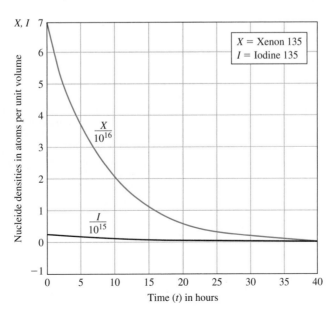

FIGURE P3.21
Nuclear reactor
response.

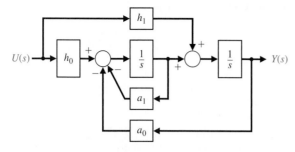

FIGURE P3.22   Model of second-order system.

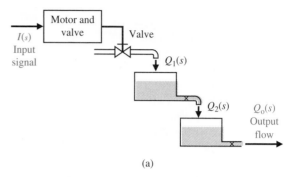

(a)

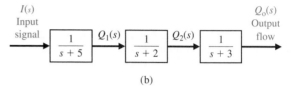

(b)

FIGURE P3.24   A two-tank system with the motor current controlling the output flow rate. (a) Physical diagram. (b) Block diagram.

P3.22   Two equivalent block diagram models for a fourth-order equation (Equation (3.46)) are shown in Figures 3.11 and 3.13. Another alternative structure is shown in Figure P3.22. In this case, the system is second order and the input–output transfer function is

$$G(s) = \frac{Y(s)}{U(s)} = \frac{b_1 s + b_0}{s^2 + a_1 s + a_0}.$$

(a) Verify that the block diagram model of Figure P3.22 is in fact a model of $G(s)$. (b) Show that the vector differential equation representing the block diagram model of Figure P3.22 is

$$\dot{\mathbf{x}} = \begin{bmatrix} 0 & 1 \\ -a_0 & -a_1 \end{bmatrix} \mathbf{x} + \begin{bmatrix} h_1 \\ h_0 \end{bmatrix} u(t),$$

where $h_1 = b_1$, and $h_0 = b_0 - b_1 a_1$.

P3.23   Determine the state variable matrix differential equation for the circuit shown in Figure P3.23. The state variables are $x_1 = i$, $x_2 = v_1$, and $x_3 = v_2$. The output variable is $v_0(t)$.

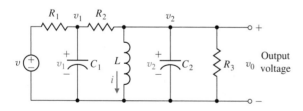

FIGURE P3.23   RLC circuit.

P3.24   The two-tank system shown in Figure P3.24(a) is controlled by a motor adjusting the input valve and ultimately varying the output flow rate. The system has the transfer function

$$\frac{Y(s)}{R(s)} = T(s) = \frac{1}{s^3 + 10s^2 + 31s + 30}$$

for the block diagram shown in Figure P3.24(b). Obtain a block diagram model and the matrix differential equation for the following models: (a) phase variables, (b) input feedforward, (c) physical state variables, and (d) decoupled state variables.

P3.25   It is desirable to use well-designed controllers to maintain building temperature with solar collector space-heating systems. One solar heating system can be described by [10]

$$\frac{dx_1}{dt} = 3x_1 + u_1 + u_2,$$

and

$$\frac{dx_2}{dt} = 2x_2 + u_2 + d,$$

where $x_1 =$ temperature deviation from desired equilibrium, and $x_2 =$ temperature of the storage material (such as a water tank). Also, u_1 and u_2 are the respective flow rates of conventional and solar heat, where the transport medium is forced air. A solar disturbance on the storage temperature (such as overcast skies) is represented by d. Write the matrix equations and solve for the system response when $u_1 = 0$, $u_2 = 1$, and $d = 1$, with zero initial conditions.

P3.26   A system has the following differential equation:

$$\dot{\mathbf{x}} = \begin{bmatrix} -1 & 0 \\ 2 & -3 \end{bmatrix} \mathbf{x} + \begin{bmatrix} 0 \\ 1 \end{bmatrix} r(t).$$

Determine $\Phi(t)$ and its transform $\Phi(s)$ for the system.

P3.27 A system has a block diagram as shown in Figure P3.27. Determine the state variable differential equation and the state transition matrix $\Phi(s)$.

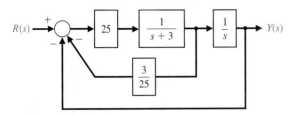

FIGURE P3.27 Feedback system.

P3.28 A gyroscope with a single degree of freedom is shown in Figure P3.28. Gyroscopes sense the angular motion of a system and are used in automatic flight control systems. The gimbal moves about the output axis OB. The input is measured around the input axis OA. The equation of motion about the output axis is obtained by equating the rate of change of angular momentum to the sum of torques. Obtain a state-space representation of the gyro system.

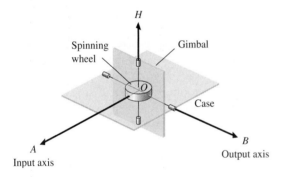

FIGURE P3.28 Gyroscope.

P3.29 A two-mass system is shown in Figure P3.29. The rolling friction constant is b. Determine the matrix differential equation when the output variable is $y_2(t)$.

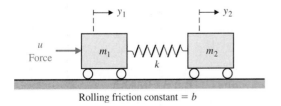

FIGURE 3.29 Two-mass system.

P3.30 There has been considerable engineering effort directed at finding ways to perform manipulative operations in space—for example, assembling a space station and acquiring target satellites. To perform such tasks, space shuttles carry a remote manipulator system (RMS) in the cargo bay [4, 12, 26]. The RMS has proven its effectiveness on recent shuttle missions, but now a new design approach is being considered—a manipulator with inflatable arm segments. Such a design might reduce manipulator weight by a factor of four while producing a manipulator that, prior to inflation, occupies only one-eighth as much space in the shuttle's cargo bay as the present RMS.

The use of an RMS for constructing a space structure in the shuttle bay is shown in Figure P3.30(a), and a model of the flexible RMS arm is shown in Figure P3.30(b), where J is the inertia of the drive motor and L is the distance to the center of gravity of the load component. Derive the state equations for this system.

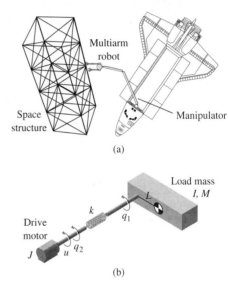

(a)

(b)

FIGURE P3.30 Remote manipulator system.

P3.31 Obtain the state equations for the two-input and one-output circuit shown in Figure P3.31, where the output is i_2.

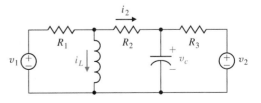

FIGURE P3.31 Two-input *RLC* circuit.

P3.32 Extenders are robot manipulators that extend (that is, increase) the strength of the human arm in load-maneuvering tasks (Figure P3.32) [23, 27]. The system is represented by the transfer function

$$\frac{Y(s)}{U(s)} = G(s) = \frac{30}{s^2 + 4s + 3},$$

where $U(s)$ is the force of the human hand applied to the robot manipulator, and $Y(s)$ is the force of the robot manipulator applied to the load. Determine a state variable model and the state transition matrix for the system.

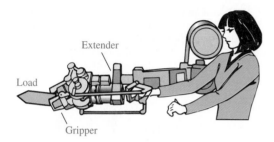

FIGURE P3.32   Extender for increasing the strength of the human arm in load maneuvering tasks.

P3.33   A drug taken orally is ingested at a rate r. The mass of the drug in the gastrointestinal tract is denoted by m_1 and in the bloodstream by m_2. The rate of change of the mass of the drug in the gastrointestinal tract is equal to the rate at which the drug is ingested minus the rate at which the drug enters the bloodstream, a rate that is taken to be proportional to the mass present. The rate of change of the mass in the bloodstream is proportional to the amount coming from the gastrointestinal tract minus the rate at which mass is lost by metabolism, which is proportional to the mass

present in the blood. Develop a state space representation of this system.

For the special case where the coefficients of A are equal to 1 (with the appropriate sign), determine the response when $m_1(0) = 1$ and $m_2(0) = 0$. Plot the state variables versus time and on the $x_1 - x_2$ state plane.

P3.34   The dynamics of a rocket are represented by

$$\frac{Y(s)}{U(s)} = G(s) = \frac{1}{s^2},$$

and state variable feedback is used where $x_1 = y(t)$, $x_2 = \dot{y}(t)$, and $u = -x_2 - 0.5x_1$. Determine the roots of the characteristic equation of this system and the response of the system when the initial conditions are $x_1(0) = 0$ and $x_2(0) = 1$. The input $U(s)$ is the applied torques, and $Y(s)$ is the rocket attitude.

P3.35   A system has the transfer function

$$\frac{Y(s)}{R(s)} = T(s) = \frac{8}{s^3 + 7s^2 + 14s + 8}.$$

(a) Construct a state-space representation of the system.
(b) Determine the element $\phi_{11}(t)$ of the state transition matrix for this system.

P3.36   Determine a state-space representation for the system shown in Figure P3.36. The motor inductance is negligible, the motor constant is $K_m = 10$, the back electromagnetic force constant is $K_b = 0.0706$, the motor friction is negligible. The motor and valve inertia is $J = 0.006$, and the area of the tank is 50 m². Note that the motor is controlled by the armature current i_a. Let $x_1 = h$, $x_2 = \theta$, and $x_3 = d\theta/dt$. Assume that $q_1 = 80\theta$, where θ is the shaft angle. The output flow is $q_0 = 50h(t)$.

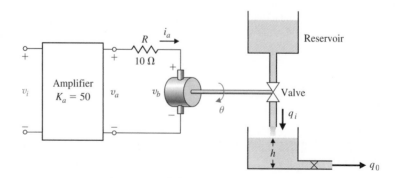

FIGURE P3.36
One-tank system.

ADVANCED PROBLEMS

AP3.1 Consider the electromagnetic suspension system shown in Figure AP3.1. An electromagnet is located at the upper part of the experimental system. Using the electromagnetic force f, we want to suspend the iron ball. Note that this simple electromagnetic suspension system is essentially unworkable. Hence feedback control is indispensable. As a gap sensor, a standard induction probe of the type of eddy current is placed below the ball [25].

Assume that the state variables are $x_1 = x$, $x_2 = dx/dt$, and $x_3 = i$. The electromagnet has an inductance $L = 0.508\,\text{H}$ and a resistance $R = 23.2\,\Omega$. Use a Taylor series approximation for the electromagnetic force. The current is $i_1 = I_0 + i$, where $I_0 = 1.06\,\text{A}$ is the operating point and i is the variable. The mass m is equal to 1.75 kg. The gap is $x_g = X_0 + x$, where $X_0 = 4.36\,\text{mm}$ is the operating point and x is the variable. The electromagnetic force is $f = k(i_1/x_g)^2$, where $k = 2.9 \times 10^{-4}\,\text{N}\,\text{m}^2/\text{A}^2$. Determine the matrix differential equation and the equivalent transfer function $X(s)/V(s)$.

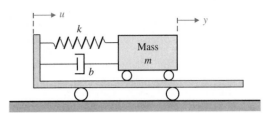

FIGURE AP3.2 Mass on cart.

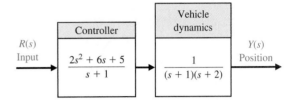

FIGURE AP3.3 Position control.

AP3.4 Front suspensions have become standard equipment on mountain bikes. Replacing the rigid fork that attaches the bicycle's front tire to its frame, such suspensions absorb bump impact energy, shielding both frame and rider from jolts. Commonly used forks, however, use only one spring constant and treat bump impacts at high and low speeds—impacts that vary greatly in severity—essentially the same.

A suspension system with multiple settings that are adjustable while the bike is in motion would be attractive. One air and coil spring with an oil damper is available that permits an adjustment of the damping constant to the terrain as well as to the rider's weight [20]. The suspension system model is shown in Figure AP3.4, where b is adjustable. Select the appropriate

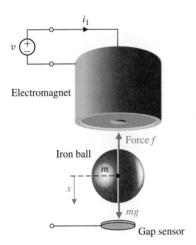

FIGURE AP3.1 Electromagnetic suspension system.

AP3.2 Consider the mass m mounted on a massless cart, as shown in Figure AP3.2. Determine the transfer function $Y(s)/U(s)$, and use the transfer function to obtain a state-space representation of the system.

AP3.3 The control of an autonomous vehicle motion from one point to another point depends on accurate control of the position of the vehicle [18]. The control of the autonomous vehicle position $Y(s)$ is obtained by the system shown in Figure AP3.3. Obtain a state variable representation of the system.

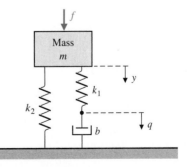

FIGURE AP3.4 Shock absorber.

value for b so that the bike accommodates (a) a large bump at high speeds and (b) a small bump at low speeds. Assume that $k_2 = 1$ and $k_1 = 2$.

AP3.5   Figure AP3.5 shows a mass M suspended from another mass m by means of a light rod of length L. Obtain the state variable differential matrix equation using a linear model assuming a small angle for θ.

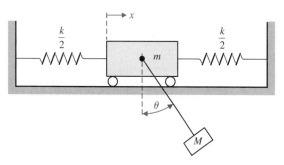

FIGURE AP3.5   Mass suspended from cart.

AP3.6   Consider a crane moving in the x direction while the mass m moves in the z direction, as shown in Figure AP3.6. The trolley motor and the hoist motor are very powerful with respect to the mass of the trolley, the hoist wire, and the load m. Consider the input control variables as the distances D and R. Also assume that $\theta < 50°$. Determine a linear model, and describe the state variable differential equation.

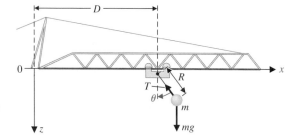

FIGURE AP3.6
A crane moving in the x-direction while the mass moves in the z-direction.

AP3.7   Consider the single-input, single-output system described by

$$\dot{\mathbf{x}}(t) = \mathbf{A}\mathbf{x}(t) + \mathbf{B}u(t)$$
$$y(t) = \mathbf{C}\mathbf{x}(t)$$

where

$$\mathbf{A} = \begin{bmatrix} -1 & 1 \\ 0 & 0 \end{bmatrix}, \mathbf{B} = \begin{bmatrix} 0 \\ 1 \end{bmatrix}, \mathbf{C} = [2 \quad 1].$$

Assume that the input is a linear combination of the states, that is,

$$u(t) = -\mathbf{K}\mathbf{x}(t) + r(t),$$

where $r(t)$ is the reference input. The matrix $\mathbf{K} = [K_1 \quad K_2]$ is known as the gain matrix. Substituting $u(t)$ into the state variable equation gives the closed-loop system

$$\dot{\mathbf{x}}(t) = [\mathbf{A} - \mathbf{B}\mathbf{K}]\mathbf{x}(t) + \mathbf{B}r(t)$$

$$y(t) = \mathbf{C}\mathbf{x}(t)$$

The design process involves finding $\mathbf{K}$ so that the eigenvalues of $\mathbf{A}-\mathbf{B}\mathbf{K}$ are at desired locations in the left-half plane. Compute the characteristic polynomial associated with the closed-loop system and determine values of $\mathbf{K}$ so that the closed-loop eigenvalues are in the left-half plane.

DESIGN PROBLEMS

CDP3.1   The traction drive uses the capstan drive system shown in Figure CDP2.1. Neglect the effect of the motor inductance and determine a state variable model for the system. The parameters are given in Table CDP2.1. The friction of the slide is negligible.

DP3.1   A spring-mass-damper system, as shown in Figure 3.3, is used as a shock absorber for a large high-performance motorcycle. The original parameters selected are $m = 1$ kg, $b = 9$ N s/m, and $k = 20$ N/m. (a) Determine the system matrix, the characteristic roots, and the transition matrix $\Phi(t)$. The harsh initial conditions are assumed to be $y(0) = 1$ and $dy/dt|_{t=0} = 2$. (b) Plot the response of $y(t)$ and dy/dt for the first two seconds. (c) Redesign the shock absorber by changing the spring constant and the damping constant in order to reduce the effect of a high rate of acceleration force

d^2y/dt^2 on the rider. The mass must remain constant at 1 kg.

DP3.2   A system has the state variable matrix equation in phase variable form

$$\dot{\mathbf{x}} = \begin{bmatrix} 0 & 1 \\ -a & -b \end{bmatrix} \mathbf{x} + \begin{bmatrix} 0 \\ d \end{bmatrix} u(t),$$

and $y = 10x_1$. It is desired that the canonical diagonal form of the differential equation be

$$\dot{\mathbf{z}} = \begin{bmatrix} -3 & 0 \\ 0 & -1 \end{bmatrix} \mathbf{z} + \begin{bmatrix} 1 \\ 1 \end{bmatrix} u,$$

$$y = [-5 \quad 5]\mathbf{z}.$$

Determine the parameters a, b, and d to yield the required diagonal matrix differential equation.

DP3.3   An aircraft arresting gear is used on an aircraft carrier as shown in Figure DP3.3. The linear model of each energy absorber has a drag force $f_D = K_D \dot{x}_3$. It is desired to halt the airplane within 30 m after engaging the arresting cable [15]. The speed of the aircraft on landing is 60 m/s. Select the required constant K_D, and plot the response of the state variables.

DP3.4   The Mile-High Bungi Jumping Company wants you to design a bungi jumping system (i.e., a cord) so that the jumper cannot hit the ground when his or her mass is less than 100 kg, but greater than 50 kg. Also, the company wants a hang time (the time a jumper is moving up

and down) greater than 25 seconds, but less than 40 seconds. Determine the characteristics of the cord. The jumper stands on a platform 90 m above the ground, and the cord will be attached to a fixed beam secured 10 m above the platform. Assume that the jumper is 2 m tall and the cord is attached at the waist (1 m high).

DP3.5   Consider the single-input, single-output system described by

$$\dot{\mathbf{x}}(t) = \mathbf{A}\mathbf{x}(t) + \mathbf{B}u(t)$$
$$y(t) = \mathbf{C}\mathbf{x}(t)$$

where

$$\mathbf{A} = \begin{bmatrix} 0 & 1 \\ -2 & 3 \end{bmatrix}, \mathbf{B} = \begin{bmatrix} 0 \\ 1 \end{bmatrix}, \mathbf{C} = [1 \quad 0].$$

Assume that the input is a linear combination of the states, that is,

$$u(t) = -\mathbf{K}\mathbf{x}(t) + r(t),$$

where $r(t)$ is the reference input. Determine $\mathbf{K} = [K_1 \quad K_2]$ so that the closed-loop system

$$\dot{\mathbf{x}}(t) = [\mathbf{A} - \mathbf{BK}]\mathbf{x}(t) + \mathbf{B}r(t)$$
$$y(t) = \mathbf{C}\mathbf{x}(t)$$

possesses closed-loop eigenvalues at r_1 and r_2. Note that if $r_1 = \sigma + j\omega$ is a complex number, then $r_2 = \sigma - j\omega$ is its complex conjugate.

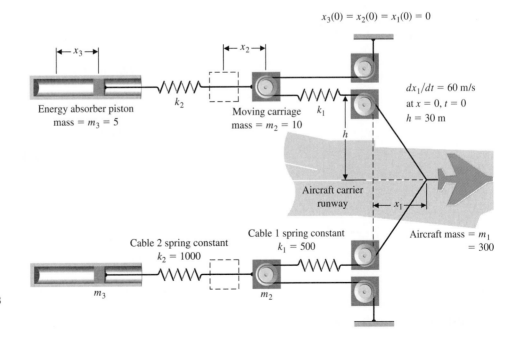

$x_3(0) = x_2(0) = x_1(0) = 0$

$dx_1/dt = 60$ m/s
at $x = 0$, $t = 0$
$h = 30$ m

Energy absorber piston
mass = $m_3 = 5$

Moving carriage
mass = $m_2 = 10$

Aircraft carrier runway

Cable 2 spring constant
$k_2 = 1000$

Cable 1 spring constant
$k_1 = 500$

Aircraft mass = $m_1 = 300$

FIGURE DP3.3
Aircraft arresting gear.

COMPUTER PROBLEMS

CP3.1 Determine a state variable representation for the following transfer functions (without feedback) using the **SS** function:

(a) $G(s) = \dfrac{1}{s + 25}$

(b) $G(s) = \dfrac{3s^2 + 10s + 3}{s^2 + 8s + 5}$

(c) $G(s) = \dfrac{s + 10}{s^3 + 3s^2 + 3s + 1}$

CP3.2 Determine a transfer function representation for the following state variable models using the **tf** function:

(a) $\mathbf{A} = \begin{bmatrix} 0 & 1 \\ 2 & 4 \end{bmatrix}, \mathbf{B} = \begin{bmatrix} 0 \\ 1 \end{bmatrix}, \mathbf{C} = [1 \quad 0]$

(b) $\mathbf{A} = \begin{bmatrix} 1 & 1 & 0 \\ -2 & 0 & 4 \\ 6 & 2 & 10 \end{bmatrix}, \mathbf{B} = \begin{bmatrix} -1 \\ 0 \\ 1 \end{bmatrix}, \mathbf{C} = [0 \ 1 \ 0]$

(c) $\mathbf{A} = \begin{bmatrix} 0 & 1 \\ -1 & -2 \end{bmatrix}, \mathbf{B} = \begin{bmatrix} 0 \\ 1 \end{bmatrix}, \mathbf{C} = [-2 \quad 1].$

CP3.3 Consider the circuit shown in Figure CP3.3. Determine the transfer function $V_0(s)/V_{in}(s)$. Assume an ideal op-amp.

(a) Determine the state variable representation when $R_1 = 1\,k\Omega$, $R_2 = 10\,k\Omega$, $C_1 = 0.5\,mF$, and $C_2 = 0.1\,mF$.

(b) Using the state variable representation from part (a), plot the unit step response with the **step** function.

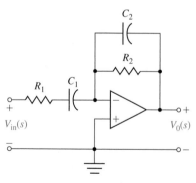

FIGURE CP3.3 An op-amp circuit.

CP3.4 Consider the system

$$\dot{\mathbf{x}} = \begin{bmatrix} 0 & 1 & 0 \\ 0 & 0 & 1 \\ -3 & -2 & -5 \end{bmatrix} \mathbf{x} + \begin{bmatrix} 0 \\ 0 \\ 1 \end{bmatrix} u,$$
$$y = [1 \quad 0 \quad 0]\mathbf{x}.$$

(a) Using the **tf** function, determine the transfer function $Y(s)/U(s)$.

(b) Plot the response of the system to the initial condition $\mathbf{x}(0) = [0 \quad -1 \quad 1]^T$ for $0 \le t \le 10$.

(c) Compute the state transition matrix using the **expm** function, and determine $\mathbf{x}(t)$ at $t = 10$ for the initial condition given in part (b). Compare the result with the system response obtained in part (b).

CP3.5 Consider the two systems

$$\dot{\mathbf{x}}_1 = \begin{bmatrix} 0 & 1 & 0 \\ 0 & 0 & 1 \\ -4 & -5 & -8 \end{bmatrix} \mathbf{x}_1 + \begin{bmatrix} 0 \\ 0 \\ 4 \end{bmatrix} u,$$
$$y = [1 \quad 0 \quad 0]\mathbf{x}_1 \qquad (1)$$

and

$$\dot{\mathbf{x}}_2 = \begin{bmatrix} 0.5000 & 0.5000 & 0.7071 \\ -0.5000 & -0.5000 & 0.7071 \\ -6.3640 & -0.7071 & -8.000 \end{bmatrix} \mathbf{x}_2 + \begin{bmatrix} 0 \\ 0 \\ 4 \end{bmatrix} u,$$
$$y = [0.7071 \quad -0.7071 \quad 0]\mathbf{x}_2. \qquad (2)$$

(a) Using the **tf** function, determine the transfer function $Y(s)/U(s)$ for system (1).

(b) Repeat part (a) for system (2).

(c) Compare the results in parts (a) and (b) and comment.

CP3.6 Consider the closed-loop control system in Figure CP3.6.

(a) Determine a state variable representation of the controller.

(b) Repeat part (a) for the process.

(c) With the controller and process in state variable form, use the **series** and **feedback** functions to compute a closed-loop system representation in state variable form and plot the closed-loop system impulse response.

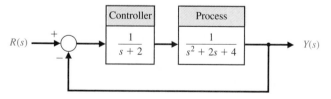

FIGURE CP3.6 A closed-loop feedback control system.

CP3.7 Consider the following system:

$$\dot{\mathbf{x}} = \begin{bmatrix} 0 & 1 \\ -2 & -3 \end{bmatrix} \mathbf{x} + \begin{bmatrix} 0 \\ 1 \end{bmatrix} u$$

$$y = \begin{bmatrix} 1 & 0 \end{bmatrix} \mathbf{x}$$

with

$$\mathbf{x}(0) = \begin{pmatrix} 1 \\ 0 \end{pmatrix}.$$

Using the lsim function obtain and plot the system response (for $x_1(t)$ and $x_2(t)$) when $u(t) = 0$.

TERMS AND CONCEPTS

Canonical form A fundamental or basic form of the state variable model representation, including phase variable canonical form, input feedforward canonical form, diagonal canonical form, and Jordan canonical form.

Diagonal canonical form A decoupled canonical form displaying the n distinct system poles on the diagonal of the state variable representation **A** matrix.

Discrete-time approximation An approximation used to obtain the time response of a system based on the division of the time into small increments Δt.

Euler's method A first-order explicit integration method utilized to obtain numerical solutions of differential equations.

Fundamental matrix *See* Transition matrix.

Input feedforward canonical form A canonical form described by n feedback loops involving the a_n coefficients of the nth order denominator polynomial of the transfer function and feedforward loops obtained by feeding forward the input signal.

Jordan canonical form A block diagonal canonical form for systems that do not possess distinct system poles.

Matrix exponential function An important matrix function, defined as $e^{\mathbf{A}t} = \mathbf{I} + \mathbf{A}t + (\mathbf{A}t)^2/2! + \cdots + (\mathbf{A}t)^k/k! + \cdots$, that plays a role in the solution of linear constant coefficient differential equations.

Output equation The algebraic equation that relates the state vector **x** and the inputs **u** to the outputs **y** through the relationship $\mathbf{y} = \mathbf{Cx} + \mathbf{Du}$.

Phase variable canonical form A canonical form described by n feedback loops involving the a_n coefficients of the

nth order denominator polynomial of the transfer function and m feedforward loops involving the b_m coefficients of the mth order numerator polynomial of the transfer function.

Phase variables The state variables associated with the phase variable canonical form.

Physical variables The state variables representing the physical variables of the system.

State differential equation The differential equation for the state vector: $\dot{\mathbf{x}} = \mathbf{Ax} + \mathbf{Bu}$.

State of a system A set of numbers such that the knowledge of these numbers and the input function will, with the equations describing the dynamics, provide the future state of the system.

State-space representation A time-domain model comprising the state differential equation $\dot{\mathbf{x}} = \mathbf{Ax} + \mathbf{Bu}$ and the output equation, $\mathbf{y} = \mathbf{Cx} + \mathbf{Du}$.

State variable feedback The use of a control signal formed as a direct function of all the state variables.

State variables The set of variables that describe the system.

State vector The vector containing all n state variables, $x_1, x_2, \ldots, x_n$.

Time domain The mathematical domain that incorporates the time response and the description of a system in terms of time t.

Time-varying system A system for which one or more parameters may vary with time.

Transition matrix $\Phi(t)$ The matrix exponential function that describes the unforced response of the system.

Feedback Control System Characteristics

P R E V I E W

In this chapter, we explore the role of error signals to characterize feedback control system performance. The areas of interest include the reduction of sensitivity to model uncertainties, disturbance rejection, measurement noise attenuation, steady-state errors and transient response characteristics. The error signal is used to control the process by negative feedback. Generally speaking, the goal is to minimize the error signal. We discuss the sensitivity of a system to parameter changes, since it is desirable to minimize the effects of parameter variations and uncertainties. We also wish to diminish the effect of unwanted disturbances and measurement noise on the ability of the system to track a desired input. We then describe the transient and steady-state performance of a feedback system and show how this performance can be readily improved with feedback. Of course, the benefits of a control system come with an attendant cost. The chapter concludes with a system performance analysis of the Sequential Design Example: Disk Drive Read System.

DESIRED OUTCOMES

Upon completion of Chapter 4, students should:

❏   Be aware of the central role of error signals in analysis of control systems.

❏   Recognize the improvements afforded by feedback control in reducing system sensitivity to parameter changes, disturbance rejection, and measurement noise attenuation.

❏   Understand the differences between controlling the transient response and the steady-state response of a system.

❏   Have a sense of the benefits and costs of feedback in the control design process.

4.1 INTRODUCTION

A control system is defined as an interconnection of components forming a system that will provide a desired system response. Because the desired system response is known, a signal proportional to the error between the desired and the actual response is generated. The use of this signal to control the process results in a closed-loop sequence of operations that is called a feedback system. This closed-loop sequence of operations is shown in Figure 4.1. The introduction of feedback to improve the control system is often necessary. It is interesting that this is also the case for systems in nature, such as biological and physiological systems; feedback is inherent in these systems. For example, the human heartrate control system is a feedback control system.

To illustrate the characteristics and advantages of introducing feedback, we will consider a single-loop feedback system. Although many control systems are multiloop, a single-loop system is illustrative. A thorough comprehension of the benefits of feedback can best be obtained from the single-loop system and then extended to multiloop systems.

A system without feedback, often called an open-loop system, is shown in Figure 4.2. The disturbance, $T_d(s)$, directly influences the output, $Y(s)$. In the absence of feedback, the control system is highly sensitive to disturbances and to changes is parameters in $G(s)$.

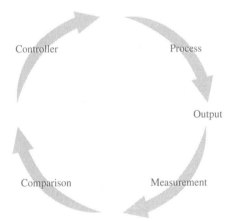

FIGURE 4.1
A closed-loop system.

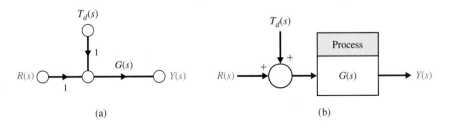

FIGURE 4.2
An open-loop system with a disturbance input, $T_d(s)$. (a) Signal-flow graph. (b) Block diagram.

> **An open-loop (direct) system operates without feedback and directly generates the output in response to an input signal.**

By contrast, a closed-loop, negative feedback control system is shown in Figure 4.3.

> **A closed-loop system uses a measurement of the output signal and a comparison with the desired output to generate an error signal that is used by the controller to adjust the actuator.**

The two forms of control systems are shown in both block diagram and signal-flow graph form. Despite the cost and increased system complexity, closed-loop feedback control has the following advantages:

❑ Decreased sensitivity of the system to variations in the parameters of the process.

❑ Improved rejection of the disturbances.

❑ Improved measurement noise attenuation.

❑ Improved reduction of the steady-state error of the system.

❑ Easy control and adjustment of the transient response of the system.

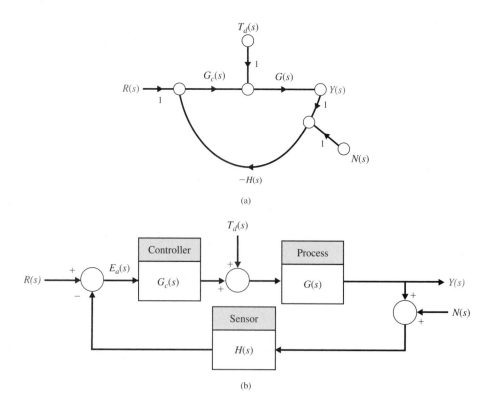

FIGURE 4.3
A closed-loop control system. (a) Signal-flow graph. (b) Block diagram.

In this chapter, we examine how the application of feedback can result in the benefits listed above. Using the notion of a tracking error signal, it will be readily apparent that it is possible to utilize feedback with a controller in the loop to improve system performance.

4.2 ERROR SIGNAL ANALYSIS

The closed-loop feedback control system shown in Figure 4.3 has three inputs—$R(s)$, $T_d(s)$, and $N(s)$—and one output, $Y(s)$. The signals $T_d(s)$ and $N(s)$ are the disturbance and measurement noise signals, respectively. Define the **tracking error** as

$$E(s) = R(s) - Y(s). \tag{4.1}$$

For ease of discussion, we will consider a unity feedback system, that is, $H(s) = 1$, in Figure 4.3. In Section 5.5 of the following chapter, the influence of a nonunity feedback element in the loop is considered.

After some block diagram manipulation, we find that the output is given by

$$Y(s) = \frac{G_c(s)G(s)}{1 + G_c(s)G(s)}R(s) + \frac{G(s)}{1 + G_c(s)G(s)}T_d(s) - \frac{G_c(s)G(s)}{1 + G_c(s)G(s)}N(s). \tag{4.2}$$

Therefore, with $E(s) = R(s) - Y(s)$, we have

$$E(s) = \frac{1}{1 + G_c(s)G(s)}R(s) - \frac{G(s)}{1 + G_c(s)G(s)}T_d(s) + \frac{G_c(s)G(s)}{1 + G_c(s)G(s)}N(s). \tag{4.3}$$

Define the function

$$L(s) = G_c(s)G(s).$$

The function, $L(s)$, is known as the **loop gain** and plays a fundamental role in control system analysis [12]. In terms of $L(s)$ the tracking error is given by

$$E(s) = \frac{1}{1 + L(s)}R(s) - \frac{G(s)}{1 + L(s)}T_d(s) + \frac{L(s)}{1 + L(s)}N(s). \tag{4.4}$$

We can define the function [41]

$$F(s) = 1 + L(s).$$

Then, in terms of $F(s)$, we define the **sensitivity function** as

$$S(s) = \frac{1}{F(s)} = \frac{1}{1 + L(s)}. \tag{4.5}$$

Similarly, in terms of the loop gain, we define the **complementary sensitivity function** as

$$C(s) = \frac{L(s)}{1 + L(s)}. \tag{4.6}$$

In terms of the functions $S(s)$ and $C(s)$, we can write the tracking error as

$$E(s) = S(s)R(s) - S(s)G(s)T_d(s) + C(s)N(s). \tag{4.7}$$

Examining Equation (4.7), we see that (for a given $G(s)$), if we want to minimize the tracking error, we want both $S(s)$ and $C(s)$ to be small. Remember that $S(s)$ and $C(s)$ are both functions of the controller, $G_c(s)$, which the control design engineer must select. However, the following special relationship between $S(s)$ and $C(s)$ holds

$$S(s) + C(s) = 1. \tag{4.8}$$

We cannot simultaneously make $S(s)$ and $C(s)$ small. Obviously, design compromises must be made.

To analyze the tracking error equation, we need to understand what it means for a transfer function to be "large" or to be "small." The discussion of magnitude of a transfer function is the subject of Chapters 8 and 9 on frequency response methods. However, for our purposes here, we describe the magnitude of the loop gain $L(s)$ by considering the magnitude $|L(j\omega)|$ over the range of frequencies, ω, of interest.

Considering the tracking error in Equation (4.4), it is evident that, for a given $G(s)$, to reduce the influence of the disturbance, $T_d(s)$, on the tracking error, $E(s)$, we desire $L(s)$ to be large over the range of frequencies that characterize the disturbances. That way, the transfer function $G(s)/(1 + L(s))$ will be small, thereby reducing the influence of $T_d(s)$. Since $L(s) = G_c(s)G(s)$, this implies that we need to design the controller $G_c(s)$ to have a large magnitude. Conversely, to attenuate the measurement noise, $N(s)$, and reduce the influence on the tracking error, we desire $L(s)$ to be small over the range of frequencies that characterize the measurement noise. The transfer function $L(s)/(1 + L(s))$ will be small, thereby reducing the influence of $N(s)$. Again, since $L(s) = G_c(s)\, G(s)$, that implies that we need to design the controller $G_c(s)$ to have a small magnitude. Fortunately, the apparent conflict between wanting to make $G_c(s)$ large to reject disturbances and the wanting to make $G_c(s)$ small to attenuate measurement noise can be addressed in the design phase by making the loop gain, $L(s)$, large at low frequencies (generally associated with the frequency range of disturbances), and making $L(s)$ small at high frequencies (generally associated with measurement noise).

More discussion on disturbance rejection and measurement noise attenuation follows in the subsequent sections. Next, we discuss how we can use feedback to reduce the sensitivity of the system to variations and uncertainty in parameters in the process, $G(s)$. This is accomplished by analyzing the tracking error in Equation (4.2) when $T_d(s) = N(s) = 0$.

4.3 SENSITIVITY OF CONTROL SYSTEMS TO PARAMETER VARIATIONS

A process, represented by the transfer function $G(s)$, whatever its nature, is subject to a changing environment, aging, ignorance of the exact values of the process parameters, and other natural factors that affect a control process. In the open-loop system, all these errors and changes result in a changing and inaccurate output. However, a closed-loop system senses the change in the output due to the process changes and attempts to correct the output. The **sensitivity** of a control system to parameter variations is of prime importance. A primary advantage of a closed-loop feedback control system is its ability to reduce the system's sensitivity [1–4, 18].

For the closed-loop case, if $G_c(s)G(s) \gg 1$ for all complex frequencies of interest, we can use Equation (4.2) to obtain (letting $T_d(s) = 0$ and $N(s) = 0$)

$$Y(s) \cong R(s).$$

The output is approximately equal to the input. However, the condition $G_c(s)G(s) \gg 1$ may cause the system response to be highly oscillatory and even unstable. But the fact that increasing the magnitude of the loop gain reduces the effect of $G(s)$ on the output is an exceedingly useful result. Therefore, the first advantage of a feedback system is that the effect of the variation of the parameters of the process, $G(s)$, is reduced.

Suppose the process (or plant) $G(s)$ undergoes a change such that the true plant model is $G(s) + \Delta G(s)$. The change in the plant may be due to a changing external environment or natural aging, or it may just represent the uncertainty in certain plant parameters. We consider the effect on the tracking error $E(s)$ due to $\Delta G(s)$. Relying on the principle of superposition, we can let $T_d(s) = N(s) = 0$ and consider only the reference input $R(s)$. From Equation (4.3), it follows that

$$E(s) + \Delta E(s) = \frac{1}{1 + G_c(s)(G(s) + \Delta G(s))} R(s).$$

Then the change in the tracking error is

$$\Delta E(s) = \frac{-G_c(s)\,\Delta G(s)}{(1 + G_c(s)G(s) + G_c(s)\,\Delta G(s))(1 + G_c(s)G(s))} R(s).$$

Since we usually find that $G_c(s)G(s) \gg G_c(s)\,\Delta G(s)$, we have

$$\Delta E(s) \approx \frac{-G_c(s)\,\Delta G(s)}{(1 + L(s))^2} R(s).$$

We see that the change in the tracking error is reduced by the factor $1 + L(s)$, which is generally greater than 1 over the range of frequencies of interest.

For large $L(s)$, we have $1 + L(s) \approx L(s)$, and we can approximate the change in the tracking error by

$$\Delta E(s) \approx -\frac{1}{L(s)} \frac{\Delta G(s)}{G(s)} R(s). \tag{4.9}$$

Larger magnitude $L(s)$ translates into smaller changes in the tracking error (that is, **reduced sensitivity** to changes in $\Delta G(s)$ in the process). Also, larger $L(s)$ implies smaller sensitivity, $S(s)$. The question arises, how do we define sensitivity? Since our goal is to reduce system sensitivity, it makes sense to formally define the term.

The **system sensitivity** is defined as the ratio of the percentage change in the system transfer function to the percentage change of the process transfer function. The system transfer function is

$$T(s) = \frac{Y(s)}{R(s)}, \qquad (4.10)$$

and therefore the sensitivity is defined as

$$S = \frac{\Delta T(s)/T(s)}{\Delta G(s)/G(s)}. \qquad (4.11)$$

In the limit, for small incremental changes, Equation (4.11) becomes

$$\boxed{S = \frac{\partial T/T}{\partial G/G} = \frac{\partial \ln T}{\partial \ln G}.} \qquad (4.12)$$

> **System sensitivity is the ratio of the change in the system transfer function to the change of a process transfer function (or parameter) for a small incremental change.**

The sensitivity of the open-loop system to changes in the plant $G(s)$ is equal to 1. The sensitivity of the closed-loop is readily obtained by using Equation (4.12). The system transfer function of the closed-loop system is

$$T(s) = \frac{G_c(s)G(s)}{1 + G_c(s)G(s)}.$$

Therefore, the sensitivity of the feedback system is

$$S_G^T = \frac{\partial T}{\partial G} \cdot \frac{G}{T} = \frac{G_c}{(1 + G_c G)^2} \cdot \frac{G}{G G_c/(1 + G_c G)}$$

or

$$\boxed{S_G^T = \frac{1}{1 + G_c(s)G(s)}.} \qquad (4.13)$$

We find that the sensitivity of the system may be reduced below that of the open-loop system by increasing $L(s) = G_c(s)G(s)$ over the frequency range of interest. Note that S_G^T in Equation (4.12) is exactly the same as the sensitive function $S(s)$ given in Equation (4.5). In fact, these functions are one and the same.

Often, we seek to determine S_α^T, where α is a parameter within the transfer function of a block G. Using the chain rule, we find that

$$S_\alpha^T = S_G^T S_\alpha^G. \qquad (4.14)$$

Very often, the transfer function of the system $T(s)$ is a fraction of the form [1]

$$T(s, \alpha) = \frac{N(s, \alpha)}{D(s, \alpha)}, \tag{4.15}$$

where α is a parameter that may be subject to variation due to the environment. Then we may obtain the sensitivity to α by rewriting Equation (4.11) as

$$S_\alpha^T = \frac{\partial \ln T}{\partial \ln \alpha} = \frac{\partial \ln N}{\partial \ln \alpha}\bigg|_{\alpha_0} - \frac{\partial \ln D}{\partial \ln \alpha}\bigg|_{\alpha_0} = S_\alpha^N - S_\alpha^D, \tag{4.16}$$

where α_0 is the nominal value of the parameter.

An important advantage of feedback control systems is the ability to reduce the effect of the variation of parameters of a control system by adding a feedback loop. To obtain highly accurate open-loop systems, the components of the open-loop, $G(s)$, must be selected carefully in order to meet the exact specifications. However, a closed-loop system allows $G(s)$ to be less accurately specified, because the sensitivity to changes or errors in $G(s)$ is reduced by the loop gain $L(s)$. This benefit of closed-loop systems is a profound advantage for the electronic amplifiers of the communication industry. A simple example will illustrate the value of feedback for reducing sensitivity.

EXAMPLE 4.1 Feedback amplifier

An amplifier used in many applications has a gain $-K_a$, as shown in Figure 4.4(a). The output voltage is

$$v_0 = -K_a v_{\text{in}}. \tag{4.17}$$

We often add feedback using a potentiometer R_p, as shown in Figure 4.4(b). The transfer function of the amplifier without feedback is

$$T = -K_a, \tag{4.18}$$

and the sensitivity to changes in the amplifier gain is

$$S_{K_a}^T = 1. \tag{4.19}$$

The block diagram model of the amplifier with feedback is shown in Figure 4.5, where

$$\beta = \frac{R_2}{R_1} \tag{4.20}$$

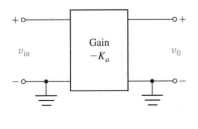

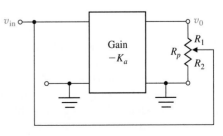

FIGURE 4.4
(a) Open-loop amplifier.
(b) Amplifier with feedback.

(a)   (b)

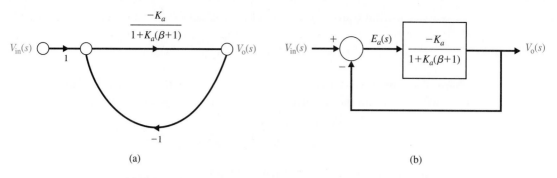

FIGURE 4.5   Block diagram model of feedback amplifier assuming $R_p \gg R_0$ of the amplifier.

and

$$R_p = R_1 + R_2. \tag{4.21}$$

The closed-loop transfer function of the feedback amplifier is

$$T = \frac{-K_a}{1 + K_a\beta}. \tag{4.22}$$

The sensitivity of the closed-loop feedback amplifier is

$$S_{K_a}^T = S_G^T S_{K_a}^G = \frac{1}{1 + K_a\beta}. \tag{4.23}$$

If K_a is large, the sensitivity is low. For example, if

$$K_a = 10^4$$

and

$$\beta = 0.1, \tag{4.24}$$

we have

$$S_{K_a}^T = \frac{1}{1 + 10^3}, \tag{4.25}$$

or the magnitude is one-thousandth of the magnitude of the open-loop amplifier.

We shall return to the concept of sensitivity in subsequent chapters. These chapters will emphasize the importance of sensitivity in the design and analysis of control systems. ∎

4.4 DISTURBANCE SIGNALS IN A FEEDBACK CONTROL SYSTEM

An important effect of feedback in a control system is the control and partial elimination of the effect of disturbance signals. A **disturbance signal** is an unwanted input signal that affects the output signal. Many control systems are subject to extraneous disturbance signals that cause the system to provide an inaccurate output. Electronic amplifiers have inherent noise generated within the integrated circuits or transistors;

radar antennas are subjected to wind gusts; and many systems generate unwanted distortion signals due to nonlinear elements. The benefit of feedback systems is that the effect of distortion, noise, and unwanted disturbances can be effectively reduced.

Disturbance Rejection

When $R(s) = N(s) = 0$, it follows from Equation (4.4) that

$$E(s) = -S(s)G(s)T_d(s) = -\frac{G(s)}{1 + L(s)}T_d(s).$$

For a fixed $G(s)$ and a given $T_d(s)$, as the loop gain $L(s)$ increases, the effect of $T_d(s)$ on the tracking error decreases. In other words, the sensitivity function $S(s)$ is small when the loop gain is large. We say that large loop gain leads to good **disturbance rejection**. More precisely, for good disturbance rejection, we require a large loop gain over the frequencies of interest associated with the expected disturbance signals.

In practice, the disturbance signals are often low frequency. When that is the case, we say that we want the loop gain to be large at low frequencies. This is equivalent to stating that we want to design the controller $G_c(s)$ so that the sensitivity function $S(s)$ is small at low frequencies.

As a specific example of a system with an unwanted disturbance, let us consider again the speed control system for a steel rolling mill. The rolls, which process steel, are subjected to large load changes or disturbances. As a steel bar approaches the rolls (see Figure 4.6), the rolls are empty. However, when the bar engages in the rolls, the load on the rolls increases immediately to a large value. This loading effect can be approximated by a step change of disturbance torque. Alternatively, the response can be seen from the speed–torque curves of a typical motor, as shown in Figure 4.8.

The transfer function model of an armature-controlled DC motor with a load torque disturbance was determined in Example 2.3 and is shown in Figure 4.7, where it is assumed that L_a is negligible. Let $R(s) = 0$ and examine $E(s) = -\omega(s)$, for a disturbance $T_d(s)$.

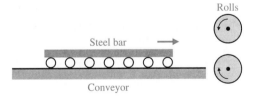

FIGURE 4.6
Steel rolling mill.

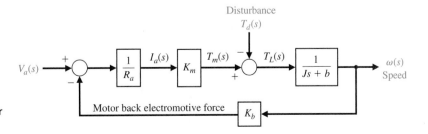

FIGURE 4.7
Open-loop speed control system (without tachometer feedback).

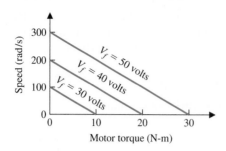

FIGURE 4.8
Motor
speed–torque
curves.

The change in speed due to the load disturbance is then

$$E(s) = -\omega(s) = \frac{1}{Js + b + K_m K_b / R_a} T_d(s). \qquad (4.26)$$

The steady-state error in speed due to the load torque, $T_d(s) = D/s$, is found by using the final-value theorem. Therefore, for the open-loop system, we have

$$\lim_{t \to \infty} E(t) = \lim_{s \to 0} sE(s) = \lim_{s \to 0} s \frac{1}{Js + b + K_m K_b / R_a} \left(\frac{D}{s} \right)$$

$$= \frac{D}{b + K_m K_b / R_a} = -\omega_0(\infty). \qquad (4.27)$$

The closed-loop speed control system is shown in block diagram form in Figure 4.9. The closed-loop system is shown in signal-flow graph and block diagram form in Figure 4.10, where $G_1(s) = K_a K_m / R_a$, $G_2(s) = 1/(Js + b)$, and $H(s) = K_t + K_b / K_a$. The error, $E(s) = -\omega(s)$, of the closed-loop system of Figure 4.10 is:

$$E(s) = -\omega(s) = \frac{G_2(s)}{1 + G_1(s)G_2(s)H(s)} T_d(s). \qquad (4.28)$$

Then, if $G_1 G_2 H(s)$ is much greater than 1 over the range of s, we obtain the approximate result

$$E(s) \approx \frac{1}{G_1(s)H(s)} T_d(s). \qquad (4.29)$$

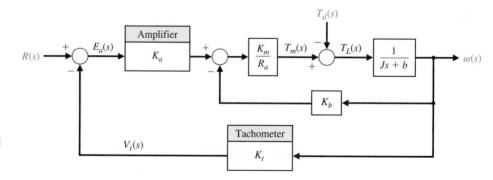

FIGURE 4.9
Closed-loop speed
tachometer control
system.

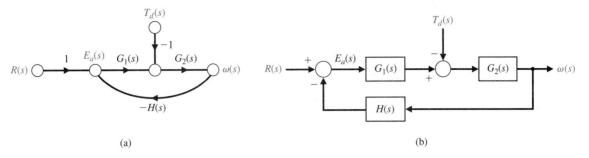

(a)   (b)

FIGURE 4.10   Closed-loop system. (a) Signal-flow graph model. (b) Block diagram model.

Therefore, if $G_1(s)H(s)$ is made sufficiently large, the effect of the disturbance can be decreased by closed-loop feedback. Note that

$$G_1(s)H(s) = \frac{K_a K_m}{R_a}\left(K_t + \frac{K_b}{K_a}\right) \approx \frac{K_a K_m K_t}{R_a},$$

since $K_a \gg K_b$. Thus, we strive to obtain a large amplifier gain, K_a, and keep $R_a < 2\ \Omega$. The error for the system shown in Figure 4.10 is

$$E(s) = R(s) - \omega(s),$$

and $R(s) = \omega_d(s)$, the desired speed. For calculation ease, we let $R(s) = 0$ and examine $\omega(s)$.

To determine the output for the speed control system of Figure 4.9, we must consider the load disturbance when the input $R(s) = 0$. This is written as

$$\omega(s) = \frac{-1/(Js + b)}{1 + (K_t K_a K_m/R_a)[1/(Js + b)] + (K_m K_b/R_a)[1/(Js + b)]}T_d(s)$$

$$= \frac{-1}{Js + b + (K_m/R_a)(K_t K_a + K_b)}T_d(s). \tag{4.30}$$

The steady-state output is obtained by utilizing the final-value theorem, and we have

$$\lim_{t \to \infty} \omega(t) = \lim_{s \to 0}(s\omega(s)) = \frac{-1}{b + (K_m/R_a)(K_t K_a + K_b)}D; \tag{4.31}$$

when the amplifier gain K_a is sufficiently high, we have

$$\omega(\infty) \approx \frac{-R_a}{K_a K_m K_t}D = \omega_c(\infty). \tag{4.32}$$

The ratio of closed-loop to open-loop steady-state speed output due to an undesired disturbance is

$$\frac{\omega_c(\infty)}{\omega_0(\infty)} = \frac{R_a b + K_m K_b}{K_a K_m K_t} \tag{4.33}$$

and is usually less than 0.02.

This advantage of a feedback speed control system can also be illustrated by considering the speed–torque curves for the closed-loop system, which are shown in

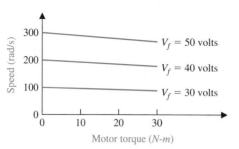

FIGURE 4.11
The speed-torque curves for the closed-loop system.

Figure 4.11. The improvement of the feedback system is evidenced by the almost horizontal curves, which indicate that the speed is almost independent of the load torque.

Measurement Noise Attenuation

When $R(s) = T_d(s) = 0$, it follows from Equation (4.4) that

$$E(s) = C(s)N(s) = \frac{L(s)}{1 + L(s)}N(s).$$

As the loop gain $L(s)$ decreases, the effect of $N(s)$ on the tracking error decreases. In other words, the complementary sensitivity function $C(s)$ is small when the loop gain $L(s)$ is small. If we design $G_c(s)$ such that $L(s) \ll 1$, then the noise is attenuated because

$$C(s) \approx L(s).$$

We say that small loop gain leads to good noise attenuation. More precisely, for effective **measurement noise attenuation**, we need a small loop gain over the frequencies associated with the expected noise signals.

In practice, measurement noise signals are often high frequency. Thus we want the loop gain to be low at high frequencies. This is equivalent to a small complementary sensitivity function at high frequencies. The separation of disturbances (at low frequencies) and measurement noise (at high frequencies) is very fortunate because it gives the control system designer a way to approach the design process: the controller should be high gain at low frequencies and low gain at high frequencies. Remember that by low and high we mean that the loop gain magnitude is low/high at the various high/low frequencies. It is not always the case that the disturbances are low frequency or that the measurement noise is high frequency. For example, an astronaut running on a treadmill on a space station may impart disturbances to the spacecraft at high frequencies. If the frequency separation does not exist, the design process usually becomes more involved (for example, we may have to use notch filters to reject disturbances at known high frequencies). A noise signal that is prevalent in many systems is the noise generated by the measurement sensor. This noise, $N(s)$, can be represented as shown in Figure 4.3. The effect of the noise on the output is

$$Y(s) = \frac{-G_c(s)G(s)}{1 + G_c(s)G(s)}N(s), \qquad (4.34)$$

which is approximately

$$Y(s) \simeq -N(s), \tag{4.35}$$

for large loop gain $L(s) = G_c(s)G(s)$. This is consistent with the earlier discussion that smaller loop gain leads to measurement noise attentuation. Clearly, the designer must shape the loop gain appropriately.

The equivalency of sensitivity, S_G^T, and the response of the closed-loop system tracking error to a reference input can be illustrated by considering Figure 4.3. The sensitivity of the system to $G(s)$ is

$$S_G^T = \frac{1}{1 + G_c(s)G(s)} = \frac{1}{1 + L(s)}. \tag{4.36}$$

The effect of the reference on the tracking error (with $T_d(s) = 0$ and $N(s) = 0$) is

$$\frac{E(s)}{R(s)} = \frac{1}{1 + G_c(s)G(s)} = \frac{1}{1 + L(s)}. \tag{4.37}$$

In both cases, we find that the undesired effects can be alleviated by increasing the loop gain. Feedback in control systems primarily reduces the sensitivity of the system to parameter variations and the effect of disturbance inputs. Note that the measures taken to reduce the effects of parameter variations or disturbances are equivalent, and fortunately, they reduce simultaneously. As a final illustration, consider the effect of the noise on the tracking error:

$$\frac{E(s)}{T_d(s)} = \frac{G_c(s)G(s)}{1 + G_c(s)G(s)} = \frac{L(s)}{1 + L(s)}. \tag{4.38}$$

We find that the undesired effects of measurement noise can be alleviated by decreasing the loop gain. Keeping in mind the relationship

$$S(s) + C(s) = 1,$$

the trade-off in the design process is evident.

4.5 CONTROL OF THE TRANSIENT RESPONSE

One of the most important characteristics of control systems is their transient response. The **transient response** is the response of a system as a function of time. Because the purpose of control systems is to provide a desired response, the transient response of control systems often must be adjusted until it is satisfactory. If an open-loop control system does not provide a satisfactory response, then the process, $G(s)$, must be replaced with a more suitable process. By contrast, a closed-loop system can often be adjusted to yield the desired response by adjusting the feedback loop parameters. It is often possible to alter the response of an open-loop system by inserting a suitable cascade controller, $G_c(s)$, preceding the process, $G(s)$, as shown in Figure 4.12. Then it is necessary to design the cascade transfer function, $G_c(s)G(s)$, so that the resulting transfer function provides the desired transient response.

FIGURE 4.12
Cascade controller
system (without
feedback).
(a) Signal-flow graph.
(b) Block diagram

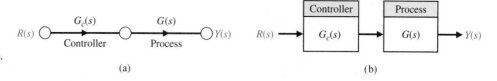

(a)   (b)

To make this concept more comprehensible, consider a specific control system, which may be operated in an open- or closed-loop manner. A speed control system, as shown in Figure 4.13, is often used in industrial processes to move materials and products. Several important speed control systems are used in steel mills for rolling the steel sheets and moving the steel through the mill [19]. The transfer function of the open-loop system (without feedback) was obtained in Equation (2.70). For $\omega(s)/V_a(s)$, we have

$$\frac{\omega(s)}{V_a(s)} = G(s) = \frac{K_1}{\tau_1 s + 1}, \tag{4.39}$$

where

$$K_1 = \frac{K_m}{R_a b + K_b K_m} \quad \text{and} \quad \tau_1 = \frac{R_a J}{R_a b + K_b K_m}.$$

In the case of a steel mill, the inertia of the rolls is quite large, and a large armature-controlled motor is required. If the steel rolls are subjected to a step command for a speed change of

$$V_a(s) = \frac{k_2 E}{s}, \tag{4.40}$$

the output response is

$$\omega(s) = G(s)V_a(s). \tag{4.41}$$

The transient speed change is then

$$\omega(t) = K_1(k_2 E)(1 - e^{-t/\tau_1}). \tag{4.42}$$

If this transient response is too slow, we must choose another motor with a different time constant τ_1, if possible. However, because τ_1 is dominated by the load inertia, J, it may not be possible to achieve much alteration of the transient response.

FIGURE 4.13
Open-loop speed
control system
(without feedback).

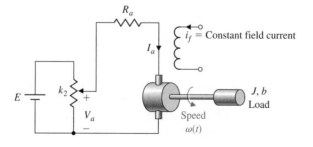

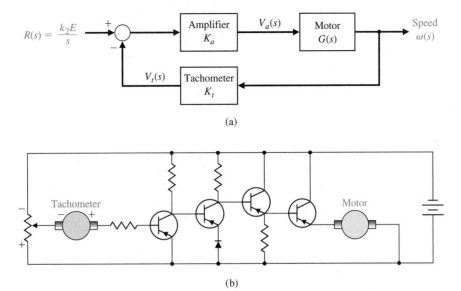

FIGURE 4.14
(a) Closed-loop speed control system.
(b) Transistorized closed-loop speed control system.

A closed-loop speed control system is easily obtained by using a tachometer to generate a voltage proportional to the speed, as shown in Figure 4.14(a). This voltage is subtracted from the potentiometer voltage and amplified as shown in Figure 4.14(a). A practical transistor amplifier circuit for accomplishing this feedback in low-power applications is shown in Figure 4.14(b) [1, 5, 7]. The closed-loop transfer function is

$$\frac{\omega(s)}{R(s)} = \frac{K_a G(s)}{1 + K_a K_t G(s)}$$

$$= \frac{K_a K_1}{\tau_1 s + 1 + K_a K_t K_1} = \frac{K_a K_1/\tau_1}{s + (1 + K_a K_t K_1)/\tau_1}. \tag{4.43}$$

The amplifier gain, K_a, may be adjusted to meet the required transient response specifications. Also, the tachometer gain constant, K_t, may be varied, if necessary.

The transient response to a step change in the input command is then

$$\omega(t) = \frac{K_a K_1}{1 + K_a K_t K_1}(k_2 E)(1 - e^{-pt}), \tag{4.44}$$

where $p = (1 + K_a K_t K_1)/\tau_1$. Because the load inertia is assumed to be very large, we alter the response by increasing K_a. Thus, we have the approximate response

$$\omega(t) \approx \frac{1}{K_t}(k_2 E)\left[1 - \exp\left(\frac{-(K_a K_t K_1)t}{\tau_1}\right)\right]. \tag{4.45}$$

For a typical application, the open-loop pole might be $1/\tau_1 = 0.10$, whereas the closed-loop pole could be at least $(K_a K_t K_1)/\tau_1 = 10$, a factor of one hundred in the improvement of the speed of response. To attain the gain $K_a K_t K_1$, the amplifier gain K_a must be reasonably large, and the armature voltage signal to the

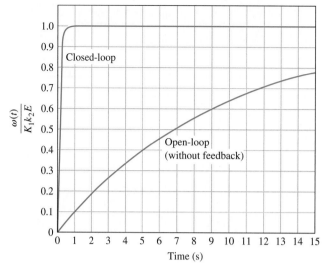

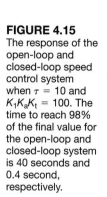

FIGURE 4.15
The response of the open-loop and closed-loop speed control system when $\tau = 10$ and $K_1 K_a K_t = 100$. The time to reach 98% of the final value for the open-loop and closed-loop system is 40 seconds and 0.4 second, respectively.

motor and its associated torque signal must be larger for the closed-loop than for the open-loop operation. Therefore, a higher-power motor will be required to avoid saturation of the motor. The responses of the closed-loop system and the open-loop system are shown in Figure 4.15. Note the rapid response of the closed-loop system.

While we are considering this speed control system, it will be worthwhile to determine the sensitivity of the open- and closed-loop systems. As before, the sensitivity of the open-loop system to a variation in the motor constant or the potentiometer constant k_2 is unity. The sensitivity of the closed-loop system to a variation in K_m is

$$S_{K_m}^T = S_G^T S_{K_m}^G \approx \frac{[s + (1/\tau_1)]}{s + (K_a K_t K_1 + 1)/\tau_1}.$$

Using the typical values given in the previous paragraph, we have

$$S_{K_m}^T \approx \frac{(s + 0.10)}{s + 10}.$$

We find that the sensitivity is a function of s and must be evaluated for various values of frequency. This type of frequency analysis is straightforward but will be deferred until a later chapter. However, it is clearly seen that at a specific low frequency—for example, $s = j\omega = j1$—the magnitude of the sensitivity is approximately $|S_{K_m}^T| \cong 0.1$.

4.6 STEADY-STATE ERROR

A feedback control system is valuable because it provides the engineer with the ability to adjust the transient response. In addition, as we have seen, the sensitivity of the system and the effect of disturbances can be reduced significantly. However, as a further requirement, we must examine and compare the final steady-state error for an open-loop and a closed-loop system. The **steady-state error** is the error after the transient response has decayed, leaving only the continuous response.

The error of the open-loop system shown in Figure 4.2 is

$$E_0(s) = R(s) - Y(s) = (1 - G(s))R(s), \tag{4.46}$$

when $T_d(s) = 0$. Figure 4.3 shows the closed-loop system. When $T_d(s) = 0$ and $N(s) = 0$, and we let $H(s) = 1$, the tracking error is given by (Equation 4.3)

$$E_c(s) = \frac{1}{1 + G_c(s)G(s)} R(s). \tag{4.47}$$

To calculate the steady-state error, we use the final-value theorem

$$\lim_{t \to \infty} e(t) = \lim_{s \to 0} sE(s). \tag{4.48}$$

Therefore, using a unit step input as a comparable input, we obtain for the open-loop system

$$e_o(\infty) = \lim_{s \to 0} s(1 - G(s))\left(\frac{1}{s}\right) = \lim_{s \to 0} (1 - G(s)) = 1 - G(0). \tag{4.49}$$

For the closed-loop system we have

$$e_c(\infty) = \lim_{s \to 0} s\left(\frac{1}{1 + G_c(s)G(s)}\right)\left(\frac{1}{s}\right) = \frac{1}{1 + G_c(0)G(0)}. \tag{4.50}$$

The value of $G(s)$ when $s = 0$ is often called the DC gain and is normally greater than one. Therefore, the open-loop system will usually have a steady-state error of significant magnitude. By contrast, the closed-loop system with a reasonably large DC loop gain $L(0) = G_c(0)G(0)$ will have a small steady-state error. In Chapter 5, we discuss steady-state error in much greater detail.

Upon examination of Equation (4.49), we note that the open-loop control system can possess a zero steady-state error by simply adjusting and calibrating the system's DC gain, $G(0)$, so that $G(0) = 1$. Therefore, we may logically ask, What is the advantage of the closed-loop system in this case? To answer this question, we return to the concept of the sensitivity of the system to parameter changes. In the open-loop system, we may calibrate the system so that $G(0) = 1$, but during the operation of the system, it is inevitable that the parameters of $G(s)$ will change due to environmental changes and that the DC gain of the system will no longer be equal to 1. Because it is an open-loop system, the steady-state error will not equal zero until the system is maintained and recalibrated. By contrast, the closed-loop feedback system continually monitors the steady-state error and provides an actuating signal to reduce the steady-state error. Because systems are susceptible to parameter drift, environmental effects, and calibration errors, negative feedback provides benefits. An example of an ingenious feedback control system is shown in Figure 4.16.

The advantage of the closed-loop system is that it reduces the steady-state error resulting from parameter changes and calibration errors. This may be illustrated by

FIGURE 4.16 The DLR German Aerospace Center is developing an advanced robotic hand. The final goal—fully autonomous operation—has not yet been acheived. Currently, the control is accomplished via a telemanipulation system consisting of a lightweight robot with a four-fingered articulated hand mounted on a mobile platform. The hand operator receives stereo video feedback and force feedback. This information is employed in conjunction with a data glove equipped with force feedback and an input device to control the robot. (Used with permission. Credit: DLR Institute of Robotics and Mechatronics.)

an example. Consider a unity feedback system with a process transfer function

$$G(s) = \frac{K}{\tau s + 1}, \tag{4.51}$$

which could represent a thermal control process, a voltage regulator, or a water-level control process. For a specific setting of the desired input variable, which may be represented by the normalized unit step input function, we have $R(s) = 1/s$. Then the steady-state error of the open-loop system is, as in Equation (4.49),

$$e_0(\infty) = 1 - G(0) = 1 - K \tag{4.52}$$

when a consistent set of dimensional units is utilized for $R(s)$ and K. The error for the closed-loop system is

$$E_c(s) = R(s) - T(s)R(s)$$

where $T(s) = G_c(s)G(s)/(1 + G_c(s)G(s))$. The steady-state error is

$$e_c(\infty) = \lim_{s \to 0} s\{1 - T(s)\}\frac{1}{s} = 1 - T(0).$$

When $G_c(s) = 1/(\tau_1 s + 1)$, we obtain $G_c(0) = 1$ and $G(0) = K$. Then we have

$$e_c(\infty) = 1 - \frac{K}{1 + K} = \frac{1}{1 + K}. \tag{4.53}$$

For the open-loop system, we would calibrate the system so that $K = 1$ and the steady-state error is zero. For the closed-loop system, we would set a large gain K. If $K = 100$, the closed-loop system steady-state error is $e_c(\infty) = 1/101$.

If the calibration of the gain setting drifts or changes by $\Delta K/K = 0.1$ (a 10% change), the open-loop steady-state error is $\Delta e_o(\infty) = 0.1$. Then the percent change from the calibrated setting is

$$\frac{\Delta e_o(\infty)}{|r(t)|} = \frac{0.10}{1}, \tag{4.54}$$

or 10%. By contrast, the steady-state error of the closed-loop system, with $\Delta K/K = 0.1$, is $e_c(\infty) = 1/91$ if the gain decreases. Thus, the change is

$$\Delta e_c(\infty) = \frac{1}{101} - \frac{1}{91}, \tag{4.55}$$

and the relative change is

$$\frac{\Delta e_c(\infty)}{|r(t)|} = 0.0011, \tag{4.56}$$

or 0.11%. This is a significant improvement, since the closed-loop relative change is two orders of magnitude lower than that of the open-loop system.

4.7 THE COST OF FEEDBACK

Adding feedback to a control system results in the advantages outlined in the previous sections. Naturally, however, these advantages have an attendant cost. The first cost of feedback is an increased number of **components** and **complexity** in the system. To add the feedback, it is necessary to consider several feedback components; the measurement component (sensor) is the key one. The sensor is often the most expensive component in a control system. Furthermore, the sensor introduces noise and inaccuracies into the system.

The second cost of feedback is the **loss of gain**. For example, in a single-loop system, the open-loop gain is $G_c(s)G(s)$ and is reduced to $G_c(s)G(s)/(1 + G_c(s)G(s))$ in a unity negative feedback system. The closed-loop gain is smaller by a factor of $1/(1 + G_c(s)G(s))$, which is exactly the factor that reduces the sensitivity of the system to parameter variations and disturbances. Usually, we have extra open-loop gain available, and we are more than willing to trade it for increased control of the system response.

We should note that it is the gain of the input–output transmittance that is reduced. The control system does retain the substantial power gain of a power amplifier and actuator, which is fully utilized in the closed-loop system.

The final cost of feedback is the introduction of the possibility of **instability**. Whereas the open-loop system is stable, the closed-loop system may not be always stable. The question of the stability of a closed-loop system is deferred until Chapter 6, where it can be treated more completely.

The addition of feedback to dynamic systems causes more challenges for the designer. However, for most cases, the advantages far outweigh the disadvantages, and a feedback system is desirable. Therefore, it is necessary to consider the additional complexity and the problem of stability when designing a control system.

Clearly, we want the output of the system, $Y(s)$, to equal the input, $R(s)$. However, upon reflection, we might ask, Why not simply set the transfer function $G(s) = Y(s)/R(s)$ equal to 1? (See Figure 4.2, assuming $T_d(s) = 0$.) The answer to this question becomes apparent once we recall that the process (or plant) $G(s)$ was necessary to provide the desired output; that is, the transfer function $G(s)$ represents a real process and possesses dynamics that cannot be neglected. If we set

$G(s)$ equal to 1, we imply that the output is directly connected to the input. We must recall that a specific output (such as temperature, shaft rotation, or engine speed), is desired, whereas the input can be a potentiometer setting or a voltage. The process $G(s)$ is necessary to provide the physical process between $R(s)$ and $Y(s)$. Therefore, a transfer function $G(s) = 1$ is unrealizable, and we must settle for a practical transfer function.

4.8 DESIGN EXAMPLES

In this section we present three illustrative examples: the English Channel boring machine, the Mars rover, and a blood pressure control problem during anesthesia. The English Channel boring machine example focuses on the closed-loop system response to disturbances. The Mars rover example highlights the advantages of closed-loop feedback control in decreasing system sensitivity to plant changes. The final example on blood pressure control is a more in-depth look at the control design problem. Since patient models in the form of transfer functions are difficult to obtain from basic biological and physical principles, a different approach using measured data is discussed. The positive impact of closed-loop feedback control is illustrated in the context of design.

EXAMPLE 4.2   **English Channel boring machines**

The construction of the tunnel under the English Channel from France to Great Britain began in December 1987. The first connection of the boring tunnels from each country was achieved in November 1990. The tunnel is 23.5 miles long and is bored 200 feet below sea level. The tunnel, completed in 1992 at a total cost of $14 billion, accommodates 50 train trips daily. This construction is a critical link between Europe and Great Britain, making it possible for a train to travel from London to Paris in three hours.

The machines, operating from both ends of the channel, bored toward the middle. To link up accurately in the middle of the channel, a laser guidance system kept the machines precisely aligned. A model of the boring machine control is shown in Figure 4.17, where $Y(s)$ is the actual angle of direction of travel of the boring machine and $R(s)$ is the desired angle. The effect of load on the machine is represented by the disturbance, $T_d(s)$.

The design objective is to select the gain K so that the response to input angle changes is desirable while we maintain minimal error due to the disturbance. The

FIGURE 4.17
A block diagram model of a boring machine control system.

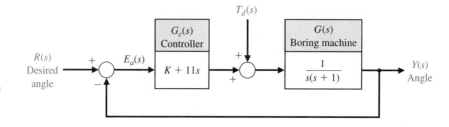

output due to the two inputs is

$$Y(s) = \frac{K + 11s}{s^2 + 12s + K} R(s) + \frac{1}{s^2 + 12s + K} T_d(s). \qquad (4.57)$$

Thus, to reduce the effect of the disturbance, we wish to set the gain greater than 10. When we select $K = 100$ and let the disturbance be zero, we have the step response for a unit step input $r(t)$, as shown in Figure 4.18(a). When the input $r(t) = 0$ and we determine the response to the unit step disturbance, we obtain $y(t)$ as shown in Figure 4.18(b). The effect of the disturbance is quite small. If we set the gain K equal to 20, we obtain the responses of $y(t)$ due to a unit step input $r(t)$ and disturbance $T_d(t)$ displayed

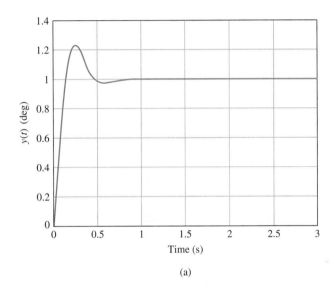

(a)

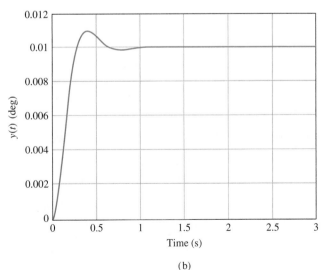

FIGURE 4.18
The response $y(t)$ to (a) a unit input step $r(t)$ and (b) a unit disturbance step input with $T_d(s) = 1/s$ for $K = 100$.

(b)

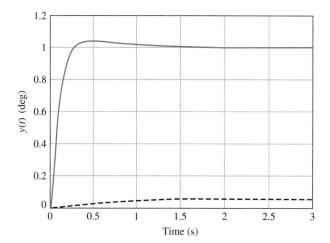

FIGURE 4.19
The response $y(t)$ for a unit step input (solid line) and for a unit step disturbance (dashed line) for $K = 20$.

together in Figure 4.19. Since the overshoot of the response is small (less than 4%) and the steady state is attained in 2 seconds, we would prefer that $K = 20$. The results are summarized in Table 4.1.

The steady-state error of the system to a unit step input $R(s) = 1/s$ is

$$\lim_{t \to \infty} e(t) = \lim_{s \to 0} s \frac{1}{1 + \dfrac{K + 11s}{s(s + 1)}\left(\dfrac{1}{s}\right)} = 0. \tag{4.58}$$

The steady-state value of $y(t)$ when the disturbance is a unit step, $T_d(s) = 1/s$, and the desired value is $r(t) = 0$ is

$$\lim_{t \to \infty} y(t) = \lim_{s \to 0}\left[\frac{1}{s(s + 12) + K}\right] = \frac{1}{K}. \tag{4.59}$$

Thus, the steady-state value is 0.01 and 0.05 for $K = 100$ and 20, respectively.

Finally, we examine the sensitivity of the system to a change in the process $G(s)$ using Equation (4.12). Then

$$S_G^T = \frac{s(s + 1)}{s(s + 12) + K}. \tag{4.60}$$

Table 4.1   Response of the Boring System for Two Gains

Gain K	Overshoot of response to $r(t)$ = step	Time for response to $r(t)$ = step to reach steady state (2% criterion)	Steady-state response $y(t)$ for unit step disturbance with $r(t)$ = 0	Steady-state error of response to $r(t)$ = step with zero disturbance
100	22%	0.7s	0.01	0
20	4%	1.0s	0.05	0

For low frequencies ($|s| < 1$), the sensitivity can be approximated by

$$S_G^T \simeq \frac{s}{K},$$ (4.61)

where $K \geq 20$. Thus, the sensitivity of the system is reduced by increasing the gain, K. In this case, we choose $K = 20$ for a reasonable design compromise. ■

EXAMPLE 4.3   Mars rover vehicle

The solar-powered Mars rover named *Sojourner* landed on Mars on July 4, 1997, and was deployed on its journey on July 5, 1997. The rover was controlled by operators on Earth using controls on the rover [21, 22]. The Mars rovers, aptly dubbed *Spirit* and *Opportunity*, are known as the twin Mars Exploration Rovers and landed on the planet in 2004. These new rovers differ in size and capability from the *Sojourner* rover. *Sojourner* was about 65 cm (2 ft) long and weighed 10 kg (22 lb), while *Spirit* and *Opportunity* are each 1.6 m (5.2 ft) long and weigh 174 kg (384 lbs). *Sojourner* traveled a total distance of about 100 m during its 12 weeks of activity on Mars. *Spirit* has traveled over 7 km and *Opportunity* has traveled over 10 km. Each vehicle has lasted many times longer than originally planned. The Mars Exploration Rovers are more autonomous; each carries its own telecommunications equipment, camera, and computers, whereas the *Sojourner* housed most of its equipment on the lander left at the base site. The solar-powered Mars rover *Spirit* is shown in Figure 4.20. The vehicle is controlled from Earth by sending it path commands, $r(t)$.

A very simplified model of a rover is depicted in Figure 4.21. The system may be operated without feedback, as shown in Figure 4.21(a), or with feedback, as shown in Figure 4.21(b). The goal is to operate the rover with modest effects from disturbances such as rocks and with low sensitivity to changes in the gain K.

The transfer function for the open-loop system is

$$T_o(s) = \frac{Y(s)}{R(s)} = \frac{K}{s^2 + 4s + 5},$$ (4.62)

FIGURE 4.20
Mars Exploration Rovers are significantly more capable than their predecessor, the Mars Pathfinder *Sojourner*. (Courtesy of NASA.)

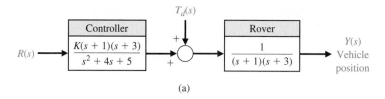

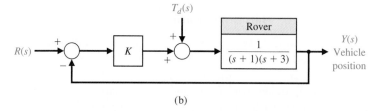

FIGURE 4.21
Control system for
the rover. (a) Open-
loop (without
feedback).
(b) Closed-loop
with feedback.

and the transfer function for the closed-loop system is

$$T_c(s) = \frac{Y(s)}{R(s)} = \frac{K}{s^2 + 4s + 3 + K}.$$ (4.63)

Then, for $K = 2$,

$$T(s) = T_o(s) = T_c(s) = \frac{2}{s^2 + 4s + 5}.$$

Hence, we can compare the sensitivity of the open-loop and closed-loop systems for the same transfer function.

The sensitivity for the open-loop system is

$$S_K^{T_o} = \frac{dT_o}{dK}\frac{K}{T_o} = 1,$$ (4.64)

and the sensitivity for the closed-loop system is

$$S_K^{T_c} = \frac{dT_c}{dK}\frac{K}{T_c} = \frac{s^2 + 4s + 3}{s^2 + 4s + 3 + K}.$$ (4.65)

To examine the effect of the sensitivity at low frequencies, we let $s = j\omega$ to obtain

$$S_K^{T_c} = \frac{(3 - \omega^2) + j4\omega}{(3 + K - \omega^2) + j4\omega}.$$ (4.66)

For $K = 2$, the sensitivity at low frequencies, $\omega < 0.1$, is $|S_K^{T_c}| \simeq 0.6$.

A frequency plot of the magnitude of the sensitivity is shown in Figure 4.22. Note that the sensitivity for low frequencies is

$$|S_K^{T_c}| < 0.8, \quad \text{for} \quad \omega \le 1.$$

The effect of the disturbance can be determined by setting $R(s) = 0$ and letting $T_d(s) = 1/s$. Then, for the open-loop system, we have the steady-state value

$$y(\infty) = \lim_{s \to 0} s\left\{\frac{1}{(s + 1)(s + 3)}\right\}\frac{1}{s} = \frac{1}{3}.$$ (4.67)

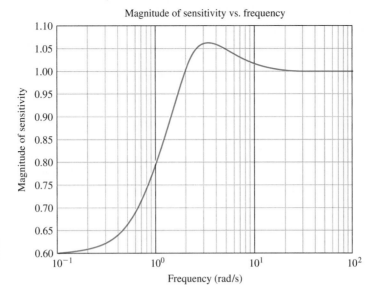

FIGURE 4.22
The magnitude of the sensitivity of the closed-loop system for the Mars rover vehicle.

As shown in Section 4.4, the output of the closed-loop system with a unit step disturbance, $T_d(s) = 1/s$, is

$$y(\infty) = \lim_{s \to 0} s \left\{ \frac{1}{(s^2 + 4s + 3 + K)} \right\} \frac{1}{s} = \frac{1}{3 + K}. \tag{4.68}$$

When $K = 2$, $y(\infty) = 1/5$. Because we seek to minimize the effect of the disturbance, it is clear that a larger value of K would be desirable. An increased value of K, such as $K = 50$, will further reduce the effect of the disturbance as well as reduce the magnitude of the sensitivity (Equation 4.66). However, as we increase K beyond $K = 50$, the transient performance of the system for the ramp input, $r(t)$, begins to deteriorate. ∎

EXAMPLE 4.4   Blood pressure control during anesthesia

The objectives of anethesia are to eliminate pain, awareness, and natural reflexes so that surgery can be conducted safely. Before about 150 years ago, alcohol, opium and cannabis were used to achieve these goals, but they proved inadequate [24]. Pain relief was insufficient both in magnitude and duration; too little pain medication and the patient felt great pain, too much medication and the patient died or became comatose. In the 1850s ether was used successfully in the United States in tooth extractions, and shortly thereafter other means of achieving unconsciousness safely were developed, including the use of chloroform and nitrous oxide.

In a modern operating room, the depth of anesthesia is the responsibility of the anesthetist. Many vital parameters, such as blood pressure, heart rate, temperature, blood oxygenation, and exhaled carbon dioxide, are controlled within acceptable bounds by the anesthetist. Of course, to ensure patient safety, adequate anesthesia must be maintained during the entire surgical procedure. Any assistance that the anesthetist can obtain automatically will increase the safety margins by freeing the anesthetist to attend to other functions not easily automated. This is an example of human computer interaction for the overall control of a process. Clearly, patient safety is the ultimate objective. Our control goal then is to develop an automated system to regulate the depth of anesthesia. This function is amenable to automatic control and in fact is in routine use in clinical applications [25, 26].

We consider how to measure the depth of anesthesia. Many anesthetists regard mean arterial pressure (MAP) as the most reliable measure of the depth of anesthesia [27]. The level of the MAP serves as a guide for the delivery of inhaled anesthesia. Based on clinical experience and the procedures followed by the anesthetist, we determine that the variable to be controlled is the mean arterial pressure.

The elements of the control system design process emphasized in this example are illustrated in Figure 4.23. From the control system design perspective, the control goal can be stated in more concrete terms:

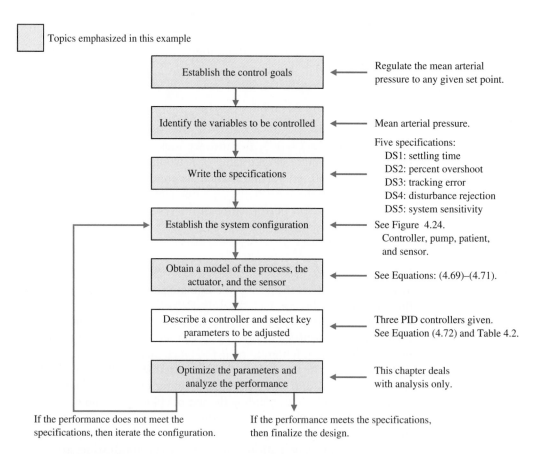

FIGURE 4.23   Elements of the control system design process emphasized in the blood pressure control example.

Control Goal

Regulate the mean arterial pressure to any desired set-point and maintain the prescribed set-point in the presence of unwanted disturbances.

Associated with the stated control goal, we identify the variable to be controlled:

Variable to Be Controlled

Mean arterial pressure (MAP).

Because it is our desire to develop a system that will be used in clinical applications, it is essential to establish realistic design specifications. In general terms the control system should have minimal complexity while satisfying the control specifications. Minimal complexity translates into increased system reliability and decreased cost.

The closed-loop system should respond rapidly and smoothly to changes in the MAP set-point (made by the anesthetist) without excessive overshoot. The closed-loop system should minimize the effects of unwanted disturbances. There are two important categories of disturbances: surgical disturbances, such as skin incisions and measurement errors, such as calibration errors and random stochastic noise. For example, a skin incision can increase the MAP rapidly by 10 mmHg [27]. Finally, since we want to apply the same control system to many different patients and we cannot (for practical reasons) have a separate model for each patient, we must have a closed-loop system that is insensitive to changes in the process parameters (that is, it meets the specifications for many different people).

Based on clinical experience [25], we can explicitly state the control specifications as follows:

Control Design Specifications

DS1   Settling time less than 20 minutes for a 10% step change from the MAP set-point.

DS2   Percent overshoot less than 15% for a 10% step change from the MAP set-point.

DS3   Zero steady-state tracking error to a step change from the MAP set-point.

DS4   Zero steady-state error to a step surgical disturbance input (of magnitude $|d(t)| \leq 50$) with a maximum response less than $\pm 5\%$ of the MAP set-point.

DS5   Minimum sensitivity to process parameter changes.

We cover the notion of percent overshoot (DS1) and settling time (DS2) more thoroughly in Chapter 5. They fall more naturally in the category of system performance. The remaining three design specifications, DS3–DS5, covering steady-state tracking errors (DS3), disturbance rejection (DS4), and system sensitivity to parameter changes (DS5) are the main topics of this chapter. The last specification, DS5, is somewhat vague; however, this is a characteristic of many real-world specifications. In the system configuration, Figure 4.24, we identify the major system elements as the controller, anesthesia pump/vaporizer, sensor, and patient.

The system input $R(s)$ is the desired mean arterial pressure change, and the output $Y(s)$ is the actual pressure change. The difference between the desired and the measured blood pressure change forms a signal used by the controller to determine value settings to the pump/vaporizer that delivers anesthesia vapor to the patient.

The model of the pump/vaporizer depends directly on the mechanical design. We will assume a simple pump/vaporizer, where the rate of change of the output

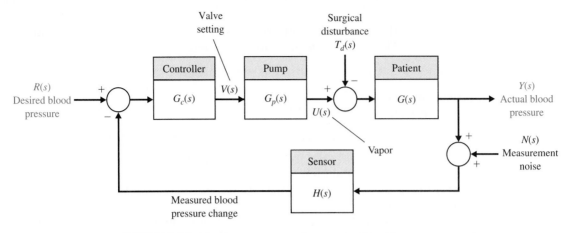

FIGURE 4.24   Blood pressure control system configuration.

vapor is equal to the input valve setting, or

$$\dot{u}(t) = v(t).$$

The transfer function of the pump is thus given by

$$G_p(s) = \frac{U(s)}{V(s)} = \frac{1}{s}. \qquad (4.69)$$

This is equivalent to saying that, from an input/output perspective, the pump has the impulse response

$$h(t) = 1 \quad t \geq 0.$$

Developing an accurate model of a patient is much more involved. Because the physiological systems in the patient (especially in a sick patient) are not easily modeled, a modeling procedure based on knowledge of the underlying physical processes is not practical. Even if such a model could be developed, it would, in general, be a nonlinear, time-varying, multi-input, multi-output model. This type of model is not directly applicable here in our linear, time-invariant, single-input, single-output system setting.

On the other hand, if we view the patient as a system and take an input/output perspective, we can use the familiar concept of an impulse response. Then if we restrict ourselves to small changes in blood pressure from a given set-point (such as 100 mmHg), we might make the case that in a small region around the set-point the patient behaves in a linear time-invariant fashion. This approach fits well into our requirement to maintain the blood pressure around a given set-point (or baseline). The impulse response approach to modeling the patient response to anesthesia has been used successfully in the past [28].

Suppose that we take a black-box approach and obtain the impulse response in Figure 4.25 for a hypothetical patient. Notice that the impulse response initially has a time delay. This reflects the fact that it takes a finite amount of time for the patient MAP to respond to the infusion of anesthesia vapor. We ignore the time-delay in

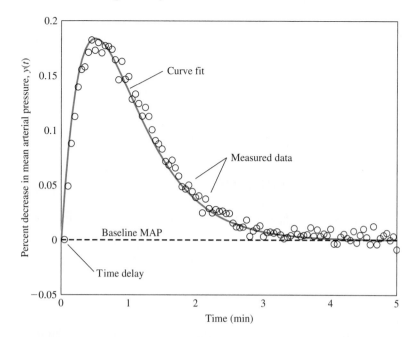

FIGURE 4.25
Mean arterial
pressure (MAP)
impulse response
for a hypothetical
patient.

our design and analysis, but we do so with caution. In subsequent chapters we will learn to handle time delays. We keep in mind that the delay does exist and should be considered in the analysis at some point.

A reasonable fit of the data shown in Figure 4.25 is given by

$$y(t) = te^{-pt} \qquad t \geq 0,$$

where $p = 2$ and time (t) is measured in minutes. Different patients are associated with different values of the parameter p. The corresponding transfer function is

$$G(s) = \frac{1}{(s + p)^2}. \tag{4.70}$$

For the sensor we assume a perfect noise-free measurement and

$$H(s) = 1. \tag{4.71}$$

Therefore, we have a unity feedback system.

A good controller for this application is a proportional-integral-derivative (PID) controller:

$$G_c(s) = K_P + sK_D + \frac{K_I}{s} = \frac{K_D s^2 + K_P s + K_I}{s}, \tag{4.72}$$

where K_P, K_D, and K_I are the controller gains to be determined to satisfy all design specifications. The selected key parameters are as follows:

Select Key Tuning Parameters
        Controller gains K_P, K_D, and K_I.

We begin the analysis by considering the steady-state errors. The tracking error (shown in Figure 4.24 with $T_d(s) = 0$ and $N(s) = 0$) is

$$E(s) = R(s) - Y(s) = \frac{1}{1 + G_c(s)G_p(s)G(s)} R(s),$$

or

$$E(s) = \frac{s^4 + 2ps^3 + p^2 s^2}{s^4 + 2ps^3 + (p^2 + K_D)s^2 + K_P s + K_I} R(s).$$

Using the final-value theorem, we determine that the steady-state tracking error is

$$\lim_{s \to 0} sE(s) = \lim_{s \to 0} \frac{R_0(s^4 + 2ps^3 + p^2 s^2)}{s^4 + 2ps^3 + (p^2 + K_D)s^2 + K_P s + K_I} = 0,$$

where $R(s) = R_0/s$ is a step input of magnitude R_0. Therefore,

$$\lim_{t \to \infty} e(t) = 0.$$

With a PID controller, we expect a zero steady-state tracking error (to a step input) for any nonzero values of K_P, K_D, and K_I. As we will see in Chapter 5, the integral term, K_I/s, in the PID controller is the reason that the steady-state error to a unit step is zero. Thus design specification DS3 is satisfied.

When considering the effect of a step disturbance input, we let $R(s) = 0$ and $N(s) = 0$. We want the steady-state output $Y(s)$ to be zero for a step disturbance. The transfer function from the disturbance $T_d(s)$ to the output $Y(s)$ is

$$Y(s) = \frac{-G(s)}{1 + G_c(s)G_p(s)G(s)} T_d(s)$$

$$= \frac{-s^2}{s^4 + 2ps^3 + (p^2 + K_D)s^2 + K_P s + K_I} T_d(s).$$

When

$$T_d(s) = \frac{D_0}{s},$$

we find that

$$\lim_{s \to 0} sY(s) = \lim_{s \to 0} \frac{-D_0 s^2}{s^4 + 2ps^3 + (p^2 + K_D)s^2 + K_P s + K_I} = 0.$$

Therefore,

$$\lim_{t \to \infty} y(t) = 0.$$

Thus a step disturbance of magnitude D_0 will produce no output in the steady-state, as desired.

The sensitivity of the closed-loop transfer function to changes in p is given by

$$S_p^T = S_G^T S_p^G.$$

We compute S_p^G as follows:

$$S_p^G = \frac{\partial G(s)}{\partial p} \cdot \frac{p}{G(s)} = \frac{-2p}{s + p},$$

and

$$S_G^T = \frac{1}{1 + G_c(s)G_p(s)G(s)} = \frac{s^2(s + p)^2}{s^4 + 2ps^3 + (p^2 + K_D)s^2 + K_Ps + K_I}.$$

Therefore,

$$S_p^T = S_G^T S_p^G = -\frac{2p(s + p)s^2}{s^4 + 2ps^3 + (p^2 + K_D)s^2 + K_Ps + K_I}. \tag{4.73}$$

We must evaluate the sensitivity function S_p^T, at various values of frequency. For low frequencies we can approximate the system sensitivity S_p^T by

$$S_p^T \approx \frac{2p^2s^2}{K_I}.$$

So at low frequencies and for a given p we can reduce the system sensitivity to variations in p by increasing the PID gain, K_I. Suppose that three PID gain sets have been proposed, as shown in Table 4.2. With $p = 2$ and the PID gains given as the cases 1–3 in Table 4.2, we can plot the magnitude of the sensitivity S_p^T as a function of frequency for each PID controller. The result is shown in Figure 4.26. We see that by using the PID 3 controller with the gains $K_P = 6$, $K_D = 4$, and $K_I = 4$, we have the smallest system sensitivity (at low frequencies) to changes in the process parameter, p. PID 3 is the controller with the largest gain K_I. As the frequency increases we see in Figure 4.26 that the sensitivity increases, and that PID 3 has the highest peak sensitivity.

Now we consider the transient response. Suppose we want to reduce the MAP by a 10% step change. The associated input is

$$R(s) = \frac{R_0}{s} = \frac{10}{s}.$$

The step response for each PID controller is shown in Figure 4.27. PID 1 and PID 2 meet the settling time and overshoot specifications; however PID 3 has excessive overshoot. The overshoot is the amount the system output exceeds the desired steady-state response. In this case the desired steady-state response is a 10% decrease in the baseline MAP. When a 15% overshoot is realized, the MAP is decreased by

Table 4.2   PID Controller Gains and System Performance Results

PID	K_P	K_D	K_I	Input response overshoot (%)	Settling time (min)	Disturbance response overshoot (%)
1	6	4	1	14.0	10.9	5.25
2	5	7	2	14.2	8.7	4.39
3	6	4	4	39.7	11.1	5.16

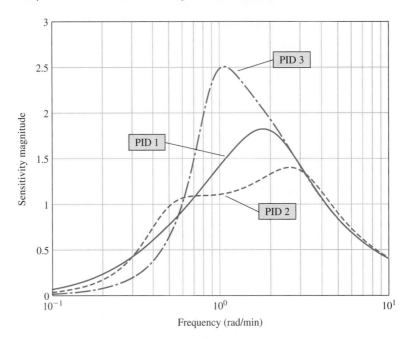

FIGURE 4.26
System sensitivity
to variations in the
parameter p.

11.5%, as illustrated in Figure 4.27. The settling time is the time required for the system output to settle within a certain percentage (for example, 2%) of the desired steady-state output amplitude. We cover the notions of overshoot and settling time more thoroughly in Chapter 5. The overshoot and settling times are summarized in Table 4.2.

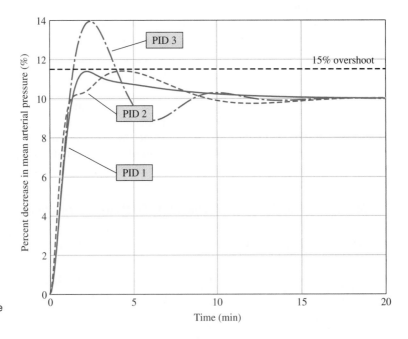

FIGURE 4.27
Mean arterial
pressure (MAP)
step input response
with $R(s) = 10/s$.

We conclude the analysis by considering the disturbance response. From previous analysis we know that the transfer function from the disturbance input $T_d(s)$ to the output $Y(s)$ is

$$Y(s) = \frac{-G(s)}{1 + G_c(s)G_p(s)G(s)}T_d(s)$$

$$= \frac{-s^2}{s^4 + 2ps^3 + (p^2 + K_D)s^2 + K_Ps + K_I}T_d(s).$$

To investigate design specification DS4, we compute the disturbance step response with

$$T_d(s) = \frac{D_0}{s} = \frac{50}{s}.$$

This is the maximum magnitude disturbance ($|T_d(t)| = D_0 = 50$). Since any step disturbance of smaller magnitude (that is, $|T_d(t)| = D_0 < 50$) will result in a smaller maximum output response, we need only to consider the maximum magnitude step disturbance input when determining whether design specification DS4 is satisfied.

The unit step disturbance for each PID controller is shown in Figure 4.28. Controller PID 2 meets design specification DS4 with a maximum response less than ±5% of the MAP set-point, while controllers PID 1 and 3 nearly meet the specification. The peak output values for each controller are summarized in Table 4.2.

In summary, given the three PID controllers, we would select PID 2 as the controller of choice. It meets all the design specifications while providing a reasonable insensitivity to changes in the plant parameter. ∎

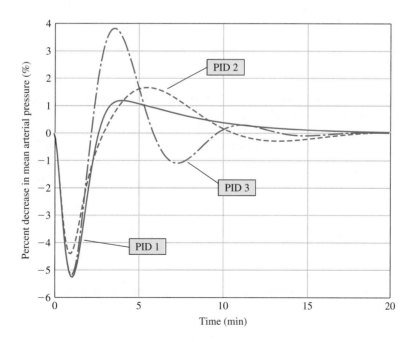

FIGURE 4.28
Mean arterial pressure (MAP) disturbance step response.

4.9 CONTROL SYSTEM CHARACTERISTICS USING CONTROL DESIGN SOFTWARE

In this section, the advantages of feedback will be illustrated with two examples. In the first example, we will introduce feedback control to a speed tachometer system in an effort to reject disturbances. The tachometer speed control system example can be found in Section 4.5. The reduction in system sensitivity to process variations, adjustment of the transient response, and reduction in steady-state error will be demonstrated using the English Channel boring machine example of Section 4.8.

EXAMPLE 4.5   **Speed control system**

The open-loop block diagram description of the armature-controlled DC motor with a load torque disturbance $T_d(s)$ is shown in Figure 4.7. The values for the various parameters (taken from Figure 4.7) are given in Table 4.3. We have two inputs to our system, $V_a(s)$ and $T_d(s)$. Relying on the principle of superposition, which applies to our linear system, we consider each input separately. To investigate the effects of disturbances on the system, we let $V_a(s) = 0$ and consider only the disturbance $T_d(s)$. Conversely, to investigate the response of the system to a reference input, we let $T_d(s) = 0$ and consider only the input $V_a(s)$.

The closed-loop speed tachometer control system block diagram is shown in Figure 4.9. The values for K_a and K_t are given in Table 4.3.

If our system displays good disturbance rejection, then we expect the disturbance $T_d(s)$ to have a small effect on the output $\omega(s)$. Consider the open-loop system in Figure 4.11 first. We can compute the transfer function from $T_d(s)$ to $\omega(s)$ and evaluate the output response to a unit step disturbance (that is, $T_d(s) = 1/s$). The time response to a unit step disturbance is shown in Figure 4.29(a). The script shown in Figure 4.29(b) is used to analyze the open-loop speed tachometer system.

The open-loop transfer function (from Equation (4.26)) is

$$\frac{\omega(s)}{T_d(s)} = \frac{-1}{2s + 1.5} = \text{sys\_o},$$

where sys_o represents the open-loop transfer function in the script. Since our desired value of $\omega(t)$ is zero (remember that $V_a(s) = 0$), the steady-state error is just the final value of $\omega(t)$, which we denote by $\omega_o(t)$ to indicate open-loop. The steady-state error, shown on the plot in Figure 4.29(a), is approximately the value of the speed when $t = 7$ seconds. We can obtain an approximate value of the steady-state error by looking at the last value in the output vector $\mathbf{y}_o$, which we computed in the process of generating the plot in Figure 4.29(a). The approximate steady-state value of ω_o is

$$\omega_o(\infty) \approx \omega_o(7) = -0.66 \text{ rad/s}.$$

The plot verifies that we have reached steady state.

Table 4.3   Tachometer Control System Parameters

R_a	K_m	J	b	K_b	K_a	K_t
$1\ \Omega$	$10\ \text{Nm/A}$	$2\ \text{kg m}^2$	$0.5\ \text{Nm s}$	$0.1\ \text{Vs}$	54	$1\ \text{Vs}$

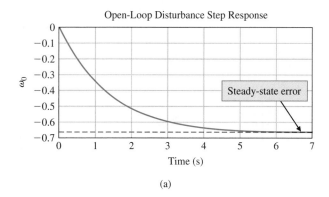

FIGURE 4.29
Analysis of the
open-loop speed
control system.
(a) Response.
(b) m-file script.

In a similar fashion, we begin the closed-loop system analysis by computing the closed-loop transfer function from $T_d(s)$ to $\omega(s)$ and then generating the time-response of $\omega(t)$ to a unit step disturbance input. The output response and the script cltach.m are shown in Figure 4.30. The closed-loop transfer function from the disturbance input (from Equation (4.30)) is

$$\frac{\omega(s)}{T_d(s)} = \frac{-1}{2s + 541.5} = \text{sys\_c}.$$

As before, the steady-state error is just the final value of $\omega(t)$, which we denote by $\omega_c(t)$ to indicate that it is a closed-loop. The steady-state error is shown on the plot in Figure 4.30(a). We can obtain an approximate value of the steady-state error by looking at the last value in the output vector $\mathbf{y}_c$, which we computed in the process of generating the plot in Figure 4.30(a). The approximate steady-state value of ω is

$$\omega_c(\infty) \approx \omega_c(0.02) = -0.002 \text{ rad/s}.$$

We generally expect that $\omega_c(\infty)/\omega_o(\infty) < 0.02$. In this example, the ratio of closed-loop to open-loop steady-state speed output due to a unit step disturbance input is

$$\frac{\omega_c(\infty)}{\omega_o(\infty)} = 0.003.$$

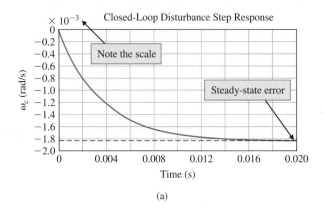

(a)

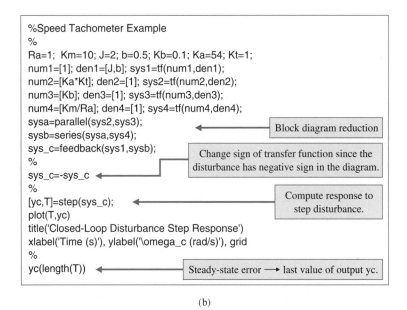

(b)

FIGURE 4.30
Analysis of the
closed-loop speed
control system.
(a) Response.
(b) m-file script.

We have achieved a remarkable improvement in disturbance rejection. It is clear that the addition of the negative feedback loop reduced the effect of the disturbance on the output. This demonstrates the disturbance rejection property of closed-loop feedback systems. ∎

EXAMPLE 4.6   **English Channel boring machines**

The block diagram description of the English Channel boring machines is shown in Figure 4.17. The transfer function of the output due to the two inputs is (Equation (4.57))

$$Y(s) = \frac{K + 11s}{s^2 + 12s + K} R(s) + \frac{1}{s^2 + 12s + K} T_d(s).$$

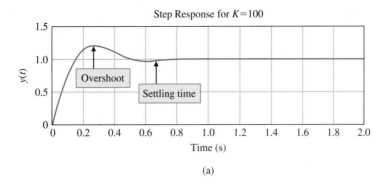

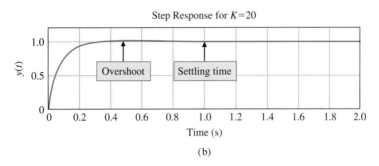

FIGURE 4.31
The response to a
step input when
(a) $K = 100$ and
(b) $K = 20$.
(c) m-file script.

The effect of the control gain, K, on the transient response is shown in Figure 4.31 along with the script used to generate the plots. Comparing the two plots in parts (a) and (b), it is apparent that decreasing K decreases the overshoot. Although it is not as obvious from the plots in Figure 4.31, it is also true that decreasing K increases the settling time. This can be verified by taking a closer look at the data used to generate the plots. This example demonstrates how the transient response can be altered by feedback control gain, K. Based on our analysis thus far, we would prefer to use $K = 20$. Other considerations must be taken into account before we can establish the final design.

Before making the final choice of K, it is important to consider the system response to a unit step disturbance, as shown in Figure 4.32. We see that increasing K reduces the

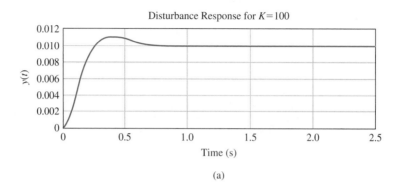

(a)

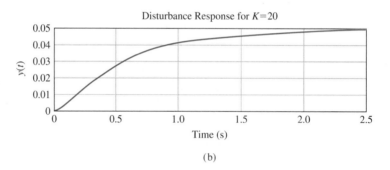

(b)

```
% Response to a Disturbance Td(s)=1/s for K=20 and K=100
%
numg=[1]; deng=[1 1 0];
sysg=tf(numg,deng);
K1=100; K2=20;
num1=[11 K1]; num2=[11 K2]; den=[0 1];
sys1=tf(num1,den); sys2=tf(num2,den);
%
sysa=feedback(sysg,sys1); sysa=minreal(sysa);          Closed-loop
sysb=feedback(sysg,sys2); sysb=minreal(sysb);        transfer functions.
%
t=[0:0.01:2.5];
[y1,t]=step(sysa,t); [y2,t]=step(sysb,t);
subplot(211),plot(t,y1), title('Disturbance Response for K=100')
xlabel('Time (s)'),ylabel('y(t)'), grid
subplot(212),plot(t,y2), title('Disturbance Response for K=20')
xlabel('Time (s)'),ylabel('y(t)'), grid          Create subplots with
                                                   x and y labels.
```

FIGURE 4.32
The response to a step disturbance when (a) K = 100 and (b) K = 20. (c) m-file script.

(c)

Table 4.4   Response of the Boring Machine Control System for $K = 20$ and $K = 100$

	$K = 20$	$K = 100$
Step Response		
Overshoot	4%	22%
T_s	1.0 s	0.7 s
Disturbance Response		
e_{ss}	5%	1%

steady-state response of $y(t)$ to the step disturbance. The steady-state value of $y(t)$ is 0.05 and 0.01 for $K = 20$ and 100, respectively. The steady-state errors, percent overshoot, and settling times (2% criteria) are summarized in Table 4.4. The steady-state values are predicted from the final-value theorem for a unit disturbance input as follows:

$$\lim_{t \to \infty} y(t) = \lim_{s \to 0} s\left\{ \frac{1}{s(s + 12) + K} \right\} \frac{1}{s} = \frac{1}{K}.$$

If our only design consideration is disturbance rejection, we would prefer to use $K = 100$.

We have just experienced a very common trade-off situation in control system design. In this particular example, increasing K leads to better disturbance rejection, whereas decreasing K leads to better performance (that is, less overshoot). The final decision on how to choose K rests with the designer. Although control design software can certainly assist in the control system design, it cannot replace the engineer's decision-making capability and intuition.

The final step in the analysis is to look at the system sensitivity to changes in the process. The system sensitivity is given by (Equation 4.60),

$$S_G^T = \frac{s(s + 1)}{s(s + 12) + K}.$$

We can compute the values of $S_G^T(s)$ for different values of s and generate a plot of the system sensitivity. For low frequencies, we can approximate the system sensitivity by

$$S_G^T \simeq \frac{s}{K}.$$

Increasing the gain K reduces the system sensitivity. The system sensitivity plots when $s = j\omega$ are shown in Figure 4.33 for $K = 20$. ∎

4.10   SEQUENTIAL DESIGN EXAMPLE: DISK DRIVE READ SYSTEM

The design of a disk drive system is an exercise in compromise and optimization. The disk drive must accurately position the head reader while being able to reduce the effects of parameter changes and external shocks and vibrations. The mechanical arm and flexure will resonate at frequencies that may be caused by excitations such as a shock to a notebook computer. Disturbances to the operation of the disk drive include

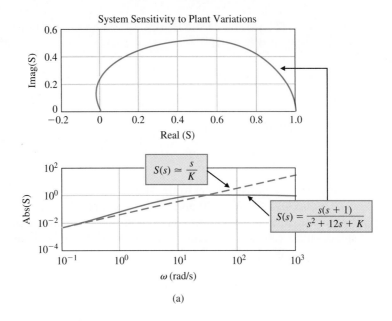

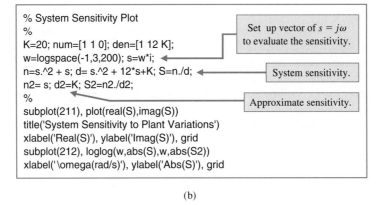

FIGURE 4.33
(a) System
sensitivity to plant
variations (s = jω).
(b) m-file script.

physical shocks, wear or wobble in the spindle bearings, and parameter changes due to component changes. In this section, we will examine the performance of the disk drive system in response to disturbances and changes in system parameters. In addition, we examine the steady-state error of the system for a step command and the transient response as the amplifier gain K_a is adjusted. Thus, in this section, we are carrying out the last two steps of the design process shown in Figure 1.15.

Let us consider the system shown in Figure 4.34. This closed-loop system uses an amplifier with a variable gain as the controller. Using the parameters specified in Table 2.10, we obtain the transfer functions as shown in Figure 4.35. First, we will determine the steady states for a unit step input, $R(s) = 1/s$, when $T_d(s) = 0$. When $H(s) = 1$, we obtain

$$E(s) = R(s) - Y(s) = \frac{1}{1 + K_a G_1(s) G_2(s)} R(s).$$

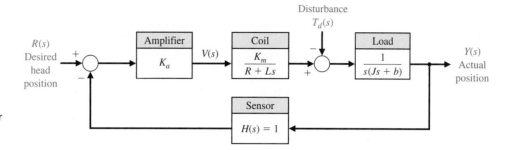

FIGURE 4.34
Control system for disk drive head reader.

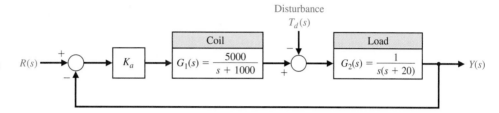

FIGURE 4.35
Disk drive head control system with the typical parameters of Table 2.10.

Therefore,

$$\lim_{t \to \infty} e(t) = \lim_{s \to 0} s \left[\frac{1}{1 + K_a G_1(s) G_2(s)} \right] \frac{1}{s}. \tag{4.74}$$

Then the steady-state error is $e(\infty) = 0$ for a step input. This performance is obtained in spite of changes in the system parameters.

Now let us determine the transient performance of the system as K_a is adjusted. The closed-loop transfer function (with $T_d(s) = 0$) is

$$T(s) = \frac{Y(s)}{R(s)} = \frac{K_a G_1(s) G_2(s)}{1 + K_a G_1(s) G_2(s)}$$

$$= \frac{5000 K_a}{s^3 + 1020 s^2 + 20000 s + 5000 K_a}. \tag{4.75}$$

Using the script shown in Figure 4.36(a), we obtain the response of the system for $K_a = 10$ and $K_a = 80$, shown in Figure 4.36(b). Clearly, the system is faster in responding to the command input when $K_a = 80$, but the response is unacceptably oscillatory.

Now let us determine the effect of the disturbance $T_d(s) = 1/s$ when $R(s) = 0$. We wish to decrease the effect of the disturbance to an insignificant level. Using the system of Figure 4.35, we obtain the response $Y(s)$ for the input $T_d(s)$ when $K_a = 80$ as

$$Y(s) = \frac{G_2(s)}{1 + K_a G_1(s) G_2(s)} T_d(s). \tag{4.76}$$

Using the script shown in Figure 4.37(a), we obtain the response of the system when $K_a = 80$ and $T_d(s) = 1/s$, as shown in Figure 4.37(b). In order to further reduce the

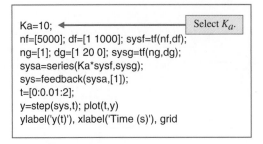

```
Ka=10;          ◄──────────────  Select Kₐ.
nf=[5000]; df=[1 1000]; sysf=tf(nf,df);
ng=[1]; dg=[1 20 0]; sysg=tf(ng,dg);
sysa=series(Ka*sysf,sysg);
sys=feedback(sysa,[1]);
t=[0:0.01:2];
y=step(sys,t); plot(t,y)
ylabel('y(t)'), xlabel('Time (s)'), grid
```

(a)

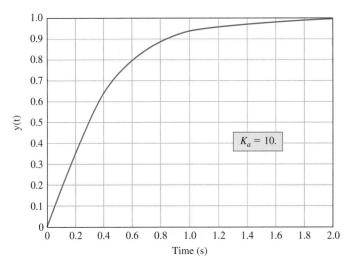

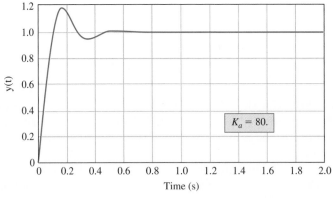

FIGURE 4.36
Closed-loop
response. (a) m-file
script. (b) Step
response for
$K_a = 10$ and
$K_a = 80$.

(b)

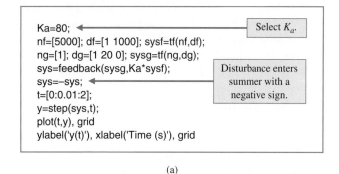

```
Ka=80;                                          Select K_a.
nf=[5000]; df=[1 1000]; sysf=tf(nf,df);
ng=[1]; dg=[1 20 0]; sysg=tf(ng,dg);
sys=feedback(sysg,Ka*sysf);                     Disturbance enters
sys=-sys;                                        summer with a
t=[0:0.01:2];                                    negative sign.
y=step(sys,t);
plot(t,y), grid
ylabel('y(t)'), xlabel('Time (s)'), grid
```

(a)

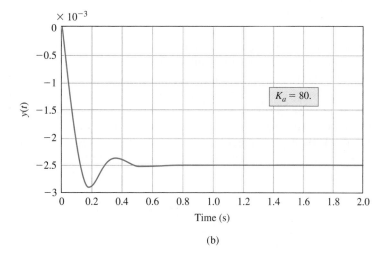

FIGURE 4.37
Disturbance step
response. (a) m-file
script.
(b) Disturbance
response for
$K_a = 80$.

(b)

effect of the disturbance, we would need to raise K_a above 80. However, the response to a step command $r(t) = 1, t > 0$ is unacceptably oscillatory. In the next chapter, we attempt to determine the best value for K_a, given our requirement for a quick, yet nonoscillatory response.

4.11 SUMMARY

The fundamental reasons for using feedback, despite its cost and additional complexity, are as follows:

1. Decrease in the sensitivity of the system to variations in the parameters of the process.
2. Improvement in the rejection of the disturbances.
3. Improvement in the attenuation of measurement noise.
4. Improvement in the reduction of the steady-state error of the system.
5. Ease of control and adjustment of the transient response of the system.

The loop gain $L(s) = G_c(s)G(s)$ plays a fundamental role in control system analysis. Associated with the loop gain we can define the sensitivity and complementary sensitivity functions as

$$S(s) = \frac{1}{1 + L(s)} \text{ and } C(s) = \frac{L(s)}{1 + L(s)},$$

respectively. The tracking error is given by

$$E(s) = S(s)R(s) - S(s)G(s)T_d(s) + C(s)N(s).$$

In order to minimize the tracking error, $E(s)$, we desire to make $S(s)$ and $C(s)$ small. Because the sensitivity and complementary sensitivity functions satisfy the constraint

$$S(s) + C(s) = 1,$$

we are faced with the fundamental trade-off in control system design between rejecting disturbances and reducing sensitivity to plant changes on the one hand, and attenuating measurement noise on the other hand.

The benefits of feedback can be illustrated by considering the system shown in Figure 4.38(a). This system can be considered for several values of gain K. Table 4.5 summarizes the results of the system operated as an open-loop system (with the feedback path disconnected) and for several values of gain, K, with the feedback connected. It is clear that the rise time and sensitivity of the system are reduced as

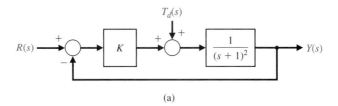

(a)

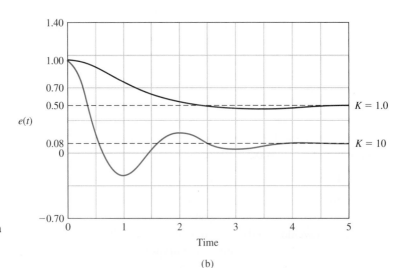

FIGURE 4.38
(a) A single-loop feedback control system. (b) The error response for a unit step disturbance when $R(s) = 0$.

(b)

Table 4.5 System Response of the System Shown in Figure 4.38(a)

	Open Loop*	Closed Loop		
	K = 1	K = 1	K = 8	K = 10
Rise time (s) (10% to 90% of final value)	3.35	1.52	0.45	0.38
Percent overshoot (%)	0	4.31	33	40
Final value of $y(t)$ due to a disturbance, $T_d(s) = 1/s$	1.0	0.50	0.11	0.09
Percent steady-state error for unit step input	0	50%	11%	9%
Percent change in steady-state error due to 10% decrease in K	10%	5.3%	1.2%	0.9%

*Response only when $K = 1$ exactly.

the gain is increased. Also, the feedback system demonstrates excellent reduction of the steady-state error as the gain is increased. Finally, Figure 4.38(b) shows the response for a unit step disturbance (when $R(s) = 0$) and shows how a larger gain will reduce the effect of the disturbance.

Feedback control systems possess many beneficial characteristics. Thus, it is not surprising that there is a multitude of feedback control systems in industry, government, and nature.

EXERCISES

E4.1   A closed-loop system is used to track the sun to obtain maximum power from a photovoltaic array. The tracking system may be represented by Figure 4.3 with $H(s) = 1$ and

$$G(s) = \frac{100}{\tau s + 1},$$

where $\tau = 3$ seconds nominally. (a) Calculate the sensitivity of this system for a small change in τ. (b) Calculate the time constant of the closed-loop system response.

Answers: $S = -3s/(3s + 101); \tau_c = 3/101$ seconds

E4.2   A digital audio system is designed to minimize the effect of disturbances and noise as shown in Figure E4.2. As an approximation, we may represent $G(s) = K_2$. (a) Calculate the sensitivity of the system due to K_2. (b) Calculate the effect of the disturbance

noise $T_d(s)$ on V_o. (c) What value would you select for K_1 to minimize the effect of the disturbance?

E4.3   A robotic arm and camera could be used to pick fruit, as shown in Figure E4.3(a). The camera is used to close the feedback loop to a microcomputer, which controls the arm [8, 9]. The transfer function for the process is

$$G(s) = \frac{K}{(s + 4)^2}.$$

(a) Calculate the expected steady-state error of the gripper for a step command A as a function of K. (b) Name a possible disturbance signal for this system.

Answers: (a) $e_{ss} = \dfrac{A}{1 + K/16}$

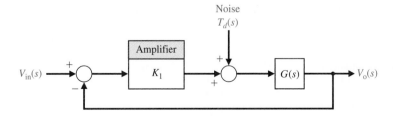

FIGURE E4.2
Digital audio
system.

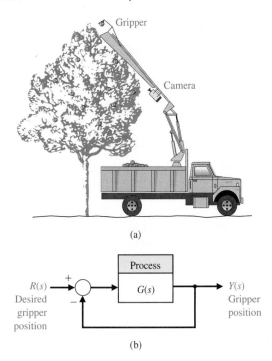

(a)

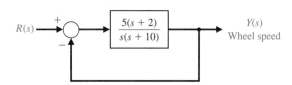

(b)

FIGURE E4.3   Robot fruit picker.

E4.4   A magnetic disk drive requires a motor to position a read/write head over tracks of data on a spinning disk, as shown in Figure E4.4. The motor and head may be represented by the transfer function

$$G(s) = \frac{10}{s(\tau s + 1)},$$

where $\tau = 0.001$ second. The controller takes the difference of the actual and desired positions and generates an error. This error is multiplied by an amplifier K. (a) What is the steady-state position error for a step change in the desired input? (b) Calculate the required K in order to yield a steady-state error of 0.1 mm for a ramp input of 10 cm/s.

Answers: $e_{ss} = 0; K = 100$

E4.5   Most people have experienced an out-of-focus slide projector. A projector with an automatic focus adjusts for variations in slide position and temperature disturbances [11]. Draw the block diagram of an autofocus system, and describe how the system works. An unfocused slide projection is a visual example of steady-state error.

E4.6   Four-wheel drive automobiles are popular in regions where winter road conditions are often slippery due to snow and ice. A four-wheel drive vehicle with antilock brakes uses a sensor to keep each wheel rotating to maintain traction. One system is shown in Figure E4.6. Find the closed-loop response of this system as it attempts to maintain a constant speed of the wheel. Determine the response when $R(s) = A/s$.

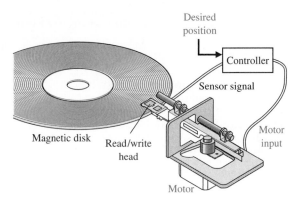

FIGURE E4.4   Disk drive control.

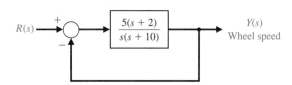

FIGURE E4.6   Four-wheel drive auto.

E4.7   Submersibles with clear plastic hulls have the potential to revolutionize underwater leisure. One small submersible vehicle has a depth-control system as illustrated in Figure E4.7.

(a) Determine the closed-loop transfer function $T(s) = Y(s)/R(s)$.

(b) Determine the sensitivity $S_{K_1}^T$ and S_K^T.

(c) Determine the steady-state error due to a disturbance $T_d(s) = 1/s$.

(d) Calculate the response $y(t)$ for a step input $R(s) = 1/s$ when $K = K_2 = 1$ and $1 < K_1 < 10$. Select K_1 for the fastest response.

E4.8   Consider the feedback control system shown in Figure E4.8. (a) Determine the steady-state error for a step input in terms of the gain, K. (b) Determine the overshoot for the step response for $40 \leq K \leq 400$. (c) Plot the overshoot and the steady-state error versus K.

E4.9   Consider the closed-loop system in Figure E4.9, where

$$G(s) = \frac{K}{s + 10} \quad \text{and} \quad H(s) = \frac{14}{s^2 + 5s + 6}.$$

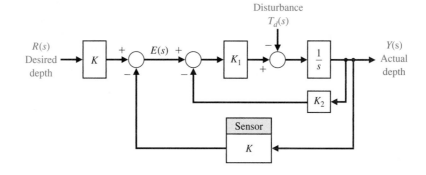

FIGURE E4.7
Depth control
system.

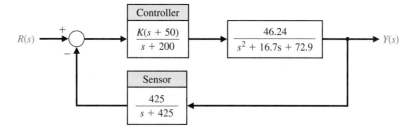

FIGURE E4.8
Feedback control
system.

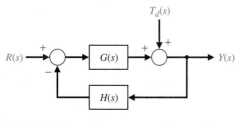

FIGURE E4.9   Closed-loop system with nonunity
feedback.

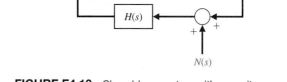

FIGURE E4.10   Closed-loop system with nonunity
feedback and measurement noise.

(a) Compute the transfer function $T(s) = Y(s)/R(s)$.
(b) Define the tracking error to be $E(s) = R(s) - Y(s)$. Compute $E(s)$ and determine the steady-state tracking error due to a unit step input, that is, let $R(s) = 1/s$.
(c) Compute the transfer function $Y(s)/T_d(s)$ and determine the steady-state error of the output due to a unit step disturbance input, that is, let $T_d(s) = 1/s$.
(d) Compute the sensitivity S_K^T.

E4.10   In Figure E4.10, consider the closed-loop system with measurement noise $N(s)$, where

$$G(s) = \frac{100}{s + 100}, \quad G_c(s) = K_1, \quad \text{and} \quad H(s) = \frac{K_2}{s + 5}.$$

In the following analysis, the tracking error is defined to be $E(s) = R(s) - Y(s)$:

(a) Compute the transfer function $T(s) = Y(s)/R(s)$ and determine the steady-state tracking error due

to a unit step response, that is, let $R(s) = 1/s$ and assume that $N(s) = 0$.
(b) Compute the transfer function $Y(s)/N(s)$ and determine the steady-state tracking error due to a unit step disturbance response, that is, let $N(s) = 1/s$ and assume that $R(s) = 0$. Remember, in this case, the desired output is zero.
(c) If the goal is to track the input while rejecting the measurement noise (in other words, while minimizing the effect of $N(s)$ on the output), how would you select the parameters K_1 and K_2?

E4.11   A closed-loop system is used in a high-speed steel rolling mill to control the accuracy of the steel strip thickness. The transfer function for the process shown in Figure E4.11 can be represented as

$$G(s) = \frac{1}{s(s + 25)}.$$

Calculate the sensitivity of the closed-loop transfer function to changes in the controller gain K.

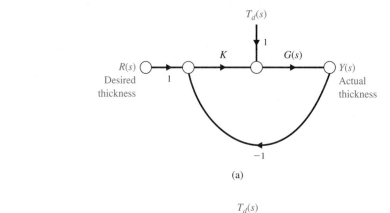

(a)

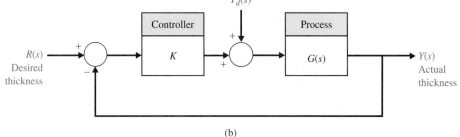

(b)

FIGURE E4.11
Control system for
a steel rolling mill.
(a) Signal flow
graph. (b) Block
diagram.

E4.12 Consider the unity feedback system shown in Figure E4.12. The system has two parameters, the controller gain K and the constant K_1 in the process.

a.   Calculate the sensitivity of the closed-loop transfer function to changes in K_1.

b.   How would you select a value for K to minimize the effects of external disturbances, $T_d(s)$?

E4.13 Reconsider the unity feedback system discussed in E4.12. This time select $K = 120$ and $K_1 = 10$. The closed-loop system is depicted in Figure E4.13.

a.   Calculate the steady-state error of the closed-loop system due to a unit step input, $R(s) = 1/s$, with $T_d(s) = 0$. Recall that the tracking error is defined as $E(s) = R(s) - Y(s)$.

b.   Calculate the steady-state response, $y_{ss} = \lim_{t \to \infty} y(t)$, when $T_d(s) = 1/s$ and $R(s) = 0$.

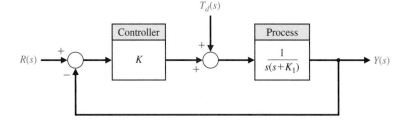

FIGURE E4.12
Closed-loop
feedback system
with two
parameters, K and
K_1.

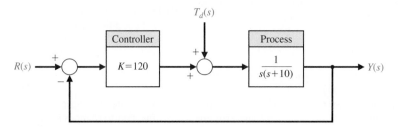

FIGURE E4.13
Closed-loop
feedback system
with $K = 120$ and
$K_1 = 10$.

PROBLEMS

P4.1 The open-loop transfer function of a fluid-flow system can be written as

$$G(s) = \frac{\Delta Q_2(s)}{\Delta Q_1(s)} = \frac{1}{\tau s + 1},$$

where $\tau = RC$, R is a constant equivalent to the resistance offered by the orifice so that $1/R = \frac{1}{2}kH_0^{-1/2}$, and C = the cross-sectional area of the tank. Since $\Delta H = R\,\Delta Q_2$, we have the following for the transfer function relating the head to the input change:

$$G_1(s) = \frac{\Delta H(s)}{\Delta Q_1(s)} = \frac{R}{RCs + 1}.$$

For a closed-loop feedback system, a float-level sensor and valve may be used as shown in Figure P4.1. Assuming the float is a negligible mass, the valve is controlled so that a reduction in the flow rate, ΔQ_1, is proportional to an increase in head, ΔH, or $\Delta Q_1 = -K\Delta H$. Draw a closed-loop flow graph or block diagram. Determine and compare the open-loop and closed-loop systems for (a) sensitivity to changes in the equivalent coefficient R and the feedback coefficient K, (b) the ability to reduce the effects of a disturbance in the level $\Delta H(s)$, and (c) the steady-state error of the level (head) for a step change of the input $\Delta Q_1(s)$.

P4.2 It is important to ensure passenger comfort on ships by stabilizing the ship's oscillations due to waves [13]. Most ship stabilization systems use fins or hydrofoils projecting into the water to generate a stabilization torque on the ship. A simple diagram of a ship stabilization system is shown in Figure P4.2. The rolling motion of a ship can be regarded as an oscillating pendulum with a deviation from the vertical of θ degrees and a typical period of 3 seconds. The transfer function of a typical ship is

$$G(s) = \frac{\omega_n^2}{s^2 + 2\zeta\omega_n s + \omega_n^2},$$

where $\omega_n = 3$ rad/s and $\zeta = 0.20$. With this low damping factor ζ, the oscillations continue for several cycles, and the rolling amplitude can reach $18°$ for the expected amplitude of waves in a normal sea. Determine and compare the open-loop and closed-loop system for (a) sensitivity to changes in the actuator constant K_a and the roll sensor K_1, and (b) the ability to reduce the effects of step disturbances of the waves. Note that the desired roll $\theta_d(s)$ is zero degrees.

P4.3 One of the most important variables that must be controlled in industrial and chemical systems is temperature. A simple representation of a thermal control system is shown in Figure P4.3 [14]. The temperature $\mathcal{T}$ of the process is controlled by the heater with a resistance R. An approximate representation of the dynamic linearly relates the heat loss from the process to the temperature difference $\mathcal{T} - \mathcal{T}_e$. This relation holds if the temperature difference is relatively small and the energy storage of the heater and the vessel walls is negligible. Also, it is assumed that the voltage e_h applied to the heater is proportional to $e_{desired}$ or $e_h = kE_b = k_a E_b e(t)$, where k_a is the constant of the

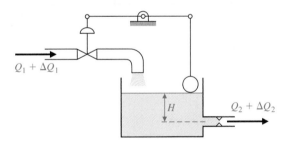

FIGURE P4.1 Tank level control.

$Q_1 + \Delta Q_1$

H

$Q_2 + \Delta Q_2$

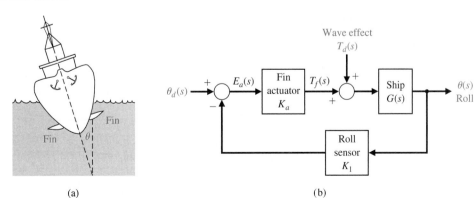

FIGURE P4.2
Ship stabilization system. The effect of the waves is a torque $T_d(s)$ on the ship

(a)

Wave effect
$T_d(s)$

$\theta_d(s)$ $E_a(s)$ Fin actuator K_a $T_f(s)$ Ship $G(s)$ $\theta(s)$ Roll

Roll sensor K_1

(b)

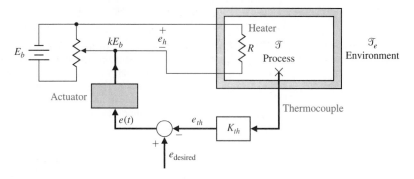

FIGURE P4.3
Temperature control
system.

actuator. Then the linearized open-loop response of
the system is

$$\mathcal{T}(s) = \frac{k_1 k_a E_b}{\tau s + 1} E(s) + \frac{\mathcal{T}_e(s)}{\tau s + 1},$$

where

$\tau = MC/(\rho A)$,
M = mass in tank,
A = surface area of tank,
ρ = heat transfer constant,
C = specific heat constant,
k_1 = a dimensionality constant, and
e_{th} = output voltage of thermocouple.

Determine and compare the open-loop and closed-
loop systems for (a) sensitivity to changes in the constant $K = k_1 k_a E_b$; (b) the ability to reduce the
effects of a step disturbance in the environmental
temperature $\Delta \mathcal{T}_e(s)$; and (c) the steady-state error of
the temperature controller for a step change in the
input, $e_{desired}$.

P4.4 A control system has two forward paths, as shown in
Figure P4.4. (a) Determine the overall transfer function
$T(s) = Y(s)/R(s)$. (b) Calculate the sensitivity, S_G^T,
using Equation (4.16). (c) Does the sensitivity depend
on $U(s)$ or $M(s)$?

P4.5 Large microwave antennas have become increasingly important for radio astronomy and satellite
tracking. A large antenna with a diameter of 60 ft, for
example, is subject to large wind gust torques. A proposed antenna is required to have an error of less
than $0.10°$ in a 35 mph wind. Experiments show that
this wind force exerts a maximum disturbance at the
antenna of 200,000 ft lb at 35 mph, or the equivalent
to 10 volts at the input $T_d(s)$ to the amplidyne. One
problem of driving large antennas is the form of the
system transfer function that possesses a structural
resonance. The antenna servosystem is shown in
Figure P4.5. The transfer function of the antenna,
drive motor, and amplidyne is approximated by

$$G(s) = \frac{\omega_n^2}{s(s^2 + 2\zeta\omega_n s + \omega_n^2)},$$

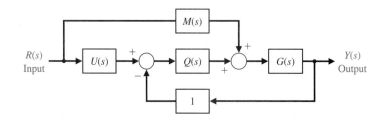

FIGURE P4.4
Two-path system.

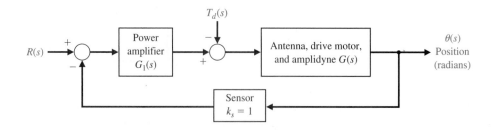

FIGURE P4.5
Antenna control
system.

where $\zeta = 0.707$ and $\omega_n = 15$. The transfer function of the power amplifier is approximately

$$G_1(s) = \frac{k_a}{\tau s + 1},$$

where $\tau = 0.15$ second. (a) Determine the sensitivity of the system to a change of the parameter k_a. (b) The system is subjected to a disturbance $T_d(s) = 10/s$. Determine the required magnitude of k_a in order to maintain the steady-state error of the system less than $0.10°$ when the input $R(s)$ is zero. (c) Determine the error of the system when subjected to a disturbance $T_d(s) = 10/s$ when it is operating as an open-loop system ($k_s = 0$) with $R(s) = 0$.

P4.6 An automatic speed control system will be necessary for passenger cars traveling on the automatic highways of the future. A model of a feedback speed control system for a standard vehicle is shown in Figure P4.6. The load disturbance due to a percent grade $\Delta T_d(s)$ is also shown. The engine gain K_e varies within the range of 10 to 1000 for various models of automobiles. The engine time constant τ_e is 20 seconds. (a) Determine the sensitivity of the system to changes in the engine gain K_e. (b) Determine the effect of the load torque on the speed. (c) Determine the constant percent grade $\Delta T_d(s) = \Delta d/s$ for which the vehicle stalls (velocity $V(s) = 0$) in terms of the gain factors. Note that since the grade is constant, the steady-state solution is sufficient. Assume that

$R(s) = 30/s$ km/hr and that $K_e K_1 \gg 1$. When $K_g/K_1 = 2$, what percent grade Δd would cause the automobile to stall?

P4.7 A robot uses feedback to control the orientation of each joint axis. The load effect varies due to varying load objects and the extended position of the arm. The system will be deflected by the load carried in the gripper. Thus, the system may be represented by Figure P4.7, where the load torque is $T_d(s) = D/s$. Assume $R(s) = 0$ at the index position. (a) What is the effect of $T_d(s)$ on $Y(s)$? (b) Determine the sensitivity of the closed loop to k_2. (c) What is the steady-state error when $R(s) = 1/s$ and $T_d(s) = 0$?

P4.8 Extreme temperature changes result in many failures of electronic circuits [1]. Temperature control feedback systems reduce the change of temperature by using a heater to overcome outdoor low temperatures. A block diagram of one system is shown in Figure P4.8. The effect of a drop in environmental temperature is a step decrease in $T_d(s)$. The actual temperature of the electronic circuit is $Y(s)$. The dynamics of the electronic circuit temperature change are represented by the transfer function.

$$G(s) = \frac{200}{s^2 + 20s + 200}.$$

(a) Determine the sensitivity of the system to K. (b) Obtain the effect of the disturbance $T_d(s)$ on the output $Y(s)$.

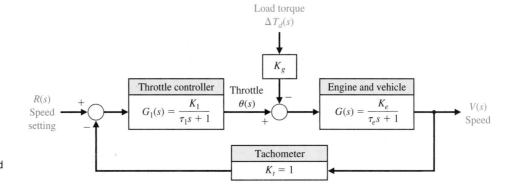

FIGURE P4.6
Automobile speed control.

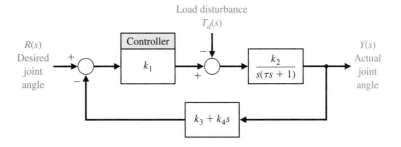

FIGURE P4.7
Robot control system.

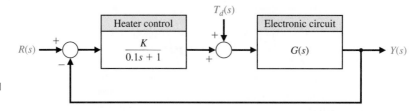

FIGURE P4.8
Temperature control
system.

P4.9 A useful unidirectional sensing device is the pho-
toemitter sensor [15]. A light source is sensitive to the
emitter current flowing and alters the resistance of the
photosensor. Both the light source and the photocon-
ductor are packaged in a single four-terminal device.
This device provides a large gain and total isolation.
A feedback circuit utilizing this device is shown in
Figure P4.9(a), and the nonlinear resistance–current
characteristic is shown in Figure P4.9(b) for the
Raytheon CK1116. The resistance curve can be represented by the equation

$$\log_{10} R = \frac{0.175}{(i - 0.005)^{1/2}},$$

where i is the lamp current. The normal operating
point is obtained when $v_o = 35$ V, and $v_{in} = 2.0$ V.

(a) Determine the closed-loop transfer function of the
system. (b) Determine the sensitivity of the system to
changes in the gain, K.

P4.10 For a paper processing plant, it is important to
maintain a constant tension on the continuous sheet
of paper between the wind-off and wind-up rolls. The
tension varies as the widths of the rolls change, and an
adjustment in the take-up motor speed is necessary, as
shown in Figure P4.10. If the wind-up motor speed is
uncontrolled, as the paper transfers from the wind-off
roll to the wind-up roll, the velocity v_0 decreases and
the tension of the paper drops [10, 14]. The three-
roller and spring combination provides a measure of
the tension of the paper. The spring force is equal to
$k_1 y$, and the linear differential transformer, rectifier,
and amplifier may be represented by $e_0 = -k_2 y$.

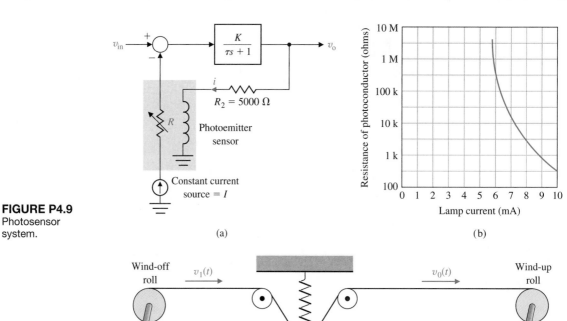

FIGURE P4.9
Photosensor
system.

(a)   (b)

FIGURE P4.10
Paper tension
control.

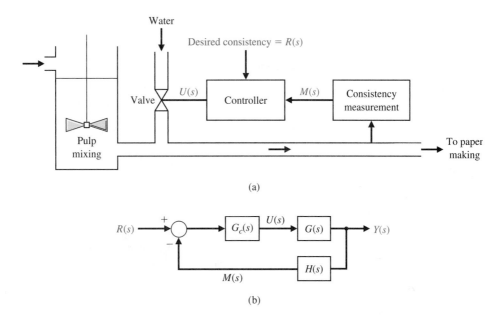

(a)

(b)

FIGURE P4.11
Paper-making
control.

$M(s)$

Therefore, the measure of the tension is described by the relation $2T(s) = k_1y$, where y is the deviation from the equilibrium condition, and $T(s)$ is the vertical component of the deviation in tension from the equilibrium condition. The time constant of the motor is $\tau = L_a/R_a$, and the linear velocity of the wind-up roll is twice the angular velocity of the motor, that is, $v_0(t) = 2\omega_0(t)$. The equation of the motor is then

$$E_0(s) = \frac{1}{K_m}[\tau s\omega_0(s) + \omega_0(s)] + k_3\Delta T(s),$$

where ΔT = a tension disturbance. (a) Draw the closed-loop block diagram for the system, including the disturbance $\Delta T(s)$. (b) Add the effect of a disturbance in the wind-off roll velocity $\Delta V_1(s)$ to the block diagram. (c) Determine the sensitivity of the system to the motor constant K_m. (d) Determine the steady-state error in the tension when a step disturbance in the input velocity, $\Delta V_1(s) = A/s$, occurs.

P4.11 One important objective of the paper-making process is to maintain uniform consistency of the stock output as it progresses to drying and rolling. A diagram of the thick stock consistency dilution control system is shown in Figure P4.11(a). The amount of water added determines the consistency. The block diagram of the system is shown in Figure P4.11(b). Let $H(s) = 1$ and

$$G_c(s) = \frac{K}{10s + 1}, \qquad G(s) = \frac{1}{2s + 1}.$$

Determine (a) the closed-loop transfer function $T(s) = Y(s)/R(s)$, (b) the sensitivity S_K^T, and (c) the steady-state error for a step change in the desired consistency $R(s) = A/s$. (d) Calculate the value of K required for an allowable steady-state error of 2%.

P4.12 Two feedback systems are shown in Figures P4.12(a) and (b). (a) Evaluate the closed-loop transfer functions T_1 and T_2 for each system. (b) Show that $T_1 = T_2 = 100$ when $K_1 = K_2 = 100$. (c) Compare the sensitivities of the two systems with respect to the parameter K_1 for the nominal values of $K_1 = K_2 = 100$.

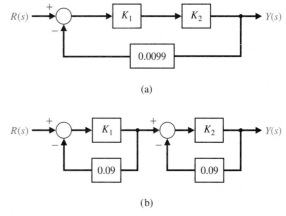

(a)

(b)

FIGURE P4.12 Two feedback systems.

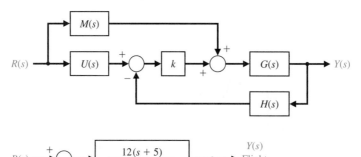

FIGURE P4.13
Closed-loop
system.

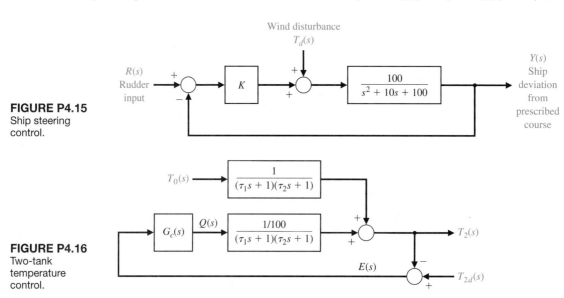

FIGURE P4.14
Hypersonic airplane
speed control.

P4.13 One form of a closed-loop transfer function is

$$T(s) = \frac{G_1(s) + kG_2(s)}{G_3(s) + kG_4(s)}.$$

(a) Use Equation (4.16) to show that [1]

$$S_k^T = \frac{k(G_2G_3 - G_1G_4)}{(G_3 + kG_4)(G_1 + kG_2)}.$$

(b) Determine the sensitivity of the system shown in Figure P4.13, using the equation verified in part (a).

P4.14 A proposed hypersonic plane would climb to 100,000 feet, fly 3800 miles per hour, and cross the Pacific in 2 hours. Control of the aircraft speed could be represented by the model in Figure P4.14. Find the sensitivity of the closed-loop transfer function $T(s)$ to a small change in the parameter a.

P4.15 The steering control of a modern ship may be represented by the system shown in Figure P4.15 [16, 20]. Find the steady-state effect of a constant wind force represented by $T_d(s) = 1/s$ for $K = 5$ and $K = 25$. (a) Assume that the rudder input $R(s)$ is zero, without any disturbance, and has not been adjusted. (b) Show that the rudder can then be used to bring the ship deviation back to zero.

P4.16 Figure P4.16 shows the model of a two-tank system containing a heated liquid, where T_0 is the temperature of the fluid flowing into the first tank and T_2 is the temperature of the liquid flowing out of the second tank. The system of two tanks has a heater in the first tank with a controllable heat input Q. The time constants are $\tau_1 = 10$ s and $\tau_2 = 50$ s. (a) Determine $T_2(s)$ in terms of $T_0(s)$ and $T_{2d}(s)$. (b) If $T_{2d}(s)$, the desired output temperature, is changed instantaneously from $T_{2d}(s) = A/s$ to $T_{2d}(s) = 2A/s$, where

FIGURE P4.15
Ship steering
control.

FIGURE P4.16
Two-tank
temperature
control.

$T_0(s) = A/s$, determine the transient response of $T_2(t)$ when $G_c(s) = K = 500$. (c) Find the steady-state error e_{ss} for the system of part (b), where $E(s) = T_{2d}(s) - T_2(s)$.

P4.17 A robot gripper, shown in part (a) of Figure P4.17, is to be controlled so that it closes to an angle θ by using a DC motor control system, as shown in part (b).

The model of the control system is shown in part (c), where $K_m = 30$, $R_f = 1\,\Omega$, $K_f = K_i = 1$, $J = 0.1$, and $b = 1$. (a) Determine the response $\theta(t)$ of the system to a step change in $\theta_d(t)$ when $K = 20$. (b) Assuming $\theta_d(t) = 0$, find the effect of a load disturbance $T_d(s) = A/s$. (c) Determine the steady-state error e_{ss} when the input is $r(t) = t, t > 0$. (Assume that $T_d(s) = 0$.)

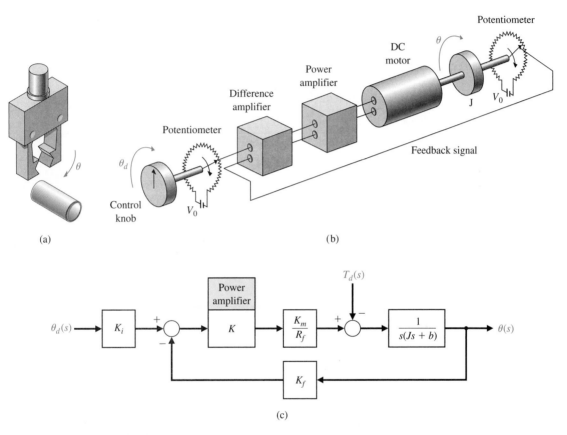

FIGURE P4.17 Robot gripper control.

ADVANCED PROBLEMS

AP4.1 A tank level regulator control is shown in Figure AP4.1(a). It is desired to regulate the level h in response to a disturbance change q_3. The block diagram shows small variable changes about the equilibrium conditions so that the desired $h_d(t) = 0$. Determine the equation for the error $E(s)$, and determine the steady-state error for a unit step disturbance when (a) $G(s) = K$ and (b) $G(s) = K/s$.

AP4.2 The shoulder joint of a robotic arm uses a DC motor with armature control and a set of gears on the output shaft. The model of the system is shown in Figure AP4.2 with a disturbance torque $T_d(s)$ which represents the effect of the load. Determine the steady-state error when the desired angle input is a step so that $\theta_d(s) = A/s$, $G_c(s) = K$, and the disturbance input is zero. When $\theta_d(s) = 0$ and the load

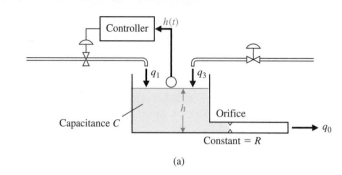

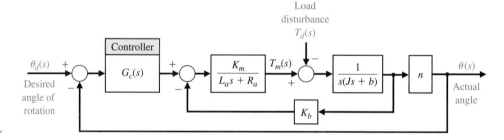

FIGURE AP4.1
A tank level regulator.

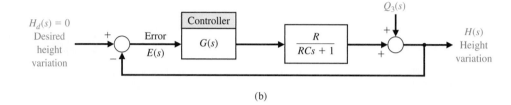

FIGURE AP4.2
Robot joint control.

effect is $T_d(s) = M/s$, determine the steady-state error when (a) $G_c(s) = K$ and (b) $G_c(s) = K/s$.

AP4.3   A machine tool is designed to follow a desired path so that

$$r(t) = (1 - t)u(t),$$

where $u(t)$ is the unit step function. The machine tool control system is shown in Figure AP4.3.

(a) Determine the steady-state error when $r(t)$ is the desired path as given and $T_d(s) = 0$.
(b) Plot the error $e(t)$ for the desired path for part (a) for $0 < t \leq 10$ seconds.
(c) If the desired input is $r(t) = 0$, find the steady-state error when $T_d(s) = 1/s$.
(d) Plot the error $e(t)$ for part (c) for $0 < t \leq 10$ seconds.

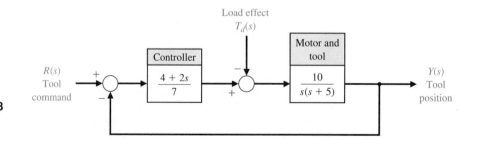

FIGURE AP4.3
Machine tool feedback.

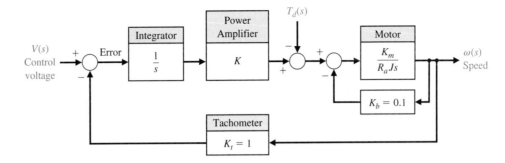

FIGURE AP4.4
DC motor with
feedback.

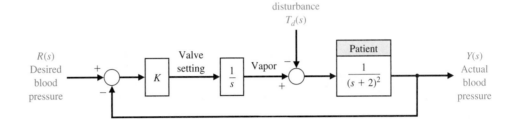

FIGURE AP4.5
Blood pressure
control.

AP4.4 An armature-controlled DC motor with tachometer feedback is shown in Figure AP4.4. Assume that $K_m = 10$, $J = 1$, and $R = 1$.

(a) Determine the required gain, K, to restrict the steady-state error to a ramp input ($v(t) = t$ for $t > 0$) to 0.1 (assume that $T_d(s) = 0$).

(b) For the gain selected in part (a), determine and plot the error, $e(t)$, due to a ramp disturbance for $0 \le t \le 5$ seconds.

AP4.5 A system that controls the mean arterial pressure during anesthesia has been designed and tested [12]. The level of arterial pressure is postulated to be a proxy for depth of anesthesia during surgery. A block diagram of the system is shown in Figure AP4.5, where the impact of surgery is represented by the disturbance $T_d(s)$.

(a) Determine the steady-state error due to a disturbance $T_d(s) = 1/s$ (let $R(s) = 0$).

(b) Determine the steady-state error for a ramp input $r(t) = t$, $t > 0$ (let $T_d(s) = 0$).

(c) Select a suitable value of K less than or equal to 10, and plot the response $y(t)$ for a unit step disturbance input (assume $r(t) = 0$).

AP4.6 A useful circuit, called a lead network, which we discuss in Chapter 10, is shown in Figure AP4.6.

(a) Determine the transfer function $G(s) = V_0(s)/V(s)$.

(b) Determine the sensitivity of $G(s)$ with respect to the capacitance C.

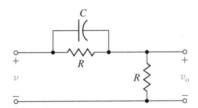

FIGURE AP4.6 A lead network.

(c) Determine and plot the transient response $v_0(t)$ for a step input $V(s) = 1/s$.

AP4.7 A feedback control system with sensor noise and a disturbance input is shown in Figure AP4.7. The goal is to reduce the effects of the noise and the disturbance. Let $R(s) = 0$.

(a) Determine the effect of the disturbance on $Y(s)$.

(b) Determine the effect of the noise on $Y(s)$.

(c) Select the best value for K when $1 \le K \le 100$ so that the effect of steady-state error due to the disturbance and the noise is minimized. Assume $T_d(s) = A/s$, and $N(s) = B/s$.

AP4.8 The block diagram of a machine-tool control system is shown in Figure AP4.8.

(a) Determine the transfer function $T(s) = Y(s)/R(s)$.

(b) Determine the sensitivity S_b^T.

(c) Select K when $1 \le K \le 50$ so that the effects of the disturbance and S_b^T are minimized

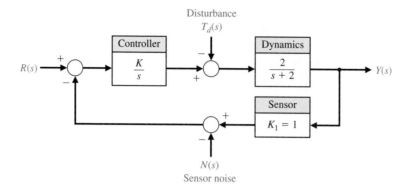

FIGURE AP4.7
Feedback system
with noise.

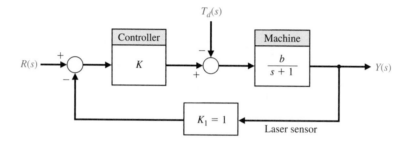

FIGURE AP4.8
Machine-tool
control.

DESIGN PROBLEMS

CDP4.1   A capstan drive for a table slide is described in CDP2.1. The position of the slide x is measured with a capacitance gauge, as shown in Figure CDP4.1, which is very linear and accurate. Sketch the model of the feedback system and determine the response of the system when the controller is an amplifier and $H(s) = 1$. Determine the step response for several selected values of the amplifier gain $G_c(s) = K_a$.

DP4.1   A closed-loop speed control system is subjected to a disturbance due to a load, as shown in Figure DP4.1. The desired speed is $\omega_d(t) = 100$ rad/s, and the load disturbance is a unit step input $T_d(s) = 1/s$. Assume that the speed has attained the no-load speed of 100 rad/s and is in steady state. (a) Determine the steady-state effect of the load disturbance, and (b) plot $\omega(t)$ for the step disturbance for selected values of gain so

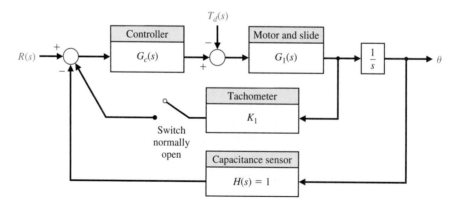

FIGURE CDP4.1
The model of the
feedback system
with a capacitance
measurement
sensor. The
tachometer may be
mounted on the
motor (optional),
and the switch will
normally be open.

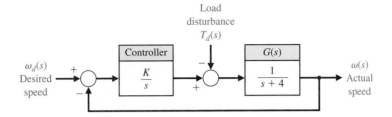

FIGURE DP4.1
Speed control
system.

that $10 \leq K \leq 25$. Determine a suitable value for the gain K.

DP4.2 The control of the roll angle of an airplane is achieved by using the torque developed by the ailerons. A linear model of the roll control system for a small experimental aircraft is shown in Figure DP4.2, where $q(t)$ is the flow of fluid into a hydraulic cylinder and

$$G(s) = \frac{1}{s^2 + 4s + 9}.$$

The goal is to maintain a small roll angle θ due to disturbances. Select an appropriate gain KK_1 that will reduce the effect of the disturbance while attaining a desirable transient response to a step disturbance, with $\theta_d(t) = 0$. To obtain a desirable transient response, let $KK_1 < 35$.

DP4.3 The speed control system of Figure DP4.1 is altered so that $G(s) = 1/(s + 5)$ and the feedback is K_1, as shown in Figure DP4.3.

(a) Determine the range of K_1 allowable so that the steady state is $e_{ss} \leq 1\%$.

(b) Determine a suitable value for K_1 and K so that the magnitude of the steady-state error to a wind disturbance $T_d(t) = 2t$ mrad/s, $0 \leq t < 5$ s, is less than 0.1 mrad.

DP4.4 Lasers have been used in eye surgery for more than 25 years. They can cut tissue or aid in coagulation

[17]. The laser allows the ophthalmologist to apply heat to a location in the eye in a controlled manner. Many procedures use the retina as a laser target. The retina is the thin sensory tissue that rests on the inner surface of the back of the eye and is the actual transducer of the eye, converting light energy into electrical pulses. On occasion, this layer will detach from the wall, resulting in death of the detached area from lack of blood and leading to partial or total blindness in that eye. A laser can be used to "weld" the retina into its proper place on the inner wall.

Automated control of position enables the ophthalmologist to indicate to the controller where lesions should be inserted. The controller then monitors the retina and controls the laser's position so that each lesion is placed at the proper location. A wide-angle video-camera system is required to monitor the movement of the retina, as shown in Figure DP4.4(a). If the eye moves during the irradiation, the laser must be either redirected or turned off. The position-control system is shown in Figure DP4.4(b). Select an appropriate gain for the controller so that the transient response to a step change in $r(t)$ is satisfactory and the effect of the disturbance due to noise in the system is minimized. Also, ensure that the steady-state error for a step input command is zero. To ensure acceptable transient response, require that $K < 10$.

FIGURE DP4.2
Control of the roll
angle of an
airplane.

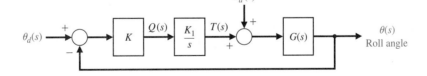

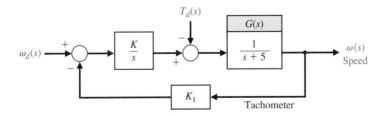

FIGURE DP4.3
Speed control
system.

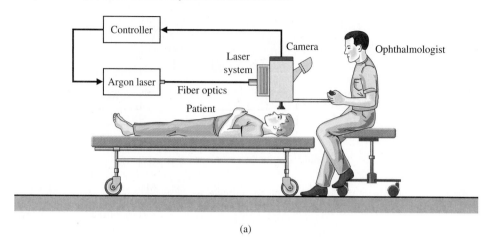

(a)

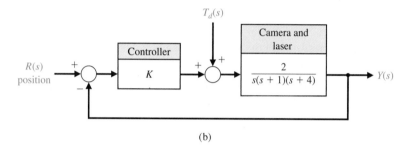

FIGURE DP4.4
Laser eye surgery
system.

(b)

DP4.5 An op-amp circuit can be used to generate a short pulse. The circuit shown in Figure DP4.5 can generate the pulse $v_0(t) = 5e^{-100t}$, $t > 0$, when the input $v(t)$ is a unit step [6]. Select appropriate values for the resistors and capacitors. Assume an ideal op-amp.

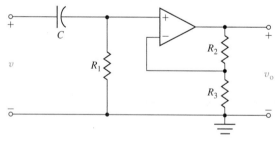

FIGURE DP4.5   Op-map circuit.

DP4.6 A hydrobot is under consideration for remote exploration under the ice of Europa, a moon of the giant planet Jupiter. Figure DP4.6(a) shows one artistic version of the mission. The hydrobot is a self-propelled underwater vehicle that would analyze the chemical composition of the water in a search for signs of life. An important aspect of the vehicle is a controlled vertical descent to depth in the presence of underwater currents. A simplified control feedback system is shown in Figure DP4.6(b). The parameter $J > 0$ is the pitching moment of inertia. (a) Suppose that $G_c(s) = K$. For what range of K is the system stable? (b) What is the steady-state error to a unit step disturbance when $G_c(s) = K$? (c) Suppose that $G_c(s) = K_p + K_D s$. For what range of K_p and K_D is the system stable? (d) What is the steady-state error to a unit step disturbance when $G_c(s) = K_p + K_D s$?

(a)

FIGURE DP4.6
(a) Europa
exploration under
the ice. (Used with
permission. Credit:
NASA.)
(b) Feedback
system.

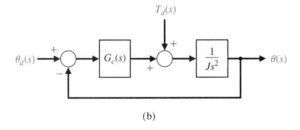

(b)

COMPUTER PROBLEMS

CP4.1 Consider the transfer function (without feedback)

$$G(s) = \frac{5}{s^2 + 2s + 20}.$$

When the input is a unit step, the desired steady-state value of the output is one. Using the step function, show that the steady-state error to a unit step input is 0.75.

CP4.2 Consider a unity feedback system with

$$G(s) = \frac{75}{s^2 + 2s + 10}.$$

Obtain the step response and determine the percent overshoot. What is the steady-state error?

CP4.3 Consider the closed-loop transfer function

$$T(s) = \frac{5K}{s^2 + 15s + K}.$$

Obtain the family of step responses for $K = 10, 200$, and 500. Co-plot the responses and develop a table of results that includes the percent overshoot, settling time, and steady-state error.

CP4.4 Consider the closed-loop control system shown in Figure CP4.4. Develop an m-file script to assist in the search for a value of k so that the percent overshoot to a unit step input is greater than 1%, but less than 10%. The script should compute the closed-loop

FIGURE CP4.4
A closed-loop
negative feedback
control system.

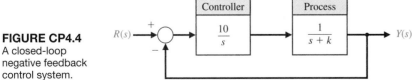

transfer function $T(s) = Y(s)/R(s)$ and generate the step response. Verify graphically that the steady-state error to a unit step input is zero.

CP4.5 Consider the closed-loop control system shown in Figure CP4.5. The controller gain is $K = 2$. The nominal value of the plant parameter is $a = 1$. The nominal value is used for design purposes only, since in reality the value is not precisely known. The objective of our analysis is to investigate the sensitivity of the closed-loop system to the parameter a.

(a) When $a = 1$, show analytically that the steady-state value of $y(t)$ is equal to 2 when $r(t)$ is a unit step. Verify that the unit step response is within 2% of the final value after 4 seconds.

(b) The sensitivity of the system to changes in the parameter a can be investigated by studying the effects of parameter changes on the transient response. Plot the unit step response for $a = 0.5, 2$, and 5. Discuss the results.

CP4.6 Consider the torsional mechanical system in Figure CP4.6(a). The torque due to the twisting of the shaft is $-k\theta$; the damping torque due to the braking device is $-b\theta$; the disturbance torque is $t_d(t)$; the

input torque is $r(t)$; and the moment of inertia of the mechanical system is J. The transfer function of the torsional mechanical system is

$$G(s) = \frac{1/J}{s^2 + (b/J)s + k/J}.$$

A closed-loop control system for the system is shown in Figure CP4.6(b). Suppose the desired angle $\theta_d = 0°, k = 5, b = 0.9$, and $J = 1$.

(a) Determine the open-loop response $\theta(t)$ of the system for a unit step disturbance (set $r(t) = 0$).

(b) With the controller gain $K_0 = 50$, determine the closed-loop response, $\theta(t)$ to a unit step disturbance.

(c) Plot the open-loop versus the closed-loop response to the disturbance input. Discuss your results and make an argument for using closed-loop feedback control to improve the disturbance rejection properties of the system.

CP4.7 A negative feedback control system is depicted in Figure CP4.7. Suppose that our design objective is to find a controller $G_c(s)$ of minimal complexity such that our closed-loop system can track a unit step input with a steady-state error of zero.

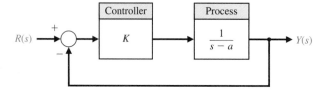

FIGURE CP4.5
A closed-loop control system with uncertain parameter *a*.

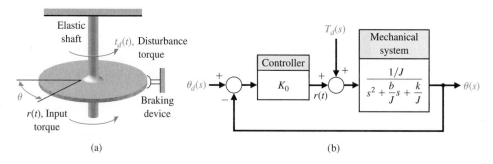

FIGURE CP4.6
(a) A torsional mechanical system.
(b) The torsional mechanical system feedback control system.

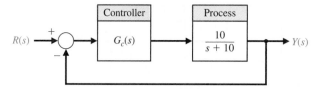

FIGURE CP4.7
A simple single-loop feedback control system.

(a) As a first try, consider a simple proportional controller

$$G_c(s) = K,$$

where K is a fixed gain. Let $K = 2$. Plot the unit step response and determine the steady-state error from the plot.

(b) Now consider a more complex controller

$$G_c(s) = K_0 + \frac{K_1}{s},$$

where $K_0 = 2$ and $K_1 = 20$. This controller is known as a proportional, integral (PI) controller. Plot the unit step response, and determine the steady-state error from the plot.

(c) Compare the results from parts (a) and (b), and discuss the trade-off between controller complexity and steady-state tracking error performance.

CP4.8 Consider the closed-loop system in Figure CP4.8, whose transfer function is

$$G(s) = \frac{10s}{s + 100} \quad \text{and} \quad H(s) = \frac{5}{s + 50}.$$

(a) Obtain the closed-loop transfer function $T(s) = Y(s)/R(s)$ and the unit step response; that is, let $R(s) = 1/s$ and assume that $N(s) = 0$.

(b) Obtain the disturbance response when

$$N(s) = \frac{10}{s^2 + 100}$$

is a sinusoidal input of frequency $\omega = 10$ rad/s. Assume that $R(s) = 0$.

(c) In the steady-state, what is the frequency and peak magnitude of the disturbance response from part (b)?

CP4.9 Consider the closed-loop system is depicted in Figure CP4.9. The controller gain K can be modified to meet the design specifications.

(a) Determine the closed-loop transfer function $T(s) = Y(s)/R(s)$.

(b) Plot the response of the closed-loop system for $K = 5, 10,$ and 50.

(c) When the controller gain is $K = 10$, determine the steady-state value of $y(t)$ when the disturbance is a unit step, that is, when $T_d(s) = 1/s$ and $R(s) = 0$.

CP4.10 Consider the non-unity feedback system is depicted in Figure CP4.10.

(a) Determine the closed-loop transfer function $T(s) = Y(s)/R(s)$.

(b) For $K = 10, 12,$ and 15, plot the unit step responses. Determine the steady-state error errors and the settling times from the plots.

For parts (a) and (b), develop an m-file that computes the closed-loop transfer function and generates the plots for varying K.

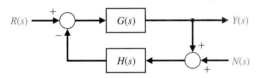

FIGURE CP4.8 Closed-loop system with nonunity feedback and measurement noise.

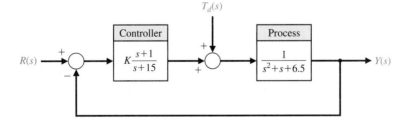

FIGURE CP4.9
Closed-loop feedback system with external disturbances.

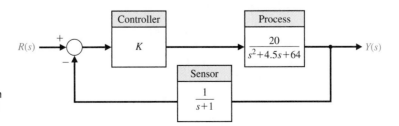

FIGURE CP4.10
Closed-loop system with a sensor in the feedback loop.

TERMS AND CONCEPTS

Closed-loop system   A system with a measurement of the output signal and a comparison with the desired output to generate an error signal that is applied to the actuator.

Complexity   A measure of the structure, intricateness, or behavior of a system that characterizes the relationships and interactions between various components.

Components   The parts, subsystems, or subassemblies that comprise a total system.

Direct system   See Open-loop system.

Disturbance signal   An unwanted input signal that affects the system's output signal.

Error signal   The difference between the desired output $R(s)$ and the actual output $Y(s)$. Therefore, $E(s) = R(s) - Y(s)$.

Instability   An attribute of a system that describes a tendency of the system to depart from the equilibrium condition when initially displaced.

Loss of gain   A reduction in the amplitude of the ratio of the output signal to the input signal through a system, usually measured in decibels.

Open-loop system   A system without feedback that directly generates the output in response to an input signal.

Steady-state error   The error when the time period is large and the transient response has decayed, leaving the continuous response.

System sensitivity   The ratio of the change in the system transfer function to the change of a process transfer function (or parameter) for a small incremental change.

Transient response   The response of a system as a function of time.

The Performance of Feedback Control Systems

PREVIEW

The ability to adjust the transient and steady-state response of a control system is a beneficial outcome of the design of control systems. In this chapter, we introduce the time-domain performance specifications and we use key input signals to test the response of the control system. The correlation between the system performance and the location of the transfer function poles and zeros is discussed. We will develop relationships between the performance specifications and the natural frequency and damping ratio for second-order systems. Relying on the notion of dominant poles, we can extrapolate the ideas associated with second-order systems to those of higher order. The concept of a performance index will be considered. We will present a set of popular quantitative performance indices that adequately represent the performance of the control system. The chapter concludes with a performance analysis of the Sequential Design Example: Disk Drive Read System.

DESIRED OUTCOMES

Upon completion of Chapter 5, students should:

❑   Be aware of key test signals used in controls and of the resulting transient response characteristics of second-order systems to test signal inputs.

❑   Recognize the direct relationship between the pole locations of second-order systems and the transient response.

❑   Be familiar with the design formulas that relate the second-order pole locations to percent overshoot, settling time, rise time, and time to peak.

❑   Be aware of the impact of a zero and a third pole on the second-order system response.

❑   Gain a sense of optimal control as measured with performance indices.

5.1 INTRODUCTION

The ability to adjust the transient and steady-state performance is a distinct advantage of feedback control systems. To analyze and design a control system, we must define and measure its performance. Based on the desired performance of the control system, the system parameters may be adjusted to provide the desired response. Because control systems are inherently dynamic, their performance is usually specified in terms of both the transient response and the steady-state response. The **transient response** is the response that disappears with time. The **steady-state response** is the response that exists for a long time following an input signal initiation.

The **design specifications** for control systems normally include several time-response indices for a specified input command, as well as a desired steady-state accuracy. In the course of any design, the specifications are often revised to effect a compromise. Therefore, specifications are seldom a rigid set of requirements, but rather a first attempt at listing a desired performance. The effective compromise and adjustment of specifications are graphically illustrated in Figure 5.1. The parameter p may minimize the performance measure M_2 if we select p as a very small value. However, this results in large measure M_1, an undesirable situation. If the performance measures are equally important, the crossover point at p_{min} provides the best compromise. This type of compromise is normally encountered in control system design. It is clear that if the original specifications called for both M_1 and M_2 to be zero, the specifications could not be simultaneously met; they would then have to be altered to allow for the compromise resulting with p_{min} [1, 12, 17, 23].

The specifications, which are stated in terms of the measures of performance, indicate the quality of the system to the designer. In other words, the performance measures help to answer the question, How well does the system perform the task for which it was designed?

5.2 TEST INPUT SIGNALS

The time-domain performance specifications are important indices because control systems are inherently time-domain systems. That is, the system transient or time performance is the response of prime interest for control systems. It is necessary to

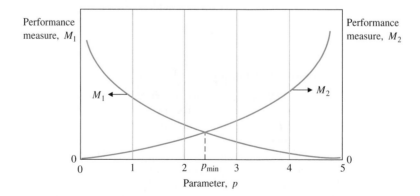

FIGURE 5.1
Two performance measures versus parameter p.

determine initially whether the system is stable; we can achieve this goal by using the techniques of ensuing chapters. If the system is stable, the response to a specific input signal will provide several measures of the performance. However, because the actual input signal of the system is usually unknown, a standard **test input signal** is normally chosen. This approach is quite useful because there is a reasonable correlation between the response of a system to a standard test input and the system's ability to perform under normal operating conditions. Furthermore, using a standard input allows the designer to compare several competing designs. Many control systems experience input signals that are very similar to the standard test signals.

The standard test input signals commonly used are the step input, the ramp input, and the parabolic input. These inputs are shown in Figure 5.2. The equations representing these test signals are given in Table 5.1, where the Laplace transform can be obtained by using Table 2.3 and a more complete list of Laplace transform pairs can be found at the MCS website. The ramp signal is the integral of the step input, and the parabola is simply the integral of the ramp input. A **unit impulse** function is also useful for test signal purposes. The unit impulse is based on a rectangular function

$$f_\epsilon(t) = \begin{cases} 1/\epsilon, & -\dfrac{\epsilon}{2} \le t \le \dfrac{\epsilon}{2}; \\ 0, & \text{otherwise,} \end{cases}$$

where $\epsilon > 0$. As ϵ approaches zero, the function $f_\epsilon(t)$ approaches the unit impulse function $\delta(t)$, which has the following properties:

$$\int_{-\infty}^{\infty} \delta(t)\, dt = 1 \quad \text{and} \quad \int_{-\infty}^{\infty} \delta(t - a)g(t)\, dt = g(a). \tag{5.1}$$

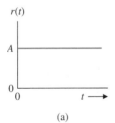

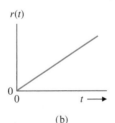

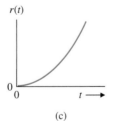

FIGURE 5.2
Test input signals:
(a) step, (b) ramp,
and (c) parabolic.

(a)   (b)   (c)

Table 5.1   Test Signal Inputs

Test Signal	r(t)	R(s)
Step	$r(t) = A, t > 0$ $= 0, t < 0$	$R(s) = A/s$
Ramp	$r(t) = At, t > 0$ $= 0, t < 0$	$R(s) = A/s^2$
Parabolic	$r(t) = At^2, t > 0$ $= 0, t < 0$	$R(s) = 2A/s^3$

FIGURE 5.3
Open-loop control
system.

(a)   (b)

The impulse input is useful when we consider the convolution integral for the output $y(t)$ in terms of an input $r(t)$, which is written as

$$y(t) = \int_{-\infty}^{t} g(t - \tau)r(\tau)\, d\tau = \mathcal{L}^{-1}\{G(s)R(s)\}. \tag{5.2}$$

This relationship is shown in block diagram form in Figure 5.3. If the input is a unit impulse function, we have

$$y(t) = \int_{-\infty}^{t} g(t - \tau)\delta(\tau)\, d\tau. \tag{5.3}$$

The integral has a value only at $\tau = 0$; therefore,

$$y(t) = g(t),$$

the impulse response of the system $G(s)$. The impulse response test signal can often be used for a dynamic system by subjecting the system to a large-amplitude, narrow-width pulse of area A.

The standard test signals are of the general form

$$r(t) = t^n, \tag{5.4}$$

and the Laplace transform is

$$R(s) = \frac{n!}{s^{n+1}}. \tag{5.5}$$

Hence, the response to one test signal may be related to the response of another test signal of the form of Equation (5.4). The step input signal is the easiest to generate and evaluate and is usually chosen for performance tests.

Consider the response of the system shown in Figure 5.3 for a unit step input when

$$G(s) = \frac{9}{s + 10}.$$

Then the output is

$$Y(s) = \frac{9}{s(s + 10)},$$

the response during the transient period is

$$y(t) = 0.9(1 - e^{-10t}),$$

and the steady-state response is

$$y(\infty) = 0.9.$$

If the error is $E(s) = R(s) - Y(s)$, then the steady-state error is

$$e_{ss} = \lim_{s \to 0} sE(s) = 0.1.$$

5.3 PERFORMANCE OF SECOND-ORDER SYSTEMS

Let us consider a single-loop second-order system and determine its response to a unit step input. A closed-loop feedback control system is shown in Figure 5.4. The closed-loop system is

$$Y(s) = \frac{G(s)}{1 + G(s)} R(s). \tag{5.6}$$

We may rewrite Equation (5.6) as

$$Y(s) = \frac{\omega_n^2}{s^2 + 2\zeta\omega_n s + \omega_n^2} R(s). \tag{5.7}$$

With a unit step input, we obtain

$$Y(s) = \frac{\omega_n^2}{s(s^2 + 2\zeta\omega_n s + \omega_n^2)}, \tag{5.8}$$

for which the transient output, as obtained from the Laplace transform table in Table 2.3, is

$$y(t) = 1 - \frac{1}{\beta} e^{-\zeta\omega_n t} \sin(\omega_n \beta t + \theta), \tag{5.9}$$

where $\beta = \sqrt{1 - \zeta^2}$, $\theta = \cos^{-1}\zeta$, and $0 < \zeta < 1$. The transient response of this second-order system for various values of the damping ratio ζ is shown in Figure 5.5.

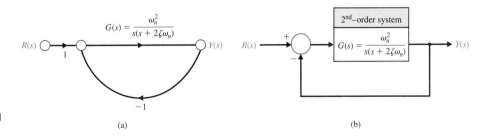

FIGURE 5.4
Second-order
closed-loop control
system.

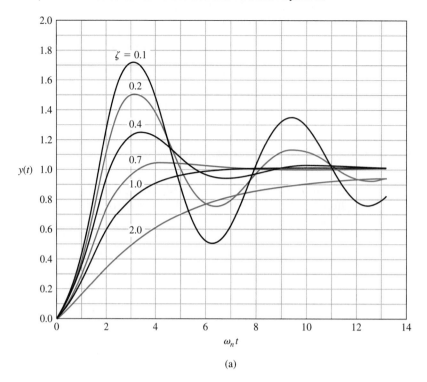

(a)

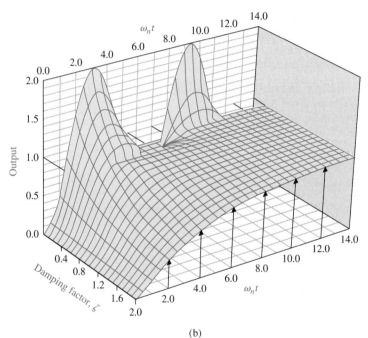

(b)

FIGURE 5.5
(a) Transient response of a second-order system (Equation 5.9) for a step input. (b) The transient response of a second-order system (Equation 5.9) for a step input as a function of ζ and $\omega_n t$. (Courtesy of Professor R. Jacquot, University of Wyoming.)

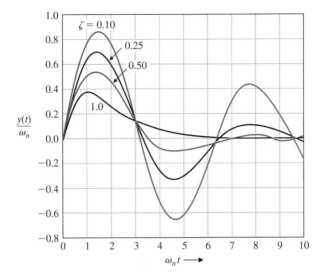

FIGURE 5.6
Response of a second-order system for an impulse function input.

As ζ decreases, the closed-loop roots approach the imaginary axis, and the response becomes increasingly oscillatory. The response as a function of ζ and time is also shown in Figure 5.5(b) for a step input.

The Laplace transform of the unit impulse is $R(s) = 1$, and therefore the output for an impulse is

$$Y(s) = \frac{\omega_n^2}{s^2 + 2\zeta\omega_n s + \omega_n^2}, \tag{5.10}$$

which is $T(s) = Y(s)/R(s)$, the transfer function of the closed-loop system. The transient response for an impulse function input is then

$$y(t) = \frac{\omega_n}{\beta} e^{-\zeta\omega_n t} \sin(\omega_n \beta t), \tag{5.11}$$

which is the derivative of the response to a step input. The impulse response of the second-order system is shown in Figure 5.6 for several values of the damping ratio ζ. The designer is able to select several alternative performance measures from the transient response of the system for either a step or impulse input.

Standard performance measures are usually defined in terms of the step response of a system as shown in Figure 5.7. The swiftness of the response is measured by the **rise time** T_r and the **peak time** T_p. For underdamped systems with an overshoot, the 0–100% rise time is a useful index. If the system is overdamped, then the peak time is not defined, and the 10–90% rise time T_{r_1} is normally used. The similarity with which the actual response matches the step input is measured by the percent overshoot and settling time T_s. The **percent overshoot** is defined as

$$P.O. = \frac{M_{P_t} - fv}{fv} \times 100\% \tag{5.12}$$

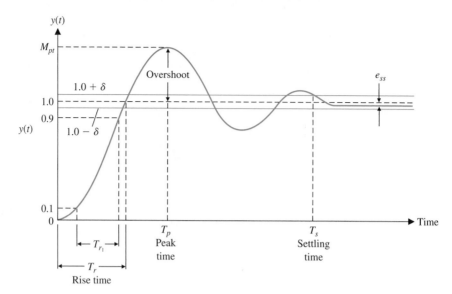

FIGURE 5.7
Step response of a
control system
(Equation 5.9).

for a unit step input, where M_{pt} is the peak value of the time response, and fv is the final value of the response. Normally, fv is the magnitude of the input, but many systems have a final value significantly different from the desired input magnitude. For the system with a unit step represented by Equation (5.8), we have $fv = 1$.

The **settling time**, T_s, is defined as the time required for the system to settle within a certain percentage δ of the input amplitude. This band of $\pm\delta$ is shown in Figure 5.7. For the second-order system with closed-loop damping constant $\zeta\omega_n$ and a response described by Equation (5.9), we seek to determine the time T_s for which the response remains within 2% of the final value. This occurs approximately when

$$e^{-\zeta\omega_n T_s} < 0.02,$$

or

$$\zeta\omega_n T_s \cong 4.$$

Therefore, we have

$$\boxed{T_s = 4\tau = \frac{4}{\zeta\omega_n}.} \tag{5.13}$$

Hence, we will define the settling time as four time constants (that is, $\tau = 1/\zeta\omega_n$) of the dominant roots of the characteristic equation. The steady-state error of the system may be measured on the step response of the system as shown in Figure 5.7.

The transient response of the system may be described in terms of two factors:

1. The swiftness of response, as represented by the rise time and the peak time.

2. The closeness of the response to the desired response, as represented by the overshoot and settling time.

As nature would have it, these are contradictory requirements; thus, a compromise must be obtained. To obtain an explicit relation for M_{pt} and T_p as a function of ζ, one can differentiate Equation (5.9) and set it equal to zero. Alternatively, one can utilize the differentiation property of the Laplace transform, which may be written as

$$\mathcal{L}\left\{\frac{dy(t)}{dt}\right\} = sY(s)$$

when the initial value of $y(t)$ is zero. Therefore, we may acquire the derivative of $y(t)$ by multiplying Equation (5.8) by s and thus obtaining the right side of Equation (5.10). Taking the inverse transform of the right side of Equation (5.10), we obtain Equation (5.11), which is equal to zero when $\omega_n \beta t = \pi$. Thus, we find that the peak time relationship for this second-order system is

$$T_p = \frac{\pi}{\omega_n \sqrt{1 - \zeta^2}}, \tag{5.14}$$

and the peak response is

$$M_{pt} = 1 + e^{-\zeta \pi / \sqrt{1 - \zeta^2}}. \tag{5.15}$$

Therefore, the percent overshoot is

$$P.O. = 100 e^{-\zeta \pi / \sqrt{1 - \zeta^2}}. \tag{5.16}$$

The percent overshoot versus the damping ratio, ζ, is shown in Figure 5.8. Also, the normalized peak time, $\omega_n T_p$, is shown versus the damping ratio, ζ, in Figure 5.8. The percent overshoot versus the damping ratio is listed in Table 5.2 for selected values of

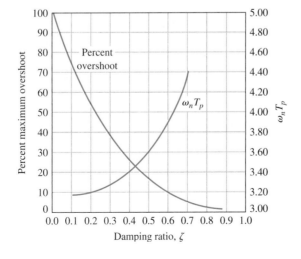

FIGURE 5.8
Percent overshoot and normalized peak time versus damping ratio ζ for a second-order system (Equation 5.8).

Table 5.2   Percent Peak Overshoot Versus Damping Ratio for a Second-Order System

Damping ratio	0.9	0.8	0.7	0.6	0.5	0.4	0.3
Percent overshoot	0.2	1.5	4.6	9.5	16.3	25.4	37.2

the damping ratio. Again, we are confronted with a necessary compromise between the swiftness of response and the allowable overshoot.

The swiftness of step response can be measured as the time it takes to rise from 10% to 90% of the magnitude of the step input. This is the definition of the rise time, T_{r1}, shown in Figure 5.7. The normalized rise time, $\omega_n T_{r1}$, versus $\zeta (0.05 \le \zeta \le 0.95)$ is shown in Figure 5.9. Although it is difficult to obtain exact analytic expressions for T_{r1}, we can utilize the linear approximation

$$T_{r1} = \frac{2.16\zeta + 0.60}{\omega_n},$$ (5.17)

which is accurate for $0.3 \le \zeta \le 0.8$. This linear approximation is shown in Figure 5.9.

The swiftness of a response to a step input as described by Equation (5.17) is dependent on ζ and ω_n. For a given ζ, the response is faster for larger ω_n, as shown in Figure 5.10. Note that the overshoot is independent of ω_n.

For a given ω_n, the response is faster for lower ζ, as shown in Figure 5.11. The swiftness of the response, however, will be limited by the overshoot that can be accepted.

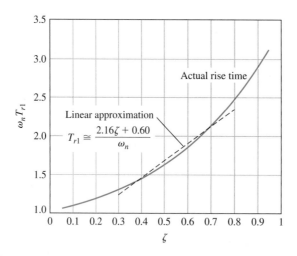

FIGURE 5.9
Normalized rise time, T_{r1}, versus ζ for a second-order system.

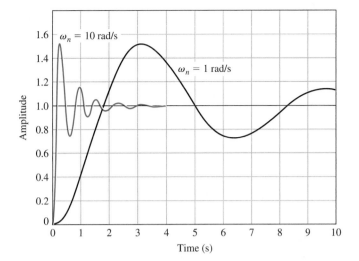

FIGURE 5.10
The step response
for $\zeta = 0.2$ for
$\omega_n = 1$ and
$\omega_n = 10$.

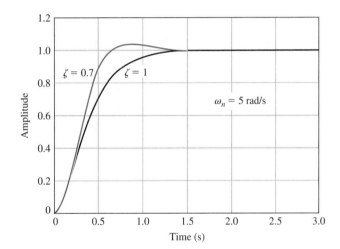

FIGURE 5.11
The step response
for $\omega_n = 5$ with
$\zeta = 0.7$ and $\zeta = 1$.

5.4 EFFECTS OF A THIRD POLE AND A ZERO ON THE SECOND-ORDER SYSTEM RESPONSE

The curves presented in Figure 5.8 are exact only for the second-order system of Equation (5.8). However, they provide a remarkably good source of data because many systems possess a dominant pair of roots, and the step response can be estimated by utilizing Figure 5.8. This approach, although an approximation, avoids the evaluation of the inverse Laplace transformation in order to determine the percent overshoot and other performance measures. For example, for a third-order system with a closed-loop transfer function

$$T(s) = \frac{1}{(s^2 + 2\zeta s + 1)(\gamma s + 1)},$$
(5.18)

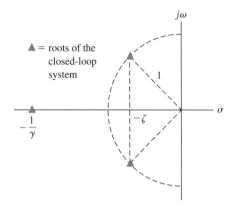

FIGURE 5.12
An s-plane diagram of a third-order system.

the s-plane diagram is shown in Figure 5.12. This third-order system is normalized with $\omega_n = 1$. It was ascertained experimentally that the performance (as indicated by the percent overshoot, $P.O.$, and the settling time, T_s), was adequately represented by the second-order system curves when [4]

$$|1/\gamma| \geq 10|\zeta\omega_n|.$$

In other words, the response of a third-order system can be approximated by the **dominant roots** of the second-order system as long as the real part of the dominant roots is less than one tenth of the real part of the third root [17, 23].

Using a computer simulation, we can determine the response of a system to a unit step input when $\zeta = 0.45$. When $\gamma = 2.25$, we find that the response is over-damped because the real part of the complex poles is -0.45, whereas the real pole is equal to -0.444. The settling time (to within 2% of the final value) is found via the simulation to be 9.6 seconds. If $\gamma = 0.90$ or $1/\gamma = 1.11$ is compared with $\zeta\omega_n = 0.45$ of the complex poles, the overshoot is 12% and the settling time is 8.8 seconds. If the complex roots were dominant, we would expect the overshoot to be 20% and the settling time to be $4/\zeta\omega_n = 8.9$ seconds. The results are summarized in Table 5.3.

The performance measures of Figure 5.8 are correct only for a transfer function without finite zeros. If the transfer function of a system possesses finite zeros and they are located relatively near the dominant complex poles, then the zeros will materially affect the transient response of the system [5].

Table 5.3   Effect of a Third Pole (Equation 5.18) for $\zeta = 0.45$

γ	$\dfrac{1}{\gamma}$	Percent Overshoot	Settling Time*
2.25	0.444	0	9.63
1.5	0.666	3.9	6.3
0.9	1.111	12.3	8.81
0.4	2.50	18.6	8.67
0.05	20.0	20.5	8.37
0∞	20.5	8.24	

*Note: Settling time is normalized time, $\omega_n T_s$ and uses a 2% criterion.

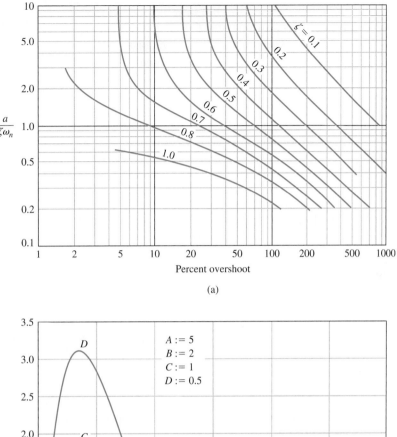

(a)

(b)

FIGURE 5.13 (a) Percent overshoot as a function of ζ and ω_ν when a second-order transfer function contains a zero. Redrawn with permission from R. N. Clark, *Introduction to Automatic Control Systems* (New York: Wiley, 1962). (b) The response for the second-order transfer function with a zero for four values of the ratio $a/\zeta\omega_n$: $A = 5$, $B = 2$, $C = 1$, and $D = 0.5$ when $\zeta = 0.45$.

Table 5.4 The Response of a Second-Order System with a Zero and $\zeta = 0.45$

$a/\zeta\omega_n$	Percent Overshoot	Settling Time	Peak Time
5	23.1	8.0	3.0
2	39.7	7.6	2.2
1	89.9	10.1	1.8
0.5	210.0	10.3	1.5

Note: Time is normalized as $\omega_n t$, and settling time is based on a 2% criterion.

The transient response of a system with one zero and two poles may be affected by the location of the zero [5]. The percent overshoot for a step input as a function of $a/\zeta\omega_n$, when $\zeta \leq 1$, is given in Figure 5.13(a) for the system transfer function

$$T(s) = \frac{(\omega_n^2/a)(s + a)}{s^2 + 2\zeta\omega_n s + \omega_n^2}.$$

The actual transient response for a step input is shown in Figure 5.13(b) for selected values of $a/\zeta\omega_n$. The actual response for these selected values is summarized in Table 5.4 when $\zeta = 0.45$.

The correlation of the time-domain response of a system with the s-plane location of the poles of the closed-loop transfer function is very useful for selecting the specifications of a system. To illustrate clearly the utility of the s-plane, let us consider a simple example.

EXAMPLE 5.1   **Parameter selection**

A single-loop feedback control system is shown in Figure 5.14. We select the gain K and the parameter p so that the time-domain specifications will be satisfied. The transient response to a step should be as fast as is attainable while retaining an overshoot of less than 5%. Furthermore, the settling time to within 2% of the final value should be less than 4 seconds. The damping ratio, ζ, for an overshoot of 4.3% is 0.707. This damping ratio is shown graphically as a line in Figure 5.15. Because the settling time is

$$T_s = \frac{4}{\zeta\omega_n} \leq 4 \text{ s},$$

FIGURE 5.14
Single-loop feedback control system.

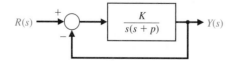

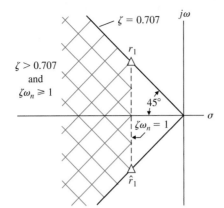

FIGURE 5.15
Specifications and
root locations on
the s-plane.

we require that the real part of the complex poles of $T(s)$ be

$$\zeta\omega_n \geq 1.$$

This region is also shown in Figure 5.15. The region that will satisfy both time-domain requirements is shown cross-hatched on the s-plane of Figure 5.15.

When the closed-loop roots are $r_1 = -1 + j1$ and $\hat{r}_1 = -1 - j1$, we have $T_s = 4$ s and an overshoot of 4.3%. Therefore, $\zeta = 1/\sqrt{2}$ and $\omega_n = 1/\zeta = \sqrt{2}$. The closed-loop transfer function is

$$T(s) = \frac{G(s)}{1 + G(s)} = \frac{K}{s^2 + ps + K} = \frac{\omega_n^2}{s^2 + 2\zeta\omega_n s + \omega_n^2}.$$

Hence, we require that $K = \omega_n^2 = 2$ and $p = 2\zeta\omega_n = 2$. A full comprehension of the correlation between the closed-loop root location and the system transient response is important to the system analyst and designer. Therefore, we shall consider the matter more completely in the following sections. ■

EXAMPLE 5.2   **Dominant poles of $T(s)$**

Consider a system with a closed-loop transfer function

$$\frac{Y(s)}{R(s)} = T(s) = \frac{\frac{\omega_n^2}{a}(s + a)}{(s^2 + 2\zeta\omega_n s + \omega_n^2)(1 + \tau s)}.$$

Both the zero and the real pole may affect the transient response. If $a \gg \zeta\omega_n$ and $\tau \ll 1/\zeta\omega_n$, then the pole and zero will have little effect on the step response. Assume that we have

$$T(s) = \frac{62.5(s + 2.5)}{(s^2 + 6s + 25)(s + 6.25)}.$$

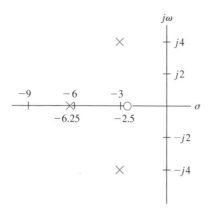

FIGURE 5.16
The poles and zeros on the s-plane for a third-order system.

Note that the DC gain is equal to 1 ($T(0) = 1$), and we expect a zero steady-state error for a step input. We have $\zeta\omega_n = 3, \tau = 0.16$, and $a = 2.5$. The poles and the zero are shown on the s-plane in Figure 5.16. As a first approximation, we neglect the real pole and obtain

$$T(s) \approx \frac{10(s + 2.5)}{s^2 + 6s + 25}.$$

We now have $\zeta = 0.6$ and $\omega_n = 5$ for dominant poles with one accompanying zero for which $a/(\zeta\omega_n) = 0.833$. Using Figure 5.13(a), we find that the percent overshoot is 55%. We expect the settling time to within 2% of the final value to be

$$T_s = \frac{4}{\zeta\omega_n} = \frac{4}{0.6(5)} = 1.33 \text{ s.}$$

Using a computer simulation for the actual third-order system, we find that the percent overshoot is equal to 38% and the settling time is 1.6 seconds. Thus, the effect of the third pole of $T(s)$ is to dampen the overshoot and increase the settling time (hence the real pole cannot be neglected). ■

The damping ratio plays a fundamental role in closed-loop system performance. As seen in the design formulas for settling time, percent overshoot, peak time, and rise time, the damping ratio is a key factor in determining the overall performance. In fact, for second-order systems, the damping ratio is the only factor determining the value of the percent overshoot to a step input. As it turns out, the damping ratio can be estimated from the response of a system to a step input [14]. The step response of a second-order system for a unit step input is given in Equation (5.9), which is

$$y(t) = 1 - \frac{1}{\beta}e^{-\zeta\omega_n t} \sin(\omega_n\beta t + \theta),$$

where $\beta = \sqrt{1 - \zeta^2}$, and $\theta = \cos^{-1}\zeta$. Hence, the frequency of the damped sinusoidal term for $\zeta < 1$ is

$$\omega = \omega_n(1 - \zeta^2)^{1/2} = \omega_n\beta,$$

and the number of cycles in 1 second is $\omega/(2\pi)$.

The time constant for the exponential decay is $\tau = 1/(\zeta\omega_n)$ in seconds. The number of cycles of the damped sinusoid during one time constant is

$$(\text{cycles/time}) \times \tau = \frac{\omega}{2\pi\zeta\omega_n} = \frac{\omega_n\beta}{2\pi\zeta\omega_n} = \frac{\beta}{2\pi\zeta}.$$

Assuming that the response decays in n visible time constants, we have

$$\text{cycles visible} = \frac{n\beta}{2\pi\zeta}. \tag{5.19}$$

For the second-order system, the response remains within 2% of the steady-state value after four time constants (4τ). Hence, $n = 4$, and

$$\text{cycles visible} = \frac{4\beta}{2\pi\zeta} = \frac{4(1 - \zeta^2)^{1/2}}{2\pi\zeta} \simeq \frac{0.55}{\zeta} \tag{5.20}$$

for $0.2 \le \zeta \le 0.6$.

As an example, examine the response shown in Figure 5.5(a) for $\zeta = 0.4$. Use $y(t) = 0$ as the first minimum point and count 1.4 cycles visible (until the response settles with 2% of the final value). Then we estimate

$$\zeta = \frac{0.55}{\text{cycles}} = \frac{0.55}{1.4} = 0.39.$$

We can use this approximation for systems with dominant complex poles so that

$$T(s) \approx \frac{\omega_n^2}{s^2 + 2\zeta\omega_n s + \omega_n^2}.$$

Then we are able to estimate the damping ratio ζ from the actual system response of a physical system.

An alternative method of estimating ζ is to determine the percent overshoot for the step response and use Figure 5.8 to estimate ζ. For example, we determine an overshoot of 25% for $\zeta = 0.4$ from the response of Figure 5.5(a). Using Figure 5.8, we estimate that $\zeta = 0.4$, as expected.

5.5 THE s-PLANE ROOT LOCATION AND THE TRANSIENT RESPONSE

The transient response of a closed-loop feedback control system can be described in terms of the location of the poles of the transfer function. The closed-loop transfer function is written in general as

$$T(s) = \frac{Y(s)}{R(s)} = \frac{\sum P_i(s) \Delta_i(s)}{\Delta(s)},$$

where $\Delta(s) = 0$ is the characteristic equation of the system. For the single-loop system of Figure 5.4, the characteristic equation reduces to $1 + G(s) = 0$. It is the poles and zeros of $T(s)$ that determine the transient response. However, for a closed-loop system, the poles of $T(s)$ are the roots of the characteristic equation $\Delta(s) = 0$ and the poles of $\Sigma P_i(s) \Delta_i(s)$. The output of a system (with gain = 1) without repeated roots and a unit step input can be formulated as a partial fraction expansion as

$$Y(s) = \frac{1}{s} + \sum_{i=1}^{M} \frac{A_i}{s + \sigma_i} + \sum_{k=1}^{N} \frac{B_k s + C_k}{s^2 + 2\alpha_k s + (\alpha_k^2 + \omega_k^2)}, \qquad (5.21)$$

where the A_i, B_k, and C_k are constants. The roots of the system must be either $s = -\sigma_i$ or complex conjugate pairs such as $s = -\alpha_k \pm j\omega_k$. Then the inverse transform results in the transient response as the sum of terms

$$y(t) = 1 + \sum_{i=1}^{M} A_i e^{-\sigma_i t} + \sum_{k=1}^{N} D_k e^{-\alpha_k t} \sin(\omega_k t + \theta_k), \qquad (5.22)$$

where D_k is a constant and depends on B_k, C_k, α_k, and ω_k. The transient response is composed of the steady-state output, exponential terms, and damped sinusoidal terms. For the response to be stable—that is, bounded for a step input—the real part of the roots, $-\sigma_i$ and $-\alpha_k$, must be in the left-hand portion of the s-plane. The impulse response for various root locations is shown in Figure 5.17. The information imparted by the location of the roots is graphic indeed, and usually well worth the effort of determining the location of the roots in the s-plane.

It is important for the control system analyst to understand the complete relationship of the complex-frequency representation of a linear system, the poles and zeros of its transfer function, and its time-domain response to step and other inputs. In such areas as signal processing and control, many of the analysis and design calculations are done in the complex-frequency plane, where a system model is

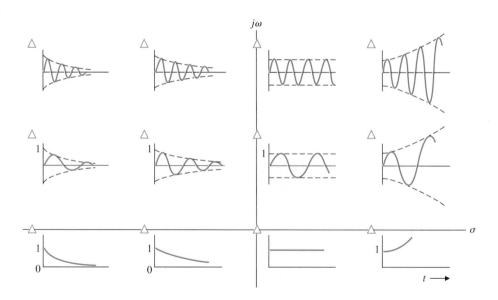

FIGURE 5.17
Impulse response for various root locations in the s-plane. (The conjugate root is not shown.)

represented in terms of the poles and zeros of its transfer function $T(s)$. On the other hand, system performance is often analyzed by examining time-domain responses, particularly when dealing with control systems.

The capable system designer will envision the effects on the step and impulse responses of adding, deleting, or moving poles and zeros of $T(s)$ in the s-plane. Likewise, the designer should visualize the necessary changes for the poles and zeros of $T(s)$, in order to effect desired changes in the model's step and impulse responses.

An experienced designer is aware of the effects of zero locations on system response. The poles of $T(s)$ determine the particular response modes that will be present, and the zeros of $T(s)$ establish the relative weightings of the individual mode functions. For example, moving a zero closer to a specific pole will reduce the relative contribution of the mode function corresponding to the pole.

A computer program can be developed to allow a user to specify arbitrary sets of poles and zeros for the transfer function of a linear system. Then the computer will evaluate and plot the system's impulse and step responses individually. It will also display them in reduced form along with the pole–zero plot.

Once the program has been run for a set of poles and zeros, the user can modify the locations of one or more of them. Plots may then be presented showing the old and new poles and zeros in the complex plane and the old and new impulse and step responses.

5.6 THE STEADY-STATE ERROR OF FEEDBACK CONTROL SYSTEMS

One of the fundamental reasons for using feedback, despite its cost and increased complexity, is the attendant improvement in the reduction of the steady-state error of the system. As illustrated in Section 4.6, the steady-state error of a stable closed-loop system is usually several orders of magnitude smaller than the error of an open-loop system. The system actuating signal, which is a measure of the system error, is denoted as $E_a(s)$. Consider the closed-loop feedback system shown in Figure 5.18. According to the discussions in Chapter 4, we know from Equation (4.3) that with $N(s) = 0$, $T_d(s) = 0$, the tracking error is

$$E(s) = \frac{1}{1 + G_c(s)G(s)} R(s).$$

Using the final value theorem and computing the steady-state tracking error yields

$$\lim_{t \to \infty} e(t) = e_{ss} = \lim_{s \to \infty} s \frac{1}{1 + G_c(s)G(s)} R(s). \tag{5.23}$$

It is useful to determine the steady-state error of the system for the three standard test inputs for the unity feedback system. Later in this section we will consider steady-state tracking errors for non-unity feedback systems.

Step Input.   The steady-state error for a step input of magnitude A is therefore

$$e_{ss} = \lim_{s \to 0} \frac{s(A/s)}{1 + G_c(s)G(s)} = \frac{A}{1 + \lim_{s \to 0} G_c(s)G(s)}.$$

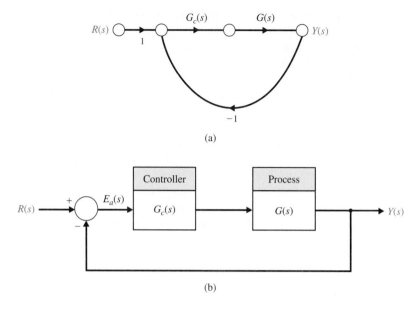

FIGURE 5.18
Closed-loop control system with unity feedback.

It is the form of the loop transfer function $G_c(s)G(s)$ that determines the steady-state error. The loop transfer function is written in general form as

$$G_c(s)G(s) = \frac{K \prod_{i=1}^{M}(s + z_i)}{s^N \prod_{k=1}^{Q}(s + p_k)}, \tag{5.24}$$

where $\prod$ denotes the product of the factors and $z_i \neq 0$, $p_k \neq 0$ for any $1 \leq i \leq M$ and $i \leq h \leq Q$. Therefore, the loop transfer function as s approaches zero depends on the number of integrations, N. If N is greater than zero, then $\lim_{s \to 0} G_c(s)G(s)$ approaches infinity, and the steady-state error approaches zero. The number of integrations is often indicated by labeling a system with a **type number** that simply is equal to N.

Consequently, for a type-zero system, $N = 0$, the steady-state error is

$$e_{ss} = \frac{A}{1 + G_c(0)G(0)} = \frac{A}{1 + K \prod_{i=1}^{M} z_i / \prod_{k=1}^{Q} p_k}. \tag{5.25}$$

The constant $G_c(0)G(0)$ is denoted by K_p, the **position error constant**, and is given by

$$\boxed{K_p = \lim_{s \to 0} G_c(s)G(s).}$$

The steady-state tracking error for a step input of magnitude A is thus given by

$$e_{ss} = \frac{A}{1 + K_p}. \tag{5.26}$$

Hence, the steady-state error for a unit step input with one integration or more, $N \geq 1$, is zero because

$$e_{ss} = \lim_{s \to 0} \frac{A}{1 + K \prod z_i / (s^N \prod p_k)}$$

$$= \lim_{s \to 0} \frac{A s^N}{s^N + K \prod z_i / \prod p_k} = 0. \tag{5.27}$$

Ramp Input.   The steady-state error for a ramp (velocity) input with a slope A is

$$e_{ss} = \lim_{s \to 0} \frac{s(A/s^2)}{1 + G_c(s)G(s)} = \lim_{s \to 0} \frac{A}{s + sG_c(s)G(s)} = \lim_{s \to 0} \frac{A}{sG_c(s)G(s)}. \tag{5.28}$$

Again, the steady-state error depends upon the number of integrations, N. For a type-zero system, $N = 0$, the steady-state error is infinite. For a type-one system, $N = 1$, the error is

$$e_{ss} = \lim_{s \to 0} \frac{A}{sK \prod (s + z_i) / [s \prod (s + p_k)]},$$

or

$$e_{ss} = \frac{A}{K \prod z_i / \prod p_k} = \frac{A}{K_v}, \tag{5.29}$$

where K_v is designated the **velocity error constant**. The velocity error constant is computed as

$$K_v = \lim_{s \to 0} sG_c(s)G(s).$$

When the transfer function possesses two or more integrations, $N \geq 2$, we obtain a steady-state error of zero. When $N = 1$, a steady-state error exists. However, the steady-state velocity of the output is equal to the velocity of input, as we shall see shortly.

Acceleration Input.   When the system input is $r(t) = At^2/2$, the steady-state error is

$$e_{ss} = \lim_{s \to 0} \frac{s(A/s^3)}{1 + G_c(s)G(s)} = \lim_{s \to 0} \frac{A}{s^2 G_c(s)G(s)}. \tag{5.30}$$

Table 5.5 Summary of Steady-State Errors

Number of Integrations in $G_c(s)G(s)$, Type Number	Step, $r(t) = A$, $R(s) = A/s$	Input Ramp, At, A/s^2	Parabola, $At^2/2$, A/s^3
0	$e_{ss} = \dfrac{A}{1 + K_p}$	Infinite	Infinite
1	$e_{ss} = 0$	$\dfrac{A}{K_v}$	Infinite
2	$e_{ss} = 0$	0	$\dfrac{A}{K_a}$

The steady-state error is infinite for one integration. For two integrations, $N = 2$, and we obtain

$$e_{ss} = \frac{A}{K \prod z_i / \prod p_k} = \frac{A}{K_a}, \tag{5.31}$$

where K_a is designated the **acceleration error constant**. The acceleration error constant is

$$K_a = \lim_{s \to 0} s^2 G_c(s)G(s).$$

When the number of integrations equals or exceeds three, then the steady-state error of the system is zero.

Control systems are often described in terms of their type number and the error constants, K_p, K_v, and K_a. Definitions for the error constants and the steady-state error for the three inputs are summarized in Table 5.5. The usefulness of the error constants can be illustrated by considering a simple example.

EXAMPLE 5.3   Mobile robot steering control

A mobile robot may be designed as an assisting device or servant for a severely disabled person [8]. The steering control system for such a robot can be represented by the block diagram shown in Figure 5.19. The steering controller is

$$G_c(s) = K_1 + K_2/s. \tag{5.32}$$

FIGURE 5.19
Block diagram of steering control system for a mobile robot.

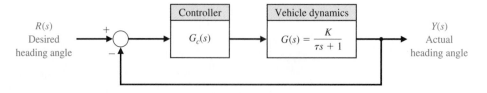

Therefore, the steady-state error of the system for a step input when $K_2 = 0$ and $G_c(s) = K_1$ is

$$e_{ss} = \frac{A}{1 + K_p}, \qquad (5.33)$$

where $K_p = KK_1$. When K_2 is greater than zero, we have a type-1 system,

$$G_c(s) = \frac{K_1 s + K_2}{s},$$

and the steady-state error is zero for a step input.

If the steering command is a ramp input, the steady-state error is

$$e_{ss} = \frac{A}{K_v}, \qquad (5.34)$$

where

$$K_v = \lim_{s \to 0} s G_c(s) G(s) = K_2 K.$$

The transient response of the vehicle to a triangular wave input when $G_c(s) = (K_1 s + K_2)/s$ is shown in Figure 5.20. The transient response clearly shows the effect of the steady-state error, which may not be objectionable if K_v is sufficiently large. Note that the output attains the desired velocity as required by the input, but it exhibits a steady-state error. ∎

The control system's error constants, K_p, K_v, and K_a, describe the ability of a system to reduce or eliminate the steady-state error. Therefore, they are utilized as numerical measures of the steady-state performance. The designer determines the error constants for a given system and attempts to determine methods of increasing the error constants while maintaining an acceptable transient response. In the case of the steering control system, we want to increase the gain factor KK_2 in order to increase K_v and reduce the steady-state error. However, an increase in KK_2 results

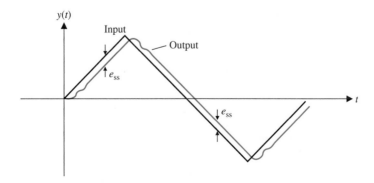

FIGURE 5.20
Triangular wave
response.

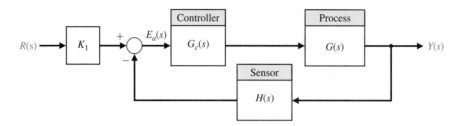

FIGURE 5.21
A nonunity
feedback system.

in an attendant decrease in the system's damping ratio ζ and therefore a more oscillatory response to a step input. Thus, we want a compromise that provides the largest K_v based on the smallest ζ allowable.

In the preceding discussions, we assumed that we had a unity feedback system where $H(s) = 1$. Now we consider nonunity feedback systems. A general feedback system with nonunity feedback is shown in Figure 5.21. For a system in which the feedback is not unity, the units of the output $Y(s)$ are usually different from the output of the sensor. For example, a speed control system is shown in Figure 5.22, where $H(s) = K_2$. The constants K_1 and K_2 account for the conversion of one set of units to another set of units (here we convert rad/s to volts). We can select K_1, and thus we set $K_1 = K_2$ and move the block for K_1 and K_2 past the summing node. Then we obtain the equivalent block diagram shown in Figure 5.23. Thus, we obtain a unity feedback system as desired.

Let us return to the system of Figure 5.21 with $H(s)$. In this case, suppose

$$H(s) = \frac{K_2}{\tau s + 1}$$

which has a DC gain of

$$\lim_{s \to 0} H(s) = K_2.$$

The factor K_2 is a conversion-of-units factor. If we set $K_2 = K_1$, then the system is transformed to that of Figure 5.23 for the steady-state calculation. To see this, consider error of the system $E(s)$, where

$$E(s) = R(s) - Y(s) = [1 - T(s)]R(s), \tag{5.35}$$

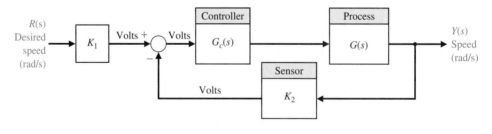

FIGURE 5.22
A speed control
system.

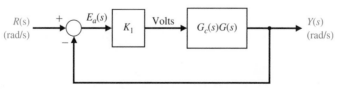

FIGURE 5.23
The speed control
system of
Figure 5.22 with
$K_1 = K_2$.

since $Y(s) = T(s)R(s)$. Note that

$$T(s) = \frac{K_1 G_c(s)G(s)}{1 + H(s)G_c(s)G(s)} = \frac{(\tau s + 1)K_1 G_c(s)G(s)}{\tau s + 1 + K_1 G_c(s)G(s)},$$

and therefore,

$$E(s) = \frac{1 + \tau s(1 - K_1 G_c(s)G(s))}{\tau s + 1 + K_1 G_c(s)G(s)} R(s).$$

Then the steady-state error for a unit step input is

$$e_{ss} = \lim_{s \to 0} s\, E(s) = \frac{1}{1 + K_1 \lim_{s \to 0} G_c(s)G(s)}. \tag{5.36}$$

We assume here that

$$\lim_{s \to 0} sG_c(s)G(s) = 0.$$

EXAMPLE 5.4   Steady-state error

Let us determine the appropriate value of K_1 and calculate the steady-state error for a unit step input for the system shown in Figure 5.21 when

$$G_c(s) = 40 \quad \text{and} \quad G(s) = \frac{1}{s + 5},$$

and

$$H(s) = \frac{20}{s + 10}.$$

We can rewrite $H(s)$ as

$$H(s) = \frac{2}{0.1s + 1}.$$

Selecting $K_1 = K_2 = 2$, we can use Equation (5.36) to determine

$$e_{ss} = \frac{1}{1 + K_1 \lim_{s \to 0} G_c(s)G(s)} = \frac{1}{1 + 2(40)(1/5)} = \frac{1}{17},$$

or 5.9% of the magnitude of the step input. ■

EXAMPLE 5.5   Feedback system

Let us consider the system of Figure 5.24, where we assume we cannot insert a gain K_1 following $R(s)$ as we did for the system of Figure 5.21. Then the actual error is given by Equation (5.35), which is

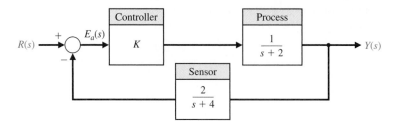

FIGURE 5.24
A system with a feedback H(s).

$$E(s) = [1 - T(s)]R(s).$$

Let us determine an appropriate gain K so that the steady-state error to a step input is minimized. The steady-state error is

$$e_{ss} = \lim_{s \to 0} s[1 - T(s)]\frac{1}{s},$$

where

$$T(s) = \frac{G_c(s)G(s)}{1 + G_c(s)G(s)H(s)} = \frac{K(s + 4)}{(s + 2)(s + 4) + 2K}.$$

Then we have

$$T(0) = \frac{4K}{8 + 2K}.$$

The steady-state error for a unit step input is

$$e_{ss} = 1 - T(0).$$

Thus, to achieve a zero steady-state error, we require that

$$T(0) = \frac{4K}{8 + 2K} = 1,$$

or $8 + 2K = 4K$. Thus, $K = 4$ will yield a zero steady-state error. ■

The determination of the steady-state error is simpler for unity feedback systems. However, it is possible to extend the notion of error constants to nonunity feedback systems by first appropriately rearranging the block diagram to obtain an equivalent unity feedback system. Remember that the underlying system must be stable, otherwise our use of the final value theorem will be compromised. Consider the nonunity feedback system in Figure 5.21 and assume that $K_1 = 1$. The closed-loop transfer function is

$$\frac{Y(s)}{R(s)} = T(s) = \frac{G_c(s)G(s)}{1 + H(s)G_c(s)G(s)}.$$

By manipulating the block diagram appropriately we can obtain the equivalent unity feedback system with

$$\frac{Y(s)}{R(s)} = T(s) = \frac{Z(s)}{1 + Z(s)} \text{ where } Z(s) = \frac{G_c(s)G(s)}{1 + G_c(s)G(s)(H(s) - 1)}.$$

The loop transfer function of the equivalent unity feedback system is $Z(s)$. It follows that the error constants for nonunity feedback systems are given as:

$$K_p = \lim_{s \to 0} Z(s), \; K_v = \lim_{s \to 0} sZ(s), \text{ and } K_a = \lim_{s \to 0} s^2 Z(s).$$

Note that when $H(s) = 1$, then $Z(s) = G_c(s)G(s)$ and we maintain the unity feedback error constants. For example, when $H(s) = 1$, then $K_p = \lim_{s \to 0} Z(s) = \lim_{s \to 0} G_c(s)G(s)$, as expected.

5.7 PERFORMANCE INDICES

Increasing emphasis on the mathematical formulation and measurement of control system performance can be found in the recent literature on automatic control. Modern control theory assumes that the systems engineer can specify quantitatively the required system performance. Then a performance index can be calculated or measured and used to evaluate the system's performance. A quantitative measure of the performance of a system is necessary for the operation of modern adaptive control systems, for automatic parameter optimization of a control system, and for the design of optimum systems.

Whether the aim is to improve the design of a system or to design a control system, a performance index must be chosen and measured.

> **A performance index is a quantitative measure of the performance**
> **of a system and is chosen so that emphasis is given**
> **to the important system specifications.**

A system is considered an **optimum control system** when the system parameters are adjusted so that the index reaches an extremum, commonly a minimum value. To be useful, a performance index must be a number that is always positive or zero. Then the best system is defined as the system that minimizes this index.

A suitable performance index is the integral of the square of the error, ISE, which is defined as

$$\text{ISE} = \int_0^T e^2(t) \, dt. \tag{5.37}$$

The upper limit T is a finite time chosen somewhat arbitrarily so that the integral approaches a steady-state value. It is usually convenient to choose T as the settling time T_s. The step response for a specific feedback control system is shown in Figure 5.25(b), and the error in Figure 5.25(c). The error squared is shown in Figure 5.25(d), and the integral of the error squared in Figure 5.25(e). This criterion will discriminate between excessively overdamped and excessively underdamped systems. The minimum value of the integral occurs for a compromise value of the damping. The performance index of Equation (5.37) is easily adapted for practical measurements because a squaring circuit is readily obtained. Furthermore, the squared error is mathematically convenient for analytical and computational purposes.

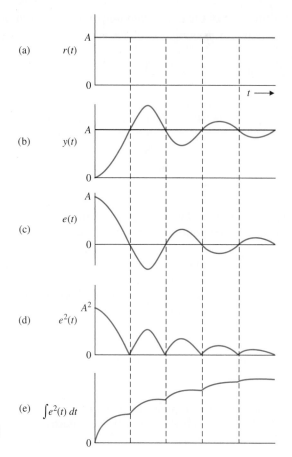

(a) $r(t)$

(b) $y(t)$

(c) $e(t)$

(d) $e^2(t)$

(e) $\int e^2(t)\,dt$

FIGURE 5.25
The calculation of
the integral squared
error.

Another readily instrumented performance criterion is the integral of the absolute magnitude of the error, IAE, which is written as

$$\text{IAE} = \int_0^T |e(t)|\,dt. \qquad (5.38)$$

This index is particularly useful for computer simulation studies.

To reduce the contribution of the large initial error to the value of the performance integral, as well as to emphasize errors occurring later in the response, the following index has been proposed [6]:

$$\text{ITAE} = \int_0^T t|e(t)|\,dt. \qquad (5.39)$$

This performance index is designated the integral of time multiplied by absolute error, ITAE. Another similar index is the integral of time multiplied by the squared error, or

$$\text{ITSE} = \int_0^T te^2(t)\,dt. \qquad (5.40)$$

The performance index ITAE provides the best selectivity of the performance indices; that is, the minimum value of the integral is readily discernible as the system parameters are varied. The general form of the performance integral is

$$I = \int_0^T f(e(t), r(t), y(t), t) \, dt, \tag{5.41}$$

where f is a function of the error, input, output, and time. We can obtain numerous indices based on various combinations of the system variables and time. Note that the minimization of IAE or ISE is often of practical significance. For example, the minimization of a performance index can be directly related to the minimization of fuel consumption for aircraft and space vehicles.

Performance indices are useful for the analysis and design of control systems. Two examples will illustrate the utility of this approach.

EXAMPLE 5.6   Performance criteria

A single-loop feedback control system is shown in Figure 5.26, where the natural frequency is the normalized value, $\omega_n = 1$. The closed-loop transfer function is then

$$T(s) = \frac{1}{s^2 + 2\zeta s + 1}. \tag{5.42}$$

Three performance indices—ISE, ITAE, and ITSE—calculated for various values of the damping ratio ζ and for a step input are shown in Figure 5.27. These curves show the selectivity of the ITAE index in comparison with the ISE index. The value of the damping ratio ζ selected on the basis of ITAE is 0.7. For a second-order system, this results in a swift response to a step with a 4.6% overshoot. ∎

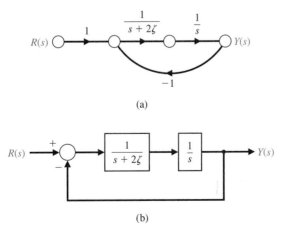

(a)

FIGURE 5.26
Single-loop feedback control system. (a) Signal-flow graph. (b) Block diagram model.

(b)

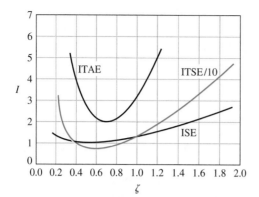

FIGURE 5.27
Three performance criteria for a second-order system.

EXAMPLE 5.7   Space telescope control system

The signal-flow graph and block diagram of a space telescope pointing control system are shown in Figure 5.28 [11]. We desire to select the magnitude of the gain, K_3, to minimize the effect of the disturbance, $T_d(s)$. In this case, the disturbance is equivalent to an initial attitude error. The closed-loop transfer function

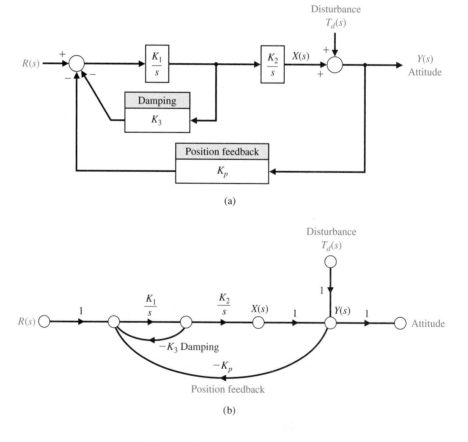

FIGURE 5.28
A space telescope pointing control system. (a) Block diagram. (b) Signal-flow graph.

for the disturbance is obtained by using Mason's signal-flow gain formula as follows:

$$\frac{Y(s)}{T_d(s)} = \frac{P_1(s)\,\Delta_1(s)}{\Delta(s)}$$

$$= \frac{1\cdot(1 + K_1K_3s^{-1})}{1 + K_1K_3s^{-1} + K_1K_2K_ps^{-2}} \qquad (5.43)$$

$$= \frac{s(s + K_1K_3)}{s^2 + K_1K_3s + K_1K_2K_p}.$$

Typical values for the constants are $K_1 = 0.5$ and $K_1K_2K_p = 2.5$. Then the natural frequency of the vehicle is $f_n = \sqrt{2.5}/(2\pi) = 0.25$ cycles/s. For a unit step disturbance, the minimum ISE can be analytically calculated. The attitude is

$$y(t) = \frac{\sqrt{10}}{\beta}\left[e^{-0.25K_3t}\sin\left(\frac{\beta}{2}t + \psi\right)\right], \qquad (5.44)$$

where $\beta = \sqrt{10 - K_3^2/4}$. Squaring $y(t)$ and integrating the result, we have

$$I = \int_0^\infty \frac{10}{\beta^2}e^{-0.5K_3t}\sin^2\left(\frac{\beta}{2}t + \psi\right)dt$$

$$= \int_0^\infty \frac{10}{\beta^2}e^{-0.5K_3t}\left(\frac{1}{2} - \frac{1}{2}\cos(\beta t + 2\psi)\right)dt \qquad (5.45)$$

$$= \frac{1}{K_3} + 0.1K_3.$$

Differentiating I and equating the result to zero, we obtain

$$\frac{dI}{dK_3} = -K_3^{-2} + 0.1 = 0. \qquad (5.46)$$

Therefore, the minimum ISE is obtained when $K_3 = \sqrt{10} = 3.2$. This value of K_3 corresponds to a damping ratio ζ of 0.50. The values of ISE and IAE for this system are plotted in Figure 5.29. The minimum for the IAE performance index is obtained when $K_3 = 4.2$ and $\zeta = 0.665$. While the ISE criterion is not as selective as the IAE criterion, it is clear that it is possible to solve analytically for the minimum value of ISE. The minimum of IAE is obtained by measuring the actual value of IAE for several values of the parameter of interest. ∎

A control system is optimum when the selected performance index is minimized. However, the optimum value of the parameters depends directly on the definition of optimum, that is, the performance index. Therefore, in Examples 5.6

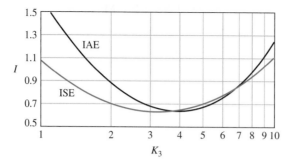

FIGURE 5.29
The performance indices of the telescope control system versus K_3.

and 5.7, we found that the optimum setting varied for different performance indices.

The coefficients that will minimize the ITAE performance criterion for a step input have been determined for the general closed-loop transfer function [6]

$$T(s) = \frac{Y(s)}{R(s)} = \frac{b_0}{s^n + b_{n-1}s^{n-1} + \cdots + b_1 s + b_0}. \tag{5.47}$$

This transfer function has a steady-state error equal to zero for a step input. Note that the transfer function has n poles and no zeros. The optimum coefficients for the ITAE criterion are given in Table 5.6. The responses using optimum coefficients for a step input are given in Figure 5.30 for ISE, IAE, and ITAE. The responses are provided for normalized time $\omega_n t$. Other standard forms based on different performance indices are available and can be useful in aiding the designer to determine the range of coefficients for a specific problem. A final example will illustrate the utility of the standard forms for ITAE.

EXAMPLE 5.8   **Two-camera control**

A very accurate and rapidly responding control system is required for a system that allows live actors to appear as if they are performing inside of complex miniature sets. The two-camera system is shown in Figure 5.31(a), where one camera is trained on the actor and the other on the miniature set. The challenge is to obtain rapid and accurate coordination of the two cameras by using sensor information from the

Table 5.6   The Optimum Coefficients of $T(s)$ Based on the ITAE Criterion for a Step Input

$$s + \omega_n$$
$$s^2 + 1.4\omega_n s + \omega_n^2$$
$$s^3 + 1.75\omega_n s^2 + 2.15\omega_n^2 s + \omega_n^3$$
$$s^4 + 2.1\omega_n s^3 + 3.4\omega_n^2 s^2 + 2.7\omega_n^3 s + \omega_n^4$$
$$s^5 + 2.8\omega_n s^4 + 5.0\omega_n^2 s^3 + 5.5\omega_n^3 s^2 + 3.4\omega_n^4 s + \omega_n^5$$
$$s^6 + 3.25\omega_n s^5 + 6.60\omega_n^2 s^4 + 8.60\omega_n^3 s^3 + 7.45\omega_n^4 s^2 + 3.95\omega_n^5 s + \omega_n^6$$

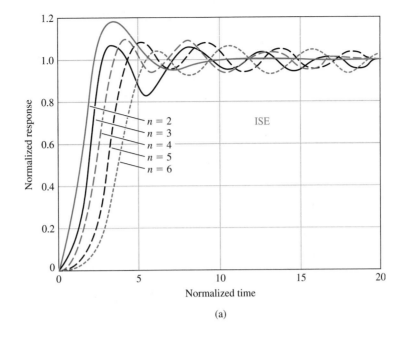

(a)

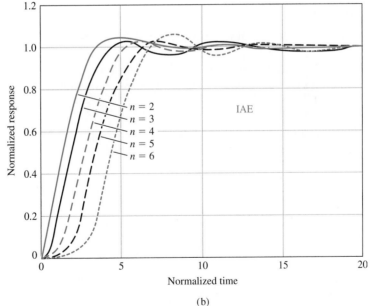

FIGURE 5.30
Step responses of a
normalized transfer
function using
optimum
coefficients for
(a) ISE, (b) IAE, and
(c) ITAE. The
response is for
normalized time,
$\omega_n t$.

(b)

foreground camera to control the movement of the background camera. The block
diagram of the background camera system is shown in Figure 5.31(b) for one axis of
movement of the background camera. The closed-loop transfer function is

$$T(s) = \frac{K_a K_m \omega_0^2}{s^3 + 2\zeta\omega_0 s^2 + \omega_0^2 s + K_a K_m \omega_0^2}. \tag{5.48}$$

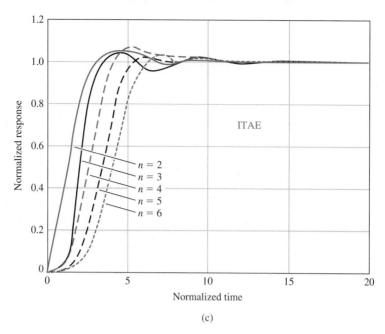

FIGURE 5.30

(*Continued*)

(c)

The standard form for a third-order system given in Table 5.6 requires that

$$2\zeta\omega_0 = 1.75\omega_n, \qquad \omega_0^2 = 2.15\omega_n^2, \quad \text{and} \quad K_a K_m \omega_0^2 = \omega_n^3.$$

Examining Figure 5.30(c) for $n = 3$, we estimate that the settling time is approximately 8 seconds (normalized time). Therefore, we estimate that

$$\omega_n T_s = 8.$$

Because a rapid response is required, a large ω_n will be selected so that the settling time will be less than 1 second. Thus, ω_n will be set equal to 10 rad/s. Then, for an ITAE system, it is necessary that the parameters of the camera dynamics be

$$\omega_0 = 14.67 \text{ rad/s}$$

and

$$\zeta = 0.597.$$

The amplifier and motor gain are required to be

$$K_a K_m = \frac{\omega_n^3}{\omega_0^2} = \frac{\omega_n^3}{2.15\omega_n^2} = \frac{\omega_n}{2.15} = 4.65.$$

Then the closed-loop transfer function is

$$T(s) = \frac{1000}{s^3 + 17.5s^2 + 215s + 1000}$$

$$= \frac{1000}{(s + 7.08)(s + 5.21 + j10.68)(s + 5.21 - j10.68)}. \qquad (5.49)$$

FIGURE 5.31
The foreground camera, which may be either a film or video camera, is trained on the blue cyclorama stage. The electronic servocontrol installation permits the slaving, by means of electronic servodevices, of the two cameras. The background camera reaches into the miniature set with a periscope lens and instantaneously reproduces all movements of the foreground camera in the scale of the miniature. The video control installation allows the composite image to be monitored and recorded live. (Part (a) reprinted with permission from *Electronic Design* 24, 11, May 24, 1976. Copyright © Hayden Publishing Co., Inc., 1976.)

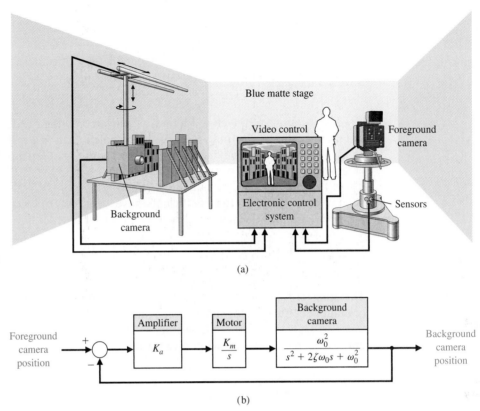

(a)

(b)

FIGURE 5.32
The closed-loop roots of a minimum ITAE system.

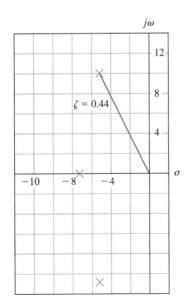

Table 5.7   The Optimum Coefficients of $T(s)$ Based on the ITAE Criterion for a Ramp Input

$$s^2 + 3.2\omega_n s + \omega_n^2$$

$$s^3 + 1.75\omega_n s^2 + 3.25\omega_n^2 s + \omega_n^3$$

$$s^4 + 2.41\omega_n s^3 + 4.93\omega_n^2 s^2 + 5.14\omega_n^3 s + \omega_n^4$$

$$s^5 + 2.19\omega_n s^4 + 6.50\omega_n^2 s^3 + 6.30\omega_n^3 s^2 + 5.24\omega_n^4 s + \omega_n^5$$

The locations of the closed-loop roots dictated by the ITAE system are shown in Figure 5.32. The damping ratio of the complex roots is $\zeta = 0.44$. However, the complex roots do not dominate. The actual response to a step input using a computer simulation showed the overshoot to be only 2% and the settling time (to within 2% of the final value) to be equal to 0.75 second.

For a ramp input, the coefficients have been determined that minimize the ITAE criterion for the general closed-loop transfer function [6]

$$T(s) = \frac{b_1 s + b_0}{s^n + b_{n-1} s^{n-1} + \cdots + b_1 s + b_0}. \tag{5.50}$$

This transfer function has a steady-state error equal to zero for a ramp input. The optimum coefficients for this transfer function are given in Table 5.7. The transfer function, Equation (5.50), implies that the process $G(s)$ has two or more pure integrations, as required to provide zero steady-state error. ■

5.8 THE SIMPLIFICATION OF LINEAR SYSTEMS

It is quite useful to study complex systems with high-order transfer functions by using lower-order approximate models. For example, a fourth-order system could be approximated by a second-order system leading to a use of the performance indices in Figure 5.8. Several methods are available for reducing the order of a systems transfer function.

One relatively simple way to delete a certain insignificant pole of a transfer function is to note a pole that has a negative real part that is much more negative than the other poles. Thus, that pole is expected to affect the transient response insignificantly.

For example, if we have a system with transfer function

$$G(s) = \frac{K}{s(s + 2)(s + 30)},$$

we can safely neglect the impact of the pole at $s = -30$. However, we must retain the steady-state response of the system, so we reduce the system to

$$G(s) = \frac{(K/30)}{s(s + 2)}.$$

A more sophisticated approach attempts to match the frequency response of the reduced-order transfer function with the original transfer function frequency response as closely as possible. Although frequency response methods are covered in Chapter 8, the associated approximation method strictly relies on algebraic manipulation and is presented here. We will let the high-order system be described by the transfer function

$$G_H(s) = K\frac{a_m s^m + a_{m-1}s^{m-1} + \cdots + a_1 s + 1}{b_n s^n + b_{n-1}s^{n-1} + \cdots + b_1 s + 1}, \tag{5.51}$$

in which the poles are in the left-hand s-plane and $m \le n$. The lower-order approximate transfer function is

$$G_L(s) = K\frac{c_p s^p + \cdots + c_1 s + 1}{d_g s^g + \cdots + d_1 s + 1}, \tag{5.52}$$

where $p \le g < n$. Notice that the gain constant, K, is the same for the original and approximate system; this ensures the same steady-state response. The method outlined in Example 5.9 is based on selecting c_i and d_i in such a way that $G_L(s)$ has a frequency response (see Chapter 8) very close to that of $G_H(s)$. This is equivalent to stating that $G_H(j\omega)/G_L(j\omega)$ is required to deviate the least amount from unity for various frequencies. The c and d coefficients are obtained by using the equations

$$M^{(k)}(s) = \frac{d^k}{ds^k}M(s) \tag{5.53}$$

and

$$\Delta^{(k)}(s) = \frac{d^k}{ds^k}\Delta(s), \tag{5.54}$$

where $M(s)$ and $\Delta(s)$ are the numerator and denominator polynomials of $G_H(s)/G_L(s)$, respectively. We also define

$$M_{2q} = \sum_{k=0}^{2q} \frac{(-1)^{k+q}M^{(k)}(0)M^{(2q-k)}(0)}{k!(2q-k)!}, \qquad q = 0, 1, 2\ldots \tag{5.55}$$

and an analogous equation for Δ_{2q}. The solutions for the c and d coefficients are obtained by equating

$$M_{2q} = \Delta_{2q} \tag{5.56}$$

for $q = 1, 2, \ldots$ up to the number required to solve for the unknown coefficients. Let us consider an example to clarify the use of these equations.

EXAMPLE 5.9   **A simplified model**

Consider the third-order system

$$G_H(s) = \frac{6}{s^3 + 6s^2 + 11s + 6} = \frac{1}{1 + \dfrac{11}{6}s + s^2 + \dfrac{1}{6}s^3}. \tag{5.57}$$

Using the second-order model

$$G_L(s) = \frac{1}{1 + d_1s + d_2s^2}, \tag{5.58}$$

we determine that

$$M(s) = 1 + d_1s + d_2s^2, \quad \text{and} \quad \Delta(s) = 1 + \frac{11}{6}s + s^2 + \frac{1}{6}s^3.$$

Then we know that

$$M^{(0)}(s) = 1 + d_1s + d_2s^2, \tag{5.59}$$

and $M^{(0)}(0) = 1$. Similarly, we have

$$M^{(1)} = \frac{d}{ds}(1 + d_1s + d_2s^2) = d_1 + 2d_2s. \tag{5.60}$$

Therefore, $M^{(1)}(0) = d_1$. Continuing this process, we find that

$$M^{(0)}(0) = 1 \qquad \Delta^{(0)}(0) = 1,$$
$$M^{(1)}(0) = d_1 \qquad \Delta^{(1)}(0) = \frac{11}{6},$$
$$M^{(2)}(0) = 2d_2 \qquad \Delta^{(2)}(0) = 2, \tag{5.61}$$

and

$$M^{(3)}(0) = 0 \qquad \Delta^{(3)}(0) = 1.$$

We now equate $M_{2q} = \Delta_{2q}$ for $q = 1$ and 2. We find that, for $q = 1$,

$$M_2 = (-1)\frac{M^{(0)}(0)M^{(2)}(0)}{2} + \frac{M^{(1)}(0)M^{(1)}(0)}{1} + (-1)\frac{M^{(2)}(0)M^{(0)}(0)}{2}$$
$$= -d_2 + d_1{}^2 - d_2 = -2d_2 + d_1{}^2. \tag{5.62}$$

Since the equation for Δ_2 is similar, we have

$$\Delta_2 = (-1)\frac{\Delta^{(0)}(0)\,\Delta^{(2)}(0)}{2} + \frac{\Delta^{(1)}(0)\,\Delta^{(1)}(0)}{1} + (-1)\frac{\Delta^{(2)}(0)\,\Delta^{(0)}(0)}{2}$$
$$= -1 + \frac{121}{36} - 1 = \frac{49}{36}. \tag{5.63}$$

Equation (5.56) with $q = 1$ requires that $M_2 = \Delta_2$; therefore,

$$-2d_2 + d_1^2 = \frac{49}{36}. \tag{5.64}$$

Completing the process for $M_4 = \Delta_4$, we obtain

$$d_2^2 = \frac{7}{18}. \tag{5.65}$$

Solving Equations (5.64) and (5.65) yields $d_1 = 1.615$ and $d_2 = 0.624$. (The other sets of solutions are rejected because they lead to unstable poles.) The lower-order system transfer function is

$$G_L(s) = \frac{1}{1 + 1.615s + 0.624s^2} = \frac{1.60}{s^2 + 2.590s + 1.60}. \tag{5.66}$$

It is interesting to see that the poles of $H(s)$ are $s = -1, -2, -3$, whereas the poles of $G_L(s)$ are $s = -1.024$ and -1.565. Because the lower-order model has two poles, we estimate that we would obtain a slightly overdamped step response with a settling time to within 2% of the final value in approximately 3 seconds. ∎

It is sometimes desirable to retain the dominant poles of the original system, $G_H(s)$, in the low-order model. This can be accomplished by specifying the denominator of $G_L(s)$ to be the dominant poles of $G_H(s)$ and allowing the numerator of $G_L(s)$ to be subject to approximation.

Another novel and useful method for reducing the order is the Routh approximation method based on the idea of truncating the Routh table used to determine stability. The Routh approximants can be computed by a finite recursive algorithm that is suited for programming on a digital computer [22].

A robot named Domo was developed to investigate robot manipulation in unstructured environments [25–26]. The robot shown in Figure 5.33 has 29 degrees of freedom, making it a very complex system. Domo employs two six-degree-of-freedom arms and hands with compliant and force-sensitive actuators coupled with a behavior-based system architecture to achieve robotic manipulation tasks in human environments. Designing a controller to control the motion of the arm and hands would require significant model reduction and approximation before the methods of design discussed in the subsequent chapters (e.g., root locus design methods) could be successfully applied.

5.9 DESIGN EXAMPLES

In this section we present two illustrative examples. The first example is a simplified view of the Hubble space telescope pointing control problem. The Hubble space telescope problem highlights the process of computing controller gains to achieve desired percent overshoot specifications, as well as meeting steady-state error specifications. The second example considers the control of the bank angle of an airplane. The airplane attitude motion control example represents a more in-depth look at the control design problem. Here we consider a complex fourth-order model of

FIGURE 5.33
An upper-torso
humanoid robot
named Domo helps
researchers
investigate robot
manipulation in
unstructured
environments.
(Photo courtesy of
Aaron Edsinger, MIT
Humanoid Robotics
Group.)

the lateral dynamics of the aircraft motion that is approximated by a second-order
model using the approximation methods of Section 5.8. The simplified model can be
used to gain insight into the controller design and the impact of key controller parameters on the transient performance.

EXAMPLE 5.10   **Hubble space telescope control**

The orbiting Hubble space telescope is the most complex and expensive scientific instrument that has ever been built. Launched to 380 miles above the earth on April 24,
1990, the telescope has pushed technology to new limits. The telescope's 2.4 meter
(94.5-inch) mirror has the smoothest surface of any mirror made, and its pointing system can center it on a dime 400 miles away [21]. The telescope had a spherical aber-
ration that was largely corrected during space missions in 1993 and 1997 [24].
Consider the model of the telescope-pointing system shown in Figure 5.34.

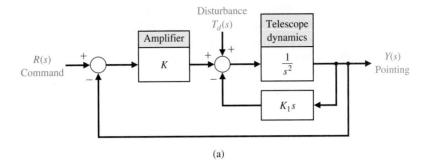

(a)

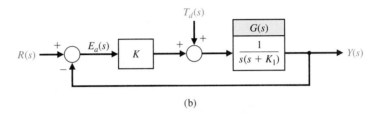

(b)

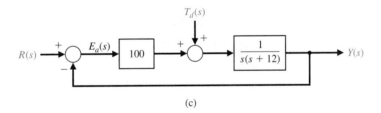

(c)

FIGURE 5.34
(a) The Hubble
telescope pointing
system, (b) reduced
block diagram,
(c) system design,
and (d) system
response to a unit
step input
command and a
unit step
disturbance input.

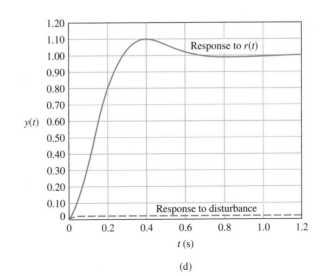

(d)

The goal of the design is to choose K_1 and K so that (1) the percent overshoot of the output to a step command, $r(t)$, is less than or equal to 10%, (2) the steady-state error to a ramp command is minimized, and (3) the effect of a step disturbance is reduced. Since the system has an inner loop, block diagram reduction can be used to obtain the simplified system of Figure 5.34(b).

The output due to the two inputs of the system of Figure 5.34(b) is given by

$$Y(s) = T(s)R(s) + [T(s)/K]T_d(s), \tag{5.67}$$

where

$$T(s) = \frac{KG(s)}{1 + KG(s)} = \frac{L(s)}{1 + L(s)}.$$

The error is

$$E(s) = \frac{1}{1 + L(s)}R(s) - \frac{G(s)}{1 + L(s)}T_d(s). \tag{5.68}$$

First, let us select K and K_1 to meet the percent overshoot requirement for a step input, $R(s) = A/s$. Setting $T_d(s) = 0$, we have

$$Y(s) = \frac{KG(s)}{1 + KG(s)}R(s)$$

$$= \frac{K}{s(s + K_1) + K}\left(\frac{A}{s}\right) = \frac{K}{s^2 + K_1 s + K}\left(\frac{A}{s}\right). \tag{5.69}$$

To set the overshoot less than 10%, we select $\zeta = 0.6$ by examining Figure 5.8 or using Equation (5.16) to determine that the overshoot will be 9.5% for $\zeta = 0.6$. We next examine the steady-state error for a ramp, $r(t) = Bt, t \geq 0$, using (Equation 5.28):

$$e_{ss} = \lim_{s \to 0}\left\{\frac{B}{sKG(s)}\right\} = \frac{B}{K/K_1}. \tag{5.70}$$

The steady-state error due to a unit step disturbance is equal to $-1/K$. (The student should show this.) The transient response of the error due to the step disturbance input can be reduced by increasing K (see Equation 5.68). In summary, we seek a large K and a large value of K/K_1 to obtain a low steady-state error for the ramp input (see Equation 5.70). However, we also require $\zeta = 0.6$ to limit the overshoot.

For our design, we need to select K. With $\zeta = 0.6$, the characteristic equation of the system is

$$s^2 + 2\zeta\omega_n s + \omega_n^2 = s^2 + 2(0.6)\omega_n s + K. \tag{5.71}$$

Therefore, $\omega_n = \sqrt{K}$, and the second term of the denominator of Equation (5.69) requires $K_1 = 2(0.6)\omega_n$. Then $K_1 = 1.2\sqrt{K}$, so the ratio K/K_1 becomes

$$\frac{K}{K_1} = \frac{K}{1.2\sqrt{K}} = \frac{\sqrt{K}}{1.2}.$$

Selecting $K = 25$, we have $K_1 = 6$ and $K/K_1 = 4.17$. If we select $K = 100$, we have $K_1 = 12$ and $K/K_1 = 8.33$. Realistically, we must limit K so that the system's operation remains linear. Using $K = 100$, we obtain the system shown in Figure 5.34(c). The responses of the system to a unit step input command and a unit step disturbance input are shown in Figure 5.34(d). Note how the effect of the disturbance is relatively insignificant.

Finally, we note that the steady-state error for a ramp input (see Equation 5.70) is

$$e_{ss} = \frac{B}{8.33} = 0.12B.$$

This design, using $K = 100$, is an excellent system. ∎

EXAMPLE 5.11   **Attitude control of an airplane**

Each time we fly on a commercial airliner, we experience first-hand the benefits of automatic control systems. These systems assist pilots by improving the handling qualities of the aircraft over a wide range of flight conditions and by providing pilot relief (for such emergencies as going to the restroom) during extended flights. The special relationship between flight and controls began in the early work of the Wright brothers. Using wind tunnels, the Wright brothers applied systematic design techniques to make their dream of powered flight a reality. This systematic approach to design contributed to their success.

Another significant aspect of their approach was their emphasis on flight controls; the brothers insisted that their aircraft be pilot-controlled. Observing birds control their rolling motion by twisting their wings, the Wright brothers built aircraft with mechanical mechanisms that twisted their airplane wings. Today we no longer use wing warping as a mechanism for performing a roll maneuver; instead we control rolling motion by using ailerons, as shown in Figure 5.35. The Wright brothers also used elevators (located forward) for longitudinal

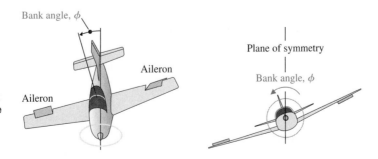

FIGURE 5.35
Control of the bank angle of an airplane using differential deflections of the ailerons.

control (pitch motion) and rudders for lateral control (yaw motion). Today's aircraft still use both elevators and rudders, although the elevators are generally located on the tail (rearward).

The first controlled, powered, unassisted take-off flight occurred in 1903 with the *Wright Flyer I* (a.k.a. *Kitty Hawk*). The first practical airplane, the *Flyer III*, could fly figure eights and stay aloft for half an hour. Three-axis flight control was a major (and often overlooked) contribution of the Wright brothers. A concise historical perspective is presented in Stevens and Lewis [27]. The continuing desire to fly faster, lighter, and longer fostered further developments in automatic flight control. Today's challenge is to develop a single-stage-to-orbit aircraft/spacecraft that can take off and land on a standard runway.

The main topic of this chapter is control of the automatic rolling motion of an airplane. The elements of the design process emphasized in this chapter are illustrated in Figure 5.36.

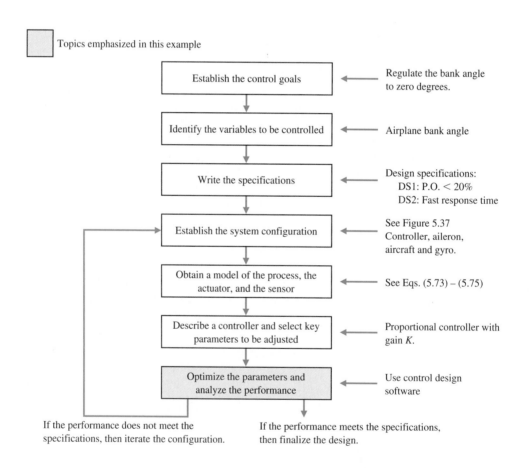

FIGURE 5.36   Elements of the control system design process emphasized in the airplane attitude control example.

We begin by considering the model of the lateral dynamics of an airplane moving along a steady, wings-level flight path. By lateral dynamics, we mean the attitude motion of the aircraft about the forward velocity. An accurate mathematical model describing the motion (translational and rotational) of an aircraft is a complicated set of highly nonlinear, time-varying, coupled differential equations. A good description of the process of developing such a mathematical model appears in Etkin and Reid [28].

For our purposes a simplified dynamic model is required for the autopilot design process. A simplified model might consist of a transfer function describing the input/output relationship between the aileron deflection and the aircraft bank angle. Obtaining such a transfer function would require many prudent simplifications to the original high-fidelity, nonlinear mathematical model.

Suppose we have a rigid aircraft with a plane of symmetry. The airplane is assumed to be cruising at subsonic or low supersonic (Mach < 3) speeds. This allows us to make a flat-earth approximation. We ignore any rotor gyroscopic effects due to spinning masses on the aircraft (such as propellors or turbines). These assumptions allow us to decouple the longitudinal rotational (pitching) motion from the lateral rotational (rolling and yawing) motion.

Of course, we also need to consider a linearization of the nonlinear equations of motion. To accomplish this, we consider only steady-state flight conditions such as

- ❑ Steady, wings-level flight
- ❑ Steady, level turning flight
- ❑ Steady, symmetric pull-up
- ❑ Steady roll.

For this example we assume that the airplane is flying at low speed in a steady, wings-level attitude, and we want to design an autopilot to control the rolling motion. We can state the control goal as follows:

Control Goal
> Regulate the airplane bank angle to zero degrees (steady, wings level) and maintain the wings-level orientation in the presence of unpredictable external disturbances.

We identify the variable to be controlled as

Variable to Be Controlled
> Airplane bank angle (denoted by ϕ).

Defining system specifications for aircraft control is complicated, so we do not attempt it here. It is a subject in and of itself, and many engineers have spent significant efforts developing good, practical design specifications. The goal is to design a control system such that the dominant closed-loop system poles have satisfactory natural frequency and damping [27]. We must define satisfactory and choose test input signals on which to base our analysis.

The Cooper–Harper pilot opinion ratings provide a way to correlate the feel of the airplane with control design specifications [29]. These ratings address the handling qualities issues. Many flying qualities requirements are specified by government agencies, such as the United States Air Force [30]. The USAF MIL-F-8785C is a source of time-domain control system design specifications.

For example we might design an autopilot control system for an aircraft in steady, wings-level flight to achieve a 20% overshoot to a step input with minimal oscillatory motion and rapid response time (that is, a short time-to-peak). Subsequently we implement the controller in the aircraft control system and conduct flight tests or high-fidelity computer simulations, after which the pilots tell us whether they liked the performance of the aircraft. If the overall performance was not satisfactory, we change the time-domain specification (in this case a percent overshoot specification) and redesign until we achieve a feel and performance that pilots (and ultimately passengers) will accept. Despite the simplicity of this approach and many years of research, precise-control system design specifications that provide acceptable airplane flying characteristics in all cases are still not available [27].

The control design specifications given in this example may seem somewhat contrived. In reality the specifications would be much more involved and, in many ways, less precisely known. But recall in Chapter 1 we discussed the fact that we must begin the design process somewhere. With that approach in mind, we select simple design specifications and begin the iterative design process. The design specifications are

Control Design Specifications

DS1   Percent overshoot less than 20% for a unit step input.

DS2   Fast response time as measured by time-to-peak.

By making the simplifying assumptions discussed above and linearizing about the steady, wings-level flight condition, we can obtain a transfer function model describing the bank angle output, $\phi(s)$, to the aileron deflection input, $\delta_a(s)$. The transfer function has the form

$$\frac{\phi(s)}{\delta_a(s)} = \frac{k(s - c_0)(s^2 + b_1 s + b_0)}{s(s + d_0)(s + e_0)(s^2 + f_1 s + f_0)}. \tag{5.72}$$

The lateral (roll/yaw) motion has three main modes: Dutch roll mode, spiral mode, and roll subsidence mode. The Dutch roll mode, which gets its name from its similarities to the motion of an ice speed skater, is characterized by a rolling and yawing motion. The airplane center of mass follows nearly a straightline path, and a rudder impulse can excite this mode. The spiral mode is characterized by a mainly yawing motion with some roll motion. This is a weak mode, but it can cause an airplane to enter a steep spiral dive. The roll subsidence motion is almost a pure roll motion. This is the motion we are concerned with for our autopilot design. The denominator of the transfer function in Equation (5.72) shows two first-order modes (spiral and roll subsidence modes) and a second-order mode (Dutch roll mode).

In general the coefficients c_0, b_0, b_1, d_0, e_0, f_0, f_1 and the gain k are complicated functions of stability derivatives. The stability derivatives are functions of the flight conditions and the aircraft configuration; they differ for different aircraft types. The coupling between the roll and yaw is included in Equation (5.72).

In the transfer function in Equation (5.72), the pole at $s = -d_0$ is associated with the spiral mode. The pole at $s = -e_0$ is associated with the roll subsidence mode. Generally, $e_0 \gg d_0$. For an F-16 flying at 500 ft/s in steady, wings-level flight, we have $e_0 = 3.57$ and $d_0 = 0.0128$ [27]. The complex conjugate poles given by the term $s^2 + f_1 s + f_0$ represent the Dutch roll motion.

For low angles of attack (such as with steady, wings-level flight), the Dutch roll mode generally cancels out of the transfer function with the $s^2 + b_1 s + b_0$ term. This is an approximation, but it is consistent with our other simplifying assumptions. Also, we can ignore the spiral mode since it is essentially a yaw motion only weakly coupled to the roll motion. The zero at $s = c_0$ represents a gravity effect that causes the aircraft to sideslip as it rolls. We assume that this effect is negligible, since it is most pronounced in a slow roll maneuver in which the sideslip is allowed to build up, and we assume that the aircraft sideslip is small or zero. Therefore we can simplify the transfer function in Eq. (5.72) to obtain a single-degree-of-freedom approximation:

$$\frac{\phi(s)}{\delta_a(s)} = \frac{k}{s(s + e_0)}. \tag{5.73}$$

For our aircraft we select $e_0 = 1.4$ and $k = 11.4$. The associated time-constant of the roll subsidence is $\tau = 1/e_0 = 0.7\,\text{s}$. These values represent a fairly fast rolling motion response.

For the aileron actuator model, we typically use a simple first-order system model,

$$\frac{\delta_a(s)}{e(s)} = \frac{p}{s + p}, \tag{5.74}$$

where $e(s) = \phi_d(s) - \phi(s)$. In this case we select $p = 10$. This corresponds to a time constant of $\tau = 1/p = 0.1$ s. This is a typical value consistent with a fast response. We need to have an actuator with a fast response so that the dynamics of the actively controlled airplane will be the dominant component of the system response. A slow actuator is akin to a time delay that can cause performance and stability problems.

For a high-fidelity simulation, we would need to develop an accurate model of the gyro dynamics. The gyro, typically an integrating gyro, is usually characterized by a very fast response. To remain consistent with our other simplifying assumptions, we ignore the gyro dynamics in the design process. This means we assume that the sensor measures the bank angle precisely. The gyro model is given by a unity transfer function,

$$K_g = 1. \tag{5.75}$$

Thus our physical system model is given by Equations (5.73), (5.74), and (5.75).

The controller we select for this design is a proportional controller,

$$G_c(s) = K.$$

The system configuration is shown in Figure 5.37. The select key parameter is as follows:

Select Key Tuning Parameter
    Controller gain K.

The closed-loop transfer function is

$$T(s) = \frac{\phi(s)}{\phi_d(s)} = \frac{114K}{s^3 + 11.4s^2 + 14s + 114K}. \tag{5.76}$$

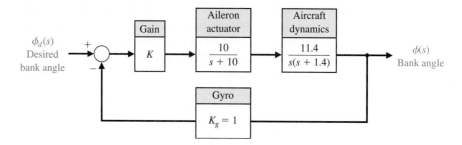

FIGURE 5.37
Bank angle control autopilot.

We want to determine analytically the values of K that will give us the desired response, namely, a percent overshoot less than 20% and a fast time-to-peak. The analytic analysis would be simpler if our closed-loop system were a second-order system (since we have valuable relationships between settling time, percent overshoot, natural frequency and damping ratio); however we have a third-order system, given by $T(s)$ in Equation (5.76). We could consider approximating the third-order transfer function by a second-order transfer function—this is sometimes a very good engineering approach to analysis. There are many methods available to obtain approximate transfer functions. Here we use the algebraic method described in Section 5.8 that attempts to match the frequency response of the approximate system as closely as possible to the actual system.

Our transfer function can be rewritten as

$$T(s) = \frac{1}{1 + \frac{14}{114K}s + \frac{11.4}{114K}s^2 + \frac{1}{114K}s^3},$$

by factoring the constant term out of the numerator and denominator. Suppose our approximate transfer function is given by the second-order system

$$G_L(s) = \frac{1}{1 + d_1s + d_2s^2}.$$

The objective is to find appropriate values of d_1 and d_2. As in Section 5.8, we define $M(s)$ and $\Delta(s)$ as the numerator and denominator of $T(s)/G_L(s)$. We also define

$$M_{2q} = \sum_{k=0}^{2q} \frac{(-1)^{k+q}M^{(k)}(0)M^{(2q-k)}(0)}{k!(2q-k)!}, \quad q = 1, 2, \ldots, \tag{5.77}$$

and

$$\Delta_{2q} = \sum_{k=0}^{2q} \frac{(-1)^{k+q}\Delta^{(k)}(0)\Delta^{(2q-k)}(0)}{k!(2q-k)!}, \quad q = 1, 2, \ldots. \tag{5.78}$$

Then, forming the set of algebraic equations

$$M_{2q} = \Delta_{2q}, \quad q = 1, 2, \ldots, \tag{5.79}$$

we can solve for the unknown parameters of the approximate function. The index q is incremented until sufficient equations are obtained to solve for the unknown coefficients of the approximate function. In this case, $q = 1, 2$ since we have two parameters d_1 and d_2 to compute.

We have

$$M(s) = 1 + d_1 s + d_2 s^2$$

$$M^{(1)}(s) = \frac{dM}{ds} = d_1 + 2d_2 s$$

$$M^{(2)}(s) = \frac{d^2 M}{ds^2} = 2d_2$$

$$M^{(3)}(s) = M^4(s) = \cdots = 0.$$

Thus evaluating at $s = 0$ yields

$$M^{(1)}(0) = d_1$$

$$M^{(2)}(0) = 2d_2$$

$$M^{(3)}(0) = M^{(4)}(0) = \cdots = 0.$$

Similarly,

$$\Delta(s) = 1 + \frac{14}{114K} s + \frac{11.4}{114K} s^2 + \frac{s^3}{114K}$$

$$\Delta^{(1)}(s) = \frac{d\Delta}{ds} = \frac{14}{114K} + \frac{22.8}{114K} s + \frac{3}{114K} s^2$$

$$\Delta^{(2)}(s) = \frac{d^2\Delta}{ds^2} = \frac{22.8}{114K} + \frac{6}{114K} s$$

$$\Delta^{(3)}(s) = \frac{d^3\Delta}{ds^3} = \frac{6}{114K}$$

$$\Delta^{(4)}(s) = \Delta^5(s) = \cdots = 0.$$

Evaluating at $s = 0$, it follows that

$$\Delta^{(1)}(0) = \frac{14}{114K},$$

$$\Delta^{(2)}(0) = \frac{22.8}{114K},$$

$$\Delta^{(3)}(0) = \frac{6}{114K},$$

$$\Delta^{(4)}(0) = \Delta^{(5)}(0) = \cdots = 0.$$

Using Equation (5.77) for $q = 1$ and $q = 2$ yields

$$M_2 = -\frac{M(0)M^{(2)}(0)}{2} + \frac{M^{(1)}(0)M^{(1)}(0)}{1} - \frac{M^{(2)}(0)M(0)}{2} = -2d_2 + d_1^2;$$

and

$$M_4 = \frac{M(0)M^{(4)}(0)}{0!\,4!} - \frac{M^{(1)}(0)M^{(3)}(0)}{1!\,3!} + \frac{M^{(2)}(0)M^{(2)}(0)}{2!\,2!}$$
$$- \frac{M^{(3)}(0)M^{(1)}(0)}{3!\,1!} + \frac{M^{(4)}(0)M(0)}{4!\,0!} = d_2^2.$$

Similarly using Equation (5.78), we find that

$$\Delta_2 = \frac{-22.8}{114K} + \frac{196}{(114K)^2} \quad \text{and} \quad \Delta_4 = \frac{101.96}{(114K)^2}.$$

Thus forming the set of algebraic equations in Equation (5.79),

$$M_2 = \Delta_2 \quad \text{and} \quad M_4 = \Delta_4,$$

we obtain

$$-2d_2 + d_1{}^2 = \frac{-22.8}{114K} + \frac{196}{(114K)^2} \quad \text{and} \quad d_2{}^2 = \frac{101.96}{(114K)^2}.$$

Solving for d_1 and d_2 yields

$$d_1 = \frac{\sqrt{196 - 296.96K}}{114K}, \tag{5.80}$$

$$d_2 = \frac{10.097}{114K}, \tag{5.81}$$

where we always choose the positive values of d_1 and d_2 so that $G_L(s)$ has poles in the left half-plane. Thus (after some manipulation) the approximate transfer function is

$$G_L(s) = \frac{11.29K}{s^2 + \sqrt{1.92 - 2.91K}\,s + 11.29K}. \tag{5.82}$$

We require that $K < 0.65$ so that the coefficient of the s term remains a real number (we do not want to have a transfer function with complex valued parameters). Our desired second-order transfer function can be written as

$$G_L(s) = \frac{\omega_n^2}{s^2 + 2\zeta\omega_n s + \omega_n^2} \tag{5.83}$$

Comparing coefficients in Equations (5.82) and (5.83) yields

$$\omega_n^2 = 11.29K \quad \text{and} \quad \zeta^2 = \frac{0.043}{K} - 0.065. \tag{5.84}$$

The design specification that the percent overshoot *P.O.* is to be less than 20% implies that we want $\zeta \geq 0.45$. This follows from solving Equation (5.16)

$$P.O. = 100\,e^{-\pi\zeta/\sqrt{1-\zeta^2}}$$

for ζ. Setting $\zeta = 0.45$ in Equation (5.84) and solving for K yields

$$K = 0.16.$$

With $K = 0.16$ we compute

$$\omega_n = \sqrt{11.29K} = 1.34.$$

Then we can estimate the time-to-peak T_p from Equation (5.14) to be

$$T_p = \frac{\pi}{\omega_n\sqrt{1 - \zeta^2}} = 2.62\text{s}.$$

We might be tempted at this point to select $\zeta > 0.45$ so that we reduce the percent overshoot even further than 20%. What happens if we decide to try this approach? From Equation (5.84) we see that K decreases as ζ increases. Then, since

$$\omega_n = \sqrt{11.29K},$$

as K decreases, then ω_n also decreases. But the time-to-peak, given by

$$T_p = \frac{\pi}{\omega_n\sqrt{1 - \zeta^2}},$$

increases as ω_n decreases. Since our goal is to meet the specification of percent overshoot less than 20% while minimizing the time-to-peak, we use the initial selection of $\zeta = 0.45$ so that we do not increase T_p unnecessarily.

The second-order system approximation has allowed us to gain insight into the relationship between the parameter K and the system response, as measured by percent overshoot and time-to-peak. Of course, the gain $K = 0.16$ is only a starting point in the design because we in fact have a third-order system and must consider the effect of the third pole (which we have ignored so far).

A comparison of the third-order aircraft model in Equation (5.76) with the second-order approximation in Equation (5.82) for a unit step input is shown in Figure 5.38. The step response of the second-order system is a good approximation of the original system step response, so we would expect that the analytic analysis using the simpler second-order system to provide accurate indications of the relationship between K and the percent overshoot and time-to-peak.

With the second-order approximation, we estimate that with $K = 0.16$ the percent overshoot *P.O.* = 20% and the time-to-peak $T_p = 2.62$ seconds. As shown in

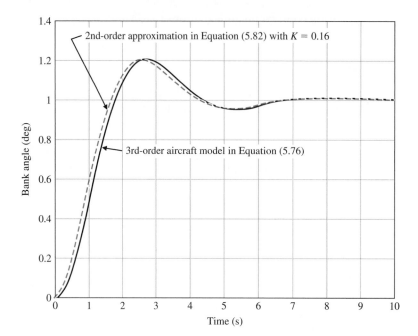

FIGURE 5.38
Step response
comparison of
third-order aircraft
model versus
second-order
approximation.

Figure 5.39 the percent overshoot of the original third-order system is $P.O. = 20.5\%$ and the time-to-peak $T_p = 2.73$ s. Thus, we see that that analytic analysis using the approximate system is an excellent predictor of the actual response. For comparison purposes, we select two variations in the gain and observe the response. For $K = 0.1$, the percent overshoot is 9.5% and the time-to-peak $T_p = 3.74$ s. For $K = 0.2$, the percent overshoot is 26.5% and the time-to-peak $T_p = 2.38$ s. So as predicted, as K decreases the damping ratio increases, leading to a reduction in the percent overshoot. Also as

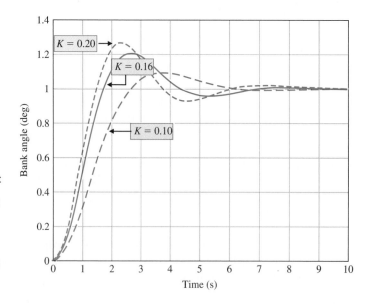

FIGURE 5.39
Step response of
the 3rd-order aircraft
model with
$K = 0.10, 0.16$, and
0. 20 showing that,
as predicted, as K
decreases percent
overshoot
decreases while the
time-to-peak
increases.

Table 5.8   Performance Comparison for $K = 0.10$, 0.16, and 0.20.

K	P.O. (%)	T_p(s)
0.10	9.5	3.74
0.16	20.5	2.73
0.20	26.5	2.38

predicted, as the percent overshoot decreases the time-to-peak increases. The results are summarized in Table 5.8. ■

5.10 SYSTEM PERFORMANCE USING CONTROL DESIGN SOFTWARE

In this section, we will investigate time-domain performance specifications given in terms of transient response to a given input signal and the resulting steady-state tracking errors. We conclude with a discussion of the simplification of linear systems. The function introduced in this section is impulse. We will revisit the lsim function (introduced in Chapter 3) and see how these functions are used to simulate a linear system.

Time-Domain Specifications.   Time-domain performance specifications are generally given in terms of the transient response of a system to a given input signal. Because the actual input signals are generally unknown, a standard test input signal is used. Consider the second-order system shown in Figure 5.4. The closed-loop output is

$$Y(s) = \frac{\omega_n^2}{s^2 + 2\zeta\omega_n s + \omega_n^2} R(s). \tag{5.87}$$

We have already discussed the use of the step function to compute the step response of a system. Now we address another important test signal: the impulse. The impulse response is the time derivative of the step response. We compute the impulse response with the impulse function shown in Figure 5.40.

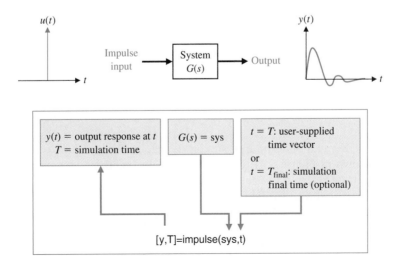

FIGURE 5.40
The **impulse** function.

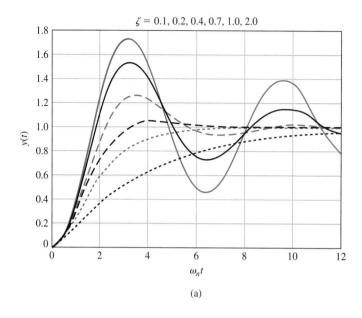

$\zeta = 0.1, 0.2, 0.4, 0.7, 1.0, 2.0$

$\omega_n t$

(a)

```
%Compute step response for a second-order system
%Duplicate Figure 5.5 (a)
%
t=[0:0.1:12]; num=[1];
zeta1=0.1; den1=[1 2*zeta1 1]; sys1=tf(num,den1);
zeta2=0.2; den2=[1 2*zeta2 1]; sys2=tf(num,den2);
zeta3=0.4; den3=[1 2*zeta3 1]; sys3=tf(num,den3);
zeta4=0.7; den4=[1 2*zeta4 1]; sys4=tf(num,den4);
zeta5=1.0; den5=[1 2*zeta5 1]; sys5=tf(num,den5);
zeta6=2.0; den6=[1 2*zeta6 1]; sys6=tf(num,den6);
%
[y1,T1]=step(sys1,t); [y2,T2]=step(sys2,t);          ◄── Compute
[y3,T3]=step(sys3,t); [y4,T4]=step(sys4,t);                step
[y5,T5]=step(sys5,t); [y6,T6]=step(sys6,t);             response.
%
plot(T1,y1,T2,y2,T3,y3,T4,y4,T5,y5,T6,y6)            ◄── Generate plot
xlabel(' \omega_n t'), ylabel('y(t)')                      and labels.
title('\zeta = 0.1, 0.2, 0.4, 0.7, 1.0, 2.0'), grid
```

FIGURE 5.41
(a) Response of a
second-order
system to a step
input. (b) m-file
script.

(b)

We can obtain a plot similar to that of Figure 5.5(a) with the step function, as shown in Figure 5.41. Using the impulse function, we can obtain a plot similar to that of Figure 5.6. The response of a second-order system for an impulse function input is shown in Figure 5.42. In the script, we set $\omega_n = 1$, which is equivalent to computing the step response versus $\omega_n t$. This gives us a more general plot valid for any $\omega_n > 0$.

In many cases, it may be necessary to simulate the system response to an arbitrary but known input. In these cases, we use the lsim function. The lsim function is

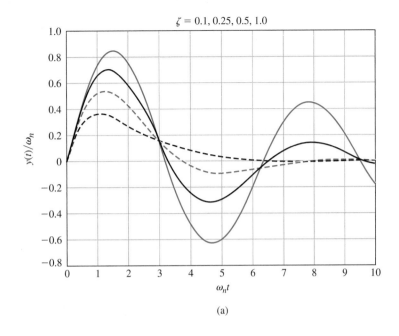

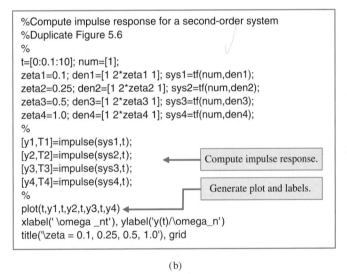

```
%Compute impulse response for a second-order system
%Duplicate Figure 5.6
%
t=[0:0.1:10]; num=[1];
zeta1=0.1; den1=[1 2*zeta1 1]; sys1=tf(num,den1);
zeta2=0.25; den2=[1 2*zeta2 1]; sys2=tf(num,den2);
zeta3=0.5; den3=[1 2*zeta3 1]; sys3=tf(num,den3);
zeta4=1.0; den4=[1 2*zeta4 1]; sys4=tf(num,den4);
%
[y1,T1]=impulse(sys1,t);
[y2,T2]=impulse(sys2,t);          Compute impulse response.
[y3,T3]=impulse(sys3,t);
[y4,T4]=impulse(sys4,t);          Generate plot and labels.
%
plot(t,y1,t,y2,t,y3,t,y4)
xlabel(' \omega _nt'), ylabel('y(t)/\omega_n')
title('\zeta = 0.1, 0.25, 0.5, 1.0'), grid
```

(b)

FIGURE 5.42
(a) Response of a second-order system to an impulse function input. (b) m-file script.

shown in Figure 5.43. We studied the lsim function in Chapter 3 for use with state-variable models; however, now we consider the use of lsim with transfer function models. An example of the use of lsim is given in Example 5.12.

EXAMPLE 5.12   **Mobile robot steering control**

The block diagram for a steering control system for a mobile robot is shown in Figure 5.19. Suppose the transfer function of the steering controller is

$$G_c(s) = K_1 + \frac{K_2}{s}.$$

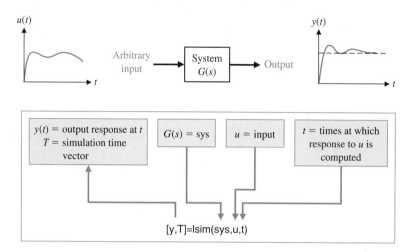

FIGURE 5.43
The **lsim** function.

When the input is a ramp, the steady-state error is

$$e_{ss} = \frac{A}{K_v},$$ (5.86)

where

$$K_v = K_2 K.$$

The effect of the controller constant, K_2, on the steady-state error is evident from Equation (5.86). Whenever K_2 is large, the steady-state error is small.

We can simulate the closed-loop system response to a ramp input using the lsim function. The controller gains, K_1 and K_2, and the system gain K can be represented symbolically in the script so that various values can be selected and simulated. The results are shown in Figure 5.44 for $K_1 = K = 1$, $K_2 = 2$, and $\tau = 1/10$. ∎

Simplification of Linear Systems.   It may be possible to develop a lower-order approximate model that closely matches the input–output response of a high-order model. A procedure for approximating transfer functions is given in Section 5.8. We can use computer simulation to compare the approximate model to the actual model, as illustrated in the following example.

EXAMPLE 5.13   **A simplified model**

Consider the third-order system

$$G_H(s) = \frac{6}{s^3 + 6s^2 + 11s + 6}.$$

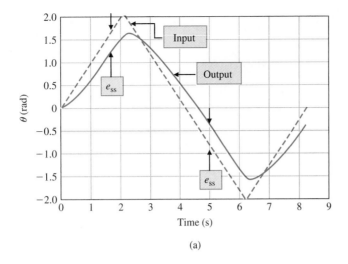

(a)

```
%Compute the response of the Mobile Robot Control
%System to a triangular wave input
%
numg=[10 20]; deng=[1 10 0]; sysg=tf(numg,deng);      ← G(s)G_c(s)
[sys]=feedback(sysg, [1]);
t=[0:0.1:8.2]';
v1=[0:0.1:2]';v2=[2:-0.1:-2]';v3=[-2:0.1:0]';          ← Compute triangular
u=[v1;v2;v3];                                             wave input.
[y,T]=lsim(sys,u,t);                                   ← Linear simulation.
plot(T,y,t,u,'--'),
xlabel('Time (s)'), ylabel('\theta (rad)'), grid
```

FIGURE 5.44
(a) Transient response of the mobile robot steering control system to a ramp input. (b) m-file script.

(b)

A second-order approximation (see Example 5.9) is

$$G_L(s) = \frac{1.60}{s^2 + 2.590s + 1.60}.$$

A comparison of their respective step responses is given in Figure 5.45. ∎

5.11 SEQUENTIAL DESIGN EXAMPLE: DISK DRIVE READ SYSTEM

In Section 4.10, we considered the response of the closed-loop reader head control system. Let us further consider the system shown in Figure 4.35. In this section, we further consider the design process. We will specify the desired performance for the system. Then we will attempt to adjust the amplifier gain K_a in order to obtain the best performance possible.

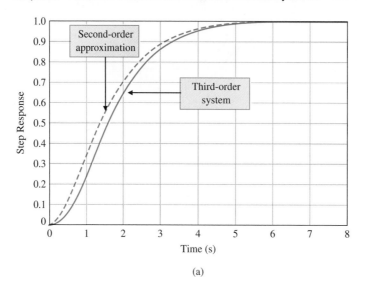

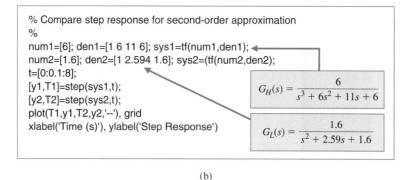

FIGURE 5.45
(a) Step response comparison for an approximate transfer function versus the actual transfer function. (b) m-file script.

(b)

Table 5.9   Specifications for the Transient Response

Performance Measure	Desired Value
Percent overshoot	Less than 5%
Settling time	Less than 250 ms
Maximum value of response to a unit step disturbance	Less than 5×10^{-3}

Our goal is to achieve the fastest response to a step input $r(t)$ while (1) limiting the overshoot and oscillatory nature of the response and (2) reducing the effect of a disturbance on the output position of the read head. The specifications are summarized in Table 5.9.

Let us consider the second-order model of the motor and arm, which neglects the effect of the coil inductance. We then have the closed-loop system shown in Figure 5.46. Then the output when $T_d(s) = 0$ is

FIGURE 5.46
Control system model with a second-order model of the motor and load.

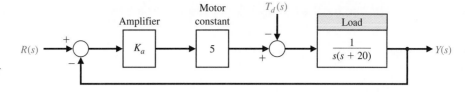

$$Y(s) = \frac{5K_a}{s(s + 20) + 5K_a} R(s)$$

$$= \frac{5K_a}{s^2 + 20s + 5K_a} R(s)$$

$$= \frac{\omega_n^2}{s^2 + 2\zeta\omega_n s + \omega_n^2} R(s). \tag{5.87}$$

Therefore, $\omega_n^2 = 5K_a$, and $2\zeta\omega_n = 20$. We then determine the response of the system as shown in Figure 5.47. Table 5.10 shows the performance measures for selected values of K_a.

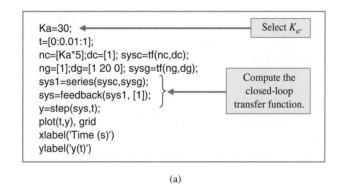

```
Ka=30;                                          ◄———  Select K_a.
t=[0:0.01:1];
nc=[Ka*5];dc=[1]; sysc=tf(nc,dc);
ng=[1];dg=[1 20 0]; sysg=tf(ng,dg);
sys1=series(sysc,sysg);                   Compute the
sys=feedback(sys1, [1]);              ◄    closed-loop
y=step(sys,t);                            transfer function.
plot(t,y), grid
xlabel('Time (s)')
ylabel('y(t)')
```

(a)

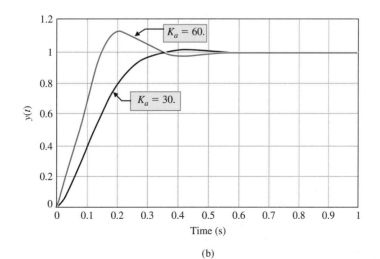

FIGURE 5.47
Response of the system to a unit step input,
$r(t) = 1, t > 0.$
(a) m-file script.
(b) Response for $K_a = 30$ and 60.

(b)

Table 5.10   Response for the Second-Order Model for a Step Input

K_a	20	30	40	60	80
Percent overshoot	0	1.2%	4.3%	10.8%	16.3%
Settling time (s)	0.55	0.40	0.40	0.40	0.40
Damping ratio	1	0.82	0.707	0.58	0.50
Maximum value of the response $y(t)$ to a unit disturbance	-10×10^{-3}	-6.6×10^{-3}	-5.2×10^{-3}	-3.7×10^{-3}	-2.9×10^{-3}

When K_a is increased to 60, the effect of a disturbance is reduced by a factor of 2. We can show this by plotting the output, $y(t)$, as a result of a unit step disturbance input, as shown in Figure 5.48. Clearly, if we wish to meet our goals with this system, we need to select a compromise gain. In this case, we select $K_a = 40$ as the best compromise. However, this compromise does not meet all the specifications. In the next chapter, we consider again the design process and change the configuration of the control system.

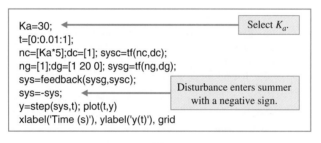

```
Ka=30;                                          Select Ka.
t=[0:0.01:1];
nc=[Ka*5];dc=[1]; sysc=tf(nc,dc);
ng=[1];dg=[1 20 0]; sysg=tf(ng,dg);
sys=feedback(sysg,sysc);
sys=-sys;                          Disturbance enters summer
y=step(sys,t); plot(t,y)            with a negative sign.
xlabel('Time (s)'), ylabel('y(t)'), grid
```

(a)

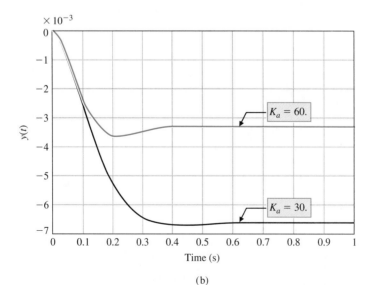

(b)

FIGURE 5.48
Response of the system to a unit step disturbance, $T_d(s) = 1/s$.
(a) m-file script.
(b) Response for $K_a = 30$ and 60.

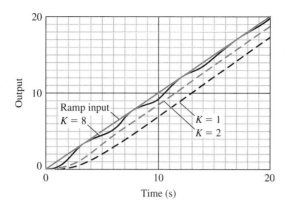

FIGURE 5.49
The response of a feedback system to a ramp input with $K = 1, 2,$ and 8 when $G(s) = K/[s(s + 1)(s + 3)]$. The steady-state error is reduced as K is increased, but the response becomes oscillatory at $K = 8$.

5.12 SUMMARY

In this chapter, we have considered the definition and measurement of the performance of a feedback control system. The concept of a performance measure or index was discussed, and the usefulness of standard test signals was outlined. Then, several performance measures for a standard step input test signal were delineated. For example, the overshoot, peak time, and settling time of the response of the system under test for a step input signal were considered. The fact that the specifications on the desired response are often contradictory was noted, and the concept of a design compromise was proposed. The relationship between the location of the s-plane root of the system transfer function and the system response was discussed. A most important measure of system performance is the steady-state error for specific test input signals. Thus, the relationship of the steady-state error of a system in terms of the system parameters was developed by utilizing the final-value theorem. The capability of a feedback control system is demonstrated in Figure 5.49. Finally, the utility of an integral performance index was outlined, and several design examples that minimized a system's performance index were completed. Thus, we have been concerned with the definition and usefulness of quantitative measures of the performance of feedback control systems.

EXERCISES

E5.1 A motor control system for a computer disk drive must reduce the effect of disturbances and parameter variations, as well as reduce the steady-state error. We want to have no steady-state error for the head-positioning control system, which is of the form shown in Figure 5.18. (a) What type number is required? (How many integrations?) (b) If the input is a ramp signal, and we want to achieve a zero steady-state error, what type number is required?

E5.2 The engine, body, and tires of a racing vehicle affect the acceleration and speed attainable [11]. The speed control of the car is represented by the model shown in Figure E5.2. (a) Calculate the steady-state error of the car to a step command in speed. (b) Calculate overshoot of the speed to a step command.

Answer: (a) $e_{ss} = A/11$; (b) $P.O. = 33\%$

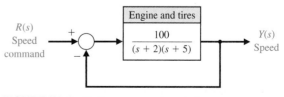

FIGURE E5.2 Racing car speed control.

E5.3 For years, Amtrak has struggled to attract passengers on its routes in the Midwest, using technology developed decades ago. During the same time, foreign railroads were developing new passenger rail systems that could profitably compete with air travel. Two of these systems, the French TGV and the Japanese Shinkansen, reach speeds of 160 mph [20]. The Transrapid-06, a U.S. experimental magnetic levitation train, is shown in Figure E5.3(a).

The use of magnetic levitation and electromagnetic propulsion to provide contactless vehicle movement makes the Transrapid-06 technology radically different from the existing Metroliner. The underside of the TR-06 carriage (where the wheel trucks would be on a conventional car) wraps around a guideway. Magnets on the bottom of the guideway attract electromagnets on the "wraparound," pulling it up toward the guideway. This suspends the vehicles about one centimeter above the guideway. (See Problem P2.27.)

The levitation control is represented by Figure E5.3(b). (a) Using Table 5.6 for a step input, select K so that the system provides an optimum ITAE response. (b) Using Figure 5.8, determine the expected overshoot to a step input of $I(s)$.

Answer: $K = 100$; 4.6%

E5.4 A feedback system with negative unity feedback has a plant transfer function

$$G(s) = \frac{2(s + 8)}{s(s + 4)}.$$

(a) Determine the closed-loop transfer function $T(s) = Y(s)/R(s)$. (b) Find the time response, $y(t)$, for a step input $r(t) = A$ for $t > 0$. (c) Using Figure 5.13(a), determine the overshoot of the response. (d) Using the final-value theorem, determine the steady-state value of $y(t)$.

Answer: (b) $y(t) = 1 - 1.07e^{-3t}\sin\left(\sqrt{7}t + 1.2\right)$

E5.5 A low-inertia plotter is shown in Figure E5.5(a). This system may be represented by the block diagram shown in Figure E5.5(b) [18]. (a) Calculate the

(a)

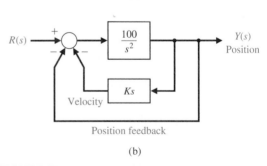

(b)

FIGURE E5.5 (a) The Hewlett-Packard x-y-plotter. (Courtesy of Hewlett-Packard Co.) (b) Block diagram of plotter.

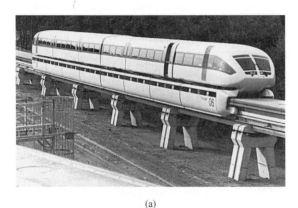

(a)

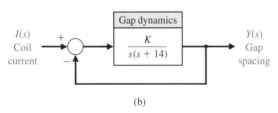

(b)

FIGURE E5.3 Levitated train control.

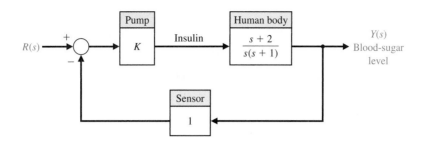

FIGURE 5.6
Blood-sugar level
control.

steady-state error for a ramp input. (b) Select a value of K that will result in zero overshoot to a step input. Provide the most rapid response that is attainable.

Plot the poles and zeros of this system and discuss the dominance of the complex poles. What overshoot for a step input do you expect?

E5.6 Effective control of insulin injections can result in better lives for diabetic persons. Automatically controlled insulin injection by means of a pump and a sensor that measures blood sugar can be very effective. A pump and injection system has a feedback control as shown in Figure E5.6. Calculate the suitable gain K so that the overshoot of the step response due to the drug injection is approximately 7%. $R(s)$ is the desired blood-sugar level and $Y(s)$ is the actual blood-sugar level. (*Hint:* Use Figure 5.13a.)

Answer: $K = 1.67$

E5.7 A control system for positioning the head of a floppy disk drive has the closed-loop transfer function

$$T(s) = \frac{0.313(s + 0.8)}{(s + 0.6)(s^2 + 4s + 5)}.$$

Plot the poles and zeros of this system and discuss the dominance of the complex poles. What overshoot for a step input do you expect?

E5.8 A unity negative feedback control system has the plant transfer function

$$G(s) = \frac{K}{s(s + \sqrt{2K})}.$$

(a) Determine the percent overshoot and settling time (using a 2% settling criterion) due to a unit step input.
(b) For what range of K is the settling time less than 1 second?

E5.9 A second-order control system has the closed-loop transfer function $T(s) = Y(s)/R(s)$. The system specifications for a step input follow:

(1) Percent overshoot $P.O. \leq 5\%$.
(2) Settling time $T_s < 4\text{s}$.
(3) Peak time $T_p < 1\text{s}$.

Show the permissible area for the poles of $T(s)$ in order to achieve the desired response. Use a 2% settling criterion to determine settling time.

E5.10 A system with unity feedback is shown in Figure E5.10. Determine the steady-state error for a step and a ramp input when

$$G(s) = \frac{6(s + 5)}{s(s + 1)(s + 3)(s + 10)}.$$

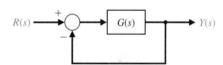

FIGURE E5.10 Unity feedback system.

E5.11 We are all familiar with the Ferris wheel featured at state fairs and carnivals. George Ferris was born in Galesburg, Illinois, in 1859; he later moved to Nevada and then graduated from Rensselaer Polytechnic Institute in 1881. By 1891, Ferris had considerable experience with iron, steel, and bridge construction. He conceived and constructed his famous wheel for the 1893 Columbian Exposition in Chicago [9]. To avoid upsetting passengers, set a requirement that the steady-state speed must be controlled to within 5% of the desired speed for the system shown in Figure E5.11.

(a) Determine the required gain K to achieve the steady-state requirement.
(b) For the gain of part (a), determine and plot the error $e(t)$ for a disturbance $T_d(s) = 1/s$. Does the speed change more than 5%? (Set $R(s) = 0$ and recall that $E(s) = R(s) - T(s)$.)

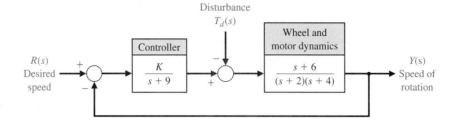

FIGURE E5.11
Speed control of a
Ferris wheel.

E5.12 For the system with unity feedback shown in Figure E5.10, determine the steady-state error for a step and a ramp input when

$$G(s) = \frac{10}{s^2 + 14s + 50}.$$

Answer: $e_{ss} = 0.83$ for a step and $e_{ss} = \infty$ for a ramp.

E5.13 A feedback system is shown in Figure E5.13.

(a) Determine the steady-state error for a unit step when $K = 0.4$ and $G_p(s) = 1$.
(b) Select an appropriate value for $G_p(s)$ so that the steady-state error is equal to zero for the unit step input.

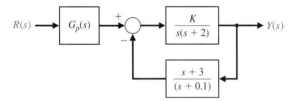

FIGURE E5.13 Feedback system.

E5.14 A closed-loop control system has a transfer function $T(s)$ as follows:

$$\frac{Y(s)}{R(s)} = T(s) = \frac{500}{(s + 10)(s^2 + 10s + 50)}.$$

Plot $y(t)$ for a step input $R(s)$ when (a) the actual $T(s)$ is used, and (b) using the relatively dominant complex poles. Compare the results.

E5.15 A second-order system is

$$\frac{Y(s)}{R(s)} = T(s) = \frac{(10/z)(s + z)}{(s + 1)(s + 8)}.$$

Consider the case where $1 < z < 8$. Obtain the partial fraction expansion, and plot $y(t)$ for a step input $r(t)$ for $z = 2, 4,$ and 6.

E5.16 A closed-loop control system transfer function $T(s)$ has two dominant complex conjugate poles. Sketch the region in the left-hand s-plane where the complex poles should be located to meet the given specifications.

(a) $0.6 \le \zeta \le 0.8,$ $\omega_n \le 10$
(b) $0.5 \le \zeta \le 0.707,$ $\omega_n \ge 10$
(c) $\zeta \ge 0.5,$ $5 \le \omega_n \le 10$
(d) $\zeta \le 0.707,$ $5 \le \omega_n \le 10$
(e) $\zeta \ge 0.6,$ $\omega_n \le 6$

E5.17 A system is shown in Figure E5.17(a). The response to a unit step, when $K = 1$, is shown in Figure E5.17(b). Determine the value of K so that the steady-state error is equal to zero.

Answer: $K = 1.25$.

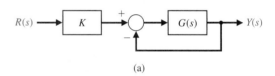

(a)

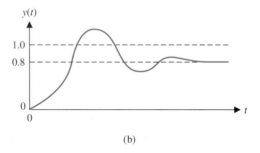

(b)

FIGURE 5.17 Feedback system with prefilter.

E5.18 A second-order system has the closed-loop transfer function

$$T(s) = \frac{Y(s)}{R(s)} = \frac{\omega_n^2}{s^2 + 2\zeta\omega_n s + \omega_n^2} = \frac{7}{s^2 + 3.175s + 7}.$$

(a) Determine the percent overshoot *P.O.*, the time to peak T_p, and the settling time T_s of the unit step response, $R(s) = 1/s$. To compute the settling time, use a 2% criterion.

(b) Obtain the system response to a unit step and verify the results in part (a).

E5.19 Consider the closed-loop system in Figure E5.19, where

$$G_c(s)G(s) = \frac{s+1}{s^2 + 03s} \text{ and } H(s) = K_a.$$

(a) Determine the closed-loop transfer function $T(s) = Y(s)/R(s)$.

(b) Determine the steady-state error of the closed-loop system response to a unit ramp input, $R(s) = 1/s^2$.

(c) Select a value for K_a so that the steady-state error of the system response to a unit step input, $R(s) = 1/s$, is zero.

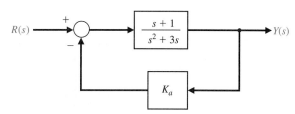

FIGURE E5.19 Nonunity closed-loop feedback control system with parameter K_a.

PROBLEMS

P5.1 An important problem for television systems is the jumping or wobbling of the picture due to the movement of the camera. This effect occurs when the camera is mounted in a moving truck or airplane. The Dynalens system has been designed to reduce the effect of rapid scanning motion; see

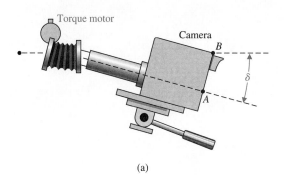

(a)

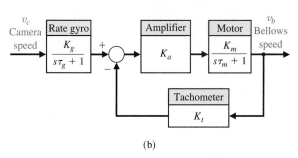

(b)

FIGURE P5.1 Camera wobble control.

Figure P5.1. A maximum scanning motion of 25°/s is expected. Let $K_g = K_t = 1$ and assume that τ_g is negligible. (a) Determine the error of the system $E(s)$. (b) Determine the necessary loop gain $K_a K_m K_t$ when a 1°/s steady-state error is allowable. (c) The motor time constant is 0.40 s. Determine the necessary loop gain so that the settling time (to within 2% of the final value of v_b) is less than or equal to 0.03 s.

P5.2 A specific closed-loop control system is to be designed for an underdamped response to a step input. The specifications for the system are as follows:

$$10\% < \text{percent overshoot} < 20\%,$$
$$\text{Settling time} < 0.6 \text{ s}.$$

(a) Identify the desired area for the dominant roots of the system. (b) Determine the smallest value of a third root r_3 if the complex conjugate roots are to represent the dominant response. (c) The closed-loop system transfer function $T(s)$ is third-order, and the feedback has a unity gain. Determine the forward transfer function $G(s) = Y(s)/E(s)$ when the settling time to within 2% of the final value is 0.6 s and the percent overshoot is 20%.

P5.3 A laser beam can be used to weld, drill, etch, cut, and mark metals, as shown in Figure P5.3(a) [16]. Assume we have a work requirement for an accurate laser to mark a parabolic path with a closed-loop control system, as shown in Figure P5.3(b). Calculate the necessary gain to result in a steady-state error of 5 mm for $r(t) = t^2$ cm.

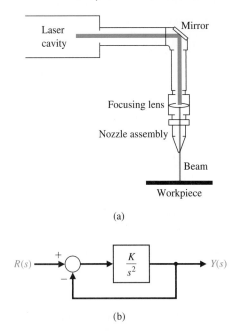

(a)

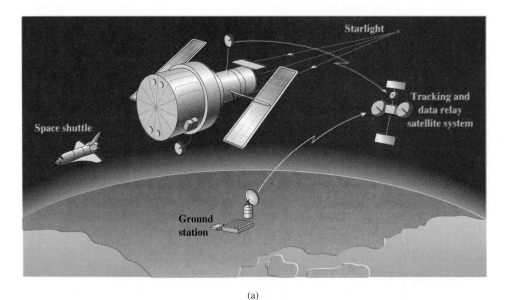

(b)

FIGURE P5.3   Laser beam control.

P5.4   The open-loop transfer function of a unity negative feedback system (see Figure E5.10) is

$$G(s) = \frac{K}{s(s + 2)}.$$

A system response to a step input is specified as follows:

peak time T_p = 1.1 s,
percent overshoot $P.O.$ = 5%.

(a) Determine whether both specifications can be met simultaneously. (b) If the specifications cannot be met simultaneously, determine a compromise value for K so that the peak time and percent overshoot specifications are relaxed by the same percentage.

P5.5   A space telescope is to be launched to carry out astronomical experiments [9]. The pointing control system is desired to achieve 0.01 minute of arc and track solar objects with apparent motion up to 0.21 arc minute per second. The system is illustrated in Figure P5.5(a). The control system is shown in

(a)

FIGURE P5.5
(a) The space telescope. (b) The space telescope pointing control system.

(b)

Figure P5.5(b). Assume that $\tau_1 = 1$ second and $\tau_2 = 0$ (an approximation). (a) Determine the gain $K = K_1 K_2$ required so that the response to a step command is as rapid as reasonable with an overshoot of less than 5%. (b) Determine the steady-state error of the system for a step and a ramp input. (c) Determine the value of $K_1 K_2$ for an ITAE optimal system for (1) a step input and (2) a ramp input.

P5.6   A robot is programmed to have a tool or welding torch follow a prescribed path [8, 13]. Consider a robot tool that is to follow a sawtooth path, as shown in Figure P5.6(a). The transfer function of the plant is

$$G(s) = \frac{75(s + 1)}{s(s + 5)(s + 20)}$$

for the closed-loop system shown in Figure 5.6(b). Calculate the steady-state error.

P5.7   Astronaut Bruce McCandless II took the first untethered walk in space on February 7, 1984, using the gas-jet propulsion device illustrated in Figure P5.7(a). The

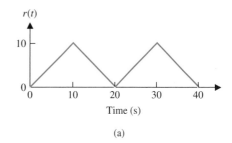

(a)

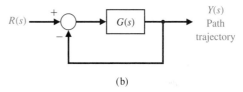

(b)

FIGURE P5.6   Robot path control.

(a)

FIGURE P5.7
Astronaut Bruce McCandless II is shown a few meters away from the earth-orbiting space shuttle. He used a nitrogen-propelled hand-controlled device called the manned maneuvering unit. (Courtesy of National Aeronautics and Space Administration.) (b) Block diagram of controller.

(b)

controller can be represented by a gain K_2, as shown in Figure P5.7(b). The moment of inertia of the equipment and man is 25 kg m². (a) Determine the necessary gain K_3 to maintain a steady-state error equal to 1 cm when the input is a ramp $r(t) = t$ (meters). (b) With this gain K_3, determine the necessary gain $K_1 K_2$ in order to restrict the percent overshoot to 10%. (c) Determine analytically the gain $K_1 K_2$ in order to minimize the ISE performance index for a step input.

P5.8 Photovoltaic arrays (solar cells) generate a DC voltage that can be used to drive DC motors or that can be converted to AC power and added to the distribution network. It is desirable to maintain the power out of the array at its maximum available as the solar incidence changes during the day. One such closed-loop system is shown in Figure P5.8. The transfer function for the process is

$$G(s) = \frac{K}{s + 10},$$

where $K = 20$. Find (a) the time constant of the closed-loop system and (b) the settling time to within 2% of the final value of the system when disturbances such as clouds occur.

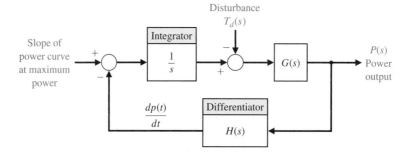

FIGURE P5.8
Solar cell control.

P5.9 The antenna that receives and transmits signals to the *Telstar* communication satellite is the largest horn antenna ever built. The microwave antenna is 177 ft long, weighs 340 tons, and rolls on a circular track. A photo of the antenna is shown in Figure P5.9. The *Telstar* satellite is 34 inches in diameter and moves about 16,000 mph at an altitude of 2500 miles. The

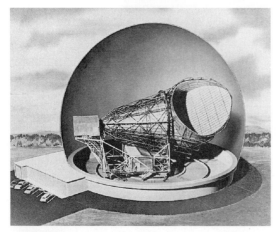

FIGURE 5.9   A model of the antenna for the Telstar System at Andover, Maine. (Photo courtesy of Bell Telephone Laboratories, Inc.)

antenna must be positioned accurately to 1/10 of a degree, because the microwave beam is 0.2° wide and highly attenuated by the large distance. If the antenna is following the moving satellite, determine the K_v necessary for the system.

P5.10 A speed control system of an armature-controlled DC motor uses the back emf voltage of the motor as a feedback signal. (a) Draw the block diagram of this system (see Equation (2.69)). (b) Calculate the steady-state error of this system to a step input command setting the speed to a new level. Assume that $R_a = L_a = J = b = 1$, the motor constant is $K_m = 1$, and $K_b = 1$. (c) Select a feedback gain for the back emf signal to yield a step response with an overshoot of 15%.

P5.11 A simple unity feedback control system has a process transfer function

$$\frac{Y(s)}{E(s)} = G(s) = \frac{K}{s}.$$

The system input is a step function with an amplitude A. The initial condition of the system at time t_0 is $y(t_0) = Q$, where $y(t)$ is the output of the system. The performance index is defined as

$$I = \int_0^\infty e^2(t) \, dt.$$

(a) Show that $I = (A - Q)^2/(2K)$. (b) Determine the gain K that will minimize the performance index I. Is this gain a practical value? (c) Select a practical value of gain and determine the resulting value of the performance index.

P5.12   Train travel between cities will increase as trains are developed that travel at high speeds, making the travel time from city center to city center equivalent to airline travel time. The Japanese National Railway has a train called the Bullet Express that travels between Tokyo and Osaka on the Tokaido line. This train travels the 320 miles in 3 hours and 10 minutes, an average speed of 101 mph [20]. This speed will be increased as new systems are used, such as magnetically levitated systems to float vehicles above an aluminum guideway. To maintain a desired speed, a speed control system is proposed that yields a zero steady-state error to a ramp input. A third-order system is sufficient. Determine the optimum system transfer function $T(s)$ for an ITAE performance criterion. Estimate the settling time (with a 2% criterion) and overshoot for a step input when $\omega_n = 10$.

P5.13   We want to approximate a fourth-order system by a lower-order model. The transfer function of the original system is

$$H(s) = \frac{s^3 + 7s^2 + 24s + 24}{s^4 + 10s^3 + 35s^2 + 50s + 24}$$

$$= \frac{s^3 + 7s^2 + 24s + 24}{(s + 1)(s + 2)(s + 3)(s + 4)}.$$

Show that if we obtain a second-order model by the method of Section 5.8, and we do not specify the poles and the zero of $L(s)$, we have

$$L(s) = \frac{0.2917s + 1}{0.399s^2 + 1.375s + 1}$$

$$= \frac{0.731(s + 3.428)}{(s + 1.043)(s + 2.4)}.$$

P5.14   For the original system of Problem P5.13, we want to find the lower-order model when the poles of the second-order model are specified as -1 and -2 and the model has one unspecified zero. Show that this low-order model is

$$L(s) = \frac{0.986s + 2}{s^2 + 3s + 2} = \frac{0.986(s + 2.028)}{(s + 1)(s + 2)}.$$

P5.15   A magnetic amplifier with a low-output impedance is shown in Figure P5.15 in cascade with a low-pass filter and a preamplifier. The amplifier has a high-input impedance and a gain of 1 and is used for adding the signals as shown. Select a value for the capacitance C so that the transfer function $V_0(s)/V_{in}(s)$ has a damping ratio of $1/\sqrt{2}$. The time constant of the magnetic amplifier is equal to 1 second, and the gain is $K = 10$. Calculate the settling time (with a 2% criterion) of the resulting system.

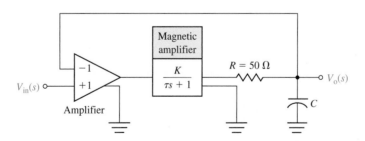

FIGURE P5.15
Feedback amplifier.

P5.16   Electronic pacemakers for human hearts regulate the speed of the heart pump. A proposed closed-loop system that includes a pacemaker and the measurement of the heart rate is shown in Figure P5.16 [2, 3]. The transfer function of the heart pump and the pacemaker is found to be

$$G(s) = \frac{K}{s(s/12 + 1)}.$$

Design the amplifier gain to yield a system with a settling time to a step disturbance of less than 1 second. The overshoot to a step in desired heart rate should be less than 10%. (a) Find a suitable range of K. (b) If the nominal value of K is $K = 10$, find the sensitivity of the system to small changes in K. (c) Evaluate the sensitivity of part (b) at DC (set $s = 0$). (d) Evaluate the magnitude of the sensitivity at the normal heart rate of 60 beats/minute.

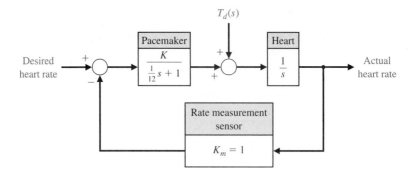

FIGURE P5.16
Heart pacemaker.

P5.17 Consider the original third-order system given in Example 5.9. Determine a first-order model with one pole unspecified and no zeros that will represent the third-order system.

P5.18 A closed-loop control system with negative unity feedback has a plant with a transfer function

$$G(s) = \frac{8}{s(s^2 + 6s + 12)}.$$

(a) Determine the closed-loop transfer function $T(s)$.
(b) Determine a second-order approximation for $T(s)$ using the method of Section 5.10. (c) Plot the response of $T(s)$ and the second-order approximation to a unit step input and compare the results.

P5.19 A system is shown in Figure P5.19.

(a) Determine the steady-state error for a unit step input in terms of K and K_1, where $E(s) = R(s) - Y(s)$.
(b) Select K_1 so that the steady-state error is zero.

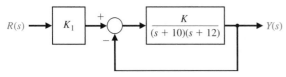

FIGURE P5.19 System with pregain, K_1.

P5.20 Consider the closed-loop system in Figure P5.20. Determine values of the parameters k and a so that the following specifications are satisfied:

(a) The steady-state error to a unit step input is zero.

(b) The closed-loop system has a percent overshoot of less than 5%.

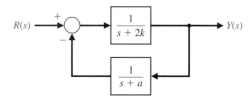

FIGURE P5.20   Closed-loop system with parameters k and a.

P5.21 Consider the closed-loop system in Figure P5.21, where

$$G_c(s)G(s) = \frac{2}{s + 0.2K} \text{ and } H(s) = \frac{2}{2s + \tau}.$$

(a) If $\tau = 2.43$, determine the value of K such that the steady-state error of the closed-loop system response to a unit step input, $R(s) = 1/s$, is zero.
(b) Determine the percent overshoot $P.O.$ and the time to peak T_p of the unit step response when K is as in part (a).

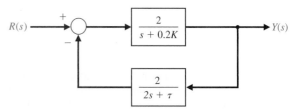

FIGURE P5.21   Nonunity closed-loop feedback control system.

ADVANCED PROBLEMS

AP5.1 A closed-loop transfer function is

$$T(s) = \frac{Y(s)}{R(s)} = \frac{96(s + 3)}{(s + 8)(s^2 + 8s + 36)}.$$

(a) Determine the steady-state error for a unit step input $R(s) = 1/s$.

(b) Assume that the complex poles dominate, and determine the overshoot and settling time to within 2% of the final value.

(c) Plot the actual system response, and compare it with the estimates of part (b).

AP5.2 A closed-loop system is shown in Figure AP5.2. Plot the response to a unit step input for the system for $\tau_z = 0, 0.05, 0.1,$ and 0.5. Record the percent overshoot, rise time, and settling time (with a 2% criterion) as τ_z varies. Describe the effect of varying τ_z. Compare the location of the zero $-1/\tau_z$ with the location of the closed-loop poles.

AP5.3 A closed-loop system is shown in Figure AP5.3. Plot the response to a unit step input for the system with $\tau_p = 0, 0.5, 2,$ and 5. Record the percent overshoot, rise time, and settling time (with a 2% criterion) as τ_p varies. Describe the effect of varying τ_p. Compare the location of the open-loop pole $-1/\tau_p$ with the location of the closed-loop poles.

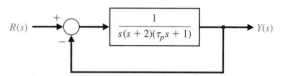

FIGURE AP5.3 System with a variable pole in the process.

AP5.4 The speed control of a high-speed train is represented by the system shown in Figure AP5.4 [20]. Determine the equation for steady-state error for K for a unit step input $r(t)$. Consider the three values for K equal to 1, 10, and 100.

(a) Determine the steady-state error.

(b) Determine and plot the response $y(t)$ for (i) a unit step input $R(s) = 1/s$ and (ii) a unit step disturbance input $T_d(s) = 1/s$.

(c) Create a table showing overshoot, settling time (with a 2% criterion), e_{ss} for $r(t)$, and $|y/t_d|_{max}$ for the three values of K. Select the best compromise value.

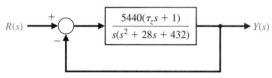

FIGURE AP5.2 System with a variable zero.

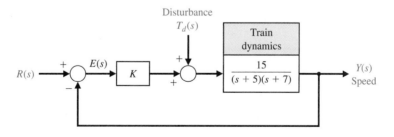

FIGURE AP5.4
Speed control.

AP5.5 A system with a controller is shown in Figure AP5.5. The zero of the controller may be varied. Let $\alpha = 0, 10, 100$.

(a) Determine the steady-state error for a step input $r(t)$ for $\alpha = 0$ and $\alpha \neq 0$.

(b) Plot the response of the system to a step input disturbance for the three values of α. Compare the results and select the best value of the three values of α.

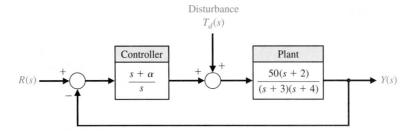

FIGURE AP5.5
System with control
parameter α.

AP5.6   The block diagram model of an armature-current-controlled DC motor is shown in Figure AP5.6.

(a) Determine the steady-state tracking error to a ramp input $r(t) = t, t \geq 0$, in terms of K, K_b, and K_m.

(b) Let $K_m = 10$ and $K_b = 0.05$, and select K so that steady-state tracking error is equal to 1.
(c) Plot the response to a unit step input and a unit ramp input for 20 seconds. Are the responses acceptable?

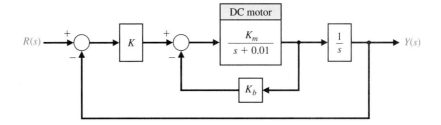

FIGURE AP5.6
DC motor control.

AP5.7   Consider the closed-loop system in Figure AP5.7 with transfer functions

$$G_c(s) = \frac{100}{s + 100} \quad \text{and} \quad G(s) = \frac{K}{s(s + 50)},$$

where

$$1000 \leq K \leq 5000.$$

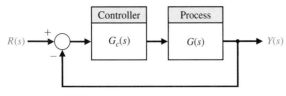

FIGURE AP5.7   Closed-loop system with unity feedback.

(a) Assume that the complex poles dominate and estimate the settling time and percent overshoot to a unit step input for $K = 1000, 2000, 3000, 4000,$ and 5000.
(b) Determine the actual settling time and percent overshoot to a unit step for the values of K in part (a).
(c) Co-plot the results of (a) and (b) and comment.

AP5.8   A unity negative feedback system (as shown in Figure E5.10) has the open-loop transfer function

$$G(s) = \frac{K(s^2 + 120s + 110)}{s^2 + 5s + 6}.$$

Determine the gain K that minimizes the damping ratio ζ of the closed-loop system poles. What is the minimum damping ratio?

DESIGN PROBLEMS

CDP5.1   The capstan drive system of the previous problems (see CDP1.1–CDP4.1) has a disturbance due to changes in the part that is being machined as material is removed. The controller is an amplifier $G_c(s) = K_a$. Evaluate the effect of a unit step disturbance, and determine the best value of the amplifier gain so that the overshoot to a step command $r(t) = A, t > 0$ is less than 5%, while reducing the effect of the disturbance as much as possible.

DP5.1   The roll control autopilot of a jet fighter is shown in Figure DP5.1. The goal is to select a suitable K so that the response to a unit step command $\phi_d(t) = A, t \geq 0$,

will provide a response $\phi(t)$ that is a fast response and has an overshoot of less than 20%. (a) Determine the closed-loop transfer function $\phi(s)/\phi_d(s)$. (b) Determine the roots of the characteristic equation for $K = 0.7, 3$, and 6. (c) Using the concept of dominant roots, find the expected overshoot and peak time for the approximate second-order system. (d) Plot the actual response and compare with the approximate results of part (c). (e) Select the gain K so that the percentage overshoot is equal to 16%. What is the resulting peak time?

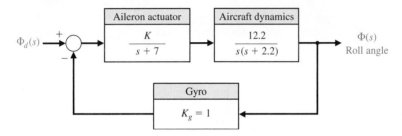

FIGURE DP5.1
Roll angle control.

DP5.2   The design of the control for a welding arm with a long reach requires the careful selection of the parameters [13]. The system is shown in Figure DP5.2, where $\zeta = 0.2$, and the gain K and the natural frequency ω_n can be selected. (a) Determine K and ω_n so that the response to a unit step input achieves a peak

time for the first overshoot (above the desired level of 1) that is less than or equal to 1 second and the overshoot is less than 5%. (*Hint:* Try $0.1 < K/\omega_n < 0.3$.) (b) Plot the response of the system designed in part (a) to a step input.

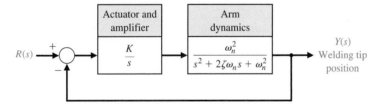

FIGURE DP5.2
Welding tip position control.

DP5.3   Active suspension systems for modern automobiles provide a comfortable firm ride. The design of an active suspension system adjusts the valves of the shock absorber so that the ride fits the conditions. A small electric motor, as shown in Figure DP5.3, changes the valve settings [15]. Select a design value

for K and the parameter q in order to satisfy the ITAE performance for a step command $R(s)$ and a settling time (with a 2% criterion) for the step response of less than or equal to 0.5 second. Upon completion of your design, predict the resulting overshoot for a step input.

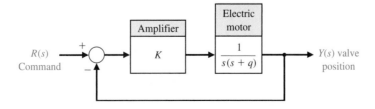

FIGURE DP5.3
Active suspension system.

DP5.4   The space satellite shown in Figure DP5.4(a) uses a control system to readjust its orientation, as shown in Figure DP5.4(b).

(a) Determine a second-order model for the closed-loop system.

(b) Using the second-order model, select a gain K so that the percent overshoot is less than 15% and the steady-state error to a step is less than 12%.

(c) Verify your design by determining the actual performance of the third-order system.

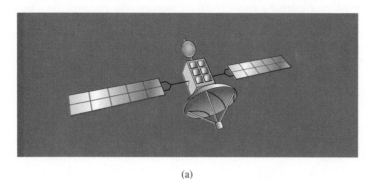

(a)

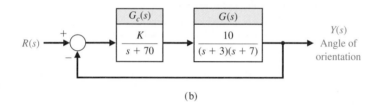

FIGURE DP5.4
Control of a space satellite.

(b)

DP5.5   A deburring robot can be used to smooth off machined parts by following a preplanned path (input command signal). In practice, errors occur due to robot inaccuracy, machining errors, large tolerances, and tool wear. These errors can be eliminated using force feedback to modify the path online [9, 13].

While force control has been able to address the problem of accuracy, it has been more difficult to solve the contact stability problem. In fact, by closing the force loop and introducing a compliant wrist force sensor (the most common type of force control), one can add to the stability problem.

A model of a robot deburring system is shown in Figure DP5.5. Determine the region of stability for the system for K_1 and K_2. Assume both adjustable gains are greater than zero.

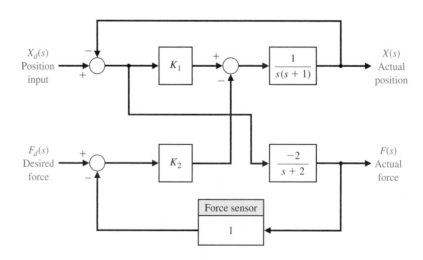

FIGURE DP5.5
Deburring robot.

DP5.6 The model for a position control system using a DC motor is shown in Figure DP5.6. The goal is to select K_1 and K_2 so that the peak time is 0.2 second and the overshoot *P.O.* for a step input is negligible ($1\% < P.O. < 4\%$).

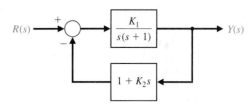

FIGURE DP5.6 Position control robot.

COMPUTER PROBLEMS

CP5.1 Consider the closed-loop transfer function

$$T(s) = \frac{6}{s^2 + 7s + 12}.$$

Obtain the impulse response analytically and compare the result to one obtained using the impulse function.

CP5.2 A unity negative feedback system has the open-loop transfer function

$$G(s) = \frac{s + 7}{s^2(s + 10)}.$$

Using lsim, obtain the response of the closed-loop system to a unit ramp input,

$$R(s) = 1/s^2.$$

Consider the time interval $0 \le t \le 25$. What is the steady-state error?

CP5.3 A working knowledge of the relationship between the pole locations of the second-order system shown in Figure CP5.3 and the transient response is important in control design. With that in mind, consider the following four cases:

1. $\omega_n = 2$, $\zeta = 0$,
2. $\omega_n = 2$, $\zeta = 0.1$,
3. $\omega_n = 1$, $\zeta = 0$,
4. $\omega_n = 1$, $\zeta = 0.2$.

Using the impulse and subplot functions, create a plot containing four subplots, with each subplot depicting the impulse response of one of the four cases listed. Compare the plot with Figure 5.17 in Section 5.5, and discuss the results.

$$R(s) \longrightarrow \boxed{\dfrac{\omega_n^2}{s^2 + 2\zeta\omega_n s + \omega_n^2}} \longrightarrow Y(s)$$

FIGURE CP5.3 A simple second-order system.

CP5.4 Consider the control system shown in Figure CP5.4.

(a) Show analytically that the expected percent overshoot of the closed-loop system response to a unit step input is about 50%.

(b) Develop an m-file to plot the unit step response of the closed-loop system and estimate the percent overshoot from the plot. Compare the result with part (a).

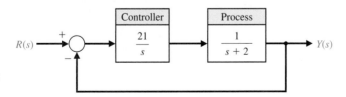

FIGURE CP5.4
A negative feedback control system.

CP5.5 The open-loop transfer function of a unity negative feedback system is

$$G(s) = \frac{50}{s(s + 10)}.$$

Develop an m-file to plot the unit step response and determine the values of peak overshoot M_p time to peak T_p and settling time T_s (with a 2% criterion) from the plot.

CP5.6 An autopilot designed to hold an aircraft in straight and level flight is shown in Figure CP5.6.

(a) Suppose the controller is a constant gain controller given by $G_c(s) = 2$. Using the lsim function, compute and plot the ramp response for $\theta_d(t) = at$, where $a = 0.5°/s$. Determine the attitude error after 10 seconds.

(b) If we increase the complexity of the controller, we can reduce the steady-state tracking error. With this objective in mind, suppose we replace the constant gain controller with the more sophisticated controller

$$G_c(s) = K_1 + \frac{K_2}{s} = 2 + \frac{1}{s}.$$

This type of controller is known as a proportional, integral (PI) controller. Repeat the simulation of part (a) with the PI controller, and compare the steady-state tracking errors of the constant gain controller versus the PI controller.

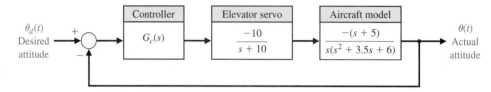

FIGURE CP5.6
An aircraft autopilot
block diagram.

CP5.7 The block diagram of a rate loop for a missile autopilot is shown in Figure CP5.7. Using the analytic formulas for second-order systems, predict M_{pt}, T_p, and T_s for the closed-loop system due to a unit step input. Compare the predicted results with the actual unit step response obtained with the **step** function. Explain any differences.

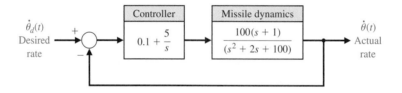

FIGURE CP5.7
A missile rate loop
autopilot.

CP5.8 Develop an m-file that can be used to analyze the closed-loop system in Figure CP5.8. Drive the system with a step input and display the output on a graph. What is the settling time and the percent overshoot?

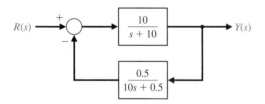

FIGURE CP5.8 Nonunity feedback system.

CP5.9 Develop an m-file to simulate the response of the system in Figure CP5.9 to a ramp input $R(s) = 1/s^2$.

What is the steady-state error? Display the output on an x-y graph.

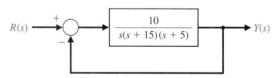

FIGURE CP5.9 Closed-loop system for Simulink simulation.

CP5.10 Consider the closed-loop system in Figure CP5.10. Develop an m-file to accomplish the following tasks:

(a) Determine the closed-loop transfer function $T(s) = Y(s)/R(s)$.

(b) Plot the closed-loop system response to an impulse input $R(s) = 1$, a unit step input $R(s) = 1/s$, and a unit ramp input $R(s) = 1/s^2$. Use the **subplot** function to display the three system responses.

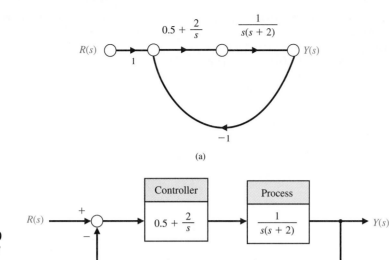

FIGURE CP5.10
A single loop unity feedback system. (a) Signal flow graph. (b) Block diagram.

CP5.11 A closed-loop transfer function is given by

$$T(s) = \frac{Y(s)}{R(s)} = \frac{77(s + 2)}{(s + 7)(s^2 + 4s + 22)}.$$

(a) Obtain the response of the closed-loop transfer function $T(s) = Y(s)/R(s)$ to a unit step input.

What is the settling time T_s (use a 2% criterion) and percent overshoot $P.O.$?

(b) Neglecting the real pole at $s = -7$, determine the settling time T_s and percent overshoot $P.O.$. Compare the results with the actual system response in part (a). What conclusions can be made regarding neglecting the pole?

TERMS AND CONCEPTS

Acceleration error constant, K_a The constant evaluated as $\lim_{s \to 0}[s^2 G_c(s)G(s)]$. The steady-state error for a parabolic input, $r(t) = At^2/2$, is equal to A/K_a.

Design specifications A set of prescribed performance criteria.

Dominant roots The roots of the characteristic equation that cause the dominant transient response of the system.

Optimum control system A system whose parameters are adjusted so that the performance index reaches an extremum value.

Peak time The time for a system to respond to a step input and rise to a peak response.

Percent overshoot The amount by which the system output response proceeds beyond the desired response.

Performance index A quantitative measure of the performance of a system.

Position error constant, K_p The constant evaluated as $\lim_{s \to 0} G_c(s)G(s)$. The steady-state error for a step input (of magnitude A) is equal to $A/(1 + K_p)$.

Rise time The time for a system to respond to a step input and attain a response equal to a percentage of the magnitude of the input. The 0–100% rise time, T_r, measures the time to 100% of the magnitude of the input. Alternatively, T_{r_1} measures the time from 10% to 90% of the response to the step input.

Settling time The time required for the system output to settle within a certain percentage of the input amplitude.

Steady-state response The constituent of the system response that exists a long time following any signal initiation.

Test input signal An input signal used as a standard test of a system's ability to respond adequately.

Transient response   The constituent of the system response that disappears with time.

Type number   The number N of poles of the transfer function, $G_c(s)G(s)$, at the origin. $G_c(s)G(s)$ is the loop transfer function.

Unit impulse   A test input consisting of an impulse of infinite amplitude and zero width, and having an area of unity. The unit impulse is used to determine the impulse response.

Velocity error constant, K_v   The constant evaluated as $\lim\limits_{s \to 0}[sG_c(s)G(s)]$. The steady-state error for a ramp input (of slope A) for a system is equal to A/K_v.

The Stability of Linear Feedback Systems

P R E V I E W

Stability of closed-loop feedback systems is central to control system design. A stable system should exhibit a bounded output if the corresponding input is bounded. This is known as bounded-input–bounded-output stability and is one of the main topics of this chapter. The stability of a feedback system is directly related to the location of the roots of the characteristic equation of the system transfer function and to the location of the eigenvalues of the system matrix for a system in state variable format. The Routh–Hurwitz method is introduced as a useful tool for assessing system stability. The technique allows us to compute the number of roots of the characteristic equation in the right half plane without actually computing the values of the roots. This gives us a design method for determining values of certain system parameters that will lead to closed-loop stability. For stable systems, we will introduce the notion of relative stability, which allows us to characterize the degree of stability. The chapter concludes with a stabilizing controller design based on the Routh–Hurwitz method for the Sequential Design Example: Disk Drive Read System.

DESIRED OUTCOMES

Upon completion of Chapter 6, students should:

- ❏ Understand the concept of stability of dynamic systems.
- ❏ Be aware of the key concepts of absolute and relative stability.
- ❏ Be familiar with the notion of bounded-input, bounded-output stability.
- ❏ Understand the relationship of the s-plane pole locations (for transfer function models) and of the eigenvalue locations (for state variable models) to system stability.
- ❏ Know how to construct a Routh array and be able to employ the Routh–Hurwitz stability criterion to determine stability.

6.1 THE CONCEPT OF STABILITY

When considering the design and analysis of feedback control systems, **stability** is of the utmost importance. From a practical point of view, a closed-loop feedback system that is unstable is of little value. As with all such general statements, there are exceptions; but for our purposes, we will declare that all our control designs must result in a closed-loop stable system. Many physical systems are inherently open-loop unstable, and some systems are even designed to be open-loop unstable. Most modern fighter aircraft are open-loop *unstable by design*, and without active feedback control assisting the pilot, they cannot fly. Active control is introduced by engineers to stabilize the unstable system—that is, the aircraft—so that other considerations, such as transient performance, can be addressed. Using feedback, we can stabilize unstable systems and then with a judicious selection of controller parameters, we can adjust the transient performance. For open-loop stable systems, we still use feedback to adjust the closed-loop performance to meet the design specifications. These specifications take the form of steady-state tracking errors, percent overshoot, settling time, time to peak, and the other indices discussed in Chapters 4 and 5.

We can say that a closed-loop feedback system is either stable or it is not stable. This type of stable/not stable characterization is referred to as **absolute stability**. A system possessing absolute stability is called a stable system—the label of absolute is dropped. Given that a closed-loop system is stable, we can further characterize the degree of stability. This is referred to as **relative stability**. The pioneers of aircraft design were familiar with the notion of relative stability—the more stable an aircraft was, the more difficult it was to maneuver (that is, to turn). One outcome of the relative instability of modern fighter aircraft is high maneuverability. A fighter aircraft is less stable than a commercial transport, hence it can maneuver more quickly. In fact, the motions of a fighter aircraft can be quite violent to the "passengers." As we will discuss later in this section, we can determine that a system is stable (in the absolute sense) by determining that all transfer function poles lie in the left-half s-plane, or equivalently, that all the eigenvalues of the system matrix $\mathbf{A}$ lie in the left-half s-plane. Given that all the poles (or eigenvalues) are in the left-half s-plane, we investigate relative-stability by examining the relative locations of the poles (or eigenvalues).

A **stable system** is defined as a system with a bounded (limited) system response. That is, if the system is subjected to a bounded input or disturbance and the response is bounded in magnitude, the system is said to be stable.

> **A stable system is a dynamic system with a bounded response to a bounded input.**

The concept of stability can be illustrated by considering a right circular cone placed on a plane horizontal surface. If the cone is resting on its base and is tipped slightly, it returns to its original equilibrium position. This position and response are said to be stable. If the cone rests on its side and is displaced slightly, it rolls with no tendency to leave the position on its side. This position is designated as the neutral stability. On the other hand, if the cone is placed on its tip and released, it falls onto its side. This position is said to be unstable. These three positions are illustrated in Figure 6.1.

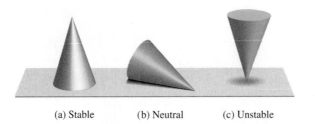

FIGURE 6.1
The stability of a
cone.

    (a) Stable   (b) Neutral   (c) Unstable

The stability of a dynamic system is defined in a similar manner. The response to a displacement, or initial condition, will result in either a decreasing, neutral, or increasing response. Specifically, it follows from the definition of stability that a linear system is stable if and only if the absolute value of its impulse response $g(t)$, integrated over an infinite range, is finite. That is, in terms of the convolution integral Equation (5.2) for a bounded input, $\int_0^\infty |g(t)| \, dt$ must be finite.

The location in the s-plane of the poles of a system indicates the resulting transient response. The poles in the left-hand portion of the s-plane result in a decreasing response for disturbance inputs. Similarly, poles on the $j\omega$-axis and in the right-hand plane result in a neutral and an increasing response, respectively, for a disturbance input. This division of the s-plane is shown in Figure 6.2. Clearly, the poles of desirable dynamic systems must lie in the left-hand portion of the s-plane [1–3].

A common example of the potential destabilizing effect of feedback is that of feedback in audio amplifier and speaker systems used for public address in auditoriums. In this case, a loudspeaker produces an audio signal that is an amplified version of the sounds picked up by a microphone. In addition to other audio inputs, the sound coming from the speaker itself may be sensed by the microphone. The strength of this particular signal depends upon the distance between the loudspeaker and the microphone. Because of the attenuating properties of air, a larger distance will cause a weaker signal to reach the microphone. Due to the finite propagation speed of sound waves, there will also be a time delay between the signal produced by the loudspeaker and the signal sensed by the microphone. In this case, the output from the feedback path is added to the external input. This is an example of positive feedback.

As the distance between the loudspeaker and the microphone decreases, we find that if the microphone is placed too close to the speaker, then the system will be unstable. The result of this instability is an excessive amplification and distortion of audio signals and an oscillatory squeal.

Another example of an unstable system is shown in Figure 6.3. The first bridge across the Tacoma Narrows at Puget Sound, Washington, was opened to traffic on July 1, 1940. The bridge was found to oscillate whenever the wind blew. After four

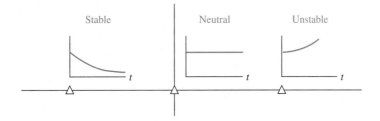

FIGURE 6.2
Stability in the
s-plane.

(a)

FIGURE 6.3
Tacoma Narrows
Bridge (a) as
oscillation begins
(b) at catastrophic
failure.

(b)

months, on November 7, 1940, a wind produced an oscillation that grew in amplitude until the bridge broke apart. Figure 6.3(a) shows the condition at the beginning of oscillation; Figure 6.3(b) shows the catastrophic failure [5].

In terms of linear systems, we recognize that the stability requirement may be defined in terms of the location of the poles of the closed-loop transfer function. The closed-loop system transfer function is written as

$$T(s) = \frac{p(s)}{q(s)} = \frac{K\prod_{i=1}^{M}(s + z_i)}{s^N \prod_{k=1}^{Q}(s + \sigma_k) \prod_{m=1}^{R}[s^2 + 2\alpha_m s + (\alpha_m^2 + \omega_m^2)]}, \tag{6.1}$$

where $q(s) = \Delta(s) = 0$ is the characteristic equation whose roots are the poles of the closed-loop system. The output response for an impulse function input (when $N = 0$) is then

$$y(t) = \sum_{k=1}^{Q} A_k e^{-\sigma_k t} + \sum_{m=1}^{R} B_m \left(\frac{1}{\omega_m}\right) e^{-\alpha_m t} \sin(\omega_m t + \theta_m), \tag{6.2}$$

where A_k and B_m are constants that depend on σ_k, z_i, α_m, K, and ω_m. To obtain a bounded response, the poles of the closed-loop system must be in the left-hand portion of the s-plane. Thus, **a necessary and sufficient condition for a feedback system to be stable is that all the poles of the system transfer function have negative real parts**. A system is stable if all the poles of the transfer function are in the left-hand s-plane. A system is not stable if not all the roots are in the left-hand plane. If the characteristic equation has simple roots on the imaginary axis ($j\omega$-axis) with all other roots in the left half-plane, the steady-state output will be sustained oscillations for a bounded input, unless the input is a sinusoid (which is bounded) whose frequency is equal to the magnitude of the $j\omega$-axis roots. For this case, the output becomes unbounded. Such a system is called **marginally stable**, since only certain bounded inputs (sinusoids of the frequency of the poles) will cause the output to become unbounded. For an unstable system, the characteristic equation has at least one root in the right half of the s-plane or repeated $j\omega$ roots; for this case, the output will become unbounded for any input.

For example, if the characteristic equation of a closed-loop system is

$$(s + 10)(s^2 + 16) = 0,$$

then the system is said to be marginally stable. If this system is excited by a sinusoid of frequency $\omega = 4$, the output becomes unbounded.

To ascertain the stability of a feedback control system, we could determine the roots of the characteristic polynomial $q(s)$. However, we are first interested in determining the answer to the question, Is the system stable? If we calculate the roots of the characteristic equation in order to answer this question, we have determined much more information than is necessary. Therefore, several methods have been developed that provide the required yes or no answer to the stability question. The three approaches to the question of stability are (1) the s-plane approach, (2) the frequency plane ($j\omega$) approach, and (3) the time-domain approach. The real frequency ($j\omega$) approach is outlined in Chapter 9, and the discussion of the time-domain approach is considered in Section 6.4.

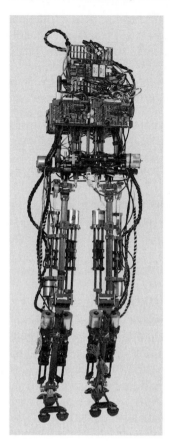

FIGURE 6.4
The M2 robot is more energy-efficient but less stable than many other designs that are well-balanced but consume much more power. (Courtesy of Professor Gill Pratt, Olin College.)

There are about one million robots in service throughout the world [10]. As the capability of robots increases, it is reasonable to assume that the numbers in service will continue to rise. Especially interesting are robots with human characteristics, particularly those that can walk upright. A class of robots that utilize series-elastic actuators as mechanical muscles emerged in the late 1990s. The M2 robot depicted in Figure 6.4 is more energy-efficient but less stable than many other designs that are well-balanced but consume much more power [22]. Examining the M2 robot in Figure 6.4, one can imagine that it is not inherently stable and that active control is required to keep it upright during the walking motion. In the next sections we present the Routh–Hurwitz stability criterion to investigate system stability by analyzing the characteristic equation without direct computation of the roots.

6.2 THE ROUTH–HURWITZ STABILITY CRITERION

The discussion and determination of stability has occupied the interest of many engineers. Maxwell and Vyshnegradskii first considered the question of stability of dynamic systems. In the late 1800s, A. Hurwitz and E. J. Routh independently published a method of investigating the stability of a linear system [6, 7]. The Routh–Hurwitz stability method provides an answer to the question of stability by considering the characteristic equation of the system. The characteristic equation in the Laplace variable is written as

$$\Delta(s) = q(s) = a_n s^n + a_{n-1} s^{n-1} + \cdots + a_1 s + a_0 = 0. \tag{6.3}$$

To ascertain the stability of the system, it is necessary to determine whether any one of the roots of $q(s)$ lies in the right half of the s-plane. If Equation (6.3) is written in factored form, we have

$$a_n(s - r_1)(s - r_2) \cdots (s - r_n) = 0, \tag{6.4}$$

where $r_i = i$th root of the characteristic equation. Multiplying the factors together, we find that

$$\begin{aligned}
q(s) = a_n s^n &- a_n(r_1 + r_2 + \cdots + r_n) s^{n-1} \\
&+ a_n(r_1 r_2 + r_2 r_3 + r_1 r_3 + \cdots) s^{n-2} \\
&- a_n(r_1 r_2 r_3 + r_1 r_2 r_4 \cdots) s^{n-3} + \cdots \\
&+ a_n(-1)^n r_1 r_2 r_3 \cdots r_n = 0.
\end{aligned} \tag{6.5}$$

In other words, for an nth-degree equation, we obtain

$$\begin{aligned}
q(s) = a_n s^n &- a_n \text{ (sum of all the roots) } s^{n-1} \\
&+ a_n \text{ (sum of the products of the roots taken 2 at a time) } s^{n-2} \\
&- a_n \text{ (sum of the products of the roots taken 3 at a time) } s^{n-3} \\
&+ \cdots + a_n(-1)^n \text{ (product of all } n \text{ roots)} = 0.
\end{aligned} \tag{6.6}$$

Examining Equation (6.5), we note that all the coefficients of the polynomial must have the same sign if all the roots are in the left-hand plane. Also, it is necessary that all the coefficients for a stable system be nonzero. These requirements are necessary but not sufficient. That is, we immediately know the system is unstable if they are not satisfied; yet if they are satisfied, we must proceed further to ascertain the stability of the system. For example, when the characteristic equation is

$$q(s) = (s + 2)(s^2 - s + 4) = (s^3 + s^2 + 2s + 8), \tag{6.7}$$

the system is unstable, and yet the polynomial possesses all positive coefficients.

The **Routh–Hurwitz criterion** is a necessary and sufficient criterion for the stability of linear systems. The method was originally developed in terms of determinants, but we shall use the more convenient array formulation.

The Routh–Hurwitz criterion is based on ordering the coefficients of the characteristic equation

$$a_n s^n + a_{n-1} s^{n-1} + a_{n-2} s^{n-2} + \cdots + a_1 s + a_0 = 0 \tag{6.8}$$

into an array or schedule as follows [4]:

s^n	a_n	a_{n-2}	$a_{n-4} \cdots$
s^{n-1}	a_{n-1}	a_{n-3}	$a_{n-5} \cdots$

Further rows of the schedule are then completed as

$$
\begin{array}{c|ccc}
s^n & a_n & a_{n-2} & a_{n-4} \\
s^{n-1} & a_{n-1} & a_{n-3} & a_{n-5} \\
s^{n-2} & b_{n-1} & b_{n-3} & b_{n-5} \\
s^{n-3} & c_{n-1} & c_{n-3} & n_{n-5} \\
\vdots & \vdots & \vdots & \vdots \\
s^0 & h_{n-1} &
\end{array}
$$

where

$$
b_{n-1} = \frac{a_{n-1}a_{n-2} - a_n a_{n-3}}{a_{n-1}} = \frac{-1}{a_{n-1}}\begin{vmatrix} a_n & a_{n-2} \\ a_{n-1} & a_{n-3} \end{vmatrix},
$$

$$
b_{n-3} = -\frac{1}{a_{n-1}}\begin{vmatrix} a_n & a_{n-4} \\ a_{n-1} & a_{n-5} \end{vmatrix},
$$

$$
c_{n-1} = \frac{-1}{b_{n-1}}\begin{vmatrix} a_{n-1} & a_{n-3} \\ b_{n-1} & b_{n-3} \end{vmatrix},
$$

and so on. The algorithm for calculating the entries in the array can be followed on a determinant basis or by using the form of the equation for b_{n-1}.

The Routh–Hurwitz criterion states that the number of roots of $q(s)$ with positive real parts is equal to the number of changes in sign of the first column of the Routh array. This criterion requires that there be no changes in sign in the first column for a stable system. This requirement is both necessary and sufficient.

Four distinct cases or configurations of the first column array must be considered, and each must be treated separately and requires suitable modifications of the array calculation procedure: (1) No element in the first column is zero; (2) there is a zero in the first column, but some other elements of the row containing the zero in the first column are nonzero; (3) there is a zero in the first column, and the other elements of the row containing the zero are also zero; and (4) as in the third case, but with repeated roots on the $j\omega$-axis.

To illustrate this method clearly, several examples will be presented for each case.

Case 1.   No element in the first column is zero.

EXAMPLE 6.1   **Second-order system**

The characteristic polynomial of a second-order system is

$$
q(s) = a_2 s^2 + a_1 s + a_0.
$$

The Routh array is written as

$$
\begin{array}{c|cc}
s^2 & a_2 & a_0 \\
s^1 & a_1 & 0 \\
s^0 & b_1 & 0
\end{array},
$$

where

$$b_1 = \frac{a_1 a_0 - (0)a_2}{a_1} = \frac{-1}{a_1}\begin{vmatrix} a_2 & a_0 \\ a_1 & 0 \end{vmatrix} = a_0.$$

Therefore, the requirement for a stable second-order system is simply that all the coefficients be positive or all the coefficients be negative. ∎

EXAMPLE 6.2   **Third-order system**

The characteristic polynomial of a third-order system is

$$q(s) = a_3 s^3 + a_2 s^2 + a_1 s + a_0.$$

The Routh array is

$$
\begin{array}{c|cc}
s^3 & a_3 & a_1 \\
s^2 & a_2 & a_0 \\
s^1 & b_1 & 0 \\
s^0 & c_1 & 0
\end{array},
$$

where

$$b_1 = \frac{a_2 a_1 - a_0 a_3}{a_2} \quad \text{and} \quad c_1 = \frac{b_1 a_0}{b_1} = a_0.$$

For the third-order system to be stable, it is necessary and sufficient that the coefficients be positive and $a_2 a_1 > a_0 a_3$. The condition when $a_2 a_1 = a_0 a_3$ results in a marginal stability case, and one pair of roots lies on the imaginary axis in the s-plane. This marginal case is recognized as Case 3 because there is a zero in the first column when $a_2 a_1 = a_0 a_3$. It will be discussed under Case 3.

As a final example of characteristic equations that result in no zero elements in the first row, let us consider the polynomial

$$q(s) = \left(s - 1 + j\sqrt{7}\right)\left(s - 1 - j\sqrt{7}\right)(s + 3) = s^3 + s^2 + 2s + 24. \quad (6.9)$$

The polynomial satisfies all the necessary conditions because all the coefficients exist and are positive. Therefore, utilizing the Routh array, we have

$$
\begin{array}{c|cc}
s^3 & 1 & 2 \\
s^2 & 1 & 24 \\
s^1 & -22 & 0 \\
s^0 & 24 & 0
\end{array}.
$$

Because two changes in sign appear in the first column, we find that two roots of $q(s)$ lie in the right-hand plane, and our prior knowledge is confirmed. ∎

Case 2.   **There is a zero in the first column, but some other elements of the row containing the zero in the first column are nonzero.** If only one element in the array is zero, it may be replaced with a small positive number, ϵ, that is allowed to approach zero after completing the array. For example, consider the following characteristic polynomial:

$$q(s) = s^5 + 2s^4 + 2s^3 + 4s^2 + 11s + 10. \tag{6.10}$$

The Routh array is then

s^5	1	2	11
s^4	2	4	10
s^3	ϵ	6	0
s^2	c_1	10	0'
s^1	d_1	0	0
s^0	10	0	0

where

$$c_1 = \frac{4\epsilon - 12}{\epsilon} = \frac{-12}{\epsilon} \quad \text{and} \quad d_1 = \frac{6c_1 - 10\epsilon}{c_1} \to 6.$$

There are two sign changes due to the large negative number in the first column, $c_1 = -12/\epsilon$. Therefore, the system is unstable, and two roots lie in the right half of the plane.

EXAMPLE 6.3   **Unstable system**

As a final example of the type of Case 2, consider the characteristic polynomial

$$q(s) = s^4 + s^3 + s^2 + s + K, \tag{6.11}$$

where we desire to determine the gain K that results in marginal stability. The Routh array is then

s^4	1	1	K
s^3	1	1	0
s^2	ϵ	K	0,
s^1	c_1	0	0
s^0	K	0	0

where

$$c_1 = \frac{\epsilon - K}{\epsilon} \to \frac{-K}{\epsilon}.$$

Therefore, for any value of K greater than zero, the system is unstable. Also, because the last term in the first column is equal to K, a negative value of K will result in an unstable system. Consequently, the system is unstable for all values of gain K. ∎

Case 3.   There is a zero in the first column, and the other elements of the row containing the zero are also zero. Case 3 occurs when all the elements in one row are zero or when the row consists of a single element that is zero. This condition occurs when the polynomial contains singularities that are symmetrically located about the origin of the s-plane. Therefore, Case 3 occurs when factors such as $(s + \sigma)(s - \sigma)$

or $(s + j\omega)(s - j\omega)$ occur. This problem is circumvented by utilizing the **auxiliary polynomial**, $U(s)$, which immediately precedes the zero entry in the Routh array. The order of the auxiliary polynomial is always even and indicates the number of symmetrical root pairs.

To illustrate this approach, let us consider a third-order system with the characteristic polynomial

$$q(s) = s^3 + 2s^2 + 4s + K, \tag{6.12}$$

where K is an adjustable loop gain. The Routh array is then

$$
\begin{array}{c|cc}
s^3 & 1 & 4 \\
s^2 & 2 & K \\
s^1 & \dfrac{8 - K}{2} & 0 \\
s^0 & K & 0
\end{array}
$$

For a stable system, we require that

$$0 < K < 8.$$

When $K = 8$, we have two roots on the $j\omega$-axis and a marginal stability case. Note that we obtain a row of zeros (Case 3) when $K = 8$. The auxiliary polynomial, $U(s)$, is the equation of the row preceding the row of zeros. The equation of the row preceding the row of zeros is, in this case, obtained from the s^2-row. We recall that this row contains the coefficients of the even powers of s, and therefore we have

$$U(s) = 2s^2 + Ks^0 = 2s^2 + 8 = 2(s^2 + 4) = 2(s + j2)(s - j2). \tag{6.13}$$

To show that the auxiliary polynomial, $U(s)$, is indeed a factor of the characteristic polynomial, we divide $q(s)$ by $U(s)$ to obtain

$$
\begin{array}{r}
\frac{1}{2}s + 1 \\
2s^2 + 8 \overline{)s^3 + 2s^2 + 4s + 8} \\
\underline{s^3 \qquad\quad + 4s} \\
2s^2 \quad + 8 \\
\underline{2s^2 \quad + 8}
\end{array}
$$

When $K = 8$, the factors of the characteristic polynomial are

$$q(s) = (s + 2)(s + j2)(s - j2). \tag{6.14}$$

The marginal case response is an unacceptable oscillation.

Case 4. Repeated roots of the characteristic equation on the $j\omega$-axis. If the $j\omega$-axis roots of the characteristic equation are simple, the system is neither stable nor unstable; it is instead called marginally stable, since it has an undamped sinusoidal mode. If the $j\omega$-axis roots are repeated, the system response will be unstable with a form $t \sin(\omega t + \phi)$. The Routh–Hurwitz criteria will not reveal this form of instability [21].

Consider the system with a characteristic polynomial

$$q(s) = (s + 1)(s + j)(s - j)(s + j)(s - j) = s^5 + s^4 + 2s^3 + 2s^2 + s + 1.$$

The Routh array is

$$\begin{array}{c|ccc} s^5 & 1 & 2 & 1 \\ s^4 & 1 & 2 & 1 \\ s^3 & \epsilon & \epsilon & 0 \\ s^2 & 1 & 1 & \\ s^1 & \epsilon & 0 & \\ s^0 & 1 & & \end{array},$$

where $\epsilon \rightarrow 0$. Note the absence of sign changes, a condition that falsely indicates that the system is marginally stable. The impulse response of the system increases with time as $t \sin(t + \phi)$. The auxiliary polynomial at the s^2 line is $s^2 + 1$, and the auxiliary polynomial at the s^4 line is $s^4 + 2s^2 + 1 = (s^2 + 1)^2$, indicating the repeated roots on the $j\omega$-axis.

EXAMPLE 6.4 **Fifth-order system with roots on the $j\omega$-axis**

Consider the characteristic polynomial

$$q(s) = s^5 + s^4 + 4s^3 + 24s^2 + 3s + 63. \tag{6.15}$$

The Routh array is

$$\begin{array}{c|ccc} s^5 & 1 & 4 & 3 \\ s^4 & 1 & 24 & 63 \\ s^3 & -20 & -60 & 0. \\ s^2 & 21 & 63 & 0 \\ s^1 & 0 & 0 & 0 \end{array}$$

Therefore, the auxiliary polynomial is

$$U(s) = 21s^2 + 63 = 21(s^2 + 3) = 21\left(s + j\sqrt{3}\right)\left(s - j\sqrt{3}\right), \tag{6.16}$$

which indicates that two roots are on the imaginary axis. To examine the remaining roots, we divide by the auxiliary polynomial to obtain

$$\frac{q(s)}{s^2 + 3} = s^3 + s^2 + s + 21.$$

Establishing a Routh array for this equation, we have

$$\begin{array}{c|cc} s^3 & 1 & 1 \\ s^2 & 1 & 21 \\ s^1 & -20 & 0^{\cdot} \\ s^0 & 21 & 0 \end{array}$$

The two changes in sign in the first column indicate the presence of two roots in the right-hand plane, and the system is unstable. The roots in the right-hand plane are $s = +1 \pm j\sqrt{6}$. ∎

EXAMPLE 6.5   Welding control

Large welding robots are used in today's auto plants. The welding head is moved to different positions on the auto body, and a rapid, accurate response is required. A block diagram of a welding head positioning system is shown in Figure 6.5. We desire to determine the range of K and a for which the system is stable. The characteristic equation is

$$1 + G(s) = 1 + \frac{K(s + a)}{s(s + 1)(s + 2)(s + 3)} = 0.$$

Therefore, $q(s) = s^4 + 6s^3 + 11s^2 + (K + 6)s + Ka = 0$. Establishing the Routh array, we have

$$
\begin{array}{c|ccc}
s^4 & 1 & 11 & Ka \\
s^3 & 6 & K + 6 & \\
s^2 & b_3 & Ka & \\
s^1 & c_3 & & \\
s^0 & Ka & &
\end{array}
\quad ,
$$

where

$$b_3 = \frac{60 - K}{6} \quad \text{and} \quad c_3 = \frac{b_3(K + 6) - 6Ka}{b_3}.$$

The coefficient c_3 sets the acceptable range of K and a, while b_3 requires that K be less than 60. Requiring $c_3 \geq 0$, we obtain

$$(K - 60)(K + 6) + 36Ka \leq 0.$$

The required relationship between K and a is then

$$a \leq \frac{(60 - K)(K + 6)}{36K}$$

when a is positive. Therefore, if $K = 40$, we require $a \leq 0.639$. ∎

The general form of the characteristic equation of an nth-order system is

$$s^n + a_{n-1}s^{n-1} + a_{n-2}s^{n-2} + \cdots + a_1 s + \omega_n{}^n = 0.$$

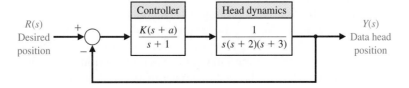

FIGURE 6.5
Welding head
position control.

R(s) Desired position → Controller $\dfrac{K(s + a)}{s + 1}$ → Head dynamics $\dfrac{1}{s(s + 2)(s + 3)}$ → Y(s) Data head position

Table 6.1   The Routh–Hurwitz Stability Criterion

n	Characteristic Equation	Criterion
2	$s^2 + bs + 1 = 0$	$b > 0$
3	$s^3 + bs^2 + cs + 1 = 0$	$bc - 1 > 0$
4	$s^4 + bs^3 + cs^2 + ds + 1 = 0$	$bcd - d^2 - b^2 > 0$
5	$s^5 + bs^4 + cs^3 + ds^2 + es + 1 = 0$	$bcd + b - d^2 - b^2e > 0$
6	$s^6 + bs^5 + cs^4 + ds^3 + es^2 + fs + 1 = 0$	$(bcd + bf - d^2 - b^2e)e + b^2c - bd - bc^2f - f^2 + bfe + cdf > 0$

Note: The equations are normalized by $(\omega_n)^n$.

We divide through by $\omega_n{}^n$ and use $\overset{*}{s} = s/\omega_n$ to obtain the normalized form of the characteristic equation:

$$\overset{*}{s}{}^n + b\overset{*}{s}{}^{n-1} + c\overset{*}{s}{}^{n-2} + \cdots + 1 = 0.$$

For example, we normalize

$$s^3 + 5s^2 + 2s + 8 = 0$$

by dividing through by $8 = \omega_n{}^3$, obtaining

$$\frac{s^3}{\omega_n{}^3} + \frac{5}{2}\frac{s^2}{\omega_n{}^2} + \frac{2}{4}\frac{s}{\omega_n} + 1 = 0,$$

or

$$\overset{*}{s}{}^3 + 2.5\overset{*}{s}{}^2 + 0.5\overset{*}{s} + 1 = 0,$$

where $\overset{*}{s} = s/\omega_n$. In this case, $b = 2.5$ and $c = 0.5$. Using this normalized form of the characteristic equation, we summarize the stability criterion for up to a sixth-order characteristic equation, as provided in Table 6.1. Note that $bc = 1.25$ and the system is stable.

6.3 THE RELATIVE STABILITY OF FEEDBACK CONTROL SYSTEMS

The verification of stability using the Routh–Hurwitz criterion provides only a partial answer to the question of stability. The Routh–Hurwitz criterion ascertains the absolute stability of a system by determining whether any of the roots of the characteristic equation lie in the right half of the s-plane. However, if the system satisfies the Routh–Hurwitz criterion and is absolutely stable, it is desirable to determine the **relative stability**; that is, it is necessary to investigate the relative damping of each root of the characteristic equation. The relative stability of a system can be defined as the property that is measured by the relative real part of each root or pair of roots. Thus, root r_2 is relatively more stable than the roots $r_1, \hat{r}_1$, as shown in Figure 6.6. The relative stability of a system can also be defined in terms of the relative damping coefficients ζ of each complex root pair and, therefore, in terms of the speed of response and overshoot instead of settling time.

Hence, the investigation of the relative stability of each root is clearly necessary because, as we found in Chapter 5, the location of the closed-loop poles in the s-plane determines the performance of the system. Thus, it is imperative that we

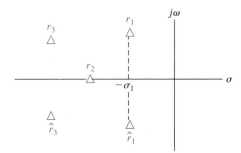

FIGURE 6.6
Root locations in
the s-plane.

reexamine the characteristic polynomial $q(s)$ and consider several methods for the determination of relative stability.

Because the relative stability of a system is dictated by the location of the roots of the characteristic equation, a first approach using an s-plane formulation is to extend the Routh–Hurwitz criterion to ascertain relative stability. This can be simply accomplished by utilizing a change of variable, which shifts the s-plane axis in order to utilize the Routh–Hurwitz criterion. Examining Figure 6.6, we notice that a shift of the vertical axis in the s-plane to $-\sigma_1$ will result in the roots r_1, $\hat{r}_1$ appearing on the shifted axis. The correct magnitude to shift the vertical axis must be obtained on a trial-and-error basis. Then, without solving the fifth-order polynomial $q(s)$, we may determine the real part of the dominant roots r_1, $\hat{r}_1$.

EXAMPLE 6.6   **Axis shift**

Consider the simple third-order characteristic equation

$$q(s) = s^3 + 4s^2 + 6s + 4. \tag{6.17}$$

As a first try, let $s_n = s + 2$ and note that we obtain a Routh array without a zero occurring in the first column. However, upon setting the shifted variable s_n equal to $s + 1$, we obtain

$$(s_n - 1)^3 + 4(s_n - 1)^2 + 6(s_n - 1) + 4 = s_n^3 + s_n^2 + s_n + 1. \tag{6.18}$$

Then the Routh array is established as

$$
\begin{array}{c|cc}
s_n^3 & 1 & 1 \\
s_n^2 & 1 & 1 \\
s_n^1 & 0 & 0 \\
s_n^0 & 1 & 0
\end{array}
$$

There are roots on the shifted imaginary axis that can be obtained from the auxiliary polynomial

$$U(s_n) = s_n^2 + 1 = (s_n + j)(s_n - j) = (s + 1 + j)(s + 1 - j). \tag{6.19} \ \blacksquare$$

The shifting of the s-plane axis to ascertain the relative stability of a system is a very useful approach, particularly for higher-order systems with several pairs of closed-loop complex conjugate roots.

6.4 THE STABILITY OF STATE VARIABLE SYSTEMS

The stability of a system modeled by a state variable flow graph model can be readily ascertained. The stability of a system with an input–output transfer function $T(s)$ can be determined by examining the denominator polynomial of $T(s)$. Therefore, if the transfer function is written as

$$T(s) = \frac{p(s)}{q(s)},$$

where $p(s)$ and $q(s)$ are polynomials in s, then the stability of the system is represented by the roots of $q(s)$. The polynomial $q(s)$, when set equal to zero, is called the characteristic equation. The roots of the characteristic equation must lie in the left-hand s-plane for the system to exhibit a stable time response. Therefore, to ascertain the stability of a system represented by a transfer function, we investigate the characteristic equation and utilize the Routh–Hurwitz criterion. If the system we are investigating is represented by a signal-flow graph state model, we obtain the characteristic equation by evaluating the flow graph determinant. If the system is represented by a block diagram model we obtain the characteristic equation using the block diagram reduction methods. As an illustration of these methods, let us investigate the stability of the system of Example 3.2.

EXAMPLE 6.7   **Stability of a system**

The transfer function $T(s)$ examined in Example 3.2 is

$$T(s) = \frac{2s^2 + 8s + 6}{s^3 + 8s^2 + 16s + 6}. \qquad (6.20)$$

The characteristic polynomial for this system is

$$q(s) = s^3 + 8s^2 + 16s + 6. \qquad (6.21)$$

This characteristic polynomial is also readily obtained from either the flow graph model or block diagram model shown in Figure 3.11 or the ones shown in Figure 3.13. Using the Routh–Hurwitz criterion, we find that the system is stable and that all the roots of $q(s)$ lie in the left-hand s-plane. ∎

We often determine the flow graph or block diagram model directly from a set of state differential equations. We can use the flow graph directly to determine the stability of the system by obtaining the characteristic equation from the flow graph determinant $\Delta(s)$. Similarly, we can use block diagram reduction to define the characteristic equation. An illustration of these approaches will aid in comprehending these methods.

EXAMPLE 6.8   **Stability of a second-order system**

A second-order system is described by the two first-order differential equations

$$\dot{x}_1 = -3x_1 + x_2 \quad \text{and} \quad \dot{x}_2 = +1x_2 - Kx_1 + Ku,$$

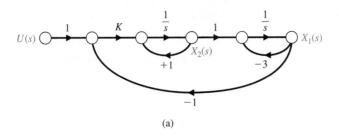

(a)

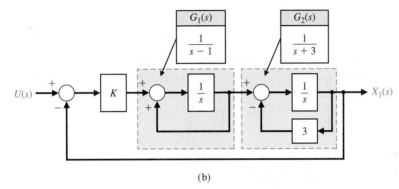

FIGURE 6.7
(a) Flow graph
model for state
variable equations
of Example 6.8.
(b) Block diagram
model.

(b)

where the dot notation implies the first derivative and $u(t)$ is the input. The flow graph model of this set of differential equations is shown in Figure 6.7(a) and the block diagram model is shown in Figure 6.7(b).

Using Mason's signal-flow gain formula, we note three loops:

$$L_1 = s^{-1}, \qquad L_2 = -3s^{-1}, \quad \text{and} \quad L_3 = -Ks^{-2},$$

where L_1 and L_2 do not share a common node. Therefore, the determinant is

$$\Delta = 1 - (L_1 + L_2 + L_3) + L_1 L_2 = 1 - (s^{-1} - 3s^{-1} - Ks^{-2}) + (-3s^{-2}).$$

We multiply by s^2 to obtain the characteristic equation

$$s^2 + 2s + (K - 3) = 0.$$

Since all coefficients must be positive, we require $K > 3$ for stability. A similar analysis can be undertaken using the block diagram. Closing the two feedback loops yields the two transfer functions

$$G_1(s) = \frac{1}{s - 1} \quad \text{and} \quad G_2(s) = \frac{1}{s + 3},$$

as illustrated in Figure 6.7(b). The closed loop transfer function is thus

$$T(s) = \frac{KG_1(s)G_2(s)}{1 + KG_1(s)G_2(s)}.$$

Therefore, the characteristic equation is

$$\Delta(s) = 1 + KG_1(s)G_2(s) = 0,$$

or

$$\Delta(s) = (s - 1)(s + 3) + K = s^2 + 2s + (K - 3) = 0.$$

This confirms the results obtained using signal-flow graph techniques. ∎

A method of obtaining the characteristic equation directly from the vector differential equation is based on the fact that the solution to the unforced system is an exponential function. The vector differential equation without input signals is

$$\dot{\mathbf{x}} = \mathbf{A}\mathbf{x}, \tag{6.22}$$

where $\mathbf{x}$ is the state vector. The solution is of exponential form, and we can define a constant λ such that the solution of the system for one state can be of the form $x_i(t) = k_i e^{\lambda_i t}$. The λ_i are called the characteristic roots or eigenvalues of the system, which are simply the roots of the characteristic equation. If we let $\mathbf{x} = \mathbf{k}e^{\lambda t}$ and substitute into Equation (6.22), we have

$$\lambda \mathbf{k}e^{\lambda t} = \mathbf{A}\mathbf{k}e^{\lambda t}, \tag{6.23}$$

or

$$\lambda \mathbf{x} = \mathbf{A}\mathbf{x}. \tag{6.24}$$

Equation (6.24) can be rewritten as

$$(\lambda \mathbf{I} - \mathbf{A})\mathbf{x} = \mathbf{0}, \tag{6.25}$$

where $\mathbf{I}$ equals the identity matrix and $\mathbf{0}$ equals the null matrix. This set of simultaneous equations has a nontrivial solution if and only if the determinant vanishes—that is, only if

$$\det(\lambda \mathbf{I} - \mathbf{A}) = 0. \tag{6.26}$$

The nth-order equation in λ resulting from the evaluation of this determinant is the characteristic equation, and the stability of the system can be readily ascertained. Let us consider again the third-order system described in Example 3.3 to illustrate this approach.

EXAMPLE 6.9 **Closed epidemic system**

The vector differential equation of the epidemic system is given in Equation (3.63) and repeated here as

$$\frac{d\mathbf{x}}{dt} = \begin{bmatrix} -\alpha & -\beta & 0 \\ \beta & -\gamma & 0 \\ \alpha & \gamma & 0 \end{bmatrix} \mathbf{x} + \begin{bmatrix} 1 & 0 \\ 0 & 1 \\ 0 & 0 \end{bmatrix} \begin{bmatrix} u_1 \\ u_2 \end{bmatrix}.$$

The characteristic equation is then

$$\det(\lambda \mathbf{I} - \mathbf{A}) = \det \left\{ \begin{bmatrix} \lambda & 0 & 0 \\ 0 & \lambda & 0 \\ 0 & 0 & \lambda \end{bmatrix} - \begin{bmatrix} -\alpha & -\beta & 0 \\ \beta & -\gamma & 0 \\ \alpha & \gamma & 0 \end{bmatrix} \right\}$$

$$= \det \begin{bmatrix} \lambda + \alpha & \beta & 0 \\ -\beta & \lambda + \gamma & 0 \\ -\alpha & -\gamma & \lambda \end{bmatrix}$$

$$= \lambda[(\lambda + \alpha)(\lambda + \gamma) + \beta^2]$$
$$= \lambda[\lambda^2 + (\alpha + \gamma)\lambda + (\alpha\gamma + \beta^2)] = 0.$$

Thus, we obtain the characteristic equation of the system, and it is similar to that obtained in Equation (3.65) by flow graph methods. The additional root $\lambda = 0$ results from the definition of x_3 as the integral of $\alpha x_1 + \gamma x_2$, and x_3 does not affect the other state variables. Thus, the root $\lambda = 0$ indicates the integration connected with x_3. The characteristic equation indicates that the system is marginally stable when $\alpha + \gamma > 0$ and $\alpha\gamma + \beta^2 > 0$. ∎

As another example, consider again the inverted pendulum described in Example 3.4. The system matrix is

$$\mathbf{A} = \begin{bmatrix} 0 & 1 & 0 & 0 \\ 0 & 0 & -mg/M & 0 \\ 0 & 0 & 0 & 1 \\ 0 & 0 & g/l & 0 \end{bmatrix}.$$

The characteristic equation can be obtained from the determinant of $(\lambda\mathbf{I} - \mathbf{A})$ as follows:

$$\det\begin{bmatrix} \lambda & -1 & 0 & 0 \\ 0 & \lambda & mg/M & 0 \\ 0 & 0 & \lambda & -1 \\ 0 & 0 & -g/l & \lambda \end{bmatrix} = \lambda\left[\lambda\left(\lambda^2 - \frac{g}{l}\right)\right] = \lambda^2\left(\lambda^2 - \frac{g}{l}\right) = 0.$$

The characteristic equation indicates that there are two roots at $\lambda = 0$: a root at $\lambda = +\sqrt{g/l}$ and a root at $\lambda = -\sqrt{g/l}$. Hence, the system is unstable, because there is a root in the right-hand plane at $\lambda = +\sqrt{g/l}$. The two roots at $\lambda = 0$ will also result in an unbounded response.

6.5 DESIGN EXAMPLES

In this section we present two illustrative examples. The first example is a tracked vehicle control problem. In this first example, stability issues are addressed employing the Routh-Hurwitz stability criterion and the outcome is the selection of two key system parameters. The second example illustrates the stability problem robot-controlled motorcycle and how Routh-Hurwitz can be used in the selection of controller gains during the design process. The robot-controlled motorcycle example highlights the design process with special attention to the impact of key controller parameters on stability.

EXAMPLE 6.10   Tracked vehicle turning control

The design of a turning control for a tracked vehicle involves the selection of two parameters [8]. In Figure 6.8, the system shown in part (a) has the model shown in part (b). The two tracks are operated at different speeds in order to turn the vehicle.

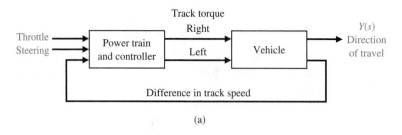

(a)

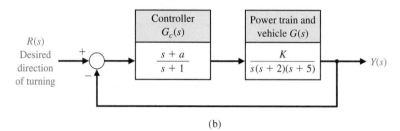

(b)

FIGURE 6.8
(a) Turning control system for a two-track vehicle.
(b) Block diagram.

We must select K and a so that the system is stable and the steady-state error for a ramp command is less than or equal to 24% of the magnitude of the command.

The characteristic equation of the feedback system is

$$1 + G_c G(s) = 0,$$

or

$$1 + \frac{K(s + a)}{s(s + 1)(s + 2)(s + 5)} = 0. \tag{6.27}$$

Therefore, we have

$$s(s + 1)(s + 2)(s + 5) + K(s + a) = 0,$$

or

$$s^4 + 8s^3 + 17s^2 + (K + 10)s + Ka = 0. \tag{6.28}$$

To determine the stable region for K and a, we establish the Routh array as

$$
\begin{array}{c|cccc}
s^4 & 1 & 17 & Ka \\
s^3 & 8 & K + 10 & 0 \\
s^2 & b_3 & Ka & \\
s^1 & c_3 & & \\
s^0 & Ka & &
\end{array} \quad ,
$$

where

$$b_3 = \frac{126 - K}{8} \quad \text{and} \quad c_3 = \frac{b_3(K + 10) - 8Ka}{b_3}.$$

For the elements of the first column to be positive, we require that Ka, b_3, and c_3 be positive. Therefore, we require that

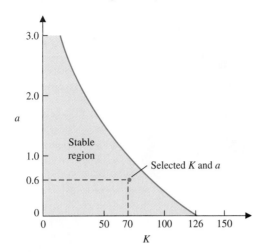

FIGURE 6.9
The stable region.

$$K < 126,$$

$$Ka > 0, \quad \text{and}$$

$$(K + 10)(126 - K) - 64Ka > 0. \tag{6.29}$$

The region of stability for $K > 0$ is shown in Figure 6.9. The steady-state error to a ramp input $r(t) = At, t > 0$ is

$$e_{ss} = A/K_v,$$

where

$$K_v = \lim_{s \to 0} sG_cG = Ka/10.$$

Therefore, we have

$$e_{ss} = \frac{10A}{Ka}. \tag{6.30}$$

When e_{ss} is equal to 23.8% of A, we require that $Ka = 42$. This can be satisfied by the selected point in the stable region when $K = 70$ and $a = 0.6$, as shown in Figure 6.9. Another acceptable design would be attained when $K = 50$ and $a = 0.84$. We can calculate a series of possible combinations of K and a that can satisfy $Ka = 42$ and that lie within the stable region, and all will be acceptable design solutions. However, not all selected values of K and a will lie within the stable region. Note that K cannot exceed 126. ∎

EXAMPLE 6.11   **Robot-controlled motorcycle**

Consider the robot-controlled motorcycle shown in Figure 6.10. The motorcycle will move in a straight line at constant forward speed v. Let ϕ denote the angle between the plane of symmetry of the motorcycle and the vertical. The desired angle ϕ_d is equal to zero:

$$\phi_d(s) = 0.$$

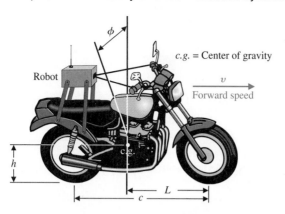

FIGURE 6.10
The robot-
controlled
motorcycle.

The design elements highlighted in this example are illustrated in Figure 6.11. Using the Routh–Hurwitz stability criterion will allow us to get to the heart of the matter, that is, to develop a strategy for computing the controller gains while ensuring closed-loop stability.

The control goal is

Control Goal
Control the motorcycle in the vertical position, and maintain the prescribed position in the presence of disturbances.

The variable to be controlled is

Variable to Be Controlled
The motorcycle position from vertical (ϕ).

Since our focus here is on stability rather than transient response characteristics, the control specifications will be related to stability only; transient performance is an issue that we need to address once we have investigated all the stability issues. The control design specification is

Design Specification
DS1 The closed-loop system must be stable.

The main components of the robot-controlled motorcycle are the motorcycle and robot, the controller, and the feedback measurements. The main subject of the chapter is not modeling, so we do not concentrate on developing the motorcycle dynamics model. We rely instead on the work of others (see [25]). The motorcycle model is given by

$$G(s) = \frac{1}{s^2 - \alpha_1}, \tag{6.31}$$

where $\alpha_1 = g/h$, $g = 9.806$ m/s^2, and h is the height of the motorcycle center of gravity above the ground (see Figure 6.10). The motorcycle is unstable with poles at $s = \pm\sqrt{\alpha_1}$. The controller is given by

$$G_c(s) = \frac{\alpha_2 + \alpha_3 s}{\tau s + 1}, \tag{6.32}$$

Topics emphasized in this example

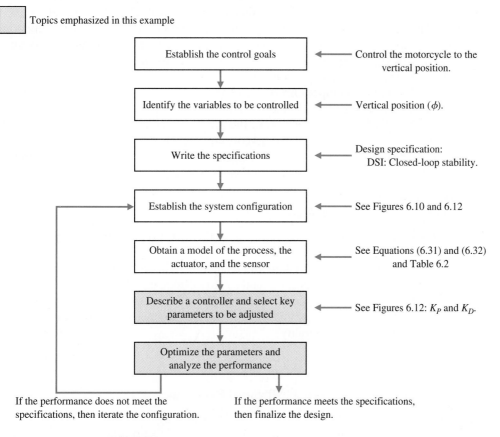

FIGURE 6.11   Elements of the control system design process emphasized in this robot-controlled motorcycle example.

where

$$\alpha_2 = v^2/(hc)$$

and

$$\alpha_3 = vL/(hc).$$

The forward speed of the motorcycle is denoted by v, and c denotes the wheel-base (the distance between the wheel centers). The length, L, is the horizontal distance between the front wheel axle and the motorcycle center of gravity. The time-constant of the controller is denoted by τ. This term represents the speed of response of the controller; smaller values of τ indicate an increased speed of response. Many simplifying assumptions are necessary to obtain the simple transfer function models in Equations (6.31) and (6.32).

Control is accomplished by turning the handlebar. The front wheel rotation about the vertical is not evident in the transfer functions. Also, the transfer functions assume a constant forward speed v which means that we must have another control system at work regulating the forward speed. Nominal motorcycle and robot controller parameters are given in Table 6.2.

Assembling the components of the feedback system gives us the system configuration shown in Figure 6.12. Examination of the configuration reveals that the robot controller block is a function of the physical system (h, c, and L), the operating conditions (v), and the robot time-constant (τ). No parameters need adjustment unless we physically change the motorcycle parameters and/or speed. In fact, in this example the parameters we want to adjust are in the feedback loop:

Select Key Tuning Parameters
Feedback gains K_P and K_D.

The key tuning parameters are not always in the forward path; in fact they may exist in any subsystem in the block diagram.

We want to use the Routh–Hurwitz technique to analyze the closed-loop system stability. What values of K_P and K_D lead to closed-loop stability? A related question that we can pose is, given specific values of K_P and K_D for the nominal system (that is, nominal values of α_1, α_2, α_3, and τ), how can the parameters themselves vary while still retaining closed-loop stability?

Table 6.2	Physical Parameters
τ	0.2 s
α_1	9 1/s^2
α_2	2.7 1/s^2
α_3	1.35 1/s
h	1.09 m
V	2.0 m/s
L	1.0 m
c	1.36 m

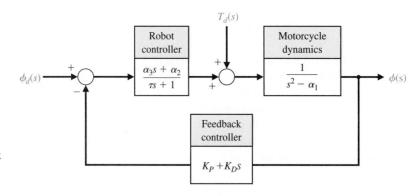

FIGURE 6.12
The robot-controlled motorcycle feedback system block diagram.

The closed-loop transfer function from $\phi_d(s)$ to $\phi(s)$ is

$$T(s) = \frac{\alpha_2 + \alpha_3 s}{\Delta(s)},$$

where

$$\Delta(s) = \tau s^3 + (1 + K_D\alpha_3)s^2 + (K_D\alpha_2 + K_P\alpha_3 - \tau\alpha_1)s + K_P\alpha_2 - \alpha_1.$$

The characteristic equation is

$$\Delta(s) = 0.$$

The question that we need to answer is for what values of K_P and K_D does the characteristic equation $\Delta(s) = 0$ have all roots in the left half-plane?

We can set up the following Routh array:

$$
\begin{array}{c|cc}
s^3 & \tau & K_D\alpha_2 + K_P\alpha_3 - \tau\alpha_1 \\
s^2 & 1 + K_D\alpha_3 & K_P\alpha_2 - \alpha_1 \\
s & a & \\
1 & K_P\alpha_2 - \alpha_1 &
\end{array}
$$

where

$$a = \frac{(1 + K_D\alpha_3)(K_D\alpha_2 + K_P\alpha_3 - \tau\alpha_1) - \tau(\alpha_2 K_P - \alpha_1)}{1 + K_D\alpha_3}.$$

By inspecting column 1, we determine that for stability we require

$$\tau > 0, \ K_D > -1/\alpha_3, \ K_P > \alpha_1/\alpha_2, \ \text{and} \ a > 0.$$

Choosing $K_D > 0$ satisfies the second inequality (note that $\alpha_3 > 0$). In the event $\tau = 0$, we would reformulate the characteristic equation and rework the Routh array.

The computational difficulty arises in determining the conditions on K_P and K_D such that $a > 0$. We find that $a > 0$ implies that the following relationship must be satisfied:

$$f = \alpha_2\alpha_3 K_D^2 + (\alpha_2 - \tau\alpha_1\alpha_3 + \alpha_3^2 K_P)K_D + (\alpha_3 - \tau\alpha_2)K_P > 0. \qquad (6.33)$$

Using the nominal values of the parameters α_1, α_2, α_3, and τ (see Table 6.2), the stability region is shown in Figure 6.13. For all $K_D > 0$ and $K_P > 3.33$, the function $f > 0$, hence $a > 0$. Taking into account all the inequalities, a valid region for selecting the gains is $K_D > 0$ and $K_P > \alpha_1/\alpha_2 = 3.33$.

Selecting any point (K_P, K_D) in the stability region yields a valid (that is, stable) set of gains for the feedback loop. For example, selecting

$$K_P = 10 \text{ and } K_D = 5$$

yields a stable closed-loop system. The closed-loop poles are

$$s_1 = -35.2477, \ s_2 = -2.4674, \text{ and } s_3 = -1.0348.$$

Since all the poles have negative real parts, we know the system response to any bounded input will be bounded.

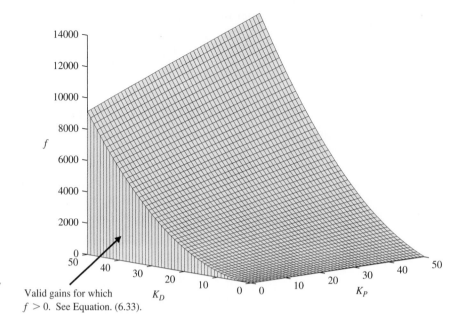

FIGURE 6.13
Region of valid
gains (K_D, K_P) for
which the inequality
in Equation. (6.33)
is satisfied.

Valid gains for which
$f > 0$. See Equation. (6.33).

For this robot-controlled motorcycle, we do not expect to have to respond to nonzero command inputs (that is, $\phi_d \neq 0$) since we want the motorcycle to remain upright, and we certainly want to remain upright in the presence of external disturbances.. The transfer function for the disturbance $T_d(s)$ to the output $\phi(s)$ without feedback is

$$\phi(s) = \frac{1}{s^2 - \alpha_1} T_d(s).$$

The characteristic equation is

$$q(s) = s^2 - \alpha_1 = 0.$$

The system poles are

$$s_1 = -\sqrt{\alpha_1} \text{ and } s_2 = +\sqrt{\alpha_1}.$$

Thus we see that the motorcycle is unstable; it possesses a pole in the right half-plane. Without feedback control, any external disturbance will result in the motorcycle falling over. Clearly the need for a control system (usually provided by the human rider) is necessary. With the feedback and robot controller in the loop, the closed-loop transfer function from the disturbance to the output is

$$\frac{\phi(s)}{T_d(s)} = \frac{\tau s + 1}{\tau s^3 + (1 + K_D \alpha_3)s^2 + (K_D \alpha_2 + K_P \alpha_3 - \tau \alpha_1)s + K_P \alpha_2 - \alpha_1}.$$

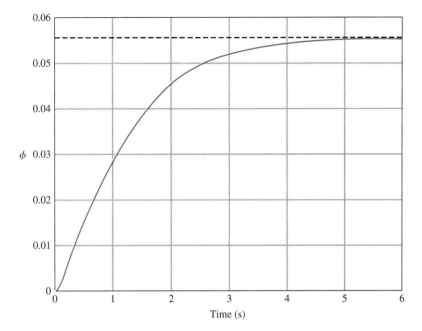

FIGURE 6.14
Disturbance
response with
K_P = 10 and
K_D = 5.

The response to a step disturbance

$$T_d(s) = \frac{1}{s},$$

is shown in Figure 6.14; the response is stable. The control system manages to keep the motorcycle upright, although it is tilted at about ϕ = 0.055 rad = 3.18 deg.

It is important to give the robot the ability to control the motorcycle over a wide range of forward speeds. Is it possible for the robot, with the feedback gains as selected (K_P = 10 and K_D = 5), to control the motorcycle as the velocity varies? From experience we know that at slower speeds a bicycle becomes more difficult to control. We expect to see the same characteristics in the stability analysis of our system. Whenever possible, we try to relate the engineering problem at hand to real-life experiences. This helps to develop intuition that can be used as a reasonableness check on our solution.

A plot of the roots of the characteristic equation as the forward speed v varies is shown in Figure 6.15. The data in the plot were generated using the nominal values of the feedback gains, K_P = 10 and K_D = 5. We selected these gains for the case where v = 2 m/s. Figure 6.15 shows that as v increases, the roots of the characteristic equation remain stable (that is, in the left half-plane) with all points negative. But as the motorcycle forward speed decreases, the roots move toward zero, with one root becoming positive at v = 1.15 m/s. At the point where one root is positive, the motorcycle is unstable. ∎

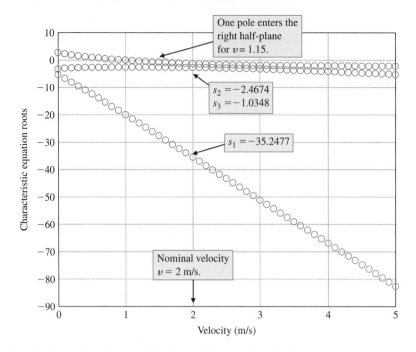

FIGURE 6.15
Roots of the characteristic equation as the motorcycle velocity varies.

6.6 SYSTEM STABILITY USING CONTROL DESIGN SOFTWARE

This section begins with a discussion of the Routh–Hurwitz stability method. We will see how the computer can assist us in the stability analysis by providing an easy and accurate method for computing the poles of the characteristic equation. For the case of the characteristic equation as a function of a single parameter, it will be possible to generate a plot displaying the movement of the poles as the parameter varies. The section concludes with an example.

The function introduced in this section is the function for, which is used to repeat a number of statements a specific number of times.

Routh–Hurwitz Stability. As stated earlier, the Routh–Hurwitz criterion is a necessary and sufficient criterion for stability. Given a characteristic equation with fixed coefficients, we can use Routh–Hurwitz to determine the number of roots in the right half-plane. For example, consider the characteristic equation

$$q(s) = s^3 + s^2 + 2s + 24 = 0$$

associated with the closed-loop control system shown in Figure 6.16. The corresponding Routh–Hurwitz array is shown in Figure 6.17. The two sign changes in the first column indicate that there are two roots of the characteristic polynomial in the right half-plane; hence, the closed-loop system is unstable. We can verify the Routh–Hurwitz result by directly computing the roots of the characteristic equation, as shown in Figure 6.18, using the pole function. Recall that the pole function computes the system poles.

Whenever the characteristic equation is a function of a single parameter, the Routh–Hurwitz method can be utilized to determine the range of values that the

FIGURE 6.16
Closed-loop control system with $T(s) = Y(s)/R(s) = 1/(s^3 + s^2 + 2s + 24)$.

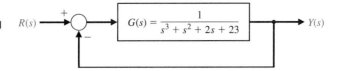

FIGURE 6.17
Routh array for the closed-loop control system with $T(s) = Y(s)/R(s) = 1/(s^3 + s^2 + 2s + 24)$.

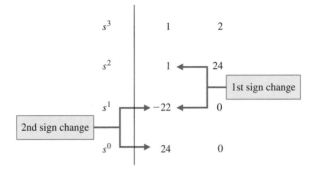

parameter may take while maintaining stability. Consider the closed-loop feedback system in Figure 6.19. The characteristic equation is

$$q(s) = s^3 + 2s^2 + 4s + K = 0.$$

Using a Routh–Hurwitz approach, we find that we require $0 < K < 8$ for stability (see Equation 6.12). We can verify this result graphically. As shown in Figure 6.20(a), we establish a vector of values for K at which we wish to compute the roots of the characteristic equation. Then using the roots function, we calculate and plot the roots of the characteristic equation, as shown in Figure 6.20(b). It can be seen that as K increases, the roots of the characteristic equation move toward the right half-plane as the gain tends toward $K = 8$, and eventually into the right half-plane when $K > 8$.

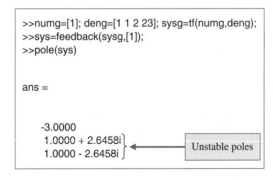

```
>>numg=[1]; deng=[1 1 2 23]; sysg=tf(numg,deng);
>>sys=feedback(sysg,[1]);
>>pole(sys)


ans =


   -3.0000
    1.0000 + 2.6458i
    1.0000 - 2.6458i
```

FIGURE 6.18
Using the **pole** function to compute the closed-loop control system poles of the system shown in Figure 6.16.

FIGURE 6.19
Closed-loop control system with $T(s) = Y(s)/R(s) = K/(s^3 + 2s^2 + 4s + K)$.

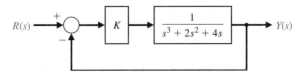

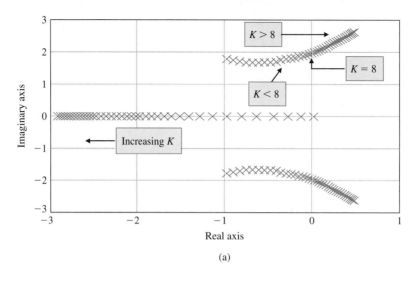

(a)

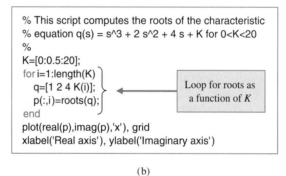

(b)

FIGURE 6.20
(a) Plot of root locations of $q(s) = s^3 + 2s^2 + 4s + K$ for $0 \le K \le 20$.
(b) m-file script.

The script in Figure 6.20 contains the for function. This function provides a mechanism for repeatedly executing a series of statements a given number of times. The for function connected to an end statement sets up a repeating calculation loop. Figure 6.21 describes the for function format and provides an illustrative example of its usefulness. The example sets up a loop that repeats ten times. During the ith iteration, where $1 \le i \le 10$, the ith element of the vector **a** is set equal to 20, and the scalar b is recomputed.

The Routh–Hurwitz method allows us to make definitive statements regarding absolute stability of a linear system. The method does not address the issue of relative stability, which is directly related to the location of the roots of the characteristic equation. Routh–Hurwitz tells us how many poles lie in the right half-plane, but not the specific location of the poles. With control design software, we can easily calculate the poles explicitly, thus allowing us to comment on the relative stability.

EXAMPLE 6.12   Tracked vehicle control

The block diagram of the control system for the two-track vehicle is shown in Figure 6.8. The design objective is to find a and K such that the system is stable and the steady-state error for a ramp input is less than or equal to 24% of the command.

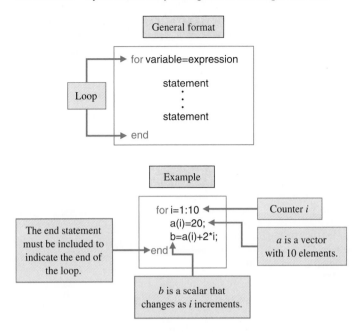

FIGURE 6.21
The **for** function
and an illustrative
example.

We can use the Routh–Hurwitz method to aid in the search for appropriate values of a and K. The closed-loop characteristic equation is

$$q(s) = s^4 + 8s^3 + 17s^2 + (K + 10)s + aK = 0.$$

Using the Routh array, we find that, for stability, we require that

$$K < 126, \frac{126 - K}{8}(K + 10) - 8aK > 0, \quad \text{and} \quad aK > 0.$$

For positive K, it follows that we can restrict our search to $0 < K < 126$ and $a > 0$. Our approach will be to use the computer to help find a parameterized a versus K region in which stability is assured. Then we can find a set of (a, K) belonging to the stable region such that the steady-state error specification is met. This procedure, shown in Figure 6.22, involves selecting a range of values for a and K and computing the roots of the characteristic polynomial for specific values of a and K. For each value of K, we find the first value of a that results in at least one root of the characteristic equation in the right half-plane. The process is repeated until the entire selected range of a and K is exhausted. The plot of the (a, K) pairs defines the separation between the stable and unstable regions. The region to the left of the plot of a versus K in Figure 6.22 is the stable region.

If we assume that $r(t) = At, t > 0$, then the steady-state error is

$$e_{ss} = \lim_{s \to 0} s \cdot \frac{s(s + 1)(s + 2)(s + 5)}{s(s + 1)(s + 2)(s + 5) + K(s + a)} \cdot \frac{A}{s^2} = \frac{10A}{aK},$$

where we have used the fact that

$$E(s) = \frac{1}{1 + G_c G(s)} R(s) = \frac{s(s + 1)(s + 2)(s + 5)}{s(s + 1)(s + 2)(s + 5) + K(s + a)} R(s).$$

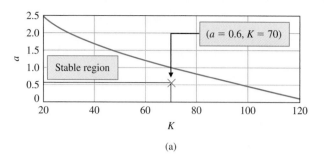

(a)

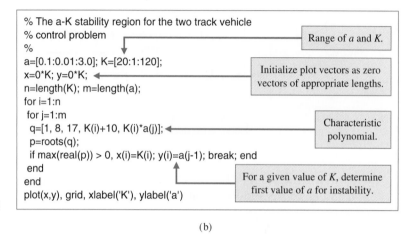

(b)

FIGURE 6.22
(a) Stability region for a and K for two-track vehicle turning control. (b) m-file script.

Given the steady-state specification, $e_{ss} < 0.24A$, we find that the specification is satisfied when

$$\frac{10A}{aK} < 0.24A,$$

or

$$aK > 41.67. \qquad (6.34)$$

Any values of a and K that lie in the stable region in Figure 6.22 and satisfy Equation (6.34) will lead to an acceptable design. For example, $K = 70$ and $a = 0.6$ will satisfy all the design requirements. The closed-loop transfer function (with $a = 0.6$ and $K = 70$) is

$$T(s) = \frac{70s + 42}{s^4 + 8s^3 + 17s^2 + 80s + 42}.$$

The associated closed-loop poles are

$$s = -7.0767,$$
$$s = -0.5781,$$
$$s = -0.1726 + 3.1995i, \quad \text{and}$$
$$s = -0.1726 - 3.1995i.$$

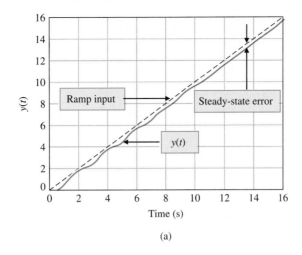

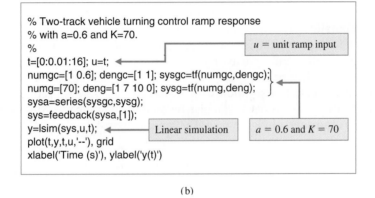

FIGURE 6.23
(a) Ramp response
for a = 0.6 and
K = 70 for two-
track vehicle
turning control.
(b) m-file script.

(b)

The corresponding unit ramp input response is shown in Figure 6.23. The steady-state error is less than 0.24, as desired. ∎

The Stability of State Variable Systems. Now let us turn to determining the stability of systems described in state variable form. Suppose we have a system in state-space form as in Equation (6.22). The stability of the system can be evaluated with the **characteristic equation** associated with the system matrix **A**. The characteristic equation is

$$\det(s\mathbf{I} - \mathbf{A}) = 0. \tag{6.35}$$

The left-hand side of the characteristic equation is a polynomial in s. If all of the roots of the characteristic equation have negative real parts (i.e., $\text{Re}(s_i) < 0$), then the system is stable.

When the system model is given in state variable form, we must calculate the characteristic polynomial associated with the **A** matrix. In this regard, we have several options. We can calculate the characteristic equation directly from Equation (6.35) by manually computing the determinant of $s\mathbf{I} - \mathbf{A}$. Then, we can compute the roots using the roots function to check for stability, or alternatively, we can use

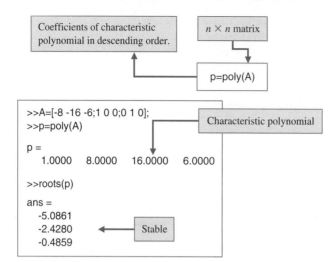

FIGURE 6.24
Computing the characteristic polynomial of **A** with the **poly** function.

the Routh–Hurwitz method to detect any unstable roots. Unfortunately, the manual computations can become lengthy, especially if the dimension of **A** is large. We would like to avoid this manual computation if possible. As it turns out, the computer can assist in this endeavor.

The poly function described in Section 2.9 can be used to compute the characteristic equation associated with **A**. Recall that poly is used to form a polynomial from a vector of roots. It can also be used to compute the characteristic equation of **A**, as illustrated in Figure 6.24. The input matrix **A** is

$$A = \begin{bmatrix} -8 & -16 & -6 \\ 1 & 0 & 0 \\ 0 & 1 & 0 \end{bmatrix},$$

and the associated characteristic polynomial is

$$s^3 + 8s^2 + 16s + 6 = 0.$$

If **A** is an $n \times n$ matrix, poly(**A**) is an $n + 1$ element row vector whose elements are the coefficients of the characteristic equation $\det(s\mathbf{I} - \mathbf{A}) = 0$.

EXAMPLE 6.13   **Stability region for an unstable process**

A jump-jet aircraft has a control system as shown in Figure 6.25 [17]. Assume that $z > 0$ and $p > 0$. The system is open-loop unstable (without feedback), since the characteristic equation of the process and controller is

$$s(s - 1)(s + p) = s[s^2 + (p - 1)s - p] = 0.$$

FIGURE 6.25
Control system for jump-jet aircraft. Assume that $z > 0$ and $p > 0$.

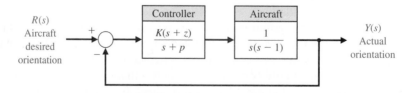

Note that since one term within the bracket has a negative coefficient, the characteristic equation has at least one root in the right-hand s-plane. The characteristic equation of the closed-loop system is

$$s^3 + (p - 1)s^2 + (K - p)s + Kz = 0.$$

The goal is to determine the region of stability for K, p, and z. The Routh array is

$$
\begin{array}{c|cc}
s^3 & 1 & K - p \\
s^2 & p - 1 & Kz \\
s^1 & b_2 & \\
s^0 & Kz &
\end{array}
$$

where

$$b_2 = \frac{(p - 1)(K - p) - Kz}{p - 1}.$$

From the Routh–Hurwitz criterion, we find that we require $Kz > 0$ and $p > 1$. Setting $b_2 > 0$, we have

$$(p - 1)(K - p) - Kz = K[(p - 1) - z] - p(p - 1) > 0.$$

Consider two cases:

1. $z \geq p - 1$: there is no $0 < K < \infty$ that leads to stability.
2. $z < p - 1$: any $0 < K < \infty$ satisfying the stability condition for a given p and z will result in stability:

$$K > \frac{p(p - 1)}{(p - 1) - z} \tag{6.36}$$

The stability conditions can be depicted graphically. The m-file script used to generate a three-dimensional stability surface is shown in Figure 6.26. This script uses mesh to create the three-dimensional surface and meshgrid to generate arrays for use with the mesh surface.

The three-dimensional plot of the stability region for K, p, and z is shown in Figure 6.27. One acceptable stability point is $z = 1$, $p = 10$, and $K = 15$. ∎

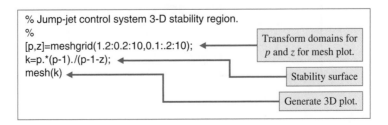

```
% Jump-jet control system 3-D stability region.
%
[p,z]=meshgrid(1.2:0.2:10,0.1:.2:10);
k=p.*(p-1)./(p-1-z);
mesh(k)
```

Transform domains for p and z for mesh plot.

Stability surface

Generate 3D plot.

FIGURE 6.26
m-file script for stability region.

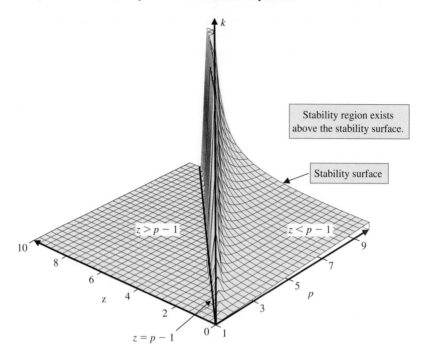

FIGURE 6.27
The three-dimensional region of stability lies above the surface shown.

6.7 SEQUENTIAL DESIGN EXAMPLE: DISK DRIVE READ SYSTEM

 In Section 5.11, we examined the design of the head reader system with an adjustable gain K_a. In this section, we will examine the stability of the system as K_a is adjusted and then reconfigure the system.

Let us consider the system as shown in Figure 6.28. This is the same system with a model of the motor and load as considered in Chapter 5, except that the velocity

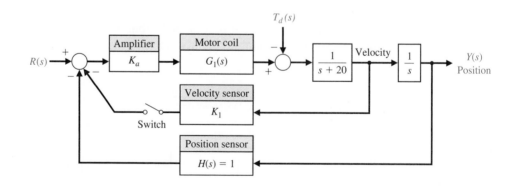

FIGURE 6.28
The closed-loop disk drive head system with an optional velocity feedback.

feedback sensor was added, as shown in Figure 6.28. Initially, we consider the case where the switch is open. Then the closed-loop transfer function is

$$\frac{Y(s)}{R(s)} = \frac{K_a G_1(s) G_2(s)}{1 + K_a G_1(s) G_2(s)}, \tag{6.37}$$

where

$$G_1(s) = \frac{5000}{s + 1000}$$

and

$$G_2(s) = \frac{1}{s(s + 20)}.$$

The characteristic equation is

$$s(s + 20)(s + 1000) + 5000K_a = 0, \tag{6.38}$$

or

$$s^3 + 1020s^2 + 20000s + 5000K_a = 0.$$

We use the Routh array

s^3	1	20000
s^2	1020	$5000K_a$
s^1	b_1	
s^0	$5000K_a$	

where

$$b_1 = \frac{(20000)1020 - 5000K_a}{1020}.$$

The case $b_1 = 0$ results in marginal stability when $K_a = 4080$. Using the auxiliary equation, we have

$$1020s^2 + 5000(4080) = 0,$$

or the roots of the $j\omega$-axis are $s = \pm j141.4$. In order for the system to be stable, $K_a < 4080$.

Now let us add the velocity feedback by closing the switch in the system of Figure 6.28. The closed-loop transfer function for the system is then

$$\frac{Y(s)}{R(s)} = \frac{K_a G_1(s) G_2(s)}{1 + [K_a G_1(s) G_2(s)](1 + K_1 s)}, \tag{6.39}$$

since the feedback factor is equal to $1 + K_1 s$, as shown in Figure 6.29.

FIGURE 6.29
Equivalent system
with the velocity
feedback switch
closed.

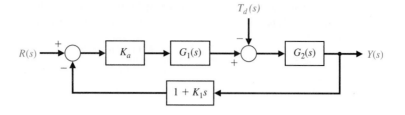

The characteristic equation is then

$$1 + [K_a G_1(s)G_2(s)](1 + K_1 s) = 0,$$

or

$$s(s + 20)(s + 1000) + 5000K_a(1 + K_1 s) = 0.$$

Therefore, we have

$$s^3 + 1020s^2 + [20000 + 5000K_a K_1]s + 5000K_a = 0.$$

Then the Routh array is

$$
\begin{array}{c|cc}
s^3 & 1 & 20000 + 5000K_a K_1 \\
s^2 & 1020 & 5000K_a \\
s^1 & b_1 & \\
s^0 & 5000K_a &
\end{array}
\text{,}
$$

where

$$b_1 = \frac{1020\,(20000 + 5000K_a K_1) - 5000K_a}{1020}.$$

To guarantee stability, it is necessary to select the pair (K_a, K_1) such that $b_1 > 0$, where $K_a > 0$. When $K_1 = 0.05$ and $K_a = 100$, we can determine the system response using the script shown in Figure 6.30. The settling time (with a 2% criterion) is approximately 260 ms, and the percent overshoot is zero. The system performance is summarized in Table 6.3. The performance specifications are nearly satisfied, and some iteration of K_1 is necessary to obtain the desired 250 ms settling time.

Table 6.3   Performance of the Disk Drive System Compared to the Specifications

Performance Measure	Desired Value	Actual Response
Percent overshoot	Less than 5%	0%
Settling time	Less than 250 ms	260 ms
Maximum response to a unit disturbance	Less than 5×10^{-3}	2×10^{-3}

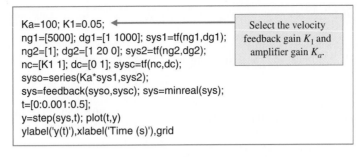

```
Ka=100; K1=0.05;
ng1=[5000]; dg1=[1 1000]; sys1=tf(ng1,dg1);
ng2=[1]; dg2=[1 20 0]; sys2=tf(ng2,dg2);
nc=[K1 1]; dc=[0 1]; sysc=tf(nc,dc);
syso=series(Ka*sys1,sys2);
sys=feedback(syso,sysc); sys=minreal(sys);
t=[0:0.001:0.5];
y=step(sys,t); plot(t,y)
ylabel('y(t)'),xlabel('Time (s)'),grid
```

> Select the velocity
> feedback gain K_1 and
> amplifier gain K_a.

(a)

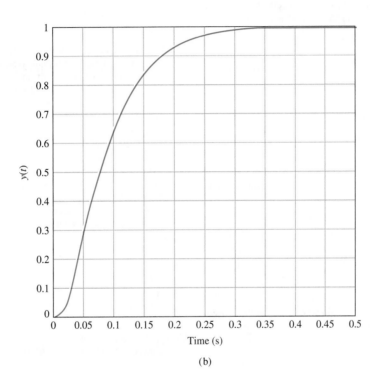

(b)

FIGURE 6.30
Response of the
system with
velocity feedback.
(a) m-file script.
(b) Response with
$K_a = 100$ and
$K_1 = 0.05$.

6.8 SUMMARY

In this chapter, we have considered the concept of the stability of a feedback control
system. A definition of a stable system in terms of a bounded system response was
outlined and related to the location of the poles of the system transfer function in
the s-plane.

The Routh–Hurwitz stability criterion was introduced, and several examples
were considered. The relative stability of a feedback control system was also considered in terms of the location of the poles and zeros of the system transfer function in
the s-plane. The stability of state variable systems was considered.

EXERCISES

E6.1 A system has a characteristic equation $s^3 + 3Ks^2 + (2 + K)s + 5 = 0$. Determine the range of K for a stable system.

Answer: K > 0.63

E6.2 A system has a characteristic equation $s^3 + 9s^2 + 2s + 24 = 0$. Using the Routh–Hurwitz criterion, show that the system is unstable.

E6.3 Find the roots of the characteristic equation $s^4 + 9.5s^3 + 30.5s^2 + 37s + 12 = 0$.

E6.4 A control system has the structure shown in Figure E6.4. Determine the gain at which the system will become unstable.

Answer: 0 < K < 1.5

E6.5 A unity feedback system has a loop transfer function

$$L(s) = \frac{K}{(s + 1)(s + 3)(s + 6)},$$

where $K = 20$. Find the roots of the closed-loop system's characteristic equation.

E6.6 For the feedback system of Exercise E6.5, find the value of K when two roots lie on the imaginary axis. Determine the value of the three roots.

Answer: s = −10, ±j5.2

E6.7 A negative feedback system has a loop transfer function

$$L(s) = \frac{K(s + 2)}{s(s - 1)}.$$

(a) Find the value of the gain when the ζ of the closed-loop roots is equal to 0.707. (b) Find the value of the gain when the closed-loop system has two roots on the imaginary axis.

E6.8 Designers have developed small, fast, vertical-take-off fighter aircraft that are invisible to radar (stealth aircraft). This aircraft concept uses quickly turning jet nozzles to steer the airplane [22]. The control system for the heading or direction control is shown in Figure E6.8. Determine the maximum gain of the system for stable operation.

E6.9 A system has a characteristic equation

$$s^3 + 2s^2 + (K + 1)s + 6 = 0.$$

Find the range of K for a stable system.

Answer: K > 2

E6.10 We all use our eyes and ears to achieve balance. Our orientation system allows us to sit or stand in a desired position even while in motion. This orientation system is primarily run by the information received in the inner ear, where the semicircular canals sense angular acceleration and the otoliths measure linear acceleration. But these acceleration measurements need to be supplemented by visual signals. Try the following experiment: (a) Stand with one foot in front of another, with your hands resting on your hips and your elbows bowed outward. (b) Close your eyes. Did you experience a low-frequency oscillation that grew until you lost balance? Is this orientation position stable with and without the use of your eyes?

E6.11 A system with a transfer function $Y(s)/R(s)$ is

$$\frac{Y(s)}{R(s)} = \frac{24(s + 1)}{s^4 + 6s^3 + 2s^2 + s + 3}.$$

Determine the steady-state error to a unit step input. Is the system stable?

E6.12 By using magnetic bearings, a rotor is supported contactless. The technique of contactless support for

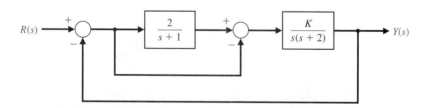

FIGURE E6.4
Feedforward system.

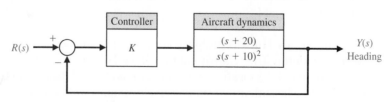

FIGURE E6.8
Aircraft heading control.

rotors becomes more important in light and heavy industrial applications [14]. The matrix differential equation for a magnetic bearing system is

$$\dot{\mathbf{x}} = \begin{bmatrix} 0 & 1 & 0 \\ -3 & -1 & 0 \\ -2 & -1 & -2 \end{bmatrix} \mathbf{x},$$

where $\mathbf{x}^T = [y, dy/dt, i]$, y = bearing gap, and i is the electromagnetic current. Determine whether the system is stable.

Answer: The system is stable.

E6.13 A system has a characteristic equation

$$q(s) = s^6 + 9s^5 + 31.25s^4 + 61.25s^3$$
$$+ 67.75s^2 + 14.75s + 15 = 0.$$

(a) Determine whether the system is stable, using the Routh–Hurwitz criterion. (b) Determine the roots of the characteristic equation.

Answer: (a) The system is marginally stable. (b) $s = -3, -4, -1 \pm 2j, \pm 0.5j$

E6.14 A system has a characteristic equation

$$q(s) = s^4 + 9s^3 + 45s^2 + 87s + 50 = 0.$$

(a) Determine whether the system is stable, using the Routh–Hurwitz criterion. (b) Determine the roots of the characteristic equation.

E6.15 The matrix differential equation of a state variable model of a system has

$$\mathbf{A} = \begin{bmatrix} 0 & 1 & -1 \\ -6 & -11 & 6 \\ -6 & -11 & 5 \end{bmatrix}.$$

(a) Determine the characteristic equation. (b) Determine whether the system is stable. (c) Determine the roots of the characteristic equation.

Answer: (a) $q(s) = s^3 + 6s^2 + 11s + 6 = 0$

E6.16 A system has a characteristic equation

$$q(s) = s^3 + 20s^2 + 5s + 100 = 0.$$

(a) Determine whether the system is stable, using the Routh–Hurwitz criterion. (b) Determine the roots of the characteristic equation.

E6.17 Determine whether the systems with the following characteristic equations are stable or unstable:
(a) $s^3 - 4s^2 + 6s + 100 = 0$,
(b) $s^4 - 6s^3 - s^2 - 17s - 6 = 0$, and
(c) $s^2 + 6s + 3 = 0$.

E6.18 Find the roots of the following polynomials:
(a) $s^3 + 5s^2 + 8s + 4 = 0$ and
(b) $s^3 + 9s^2 + 27s + 27 = 0$.

E6.19 A system has the characteristic equation

$$q(s) = s^3 + 10s^2 + 29s + K = 0.$$

Shift the vertical axis to the right by 2 by using $s = s_n - 2$, and determine the value of gain K so that the complex roots are $s = -2 \pm j$.

E6.20 A system has a transfer function $Y(s)/R(s) = T(s) = 1/s$. (a) Is this system stable? (b) If $r(t)$ is a unit step input, determine the response $y(t)$.

E6.21 A system is represented by Equation (6.22) where

$$\mathbf{A} = \begin{bmatrix} 0 & 1 & 0 \\ 0 & 0 & 1 \\ -6 & -k & -3 \end{bmatrix}.$$

Find the range of k where the system is stable.

E6.22 Consider the system represented in state variable form

$$\dot{\mathbf{x}} = \mathbf{Ax} + \mathbf{B}u$$
$$y = \mathbf{Cx} + \mathbf{D}u,$$

where

$$\mathbf{A} = \begin{bmatrix} 0 & 1 & 0 \\ 0 & 0 & 1 \\ -k & -k & -k \end{bmatrix}, \mathbf{B} = \begin{bmatrix} 0 \\ 0 \\ 1 \end{bmatrix}$$

$$\mathbf{C} = [1 \quad 0 \quad 0], \mathbf{D} = [0].$$

(a) What is the system transfer function? (b) For what values of k is the system stable?

E6.23 A closed-loop feedback system is shown in Figure E6.23. For what range of values of the parameters K and p is the system stable?

E6.24 Consider the closed-loop system in Figure E6.24, where

$$G(s) = \frac{10}{s - 10} \quad \text{and} \quad G_c(s) = \frac{1}{2s + K}.$$

(a) Determine the characteristic equation associated with the closed-loop system.
(b) Determine the values of K for which the closed-loop system is stable.

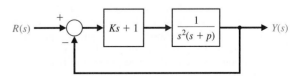

FIGURE E6.23 Closed-loop system with parameters K and p.

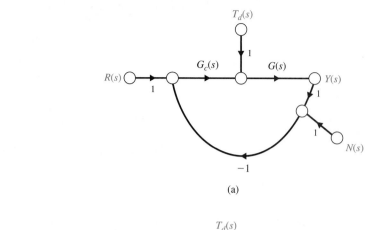

(a)

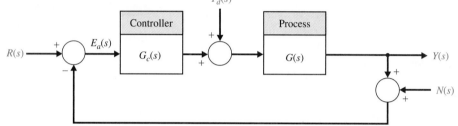

FIGURE E6.24
Closed-loop
feedback control
system with
parameter K.

PROBLEMS

P6.1 Utilizing the Routh–Hurwitz criterion, determine the stability of the following polynomials:

(a) $s^2 + 5s + 2$

(b) $s^3 + 4s^2 + 8s + 4$

(c) $s^3 + 2s^2 - 4s + 20$

(d) $s^4 + s^3 + 2s^2 + 10s + 8$

(e) $s^4 + s^3 + 3s^2 + 2s + K$

(f) $s^5 + s^4 + 2s^3 + s + 6$

(g) $s^5 + s^4 + 2s^3 + s^2 + s + K$

Determine the number of roots, if any, in the right-hand plane. If it is adjustable, determine the range of K that results in a stable system.

P6.2 An antenna control system was analyzed in Problem 4.5, and it was determined that, to reduce the effect of wind disturbances, the gain of the magnetic amplifier, k_a, should be as large as possible. (a) Determine the limiting value of gain for maintaining a stable system. (b) We want to have a system settling time equal to 1.5 seconds. Using a shifted axis and the Routh–Hurwitz criterion, determine the value of the gain that satisfies this requirement. Assume that the complex roots of the closed-loop system dominate the transient response. (Is this a valid approximation in this case?)

P6.3 Arc welding is one of the most important areas of application for industrial robots [11]. In most manufacturing welding situations, uncertainties in dimensions of the part, geometry of the joint, and the welding process itself require the use of sensors for maintaining weld quality. Several systems use a vision system to measure the geometry of the puddle of melted metal, as shown in Figure P6.3. This system uses a constant rate of feeding the wire to be melted. (a) Calculate the maximum value for K for the system that will result in a stable system. (b) For half of the maximum value of K found in part (a), determine the roots of the characteristic equation. (c) Estimate the overshoot of the system of part (b) when it is subjected to a step input.

P6.4 A feedback control system is shown in Figure P6.4. The controller and process transfer functions are given by

$$G_c(s) = K \text{ and } G(s) = \frac{s + 40}{s(s + 10)}$$

and the feedback transfer function is $H(s) = 1/(s + 20)$.

(a) Determine the limiting value of gain K for a stable

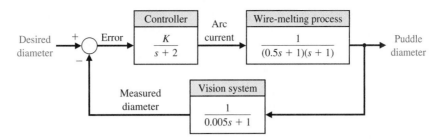

FIGURE P6.3
Welder control.

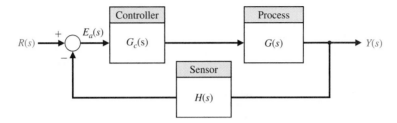

FIGURE P6.4
Nonunity feedback system.

system. (b) For the gain that results in marginal stability, determine the magnitude of the imaginary roots. (c) Reduce the gain to half the magnitude of the marginal value and determine the relative stability of the system (1) by shifting the axis and using the Routh–Hurwitz criterion and (2) by determining the root locations. Show the roots are between −1 and −2.

P6.5 Determine the relative stability of the systems with the following characteristic equations (1) by shifting the axis in the s-plane and using the Routh–Hurwitz criterion, and (2) by determining the location of the complex roots in the s-plane:

(a) $s^3 + 3s^2 + 4s + 2 = 0$.

(b) $s^4 + 9s^3 + 30s^2 + 42s + 20 = 0$.

(c) $s^3 + 19s^2 + 110s + 200 = 0$.

P6.6 A unity-feedback control system is shown in Figure P6.6. Determine the relative stability of the

system with the following transfer functions by locating the complex roots in the s-plane:

(a) $G_c(s)G(s) = \dfrac{10s + 2}{s^2(s + 1)}$

(b) $G_c(s)G(s) = \dfrac{24}{s(s^3 + 10s^2 + 35s + 50)}$

(c) $G_c(s)G(s) = \dfrac{(s + 2)(s + 3)}{s(s + 4)(s + 6)}$

P6.7 The linear model of a phase detector (phase-lock loop) can be represented by Figure P6.7 [9]. The phase-lock systems are designed to maintain zero difference in phase between the input carrier signal and a local voltage-controlled oscillator. Phase-lock loops find application in color television, missile tracking, and

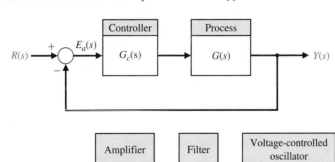

FIGURE P6.6
Unity feedback system.

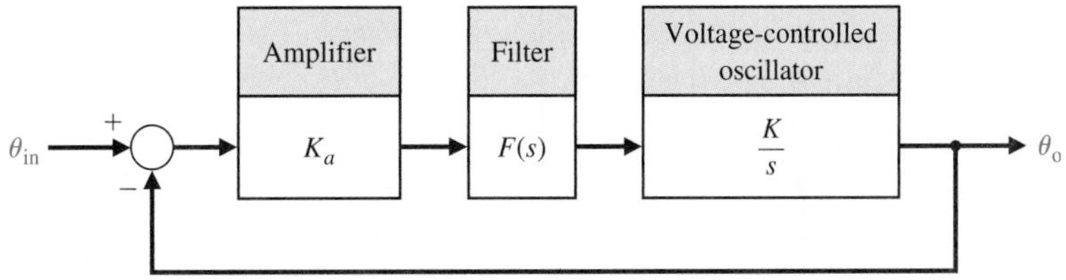

FIGURE P6.7
Phase-lock loop system.

space telemetry. The filter for a particular application is chosen as

$$F(s) = \frac{10(s + 10)}{(s + 1)(s + 100)}.$$

We want to minimize the steady-state error of the system for a ramp change in the phase information signal. (a) Determine the limiting value of the gain $K_a K = K_v$ in order to maintain a stable system. (b) A steady-state error equal to $1°$ is acceptable for a ramp signal of 100 rad/s. For that value of gain K_v, determine the location of the roots of the system.

P6.8   A very interesting and useful velocity control system has been designed for a wheelchair control system. We want to enable people paralyzed from the neck down to drive themselves in motorized wheelchairs. A proposed system utilizing velocity sensors mounted in a headgear is shown in Figure P6.8. The headgear sensor provides an output proportional to the magnitude of the head movement. There is a sensor mounted at $90°$ intervals so that forward, left, right, or reverse can be commanded. Typical values for the time constants are $\tau_1 = 0.5s$, $\tau_3 = 1s$, and $\tau_4 = \frac{1}{4}s$.
(a) Determine the limiting gain $K = K_1 K_2 K_3$ for a stable system.
(b) When the gain K is set equal to one-third of the limiting value, determine whether the settling time (to within 2% of the final value of the system) is less than 4 s.
(c) Determine the value of gain that results in a system with a settling time of 4 s. Also, obtain the value of the roots of the characteristic equation when the settling time is equal to 4 s.

P6.9   A cassette tape storage device has been designed for mass-storage [1]. It is necessary to control the velocity of the tape accurately. The speed control of the tape

drive is represented by the system shown in Figure P6.9.
(a) Determine the limiting gain for a stable system.
(b) Determine a suitable gain so that the overshoot to a step command is approximately 5%.

P6.10   Robots can be used in manufacturing and assembly operations that require accurate, fast, and versatile manipulation [10, 11]. The open-loop transfer function of a direct-drive arm may be approximated by

$$G(s)H(s) = \frac{K(s + 10)}{s(s + 3)(s^2 + 4s + 8)}.$$

(a) Determine the value of gain K when the system oscillates. (b) Calculate the roots of the closed-loop system for the K determined in part (a).

P6.11   A feedback control system has a characteristic equation

$$s^3 + (1 + K)s^2 + 10s + (5 + 15K) = 0.$$

The parameter K must be positive. What is the maximum value K can assume before the system becomes unstable? When K is equal to the maximum value, the system oscillates. Determine the frequency of oscillation.

P6.12   A feedback control system has a characteristic equation

$$s^6 + 5s^5 + 14s^4 + 40s^3 + 64s^2 + 80s + 96 = 0.$$

Determine whether the system is stable, and determine the values of the roots.

P6.13   The stability of a motorcycle and rider is an important area for study because many motorcycle designs result in vehicles that are difficult to control [12, 13]. The handling characteristics of a motorcycle must include a model of the rider as well as one of the vehicle. The dynamics of one motorcycle and

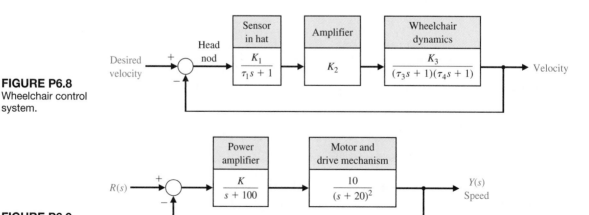

FIGURE P6.8
Wheelchair control system.

FIGURE P6.9
Tape drive control.

rider can be represented by an open-loop transfer function (Figure P6.4)

$$GH(s) = \frac{K(s^2 + 30s + 1125)}{s(s + 20)(s^2 + 10s + 125)(s^2 + 60s + 3400)}.$$

(a) As an approximation, calculate the acceptable range of K for a stable system when the numerator polynomial (zeros) and the denominator polynomial $(s^2 + 60s + 3400)$ are neglected. (b) Calculate the actual range of acceptable K, account for all zeros and poles.

P6.14 A system has a transfer function

$$T(s) = \frac{1}{s^3 + 5s^2 + 20s + 6}.$$

(a) Determine whether the system is stable. (b) Determine the roots of the characteristic equation. (c) Plot the response of the system to a unit step input.

P6.15 On July 16, 1993, the elevator in Yokohama's 70-story Landmark Tower, operating at a peak speed of 45 km/hr (28 mph), was inaugurated as the fastest super-fast elevator. To reach such a speed without inducing discomfort in passengers, the elevator accelerates for longer periods, rather than more precipitously. Going up, it reaches full speed only at the 27th floor; it begins decelerating 15 floors later. The result is a peak acceleration similar to that of other skyscraper elevators—a bit less than a tenth of the force of gravity. Admirable ingenuity has gone into making this safe and comfortable. Special ceramic brakes had to be developed; iron ones would melt. Computer-controlled systems damp out vibrations. The lift has been streamlined to reduce the wind noise as it speeds up and down [20]. One proposed control system for the elevator's vertical position is shown in Figure P6.15. Determine the range of K for a stable system.

P6.16 Consider the case of rabbits and foxes in Australia. The number of rabbits is x_1 and, if left alone, it would

grow indefinitely (until the food supply was exhausted) so that

$$\dot{x}_1 = kx_1.$$

However, with foxes present on the continent, we have

$$\dot{x}_1 = kx_1 - ax_2,$$

where x_2 is the number of foxes. Now, if the foxes must have rabbits to exist, we have

$$\dot{x}_2 = -hx_2 + bx_1.$$

Determine whether this system is stable and thus decays to the condition $x_1(t) = x_2(t) = 0$ at $t = \infty$. What are the requirements on $a, b, h,$ and k for a stable system? What is the result when k is greater than h?

P6.17 The goal of vertical takeoff and landing (VTOL) aircraft is to achieve operation from relatively small airports and yet operate as a normal aircraft in level flight [17]. An aircraft taking off in a form similar to a missile (on end) is inherently unstable (see Example 3.4 for a discussion of the inverted pendulum). A control system using adjustable jets can control the vehicle, as shown in Figure P6.17. (a) Determine the range of gain for which the system is stable. (b) Determine the gain K for which the system is marginally stable and the roots of the characteristic equation for this value of K.

P6.18 A vertical-liftoff vehicle is shown in Figure P6.18(a). The four engines swivel for liftoff. The control system for aircraft altitude is shown in Figure P6.18(b). (a) For $K = 1$, determine whether the system is stable. (b) Determine a range of stability, if any, for $K > 0$.

P6.19 Consider the system described in state variable form by

$$\dot{\mathbf{x}}(t) = \mathbf{A}\mathbf{x}(t) + \mathbf{B}u(t)$$
$$y(t) = \mathbf{C}\mathbf{x}(t)$$

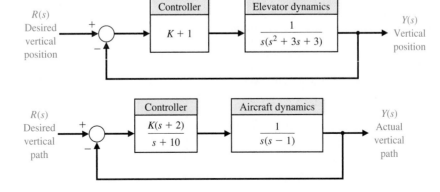

FIGURE P6.15
Elevator control system.

FIGURE P6.17
Control of a jump-jet aircraft.

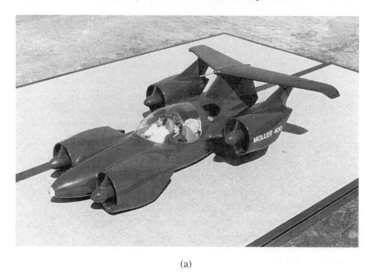

(a)

FIGURE P6.18
(a) Vertical-takeoff
aircraft. (Courtesy
of Moller
International.)
(b) Control system.

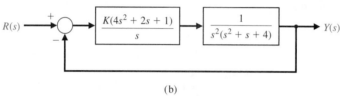

(b)

where

$$\mathbf{A} = \begin{bmatrix} 0 & 1 \\ -k_1 & -k_2 \end{bmatrix}, \mathbf{B} = \begin{bmatrix} 0 \\ 1 \end{bmatrix}, \text{ and } \mathbf{C} = \begin{bmatrix} 1 & -1 \end{bmatrix},$$

and where $k_1 \neq k_2$ and both k_1 and k_2 are real numbers.

(a) Compute the state transition matrix $\Phi(t, 0)$.
(b) Compute the eigenvalues of the system matrix $\mathbf{A}$.
(c) Compute the roots of the characteristic polynomial. (d) Discuss the results of parts (a)–(c) in terms of stability of the system.

ADVANCED PROBLEMS

AP6.1 A teleoperated control system incorporates both a person (operator) and a remote machine. The normal teleoperation system is based on a one-way link to the machine and limited feedback to the operator. However, two-way coupling using bilateral information exchange enables better operation [19]. In the case of remote control of a robot, force feedback plus position feedback is useful. The characteristic equation for a teleoperated system, as shown in Figure AP6.1, is

$$s^4 + 20s^3 + K_1 s^2 + 4s + K_2 = 0,$$

where K_1 and K_2 are feedback gain factors. Determine and plot the region of stability for this system for K_1 and K_2.

AP6.2 Consider the case of a navy pilot landing an aircraft on an aircraft carrier. The pilot has three basic tasks. The first task is guiding the aircraft's approach

to the ship along the extended centerline of the runway. The second task is maintaining the aircraft on the correct glideslope. The third task is maintaining the correct speed. A model of a lateral position control system is shown in Figure AP6.2. Determine the range of stability for $K \geq 0$.

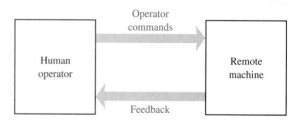

FIGURE AP6.1 Model of a teleoperated machine.

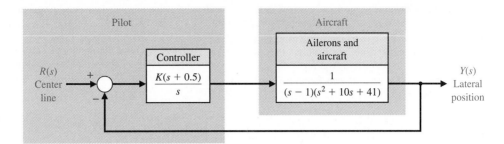

FIGURE AP6.2
Lateral position
control for landing
on an aircraft
carrier.

AP6.3 A control system is shown in Figure AP6.3. We want the system to be stable and the steady-state error for a unit step input to be less than or equal to 0.05 (5%). (a) Determine the range of α that satisfies the error requirement. (b) Determine the range of α that

satisfies the stability requirement. (c) Select an α that meets both requirements.

AP6.4 A bottle-filling line uses a feeder screw mechanism, as shown in Figure AP6.4. The tachometer feedback is

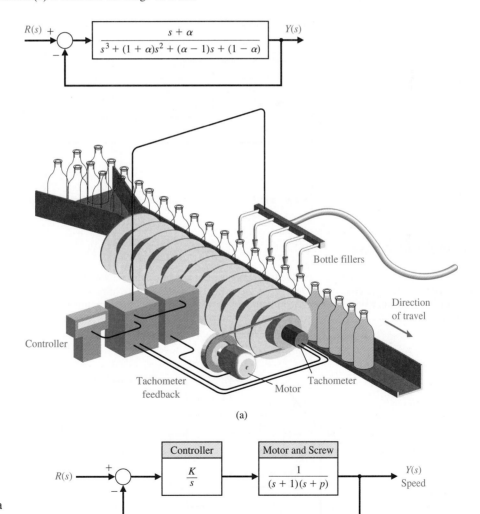

FIGURE AP6.3
Third-order unity
feedback system.

FIGURE AP6.4
Speed control of a
bottle-filling line.
(a) System layout.
(b) Block diagram.

used to maintain accurate speed control. Determine and plot the range of K and p that permits stable operation.

AP6.5   Consider the closed-loop system in Figure AP6.5. Suppose that all gains are positive, that is, $K_1 > 0$, $K_2 > 0$, $K_3 > 0$, $K_4 > 0$, and $K_5 > 0$.

(a) Determine the closed-loop transfer function $T(s) = Y(s)/R(s)$.

(b) Obtain the conditions on selecting the gains K_1, K_2, K_3, K_4, and K_5, so that the closed-loop system is guaranteed to be stable.

(c) Using the results of part (b), select values of the five gains so that the closed-loop system is stable, and plot the unit step response.

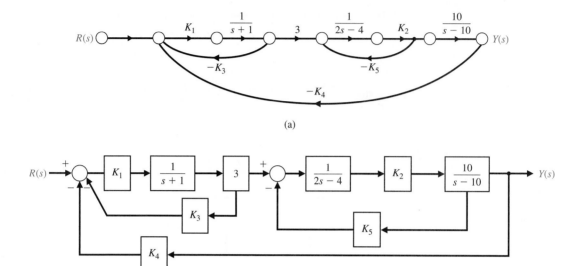

(a)

(b)

FIGURE AP6.5   Multiloop feedback control system. (a) Signal flow graph. (b) Block diagram.

DESIGN PROBLEMS

CP6.1   The capstan drive system of problem CDP5.1 uses the amplifier as the controller. Determine the maximum value of the gain K_a before the system becomes unstable.

DP6.1   The control of the spark ignition of an automotive engine requires constant performance over a wide range of parameters [15]. The control system is shown in Figure DP6.1, with a controller gain K to be selected.

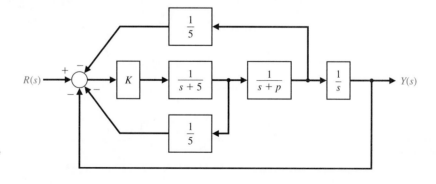

FIGURE DP6.1
Automobile engine
control.

The parameter p is equal to 2 for many autos but can equal zero for those with high performance. Select a gain K that will result in a stable system for both values of p.

DP6.2 An automatically guided vehicle on Mars is represented by the system in Figure DP6.2. The system has a steerable wheel in both the front and back of the vehicle, and the design requires that $H(s) = Ks + 1$. Determine (a) the value of K required for stability, (b) the value of K when one root of the characteristic equation is equal to $s = -5$, and (c) the value of the two remaining roots for the gain selected in part (b). (d) Find the response of the system to a step command for the gain selected in part (b).

DP6.3 A unity negative feedback system with

$$G_c(s)G(s) = \frac{K(s + 2)}{s(1 + \tau s)(1 + 2s)}$$

has two parameters to be selected. (a) Determine and plot the regions of stability for this system. (b) Select τ and K so that the steady-state error to a ramp input is less than or equal to 25% of the input magnitude. (c) Determine the percent overshoot for a step input for the design selected in part (b).

DP6.4 The attitude control system of a space shuttle rocket is shown in Figure DP6.4 [18]. (a) Determine

the range of gain K and parameter m so that the system is stable, and plot the region of stability. (b) Select the gain and parameter values so that the steady-state error to a ramp input is less than or equal to 10% of the input magnitude. (c) Determine the percent overshoot for a step input for the design selected in part (b).

DP6.5 A traffic control system is designed to control the distance between vehicles, as shown in Figure DP6.5 [15]. (a) Determine the range of gain K for which the system is stable. (b) If K_m is the maximum value of K so that the characteristic roots are on the $j\omega$-axis, then let $K = K_m/N$, where $6 < N < 7$. We want the peak time to be less than 2 seconds and the percent overshoot to be less than 18%. Determine an appropriate value for N.

DP6.6 Consider the single-input, single-output system as described by

$$\dot{\mathbf{x}}(t) = \mathbf{A}\mathbf{x}(t) + \mathbf{B}u(t)$$
$$y(t) = \mathbf{C}\mathbf{x}(t)$$

where

$$\mathbf{A} = \begin{bmatrix} 0 & 1 \\ 2 & -2 \end{bmatrix}, \mathbf{B} = \begin{bmatrix} 0 \\ 1 \end{bmatrix}, \mathbf{C} = \begin{bmatrix} 1 & 0 \end{bmatrix}.$$

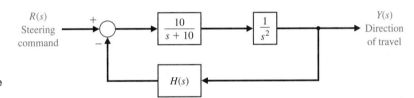

FIGURE DP6.2
Mars guided vehicle control.

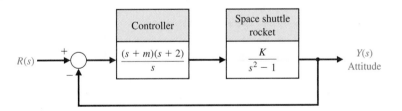

FIGURE DP6.4
Shuttle attitude control.

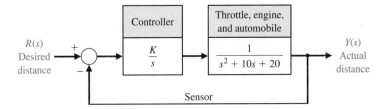

FIGURE DP6.5
Traffic distance control.

Assume that the input is a linear combination of the states, that is,

$$u(t) = -\mathbf{K}\mathbf{x}(t) + r(t),$$

where $r(t)$ is the reference input. The matrix $\mathbf{K} = [K_1 \quad K_2]$ is known as the gain matrix. If you substitute $u(t)$ into the state variable equation you will obtain the closed-loop system

$$\dot{\mathbf{x}}(t) = [\mathbf{A} - \mathbf{B}\mathbf{K}]\mathbf{x}(t) + \mathbf{B}r(t)$$
$$y(t) = \mathbf{C}\mathbf{x}(t)$$

For what values of $\mathbf{K}$ is the closed-loop system stable? Determine the region of the left half-plane where the desired closed-loop eigenvalues should be placed so that the percent overshoot to a unit step input, $R(s) = 1/s$, is less than $P.O. < 5\%$ and the settling time is less than $T_s < 4s$. Select a gain matrix, $\mathbf{K}$, so that the system step response meets the specifications $P.O. < 5\%$ and $T_s < 4s$.

DP6.7   Consider the feedback control system in Figure DP6.7. The system has an inner loop and an outer loop.

The inner loop must be stable and have a quick speed of response. (a) Consider the inner loop first. Determine the range of K_1 resulting in a stable inner loop. That is, the transfer function $Y(s)/U(s)$ must be stable. (b) Select the value of K_1 in the stable range leading to the fastest step response. (c) For the value of K_1 selected in (b), determine the range of K_2 such that the closed-loop system $T(s) = Y(s)/R(s)$ is stable.

DP6.8   Consider the feedback system shown in Figure DP6.8. The process transfer function is marginally stable. The controller is the proportional-derivative (PD) controller

$$G_c(s) = K_P + K_D s.$$

Determine if it is possible to find values of K_P and K_D such that the closed-loop system is stable. If so, obtain values of the controller parameters such that the steady-state tracking error $E(s) = R(s) - Y(s)$ to a unit step input $R(s) = 1/s$ is $e_{ss} = \lim_{t \to \infty} e(t) \le 0.1$ and the damping of the closed-loop system is $\zeta = \sqrt{2}/2$.

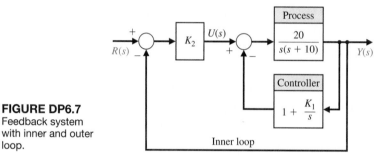

FIGURE DP6.7
Feedback system with inner and outer loop.

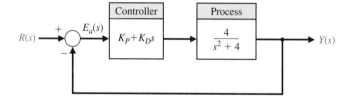

FIGURE DP6.8
A marginally stable plant with a PD controller in the loop.

 COMPUTER PROBLEMS

CP6.1   Determine the roots of the following characteristic equations:

(a) $q(s) = s^3 + 3s^2 + 5s + 7 = 0$.

(b) $q(s) = s^4 + 3s^3 + 4s^2 + 4s + 10 = 0$.

(c) $q(s) = s^4 + 2s^2 + 1 = 0$.

CP6.2   Consider a unity negative feedback system with

$$G_c(s) = K \text{ and } G(s) = \frac{s^2 - s + 2}{s^2 + 2s + 1}.$$

Develop an m-file to compute the roots of the closed-loop transfer function characteristic polynomial for $K = 1, 2,$ and 5. For which values of K is the closed-loop system stable?

CP6.3   A unity negative feedback system has the loop transfer function

$$G_c(s)G(s) = \frac{s + 1}{s^3 + 4s^2 + 6s + 10}.$$

Develop an m-file to determine the closed-loop transfer function and show that the roots of the characteristic equation are $s_1 = -2.89$ and $s_{2,3} = -0.55 \pm j1.87$.

CP6.4 Consider the closed-loop transfer function

$$T(s) = \frac{1}{s^5 + 2s^4 + 2s^3 + 4s^2 + s + 2}.$$

(a) Using the Routh–Hurwitz method, determine whether the system is stable. If it is not stable, how many poles are in the right half-plane? (b) Compute the poles of $T(s)$ and verify the result in part (a). (c) Plot the unit step response, and discuss the results.

CP6.5 A "paper-pilot" model is sometimes utilized in aircraft control design and analysis to represent the pilot in the loop. A block diagram of an aircraft with a pilot "in the loop" is shown in Figure CP6.5. The variable τ represents the pilot's time delay. We can represent a slower pilot with $\tau = 0.5$ and a faster pilot with $\tau = 0.25$. The remaining variables in the pilot model are assumed to be $K = 1, \tau_1 = 2$, and $\tau_2 = 0.5$. Develop an m-file to compute the closed-loop system poles for the fast and slow pilots. Comment on the results. What is the maximum pilot time delay allowable for stability?

CP6.6 Consider the feedback control system in Figure CP6.6. Using the for function, develop an m-file script

to compute the closed-loop transfer function poles for $0 \le K \le 5$ and plot the results denoting the poles with the "×" symbol. Determine the maximum range of K for stability with the Routh–Hurwitz method. Compute the roots of the characteristic equation when K is the minimum value allowed for stability.

CP6.7 Consider a system in state variable form:

$$\dot{x} = \begin{bmatrix} 0 & 1 & 0 \\ 0 & 0 & 1 \\ -10 & -15 & -10 \end{bmatrix} x + \begin{bmatrix} 0 \\ 0 \\ 10 \end{bmatrix} u,$$

$$y = [1 \quad 1 \quad 0]x.$$

(a) Compute the characteristic equation using the poly function. (b) Compute the roots of the characteristic equation, and determine whether the system is stable. (c) Obtain the response plot of $y(t)$ when $u(t)$ is a unit step and when the system has zero initial conditions.

CP6.8 Consider the feedback control system in Figure CP6.8. (a) Using the Routh–Hurwitz method, determine the range of K_1 resulting in closed-loop stability. (b) Develop an m-file to plot the pole locations as a function of $0 < K_1 < 30$ and comment on the results.

CP6.9 Consider a system represented in state variable form

$$\dot{x} = Ax + Bu$$
$$y = Cx + Du,$$

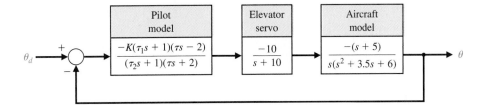

FIGURE CP6.5
An aircraft with a pilot in the loop.

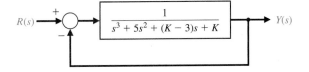

FIGURE CP6.6
A single-loop feedback control system with parameter K.

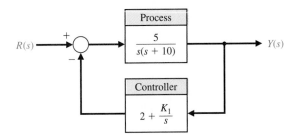

FIGURE CP6.8
Nonunity feedback system with parameter K_1.

where

$$\mathbf{A} = \begin{bmatrix} 0 & 1 & 0 \\ 2 & 0 & 1 \\ -k & -3 & -2 \end{bmatrix}, \mathbf{B} = \begin{bmatrix} -1 \\ 0 \\ 1 \end{bmatrix},$$

$$\mathbf{C} = \begin{bmatrix} 1 & 2 & 0 \end{bmatrix}, \mathbf{D} = \begin{bmatrix} 0 \end{bmatrix}.$$

(a) For what values of k is the system stable?

(b) Develop an m-file to plot the pole locations as a function of $0 < k < 10$ and comment on the results.

TERMS AND CONCEPTS

Absolute stability   A system description that reveals whether a system is stable or not stable without consideration of other system attributes such as degree of stability.

Auxiliary polynomial   The equation that immediately precedes the zero entry in the Routh array.

Marginally stable   A system is marginally stable if and only if the zero input response remains bounded as $t \to \infty$.

Relative stability   The property that is measured by the relative real part of each root or pair of roots of the characteristic equation.

Routh–Hurwitz criterion   A criterion for determining the stability of a system by examining the characteristic equation of the transfer function. The criterion states that the number of roots of the characteristic equation with positive real parts is equal to the number of changes of sign of the coefficients in the first column of the Routh array.

Stability   A performance measure of a system. A system is stable if all the poles of the transfer function have negative real parts.

Stable system   A dynamic system with a bounded system response to a bounded input.

The Root Locus Method

P R E V I E W

The performance of a feedback system can be described in terms of the location of the roots of the characteristic equation in the *s*-plane. A graph showing how the roots of the characteristic equation move around the *s*-plane as a single parameter varies is known as a root locus plot. The root locus is a powerful tool for designing and analyzing feedback control systems. We will discuss practical techniques for obtaining a sketch of a root locus plot by hand. We also consider computer-generated root locus plots and illustrate their effectiveness in the design process. We will show that it is possible to use root locus methods for controller design when more than one parameter varies. This is important because we know that the response of a closed-loop feedback system can be adjusted to achieve the desired performance by judicious selection of one or more controller parameters. The popular PID controller is introduced as a practical controller structure with three adjustable parameters. We will also define a measure of sensitivity of a specified root to a small incremental change in a system parameter. The chapter concludes with a controller design based on root locus methods for the Sequential Design Example: Disk Drive Read System.

DESIRED OUTCOMES

Upon completion of Chapter 7, students should:

❑ Understand the powerful concept of the root locus and its role in control system design.
❑ Know how to sketch a root locus and also how to obtain a computer-generated root locus plot.
❑ Be familiar with the PID controller as a key element of many feedback systems in use today.
❑ Recognize the role of root locus plots in parameter design and system sensitivity analysis.
❑ Be capable of designing a controller to meet desired specifications using root locus methods.

7.1 INTRODUCTION

The relative stability and the transient performance of a closed-loop control system are directly related to the location of the closed-loop roots of the characteristic equation in the s-plane. It is frequently necessary to adjust one or more system parameters in order to obtain suitable root locations. Therefore, it is worthwhile to determine how the roots of the characteristic equation of a given system migrate about the s-plane as the parameters are varied; that is, it is useful to determine the **locus** of roots in the s-plane as a parameter is varied. The **root locus method** was introduced by Evans in 1948 and has been developed and utilized extensively in control engineering practice [1–3]. The root locus technique is a graphical method for sketching the locus of roots in the s-plane as a parameter is varied. In fact, the root locus method provides the engineer with a measure of the sensitivity of the roots of the system to a variation in the parameter being considered. The root locus technique may be used to great advantage in conjunction with the Routh–Hurwitz criterion.

The root locus method provides graphical information, and therefore an approximate sketch can be used to obtain qualitative information concerning the stability and performance of the system. Furthermore, the locus of roots of the characteristic equation of a multiloop system may be investigated as readily as for a single-loop system. If the root locations are not satisfactory, the necessary parameter adjustments often can be readily ascertained from the root locus [4].

7.2 THE ROOT LOCUS CONCEPT

The dynamic performance of a closed-loop control system is described by the closed-loop transfer function

$$T(s) = \frac{Y(s)}{R(s)} = \frac{p(s)}{q(s)}, \tag{7.1}$$

where $p(s)$ and $q(s)$ are polynomials in s. The roots of the characteristic equation $q(s)$ determine the modes of response of the system. In the case of the simple single-loop system shown in Figure 7.1, we have the characteristic equation

$$\boxed{1 + KG(s) = 0,} \tag{7.2}$$

where K is a variable parameter. The characteristic roots of the system must satisfy Equation (7.2), where the roots lie in the s-plane. Because s is a complex variable, Equation (7.2) may be rewritten in polar form as

$$|KG(s)| \underline{/KG(s)} = -1 + j0, \tag{7.3}$$

FIGURE 7.1
Closed-loop
control system with
a variable
parameter K.

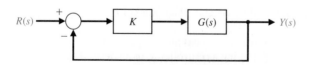

and therefore it is necessary that

$$\boxed{|KG(s)| = 1}$$

and

$$\boxed{\underline{/KG(s)} = 180° + k360°,} \tag{7.4}$$

where $k = 0, \pm1, \pm2, \pm3, \ldots$.

> **The root locus is the path of the roots of the characteristic equation traced out in the s-plane as a system parameter is changed.**

The simple second-order system considered in the previous chapters is shown in Figure 7.2. The characteristic equation representing this system is

$$\Delta(s) = 1 + KG(s) = 1 + \frac{K}{s(s + 2)} = 0,$$

or, alternatively,

$$\Delta(s) = s^2 + 2s + K = s^2 + 2\zeta\omega_n s + \omega_n^2 = 0. \tag{7.5}$$

The locus of the roots as the gain K is varied is found by requiring that

$$|KG(s)| = \left|\frac{K}{s(s + 2)}\right| = 1 \tag{7.6}$$

and

$$\underline{/KG(s)} = \pm180°, \pm540°, \ldots \tag{7.7}$$

The gain K may be varied from zero to an infinitely large positive value. For a second-order system, the roots are

$$s_1, s_2 = -\zeta\omega_n \pm \omega_n\sqrt{\zeta^2 - 1}, \tag{7.8}$$

and for $\zeta < 1$, we know that $\theta = \cos^{-1}\zeta$. Graphically, for two open-loop poles as shown in Figure 7.3, the locus of roots is a vertical line for $\zeta \leq 1$ in order to satisfy the angle requirement, Equation (7.7). For example, as shown in Figure 7.4, at a root s_1, the angles are

$$\underline{/\frac{K}{s(s + 2)}}\bigg|_{s=s_1} = -\underline{/s_1} - \underline{/(s_1 + 2)} = -[(180° - \theta) + \theta] = -180°. \tag{7.9}$$

FIGURE 7.2
Unity feedback control system. The gain K is a variable parameter.

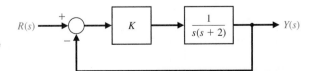

$R(s) \longrightarrow \boxed{K} \longrightarrow \boxed{\dfrac{1}{s(s + 2)}} \longrightarrow Y(s)$

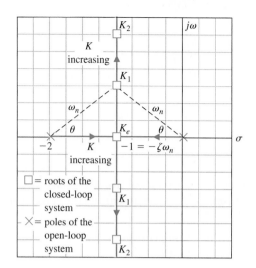

FIGURE 7.3
Root locus for a second-order system when $K_e < K_1 < K_2$. The locus is shown as heavy lines, with arrows indicating the direction of increasing K. Note that roots of the characteristic equation are denoted by "□" on the root locus.

This angle requirement is satisfied at any point on the vertical line that is a perpendicular bisector of the line 0 to −2. Furthermore, the gain K at the particular points is found by using Equation (7.6) as

$$\left|\frac{K}{s(s+2)}\right|_{s=s_1} = \frac{K}{|s_1||s_1+2|} = 1, \tag{7.10}$$

and thus

$$K = |s_1||s_1+2|, \tag{7.11}$$

where $|s_1|$ is the magnitude of the vector from the origin to s_1, and $|s_1+2|$ is the magnitude of the vector from −2 to s_1.

For a multiloop closed-loop system, we found in Section 2.7 that by using Mason's signal-flow gain formula, we had

$$\Delta(s) = 1 - \sum_{n=1}^{N} L_n + \sum_{\substack{n,m \\ \text{nontouching}}} L_n L_m - \sum_{\substack{n,m,p \\ \text{nontouching}}} L_n L_m L_p + \cdots, \tag{7.12}$$

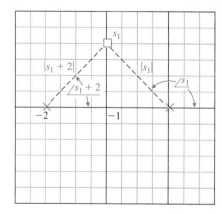

FIGURE 7.4
Evaluation of the angle and gain at s_1 for gain $K = K_1$.

where L_n equals the value of the nth self-loop transmittance. Hence, we have a characteristic equation, which may be written as

$$q(s) = \Delta(s) = 1 + F(s). \tag{7.13}$$

To find the roots of the characteristic equation, we set Equation (7.13) equal to zero and obtain

$$1 + F(s) = 0. \tag{7.14}$$

Equation (7.14) may be rewritten as

$$F(s) = -1 + j0, \tag{7.15}$$

and the roots of the characteristic equation must also satisfy this relation.

In general, the function $F(s)$ may be written as

$$F(s) = \frac{K(s + z_1)(s + z_2)(s + z_3) \cdots (s + z_M)}{(s + p_1)(s + p_2)(s + p_3) \cdots (s + p_n)}.$$

Then the magnitude and angle requirement for the root locus are

$$|F(s)| = \frac{K|s + z_1||s + z_2| \cdots}{|s + p_1||s + p_2| \cdots} = 1 \tag{7.16}$$

and

$$\underline{/F(s)} = \underline{/s + z_1} + \underline{/s + z_2} + \cdots$$
$$- (\underline{/s + p_1} + \underline{/s + p_2} + \cdots) = 180° + k360°, \tag{7.17}$$

where k is an integer. The magnitude requirement, Equation (7.16), enables us to determine the value of K for a given root location s_1. A test point in the s-plane, s_1, is verified as a root location when Equation (7.17) is satisfied. All angles are measured in a counterclockwise direction from a horizontal line.

To further illustrate the root locus procedure, let us consider again the second-order system of Figure 7.5(a). The effect of varying the parameter a can

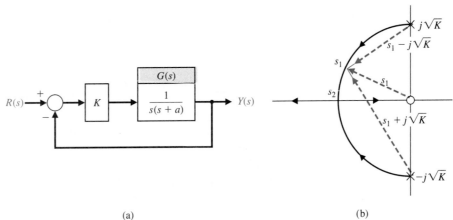

FIGURE 7.5
(a) Single-loop system. (b) Root locus as a function of the parameter a, where $a > 0$.

(a)

(b)

be effectively portrayed by rewriting the characteristic equation for the root locus form with a as the multiplying factor in the numerator. Then the characteristic equation is

$$1 + KG(s) = 1 + \frac{K}{s(s + a)} = 0,$$

or, alternatively,

$$s^2 + as + K = 0.$$

Dividing by the factor $s^2 + K$, we obtain

$$1 + \frac{as}{s^2 + K} = 0. \tag{7.18}$$

Then the magnitude criterion is satisfied when

$$\frac{a|s_1|}{|s_1^2 + K|} = 1 \tag{7.19}$$

at the root s_1. The angle criterion is

$$\angle s_1 - \left(\angle s_1 + j\sqrt{K} + \angle s_1 - j\sqrt{K} \right) = \pm 180°, \pm 540°, \dots.$$

In principle, we could construct the root locus by determining the points in the s-plane that satisfy the angle criterion. In the next section, we will develop a multistep procedure to sketch the root locus. The root locus for the characteristic equation in Equation (7.18) is shown in Figure 7.5(b). Specifically at the root s_1, the magnitude of the parameter a is found from Equation (7.19) as

$$a = \frac{|s_1 - j\sqrt{K}||s_1 + j\sqrt{K}|}{|s_1|}. \tag{7.20}$$

The roots of the system merge on the real axis at the point s_2 and provide a critically damped response to a step input. The parameter a has a magnitude at the critically damped roots, $s_2 = \sigma_2$, equal to

$$a = \frac{|\sigma_2 - j\sqrt{K}||\sigma_2 + j\sqrt{K}|}{\sigma_2} = \frac{1}{\sigma_2}(\sigma_2^2 + K) = 2\sqrt{K}, \tag{7.21}$$

where σ_2 is evaluated from the s-plane vector lengths as $\sigma_2 = \sqrt{K}$. As a increases beyond the critical value, the roots are both real and distinct; one root is larger than σ_2, and one is smaller.

In general, we desire an orderly process for locating the locus of roots as a parameter varies. In the next section, we will develop such an orderly approach to sketching a root locus diagram.

7.3 THE ROOT LOCUS PROCEDURE

The roots of the characteristic equation of a system provide a valuable insight concerning the response of the system. To locate the roots of the characteristic equation in a graphical manner on the s-plane, we will develop an orderly procedure of seven steps that facilitates the rapid sketching of the locus.

Step 1: Prepare the root locus sketch. Begin by writing the characteristic equation as

$$1 + F(s) = 0. \tag{7.22}$$

Rearrange the equation, if necessary, so that the parameter of interest, K, appears as the multiplying factor in the form,

$$1 + KP(s) = 0. \tag{7.23}$$

We are usually interested in determining the locus of roots as K varies as

$$0 \le K \le \infty.$$

Factor $P(s)$, and write the polynomial in the form of poles and zeros as follows:

$$1 + K \frac{\displaystyle\prod_{i=1}^{M}(s + z_i)}{\displaystyle\prod_{j=1}^{n}(s + p_j)} = 0. \tag{7.24}$$

Locate the poles $-p_i$ and zeros $-z_i$ on the s-plane with selected symbols. By convention, we use 'x' to denote poles and 'o' to denote zeros.

Rewriting Equation (7.24), we have

$$\prod_{j=1}^{n}(s + p_j) + K \prod_{i=1}^{M}(s + z_i) = 0. \tag{7.25}$$

Note that Equation (7.25) is another way to write the characteristic equation. When $K = 0$, the roots of the characteristic equation are the poles of $P(s)$. To see this, consider Equation (7.25) with $K = 0$. Then, we have

$$\prod_{j=1}^{n}(s + p_j) = 0.$$

When solved, this yields the values of s that coincide with the poles of $P(s)$. Conversely, as $K \to \infty$, the roots of the characteristic equation are the zeros of $P(s)$. To see this, first divide Equation (7.25) by K. Then, we have

$$\frac{1}{K} \prod_{j=1}^{n}(s + p_j) + \prod_{j=1}^{M}(s + z_j) = 0,$$

which, as $K \to \infty$, reduces to

$$\prod_{j=1}^{M} (s + z_j) = 0.$$

When solved, this yields the values of s that coincide with the zeros of $P(s)$. Therefore, we note that **the locus of the roots of the characteristic equation $1 + KP(s) = 0$ begins at the poles of $P(s)$ and ends at the zeros of $P(s)$ as K increases from zero to infinity.** For most functions $P(s)$ that we will encounter, several of the zeros of $P(s)$ lie at infinity in the s-plane. This is because most of our functions have more poles than zeros. With n poles and M zeros and $n > M$, we have $n - M$ branches of the root locus approaching the $n - M$ zeros at infinity.

Step 2: Locate the segments of the real axis that are root loci. **The root locus on the real axis always lies in a section of the real axis to the left of an odd number of poles and zeros.** This fact is ascertained by examining the angle criterion of Equation (7.17). These two useful steps in plotting a root locus will be illustrated by a suitable example.

EXAMPLE 7.1   Second-order system

A single-loop feedback control system possesses the characteristic equation

$$1 + GH(s) = 1 + \frac{K\left(\frac{1}{2}s + 1\right)}{\frac{1}{4}s^2 + s} = 0. \tag{7.26}$$

STEP 1: The characteristic equation can be written as

$$1 + K\frac{2(s + 2)}{s^2 + 4s} = 0,$$

where

$$P(s) = \frac{2(s + 2)}{s^2 + 4s}.$$

The transfer function, $P(s)$, is rewritten in terms of poles and zeros as

$$1 + K\frac{2(s + 2)}{s(s + 4)} = 0, \tag{7.27}$$

and the multiplicative gain parameter is K. To determine the locus of roots for the gain $0 \le K \le \infty$, we locate the poles and zeros on the real axis as shown in Figure 7.6(a).

FIGURE 7.6
(a) The zero and poles of a second-order system, (b) the root locus segments, and (c) the magnitude of each vector at s_1.

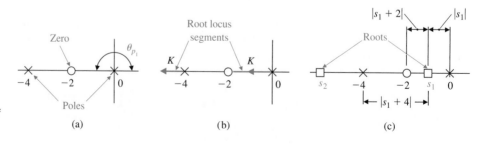

STEP 2: The angle criterion is satisfied on the real axis between the points 0 and −2, because the angle from pole p_1 at the origin is $180°$, and the angle from the zero and pole p_2 at $s = −4$ is zero degrees. The locus begins at the pole and ends at the zeros, and therefore the locus of roots appears as shown in Figure 7.6(b), where the direction of the locus as K is increasing ($K\uparrow$) is shown by an arrow. We note that because the system has two real poles and one real zero, the second locus segment ends at a zero at negative infinity. To evaluate the gain K at a specific root location on the locus, we use the magnitude criterion, Equation (7.16). For example, the gain K at the root $s = s_1 = −1$ is found from (7.16) as

$$\frac{(2K)|s_1 + 2|}{|s_1||s_1 + 4|} = 1$$

or

$$K = \frac{|-1||-1 + 4|}{2|-1 + 2|} = \frac{3}{2}. \tag{7.28}$$

This magnitude can also be evaluated graphically, as shown in Figure 7.6(c). For the gain of $K = \frac{3}{2}$, one other root exists, located on the locus to the left of the pole at −4. The location of the second root is found graphically to be located at $s = −6$, as shown in Figure 7.6(c).

Now, we determine the number of separate loci, SL. Because the loci begin at the poles and end at the zeros, the **number of separate loci is equal to the number of poles** since the number of poles is greater than or equal to the number of zeros. Therefore, as we found in Figure 7.6, the number of separate loci is equal to two because there are two poles and one zero.

Note that the **root loci must be symmetrical with respect to the horizontal real axis** because the complex roots must appear as pairs of complex conjugate roots. ∎

We now return to developing a general list of root locus steps.

Step 3: The loci proceed to the zeros at infinity along asymptotes centered at σ_A and with angles ϕ_A. When the number of finite zeros of $P(s)$, M, is less than the number of poles n by the number $N = n − M$, then N sections of loci must end at zeros at infinity. These sections of loci proceed to the zeros at infinity along **asymptotes** as K approaches infinity. These linear **asymptotes are centered** at a point on the real axis given by

$$\sigma_A = \frac{\sum \text{poles of } P(s) - \sum \text{zeros of } P(s)}{n - M} = \frac{\sum_{j=1}^{n}(-p_j) - \sum_{i=1}^{M}(-z_i)}{n - M}. \tag{7.29}$$

The **angle of the asymptotes** with respect to the real axis is

$$\phi_A = \frac{2k + 1}{n - M}180°, \quad k = 0, 1, 2, \ldots, (n - M - 1), \tag{7.30}$$

where k is an integer index [3]. The usefulness of this rule is obvious for sketching the approximate form of a root locus. Equation (7.30) can be readily derived by considering a point on a root locus segment at a remote distance from the finite poles and zeros in the s-plane. The net phase angle at this remote point is $180°$, because it is a point on a root locus segment. The finite poles and zeros of $P(s)$ are a great distance from the remote point, and so the angles from each pole and zero, ϕ, are essentially equal, and therefore the net angle is simply $(n - M)\phi$, where n and M are the number of finite poles and zeros, respectively. Thus, we have

$$(n - M)\phi = 180°,$$

or, alternatively,

$$\phi = \frac{180°}{n - M}.$$

Accounting for all possible root locus segments at remote locations in the s-plane, we obtain Equation (7.30).

The center of the linear asymptotes, often called the **asymptote centroid**, is determined by considering the characteristic equation in Equation (7.24). For large values of s, only the higher-order terms need be considered, so that the characteristic equation reduces to

$$1 + \frac{Ks^M}{s^n} = 0.$$

However, this relation, which is an approximation, indicates that the centroid of $n - M$ asymptotes is at the origin, $s = 0$. A better approximation is obtained if we consider a characteristic equation of the form

$$1 + \frac{K}{(s - \sigma_A)^{n-M}} = 0$$

with a centroid at σ_A.

The centroid is determined by considering the first two terms of Equation (7.24), which may be found from the relation

$$1 + \frac{K \prod_{i=1}^{M}(s + z_i)}{\prod_{j=1}^{n}(s + p_j)} = 1 + K \frac{s^M + b_{M-1}s^{M-1} + \cdots + b_0}{s^n + a_{n-1}s^{n-1} + \cdots + a_0}.$$

From Chapter 6, especially Equation (6.5), we note that

$$b_{M-1} = \sum_{i=1}^{M} z_i \quad \text{and} \quad a_{n-1} = \sum_{j=1}^{n} p_j.$$

Considering only the first two terms of this expansion, we have

$$1 + \frac{K}{s^{n-M} + (a_{n-1} - b_{M-1})s^{n-M-1}} = 0.$$

The first two terms of

$$1 + \frac{K}{(s - \sigma_A)^{n-M}} = 0$$

are

$$1 + \frac{K}{s^{n-M} - (n - M)\sigma_A s^{n-M-1}} = 0.$$

Equating the term for s^{n-M-1}, we obtain

$$a_{n-1} - b_{M-1} = -(n - M)\sigma_A,$$

or

$$\sigma_A = \frac{\sum_{i=1}^{n}(-p_i) - \sum_{i=1}^{M}(-z_i)}{n - M}$$

which is Equation (7.29).

For example, reexamine the system shown in Figure 7.2 and discussed in Section 7.2. The characteristic equation is written as

$$1 + \frac{K}{s(s + 2)} = 0.$$

Because $n - M = 2$, we expect two loci to end at zeros at infinity. The asymptotes of the loci are located at a center

$$\sigma_A = \frac{-2}{2} = -1$$

and at angles of

$$\phi_A = 90° \text{ (for } k = 0) \quad \text{and} \quad \phi_A = 270° \text{ (for } k = 1).$$

The root locus is readily sketched, and the locus shown in Figure 7.3 is obtained. An example will further illustrate the process of using the asymptotes.

EXAMPLE 7.2   **Fourth-order system**

A single-loop feedback control system has a characteristic equation as follows:

$$1 + GH(s) = 1 + \frac{K(s + 1)}{s(s + 2)(s + 4)^2}. \tag{7.31}$$

We wish to sketch the root locus in order to determine the effect of the gain K. The poles and zeros are located in the s-plane, as shown in Figure 7.7(a). The root loci on the real axis must be located to the left of an odd number of poles and zeros; they are shown as heavy lines in Figure 7.7(a). The intersection of the asymptotes is

$$\sigma_A = \frac{(-2) + 2(-4) - (-1)}{4 - 1} = \frac{-9}{3} = -3. \tag{7.32}$$

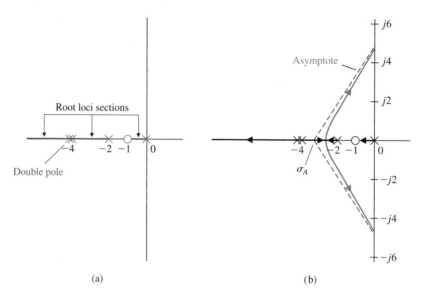

FIGURE 7.7
A fourth-order
system with (a) a
zero and (b) root
locus.

The angles of the asymptotes are

$$\phi_A = +60° \quad (k = 0),$$
$$\phi_A = 180° \quad (k = 1), \text{and}$$
$$\phi_A = 300° \quad (k = 2),$$

where there are three asymptotes, since $n - M = 3$. Also, we note that the root loci must begin at the poles; therefore, two loci must leave the double pole at $s = -4$. Then with the asymptotes sketched in Figure 7.7(b), we may sketch the form of the root locus as shown in Figure 7.7(b). The actual shape of the locus in the area near σ_A would be graphically evaluated, if necessary. ■

We now proceed to develop more steps for the process of determining the root loci.

Step 4: Determine where the locus crosses the imaginary axis (if it does so), using the Routh–Hurwitz criterion. **The actual point at which the root locus crosses the imaginary axis is readily evaluated by using the criterion.**

Step 5: Determine the breakaway point on the real axis (if any). The root locus in Example 7.2 left the real axis at a **breakaway point**. The locus breakaway from the real axis occurs where the net change in angle caused by a small displacement is zero. The locus leaves the real axis where there is a multiplicity of roots (typically, two). The breakaway point for a simple second-order system is shown in Figure 7.8(a) and, for a special case of a fourth-order system, is shown in Figure 7.8(b). In general, due to the phase criterion, **the tangents to the loci at the breakaway point are equally spaced over 360°. Therefore, in Figure 7.8(a), we find that the two loci at the breakaway point are spaced 180° apart, whereas in Figure 7.8(b), the four loci are spaced 90° apart.**

The breakaway point on the real axis can be evaluated graphically or analytically. The most straightforward method of evaluating the breakaway point involves

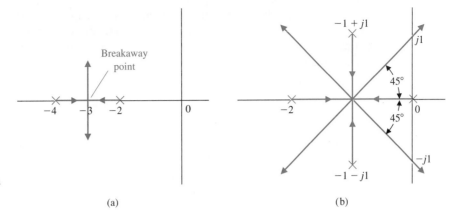

FIGURE 7.8
Illustration of the breakaway point (a) for a simple second-order system and (b) for a fourth-order system.

(a)   (b)

the rearranging of the characteristic equation to isolate the multiplying factor K. Then the characteristic equation is written as

$$p(s) = K. \tag{7.33}$$

For example, consider a unity feedback closed-loop system with an open-loop transfer function

$$G(s) = \frac{K}{(s + 2)(s + 4)},$$

which has the characteristic equation

$$1 + G(s) = 1 + \frac{K}{(s + 2)(s + 4)} = 0. \tag{7.34}$$

Alternatively, the equation may be written as

$$K = p(s) = -(s + 2)(s + 4). \tag{7.35}$$

The root loci for this system are shown in Figure 7.8(a). We expect the breakaway point to be near $s = \sigma = -3$ and plot $p(s)|_{s=\sigma}$ near that point, as shown in Figure 7.9. In this case, $p(s)$ equals zero at the poles $s = -2$ and $s = -4$. The plot of $p(s)$ versus $s - \sigma$ is symmetrical, and the maximum point occurs at $s = \sigma = -3$, the breakaway point.

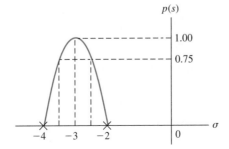

FIGURE 7.9
A graphical evaluation of the breakaway point.

Analytically, the very same result may be obtained by determining the maximum of $K = p(s)$. To find the maximum analytically, we differentiate, set the differentiated polynomial equal to zero, and determine the roots of the polynomial. Therefore, we may evaluate

$$\frac{dK}{ds} = \frac{dp(s)}{ds} = 0 \tag{7.36}$$

in order to find the breakaway point. Equation (7.36) is an analytical expression of the graphical procedure outlined in Figure 7.9 and will result in an equation of only one degree less than the total number of poles and zeros $n + M - 1$.

The proof of Equation (7.36) is obtained from a consideration of the characteristic equation

$$1 + F(s) = 1 + \frac{KY(s)}{X(s)} = 0,$$

which may be written as

$$X(s) + KY(s) = 0. \tag{7.37}$$

For a small increment in K, we have

$$X(s) + (K + \Delta K)Y(s) = 0.$$

Dividing by $X(s) + KY(s)$ yields

$$1 + \frac{\Delta KY(s)}{X(s) + KY(s)} = 0. \tag{7.38}$$

Because the denominator is the original characteristic equation, a multiplicity m of roots exists at a breakaway point, and

$$\frac{Y(s)}{X(s) + KY(s)} = \frac{C_i}{(s - s_i)^m} = \frac{C_i}{(\Delta s)^m}. \tag{7.39}$$

Then we may write Equation (7.38) as

$$1 + \frac{\Delta KC_i}{(\Delta s)^m} = 0, \tag{7.40}$$

or, alternatively,

$$\frac{\Delta K}{\Delta s} = \frac{-(\Delta s)^{m-1}}{C_i}. \tag{7.41}$$

Therefore, as we let Δs approach zero, we obtain

$$\frac{dK}{ds} = 0 \tag{7.42}$$

at the breakaway points.

Now, considering again the specific case where

$$G(s) = \frac{K}{(s + 2)(s + 4)},$$

we obtain

$$p(s) = K = -(s + 2)(s + 4) = -(s^2 + 6s + 8). \tag{7.43}$$

Then, when we differentiate, we have

$$\frac{dp(s)}{ds} = -(2s + 6) = 0, \tag{7.44}$$

or the breakaway point occurs at $s = -3$. A more complicated example will illustrate the approach and demonstrate the use of the graphical technique to determine the breakaway point.

EXAMPLE 7.3   Third-order system

A feedback control system is shown in Figure 7.10. The characteristic equation is

$$1 + G(s)H(s) = 1 + \frac{K(s + 1)}{s(s + 2)(s + 3)} = 0. \tag{7.45}$$

The number of poles n minus the number of zeros M is equal to 2, and so we have two asymptotes at $\pm 90°$ with a center at $\sigma_A = -2$. The asymptotes and the sections of loci on the real axis are shown in Figure 7.11(a). A breakaway point occurs between $s = -2$ and $s = -3$. To evaluate the breakaway point, we rewrite the characteristic equation so that K is separated; thus,

$$s(s + 2)(s + 3) + K(s + 1) = 0,$$

or

$$p(s) = \frac{-s(s + 2)(s + 3)}{s + 1} = K. \tag{7.46}$$

Then, evaluating $p(s)$ at various values of s between $s = -2$ and $s = -3$, we obtain the results of Table 7.1, as shown in Figure 7.11(b). Alternatively, we differentiate

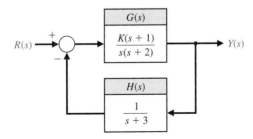

FIGURE 7.10
Closed-loop
system.

Table 7.1

$p(s)$	0	0.411	0.419	0.417	+0.390	0
s	−2.00	−2.40	−2.46	−2.50	−2.60	−3.0

Equation (7.46) and set it equal to zero to obtain

$$\frac{d}{ds}\left(\frac{-s(s+2)(s+3)}{(s+1)}\right) = \frac{(s^3 + 5s^2 + 6s) - (s+1)(3s^2 + 10s + 6)}{(s+1)^2} = 0$$

$$2s^3 + 8s^2 + 10s + 6 = 0. \qquad (7.47)$$

Now to locate the maximum of $p(s)$, we locate the roots of Equation (7.47) to obtain $s = -2.46, -0.77 \pm 0.79j$. The only value of s on the real axis in the interval $s = -2$ to $s = -3$ is $s = -2.46$; hence this must be the breakaway point. It is evident from this one example that the numerical evaluation of $p(s)$ near the expected breakaway point provides an effective method of evaluating the breakaway point. ∎

Step 6: Determine the angle of departure of the locus from a pole and the angle of arrival of the locus at a zero, using the phase angle criterion. The **angle of locus departure from a pole is the difference between the net angle due to all other poles and zeros and the criterion angle of ±180° (2k + 1)**, and similarly for the locus angle of arrival at a zero. The angle of departure (or arrival) is particularly of interest for complex poles (and zeros) because the information is helpful in completing the root locus. For example, consider the third-order open-loop transfer function

$$F(s) = G(s)H(s) = \frac{K}{(s+p_3)(s^2 + 2\zeta\omega_n s + \omega_n^2)}. \qquad (7.48)$$

The pole locations and the vector angles at one complex pole $-p_1$ are shown in Figure 7.12(a). The angles at a test point s_1, an infinitesimal distance from $-p_1$, must

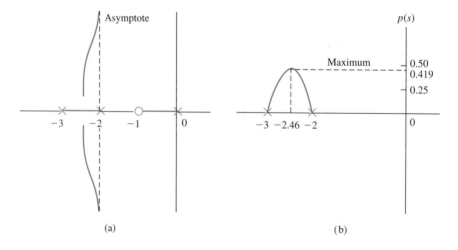

FIGURE 7.11
Evaluation of the (a) asymptotes and (b) breakaway point.

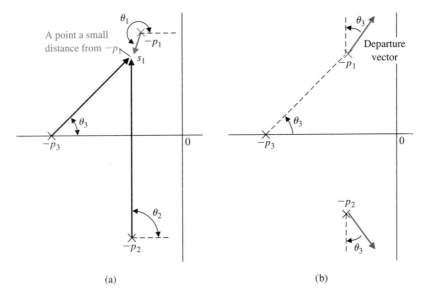

FIGURE 7.12
Illustration of the
angle of departure.
(a) Test point
infinitesimal
distance from $-p_1$.
(b) Actual departure
vector at $-p_1$.

(a)   (b)

meet the angle criterion. Therefore, since $\theta_2 = 90°$, we have

$$\theta_1 + \theta_2 + \theta_3 = \theta_1 + 90° + \theta_3 = +180°,$$

or the angle of departure at pole p_1 is

$$\theta_1 = 90° - \theta_3,$$

as shown in Figure 7.12(b). The departure at pole $-p_2$ is the negative of that at $-p_1$, because $-p_1$ and $-p_2$ are complex conjugates. Another example of a departure angle is shown in Figure 7.13. In this case, the departure angle is found from

$$\theta_2 - (\theta_1 + \theta_3 + 90°) = 180° + k360°.$$

Since $\theta_2 - \theta_3 = \gamma$ in the diagram, we find that the departure angle is $\theta_1 = 90° + \gamma$.

Step 7: The final step in the root locus sketching procedure is to complete the sketch. This entails sketching in all sections of the locus not covered in the previous

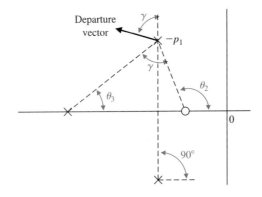

FIGURE 7.13
Evaluation of the
angle of departure.

six steps. If a more detailed root locus is required, we recommend using a computer-aided tool. (See Section 7.8.)

In some situation, we may want to determine a root location s_x and the value of the parameter K_x at that root location. Determine the root locations that satisfy the phase criterion at the root s_x, $x = 1, 2, \ldots, n$, using the phase criterion. The phase criterion, given in Equation (17.17), is

$$\angle P(s) = 180° + k360°, \quad \text{and} \quad k = 0, \pm1, \pm2, \ldots.$$

To determine the parameter value K_x at a specific root s_x, we use the magnitude requirement (Equation 7.16). The magnitude requirement at s_x is

$$K_x = \left. \frac{\displaystyle\prod_{j=1}^{n} |s + p_i|}{\displaystyle\prod_{i=1}^{M} |s + z_i|} \right|_{s=s_x}.$$

It is worthwhile at this point to summarize the seven steps utilized in the root locus method (Table 7.2) and then illustrate their use in a complete example.

Table 7.2 Seven Steps for Sketching a Root Locus

Step	Related Equation or Rule
1. Prepare the root locus sketch.	
(a) Write the characteristic equation so that the parameter of interest, K, appears as a multiplier.	$1 + KP(s) = 0.$
(b) Factor $P(s)$ in terms of n poles and M zeros.	$1 + K\dfrac{\displaystyle\prod_{i=1}^{M}(s + z_i)}{\displaystyle\prod_{j=1}^{n}(s + p_j)} = 0.$
(c) Locate the open-loop poles and zeros of $P(s)$ in the s-plane with selected symbols.	$\times$ = poles, $\bigcirc$ = zeros Locus begins at a pole and ends at a zero.
(d) Determine the number of separate loci, SL.	$SL = n$ when $n \geq M$; n = number of finite poles, M = number of finite zeros.
(e) The root loci are symmetrical with respect to the horizontal real axis.	
2. Locate the segments of the real axis that are root loci.	Locus lies to the left of an odd number of poles and zeros.
3. The loci proceed to the zeros at infinity along asymptotes centered at σ_A and with angles ϕ_A.	$\sigma_A = \dfrac{\sum(-p_j) - \sum(-z_i)}{n - M}.$ $\phi_A = \dfrac{2k + 1}{n - M}180°, k = 0, 1, 2, \ldots (n - M - 1).$
4. Determine the points at which the locus crosses the imaginary axis (if it does so).	Use Routh–Hurwitz criterion (see Section 6.2).
5. Determine the breakaway point on the real axis (if any).	a) Set $K = p(s)$. b) Determine roots of $dp(s)/ds = 0$ or use graphical method to find maximum of $p(s)$. $\angle P(s) = 180° + k360°$ at $s = -p_j$ or $-z_i$.
6. Determine the angle of locus departure from complex poles and the angle of locus arrival at complex zeros, using the phase criterion.	
7. Complete the root locus sketch.	

EXAMPLE 7.4 **Fourth-order system**

1. (a). We desire to plot the root locus for the characteristic equation of a system as K varies for $K > 0$ when

$$1 + \frac{K}{s^4 + 12s^3 + 64s^2 + 128s} = 0.$$

   (b) Determining the poles, we have

$$1 + \frac{K}{s(s + 4)(s + 4 + j4)(s + 4 - j4)} = 0 \qquad (7.49)$$

   as K varies from zero to infinity. This system has no finite zeros.

   (c) The poles are located on the s-plane as shown in Figure 7.14(a).

   (d) Because the number of poles n is equal to 4, we have four separate loci.

   (e) The root loci are symmetrical with respect to the real axis.

2. A segment of the root locus exists on the real axis between $s = 0$ and $s = -4$.

3. The angles of the asymptotes are

$$\phi_A = \frac{(2k + 1)}{4} 180°, \qquad k = 0, 1, 2, 3;$$

$$\phi_A = +45°, 135°, 225°, 315°.$$

   The center of the asymptotes is

$$\sigma_A = \frac{-4 - 4 - 4}{4} = -3.$$

   Then the asymptotes are drawn as shown in Figure 7.14(a).

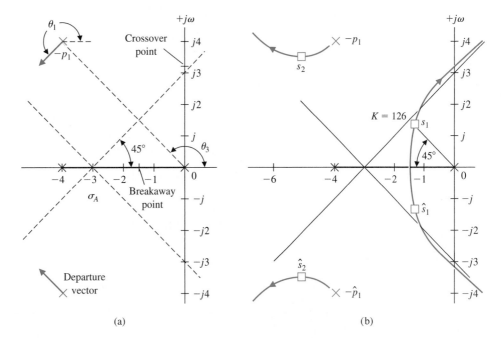

FIGURE 7.14
The root locus for
Example 7.4.
Locating (a) the
poles and (b) the
asymptotes.

4. The characteristic equation is rewritten as

$$s(s + 4)(s^2 + 8s + 32) + K = s^4 + 12s^3 + 64s^2 + 128s + K = 0. \qquad (7.50)$$

Therefore, the Routh array is

$$
\begin{array}{c|ccc}
s^4 & 1 & 64 & K \\
s^3 & 12 & 128 & \\
s^2 & b_1 & K & \\
s^1 & c_1 & & \\
s^0 & K & &
\end{array} \quad ,
$$

where

$$b_1 = \frac{12(64) - 128}{12} = 53.33 \quad \text{and} \quad c_1 = \frac{53.33(128) - 12K}{53.33}.$$

Hence, the limiting value of gain for stability is $K = 568.89$, and the roots of the auxiliary equation are

$$53.33s^2 + 568.89 = 53.33(s^2 + 10.67) = 53.33(s + j3.266)(s - j3.266). \qquad (7.51)$$

The points where the locus crosses the imaginary axis are shown in Figure 7.14(a). Therefore, when $K = 568.89$, the root locus crosses the $j\omega$-axis at $s = \pm j3.266$.

5. The breakaway point is estimated by evaluating

$$K = p(s) = -s(s + 4)(s + 4 + j4)(s + 4 - j4)$$

between $s = -4$ and $s = 0$. We expect the breakaway point to lie between $s = -3$ and $s = -1$, so we search for a maximum value of $p(s)$ in that region. The resulting values of $p(s)$ for several values of s are given in Table 7.3. The maximum of $p(s)$ is found to lie at approximately $s = -1.577$, as indicated in the table. A more accurate estimate of the breakaway point is normally not necessary. The breakaway point is then indicated on Figure 7.14(a).

6. The angle of departure at the complex pole p_1 can be estimated by utilizing the angle criterion as follows:

$$\theta_1 + 90° + 90° + \theta_3 = 180° + k360°.$$

Here, θ_3 is the angle subtended by the vector from pole p_3. The angles from the pole at $s = -4$ and $s = -4 - j4$ are each equal to 90°. Since $\theta_3 = 135°$, we find that

$$\theta_1 = -135° \equiv +225°,$$

as shown in Figure 7.14(a).

7. Complete the sketch as shown in Figure 7.14(b).

Table 7.3

$p(s)$	0	51.0	68.44	80.0	83.57	75.0	0
s	-4.0	-3.0	-2.5	-2.0	-1.577	-1.0	0

Using the information derived from the seven steps of the root locus method, the complete root locus sketch is obtained by filling in the sketch as well as possible by visual inspection. The root locus for this system is shown in Figure 7.14(b). When the complex roots near the origin have a damping ratio of $\zeta = 0.707$, the gain K can be determined graphically as shown in Figure 7.14(b). The vector lengths to the root location s_1 from the open-loop poles are evaluated and result in a gain at s_1 of

$$K = |s_1||s_1 + 4||s_1 - p_1||s_1 - \hat{p}_1| = (1.9)(2.9)(3.8)(6.0) = 126. \quad (7.52)$$

The remaining pair of complex roots occurs at s_2 and $\hat{s}_2$, when $K = 126$. The effect of the complex roots at s_2 and $\hat{s}_2$ on the transient response will be negligible compared to the roots s_1 and $\hat{s}_1$. This fact can be ascertained by considering the damping of the response due to each pair of roots. The damping due to s_1 and $\hat{s}_1$ is

$$e^{-\zeta_1 \omega_{n_1} t} = e^{-\sigma_1 t},$$

and the damping factor due to s_2 and $\hat{s}_2$ is

$$e^{-\zeta_2 \omega_{n_2} t} = e^{-\sigma_2 t},$$

where σ_2 is approximately five times as large as σ_1. Therefore, the transient response term due to s_2 will decay much more rapidly than the transient response term due to s_1. Thus, the response to a unit step input may be written as

$$y(t) = 1 + c_1 e^{-\sigma_1 t} \sin(\omega_1 t + \theta_1) + c_2 e^{-\sigma_2 t} \sin(\omega_2 t + \theta_2)$$

$$\approx 1 + c_1 e^{-\sigma_1 t} \sin(\omega_1 t + \theta_1). \quad (7.53)$$

The complex conjugate roots near the origin of the s-plane relative to the other roots of the closed-loop system are labeled the **dominant roots** of the system because they represent or dominate the transient response. The relative dominance of the complex roots, in a third-order system with a pair of complex conjugate roots, is determined by the ratio of the real root to the real part of the complex roots and will result in approximate dominance for ratios exceeding 5.

The dominance of the second term of Equation (7.53) also depends upon the relative magnitudes of the coefficients c_1 and c_2. These coefficients, which are the residues evaluated at the complex roots, in turn depend upon the location of the zeros in the s-plane. Therefore, the concept of dominant roots is useful for estimating the response of a system, but must be used with caution and with a comprehension of the underlying assumptions. ∎

EXAMPLE 7.5   **Automatic self-balancing scale**

The analysis and design of a control system can be accomplished by using the Laplace transform, a signal-flow diagram or block diagram, the s-plane, and the root locus method. At this point, it will be worthwhile to examine a control system and select suitable parameter values based on the root locus method.

Figure 7.15 shows an automatic self-balancing scale in which the weighing operation is controlled by the physical balance function through an electrical feedback loop [5]. The balance is shown in the equilibrium condition, and x is the travel of the counterweight W_c from an unloaded equilibrium condition. The weight W to be

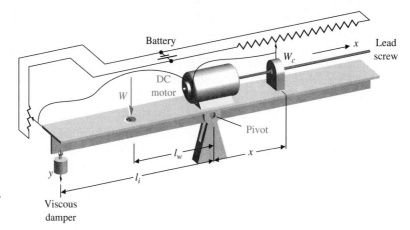

FIGURE 7.15
An automatic self-balancing scale. (Reprinted with permission from J. H. Goldberg, *Automatic Controls*, Allyn and Bacon, Boston, 1964.)

measured is applied 5 cm from the pivot, and the length l_i of the beam to the viscous damper is 20 cm. We desire to accomplish the following:

1. Select the parameters and the specifications of the feedback system.
2. Obtain a model representing the system.
3. Select the gain K based on a root locus diagram.
4. Determine the dominant mode of response.

An inertia of the beam equal to 0.05 kg m^2 will be chosen. We must select a battery voltage that is large enough to provide a reasonable position sensor gain, so we will choose $E_b = 24$ volts. We will use a lead screw of 20 turns/cm and a potentiometer for x equal to 6 cm in length. Accurate balances are required; therefore, an input potentiometer 0.5 cm in length for y will be chosen. A reasonable viscous damper will be chosen with a damping constant $b = 10\sqrt{3}\ N/(m/s)$. Finally, a counterweight W_c is chosen so that the expected range of weights W can be balanced. The parameters of the system are selected as listed in Table 7.4.

Specifications.   A rapid and accurate response resulting in a small steady-state weight measurement error is desired. Therefore, we will require that the system be at least a type one so that a zero measurement error is obtained. An underdamped response to a step change in the measured weight W is satisfactory, so a dominant response with $\zeta = 0.5$ will be specified. We want the settling time to be less than 2

Table 7.4   Self-Balancing Scale Parameters

$W_c = 2$ N	Lead screw gain $K_s = \dfrac{1}{4000\pi}$ m/rad.
$I = 0.05$ kg m^2	
$l_w = 5$ cm	Input potentiometer gain $K_i = 4800$ V/m.
$l_i = 20$ cm	
$b = 10\sqrt{3}$ N m/s	Feedback potentiometer gain $K_f = 400$ V/m.

Table 7.5   Specifications

Steady-state error	$K_p = \infty$, $e_{ss} = 0$ for a step input
Underdamped response	$\zeta = 0.5$
Settling time (2% criterion)	Less than 2 seconds

seconds in order to provide a rapid weight-measuring device. The settling time must be within 2% of the final value of the balance following the introduction of a weight to be measured. The specifications are summarized in Table 7.5.

The derivation of a model of the electromechanical system may be accomplished by obtaining the equations of motion of the balance. For small deviations from balance, the deviation angle is

$$\theta \approx \frac{y}{l_i}. \tag{7.54}$$

The motion of the beam about the pivot is represented by the torque equation

$$I\frac{d^2\theta}{dt^2} = \sum \text{torques}.$$

Therefore, in terms of the deviation angle, the motion is represented by

$$I\frac{d^2\theta}{dt^2} = l_w W - xW_c - l_i^2 b\frac{d\theta}{dt}. \tag{7.55}$$

The input voltage to the motor is

$$v_m(t) = K_i y - K_f x. \tag{7.56}$$

The lead screw motion and transfer function of the motor are described by

$$X(s) = K_s \theta_m(s) \quad \text{and} \quad \frac{\theta_m(s)}{V_m(s)} = \frac{K_m}{s(\tau s + 1)}, \tag{7.57}$$

where τ will be negligible with respect to the time constants of the overall system, and θ_m is the output shaft rotation. A signal-flow graph and block diagram representing Equations (7.54) through (7.57) is shown in Figure 7.16. Examining the forward path from W to $X(s)$, we find that the system is a type one due to the integration preceding $Y(s)$. Therefore, the steady-state error of the system is zero.

The closed-loop transfer function of the system is obtained by utilizing Mason's signal-flow gain formula and is found to be

$$\frac{X(s)}{W(s)} = \frac{l_w l_i K_i K_m K_s/(Is^3)}{1 + l_i^2 b/(Is) + (K_m K_s K_f/s) + l_i K_i K_m K_s W_c/(Is^3) + l_i^2 b K_m K_s K_f/(Is^2)}, \tag{7.58}$$

where the numerator is the path factor from W to X, the second term in the denominator is the loop L_1, the third term is the loop factor L_2, the fourth term is the loop

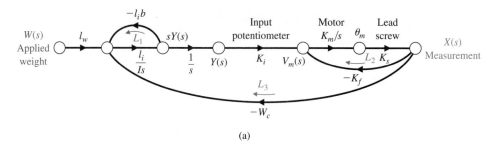

(a)

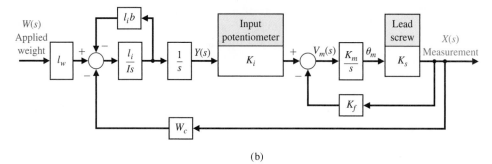

FIGURE 7.16
Model of the
automatic self-
balancing scale.
(a) Signal-flow
graph. (b) Block
diagram.

(b)

L_3, and the fifth term is the two nontouching loops $L_1 L_2$. Therefore, the closed-loop transfer function is

$$\frac{X(s)}{W(s)} = \frac{l_w l_i K_i K_m K_s}{s(Is + l_i^2 b)(s + K_m K_s K_f) + W_c K_m K_s K_i l_i}. \tag{7.59}$$

The steady-state gain of the system is then

$$\lim_{t \to \infty} \frac{x(t)}{|W|} = \lim_{s \to 0} \frac{X(s)}{W(s)} = \frac{l_w}{W_c} = 2.5 \text{ cm/kg} \tag{7.60}$$

when $W(s) = |W|/s$. To obtain the root locus as a function of the motor constant K_m, we substitute the selected parameters into the characteristic equation, which is the denominator of Equation (7.59). Therefore, we obtain the following characteristic equation:

$$s\left(s + 8\sqrt{3}\right)\left(s + \frac{K_m}{10\pi}\right) + \frac{96 K_m}{10\pi} = 0. \tag{7.61}$$

Rewriting the characteristic equation in root locus form, we first isolate K_m as follows:

$$s^2\left(s + 8\sqrt{3}\right) + s\left(s + 8\sqrt{3}\right)\frac{K_m}{10\pi} + \frac{96 K_m}{10\pi} = 0. \tag{7.62}$$

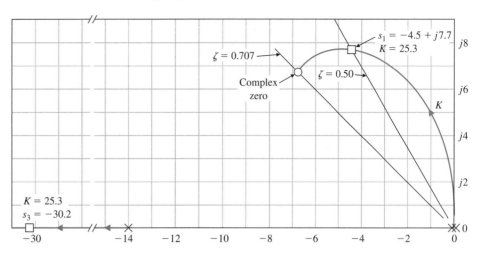

FIGURE 7.17
Root locus as K_m varies (only upper halfplane shown). One locus leaves the two poles at the origin and goes to the two complex zeros as K increases. The other locus is to the left of the pole at $s = -14$.

Then, rewriting Equation (7.62) in root locus form, we have

$$1 + KP(s) = 1 + \frac{K_m/(10\pi)\left[s\left(s + 8\sqrt{3}\right) + 96\right]}{s^2\left(s + 8\sqrt{3}\right)} = 0$$

$$= 1 + \frac{K_m/(10\pi)(s + 6.93 + j6.93)(s + 6.93 - j6.93)}{s^2\left(s + 8\sqrt{3}\right)}. \qquad (7.63)$$

The root locus as K_m varies is shown in Figure 7.17. The dominant roots can be placed at $\zeta = 0.5$ when $K = 25.3 = K_m/10\pi$. To achieve this gain,

$$K_m = 795\frac{\text{rad/s}}{\text{volt}} = 7600\frac{\text{rpm}}{\text{volt}}, \qquad (7.64)$$

an amplifier would be required to provide a portion of the required gain. The real part of the dominant roots is less than -4; therefore, the settling time, $4/\sigma$, is less than 1 second, and the settling time requirement is satisfied. The third root of the characteristic equation is a real root at $s = -30.2$, and the underdamped roots clearly dominate the response. Therefore, the system has been analyzed by the root locus method and a suitable design for the parameter K_m has been achieved. The efficiency of the s-plane and root locus methods is clearly demonstrated by this example. ∎

7.4   PARAMETER DESIGN BY THE ROOT LOCUS METHOD

Originally, the root locus method was developed to determine the locus of roots of the characteristic equation as the system gain, K, is varied from zero to infinity. However, as we have seen, the effect of other system parameters may be readily investigated by using the root locus method. Fundamentally, the root locus method is concerned with a characteristic equation (Equation 7.22), which may be written as

$$1 + F(s) = 0. \tag{7.65}$$

Then the standard root locus method we have studied may be applied. The question arises: How do we investigate the effect of two parameters, α and β? It appears that the root locus method is a single-parameter method; fortunately, it can be readily extended to the investigation of two or more parameters. This method of **parameter design** uses the root locus approach to select the values of the parameters.

The characteristic equation of a dynamic system may be written as

$$a_n s^n + a_{n-1} s^{n-1} + \cdots + a_1 s + a_0 = 0. \tag{7.66}$$

Hence, the effect of the coefficient a_1 may be ascertained from the root locus equation

$$1 + \frac{a_1 s}{a_n s^n + a_{n-1} s^{n-1} + \cdots + a_2 s^2 + a_0} = 0. \tag{7.67}$$

If the parameter of interest, α, does not appear solely as a coefficient, the parameter may be isolated as

$$a_n s^n + a_{n-1} s^{n-1} + \cdots + (a_{n-q} - \alpha)s^{n-q} + \alpha s^{n-q} + \cdots + a_1 s + a_0 = 0. \tag{7.68}$$

For example, a third-order equation of interest might be

$$s^3 + (3 + \alpha)s^2 + 3s + 6 = 0. \tag{7.69}$$

To ascertain the effect of the parameter α, we isolate the parameter and rewrite the equation in root locus form, as shown in the following steps:

$$s^3 + 3s^2 + \alpha s^2 + 3s + 6 = 0; \tag{7.70}$$

$$1 + \frac{\alpha s^2}{s^3 + 3s^2 + 3s + 6} = 0. \tag{7.71}$$

Then, to determine the effect of two parameters, we must repeat the root locus approach twice. Thus, for a characteristic equation with two variable parameters, α and β, we have

$$a_n s^n + a_{n-1} s^{n-1} + \cdots + (a_{n-q} - \alpha)s^{n-q} + \alpha s^{n-q} + \cdots$$
$$+ (a_{n-r} - \beta)s^{n-r} + \beta s^{n-r} + \cdots + a_1 s + a_0 = 0. \tag{7.72}$$

The two variable parameters have been isolated, and the effect of α will be determined. Then, the effect of β will be determined. For example, for a certain third-order characteristic equation with α and β as parameters, we obtain

$$s^3 + s^2 + \beta s + \alpha = 0. \tag{7.73}$$

In this particular case, the parameters appear as the coefficients of the characteristic equation. The effect of varying β from zero to infinity is determined from the root locus equation

$$1 + \frac{\beta s}{s^3 + s^2 + \alpha} = 0. \tag{7.74}$$

We note that the denominator of Equation (7.74) is the characteristic equation of the system with $\beta = 0$. Therefore, we must first evaluate the effect of varying α from zero to infinity by using the equation

$$s^3 + s^2 + \alpha = 0,$$

rewritten as

$$1 + \frac{\alpha}{s^2(s + 1)} = 0, \tag{7.75}$$

where β has been set equal to zero in Equation (7.73). Then, upon evaluating the effect of α, a value of α is selected and used with Equation (7.74) to evaluate the effect of β. This two-step method of evaluating the effect of α and then β may be carried out as two root locus procedures. First, we obtain a locus of roots as α varies, and we select a suitable value of α; the results are satisfactory root locations. Then, we obtain the root locus for β by noting that the poles of Equation (7.74) are the roots evaluated by the root locus of Equation (7.75). A limitation of this approach is that we will not always be able to obtain a characteristic equation that is linear in the parameter under consideration (for example, α).

To illustrate this approach effectively, let us obtain the root locus for α and then β for Equation (7.73). A sketch of the root locus as α varies for Equation (7.75) is shown in Figure 7.18(a), where the roots for two values of gain α are shown. If the gain α is selected as α_1, then the resultant roots of Equation (7.75) become the poles of Equation (7.74). The root locus of Equation (7.74) as β varies is shown in Figure 7.18(b), and a suitable β can be selected on the basis of the desired root locations.

Using the root locus method, we will further illustrate this parameter design approach by a specific design example.

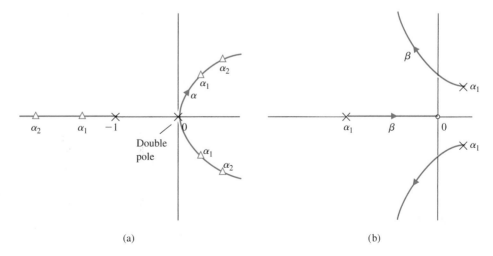

FIGURE 7.18
Root loci as a function of α and β.
(a) Loci as α varies.
(b) Loci as β varies for one value of $\alpha = \alpha_1$.

(a)   (b)

EXAMPLE 7.6   **Welding head control**

A welding head for an auto body requires an accurate control system for positioning the welding head [4]. The feedback control system is to be designed to satisfy the following specifications:

1. Steady-state error for a ramp input $\leq 35\%$ of input slope

2. Damping ratio of dominant roots ≥ 0.707

3. Settling time to within 2% of the final value ≤ 3 seconds

The structure of the feedback control system is shown in Figure 7.19, where the amplifier gain K_1 and the derivative feedback gain K_2 are to be selected. The steady-state error specification can be written as

$$e_{ss} = \lim_{t \to \infty} e(t) = \lim_{s \to 0} sE(s) = \lim_{s \to 0} \frac{s(|R|/s^2)}{1 + G_2(s)}, \tag{7.76}$$

where $G_2(s) = G(s)/(1 + G(s)H_1(s))$. Therefore, the steady-state error requirement is

$$\frac{e_{ss}}{|R|} = \frac{2 + K_1K_2}{K_1} \leq 0.35. \tag{7.77}$$

Thus, we will select a small value of K_2 to achieve a low value of steady-state error. The damping ratio specification requires that the roots of the closed-loop system be below the line at $45°$ in the left-hand s-plane. The settling time specification can be rewritten in terms of the real part of the dominant roots as

$$T_s = \frac{4}{\sigma} \leq 3 \text{ s}. \tag{7.78}$$

Therefore, it is necessary that $\sigma \geq {}^4\!/_3$; this area in the left-hand s-plane is indicated along with the ζ-requirement in Figure 7.20. Note that $\sigma \geq {}^4\!/_3$ implies that we want the dominant roots to lie to the left of the line defined by $\sigma = -{}^4\!/_3$. To satisfy the specifications, all the roots must lie within the shaded area of the left-hand plane.

The parameters to be selected are $\alpha = K_1$ and $\beta = K_2K_1$. The characteristic equation is

$$1 + GH(s) = s^2 + 2s + \beta s + \alpha = 0. \tag{7.79}$$

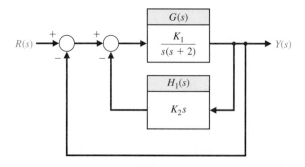

FIGURE 7.19
Block diagram of welding head control system.

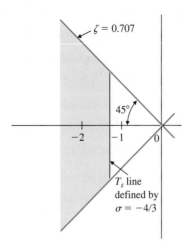

FIGURE 7.20
A region in the
s-plane for desired
root location.

The locus of roots as $\alpha = K_1$ varies (set $\beta = 0$) is determined from the equation

$$1 + \frac{\alpha}{s(s + 2)} = 0, \qquad (7.80)$$

as shown in Figure 7.21(a). For a gain of $K_1 = \alpha = 20$, the roots are indicated on the locus. Then the effect of varying $\beta = 20K_2$ is determined from the locus equation

$$1 + \frac{\beta s}{s^2 + 2s + \alpha} = 0, \qquad (7.81)$$

where the poles of this root locus are the roots of the locus of Figure 7.21(a). The root locus for Equation (7.81) is shown in Figure 7.21(b), and roots with $\zeta = 0.707$ are obtained when $\beta = 4.3 = 20K_2$ or when $K_2 = 0.215$. The real part of these roots is

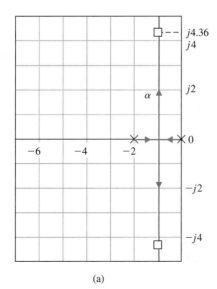

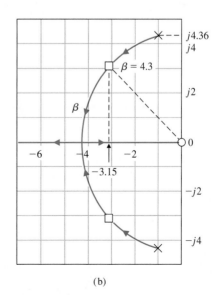

FIGURE 7.21
Root loci as a
function of (a) α
and (b) β.

(a)

(b)

$\sigma = -3.15$; therefore, the time to settle (to within 2% of the final value) is equal to 1.27 seconds, which is considerably less than the specification of 3 seconds. ∎

We can extend the root locus method to more than two parameters by extending the number of steps in the method outlined in this section. Furthermore, a family of root loci can be generated for two parameters in order to determine the total effect of varying two parameters. For example, let us determine the effect of varying α and β of the following characteristic equation:

$$s^3 + 3s^2 + 2s + \beta s + \alpha = 0. \tag{7.82}$$

The root locus equation as a function of α is (set $\beta = 0$)

$$1 + \frac{\alpha}{s(s+1)(s+2)} = 0. \tag{7.83}$$

The root locus as a function of β is

$$1 + \frac{\beta s}{s^3 + 3s^2 + 2s + \alpha} = 0. \tag{7.84}$$

The root locus for Equation (7.83) as a function of α is shown in Figure 7.22 (unbroken lines). The roots of this locus, indicated by slashes, become the poles for the locus of Equation (7.84). Then the locus of Equation (7.84) is continued on Figure 7.22 (dotted lines), where the locus for β is shown for several selected values of α. This family of loci, often called **root contours**, illustrates the effect of α and β on the roots of the characteristic equation of a system [3].

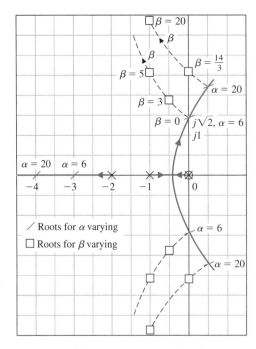

FIGURE 7.22
Two-parameter root locus. The loci for α varying are solid; the loci for β varying are dashed.

7.5 SENSITIVITY AND THE ROOT LOCUS

One of the prime reasons for the use of negative feedback in control systems is the reduction of the effect of parameter variations. The effect of parameter variations, as we found in Section 4.3, can be described by a measure of the **sensitivity** of the system performance to specific parameter changes. In Section 4.3, we defined the **logarithmic sensitivity** originally suggested by Bode as

$$S_K^T = \frac{\partial \ln T}{\partial \ln K} = \frac{\partial T/T}{\partial K/K}, \tag{7.85}$$

where the system transfer function is $T(s)$ and the parameter of interest is K.

In recent years, there has been an increased use of the pole–zero (s-plane) approach. Therefore, it has become useful to define a sensitivity measure in terms of the positions of the roots of the characteristic equation [7–9]. Because these roots represent the dominant modes of transient response, the effect of parameter variations on the position of the roots is an important and useful measure of the sensitivity. The **root sensitivity** of a system $T(s)$ can be defined as

$$\boxed{S_K^{r_i} = \frac{\partial r_i}{\partial \ln K} = \frac{\partial r_i}{\partial K/K},} \tag{7.86}$$

where r_i equals the ith root of the system, so that

$$T(s) = \frac{K_1 \displaystyle\prod_{j=1}^{M}(s + z_j)}{\displaystyle\prod_{i=1}^{n}(s + r_i)} \tag{7.87}$$

and K is a parameter affecting the roots. The root sensitivity relates the changes in the location of the root in the s-plane to the change in the parameter. The root sensitivity is related to the logarithmic sensitivity by the relation

$$S_K^T = \frac{\partial \ln K_1}{\partial \ln K} - \sum_{i=1}^{n} \frac{\partial r_i}{\partial \ln K} \cdot \frac{1}{s + r_i} \tag{7.88}$$

when the zeros of $T(s)$ are independent of the parameter K, so that

$$\frac{\partial z_j}{\partial \ln K} = 0.$$

This logarithmic sensitivity can be readily obtained by determining the derivative of $T(s)$, Equation (7.87), with respect to K. For this particular case, when the gain of the system is independent of the parameter K, we have

$$S_K^T = -\sum_{i=1}^{n} S_K^{r_i} \cdot \frac{1}{s + r_i}, \tag{7.89}$$

and the two sensitivity measures are directly related.

The evaluation of the root sensitivity for a control system can be readily accomplished by utilizing the root locus methods of the preceding section. The root sensitivity $S_K^{r_i}$ may be evaluated at root $-r_i$ by examining the root contours for the parameter K. We can change K by a small finite amount ΔK and determine the modified root $-(r_i + \Delta r_i)$ at $K + \Delta K$. Then, using Equation (7.86), we have

$$S_K^{r_i} \approx \frac{\Delta r_i}{\Delta K / K}. \qquad (7.90)$$

Equation (7.90) is an approximation that approaches the actual value of the sensitivity as $\Delta K \to 0$. An example will illustrate the process of evaluating the root sensitivity.

EXAMPLE 7.7   **Root sensitivity of a control system**

The characteristic equation of the feedback control system shown in Figure 7.23 is

$$1 + \frac{K}{s(s + \beta)} = 0,$$

or, alternatively,

$$s^2 + \beta s + K = 0. \qquad (7.91)$$

The gain K will be considered to be the parameter α. Then the effect of a change in each parameter can be determined by utilizing the relations

$$\alpha = \alpha_0 \pm \Delta \alpha \quad \text{and} \quad \beta = \beta_0 \pm \Delta \beta,$$

where α_0 and β_0 are the nominal or desired values for the parameters α and β, respectively. We shall consider the case when the nominal pole value is $\beta_0 = 1$ and the desired gain is $\alpha_0 = K = 0.5$. Then the root locus can be obtained as a function of $\alpha = K$ by utilizing the root locus equation

$$1 + \frac{K}{s(s + \beta_0)} = 1 + \frac{K}{s(s + 1)} = 0, \qquad (7.92)$$

as shown in Figure 7.24. The nominal value of gain $K = \alpha_0 = 0.5$ results in two complex roots, $-r_1 = -0.5 + j0.5$ and $-r_2 = -\hat{r}_1$, as shown in Figure 7.24. To evaluate the effect of unavoidable changes in the gain, the characteristic equation with $\alpha = \alpha_0 \pm \Delta \alpha$ becomes

$$s^2 + s + \alpha_0 \pm \Delta \alpha = s^2 + s + 0.5 \pm \Delta \alpha. \qquad (7.93)$$

Therefore, the effect of changes in the gain can be evaluated from the root locus of Figure 7.24. For a 20% change in α, we have $\Delta \alpha = \pm 0.1$. The root locations for a

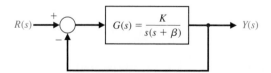

FIGURE 7.23
A feedback control system.

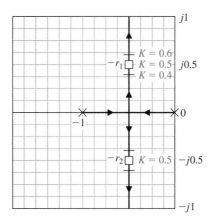

FIGURE 7.24
The root locus
for K.

gain $\alpha = 0.4$ and $\alpha = 0.6$ are readily determined by root locus methods, and the root locations for $\Delta\alpha = \pm 0.1$ are shown in Figure 7.24. When $\alpha = K = 0.6$, the root in the second quadrant of the s-plane is

$$(-r_1) + \Delta r_1 = -0.5 + j0.59,$$

and the change in the root is $\Delta r_1 = +j0.09$. When $\alpha = K = 0.4$, the root in the second quadrant is

$$-(r_1) + \Delta r_1 = -0.5 + j0.387,$$

and the change in the root is $-\Delta r = -j0.11$. Thus, the root sensitivity for r_1 is

$$S_{K+}^{r_1} = \frac{\Delta r_1}{\Delta K / K} = \frac{+j0.09}{+0.2} = j0.45 = 0.45 \underline{/+90°} \tag{7.94}$$

for positive changes of gain. For negative increments of gain, the sensitivity is

$$S_{K-}^{r_1} = \frac{\Delta r_1}{\Delta K / K} = \frac{-j0.11}{+0.2} = -j0.55 = 0.55 \underline{/-90°}.$$

For infinitesimally small changes in the parameter K, the sensitivity will be equal for negative or positive increments in K. The angle of the root sensitivity indicates the direction the root moves as the parameter varies. The angle of movement for $+\Delta\alpha$ is always 180° from the angle of movement for $-\Delta\alpha$ at the point $\alpha = \alpha_0$.

The pole β varies due to environmental changes, and it may be represented by $\beta = \beta_0 + \Delta\beta$, where $\beta_0 = 1$. Then the effect of variation of the poles is represented by the characteristic equation

$$s^2 + s + \Delta\beta s + K = 0,$$

or, in root locus form,

$$1 + \frac{\Delta\beta s}{s^2 + s + K} = 0. \tag{7.95}$$

The denominator of the second term is the unchanged characteristic equation when $\Delta\beta = 0$. The root locus for the unchanged system ($\Delta\beta = 0$) is shown in Figure 7.24 as a function of K. For a design specification requiring $\zeta = 0.707$, the complex roots lie at

$$-r_1 = -0.5 + j0.5 \quad \text{and} \quad -r_2 = -\hat{r}_1 = -0.5 - j0.5.$$

Then, because the roots are complex conjugates, the root sensitivity for r_1 is the conjugate of the root sensitivity for $\hat{r}_1 = r_2$. Using the parameter root locus techniques discussed in the preceding section, we obtain the root locus for $\Delta\beta$ as shown in Figure 7.25. We are normally interested in the effect of a variation for the parameter so that $\beta = \beta_0 \pm \Delta\beta$, for which the locus as $\Delta\beta$ decreases is obtained from the root locus equation

$$1 + \frac{-(\Delta\beta)s}{s^2 + s + K} = 0.$$

We note that the equation is of the form

$$1 - \Delta\beta P(s) = 0.$$

Comparing this equation with Equation (7.23) in Section 7.3, we find that the sign preceding the gain $\Delta\beta$ is negative in this case. In a manner similar to the development of the root locus method in Section 7.3, we require that the root locus satisfy the equations

$$|\Delta\beta P(s)| = 1 \quad \text{and} \quad \underline{/P(s)} = 0° \pm k360°,$$

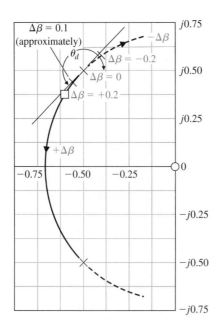

FIGURE 7.25
The root locus for
the parameter β.

where k is an integer. The locus of roots follows a zero-degree locus in contrast with the 180° locus considered previously. However, the root locus rules of Section 7.3 may be altered to account for the zero-degree phase angle requirement, and then the root locus may be obtained as in the preceding sections. Therefore, to obtain the effect of reducing β, we determine the zero-degree locus in contrast to the 180° locus, as shown by a dotted locus in Figure 7.25. To find the effect of a 20% change of the parameter β, we evaluate the new roots for $\Delta\beta = \pm 0.20$, as shown in Figure 7.25. The root sensitivity is readily evaluated graphically and, for a positive change in β, is

$$S_{\beta+}^{r_1} = \frac{\Delta r_1}{\Delta\beta/\beta} = \frac{0.16\underline{/-128°}}{0.20} = 0.80\underline{/-128°}.$$

The root sensitivity for a negative change in β is

$$S_{\beta-}^{r_1} = \frac{\Delta r_1}{\Delta\beta/\beta} = \frac{0.125\underline{/39°}}{0.20} = 0.625\underline{/+39°}.$$

As the percentage change $\Delta\beta/\beta$ decreases, the sensitivity measures $S_{\beta+}^{r_1}$ and $S_{\beta-}^{r_1}$ will approach equality in magnitude and a difference in angle of 180°. Thus, for small changes when $\Delta\beta/\beta \leq 0.10$, the sensitivity measures are related as

$$|S_{\beta+}^{r_1}| = |S_{\beta-}^{r_1}|$$

and

$$\underline{/S_{\beta+}^{r_1}} = 180° + \underline{/S_{\beta-}^{r_1}}. \tag{7.95}$$

Often, the desired root sensitivity measure is desired for small changes in the parameter. When the relative change in the parameter is of the order $\Delta\beta/\beta = 0.10$, we can estimate the increment in the root change by approximating the root locus with the line at the angle of departure θ_d. This approximation is shown in Figure 7.25 and is accurate for only relatively small changes in $\Delta\beta$. However, the use of this approximation allows the analyst to avoid sketching the complete root locus diagram. Therefore, for Figure 7.25, the root sensitivity may be evaluated for $\Delta\beta/\beta = 0.10$ along the departure line, and we obtain

$$S_{\beta+}^{r_1} = \frac{0.075\underline{/-132°}}{0.10} = 0.075\underline{/-132°}. \tag{7.96}$$

The root sensitivity measure for a parameter variation is useful for comparing the sensitivity for various design parameters and at different root locations. Comparing Equation (7.96) for β with Equation (7.94) for α, we find (a) that the sensitivity for β is greater in magnitude by approximately 50% and (b) that the angle for $S_{\beta-}^{r_1}$ indicates that the approach of the root toward the $j\omega$-axis is more sensitive for changes in β. Therefore, the tolerance requirements for β would be more stringent than for α. This information provides the designer with a comparative measure of the required tolerances for each parameter. ■

EXAMPLE 7.8   **Root sensitivity to a parameter**

A unity feedback control system has a forward transfer function

$$G(s) = \frac{20.7(s + 3)}{s(s + 2)(s + \beta)},$$

where $\beta = \beta_0 + \Delta\beta$ and $\beta_0 = 8$. The characteristic equation, as a function of $\Delta\beta$, is

$$s(s + 2)(s + 8 + \Delta\beta) + 20.7(s + 3) = 0,$$

or

$$s(s + 2)(s + 8) + \Delta\beta s(s + 2) + 20.7(s + 3) = 0.$$

When $\Delta\beta = 0$, the roots are

$$-r_1 = -2.36 + j2.48, \qquad -r_2 = \hat{r}_1, \quad \text{and} \quad -r_3 = -5.27.$$

The root locus for $\Delta\beta$ is determined by using the root locus equation

$$1 + \frac{\Delta\beta s(s + 2)}{(s + r_1)(s + \hat{r}_1)(s + r_3)} = 0. \tag{7.97}$$

The roots and zeros of Equation (7.97) are shown in Figure 7.26. The angle of departure at r_1 is evaluated from the angles as follows:

$$180° = -(\theta_d + 90° + \theta_{p3}) + (\theta_{z_1} + \theta_{z_2})$$
$$= -(\theta_d + 90° + 40°) + (133° + 98°).$$

Therefore, $\theta_d = -80°$ and the locus is approximated near $-r_1$ by the line at an angle of θ_d. For a change of $\Delta r_1 = 0.2\underline{/-80°}$ along the departure line, the $+\Delta\beta$ is evaluated by determining the vector lengths from the poles and zeros. Then we have

$$+\Delta\beta = \frac{4.8(3.75)(0.2)}{(3.25)(2.3)} = 0.48.$$

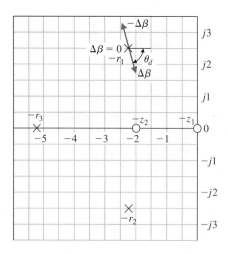

FIGURE 7.26
Pole and zero
diagram for the
parameter β.

Therefore, the sensitivity at r_1 is

$$S_\beta^{r_1} = \frac{\Delta r_1}{\Delta \beta / \beta} = \frac{0.2 \underline{/-80°}}{0.48/8} = 3.34 \underline{/-80°},$$

which indicates that the root is quite sensitive to this 6% change in the parameter β. For comparison, it is worthwhile to determine the sensitivity of the root $-r_1$ to a change in the zero $s = -3$. Then the characteristic equation is

$$s(s + 2)(s + 8) + 20.7(s + 3 + \Delta\gamma) = 0,$$

or

$$1 + \frac{20.7 \, \Delta\gamma}{(s + r_1)(s + \hat{r}_1)(s + r_3)} = 0. \tag{7.98}$$

The pole–zero diagram for Equation (7.98) is shown in Figure 7.27. The angle of departure at root $-r_1$ is $180° = -(\theta_d + 90° + 40°)$, or

$$\theta_d = +50°.$$

For a change of $r_1 = 0.2 \underline{/+50°}$, the $\Delta\gamma$ is positive. Obtaining the vector lengths, we find that

$$|\Delta\gamma| = \frac{5.22(4.18)(0.2)}{20.7} = 0.21.$$

Therefore, the sensitivity at r_1 for $+\Delta\gamma$ is

$$S_\gamma^{r_1} = \frac{\Delta r_1}{\Delta\gamma / \gamma} = \frac{0.2 \underline{/+50°}}{0.21/3} = 2.84 \underline{/+50°}.$$

Thus, we find that the magnitude of the root sensitivity for the pole β and the zero γ is approximately equal. However, the sensitivity of the system to the pole can be considered to be less than the sensitivity to the zero because the angle of the sensitivity, $S_\gamma^{r_1}$, is equal to $+50°$ and the direction of the root change is toward the $j\omega$-axis.

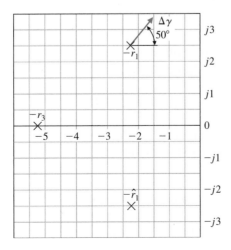

FIGURE 7.27
Pole–zero diagram for the parameter γ.

Evaluating the root sensitivity in the manner of the preceding paragraphs, we find that the sensitivity for the pole $s = -\delta_0 = -2$ is

$$S_{\delta-}^{r_1} = 2.1 \underline{/+27°}.$$

Thus, for the parameter δ, the magnitude of the sensitivity is less than for the other parameters, but the direction of the change of the root is more important than for β and γ. ∎

To utilize the root sensitivity measure for the analysis and design of control systems, a series of calculations must be performed; they will determine the various selections of possible root configurations and the zeros and poles of the open-loop transfer function. Therefore, the root sensitivity measure as a design technique is somewhat limited by two things: the relatively large number of calculations required and the lack of an obvious direction for adjusting the parameters in order to provide a minimized or reduced sensitivity. However, the root sensitivity measure can be utilized as an analysis measure, which permits the designer to compare the sensitivity for several system designs based on a suitable method of design. The root sensitivity measure is a useful index of the system's sensitivity to parameter variations expressed in the s-plane. The weakness of the sensitivity measure is that it relies on the ability of the root locations to represent the performance of the system. As we have seen in the preceding chapters, the root locations represent the performance quite adequately for many systems, but due consideration must be given to the location of the zeros of the closed-loop transfer function and the dominancy of the pertinent roots. The root sensitivity measure is a suitable measure of system performance sensitivity and can be used reliably for system analysis and design.

7.6 THREE-TERM (PID) CONTROLLERS

One form of controller widely used in industrial process control is called a three-term, or **PID controller**. This controller has a transfer function

$$G_c(s) = K_p + \frac{K_I}{s} + K_D s.$$

The controller provides a proportional term, an integration term, and a derivative term [4, 10]. The equation for the output in the time domain is

$$u(t) = K_p e(t) + K_I \int e(t)\, dt + K_D \frac{de(t)}{dt}.$$

The three-mode controller is also called a PID controller because it contains a proportional, an integral, and a derivative term. The transfer function of the derivative term is actually

$$G_d(s) = \frac{K_D s}{\tau_d s + 1},$$

but τ_d is usually much smaller than the time constants of the process itself, so it may be neglected.

If we set $K_D = 0$, then we have the **proportional plus integral (PI) controller**

$$G_c(s) = K_p + \frac{K_I}{s}.$$

When $K_I = 0$, we have

$$G_c(s) = K_p + K_D s,$$

which is called a **proportional plus derivative (PD) controller**.

Many industrial processes are controlled using proportional–integral–derivative (PID) controllers. The popularity of PID controllers can be attributed partly to their good performance in a wide range of operating conditions and partly to their functional simplicity, which allows engineers to operate them in a simple, straightforward manner. To implement such a controller, three parameters must be determined for the given process: proportional gain, integral gain, and derivative gain [10].

The PID controller can also be viewed as a cascade of the PI and the PD controllers. Consider the PI controller

$$G_{PI}(s) = \hat{K}_P + \frac{\hat{K}_I}{s}$$

and the PD controller

$$G_{PD}(s) = \overline{K}_P + \overline{K}_D s,$$

where $\hat{K}_P$ and $\hat{K}_I$ are the PI controller gains and $\overline{K}_P$ and $\overline{K}_D$ are the PD controller gains. Cascading the two controllers (that is, placing them in series) yields

$$G_c(s) = G_{PI}(s)G_{PD}(s)$$

$$= \left(\hat{K}_P + \frac{\hat{K}_I}{s} \right)(\overline{K}_P + \overline{K}_D s)$$

$$= (\overline{K}_P \hat{K}_P + \hat{K}_I \overline{K}_D) + \hat{K}_P \overline{K}_D s + \frac{\hat{K}_I \overline{K}_D}{s}$$

$$= K_P + K_D s + \frac{K_I}{s},$$

where we have the following relationships between the PI and PD controller gains and the PID controller gains

$$K_P = \overline{K}_P \hat{K}_P + \hat{K}_I \overline{K}_D$$
$$K_D = \hat{K}_P \overline{K}_D$$
$$K_I = \hat{K}_I \overline{K}_D.$$

Consider the PID controller

$$G_c(s) = K_P + \frac{K_I}{s} + K_D s = \frac{K_D s^2 + K_P s + K_I}{s}$$

$$= \frac{K_D(s^2 + as + b)}{s} = \frac{K_D(s + z_1)(s + z_2)}{s},$$

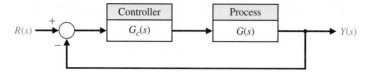

FIGURE 7.28
Closed-loop system
with a controller.

where $a = K_P/K_D$ and $b = K_I/K_D$. Therefore, a PID controller introduces a transfer function with one pole at the origin and two zeros that can be located anywhere in the left-hand s-plane.

Recall that a root locus begins at the poles and ends at the zeros. If we have a system, as shown in Figure 7.28, with

$$G(s) = \frac{1}{(s+2)(s+3)},$$

and we use a PID controller with complex zeros $-z_1$ and $-z_2$, where $-z_1 = -3 + j1$ and $-z_2 = -\hat{z}_1$, we can plot the root locus as shown in Figure 7.29. As the gain, K_D, of the controller is increased, the complex roots approach the zeros. The closed-loop transfer function is

$$T(s) = \frac{G(s)G_c(s)}{1 + G(s)G_c(s)}$$

$$= \frac{K_D(s+z_1)(s+\hat{z}_1)}{(s+r_2)(s+r_1)(s+\hat{r}_1)}.$$

The response of this system will be attractive. The percent overshoot to a step will be less than 2%, and the steady-state error for a step input will be zero. The settling

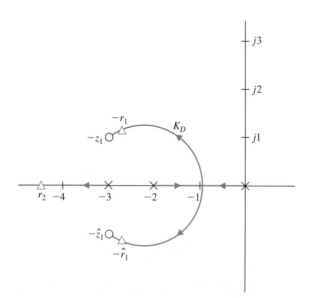

FIGURE 7.29
Root locus for plant
with a PID
controller with
complex zeros.

time will be approximately 1 second. If a shorter settling time is desired, then we select z_1 and z_2 to lie further left in the left-hand s-plane and set K_D to drive the roots near the complex zeros.

We will use the PD controller later in this chapter to control the hard disk drive sequential design problem (see Section 7.9).

7.7 DESIGN EXAMPLES

In this section we present three illustrative examples. The first example is a laser manipulator control system. Here the root locus method is used to show how the closed-loop system poles move in the s-plane as the proportional controller amplifier gain varies. The second example considers a simplified robotic replication facility. In the example, the system is represented by a fifth-order transfer function model. The feedback control strategy employs a velocity feedback coupled with a controller in the forward loop. Root locus design methods are used to select the two feedback controller gains. In the final example, the automatic control of the velocity of an automobile is considered. In this example, the root locus method is extended from one parameter to three parameters as the three gains of a PID controller are determined. The design process is emphasized, including considering the control goals and associated variables to be controlled, the design specifications, and the PID controller design using root locus methods.

EXAMPLE 7.9   **Laser manipulator control system**

Lasers can be used to drill the hip socket for the appropriate insertion of an artificial hip joint. The use of lasers for surgery requires high accuracy for position and velocity response. Let us consider the system shown in Figure 7.30, which uses a DC motor manipulator for the laser. The amplifier gain K must be adjusted so that the steady-state error for a ramp input, $r(t) = At$ (where $A = 1$ mm/s), is less than or equal to 0.1 mm, while a stable response is maintained.

To obtain the steady-state error required and a good response, we select a motor with a field time constant $\tau_1 = 0.1$ s and a motor-plus-load time constant $\tau_2 = 0.2$ s. We then have

$$T(s) = \frac{KG(s)}{1 + KG(s)} = \frac{K}{s(\tau_1 s + 1)(\tau_2 s + 1) + K}$$

$$= \frac{K}{0.02s^3 + 0.3s^2 + s + K} = \frac{50K}{s^3 + 15s^2 + 50s + 50K}. \qquad (7.99)$$

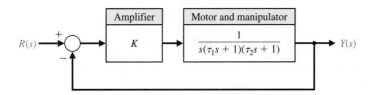

FIGURE 7.30
Laser manipulator
control system.

The steady-state error for a ramp, $R(s) = A/s^2$, from Equation (5.29), is

$$e_{ss} = \frac{A}{K_v} = \frac{A}{K}.$$

Since we desire $e_{ss} = 0.1$ mm (or less) and $A = 1$ mm, we require $K = 10$ (or greater).

To ensure a stable system, we obtain the characteristic equation from Equation (7.99) as

$$s^3 + 15s^2 + 50s + 50K = 0.$$

Establishing the Routh array, we have

$$
\begin{array}{c|cc}
s^3 & 1 & 50 \\
s^2 & 15 & 50K \\
s^1 & b_1 & 0 \\
s_0 & 50K &
\end{array}
\quad ,
$$

where

$$b_1 = \frac{750 - 50K}{15}.$$

Therefore, the system is stable for

$$0 \le K \le 15.$$

The characteristic equation can be written as

$$1 + K \frac{50}{s^3 + 15s^2 + 50s} = 0.$$

The root locus for $K > 0$ is shown in Figure 7.31. Using $K = 10$ results in a stable system that also satisfies the steady-state tracking error specification. The roots at $K = 10$ are $-r_2 = -13.98$, $-r_1 = -0.51 + j5.96$, and $-\hat{r}_1$. The ζ of the complex roots is 0.085 and $\zeta\omega_n = 0.51$. Thus, assuming that the complex roots are dominant, we expect (using Equation 5.16 and 5.13) a step input to have an overshoot of 76% and a settling time (to within 2% of the final value) of

$$T_s = \frac{4}{\zeta\omega_n} = \frac{4}{0.51} = 7.8 \text{ s.}$$

Plotting the actual system response, we find that the overshoot is 70% and the settling time is 7.5 seconds. Thus, the complex roots are essentially dominant. The system response to a step input is highly oscillatory and cannot be tolerated for laser surgery. The command signal must be limited to a low-velocity ramp signal. The response to a ramp signal is shown in Figure 7.32. ∎

EXAMPLE 7.10   Robot control system

The concept of robot replication is relatively easy to grasp. The central idea is that robots replicate themselves and develop a factory that automatically produces robots. An example of a robot replication facility is shown in Figure 7.33. To achieve

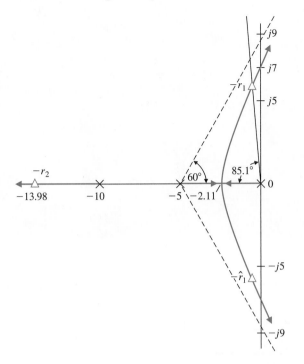

FIGURE 7.31
Root locus for a
laser control
system.

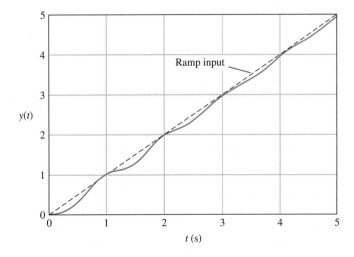

FIGURE 7.32
The response to a
ramp input for a
laser control
system.

the rapid and accurate control of a robot, it is important to keep the robotic arm stiff
and yet lightweight [6].

   The specifications for controlling the motion of the arm are (1) a settling time to
within 2% of the final value of less than 2 seconds, (2) a percent overshoot of less
than 10% for a step input, and (3) a steady-state error of zero for a step input.

   The block diagram of the proposed system with a controller is shown in
Figure 7.34. The configuration proposes the use of velocity feedback as well as

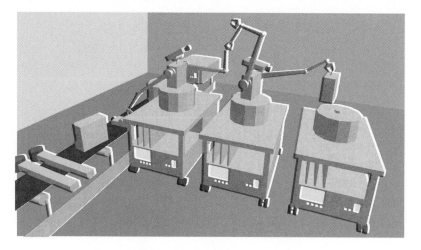

FIGURE 7.33
A robot replication facility.

FIGURE 7.34
Proposed configuration for control of the lightweight robotic arm.

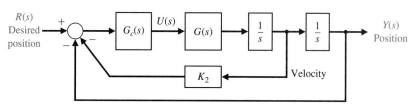

the use of a controller $G_c(s)$. The transfer function of the arm is

$$\frac{Y(s)}{U(s)} = \frac{1}{s^2}G(s)$$

where

$$G(s) = \frac{(s^2 + 4s + 10004)(s^2 + 12s + 90036)}{(s + 10)(s^2 + 2s + 2501)(s^2 + 6s + 22509)}.$$

The complex zeros are located at

$$s = -2 \pm j100 \quad \text{and} \quad s = -6 \pm j300.$$

The complex poles are located at

$$s = -1 \pm j50 \quad \text{and} \quad s = -3 \pm j150.$$

A sketch of the root locus when $K_2 = 0$ and the controller is an adjustable gain, $G_c(s) = K_1$, is shown in Figure 7.35. The system is unstable since two roots of the characteristic equation appear in the right-hand s-plane for $K_1 > 0$.

It is clear that we need to introduce the use of velocity feedback by setting K_2 to a positive magnitude. Then we have $H(s) = 1 + K_2s$; therefore, the loop transfer function is

$$\frac{1}{s^2}G_c(s)G(s)H(s) = \frac{K_1K_2\left(s + \frac{1}{K_2}\right)(s^2 + 4s + 10004)(s^2 + 12s + 90036)}{s^2(s + 10)(s^2 + 2s + 2501)(s^2 + 6s + 22509)},$$

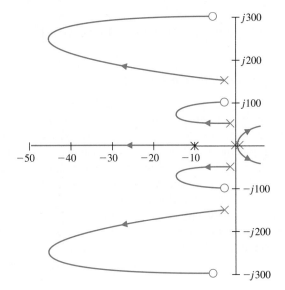

FIGURE 7.35
Root locus of the
system if
$K_2 = 0$, K_1 is varied
from $K_1 = 0$ to
$K_1 = \infty$, and
$G_c(s) = K_1$.

where K_1 is the gain of $G_c(s)$. We now have available two parameters, K_1 and K_2, that we may adjust. We select $5 < K_2 < 10$ in order to place the adjustable zero near the origin.

When $K_2 = 5$ and K_1 is varied, we obtain the root locus sketched in Figure 7.36. When $K_1 = 0.8$ and $K_2 = 5$, we obtain a step response with a percent overshoot of 12% and a settling time of 1.8 seconds. This is the optimum achievable response. If we try $K_2 = 7$ or $K_2 = 4$, the overshoot will be larger than desired. Therefore, we have achieved the best performance with this system. If we desired to continue the design process, we would use a controller $G_c(s)$ with a pole and zero in addition to retaining the velocity feedback with $K_2 = 5$.

One possible selection of a controller is

$$G_c(s) = \frac{K_1(s + z)}{s + p}.$$

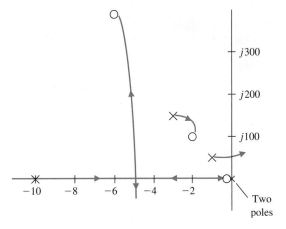

FIGURE 7.36
Root locus for the
robot controller with
a zero inserted at
$s = -0.2$ with
$G_c(s) = K_1$.

If we select $z = 1$ and $p = 5$, then, when $K_1 = 5$, we obtain a step response with an overshoot of 8% and a settling time of 1.6 seconds. ■

EXAMPLE 7.11   **Automobile velocity control**

The automotive electronics market is expected to reach \$52 billion by 2010. It is predicted that there will be growth of about 5% up to the year 2010 in electronic braking, steering, and driver information in North America alone [32]. Much of the additional computing power will be used for new technology for smart cars and smart roads, such as IVHS (intelligent vehicle/highway systems) [14, 33]. New systems on-board the automobile will support semi-autonomous automobiles, safety enhancements, emission reduction, and other features including intelligent cruise control, and brake by wire systems eliminating the hydraulics [34].

The term IVHS refers to a varied assortment of electronics that provides real-time information on accidents, congestion, and roadside services to drivers and traffic controllers. IVHS also encompasses devices that make vehicles more autonomous: collision-avoidance systems and lane-tracking technology that alert drivers to impending disasters and allow a car to drive itself.

An example of an automated highway system is shown in Figure 7.37. A velocity control system for maintaining the velocity between vehicles is shown in Figure 7.38. The output $Y(s)$ is the relative velocity of the two automobiles; the input $R(s)$ is the desired relative velocity between the two vehicles. Our design goal is to develop a controller that can maintain the prescribed velocity between the vehicles and maneuver the active vehicle (in this case the rearward automobile) as commanded. The elements of the design process emphasized in this example are depicted in Figure 7.39.

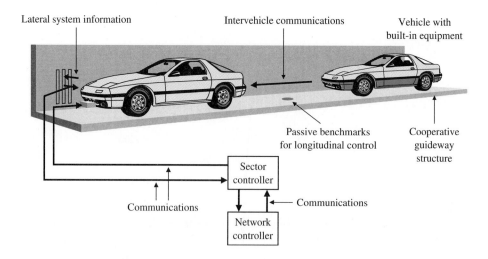

FIGURE 7.37
Automated
highway system.

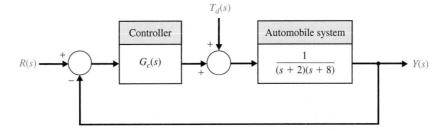

FIGURE 7.38
Vehicle velocity
control system.

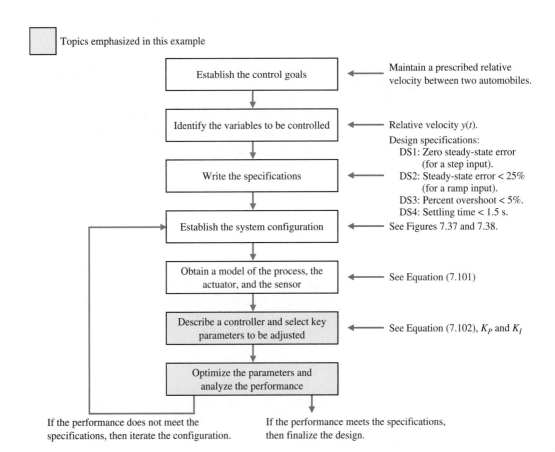

FIGURE 7.39   Elements of the control system design process emphasized in the automobile velocity control example.

The control goal is

Control Goal
   Maintain the prescribed velocity between the two vehicles, and maneuver the active vehicle as commanded.

The variable to be controlled is the relative velocity between the two vehicles:

Variable to Be Controlled
  The relative velocity between vehicles, denoted by $y(t)$.

The design specifications are

Design Specifications
  DS1   Zero steady-state error to a step input.
  DS2   Steady-state error due to a ramp input of less than 25% of the input magnitude.
  DS3   Percent overshoot less than 5% to a step input.
  DS4   Settling time less than 1.5 seconds to a step input (using a 2% criterion to establish settling time).

From the design specifications and knowledge of the open-loop system, we find that we need a type 1 system to guarantee a zero steady-state error to a step input. The open-loop system transfer function is a type 0 system; therefore, the controller needs to increase the system type by at least 1. A type 1 controller (that is, a controller with one integrator) satisfies DS1. To meet DS2 we need to have the velocity error constant (see Equation (5.29))

$$K_v = \lim_{s \to 0} sG_c(s)G(s) \geq \frac{1}{0.25} = 4, \qquad (7.100)$$

where

$$G(s) = \frac{1}{(s + 2)(s + 8)}, \qquad (7.101)$$

and $G_c(s)$ is the controller (yet to be specified).
  The percent overshoot specification DS3 allows us to define a target damping ratio (see Figure 5.8):

$$P.O. \leq 5\% \quad \text{implies} \quad \zeta \geq 0.69.$$

Similarly from the settling time specification DS4 we have (see Equation (5.13))

$$T_s \approx \frac{4}{\zeta\omega_n} \leq 1.5.$$

Solving for $\zeta\omega_n$ yields $\zeta\omega_n \geq 2.6$.
  The desired region for the poles of the closed-loop transfer function is shown in Figure 7.40. Using a proportional controller $G_c(s) = K_P$, is not reasonable, because DS2 cannot be satisfied. We need at least one pole at the origin to track a ramp input. Consider the PI controller

$$G_c(s) = \frac{K_P s + K_I}{s} = K_P \frac{s + \dfrac{K_I}{K_P}}{s}. \qquad (7.102)$$

The question is where to place the zero at $s = -K_I/K_P$.

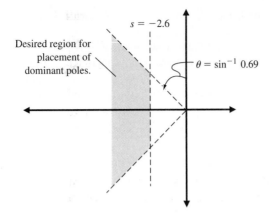

FIGURE 7.40
Desired region in the complex plane for locating the dominant system poles.

We ask for what values of K_P and K_I is the system stable. The closed-loop transfer function is

$$T(s) = \frac{K_P s + K_I}{s^3 + 10s^2 + (16 + K_P)s + K_I}.$$

The corresponding Routh array is

$$
\begin{array}{c|cc}
s^3 & 1 & 16 + K_P \\
s^2 & 10 & K_I \\
s & \dfrac{10(K_P + 16) - K_I}{10} & 0 \\
1 & K_I
\end{array}
$$

The first requirement for stability (from column one, row four) is

$$K_I > 0. \tag{7.103}$$

From the first column, third row, we have the inequality

$$K_P > \frac{K_I}{10} - 16. \tag{7.104}$$

It follows from DS2 that

$$K_v = \lim_{s \to 0} sG_c(s)G(s) = \lim_{s \to 0} s \frac{K_P\left(s + \dfrac{K_I}{K_P}\right)}{s} \frac{1}{(s + 2)(s + 8)} = \frac{K_I}{16} > 4.$$

Therefore, the integral gain must satisfy

$$K_I > 64. \tag{7.105}$$

If we select $K_I > 64$, then the inequality in Equation (7.103) is satisfied. The valid region for K_P is then given by Equation (7.104), where $K_I > 64$.

We need to consider DS4. Here we want to have the dominant poles to the left of the $s = -2.6$ line. We know from our experience sketching the root locus that since we have three poles (at $s = 0, -2,$ and -8) and one zero (at $s = -K_I/K_P$), we expect two branches of the loci to go to infinity along two asymptotes at $\phi = -90°$ and $+90°$ centered at

$$\sigma_A = \frac{\sum(-p_i) - \sum(-z_i)}{n_p - n_z},$$

where $n_p = 3$ and $n_z = 1$. In our case

$$\sigma_A = \frac{-2 - 8 - \left(-\dfrac{K_I}{K_P}\right)}{2} = -5 + \frac{1}{2}\frac{K_I}{K_P}.$$

We want to have $\alpha < -2.6$ so that the two branches will bend into the desired regions. Therefore,

$$-5 + \frac{1}{2}\frac{K_I}{K_P} < -2.6,$$

or

$$\frac{K_I}{K_P} < 4.7. \tag{7.106}$$

So as a first design, we can select K_P and K_I such that

$$K_I > 64, \quad K_P > \frac{K_I}{10} - 16, \quad \text{and} \quad \frac{K_I}{K_P} < 4.7.$$

Suppose we choose $K_I/K_P = 2.5$. Then the closed-loop characteristic equation is

$$1 + K_P\frac{s + 2.5}{s(s + 2)(s + 8)} = 0.$$

The root locus is shown in Figure 7.41. To meet the $\zeta = 0.69$ (which evolved from DS3), we need to select $K_P < 30$. We selected the value at the boundary of the performance region (see Figure 7.41) as carefully as possible.

Selecting $K_P = 26$, we have $K_I/K_P = 2.5$ which implies $K_I = 65$. This satisfies the steady-state tracking error specification (DS2) since $K_I = 65 > 64$.

The resulting PI controller is

$$G_c(s) = 26 + \frac{65}{s}. \tag{7.107}$$

The step response is shown in Figure 7.42.

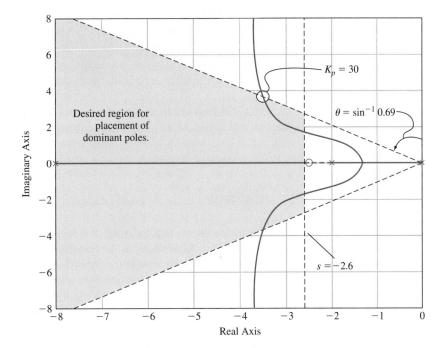

FIGURE 7.41
Root locus for
$K_I/K_P = 2.5$.

In the figure: "Desired region for placement of dominant poles.", $K_p = 30$, $\theta = \sin^{-1} 0.69$, $s = -2.6$

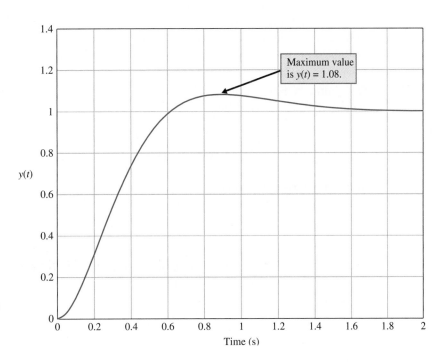

FIGURE 7.42
Automobile velocity
control using the PI
controller in
Eq. (7.107).

In the figure: "Maximum value is $y(t) = 1.08$."

The percent overshoot is $P.O. = 8\%$, and the settling time is $T_s = 1.45$ s. The percent overshoot specification is not precisely satisfied, but the controller in Equation (7.107) represents a very good first design. We can iteratively refine it. Even though the closed-loop poles lie in the desired region, the response does not exactly meet the specifications because the controller zero influences the response. The closed-loop system is a third-order system and does not have the performance of a second-order system. We might consider moving the zero to $s = -2$ (by choosing $K_I/K_P = 2$) so that the pole at $s = -2$ is cancelled and the resulting system is a second-order system. ■

7.8 THE ROOT LOCUS USING CONTROL DESIGN SOFTWARE

An approximate root locus sketch can be obtained by applying the orderly procedure summarized in Table 7.2. Alternatively, we can use control design software to obtain an accurate root locus plot. However, we should not be tempted to rely solely on the computer for obtaining root locus plots while neglecting the manual steps in developing an approximate root locus. The fundamental concepts behind the root locus method are embedded in the manual steps, and it is essential to understand their application fully.

The section begins with a discussion on obtaining a computer-generated root locus plot. This is followed by a discussion of the connections between the partial fraction expansion, dominant poles, and the closed-loop system response. Root sensitivity is covered in the final paragraphs.

The functions covered in this section are rlocus, rlocfind, and residue. The functions rlocus and rlocfind are used to obtain root locus plots, and the residue function is utilized for partial fraction expansions of rational functions.

Obtaining a Root Locus Plot.   Consider the closed-loop control system in Figure 7.10. The closed-loop transfer function is

$$T(s) = \frac{Y(s)}{R(s)} = \frac{K(s + 1)(s + 3)}{s(s + 2)(s + 3) + K(s + 1)}.$$

The characteristic equation can be written as

$$1 + K\frac{s + 1}{s(s + 2)(s + 3)} = 0. \tag{7.108}$$

The form of the characteristic equation in Equation (7.108) is necessary to use the rlocus function for generating root locus plots. The general form of the characteristic equation necessary for application of the rlocus function is

$$1 + KG(s) = 1 + K\frac{p(s)}{q(s)} = 0, \tag{7.109}$$

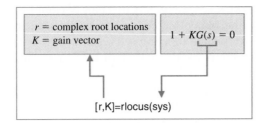

FIGURE 7.43
The **rlocus**
function.

where K is the parameter of interest to be varied from $0 < K < \infty$. The rlocus function is shown in Figure 7.43, where we define the transfer function object sys $= G(s)$. The steps to obtaining the root locus plot associated with Equation (7.108), along with the associated root locus plot, are shown in Figure 7.44. Invoking the rlocus function without left-hand arguments results in an automatic generation of the root locus plot. When invoked with left-hand arguments, the rlocus function returns a matrix of root locations and the associated gain vector.

The steps to obtain a computer-generated root locus plot are as follows:

1. Obtain the characteristic equation in the form given in Equation (7.109), where K is the parameter of interest.

2. Use the rlocus function to generate the plots.

Referring to Figure 7.44, we can see that as K increases, two branches of the root locus break away from the real axis. This means that, for some values of K, the closed-loop system characteristic equation will have two complex roots. Suppose we

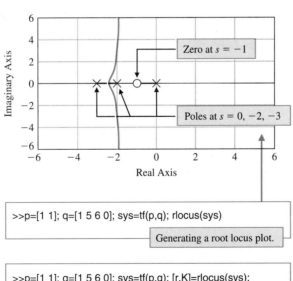

FIGURE 7.44
The root locus for
the characteristic
equation,
Equation (7.108).

want to find the value of K corresponding to a pair of complex roots. We can use the rlocfind function to do this, but only after a root locus has been obtained with the rlocus function. Executing the rlocfind function will result in a cross-hair marker appearing on the root locus plot. We move the cross-hair marker to the location on the locus of interest and hit the enter key. The value of the parameter K and the value of the selected point will then be displayed in the command display. The use of the rlocfind function is illustrated in Figure 7.45.

Control design software packages may respond differently when interacting with plots, such as with the rlocfind function on the root locus. The response of rlocfind in Figure 7.45 corresponds to MATLAB. Refer to the companion website for more information on other control design software applications.

Continuing our third-order root locus example, we find that when $K = 20.5775$, the closed-loop transfer function has three poles and two zeros, at

$$\text{poles: } s = \begin{pmatrix} -2.0505 + j4.3227 \\ -2.0505 - j4.3227 \\ -0.8989 \end{pmatrix}; \quad \text{zeros: } s = \begin{pmatrix} -1 \\ -3 \end{pmatrix}.$$

Considering the closed-loop pole locations only, we would expect that the real pole at $s = -.8989$ would be the dominant pole. To verify this, we can study the closed-loop system response to a step input, $R(s) = 1/s$. For a step input, we have

$$Y(s) = \frac{20.5775(s + 1)(s + 3)}{s(s + 2)(s + 3) + 20.5775(s + 1)} \cdot \frac{1}{s}. \quad (7.110)$$

Generally, the first step in computing $y(t)$ is to expand Equation (7.110) in a partial fraction expansion. The residue function can be used to expand Equation (7.110), as shown in Figure 7.46. The residue function is described in Figure 7.47.

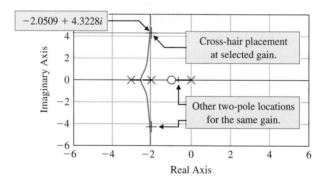

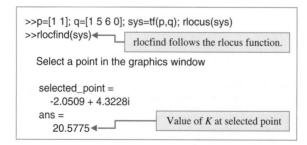

FIGURE 7.45
Using the **rlocfind** function.

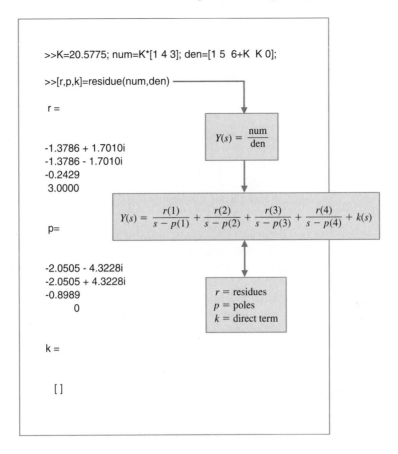

>>K=20.5775; num=K*[1 4 3]; den=[1 5 6+K K 0];

>>[r,p,k]=residue(num,den)

r =

-1.3786 + 1.7010i
-1.3786 - 1.7010i
-0.2429
3.0000

$$Y(s) = \frac{num}{den}$$

$$Y(s) = \frac{r(1)}{s - p(1)} + \frac{r(2)}{s - p(2)} + \frac{r(3)}{s - p(3)} + \frac{r(4)}{s - p(4)} + k(s)$$

p=

-2.0505 - 4.3228i
-2.0505 + 4.3228i
-0.8989
0

r = residues
p = poles
k = direct term

k =

[ ]

FIGURE 7.46
Partial fraction
expansion of
Equation (7.110).

The partial fraction expansion of Equation (7.110) is

$$Y(s) = \frac{-1.3786 + j1.7010}{s + 2.0505 + j4.3228} + \frac{-1.3786 - j1.7010}{s + 2.0505 - j4.3228} + \frac{-0.2429}{s + 0.8989} + \frac{3}{s}.$$

Comparing the residues, we see that the coefficient of the term corresponding to the pole at $s = -0.8989$ is considerably smaller than the coefficient of the terms corresponding to the complex-conjugate poles at $s = -2.0505 \pm j4.3227$. From this, we expect that the influence of the pole at $s = -0.8989$ on the output response $y(t)$ is not dominant. The settling time (to within 2% of the final value) is then predicted by considering the complex-conjugate poles. The poles at $s = -2.0505 \pm j4.3227$ correspond to a damping of $\zeta = 0.4286$ and a natural frequency of $\omega_n = 4.7844$. Thus, the settling time is predicted to be

$$T_s \simeq \frac{4}{\zeta \omega_n} = 1.95 \text{ s}.$$

Using the step function, as shown in Figure 7.48, we find that $T_s = 1.6$ s. Hence, our approximation of settling time $T_s \simeq 1.95$ is a fairly good approximation. The percent overshoot can be predicted using Figure 5.13 since the zero of $T(s)$ at $s = -3$ will impact the system response. Using Figure 5.13, we predict an overshoot of 60%. As can be seen in Figure 7.48, the actual overshoot is 50%.

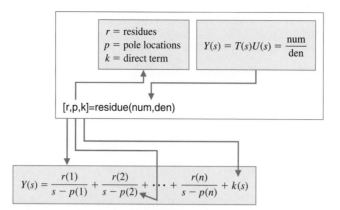

FIGURE 7.47
The **residue** function.

When using the **step** function, we can right-click on the figure to access the pull-down menu, which allows us to determine the step response settling time and peak response, as illustrated in Figure 7.48. On the pull-down menu select "Characteristics" and select "Settling Time." A dot will appear on the figure at the settling point. Place the cursor over the dot to determine the settling time.

In this example, the role of the system zeros on the transient response is illustrated. The proximity of the zero at $s = -1$ to the pole at $s = -0.8989$ reduces the impact of that pole on the transient response. The main contributors to the transient response are the complex-conjugate poles at $s = -2.0505 \pm j4.3228$ and the zero at $s = -3$.

There is one final point regarding the **residue** function: We can convert the partial fraction expansion back to the polynomials num/den, given the residues r, the pole locations p, and the direct terms k, with the command shown in Figure 7.49.

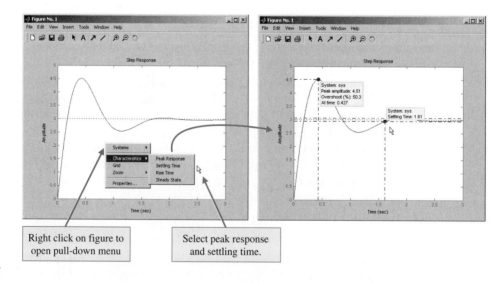

FIGURE 7.48
Step response for the closed-loop system in Figure 7.10 with $K = 20.5775$.

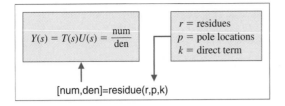

FIGURE 7.49
Converting a partial fraction expansion back to a rational function.

Sensitivity and the Root Locus. The roots of the characteristic equation play an important role in defining the closed-loop system transient response. The effect of parameter variations on the roots of the characteristic equation is a useful measure of sensitivity. The root sensitivity is defined to be

$$\frac{\partial r_i}{\partial K/K}. \tag{7.111}$$

We can use Equation (7.111) to investigate the sensitivity of the roots of the characteristic equation to variations in the parameter K. If we change K by a small finite amount ΔK, and evaluate the modified root $r_i + \Delta r_i$, it follows that

$$S_K^{r_i} \approx \frac{\Delta r_i}{\Delta K/K}. \tag{7.112}$$

The quantity $S_K^{r_i}$ is a complex number. Referring back to the third-order example of Figure 7.10 (Equation 7.108), if we change K by a factor of 5%, we find that the dominant complex-conjugate pole at $s = -2.0505 + j4.3228$ changes by

$$\Delta r_i = -0.0025 - j0.1168$$

when K changes from $K = 20.5775$ to $K = 21.6064$. From Equation (7.112), it follows that

$$S_K^{r_i} = \frac{-0.0025 - j0.1168}{1.0289/20.5775} = -0.0494 - j2.3355.$$

The sensitivity $S_K^{r_i}$ can also be written in the form

$$S_K^{r_i} = 2.34 \underline{/268.79°}.$$

The magnitude and direction of $S_K^{r_i}$ provides a measure of the root sensitivity. The script used to perform these sensitivity calculations is shown in Figure 7.50.

The root sensitivity measure may be useful for comparing the sensitivity for various system parameters at different root locations.

7.9 SEQUENTIAL DESIGN EXAMPLE: DISK DRIVE READ SYSTEM

In Chapter 6, we introduced a new configuration for the control system using velocity feedback (see Section 6.7). In this chapter, we will use the PID controller to obtain a desirable response. We will proceed with our model and then select a controller. Finally, we will optimize the parameters and analyze the performance. In this chapter, we will use the root locus method in the selection of the controller parameters.

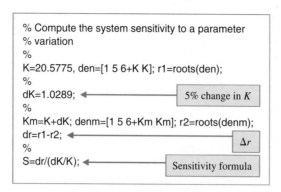

FIGURE 7.50
Sensitivity
calculations for the
root locus for a 5%
change in
$K = 20.5775$.

We use the root locus to select the controller gains. The PID controller introduced in this chapter is

$$G_c(s) = K_P + \frac{K_I}{s} + K_D s.$$

Since the process model $G_1(s)$ already possesses an integration, we set $K_I = 0$. Then we have the PD controller

$$G_c(s) = K_P + K_D s,$$

and our goal is to select K_P and K_D in order to meet the specifications. The system is shown in Figure 7.51. The closed-loop transfer function of the system is

$$\frac{Y(s)}{R(s)} = T(s) = \frac{G_c(s)G_1(s)G_2(s)}{1 + G_c(s)G_1(s)G_2(s)H(s)},$$

where $H(s) = 1$.

In order to obtain the root locus as a function of a parameter, we write $G_c(s)G_1(s)G_2(s)H(s)$ as

$$G_cG_1G_2H(s) = \frac{5000(K_P + K_D s)}{s(s + 20)(s + 1000)} = \frac{5000K_D(s + z)}{s(s + 20)(s + 1000)},$$

where $z = K_P/K_D$. We use K_P to select the location of the zero z and then sketch the locus as a function of K_D. Based on the insight developed in Section 6.7, we select $z = 1$ so that

$$G_cG_1G_2H(s) = \frac{5000K_D(s + 1)}{s(s + 20)(s + 1000)}.$$

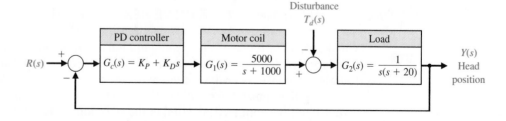

FIGURE 7.51
Disk drive control
system with a PD
controller.

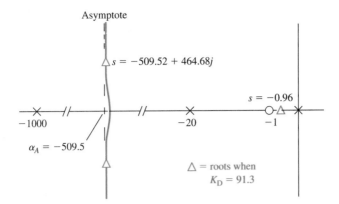

FIGURE 7.52
Sketch of the root
locus.

Table 7.6 Disk Drive Control System Specifications and Actual Design Performance

Performance Measure	Desired Value	Actual Response
Percent overshoot	Less than 5%	0%
Settling time	Less than 250 ms	20 ms
Maximum response to a unit disturbance	Less than 5×10^{-3}	2×10^{-3}

The number of poles minus the number of zeros is 2, and we expect asymptotes at $\phi_A = \pm 90°$ with a centroid

$$\sigma_A = \frac{-1020 + 1}{2} = -509.5,$$

as shown in Figure 7.52. We can quickly sketch the root locus, as shown in Figure 7.52. We use the computer-generated root locus to determine the root values for various values of K_D. When $K_D = 91.3$, we obtain the roots shown in Figure 7.52. Then, obtaining the system response, we achieve the actual response measures as listed in Table 7.6. As designed, the system meets all the specifications. It takes the system a settling time of 20 ms to "practically" reach the final value. In reality, the system drifts very slowly toward the final value after quickly achieving 97% of the final value.

7.10 SUMMARY

The relative stability and the transient response performance of a closed-loop control system are directly related to the location of the closed-loop roots of the characteristic equation. Therefore, we have investigated the movement of the characteristic roots on the s-plane as the system parameters are varied by using the root locus method. The root locus method, a graphical technique, can be used to obtain an approximate sketch in order to analyze the initial design of a system and determine suitable alterations of the system structure and the parameter values. A computer is commonly used to calculate several accurate roots at important points on the locus. A summary of fifteen typical root locus diagrams is shown in Table 7.7.

Table 7.7   Root Locus Plots for Typical Transfer Functions

G(s)	Root Locus	G(s)	Root Locus
1. $\dfrac{K}{s\tau_1 + 1}$		4. $\dfrac{K}{s}$	
2. $\dfrac{K}{(s\tau_1 + 1)(s\tau_2 + 1)}$		5. $\dfrac{K}{s(s\tau_1 + 1)}$	
3. $\dfrac{K}{(s\tau_1 + 1)(s\tau_2 + 1)(s\tau_3 + 1)}$		6. $\dfrac{K}{s(s\tau_1 + 1)(s\tau_2 + 1)}$	

Table 7.7   (continued)

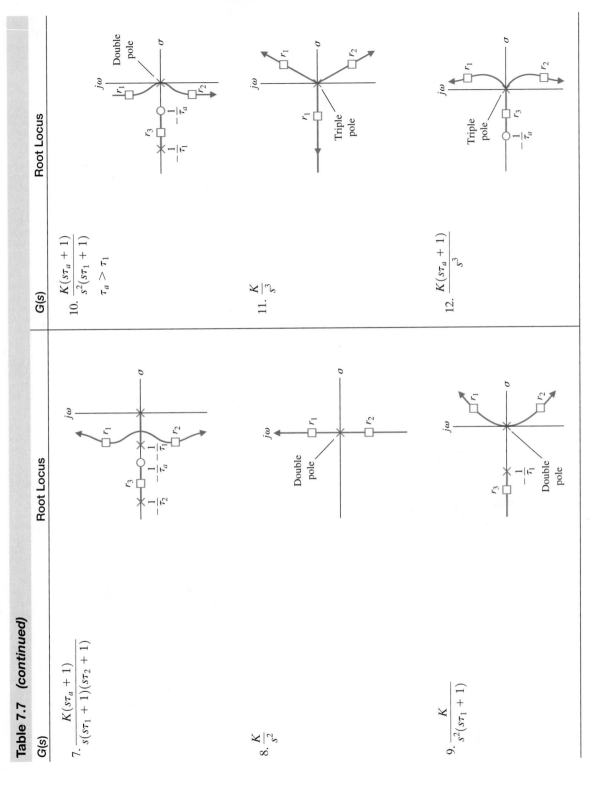

G(s)	Root Locus
7. $\dfrac{K(s\tau_a + 1)}{s(s\tau_1 + 1)(s\tau_2 + 1)}$	
8. $\dfrac{K}{s^2}$	
9. $\dfrac{K}{s^2(s\tau_1 + 1)}$	
10. $\dfrac{K(s\tau_a + 1)}{s^2(s\tau_1 + 1)}$ $\tau_a > \tau_1$	
11. $\dfrac{K}{s^3}$	
12. $\dfrac{K(s\tau_a + 1)}{s^3}$	

Table 7.7 (continued)

G(s)	Root Locus	G(s)	Root Locus

13. $\dfrac{K(s\tau_a + 1)(s\tau_b + 1)}{s^3}$

14. $\dfrac{K(s\tau_a + 1)(s\tau_b + 1)}{s(s\tau_1 + 1)(s\tau_2 + 1)(s\tau_3 + 1)(s\tau_4 + 1)}$

15. $\dfrac{K(s\tau_a + 1)}{s^2(s\tau_1 + 1)(s\tau_2 + 1)}$

Furthermore, we extended the root locus method for the design of several para-meters for a closed-loop control system. Then the sensitivity of the characteristic roots was investigated for undesired parameter variations by defining a root sensitivity measure. It is clear that the root locus method is a powerful and useful approach for the analysis and design of modern control systems and will continue to be one of the most important procedures of control engineering.

EXERCISES

E7.1 Let us consider a device that consists of a ball rolling on the inside rim of a hoop [11]. This model is similar to the problem of liquid fuel sloshing in a rocket. The hoop is free to rotate about its horizontal principal axis as shown in Figure E7.1. The angular position of the hoop may be controlled via the torque T applied to the hoop from a torque motor attached to the hoop drive shaft. If negative feedback is used, the system characteristic equation is

$$1 + \frac{Ks(s + 4)}{s^2 + 2s + 2} = 0.$$

(a) Sketch the root locus. (b) Find the gain when the roots are both equal. (c) Find these two equal roots. (d) Find the settling time of the system when the roots are equal.

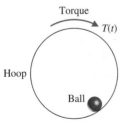

FIGURE E7.1 Hoop rotated by motor.

E7.2 A tape recorder has a speed control system so that $H(s) = 1$ with negative feedback and

$$G_c(s)G(s) = \frac{K}{s(s + 2)(s^2 + 4s + 5)}.$$

(a) Sketch a root locus for K, and show that the dominant roots are $s = -0.35 \pm j0.80$ when $K = 6.5$.
(b) For the dominant roots of part (a), calculate the settling time and overshoot for a step input.

E7.3 A control system for an automobile suspension tester has negative unity feedback and a process [12]

$$G_c(s)G(s) = \frac{K(s^2 + 4s + 8)}{s^2(s + 4)}.$$

We desire the dominant roots to have a ζ equal to 0.5. Using the root locus, show that $K = 7.35$ is required and the dominant roots are $s = -1.3 \pm j2.2$.

E7.4 Consider a unity feedback system with

$$G_c(s)G(s) = \frac{K(s + 1)}{s^2 + 4s + 5}.$$

(a) Find the angle of departure of the root locus from the complex poles. (b) Find the entry point for the root locus as it enters the real axis.

Answers: $\pm 225°$; -2.4

E7.5 Consider a unity feedback system with a loop transfer function

$$G_c(s)G(s) = \frac{s^2 + 2s + 10}{s^4 + 38s^3 + 515s^2 + 2950s + 6000}.$$

(a) Find the breakaway points on the real axis. (b) Find the asymptote centroid. (c) Find the values of K at the breakaway points.

E7.6 One version of a space station is shown in Figure E7.6 [30]. It is critical to keep this station in the proper orientation toward the sun and the Earth for generating power and communications. The orientation controller may be represented by a unity feedback system with an actuator and controller, such as

$$G_c(s)G(s) = \frac{20K}{s(s^2 + 20s + 100)}.$$

Sketch the root locus of the system as K increases. Find the value of K that results in an unstable system.

Answers: $K = 100$

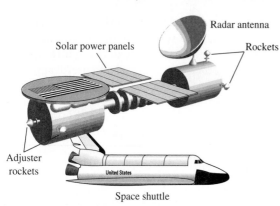

Radar antenna

Solar power panels

Rockets

Adjuster rockets

United States

Space shuttle

FIGURE E7.6   Space station.

E7.7   The elevator in a modern office building travels at a top speed of 25 feet per second and is still able to stop within one-eighth of an inch of the floor outside. The loop transfer function of the unity feedback elevator position control is

$$G_c(s)G(s) = \frac{K(s + 10)}{s(s + 5)(s + 6)(s + 8)}.$$

Determine the gain K when the complex roots have a ζ equal to 0.8.

E7.8   Sketch the root locus for a unity feedback system with

$$G_c(s)G(s) = \frac{K(s + 1)}{s^2(s + 9)}.$$

(a) Find the gain when all three roots are real and equal. (b) Find the roots when all the roots are equal as in part (a).

Answers: $K = 27; s = -3$

E7.9   The world's largest telescope is located in Hawaii. The primary mirror has a diameter of 10 m and consists of a mosaic of 36 hexagonal segments with the orientation of each segment actively controlled. This unity feedback system for the mirror segments has the loop transfer function

$$G_c(s)G(s) = \frac{K}{s(s^2 + 2s + 5)}.$$

(a) Find the asymptotes and draw them in the s-plane.
(b) Find the angle of departure from the complex poles.
(c) Determine the gain when two roots lie on the imaginary axis. (d) Sketch the root locus.

E7.10   A unity feedback system has the loop transfer function

$$KG(s) = \frac{K(s + 2)}{s(s + 1)}.$$

(a) Find the breakaway and entry points on the real axis.
(b) Find the gain and the roots when the real part of the complex roots is located at -2. (c) Sketch the locus.

Answers: (a) $-0.59, -3.41$; (b) $K = 3, s = -2 \pm j\sqrt{2}$

E7.11   A robot force control system with unity feedback has a loop transfer function [6]

$$KG(s) = \frac{K(s + 2.5)}{(s^2 + 2s + 2)(s^2 + 4s + 5)}.$$

(a) Find the gain K that results in dominant roots with a damping ratio of 0.707. Sketch the root locus.
(b) Find the actual percent overshoot and peak time for the gain K of part (a).

E7.12   A unity feedback system has a loop transfer function

$$KG(s) = \frac{K(s + 1)}{s(s^2 + 6s + 18)}.$$

(a) Sketch the root locus for $K > 0$. (b) Find the roots when $K = 10$ and 20. (c) Compute the 0–100% rise time, percent overshoot, and settling time (with a 2% criterion) of the system for a unit step input when $K = 10$ and 20.

E7.13   A unity feedback system has a loop transfer function

$$G_c(s)G(s) = \frac{4(s + z)}{s(s + 1)(s + 3)}.$$

(a) Draw the root locus as z varies from 0 to 100.
(b) Using the root locus, estimate the percent overshoot and settling time (with a 2% criterion) of the system at $z = 0.6, 2,$ and 4 for a step input. (c) Determine the actual overshoot and settling time at $z = 0.6, 2,$ and 4.

E7.14   A unity feedback system has the loop transfer function

$$G_c(s)G(s) = \frac{K(s + 10)}{s(s + 5)}.$$

(a) Determine the breakaway and entry points of the root locus and sketch the root locus for $K > 0$.
(b) Determine the gain K when the two characteristic roots have a ζ of $1/\sqrt{2}$. (c) Calculate the roots.

E7.15   (a) Plot the root locus for a unity feedback system with loop transfer function

$$G_c(s)G(s) = \frac{K(s + 10)(s + 2)}{s^3}.$$

(b) Calculate the range of K for which the system is stable. (c) Predict the steady-state error of the system for a ramp input.

Answers: (a) $K > 1.67$; (b) $e_{ss} = 0$

E7.16 A negative unity feedback system has a loop transfer function

$$G_c(s)G(s) = \frac{Ke^{-sT}}{s + 1},$$

where $T = 0.1$ s. Show that an approximation for the time delay is

$$e^{-sT} \approx \frac{\frac{2}{T} - s}{\frac{2}{T} + s}.$$

Using

$$e^{-0.1s} = \frac{20 - s}{20 + s},$$

obtain the root locus for the system for $K > 0$. Determine the range of K for which the system is stable.

E7.17 A control system, as shown in Figure E7.17, has a process

$$G(s) = \frac{1}{s(s - 1)}.$$

(a) When $G_c(s) = K$, show that the system is always unstable by sketching the root locus. (b) When

$$G_c(s) = \frac{K(s + 2)}{s + 20},$$

sketch the root locus and determine the range of K for which the system is stable. Determine the value of K and the complex roots when two roots lie on the $j\omega$-axis.

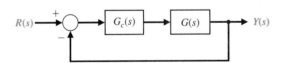

FIGURE E7.17 Feedback system.

E7.18 A closed-loop negative unity feedback system is used to control the yaw of the A-6 Intruder attack jet. When the loop transfer function is

$$G_c(s)G(s) = \frac{K}{s(s + 3)(s^2 + 2s + 2)},$$

determine (a) the root locus breakaway point and (b) the value of the roots on the $j\omega$-axis and the gain required for those roots. Sketch the root locus.

Answers: (a) Breakaway: $s = -2.29$ (b) $j\omega$-axis: $s = \pm j1.09$, $K = 8$

E7.19 A unity feedback system has a loop transfer function

$$G_c(s)G(s) = \frac{K}{s(s + 3)(s^2 + 6s + 64)}.$$

(a) Determine the angle of departure of the root locus at the complex poles. (b) Sketch the root locus. (c) Determine the gain K when the roots are on the $j\omega$-axis and determine the location of these roots.

E7.20 A unity feedback system has a loop transfer function

$$G_c(s)G(s) = \frac{K(s + 1)}{s(s - 1)(s + 4)}.$$

(a) Determine the range of K for stability. (b) Sketch the root locus. (c) Determine the maximum ζ of the stable complex roots.

Answers: (a) $K > 6$; (b) $\zeta = 0.2$

E7.21 A unity feedback system has a loop transfer function

$$G_c(s)G(s) = \frac{Ks}{s^3 + 5s^2 + 10}.$$

Sketch the root locus. Determine the gain K when the complex roots of the characteristic equation have a ζ approximately equal to 0.66.

E7.22 A high-performance missile for launching a satellite has a unity feedback system with a loop transfer function

$$G_c(s)G(s) = \frac{K(s^2 + 20)(s + 1)}{(s^2 - 2)(s + 10)}.$$

Sketch the root locus as K varies from $0 < K < \infty$.

E7.23 A unity feedback system has a loop transfer function

$$G_c(s)G(s) = \frac{4(s^2 + 1)}{s(s + a)}.$$

Sketch the root locus for $0 \le a < \infty$.

E7.24 Consider the system represented in state variable form

$$\dot{x} = Ax + Bu$$
$$y = Cx + Du,$$

where

$$A = \begin{bmatrix} 0 & 1 \\ -2 & -k \end{bmatrix}, B = \begin{bmatrix} 0 \\ 1 \end{bmatrix},$$
$$C = [1 \quad 0], \text{ and } D = [0].$$

Determine the characteristic equation and then sketch the root locus as $0 < k < \infty$.

E7.25 A closed-loop feedback system is shown in Figure E7.25. For what range of values of the parameters K is the system stable? Sketch the root locus as $0 < K < \infty$.

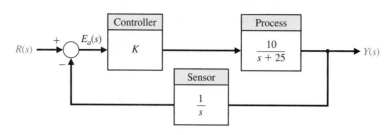

FIGURE 7.25
Nonunity feedback
system with
parameter K.

E7.26 Consider the signle-input, single-output system is
described by

$$\dot{x}(t) = Ax(t) + Bu(t)$$
$$y(t) = Cx(t)$$

where

$$A = \begin{bmatrix} 0 & 1 \\ 3 - K & -2 - K \end{bmatrix}, B = \begin{bmatrix} 0 \\ 1 \end{bmatrix}, C = [1 \quad -1].$$

Compute the characteristic polynomial and plot the
root locus as $0 \le K < \infty$. For what values of K is the
system stable?

E7.27 Consider the unity feedback system in Figure
E7.27. Sketch the root locus as $0 \le p < \infty$.

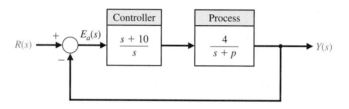

FIGURE 7.27
Unity feedback
system with
parameter p.

PROBLEMS

P7.1 Sketch the root locus for the following loop transfer
functions of the system shown in Figure P7.1 when
$0 < K < \infty$:

(a) $G_c(s)G(s) = \dfrac{K}{s(s + 10)(s + 8)}$

(b) $G_c(s)G(s) = \dfrac{K}{(s^2 + 2s + 2)(s + 1)}$

(c) $G_c(s)G(s) = \dfrac{K(s + 5)}{s(s + 2)(s + 7)}$

(d) $G_c(s)G(s) = \dfrac{K(s^2 + 4s + 8)}{s^2(s + 7)}$

P7.2 The linear model of a phase detector was presented in
Problem 6.7. Sketch the root locus as a function of the
gain $K_v = K_aK$. Determine the value of K_v attained
if the complex roots have a damping ratio equal to
0.60 [13].

P7.3 A unity feedback system has the loop transfer function

$$G_c(s)G(s) = \dfrac{K}{s(s + 2)(s + 5)}.$$

Find (a) the breakaway point on the real axis and the
gain K for this point, (b) the gain and the roots when
two roots lie on the imaginary axis, and (c) the roots
when $K = 6$. (d) Sketch the root locus.

P7.4 The analysis of a large antenna was presented in
Problem 4.5. Sketch the root locus of the system as
$0 < k_a < \infty$. Determine the maximum allowable
gain of the amplifier for a stable system.

P7.5 Automatic control of helicopters is necessary
because, unlike fixed-wing aircraft which possess a fair
degree of inherent stability, the helicopter is quite
unstable. A helicopter control system that utilizes an
automatic control loop plus a pilot stick control is
shown in Figure P7.5. When the pilot is not using the
control stick, the switch may be considered to be open.

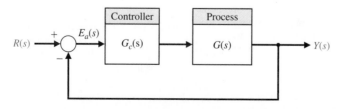

FIGURE P7.1

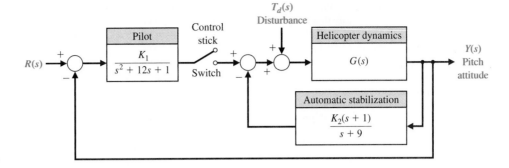

FIGURE P7.5
Helicopter control.

The dynamics of the helicopter are represented by the transfer function

$$G(s) = \frac{25(s + 0.03)}{(s + 0.4)(s^2 - 0.36s + 0.16)}.$$

(a) With the pilot control loop open (hands-off control), sketch the root locus for the automatic stabilization loop. Determine the gain K_2 that results in a damping for the complex roots equal to $\zeta = 0.707$. (b) For the gain K_2 obtained in part (a), determine the steady-state error due to a wind gust $T_d(s) = 1/s$. (c) With the pilot loop added, draw the root locus as K_1 varies from zero to ∞ when K_2 is set at the value calculated in part (a). (d) Recalculate the steady-state error of part (b) when K_1 is equal to a suitable value based on the root locus.

P7.6 An attitude control system for a satellite vehicle within the earth's atmosphere is shown in Figure P7.6. The transfer functions of the system are

$$G(s) = \frac{K(s + 0.20)}{(s + 0.90)(s - 0.60)(s - 0.10)}$$

and

$$G_c(s) = \frac{(s + 2 + j1.5)(s + 2 - j1.5)}{s + 4.0}.$$

(a) Draw the root locus of the system as K varies from 0 to ∞. (b) Determine the gain K that results in a system with a settling time (with a 2% criterion) less than 12 seconds and a damping ratio for the complex roots greater than 0.50.

P7.7 The speed control system for an isolated power system is shown in Figure P7.7. The valve controls the steam flow input to the turbine in order to account for load changes $\Delta L(s)$ within the power distribution network. The equilibrium speed desired results in a generator frequency equal to 60 cps. The effective rotary inertia J is equal to 4000 and the friction constant b is equal to 0.75. The steady-state speed regulation factor R is represented by the equation $R \approx (\omega_0 - \omega_r)/\Delta L$, where ω_r equals the speed at rated load and ω_0 equals the speed at no load. We want to obtain a very small R, usually less than 0.10. (a) Using root locus techniques, determine the regulation R attainable when

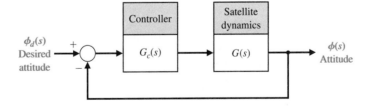

FIGURE P7.6
Satellitte attitude
control.

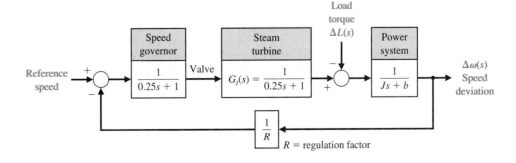

FIGURE P7.7
Power system
control.

the damping ratio of the roots of the system must be greater than 0.60. (b) Verify that the steady-state speed deviation for a load torque change $\Delta L(s) = \Delta L/s$ is, in fact, approximately equal to $R\Delta L$ when $R \leq 0.1$.

P7.8 Consider again the power control system of Problem P7.7 when the steam turbine is replaced by a hydroturbine. For hydroturbines, the large inertia of the water used as a source of energy causes a considerably larger time constant. The transfer function of a hydroturbine may be approximated by

$$G_t(s) = \frac{-\tau s + 1}{(\tau/2)s + 1},$$

where $\tau = 1$ second. With the rest of the system remaining as given in Problem P7.7, repeat parts (a) and (b) of Problem P7.7.

P7.9 The achievement of safe, efficient control of the spacing of automatically controlled guided vehicles is an important part of the future use of the vehicles in a manufacturing plant [14, 15]. It is important that the system eliminate the effects of disturbances (such as oil on the floor) as well as maintain accurate spacing between vehicles on a guideway. The system can be represented by the block diagram of Figure P7.9. The vehicle dynamics can be represented by

$$G(s) = \frac{(s + 0.1)(s^2 + 2s + 289)}{s(s - 0.4)(s + 0.8)(s^2 + 1.45s + 361)}.$$

(a) Sketch the root locus of the system. (b) Determine all the roots when the loop gain $K = K_1K_2$ is equal to 4000.

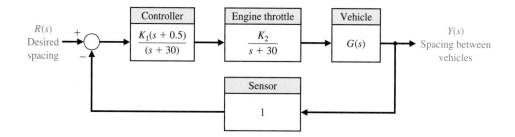

FIGURE P7.9
Guided vehicle control.

P7.10 New concepts in passenger airliner design will have the range to cross the Pacific in a single flight and the efficiency to make it economical [16, 31]. These new designs will require the use of temperature-resistant, lightweight materials and advanced control systems. Noise control is an important issue in modern aircraft designs since most airports have strict noise level requirements. One interesting concept is the Boeing Sonic Cruiser depicted in Figure P7.10(a). It would seat 200 to 250 passengers and cruise at just below the speed of sound.

The flight control system must provide good handling characteristics and comfortable flying conditions. An automatic control system can be designed for the next generation passenger aircraft.

The desired characteristics of the dominant roots of the control system shown in Figure P7.10(b) have a $\zeta = 0.707$. The characteristics of the aircraft are $\omega_n = 2.5$, $\zeta = 0.30$, and $\tau = 0.1$. The gain factor K_1, however, will vary over the range 0.02 at medium-weight cruise conditions to 0.20 at lightweight descent conditions. (a) Sketch the root locus as a function of the loop gain K_1K_2. (b) Determine the gain K_2 necessary to yield roots with $\zeta = 0.707$ when the aircraft is in the medium-cruise condition. (c) With the gain K_2 as found in part (b), determine the ζ of the roots when the gain K_1 results from the condition of light descent.

P7.11 A computer system requires a high-performance magnetic tape transport system [17]. The environmental conditions imposed on the system result in a severe test of control engineering design. A direct-drive DC motor system for the magnetic tape reel system is shown in Figure P7.11, where r equals the reel radius, and J equals the reel and rotor inertia. A complete reversal of the tape reel direction is required in 6 ms, and the tape reel must follow a step command in 3 ms or less. The tape is normally operating at a speed of 100 in/s. The motor and components selected for this system possess the following characteristics:

$K_b = 0.40$	$r = 0.2$
$K_p = 1$	$K_1 = 2.0$
$\tau_1 = \tau_a = 1$ ms	K_2 is adjustable.
$K_T/(LJ) = 2.0$	

The inertia of the reel and motor rotor is 2.5×10^{-3} when the reel is empty, and 5.0×10^{-3} when the reel is full. A series of photocells is used as an error-sensing device. The time constant of the motor is $L/R = 0.5$ ms. (a) Sketch the root locus for the system when $K_2 = 10$ and $J = 5.0 \times 10^{-3}$, $0 < K_a < \infty$. (b) Determine the gain K_a that results in a well-damped system so that the ζ of all the roots is greater than or equal to 0.60. (c) With the K_a determined from part (b), sketch a root locus for $0 < K_2 < \infty$.

(a)

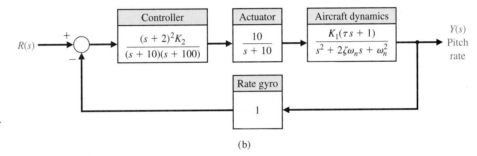

(b)

FIGURE P7.10
(a) A passenger jet aircraft of the future. (™ and © Boeing. Used under license.) (b) Control system.

P7.12 A precision speed control system (Figure P7.12) is required for a platform used in gyroscope and inertial system testing where a variety of closely controlled speeds is necessary. A direct-drive DC torque motor system was utilized to provide (1) a speed range of 0.01°/s to 600°/s, and (2) 0.1% steady-state error maximum for a step input. The direct-drive DC torque motor avoids the use of a gear train with its attendant backlash and friction. Also, the direct-drive motor has a high-torque capability, high efficiency, and low motor time constants. The motor gain constant is nominally $K_m = 1.8$, but is subject to variations up to 50%. The amplifier gain K_a is normally greater than 10 and subject to a variation of 10%. (a) Determine the minimum loop gain necessary to satisfy the steady-state error requirement. (b) Determine the limiting value of gain for stability. (c) Sketch the root locus as K_a varies from 0 to ∞. (d) Determine the roots when $K_a = 40$, and estimate the response to a step input.

P7.13 A unity feedback system has the loop transfer function

$$G_c(s)G(s) = \frac{K}{s(s + 3)(s^2 + 4s + 7.84)}.$$

(a) Find the breakaway point on the real axis and the gain for this point. (b) Find the gain to provide two complex roots nearest the $j\omega$-axis with a damping ratio of 0.707. (c) Are the two roots of part (b) dominant? (d) Determine the settling time (with a 2% criterion) of the system when the gain of part (b) is used.

P7.14 The loop transfer function of a single-loop negative feedback system is

$$G_c(s)G(s) = \frac{K(s + 2)(s + 3)}{s^2(s + 1)(s + 10)(s + 50)}.$$

This system is called **conditionally stable** because it is stable only for a range of the gain K such that $k_1 < K < k_2$. Using the Routh–Hurwitz criteria and the root locus method, determine the range of the gain for which the system is stable. Sketch the root locus for $0 < K < ∞$.

P7.15 Let us again consider the stability and ride of a rider and high performance motorcycle as outlined in Problem 6.13. The dynamics of the motorcycle and rider can be represented by the loop transfer function

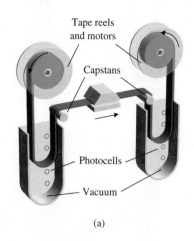

(a)

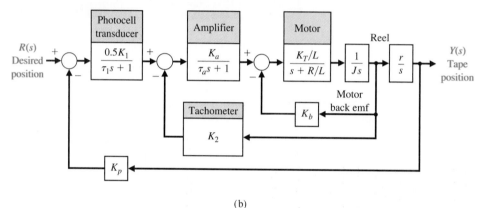

FIGURE P7.11
(a) Tape control
system. (b) Block
diagram.

(b)

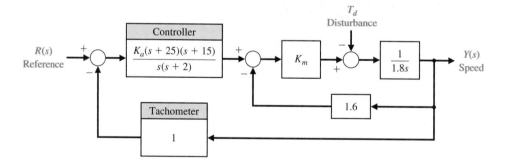

FIGURE P7.12
Speed control.

$$G_c(s)G(s) = \frac{K(s^2 + 30s + 625)}{s(s + 20)(s^2 + 20s + 200)(s^2 + 60s + 3400)}.$$

Sketch the root locus for the system. Determine the ζ of the dominant roots when $K = 3 \times 10^4$.

P7.16 Control systems for maintaining constant tension on strip steel in a hot strip finishing mill are called "loopers." A typical system is shown in Figure P7.16.

The looper is an arm 2 to 3 feet long with a roller on the end; it is raised and pressed against the strip by a motor [18]. The typical speed of the strip passing the looper is 2000 ft/min. A voltage proportional to the looper position is compared with a reference voltage and integrated where it is assumed that a change in looper position is proportional to a change in the steel strip tension. The time constant τ of the filter is negligible relative to the other time constants in the system. (a) Sketch the

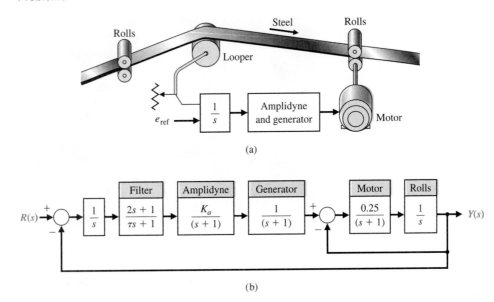

(a)

FIGURE P7.16
Steel mill control
system.

(b)

root locus of the control system for $0 < K_a < \infty$. (b) Determine the gain K_a that results in a system whose roots have a damping ratio of $\zeta = 0.707$ or greater. (c) Determine the effect of τ as τ increases from a negligible quantity.

P7.17 Consider again the vibration absorber discussed in Problems 2.2 and 2.10 as a design problem. Using the root locus method, determine the effect of the parameters M_2 and k_{12}. Determine the specific values of the parameters M_2 and k_{12} so that the mass M_1 does not vibrate when $F(t) = a\sin(\omega_0 t)$. Assume that $M_1 = 1, k_1 = 1,$ and $b = 1$. Also assume that $k_{12} < 1$ and that the term k_{12}^2 may be neglected.

P7.18 A feedback control system is shown in Figure P7.18. The filter $G_c(s)$ is often called a compensator, and the design problem involves selecting the parameters α and β. Using the root locus method, determine the effect of varying the parameters. Select a suitable filter so that the time to settle (to within 2% of the final value) is less than 4 seconds and the damping ratio of the dominant roots is greater than 0.60.

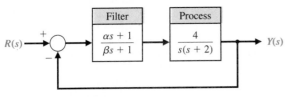

FIGURE P7.18 Filter design.

P7.19 In recent years, many automatic control systems for guided vehicles in factories have been installed. One system uses a guidance cable embedded in the floor to guide the vehicle along the desired lane [10, 15]. An error detector, composed of two coils mounted on the front of the cart, senses a magnetic field produced by the current in the guidance cable. An example of a guided vehicle in a factory is shown in Figure P7.19(a). We have

$$G(s) = \frac{s^2 + 3.6s + 81}{s(s+1)(s+5)}$$

and K_a is the amplifier gain. (a) Sketch a root locus and determine a suitable gain K_a so that the damping ratio of the complex roots is 0.707. (b) Determine the root sensitivity of the system for the complex root r_1 as a function of (1) K_a and (2) the pole of $G(s)$ at $s = -1$.

P7.20 Determine the root sensitivity for the dominant roots of the design for Problem 7.18 for the gain $K = 4\alpha/\beta$ and the pole $s = -2$.

P7.21 Determine the root sensitivity of the dominant roots of the power system of Problem P7.7. Evaluate the sensitivity for variations of (a) the poles at $s = -4$, and (b) the feedback gain, $1/R$.

P7.22 Determine the root sensitivity of the dominant roots of Problem P7.1(a) when K is set so that the damping ratio of the unperturbed roots is 0.707. Evaluate and compare the sensitivity as a function of the poles and zeros of $G_c(s)G(s)$.

(a)

FIGURE P7.19
(a) An automatically
guided vehicle.
(Photo courtesy of
Control Engineering
Corporation.)
(b) Block diagram.

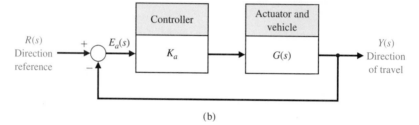

(b)

P7.23 Repeat Problem P7.22 for the loop transfer function $G_c(s)G(s)$ of Problem P7.1(c).

P7.24 For systems of relatively high degree, the form of the root locus can often assume an unexpected pattern. The root loci of four different feedback systems of third order or higher are shown in Figure P7.24. The open-loop poles and zeros of $KG(s)$ are shown, and the form of the root loci as K varies from zero to infinity is presented. Verify the diagrams of Figure P7.24 by constructing the root loci.

P7.25 Solid-state integrated electronic circuits are composed of distributed R and C elements. Therefore, feedback electronic circuits in integrated circuit form must be investigated by obtaining the transfer function of the distributed RC networks. It has been shown that the slope of the attenuation curve of a distributed RC network is $10n$ dB/decade, where n is the order of the RC filter [13]. This attenuation is in contrast with the normal $20n$ dB/decade for the lumped parameter circuits. (The concept of the slope of an attenuation curve is considered in Chapter 8. If it is unfamiliar, reexamine this problem after studying Chapter 8.) An interesting case arises when the distributed RC network occurs in a series-to-shunt feedback path of a transistor amplifier. Then the loop transfer function may be written as

$$G_c(s)G(s) = \frac{K(s - 1)(s + 3)^{1/2}}{(s + 1)(s + 2)^{1/2}}.$$

(a) Using the root locus method, determine the locus of roots as K varies from zero to infinity. (b) Calculate the gain at borderline stability and the frequency of oscillation for this gain.

P7.26 A single-loop negative feedback system has a loop transfer function

$$G_c(s)G(s) = \frac{K(s + 2)^2}{s(s^2 + 1)(s + 8)}.$$

(a) Sketch the root locus for $0 \leq K \leq \infty$ to indicate the significant features of the locus. (b) Determine the range of the gain K for which the system is stable. (c) For what value of K in the range $K \geq 0$ do purely imaginary roots exist? What are the values of these roots? (d) Would the use of the dominant roots approximation for an estimate of settling time be justified in this case for a large magnitude of gain ($K > 50$)?

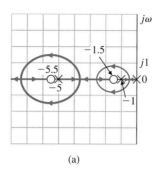

(a)

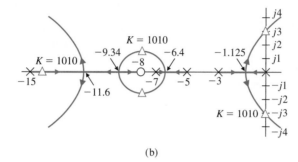

(b)

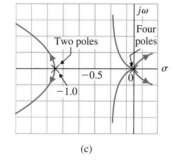

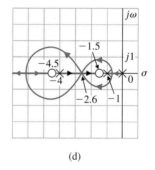

FIGURE P7.24
Root loci of four
systems.

(c)   (d)

P7.27 A unity negative feedback system has a loop transfer function

$$G_c(s)G(s) = \frac{K(s^2 + 0.105625)}{s(s^2 + 1)}$$

$$= \frac{K(s + j0.325)(s - j0.325)}{s(s^2 + 1)}.$$

Sketch the root locus as a function of K. Carefully calculate where the segments of the locus enter and leave the real axis.

P7.28 To meet current U.S. emissions standards for automobiles, hydrocarbon (HC) and carbon monoxide (CO) emissions are usually controlled by a catalytic converter in the automobile exhaust. Federal standards

for nitrogen oxides (NO_x) emissions are met mainly by exhaust-gas recirculation (EGR) techniques. However, as NO_x emissions standards were tightened from the current limit of 2.0 grams per mile to 1.0 gram per mile, these techniques alone were no longer sufficient.

Although many schemes are under investigation for meeting the emissions standards for all three emissions, one of the most promising employs a three-way catalyst—for HC, CO, and NO_x emissions—in conjunction with a closed-loop engine-control system. The approach is to use a closed-loop engine control, as shown in Figure P7.28 [19, 23]. The exhaust-gas sensor gives an indication of a rich or lean exhaust and compares it to a reference. The difference signal is processed by the controller, and the output of the controller modulates the vacuum level in the carburetor

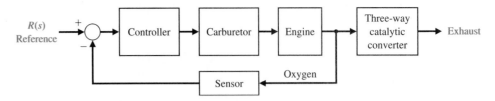

FIGURE P7.28
Auto engine control.

to achieve the best air–fuel ratio for proper operation of the catalytic converter. The loop transfer function is represented by

$$L(s) = \frac{Ks^2 + 12s + 20}{s^3 + 10s^2 + 25s}.$$

Calculate the root locus as a function of K. Carefully calculate where the segments of the locus enter and leave the real axis. Determine the roots when $K = 2$. Predict the step response of the system when $K = 2$.

P7.29   A unity feedback control system has a transfer function

$$G_c(s)G(s) = \frac{K(s^2 + 8s + 25)}{s^2(s + 4)}.$$

We desire the dominant roots to have a damping ratio equal to 0.707. Find the gain K when this condition is satisfied. Show that the complex roots are $s = -4 \pm j4$ at this gain.

P7.30   An *RLC* network is shown in Figure P7.30. The nominal values (normalized) of the network elements are $L - C = 1$ and $R = 2.5$. Show that the root sensitivity of the two roots of the input impedance $Z(s)$ to a change in R is different by a factor of 4.

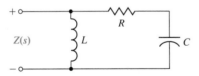

FIGURE P7.30   *RLC* network.

P7.31   The development of high-speed aircraft and missiles requires information about aerodynamic parameters prevailing at very high speeds. Wind tunnels are used to test these parameters. These wind tunnels are constructed by compressing air to very high pressures and releasing it through a valve to create a wind. Since the air pressure drops as the air escapes, it is necessary to open the valve wider to maintain a constant wind speed. Thus, a control system is needed to adjust the valve to maintain a constant wind speed. The loop transfer function for a unity feedback system is

$$G_c(s)G(s) = \frac{K(s + 4)}{s(s + 0.16)(s + p)(s - \bar{p})},$$

where $p = 7.3 + 9.7831j$. Sketch the root locus and show the location of the roots for $K = 326$ and $K = 1350$.

P7.32   A mobile robot suitable for nighttime guard duty is available. This guard never sleeps and can tirelessly patrol large warehouses and outdoor yards. The steering control system for the mobile robot has a unity feedback with the loop transfer function

$$G_c(s)G(s) = \frac{K(s + 1)(s + 5)}{s(s + 1.5)(s + 2)}.$$

(a) Find K for all breakaway and entry points on the real axis. (b) Find K when the damping ratio of the complex roots is 0.707. (c) Find the minimum value of the damping ratio for the complex roots and the associated gain K. (d) Find the overshoot and the time to settle (to within 2% of the final value) for a unit step input for the gain, K, determined in parts (b) and (c).

P7.33   The Bell-Boeing V-22 Osprey Tiltrotor is both an airplane and a helicopter. Its advantage is the ability to rotate its engines to 90° from a vertical position for takeoffs and landings as shown in Figure P7.33(a), and then to switch the engines to a horizontal position for cruising as an airplane [20]. The altitude control system in the helicopter mode is shown in Figure P7.33(b). (a) Determine the root locus as K varies and determine the range of K for a stable system. (b) For $K = 280$, find the actual $y(t)$ for a unit step input $r(t)$ and the percentage overshoot and settling time (with a 2% criterion). (c) When $K = 280$ and $r(t) = 0$, find $y(t)$ for a unit step disturbance, $T_d(s) = 1/s$. (d) Add a prefilter between $R(s)$ and the summing node so that

$$G_p(s) = \frac{0.5}{s^2 + 1.5s + 0.5},$$

and repeat part (b).

P7.34   The fuel control for an automobile uses a diesel pump that is subject to parameter variations. A unity negative feedback has a loop transfer function

$$G_c(s)G(s) = \frac{K(s + 2)}{(s + 1)(s + 2.5)(s + 4)(s + 10)}.$$

(a) Sketch the root locus as K varies from 0 to 2000. (b) Find the roots for K equal to 400, 500, and 600. (c) Predict how the percent overshoot to a step will vary for the gain K, assuming dominant roots. (d) Find

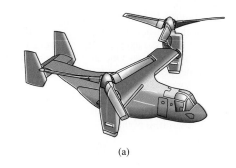

(a)

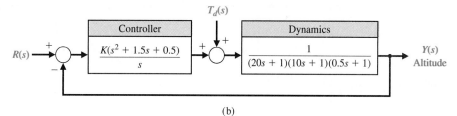

(b)

FIGURE P7.33
(a) Osprey Tiltrotor
aircraft. (b) Its
control system.

the actual time response for a step input for all three gains and compare the actual overshoot with the predicted overshoot.

P7.35 A powerful electrohydraulic forklift can be used to lift pallets weighing several tons on top of 35-foot scaffolds at a construction site. The negative unity feedback system has a loop transfer function

$$G_c(s)G(s) = \frac{K(s + 1)^2}{s(s^2 + 1)}.$$

(a) Sketch the root locus for $K > 0$. (b) Find the gain K when two complex roots have a ζ of 0.707, and calculate all three roots. (c) Find the entry point of the root locus at the real axis. (d) Estimate the expected overshoot to a step input, and compare it with the actual overshoot determined from a computer program.

P7.36 A microrobot with a high-performance manipulator has been designed for testing very small particles, such as simple living cells [6]. The single-loop unity negative feedback system has a loop transfer function

$$G_c(s)G(s) = \frac{K(s + 1)(s + 2)(s + 3)}{s^3(s - 1)}.$$

(a) Sketch the root locus for $K > 0$. (b) Find the gain and roots when the characteristic equation has two imaginary roots. (c) Determine the characteristic roots when $K = 20$ and $K = 100$. (d) For $K = 20$, estimate the percent overshoot to a step input, and compare the estimate to the actual overshoot determined from a computer program.

P7.37 Identify the parameters K, a, and b of the system shown in Figure P7.37. The system is subject to a unit step input, and the output response has an overshoot but ultimately attains the final value of 1. When the closed-loop system is subjected to a ramp input, the output response follows the ramp input with a finite steady-state error. When the gain is doubled to $2K$, the output response to an impulse input is a pure sinusoid with a period of 0.314 second. Determine K, a, and b.

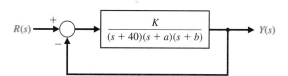

FIGURE P7.37 Feedback system.

P7.38 A unity feedback system has the loop transfer function

$$G_c(s)G(s) = \frac{K(s + 1)}{s(s - 3)}.$$

This system is open-loop unstable. (a) Determine the range of K so that the closed-loop system is stable. (b) Sketch the root locus. (c) Determine the roots for $K = 10$. (d) For $K = 10$, predict the percent overshoot for a step input using Figure 5.13. (e) Determine the actual overshoot by plotting the response.

P7.39 High-speed trains for U.S. railroad tracks must traverse twists and turns. In conventional trains, the axles are fixed in steel frames called trucks. The trucks pivot as the train goes into a curve, but the fixed axles stay parallel to each other, even though the front axle tends to go in a different direction from the rear axle [24]. If the train is going fast, it may jump the tracks. One solution uses axles that pivot independently. To counterbalance the strong centrifugal forces in a curve, the train also has a computerized hydraulic system that tilts each car as it rounds a turn. On-board sensors calculate the train's speed and the sharpness of the curve

and feed this information to hydraulic pumps under the floor of each car. The pumps tilt the car up to eight degrees, causing it to lean into the curve like a race car on a banked track.

The tilt control system is shown in Figure P7.39. Sketch the root locus, and determine the value of K when the complex roots have maximum damping. Predict the response of this system to a step input $R(s)$.

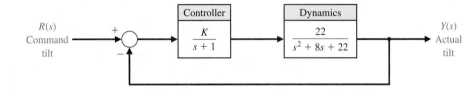

FIGURE P7.39
Tilt control for a
high-speed train.

ADVANCED PROBLEMS

AP7.1 The top view of a high-performance jet aircraft is shown in Figure AP7.1(a) [20]. Sketch the root locus and determine the gain K so that the ζ of the complex poles near the $j\omega$-axis is the maximum achievable. Evaluate the roots at this K and predict the response to a step input. Determine the actual response and compare it to the predicted response.

AP7.2 A magnetically levitated high-speed train "flies" on an air gap above its rail system [24]. The air gap control system has a unity feedback system with a

loop transfer function

$$G_c(s)G(s) = \frac{K(s+1)(s+3)}{s(s-1)(s+4)(s+8)}.$$

The goal is to select K so that the response for a unit step input is reasonably damped and the settling time is less than 3 seconds. Sketch the root locus, and select K so that all of the complex roots have a ζ greater than 0.6. Determine the actual response for the selected K and the percent overshoot.

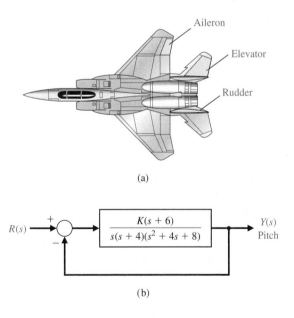

(a)

(b)

FIGURE AP7.1
(a) High-performance aircraft. (b) Pitch control system.

AP7.3 A compact disc player for portable use requires a good rejection of disturbances and an accurate position of the optical reader sensor. The position control system uses unity feedback and a loop transfer function

$$G_c(s)G(s) = \frac{10}{s(s + 1)(s + p)}.$$

The parameter p can be chosen by selecting the appropriate DC motor. Sketch the root locus as a function of p. Select p so that the ζ of the complex roots of the characteristic equation is approximately $1/\sqrt{2}$.

AP7.4 A remote manipulator control system has unity feedback and a loop transfer function

$$G_c(s)G(s) = \frac{(s + \alpha)}{s^3 + (1 + \alpha)s^2 + (\alpha - 1)s + 1 - \alpha}.$$

We want the steady-state position error for a step input to be less than or equal to 10% of the magnitude of the input. Sketch the root locus as a function of the parameter α. Determine the range of α required for the desired steady-state error. Locate the roots for the allowable value of α to achieve the required steady-state error, and estimate the step response of the system.

AP7.5 A unity feedback system has a loop transfer function

$$G_c(s)G(s) = \frac{K}{s^3 + 10s^2 + 7s - 18}.$$

(a) Sketch the root locus and determine K for a stable system with complex roots with ζ equal to $1/\sqrt{2}$.
(b) Determine the root sensitivity of the complex roots of part (a).
(c) Determine the percent change in K (increase or decrease) so that the roots lie on the $j\omega$-axis.

AP7.6 A unity feedback system has a loop transfer function

$$G_c(s)G(s) = \frac{K(s^2 + 2s + 5)}{s^3 + 2s^2 + 2s + 1}.$$

Sketch the root locus for $K > 0$, and select a value for K that will provide a closed step response with settling time less than 2 second.

AP7.7 A feedback system with positive feedback is shown in Figure AP7.7. The root locus for $K > 0$ must meet the condition

$$KG(s) = 1\underline{/\pm k360°}$$
$$\text{for } k = 0, 1, 2, \dots.$$

Sketch the root locus for $0 < K < \infty$.

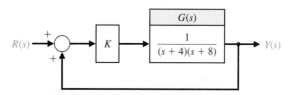

FIGURE AP7.7 A closed-loop system with positive feedback.

AP7.8 A position control system for a DC motor is shown in Figure AP7.8. Obtain the root locus for the velocity feedback constant K, and select K so that all the roots of the characteristic equation are real (two are equal and real). Estimate the step response of the system for the K selected. Compare the estimate with the actual response.

AP7.9 A control system is shown in Figure AP7.9. Sketch the root loci for the following transfer functions $G_c(s)$:
(a) $G_c(s) = K$
(b) $G_c(s) = K(s + 3)$

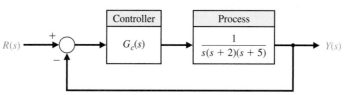

FIGURE AP7.8
A position control system with velocity feedback.

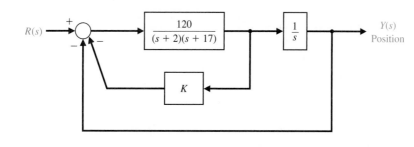

FIGURE AP7.9
A unity feedback control system.

(c) $G_c(s) = \dfrac{K(s + 1)}{s + 20}$

(d) $G_c(s) = \dfrac{K(s + 1)(s + 4)}{s + 10}$

AP7.10   A feedback system is shown in Figure AP7.10. Sketch the root locus as K varies when $K \geq 0$. Determine a value for K that will provide a step response with an overshoot less than 5% and a settling time (with a 2% criterion) less than 2.5 seconds.

AP7.11   A control system is shown in Figure AP7.11. Sketch the root locus, and select a gain K so that the step response of the system has an overshoot of less than 20% and the settling time (with a 2% criterion) is less than 5 seconds.

AP7.12   A control system with PI control is shown in Figure AP7.12. (a) Let $K_I/K_P = 0.2$ and determine K_P so that the complex roots have maximum damping ratio. (b) Predict the step response of the system with K_P set to the value determined in part (a).

AP7.13   The feedback system shown in Figure AP7.13 has two unknown parameters K_1 and K_2. The process transfer function is unstable. Sketch the root locus for $0 \leq K_1, K_2 < \infty$. What is the fastest settling time that you would expect of the closed-loop system in response to a unit step input $R(s) = 1/s$? Explain.

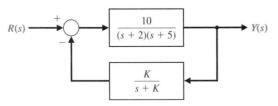

FIGURE AP7.10   A nonunity feedback control system.

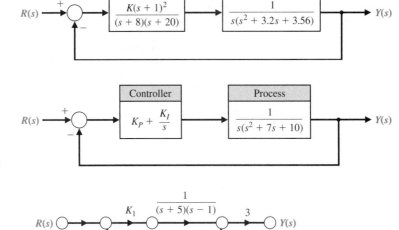

FIGURE AP7.11
A control system
with parameter K.

FIGURE AP7.12
A control system
with a PI controller.

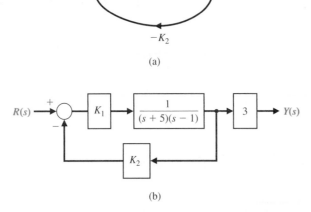

(a)

(b)

FIGURE AP7.13
An unstable plant
with two
parameters K_1 and
K_2.

DESIGN PROBLEMS

CDP7.1   The drive motor and slide system uses the output of a tachometer mounted on the shaft of the motor as shown in Figure CDP4.1 (switch-closed option). The output voltage of the tachometer is $v_T = K_1\theta$. Use the velocity feedback with the adjustable gain K_1. Select the best values for the gain K_1 and the amplifier gain K_a so that the transient response to a step input has an overshoot less than 5% and a settling time (to within 2% of the final value) less than 300 ms.

DP7.1   A high-performance aircraft, shown in Figure DP7.1(a), uses the ailerons, rudder, and elevator to steer through a three-dimensional flight path [20]. The pitch rate control system for a fighter aircraft at 10,000 m and Mach 0.9 can be represented by the system in Figure DP7.1(b), where

$$G(s) = \frac{-18(s + 0.015)(s + 0.45)}{(s^2 + 1.2s + 12)(s^2 + 0.01s + 0.0025)}.$$

(a) Sketch the root locus when the controller is a gain, so that $G_c(s) = K$, and determine K when ζ for the roots with $\omega_n > 2$ is larger than 0.15 (seek a maximum ζ). (b) Plot the response $q(t)$ for a step input $r(t)$ with K as in (a). (c) A designer suggests an anticipatory

controller with $G_c(s) = K_1 + K_2s = K(s + 2)$. Sketch the root locus for this system as K varies and determine a K so that the ζ of all the closed-loop roots is >0.8. (d) Plot the response $q(t)$ for a step input $r(t)$ with K as in (c).

DP7.2   A large helicopter uses two tandem rotors rotating in opposite directions, as shown in Figure P7.33(a). The controller adjusts the tilt angle of the main rotor and thus the forward motion as shown in Figure DP7.2. The helicopter dynamics are represented by

$$G(s) = \frac{10}{s^2 + 4.5s + 9},$$

and the controller is selected as

$$G_c(s) = K_1 + \frac{K_2}{s} = \frac{K(s + 1)}{s}.$$

(a) Sketch the root locus of the system and determine K when ζ of the complex roots is equal to 0.6. (b) Plot the response of the system to a step input $r(t)$ and find the settling time (with a 2% criterion) and overshoot for the system of part (a). What is the steady-state

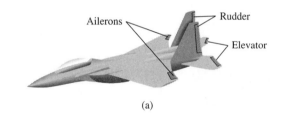

Ailerons   Rudder

Elevator

(a)

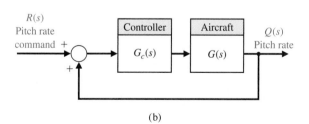

$R(s)$
Pitch rate command

Controller $G_c(s)$

Aircraft $G(s)$

$Q(s)$
Pitch rate

FIGURE DP7.1
(a) High-performance aircraft. (b) Pitch rate control system.

(b)

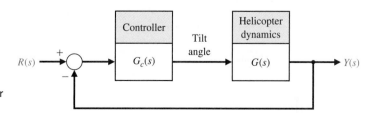

Controller $G_c(s)$

Tilt angle

Helicopter dynamics $G(s)$

$R(s)$

$Y(s)$

FIGURE DP7.2
Two-rotor helicopter velocity control.

error for a step input? (c) Repeat parts (a) and (b) when the ζ of the complex roots is 0.41. Compare the results with those obtained in parts (a) and (b).

DP7.3   The vehicle Rover has been designed for maneuvering at 0.25 mph over Martian terrain. Because Mars is 189 million miles from Earth and it would take up to 40 minutes each way to communicate with Earth [22, 27], Rover must act independently and reliably. Resembling a cross between a small flatbed truck and an elevated jeep, Rover is constructed of three articulated sections, each with its own two independent, axle-bearing, one-meter conical wheels. A pair of sampling arms—one for chipping and drilling, the other for manipulating fine objects—extend from its front end like pincers. The control of the arms can be represented by the system shown in Figure DP7.3. (a) Sketch the root locus for K and identify the roots for $K = 4.1$ and 41. (b) Determine the gain K that results in an overshoot to a step of approximately 1%. (c) Determine the gain that minimizes the settling time (with a 2% criterion) while maintaining an overshoot of less than 1%.

DP7.4   A welding torch is remotely controlled to achieve high accuracy while operating in changing and hazardous environments [21]. A model of the welding arm position control is shown in Figure DP7.4, with the disturbance representing the environmental changes. (a) With $T_d(s) = 0$, select K_1 and K to provide high-quality performance of the position control system. Select a set of performance criteria, and examine the results of your design. (b) For the system in part (a), let $R(s) = 0$ and determine the effect of a unit step $T_d(s) = 1/s$ by obtaining $y(t)$.

DP7.5   A high-performance jet aircraft with an autopilot control system has a unity feedback and control system, as shown in Figure DP7.5. Sketch the root locus, and predict the step response of the system. Determine the actual response of the system, and compare it to the predicted response.

DP7.6   A system to aid and control the walk of a partially disabled person could use automatic control of the walking motion [25]. One model of a system that is open-loop unstable is shown in Figure DP7.6. Using the root locus, select K for the maximum achievable ζ

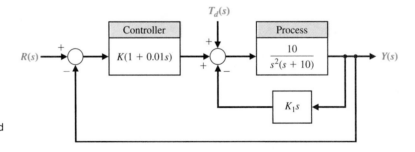

FIGURE DP7.3
Mars vehicle robot control system.

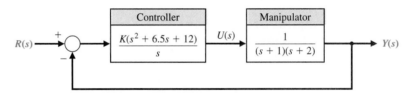

FIGURE DP7.4
Remotely controlled welder.

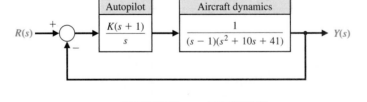

FIGURE DP7.5
High-performance jet aircraft.

FIGURE DP7.6
Automatic control of walking motion.

of the complex roots. Predict the step response of the system, and compare it with the actual step response.

DP7.7 Most commercial op-amps are designed to be unity-gain stable [26]. That is, they are stable when used in a unity-gain configuration. To achieve higher bandwidth, some op-amps relax the requirement to be unity-gain stable. One such amplifier has a DC gain of 10^5 and a bandwidth of 10 kHz. The amplifier, $G(s)$, is connected in the feedback circuit shown in Figure DP7.7(a). The amplifier is represented by the model shown in Figure DP7.7(b), where $K_a = 10^5$. Sketch the root locus of the system for K. Determine the minimum value of the DC gain of the closed-loop amplifier for stability. Select a DC gain and the resistors R_1 and R_2.

DP7.8 A robotic arm actuated at the elbow joint is shown in Figure DP7.8(a), and the control system for the actuator is shown in Figure DP7.8(b). Plot the root locus for $K \geq 0$. Select $G_p(s)$ so that the steady-state error for a step input is equal to zero. Using the $G_p(s)$ selected, plot $y(t)$ for K equal to 1, 1.5, and 2.85. Record the rise time, settling time (with a 2% criterion), and percent overshoot for the three gains. We wish to limit the overshoot to less than 6% while achieving the shortest rise time possible. Select the best system for $1 \leq K \leq 2.85$.

DP7.9 The four-wheel-steering automobile has several benefits. The system gives the driver a greater degree of control over the automobile. The driver gets a more forgiving vehicle over a wide variety of conditions.

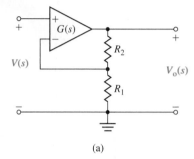

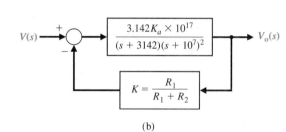

FIGURE DP7.7
(a) Op-amp circuit.
(b) Control system.

(a)   (b)

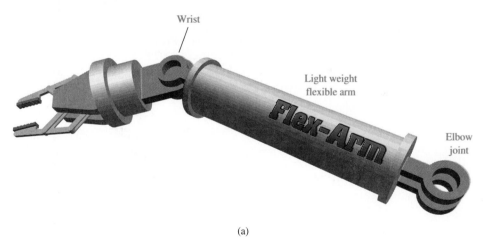

Wrist

Light weight
flexible arm

Elbow
joint

(a)

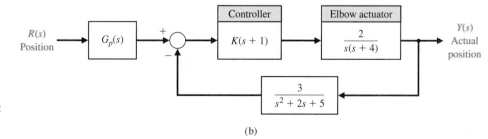

(b)

FIGURE DP7.8
(a) A robotic arm actuated at the joint elbow. (b) Its control system.

The system enables the driver to make sharp, smooth lane transitions. It also prevents yaw, which is the swaying of the rear end during sudden movements. Furthermore, the four-wheel-steering system gives a car increased maneuverability. This enables the driver to park the car in extremely tight quarters. With additional closed-loop computer operating systems, a car could be prevented from sliding out of control in abnormal icy or wet road conditions.

The system works by moving the rear wheels relative to the front-wheel-steering angle. The control system takes information about the front wheels' steering angle and passes it to the actuator in the back. This actuator then moves the rear wheels appropriately.

When the rear wheels are given a steering angle relative to the front ones, the vehicle can vary its lateral acceleration response according to the loop transfer function

$$G_c(s)G(s) = K\frac{1 + (1 + \lambda)T_1 s + (1 + \lambda)T_2 s^2}{s[1 + (2\zeta/\omega_n)s + (1/\omega_n^2)s^2]},$$

where $\lambda = 2q/(1 - q)$, and q is the ratio of rear wheel angle to front wheel steering angle [14]. We will assume that $T_1 = T_2 = 1$ second and $\omega_n = 4$. Design a unity feedback system, selecting an appropriate set of parameters (λ, K, ζ) so that the steering control response is rapid and yet will yield modest overshoot characteristics. In addition, q must be between 0 and 1.

DP7.10 A pilot crane control is shown in Figure DP7.10(a). The trolley is moved by an input $F(t)$ in order to control $x(t)$ and $\phi(t)$ [13]. The model of the pilot crane control is shown in Figure DP7.10(b).

Design a controller that will achieve control of the desired variables when $G_c(s) = K$.

DP7.11 A rover vehicle designed for use on other planets and moons is shown in Figure DP7.11(a) [21]. The block diagram of the steering control is shown in Figure DP7.11(b), where

$$G(s) = \frac{s + 1.5}{(s + 1)(s + 2)(s + 4)(s + 10)}.$$

(a)

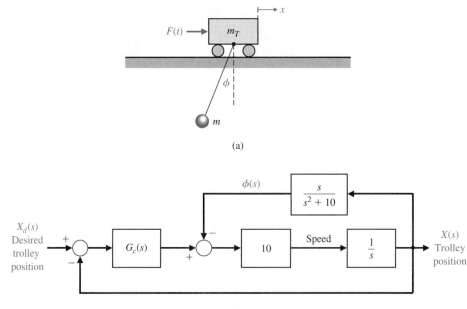

(b)

FIGURE DP7.11   (a) Planetary rover vehicle. (b) Steering control system.

FIGURE DP7.10
(a) Pilot crane control system.
(b) Block diagram.

(a) When $G_c(s) = K$, sketch the root locus as K varies from 0 to 1000. Find the roots for K equal to 100, 300, and 600. (b) Predict the overshoot, settling time (with a 2% criterion), and steady-state error for a step input, assuming dominant roots. (c) Determine the actual time response for a step input for the three values of the gain K, and compare the actual results with the predicted results.

DP7.12 The automatic control of an airplane is one example that requires multiple-variable feedback methods. In this system, the attitude of an aircraft is controlled by three sets of surfaces: elevators, a rudder, and ailerons, as shown in Figure DP7.12(a). By manipulating these surfaces, a pilot can set the aircraft on a desired flight path [20].

An autopilot, which will be considered here, is an automatic control system that controls the roll angle ϕ by adjusting aileron surfaces. The deflection of the aileron surfaces by an angle θ generates a torque due to air pressure on these surfaces. This causes a rolling motion of the aircraft. The aileron surfaces are controlled by a hydraulic actuator with a transfer function $1/s$.

The actual roll angle ϕ is measured and compared with the input. The difference between the desired roll angle ϕ_d and the actual angle ϕ will drive the hydraulic actuator, which in turn adjusts the deflection of the aileron surface.

A simplified model where the rolling motion can be considered independent of other motions is assumed, and its block diagram is shown in Figure DP7.12(b). Assume that $K_1 = 1$ and that the roll rate $\dot{\phi}$ is fed back using a rate gyro. The step response desired has an overshoot less than 10% and a settling time (with a 2% criterion) less than 9 seconds. Select the parameters K_a and K_2.

DP7.13 Consider the feedback system shown in Figure DP7.13. The process transfer function is marginally stable. The controller is the proportional-derivative (PD) controller

$$G_c(s) = K_P + K_D s.$$

(a) Determine the characteristic equation of the closed-loop system.
(b) Let $\tau = K_P/K_D$. Write the characteristic equation in the form

$$\Delta(s) = 1 + K_D \frac{n(s)}{d(s)}.$$

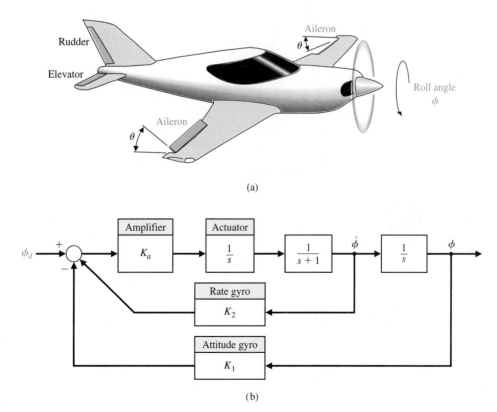

FIGURE DP7.12
(a) An airplane with a set of ailerons.
(b) The block diagram for controlling the roll rate of the airplane.

(a)

(b)

(c) Plot the root locus for $0 \le K_D < \infty$ when $\tau = 6$.

(d) What is the effect on the root locus when $0 < \tau < \sqrt{10}$?

(e) Design the PD controller to meet the following specifications:

    (i) $P.O. < 5\%$

    (ii) $T_s < 1$ s

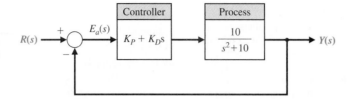

FIGURE DP7.13
A marginally stable plant with a PD controller in the loop.

COMPUTER PROBLEMS

CP7.1 Using the rlocus function, obtain the root locus for the following transfer functions of the system shown in Figure CP7.1 when $0 < K < \infty$:

(a) $G(s) = \dfrac{10}{s^3 + 14s^2 + 43s + 30}$,

(b) $G(s) = \dfrac{s + 20}{s^2 + 5s + 20}$,

(c) $G(s) = \dfrac{s^2 + s + 1}{s(s^2 + 5s + 10)}$,

(d) $G(s) = \dfrac{s^5 + 4s^4 + 6s^3 + 8s^2 + 6s + 4}{s^6 + 2s^5 + 2s^4 + s^3 + s^2 + 10s + 1}$.

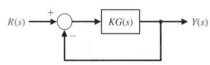

FIGURE CP7.1 A single-loop feedback system with parameter K.

CP7.2 A unity negative feedback system has the loop transfer function

$$KG(s) = K \frac{s^2 - 2s + 2}{s(s^2 + 3s + 2)}.$$

Develop an m-file to plot the root locus and show with the rlocfind function that the maximum value of K for a stable system is $K = 0.79$.

CP7.3 Compute the partial fraction expansion of

$$Y(s) = \frac{s + 2}{s(s^2 + 6s + 5)}$$

and verify the result using the residue function.

CP7.4 A unity negative feedback system has the loop transfer function

$$G_c(s)G(s) = \frac{(1 + p)s - p}{s^2 + 3s + 6}.$$

Develop an m-file to obtain the root locus as p varies; $0 < p < \infty$. For what values of p is the closed-loop stable?

CP7.5 Consider the feedback system shown in Figure CP7.1, where

$$G(s) = \frac{s + 1}{s^2}.$$

For what value of K is $\zeta = 0.707$ for the dominant closed-loop poles?

CP7.6 Consider the feedback control system in Figure CP7.6. We have three potential controllers for our system:

    1. $G_c(s) = K$ (proportional controller)

    2. $G_c(s) = K/s$ (integral controller)

    3. $G_c(s) = K(1 + 1/s)$ (proportional, integral (PI) controller)

The design specifications are $T_s \le 10$ seconds and $P.O. \le 10\%$ for a unit step input.

(a) For the proportional controller, develop an m-file to sketch the root locus for $0 < K < \infty$, and determine the value of K so that the design specifications are satisfied.

(b) Repeat part (a) for the integral controller.

(c) Repeat part (a) for the PI controller.

(d) Co-plot the unit step responses for the closed-loop systems with each controller designed in parts (a)–(c).

FIGURE CP7.6
A single-loop feedback control system with controller $G_c(s)$.

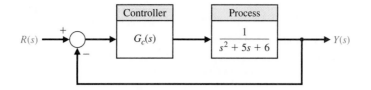

FIGURE CP7.7
A spacecraft attitude control system with a proportional-derivative controller.

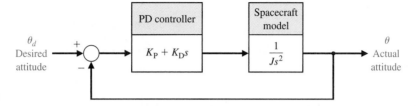

(e) Compare and contrast the three controllers obtained in parts (a)–(c), concentrating on the steady-state errors and transient performance.

CP7.7 Consider the spacecraft single-axis attitude control system shown in Figure CP7.7. The controller is known as a proportional-derivative (PD) controller. Suppose that we require the ratio of $K_p/K_D = 5$. Then, develop an m-file using root locus methods find the values of K_D/J and K_p/J so that the settling time T_s is less than or equal to 4 seconds, and the peak overshoot *P.O.* is less than or equal to 10% for a unit step input. Use a 2% criterion in determining the settling time.

CP7.8 Consider the feedback control system in Figure CP7.8. Develop an m-file to plot the root locus for $0 < K < \infty$. Find the value of K resulting in a damping ratio of the closed-loop poles equal to 0.707.

CP7.9 Consider the system represented in state variable form

$$\dot{\mathbf{x}} = \mathbf{A}\mathbf{x} + \mathbf{B}u$$
$$y = \mathbf{C}\mathbf{x} + \mathbf{D}u,$$

where

$$\mathbf{A} = \begin{bmatrix} 0 & 1 & 0 \\ 0 & 0 & 1 \\ -1 & -5 & -2-k \end{bmatrix}, \mathbf{B} = \begin{bmatrix} 1 \\ 0 \\ 4 \end{bmatrix},$$

$$\mathbf{C} = \begin{bmatrix} 1 & -9 & 12 \end{bmatrix}, \quad \text{and} \quad \mathbf{D} = \begin{bmatrix} 0 \end{bmatrix}.$$

(a) Determine the characteristic equation. (b) Using the Routh–Hurwitz criterion, determine the values of k for which the system is stable. (c) Develop an m-file to plot the root locus and compare the results to those obtained in (b).

FIGURE CP7.8
Unity feedback system with parameter K.

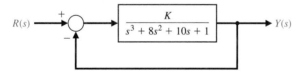

TERMS AND CONCEPTS

Angle of departure   The angle at which a locus leaves a complex pole in the s-plane.

Angle of the asymptotes   The angle ϕ_A that the asymptote makes with respect to the real axis.

Asymptote   The path the root locus follows as the parameter becomes very large and approaches infinity. The number of asymptotes is equal to the number of poles minus the number of zeros.

Asymptote centroid   The center σ_A of the linear asymptotes.

Breakaway point   The point on the real axis where the locus departs from the real axis of the s-plane.

Dominant roots   The roots of the characteristic equation that represent or dominate the closed-loop transient response.

Locus   A path or trajectory that is traced out as a parameter is changed.

Logarithmic sensitivity   A measure of the sensitivity of the system performance to specific parameter changes, given by $S_K^T(s) = \dfrac{\partial T(s)/T(s)}{\partial K/K}$, where $T(s)$ is the system transfer function and K is the parameter of interest.

Number of separate loci   Equal to the number of poles of the transfer function, assuming that the number of poles is greater than or equal to the number of zeros of the transfer function.

Parameter design   A method of selecting one or two parameters using the root locus method.

PID controller   A widely used controller used in industry of the form $G_c(s) = K_p + \dfrac{K_I}{s} + K_D s$, where K_p is the proportional gain, K_I is the integral gain, and K_D is the derivative gain.

Proportional plus deriviative (PD) controller   A two-term controller of the form $G_c(s) = K_p + K_D s$, where K_p is the proportional gain and K_D is the derivative gain.

Proportional plus integral (PI) controller   A two-term controller of the form $G_c(s) = K_p + \dfrac{K_I}{s}$, where K_p is the proportional gain and K_I is the integral gain.

Root contours   The family of loci that depict the effect of varying two parameters on the roots of the characteristic equation.

Root locus   The locus or path of the roots traced out on the s-plane as a parameter is changed.

Root locus method   The method for determining the locus of roots of the characteristic equation $1 + KP(s) = 0$ as K varies from 0 to infinity.

Root locus segments on the real axis   The root locus lying in a section of the real axis to the left of an odd number of poles and zeros.

Root sensitivity   The sensitivity of the roots as a parameter changes from its normal value. The root sensitivity is given by $S_K^r = \dfrac{\partial r}{\partial K/K}$, the incremental change in the root divided by the proportional change of the parameter.

Frequency Response Methods

P R E V I E W

In previous chapters, we examined the use of test signals such as a step and a ramp signal. In this chapter, we consider the steady-state response of a system to a sinusoidal input test signal. We will see that the response of a linear constant coefficient system to a sinusoidal input signal is an output sinusoidal signal at the same frequency as the input. However, the magnitude and phase of the output signal differ from those of the input sinusoidal signal, and the amount of difference is a function of the input frequency. Thus, we will be investigating the steady-state response of the system to a sinusoidal input as the frequency varies.

We will examine the transfer function $G(s)$ when $s = j\omega$ and develop methods for graphically displaying the complex number $G(j\omega)$ as ω varies. The Bode plot is one of the most powerful graphical tools for analyzing and designing control systems, and we will cover that subject in this chapter. We will also consider polar plots and log magnitude and phase diagrams. We will develop several time-domain performance measures in terms of the frequency response of the system, as well as introduce the concept of system bandwidth. The chapter concludes with a frequency response analysis of the Sequential Design Example: Disk Drive Read System.

DESIRED OUTCOMES

Upon completion of Chapter 8, students should:

❑ Understand the powerful concept of frequency response and its role in control system design.

❑ Know how to sketch a Bode plot and also how to obtain a computer-generated Bode plot.

❑ Be familiar with log magnitude and phase diagrams.

❑ Understand performance specifications in the frequency domain and relative stability based on gain and phase margins.

❑ Be capable of designing a controller to meet desired specifications using frequency response methods.

8.1 INTRODUCTION

In preceding chapters, the response and performance of a system have been described in terms of the complex frequency variable s and the location of the poles and zeros on the s-plane. A very practical and important alternative approach to the analysis and design of a system is the **frequency response** method.

> **The frequency response of a system is defined as the steady-state response of the system to a sinusoidal input signal. The sinusoid is a unique input signal, and the resulting output signal for a linear system, as well as signals throughout the system, is sinusoidal in the steady state; it differs from the input waveform only in amplitude and phase angle.**

For example, consider the system $Y(s) = T(s)R(s)$ with $r(t) = A \sin \omega t$. We have

$$R(s) = \frac{A\omega}{s^2 + \omega^2}$$

and

$$T(s) = \frac{m(s)}{q(s)} = \frac{m(s)}{\displaystyle\prod_{i=1}^{n}(s + p_i)},$$

where $-p_i$ are assumed to be distinct poles. Then, in partial fraction form, we have

$$Y(s) = \frac{k_1}{s + p_1} + \cdots + \frac{k_n}{s + p_n} + \frac{\alpha s + \beta}{s^2 + \omega^2}.$$

Taking the inverse Laplace transform yields

$$y(t) = k_1 e^{-p_1 t} + \cdots + k_n e^{-p_n t} + \mathcal{L}^{-1}\left\{\frac{\alpha s + \beta}{s^2 + \omega^2}\right\},$$

where α and β are constants which are problem dependent. If the system is stable, then all p_i have positive real parts and

$$\lim_{t \to \infty} y(t) = \lim_{t \to \infty} \mathcal{L}^{-1}\left\{\frac{\alpha s + \beta}{s^2 + \omega^2}\right\},$$

since each exponential term $k_i e^{-p_i t}$ decays to zero as $t \to \infty$.

In the limit for $y(t)$, it can be shown, for $t \to \infty$ (the steady state),

$$y(t) = \mathcal{L}^{-1}\left[\frac{\alpha s + \beta}{s^2 + \omega^2}\right]$$

$$= \frac{1}{\omega}\left|A\omega T(j\omega)\right| \sin(\omega t + \phi)$$

$$= A|T(j\omega)| \sin(\omega t + \phi), \tag{8.1}$$

where $\phi = \underline{/T(j\omega)}$.

Thus, the steady-state output signal depends only on the magnitude and phase of $T(j\omega)$ at a specific frequency ω. Notice that the steady-state response, as described in Equation (8.1), is true only for stable systems, $T(s)$.

One advantage of the frequency response method is the ready availability of sinusoid test signals for various ranges of frequencies and amplitudes. Thus, the experimental determination of the system's frequency response is easily accomplished; it is the most reliable and uncomplicated method for the experimental analysis of a system. Often, as we shall find in Section 8.4, the unknown transfer function of a system can be deduced from the experimentally determined frequency response of a system [1, 2]. Furthermore, the design of a system in the frequency domain provides the designer with control of the bandwidth of a system, as well as some measure of the response of the system to undesired noise and disturbances.

A second advantage of the frequency response method is that the transfer function describing the sinusoidal steady-state behavior of a system can be obtained by replacing s with $j\omega$ in the system transfer function $T(s)$. The transfer function representing the sinusoidal steady-state behavior of a system is then a function of the complex variable $j\omega$ and is itself a complex function $T(j\omega)$ that possesses a magnitude and phase angle. The magnitude and phase angle of $T(j\omega)$ are readily represented by graphical plots that provide significant insight into the analysis and design of control systems.

The basic disadvantage of the frequency response method for analysis and design is the indirect link between the frequency and the time domain. Direct correlations between the frequency response and the corresponding transient response characteristics are somewhat tenuous, and in practice the frequency response characteristic is adjusted by using various design criteria that will normally result in a satisfactory transient response.

The **Laplace transform pair** was given in Section 2.4; it is written as

$$F(s) = \mathcal{L}\{f(t)\} = \int_0^\infty f(t)e^{-st}\, dt \qquad (8.2)$$

and

$$f(t) = \mathcal{L}^{-1}\{F(s)\} = \frac{1}{2\pi j}\int_{\sigma-j\infty}^{\sigma+j\infty} F(s)e^{st}\, ds, \qquad (8.3)$$

where the complex variable $s = \sigma + j\omega$. Similarly, the **Fourier transform pair** is written as

$$F(\omega) = \mathcal{F}\{f(t)\} = \int_{-\infty}^\infty f(t)e^{-j\omega t}\, dt \qquad (8.4)$$

and

$$f(t) = \mathcal{F}^{-1}\{F(\omega)\} = \frac{1}{2\pi}\int_{-\infty}^\infty F(\omega)e^{j\omega t}\, d\omega. \qquad (8.5)$$

The Fourier transform exists for $f(t)$ when

$$\int_{-\infty}^{\infty} |f(t)|\, dt < \infty.$$

The Fourier and Laplace transforms are closely related, as we can see by examining Equations (8.2) and (8.4). When the function $f(t)$ is defined only for $t \geq 0$, as is often the case, the lower limits on the integrals are the same. Then we note that the two equations differ only in the complex variable. Thus, if the Laplace transform of a function $f_1(t)$ is known to be $F_1(s)$, we can obtain the Fourier transform of this same time function by setting $s = j\omega$ in $F_1(s)$ [3].

Again we might ask, Since the Fourier and Laplace transforms are so closely related, why can't we always use the Laplace transform? Why use the Fourier transform at all? The Laplace transform permits us to investigate the s-plane location of the poles and zeros of a transfer function $T(s)$, as in Chapter 7. However, the frequency response method allows us to consider the transfer function $T(j\omega)$ and to concern ourselves with the amplitude and phase characteristics of the system. This ability to investigate and represent the character of a system by amplitude, phase equations, and curves is an advantage for the analysis and design of control systems.

If we consider the frequency response of the closed-loop system, we might have an input $r(t)$ that has a Fourier transform in the frequency domain as follows:

$$R(j\omega) = \int_{-\infty}^{\infty} r(t)e^{-j\omega t}\, dt.$$

Then the output frequency response of a single-loop control system can be obtained by substituting $s = j\omega$ in the closed-loop system relationship, $Y(s) = T(s)R(s)$, so that we have

$$Y(j\omega) = T(j\omega)R(j\omega) = \frac{G(j\omega)}{1 + G(j\omega)H(j\omega)}R(j\omega). \qquad (8.6)$$

Using the inverse Fourier transform, the output transient response would be

$$y(t) = \mathscr{F}^{-1}\{Y(j\omega)\} = \frac{1}{2\pi}\int_{-\infty}^{\infty} Y(j\omega)e^{j\omega t}\, d\omega. \qquad (8.7)$$

However, it is usually quite difficult to evaluate this inverse transform integral for all but the simplest systems, and a graphical integration may be used. Alternatively, as we will note in succeeding sections, several measures of the transient response can be related to the frequency characteristics and utilized for design purposes.

8.2 FREQUENCY RESPONSE PLOTS

The transfer function of a system $G(s)$ can be described in the frequency domain by the relation

$$G(j\omega) = G(s)|_{s=j\omega} = R(\omega) + jX(\omega), \qquad (8.8)$$

where

$$R(\omega) = \text{Re}[G(j\omega)] \quad \text{and} \quad X(\omega) = \text{Im}[G(j\omega)].$$

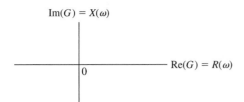

$$\text{Im}(G) = X(\omega)$$

$$\text{Re}(G) = R(\omega)$$

FIGURE 8.1
The polar plane.

See the MCS website for a review of complex numbers.
Alternatively, the transfer function can be represented by a magnitude $|G(j\omega)|$ and a phase $\phi(j\omega)$ as

$$G(j\omega) = |G(j\omega)|e^{j\phi(\omega)} = |G(j\omega)|\underline{/\phi(\omega)}, \quad (8.9)$$

where

$$\phi(\omega) = \tan^{-1}\frac{X(\omega)}{R(\omega)} \quad \text{and} \quad |G(j\omega)|^2 = [R(\omega)]^2 + [X(\omega)]^2.$$

The graphical representation of the frequency response of the system $G(j\omega)$ can utilize either Equation (8.8) or Equation (8.9). The **polar plot** representation of the frequency response is obtained by using Equation (8.8). The coordinates of the polar plot are the real and imaginary parts of $G(j\omega)$, as shown in Figure 8.1. An example of a polar plot will illustrate this approach.

EXAMPLE 8.1   **Frequency response of an *RC* filter**

A simple *RC* filter is shown in Figure 8.2. The transfer function of this filter is

$$G(s) = \frac{V_2(s)}{V_1(s)} = \frac{1}{RCs + 1}, \quad (8.10)$$

and the sinusoidal steady-state transfer function is

$$G(j\omega) = \frac{1}{j\omega(RC) + 1} = \frac{1}{j(\omega/\omega_1) + 1}, \quad (8.11)$$

where

$$\omega_1 = \frac{1}{RC}.$$

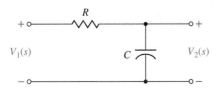

FIGURE 8.2
An *RC* filter.

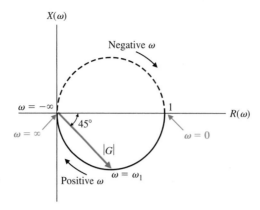

FIGURE 8.3
Polar plot for *RC* filter.

Then the polar plot is obtained from the relation

$$G(j\omega) = R(\omega) + jX(\omega)$$
$$= \frac{1 - j(\omega/\omega_1)}{(\omega/\omega_1)^2 + 1}$$
$$= \frac{1}{1 + (\omega/\omega_1)^2} - \frac{j(\omega/\omega_1)}{1 + (\omega/\omega_1)^2}. \qquad (8.12)$$

The first step is to determine $R(\omega)$ and $X(\omega)$ at the two frequencies, $\omega = 0$ and $\omega = \infty$. At $\omega = 0$, we have $R(\omega) = 1$ and $X(\omega) = 0$. At $\omega = \infty$, we have $R(\omega) = 0$ and $X(\omega) = 0$. These two points are shown in Figure 8.3. The locus of the real and imaginary parts is also shown in Figure 8.3 and is easily shown to be a circle with the center at $\left(\frac{1}{2}, 0\right)$. When $\omega = \omega_1$, the real and imaginary parts are equal in magnitude, and the angle $\phi(\omega) = -45°$. The polar plot can also be readily obtained from Equation (8.9) as

$$G(j\omega) = |G(j\omega)| \underline{/\phi(\omega)}, \qquad (8.13)$$

where

$$|G(j\omega)| = \frac{1}{[1 + (\omega/\omega_1)^2]^{1/2}} \quad \text{and} \quad \phi(\omega) = -\tan^{-1}(\omega/\omega_1).$$

Hence, when $\omega = \omega_1$, the magnitude is $|G(j\omega_1)| = 1/\sqrt{2}$ and the phase $\phi(\omega_1) = -45°$. Also, when ω approaches $+\infty$, we have $|G(j\omega)| \to 0$ and $\phi(\omega) = -90°$. Similarly, when $\omega = 0$, we have $|G(j\omega)| = 1$ and $\phi(\omega) = 0$. ∎

EXAMPLE 8.2   **Polar plot of a transfer function**

The polar plot of a transfer function is useful for investigating system stability and will be utilized in Chapter 9. Therefore, it is worthwhile to complete another example at this point. Consider a transfer function

$$G(s)|_{s=j\omega} = G(j\omega) = \frac{K}{j\omega(j\omega\tau + 1)} = \frac{K}{j\omega - \omega^2\tau}. \qquad (8.14)$$

Table 8.1

ω	0	$1/2\tau$	$1/\tau$	∞
$\lvert G(j\omega)\rvert$	∞	$4K\tau/\sqrt{5}$	$K\tau/\sqrt{2}$	0
$\phi(\omega)$	$-90°$	$-117°$	$-135°$	$-180°$

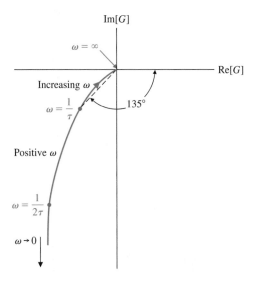

FIGURE 8.4
Polar plot for $G(j\omega) = K/(j\omega(j\omega\tau + 1))$. Note that $\omega = \infty$ at the origin.

Then the magnitude and phase angle are written as

$$\lvert G(j\omega)\rvert = \frac{K}{(\omega^2 + \omega^4\tau^2)^{1/2}} \quad \text{and} \quad \phi(\omega) = -\tan^{-1}\frac{1}{-\omega\tau}.$$

The phase angle and the magnitude are readily calculated at the frequencies $\omega = 0$, $\omega = 1/\tau$, and $\omega = +\infty$. The values of $\lvert G(\omega)\rvert$ and $\phi(\omega)$ are given in Table 8.1, and the polar plot of $G(j\omega)$ is shown in Figure 8.4.

An alternative solution uses the real and imaginary parts of $G(j\omega)$ as

$$G(j\omega) = \frac{K}{j\omega - \omega^2\tau} = \frac{K(-j\omega - \omega^2\tau)}{\omega^2 + \omega^4\tau^2} = R(\omega) + jX(\omega), \qquad (8.15)$$

where $R(\omega) = -K\omega^2\tau/M(\omega)$ and $X(\omega) = -\omega K/M(\omega)$, and where $M(\omega) = \omega^2 + \omega^4\tau^2$. Then when $\omega = \infty$, we have $R(\omega) = 0$ and $X(\omega) = 0$. When $\omega = 0$, we have $R(\omega) = -K\tau$ and $X(\omega) = -\infty$. When $\omega = 1/\tau$, we have $R(\omega) = -K\tau/2$ and $X(\omega) = -K\tau/2$, as shown in Figure 8.4.

Another method of obtaining the polar plot is to evaluate the vector $G(j\omega)$ graphically at specific frequencies, ω, along the $s = j\omega$ axis on the s-plane. We consider

$$G(s) = \frac{K/\tau}{s(s + 1/\tau)}$$

with the two poles shown on the s-plane in Figure 8.5.

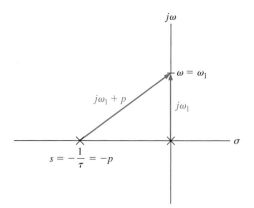

FIGURE 8.5
Two vectors on the
s-plane to evaluate
$G(j\omega_1)$.

When $s = j\omega$, we have

$$G(j\omega) = \frac{K/\tau}{j\omega(j\omega + p)},$$

where $p = 1/\tau$. The magnitude and phase of $G(j\omega)$ can be evaluated at a specific frequency, ω_1, on the $j\omega$-axis, as shown in Figure 8.5. The magnitude and the phase are, respectively,

$$|G(j\omega_1)| = \frac{K/\tau}{|j\omega_1||j\omega_1 + p|}$$

and

$$\phi(\omega) = -\angle(j\omega_1) - \angle(j\omega_1 + p) = -90° - \tan^{-1}(\omega_1/p). \quad \blacksquare$$

There are several possibilities for coordinates of a graph portraying the frequency response of a system. As we have seen, we may use a polar plot to represent the frequency response (Equation 8.8) of a system. However, the limitations of polar plots are readily apparent. The addition of poles or zeros to an existing system requires the recalculation of the frequency response, as outlined in Examples 8.1 and 8.2. (See Table 8.1.) Furthermore, calculating the frequency response in this manner is tedious and does not indicate the effect of the individual poles or zeros.

The introduction of **logarithmic plots**, often called **Bode plots**, simplifies the determination of the graphical portrayal of the frequency response. The logarithmic plots are called Bode plots in honor of H. W. Bode, who used them extensively in his studies of feedback amplifiers [4, 5]. The transfer function in the frequency domain is

$$G(j\omega) = |G(j\omega)|e^{j\phi(\omega)}. \tag{8.16}$$

The logarithm of the magnitude is normally expressed in terms of the logarithm to the base 10, so we use

$$\boxed{\text{Logarithmic gain} = 20 \log_{10}|G(j\omega)|,} \tag{8.17}$$

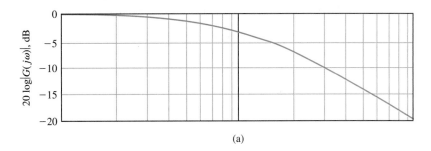

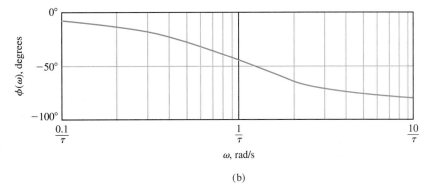

FIGURE 8.6
Bode diagram for
$G(j\omega) = 1/(j\omega\tau + 1)$:
(a) magnitude plot
and (b) phase plot.

where the units are decibels (dB). A decibel conversion table is given on the MCS website. The logarithmic gain in dB and the angle $\phi(\omega)$ can be plotted versus the frequency ω by utilizing several different arrangements. For a Bode diagram, the plot of logarithmic gain in dB versus ω is normally plotted on one set of axes, and the phase $\phi(\omega)$ versus ω on another set of axes, as shown in Figure 8.6. For example, the Bode diagram of the transfer function of Example 8.1 can be readily obtained, as we will find in the following example.

EXAMPLE 8.3 **Bode diagram of an *RC* filter**

The transfer function of Example 8.1 is

$$G(j\omega) = \frac{1}{j\omega(RC) + 1} = \frac{1}{j\omega\tau + 1}, \tag{8.18}$$

where

$$\tau = RC,$$

the time constant of the network. The logarithmic gain is

$$20 \log|G(j\omega)| = 20 \log\left(\frac{1}{1 + (\omega\tau)^2}\right)^{1/2} = -10 \log(1 + (\omega\tau)^2). \tag{8.19}$$

For small frequencies—that is, $\omega \ll 1/\tau$—the logarithmic gain is

$$20 \log|G(j\omega)| = -10 \log(1) = 0 \text{ dB}, \qquad \omega \ll 1/\tau. \tag{8.20}$$

For large frequencies—that is, $\omega \gg 1/\tau$—the logarithmic gain is

$$20 \log G(j\omega) = -20 \log(\omega\tau) \qquad \omega \gg 1/\tau, \tag{8.21}$$

and at $\omega = 1/\tau$, we have

$$20 \log|G(j\omega)| = -10 \log 2 = -3.01 \text{ dB}.$$

The magnitude plot for this network is shown in Figure 8.6(a). The phase angle of the network is

$$\phi(\omega) = -\tan^{-1}(\omega\tau). \tag{8.22}$$

The phase plot is shown in Figure 8.6(b). The frequency $\omega = 1/\tau$ is often called the **break frequency** or **corner frequency**. ■

A linear scale of frequency is not the most convenient or judicious choice, and we consider the use of a logarithmic scale of frequency. The convenience of a logarithmic scale of frequency can be seen by considering Equation (8.21) for large frequencies $\omega \gg 1/\tau$, as follows:

$$20 \log|G(j\omega)| = -20 \log(\omega\tau) = -20 \log \tau - 20 \log \omega. \tag{8.23}$$

Then, on a set of axes where the horizontal axis is $\log \omega$, the asymptotic curve for $\omega \gg 1/\tau$ is a straight line, as shown in Figure 8.7. The slope of the straight line can be ascertained from Equation (8.21). An interval of two frequencies with a ratio equal to 10 is called a **decade**, so that the range of frequencies from ω_1 to ω_2, where $\omega_2 = 10\omega_1$, is called a decade. The difference between the logarithmic gains, for $\omega \gg 1/\tau$, over a decade of frequency is

$$\begin{aligned}
20 \log|G(j\omega_1)| - 20 \log|G(j\omega_2)| &= -20 \log(\omega_1\tau) - (-20 \log(\omega_2\tau)) \\
&= -20 \log \frac{\omega_1\tau}{\omega_2\tau} \tag{8.24} \\
&= -20 \log \frac{1}{10} = +20 \text{ dB};
\end{aligned}$$

that is, the slope of the asymptotic line for this first-order transfer function is -20 dB/decade, and the slope is shown for this transfer function in Figure 8.7. Instead of using a horizontal axis of $\log \omega$ and linear rectangular coordinates, it is easier to use semilog paper with a linear rectangular coordinate for dB and a logarithmic coordinate for ω. Alternatively, we could use a logarithmic coordinate for the magnitude as well as for frequency and avoid the necessity of calculating the logarithm of the magnitude.

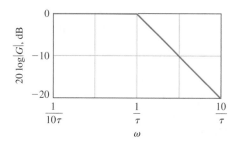

FIGURE 8.7
Asymptotic curve for $(j\omega\tau + 1)^{-1}$.

The frequency interval $\omega_2 = 2\omega_1$ is often used and is called an **octave** of frequencies. The difference between the logarithmic gains for $\omega \gg 1/\tau$, for an octave, is

$$20 \log|G(j\omega_1)| - 20 \log|G(j\omega_2)| = -20 \log \frac{\omega_1 \tau}{\omega_2 \tau}$$

$$= -20 \log \frac{1}{2} = 6.02 \text{ dB}. \qquad (8.25)$$

Therefore, the slope of the asymptotic line is -6 dB/octave.

The primary advantage of the logarithmic plot is the conversion of multiplicative factors, such as $(j\omega\tau + 1)$, into additive factors, $20 \log(j\omega\tau + 1)$, by virtue of the definition of logarithmic gain. This can be readily ascertained by considering the generalized transfer function

$$G(j\omega) = \frac{K_b \prod_{i=1}^{Q}(1 + j\omega\tau_i)}{(j\omega)^N \prod_{m=1}^{M}(1 + j\omega\tau_m) \prod_{k=1}^{R}[(1 + (2\zeta_k/\omega_{nk})j\omega + (j\omega/\omega_{nk})^2)]}. \qquad (8.26)$$

This transfer function includes Q zeros, N poles at the origin, M poles on the real axis, and R pairs of complex conjugate poles. Obtaining the polar plot of such a function would be a formidable task indeed. However, the logarithmic magnitude of $G(j\omega)$ is

$$20 \log|G(j\omega)| = 20 \log K_b + 20 \sum_{i=1}^{Q} \log|1 + j\omega\tau_i|$$

$$-20 \log|(j\omega)^N| - 20 \sum_{m=1}^{M} \log|1 + j\omega\tau_m|$$

$$-20 \sum_{k=1}^{R} \log\left|1 + \frac{2\zeta_k}{\omega_{nk}}j\omega + \left(\frac{j\omega}{\omega_{nk}}\right)^2\right|, \qquad (8.27)$$

and the Bode diagram can be obtained by adding the plot due to each individual factor. Furthermore, the separate phase angle plot is obtained as

$$\phi(\omega) = +\sum_{i=1}^{Q} \tan^{-1}(\omega\tau_i) - N(90°) - \sum_{m=1}^{M} \tan^{-1}(\omega\tau_m)$$

$$-\sum_{k=1}^{R} \tan^{-1} \frac{2\zeta_k\omega_{nk}\omega}{\omega_{nk}^2 - \omega^2}, \qquad (8.28)$$

which is simply the summation of the phase angles due to each individual factor of the transfer function.

Therefore, the four different kinds of factors that may occur in a transfer function are as follows:

1. Constant gain K_b

2. Poles (or zeros) at the origin $(j\omega)$

3. Poles (or zeros) on the real axis $(j\omega\tau + 1)$

4. Complex conjugate poles (or zeros) $[1 + (2\zeta/\omega_n)j\omega + (j\omega/\omega_n)^2]$

We can determine the logarithmic magnitude plot and phase angle for these four factors and then use them to obtain a Bode diagram for any general form of a transfer function. Typically, the curves for each factor are obtained and then added together graphically to obtain the curves for the complete transfer function. Furthermore, this procedure can be simplified by using the asymptotic approximations to these curves and obtaining the actual curves only at specific important frequencies.

Constant Gain K_b. The logarithmic gain for the **constant K_b** is

$$20 \log K_b = \text{constant in dB},$$

and the phase angle is

$$\phi(\omega) = 0.$$

The gain curve is a horizontal line on the Bode diagram.

If the gain is a negative value, $-K_b$, the logarithmic gain remains $20 \log K_b$. The negative sign is accounted for by the phase angle, $-180°$.

Poles (or Zeros) at the Origin, $(j\omega)$. A pole at the origin has a logarithmic magnitude

$$20 \log \left| \frac{1}{j\omega} \right| = -20 \log \omega \text{ dB} \tag{8.29}$$

and a phase angle

$$\phi(\omega) = -90°.$$

The slope of the magnitude curve is -20 dB/decade for a pole. Similarly, for a multiple pole at the origin, we have

$$20 \log \left| \frac{1}{(j\omega)^N} \right| = -20N \log \omega, \tag{8.30}$$

and the phase is

$$\phi(\omega) = -90°N.$$

In this case, the slope due to the multiple pole is $-20N$ dB/decade. For a zero at the origin, we have a logarithmic magnitude

$$20 \log |j\omega| = +20 \log \omega, \tag{8.31}$$

where the slope is $+20$ dB/decade and the phase angle is

$$\phi(\omega) = +90°.$$

The Bode diagram of the magnitude and phase angle of $(j\omega)^{\pm N}$ is shown in Figure 8.8 for $N = 1$ and $N = 2$.

Poles or Zeros on the Real Axis. The pole factor $(1 + j\omega\tau)^{-1}$ has been considered previously, and we found that, for a pole on the real axis,

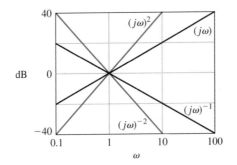

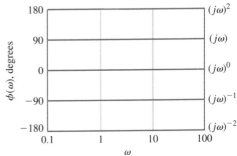

FIGURE 8.8
Bode diagram for
$(j\omega)^{\pm N}$.

$$20 \log \left| \frac{1}{1 + j\omega\tau} \right| = -10 \log(1 + \omega^2\tau^2). \qquad (8.32)$$

The asymptotic curve for $\omega \ll 1/\tau$ is $20 \log 1 = 0$ dB, and the asymptotic curve for $\omega \gg 1/\tau$ is $-20 \log(\omega\tau)$, which has a slope of -20 dB/decade. The intersection of the two asymptotes occurs when

$$20 \log 1 = 0 \text{ dB} = -20 \log(\omega\tau),$$

or when $\omega = 1/\tau$, the **break frequency**. The actual logarithmic gain when $\omega = 1/\tau$ is -3 dB for this factor. The phase angle is $\phi(\omega) = -\tan^{-1}(\omega\tau)$ for the denominator factor. The Bode diagram of a pole factor $(1 + j\omega\tau)^{-1}$ is shown in Figure 8.9.

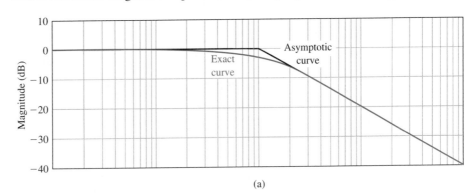

(a)

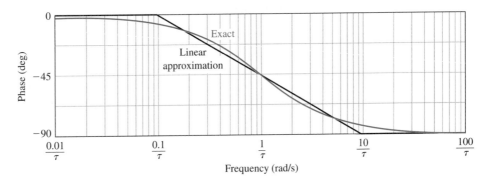

FIGURE 8.9
Bode diagram for
$(1 + j\omega\tau)^{-1}$.

(b)

The Bode diagram of a zero factor $1 + j\omega\tau$ is obtained in the same manner as that of the pole. However, the slope is positive at +20 dB/decade, and the phase angle is $\phi(\omega) = +\tan^{-1}(\omega\tau)$.

A piecewise linear approximation to the phase angle curve can be obtained as shown in Figure 8.9. This linear approximation, which passes through the correct phase at the break frequency, is within 6° of the actual phase curve for all frequencies. This approximation will provide a useful means for readily determining the form of the phase angle curves of a transfer function $G(s)$. However, often the accurate phase angle curves are required, and the actual phase curve for the first-order factor must be obtained via a computer program. The exact values of the frequency response for the pole $(1 + j\omega\tau)^{-1}$, as well as the values obtained by using the approximation for comparison, are given in Table 8.2.

Complex Conjugate Poles or Zeros $[1 + (2\zeta/\omega_n)j\omega + (j\omega/\omega_n)^2]$. The quadratic factor for a pair of complex conjugate poles can be written in normalized form as

$$[1 + j2\zeta u - u^2]^{-1}, \tag{8.33}$$

where $u = \omega/\omega_n$. Then the logarithmic magnitude for a pair of complex conjugate poles is

$$20 \log|G(j\omega)| = -10 \log((1 - u^2)^2 + 4\zeta^2 u^2), \tag{8.34}$$

and the phase angle is

$$\phi(\omega) = -\tan^{-1}\frac{2\zeta u}{1 - u^2}. \tag{8.35}$$

When $u \ll 1$, the magnitude is

$$20 \log|G(j\omega)| = -10 \log 1 = 0 \text{ dB},$$

and the phase angle approaches 0°. When $u \gg 1$, the logarithmic magnitude approaches

$$20 \log|G(j\omega)| = -10 \log u^4 = -40 \log u,$$

which results in a curve with a slope of −40 dB/decade. The phase angle, when $u \gg 1$, approaches −180°. The magnitude asymptotes meet at the 0 dB line when $u = \omega/\omega_n = 1$. However, the difference between the actual magnitude curve and the asymptotic approximation is a function of the damping ratio and must be accounted for when $\zeta < 0.707$. The Bode diagram of a quadratic factor due to a

Table 8.2

$\omega\tau$	0.10	0.50	0.76	1	1.31	2	5	10
$20 \log|(1 + j\omega\tau)^{-1}|$, dB	−0.04	−1.0	−2.0	−3.0	−4.3	−7.0	−14.2	−20.04
Asymptotic approximation, dB	0	0	0	0	−2.3	−6.0	−14.0	−20.0
$\phi(\omega)$, degrees	−5.7	−26.6	−37.4	−45.0	−52.7	−63.4	−78.7	−84.3
Linear approximation, degrees	0	−31.50	−39.5	−45.0	−50.3	−58.5	−76.5	−90.0

pair of complex conjugate poles is shown in Figure 8.10. The maximum value $M_{p\omega}$ of the frequency response occurs at the **resonant frequency** ω_r. When the damping ratio approaches zero, then ω_r approaches ω_n, the natural frequency. The resonant frequency is determined by taking the derivative of the magnitude of Equation (8.33) with respect to the normalized frequency, u, and setting it equal to zero. The resonant frequency is given by the relation

$$\omega_r = \omega_n\sqrt{1 - 2\zeta^2}, \quad \zeta < 0.707, \tag{8.36}$$

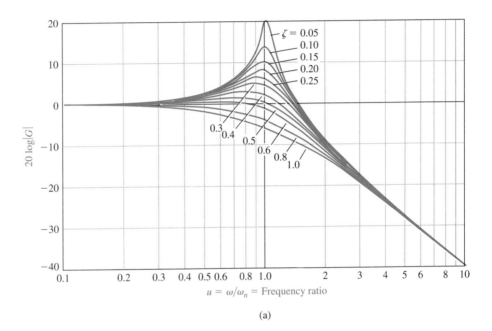

(a)

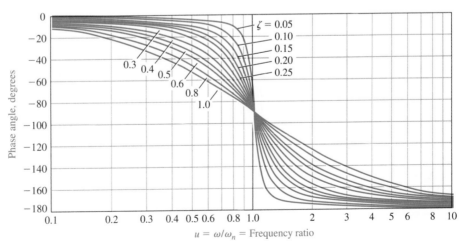

FIGURE 8.10
Bode diagram for
$G(j\omega) = [1 + (2\zeta/\omega_n)$
$j\omega + (j\omega/\omega_n)^2]^{-1}$.

(b)

and the maximum value of the magnitude $|G(j\omega)|$ is

$$M_{p\omega} = |G(j\omega_r)| = \left(2\zeta\sqrt{1 - \zeta^2}\right)^{-1}, \quad \zeta < 0.707, \qquad (8.37)$$

for a pair of complex poles. The maximum value of the frequency response, $M_{p\omega}$, and the resonant frequency ω_r are shown as a function of the damping ratio ζ for a pair of complex poles in Figure 8.11. Assuming the dominance of a pair of complex conjugate closed-loop poles, we find that these curves are useful for estimating the damping ratio of a system from an experimentally determined frequency response.

The frequency response curves can be evaluated on the s-plane by determining the vector lengths and angles at various frequencies ω along the $(s = +j\omega)$-axis. For example, considering the second-order factor with complex conjugate poles, we have

$$G(s) = \frac{1}{(s/\omega_n)^2 + 2\zeta s/\omega_n + 1} = \frac{\omega_n^2}{s^2 + 2\zeta\omega_n s + \omega_n^2}. \qquad (8.38)$$

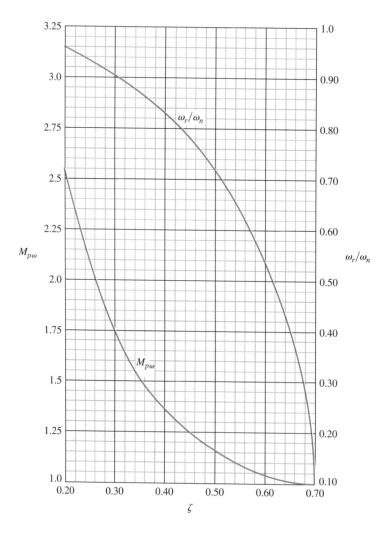

FIGURE 8.11
The maximum $M_{p\omega}$ of the frequency response and the resonant frequency ω_r versus ζ for a pair of complex conjugate poles.

The poles for varying ζ lie on a circle of radius ω_n and are shown for a particular ζ in Figure 8.12(a). The transfer function evaluated for real frequency $s = j\omega$ is written as

$$G(j\omega) = \left.\frac{\omega_n^2}{(s - s_1)(s - \hat{s}_1)}\right|_{s=j\omega} = \frac{\omega_n^2}{(j\omega - s_1)(j\omega - \hat{s}_1)}, \tag{8.39}$$

where s_1 and s_1^* are the complex conjugate poles. The vectors $j\omega - s_1$ and $j\omega - \hat{s}_1$ are the vectors from the poles to the frequency $j\omega$, as shown in Figure 8.12(a). Then the magnitude and phase may be evaluated for various specific frequencies. The magnitude is

$$|G(j\omega)| = \frac{\omega_n^2}{|j\omega - s_1||j\omega - s_1^*|}, \tag{8.40}$$

and the phase is

$$\phi(\omega) = -\angle(j\omega - s_1) - \angle(j\omega - \hat{s}_1).$$

The magnitude and phase may be evaluated for three specific frequencies, namely,

$$\omega = 0, \qquad \omega = \omega_r, \quad \text{and} \quad \omega = \omega_d,$$

as shown in Figure 8.12 in parts (b), (c), and (d), respectively. The magnitude and phase corresponding to these frequencies are shown in Figure 8.13.

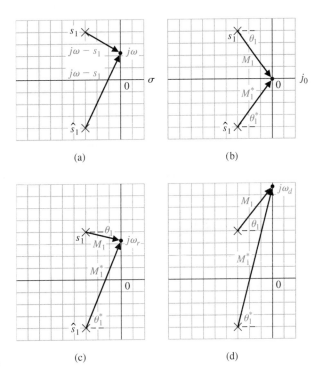

FIGURE 8.12
Vector evaluation of the frequency response for selected values of ω.

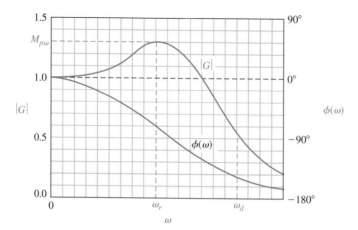

FIGURE 8.13
Bode diagram for complex conjugate poles.

EXAMPLE 8.4   Bode diagram of a twin-T network

As an example of the determination of the frequency response using the pole–zero diagram and the vectors to $j\omega$, consider the twin-T network shown in Figure 8.14 [6]. The transfer function of this network is

$$G(s) = \frac{V_o(s)}{V_{in}(s)} = \frac{(s\tau)^2 + 1}{(s\tau)^2 + 4s\tau + 1},\tag{8.41}$$

where $\tau = RC$. The zeros are at $\pm j1$, and the poles are at $-2 \pm \sqrt{3}$ in the $s\tau$-plane, as shown in Figure 8.15(a). At $\omega = 0$, we have $|G(j\omega)| = 1$ and $\phi(\omega) = 0°$. At $\omega = 1/\tau$, $|G(j\omega)| = 0$ and the phase angle of the vector from the zero at $s\tau = j1$ passes through a transition of $180°$. When ω approaches ∞, $|G(j\omega)| = 1$ and $\phi(\omega) = 0$ again. Evaluating several intermediate frequencies, we can readily obtain the frequency response, as shown in Figure 8.15(b). ∎

A summary of the asymptotic curves for basic terms of a transfer function is provided in Table 8.3.

In the previous examples, the poles and zeros of $G(s)$ have been restricted to the left-hand plane. However, a system may have zeros located in the right-hand s-plane and may still be stable. Transfer functions with zeros in the right-hand s-plane are classified as **nonminimum phase transfer functions**. If the zeros of a transfer function are all reflected about the $j\omega$-axis, there is no change in the magnitude of the transfer function, and the only difference is in the phase-shift characteristics. If the phase characteristics of the two system functions are compared, it can be readily

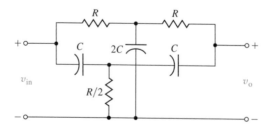

FIGURE 8.14
Twin-T network.

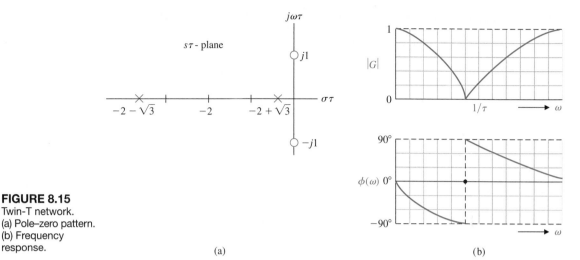

FIGURE 8.15
Twin-T network.
(a) Pole–zero pattern.
(b) Frequency
response.

(a)   (b)

shown that the net phase shift over the frequency range from zero to infinity is less for the system with all its zeros in the left-hand s-plane. Thus, the transfer function $G_1(s)$, with all its zeros in the left-hand s-plane, is called a **minimum phase transfer function**. The transfer function $G_2(s)$, with $|G_2(j\omega)| = |G_1(j\omega)|$ and all the zeros of $G_1(s)$ reflected about the $j\omega$-axis into the right-hand s-plane, is called a nonminimum phase transfer function. Reflection of any zero or pair of zeros into the right half-plane results in a nonminimum phase transfer function.

> **A transfer function is called a minimum phase transfer function if all its zeros lie in the left-hand s-plane. It is called a nonminimum phase transfer function if it has zeros in the right-hand s-plane.**

The two pole–zero patterns shown in Figures 8.16(a) and (b) have the same amplitude characteristics as can be deduced from the vector lengths. However, the phase characteristics are different for Figures 8.16(a) and (b). The minimum phase characteristic of Figure 8.16(a) and the nonminimum phase characteristic of Figure 8.16(b) are shown in Figure 8.17. Clearly, the phase shift of

$$G_1(s) = \frac{s + z}{s + p}$$

ranges over less than $80°$, whereas the phase shift of

$$G_2(s) = \frac{s - z}{s + p}$$

ranges over $180°$. The meaning of the term **minimum phase** is illustrated by Figure 8.17. The range of phase shift of a minimum phase transfer function is the least possible or minimum corresponding to a given amplitude curve, whereas the range of the nonminimum phase curve is the greatest possible for the given amplitude curve.

Table 8.3 Asymptotic Curves for Basic Terms of a Transfer Function

Term	Magnitude 20 log|G|	Phase $\phi(\omega)$
1. Gain, $G(j\omega) = K$		
2. Zero, $G(j\omega) = 1 + j\omega/\omega_1$		
3. Pole, $G(j\omega) = (1 + j\omega/\omega_1)^{-1}$		
4. Pole at the origin, $G(j\omega) = 1/j\omega$		
5. Two complex poles, $0.1 < \zeta < 1$, $G(j\omega) = (1 + j2\zeta u - u^2)^{-1}$, $u = \omega/\omega_n$		

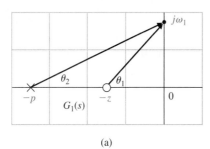
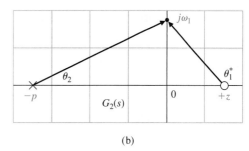

FIGURE 8.16
Pole–zero patterns giving the same amplitude response and different phase characteristics.

(a)   (b)

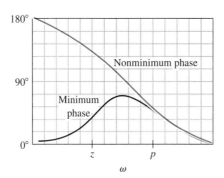

FIGURE 8.17
The phase characteristics for the minimum phase and nonminimum phase transfer function.

A particularly interesting nonminimum phase network is the **all-pass network**, which can be realized with a symmetrical lattice network [8]. A symmetrical pattern of poles and zeros is obtained as shown in Figure 8.18(a). Again, the magnitude $|G(j\omega)|$ remains constant; in this case, it is equal to unity. However, the angle varies

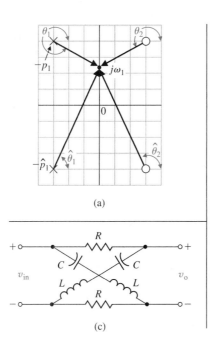
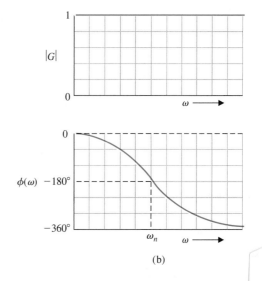

FIGURE 8.18
The all-pass network (a) pole–zero pattern, (b) frequency response, and (c) a lattice network.

(a)

(c)

(b)

from $0°$ to $-360°$. Because $\theta_2 = 180° - \theta_1$ and $\theta_2^* = 180° - \theta_1^*$, the phase is given by $\phi(\omega) = -2(\theta_1 + \theta_1^*)$. The magnitude and phase characteristic of the all-pass network is shown in Figure 8.18(b). A nonminimum phase lattice network is shown in Figure 8.18(c).

EXAMPLE 8.5   **Sketching a bode plot**

The Bode diagram of a transfer function $G(s)$, which contains several zeros and poles, is obtained by adding the plot due to each individual pole and zero. The simplicity of this method will be illustrated by considering a transfer function that possesses all the factors considered in the preceding section. The transfer function of interest is

$$G(j\omega) = \frac{5(1 + j0.1\omega)}{j\omega(1 + j0.5\omega)(1 + j0.6(\omega/50) + (j\omega/50)^2)}. \tag{8.42}$$

The factors, in order of their occurrence as frequency increases, are as follows:

1. A constant gain $K = 5$
2. A pole at the origin
3. A pole at $\omega = 2$
4. A zero at $\omega = 10$
5. A pair of complex poles at $\omega = \omega_n = 50$

First, we plot the magnitude characteristic for each individual pole and zero factor and the constant gain:

1. The constant gain is $20 \log 5 = 14$ dB, as shown in Figure 8.19.

2. The magnitude of the pole at the origin extends from zero frequency to infinite frequencies and has a slope of -20 dB/decade intersecting the 0-dB line at $\omega = 1$, as shown in Figure 8.19.

3. The asymptotic approximation of the magnitude of the pole at $\omega = 2$ has a slope of -20 dB/decade beyond the break frequency at $\omega = 2$. The asymptotic magnitude below the break frequency is 0 dB, as shown in Figure 8.19.

4. The asymptotic magnitude for the zero at $\omega = +10$ has a slope of $+20$ dB/decade beyond the break frequency at $\omega = 10$, as shown in Figure 8.19.

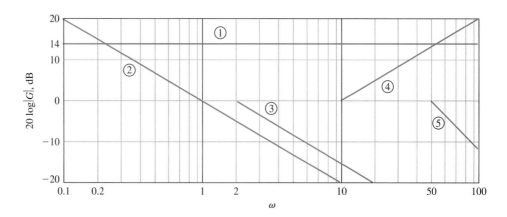

FIGURE 8.19
Magnitude asymptotes of poles and zeros used in the example.

5. The magnitude for the complex poles is -40 dB/decade. The break frequency is $\omega = \omega_n = 50$, as shown in Figure 8.19. This approximation must be corrected to the actual magnitude because the damping ratio is $\zeta = 0.3$, and the magnitude differs appreciably from the approximation, as shown in Figure 8.20.

Therefore, the total asymptotic magnitude can be plotted by adding the asymptotes due to each factor, as shown by the solid line in Figure 8.20. Examining the asymptotic curve of Figure 8.20, we note that the curve can be obtained directly by plotting each asymptote in order as frequency increases. Thus, the slope is -20 dB/decade due to $(j\omega)^{-1}$ intersecting 14 dB at $\omega = 1$. Then, at $\omega = 2$, the slope becomes -40 dB/decade due to the pole at $\omega = 2$. The slope changes to -20 dB/decade due to the zero at $\omega = 10$. Finally, the slope becomes -60 dB/decade at $\omega = 50$ due to the pair of complex poles at $\omega_n = 50$.

The exact magnitude curve is then obtained by using Table 8.2, which provides the difference between the actual and asymptotic curves for a single pole or zero. The exact magnitude curve for the pair of complex poles is obtained by utilizing Figure 8.10(a) for the quadratic factor. The exact magnitude curve for $G(j\omega)$ is shown by a dashed line in Figure 8.20.

The phase characteristic can be obtained by adding the phase due to each individual factor. Usually, the linear approximation of the phase characteristic for a single pole or zero is suitable for the initial analysis or design attempt. Thus, the individual phase characteristics for the poles and zeros are shown in Figure 8.21 and are as follows:

1. The phase of the constant gain is $0°$.

2. The phase of the pole at the origin is a constant $-90°$.

3. The linear approximation of the phase characteristic for the pole at $\omega = 2$ is shown in Figure 8.21, where the phase shift is $-45°$ at $\omega = 2$.

4. The linear approximation of the phase characteristic for the zero at $\omega = 10$ is also shown in Figure 8.21, where the phase shift is $+45°$ at $\omega = 10$.

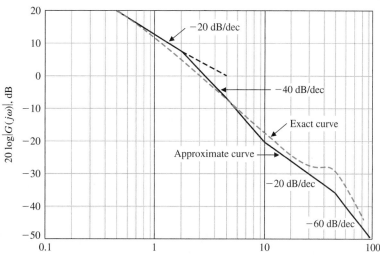

FIGURE 8.20
Magnitude
characteristic.

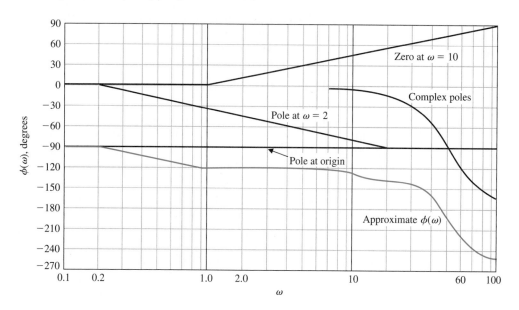

FIGURE 8.21
Phase
characteristic.

5. The actual phase characteristic for the pair of complex poles is obtained from Figure 8.10 and is shown in Figure 8.21.

   Therefore, the total phase characteristic, $\phi(\omega)$, is obtained by adding the phase due to each factor as shown in Figure 8.21. While this curve is an approximation, its usefulness merits consideration as a first attempt to determine the phase characteristic. Thus, a frequency of interest, as we shall note in the following section, is the frequency for which $\phi(\omega) = -180°$. The approximate curve indicates that a phase shift of $-180°$ occurs at $\omega = 46$. The actual phase shift at $\omega = 46$ can be readily calculated as

$$\phi(\omega) = -90° - \tan^{-1} \omega\tau_1 + \tan^{-1} \omega\tau_2 - \tan^{-1}\frac{2\zeta u}{1 - u^2}, \tag{8.43}$$

where

$$\tau_1 = 0.5, \qquad \tau_2 = 0.1, \quad 2\zeta = 0.6, \quad \text{and} \quad u = \omega/\omega_n = \omega/50.$$

Then we find that

$$\phi(46) = -90° - \tan^{-1} 23 + \tan^{-1} 4.6 - \tan^{-1} 3.55 = -175°, \tag{8.44}$$

and the approximate curve has an error of $5°$ at $\omega = 46$. However, once the approximate frequency of interest is ascertained from the approximate phase curve, the accurate phase shift for the neighboring frequencies is readily determined by using the exact phase shift relation (Equation 8.43). This approach is usually preferable to the calculation of the exact phase shift for all frequencies over several decades. In summary, we may obtain approximate curves for the magnitude and phase shift of a transfer function $G(j\omega)$ in order to determine the important frequency ranges. Then, within the relatively small important frequency ranges, the

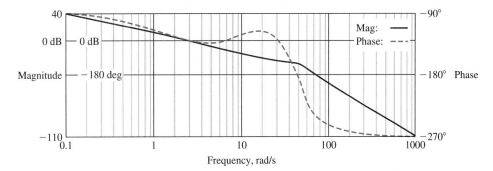

FIGURE 8.22
The Bode plot of
the $G(j\omega)$ of
Equation (8.42).

exact magnitude and phase shift can be readily evaluated by using the exact equations, such as Equation (8.43).

The frequency response of $G(j\omega)$ can be calculated and plotted using a computer program. The Bode plot for the example in this section (Equation 8.42) can be readily obtained, as shown in Figure 8.22. The plot is generated for four decades, and the 0-dB line is indicated, as well as the $-180°$ line. The data above the plot indicate that the magnitude is 34 dB and that the phase is $-92.36°$ at $\omega = 0.1$. Similarly, the data indicate that the magnitude is -43 dB and that the phase is $-243°$ at $\omega = 100$. Using the tabular data provided, we find that the magnitude is 0 dB at $\omega = 3.0$, and the phase is $-180°$ at $\omega = 50$. ∎

8.3 FREQUENCY RESPONSE MEASUREMENTS

A sine wave can be used to measure the open-loop frequency response of a control system. In practice, a plot of amplitude versus frequency and phase versus frequency will be obtained [1, 3, 6]. From these two plots, the open-loop transfer function $GH(j\omega)$ can be deduced. Similarly, the closed-loop frequency response of a control system, $T(j\omega)$, may be obtained and the actual transfer function deduced.

A device called a wave analyzer can be used to measure the amplitude and phase variations as the frequency of the input sine wave is altered. Also, a device called a transfer function analyzer can be used to measure the open-loop and closed-loop transfer functions [6].

A typical signal analyzer instrument can perform frequency response measurements from DC to 100 kHz. Built-in analysis and modeling capabilities can derive poles and zeros from measured frequency responses or construct phase and magnitude responses from user-supplied models. This device can also synthesize the frequency response of a model of a system, allowing a comparison with an actual response.

As an example of determining the transfer function from the Bode plot, let us consider the plot shown in Figure 8.23. The system is a stable circuit consisting of resistors and capacitors. Because the magnitude declines at about -20 dB/decade as ω increases between 100 and 1000, and because the phase is $-45°$ and the magnitude is -3 dB at 300 rad/s, we can deduce that one factor is a pole at $p_1 = 300$. Next, we deduce that a pair of quadratic zeros exist at $\omega_n = 2450$. This is inferred by noting

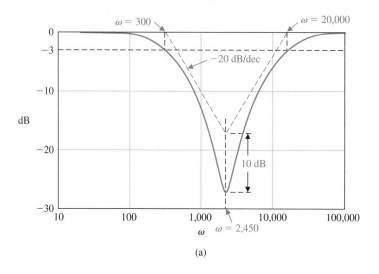

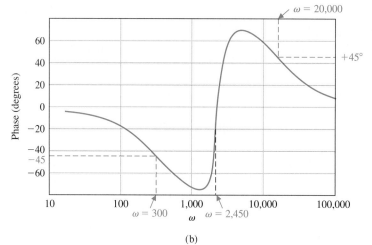

FIGURE 8.23
A Bode diagram for
a system with an
unidentified transfer
function.

that the phase changes abruptly by nearly $+180°$, passing through $0°$ at $\omega_n = 2450$. Also, the slope of the magnitude changes from -20 dB/decade to $+20$ dB/decade at $\omega_n = 2450$. Because the slope of the magnitude returns to 0 dB/decade as ω exceeds 50,000, we determine that there is a second pole as well as two zeros. This second pole is at $p_2 = 20,000$, because the magnitude is -3 dB from the asymptote and the phase is $+45°$ at this point ($-90°$ for the first pole, $+180°$ for the pair of quadratic zeros, and $-45°$ for the second pole). We sketch the asymptotes for the poles and the numerator of the proposed transfer function $T(s)$ of Equation (8.45), as shown in Figure 8.23(a). The equation is

$$T(s) = \frac{(s/\omega_n)^2 + (2\zeta/\omega_n)s + 1}{(s/p_1 + 1)(s/p_2 + 1)}. \tag{8.45}$$

The difference in magnitude from the corner frequency ($\omega_n = 2450$) of the asymptotes to the minimum response is 10 dB, which, from Equation (8.37), indicates that $\zeta = 0.16$. (Compare the plot of the quadratic zeros to the plot of the quadratic poles in Figure 8.10. Note that the plots need to be turned "upside down" for the quadratic zeros and that the phase goes from 0° to +180° instead of −180°.) Therefore, the transfer function is

$$T(s) = \frac{(s/2450)^2 + (0.32/2450)s + 1}{(s/300 + 1)(s/20000 + 1)}.$$

This frequency response is actually obtained from a bridged-T network (see Problems 2.8 and 8.3 and Figure 8.14).

8.4 PERFORMANCE SPECIFICATIONS IN THE FREQUENCY DOMAIN

We must continually ask the question: how does the frequency response of a system relate to the expected transient response of the system? In other words, given a set of time-domain (transient performance) specifications, how do we specify the frequency response? For a simple second-order system, we have already answered this question by considering the time-domain performance in terms of overshoot, settling time, and other performance criteria, such as integral squared error. For the second-order system shown in Figure 8.24, the closed-loop transfer function is

$$T(s) = \frac{\omega_n^2}{s^2 + 2\zeta\omega_n s + \omega_n^2}. \tag{8.46}$$

The frequency response of this feedback system will appear as shown in Figure 8.25. Because this is a second-order system, the damping ratio of the system is related to the maximum magnitude $M_{p\omega}$, which occurs at the frequency ω_r as shown in Figure 8.25.

At the resonant frequency ω_r a maximum value $M_{p\omega}$ of the frequency response is attained.

FIGURE 8.24
A second-order closed-loop system.

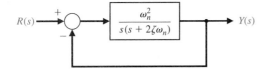

FIGURE 8.25
Magnitude characteristic of the second-order system.

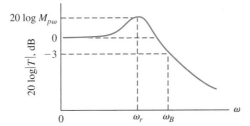

The bandwidth, ω_B, is a measure of a ability of the system to faithfully reproduce an input signal.

> **The bandwidth is the frequency ω_B at which the frequency response has declined 3 dB from its low-frequency value. This corresponds to approximately half an octave, or about $1/\sqrt{2}$ of the low-frequency value.**

The resonant frequency ω_r and the -3-dB **bandwidth** can be related to the speed of the transient response. Thus, as the bandwidth ω_B increases, the rise time of the step response of the system will decrease. Furthermore, the overshoot to a step input can be related to $M_{p\omega}$ through the damping ratio ζ. The curves of Figure 8.11 relate the resonance magnitude and frequency to the damping ratio of the second-order system. Then the step response overshoot may be estimated from Figure 5.8 or may be calculated by utilizing Equation (5.15). Thus, we find as the resonant peak $M_{p\omega}$ increases in magnitude, the overshoot to a step input increases. In general, the magnitude $M_{p\omega}$ indicates the relative stability of a system.

The bandwidth of a system ω_B, as indicated on the frequency response, can be approximately related to the natural frequency of the system. Figure 8.26 shows the normalized bandwidth ω_B/ω_n versus ζ for the second-order system of Equation (8.46). The response of the second-order system to a unit step input is of the form (see Equation (5.9))

$$y(t) = 1 + Be^{-\zeta\omega_n t}\cos(\omega_1 t + \theta). \tag{8.47}$$

The greater the magnitude of ω_n when ζ is constant, the more rapidly the response approaches the desired steady-state value. Thus, desirable frequency-domain specifications are as follows:

1. Relatively small resonant magnitudes: $M_{p\omega} < 1.5$, for example.
2. Relatively large bandwidths so that the system time constant $\tau = 1/(\zeta\omega_n)$ is sufficiently small.

FIGURE 8.26
Normalized bandwidth, ω_B/ω_n, versus ζ for a second-order system (Equation 8.46). The linear approximation $\omega_B/\omega_n = -1.19\zeta + 1.85$ is accurate for $0.3 \le \zeta \le 0.8$.

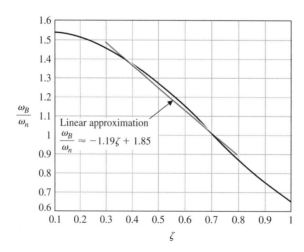

Linear approximation
$$\frac{\omega_B}{\omega_n} \approx -1.19\zeta + 1.85$$

The usefulness of these frequency response specifications and their relation to the actual transient performance depend upon the approximation of the system by a second-order pair of complex poles. This approximation was discussed in Section 7.3, and the second-order poles of $T(s)$ are called the **dominant roots**. If the frequency response is dominated by a pair of complex poles, the relationships between the frequency response and the time response discussed in this section will be valid. Fortunately, a large proportion of control systems satisfy this dominant second-order approximation in practice.

The steady-state error specification can also be related to the frequency response of a closed-loop system. As we found in Section 5.6, the steady-state error for a specific test input signal can be related to the gain and number of integrations (poles at the origin) of the loop transfer function. Therefore, for the system shown in Figure 8.24, the steady-state error for a ramp input is specified in terms of K_v, the velocity constant. The steady-state error for the system is

$$\lim_{t \to \infty} e(t) = \frac{A}{K_v},$$

where A = magnitude of the ramp input. The velocity constant for the system of Figure 8.24 without feedback is

$$K_v = \lim_{s \to 0} sG(s) = \lim_{s \to 0} s\left(\frac{\omega_n^2}{s(s + 2\zeta\omega_n)}\right) = \frac{\omega_n}{2\zeta}. \tag{8.48}$$

In Bode diagram form (in terms of time constants), the transfer function is written as

$$G(s) = \frac{\omega_n/(2\zeta)}{s(s/(2\zeta\omega_n) + 1)} = \frac{K_v}{s(\tau s + 1)}, \tag{8.49}$$

and the gain constant is K_v for this type-one system. For example, reexamining Example 8.5, we had a type-one system with a loop transfer function

$$G(j\omega) = \frac{5(1 + j\omega\tau_2)}{j\omega(1 + j\omega\tau_1)(1 + j0.6u - u^2)}, \tag{8.50}$$

where $u = \omega/\omega_n$. Therefore, in this case, we have $K_v = 5$. In general, if the loop transfer function of a feedback system is written as

$$G(j\omega) = \frac{K\displaystyle\prod_{i=1}^{M}(1 + j\omega\tau_i)}{(j\omega)^N\displaystyle\prod_{k=1}^{Q}(1 + j\omega\tau_k)}, \tag{8.51}$$

then the system is type N and the gain K is the gain constant for the steady-state error. Thus, for a type-zero system that has a loop transfer function, we have

$$G(j\omega) = \frac{K}{(1 + j\omega\tau_1)(1 + j\omega\tau_2)}. \tag{8.52}$$

In this equation, $K = K_p$ (the position error constant) that appears as the low-frequency gain on the Bode diagram.

Furthermore, the gain constant $K = K_v$ for the type-one system appears as the gain of the low-frequency section of the magnitude characteristic. Considering only the pole and gain of the type-one system of Equation (8.50), we have

$$G(j\omega) = \frac{5}{j\omega} = \frac{K_v}{j\omega}, \qquad \omega < 1/\tau_1, \qquad (8.53)$$

and the K_v is equal to the magnitude when this portion of the magnitude characteristic intersects the 0-dB line. For example, the low-frequency intersection of $K_v/j\omega$ in Figure 8.20 is equal to $\omega = 5$, as we expect.

Therefore, the frequency response characteristics represent the performance of a system quite adequately, and with some experience, they are quite useful for the analysis and design of feedback control systems.

8.5 LOG MAGNITUDE AND PHASE DIAGRAMS

There are several alternative methods for presenting the frequency response of a function $G(j\omega)$. We have seen that suitable graphical presentations of the frequency response are (1) the polar plot and (2) the Bode diagram. An alternative approach to portraying the frequency response graphically is to plot the logarithmic magnitude in dB versus the phase angle for a range of frequencies. Because this information is equivalent to that portrayed by the Bode diagram, it is normally easier to obtain the Bode diagram and transfer the information to the coordinates of the log magnitude versus phase diagram.

An illustration will best portray the use of the log-magnitude–phase diagram. This diagram for a transfer function

$$G_1(j\omega) = \frac{5}{j\omega(0.5j\omega + 1)(j\omega/6 + 1)} \qquad (8.54)$$

is shown in Figure 8.27. The numbers indicated along the curve are for values of frequency ω.

The log-magnitude–phase curve for the transfer function

$$G_2(j\omega) = \frac{5(0.1j\omega + 1)}{j\omega(0.5j\omega + 1)(1 + j0.6(\omega/50) + (j\omega/50)^2)} \qquad (8.55)$$

considered in Section 8.2 is shown in Figure 8.28. This curve is obtained most readily by utilizing the Bode diagrams of Figures 8.20 and 8.21 to transfer the frequency response information to the log magnitude and phase coordinates. The shape of the locus of the frequency response on a log-magnitude–phase diagram is particularly important as the phase approaches $-180°$ and the magnitude approaches 0 dB. The locus of Equation (8.54) and Figure 8.27 differs substantially from the locus of Equation (8.55) and Figure 8.28. Therefore, as the correlation between the shape of the locus and the transient response of a system is established, we will obtain another useful portrayal of the frequency response of a system. In Chapter 9, we will establish a stability criterion in the frequency domain for which it will be useful to utilize the log-magnitude–phase diagram to investigate the relative stability of closed-loop feedback control systems.

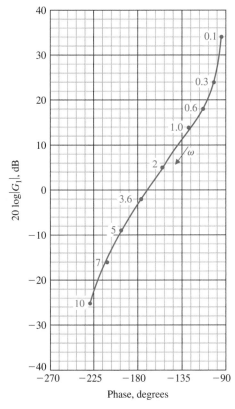

FIGURE 8.27   Log-magnitude–phase curve for $G_1(j\omega)$.

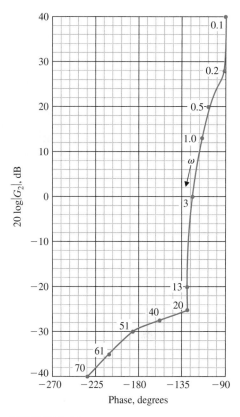

FIGURE 8.28   Log-magnitude–phase curve for $G_2(j\omega)$.

8.6 DESIGN EXAMPLES

In this section we present two illustrative examples using frequency response methods to design controllers. The first example illustrates the use of log-magnitude-phase plots, as well as open-and closed-loop Bode plots. The specific problem is to design a proportional controller gain for an engraving machine control feedback control system. The second example considers the control of one leg of a six-legged robotic device. In this example, the specifications that must be satisfied include a mix of time-domain specifications (percent overshoot and settling time) and frequency-domain specifications (bandwidth). The design process leads to a viable PID controller meeting all the specifications.

EXAMPLE 8.6   **Engraving machine control system**

The engraving machine shown in Figure 8.29(a) uses two drive motors and associated lead screws to position the engraving scribe in the x direction [7]. A separate motor is used for both the y- and z-axes, as shown. The block diagram model for the x-axis position control system is shown in Figure 8.29(b). The goal is to select an

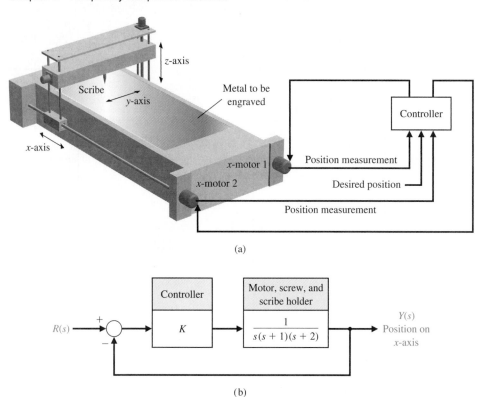

FIGURE 8.29
(a) Engraving
machine control
system. (b) Block
diagram model.

appropriate gain K, using frequency response methods, so that the time response to step commands is acceptable.

To represent the frequency response of the system, we will first obtain the open-loop and closed-loop Bode diagrams. Then we will use the closed-loop Bode diagram to predict the time response of the system and check the predicted results with the actual results.

To plot the frequency response, we arbitrarily select $K = 2$ and proceed with obtaining the Bode diagram. If the resulting system is not acceptable, we will later adjust the gain.

The frequency response of $G(j\omega)$ is partially listed in Table 8.4 and is plotted in Figure 8.30. We need the frequency response of the closed-loop transfer function

$$T(s) = \frac{2}{s^3 + 3s^2 + 2s + 2}. \tag{8.56}$$

Table 8.4   Frequency Response for $G(j\omega)$

ω	0.2	0.4	0.8	1.0	1.4	1.8
$20 \log\lvert G\rvert$	14	7	−1	−4	−9	−13
ϕ	−107°	−123°	−150.5°	−162°	−179.5°	−193°

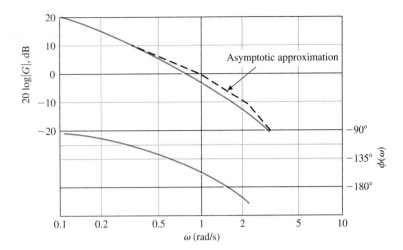

FIGURE 8.30
Bode diagram for
$G(j\omega)$.

Therefore, we let $s = j\omega$, obtaining

$$T(j\omega) = \frac{2}{(2 - 3\omega^2) + j\omega(2 - \omega^2)}. \qquad (8.57)$$

The Bode diagram of the closed-loop system is shown in Figure 8.31, where $20 \log|T(j\omega)| = 5$ dB at $\omega_r = 0.8$. Hence,

$$20 \log M_{p\omega} = 5 \quad \text{or} \quad M_{p\omega} = 1.78.$$

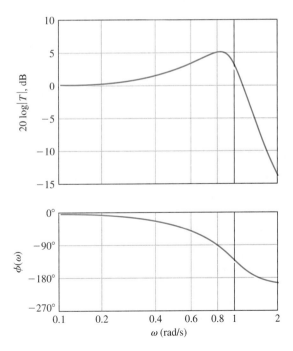

FIGURE 8.31
Bode diagram for
closed-loop
system.

If we assume that the system has dominant second-order roots, we can approximate the system with a second-order frequency response of the form shown in Figure 8.10. Since $M_{p\omega} = 1.78$, we use Figure 8.11 to estimate ζ to be 0.29. Using this ζ and $\omega_r = 0.8$, we can use Figure 8.11 to estimate $\omega_r/\omega_n = 0.91$. Therefore,

$$\omega_n = \frac{0.8}{0.91} = 0.88.$$

Since we are now approximating $T(s)$ as a second-order system, we have

$$T(s) \approx \frac{\omega_n^2}{s^2 + 2\zeta\omega_n s + \omega_n^2} = \frac{0.774}{s^2 + 0.51s + 0.774} \tag{8.58}$$

We use Figure 5.8 to predict the overshoot to a step input as 37% for $\zeta = 0.29$. The settling time (to within 2% of the final value) is estimated as

$$T_s = \frac{4}{\zeta\omega_n} = \frac{4}{(0.29)0.88} = 15.7\text{s}.$$

The actual overshoot for a step input is 34%, and the actual settling time is 17 seconds. We see that the second-order approximation is reasonable in this case and can be used to determine suitable parameters on a system. If we require a system with lower overshoot, we would reduce K to 1 and repeat the procedure. ∎

EXAMPLE 8.7   **Control of one leg of a six-legged robot**

The Ambler is a six-legged walking machine being developed at Carnegie-Mellon University [23]. An artist's conception of the Ambler is shown in Figure 8.32.

In this example we consider the control system design for position control of one leg. The elements of the design process emphasized in this example are highlighted in Figure 8.33. The mathematical model of the actuator and leg is provided. The transfer function is

$$G(s) = \frac{1}{s(s^2 + 2s + 10)}. \tag{8.59}$$

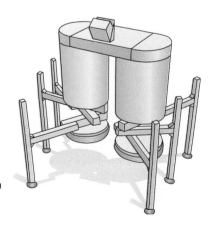

FIGURE 8.32
An artist's conception of the six-legged Ambler.

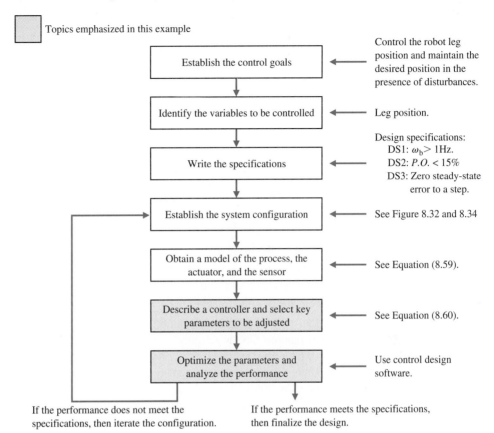

Topics emphasized in this example

Establish the control goals — Control the robot leg position and maintain the desired position in the presence of disturbances.

Identify the variables to be controlled — Leg position.

Write the specifications — Design specifications:
DS1: $\omega_b >$ 1Hz.
DS2: *P.O.* < 15%
DS3: Zero steady-state error to a step.

Establish the system configuration — See Figure 8.32 and 8.34

Obtain a model of the process, the actuator, and the sensor — See Equation (8.59).

Describe a controller and select key parameters to be adjusted — See Equation (8.60).

Optimize the parameters and analyze the performance — Use control design software.

If the performance does not meet the specifications, then iterate the configuration.

If the performance meets the specifications, then finalize the design.

FIGURE 8.33
Elements of the control system design process emphasized in this six-legged robot example.

The input is a voltage command to the actuator, and the output is the leg position (vertical position only). A block diagram of the control system is shown in Figure 8.34. The control goal is

Control Goal
Control the robot leg position and maintain the position in the presence of unwanted measurement noise.

The variable to be controlled is

Variable to Be Controlled
Leg position, $Y(s)$.

We want the leg to move to the commanded position as fast as possible but with minimal overshoot. As a practical first step, the design goal will be to produce a system that moves, albeit slowly. In other words, the control system bandwidth will initially be low.

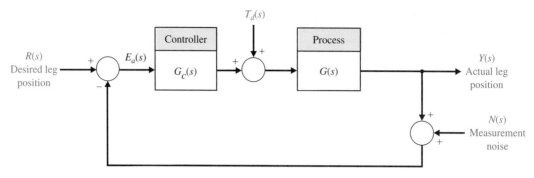

FIGURE 8.34   Control system for one leg.

The control design specifications are

Control Design Specifications

    DS1   Closed-loop bandwidth greater than 1 Hz.

    DS2   Percent overshoot less than 15% to a step input.

    DS3   Zero steady-state tracking error to a step input.

Specifications DS1 and DS2 are intended to ensure acceptable tracking performance. Design specification DS3 is actually a nonissue in our design: the actuator/leg transfer function is a type-one system so a zero steady-state tracking error to a step input is guaranteed. We simply need to ensure that $G_c(s)G(s)$ remains at least a type-one system.

    Consider the controller

$$G_c(s) = \frac{K(s^2 + as + b)}{s + c}. \tag{8.60}$$

As $c \to 0$, a PID controller is obtained with $K_P = K_a$, $K_D = K$, and $K_I = Kb$. We can let c be a parameter at this point and see if the additional freedom in selecting $c \neq 0$ is useful. It may be that we can simply set $c = 0$ and use the PID form. The key tuning parameters are

Select Key Tuning Parameters

    K, a, b, and c.

The controller in Equation (8.60) is not the only controller that we can consider. For example, we might consider

$$G_c(s) = K\frac{s + z}{s + p}, \tag{8.61}$$

where K, z, and p are the key tuning parameters. The design of the type of controller given in Equation (8.61) will be left as a design problem at the end of the chapter.

    The response of a closed-loop control system is determined predominantly by the location of the dominant poles. Our approach to the design is to determine appropriate locations for the dominant poles of the closed-loop system. We can determine the locations from the performance specifications by using second-order system approximation formulas. Once the controller parameters are obtained so that the closed-loop system has the desired dominant poles, the remaining poles are located so that their contribution to the overall response is negligible.

The bandwidth ω_B is approximately related to the natural frequency ω_n by

$$\frac{\omega_B}{\omega_n} \approx -1.1961\zeta + 1.8508 \quad (0.3 \le \zeta \le 0.8). \tag{8.62}$$

This approximation applies to second-order systems. Per specification DS1, we want

$$\omega_B = 1 \text{ Hz} = 6.28 \text{ rad/s}. \tag{8.63}$$

From the percent overshoot specification, we can determine the minimum value of ζ. Thus for $P.O. \le 15\%$, we require

$$\zeta \ge 0.52, \tag{8.64}$$

where we have used Equation (5.16) (valid for second-order systems) that

$$P.O. = 100e^{-\zeta\pi/\sqrt{1-\zeta^2}}.$$

Another useful design formula (Equation (8.37)) relates $M_{p\omega} = |T(\omega_r)|$ to the damping ratio:

$$M_{p\omega} = |T(\omega_r)| = \frac{1}{2\zeta\sqrt{1-\zeta^2}} \quad (\zeta < 0.707). \tag{8.65}$$

The relationship between the resonant frequency, ω_r the natural frequency ω_n, and the damping ratio ζ is given by (Equation (8.36))

$$\omega_r = \omega_n\sqrt{1 - 2\zeta^2} \quad (\zeta < 0.707). \tag{8.66}$$

We require $\zeta \ge 0.52$; therefore, we will design with $\zeta = 0.52$. Even though settling time is not a design specification for this problem, we usually attempt to make the system response as fast as possible while still meeting all the design specifications. From Equations (8.62) and (8.63) it follows that

$$\omega_n = \frac{\omega_B}{-1.1961\zeta + 1.8508} = 5.11 \text{ rad/s}. \tag{8.67}$$

Then with $\omega_n = 5.11$ rad/s and $\zeta = 0.52$ and using Equation (8.66) we compute

$$\omega_r = 3.46 \text{ rad/s}. \tag{8.68}$$

So if we had a second-order system, we would want to determine values of the control gains such that

$$\omega_n = 5.11 \text{ rad/s} \quad \text{and} \quad \zeta = 0.52,$$

which give

$$M_{p\omega} = 1.125 \quad \text{and} \quad \omega_r = 3.46 \text{ rad/s}.$$

Our closed-loop system is a fourth-order system and not a second-order system. So, a valid design approach would be to select K, a, b, and c so that two poles are dominant and located appropriately to meet the design specifications. This will be the approach followed here.

Another valid approach is to develop a second-order approximation of the fourth-order system. In the approximate transfer function, the parameters K, a, b, and c are left as variables. Following the approach discussed in Chapter 5, we can obtain an approximate transfer function $T_L(s)$ in such a way that the frequency response of $T_L(s)$ is very close to that of the original system.

The loop transfer function is

$$G_c(s)G(s) = \frac{K(s^2 + as + b)}{s(s^2 + 2s + 10)(s + c)},$$

and the closed-loop transfer function is

$$T(s) = \frac{G_c(s)G(s)}{1 + G_c(s)G(s)} \tag{8.69}$$

$$= \frac{K(s^2 + as + b)}{s^4 + (2 + c)s^3 + (10 + 2c + K)s^2 + (10c + Ka)s + Kb}.$$

The associated characteristic equation is

$$s^4 + (2 + c)s^3 + (10 + 2c + K)s^2 + (10c + Ka)s + Kb = 0. \tag{8.70}$$

The desired characteristic polynomial must also be fourth-order, but we want it to be composed of multiple factors, as follows:

$$P_d(s) = (s^2 + 2\zeta\omega_n s + \omega_n^2)(s^2 + d_1 s + d_0),$$

where ζ and ω_n are selected to meet the design specifications, and the roots of $s^2 + 2\zeta\omega_n s + \omega_n^2 = 0$ are the dominant roots. Conversely we want the roots of $s^2 + d_1 s + d_0 = 0$ to be the nondominant roots. The dominant roots should lie on a vertical line in the complex plane defined by the distance $s = -\zeta\omega_n$ away from the imaginary axis. Let

$$d_1 = 2\alpha\zeta\omega_n.$$

Then the roots of $s^2 + d_1 s + d_0 = 0$, when complex, lie on a vertical line in the complex plane defined by $s = -\alpha\zeta\omega_n$. By choosing $\alpha > 1$, we effectively move the roots to the left of the dominant roots. The larger we select α, the further the nondominant roots lie to the left of the dominant roots. A reasonable value of α is

$$\alpha = 12.$$

Also, if we select

$$d_0 = \alpha^2\zeta^2\omega_n^2,$$

then we obtain two real roots

$$s^2 + d_1 s + d_0 = (s + \alpha\zeta\omega_n)^2 = 0.$$

Choosing $d_0 = \alpha^2 \zeta^2 \omega_n^2$ is not required, but this seems to be a reasonable choice since we would like the contribution of the nondominant roots to the overall response to be quickly fading and nonoscillatory.

The desired characteristic polynomial is then

$$s^4 + [2\zeta\omega_n(1 + \alpha)]s^3 + [\omega_n^2(1 + \alpha\zeta^2(\alpha + 4))]s^2 \tag{8.71}$$

$$+ [2\alpha\zeta\omega_n^3(1 + \zeta^2\alpha)]s + \alpha^2\zeta^2\omega_n^4 = 0.$$

Equating the coefficients of Equations (8.70) and (8.71) yields four relationships involving K, a, b, c, and α:

$$2\zeta\omega_n(1 + \alpha) = 2 + c,$$

$$\omega_n^2(1 + \alpha\zeta^2(4 + \alpha)) = 10 + 2c + K,$$

$$2\alpha\zeta\omega_n^3(1 + \zeta^2\alpha) = 10c + Ka,$$

$$\alpha^2\zeta^2\omega_n^4 = Kb.$$

In our case $\zeta = 0.52$, $\omega_n = 5.11$, and $\alpha = 12$. Thus we obtain

$$c = 67.13$$

$$K = 1239.2$$

$$a = 5.17$$

$$b = 21.48$$

and the resulting controller is

$$G_c(s) = 1239\frac{s^2 + 5.17s + 21.48}{s + 67.13}. \tag{8.72}$$

The step response of the closed-loop system using the controller in Equation (8.72) is shown in Figure 8.35. The percent overshoot is $P.O. = 14\%$, and the settling time is $T_s = 0.96$ second.

The magnitude plot of the closed-loop system is shown in Figure 8.36. The bandwidth is $\omega_B = 27.2$ rad/s $= 4.33$ Hz. This satisfies DS1 but is larger than the $\omega_B = 1$ Hz used in the design (due to the fact that our system is not a second-order system). The higher bandwidth leads us to expect a faster settling time. The peak magnitude is $M_{p\omega} = 1.21$. We were expecting $M_{p\omega} = 1.125$.

What is the steady-state response of the closed-loop system if the input is a sinusoidal input? From our previous discussions we expect that as the input frequency increases, the magnitude of the output will decrease. Two cases are presented here. In Figure 8.37 the input frequency is $\omega = 1$ rad/s. The output magnitude is approximately equal to 1 in the steady-state. In Figure 8.38 the input frequency is $\omega = 500$ rad/s. The output magnitude is less than 0.005 in the steady-state. This verifies our intuition that the system response decreases as the input sinusoidal frequency increases.

Using simple analytic methods, we obtained an initial set of controller parameters for the mobile robot. The controller thus designed proved to satisfy the design requirements. Some fine-tuning would be necessary to meet the design specifications exactly. ∎

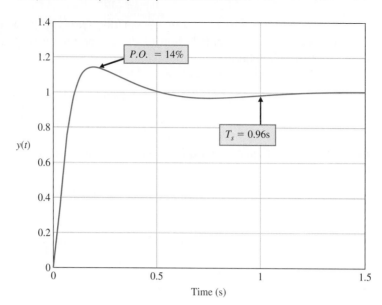

FIGURE 8.35
Step response using the controller in Equation (8.72).

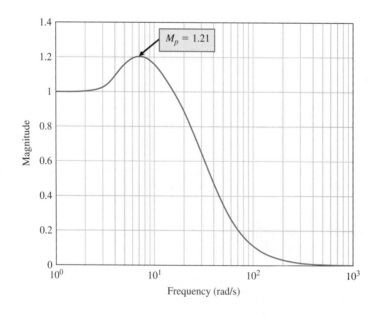

FIGURE 8.36
Magnitude plot of the closed-loop system with the controller in Equation (8.72).

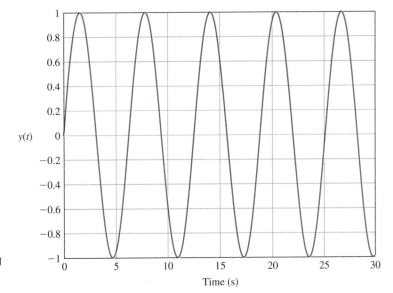

FIGURE 8.37
Output response of the closed-loop system when the input is a sinusoidal signal of frequency $\omega = 1$ rad/s.

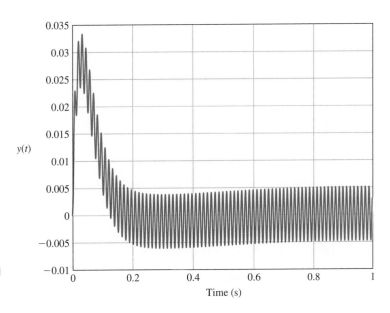

FIGURE 8.38
Output response of the closed-loop system when the input is a sinusoidal signal of frequency $\omega = 500$ rad/s.

8.7 FREQUENCY RESPONSE METHODS USING CONTROL DESIGN SOFTWARE

This section begins with an introduction to the Bode diagram and then discusses the connection between the frequency response and performance specifications in the time domain. The section concludes with an illustrative example of designing a control system in the frequency domain.

We will cover the functions **bode** and **logspace**. The **bode** function is used to generate a Bode diagram, and the **logspace** function generates a logarithmically spaced vector of frequencies utilized by the **bode** function.

Bode Diagram.   Consider the transfer function

$$G(s) = \frac{5(1 + 0.1s)}{s(1 + 0.5s)(1 + (0.6/50)s + (1/50^2)s^2)}. \tag{8.73}$$

The Bode diagram corresponding to Equation (8.73) is shown in Figure 8.39. The diagram consists of the logarithmic gain in dB versus ω in one plot and the phase $\phi(\omega)$ versus ω in a second plot. As with the root locus plots, it will be tempting to rely exclusively on control design software to obtain the Bode diagrams. The software should be treated as one tool in a tool kit that can be used to design and analyze control systems. It is essential to develop the capability to obtain approximate Bode diagrams manually. There is no substitute for a clear understanding of the underlying theory.

A Bode diagram is obtained with the **bode** function, shown in Figure 8.40. The Bode diagram is automatically generated if the **bode** function is invoked without left-hand arguments. Otherwise, the magnitude and phase characteristics are placed in the workspace through the variables *mag* and *phase*. A Bode diagram is obtained with the **plot** or **semilogx** function using mag, phase, and ω. The vector ω contains the values of the frequency in rad/s at which the Bode diagram will be calculated.

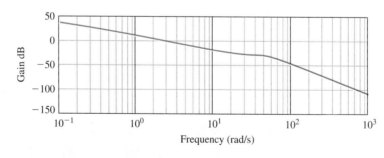

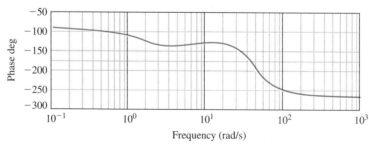

FIGURE 8.39
The Bode plot associated with Equation (8.73).

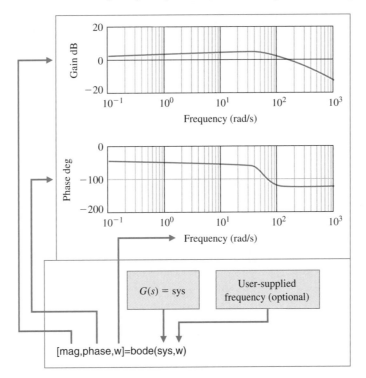

FIGURE 8.40
The **bode** function, given $G(s)$.

If ω is not specified, the bode function will automatically choose the frequency values by placing more points in regions where the frequency response is changing quickly. If the frequencies are specified explicitly, it is desirable to generate the vector ω using the logspace function. The logspace function is shown in Figure 8.41.

The Bode diagram in Figure 8.39 is generated using the script shown in Figure 8.42. The bode function automatically selected the frequency range. This range is user selectable using the logspace function. The bode function can be used with a state variable model, as shown in Figure 8.43. The use of the bode function is exactly the same as with transfer functions, except that the input is a state-space object instead of a transfer function object.

Keep in mind that our goal is to design control systems that satisfy certain performance specifications given in the time domain. Thus, we must establish a connection between the frequency response and the transient time response of a system. The relationship between specifications given in the time domain to those given in the frequency domain depends upon approximation of the system by a second-order system with the poles being the system dominant roots.

Consider the second-order system shown in Figure 8.24. The closed-loop transfer function is

$$T(s) = \frac{\omega_n^2}{s^2 + 2\zeta\omega_n s + \omega_n^2}. \tag{8.74}$$

The Bode diagram magnitude characteristic associated with the closed-loop transfer function in Equation (8.74) is shown in Figure 8.25. The relationship

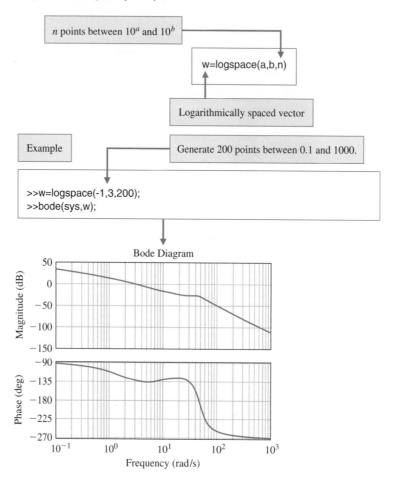

FIGURE 8.41
The **logspace** function.

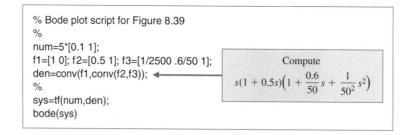

FIGURE 8.42
The script for the Bode diagram in Figure 8.39.

between the resonant frequency, ω_r, the maximum of the frequency response, $M_{p\omega}$, and the damping ratio, ζ, and the natural frequency, ω_n, is shown in Figure 8.44 (and in Figure 8.11). The information in Figure 8.44 will be quite helpful in designing control systems in the frequency domain while satisfying time-domain specifications.

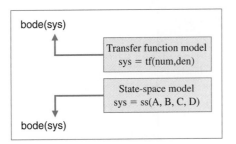

FIGURE 8.43
The **bode** function with a state variable model.

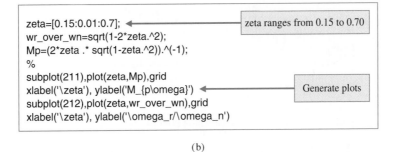

FIGURE 8.44
(a) The relationship between ($M_{p\omega}$, ω_r) and (ζ, ω_n) for a second-order system. (b) m-file script.

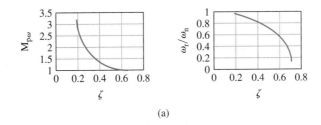

EXAMPLE 8.8   Engraving machine system

Consider the block diagram model in Figure 8.29. Our objective is to select K so that the closed-loop system has an acceptable time response to a step command. A functional block diagram describing the frequency-domain design process is shown in Figure 8.45. First, we choose $K = 2$ and then iterate K if the performance is unacceptable. The script shown in Figure 8.46 is used in the design. The value of K is defined at the command level. Then the script is executed and the closed-loop Bode diagram is generated. The values of $M_{p\omega}$ and ω_r are determined by inspection from the Bode diagram. Those values are used in conjunction with Figure 8.44 to determine the corresponding values of ζ and ω_n.

Given the damping ratio, ζ, and the natural frequency, ω_n, the settling time and percent overshoot are estimated using the formulas

$$T_s \approx \frac{4}{\zeta\omega_n}, \qquad P.O. \approx 100 \exp\frac{-\zeta\pi}{\sqrt{1-\zeta^2}}.$$

If the time-domain specifications are not satisfied, then we adjust K and iterate.

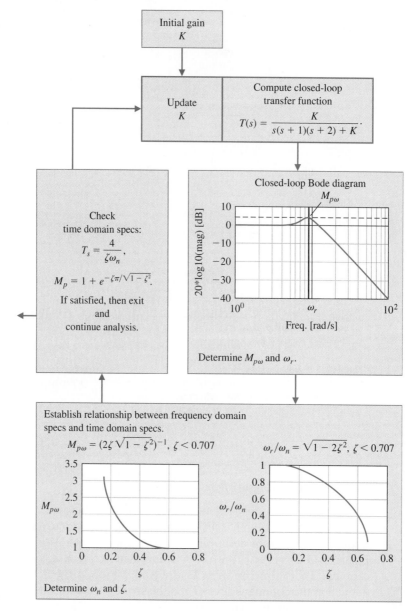

FIGURE 8.45
Frequency design functional block diagram for the engraving machine.

The values for ζ and ω_n corresponding to $K = 2$ are $\zeta = 0.29$ and $\omega_n = 0.88$. This leads to a prediction of $P.O. = 37\%$ and $T_s = 15.7$ seconds. The step response, shown in Figure 8.47, is a verification that the performance predictions are quite accurate and that the closed-loop system performs adequately.

In this example, the second-order system approximation is reasonable and leads to an acceptable design. However, the second-order approximation may not always lead directly to a good design. Fortunately, the control design software allows us to construct an interactive design facility to assist in the design process by reducing the manual computational loads while providing easy access to a host of classical and modern control tools. ■

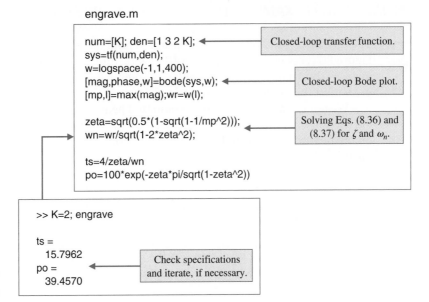

engrave.m

```
num=[K]; den=[1 3 2 K];          ◄──  Closed-loop transfer function.
sys=tf(num,den);
w=logspace(-1,1,400);
[mag,phase,w]=bode(sys,w);       ◄──  Closed-loop Bode plot.
[mp,l]=max(mag);wr=w(l);

zeta=sqrt(0.5*(1-sqrt(1-1/mp^2)));   ◄── Solving Eqs. (8.36) and
wn=wr/sqrt(1-2*zeta^2);                    (8.37) for ζ and ωₙ.

ts=4/zeta/wn
po=100*exp(-zeta*pi/sqrt(1-zeta^2))
```

```
>> K=2; engrave

ts =
   15.7962                 Check specifications
po =              ◄──     and iterate, if necessary.
   39.4570
```

FIGURE 8.46
Script for the
design of an
engraving machine.

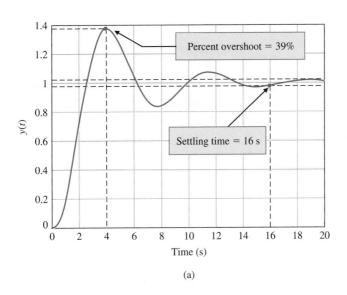

Percent overshoot = 39%

Settling time = 16 s

(a)

```
K=2; num=[K]; den=[1 3 2 K]; sys=tf(num,den);
t=[0:0.01:20];
y=step(sys,t); plot(t,y); grid
xlabel('Time (s)'), ylabel('y(t)')
```

FIGURE 8.47
(a) Engraving
machine step
response for $K = 2$.
(b) m-file script.

(b)

8.8 SEQUENTIAL DESIGN EXAMPLE: DISK DRIVE READ SYSTEM

The disk drive uses a flexure suspension to hold the reader head mount, as shown in Figure 2.71. As noted in Section 3.10, this flexure may be modeled by a spring and mass, as shown in Figure 3.40. In this chapter, we will include the effect of the flexure within the model of the motor-load system [22].

We model the flexure with the mounted head as a mass M, a spring k, and a sliding friction b, as shown in Figure 8.48. Here, we assume that the force $u(t)$ is exerted on the flexure by the arm. The transfer function of a spring-mass-damper was developed in Chapter 2, where

$$\frac{Y(s)}{U(s)} = G_3(s) = \frac{\omega_n^2}{s^2 + 2\zeta\omega_n s + \omega_n^2} = \frac{1}{1 + (2\zeta s/\omega_n) + (s/\omega_n)^2}.$$

A typical flexure and head has $\zeta = 0.3$ and a natural resonance at $f_n = 3000$ Hz. Therefore, $\omega_n = 18.85 \times 10^3$ as shown in the model of the system (see Figure 8.49).

First, we sketch the magnitude characteristics for the open-loop Bode diagram. The Bode diagram sketch is shown in Figure 8.50. Note that the actual plot has a 10-dB gain (over the asymptotic plot) at the resonance $\omega = \omega_n$, as shown in the sketch. The sketch is a plot of

$$20 \log |K(j\omega + 1)G_1(j\omega)G_2(j\omega)G_3(j\omega)|,$$

for the system of Figure 8.49 when $K = 400$. Note the resonance at ω_n. Clearly, we wish to avoid exciting this resonance.

FIGURE 8.48
Spring, mass, friction model of flexure and head.

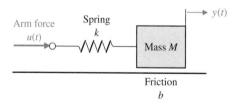

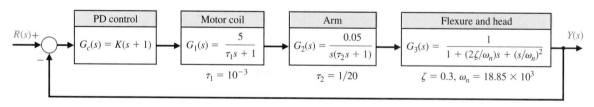

FIGURE 8.49 Disk drive head position control, including effect of flexure head mount.

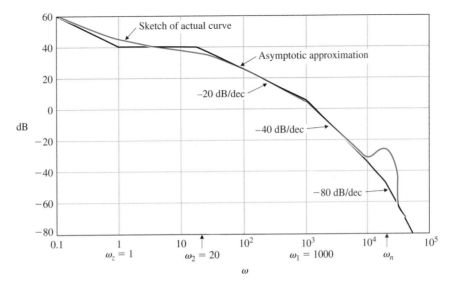

FIGURE 8.50   Sketch of the Bode diagram magnitude for the system of Figure 8.49.

Plots of the magnitude of the open-loop Bode diagram and the closed-loop Bode diagram are shown in Figure 8.51. The bandwidth of the closed-loop system is $\omega_B = 2000$ rad/s. We can estimate the settling time (with a 2% criterion) of this system using

$$T_s = \frac{4}{\zeta \omega_n},$$

where $\zeta \approx 0.8$ and $\omega_n \approx \omega_B = 2000$ rad/s. Therefore, we expect $T_s = 2.5$ ms for the system of Figure 8.49. As long as $K \leq 400$, the resonance is outside the bandwidth of the system.

8.9 SUMMARY

In this chapter, we have considered the representation of a feedback control system by its frequency response characteristics. The frequency response of a system was defined as the steady-state response of the system to a sinusoidal input signal. Several alternative forms of frequency response plots were considered. They included the polar plot of the frequency response of a system $G(j\omega)$ and logarithmic plots, often called Bode plots. The value of the logarithmic measure was also illustrated. The ease of obtaining a Bode plot for the various factors of $G(j\omega)$ was noted, and an example was considered in detail. The asymptotic approximation for sketching the Bode diagram simplifies the computation considerably. A summary of fifteen typical Bode plots is shown in Table 8.5. Several performance specifications in the frequency domain were discussed; among them were the maximum magnitude $M_{p\omega}$ and the resonant frequency ω_r. The relationship between the Bode diagram plot and the system error constants (K_p and K_v) was noted. Finally, the log-magnitude versus phase diagram was considered for graphically representing the frequency response of a system.

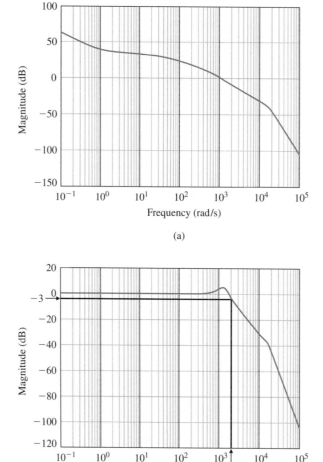

(a)

(b)

FIGURE 8.51   The magnitude Bode plot for (a) the open-loop transfer function and (b) the closed-loop system.

Table 8.5   Bode Diagram Plots for Typical Transfer Functions

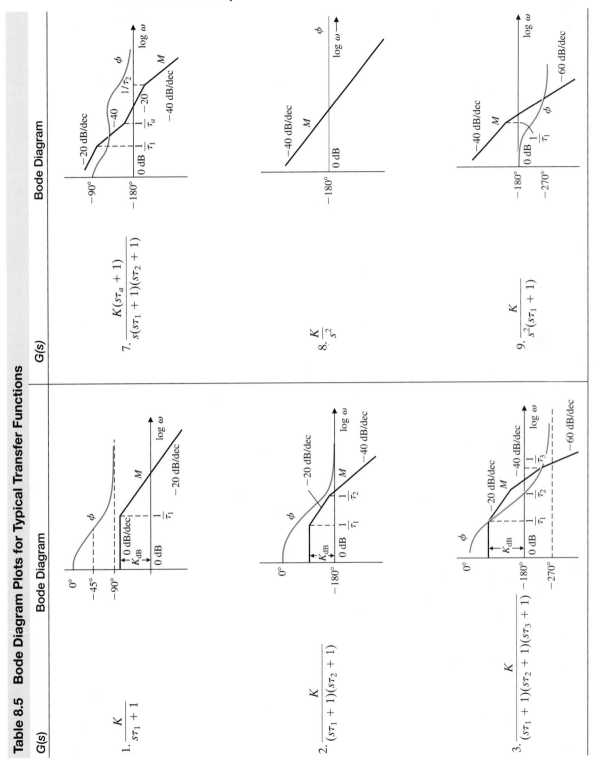

$G(s)$ Bode Diagram

1. $\dfrac{K}{s\tau_1 + 1}$

2. $\dfrac{K}{(s\tau_1 + 1)(s\tau_2 + 1)}$

3. $\dfrac{K}{(s\tau_1 + 1)(s\tau_2 + 1)(s\tau_3 + 1)}$

$G(s)$ Bode Diagram

7. $\dfrac{K(s\tau_a + 1)}{s(s\tau_1 + 1)(s\tau_2 + 1)}$

8. $\dfrac{K}{s^2}$

9. $\dfrac{K}{s^2(s\tau_1 + 1)}$

Table 8.5 *(continued)*

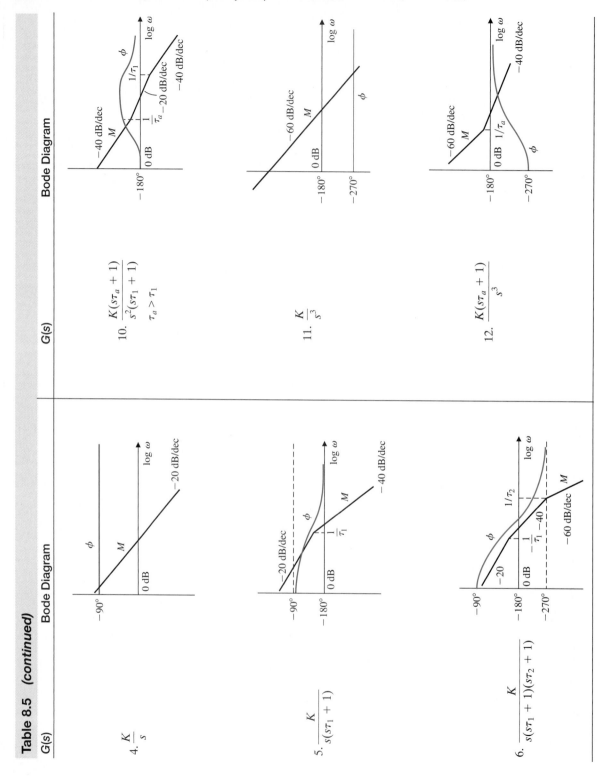

$G(s)$	Bode Diagram

$$4. \ \frac{K}{s}$$

$$5. \ \frac{K}{s(s\tau_1 + 1)}$$

$$6. \ \frac{K}{s(s\tau_1 + 1)(s\tau_2 + 1)}$$

$$10. \ \frac{K(s\tau_a + 1)}{s^2(s\tau_1 + 1)}$$
$$\tau_a > \tau_1$$

$$11. \ \frac{K}{s^3}$$

$$12. \ \frac{K(s\tau_a + 1)}{s^3}$$

Table 8.5   *(continued)*

G(s)	Bode Diagram

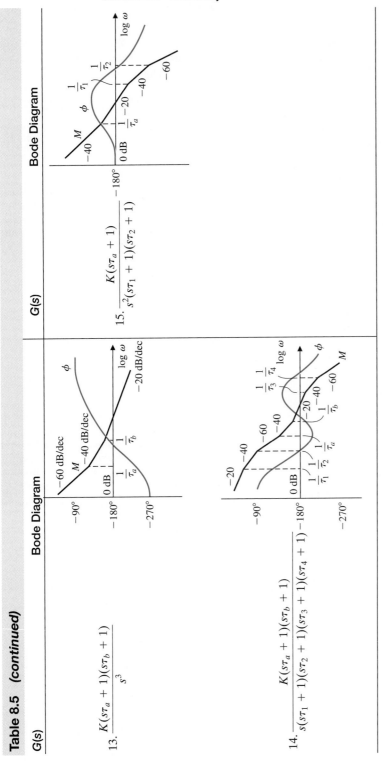

13. $\dfrac{K(s\tau_a + 1)(s\tau_b + 1)}{s^3}$

14. $\dfrac{K(s\tau_a + 1)(s\tau_b + 1)}{s(s\tau_1 + 1)(s\tau_2 + 1)(s\tau_3 + 1)(s\tau_4 + 1)}$

15. $\dfrac{K(s\tau_a + 1)}{s^2(s\tau_1 + 1)(s\tau_2 + 1)}$

EXERCISES

E8.1 Increased track densities for computer disk drives necessitate careful design of the head positioning control [1]. The loop transfer function is

$$L(s) = G_c(s)G(s) = \frac{K}{(s + 1)^2}.$$

Plot the polar plot for this system when $K = 4$. Calculate the phase and magnitude at $\omega = 0.5, 1, 2, 4,$ and ∞.

Answer: $|L(j0.5)| = 3.2$ and $\angle L(j0.5) = -53°$.

E8.2 A tendon-operated robotic hand can be implemented using a pneumatic actuator [8]. The actuator can be represented by

$$G(s) = \frac{2572}{s^2 + 386s + 15434}$$

$$= \frac{2572}{(s + 45.3)(s + 341)}.$$

Plot the frequency response of $G(j\omega)$. Show that the magnitude of $G(j\omega)$ is -15.6 dB at $\omega = 10$ and -30 dB at $\omega = 200$. Show also that the phase is $-150°$ at $\omega = 700$.

E8.3 A robotic arm has a joint-control loop transfer function

$$G_c(s)G(s) = \frac{300(s + 100)}{s(s + 10)(s + 40)}.$$

Prove that the frequency equals 28.3 rad/s when the phase angle of $(j\omega)$ is $-180°$. Find the magnitude of $G_c(j\omega)G(j\omega)$ at that frequency.

Answer: $|G_c(j28.3)G(j28.3)| = -2.5$ dB

E8.4 The frequency response for a process of the form

$$G(s) = \frac{Ks}{(s + a)(s^2 + 20s + 100)}$$

is shown in Figure E8.4. Determine K and a by examining the frequency response curves.

E8.5 The magnitude plot of a transfer function

$$G(s) = \frac{K(1 + 0.5s)(1 + as)}{s(1 + s/8)(1 + bs)(1 + s/36)}$$

is shown in Figure E8.5. Determine K, a, and b from the plot.

Answer: $K = 8$, $a = 1/4$, $b = 1/24$

E8.6 Several studies have proposed an extravehicular robot that could move around in a NASA space station and perform physical tasks at various worksites [9]. The arm is controlled by a unity feedback control with loop transfer function

$$L(s) = G_c(s)G(s) = \frac{K}{s(s/6 + 1)(s/100 + 1)}.$$

Draw the Bode diagram for $K = 10$, and determine the frequency when $20 \log |L(j\omega)|$ is 0 dB.

E8.7 Consider a system with a closed-loop transfer function

$$T(s) = \frac{Y(s)}{R(s)} = \frac{4}{(s^2 + s + 1)(s^2 + 0.4s + 4)}.$$

This system will have no steady-state error for a step input. (a) Plot the frequency response, noting the two peaks in the magnitude response. (b) Predict the time response to a step input, noting that the system has four poles and cannot be represented as a dominant second-order system. (c) Plot the step response.

E8.8 A feedback system has a loop transfer function

$$G_c(s)G(s) = \frac{100(s - 1)}{s^2 + 25s + 100}.$$

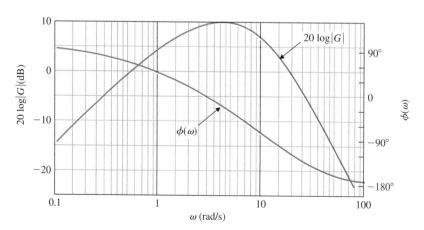

FIGURE E8.4
Bode diagram.

(a) Determine the corner frequencies (break frequencies) for the Bode plot. (b) Determine the slope of the asymptotic plot at very low frequencies and at high frequencies. (c) Sketch the Bode magnitude plot.

E8.9 The Bode diagram of a system is shown in Figure E8.9. Determine the transfer function $G(s)$.

E8.10 The dynamic analyzer shown in Figure E8.10(a) can be used to display the frequency response of a selected $G(j\omega)$ model. Also shown is a head positioning mechanism for a disk drive, which uses a linear motor to position the head. Figure E8.10(b) shows the actual frequency response of the head positioning mechanism. Estimate the poles and zeros of the device. Note $X = 1.37$ kHz at the first cursor, and $\Delta X = 1.257$ kHz to the second cursor.

E8.11 Consider the feedback control system in Figure E8.11. Sketch the Bode plot of $G(s)$ and determine

the crossover frequency, that is, the frequency when $20 \log_{10}|G(j\omega)| = 0$ dB.

E8.12 Consider the system represented in state variable form

$$\dot{\mathbf{x}} = \begin{bmatrix} 0 & 1 \\ -2 & -3 \end{bmatrix} \mathbf{x} + \begin{bmatrix} 0 \\ 5 \end{bmatrix} u$$

$$y = \begin{bmatrix} 1 & -1 \end{bmatrix} \mathbf{x} + \begin{bmatrix} 0 \end{bmatrix} u$$

(a) Determine the transfer function representation of the system. (b) Sketch the Bode plot.

E8.13 Determine the bandwidth of the feedback control system in Figure E8.13.

E8.14 Consider the nonunity feedback system in Figure E8.14, where the controller gain is $K = 2$. Sketch the Bode plot of the loop transfer function. Determine the phase of the loop transfer function when the magnitude

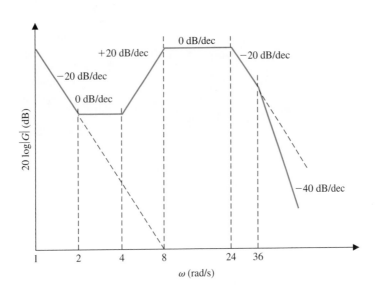

FIGURE E8.5
Bode diagram.

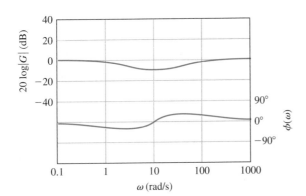

FIGURE E8.9
Bode diagram.

(a)

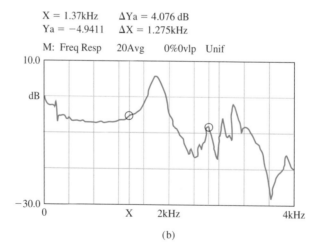

$$X = 1.37\text{kHz} \qquad \Delta Ya = 4.076 \text{ dB}$$
$$Ya = -4.9411 \qquad \Delta X = 1.275\text{kHz}$$

M: Freq Resp   20Avg   0%0vlp   Unif

(b)

FIGURE E8.10   (a) Dual-exposure photo showing the head positioner and the Signal Analyzer 3562A. (b) Frequency response. (Courtesy of Hewlett-Packard Co.)

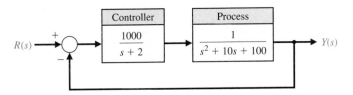

FIGURE E8.11
Unity feedback
system.

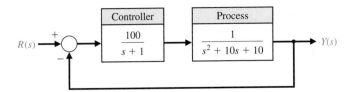

FIGURE E8.13
Third-order
feedback system.

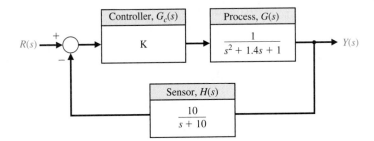

FIGURE E8.14
Nonunity feedback
system with
controller gain K.

$20 \log|L(j\omega)| = 0$ dB. Recall that the loop transfer function is $L(s) = G_c(s)G(s)H(s)$.

E8.15 Consider the single-input, single-output system described by

$$\dot{x}(t) = \mathbf{A}x(t) + \mathbf{B}u(t)$$

$$y(t) = \mathbf{C}x(t)$$

where

$$\mathbf{A} = \begin{bmatrix} 0 & 1 \\ -5 - K & -2 \end{bmatrix}, \mathbf{B} = \begin{bmatrix} 0 \\ 1 \end{bmatrix}, \mathbf{C} = [6 \quad 3].$$

Compute the bandwidth of the system for $K = 1, 2,$ and 10. As K increases, does the bandwidth increase or decrease?

PROBLEMS

P8.1 Sketch the polar plot of the frequency response for the following loop transfer functions:

(a) $G_c(s)G(s) = \dfrac{1}{(1 + 0.5s)(1 + 2s)}$

(b) $G_c(s)G(s) = \dfrac{10(s^2 + 1.4s + 1)}{(s - 1)^2}$

(c) $G_c(s)G(s) = \dfrac{s - 10}{s^2 + 6s + 10}$

(d) $G_c(s)G(s) = \dfrac{30(s + 8)}{s(s + 2)(s + 4)}$

P8.2 Sketch the Bode diagram representation of the frequency response for the transfer functions given in Problem 8.1.

P8.3 A rejection network that can be used instead of the twin-T network of Example 8.4 is the bridged-T network shown in Figure P8.3. The transfer function of this network is

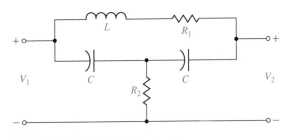

FIGURE P8.3 Bridged-T network.

$$G(s) = \frac{s^2 + \omega_n^2}{s^2 + 2(\omega_n/Q)s + \omega_n^2}$$

(can you show this?), where $\omega_n^2 = 2/LC$, $Q = \omega_n L/R_1$, and R_2 is adjusted so that $R_2 = (\omega_n L)^2/4R_1$ [3]. (a) Determine the pole–zero pattern and, using the vector approach, evaluate the approximate frequency response. (b) Compare the frequency response of the twin-T and bridged-T networks when $Q = 10$.

P8.4 A control system for controlling the pressure in a closed chamber is shown in Figure P8.4. The transfer function for the measuring element is

$$H(s) = \frac{150}{s^2 + 15s + 150}$$

and the transfer function for the valve is

$$G_1(s) = \frac{1}{(0.1s + 1)(s/20 + 1)}.$$

The controller transfer function is

$$G_c(s) = 2s + 1.$$

Obtain the frequency response characteristics for the loop transfer function

$$G_c(s)G_1(s)H(s) \cdot [1/s].$$

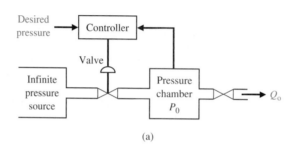

(a)

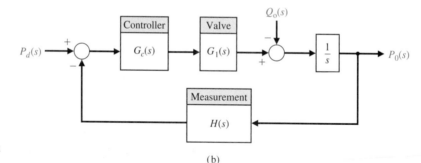

FIGURE P8.4
(a) Pressure controller. (b) Block diagram model.

(b)

P8.5 The robot industry in the United States is growing at a rate of 30% a year [8]. A typical industrial robot has six axes or degrees of freedom. A unity feedback position control system for a force-sensing joint has a loop transfer function

$$G_c(s)G(s) = \frac{K}{(1 + s/5)(1 + s)(1 + s/15)(1 + s/75)},$$

where $K = 10$. Sketch the Bode diagram of this system.

P8.6 The asymptotic log-magnitude curves for two transfer functions are given in Figure P8.6. Sketch the corresponding asymptotic phase shift curves for each system. Determine the transfer function for each system. Assume that the systems have minimum phase transfer functions.

P8.7 Driverless vehicles can be used in warehouses, airports, and many other applications. These vehicles follow a wire embedded in the floor and adjust the steerable front wheels in order to maintain proper direction, as shown in Figure P8.7(a) [10]. The sensing coils, mounted on the front wheel assembly, detect an error in the direction of travel and adjust the steering. The overall control system is shown in Figure P8.7(b). The loop transfer function is

$$L(s) = \frac{K}{s(s + \pi)^2} = \frac{K_v}{s(s/\pi + 1)^2}.$$

We want the bandwidth of the closed-loop system to exceed 2π rad/s. (a) Set $K_v = 2\pi$ and sketch the Bode diagram. (b) Using the Bode diagram, obtain the logarithmic-magnitude versus phase angle curve.

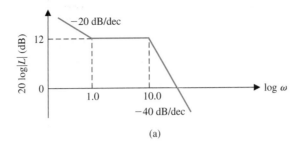

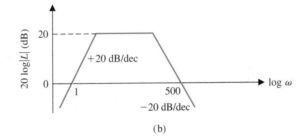

FIGURE P8.6
Log-magnitude
curves.

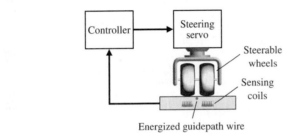

FIGURE P8.7
Steerable wheel
control.

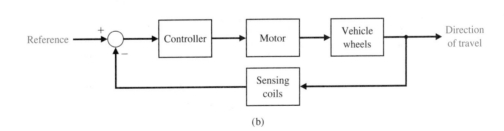

P8.8 A feedback control system is shown in Figure P8.8. The specification for the closed-loop system requires that the overshoot to a step input be less than 10%. (a) Determine the corresponding specification $M_{p\omega}$ in the frequency domain for the closed-loop transfer function

$$\frac{Y(j\omega)}{R(j\omega)} = T(j\omega).$$

(b) Determine the resonant frequency ω_r. (c) Determine the bandwidth of the closed-loop system.

P8.9 Sketch the logarithmic-magnitude versus phase angle curves for the transfer functions (a) and (b) of Problem 8.1.

P8.10 A linear actuator is used in the system shown in Figure P8.10 to position a mass M. The actual position of the mass is measured by a slide wire resistor, and thus $H(s) = 1.0$. The amplifier gain is selected so that the steady-state error of the system is less than 1% of the magnitude of the position reference $R(s)$. The

FIGURE P8.8
Second-order unity
feedback system.

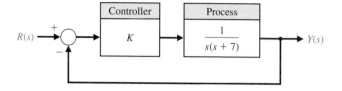

actuator has a field coil with a resistance $R_f = 0.1\ \Omega$ and $L_f = 0.2$ H. The mass of the load is 0.1 kg, and the friction is 0.2 N s/m. The spring constant is equal to 0.4 N/m. (a) Determine the gain K necessary to maintain a steady-state error for a step input less than 1%. That is, K_p must be greater than 99. (b) Sketch the Bode diagram of the loop transfer function, $L(s) = G(s)H(s)$. (c) Sketch the logarithmic magnitude versus phase angle curve for $L(j\omega)$. (d) Sketch the Bode diagram for the closed-loop transfer function, $Y(j\omega)/R(j\omega)$. Determine $M_{p\omega}$, ω_r, and the bandwidth.

P8.11   Automatic steering of a ship would be a particularly useful application of feedback control theory [20]. In the case of heavily traveled seas, it is important to maintain the motion of the ship along an accurate track. An automatic system would be more likely to maintain a smaller error from the desired heading than a helmsman who recorrects at infrequent intervals. A mathematical model of the steering system has been developed for a ship moving at a constant velocity and for small deviations from the desired track. For a large tanker, the transfer function of the ship is

$$G(s) = \frac{E(s)}{\delta(s)} = \frac{0.164(s + 0.2)(-s + 0.32)}{s^2(s + 0.25)(s - 0.009)},$$

where $E(s)$ is the Laplace transform of the deviation of the ship from the desired heading and $\delta(s)$ is the Laplace transform of the angle of deflection of the steering rudder. Verify that the frequency response of the ship, $E(j\omega)/\delta(j\omega)$, is that shown in Figure P8.11.

P8.12   The block diagram of a feedback control system is shown in Figure P8.12(a). The transfer functions of the blocks are represented by the frequency response curves shown in Figure P8.12(b). (a) When G_3 is disconnected from the system, determine the damping ratio ζ of the system. (b) Connect G_3 and determine the damping ratio ζ. Assume that the systems have minimum phase transfer functions.

P8.13   A position control system may be constructed by using an AC motor and AC components, as shown in Figure P8.13. The syncro and control transformer may be considered to be a transformer with a rotating winding. The syncro position detector rotor turns with the load through an angle θ_0. The syncro motor is energized with an AC reference voltage, for example, 115 volts, 60 Hz. The input signal or command is $R(s) = \theta_{in}(s)$ and is applied by turning the rotor of the control transformer. The AC two-phase motor operates as a result of the amplified error signal. The advantages of an AC control system are (1) freedom from DC drift effects and (2) the simplicity and accuracy of AC components. To measure the open-loop frequency response, we simply disconnect X from Y and X' from Y' and then apply a sinusoidal modulation signal generator to the $Y - Y'$ terminals and measure the response at $X - X'$. (The error $(\theta_0 - \theta_i)$ will be adjusted to zero before applying the AC generator.) The resulting frequency response of the loop transfer function $L(j\omega) = G_c(j\omega)G(j\omega)H(j\omega)$, is

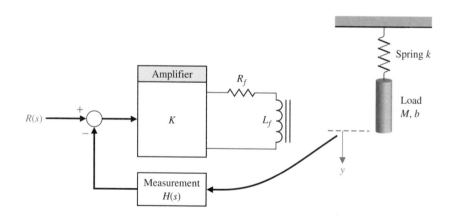

FIGURE P8.10
Linear actuator
control.

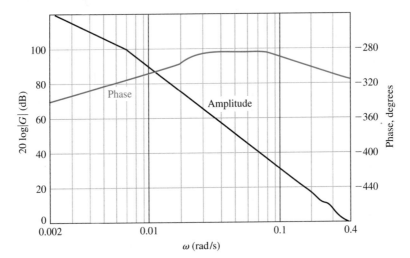

FIGURE P8.11
Frequency
response of ship
control system.

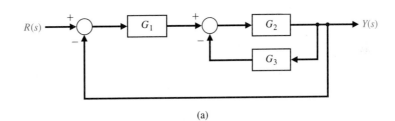

(a)

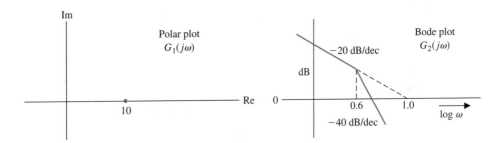

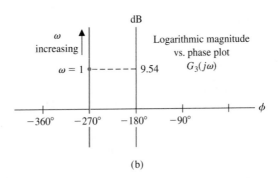

FIGURE P8.12
Feedback system.

(b)

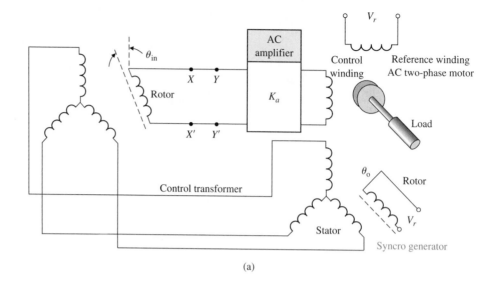

(a)

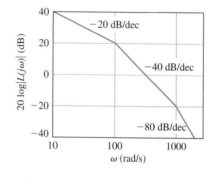

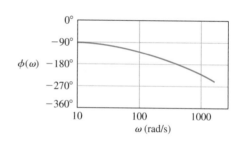

FIGURE P8.13
(a) AC motor control,
(b) Frequency
response.

(b)

shown in Figure P8.13(b). Determine the transfer function $L(j\omega)$. Assume that the system has a minimum phase transfer function.

P8.14 A bandpass amplifier may be represented by the circuit model shown in Figure P8.14 [3]. When $R_1 = R_2 = 1 \text{ k}\Omega, C_1 = 100 \text{ pF}, C_2 = 1 \, \mu\text{F}$, and $K = 100$, show that

$$G(s) = \frac{10^9 s}{(s + 1000)(s + 10^7)}.$$

(a) Sketch the Bode diagram of $G(j\omega)$. (b) Find the midband gain (in dB). (c) Find the high and low frequency −3 dB points.

P8.15 To determine the transfer function of a process $G(s)$, the frequency response may be measured using a sinusoidal input. One system yields the data in the following table:

ω, rad/s	$|G(j\omega)|$	Phase, degrees
0.1	50	−90
1	5.02	−92.4
2	2.57	−96.2
4	1.36	−100
5	1.17	−104
6.3	1.03	−110
8	0.97	−120
10	0.97	−143
12.5	0.74	−169
20	0.13	−245
31	0.026	−258

Determine the transfer function $G(s)$.

P8.16 The space shuttle has been used to repair satellites and the Hubble telescope. Figure P8.16 illustrates how

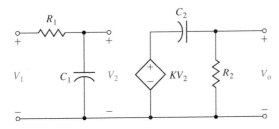

FIGURE P8.14   Bandpass amplifier.

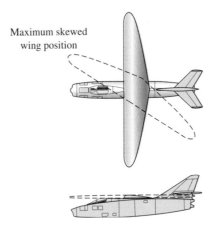

FIGURE P8.17   The Oblique Wing Aircraft, top and side views.

a crew member, with his feet strapped to the platform on the end of the shuttle's robotic arm, used his arms to stop the satellite's spin. The control system of the robotic arm has a closed-loop transfer function

$$\frac{Y(s)}{R(s)} = \frac{53.5}{s^2 + 14.1s + 53.5}.$$

(a) Determine the response $y(t)$ to a unit step input, $R(s) = 1/s$. (b) Determine the bandwidth of the system.

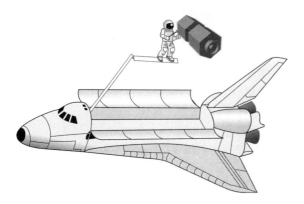

FIGURE P8.16   Satellite repair.

P8.17   The experimental Oblique Wing Aircraft (OWA) has a wing that pivots, as shown in Figure P8.17. The wing is in the normal unskewed position for low speeds and can move to a skewed position for improved supersonic flight [11]. The aircraft control system loop transfer function is

$$G_c(s)G(s) = \frac{4(0.5s + 1)}{s(2s + 1)\left[\left(\dfrac{s}{8}\right)^2 + \left(\dfrac{s}{20}\right) + 1\right]}.$$

(a) Sketch the Bode diagram. (b) Find the frequency ω_1 when the magnitude is 0 dB, and find the frequency ω_2 when the phase is $-180°$.

P8.18   Remote operation plays an important role in hostile environments, such as those in nuclear or high-temperature environments and in deep space. In spite

of the efforts of many researchers, a teleoperation system that is comparable to the human's direct operation has not been developed. Research engineers have been trying to improve teleoperations by feeding back rich sensory information acquired by the robot to the operator with a sensation of presence. This concept is called tele-existence or telepresence [9].

The tele-existence master–slave system consists of a master system with a visual and auditory sensation of presence, a computer control system, and an anthropomorphic slave robot mechanism with an arm having seven degrees of freedom and a locomotion mechanism. The operator's head movement, right arm movement, right hand movement, and other auxiliary motion are measured by the master system. A specially designed stereo visual and auditory input system mounted on the neck mechanism of the slave robot gathers visual and auditory information from the remote environment. These pieces of information are sent back to the master system and are applied to the specially designed stereo display system to evoke the sensation of presence of the operator. The locomotion control system has the loop transfer function

$$G_c(s)G(s) = \frac{12(s + 0.5)}{s^2 + 13s + 30}.$$

Obtain the Bode diagram for $G_c(j\omega)G(j\omega)$ and determine the frequency when $20 \log|G_c(j\omega)G(j\omega)|$ is very close to 0 dB.

P8.19   A DC motor controller used extensively in automobiles is shown in Figure P8.19(a). The measured plot of $\Theta(s)/I(s)$ is shown in Figure P8.19(b). Determine the transfer function of $\Theta(s)/I(s)$.

P8.20   For the successful development of space projects, robotics and automation will be a key technology. Autonomous and dexterous space robots can reduce the workload of astronauts and increase operational

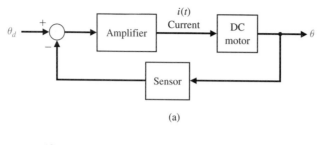

(a)

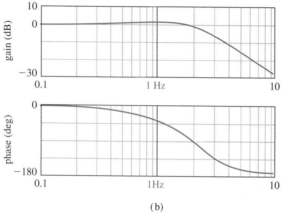

FIGURE P8.19
(a) Motor controller.
(b) Measured plot.

(b)

efficiency in many missions. Figure P8.20 shows a concept called a free-flying robot [9, 13]. A major characteristic of space robots, which clearly distinguishes them from robots operated on earth, is the lack of a fixed base. Any motion of the manipulator arm will induce reaction forces and moments in the base, which disturb its position and attitude.

The control of one of the joints of the robot can be represented by the loop transfer function

$$G_c(s)G(s) = \frac{781(s + 10)}{s^2 + 22s + 484}.$$

(a) Sketch the Bode diagram of $G_c(j\omega)G(j\omega)$. (b) Determine the maximum value of $20 \log|G_c(j\omega)G(j\omega)|$, the frequency at which it occurs, and the phase at that frequency.

P8.21   Low-altitude wind shear is a major cause of air carrier accidents in the United States. Most of these accidents have been caused by either microbursts (small-scale, low-altitude, intense thunderstorm downdrafts that impact the surface and cause strong divergent outflows of wind) or by the gust front at the leading edge of expanding thunderstorm outflows. A microburst encounter is a serious problem for either landing or departing aircraft, because the aircraft is at low altitudes and is traveling at just over 25% above its stall speed [12].

The design of the control of an aircraft encountering wind shear after takeoff may be treated as a problem of stabilizing the climb rate about a desired value of the climb rate. The resulting controller uses only climb rate information.

The standard negative unity feedback system of Figure 8.24 has a loop transfer function

$$G_c(s)G(s) = \frac{-200s^2}{s^3 + 14s^2 + 44s + 40}.$$

FIGURE P8.20   A space robot with three arms, shown capturing a satellite.

Note the negative gain in $G_c(s)G(s)$. This system represents the control system for the climb rate. Sketch the Bode diagram and determine gain (in dB) when the phase is $-180°$.

P8.22 The frequency response of a process $G(j\omega)$ is shown in Figure P8.22. Determine $G(s)$.

P8.23 The frequency response of a process $G(j\omega)$ is shown in Figure P8.23. Deduce the type number (number of integrations) for the system. Determine the transfer function of the system, $G(s)$. Calculate the error to a unit step input.

P8.24 The Bode diagram of a closed-loop film transport system is shown in Figure P8.24 [17]. Assume that the system transfer function $T(s)$ has two dominant complex conjugate poles. (a) Determine the best second-order model for the system. (b) Determine the system bandwidth. (c) Predict the percent overshoot and settling time (with a 2% criterion) for a step input.

P8.25 A unity feedback closed-loop system has a steady-state error equal to $A/10$, where the input is $r(t) = At^2/2$. The Bode plot of the magnitude and phase angle versus ω is shown in Figure P8.25 for $G(j\omega)$. Determine the transfer function $G(s)$.

P8.26 Determine the transfer function of the op-amp circuit shown in Figure P8.26. Assume an ideal op-amp. Plot the frequency response when $R = 10 \, k\Omega$, $R_1 = 9 \, k\Omega$, $R_2 = 1 \, k\Omega$, and $C = 1 \, \mu F$.

P8.27 A unity feedback system has the loop transfer function

$$L(s) = G_c(s)G(s) = \frac{K(s + 50)}{s^2 + 10s + 25}.$$

Sketch the Bode plot of the loop transfer function and indicate how the magnitude $20 \log|L(j\omega)|$ plot varies as K varies. Develop a table for $K = 0.75, 2$, and 10, and for each K determine the crossover frequency (ω_c for $20 \log|L(j\omega)| = 0$ dB), the magnitude at low frequency ($20 \log|L(j\omega)|$ for $\omega \ll 1$), and for the closed-loop system determine the bandwidth for each K.

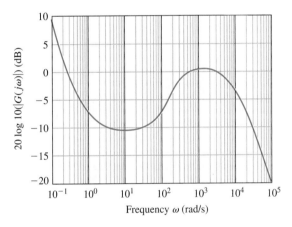

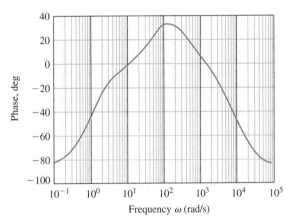

FIGURE P8.22 Bode plot of $G(s)$.

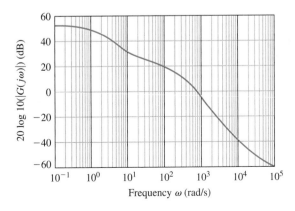

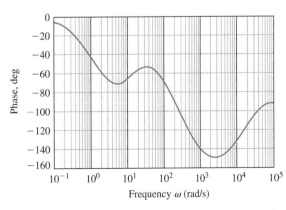

FIGURE P8.23 Frequency response of $G(j\omega)$.

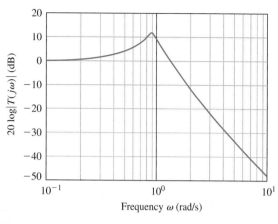

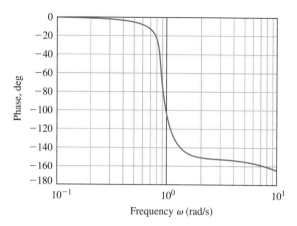

FIGURE P8.24 Bode plot of a closed-film transport system.

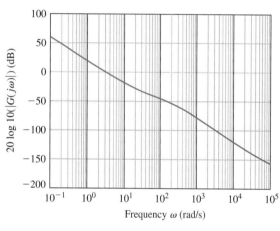

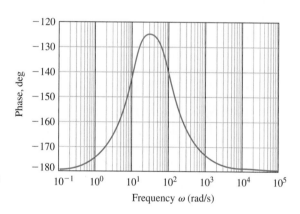

FIGURE P8.25 Bode plot of a unity feedback system.

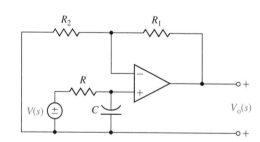

FIGURE P8.26
An op-amp circuit.

ADVANCED PROBLEMS

AP8.1 A spring-mass-damper system is shown in Figure AP8.1(a). The Bode diagram obtained by experimental means using a sinusoidal forcing function is shown in Figure AP8.1(b). Determine the numerical values of m, b, and k.

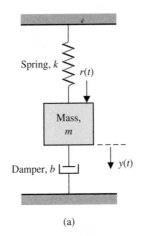

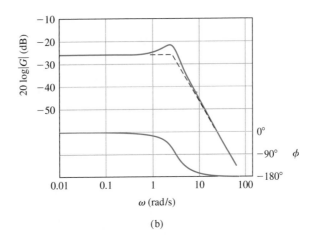

FIGURE AP8.1
A spring-mass-
damper system.

(a)

(b)

AP8.2 A system is shown in Figure AP8.2. The nominal value of the parameter b is 4.0. Determine the sensitivity S_b^T and plot $20 \log|S_b^T|$, the Bode magnitude diagram for $K = 2$.

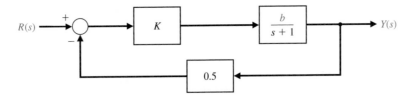

FIGURE AP8.2
System with
parameter b.

AP8.3 As an automobile moves along the road, the vertical displacements at the tires act as the motion excitation to the automobile suspension system [16]. Figure AP8.3 is a schematic diagram of a simplified automobile suspension system, for which we assume the input is sinusoidal. Determine the transfer function $X(s)/R(s)$, and sketch the Bode diagram when $M = 1$ kg, $b = 4$ N s/m, and $k = 18$ N/m.

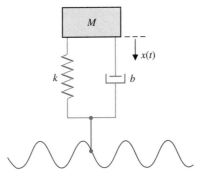

FIGURE AP8.3
Auto suspension
system model.

AP8.4   A helicopter with a load on the end of a cable is shown in Figure AP8.4(a). The position control system is shown in Figure AP8.4(b), where the visual feedback is represented by $H(s)$. Sketch the Bode diagram of $G(j\omega)H(j\omega)$.

AP8.5   A closed-loop system with unity feedback has a transfer function

$$T(s) = \frac{10(s + 1)}{s^2 + 9s + 10}.$$

(a) Determine the loop transfer function $G_c(s)\,G(s)$.
(b) Plot the log-magnitude–phase (similar to Figure 8.27), and identify the frequency points for ω equal to 1, 10, 50, 110, and 500. (c) Is the open-loop system stable? Is the closed-loop system stable?

AP8.6   Consider the spring-mass system depicted in Figure AP8.6. Develop a transfer function model to describe the motion of the mass $M = 2$ kg, when the input is $u(t)$ and the output is $x(t)$. Assume that the initial conditions are $x(0) = 0$ and $\dot{x}(0) = 0$. Determine values of k and b such that the maximum steady-state response of the system to a sinusoidal input $u(t) = \sin(\omega t)$ is less than 1 for all ω. For the values you selected for k and b, what is the frequency at which the peak response occurs?

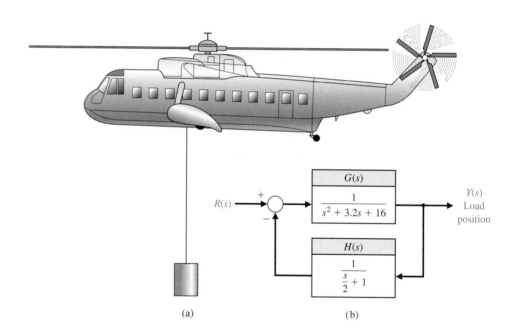

FIGURE AP8.4
A helicopter feedback control system.

(a)   (b)

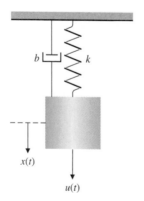

FIGURE AP8.6
Suspended spring-mass system with parameters k and b.

DESIGN PROBLEMS

CDP8.1 In this chapter, we wish to use a PD controller such that

$$G_c(s) = K(s + 2).$$

The tachometer is not used (see Figure CDP4.1). Plot the Bode diagram for the system when $K = 40$. Determine the step response of this system and estimate the overshoot and settling time (with a 2% criterion).

Figure DP8.2(a), being developed at NASA Jet Propulsion Laboratory [18]. The control system of one leg is shown in Figure DP8.2(b).

(a) Sketch the Bode diagram for $G_c(s)G(s)$ when $K = 20$. Determine (1) the frequency when the phase is $-180°$ and (2) the frequency when $20 \log|G_cG| = 0$ dB. (b) Plot the Bode diagram for the closed-loop transfer function $T(s)$ when $K = 20$.

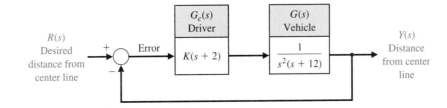

FIGURE DP8.1
Human steering control system.

DP8.1 Understanding the behavior of a human steering an automobile remains an interesting subject [14, 15, 16, 21]. The design and development of systems for four-wheel steering, active suspensions, active, independent braking, and "drive-by-wire" steering provide the engineer with considerably more freedom in altering vehicle-handling qualities than existed in the past.

The vehicle and the driver are represented by the model in Figure DP8.1, where the driver develops anticipation of the vehicle deviation from the center line. For $K = 1$, plot the Bode diagram of (a) the loop transfer function $G_c(s)G(s)$ and (b) the closed-loop transfer function $T(s)$. (c) Repeat parts (a) and (b) when $K = 50$. (d) A driver can select the gain K. Determine the appropriate gain so that $M_{p\omega} \leq 2$, and the bandwidth is the maximum attainable for the closed-loop system. (e) Determine the steady-state error of the system for a ramp input $r(t) = t$.

DP8.2 The unmanned exploration of planets such as Mars requires a high level of autonomy because of the communication delays between robots in space and their Earth-based stations. This affects all the components of the system: planning, sensing, and mechanism. In particular, such a level of autonomy can be achieved only if each robot has a perception system that can reliably build and maintain models of the environment. The perception system is a major part of the development of a complete system that includes planning and mechanism design. The target vehicle is the Spider-bot, a four-legged walking robot shown in

(c) Determine $M_{p\omega}$, ω_r, and ω_B for the closed-loop system when $K = 20$ and $K = 40$. (d) Select the best gain of the two specified in part (c) when it is desired that the overshoot of the system to a step input $r(t)$, be less than 35% and the settling time be as short as possible.

DP8.3 A table is used to position vials under a dispenser head, as shown in Figure DP8.3(a). The objective is speed, accuracy, and smooth motion in order to eliminate spilling. The position control system is shown in Figure DP8.3(b). Since we want small overshoot for a step input and yet desire a short settling time, we will limit $20 \log M_{p\omega}$ to 3 dB for $T(j\omega)$. Plot the Bode diagram for a gain K that will result in a stable system. Then adjust K until $20 \log M_{p\omega} = 3$ dB, and determine the closed-loop system bandwidth. Determine the steady-state error for the system for the gain K selected to meet the requirement for $M_{p\omega}$.

DP8.4 Anesthesia can be administered automatically by a control system. For certain operations, such as brain and eye surgery, involuntary muscle movements can be disastrous. To ensure adequate operating conditions for the surgeon, muscle relaxant drugs, which block involuntary muscle movements, are administered.

A conventional method used by anesthesiologists for muscle relaxant administration is to inject a bolus dose whose size is determined by experience and to inject supplements as required. However, an anesthesiologist may sometimes fail to maintain a steady level of relaxation, resulting in a large drug consumption by the patient. Significant improvements may be achieved by introducing the concept of

(a)

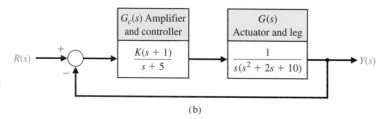

(b)

FIGURE DP8.2
(a) The Mars-bound
Spider-bot. (Photo
courtesy of NASA.)
(b) Block diagram of
the control system
for one leg.

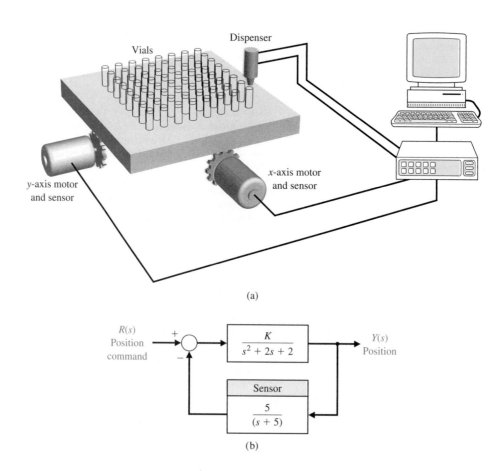

(a)

(b)

FIGURE DP8.3
Automatic table
and dispenser.

automatic control, which results in a considerable reduction in the total relaxant drug consumed [19].

A model of the anesthesia process is shown in Figure DP8.4. Select a gain K so that the bandwidth of the closed-loop system is maximized while $M_{p\omega} \leq 1.5$. Determine the bandwidth attained for your design.

DP8.5 Consider the control system depicted in Figure DP8.5(a) where the plant is a "black box" for which little is known in the way of mathematical models. The only information available on the plant is the frequency response shown in Figure DP8.5(b). Design a controller $G_c(s)$ to meet the following specifications: (i) The crossover frequency is between 10 rad/s and 50 rad/s; (ii) The magnitude of $G_c(s)G(s)$ is greater than 20 dB for $\omega < 0.1$ rad/s.

DP8.6 A single-input, single-output system is described by

$$\dot{\mathbf{x}}(t) = \begin{bmatrix} 0 & 1 \\ -1 & -p \end{bmatrix}\mathbf{x}(t) + \begin{bmatrix} K \\ 0 \end{bmatrix}u(t)$$

$$y(t) = [0 \quad 1]\mathbf{x}(t)$$

(a) Determine p and K such that the unit step response exhibits a zero steady-state error and the percent overshoot meets the requirement $P.O. \leq 5\%$.

(b) For the values of p and K determined in part (a), determine the system damping ratio ζ and the natural frequency ω_n.

(c) For the values of p and K determined in part (a), obtain the Bode plot of the system and determine the bandwidth ω_B.

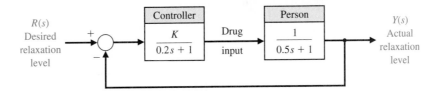

FIGURE DP8.4
Model of an anesthesia control system.

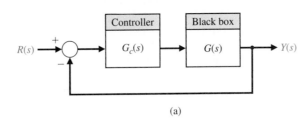

(a)

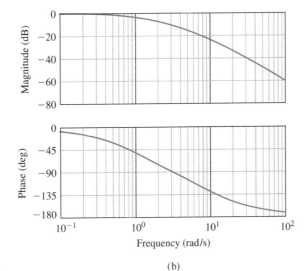

FIGURE DP8.5
(a) Feedback system with "black box" plant. (b) Frequency response plot of the "black box" represented by G(s).

(b)

(d) Using the approximate formula shown in Figure 8.26, compute the bandwidth using ζ and ω_n and compare the value to the actual bandwidth from part (c).

COMPUTER PROBLEMS

CP8.1   Consider the closed-loop transfer function

$$T(s) = \frac{25}{s^2 + s + 25}.$$

Develop an m-file to, obtain the Bode plot and verify that the resonant frequency is 5 rad/s and that the peak magnitude $M_{p\omega}$ is 14 dB.

CP8.2   For the following transfer functions, sketch the Bode plots, then verify with the bode function:

(a) $G(s) = \dfrac{1}{(s + 1)(s + 10)}$

(b) $G(s) = \dfrac{s + 10}{(s + 2)(s + 40)}$

(c) $G(s) = \dfrac{1}{s^2 + 2s + 50}$

(d) $G(s) = \dfrac{s - 7}{(s + 2)(s^2 + 12s + 50)}$

CP8.3   For each of the following transfer functions, sketch the Bode plot and determine the crossover frequency (that is, the frequency at which $20 \log_{10}|G(j\omega)| = 0$ dB):

(a) $G(s) = \dfrac{1000}{(s + 10)(s + 30)}$

(b) $G(s) = \dfrac{100}{(s + 0.2)(s^2 + s + 20)}$

(c) $G(s) = \dfrac{50(s + 100)}{(s + 1)(s + 50)}$

(d) $G(s) = \dfrac{100(s^2 + 14s + 50)}{(s + 1)(s + 2)(s + 500)}$

CP8.4   A unity negative feedback system has the loop transfer function

$$G_c(s)G(s) = \frac{50}{s(s + 5)}.$$

Determine the closed-loop system bandwidth by using the bode function to obtain the Bode plot, and estimate the bandwidth from the plot. Label the plot with the bandwidth.

CP8.5   A block diagram of a second-order system is shown in Figure CP8.5.

(a) Determine the resonant peak $M_{p\omega}$ the resonant frequency ω_r, and the bandwidth ω_B, of the system

from the closed-loop Bode plot. Generate the Bode plot with an m-file for $\omega = 0.1$ to $\omega = 1000$ rad/s using the logspace function. (b) Estimate the system damping ratio, ζ, and natural frequency ω_n, using Equations (8.36) and (8.37) in Section 8.2. (c) From the closed-loop transfer function, compute the actual ζ and ω_n and compare with your results in part (b).

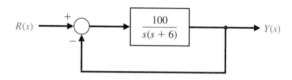

FIGURE CP8.5   A second-order feedback control system.

CP8.6   Consider the feedback system in Figure CP8.6. Obtain the Bode plots of the loop and closed-loop transfer functions using an m-file.

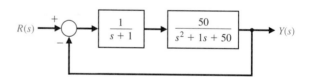

FIGURE CP8.6   Closed-loop feedback system.

CP8.7   A unity feedback system has the loop transfer function

$$G_c(s)G(s) = \frac{1}{s(s + 2p)}.$$

Generate a plot of the bandwidth versus the parameter p as $0 < p < 1$.

CP8.8   Consider the problem of controlling an inverted pendulum on a moving base, as shown in Figure CP8.8(a). The transfer function of the system is

$$G(s) = \frac{-1/(M_b L)}{s^2 - (M_b + M_s)g/(M_b L)}.$$

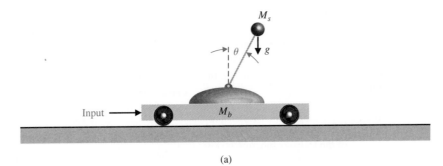

(a)

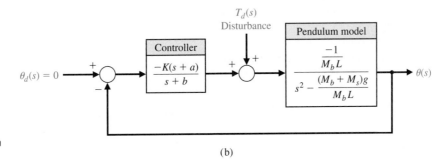

(b)

FIGURE CP8.8
(a) An inverted
pendulum on a
moving base.
(b) A block diagram
representation.

The design objective is to balance the pendulum (i.e., $\theta(t) \approx 0$) in the presence of disturbance inputs. A block diagram representation of the system is depicted in Figure CP8.8(b). Let $M_s = 10$ kg, $M_b = 100$ kg, $L = 1$ m, $g = 9.81$ m/s^2, $a = 5$, and $b = 10$. The design specifications, based on a unit step disturbance, are as follows:

1. settling time (with a 2% criterion) less than 10 seconds,
2. percent overshoot less than 40%, and
3. steady-state tracking error less than 0.1° in the presence of the disturbance.

Develop a set of interactive m-file scripts to aid in the control system design. The first script should accomplish at least the following:

1. Compute the closed-loop transfer function from the disturbance to the output with K as an adjustable parameter,
2. Draw the Bode plot of the closed-loop system.
3. Automatically compute and output $M_{p\omega}$ and ω_r.

As an intermediate step, use $M_{p\omega}$ and ω_r and Equations (8.36) and (8.37) in Section 8.2 to estimate ζ and ω_n. The second script should at least estimate the settling time and percent overshoot using ζ and ω_n as input variables.

If the performance specifications are not satisfied, change K and iterate on the design using the first two scripts. After completion of the first two steps, the final step is to test the design by simulation. The functions of the third script are as follows:

1. plot the response, $\theta(t)$, to a unit step disturbance with K as an adjustable parameter, and
2. label the plot appropriately.

Utilizing the interactive scripts, design the controller to meet the specifications using frequency response Bode methods. To start the design process, use analytic methods to compute the minimum value of K to meet the steady-state tracking error specification. Use the minimum K as the first guess in the design iteration.

CP8.9 Design a filter, $G(s)$, with the following frequency response:

1. For $\omega < 1$ rad/s, the magnitude $20 \log_{10}|G(j\omega)| < 0$ dB
2. For $1 < \omega < 1000$ rad/s, the magnitude $20 \log_{10} |G(j\omega)| \geq 0$ dB
3. For $\omega > 1000$ rad/s, the magnitude $20 \log_{10} |G(j\omega)| < 0$ dB

Try to maximize the peak magnitude as close to $\omega = 40$ rad/s as possible.

TERMS AND CONCEPTS

All-pass network   A nonminimum phase system that passes all frequencies with equal gain.

Bandwidth   The frequency at which the frequency response has declined 3 dB from its low-frequency value.

Bode plot   The logarithm of the magnitude of the transfer function is plotted versus the logarithm of ω, the frequency. The phase ϕ of the transfer function is separately plotted versus the logarithm of the frequency.

Break frequency   The frequency at which the asymptotic approximation of the frequency response for a pole (or zero) changes slope.

Corner frequency   See Break frequency.

Decade   A factor of 10 in frequency (e.g., the range of frequencies from 1 rad/s to 10 rad/s is one decade).

Decibel (dB)   The units of the logarithmic gain.

Dominant roots   The roots of the characteristic equation that represent or dominate the closed-loop transient response.

Fourier transform   The transformation of a function of time $f(t)$ into the frequency domain.

Fourier transform pair   A pair of functions, one in the time domain, denoted by $f(t)$, and the other in the frequency domain, denoted by $F(\omega)$, related by the Fourier transform as $F(\omega) = \mathscr{F}\{f(t)\}$, where $\mathscr{F}$ denotes the Fourier transform.

Frequency response   The steady-state response of a system to a sinusoidal input signal.

Laplace transform pair   A pair of functions, one in the time domain, denoted by $f(t)$, and the other in the frequency domain, denoted by $F(s)$, related by the Laplace transform as $F(s) = \mathscr{L}\{f(t)\}$, where $\mathscr{L}$ denotes the Laplace transform.

Logarithmic magnitude   The logarithm of the magnitude of the transfer function, usually expressed in units of 20 dB, thus $20 \log_{10}|G|$.

Logarithmic plot   See Bode plot.

Maximum value of the frequency response   A pair of complex poles will result in a maximum value for the frequency response occurring at the resonant frequency.

Minimum phase transfer function   All the zeros of a transfer function lie in the left-hand side of the s-plane.

Natural frequency   The frequency of natural oscillation that would occur for two complex poles if the damping were equal to zero.

Nonminimum phase transfer function   Transfer functions with zeros in the right-hand s-plane.

Polar plot   A plot of the real part of $G(j\omega)$ versus the imaginary part of $G(j\omega)$.

Resonant frequency   The frequency ω_r at which the maximum value of the frequency response of a complex pair of poles is attained.

Transfer function in the frequency domain   The ratio of the output to the input signal where the input is a sinusoid. It is expressed as $G(j\omega)$.

PREVIEW

In previous chapters, we discussed stability and developed various tools to determine stability and to assess relative stability. We continue that discussion in this chapter by showing how frequency response methods can be used to investigate stability. The important concepts of gain margin, phase margin, and bandwidth are developed in the context of Bode plots and Nyquist diagrams. A frequency response stability result—known as the Nyquist stability criterion—is presented and its use illustrated through several interesting examples. The implications of having pure time delays in the system on both stability and performance are discussed. We will see that the phase lag introduced by the time delay can destabilize an otherwise stable system. The chapter concludes with a frequency response analysis of the Sequential Design Example: Disk Drive Read System.

DESIRED OUTCOMES

Upon completion of Chapter 9, students should:

❑ Understand the Nyquist stability criterion and the role of the Nyquist plot.
❑ Be familiar with time-domain performance specifications in the frequency domain.
❑ Appreciate the importance of considering time delays in feedback control systems.
❑ Be capable of analyzing the relative stability and performance of feedback control systems using frequency response methods considering phase and gain margin, and system bandwidth.

9.1 INTRODUCTION

For a control system, it is necessary to determine whether the system is stable. Furthermore, if the system is stable, it is often necessary to investigate the relative stability. In Chapter 6, we discussed the concept of stability and several methods of determining the absolute and relative stability of a system. The Routh–Hurwitz method, discussed in Chapter 6, is useful for investigating the characteristic equation expressed in terms of the complex variable $s = \sigma + j\omega$. Then, in Chapter 7, we investigated the relative stability of a system utilizing the root locus method, which is also expressed in terms of the complex variable s. In this chapter, we are concerned with investigating the stability of a system in the real frequency domain, that is, in terms of the frequency response discussed in Chapter 8.

The frequency response of a system represents the sinusoidal steady-state response of a system and provides sufficient information for the determination of the relative stability of the system. The frequency response of a system can readily be obtained experimentally by exciting the system with sinusoidal input signals; therefore, it can be utilized to investigate the relative stability of a system when the system parameter values have not been determined. Furthermore, a frequency-domain stability criterion would be useful for determining suitable approaches to adjusting the parameters of a system in order to increase its relative stability.

A frequency domain stability criterion was developed by H. Nyquist in 1932, and it remains a fundamental approach to the investigation of the stability of linear control systems [1, 2]. The **Nyquist stability criterion** is based on a theorem in the theory of the function of a complex variable due to Cauchy. Cauchy's theorem is concerned with mapping contours in the complex s-plane, and fortunately the theorem can be understood without a formal proof requiring complex variable theory.

To determine the relative stability of a closed-loop system, we must investigate the characteristic equation of the system:

$$F(s) = 1 + L(s) = 0. \tag{9.1}$$

For the single-loop control system of Figure 9.1, $L(s) = G_c(s)G(s)H(s)$. For a multiloop system, we found in Section 2.7 that, in terms of signal-flow graphs, the characteristic equation is

$$F(s) = \Delta(s) = 1 - \Sigma L_n + \Sigma L_m L_q \ldots = 0,$$

where $\Delta(s)$ is the graph determinant. Therefore, we can represent the characteristic equation of single-loop or multiple-loop systems by Equation (9.1), where $L(s)$ is a rational function of s. To ensure stability, we must ascertain that all the zeros of $F(s)$ lie in the left-hand s-plane. Nyquist thus proposed a mapping of the right-hand s-plane into the $F(s)$-plane. Therefore, to use and understand Nyquist's criterion, we shall first consider briefly the mapping of contours in the complex plane.

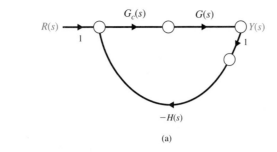

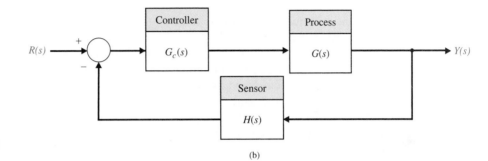

FIGURE 9.1
Single-loop
feedback control
system.

9.2 MAPPING CONTOURS IN THE *s*-PLANE

We are concerned with the mapping of contours in the *s*-plane by a function $F(s)$. A
contour map is a contour or trajectory in one plane mapped or translated into another plane by a relation $F(s)$. Since *s* is a complex variable, $s = \sigma + j\omega$, the function $F(s)$
is itself complex; it can be defined as $F(s) = u + jv$ and can be represented on a complex $F(s)$-plane with coordinates *u* and *v*. As an example, let us consider a function
$F(s) = 2s + 1$ and a contour in the *s*-plane, as shown in Figure 9.2(a). The mapping

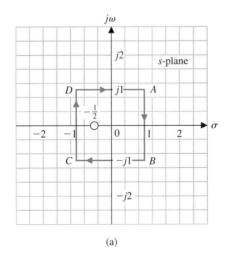

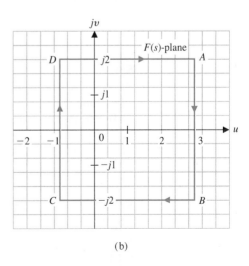

FIGURE 9.2
Mapping a square
contour by
$F(s) = 2s + 1 = 2(s + 1/2)$.

of the s-plane unit square contour to the $F(s)$-plane is accomplished through the relation $F(s)$, and so

$$u + jv = F(s) = 2s + 1 = 2(\sigma + j\omega) + 1. \qquad (9.2)$$

Therefore, in this case, we have

$$u = 2\sigma + 1 \qquad (9.3)$$

and

$$v = 2\omega. \qquad (9.4)$$

Thus, the contour has been mapped by $F(s)$ into a contour of an identical form, a square, with the center shifted by one unit and the magnitude of a side multiplied by two. This type of mapping, which retains the angles of the s-plane contour on the $F(s)$-plane, is called a **conformal mapping**. We also note that a closed contour in the s-plane results in a closed contour in the $F(s)$-plane.

The points $A, B, C,$ and D, as shown in the s-plane contour, map into the points $A, B, C,$ and D shown in the $F(s)$-plane. Furthermore, a direction of traversal of the s-plane contour can be indicated by the direction $ABCD$ and the arrows shown on the contour. Then a similar traversal occurs on the $F(s)$-plane contour as we pass $ABCD$ in order, as shown by the arrows. By convention, the area within a contour to the right of the traversal of the contour is considered to be the area enclosed by the contour. Therefore, we will assume clockwise traversal of a contour to be positive and the area enclosed within the contour to be on the right. This convention is opposite to that usually employed in complex variable theory, but is equally applicable and is generally used in control system theory. We might consider the area on the right as we walk along the contour in a clockwise direction and call this rule "clockwise and eyes right."

Typically, we are concerned with an $F(s)$ that is a rational function of s. Therefore, it will be worthwhile to consider another example of a mapping of a contour. Let us again consider the unit square contour for the function

$$F(s) = \frac{s}{s + 2}. \qquad (9.5)$$

Several values of $F(s)$ as s traverses the square contour are given in Table 9.1, and the resulting contour in the $F(s)$-plane is shown in Figure 9.3(b). The contour in the $F(s)$-plane encloses the origin of the $F(s)$-plane because the origin lies within the enclosed area of the contour in the $F(s)$-plane.

Table 9.1   Values of $F(s)$

	Point A		Point B		Point C		Point D	
$s = \sigma + j\omega$	$1 + j1$	1	$1 - j1$	$-j1$	$-1 - j1$	-1	$-1 + j1$	$j1$
$F(s) = u + jv$	$\dfrac{4 + 2j}{10}$	$\dfrac{1}{3}$	$\dfrac{4 - 2j}{10}$	$\dfrac{1 - 2j}{5}$	$-j$	-1	$+j$	$\dfrac{1 + 2j}{5}$

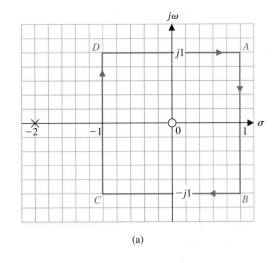
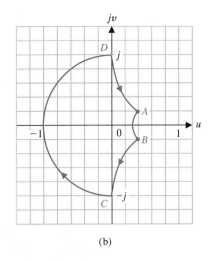

FIGURE 9.3
Mapping for
$F(s) = s/(s + 2)$.

(a)   (b)

Cauchy's theorem is concerned with mapping a function $F(s)$ that has a finite number of poles and zeros within the contour, so that we may express $F(s)$ as

$$F(s) = \frac{K\prod_{i=1}^{n}(s + z_i)}{\prod_{k=1}^{M}(s + p_k)}, \tag{9.6}$$

where $-z_i$ are the zeros of the function $F(s)$ and $-p_k$ are the poles of $F(s)$. The function $F(s)$ is the characteristic equation, and so

$$F(s) = 1 + L(s), \tag{9.7}$$

where

$$L(s) = \frac{N(s)}{D(s)}.$$

Therefore, we have

$$F(s) = 1 + L(s) = 1 + \frac{N(s)}{D(s)} = \frac{D(s) + N(s)}{D(s)} = \frac{K\prod_{i=1}^{n}(s + z_i)}{\prod_{k=1}^{M}(s + p_k)}, \tag{9.8}$$

and the poles of $L(s)$ are the poles of $F(s)$. However, it is the zeros of $F(s)$ that are the characteristic roots of the system and that indicate its response. This is clear if we recall that the output of the system is

$$Y(s) = T(s)R(s) = \frac{\sum P_k \Delta_k}{\Delta(s)} R(s) = \frac{\sum P_k \Delta_k}{F(s)} R(s), \tag{9.9}$$

where P_k and Δ_k are the path factors and cofactors as defined in Section 2.7.

Reexamining the example when $F(s) = 2(s + 1/2)$, we have one zero of $F(s)$ at $s = -1/2$, as shown in Figure 9.2. The contour that we chose (that is, the unit square) enclosed and encircled the zero once within the area of the contour. Similarly, for the function $F(s) = s/(s + 2)$, the unit square encircled the zero at the origin but did not encircle the pole at $s = -2$. The encirclement of the poles and zeros of $F(s)$ can be related to the encirclement of the origin in the $F(s)$-plane by **Cauchy's theorem**, commonly known as the **principle of the argument**, which states [3, 4]:

> If a contour Γ_s in the s-plane encircles Z zeros and P poles of $F(s)$ and does not pass through any poles or zeros of $F(s)$ and the traversal is in the clockwise direction along the contour, the corresponding contour Γ_F in the $F(s)$-plane encircles the origin of the $F(s)$-plane $N = Z - P$ times in the clockwise direction.

Thus, for the examples shown in Figures 9.2 and 9.3, the contour in the $F(s)$-plane encircles the origin once, because $N = Z - P = 1$, as we expect. As another example, consider the function $F(s) = s/(s + 1/2)$. For the unit square contour shown in Figure 9.4(a), the resulting contour in the $F(s)$ plane is shown in Figure 9.4(b). In this case, $N = Z - P = 0$, as is the case in Figure 9.4(b), since the contour Γ_F does not encircle the origin.

Cauchy's theorem can be best comprehended by considering $F(s)$ in terms of the angle due to each pole and zero as the contour Γ_s is traversed in a clockwise direction. Thus, let us consider the function

$$F(s) = \frac{(s + z_1)(s + z_2)}{(s + p_1)(s + p_2)}, \tag{9.10}$$

where $-z_i$ is a zero of $F(s)$, and $-p_k$ is a pole of $F(s)$. Equation (9.10) can be written as

$$F(s) = |F(s)| \underline{/F(s)}$$

$$= \frac{|s + z_1||s + z_2|}{|s + p_1||s + p_2|} \left(\underline{/s + z_1} + \underline{/s + z_2} - \underline{/s + p_1} - \underline{/s + p_2} \right)$$

$$= |F(s)|(\phi_{z_1} + \phi_{z_2} - \phi_{p_1} - \phi_{p_2}). \tag{9.11}$$

Now, considering the vectors as shown for a specific contour Γ_s (Figure 9.5a), we can determine the angles as s traverses the contour. Clearly, the net angle change as s traverses along Γ_s (a full rotation of 360° for ϕ_{p_1}, ϕ_{p_2} and ϕ_{z_2}) is zero degrees. However, for ϕ_{z_1} as s traverses 360° around Γ_s, the angle ϕ_{z_1} traverses a full 360° clockwise. Thus, as Γ_s is completely traversed, the net angle increase of $F(s)$ is equal to 360°, since only one zero is enclosed. If Z zeros were enclosed within Γ_s, then the net angle increase would be equal to $\phi_z = 2\pi Z$ rad. Following this reasoning, if Z zeros and P poles are encircled as Γ_s is traversed, then $2\pi Z - 2\pi P$ is the net resultant angle increase of $F(s)$. Thus, the net angle increase of Γ_F of the contour in the

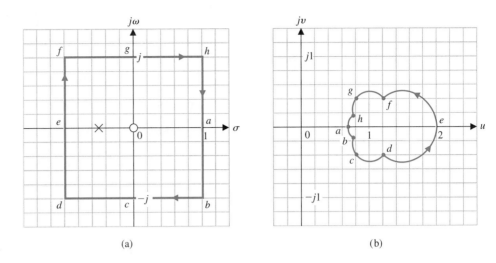

FIGURE 9.4
Mapping for
$F(s) = s/(s + 1/2)$.

(a)   (b)

$F(s)$-plane is simply

$$\phi_F = \phi_Z - \phi_P,$$

or

$$2\pi N = 2\pi Z - 2\pi P, \qquad (9.12)$$

and the net number of encirclements of the origin of the $F(s)$-plane is $N = Z - P$. Thus, for the contour shown in Figure 9.5(a), which encircles one zero, the contour Γ_F shown in Figure 9.5(b) encircles the origin once in the clockwise direction.

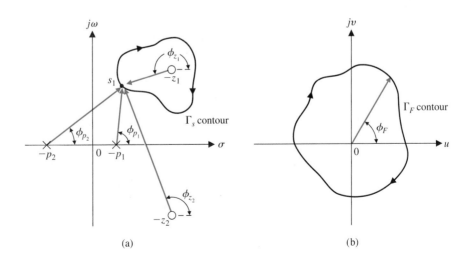

FIGURE 9.5
Evaluation of the
net angle of Γ_F.

(a)   (b)

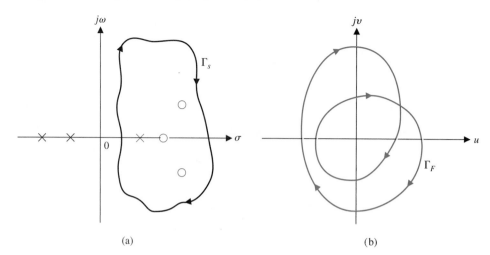

FIGURE 9.6
Example of
Cauchy's theorem
with three zeros
and one pole
within Γ_s.

(a)

(b)

As an example of the use of Cauchy's theorem, consider the pole–zero pattern shown in Figure 9.6(a) with the contour Γ_s to be considered. The contour encloses and encircles three zeros and one pole. Therefore, we obtain

$$N = 3 - 1 = +2,$$

and Γ_F completes two clockwise encirclements of the origin in the $F(s)$-plane, as shown in Figure 9.6(b).

For the pole and zero pattern shown and the contour Γ_s as shown in Figure 9.7(a), one pole is encircled and no zeros are encircled. Therefore, we have

$$N = Z - P = -1,$$

and we expect one encirclement of the origin by the contour Γ_F in the $F(s)$-plane. However, since the sign of N is negative, we find that the encirclement moves in the counterclockwise direction, as shown in Figure 9.7(b).

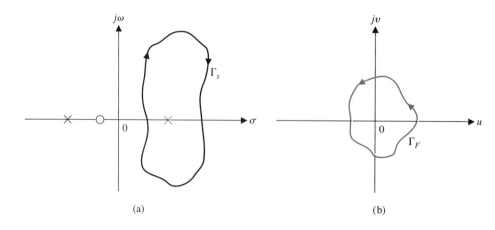

FIGURE 9.7
Example of
Cauchy's theorem
with one pole
within Γ_s.

(a)

(b)

Now that we have developed and illustrated the concept of mapping of contours through a function $F(s)$, we are ready to consider the stability criterion proposed by Nyquist.

9.3 THE NYQUIST CRITERION

To investigate the stability of a control system, we consider the characteristic equation, which is $F(s) = 0$, so that

$$F(s) = 1 + L(s) = \frac{K \prod_{i=1}^{n} (s + z_i)}{\prod_{k=1}^{M} (s + p_k)} = 0. \tag{9.13}$$

For a system to be stable, all the zeros of $F(s)$ must lie in the left-hand s-plane. Thus, we find that the roots of a stable system (the zeros of $F(s)$) must lie to the left of the $j\omega$-axis in the s-plane. Therefore, we choose a contour Γ_s in the s-plane that encloses the entire right-hand s-plane, and we determine whether any zeros of $F(s)$ lie within Γ_s by utilizing Cauchy's theorem. That is, we plot Γ_F in the $F(s)$-plane and determine the number of encirclements of the origin N. Then the number of zeros of $F(s)$ within the Γ_s contour (and therefore, the unstable zeros of $F(s)$) is

$$\boxed{Z = N + P.} \tag{9.14}$$

Thus, if $P = 0$, as is usually the case, we find that the number of unstable roots of the system is equal to N, the number of encirclements of the origin of the $F(s)$-plane.

The Nyquist contour that encloses the entire right-hand s-plane is shown in Figure 9.8. The contour Γ_s passes along the $j\omega$-axis from $-j\infty$ to $+j\infty$, and this part of the contour provides the familiar $F(j\omega)$. The contour is completed by a

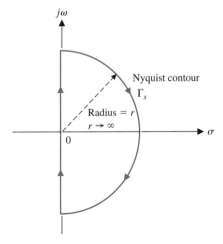

FIGURE 9.8
Nyquist contour is shown as the heavy line.

semicircular path of radius r, where r approaches infinity so this part of the contour typically maps to a point. This contour Γ_F is known as the Nyquist diagram or polar plot.

Now, the Nyquist criterion is concerned with the mapping of the characteristic equation

$$F(s) = 1 + L(s) \tag{9.15}$$

and the number of encirclements of the origin of the $F(s)$-plane. Alternatively, we may define the function

$$F'(s) = F(s) - 1 = L(s). \tag{9.16}$$

The change of functions represented by Equation (9.16) is very convenient because $L(s)$ is typically available in factored form, while $1 + L(s)$ is not. Then, the mapping of Γ_s in the s-plane will be through the function $F'(s) = L(s)$ into the $L(s)$-plane. In this case, the number of clockwise encirclements of the origin of the $F(s)$-plane becomes the number of clockwise encirclements of the -1 point in the $F'(s) = L(s)$-plane because $F'(s) = F(s) - 1$. Therefore, the **Nyquist stability criterion** can be stated as follows:

> **A feedback system is stable if and only if the contour Γ_L in the $L(s)$-plane does not encircle the $(-1, 0)$ point when the number of poles of $L(s)$ in the right-hand s-plane is zero ($P = 0$).**

When the number of poles of $L(s)$ in the right-hand s-plane is other than zero, the Nyquist criterion is stated as follows:

> **A feedback control system is stable if and only if, for the contour Γ_L, the number of counterclockwise encirclements of the $(-1, 0)$ point is equal to the number of poles of $L(s)$ with positive real parts.**

The basis for the two statements is the fact that, for the $F'(s) = L(s)$ mapping, the number of roots (or zeros) of $1 + L(s)$ in the right-hand s-plane is represented by the expression

$$Z = N + P.$$

Clearly, if the number of poles of $L(s)$ in the right-hand s-plane is zero ($P = 0$), we require for a stable system that $N = 0$, and the contour Γ_p must not encircle the -1 point. Also, if P is other than zero and we require for a stable system that $Z = 0$, then we must have $N = -P$, or P counterclockwise encirclements.

It is best to illustrate the use of the Nyquist criterion by completing several examples.

EXAMPLE 9.1   **System with two real poles**

A single-loop control system is shown in Figure 9.1, where

$$L(s) = \frac{K}{(\tau_1 s + 1)(\tau_2 s + 1)}. \tag{9.17}$$

In this case, $L(s) = G_c(s)G(s)H(s)$, and we use a contour Γ_L in the $L(s)$-plane. The contour Γ_s in the s-plane is shown in Figure 9.9(a), and the contour Γ_L is shown in Figure 9.9(b) for $\tau_1 = 1, \tau_2 = 1/10$, and $K = 100$. The magnitude and phase of $L(j\omega)$ for selected values of ω are given in Table 9.2. We use these values to obtain the polar plot of Figure 9.9(b).

The $+j\omega$-axis is mapped into the solid line, as shown in Figure 9.9. The $-j\omega$-axis is mapped into the dashed line, as shown in Figure 9.9. The semicircle with $r \to \infty$ in the s-plane is mapped into the origin of the $L(s)$-plane.

We note that the number of poles of $L(s)$ in the right-hand s-plane is zero, and thus $P = 0$. Therefore, for this system to be stable, we require $N = Z = 0$, and the contour must not encircle the -1 point in the $L(s)$-plane. Examining Figure 9.9(b) and Equation (9.17), we find that, irrespective of the value of K, the contour does not encircle the -1 point, and the system is always stable for all K greater than zero. ∎

FIGURE 9.9
Nyquist contour and mapping for $L(s) = \dfrac{100}{(s + 1)(s/10 + 1)}$.

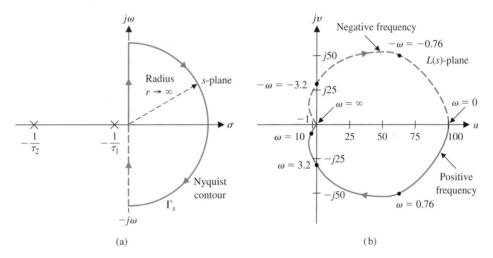

(a)   (b)

Table 9.2   Magnitude and Phase of $L(j\omega)$

ω	0	0.1	0.76	1	2	10	20	100	∞
$\lvert L(j\omega)\rvert$	100	96	79.6	70.7	50.2	6.8	2.24	0.10	0
$\angle L(j\omega)$ (degrees)	0	−5.7	−41.5	−50.7	−74.7	−129.3	−150.5	−173.7	−180

EXAMPLE 9.2   **System with a pole at the origin**

A single-loop control system is shown in Figure 9.1, where

$$L(s) = \frac{K}{s(\tau s + 1)}.$$

In this single-loop case, $L(s) = G_c(s)G(s)H(s)$, and we determine the contour Γ_L in the $L(s)$-plane. The contour Γ_s in the s-plane is shown in Figure 9.10(a), where an infinitesimal detour around the pole at the origin is effected by a small semicircle of radius ϵ, where $\epsilon \to 0$. This detour is a consequence of the condition of Cauchy's theorem, which requires that the contour cannot pass through the pole at the origin. A sketch of the contour Γ_L is shown in Figure 9.10(b). Clearly, the portion of the contour Γ_L from $\omega = 0^+$ to $\omega = +\infty$ is simply $L(j\omega)$, the real frequency polar plot. Let us consider each portion of the Nyquist contour Γ_s in detail and determine the corresponding portions of the $L(s)$-plane contour Γ_L.

(a) The Origin of the s-plane. The small semicircular detour around the pole at the origin can be represented by setting $s = \epsilon e^{j\phi}$ and allowing ϕ to vary from $-90°$ at $\omega = 0^-$ to $+90°$ at $\omega = 0^+$. Because ϵ approaches zero, the mapping for $L(s)$ is

$$\lim_{\epsilon \to 0} L(s) = \lim_{\epsilon \to 0} \frac{K}{\epsilon e^{j\phi}} = \lim_{\epsilon \to 0} \frac{K}{\epsilon} e^{-j\phi}. \tag{9.18}$$

Therefore, the angle of the contour in the $L(s)$-plane changes from $90°$ at $\omega = 0_-$ to $-90°$ at $\omega = 0_+$, passing through $0°$ at $\omega = 0$. The radius of the contour in the $L(s)$-plane for this portion of the contour is infinite, and this portion of the contour is shown in Figure 9.10(b). The points denoted by $A, B,$ and C in Figure 9.10(a) map to $A, B,$ and C, respectively, in Figure 9.10(b).

(b) The Portion from $\omega = 0_+$ to $\omega = +\infty$. The portion of the contour Γ_s from $\omega = 0_+$ to $\omega = +\infty$ is mapped by the function $L(s)$ as the real frequency polar

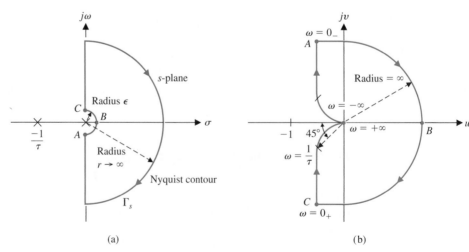

FIGURE 9.10
Nyquist contour and mapping for
$L(s) = K/(s(\tau s + 1))$.

plot because $s = j\omega$ and

$$L(s)|_{s=j\omega} = L(j\omega) \tag{9.19}$$

for this part of the contour. This results in the real frequency polar plot shown in Figure 9.10(b). When ω approaches $+\infty$, we have

$$\lim_{\omega \to +\infty} L(j\omega) = \lim_{\omega \to +\infty} \frac{K}{+j\omega(j\omega\tau + 1)}$$

$$= \lim_{\omega \to \infty} \left|\frac{K}{\tau\omega^2}\right| \underline{/-(\pi/2) - \tan^{-1}(\omega\tau)}. \tag{9.20}$$

Therefore, the magnitude approaches zero at an angle of $-180°$.

(c) *The Portion from $\omega = +\infty$ to $\omega = -\infty$.* The portion of Γ_s from $\omega = +\infty$ to $\omega = -\infty$ is mapped into the point zero at the origin of the $L(s)$-plane by the function $L(s)$. The mapping is represented by

$$\lim_{r \to \infty} L(s)|_{s=re^{j\phi}} = \lim_{r \to \infty} \left|\frac{K}{\tau r^2}\right| e^{-2j\phi} \tag{9.21}$$

as ϕ changes from $\phi = +90°$ at $\omega = +\infty$ to $\phi = -90°$ at $\omega = -\infty$. Thus, the contour moves from an angle of $-180°$ at $\omega = +\infty$ to an angle of $+180°$ at $\omega = -\infty$. The magnitude of the $L(s)$ contour when r is infinite is always zero or a constant.

(d) *The Portion from $\omega = -\infty$ to $\omega = 0_-$.* The portion of the contour Γ_s from $\omega = -\infty$ to $\omega = 0_-$ is mapped by the function $L(s)$ as

$$L(s)|_{s=-j\omega} = L(-j\omega). \tag{9.22}$$

Thus, we obtain the complex conjugate of $L(j\omega)$, and the plot for the portion of the polar plot from $\omega = -\infty$ to $\omega = 0_-$ is symmetrical to the polar plot from $\omega = +\infty$ to $\omega = 0_+$. This symmetrical polar plot is shown on the $L(s)$-plane in Figure 9.10(b).

To investigate the stability of this second-order system, we first note that the number of poles, P, within the right-hand s-plane is zero. Therefore, for this system to be stable, we require $N = Z = 0$, and the contour Γ_L must not encircle the -1 point in the $L(s)$-plane. Examining Figure 9.10(b), we find that irrespective of the value of the gain K and the time constant τ, the contour does not encircle the -1 point, and the system is always stable. As in Chapter 7, we are considering positive values of gain K. If negative values of gain are to be considered, we should use $-K$, where $K \geq 0$.

We may draw two general conclusions from this example:

1. The plot of the contour Γ_L for the range $-\infty < \omega < 0_-$ will be the complex conjugate of the plot for the range $0_+ < \omega < +\infty$, and the polar plot of $L(s) = G_c(s)G(s)H(s)$ will be symmetrical in the $L(s)$-plane about the u-axis. Therefore, **it is sufficient to construct the contour Γ_L for the frequency range $0_+ < \omega < +\infty$ in order to investigate the stability** (keeping in mind the detour around the origin).

2. The magnitude of $L(s) = G_c(s)G(s)H(s)$ as $s = re^{j\phi}$ and $r \to \infty$ will normally approach zero or a constant. ∎

EXAMPLE 9.3   **System with three poles**

Let us again consider the single-loop system shown in Figure 9.1 when

$$L(s) = G_c(s)G(s)H(s) = \frac{K}{s(\tau_1 s + 1)(\tau_2 s + 1)}. \tag{9.23}$$

The Nyquist contour Γ_s is shown in Figure 9.10(a). Again, this mapping is symmetrical for $L(j\omega)$ and $L(-j\omega)$ so that it is sufficient to investigate the $L(j\omega)$-locus. The small semicircle around the origin of the s-plane maps into a semicircle of infinite radius, as in Example 9.2. Also, the semicircle $re^{j\phi}$ in the s-plane as $r \to \infty$ maps into the point $L(s) = 0$, as we expect. Therefore, to investigate the stability of the system, it is sufficient to plot the portion of the contour Γ_L that is the real frequency polar plot $L(j\omega)$ for $0_+ < \omega < +\infty$. Thus, when $s = +j\omega$, we have

$$L(j\omega) = \frac{K}{j\omega(j\omega\tau_1 + 1)(j\omega\tau_2 + 1)}$$

$$= \frac{-K(\tau_1 + \tau_2) - jK(1/\omega)(1 - \omega^2\tau_1\tau_2)}{1 + \omega^2(\tau_1^2 + \tau_2^2) + \omega^4\tau_1^2\tau_2^2}$$

$$= \frac{K}{[\omega^4(\tau_1 + \tau_2)^2 + \omega^2(1 - \omega^2\tau_1\tau_2)^2]^{1/2}}$$

$$\times \underline{/-\tan^{-1}(\omega\tau_1) - \tan^{-1}(\omega\tau_2) - (\pi/2)}. \tag{9.24}$$

When $\omega = 0_+$, the magnitude of the locus is infinite at an angle of $-90°$ in the $L(s)$-plane. When ω approaches $+\infty$, we have

$$\lim_{\omega \to \infty} L(j\omega) = \lim_{\omega \to \infty} \left| \frac{1}{\omega^3\tau_1\tau_2} \right| \underline{/-(\pi/2) - \tan^{-1}(\omega\tau_1) - \tan^{-1}(\omega\tau_2)}$$

$$= \lim_{\omega \to \infty} \left| \frac{1}{\omega^3\tau_1\tau_2} \right| \underline{/-3\pi/2}. \tag{9.25}$$

Therefore, $L(j\omega)$ approaches a magnitude of zero at an angle of $-270°$ [30]. To approach at an angle of $-270°$, the locus must cross the u-axis in the $L(s)$-plane, as shown in Figure 9.11. Thus, it is possible to encircle the -1 point. The number of encirclements when the -1 point lies within the locus, as shown in Figure 9.11, is equal to two, and the system is unstable with two roots in the right-hand s-plane. The point where the $L(s)$-locus intersects the real axis can be found by setting the imaginary part of $L(j\omega) = u + jv$ equal to zero. We then have, from Equation (9.24),

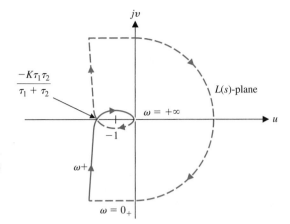

FIGURE 9.11
Nyquist diagram for
$L(s) = K/(s(\tau_1 s + 1)(\tau_2 s + 1))$. The tic mark shown to the left of the origin is the -1 point.

$$v = \frac{-K(1/\omega)(1 - \omega^2 \tau_1 \tau_2)}{1 + \omega^2(\tau_1^2 + \tau_2^2) + \omega^4 \tau_1^2 \tau_2^2} = 0. \tag{9.26}$$

Thus, $v = 0$ when $1 - \omega^2 \tau_1 \tau_2 = 0$ or $\omega = 1/\sqrt{\tau_1 \tau_2}$. The magnitude of the real part of $L(j\omega)$ at this frequency is

$$u = \frac{-K(\tau_1 + \tau_2)}{1 + \omega^2(\tau_1^2 + \tau_2^2) + \omega^4 \tau_1^2 \tau_2^2}\bigg|_{\omega^2 = 1/\tau_1 \tau_2}$$

$$= \frac{-K(\tau_1 + \tau_2)\tau_1 \tau_2}{\tau_1 \tau_2 + (\tau_1^2 + \tau_2^2) + \tau_1 \tau_2} = \frac{-K\tau_1 \tau_2}{\tau_1 + \tau_2}. \tag{9.27}$$

Therefore, the system is stable when

$$\frac{-K\tau_1 \tau_2}{\tau_1 + \tau_2} \ge -1,$$

or

$$K \le \frac{\tau_1 + \tau_2}{\tau_1 \tau_2}. \tag{9.28}$$

Consider the case where $\tau_1 = \tau_2 = 1$, so that

$$L(s) = G_c(s)G(s)H(s) = \frac{K}{s(s + 1)^2}.$$

Using Equation (9.28), we expect stability when

$$K \le 2.$$

The Nyquist diagrams for three values of K are shown in Figure 9.12. ∎

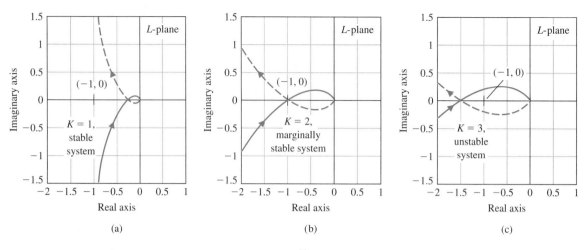

FIGURE 9.12 Nyquist plot for $L(s) = G_c(s)G(s)H(s) = \dfrac{K}{s(s+1)^2}$ when (a) $K = 1$, (b) $K = 2$, and (c) $K = 3$.

EXAMPLE 9.4   **System with two poles at the origin**

Again, let us determine the stability of the single-loop system shown in Figure 9.1 when

$$L(s) = G_c(s)G(s)H(s) = \frac{K}{s^2(\tau s + 1)}. \tag{9.29}$$

The real frequency polar plot is obtained when $s = j\omega$, and we have

$$L(j\omega) = \frac{K}{-\omega^2(j\omega\tau + 1)} = \frac{K}{[\omega^4 + \tau^2\omega^6]^{1/2}} \underline{/-\pi - \tan^{-1}(\omega\tau)}. \tag{9.30}$$

We note that the angle of $L(j\omega)$ is always $-180°$ or less, and the locus of $L(j\omega)$ is above the u-axis for all values of ω. As ω approaches 0_+, we have

$$\lim_{\omega \to 0+} L(j\omega) = \lim_{\omega \to 0+} \left|\frac{K}{\omega^2}\right| \underline{/-\pi}. \tag{9.31}$$

As ω approaches $+\infty$, we have

$$\lim_{\omega \to +\infty} L(j\omega) = \lim_{\omega \to +\infty} \frac{K}{\omega^3} \underline{/-3\pi/2}. \tag{9.32}$$

At the small semicircular detour at the origin of the s-plane where $s = \epsilon e^{j\phi}$, we have

$$\lim_{\epsilon \to 0} L(s) = \lim_{\epsilon \to 0} \frac{K}{\epsilon^2} e^{-2j\phi}, \tag{9.33}$$

where $-\pi/2 \le \phi \le \pi/2$. Thus, the contour Γ_L ranges from an angle of $+\pi\omega = 0_-$ to $-\pi$ at $\omega = 0_+$ and passes through a full circle of 2π rad as ω changes from $\omega = 0_-$ to $\omega = 0_+$. The complete contour plot of Γ_L is shown in Figure 9.13. Because the

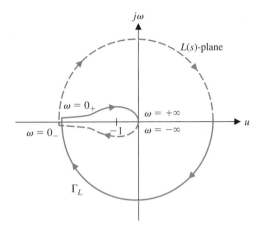

FIGURE 9.13
Nyquist contour plot for $L(s) = K/(s^2(\tau s + 1))$.

contour encircles the -1 point twice, there are two roots of the closed-loop system in the right-hand plane, and the system, irrespective of the gain K, is unstable. ∎

EXAMPLE 9.5   **System with a pole in the right-hand s-plane**

Let us consider the control system shown in Figure 9.14 and determine the stability of the system. First, let us consider the system without derivative feedback, so that $K_2 = 0$. We then have the loop transfer function

$$L(s) = G_c(s)G(s)H(s) = \frac{K_1}{s(s - 1)}. \tag{9.34}$$

Thus, the loop transfer function has one pole in the right-hand s-plane, and therefore $P = 1$. For this system to be stable, we require $N = -P = -1$, one counterclockwise encirclement of the -1 point. At the semicircular detour at the origin

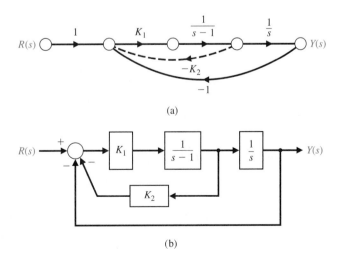

(a)

(b)

FIGURE 9.14
Second-order feedback control system. (a) Signal-flow graph. (b) Block diagram.

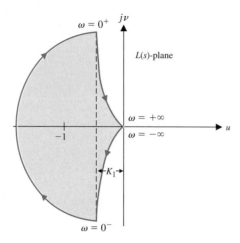

FIGURE 9.15
Nyquist diagram for
$L(s) = K_1/(s(s - 1))$.

of the s-plane, we let $s = \epsilon e^{j\phi}$ when $-\pi/2 \leq \phi \leq \pi/2$. Then, when $s = \epsilon e^{j\phi}$, we have

$$\lim_{\epsilon \to 0} L(s) = \lim_{\epsilon \to 0} \frac{K_1}{-\epsilon e^{j\phi}} = \lim_{\epsilon \to 0} \left| \frac{K_1}{\epsilon} \right| \underline{/-180° - \phi}. \qquad (9.35)$$

Therefore, this portion of the contour Γ_L is a semicircle of infinite magnitude in the left-hand $L(s)$-plane, as shown in Figure 9.15. When $s = j\omega$, we have

$$L(j\omega) = G_c(j\omega)G(j\omega)H(j\omega) = \frac{K_1}{j\omega(j\omega - 1)} = \frac{K_1}{(\omega^2 + \omega^4)^{1/2}} \underline{/(-\pi/2) - \tan^{-1}(-\omega)}$$

$$= \frac{K_1}{(\omega^2 + \omega^4)^{1/2}} \underline{/+\pi/2 + \tan^{-1}\omega}. \qquad (9.36)$$

Finally, for the semicircle of radius r as r approaches infinity, we have

$$\lim_{r \to \infty} L(s)|_{s=re^{j\phi}} = \lim_{r \to \infty} \left| \frac{K_1}{r^2} \right| e^{-2j\phi}, \qquad (9.37)$$

where ϕ varies from $\pi/2$ to $-\pi/2$ in a clockwise direction. Therefore, the contour Γ_L, at the origin of the $L(s)$-plane, varies 2π rad in a counterclockwise direction. Several important values of the $L(s)$-locus are given in Table 9.3. The contour Γ_L in the $L(s)$-plane encircles the -1 point once in the clockwise direction so $N = +1$,

Table 9.3   Values of $L(s) = G_c(s)G(s)H(s)$

s	$j0_-$	$j0_+$	$j1$	$+j\infty$	$-j\infty$
$|L|/K_1$	∞	∞	$1/\sqrt{2}$	0	0
$\underline{/L}$	$-90°$	$+90°$	$+135°$	$+180°$	$-180°$

and there is one pole $s = 1$ in the right-hand plane so $P = 1$. Hence,

$$Z = N + P = 2, \tag{9.38}$$

and the system is unstable because two roots of the characteristic equation, irrespective of the value of the gain K_1, lie in the right half of the s-plane.

Let us now consider again the system when the derivative feedback is included in the system shown in Figure 9.14 ($K_2 > 0$). Then the loop transfer function is

$$L(s) = G_c(s)G(s)H(s) = \frac{K_1(1 + K_2 s)}{s(s - 1)}. \tag{9.39}$$

The portion of the contour Γ_L when $s = \epsilon e^{j\phi}$ is the same as the system without derivative feedback, as shown in Figure 9.16. However, when $s = re^{j\phi}$ as r approaches infinity, we have

$$\lim_{r \to \infty} L(s)|_{s=re^{j\phi}} = \lim_{r \to \infty} \left| \frac{K_1 K_2}{r} \right| e^{-j\phi}, \tag{9.40}$$

and the Γ_L-contour at the origin of the $L(s)$-plane varies π rad in a counterclockwise direction. The frequency locus $L(j\omega)$ crosses the u-axis at a point determined by considering the real frequency transfer function

$$L(j\omega) = G_c(j\omega)G(j\omega)H(j\omega) = \frac{K_1(1 + K_2 j\omega)}{-\omega^2 - j\omega}$$

$$= \frac{-K_1(\omega^2 + \omega^2 K_2) + j(\omega - K_2\omega^3)K_1}{\omega^2 + \omega^4}. \tag{9.41}$$

The $L(j\omega)$-locus intersects the u-axis at a point where the imaginary part of $L(j\omega)$ is zero. Therefore,

$$\omega - K_2\omega^3 = 0$$

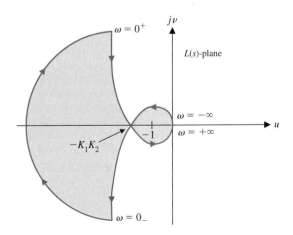

FIGURE 9.16
Nyquist diagram for
$L(s) = K_1(1 + K_2 s)/$
$(s(s - 1))$.

at this point, or $\omega^2 = 1/K_2$. The value of the real part of $L(j\omega)$ at the intersection is then

$$u\big|_{\omega^2=1/K_2} = \frac{-\omega^2 K_1(1 + K_2)}{\omega^2 + \omega^4}\bigg|_{\omega^2=1/K_2} = -K_1 K_2. \tag{9.42}$$

Therefore, when $-K_1 K_2 < -1$ or $K_1 K_2 > 1$, the contour Γ_L encircles the -1 point once in a counterclockwise direction, and therefore $N = -1$. Then the number of zeros of the system in the right-hand plane, is

$$Z = N + P = -1 + 1 = 0.$$

Thus, the system is stable when $K_1 K_2 > 1$. Often, it may be useful to utilize a computer to plot the Nyquist diagram [5]. ∎

EXAMPLE 9.6   **System with a zero in the right-hand s-plane**

Let us consider the feedback control system shown in Figure 9.1 when

$$L(s) = G_c(s)G(s)H(s) = \frac{K(s - 2)}{(s + 1)^2}.$$

We have

$$L(j\omega) = \frac{K(j\omega - 2)}{(j\omega + 1)^2} = \frac{K(j\omega - 2)}{(1 - \omega^2) + j2\omega}. \tag{9.43}$$

As ω approaches $+\infty$ on the $+j\omega$ axis, we have

$$\lim_{\omega \to +\infty} L(j\omega) = \lim_{\omega \to +\infty} \frac{K}{\omega}\angle{-\pi/2}.$$

When $\omega = \sqrt{5}$, we have $L(j\omega) = K/2$. At $\omega = 0_+$, we have $L(j\omega) = -2K$. The Nyquist diagram for $L(j\omega)/K$ is shown in Figure 9.17. $L(j\omega)$ intersects the $-1 + j0$ point when $K = 1/2$. Thus, the system is stable for the limited range of gain $0 < K \le 1/2$. When $K > 1/2$, the number of encirclements of the -1 point is $N = 1$. The number of poles of $L(s)$ in the right half s-plane is $P = 0$. Therefore, we have

$$Z = N + P = 1,$$

and the system is unstable. Examining the Nyquist diagram of Figure 9.17, which is plotted for $L(j\omega)/K$, we conclude that the system is unstable for all $K > 1/2$. ∎

9.4 RELATIVE STABILITY AND THE NYQUIST CRITERION

We discussed the relative stability of a system in terms of the s-plane in Section 6.3. For the s-plane, we defined the relative stability of a system as the property measured by the relative settling time of each root or pair of roots. Therefore, a system with a shorter settling time is considered more relatively stable. We would like to

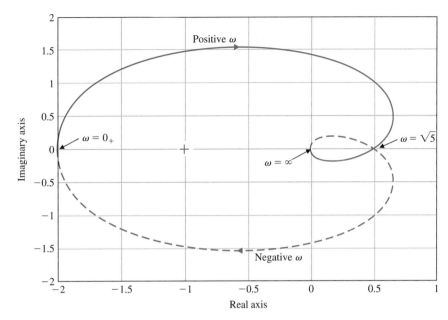

FIGURE 9.17
Nyquist diagram for
Example 9.6 for
$L(j\omega)/K$.

determine a similar measure of relative stability useful for the frequency response method. The Nyquist criterion provides us with suitable information concerning the absolute stability and, furthermore, can be utilized to define and ascertain the relative stability of a system.

The Nyquist stability criterion is defined in terms of the $(-1, 0)$ point on the polar plot or the 0-dB, $-180°$ point on the Bode diagram or log-magnitude–phase diagram. Clearly, the proximity of the $L(j\omega)$-locus to this stability point is a measure of the relative stability of a system. The polar plot for $L(j\omega)$ for several values of K and

$$L(j\omega) = G_c(j\omega)G(j\omega)H(j\omega) = \frac{K}{j\omega(j\omega\tau_1 + 1)(j\omega\tau_2 + 1)} \quad (9.44)$$

is shown in Figure 9.18. As K increases, the polar plot approaches the -1 point and eventually encircles the -1 point for a gain $K = K_3$. We determined in Section 9.3 that the locus intersects the u-axis at a point

$$u = \frac{-K\tau_1\tau_2}{\tau_1 + \tau_2}. \quad (9.45)$$

Therefore, the system has roots on the $j\omega$-axis when

$$u = -1 \quad \text{or} \quad K = \frac{\tau_1 + \tau_2}{\tau_1\tau_2}.$$

As K is decreased below this marginal value, the stability is increased, and the margin between the critical gain $K = (\tau_1 + \tau_2)/\tau_1\tau_2$ and a gain $K = K_2$ is a measure of

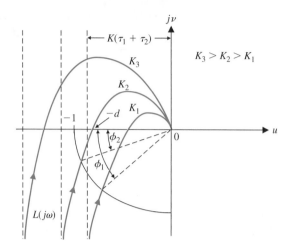

FIGURE 9.18
Polar plot for $L(j\omega)$
for three values of
gain.

the relative stability. This measure of relative stability is called the **gain margin** and is defined as **the reciprocal of the gain $|L(j\omega)|$ at the frequency at which the phase angle reaches $-180°$** (that is, $v = 0$). The gain margin is a measure of the factor by which the system gain would have to be increased for the $L(j\omega)$ locus to pass through the $u = -1$ point. Thus, for a gain $K = K_2$ in Figure 9.18, the gain margin is equal to the reciprocal of $L(j\omega)$ when $v = 0$. Because $\omega = 1/\sqrt{\tau_1\tau_2}$ when the phase shift is $-180°$, we have a gain margin equal to

$$\frac{1}{|L(j\omega)|} = \left[\frac{K_2\tau_1\tau_2}{\tau_1 + \tau_2}\right]^{-1} = \frac{1}{d}. \tag{9.46}$$

The gain margin can be defined in terms of a **logarithmic (decibel) measure** as

$$20\log\frac{1}{d} = -20\log d \text{ dB.} \tag{9.47}$$

For example, when $\tau_1 = \tau_2 = 1$, the system is stable when $K \leq 2$. Thus, when $K = K_2 = 0.5$, the gain margin is equal to

$$\frac{1}{d} = \left[\frac{K_2\tau_1\tau_2}{\tau_1 + \tau_2}\right]^{-1} = 4, \tag{9.48}$$

or, in logarithmic measure,

$$20\log 4 = 12 \text{ dB.} \tag{9.49}$$

Therefore, the gain margin indicates that the system gain can be increased by a factor of four (12 dB) before the stability boundary is reached.

> **The gain margin is the increase in the system gain when phase $= -180°$ that will result in a marginally stable system with intersection of the $-1 + j0$ point on the Nyquist diagram.**

An alternative measure of relative stability can be defined in terms of the phase angle margin between a specific system and a system that is marginally stable. The **phase margin** is defined as **the phase angle through which the $L(j\omega)$ locus must be rotated so that the unity magnitude $|L(j\omega)| = 1$ point will pass through the $(-1, 0)$ point in the $L(j\omega)$ plane**. This measure of relative stability is equal to the additional phase lag required before the system becomes unstable. This information can be determined from the Nyquist diagram shown in Figure 9.18. For a gain $K = K_2$, an additional phase angle, ϕ_2, may be added to the system before the system becomes unstable. Similarly, for the gain K_1, the phase margin is equal to ϕ_1, as shown in Figure 9.18.

> **The phase margin is the amount of phase shift of the $L(j\omega)$ at unity magnitude that will result in a marginally stable system with intersection of the $-1 + j0$ point on the Nyquist diagram.**

The gain and phase margins are easily evaluated from the Bode diagram, and because it is preferable to draw the Bode diagram in contrast to the polar plot, it is worthwhile to illustrate the relative stability measures for the Bode diagram. The critical point for stability is $u = -1$, $v = 0$ in the $L(j\omega)$-plane, which is equivalent to a logarithmic magnitude of 0 dB and a phase angle of 180° (or −180°) on the Bode diagram.

It is relatively straightforward to examine the Nyquist diagram of a minimum-phase system. Special care is required with a nonminimum-phase system, however, and the complete Nyquist diagram should be studied to determine stability.

The gain margin and phase margin can be readily calculated by utilizing a computer program, assuming the system is minimum phase. In contrast, for nonminimum-phase systems, the complete Nyquist diagram must be constructed.

The Bode diagram of

$$L(j\omega) = G_c(j\omega)G(j\omega)H(j\omega) = \frac{1}{j\omega(j\omega + 1)(0.2j\omega + 1)} \tag{9.50}$$

is shown in Figure 9.19. The phase angle when the logarithmic magnitude is 0 dB is equal to 137°. Thus, the phase margin is $180° - 137° = 43°$, as shown in Figure 9.19. The logarithmic magnitude when the phase angle is −180° is −15 dB, and therefore the gain margin is equal to 15 dB, as shown in Figure 9.19.

The frequency response of a system can be graphically portrayed on the logarithmic-magnitude–phase-angle diagram. For the log-magnitude–phase diagram, the critical stability point is the 0-dB, −180° point, and the gain margin and phase margin can be easily determined and indicated on the diagram. The log-magnitude–phase locus of

$$L_1(j\omega) = G_c(j\omega)G(j\omega)H_1(j\omega) = \frac{1}{j\omega(j\omega + 1)(0.2j\omega + 1)} \tag{9.51}$$

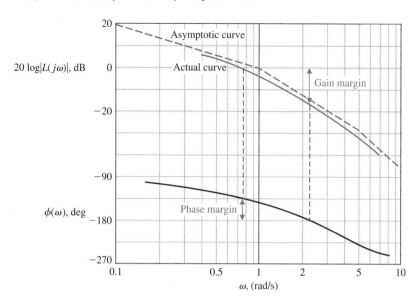

FIGURE 9.19
Bode diagram for
$L(j\omega) =$
$1/(j\omega(j\omega + 1))$
$(0.2\,j\omega + 1)$.

is shown in Figure 9.20. The indicated phase margin is 43°, and the gain margin is 15 dB. For comparison, the locus for

$$L_2(j\omega) = G_c(j\omega)G(j\omega)H_2(j\omega) = \frac{1}{j\omega(j\omega + 1)^2} \qquad (9.52)$$

is also shown in Figure 9.20. The gain margin for L_2 is equal to 5.7 dB, and the phase margin for L_2 is equal to 20°. Clearly, the feedback system $L_2(j\omega)$ is relatively less stable than the system $L_1(j\omega)$. However, the question still remains: How much less stable is the system $L_2(j\omega)$ in comparison to the system $L_1(j\omega)$? In the following, we answer this question for a second-order system, and the general usefulness of the relation that we develop will depend on the presence of dominant roots.

Let us now determine the phase margin of a second-order system and relate the phase margin to the damping ratio ζ of an underdamped system. Consider the loop-transfer function of the system shown in Figure 9.1, where

$$L(s) = G_c(s)G(s)H(s) = \frac{\omega_n^2}{s(s + 2\zeta\omega_n)}. \qquad (9.53)$$

The characteristic equation for this second-order system is

$$s^2 + 2\zeta\omega_n s + \omega_n^2 = 0.$$

Therefore, the closed-loop roots are

$$s = -\zeta\omega_n \pm j\omega_n\sqrt{1 - \zeta^2}.$$

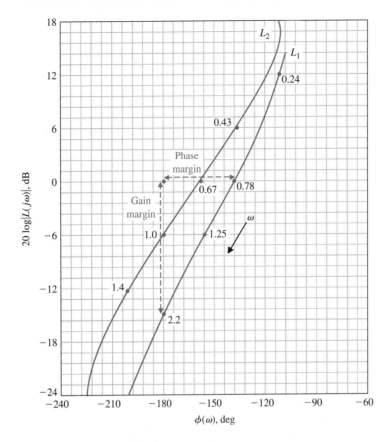

FIGURE 9.20
Log-magnitude–phase curve for L_1 and L_2.

The frequency domain form of Equation (9.53) is

$$L(j\omega) = \frac{\omega_n^2}{j\omega(j\omega + 2\zeta\omega_n)}. \qquad (9.54)$$

The magnitude of the frequency response is equal to 1 at a frequency ω_c; thus,

$$\frac{\omega_n^2}{\omega_c(\omega_c^2 + 4\zeta^2\omega_n^2)^{1/2}} = 1. \qquad (9.55)$$

Rearranging Equation (9.55), we obtain

$$(\omega_c^2)^2 + 4\zeta^2\omega_n^2(\omega_c^2) - \omega_n^4 = 0. \qquad (9.56)$$

Solving for ω_c, we find that

$$\frac{\omega_c^2}{\omega_n^2} = (4\zeta^4 + 1)^{1/2} - 2\zeta^2.$$

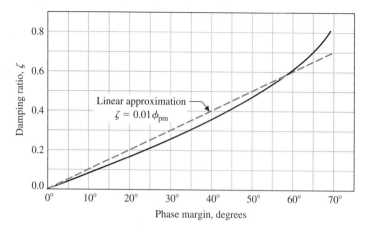

FIGURE 9.21
Damping ratio versus phase margin for a second-order system.

The phase margin for this system is

$$\phi_{pm} = 180° - 90° - \tan^{-1}\frac{\omega_c}{2\zeta\omega_n}$$

$$= 90° - \tan^{-1}\left(\frac{1}{2\zeta}[(4\zeta^4 + 1)^{1/2} - 2\zeta^2]^{1/2}\right)$$

$$= \tan^{-1}\frac{2}{[(4 + 1/\zeta^4)^{1/2} - 2]^{1/2}}. \tag{9.57}$$

Equation (9.57) is the relationship between the damping ratio ζ and the phase margin ϕ_{pm}, which provides a correlation between the frequency response and the time response. A plot of ζ versus ϕ_{pm} is shown in Figure 9.21. The actual curve of ζ versus ϕ_{pm} can be approximated by the dashed line shown in Figure 9.21. The slope of the linear approximation is equal to 0.01, and therefore an approximate linear relationship between the damping ratio and the phase margin is

$$\boxed{\zeta = 0.01\phi_{pm},} \tag{9.58}$$

where the phase margin is measured in degrees. This approximation is reasonably accurate for $\zeta \le 0.7$ and is a useful index for correlating the frequency response with the transient performance of a system. Equation (9.58) is a suitable approximation for a second-order system and may be used for higher-order systems if we can assume that the transient response of the system is primarily due to a pair of dominant underdamped roots. The approximation of a higher-order system by a dominant second-order system is a useful approximation indeed! Although it must be used with care, control engineers find this approach to be a simple, yet fairly accurate, technique of setting the specifications of a control system.

Therefore, for the system with a loop transfer function

$$L(j\omega) = \frac{1}{j\omega(j\omega + 1)(0.2j\omega + 1)}, \tag{9.59}$$

we found that the phase margin was 43°, as shown in Figure 9.19. Thus, the damping ratio is approximately

$$\zeta \approx 0.01\phi_{pm} = 0.43. \tag{9.60}$$

Then the percent overshoot to a step input for this system is approximately

$$P.O. = 22\%, \tag{9.61}$$

as obtained from Figure 5.8 for $\zeta = 0.43$.

It is feasible to develop a computer program to calculate and plot phase margin and gain margin versus the gain K for a specified $L(j\omega)$. Consider the system of Figure 9.1 with

$$L(s) = G_c(s)G(s)H(s) = \frac{K}{s(s + 4)^2}.$$

The gain for which the system is marginally stable is $K = K^* = 128$. The gain margin and the phase margin plotted versus K are shown in Figures 9.22(a) and (b), respectively. The gain margin is plotted versus the phase margin, as shown in Figure 9.22(c). Note that either the phase margin or the gain margin is a suitable measure of the performance of the system. We will normally emphasize phase margin as a frequency-domain specification.

The phase margin of a system is a quite suitable frequency response measure for indicating the expected transient performance of a system. Another useful index of performance in the frequency domain is $M_{p\omega}$, the maximum magnitude of the closed-loop frequency response, and we shall now consider this practical index.

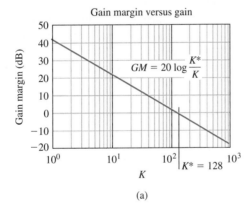

(a)

FIGURE 9.22
(a) Gain margin versus gain K.
(b) Phase margin versus gain K.
(c) Gain margin versus phase margin.

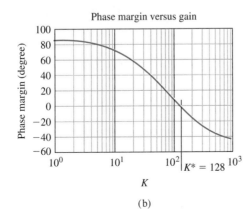

(b)

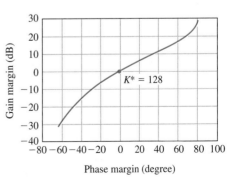

(c)

9.5 TIME-DOMAIN PERFORMANCE CRITERIA IN THE FREQUENCY DOMAIN

The transient performance of a feedback system can be estimated from the closed-loop frequency response. The **closed-loop frequency response** is the frequency response of the closed-loop transfer function $T(j\omega)$. The open- and closed-loop frequency responses for a single-loop system are related as follows:

$$\frac{Y(j\omega)}{R(j\omega)} = T(j\omega) = \frac{G_c(j\omega)G(j\omega)}{1 + G_c(j\omega)G(j\omega)H(j\omega)}. \tag{9.62}$$

The Nyquist criterion and the phase margin index are defined for the loop transfer function $L(j\omega) = G_c(j\omega)G(j\omega)H(j\omega)$. However, as we found in Section 8.2, the maximum magnitude of the closed-loop frequency response can be related to the damping ratio of a second-order system of

$$\boxed{M_{p\omega} = |T(\omega_r)| = \left(2\zeta\sqrt{1 - \zeta^2}\right)^{-1},} \qquad \zeta < 0.707. \tag{9.63}$$

This relation is graphically portrayed in Figure 8.11. Because this relationship between the closed-loop frequency response and the transient response is a useful one, we would like to be able to determine $M_{p\omega}$ from the plots completed for the investigation of the Nyquist criterion. That is, we want to be able to obtain the closed-loop frequency response (Equation 9.62) from the open-loop frequency response. Of course, we could determine the closed-loop roots of $1 + L(s)$ and plot the closed-loop frequency response. However, once we have invested all the effort necessary to find the closed-loop roots of a characteristic equation, then a closed-loop frequency response is not necessary.

The relation between the closed-loop and open-loop frequency response is illuminated on the magnitude-phase plot when considering unity feedback systems, that is, when $H(s) = 1$ in Figure 9.1. In the unity feedback case, key performance indicators such as $M_{p\omega}$ and ω_r can be determined from the magnitude-phase plot using circles of constant magnitude of the closed-loop transfer function. These circles are known as constant M-circles. If the system is not in fact a unity feedback system where $H(j\omega) = 1$, we can modify the system (see Section 5.6). For unity feedback systems, Equation (9.62) becomes

$$T(j\omega) = M(\omega)e^{j\phi(\omega)} = \frac{G_c(j\omega)G(j\omega)}{1 + G_c(j\omega)G(j\omega)}. \tag{9.64}$$

The relationship between $T(j\omega)$ and $G_c(j\omega)G(j\omega)$ is readily obtained in terms of complex variables in the $G_cG(j\omega)$-plane. The coordinates of the $G_cG(j\omega)$-plane are u and v, and we have

$$G_c(j\omega)G(j\omega) = u + jv. \tag{9.65}$$

Therefore, the magnitude of the closed-loop transfer function is

$$M(\omega) = \left|\frac{G_c(j\omega)G(j\omega)}{1 + G_c(j\omega)G(j\omega)}\right| = \left|\frac{u + jv}{1 + u + jv}\right| = \frac{(u^2 + v^2)^{1/2}}{[(1 + u)^2 + v^2]^{1/2}}. \tag{9.66}$$

Squaring Equation (9.66) and rearranging, we obtain

$$(1 - M^2)u^2 + (1 - M^2)v^2 - 2M^2u = M^2. \tag{9.67}$$

Dividing Equation (9.67) by $1 - M^2$ and adding the term $[M^2/(1 - M^2)]^2$ to both sides, we have

$$u^2 + v^2 - \frac{2M^2u}{1 - M^2} + \left(\frac{M^2}{1 - M^2}\right)^2 = \left(\frac{M^2}{1 - M^2}\right) + \left(\frac{M^2}{1 - M^2}\right)^2. \tag{9.68}$$

Rearranging, we obtain

$$\left(u - \frac{M^2}{1 - M^2}\right)^2 + v^2 = \left(\frac{M}{1 - M^2}\right)^2, \tag{9.69}$$

which is the equation of a circle on the (u, v)-plane with the center at

$$u = \frac{M^2}{1 - M^2}, \qquad v = 0.$$

The radius of the circle is equal to $|M/(1 - M^2)|$. Therefore, we can plot several circles of constant magnitude M in the $[G_c(j\omega)G(j\omega) = u + jv]$-plane. Several constant M circles are shown in Figure 9.23. The circles to the left of $u = -1/2$ are for $M > 1$, and the circles to the right of $u = -1/2$ are for $M < 1$. When $M = 1$, the circle becomes the straight line $u = -1/2$, which is evident from inspection of Equation (9.67).

The open-loop frequency response for a system is shown in Figure 9.24 for two gain values where $K_2 > K_1$. The frequency response curve for the system with gain K_1 is tangent to magnitude circle M_1 at a frequency ω_{r1}. Similarly, the frequency response curve for gain K_2 is tangent to magnitude circle M_2 at the frequency ω_{r2}. Therefore, the closed-loop frequency response magnitude curves are estimated as shown in Figure 9.25. Hence, we can obtain the closed-loop frequency response of a

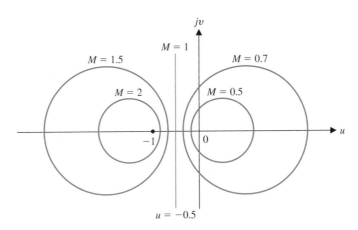

FIGURE 9.23
Constant M circles.

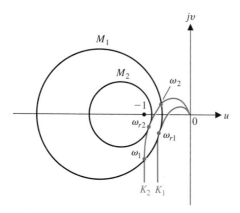

FIGURE 9.24   Polar plot of $G_c(j\omega)G(j\omega)$ for two values of a gain ($K_2 > K_1$).

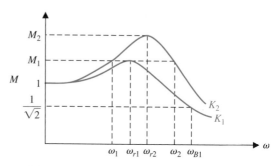

FIGURE 9.25   Closed-loop frequency response of $T(j\omega) = G_c(j\omega)G(j\omega)/(1 + G_c(j\omega)G(j\omega))$. Note that $K_2 > K_1$.

system from the $(u + jv)$-plane. If the maximum magnitude, $M_{p\omega}$, is the only information desired, then it is sufficient to read this value directly from the polar plot. The maximum magnitude of the closed-loop frequency response, $M_{p\omega}$, is the value of the M circle that is tangent to the $G_c(j\omega)G(j\omega)$-locus. The point of tangency occurs at the frequency ω_r, the resonant frequency. The complete closed-loop frequency response of a system can be obtained by reading the magnitude M of the circles that the $G_c(j\omega)G(j\omega)$-locus intersects at several frequencies. Therefore, the system with a gain $K = K_2$ has a closed-loop magnitude M_1 at the frequencies ω_1 and ω_2. This magnitude is read from Figure 9.24 and is shown on the closed-loop frequency response in Figure 9.25. The **bandwidth** for K_1 is shown as ω_{B1}.

It may be empirically shown that the crossover frequency ω_c on the open-loop Bode diagram is related to the closed-loop system bandwidth ω_B by the approximation $\omega_B = 1.6\omega_c$ for ζ in the range 0.2 to 0.8.

In a similar manner, we can obtain circles of constant closed-loop phase angles. Thus, for Equation (9.64), the angle relation is

$$\phi = \underline{/T(j\omega)} = \underline{/(u + jv)/(1 + u + jv)}$$

$$= \tan^{-1}\left(\frac{v}{u}\right) - \tan^{-1}\left(\frac{v}{1 + u}\right). \tag{9.70}$$

Taking the tangent of both sides and rearranging, we have

$$u^2 + v^2 + u - \frac{v}{N} = 0, \tag{9.71}$$

where $N = \tan \phi$. Adding the term $1/4[1 + 1/N^2]$ to both sides of the equation and simplifying, we obtain

$$\left(u + \frac{1}{2}\right)^2 + \left(v - \frac{1}{2N}\right)^2 = \frac{1}{4}\left(1 + \frac{1}{N^2}\right), \tag{9.72}$$

which is the equation of a circle with its center at $u = -1/2$ and $v = +1/(2N)$. The radius of the circle is equal to $1/2[1 + 1/N^2]^{1/2}$. Therefore, the constant phase angle curves can be obtained for various values of N in a manner similar to the M circles.

The constant M and N circles can be used for analysis and design in the polar plane. However, it is much easier to obtain the Bode diagram for a system, and it would be preferable if the constant M and N circles were translated to a logarithmic gain phase. N. B. Nichols transformed the constant M and N circles to the log-magnitude–phase diagram, and the resulting chart is called the **Nichols chart** [3, 7]. The M and N circles appear as contours on the Nichols chart shown in Figure 9.26. The coordinates of the log-magnitude–phase diagram are the same as those used in Section 8.5. However, superimposed on the log-magnitude–phase plane we find constant M and N lines. The constant M lines are given in decibels and the N lines in degrees. An example will illustrate the use of the Nichols chart to determine the closed-loop frequency response.

EXAMPLE 9.7   **Stability using the Nichols chart**

Consider a unity feedback system with a loop transfer function

$$G_c(j\omega)G(j\omega) = \frac{1}{j\omega(j\omega + 1)(0.2j\omega + 1)}. \tag{9.73}$$

The $G_c(j\omega)G(j\omega)$-locus is plotted on the Nichols chart and is shown in Figure 9.27. The maximum magnitude, $M_{p\omega}$, is equal to $+2.5$ dB and occurs at a frequency $\omega_r = 0.8$. The closed-loop phase angle at ω_r is equal to $-72°$. The 3-dB closed-loop bandwidth, where the closed-loop magnitude is -3 dB, is equal to $\omega_B = 1.33$, as shown in Figure 9.27. The closed-loop phase angle at ω_B is equal to $-142°$. ∎

EXAMPLE 9.8   **Third-order system**

Let us consider a unity feedback system with a loop transfer function

$$G_c(j\omega)G(j\omega) = \frac{0.64}{j\omega[(j\omega)^2 + j\omega + 1]}, \tag{9.74}$$

where $\zeta = 0.5$ for the complex poles. The Nichols diagram for this system is shown in Figure 9.28. The phase margin for this system as it is determined from the Nichols chart is 30°. On the basis of the phase, we use Equation (9.58) to estimate the system damping ratio as $\zeta = 0.30$. The maximum magnitude is equal to $+9$ dB occurring at a frequency $\omega_r = 0.88$. Therefore,

$$20 \log M_{p\omega} = 9 \text{ dB}, \quad \text{or} \quad M_{p\omega} = 2.8.$$

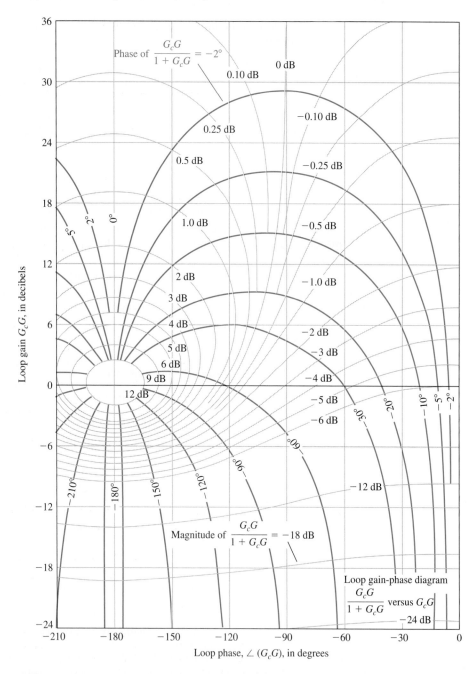

FIGURE 9.26   Nichols chart. The phase curves for the closed-loop system are shown as heavy curves.

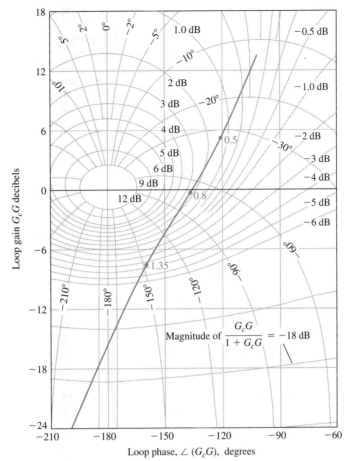

FIGURE 9.27
Nichols diagram for
$G_c(j\omega)G(j\omega) =$
$1/(j\omega(j\omega + 1)$
$(0.2j\omega + 1))$.
Three points on
curve are shown for
$\omega = 0.5, 0.8,$ and
1.35, respectively.

Solving Equation (9.63), we find that $\zeta = 0.18$. We are confronted with two conflicting damping ratios, where one is obtained from a phase margin measure and another from a peak frequency response measure. In this case, we have discovered an example in which the correlation between the frequency domain and the time domain is unclear and uncertain. This apparent conflict is caused by the nature of the $G_c(j\omega)G(j\omega)$-locus, which slopes rapidly toward the 180° line from the 0-dB axis. If we determine the roots of the characteristic equation for $1 + L(s)$, we obtain

$$q(s) = (s + 0.77)(s^2 + 0.225s + 0.826) = 0. \qquad (9.75)$$

The damping ratio of the complex conjugate roots is equal to 0.124, where the complex roots do not dominate the response of the system. Therefore, the real root will add some damping to the system, and we might estimate the damping ratio to be approximately the value determined from the $M_{p\omega}$ index; that is, $\zeta = 0.18$. A designer

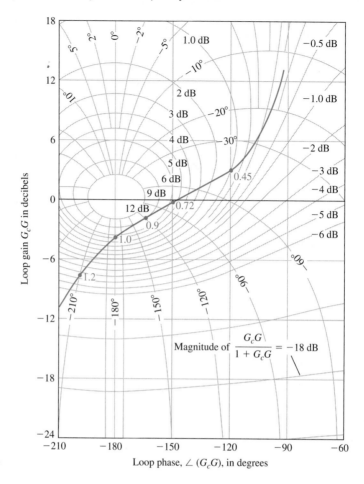

must use the frequency-domain-to-time-domain correlations with caution. However, we are usually safe if the lower value of the damping ratio resulting from the phase margin and the $M_{p\omega}$ relation is used for analysis and design purposes. ∎

The Nichols chart can be used for design purposes by altering the $G_cG(j\omega)$-locus so we can obtain a desirable phase margin and $M_{p\omega}$. The system gain K is readily adjusted to provide a suitable phase margin and $M_{p\omega}$ by inspecting the Nichols chart. For example, let us consider again Example 9.8, where

$$G_c(j\omega)G(j\omega) = \frac{K}{j\omega[(j\omega)^2 + j\omega + 1]}. \tag{9.76}$$

The $G_cG(j\omega)$-locus on the Nichols chart for $K = 0.64$ is shown in Figure 9.28. Let us determine a suitable value for K so that the system damping ratio is greater than 0.30. Examining Figure 8.11, we find that it is required that $M_{p\omega}$ be less than 1.75 (4.9 dB). From Figure 9.28, we find that the $G_cG(j\omega)$-locus will be tangent to the 4.9-dB curve if the $G_cG(j\omega)$-locus is lowered by a factor of 2.2 dB. Therefore, K should be reduced by 2.2 dB or the factor antilog(2.2/20) = 1.28. Thus, the gain K must be less than $0.64/1.28 = 0.50$ if the system damping ratio is to be greater than 0.30.

9.6 SYSTEM BANDWIDTH

The bandwidth of the closed-loop control system is an excellent measurement of the range of fidelity of response of the system. In systems where the low-frequency magnitude is 0 dB on the Bode diagram, the bandwidth is measured at the -3-dB frequency. The speed of response to a step input will be roughly proportional to ω_B, and the settling time is inversely proportional to ω_B. Thus, we seek a large bandwidth consistent with reasonable system components [12].

Consider the following two closed-loop system transfer functions:

$$T_1(s) = \frac{1}{s + 1}$$

and

$$T_2(s) = \frac{1}{5s + 1}. \tag{9.77}$$

The frequency response of the two systems is contrasted in part (a) of Figure 9.29, and the step response of the systems is shown in part (b). Also the response to a ramp is shown in part (c) of that figure. The system with the larger bandwidth provides the faster step response and higher fidelity ramp response.

Now consider the two second-order systems with closed-loop transfer functions

$$T_3(s) = \frac{100}{s^2 + 10s + 100}$$

and

$$T_4(s) = \frac{900}{s^2 + 30s + 900}. \tag{9.78}$$

Both systems have a ζ of 0.5. The frequency response of both closed-loop systems is shown in Figure 9.30(a). The natural frequency is 10 and 30 for systems T_3 and T_4, respectively. The bandwidth is 12.7 and 38.1 for systems T_3 and T_4, respectively. Both systems have a 16% overshoot, but T_4 has a peak time of 0.12 second compared to 0.36 for T_3, as shown in Figure 9.30(b). Also, note that the settling time for T_4 is 0.27 second, while the settling time for T_3 is 0.8 second. The system with a larger bandwidth provides a faster response.

9.7 THE STABILITY OF CONTROL SYSTEMS WITH TIME DELAYS

The Nyquist stability criterion has been discussed and illustrated in the previous sections for control systems whose transfer functions are rational polynomials of $j\omega$. Many control systems have a time delay within the closed loop of the system that affects the stability of the system. A **time delay** is the time interval between the start of an event at one point in a system and its resulting action at another point in

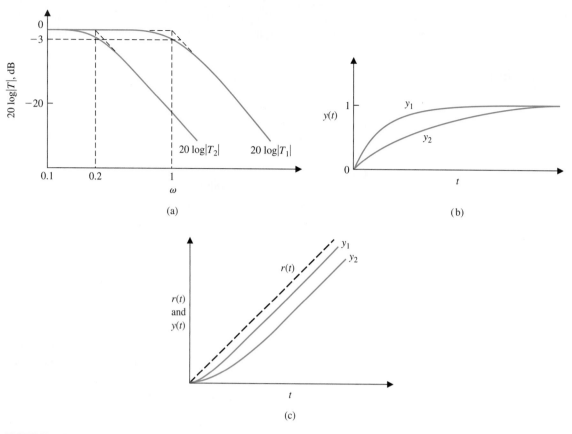

FIGURE 9.29   Response of two first-order systems.

the system. Fortunately, the Nyquist criterion can be utilized to determine the effect of the time delay on the relative stability of the feedback system. A pure time delay, without attenuation, is represented by the transfer function

$$G_d(s) = e^{-sT}, \tag{9.79}$$

where T is the delay time. The Nyquist criterion remains valid for a system with a time delay because the factor e^{-sT} does not introduce any additional poles or zeros within the contour. The factor adds a phase shift to the frequency response without altering the magnitude curve.

This type of time delay occurs in systems that have a movement of a material that requires a finite time to pass from an input or control point to an output or measured point [8, 9].

For example, a steel rolling mill control system is shown in Figure 9.31. The motor adjusts the separation of the rolls so that the thickness error is minimized. If the steel is traveling at a velocity v, then the time delay between the roll adjustment and the measurement is

$$T = \frac{d}{v}.$$

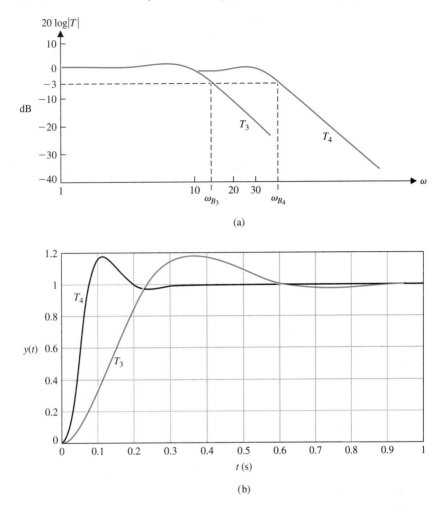

(a)

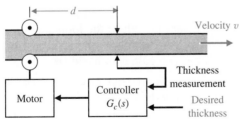

FIGURE 9.30
Response of two
second-order
systems.

(b)

FIGURE 9.31
Steel rolling mill
control system.

Therefore, to have a negligible time delay, we must decrease the distance to the measurement and increase the velocity of the flow of steel. Usually, we cannot eliminate the effect of time delay; thus, the loop transfer function is [10]

$$G_c(s)G(s)e^{-sT}. \tag{9.80}$$

However, we note that the frequency response of this system is obtained from the loop transfer function

$$L(j\omega) = G_c(j\omega)G(j\omega)e^{-j\omega T}. \qquad (9.81)$$

The usual loop transfer function is plotted on the $L(j\omega)$-plane and the stability ascertained relative to the -1 point. Alternatively, we can plot the Bode diagram including the delay factor and investigate the stability relative to the 0-dB, $-180°$ point. The delay factor $e^{-j\omega T}$ results in a phase shift

$$\boxed{\phi(\omega) = -\omega T} \qquad (9.82)$$

and is readily added to the phase shift resulting from $G_c(j\omega)G(j\omega)$. Note that the angle is in radians in Equation (9.82). An example will show the simplicity of this approach on the Bode diagram.

EXAMPLE 9.9   **Liquid level control system**

A level control system is shown in Figure 9.32(a) and the block diagram in Figure 9.32(b) [11]. The time delay between the valve adjustment and the fluid output is $T = d/v$. Therefore, if the flow rate is 5 m³/s, the cross-sectional area of the pipe is 1 m², and the distance is equal to 5 m, then we have a time delay $T = 1$ s. The loop transfer function is then

$$L(s) = G_A(s)G(s)G_f(s)e^{-sT}$$

$$= \frac{31.5}{(s + 1)(30s + 1)[(s^2/9) + (s/3) + 1]}e^{-sT}. \qquad (9.83)$$

The Bode diagram for this system is shown in Figure 9.33. The phase angle is shown both for the denominator factors alone and with the additional phase lag due to the time delay. The logarithmic gain curve crosses the 0-dB line at $\omega = 0.8$. Therefore, the phase margin of the system without the pure time delay would be 40°. However, with the time delay added, we find that the phase margin is equal to $-3°$, and the system is unstable. Consequently, the system gain must be reduced in order to provide a reasonable phase margin. To provide a phase margin of 30°, the gain would have to be decreased by a factor of 5 dB, to $K = 31.5/1.78 = 17.7$.

A time delay e^{-sT} in a feedback system introduces an additional phase lag and results in a less stable system. Therefore, as pure time delays are unavoidable in many systems, it is often necessary to reduce the loop gain in order to obtain a stable response. However, the cost of stability is the resulting increase in the steady-state error of the system as the loop gain is reduced. ∎

The systems considered by most analytical tools are described by rational functions (that is, transfer functions) or by a finite set of ordinary constant coefficient differential equations. Since the time-delay is given by e^{-sT}, where T is the delay, we see that the time delay is nonrational. It would be helpful if we could obtain a rational function approximation of the time-delay. Then it would be more convenient to incorporate the delay into the block diagram for analysis and design purposes.

The **Padé** approximation uses a series expansion of the transcendental function e^{-sT} and matches as many coefficients as possible with a series expansion of a rational function of specified order. For example, to approximate the function e^{-sT} with a first-order rational function, we begin by expanding both functions in a series (actually a Maclaurin series[1]),

[1] $f(s) = f(0) + \frac{s}{1!}\dot{f}(0) + \frac{s^2}{2!}\ddot{f}(0) + \cdots$

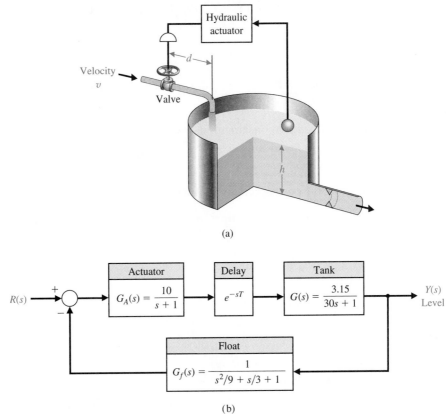

(a)

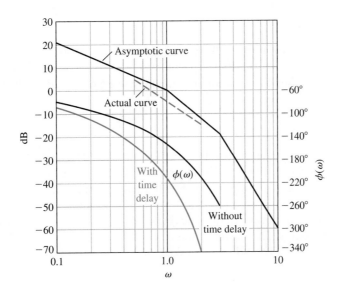

(b)

FIGURE 9.32
(a) Liquid level
control system.
(b) Block diagram.

FIGURE 9.33
Bode diagram for
level control
system.

$$e^{-sT} = 1 - sT + \frac{(sT)^2}{2!} - \frac{(sT)^3}{3!} + \frac{(sT)^4}{4!} - \frac{(sT)^5}{5!} + \cdots, \tag{9.84}$$

and

$$\frac{n_1 s + n_0}{d_1 s + d_0} = \frac{n_0}{d_0} + \left(\frac{d_0 n_1 - n_0 d_1}{d_0^2} \right) s + \left(\frac{d_1^2 n_0}{d_0^3} - \frac{d_1 n_1}{d_0^2} \right) s^2 + \cdots$$

For a first-order approximation, we want to find n_0, n_1, d_0, and d_1 such that

$$e^{-sT} \approx \frac{n_1 s + n_0}{d_1 s + d_0}.$$

Equating the corresponding coefficients of the terms in s, we obtain the relationships

$$\frac{n_0}{d_0} = 1, \quad \frac{n_1}{d_0} - \frac{n_0 d_1}{d_0^2} = -T, \quad \frac{d_1^2 n_0}{d_0^3} - \frac{d_1 n_1}{d_0^2} = \frac{T^2}{2}, \cdots$$

Solving for n_0, d_0, n_1, and d_1 yields

$$n_0 = d_0,$$

$$d_1 = \frac{d_0 T}{2},$$

$$n_1 = -\frac{d_0 T}{2}.$$

Setting $d_0 = 1$, and solving yields

$$e^{-sT} \approx \frac{n_1 s + n_0}{d_1 s + d_0} = \frac{-\frac{T}{2} s + 1}{\frac{T}{2} s + 1}. \tag{9.85}$$

A series expansion of Equation (9.85) yields

$$\frac{n_1 s + n_0}{d_1 s + d_0} = \frac{-\frac{T}{2} s + 1}{\frac{T}{2} s + 1} = 1 - Ts + \frac{T^2 s^2}{2} - \frac{T^3 s^3}{4} + \cdots. \tag{9.86}$$

Comparing Equation (9.86) to Equation (9.84), we verify that the first three terms match. So for small s, the Padé approximation is a reasonable representation of the time-delay. Higher-order rational functions can be obtained.

9.8 DESIGN EXAMPLES

In this section we present two illustrative examples. The first example is a remotely controlled reconnaissance vehicle control design. The Nichols chart is illustrated as a key element of the design of a controller gain to meet time-domain specifications. The second example considers the control of a hot ingot robot used in manufacturing. The goal is to minimize the tracking error in the presence of disturbances and a known time-delay. The design process is illustrated, leading to a PI controller that meets a mixture of time-domain and frequency-domain performance specifications.

EXAMPLE 9.10   **Remotely controlled reconnaissance vehicle**

The use of remotely controlled vehicles for reconnaissance for U.N. peacekeeping missions may be an idea whose time has come. One concept of a roving vehicle is shown in Figure 9.34(a), and a proposed speed control system is shown in Figure 9.34(b). The desired speed $R(s)$ is transmitted by radio to the vehicle; the disturbance $T_d(s)$ represents hills and rocks. The goal is to achieve good overall control with a low steady-state error and a low-overshoot response to step commands, $R(s)$ [13].

First, to achieve a low steady-state error for a unit step command, we calculate

$$e_{ss} = \lim_{s \to 0} sE(s)$$

$$= \lim_{s \to 0} s\left[\frac{R(s)}{1 + L(s)}\right]$$

$$= \frac{1}{1 + L(s)} = \frac{1}{1 + K/2},$$

where $L(s) = G_c(s)G(s)$. If we select $K = 20$, we will obtain a steady-state error of 9% of the magnitude of the input command. Using $K = 20$, we reformulate $L(s) = G_c(s)G(s)$ for Bode diagram calculations, obtaining

$$L(s) = G_c(s)G(s) = \frac{10(1 + s/2)}{(1 + s)(1 + s/2 + s^2/4)}.$$

(a)

FIGURE 9.34
(a) Remotely controlled reconnaissance vehicle. (b) Speed control system. This vehicle could be used for United Nations peacekeeping missions.

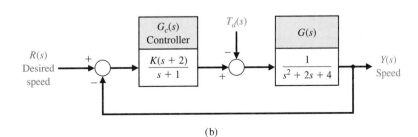

(b)

Table 9.4   Frequency Response Data for Design Example

ω	0	1.2	1.6	2.0	2.8	4	6
dB	20	18.4	17.8	16.0	10.5	2.7	−5.2
Degrees	0	−65	−86	−108	−142	−161	−170

The calculations for $0 \le \omega \le 6$ provide the data summarized in Table 9.4. The Nichols diagram for $K = 20$ is shown in Figure 9.35. Examining the Nichols chart, we find that $M_{p\omega}$ is 12 dB and the phase margin is 15 degrees. The step response of this system is underdamped, and we use Equation (9.58) and Figure 5.8 to predict an excessive overshoot of approximately 61%.

To reduce the overshoot to a step input, we can reduce the gain to achieve a predicted overshoot. To limit the overshoot to 25%, we select a desired ζ of the dominant roots as 0.4 (from Figure 5.8) and thus require $M_{p\omega} = 1.35$ (from Figure 8.11) or $20 \log M_{p\omega} = 2.6$ dB. To lower the gain, we will move the frequency response

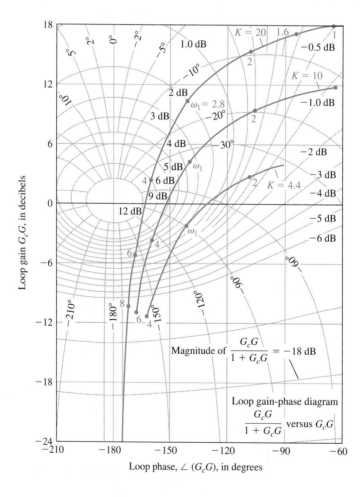

FIGURE 9.35
Nichols diagram for the design example when $K = 20$ and for two reduced gains.

vertically down on the Nichols chart, as shown in Figure 9.35. At $\omega_1 = 2.8$, we just intersect the 2.6-dB closed-loop curve. The reduction (vertical drop) in gain is equal to 13 dB, or a factor of 4.5. Thus, $K = 20/4.5 = 4.44$. For this reduced gain, the steady-state error is

$$e_{ss} = \frac{1}{1 + 4.4/2} = 0.31,$$

so that we have a 31% steady-state error.

The actual step response when $K = 4.44$, as shown in Figure 9.36, has an overshoot of 32%. If we use a gain of 10, we have an overshoot of 48% with a steady-state error of 17%. The performance of the system is summarized in Table 9.5. As a suitable compromise, we select $K = 10$ and draw the frequency response on the Nichols chart by moving the response for $K = 20$ down by $20 \log 2 = 6$ dB, as shown in Figure 9.35.

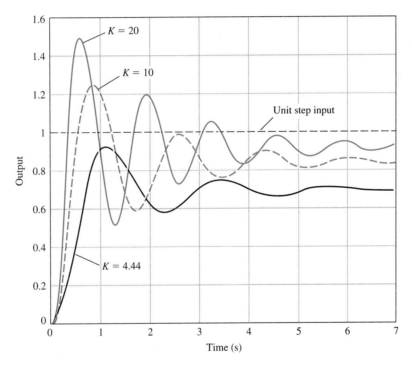

FIGURE 9.36
The response of the system for three values of K for a unit step input $r(t)$.

Table 9.5   Actual Response for Selected Gains

K	4.44	10	20
Percent overshoot	32.4	48.4	61.4
Settling time (seconds)	4.94	5.46	6.58
Peak time (seconds)	1.19	0.88	0.67
e_{ss}	31%	16.7%	9.1%

Note: Percent overshoot is defined by Equation (5.12).

Examining the Nichols chart for $K = 10$, we have $M_{p\omega} = 7$ dB, and a phase margin of 26 degrees. Thus, we estimate a ζ for the dominant roots of 0.23 which should result in an overshoot to a step input of 23%. The actual response is recorded in Table 9.5. The bandwidth of the system is $\omega_B \approx 5$. Therefore, we predict a settling time (with a 2% criterion) of

$$T_s = \frac{4}{\zeta\omega_n} = \frac{4}{(0.34)(\omega_B/1.4)} = 3.3 \text{ s},$$

since $\omega_B \approx 1.4\omega_n$ for $\zeta = 0.34$, using Figure 8.26. The actual settling time is approximately 5.4 seconds, as shown in Figure 9.36.

The steady-state effect of a unit step disturbance can be determined by using the final-value theorem with $R(s) = 0$, as follows:

$$y(\infty) = \lim_{s \to 0} s \left[\frac{G(s)}{1 + L(s)} \right] \left(\frac{1}{s} \right) = \frac{1}{4 + 2K}. \tag{9.86}$$

Thus, the unit disturbance is reduced by the factor $4 + 2K$. For $K = 10$, we have $y(\infty) = 1/24$, or the steady-state disturbance is reduced to 4% of the disturbance magnitude. Thus we have achieved a reasonable result with $K = 10$.

The best compromise design would be $K = 10$, since we achieve a compromise steady-state error of 16.7%. If the overshoot and settling time are excessive, then we need to reshape the $L(j\omega)$-locus on the Nichols chart by methods we will describe in Chapter 10. ∎

EXAMPLE 9.11   **Hot ingot robot control**

The hot ingot robot mechanism is shown in Figure 9.37. The robot picks up hot ingots and sets them in a quenching tank. A vision sensor is in place to provide a measurement of the ingot position. The controller uses the sensed position information to orient the robot over the ingot (along the x-axis). The vision sensor provides the desired position input $R(s)$ to the controller. The block diagram depiction of the closed-loop

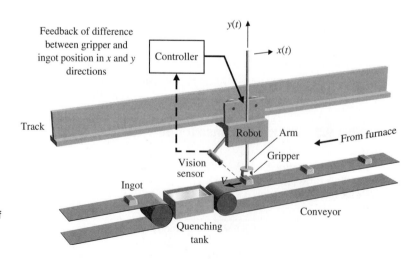

FIGURE 9.37
Artist's depiction of the hot ingot robot control system.

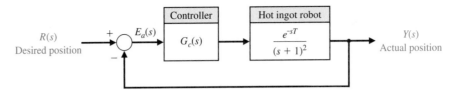

FIGURE 9.38
Hot ingot robot
control system
block diagram.

system is shown in Figure 9.38. More information on robots and robot vision systems can be found in [15, 31, 32].

The position of the robot along the track is also measured (by a sensor other than the vision sensor) and is available for feedback to the controller. We assume that the position measurement is noise free. This is not a restrictive assumption since many accurate position sensors are available today. For example some laser diode systems are self-contained (including the power supply, optics, and laser diode) and provide position accuracy of over 99.9%.

The robot dynamics are modeled as a second-order system with two poles at $s = -1$ and include a time delay of $T = \pi/4$ s. Therefore,

$$G(s) = \frac{e^{-sT}}{(s + 1)^2}, \tag{9.87}$$

where $T = \pi/4$ s. The elements of the design process emphasized in this example are highlighted in Figure 9.39. The control goal is as follows:

Control Goal
 Minimize the tracking error $E(s) = R(s) - Y(s)$ in the presence of external disturbances while accounting for the known time-delay.

To this end the following control specifications must be satisfied:

Design Specifications

 DS1   Achieve a steady-state tracking error less than 10% for a step input.

 DS2   Phase margin greater than 50° with the time-delay $T = \pi/4$ s.

 DS3   Percent overshoot less than 10% for a step input.

Our design method is first to consider a proportional controller. We will show that the design specifications cannot be simultaneously satisfied with a proportional controller; however, the feedback system with proportional control provides a useful vehicle to discuss in some detail the effects of the time-delay. In particular, we consider the effects of the time-delay on the Nyquist plot. The final design uses a PI controller, which is capable of providing adequate performance (that is, it satisfies all design specifications).

As a first try, we consider a simple proportional controller:

$$G_c(s) = K.$$

Then ignoring the time-delay for the moment, we have the loop gain

$$L(s) = G_c(s)G(s) = \frac{K}{(s + 1)^2} = \frac{K}{s^2 + 2s + 1}.$$

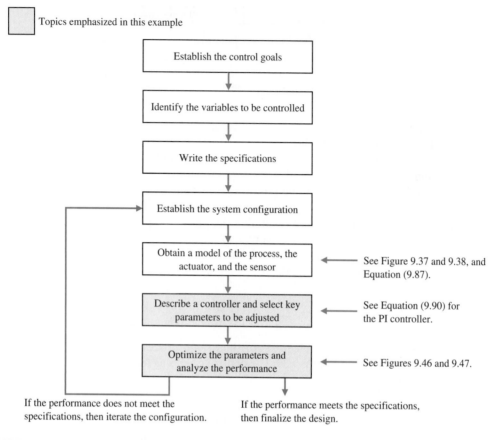

FIGURE 9.39   Elements of the control system design process emphasized in the hot ingot robot control example.

FIGURE 9.40
Hot ingot robot
control system
block diagram with
the proportional
controller and no
time-delay.

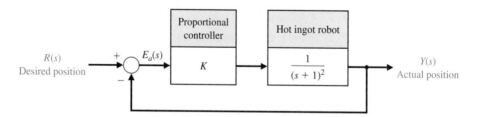

The feedback control system is shown in Figure 9.40 with a proportional controller and no time-delay. The system is a type-zero system, so we expect a nonzero steady-state tracking error to a step input (see Section 5.6 for a review of system type). The closed-loop transfer function is

$$T(s) = \frac{K}{s^2 + 2s + 1 + K}.$$

With the tracking error defined as

$$E(s) = R(s) - Y(s),$$

and with $R(s) = a/s$, where a is the input magnitude, we have

$$E(s) = \frac{s^2 + 2s + 1}{s^2 + 2s + 1 + K} \frac{a}{s}.$$

Using the final value theorem (which is possible since the system is stable for all positive values of K) yields

$$e_{ss} = \lim_{s \to 0} sE(s) = \frac{a}{1 + K}.$$

Per specification DS1, we require the steady-state tracking error be less than 10%. Therefore,

$$e_{ss} \le \frac{a}{10}.$$

Solving for the appropriate gain K yields $K \ge 9$. With $K = 9$, we obtain the Bode plot shown in Figure 9.41.

If we raise the gain above $K = 9$, we find that the crossover moves to the right (that is, ω_c increases) and the corresponding phase margin (*P.M.*) decreases. Is a *P.M.* $= 38.9°$ at $\omega = 2.8$ rad/s sufficient for stability in the presence of a time-delay of $T = \pi/4$ s? The addition of the time-delay term causes a phase lag without changing the magnitude plot. The amount of time-delay that our system can withstand while remaining stable is $\phi = -\omega T$ which implies that

$$\frac{-38.9\pi}{180} = -2.8T.$$

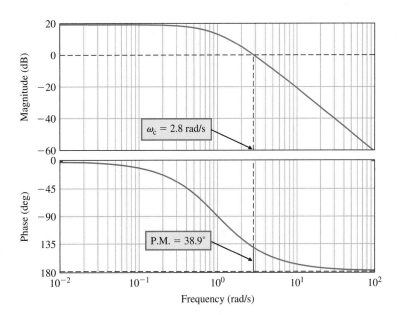

FIGURE 9.41
Bode plot with $K = 9$ and no time-delay showing gain margin *G.M.* $= \infty$ and phase margin *P.M.* $= 38.9°$.

Solving for T yields $T = 0.24$ s. Thus for time-delays less than $T = 0.24$ s, our closed-loop system remains stable. However, the time-delay $T = \pi/4$ s will cause instability. Raising the gain only exacerbates matters, since the phase margin goes down further. Lowering the gain raises the phase margin, but the steady-state tracking error exceeds the 10% limit. A more complex controller is necessary. Before proceeding, let us consider the Nyquist plot and see how it changes with the addition of the time-delay. The Nyquist plot for the system (without the time-delay)

$$L(s) = G_c(s)G(s) = \frac{K}{(s + 1)^2},$$

is shown in Figure 9.42, where we use $K = 9$. The number of open-loop poles of $G_c(s)G(s)$ in the right half-plane is $P = 0$. From Figure 9.42 we see that there are no encirclements of the -1 point, thus, $N = 0$.

By the Nyquist theorem, we know that the net number of encirclements N equals the number of zeros Z (or closed-loop system poles) in the right half-plane minus the number of open-loop poles P in the right half-plane. Therefore,

$$Z = N + P = 0.$$

Since $Z = 0$, the closed-loop system is stable. More importantly, even when the gain K is increased (or decreased), the -1 point is never encircled—the gain margin is ∞. Similarly when the time-delay is absent, the phase margin is always positive. The value of the $P.M.$ varies as K varies, but the $P.M.$ is always greater than zero.

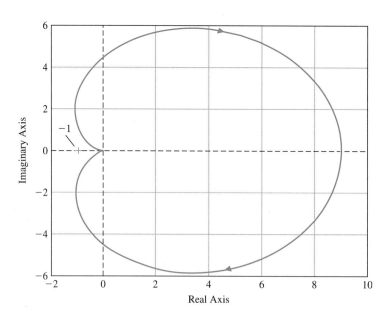

FIGURE 9.42
Nyquist plot with $K = 9$ and no time-delay showing no encirclements of the minus 1 point.

With the time-delay in the loop, we can rely on analytic methods to obtain the Nyquist plot. The loop transfer function with the time-delay is

$$L(s) = G_c(s)G(s) = \frac{K}{(s + 1)^2}e^{-sT}.$$

Using the Euler identity

$$e^{-j\omega T} = \cos(\omega T) - j\sin(\omega T),$$

and substituting $s = j\omega$ into $L(s)$ yields

$$L(j\omega) = \frac{K}{(j\omega + 1)^2}e^{-j\omega T}$$

$$= \frac{K}{\Delta}([(1 - \omega^2)\cos(\omega T) - 2\omega\sin(\omega T) - j[(1 - \omega^2)\sin(\omega T) + 2\omega\cos(\omega T)],$$

$$(9.88)$$

where

$$\Delta = (1 - \omega^2)^2 + 4\omega^2.$$

Generating a plot of $\text{Re}(L(j\omega))$ versus $\text{Im}(L(j\omega))$ for various values of ω leads to the plot shown in Figure 9.43. With $K = 9$, the number of encirclements of the -1 point is $N = 2$. Therefore, the system is unstable since $Z = N + P = 2$.

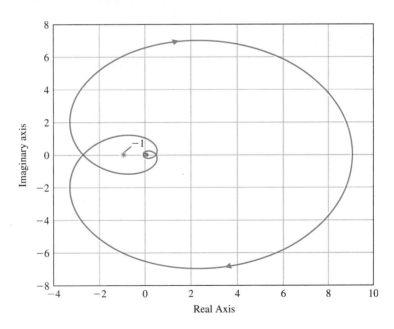

FIGURE 9.43
Nyquist plot with $K = 9$ and $T = \pi/4$ showing two encirclements of the -1 point, $N = 2$.

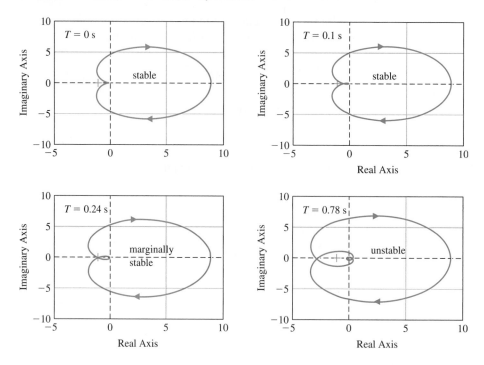

FIGURE 9.44
Nyquist plot with
$K = 9$ and various
time-delays.

Figure 9.44 shows the Nyquist plot for four values of time-delay: $T = 0$, 0.1, 0.24, and $\pi/4 = 0.78$ s. For $T = 0$ there is no possibility of an encirclement of the -1 point as K varies (see the upper left graph of Figure 9.44). We have stability (that is, $N = 0$) for $T = 0.1$ s (upper right graph), marginal stability for $T = 0.24$ s (lower left graph), and for $T = \pi/4 = 0.78$ s we have $N = 1$ (lower right graph), thus the closed-loop system is unstable.

Since we know that $T = \pi/4$ in this example, the proportional gain controller is not a viable controller. With it we cannot meet the steady-state error specifications and have a stable closed-loop system in the presence of the time-delay $T = \pi/4$. However, before proceeding with the design of a controller that meets all the specifications, let us take a closer look at the Nyquist plot with a time-delay.

Suppose we have the case where $K = 9$ and $T = 0.1$ s. The associated Nyquist plot is shown in the upper right of Figure 9.44. The Nyquist plot intersects (or crosses over) the real axis whenever the imaginary part of $G_c(j\omega)G(j\omega) = 0$ [see Equation (9.88)], or

$$(1 - \omega^2)\sin(0.1\omega) + 2\omega\cos(0.1\omega) = 0.$$

Thus we obtain the relation that describes the frequencies ω at which crossover occurs:

$$\frac{(1 - \omega^2)\tan(0.1\omega)}{2\omega} = -1. \qquad (9.89)$$

Equation (9.89) has an infinite number of solutions. The first real-axis crossing (farthest in the left half-plane) occurs when $\omega = 4.43$ rad/s.

The magnitude of $|L(j4.43)|$ is equal to 0.0484 K. For stability we require that $|L(j\omega)| < 1$ when $\omega = 4.43$ (to avoid an encirclement of the -1 point). Thus, for stability we find

$$K < \frac{1}{0.0484} = 20.67,$$

when $T = 0.1$. When $K = 9$, the closed-loop system is stable, as we already know. If the gain $K = 9$ increases by a factor of 2.3 to $K = 20.67$, we will be on the border of instability. This factor δ is the gain margin:

$$G.M. = 20 \log_{10} 2.3 = 7.2 \text{ dB}.$$

Consider the PI controller

$$G_c(s) = K_P + \frac{K_I}{s} = \frac{K_P s + K_I}{s}. \tag{9.90}$$

The loop system transfer function is

$$L(s) = G_c(s)G(s) = \frac{K_P s + K_I}{s} \frac{K}{(s + 1)^2} e^{-sT}.$$

The system type is now equal to 1; thus we expect a zero steady-state error to a step input. The steady-state error specification DS1 is satisfied. We can now concentrate on meeting specification DS3, $P.O. < 10\%$ and DS2, the requirement for stability in the presence of the time-delay $T = \pi/4$ s.

From the percent overshoot specification we can determine a desired system damping ratio. Thus we determine for $P.O. \leq 10\%$ that $\zeta \geq 0.59$. Due to the PI controller, the system now has a zero at $s = -K_I/K_P$. The zero will not affect the closed-loop system stability, but it will affect the performance. Using the approximation (valid for small ζ, $P.M.$ expressed in degrees)

$$\zeta \approx \frac{P.M.}{100},$$

we determine a good target phase margin (since we want $\zeta \geq 0.59$) to be 60%. We can rewrite the PI controller as

$$G_c(s) = K_I \frac{1 + \tau s}{s},$$

where $1/\tau = K_I/K_P$ is the break frequency of the controller. The PI controller is essentially a low-pass filter and adds phase lag to the system below the break frequency. We would like to place the break frequency below the crossover frequency so that the phase margin is not reduced significantly due to the presence of the PI zero.

The uncompensated Bode plot is shown in Figure 9.45 for

$$G(s) = \frac{9}{(s + 1)^2} e^{-sT},$$

where $T = \pi/4$. The uncompensated system phase margin is $P.M. = -88.34°$ at $\omega_c = 2.83$ rad/s. Since we want $P.M. = 60°$, we need the phase to be minus 120° at the crossover frequency. In Figure 9.45 we can estimate the phase $\phi = -120°$ at $\omega \approx 0.87$ rad/s. This is an approximate value but is sufficiently accurate for the design procedure. At $\omega = 0.87$ the magnitude is about 14.5 dB. If we want the crossover to be $\omega_c = 0.87$ rad/s, the controller needs to attenuate the system gain

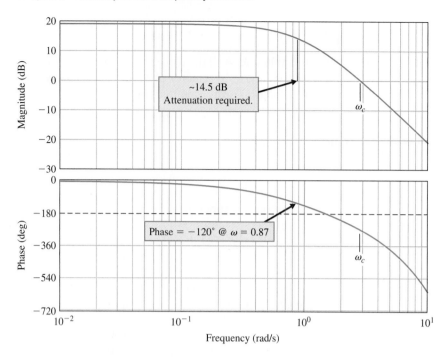

FIGURE 9.45
Uncompensated
Bode plot with
$K = 9$ and
$T = \pi/4$.

by 14.5 dB, so that the magnitude is 0 dB at $\omega_c = 0.87$. With

$$G_c(s) = K_P \frac{s + \dfrac{K_I}{K_P}}{s},$$

we can consider K_P to be the gain of the compensator (a good approximation for large ω). Therefore,

$$K_P = 10^{-(14.5/20)} = 0.188.$$

Finally we need to select K_I. Since we want the break frequency of the controller to be below the crossover frequency (so that the phase margin is not reduced significantly due to the presence of the PI zero), a good rule-of-thumb is to select $1/\tau = K_I/K_P = 0.1\omega_c$. To make the break frequency of the controller zero one decade below the crossover frequency. The final value of K_I is computed to be $K_I = 0.1\omega_c K_P = 0.0164$, where $\omega_c = 0.87$ rad/s. Thus the PI controller is

$$G_c(s) = \frac{0.188s + 0.0164}{s}. \tag{9.91}$$

The Bode plot of $G_c(s)G(s)$ is shown in Figure 9.46. The gain and phase margins are $G.M. = 5.3$ dB and $P.M. = 56.5°$.

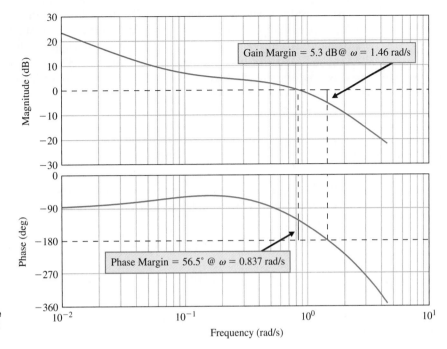

FIGURE 9.46
Compensated Bode plot with $K = 9$ and $T = \pi/4$.

Gain Margin = 5.3 dB @ $\omega = 1.46$ rad/s

Phase Margin = 56.5° @ $\omega = 0.837$ rad/s

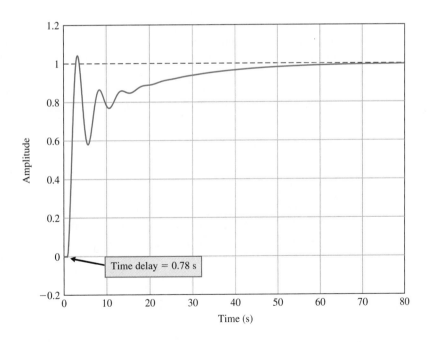

FIGURE 9.47
Hot ingot robot control step response.

Time delay = 0.78 s

We consider whether the design specifications have been met. The steady-state tracking specification (DS1) is certainly satisfied since our system is type one; the PI controller introduced an integrator. The phase margin (with the time-delay) is $P.M. = 56.5°$, so the phase margin specification, DS2, in satisfied. The unit step response is shown in Figure 9.47. The percent overshoot is approximately $P.O. \approx 4.2\%$. The target percent overshoot was 10%, so DS3 is satisfied. Overall the design specifications are satisfied.

9.9 PID CONTROLLERS IN THE FREQUENCY DOMAIN

The PID controller provides a proportional term, an integral term, and a derivative term (see Section 7.6). We then have the PID controller transfer function as

$$G_c(s) = K_P + \frac{K_I}{s} + K_D s. \tag{9.92}$$

If we set $K_D = 0$, we have the PI controller

$$G_c(s) = K_P + \frac{K_I}{s}. \tag{9.93}$$

If we set $K_I = 0$, we have the PD controller

$$G_c(s) = K_P + K_D s. \tag{9.94}$$

In general, we note that PID controllers are particularly useful for reducing the steady-state error and improving the transient response when $G(s)$ has one or two poles (or may be approximated by a second-order process).

We may use frequency response methods to represent the addition of a PID controller. The PID controller, Equation (9.92), may be rewritten as

$$G_c(s) = \frac{K_I\left(\dfrac{K_D}{K_I}s^2 + \dfrac{K_P}{K_I}s + 1\right)}{s} = \frac{K_I(\tau s + 1)\left(\dfrac{\tau}{\alpha}s + 1\right)}{s}. \tag{9.95}$$

The Bode diagram of Equation (9.95) is shown in Figure 9.48 for $\omega\tau$, $K_I = 2$, and $\alpha = 10$. The PID controller is a form of a notch (or bandstop) compensator with a variable gain, K_I. Of course, it is possible that the controller will have complex zeros and a Bode diagram that will be dependent on the ζ of the complex zeros. The contribution by the zeros to the Bode chart may be visualized by reviewing Figure 8.10 for complex poles and noting that the phase and magnitude change as ζ changes. The PID controller with complex zeros is

$$G_c(\omega) = \frac{K_I[1 + (2\zeta/\omega_n)j\omega - (\omega/\omega_n)^2]}{j\omega}. \tag{9.96}$$

Normally, we choose $0.9 > \zeta > 0.7$.

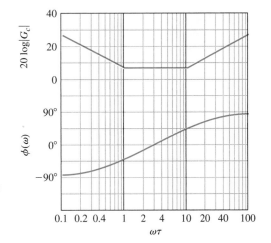

FIGURE 9.48
Bode diagram for a PID controller using the asymptomatic approximation for the magnitude curve.

9.10 STABILITY IN THE FREQUENCY DOMAIN USING CONTROL DESIGN SOFTWARE

We now approach the issue of stability using the computer as a tool. This section revisits the Nyquist diagram, the Nichols chart, and the Bode diagram in our discussions on relative stability. Two examples will illustrate the frequency-domain design approach. We will make use of the frequency response of the closed-loop transfer function $T(j\omega)$ as well as the loop transfer function $L(j\omega)$. We also present an illustrative example that shows how to deal with a time delay in the system by utilizing a Padé approximation [6]. The functions covered in this section are nyquist, nichols, margin, pade, and ngrid.

It is generally more difficult to manually generate the Nyquist plot than the Bode diagram. However, we can use the control design software to generate the Nyquist plot. The Nyquist plot is generated with the nyquist function, as shown in Figure 9.49. When nyquist is used without left-hand arguments, the Nyquist plot is

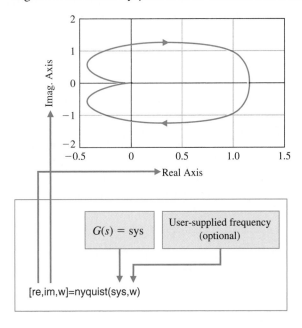

FIGURE 9.49
The **nyquist** function.

```
>>num=[0.5]; den=[1 2 1 0.5 ];
>>sys=tf(num,den);
>>nyquist(sys)
```

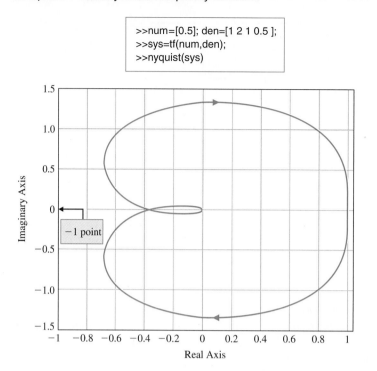

FIGURE 9.50
An example of the
nyquist function.

automatically generated; otherwise, the real and imaginary parts of the frequency response (along with the frequency vector ω) is returned. An illustration of the nyquist function is given in Figure 9.50.

As discussed in Section 9.4, relative stability measures of **gain margin** and **phase margin** can be determined from both the Nyquist plot and the Bode diagram. The gain margin is a measure of how much the system gain would have to be increased for the $L(j\omega)$ locus to pass through the $-1 + j0$ point, thus resulting in an unstable system. The phase margin is a measure of the additional phase lag required before the system becomes unstable. Gain and phase margins can be determined from both the Nyquist plot and the Bode diagram.

Consider the system shown in Figure 9.51. Relative stability can be determined from the Bode diagram using the margin function, which is shown in Figure 9.52. If the margin function is invoked without left-hand arguments, the Bode diagram is automatically generated with the gain and phase margins labeled on the diagram. This is illustrated in Figure 9.53 for the system shown in Figure 9.51.

The script to generate the Nyquist plot for the system in Figure 9.51 is shown in Figure 9.54. In this case, the number of poles of $L(s) = G_c(s)G(s)H(s)$ with positive real parts is zero, and the number of counterclockwise encirclements of -1 is zero;

FIGURE 9.51
A closed-loop
control system
example for Nyquist
and Bode with
relative stability.

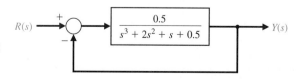

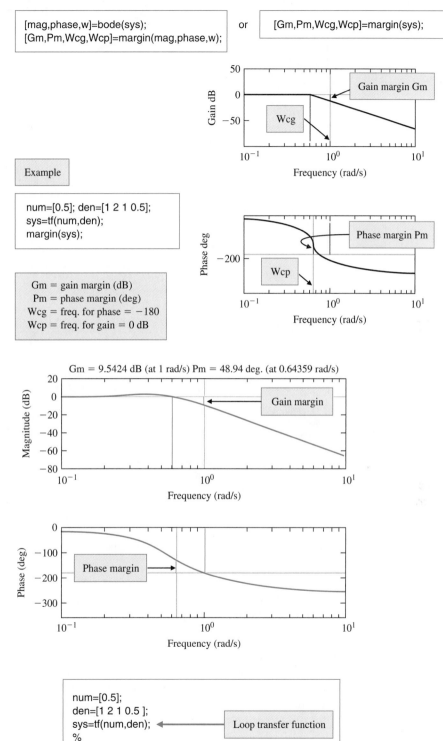

```
[mag,phase,w]=bode(sys);
[Gm,Pm,Wcg,Wcp]=margin(mag,phase,w);
```
or
```
[Gm,Pm,Wcg,Wcp]=margin(sys);
```

Example

```
num=[0.5]; den=[1 2 1 0.5];
sys=tf(num,den);
margin(sys);
```

Gm = gain margin (dB)
Pm = phase margin (deg)
Wcg = freq. for phase = −180
Wcp = freq. for gain = 0 dB

FIGURE 9.52
The **margin**
function.

Gm = 9.5424 dB (at 1 rad/s) Pm = 48.94 deg. (at 0.64359 rad/s)

FIGURE 9.53
The Bode diagram
for the system in
Figure 9.51 with the
gain margin and the
phase margin
indicated on the
plots.

```
num=[0.5];
den=[1 2 1 0.5 ];
sys=tf(num,den);          Loop transfer function
%
margin(sys)
```

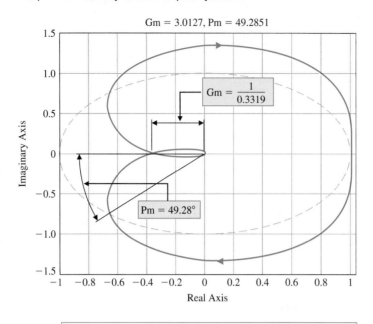

Gm = 3.0127, Pm = 49.2851

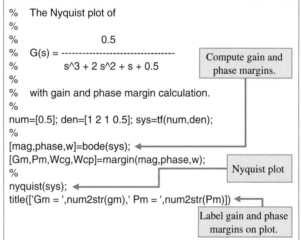

FIGURE 9.54

(a) The Nyquist plot for the system in Figure 9.51 with gain and phase margins. (b) m-file script.

hence, the closed-loop system is stable. We can also determine the gain margin and phase margin, as indicated in Figure 9.54.

Nichols Chart.   Nichols charts can be generated using the nichols function, shown in Figure 9.55. If the nichols function is invoked without left-hand arguments, the Nichols chart is automatically generated; otherwise the nichols function returns the magnitude and phase in degrees (along with the frequency ω). A Nichols chart grid is drawn on the existing plot with the ngrid function. The Nichols chart, shown in

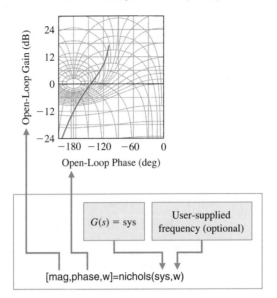

FIGURE 9.55
The **nichols** function.

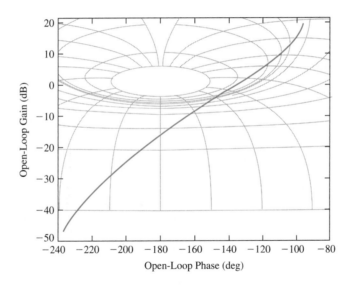

FIGURE 9.56
Nichols chart for the system of Equation (9.97).

Figure 9.56, is for the system

$$G(j\omega) = \frac{1}{j\omega(j\omega + 1)(0.2j\omega + 1)}. \tag{9.97}$$

EXAMPLE 9.12   Liquid level control system

Consider a liquid level control system described by the block diagram shown in Figure 9.32 (see Example 9.9). Note that this system has a time delay. The loop transfer function is given by

$$L(s) = \frac{31.5e^{-sT}}{(s + 1)(30s + 1)(s^2/9 + s/3 + 1)}. \tag{9.98}$$

We first change Equation (9.98) in such a way that $L(s)$ has a transfer function form with polynomials in the numerator and the denominator. To do this, we can make an approximation to e^{-sT} with the **pade** function, shown in Figure 9.57. For example, suppose our time delay is $T = 1$ s, and we want a second-order approximation $n = 2$. Using the **pade** function, we find that

$$e^{-s} \simeq \frac{s^2 - 6s + 12}{s^2 + 6s + 12}. \tag{9.99}$$

Substituting Equation (9.99) into Equation (9.98), we have

$$L(s) \simeq \frac{31.5(s^2 - 6s + 12)}{(s + 1)(30s + 1)(s^2/9 + s/3 + 1)(s^2 + 6s + 12)}.$$

Now we can build a script to investigate the relative stability of the system using the Bode diagram. Our goal is to have a phase margin of 30°. The associated script is shown in Figure 9.58. To make the script interactive, we let the gain K (now set at $K = 31.5$) be adjustable and defined outside the script at the command level. Then we set K and run the script to check the phase margin and iterate if necessary. The final selected gain is $K = 16$. Remember that we have utilized a second-order Padé approximation of the time delay in our analysis. ■

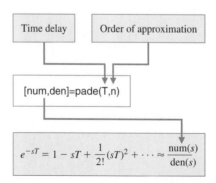

FIGURE 9.57
The **pade** function.

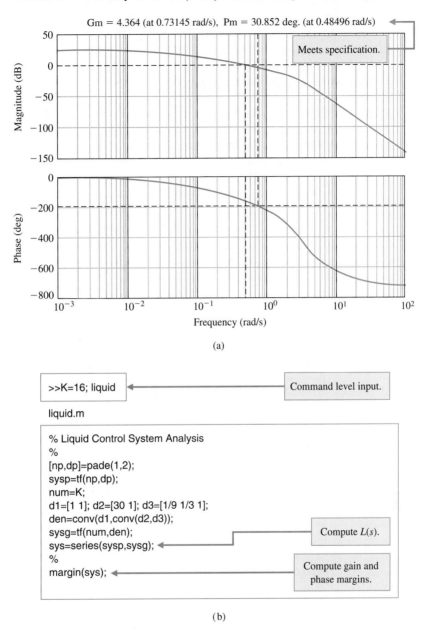

FIGURE 9.58
(a) Bode diagram for the liquid level control system.
(b) m-file script.

EXAMPLE 9.13   **Remotely controlled reconnaissance vehicle**

Consider the speed control system for a remotely controlled reconnaissance vehicle shown in Figure 9.34. The design objective is to achieve good control with a low steady-state error and a low overshoot to a step command. Building a script will allow us to perform many design iterations quickly and efficiently. First, we investigate the steady-state error specification. The steady-state error to a unit step command is

$$e_{ss} = \frac{1}{1 + K/2}. \tag{9.100}$$

The effect of the gain K on the steady-state error is clear from Equation (9.100): If $K = 20$, the error is 9% of the input magnitude; if $K = 10$, the error is 17% of the input magnitude.

Now we can investigate the overshoot specification in the frequency domain. Suppose we require that the percent overshoot is less than 50%. Solving

$$P.O. \approx 100\, exp^{-\zeta\pi/\sqrt{1-\zeta^2}} \le 50$$

for ζ yields $\zeta \ge 0.215$. Referring to Figure 8.11, we find that $M_{p\omega} \le 2.45$. We must keep in mind that the information in Figure 8.11 is for second-order systems only and can be used here only as a guideline. We now compute the closed-loop Bode diagram and check the values of $M_{p\omega}$. Any gain K for which $M_{p\omega} \le 2.45$ may be a valid gain for our design, but we will have to investigate step responses further to check the actual overshoot. The script in Figure 9.59 aids us in this task. We further investigate the gains $K = 20$, 10, and 4.44 (even though $M_{p\omega} > 2.45$ for $K = 20$).

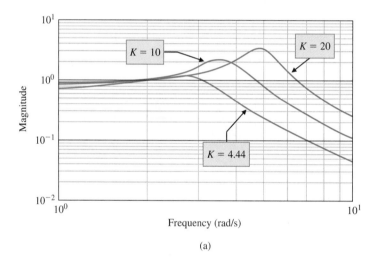

(a)

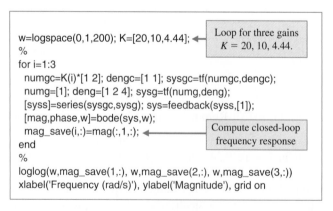

(b)

FIGURE 9.59
Remotely controlled vehicle. (a) Closed-loop system Bode diagram. (b) m-file script.

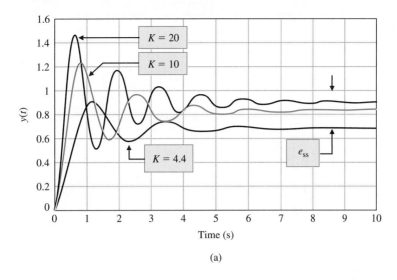

(a)

```
t=[0:0.01:10]; K=[20,10,4.44];          Loop for three gains
y=zeros(length(t), length(k));           K = 20, 10, 4.44.
%
for i=1:3
   numgc=K(i)*[1 2]; dengc=[1 1]; sysgc=tf(numgc,dengc);
   numg=[1]; deng=[1 2 4]; sysg=tf(numg,deng);
   syss=series(sysgc,sysg);
   sys=feedback(syss,[1]);
   y(:,i)=step(sys,t);                   Compute step
end                                       response.
%
plot(t,y(:,1),t,y(:,2),t,y(:,3)),grid
xlabel('Time (s)'), ylabel('y(t)')
```

(b)

FIGURE 9.60
Remotely controlled vehicle. (a) Step response, (b) m-file script.

We can plot the step responses to quantify the overshoot as shown in Figure 9.60. Additionally, we could have used a Nichols chart to aid the design process, as shown in Figure 9.61.

The results of the analysis are summarized in Table 9.5 for $K = 20$, 10, and 4.44. We choose $K = 10$ as our design gain. Then we obtain the Nyquist plot and check relative stability, as shown in Figure 9.62. The gain margin is $GM = 49.56$ dB and the phase margin is $PM = 26.11°$. ∎

9.11 SEQUENTIAL DESIGN EXAMPLE: DISK DRIVE READ SYSTEM

In this chapter, we will examine the system described in Chapter 8, using the system represented by Figure 8.49. This system includes the effect of the flexure resonance

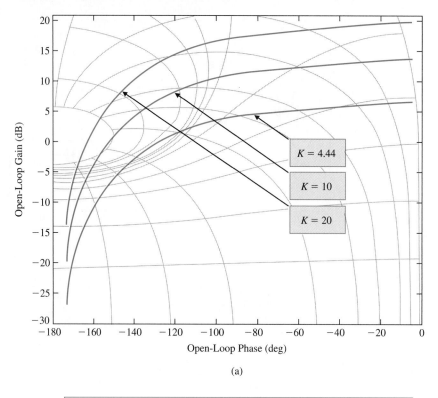

(a)

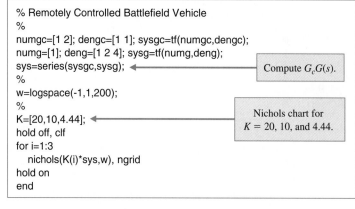

(b)

FIGURE 9.61
Remotely controlled
vehicle. (a) Nichols
chart. (b) m-file
script.

and incorporates a PD controller with a zero at $s = -1$. We will determine the system gain margin and phase margin when $K = 400$.

The Bode diagram for the system of Figure 8.49 when $K = 400$ is shown in Figure 9.63. The gain margin is 22.9 dB, and the phase margin is 37.2°. The plot of the step response of this system is shown in Figure 9.64. The settling time of this design is $T_s = 9.6$ ms.

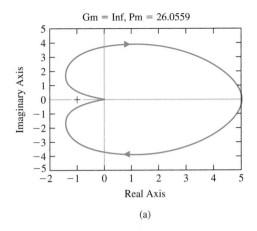

Gm = Inf, Pm = 26.0559

```
% Remotely Controlled Vehicle
% Nyquist plot for K=10
%
numgc=10*[1 2]; dengc=[1 1]; sysgc=tf(numgc,dengc);
numg=[1]; deng=[1 2 4]; sysg=tf(numg,deng);
sys=series(sysgc,sysg);
%
[Gm,Pm,Wcg,Wcp]=margin(sys);
%
nyquist(sys);
title(['Gm = ',num2str(Gm), '  Pm = ',num2str(Pm)])
```

FIGURE 9.62
(a) Nyquist chart for the remotely controlled vehicle with $K = 10$.
(b) m-file script.

(b)

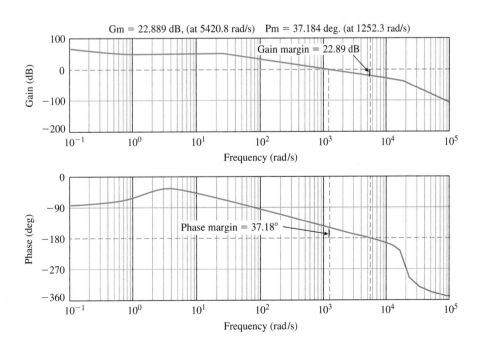

Gm = 22.889 dB, (at 5420.8 rad/s)   Pm = 37.184 deg. (at 1252.3 rad/s)

Gain margin = 22.89 dB

Phase margin = 37.18°

FIGURE 9.63
Bode diagram of the system shown in Figure 8.49.

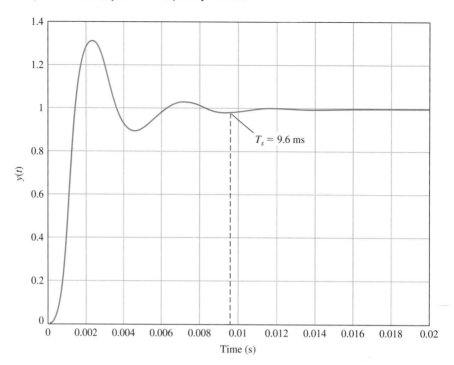

FIGURE 9.64
Response of the
system to a step
input.

9.12 SUMMARY

The stability of a feedback control system can be determined in the frequency domain by utilizing Nyquist's criterion. Furthermore, Nyquist's criterion provides us with two relative stability measures: (1) gain margin and (2) phase margin. These relative stability measures can be utilized as indices of the transient performance on the basis of correlations established between the frequency domain and the transient response. The magnitude and phase of the closed-loop system can be determined from the frequency response of the open-loop transfer function by utilizing constant magnitude and phase circles on the polar plot. Alternatively, we can utilize a log-magnitude–phase diagram with closed-loop magnitude and phase curves superimposed (called the Nichols chart) to obtain the closed-loop frequency response. A measure of relative stability, the maximum magnitude of the closed-loop frequency response, $M_{p\omega}$, is available from the Nichols chart. The frequency response, $M_{p\omega}$, can be correlated with the damping ratio of the time response and is a useful index of performance. Finally, a control system with a pure time delay can be investigated in a manner similar to that for systems without time delay. A summary of the Nyquist criterion, the relative stability measures, and the Nichols diagram is given in Table 9.6 for several transfer functions.

Table 9.6 is very useful and important to the designer and analyst of control systems. If we have the model of a process $G(s)$ and a controller $G_c(s)$, then we can determine $L(s) = G_c(s)G(s)$. With this loop transfer function, we can examine the transfer function table in column 1. This table contains fifteen typical transfer functions. For a selected transfer function, the table gives the Bode diagram, the Nichols diagram, and the root locus. With this information, the designer can determine or estimate the performance of the system and consider the addition or alteration of the controller $G_c(s)$.

Table 9.6   Transfer Function Plots for Typical Transfer Functions

$L(s)$	Polar Plot	Bode Diagram
1. $\dfrac{K}{s\tau_1 + 1}$	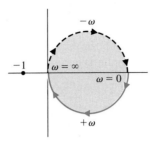	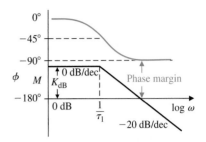
2. $\dfrac{K}{(s\tau_1 + 1)(s\tau_2 + 1)}$	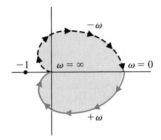	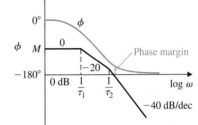
3. $\dfrac{K}{(s\tau_1 + 1)(s\tau_2 + 1)(s\tau_3 + 1)}$	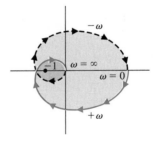	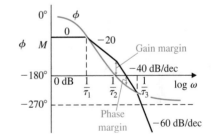
4. $\dfrac{K}{s}$	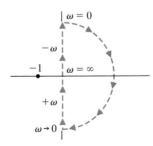	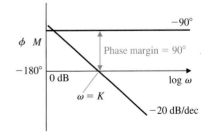

Table 9.6 *(continued)*

Nichols Diagram	Root Locus	Comments
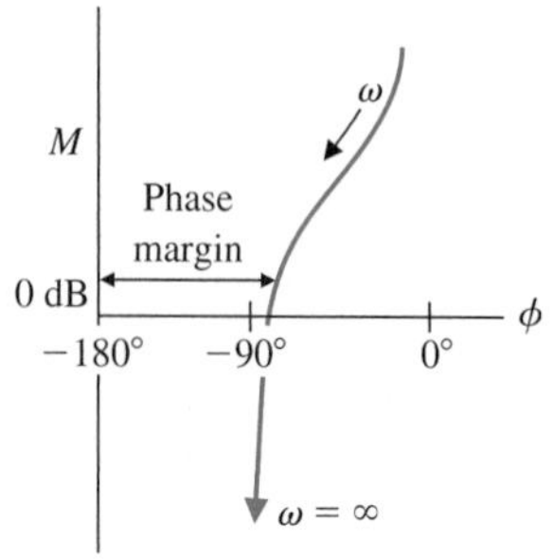	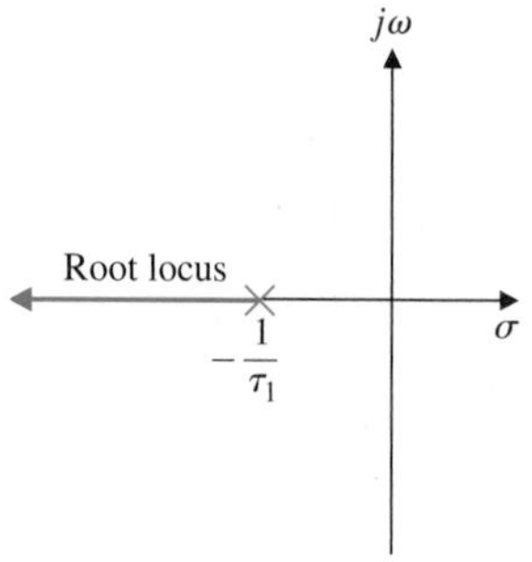	Stable; gain margin $= \infty$
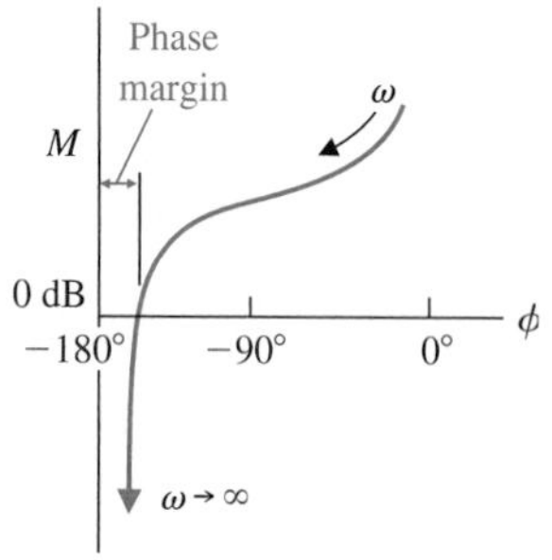	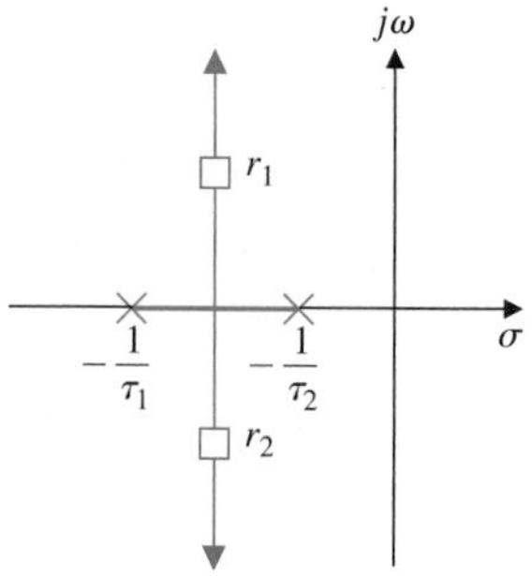	Elementary regulator; stable; gain margin $= \infty$
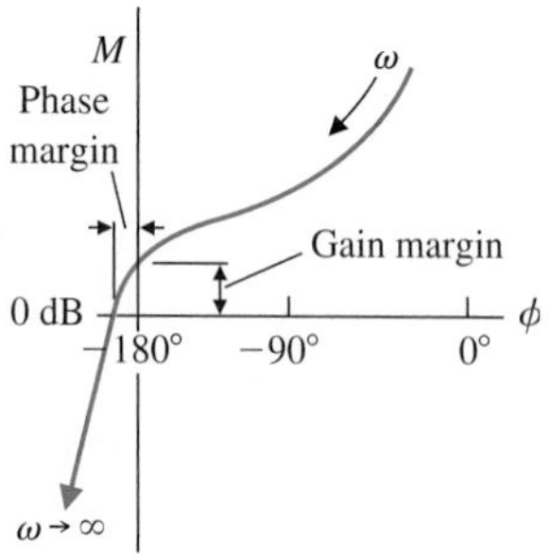	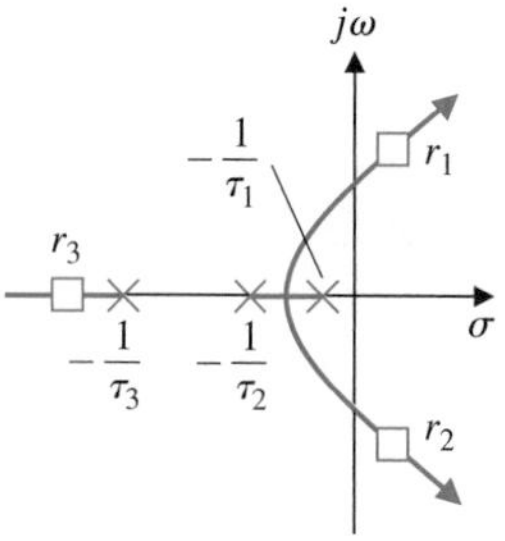	Regulator with additional energy-storage component; unstable, but can be made stable by reducing gain
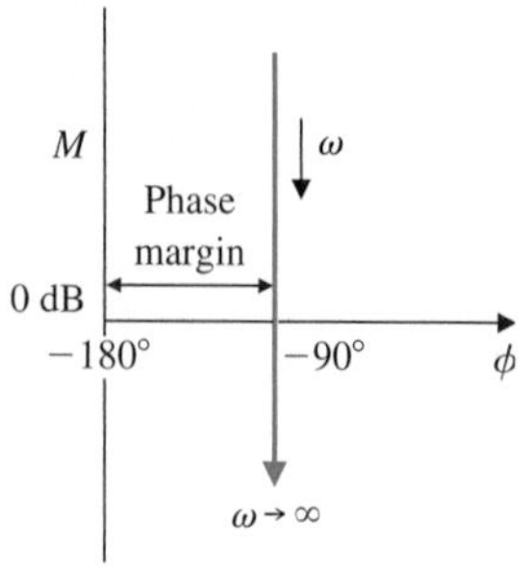	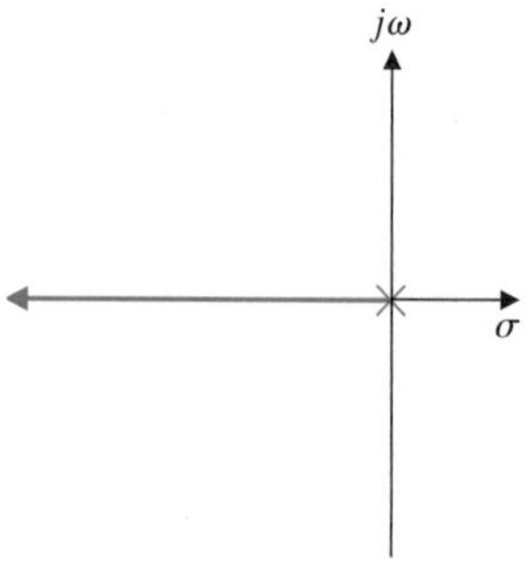	Ideal integrator; stable

(continued)

Table 9.6   *(continued)*

$L(s)$	Polar Plot	Bode Diagram
5. $\dfrac{K}{s(s\tau_1 + 1)}$	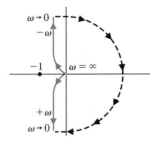	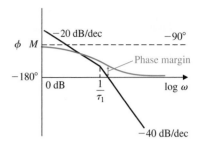
6. $\dfrac{K}{s(s\tau_1 + 1)(s\tau_2 + 1)}$	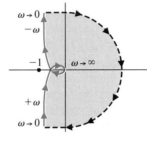	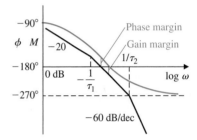
7. $\dfrac{K(s\tau_a + 1)}{(s\tau_1 + 1)(s\tau_2 + 1)}$	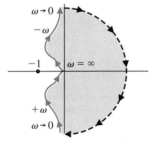	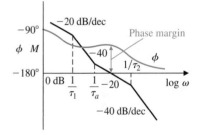
8. $\dfrac{K}{s^2}$	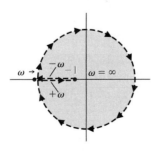	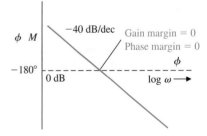

Table 9.6   (continued)

Nichols Diagram	Root Locus	Comments
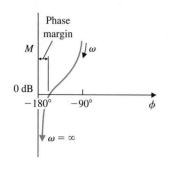	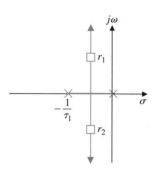	Elementary instrument servo; inherently stable; gain margin $= \infty$
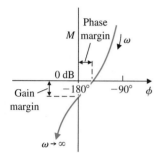	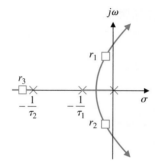	Instrument servo with field control motor or power servo with elementary Wark–Leonard drive; stable as shown, but may become unstable with increased gain
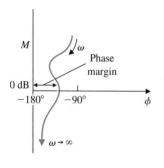	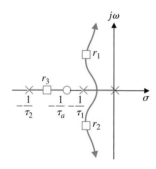	Elementary instrument servo with phase-lead (derivative) compensator; stable
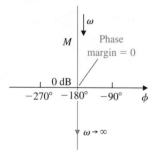	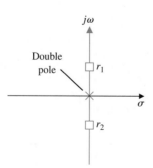	Inherently marginally stable; must be compensated

(continued)

Table 9.6 *(continued)*

$L(s)$	Polar Plot	Bode Diagram
9. $\dfrac{K}{s^2(s\tau_1 + 1)}$	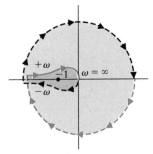	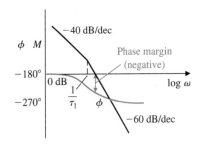
10. $\dfrac{K(s\tau_a + 1)}{s^2(s\tau_1 + 1)}$ $\tau_a > \tau_1$	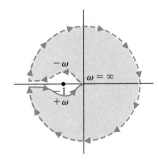	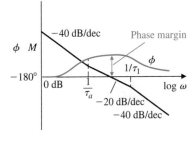
11. $\dfrac{K}{s^3}$	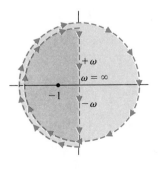	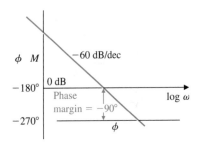
12. $\dfrac{K(s\tau_a + 1)}{s^3}$	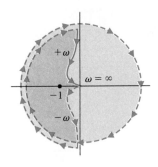	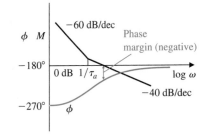

Table 9.6   *(continued)*

Nichols Diagram	Root Locus	Comments
		Inherently unstable; must be compensated
		Stable for all gains
		Inherently unstable
		Inherently unstable

(continued)

Table 9.6   *(continued)*

L(s)	Polar Plot	Bode Diagram

13. $\dfrac{K(s\tau_a + 1)(s\tau_b + 1)}{s^3}$

14. $\dfrac{K(s\tau_a + 1)(s\tau_b + 1)}{s(s\tau_1 + 1)(s\tau_2 + 1)(s\tau_3 + 1)(s\tau_4 + 1)}$

15. $\dfrac{K(s\tau_a + 1)}{s^2(s\tau_1 + 1)(s\tau_2 + 1)}$

Table 9.6 *(continued)*

Nichols Diagram	Root Locus	Comments
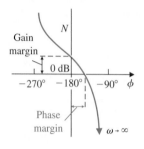	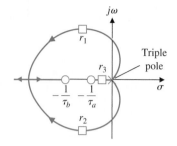	Conditionally stable; becomes unstable if gain is too low
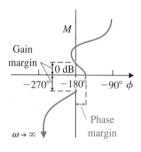	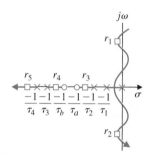	Conditionally stable; stable at low gain, becomes unstable as gain is raised, again becomes stable as gain is further increased, and becomes unstable for very high gains
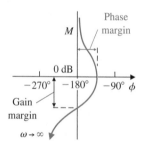	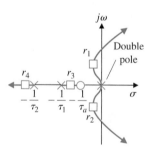	Conditionally stable; becomes unstable at high gain

EXERCISES

E9.1   A system has the loop transfer function

$$G_c(s)G(s) = \frac{2.5(1 + s/5)}{s(1 + 2s)(1 + s/7 + s^2/49)}.$$

Plot the Bode diagram. Show that the phase margin is approximately $28°$ and that the gain margin is approximately 21 dB.

E9.2   A system has the loop transfer function

$$G_c(s)G(s) = \frac{K(1 + s/6)}{s(1 + s/3)(1 + s/12)},$$

where $K = 8.1$. Show that the system crossover (0 dB) frequency is 5 rad/s and that the phase margin is $48°$.

E9.3   An integrated circuit is available to serve as a feedback system to regulate the output voltage of a power supply. The Bode diagram of the required loop transfer function $G_c(j\omega)G(j\omega)$ is shown in Figure E9.3 Estimate the gain and phase margins of the regulator.

Answer: $GM = 25$ dB, $PM = 75°$

E9.4   Consider a system with a loop transfer function

$$G_c(s)G(s) = \frac{100}{s(s + 10)}.$$

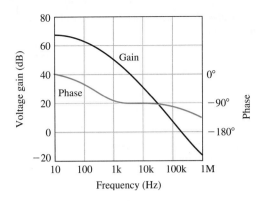

FIGURE 9.3 Power supply regulator

We wish to obtain a resonant peak $M_{p\omega} = 3.0$ dB for the closed-loop system. The peak occurs between 6 and 9 rad/s and is only 1.25 dB. Plot the Nichols chart for the range of frequency from 6 to 15 rad/s. Show that the system gain needs to be raised by 4.6 dB to 171. Determine the resonant frequency for the adjusted system.

Answer: $\omega_r = 11$ rad/s

E9.5 An integrated CMOS digital circuit can be represented by the Bode diagram shown in Figure E9.5. (a) Find the gain and phase margins of the circuit. (b) Estimate how much we would need to reduce the system gain (dB) to obtain a phase margin of 60°.

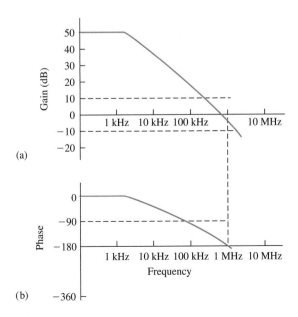

FIGURE 9.5 CMOS circuit

E9.6 A system has a loop transfer function

$$G_c(s)G(s) = \frac{K(s + 100)}{s(s + 10)(s + 40)}.$$

When $K = 500$, the system is unstable. Show that if we reduce the gain to 50, the resonant peak is 3.5 dB. Find the phase margin of the system with $K = 50$.

E9.7 A unity feedback system has a loop transfer function

$$G_c(s)G(s) = \frac{K}{s - 2}.$$

Determine the range of K for which the system is stable by drawing the polar plot.

E9.8 Consider a unity feedback system with the loop transfer function

$$G_c(s)G(s) = \frac{K}{s(s + 1)(s + 2)}.$$

(a) For $K = 4$, show that the gain margin is 3.5 dB.
(b) If we wish to achieve a gain margin equal to 16 dB, determine the value of the gain K.

Answer: (b) $K = 0.98$

E9.9 For the system of E9.8, find the phase margin of the system for $K = 5$.

E9.10 Consider the wind tunnel control system of Problem P7.31. Obtain the Bode diagram and show that the phase margin is 25° and that the gain margin is 10 dB. Also, show that the bandwidth of the closed-loop system is 6 rad/s.

E9.11 Consider a unity feedback system with the loop transfer function

$$G_c(s)G(s) = \frac{10(1 + 0.4s)}{s(1 + 2s)(1 + 0.24s + 0.04s^2)}.$$

(a) Plot the Bode diagram. (b) Find the gain margin and the phase margin.

E9.12 A closed-loop system, as shown in Figure 9.1, has $H(s) = 1$ and

$$G_c(s)G(s) = \frac{K}{s(\tau_1 s + 1)(\tau_2 s + 1)},$$

where $\tau_1 = 0.02$ and $\tau_2 = 0.2$ s. (a) Select a gain K so that the steady-state error for a ramp input is 10% of the magnitude of the ramp function A, where $r(t) = At, t \geq 0$. (b) Plot the Bode plot of $G_c(s)G(s)$, and determine the phase and gain margins. (c) Using the Nichols chart, determine the bandwidth ω_B, the resonant peak $M_{p\omega}$, and the resonant frequency ω_r of the closed-loop system.

Answer:

(a) $K = 10$
(b) $PM = 32°, GM = 15$ dB
(c) $\omega_B = 10.3, M_{p\omega} = 1.84, \omega_r = 6.5$

E9.13 A unity feedback system has a loop transfer function

$$G_c(s)G(s) = \frac{150}{s(s + 5)}.$$

(a) Find the maximum magnitude of the closed-loop frequency response using the Nichols chart. (b) Find the bandwidth and the resonant frequency of this system. (c) Use these frequency measures to estimate the overshoot of the system to a step response.

Answers: (a) 7.5 dB, (b) $\omega_B = 19$, $\omega_r = 12.6$

E9.14 A Nichols chart is given in Figure E9.14 for a system where $G_c(j\omega)G(j\omega)$ is plotted. Using the following table, find (a) the peak resonance $M_{p\omega}$ in dB; (b) the resonant frequency ω_r; (c) the 3-dB bandwidth; and (d) the phase margin of the system.

	ω_1	ω_2	ω_3	ω_4
rad/s	1	3	6	10

E9.15 Consider a unity feedback system with the loop transfer function

$$G_c(s)G(s) = \frac{100}{s + 10}.$$

Find the bandwidth of the closed-loop system.

Answers: $\omega_B = 109$

E9.16 The pure time delay e^{-sT} may be approximated by a transfer function as

$$e^{-sT} \approx \frac{1 - Ts/2}{1 + Ts/2}$$

for $0 < \omega < 2/T$. Obtain the Bode diagram for the actual transfer function and the approximation for $T = 2$ for $0 < \omega < 1$.

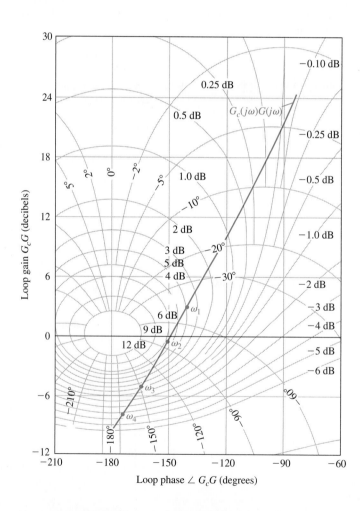

FIGURE 9.14
Nichols chart for $G_c(j\omega)G(j\omega)$.

E9.17 A unity feedback system has a loop transfer function

$$G_c(s)G(s) = \frac{K(s + 2)}{s^3 + 2s^2 + 15s}.$$

(a) Plot the Bode diagram and (b) determine the gain K required to obtain a phase margin of 30°. What is the steady-state error for a ramp input for the gain of part (b)?

E9.18 An actuator for a disk drive uses a shock mount to absorb vibrational energy at approximately 60 Hz [14]. The Bode diagram of $G_c(s)G(s)$ of the control system is shown in Figure E9.18. (a) Find the expected percent overshoot for a step input for the closed-loop

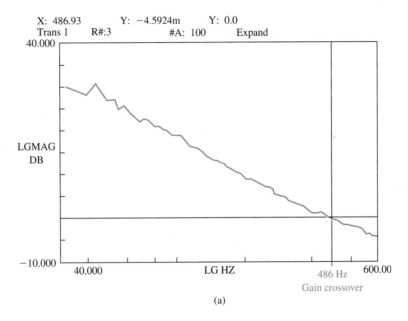

(a)

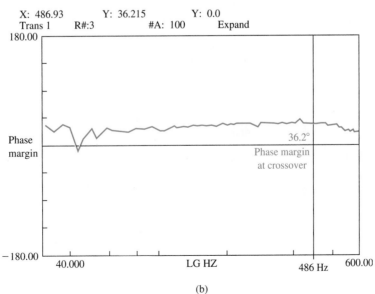

(b)

FIGURE 9.18
Bode diagram of the disk drive, $G_c(s)G(s)$.

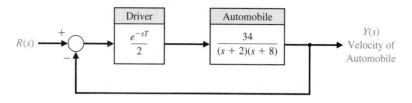

FIGURE 9.20
Automobile control
system.

system, (b) estimate the bandwidth of the closed-loop system, and (c) estimate the settling time (with a 2% criterion) of the system.

E9.19   A unity feedback system with $G_c(s) = K$ has

$$G(s) = \frac{e^{-0.1s}}{s + 10}.$$

Select a gain K so that the phase margin of the system is $50°$. Determine the gain margin for the selected gain, K.

E9.20   Consider a simple model of an automobile driver following another car on the highway at high speed. The model shown in Figure E9.20 incorporates the driver's reaction time, T. One driver has $T = 1$ s, and another has $T = 1.5$ s. Determine the time response $y(t)$ of the system for both drivers for a step change in the command signal $R(s) = -1/s$, due to the braking of the lead car.

E9.21   A unity feedback control system has a loop transfer function

$$G_c(s)G(s) = \frac{K}{s(s + 2)(s + 50)}.$$

Determine the phase margin, the crossover frequency, and the gain margin when $K = 1300$.

Answers: $PM = 16.6°$, $\omega_c = 4.9$, $GM = 4$ or 12 dB

E9.22   A unity feedback system has a loop transfer function

$$G_c(s)G(s) = \frac{K(s + 1)}{(s - 1)(s - 6)}.$$

(a) Using a Bode diagram for $K = 8$, determine the system phase margin. (b) Select a gain K so that the phase margin is at least $45°$.

E9.23   Consider again the system of E9.21 when $K = 438$. Determine the closed-loop system bandwidth, resonant frequency, and $M_{p\omega}$ using the Nichols chart.

Answers: $\omega_B = 4.25$ rad/s, $\omega_r = 2.7$, $M_{p\omega} = 1.7$

E9.24   A unity feedback system has a loop transfer function

$$G_c(s)G(s) = \frac{K}{-1 + \tau s},$$

where $K = \frac{1}{2}$ and $\tau = 1$. The polar plot for $G_c(j\omega)G(j\omega)$ is shown in Figure E9.24. Determine whether the system is stable by using the Nyquist criterion.

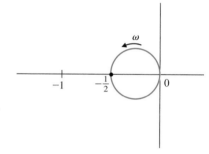

FIGURE 9.24
Polar plot for
$G_c(s)G(s) =$
$K/(-1 + \tau s)$

E9.25   A unity feedback system has a loop transfer function

$$G_c(s)G(s) = \frac{11.7}{s(1 + 0.05s)(1 + 0.1s)}.$$

Determine the phase margin and the crossover frequency.

Answers: $PM = 27.7°$, $\omega_c = 8.31$ rad/s

E9.26   For the system of E9.25, determine $M_{p\omega}$, ω_r, and ω_B for the closed-loop frequency response by using the Nichols chart.

E9.27   A unity feedback system has a loop transfer function

$$G_c(s)G(s) = \frac{K}{s(s + 6)^2}.$$

Determine the maximum gain K for which the phase margin is at least 40° and the gain margin is at least 6 dB. What are the gain margin and phase margin for this value of K?

E9.28   A unity feedback system has the loop transfer function

$$G_c(s)G(s) = \frac{K}{s(s + 0.2)}.$$

(a) Determine the phase margin of the system when $K = 0.16$. (b) Use the phase margin to estimate ζ and predict the overshoot. (c) Calculate the actual response for this second-order system, and compare the result with the part (b) estimate.

E9.29   A loop transfer function is

$$G_c(s)G(s) = \frac{1}{s + 2}.$$

Using the contour in the s-plane shown in Figure E9.29, determine the corresponding contour in the $F(s)$-plane ($B = -1 + j$).

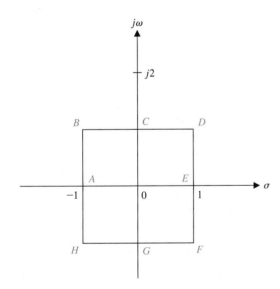

FIGURE 9.29   Contour in the s-plane.

E9.30   Consider the system represented in state variable form

$$\dot{\mathbf{x}} = \mathbf{A}\mathbf{x} + \mathbf{B}u$$
$$y = \mathbf{C}\mathbf{x} + \mathbf{D}u,$$

where

$$\mathbf{A} = \begin{bmatrix} 0 & 1 \\ -10 & -100 \end{bmatrix}, \mathbf{B} = \begin{bmatrix} 0 \\ 1 \end{bmatrix},$$

$$\mathbf{C} = [1000 \quad 0], \text{ and } \mathbf{D} = [0].$$

Sketch the Bode plot.

E9.31   A closed-loop feedback system is shown in Figure E9.31. Sketch the Bode plot and determine the phase margin.

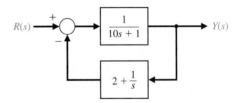

FIGURE 9.31   Nonunity feedback system.

E9.32   Consider the system described in state variable form by

$$\dot{\mathbf{x}}(t) = \mathbf{A}\mathbf{x}(t) + \mathbf{B}u(t)$$
$$y(t) = \mathbf{C}\mathbf{x}(t)$$

where

$$\mathbf{A} = \begin{bmatrix} 0 & 1 \\ -4 & -1 \end{bmatrix}, \mathbf{B} = \begin{bmatrix} 0 \\ 3.62 \end{bmatrix}, \mathbf{C} = [1 \quad 0].$$

Compute the phase margin.

E9.33   Consider the system shown in Figure E9.33. Compute the loop transfer function $L(s)$, and sketch the Bode plot. Determine the phase margin and gain margin when the controller gain $K = 5$.

FIGURE 9.33
Nonunity feedback
system with
proportional
controller K.

R(s) ⟶ + ⊘ − ⟶ Controller K ⟶ Process $\dfrac{4}{s^2 + 2.83s + 4}$ ⟶ $Y(s)$

Sensor $\dfrac{10}{s + 10}$

PROBLEMS

P9.1 For the polar plots of Problem P8.1, use the Nyquist criterion to ascertain the stability of the various systems. In each case, specify the values of N, P, and Z.

P9.2 Sketch the polar plots of the following loop transfer functions $G_c(s)G(s)$, and determine whether the system is stable by applying the Nyquist criterion:

(a) $G_c(s)G(s) = \dfrac{K}{s(s^2 + s + 4)}$.

(b) $G_c(s)G(s) = \dfrac{K(s + 2)}{s^2(s + 4)}$.

If the system is stable, find the maximum value for K by determining the point where the polar plot crosses the u-axis.

P9.3 (a) Find a suitable contour Γ_s in the s-plane that can be used to determine whether all roots of the characteristic equation have damping ratios greater than ζ_1. (b) Find a suitable contour Γ_s in the s-plane that can be used to determine whether all the roots of the characteristic equation have real parts less than $s = -\sigma_1$. (c) Using the contour of part (b) and Cauchy's theorem, determine whether the following characteristic equation has roots with real parts less than $s = -1$:

$$q(s) = s^3 + 11s^2 + 56s + 96.$$

P9.4 The polar plot of a conditionally stable system is shown in Figure P9.4 for a specific gain K. (a) Determine

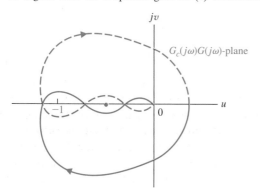

FIGURE P9.4 Polar plot of conditionally stable system.

whether the system is stable, and find the number of roots (if any) in the right-hand s-plane. The system has no poles of $G_c(s)G(s)$ in the right half-plane. (b) Determine whether the system is stable if the -1 point lies at the dot on the axis.

P9.5 A speed control for a gasoline engine is shown in Figure P9.5. Because of the restriction at the carburetor intake and the capacitance of the reduction manifold, the lag τ_t occurs and is equal to 1 second. The engine time constant τ_e is equal to $J/b = 3$ s. The speed measurement time constant is $\tau_m = 0.4$ s. (a) Determine the necessary gain K if the steady-state speed error is required to be less than 10% of the speed reference setting. (b) With the gain determined from part (a), apply the Nyquist criterion to investigate the stability of the system. (c) Determine the phase and gain margins of the system.

P9.6 A direct-drive arm is an innovative mechanical arm in which no reducers are used between motors and their loads. Because the motor rotors are directly coupled to the loads, the drive systems have no backlash, small friction, and high mechanical stiffness, which are all important features for fast and accurate positioning and dexterous handling using sophisticated torque control.

The goal of the MIT direct-drive arm project is to achieve arm speeds of 10 m/s [15]. The arm has torques of up to 660 N m (475 ft lb). Feedback and a set of position and velocity sensors are used with each motor. The frequency response of one joint of the arm is shown in Figure P9.6(a). The two poles appear at 3.7 Hz and 68 Hz. Figure P9.6(b) shows the step response with position and velocity feedback used. The time constant of the closed-loop system is 82 ms. Develop the block diagram of the drive system and prove that 82 ms is a reasonable result.

P9.7 A vertical takeoff (VTOL) aircraft is an inherently unstable vehicle and requires an automatic stabilization system. An attitude stabilization system for the K-16B U.S. Army VTOL aircraft has been designed and is shown in block diagram form in Figure P9.7 [16]. At 40 knots, the dynamics of the vehicle are approximately represented by the transfer function

$$G(s) = \frac{10}{s^2 + 0.36}.$$

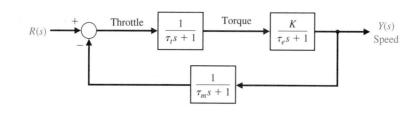

FIGURE P9.5
Engine speed control.

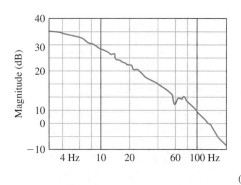
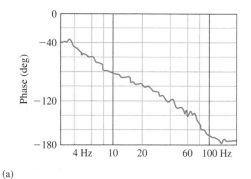

(a)

FIGURE P9.6
The MIT arm:
(a) frequency
response, and
(b) position
response.

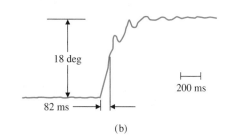

18 deg

200 ms

82 ms

(b)

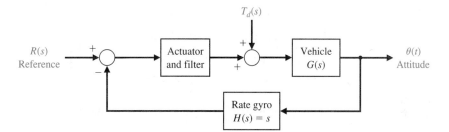

FIGURE P9.7
VTOL aircraft
stabilization
system.

The actuator and filter are represented by the transfer function

$$G_c(s) = \frac{K_1(s + 7)}{s + 3}.$$

(a) Obtain the Bode diagram of the loop transfer function $L(s) = G_c(s)G(s)H(s)$ when the gain is $K_1 = 2$. (b) Determine the gain and phase margins of this system. (c) Determine the steady-state error for a wind disturbance of $T_d(s) = 1/s$. (d) Determine the maximum amplitude of the resonant peak of the closed-loop frequency response and the frequency of the resonance. (e) Estimate the damping ratio of the system from $M_{p\omega}$ and the phase margin.

P9.8 Electrohydraulic servomechanisms are used in control systems requiring a rapid response for a large mass. An electrohydraulic servomechanism can provide an output of 100 kW or greater [17]. A photo of a servovalve and actuator is shown in Figure P9.8(a).

The output sensor yields a measurement of actuator position, which is compared with V_{in}. The error is amplified and controls the hydraulic valve position, thus controlling the hydraulic fluid flow to the actuator. The block diagram of a closed-loop electrohydraulic servomechanism using pressure feedback to obtain damping is shown in Figure P9.8(b) [17, 18]. Typical values for this system are $\tau = 0.02$ s; for the hydraulic system they are $\omega_2 = 7(2\pi)$ and $\zeta_2 = 0.05$. The structural resonance ω_1 is equal to $10(2\pi)$, and the damping is $\zeta_1 = 0.05$. The loop gain is $K_A K_1 K_2 = 1.0$. (a) Sketch the Bode diagram and determine the phase margin of the system. (b) The damping of the system can be increased by drilling a small hole in the piston so that $\zeta_2 = 0.25$. Sketch the Bode diagram and determine the phase margin of this system.

P9.9 The space shuttle, shown in Figure P9.9(a), carries large payloads into space and returns them to earth for reuse [19]. The shuttle uses elevons at the trailing

(a)

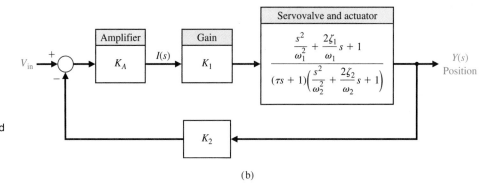

FIGURE P9.8
(a) A servovalve and
actuator (courtesy
of Moog, Inc.,
Industrial Division).
(b) Block diagram.

(b)

edge of the wing and a brake on the tail to control the flight during entry. The block diagram of a pitch rate control system is shown in Figure P9.9(b). The sensor is represented by a gain, $H(s) = 0.5$, and the vehicle by the transfer function

$$G(s) = \frac{0.30(s + 0.05)(s^2 + 1600)}{(s^2 + 0.05s + 16)(s + 70)}.$$

The controller $G_c(s)$ can be a gain or any suitable transfer function. (a) Sketch the Bode diagram of the system when $G_c(s) = 2$ and determine the stability margin. (b) Sketch the Bode diagram of the system when

$$G_c(s) = K_P + K_I/s \quad \text{and} \quad K_I/K_P = 0.5.$$

The gain K_P should be selected so that the gain margin is 10 dB.

P9.10   Machine tools are often automatically controlled as shown in Figure P9.10. These automatic systems are often called numerical machine controls [9]. On each axis, the desired position of the machine tool is

compared with the actual position and is used to actuate a solenoid coil and the shaft of a hydraulic actuator. The transfer function of the actuator (see Table 2.7) is

$$G_a(s) = \frac{X(s)}{Y(s)} = \frac{K_a}{s(\tau_a s + 1)},$$

where $K_a = 1$ and $\tau_a = 0.4$ s. The output voltage of the difference amplifier is

$$E_0(s) = K_1(X(s) - X_d(s)),$$

where $x_d(t)$ is the desired position input. The force on the shaft is proportional to the current i, so that $F = K_2 i(t)$, where $K_2 = 3.0$. The spring constant K_s is equal to 1.5, $R = 0.1$, and $L = 0.2$.

(a) Determine the gain K_1 that results in a system with a phase margin of 30°. (b) For the gain K_1 of part (a), determine $M_{p\omega}, \omega_r$, and the closed-loop system bandwidth. (c) Estimate the percent overshoot of the transient response to a step input $X_d(s) = 1/s$, and the settling time (to within 2% of the final value).

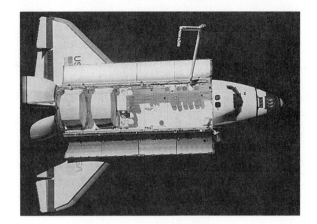

FIGURE P9.9
(a) The Earth-orbiting space shuttle against the blackness of space. The remote manipulator robot is shown with the cargo bay doors open in this top view, taken by a satellite. (b) Pitch rate control system. (Courtesy of NASA.)

(a)

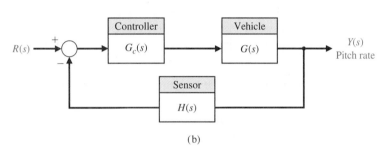

(b)

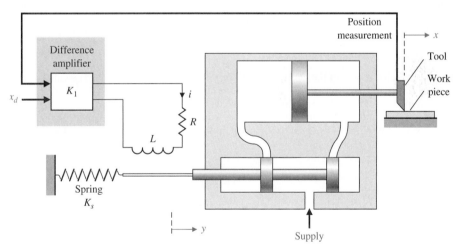

FIGURE P9.10
Machine tool control.

P9.11 A control system for a chemical concentration control system is shown in Figure P9.11. The system receives a granular feed of varying composition, and we want to maintain a constant composition of the output mixture by adjusting the feed-flow valve. The transfer function of the tank and output valve is

$$G(s) = \frac{5}{5s + 1},$$

and that of the controller is

$$G_c(s) = K_1 + \frac{K_2}{s}.$$

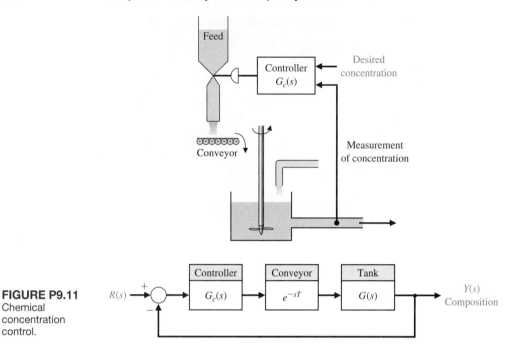

FIGURE P9.11
Chemical
concentration
control.

The transport of the feed along the conveyor requires a transport (or delay) time, $T = 1.5$ s. (a) Sketch the Bode diagram when $K_1 = K_2 = 1$, and investigate the stability of the system. (b) Sketch the Bode diagram when $K_1 = 0.1$ and $K_2 = 0.04$, and investigate the stability of the system. (c) When $K_1 = 0$, use the Nyquist criterion to calculate the maximum allowable gain K_2 for the system to remain stable.

P9.12   A simplified model of the control system for regulating the pupillary aperture in the human eye is shown in Figure P9.12 [20]. The gain K represents the pupillary gain, and τ is the pupil time constant, which is 0.6 s. The time delay T is equal to 0.15 s. The pupillary gain is equal to 3.8.

(a) Assuming the time delay is negligible, sketch the Bode diagram for the system. Determine the phase margin of the system. (b) Include the effect of the time delay by adding the phase shift due to the delay. Determine the phase margin of the system with the time delay included.

P9.13   A controller is used to regulate the temperature of a mold for plastic part fabrication, as shown in Figure P9.13. The value of the delay time is estimated as 1.2 s. (a) Using the Nyquist criterion, determine the stability of the system for $K_a = K = 1$. (b) Determine a suitable value for K_a for a stable system that will yield a phase margin greater than 50° when $K = 1$.

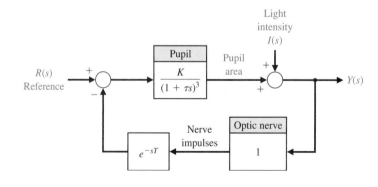

FIGURE P9.12
Human pupil
aperature control.

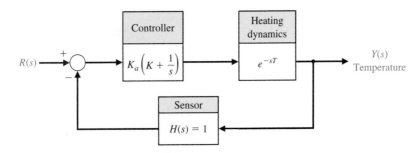

FIGURE P9.13
Temperature
controller.

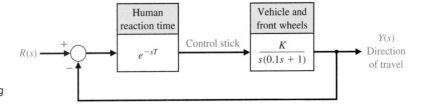

FIGURE P9.14
Automobile steering
control.

P9.14 Electronics and computers are being used to control automobiles. Figure P9.14 is an example of an automobile control system, the steering control for a research automobile. The control stick is used for steering. A typical driver has a reaction time of $T = 0.2$ s.

(a) Using the Nichols chart, determine the magnitude of the gain K that will result in a system with a peak magnitude of the closed-loop frequency response $M_{p\omega}$ less than or equal to 2 dB.
(b) Estimate the damping ratio of the system based on (1) $M_{p\omega}$ and (2) the phase margin. Compare the results and explain the difference, if any.
(c) Determine the closed-loop 3-dB bandwidth of the system.

P9.15 Consider the automatic ship-steering system discussed in Problem P8.11. The frequency response of the open-loop portion of the ship steering control system is shown in Figure P8.11. The deviation of the tanker from the straight track is measured by radar and is used to generate the error signal, as shown in Figure P9.15. This error signal is used to control the rudder angle $\delta(s)$.

(a) Is this system stable? Discuss what an unstable ship-steering system indicates in terms of the transient response of the system. Recall that the system under consideration is a ship attempting to follow a straight track.
(b) Is it possible to stabilize this system by lowering the gain of the transfer function $G(s)$?
(c) Is it possible to stabilize this system? Suggest a suitable feedback compensator?
(d) Repeat parts (a), (b), and (c) when switch S is closed.

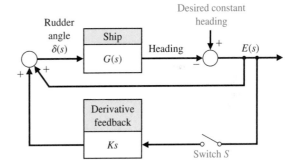

FIGURE P9.15 Automatic ship steering.

P9.16 An electric carrier that automatically follows a tape track laid out on a factory floor is shown in Figure P9.16(a) [15]. Closed-loop feedback systems are used to control the guidance and speed of the vehicle. The cart senses the tape path by means of an array of 16 phototransistors. The block diagram of the steering system is shown in Figure P9.16(b). Select a gain K so that the phase margin is approximately 30°.

P9.17 The primary objective of many control systems is to maintain the output variable at the desired or reference condition when the system is subjected to a disturbance

(a)

FIGURE P9.16
(a) An electric
carrier vehicle
(photo courtesy of
Control Engineering
Corporation).
(b) Block diagram.

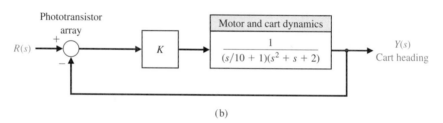

(b)

FIGURE P9.17
Chemical reactor
control.

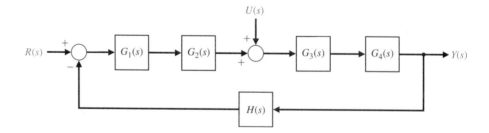

[23]. A typical chemical reactor control scheme is shown in Figure P9.17. The disturbance is represented by $U(s)$, and the chemical process by G_3 and G_4. The controller is represented by G_1 and the valve by G_2. The feedback sensor is $H(s)$ and will be assumed to be equal to 1. We will assume that G_2, G_3, and G_4 are all of the form

$$G_i(s) = \frac{K_i}{1 + \tau_i s},$$

where $\tau_3 = \tau_4 = 4$ s and $K_3 = K_4 = 0.1$. The valve constants are $K_2 = 20$ and $\tau_2 = 0.5$ s. We want to maintain a steady-state error less than 5% of the desired reference position.

(a) When $G_1(s) = K_1$, find the necessary gain to satisfy the error-constant requirement. For this condition, determine the expected overshoot to a step change in the reference signal $r(t)$.

(b) If the controller has a proportional term plus an integral term so that $G_1(s) = K_1(1 + 1/s)$, determine a suitable gain to yield a system with an overshoot less than 30%, but greater than 5%. For parts (a) and (b), use the approximation of the damping ratio as a function of phase margin that yields $\zeta = 0.01\phi_{pm}$. For these calculations, assume that $U(s) = 0$.

(c) Estimate the settling time (with a 2% criterion) of the step response of the system for the controller of parts (a) and (b).

(d) The system is expected to be subjected to a step disturbance $U(s) = A/s$. For simplicity, assume that the desired reference is $r(t) = 0$ when the system has settled. Determine the response of the system of part (b) to the disturbance.

P9.18 A model of an automobile driver attempting to steer a course is shown in Figure P9.18, where $K = 5.3$. (a) Find the frequency response and the gain and phase margins when the reaction time T is zero. (b) Find the phase margin when the reaction time is 0.1 s. (c) Find the reaction time that will cause the system to be borderline stable (phase margin = 0°).

P9.19 In the United States, billions of dollars are spent annually for solid waste collection and disposal. One system, which uses a remote control pick-up arm for collecting waste bags, is shown in Figure P9.19. The loop transfer function of the remote pick-up arm is

$$G_c(s)G(s) = \frac{0.25}{s(4s + 1)(s + 3)}.$$

(a) Plot the Nichols chart and show that the gain margin is approximately 32 dB. (b) Determine the phase margin and the $M_{p\omega}$ for the closed loop. Also, determine the closed-loop bandwidth.

P9.20 The Bell-Boeing V-22 Osprey Tiltrotor is both an airplane and a helicopter. Its advantage is the ability to rotate its engines to a vertical position, as shown in Figure P7.33(a), for takeoffs and landings and then switch the engines to a horizontal position for cruising as an airplane. The altitude control system in the helicopter mode is shown in Figure P9.20. (a) Obtain the frequency response of the system for $K = 100$. (b) Find the gain margin and the phase margin for this system. (c) Select a suitable gain K so that the phase margin is 40°. (Decrease the gain above $K = 100$.) (d) Find the response $y(t)$ of the system for the gain selected in part (c).

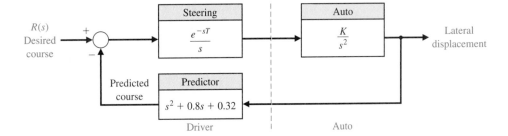

FIGURE P9.18
Automobile and driver control.

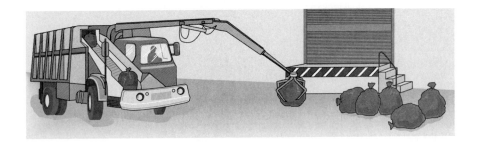

FIGURE P9.19
Waste collection system.

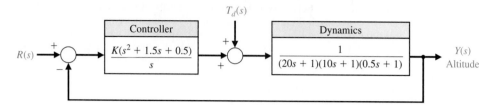

FIGURE P9.20
Tiltrotor aircraft
control.

P9.21 Consider a unity feedback system with the loop transfer function

$$G_c(s)G(s) = \frac{K}{s(s + 1)(s + 4)}.$$

(a) Sketch the Bode diagram for $K = 4$. Determine (b) the gain margin, (c) the value of K required to provide a gain margin equal to 12 dB, and (d) the value of K to yield a steady-state error of 25% of the magnitude A for the ramp input $r(t) = At, t > 0$. Can this gain be utilized and achieve acceptable performance?

P9.22 The Nichols diagram for $G_c(j\omega)G(j\omega)$ of a closed-loop system is shown in Figure P9.22. The frequency for each point on the graph is given in the following table:

Point	1	2	3	4	5	6	7	8	9
ω	1	2.0	2.6	3.4	4.2	5.2	6.0	7.0	8.0

Determine (a) the resonant frequency, (b) the bandwidth, (c) the phase margin, and (d) the gain margin. (e) Estimate the overshoot and settling time (with a 2% criterion) of the response to a step input.

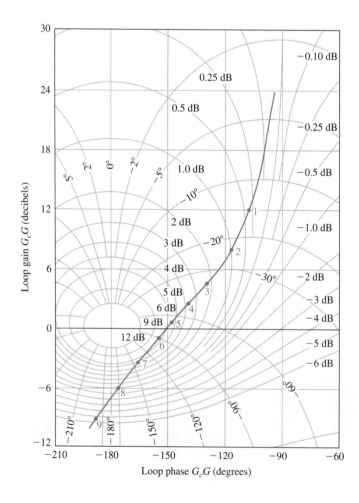

FIGURE P9.22
Nichols chart.

P9.23 A closed-loop system has a loop transfer function

$$G_c(s)G(s) = \frac{K}{s(s + 5)(s + 10)}.$$

(a) Determine the gain K so that the phase margin is 60°. (b) For the gain K selected in part (a), determine the gain margin of the system.

P9.24 A closed-loop system with unity feedback has a loop transfer function

$$G_c(s)G(s) = \frac{K(s + 20)}{s^2}.$$

(a) Determine the gain K so that the phase margin is 45°. (b) For the gain K selected in part (a), determine the gain margin. (c) Predict the bandwidth of the closed-loop system.

P9.25 A closed-loop system has the loop transfer function

$$G_c(s)G(s) = \frac{Ke^{-Ts}}{s}.$$

(a) Determine the gain K so that the phase margin is 60° when $T = 0.2$. (b) Plot the phase margin versus the time delay T for K as in part (a).

P9.26 A specialty machine shop is improving the efficiency of its surface-grinding process [22]. The existing machine is mechanically sound, but manually operated. Automating the machine will free the operator for other tasks and thus increase overall throughput of the machine shop. The grinding machine is shown in Figure P9.26(a) with all three axes automated with motors and feedback systems. The control system for the y-axis is shown in Figure P9.26(b). To achieve a low steady-state error to a ramp command, we choose $K = 10$. Sketch the Bode diagram of the open-loop system and obtain the Nichols chart plot. Determine the gain and phase margin of the system and the bandwidth of the closed-loop system. Estimate the ζ of the system and the predicted overshoot and settling time (with a 2% criterion)

P9.27 Consider the system shown in Figure P9.27. Determine the maximum value of $K = K_{max}$ for which the

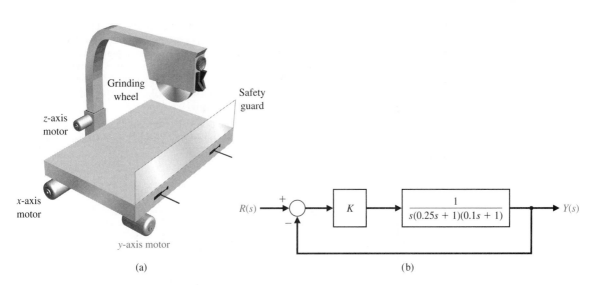

(a)   (b)

FIGURE P9.26 Surface-grinding wheel control system.

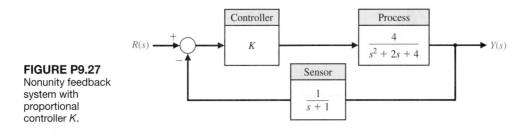

FIGURE P9.27
Nonunity feedback
system with
proportional
controller K.

closed-loop system is stable. Plot the phase margin as a function of the gain $1 \le K \le K_{max}$. Explain what happens to the phase margin as K approaches K_{max}.

P9.28 Consider the feedback system shown in Figure P9.28 with the process transfer function given as

$$G(s) = \frac{1}{s(s + 1)}.$$

The controller is the proportional controller

$$G_c(s) = K_P.$$

(a) Determine a value of K_P such that the phase margin is approximately $P.M. \approx 45°$.
(b) Using the *P.M.* obtained, predict the percent overshoot of the closed-loop system to a unit step input.
(c) Plot the step response and compare the actual percent overshoot with the predicted percent overshoot.

FIGURE P9.28
A unity feedback
system with a
proportional
controller in the
loop.

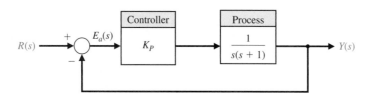

ADVANCED PROBLEMS

AP9.1 Operational spacecraft undergo substantial mass property and configuration changes during their lifetime [26]. For example, the inertias change considerably during operations. Consider the orientation control system shown in Fig. AP9.1.

(a) Plot the Bode diagram, and determine the gain and phase margins when $\omega_n{}^2 = 15{,}267$. (b) Repeat part (a) when $\omega_n{}^2 = 9500$. Note the effect of changing $\omega_n{}^2$ by 38%.

AP9.2 Anesthesia is used in surgery to induce unconsciousness. One problem with drug-induced unconsciousness is large differences in patient responsiveness. Furthermore, the patient response changes during an operation. A model of drug-induced anesthesia control is shown in Figure AP9.2. The proxy for unconsciousness is the arterial blood pressure.

(a) Plot the Bode diagram and determine the gain margin and the phase margin when $T = 0.05$ s. (b) Repeat

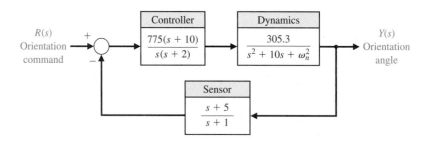

FIGURE AP9.1
Spacecraft
orientation control.

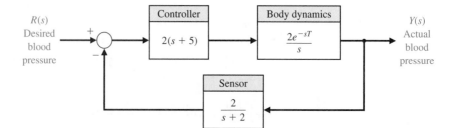

FIGURE AP9.2
Control of blood
pressure with
anesthesia.

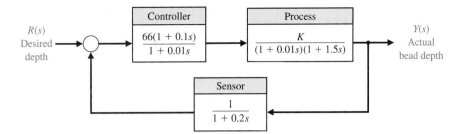

FIGURE AP9.3
Weld bead depth
control.

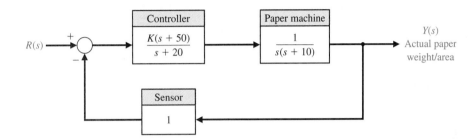

FIGURE AP9.4
Paper machine
control.

part (a) when $T = 0.1$ s. Describe the effect of the 100% increase in the time delay T. (c) Using the phase margin, predict the overshoot for a step input for parts (a) and (b).

AP9.3 Welding processes have been automated over the past decades. Weld quality features, such as final metallurgy and joint mechanics, typically are not measurable online for control. Therefore, some indirect way of controlling the weld quality is necessary. A comprehensive approach to in-process control of welding includes both geometric features of the bead (such as the cross-sectional features of width, depth, and height) and thermal characteristics (such as the heat-affected zone width and cooling rate). The weld bead depth, which is the key geometric attribute of a major class of welds, is very difficult to measure directly, but a method to estimate the depth using temperature measurement has been developed [27]. A model of the weld control system is shown in Figure AP9.3.

(a) Determine the phase margin and gain margin for the system when $K = 1$. (b) Repeat part (a) when $K = 1.5$. (c) Determine the bandwidth of the system for $K = 1$ and $K = 1.5$ by using the Nichols chart. (d) Predict the settling time (with a 2% criterion) of a step response for $K = 1$ and $K = 1.5$.

AP9.4 The control of a paper-making machine is quite complex [28]. The goal is to deposit the proper amount of fiber suspension (pulp) at the right speed and in a uniform way. Dewatering, fiber deposition, rolling,

and drying then take place in sequence. Control of the paper weight per unit area is very important. For the control system shown in Figure AP9.4, select K so that the phase margin $\geq 40°$ and the gain margin ≥ 10 dB. Plot the step response for the selected gain. Determine the bandwidth of the closed-loop system.

AP9.5 NASA is planning many Mars missions with rover vehicles. A typical rover is a solar-powered vehicle which will see where it is going with TV cameras and will measure distance to objects with laser range finders. It will be able to climb a 30° slope in dry sand and will carry a spectrometer that can determine the chemical composition of surface rocks. It will be controlled remotely from Earth.

For the model of the position control system shown in Figure AP9.5, determine the gain K that maximizes the phase margin. Determine the overshoot for a step input with the selected gain.

AP9.6 The acidity of water draining from a coal mine is often controlled by adding lime to the water. A valve controls the lime addition and a sensor is downstream. For the model of the system shown in Figure AP9.6, determine K and the distance D to maintain stability. We require $D > 2$ meters in order to allow full mixing before sensing.

AP9.7 Building elevators are limited to about 800 meters. Above that height, elevator cables become too thick and too heavy for practical use. One solution is to eliminate the cable. The key to the cordless elevator is the

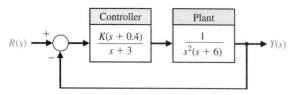

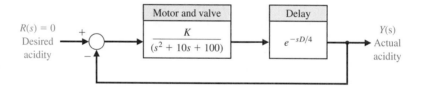

FIGURE AP9.5   Position control system of a Mars rover.

linear motor technology now being applied to the development of magnetically levitated rail transportation systems. Under consideration is a linear synchronous motor that propels a passenger car along the tracklike guideway running the length of the elevator shaft. The motor works by the interaction of an electromagnetic field from electric coils on the guideway with magnets on the car [29].

If we assume that the motor has negligible friction, the system may be represented by the model shown in Figure AP9.7. Determine K so that the phase margin of the system is 45°. For the gain K selected, determine the system bandwidth. Also calculate the maximum value of the output for a unit step disturbance for the selected gain.

AP9.8   A control system is shown in Figure AP9.8. The gain K is greater than 500 and less than 3000. Select a gain that will cause the system step response to have an overshoot of less than 18%. Plot the Nichols diagram, and calculate the phase margin.

AP9.9   Consider again the system shown in Figure AP7.12 which uses a PI controller. Let

$$\frac{K_I}{K_P} = 0.2,$$

and determine the gain K_P that provides the maximum phase margin.

AP9.10   A multiloop block diagram is shown in Figure AP9.10.

(a) Compute the transfer function $T(s) = Y(s)/R(s)$.
(b) Determine K such that the steady-state tracking error to a unit step input $R(s) = 1/s$ is zero. Plot the unit step response.
(c) Using K from part (b), compute the system bandwidth ω_b.

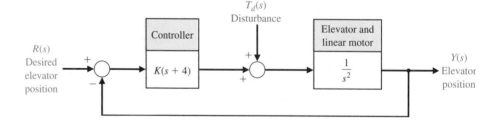

FIGURE AP9.6
Mine water acidity control.

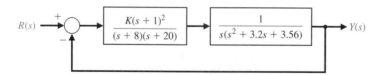

FIGURE AP9.7
Elevator position control.

FIGURE AP9.8
Gain selection.

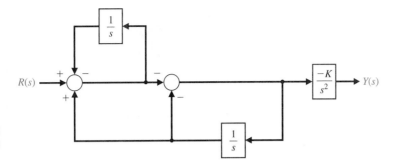

FIGURE AP 9.10
Multiloop feedback
control system.

DESIGN PROBLEMS

CDP9.1 The system of Figure CDP4.1 uses a controller 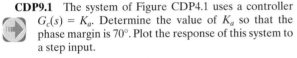 $G_c(s) = K_a$. Determine the value of K_a so that the phase margin is 70°. Plot the response of this system to a step input.

DP9.1 A mobile robot for toxic waste cleanup is shown in Figure DP9.1(a) [24]. The closed-loop speed control is represented by Figure 9.1 with $H(s) = 1$. The Nichols chart in Figure DP9.1(b) shows the plot of $G_c(j\omega)$ $G(j\omega)/K$ versus ω. The value of the frequency at the points indicated is recorded in the following table:

Point	1	2	3	4	5
ω	2	5	10	20	50

(a) Determine the gain and phase margins of the closed-loop system when $K = 1$. (b) Determine the resonant peak in dB and the resonant frequency for $K = 1$. (c) Determine the system bandwidth and estimate the settling time (with a 2% criterion) and percent overshoot of this system for a step input. (d) Determine the appropriate gain K so that the overshoot to a step input is 30%, and estimate the settling time of the system.

DP9.2 Flexible-joint robotic arms are constructed of lightweight materials and exhibit lightly damped open-loop dynamics [15]. A feedback control system for a flexible arm is shown in Figure DP9.2. Select K so that the system has maximum phase margin. Predict the overshoot for a step input based on the phase margin attained, and compare it to the actual overshoot for a step input. Determine the bandwidth of the closed-loop system. Predict the settling time (with a 2% criterion) of the system to a step input and compare it to the actual settling time. Discuss the suitability of this control system.

DP9.3 An automatic drug delivery system is used in the regulation of critical care patients suffering from cardiac failure [25]. The goal is to maintain stable patient status within narrow bounds. Consider the use of a drug delivery system for the regulation of blood pressure by the infusion of a drug. The feedback control system is shown in Figure DP9.3. Select an appropriate gain K that maintains narrow deviation for blood pressure while achieving a good dynamic response.

DP9.4 A robot tennis player is shown in Figure DP9.4(a), and a simplified control system for $\theta_2(t)$ is shown in Figure DP9.4(b). The goal of the control system is to attain the best step response while attaining a high K_v for the system. Select $K_{v1} = 0.325$ and $K_{v2} = 0.45$, and determine the phase margin, gain margin, and closed-loop bandwidth for each case. Estimate the step response for each case and select the best value for K.

DP9.5 An electrohydraulic actuator is used to actuate large loads for a robot manipulator, as shown in Figure DP9.5 [17]. The system is subjected to a step input, and we desire the steady-state error to be minimized. However, we wish to keep the overshoot less than 10%. Let $T = 0.8$ s.

(a) Select the gain K when $G_c(s) = K$, and determine the resulting overshoot, settling time (with a 2% criterion), and steady-state error. (b) Repeat part (a) when $G_c(s) = K_1 + K_2/s$ by selecting K_1 and K_2. Sketch the Nichols chart for the selected gains K_1 and K_2.

DP9.6 The physical representation of a steel strip-rolling mill is a damped-spring system [8]. The output thickness sensor is located a negligible distance from the output of the mill, and the objective is to keep the thickness as close to a reference value as possible. Any change of the input strip thickness is regarded as a disturbance. The system is a nonunity feedback system, as shown in Figure DP9.6. Depending on the maintenance of the mill, the parameter varies as $80 \le b < 300$.

Determine the phase margin and gain margin for the two extreme values of b when the normal value of

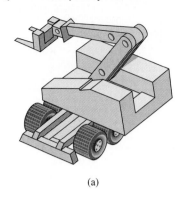

(a)

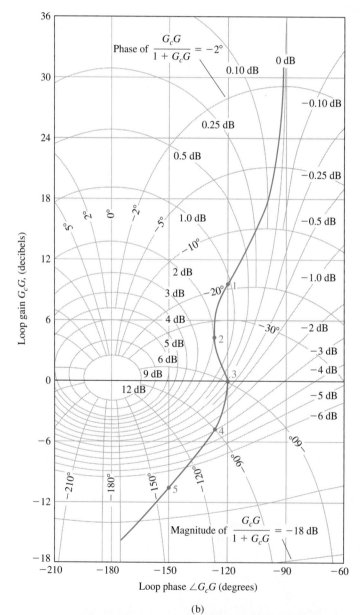

(b)

FIGURE DP9.1

(a) Mobile robot for toxic waste cleanup. (b) Nichols chart.

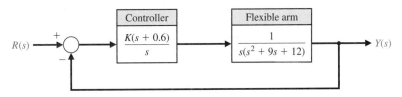

FIGURE DP9.2
Control of a flexible robot arm.

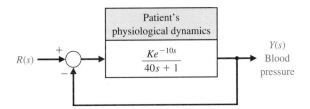

FIGURE DP9.3
Automatic drug delivery.

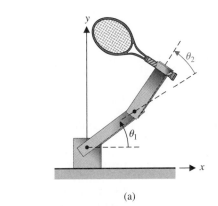

(a)

FIGURE DP9.4
(a) An articulated two-link tennis player robot.
(b) Simplified control system.

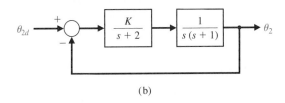

(b)

FIGURE DP9.5
Electrohydraulic actuator.

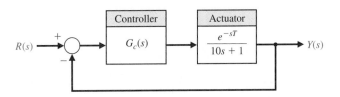

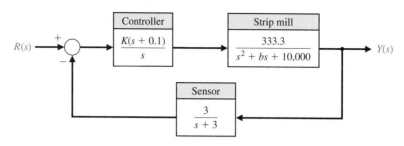

FIGURE DP9.6
Steel strip-rolling
mill.

the gain is $K = 170$. Recommend a reduced value for K so that the phase margin is greater than 40° and the gain margin is greater than 8 dB for the range of b.

DP9.7   Vehicles for lunar construction and exploration work will face conditions unlike anything found on Earth. Furthermore, they will be controlled via remote control. A block diagram of such a vehicle and the control are shown in Figure DP9.7. Select a suitable gain K when $T = 0.5$ s. The goal is to achieve a fast step response with an overshoot of less than 20%.

DP9.8   The control of a high-speed steel-rolling mill is a challenging problem. The goal is to keep the strip thickness accurate and readily adjustable. The model of the control system is shown in Figure DP9.8. Design a control system by selecting K so that the step response of the system is as fast as possible with an overshoot less than 0.5% and a settling time (with a 2% criterion) less than 4 seconds. Use the root locus to select K, and calculate the roots for the selected K. Describe the dominant root(s) of the system.

DP9.9   A two-tank system containing a heated liquid has the model shown in Figure DP9.9(a), where T_0 is the temperature of the fluid flowing into the first tank and T_2 is the temperature of the liquid flowing out of the second tank. The block diagram model is shown in Figure DP9.9(b). The system of the two tanks has a heater in tank 1 with a controllable heat input Q. The time constants are $\tau_1 = 10$ s and $\tau_2 = 50$ s.

(a) Determine $T_2(s)$ in terms of $T_0(s)$ and $T_{2d}(s)$.
(b) If $T_{2d}(s)$, the desired output temperature, is changed instantaneously from $T_{2d}(s) = A/s$ to $T_{2d}(s) = 2A/s$, determine the transient response of $T_2(t)$ when $G_c(s) = K = 500$. Assume that, prior to the abrupt temperature change, the system is at steady state.
(c) Find the steady-state error e_{ss} for the system of part (b), where $E(s) = T_{2d}(s) - T_2(s)$.
(d) Let $G_c(s) = K/s$ and repeat parts (b) and (c). Use a gain K such that the percent overshoot is less than 10%.
(e) Design a controller that will result in a system with a settling time (with a 2% criterion) of

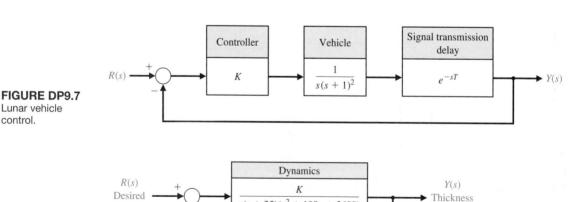

FIGURE DP9.7
Lunar vehicle
control.

FIGURE DP9.8
Steel-rolling mill
control.

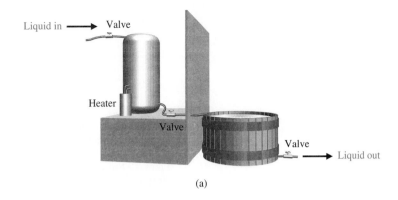

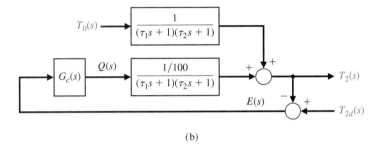

FIGURE DP9.9
Two-tank
temperature
control.

(a)

(b)

$T_s < 150$ s and a percent overshoot of less than 10%, while maintaining a zero steady-state error when

$$G_c(s) = K_P + \frac{K_I}{s}.$$

(f) Prepare a table comparing the percent overshoot, settling time, and steady-state error for the designs of parts (b) through (e).

DP9.10 Consider the system is described in state variable form by

$$\dot{\mathbf{x}}(t) = \mathbf{A}\mathbf{x}(t) + \mathbf{B}u(t)$$
$$y(t) = \mathbf{C}\mathbf{x}(t)$$

where

$$\mathbf{A} = \begin{bmatrix} 0 & 1 \\ 2 & 3 \end{bmatrix}, \mathbf{B} = \begin{bmatrix} 0 \\ 1 \end{bmatrix}, \mathbf{C} = [1 \quad 0].$$

Assume that the input is a linear combination of the states, that is,

$$u(t) = -\mathbf{K}\mathbf{x}(t) + r(t),$$

where $r(t)$ is the reference input and the gain matrix is $\mathbf{K} = [K_1 \quad K_2]$. Substituting $u(t)$ into the state variable equation yields the closed-loop system

$$\dot{\mathbf{x}}(t) = [\mathbf{A} - \mathbf{B}\mathbf{K}]\mathbf{x}(t) + \mathbf{B}r(t)$$
$$y(t) = \mathbf{C}\mathbf{x}(t)$$

(a) Obtain the characteristic equation associated with **A-BK**.
(b) Design the gain matrix **K** to meet the following specifications: (i) the closed-loop system is stable; (ii) the system bandwidth $\omega_b \geq 1$ rad/s; and (iii) the steady-state error to a unit step input $R(s) = 1/s$ is zero.

COMPUTER PROBLEMS

CP9.1 Consider a unity negative feedback control system with

$$G_c(s)G(s) = \frac{100}{s^2 + 4s + 10}.$$

Verify that the gain margin is ∞ and that the phase margin is 24°.

CP9.2 Using the nyquist function, obtain the polar plot for the following transfer functions:

(a) $G(s) = \dfrac{2}{s + 2}$;

(b) $G(s) = \dfrac{25}{s^2 + 8s + 16}$;

(c) $G(s) = \dfrac{5}{s^3 + 3s^2 + 3s + 1}$.

CP9.3   Using the nichols, and logspace functions, obtain the Nichols chart with a grid for the following transfer functions:

(a) $G(s) = \dfrac{1}{s + 0.2}$;

(b) $G(s) = \dfrac{1}{s^2 + 2s + 1}$;

(c) $G(s) = \dfrac{6}{s^3 + 6s^2 + 11s + 6}$.

Determine the approximate phase and gain margins from the Nichols charts and label the charts accordingly.

CP9.4   A negative feedback control system has the loop transfer function

$$G_c(s)G(s) = \frac{Ke^{-Ts}}{s + 1}.$$

(a) When $T = 0.1s$, find K such that the phase margin is 45° using the margin function. (b) Obtain a plot of phase margin versus T for K as in part (a), with $0 \le T \le 0.2s$.

CP9.5   Consider the paper machine control in Figure AP9.4. Develop an m-file to plot the bandwidth of the closed-loop system as K varies in the interval $1 \le K \le 50$.

CP9.6   A block diagram of the yaw acceleration control system for a bank-to-turn missile is shown in Figure CP9.6. The input is yaw acceleration command (in g's), and the output is missile yaw acceleration (in g's). The controller is specified to be a proportional, integral (PI) controller. The nominal value of b_0 is 0.5.

(a) Using the margin function, compute the phase margin, gain margin, and system crossover frequency (0 dB), assuming the nominal value of b_0.

(b) Using the gain margin from part (a), determine the maximum value of b_0 for a stable system. Verify your answer with a Routh–Hurwitz analysis of the characteristic equation.

CP9.7   An engineering laboratory has presented a plan to operate an Earth-orbiting satellite that is to be controlled from a ground station. A block diagram of the proposed system is shown in Figure CP9.7. It takes T seconds for a signal to reach the spacecraft from the ground station and the identical delay for a return signal. The proposed ground-based controller is a proportional-derivative (PD) controller, where

$$G_c(s) = K_P + K_D s.$$

(a) Assume no transmission time delay (i.e., $T = 0$), and design the controller to the following specifications: (1) percent overshoot less than 20% to a unit step input and (2) time to peak less than 30 seconds.

(b) Compute the phase margin with the controller in the loop but assuming a zero transmission time delay. Estimate the amount of allowable time delay for a stable system from the phase margin calculation.

(c) Using a second-order Padé approximation to the time delay, determine the maximum allowable delay T_{max} for system stability by developing a m-file script that employs the pade function and computes the closed-loop system poles as a function of the time delay T. Compare your answer with the one obtained in part (b).

CP9.8   Consider the system represented in state variable form

$$\dot{\mathbf{x}} = \begin{bmatrix} 0 & 1 \\ -1 & -10 \end{bmatrix}\mathbf{x} + \begin{bmatrix} 0 \\ 22 \end{bmatrix}u$$

$$y = \begin{bmatrix} 10 & 0 \end{bmatrix}\mathbf{x} + \begin{bmatrix} 0 \end{bmatrix}u$$

Using the nyquist function, obtain the polar plot.

FIGURE CP9.6
A feedback control system for the yaw acceleration control of a bank-to-turn missile.

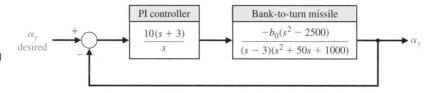

CP9.9 For the system in CP9.8, use the nichols function to obtain the Nichols chart and determine the phase margin and gain margin.

CP9.10 A closed-loop feedback system is shown in Figure CP9.10. (a) Obtain the Nyquist plot and determine the

phase margin. Assume that the time delay $T = 0$ s. (b) Compute the phase margin when $T = 0.05$ s. (c) Determine the minimum time delay that destabilizes the closed-loop system.

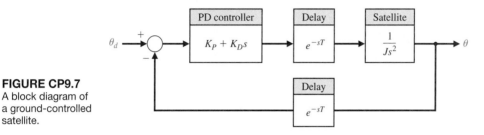

FIGURE CP9.7
A block diagram of a ground-controlled satellite.

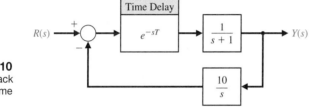

FIGURE CP9.10
Nonunity feedback system with a time delay.

TERMS AND CONCEPTS

Bandwidth The frequency at which the frequency response has declined 3 dB from its low-frequency value.

Cauchy's theorem If a contour encircles Z zeros and P poles of $F(s)$ traversing clockwise, the corresponding contour in the $F(s)$-plane encircles the origin of the $F(s)$-plane $N = Z - P$ times clockwise.

Closed-loop frequency response The frequency response of the closed-loop transfer function $T(j\omega)$.

Conformal mapping A contour mapping that retains the angles on the s-plane on the $F(s)$-plane.

Contour map A contour or trajectory in one plane is mapped into another plane by a relation $F(s)$.

Gain margin The increase in the system gain when phase $= -180°$ that will result in a marginally stable

system with intersection of the $-1 + j0$ point on the Nyquist diagram.

Logarithmic (decibel) measure A measure of the gain margin defined as $20 \log_{10}(1/d)$, where $\dfrac{1}{d} = \dfrac{1}{|L(j\omega)|}$ when the phase shift is $-180°$.

Nichols chart A chart displaying the curves for the relationship between the open-loop and closed-loop frequency response.

Nyquist stability criterion A feedback system is stable if, and only if, the contour in the $L(s)$-plane does not encircle the $(-1, 0)$ point when the number of poles of $L(s)$ in the right-hand s-plane is zero. If $L(s)$ has P poles in the right-hand plane, then the number of counterclockwise encirclements of the $(-1, 0)$ point must be equal to P for a stable system.

Phase margin   The amount of phase shift of the $L(j\omega)$ at unity magnitude that will result in a marginally stable system with intersections of the $-1 + j0$ point on the Nyquist diagram.

Principle of the argument   See Cauchy's theorem.

Time delay   A time delay T, so that events occurring at time t at one point in the system occur at another point in the system at a later time $t + T$.

CHAPTER 10

The Design of Feedback Control Systems

PREVIEW

In this chapter, we address the central issue of the design of compensators. Using the methods of the previous chapters, we develop several design techniques in the frequency domain that enable us to achieve the desired system performance. The powerful lead and lag controllers are introduced and used in several design examples. Phase-lead and phase-lag control design approaches using both root locus plots and Bode diagrams are presented. The proportional-integral (PI) controller is revisited in the context of achieving high steady-state tracking accuracies. The chapter concludes with a proportional-derivative (PD) controller design with prefiltering for the Sequential Design Example: Disk Drive Read System.

DESIRED OUTCOMES

Upon completion of Chapter 10, students should:

❑ Be familiar with the design of lead and lag compensators using root locus and Bode plot methods.

❑ Understand the value of prefilters and how to design for deadbeat response.

❑ Have a greater appreciation for the varied approaches available for control system design.

10.1 INTRODUCTION

The performance of a feedback control system is of primary importance. This subject was discussed at length in Chapter 5 and quantitative measures of performance were developed. We have found that a suitable control system is stable and that it results in an acceptable response to input commands, is less sensitive to system parameter changes, results in a minimum steady-state error for input commands, and, finally, is able to reduce the effect of undesirable disturbances. A feedback control system that provides an optimum performance without any necessary adjustments is rare indeed. Usually, we find it necessary to compromise among the many conflicting and demanding specifications and to adjust the system parameters to provide a suitable and acceptable performance when it is not possible to obtain all the desired optimum specifications.

At several points in the preceding chapters, we have considered the question of design and adjustment of the system parameters in order to provide a desirable response and performance. In Chapter 5, we defined and established several suitable measures of performance. In Chapter 6, we determined a method of investigating the stability of a control system, recognizing that a system is unacceptable unless it is stable. In Chapter 7, we used the root locus method to design a self-balancing scale and illustrated a method of parameter design by using the root locus method. Furthermore, in Chapters 8 and 9, we developed suitable measures of performance in terms of the frequency variable ω and used them to design several suitable control systems. Thus, we have been considering the problems of the design of feedback control systems as an integral part of the subjects of the preceding chapters. It is now our purpose to study the question further and to point out several significant design and compensation methods.

The preceding chapters have shown that it is often possible to adjust the system parameters in order to provide the desired system response. However, we often find that it is not sufficient to adjust a system parameter and thus obtain the desired performance. Rather, we are required to consider the structure of the system and redesign the system in order to obtain a suitable one. That is, we must examine the scheme or plan of the system and obtain a new design or plan that results in a suitable system. Thus, **the design of a control system is concerned with the arrangement, or the plan, of the system structure and the selection of suitable components and parameters**. For example, if we desire a set of performance measures to be less than some specified values, we often encounter a conflicting set of requirements. Hence, if we wish a system to have a percent overshoot less than 20% and $\omega_n T_p = 3.3$, we obtain a conflicting requirement on the system damping ratio ζ, as can be seen by examining Figure 5.8 again. If we are unable to relax these two performance requirements, we must alter the system in some way. The alteration or adjustment of a control system in order to provide a suitable performance is called **compensation**; that is, compensation is the adjustment of a system in order to make up for deficiencies or inadequacies.

In redesigning a control system to alter the system response, an additional component is inserted within the structure of the feedback system. It is this additional component or device that equalizes or compensates for the performance deficiency. The compensating device may be electric, mechanical, hydraulic, pneumatic, or some other type of device or network and is often called a **compensator**. Commonly, an electric circuit serves as a compensator in many control systems.

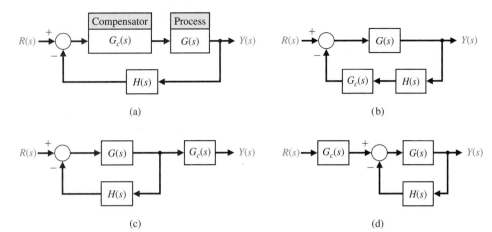

FIGURE 10.1
Types of
compensation.
(a) Cascade
compensation.
(b) Feedback
compensation.
(c) Output, or load,
compensation.
(d) Input
compensation.

> **A compensator is an additional component or circuit that is inserted into a control system to compensate for a deficient performance.**

The transfer function of a compensator is designated as $G_c(s) = E_o(s)/E_{in}(s)$, and the compensator can be placed in a suitable location within the structure of the system. Several types of compensation are shown in Figure 10.1 for a simple, single-loop feedback control system. The compensator placed in the feedforward path is called a **cascade**, or series, compensator (Figure 10.1a). Similarly, the other compensation schemes are called feedback, output (or load), and input compensation, as shown in Figures 10.1(b), (c), and (d), respectively. The selection of the compensation scheme depends upon a consideration of the specifications, the power levels at various signal nodes in the system, and the networks available for use. Usually, the output $Y(s)$ is a direct output of the process $G(s)$ and the output compensation of Figure 10.1(c) is not physically realizable. We cannot consider all the possibilities in this chapter; Chapters 11 and 12 will provide further information.

10.2   APPROACHES TO SYSTEM DESIGN

The performance of a control system can be described in terms of the time-domain performance measures or the frequency-domain performance measures. The performance of a system can be specified by requiring a certain peak time T_p, maximum overshoot, and settling-time for a step input. Furthermore, it is usually necessary to specify the maximum allowable steady-state error for several test signal inputs and disturbance inputs. These performance specifications can be defined in terms of the desirable location of the poles and zeros of the closed-loop system transfer function $T(s)$. Thus, the location of the s-plane poles and zeros of $T(s)$ can be specified. As we found in Chapter 7, the locus of the roots of the closed-loop system can be readily obtained for the variation of one system parameter. However, when the locus of roots does not result in a suitable root configuration, we must add a compensating network (Figure 10.1) to alter the locus of the roots as the parameter is varied.

Therefore, we can use the root locus method and determine a suitable compensator network transfer function so that the resultant root locus yields the desired closed-loop root configuration.

Alternatively, we can describe the performance of a feedback control system in terms of frequency performance measures. Then a system can be described in terms of the peak of the closed-loop frequency response $M_{p\omega}$, the resonant frequency ω_r, the bandwidth, and the phase margin of the system. We can add a suitable compensation network, if necessary, in order to satisfy the system specifications. The design of the network, represented by $G_c(s)$, is developed in terms of the frequency response as portrayed on the polar plane, the Bode diagram, or the Nichols chart. Because a cascade transfer function is readily accounted for on a Bode plot by adding the frequency response of the network, we usually prefer to approach the frequency response methods by utilizing the Bode diagram.

Thus, the design of a system is concerned with the alteration of the frequency response or the root locus of the system in order to obtain a suitable system performance. For frequency response methods, we are concerned with altering the system so that the frequency response of the compensated system will satisfy the system specifications. Hence, in the frequency response approach, we use compensation networks to alter and reshape the system characteristics represented on the Bode diagram and Nichols chart.

Alternatively, the design of a control system can be accomplished in the s-plane by root locus methods. For the case of the s-plane, the designer wishes to alter and reshape the root locus so that the roots of the system will lie in the desired position in the s-plane.

We have illustrated several of these approaches in the preceding chapters. In Chapter 7, we used the root locus method in considering the design of a feedback network in order to obtain a satisfactory performance. In Chapters 8 and 9, we considered the selection of the gain in order to obtain a suitable phase margin and therefore a satisfactory relative stability.

Quite often, in practice, the best and easiest way to improve the performance of a control system is to alter, if possible, the process itself. That is, if the system designer is able to specify and alter the design of the process that is represented by the transfer function $G(s)$, then the performance of the system may be readily improved. For example, to improve the transient behavior of a servomechanism position controller, we often can choose a better motor for the system. In the case of an airplane control system, we might be able to alter the aerodynamic design of the airplane and thus improve the flight transient characteristics. Thus, a control system designer should recognize that an alteration of the process may result in an improved system. However, the process is often unalterable or has been altered as much as possible and still results in unsatisfactory performance. Then the addition of compensation networks becomes useful for improving the performance of the system.

In the following sections, we will assume that the process has been improved as much as possible and that the $G(s)$ representing the process is unalterable. First, we shall consider the addition of a so-called phase-lead compensation network and describe the design of the network by root locus and frequency response techniques. Then, using both the root locus and frequency response techniques, we will describe the design of the integration compensation networks in order to obtain a suitable system performance.

10.3 CASCADE COMPENSATION NETWORKS

In this section, we will consider the design of a cascade or feedback network, as shown in Figures 10.1(a) and (b), respectively. The compensation network function $G_c(s)$ is cascaded with the specified process $G(s)$ in order to provide a suitable loop transfer function $L(s) = G_c(s)G(s)H(s)$. The compensator $G_c(s)$ can be chosen to alter either the shape of the root locus or the frequency response. In either case, the network may be chosen to have a transfer function

$$G_c(s) = \frac{K \displaystyle\prod_{i=1}^{M}(s + z_i)}{\displaystyle\prod_{j=1}^{n}(s + p_j)}. \tag{10.1}$$

Then the problem reduces to the judicious selection of the poles and zeros of the compensator. To illustrate the properties of the compensation network, we will consider a first-order compensator. The compensation approach developed on the basis of a first-order compensator can then be extended to higher-order compensators, for example, by cascading several first-order compensators.

A compensator $G_c(s)$ is used with a process $G(s)$ so that the overall loop gain can be set to satisfy the steady-state error requirement, and then $G_c(s)$ is used to adjust the system dynamics favorably without affecting the steady-state error.

Consider the first-order compensator with the transfer function

$$G_c(s) = \frac{K(s + z)}{s + p}. \tag{10.2}$$

The design problem then becomes the selection of z, p, and K in order to provide a suitable performance. When $|z| < |p|$, the network is called a **phase-lead network** and has a pole–zero s-plane configuration, as shown in Figure 10.2. If the pole was negligible, that is, $|p| \gg |z|$, and the zero occurred at the origin of the s-plane, we would have a differentiator so that

$$G_c(s) \approx \frac{K}{p}s. \tag{10.3}$$

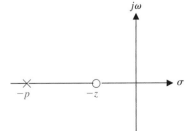

FIGURE 10.2
Pole–zero diagram
of the phase-lead
network.

Thus, a compensation network of the form of Equation (10.2) is a differentiator-type network. The differentiator network of Equation (10.3) has the frequency characteristic

$$G_c(j\omega) = j\frac{K}{p}\omega = \left(\frac{K}{p}\omega\right)e^{+j90°} \tag{10.4}$$

and a phase angle of $+90°$. Similarly, the frequency response of the differentiating network of Equation (10.2) is

$$G_c(j\omega) = \frac{K(j\omega + z)}{j\omega + p} = \frac{(Kz/p)[j(\omega/z) + 1]}{j(\omega/p) + 1} = \frac{K_1(1 + j\omega\alpha\tau)}{1 + j\omega\tau}, \tag{10.5}$$

where $\tau = 1/p$, $p = \alpha z$, and $K_1 = K/\alpha$. The frequency response of this phase-lead network is shown in Figure 10.3. The angle of the frequency characteristic is

$$\phi(\omega) = \tan^{-1}(\alpha\omega\tau) - \tan^{-1}(\omega\tau). \tag{10.6}$$

Because the zero occurs first on the frequency axis, we obtain a phase-lead characteristic, as shown in Figure 10.3. The slope of the asymptotic magnitude curve is $+20$ dB/decade.

The phase-lead compensation transfer function can be obtained with the network shown in Figure 10.4. The transfer function of this network is

$$G_c(s) = \frac{V_2(s)}{V_1(s)} = \frac{R_2}{R_2 + \dfrac{R_1/(Cs)}{R_1 + 1/(Cs)}}$$

$$= \frac{R_2}{R_1 + R_2}\frac{R_1Cs + 1}{[R_1R_2/(R_1 + R_2)]Cs + 1}. \tag{10.7}$$

Therefore, we let

$$\tau = \frac{R_1R_2}{R_1 + R_2}C \quad \text{and} \quad \alpha = \frac{R_1 + R_2}{R_2},$$

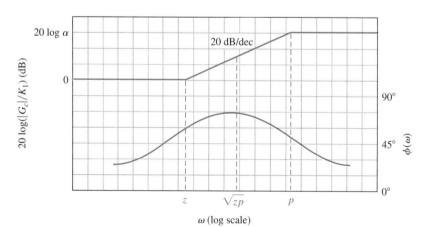

FIGURE 10.3
Bode diagram of the phase-lead network.

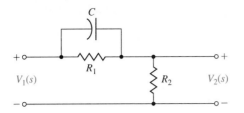

FIGURE 10.4
Phase-lead
network.

and we obtain the **phase-lead compensation** transfer function

$$G_c(s) = \frac{1 + \alpha\tau s}{\alpha(1 + \tau s)},$$ (10.8)

which is equal to Equation (10.5) when an additional cascade gain K is inserted.

The maximum value of the phase lead occurs at a frequency ω_m, where ω_m is the geometric mean of $p = 1/\tau$ and $z = 1/(\alpha\tau)$; that is, the maximum phase lead occurs halfway between the pole and zero frequencies on the logarithmic frequency scale. Therefore,

$$\omega_m = \sqrt{zp} = \frac{1}{\tau\sqrt{\alpha}}.$$

To obtain an equation for the maximum phase-lead angle, we rewrite the phase angle of Equation (10.5) as

$$\phi = \tan^{-1}\frac{\alpha\omega\tau - \omega\tau}{1 + (\omega\tau)^2\alpha}.$$ (10.9)

Then, substituting the frequency for the maximum phase angle, $\omega_m = 1/(\tau\sqrt{a})$, we have

$$\tan\phi_m = \frac{\alpha/\sqrt{\alpha} - 1/\sqrt{\alpha}}{1 + 1} = \frac{\alpha - 1}{2\sqrt{\alpha}}.$$ (10.10)

We use the trigonometric relationship $\sin\phi = \tan\phi/\sqrt{1 + \tan^2\phi}$ and obtain

$$\sin\phi_m = \frac{\alpha - 1}{\alpha + 1}.$$ (10.11)

Equation (10.11) is very useful for calculating a necessary α ratio between the pole and zero of a compensator in order to provide a required maximum phase lead. A plot of ϕ_m versus α is shown in Figure 10.5. The phase angle readily obtainable from this network is not much greater than 70°. Also, since $\alpha = (R_1 + R_2)/R_2$, there are practical limitations on the maximum value of α that we should attempt to obtain. Therefore, if we required a maximum angle greater than 70°, two cascade compensation networks would be used. Then the equivalent compensation transfer function would be $G_{c_1}(s)G_{c_2}(s)$ when the loading effect of $G_{c_2}(s)$ on $G_{c_1}(s)$ is negligible.

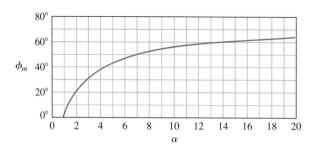

FIGURE 10.5
Maximum phase
angle ϕ_m versus α
for a phase-lead
network.

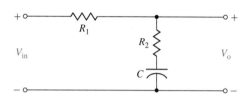

FIGURE 10.6
Phase-lag network.

It is often useful to add a cascade compensation network that provides a phase-lag characteristic. The **phase-lag network** is shown in Figure 10.6. The transfer function of the phase-lag network is

$$G_c(s) = \frac{V_o(s)}{V_{in}(s)} = \frac{R_2 + 1/(Cs)}{R_1 + R_2 + 1/(Cs)} = \frac{R_2Cs + 1}{(R_1 + R_2)Cs + 1}. \qquad (10.12)$$

When $\tau = R_2C$ and $\alpha = (R_1 + R_2)/R_2$, we have the **phase-lag compensation** transfer function

$$G_c(s) = \frac{1 + \tau s}{1 + \alpha \tau s} = \frac{1}{\alpha} \frac{s + z}{s + p}, \qquad (10.13)$$

where $z = 1/\tau$ and $p = 1/(\alpha\tau)$. In this case, because $\alpha > 1$, the pole lies closest to the origin of the s-plane, as shown in Figure 10.7. This type of compensation network is often called an integrating network because it has a frequency response like an integrator over a finite range of frequencies. The Bode diagram of the phase-lag network is obtained from the transfer function

$$G_c(j\omega) = \frac{1 + j\omega\tau}{1 + j\omega\alpha\tau} \qquad (10.14)$$

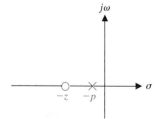

FIGURE 10.7
Pole–zero diagram
of the phase-lag
network.

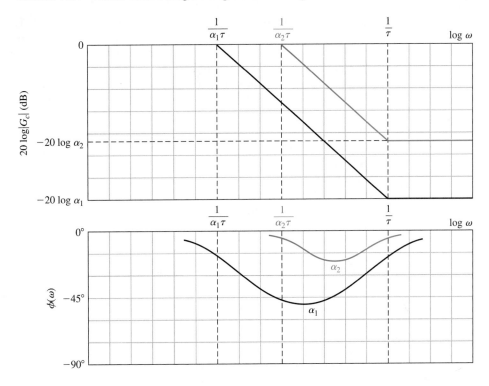

FIGURE 10.8
Bode diagram of
the phase-lag
network.

and is shown in Figure 10.8. The form of the Bode diagram of the lag network is similar to that of the phase-lead network; the difference is the resulting attenuation and phase-lag angle instead of amplification and phase-lead angle. However, note that the shapes of the diagrams of Figures 10.3 and 10.8 are similar. Therefore, we can show that the maximum phase lag occurs at $\omega_m = \sqrt{zp}$.

In the succeeding sections, we wish to utilize these compensation networks to obtain a desired system frequency response or s-plane root location. The lead network can provide a phase-lead angle and thus a satisfactory phase margin for a system. Alternatively, the phase-lead network can enable us to reshape the root locus and thus provide the desired root locations. The phase-lag network is used, not to provide a phase-lag angle, which is normally a destabilizing influence, but rather to provide an attenuation and to increase the steady-state error constant [3]. The following six sections discuss these approaches to design utilizing the phase-lead and phase-lag networks.

10.4 PHASE-LEAD DESIGN USING THE BODE DIAGRAM

The Bode diagram is used to design a suitable phase-lead network in preference to other frequency response plots. The frequency response of the cascade compensation network is added to the frequency response of the uncompensated system. That is, because the total loop transfer function of Figure 10.1(a) is $L(j\omega) = G_c(j\omega)G(j\omega)H(j\omega)$, we will first plot the Bode diagram for $G(j\omega)H(j\omega)$. Then we can examine the plot for

$G(j\omega)H(j\omega)$ and determine a suitable location for p and z of $G_c(j\omega)$ in order to satisfactorily reshape the frequency response. The uncompensated $G(j\omega)H(j\omega)$ is plotted with the desired gain to allow an acceptable steady-state error. Then the phase margin and the expected $M_{p\omega}$ are examined to find whether they satisfy the specifications. If the phase margin is not sufficient, phase lead can be added to the phase-angle curve of the system by placing the $G_c(j\omega)$ in a suitable location. To obtain maximum additional phase lead, we adjust the network so that the frequency ω_m is located at the frequency where the magnitude of the compensated magnitude curve crosses the 0-dB axis. (Recall the definition of phase margin.) The value of the added phase lead required allows us to determine the necessary value for α from Equation (10.11) or Figure 10.5. The zero $z = 1/(\alpha\tau)$ is located by noting that the maximum phase lead should occur at $\omega_m = \sqrt{zp}$, halfway between the pole and the zero. Because the total magnitude gain for the network is 20 log α, we expect a gain of 10 log α at ω_m. Thus, we determine the compensation network by completing the following steps:

1. Evaluate the uncompensated system phase margin when the error constants are satisfied.
2. Allowing for a small amount of safety, determine the necessary additional phase lead ϕ_m.
3. Evaluate α from Equation (10.11).
4. Evaluate 10 log α and determine the frequency where the uncompensated magnitude curve is equal to -10 log α dB. Because the compensation network provides a gain of 10 log α at ω_m, this frequency is the new 0-dB crossover frequency and ω_m simultaneously.
5. Calculate the pole $p = \omega_m\sqrt{a}$ and $z = p/\alpha$.
6. Draw the compensated frequency response, check the resulting phase margin, and repeat the steps if necessary. Finally, for an acceptable design, raise the gain of the amplifier in order to account for the attenuation $(1/\alpha)$.

EXAMPLE 10.1   **A lead compensator for a type-two system**

Let us consider a single-loop feedback control system as shown in Figure 10.1(a), where

$$G(s) = \frac{K_1}{s^2} \qquad (10.15)$$

and $H(s) = 1$. The uncompensated system is a type-two system and at first appears to possess a satisfactory steady-state error for both step and ramp input signals. However, the response of the uncompensated system is an undamped oscillation because

$$T(s) = \frac{Y(s)}{R(s)} = \frac{K_1}{s^2 + K_1}. \qquad (10.16)$$

Therefore, the compensation network is added so that the loop transfer function is $L(s) = G_c(s)G(s)$. The specifications for the system are

$$\text{Settling time, } T_s \leq 4 \text{ s;}$$

$$\text{System damping constant } \zeta \geq 0.45.$$

The settling time (with a 2% criterion) requirement is

$$T_s = \frac{4}{\zeta\omega_n} = 4;$$

therefore,

$$\omega_n = \frac{1}{\zeta} = \frac{1}{0.45} = 2.22.$$

Perhaps the easiest way to check the value of ω_n for the frequency response is to relate ω_n to the bandwidth ω_B, and evaluate the -3-dB bandwidth of the closed-loop system. For a closed-loop system with $\zeta = 0.45$, we estimate from Figure 8.26 that $\omega_B = 1.33\omega_n$. Therefore, we require a closed-loop bandwidth $\omega_B = 1.33(2.22) = 3.00$. The bandwidth can be checked following compensation by utilizing the Nichols chart. For the uncompensated system, the bandwidth of the system is $\omega_B = 1.33\omega_n$ and $\omega_n = \sqrt{K}$. Therefore, a loop gain equal to $K = \omega_n^2 \approx 5$ would be sufficient. To provide a suitable margin for the settling time, we will select $K = 10$ in order to draw the Bode diagram of

$$G(j\omega) = \frac{K}{(j\omega)^2}.$$

The Bode diagram of the uncompensated system is shown as solid lines in Figure 10.9.

By using Equation (9.58), the phase margin of the system is required to be approximately

$$\phi_{\text{pm}} = \frac{\zeta}{0.01} = \frac{0.45}{0.01} = 45°. \tag{10.17}$$

The phase margin of the uncompensated system is $0°$ because the double integration results in a constant $180°$ phase lag. Therefore, we must add a $45°$ phase-lead angle at the crossover (0-dB) frequency of the compensated magnitude curve. Evaluating the value of α, we have

$$\frac{\alpha - 1}{\alpha + 1} = \sin \phi_m = \sin 45°, \tag{10.18}$$

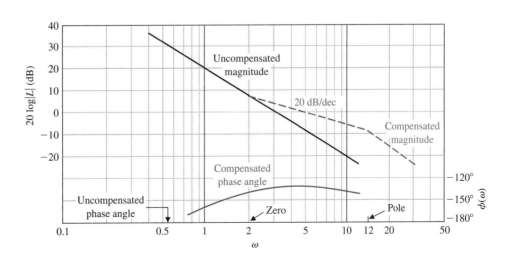

FIGURE 10.9
Bode diagram for
Example 10.1.

and thus $\alpha = 5.8$. To provide a margin of safety, we will use $\alpha = 6$. The value of $10 \log \alpha$ is then equal to 7.78 dB. Then the lead network will add an additional gain of 7.78 dB at the frequency ω_m, and we want to have ω_m equal to the compensated slope near the 0-dB axis (the dashed line) so that the new crossover is ω_m and the dashed magnitude curve is 7.78 dB above the uncompensated curve at the crossover frequency. Thus, the compensated crossover frequency is located by evaluating the frequency where the uncompensated magnitude curve is equal to -7.78 dB, which in this case is $\omega = 4.95$. Then the maximum phase-lead angle is added to $\omega = \omega_m = 4.95$, as shown in Figure 10.9. Using step 5, we determine the pole $p = \omega_m \sqrt{\alpha} = 12.0$ and the zero $z = p/\alpha = 2.0$.

The bandwidth of the compensated system can be obtained from the Nichols chart. For estimating the bandwidth, we can simply examine Figure 9.26 and note that the -3-dB line for the closed-loop system occurs when the magnitude of $G(j\omega)$ is -6 dB and the phase shift of $G(j\omega)$ is approximately $-140°$. Therefore, to estimate the bandwidth from the open-loop diagram, we will approximate the bandwidth as the frequency for which $20 \times \log|G|$ is equal to -6 dB. Thus, the bandwidth of the uncompensated system is approximately equal to $\omega_B = 4.4$, while the bandwidth of the compensated system is equal to $\omega_B = 8.4$. The lead compensation doubles the bandwidth in this case, and satisfies the specification that $\omega_B > 3.00$. Therefore, the compensation of the system is completed, and the system specifications are satisfied. The total compensated loop transfer function is

$$L(j\omega) = G_c(j\omega)G(j\omega) = \frac{10[j\omega/2.0 + 1]}{(j\omega)^2[j\omega/12.0 + 1]}. \tag{10.19}$$

The transfer function of the compensator is

$$G_c(s) = \frac{1 + \alpha\tau s}{\alpha(1 + \tau s)} = \frac{1}{6}\frac{1 + s/2.0}{1 + s/12.0}, \tag{10.20}$$

in the form of Equation (10.8). Because an attenuation of $\frac{1}{6}$ results from the passive RC network, the gain of the amplifier in the loop must be raised by a factor of 6 so that the total DC loop gain is still equal to 10, as required in Equation (10.19). When we add the compensation network Bode diagram to the uncompensated Bode diagram, as in Figure 10.9, we assume that we can raise the amplifier gain to account for this $1/\alpha$ attenuation. The pole and zero values can be read from Figure 10.9, noting that $p = \alpha z$.

The total loop transfer function is (recall that $H(s) = 1$)

$$L(s) = \frac{10(1 + s/2)}{s^2(1 + s/12)} = \frac{60(s + 2)}{s^2(s + 12)}.$$

The closed-loop transfer function is

$$T(s) = \frac{60(s + 2)}{s^3 + 12s^2 + 60s + 120} \approx \frac{60(s + 2)}{(s^2 + 6s + 20)(s + 6)},$$

and the effects of the zero at $s = -2$ and the third pole at $s = -6$ will affect the transient response. Plotting the step response, we find an overshoot of 34% and a settling time of 1.4 seconds. ∎

EXAMPLE 10.2   **A lead compensator for a second-order system**

A unity feedback control system has a loop transfer function

$$L(s) = \frac{K}{s(s + 2)},$$ (10.21)

where $L(s) = G_c(s)G(s)$ and $H(s) = 1$. We want to have a steady-state error for a ramp input equal to 5% of the velocity of the ramp. Therefore, we require that

$$K_v = \frac{A}{e_{ss}} = \frac{A}{0.05A} = 20.$$ (10.22)

Furthermore, we desire that the phase margin of the system be at least 45°. The first step is to plot the Bode diagram of the uncompensated transfer function

$$G(j\omega) = \frac{K_v}{j\omega(0.5j\omega + 1)} = \frac{20}{j\omega(0.5j\omega + 1)},$$ (10.23)

as shown in Figure 10.10(a). The frequency at which the magnitude curve crosses the 0-dB line is 6.2 rad/s, and the phase margin at this frequency is determined readily from the equation of the phase of $G(j\omega)$, which is

$$\angle G(j\omega) = \phi(\omega) = -90° - \tan^{-1}(0.5\omega).$$ (10.24)

At the crossover frequency $\omega = \omega_c = 6.2$ rad/s, we have

$$\phi(\omega) = -162°,$$ (10.25)

and therefore the phase margin is 18°. Using Equation (10.24) to evaluate the phase margin is often easier than drawing the complete phase-angle curve, which is shown in Figure 10.10(a). Thus, we need to add a phase-lead network so that the phase margin is raised to 45° at the new crossover (0-dB) frequency. Because the compensation crossover frequency is greater than the uncompensated crossover frequency, the phase lag of the uncompensated system is also greater. We shall account for this additional phase lag by attempting to obtain a maximum phase lead of $45° - 18° = 27°$, plus a small increment (10%) of phase lead to account for the added lag. Thus, we will design a compensation network with a maximum phase lead equal to $27° + 3° = 30°$. Then, calculating α, we obtain

$$\frac{\alpha - 1}{\alpha + 1} = \sin 30° = 0.5,$$ (10.26)

and therefore $\alpha = 3$.

The maximum phase lead occurs at ω_m, and this frequency will be selected so that the new crossover frequency and ω_m coincide. The magnitude of the lead network at ω_m is $10 \log \alpha = 10 \log 3 = 4.8$ dB. The compensated crossover frequency is then evaluated where the magnitude of $G(j\omega)$ is -4.8 dB, and thus $\omega_m = \omega_c = 8.4$. Drawing the compensated magnitude line so that it intersects the

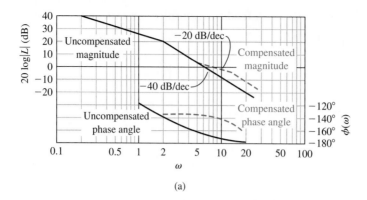

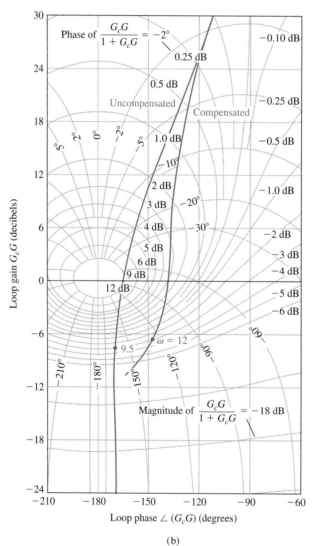

FIGURE 10.10
(a) Bode diagram for Example 10.2.
(b) Nichols diagram for Example 10.2.

0-dB axis at $\omega = \omega_c = 8.4$, we find that $z = \omega_m/\sqrt{\alpha} = 4.8$ and $p = \alpha z = 14.4$. Therefore, the compensation network is

$$G_c(s) = \frac{1}{3}\frac{1 + s/4.8}{1 + s/14.4}. \tag{10.27}$$

The total DC loop gain must be raised by a factor of three in order to account for the factor $1/\alpha = \frac{1}{3}$. Then the compensated loop transfer function is

$$L(s) = G_c(s)G(s) = \frac{20(s/4.8 + 1)}{s(0.5s + 1)(s/14.4 + 1)}. \tag{10.28}$$

To verify the final phase margin, we can evaluate the phase of $G_c(j\omega)G(j\omega)$ at $\omega = \omega_c = 8.4$ and thus obtain the phase margin. The phase angle is then

$$\phi(\omega_c) = -90° - \tan^{-1} 0.5\omega_c - \tan^{-1}\frac{\omega_c}{14.4} + \tan^{-1}\frac{\omega_c}{4.8}$$

$$= -90° - 76.5° - 30.0° + 60.2°$$

$$= -136.3°. \tag{10.29}$$

Therefore, the phase margin for the compensated system is 43.7°. If we desire to have exactly a 45° phase margin, we would repeat the steps with an increased value of α—for example, with $\alpha = 3.5$. In this case, the phase lag increased by 7° between $\omega = 6.2$ and $\omega = 8.4$, and therefore the allowance of 3° in the calculation of α was not sufficient. The step response of this system yields a 28% overshoot with a settling time of 0.75 second.

The Nichols diagram for the compensated and uncompensated system is shown in Figure 10.10(b). The reshaping of the frequency response locus is clear on this diagram. Note the increased phase margin for the compensated system as well as the reduced magnitude of $M_{p\omega}$, the maximum magnitude of the closed-loop frequency response. In this case, $M_{p\omega}$ has been reduced from an uncompensated value of +12 dB to a compensated value of approximately +3.2 dB. Also, we note that the closed-loop 3-dB bandwidth of the compensated system is equal to 12 rad/s compared with 9.5 rad/s for the uncompensated system. ∎

Looking again at Examples 10.1 and 10.2, we note that the system design is satisfactory when the asymptotic curve for the magnitude $20 \log|GG_c|$ crosses the 0-dB line with a slope of -20 dB/decade.

10.5 PHASE-LEAD DESIGN USING THE ROOT LOCUS

The design of the phase-lead compensation network can also be readily accomplished using the root locus. The phase-lead network has a transfer function

$$G_c(s) = \frac{s + 1/\alpha\tau}{s + 1/\tau} = \frac{s + z}{s + p}, \tag{10.30}$$

where α and τ are defined for the RC network in Equation (10.7). The locations of the zero and pole are selected so as to result in a satisfactory root locus for the compensated system. The specifications of the system are used to specify the desired location of the dominant roots of the system. The s-plane root locus method is as follows:

1. List the system specifications and translate them into a desired root location for the dominant roots.

2. Sketch the uncompensated root locus, and determine whether the desired root locations can be realized with an uncompensated system.

3. If a compensator is necessary, place the zero of the phase-lead network directly below the desired root location (or to the left of the first two real poles).

4. Determine the pole location so that the total angle at the desired root location is $180°$ and therefore is on the compensated root locus.

5. Evaluate the total system gain at the desired root location and then calculate the error constant.

6. Repeat the steps if the error constant is not satisfactory.

Therefore, we first locate our desired dominant root locations so that the dominant roots satisfy the specifications in terms of ζ and ω_n, as shown in Figure 10.11(a). The root locus of the uncompensated system is sketched as illustrated in Figure 10.11(b).

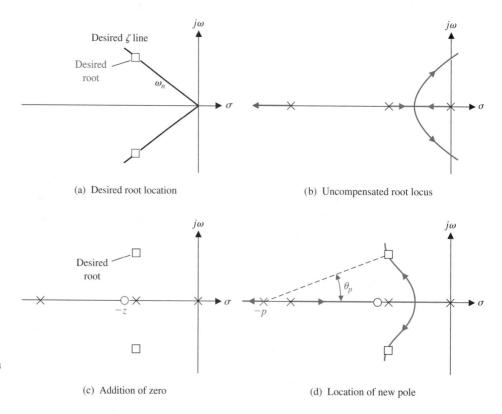

(a) Desired root location

(b) Uncompensated root locus

(c) Addition of zero

(d) Location of new pole

FIGURE 10.11
Compensation on the s-plane using a phase-lead network.

Then the zero is added to provide a phase lead by placing it to the left of the first two real poles. Some caution is necessary because the zero must not alter the dominance of the desired roots; that is, the zero should not be placed closer to the origin than the second pole on the real axis, or a real root near the origin will result and will dominate the system response. Thus, in Figure 10.11(c), we note that the desired root is directly above the second pole, and we place the zero z somewhat to the left of the second real pole.

Consequently, the real root may be near the real zero, and the coefficient of this term of the partial fraction expansion may be relatively small. Hence, the response due to this real root may have very little effect on the overall system response. Nevertheless, the designer must be continually aware that the compensated system response will be influenced by the roots and zeros of the system and that the dominant roots will not by themselves dictate the response. It is usually wise to allow for some margin of error in the design and to test the compensated system using a computer simulation.

Because the desired root is a point on the root locus when the final compensation is accomplished, we expect the algebraic sum of the vector angles to be 180° at that point. Thus, we calculate the angle θ_p from the pole of the compensator in order to result in a total angle of 180°. Then, locating a line at an angle θ_p intersecting the desired root, we are able to evaluate the compensator pole p, as shown in Figure 10.11(d).

The advantage of the root locus method is the ability of the designer to specify the location of the dominant roots and therefore the dominant transient response. The disadvantage of the method is that we cannot directly specify an error constant (for example, K_v) as in the Bode diagram approach. After the design is complete, we evaluate the gain of the system at the root location, which depends on p and z, and then calculate the error constant for the compensated system. If the error constant is not satisfactory, we must repeat the design steps and alter the location of the desired root as well as the location of the compensator pole and zero. We shall consider again Examples 10.1 and 10.2 and design a compensation network using the root locus (s-plane) approach.

EXAMPLE 10.3   Lead compensator using the root locus

Let us consider again the system of Example 10.1 where the uncompensated loop transfer function is

$$L(s) = \frac{K_1}{s^2}. \tag{10.31}$$

The characteristic equation of the uncompensated system is

$$1 + L(s) = 1 + \frac{K_1}{s^2} = 0, \tag{10.32}$$

and the root locus is the $j\omega$-axis. Therefore, we propose to compensate this system with a network

$$G_c(s) = \frac{s + z}{s + p}, \tag{10.33}$$

where $|z| < |p|$. The specifications for the system are

Settling time (with a 2% criterion), $T_s \leq 4$ s;
Percent overshoot for a step input $P.O. \leq 35\%$.

Therefore, the damping ratio should be $\zeta \geq 0.32$. The settling time requirement is

$$T_s = \frac{4}{\zeta \omega_n} = 4,$$

so $\zeta \omega_n = 1$. Thus, we will choose a desired dominant root location as

$$r_1, \hat{r}_1 = -1 \pm j2, \tag{10.34}$$

as shown in Figure 10.12 (hence, $\zeta = 0.45$).

Now we place the zero of the compensator directly below the desired location at $s = -z = -1$, as shown in Figure 10.12. Measuring the angle at the desired root, we have

$$\phi = -2(116°) + 90° = -142°.$$

Therefore, to have a total of 180° at the desired root, we evaluate the angle from the undetermined pole, θ_p, as

$$-180° = -142° - \theta_p, \tag{10.35}$$

or $\theta_p = 38°$. Then a line is drawn at an angle $\theta_p = 38°$ intersecting the desired root location and the real axis, as shown in Figure 10.12. The point of intersection with the real axis is then $s = -p = -3.6$. Therefore, the compensator is

$$G_c(s) = \frac{s + 1}{s + 3.6}, \tag{10.36}$$

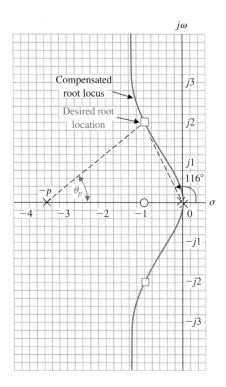

FIGURE 10.12
Phase-lead design for Example 10.3.

and the compensated loop transfer function for the system is

$$L(s) = G_c(s)G(s) = \frac{K_1(s + 1)}{s^2(s + 3.6)}. \tag{10.37}$$

The gain K_1 is evaluated by measuring the vector lengths from the poles and zeros to the root location. Hence,

$$K_1 = \frac{(2.23)^2(3.25)}{2} = 8.1. \tag{10.38}$$

Finally, the error constants of this system are evaluated. We find that this system with two open-loop integrations will result in a zero steady-state error for a step and ramp input signal. The acceleration constant is

$$K_a = \frac{8.1}{3.6} = 2.25. \tag{10.39}$$

The steady-state performance of this system is quite satisfactory, and therefore the compensation is complete. When we compare the compensation network evaluated by the s-plane method with the network obtained by using the Bode diagram approach, we find that the magnitudes of the poles and zeros are different. However, the resulting system will have the same performance, and we need not be concerned with the difference. In fact, the difference arises from the arbitrary design step (number 3), which places the zero directly below the desired root location. If we placed the zero at $s = -2.0$, we would find that the pole evaluated by the s-plane method is approximately equal to the pole evaluated by the Bode diagram approach.

The specifications for the transient response of this system were originally expressed in terms of the overshoot and the settling time of the system. These specifications were translated, on the basis of an approximation of the system by a second-order system, to an equivalent ζ and ω_n and therefore a desired root location. However, the original specifications will be satisfied only if the selected roots are dominant. The zero of the compensator and the root resulting from the addition of the compensator pole result in a third-order system with a zero. The validity of approximating this system with a second-order system without a zero is dependent upon the validity of the dominance assumption. Often, the designer will simulate the final design by using a digital computer and obtain the actual transient response of the system. In this case, a computer simulation of the system resulted in an overshoot of 46% and a settling time (to within 2% of the final value) of 3.8 seconds for a step input. These values compare moderately well with the specified values of 35% and 4 seconds, and they justify the use of the dominant root specifications. The difference in the overshoot from the specified value is due to the zero, which is not negligible. Thus, again we find that the specification of dominant roots is a useful approach but must be utilized with caution and understanding. A second attempt to obtain a compensated system with an overshoot of 30% would use a prefilter to eliminate the effect of the zero in the closed-loop transfer function, as described in Section 10.10. ∎

EXAMPLE 10.4   **Lead compensator for a type-one system**

Now, let us consider again the system of Example 10.2 and design a compensator based on the root locus approach. The system loop transfer function is

$$L(s) = \frac{K}{s(s + 2)}. \tag{10.40}$$

We want the damping ratio of the dominant roots of the system to be $\zeta = 0.45$ and the velocity error constant to be equal to 20. To satisfy the error constant requirement, the gain of the uncompensated system must be $K = 40$. When $K = 40$, the roots of the uncompensated system are

$$s^2 + 2s + 40 = (s + 1 + j6.25)(s + 1 - j6.25). \tag{10.41}$$

The damping ratio of the uncompensated roots is approximately 0.16, and therefore a compensation network must be added. To achieve a rapid settling time, we will select the real part of the desired roots as $\zeta\omega_n = 4$, and therefore $T_s = 1$ s. This implies the natural frequency of these roots is fairly large, $\omega_n = 9$; hence, the velocity constant should be reasonably large. The location of the desired roots is shown in Figure 10.13(a) for $\zeta\omega_n = 4$, $\zeta = 0.45$, and $\omega_n = 9$.

The zero of the compensator is placed at $s = -z = -4$, directly below the desired root location. Then the angle at the desired root location is

$$\phi = -116° - 104° + 90° = -130°. \tag{10.42}$$

Therefore, the angle from the undetermined pole is determined from

$$-180° = -130° - \theta_p,$$

and thus $\theta_p = 50°$. This angle is drawn to intersect the desired root location, and p is evaluated as $s = -p = -10.6$, as shown in Figure 10.13(a). The gain of the compensated system is then

$$K = \frac{9(8.25)(10.4)}{8} = 96.5. \tag{10.43}$$

The compensated system loop transfer function is then

$$L(s) = G_c(s)G(s) = \frac{96.5(s + 4)}{s(s + 2)(s + 10.6)}. \tag{10.44}$$

Therefore, the velocity constant of the compensated system is

$$K_v = \lim_{s \to 0} s[G_c(s)G(s)] = \frac{96.5(4)}{2(10.6)} = 18.2. \tag{10.45}$$

The velocity constant of the compensated system is less than the desired value of 20. Accordingly, we must repeat the design procedure for a second choice of a desired root. If we choose $\omega_n = 10$, the process can be repeated, and the resulting gain K will be increased. The compensator pole and zero location will also be altered.

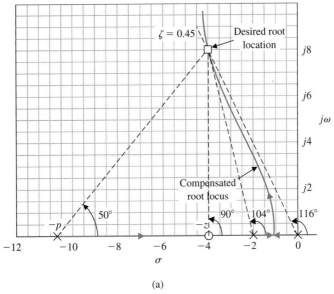

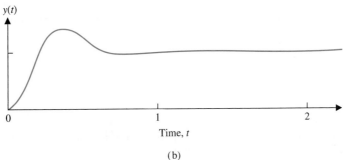

FIGURE 10.13
(a) Design of a phase-lead network on the s-plane for Example 10.4.
(b) Step response of the compensated system of Example 10.4.

Then the velocity constant can be again evaluated. We will leave it as an exercise to show that for $\omega_n = 10$, the velocity constant is $K_v = 22.7$ when $z = 4.5$ and $p = 11.6$.

Finally, for the compensation network of Equation (10.44), we have

$$G_c(s) = \frac{s + 4}{s + 10.6} = \frac{s + 1/(\alpha\tau)}{s + 1/\tau}. \tag{10.46}$$

The design of an RC-lead network to implement $G_c(s)$, as shown in Figure 10.4, follows directly from Equations (10.46) and (10.7):

$$G_c(s) = \frac{R_2}{R_1 + R_2} \frac{R_1 Cs + 1}{[R_1 R_2/(R_1 + R_2)] Cs + 1}. \tag{10.47}$$

Thus, in this case, we have

$$\frac{1}{R_1 C} = 4 \quad \text{and} \quad \alpha = \frac{R_1 + R_2}{R_2} = \frac{10.6}{4}.$$

Then, choosing $C = 1\ \mu\text{f}$, we obtain $R_1 = 250,000\ \Omega$ and $R_2 = 152,000\ \Omega$. The step response of the compensated system yields a 32% overshoot with a settling time of

0.8 second, as shown in Figure 10.13(b). As shown here, we may use a computer to verify the actual transient response. ■

The phase-lead compensation network is a useful compensator for altering the performance of a control system. The phase-lead network adds a phase-lead angle to provide an adequate phase margin for feedback systems. Using an *s*-plane design approach, we can choose the phase-lead network in order to alter the system root locus and place the roots of the system in a desired position in the *s*-plane. When the design specifications include an error constant requirement, the Bode diagram method is more suitable, because the error constant of a system designed on the *s*-plane must be ascertained following the choice of a compensator pole and zero. Therefore, the root locus method often results in an iterative design procedure when the error constant is specified. On the other hand, the root locus is a very satisfactory approach when the specifications are given in terms of overshoot and settling time, thus specifying the ζ and ω_n of the desired dominant roots in the *s*-plane. The use of a lead network compensator always extends the bandwidth of a feedback system, which may be objectionable for systems subjected to large amounts of noise. Also, lead networks are not suitable for providing high steady-state accuracy in systems requiring very high error constants. To provide large error constants, typically K_p and K_v, we must consider the use of integration-type compensation networks. This is the subject of the following section.

10.6 SYSTEM DESIGN USING INTEGRATION NETWORKS

For a large proportion of control systems, the primary objective is obtaining a high steady-state accuracy. Another goal is maintaining the transient performance of these systems within reasonable limits. As we found in Chapters 4 and 5, the steady-state accuracy of many feedback systems can be increased by increasing the amplifier gain in the forward channel. However, the resulting transient response may be totally unacceptable—even unstable. Therefore, it is often necessary to introduce a compensation network in the forward path of a feedback control system in order to provide a sufficient steady-state accuracy.

Consider the single-loop control system shown in Figure 10.14. The compensation network is chosen to provide a large error constant. With $G_p(s) = 1$, the steady-state error of this system is

$$\lim_{t \to \infty} e(t) = \lim_{s \to 0} s \frac{R(s)}{1 + G_c(s)G(s)H(s)}. \tag{10.48}$$

FIGURE 10.14
Single-loop
feedback control
system.

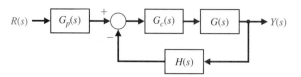

We found in Section 5.6 that the steady-state error of a system depends on the number of poles at the origin for $L(s) = G_c(s)G(s)H(s)$. A pole at the origin can be considered an integration, and therefore the steady-state accuracy of a system ultimately depends on the number of integrations in the loop transfer function $L(s) = G_c(s)G(s)H(s)$. If the steady-state accuracy is not sufficient, we will introduce an **integration-type network** $G_c(s)$ in order to compensate for the lack of integration in the uncompensated loop transfer function $G(s)H(s)$.

One widely used form of controller is the **proportional plus integral (PI) controller**, which has a transfer function

$$G_c(s) = K_p + \frac{K_I}{s}. \qquad (10.49)$$

For an example, let us consider a temperature control system where the transfer function $H(s) = 1$, and the transfer function of the heat process is [32]

$$G(s) = \frac{K_1}{(\tau_1 s + 1)(\tau_2 s + 1)}.$$

The steady-state error of the uncompensated system is then

$$\lim_{t \to \infty} e(t) = \lim_{s \to 0} s \frac{A/s}{1 + G(s)} = \frac{A}{1 + K_1}, \qquad (10.50)$$

where $R(s) = A/s$, a step input signal. To obtain a small steady-state error (less than 0.05 A, for example), the magnitude of the gain K_1 must be quite large. However, when K_1 is quite large, the transient performance of the system will very likely be unacceptable. Therefore, we must consider the addition of a compensation transfer function $G_c(s)$, as shown in Figure 10.14. To eliminate the steady-state error of this system, we might choose the compensation as

$$G_c(s) = K_P + \frac{K_I}{s} = \frac{K_P s + K_I}{s}. \qquad (10.51)$$

This PI compensation can be readily constructed by using an integrator and an amplifier and adding their output signals. The steady-state error for a step input of the system is always zero, because

$$\lim_{t \to \infty} e(t) = \lim_{s \to 0} s \frac{A/s}{1 + G_c(s)G(s)}$$

$$= \lim_{s \to 0} \frac{A}{1 + (K_P s + K_I)/s \, K_1/[(\tau_1 s + 1)(\tau_2 s + 1)]}$$

$$= 0. \qquad (10.52)$$

The transient performance can be adjusted to satisfy the system specifications by adjusting the constants K_1, K_P, and K_I. The adjustment of the transient response is perhaps best accomplished by using the root locus methods of Chapter 7 and drawing a root locus for the gain $K_P K_1$ after locating the zero $s = -K_I/K_P$ on the s-plane by the method outlined for the s-plane in the preceding section.

The addition of an integration as $G_c(s) = K_P + K_I/s$ can also be used to reduce the steady-state error for a ramp input $r(t) = t, t \geq 0$. For example, if the uncompensated system $G(s)$ possessed one integration, the additional integration due to $G_c(s)$ would result in a zero steady-state error for a ramp input. To illustrate the design of this type of integration compensator, we will consider a temperature control system in some detail.

EXAMPLE 10.5   Temperature control system

The uncompensated loop transfer function of a unity feedback temperature control system is

$$L(s) = G(s) = \frac{K_1}{(2s + 1)(0.5s + 1)} = \frac{K_1}{(s + .5)(s + 2)}, \tag{10.53}$$

where K_1 can be adjusted. To maintain zero steady-state error for a step input, we will add the PI compensation network

$$G_c(s) = K_P + \frac{K_I}{s} = K_P \frac{s + K_I/K_P}{s}. \tag{10.54}$$

Furthermore, the transient response of the system is required to have an overshoot less than or equal to 10%. Therefore, the dominant complex roots must be on (or below) the $\zeta = 0.6$ line, as shown in Figure 10.15. We will adjust the compensator zero so that the negative real part of the complex roots is $\zeta\omega_n = 0.75$, and thus the settling time (with a 2% criterion) is $T_s = 4/(\zeta\omega_n) = \frac{16}{3}$ s. Now, as in the preceding section, we will determine the location of the zero $z = -K_I/K_P$ by ensuring that the angle at the desired root is $-180°$. Therefore, the sum of the angles at the desired root is

$$-180° = -127° - 104° - 38° + \theta_z,$$

where θ_z is the angle from the undetermined zero. Consequently, we find that $\theta_z = +89°$, and the location of the zero is $z = -0.75$. Finally, to determine the gain at the desired root, we evaluate the vector lengths from the poles and zeros and obtain

$$K = K_1 K_P = \frac{1.25(1.03)1.6}{1.0} = 2.08.$$

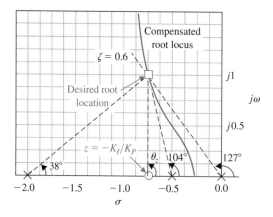

FIGURE 10.15
The s-plane design of an integration compensator.

The compensated root locus and the location of the zero are shown in Figure 10.15. Note that the zero $z = -K_I/K_P$ should be placed to the left of the pole at $s = -0.5$ to ensure that the complex roots dominate the transient response. In fact, the third root of the compensated system of Figure 10.15 can be determined as $s = -1.0$, and therefore this real root is only $\frac{4}{3}$ times the real part of the complex roots. Although complex roots dominate the response of the system, the equivalent damping of the system is somewhat less than $\zeta = 0.60$ due to the real root and zero.

The closed-loop transfer function of the system of Figure 10.14 is

$$T(s) = \frac{G_p(s)G_c(s)G(s)}{1 + G_c(s)G(s)} = \frac{2.08(s + 0.75)G_p(s)}{(s + 1)(s + r_1)(s + \hat{r}_1)}, \tag{10.55}$$

where $r_1 = -0.75 + j1$. The effect of the zero is to increase the overshoot to a step input (see Figure 5.13). If we wish to attain an overshoot of 5%, we may use a pre-filter $G_p(s)$, so that the zero is eliminated in $T(s)$ by setting

$$G_p(s) = \frac{0.75}{s + 0.75}. \tag{10.56}$$

Note that the overall DC gain (set $s = 0$) is $T(0) = 1.0$ when $G_p(s) = 1$, as obtained with the prefilter of Equation (10.56). The overshoot without the prefilter is 17.6%; with the prefilter, it is 2%. Further discussion of the use of a prefilter is provided in Section 10.10. ∎

10.7 PHASE-LAG DESIGN USING THE ROOT LOCUS

The phase-lag RC network of Figure 10.6 is an integration-type network and can be used to increase the error constant of a feedback control system. We found in Section 10.3 that the transfer function of the RC phase-lag network is of the form

$$G_c(s) = \frac{1}{\alpha}\frac{s + z}{s + p}, \tag{10.57}$$

as given in Equation (10.13), where

$$z = \frac{1}{\tau} = \frac{1}{R_2C}, \quad \alpha = \frac{R_1 + R_2}{R_2}, \quad \text{and} \quad p = \frac{1}{\alpha\tau}.$$

The steady-state error of an uncompensated unity feedback system is

$$\lim_{t\to\infty} e(t) = \lim_{s\to 0} s\left\{\frac{R(s)}{1 + G(s)}\right\}. \tag{10.58}$$

Then, for example, the velocity constant of a type-one uncompensated system is

$$K_v = \lim_{s\to 0} s\{G(s)\}, \tag{10.59}$$

as shown in Section 5.7. In general, if $G(s)$ is written as

$$G(s) = \frac{K\prod\limits_{i=1}^{M}(s + z_i)}{s\prod\limits_{j=1}^{n}(s + p_j)}, \tag{10.60}$$

we obtain the velocity constant

$$K_v = \frac{K\prod\limits_{i=1}^{M}z_i}{\prod\limits_{j=1}^{n}p_j}. \tag{10.61}$$

We will now add the integration-type phase-lag network as a compensator and determine the compensated velocity constant. If the velocity constant of the uncompensated system (Equation 10.61) is designated as $K_{v,unc}$, we have

$$K_{v,comp} = \lim_{s\to0}s\{G_c(s)G(s)\} = \lim_{s\to0}\{G_c(s)\}K_{v,unc}$$

$$= \frac{z}{p}\frac{1}{\alpha}K_{v,unc} = \frac{z}{p}\frac{K}{\alpha}\frac{\prod z_i}{\prod p_j}. \tag{10.62}$$

The gain on the compensated root locus at the desired root location will be K/α. Now, if the pole and zero of the compensator are chosen so that $|z| = \alpha|p| < 1$, the resultant K_v will be increased at the desired root location by the ratio $z/p = \alpha$. Then, for example, if $z = 0.1$ and $p = 0.01$, the velocity constant of the desired root location will be increased by a factor of 10. If the compensator pole and zero appear relatively close together on the s-plane, their effect on the location of the desired root will be negligible. Therefore, the compensator pole–zero combination near the origin of the s-plane compared to ω_n can be used to increase the error constant of a feedback system by the factor α while altering the root location very slightly. The factor α does have an upper limit, typically about 100, because the required resistors and capacitors of the network become excessively large for a higher α. For example, when $z = 0.1$ and $\alpha = 100$, we find from Equation (10.57) that

$$z = 0.1 = \frac{1}{R_2C} \quad \text{and} \quad \alpha = 100 = \frac{R_1 + R_2}{R_2}.$$

If we let $C = 10\,\mu\text{f}$, then $R_2 = 1\,\text{M}\Omega$ and $R_1 = 99\,\text{M}\Omega$. As we increase α, we increase the required magnitude of R_1. However, we should note that an attenuation α of 1000 or more may be obtained by utilizing pneumatic process controllers, which approximate a phase-lag characteristic (Figure 10.8).

The steps necessary for the design of a phase-lag network on the s-plane are as follows:

1. Obtain the root locus of the uncompensated system.

2. Determine the transient performance specifications for the system and locate suitable dominant root locations on the uncompensated root locus that will satisfy the specifications.

3. Calculate the loop gain at the desired root location and thus the system error constant.

4. Compare the uncompensated error constant with the desired error constant, and calculate the necessary increase that must result from the pole–zero ratio α of the compensator.

5. With the known ratio of the pole–zero combination of the compensator, determine a suitable location of the pole and zero of the compensator so that the compensated root locus will still pass through the desired root location. Locate the pole and zero near the origin of the s-plane in comparison to ω_n.

The fifth requirement can be satisfied if the magnitudes of the pole and zero are significantly less than ω_n of the dominant roots and they appear to merge as measured from the desired root location. The pole and zero will appear to merge at the root location if the angles from the compensator pole and zero are essentially equal as measured to the root location. One method of locating the zero and pole of the compensator is based on the requirement that the difference between the angle of the pole and the angle of the zero as measured at the desired root is less than $2°$. An example will illustrate this approach to the design of a phase-lag compensator.

EXAMPLE 10.6   Design of a phase-lag compensator

Consider the uncompensated unity feedback system of Example 10.2, where the uncompensated loop transfer function is

$$L(s) = \frac{K}{s(s + 2)}. \tag{10.63}$$

We require the damping ratio of the dominant complex roots to be 0.45, while a system velocity constant equal to 20 is attained. The uncompensated root locus is a vertical line at $s = -1$ and results in a root on the $\zeta = 0.45$ line at $s = -1 \pm j2$, as shown in Figure 10.16. Measuring the gain at this root, we have $K = (2.24)^2 = 5$. Therefore, the velocity constant of the uncompensated system is

$$K_v = \frac{K}{2} = \frac{5}{2} = 2.5.$$

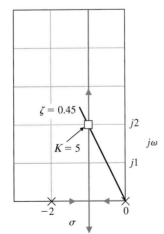

FIGURE 10.16
Root locus of the uncompensated system of Example 10.6.

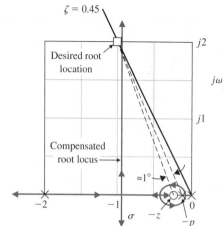

FIGURE 10.17
Root locus of the
compensated
system of Example
10.6. Note that the
actual root will differ
from the desired
root by a slight
amount. The
vertical portion of
the locus leaves the
σ axis at
$\sigma = -0.95$.

Thus, the required ratio of the zero to the pole of the compensator is

$$\left|\frac{z}{p}\right| = \alpha = \frac{K_{v,\text{comp}}}{K_{v,\text{unc}}} = \frac{20}{2.5} = 8. \tag{10.64}$$

Examining Figure 10.17, we find that we might set $z = 0.1$ and then $p = 0.1/8$. The difference of the angles from p and z at the desired root is approximately $1°$; therefore, $s = -1 \pm j2$ is still the location of the dominant roots. A sketch of the compensated root locus is shown as a heavy line in Figure 10.17. Thus, the compensated system loop transfer function is

$$L(s) = G_c(s)G(s) = \frac{5(s + 0.1)}{s(s + 2)(s + 0.0125)}, \tag{10.65}$$

where $K/\alpha = 5$, so $K = 40$, in order to account for the attenuation of the lag network. ■

EXAMPLE 10.7   **Design of a phase-lag compensator**

Let us now consider a system that is difficult to design using a phase-lead network. The loop transfer function of the uncompensated unity feedback system is

$$L(s) = \frac{K}{s(s + 10)^2}. \tag{10.66}$$

It is specified that the velocity constant of this system be equal to 20, while the damping ratio of the dominant roots is equal to 0.707. The gain necessary for a K_v of 20 is

$$K_v = 20 = \frac{K}{(10)^2},$$

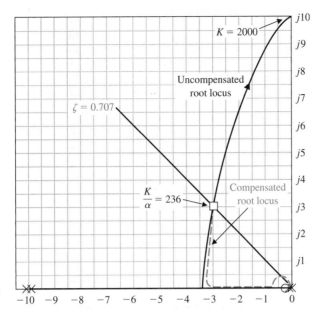

FIGURE 10.18
Design of a phase-lag compensator on the s-plane.

or $K = 2000$. However, using Routh's criterion, we find that the roots of the characteristic equation lie on the $j\omega$-axis at $\pm j10$ when $K = 2000$. The roots of the system when the K_v requirement is satisfied are a long way from satisfying the damping ratio specification, and it would be difficult to bring the dominant roots from the $j\omega$-axis to the $\zeta = 0.707$ line by using a phase-lead compensator. Therefore, we will attempt to satisfy the K_v and ζ requirements by using a phase-lag network. The uncompensated root locus of this system is shown in Figure 10.18, and the roots are shown when $\zeta = 0.707$ and $s = -2.9 \pm j2.9$. Measuring the gain at these roots, we find that $K = 236$. Therefore, the necessary ratio of the zero to the pole of the compensator (use Equation 10.64) is

$$\alpha = \left| \frac{z}{p} \right| = \frac{2000}{236} = 8.5.$$

Thus, we will choose $z = 0.1$ and $p = 0.1/9$ in order to allow a small margin of safety. Examining Figure 10.18, we find that the difference between the angle from the pole and zero of $G_c(s)$ is negligible. Therefore, the compensated system is

$$G_c(s)G(s) = \frac{236(s + 0.1)}{s(s + 10)^2(s + 0.0111)}, \tag{10.67}$$

where $K/\alpha = 236$ and $\alpha = 9$. ∎

   The design of an integration compensator to increase the error constant of an uncompensated control system is particularly illustrative using s-plane and root locus methods. We shall now turn to similarly useful methods of designing integration compensation using Bode diagrams.

10.8 PHASE-LAG DESIGN USING THE BODE DIAGRAM

The design of a phase-lag RC network suitable for compensating a feedback control system can be readily accomplished on the Bode diagram. The advantage of the Bode diagram is again apparent, for we will simply add the frequency response of the compensator to the Bode diagram of the uncompensated system in order to obtain a satisfactory system frequency response. The transfer function of the phase-lag network, written in Bode diagram form, is

$$G_c(j\omega) = \frac{1 + j\omega\tau}{1 + j\omega\alpha\tau}, \tag{10.68}$$

as we found in Equation (10.14). The Bode diagram of the phase-lag network is shown in Figure 10.8 for two values of α. On the Bode diagram, the pole and the zero of the compensator have a magnitude much smaller than the smallest pole of the uncompensated system. Thus, the phase lag is not the useful effect of the compensator; it is the attenuation $-20 \log \alpha$ that is the useful effect for compensation. The phase-lag network is used to provide an attenuation and therefore to lower the 0-dB (crossover) frequency of the system. However, at lower crossover frequencies, we usually find that the phase margin of the system is increased, and our specifications can be satisfied. The design procedure for a phase-lag network on the Bode diagram is as follows:

1. Obtain the Bode diagram of the uncompensated system with the gain adjusted for the desired error constant.

2. Determine the phase margin of the uncompensated system and, if it is insufficient, proceed with the following steps.

3. Determine the frequency where the phase margin requirement would be satisfied if the magnitude curve crossed the 0-dB line at this frequency, ω_c'. (Allow for $5°$ phase lag from the phase-lag network when determining the new crossover frequency.)

4. Place the zero of the compensator one decade below the new crossover frequency ω_c', and thus ensure only $5°$ of additional phase lag at ω_c' (see Figure 10.8) due to the lag network.

5. Measure the necessary attenuation at ω_c' to ensure that the magnitude curve crosses at this frequency.

6. Calculate α by noting that the attenuation introduced by the phase-lag network is $-20 \log \alpha$ at ω_c'.

7. Calculate the pole as $\omega_p = 1/(\alpha\tau) = \omega_z/\alpha$, and the design is completed.

An example of this design procedure will illustrate that the method is simple to carry out in practice.

EXAMPLE 10.8   **Design of a phase-lag network**

Let us consider again the unity feedback system of Example 10.6 and design a phase-lag network so that the desired phase margin is obtained. The uncompensated loop transfer function is

$$L(j\omega) = \frac{K}{j\omega(j\omega + 2)} = \frac{K_v}{j\omega(0.5j\omega + 1)}, \tag{10.69}$$

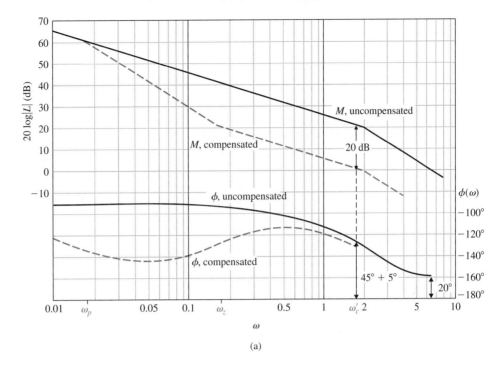

(a)

FIGURE 10.19
(a) Design of a phase-lag network on the Bode diagram for Example 10.8. (b) Time response to a step input for the uncompensated system (solid line) and the compensated system (dashed line) of Example 10.8.

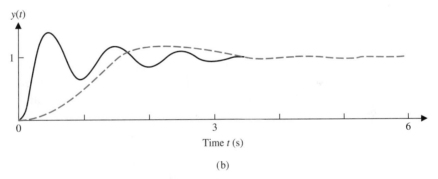

(b)

where $K_v = K/2$. We want $K_v = 20$ while a phase margin of 45° is attained. The uncompensated Bode diagram is shown as a solid line in Figure 10.19. The uncompensated system has a phase margin of 20°, and the phase margin must be increased. Allowing 5° for the phase-lag compensator, we locate the frequency ω where $\phi(\omega) = -130°$, which is to be our new crossover frequency ω'_c. In this case, we find that $\omega'_c = 1.5$, which allows for a small margin of safety. The attenuation necessary to cause ω'_c to be the new crossover frequency is equal to 20 dB. Both the compensated and uncompensated magnitude curves are an asymptotic approximation. Both the actual curves are 2 dB lower than shown. Thus, $\omega'_c = 1.5$, and the required attenuation is 20 dB.

Then we find that 20 dB $= 20 \log \alpha$, or $\alpha = 10$. Therefore, the zero is one decade below the crossover, or $\omega_z = \omega_c'/10 = 0.15$, and the pole is at $\omega_p = \omega_z/10 = 0.015$. The compensated system is then

$$G_c(j\omega)G(j\omega) = \frac{20(6.66j\omega + 1)}{j\omega(0.5j\omega + 1)(66.6j\omega + 1)}. \tag{10.70}$$

The frequency response of the compensated system is shown in Figure 10.19(a) with dashed lines. It is evident that the phase lag introduces an attenuation that lowers the crossover frequency and therefore increases the phase margin. Note that the phase angle of the lag network has almost totally disappeared at the crossover frequency ω_c'. As a final check, we numerically evaluate the phase margin and find that $\phi_{pm} = 46.8°$ at $\omega_c' = 1.58$ which is the desired result. Using the Nichols chart, we find that the closed-loop bandwidth of the system has been reduced from $\omega = 10$ rad/s for the uncompensated system to $\omega = 2.5$ rad/s for the compensated system. Due to the reduced bandwidth, we expect a slower time response to a step command.

The time response of the system is shown in Figure 10.19(b). Note that the overshoot is 25% and the peak time is 1.85 seconds. Thus, the response is within the specifications. ∎

EXAMPLE 10.9   **Design of a phase-lag compensator**

Let us consider again the unity feedback system of Example 10.7, which is

$$L(j\omega) = \frac{K}{j\omega(j\omega + 10)^2} = \frac{K_v}{j\omega(0.1j\omega + 1)^2}, \tag{10.71}$$

where $K_v = K/100$. A velocity constant of K_v equal to 20 is specified. Furthermore, a damping ratio of 0.707 for the dominant roots is required. From Figure 9.21, we estimate that a phase margin of 65° is required. The frequency response of the uncompensated system is shown in Figure 10.20. The phase margin of the uncompensated system is 0°. Allowing 5° for the lag network, we locate the frequency where the phase is $-110°$. This frequency is equal to 1.74, and therefore we will attempt to locate the new crossover frequency at $\omega_c' = 1.5$. Measuring the necessary attenuation at $\omega = \omega_c'$, we find that 23 dB is required; then $23 = 20 \log \alpha$ gives $\alpha = 14.2$. The zero of the compensator is located one decade below the crossover frequency, and thus

$$\omega_z = \frac{\omega_c'}{10} = 0.15.$$

The pole is then

$$\omega_p = \frac{\omega_z}{\alpha} = \frac{0.15}{14.2}.$$

Therefore, the compensated system is

$$G_c(j\omega)G(j\omega) = \frac{20(6.66j\omega + 1)}{j\omega(0.1j\omega + 1)^2(94.6j\omega + 1)}. \tag{10.72}$$

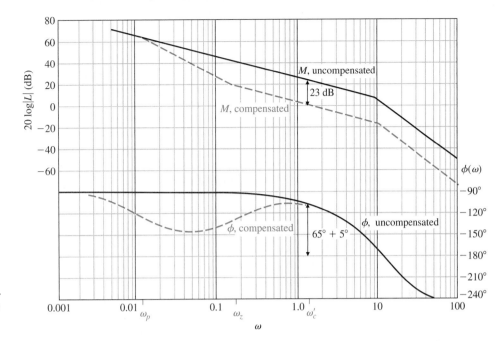

FIGURE 10.20

Design of a phase-lag network on the Bode diagram for Example 10.9.

The compensated frequency response is shown in Figure 10.20. As a final check, we evaluate the phase margin at $\omega'_c = 1.5$ and find that $\phi_{pm} = 67°$, which is within the specifications. ∎

We have seen that a phase-lag compensation network can be used to alter the frequency response of a feedback control system in order to attain satisfactory system performance. Examining both Examples 10.8 and 10.9, we note again that the system design is satisfactory when the asymptotic curve for the magnitude of the compensated system crosses the 0-dB line with a slope of -20 dB/decade. The attenuation of the phase-lag network reduces the magnitude of the crossover (0-dB) frequency to a point where the phase margin of the system is satisfactory. Thus, in contrast to the phase-lead network, the phase-lag network reduces the closed-loop bandwidth of the system as it maintains a suitable error constant.

We might ask, why not place the compensator zero more than one decade below the new crossover ω'_c (see step 4 of the design procedure) and thus ensure less than 5° of lag at ω'_c due to the compensator? This question can be answered by considering the requirements placed on the resistors and capacitors of the lag network by the values of the poles and zeros (see Equation 10.12). As the magnitudes of the pole and zero of the lag network are decreased, the magnitudes of the resistors and the capacitor required increase proportionately. The zero of the lag compensator in terms of the circuit components is $z = 1/(R_2 C)$, and the α of the network is $\alpha = (R_1 + R_2)/R_2$. Thus, considering Example 10.9, we require a zero at $\omega_z = 0.15$, which can be obtained with $C = 1\ \mu F$ and $R_2 = 6.66\ M\Omega$. However, for $\alpha = 14$, we require a resistance R_1 of $R_1 = R_2(\alpha - 1) = 88\ M\Omega$. A designer does not wish to place the zero ω_z further than one decade below ω'_c and thus require larger values of R_1, R_2, and C.

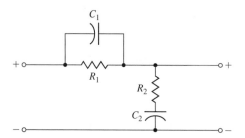

FIGURE 10.21
An *RC* lead-lag
network.

The phase-lead compensation network alters the frequency response of a network by adding a positive (leading) phase angle and therefore increases the phase margin at the crossover (0-dB) frequency. It becomes evident that a designer might wish to consider using a compensation network that provides the attenuation of a phase-lag network and the lead-phase angle of a phase-lead network. Such a network does exist. It is called a **lead-lag network** and is shown in Figure 10.21. The transfer function of this network is

$$\frac{V_2(s)}{V_1(s)} = \frac{(R_1 C_1 s + 1)(R_2 C_2 s + 1)}{R_1 R_2 C_1 C_2 s^2 + (R_1 C_1 + R_1 C_2 + R_2 C_2)s + 1}. \tag{10.73}$$

When $\alpha\tau_1 = R_1 C_1$, $\beta\tau_2 = R_2 C_2$, $\tau_1 + \tau_2 = R_1 C_1 + R_1 C_2 + R_2 C_2$, and $\tau_1\tau_2 = R_1 R_2 C_1 C_2$, we note that $\alpha\beta = 1$, and then Equation (10.73) is

$$\frac{V_2(s)}{V_1(s)} = \frac{(1 + \alpha\tau_1 s)(1 + \beta\tau_2 s)}{(1 + \tau_1 s)(1 + \tau_2 s)}, \tag{10.74}$$

where $\alpha > 1$ and $\beta < 1$. The first factors in the numerator and denominator, which are functions of τ_1, provide the phase-lead portion of the network. The second factors, which are functions of τ_2, provide the phase-lag portion of the compensation network. The parameter β is adjusted to provide suitable attenuation of the low-frequency portion of the frequency response, and the parameter α is adjusted to provide an additional phase lead at the new crossover (0-dB) frequency. Alternatively, the compensation can be designed on the *s*-plane by placing the lead pole and zero compensation in order to locate the dominant roots in a desired location. Then the phase-lag compensation is used to raise the error constant at the dominant root location by a suitable ratio $1/\beta$. The design of a phase lead-lag compensator follows the procedures already discussed. Other literature will further illustrate the utility of lead-lag compensation [2, 3, 29].

10.9 DESIGN ON THE BODE DIAGRAM USING ANALYTICAL METHODS

We will often use computers, when appropriate, to assist the designer in the selection of the parameters of a compensator. The development of algorithms for computer-aided design is an important alternative approach to the trial-and-error methods considered in earlier sections. Computer programs have been developed for the selection of suitable parameter values for compensators based on satisfaction of frequency response criteria such as the phase margin [3, 4].

An analytical technique of selecting the parameters of a lead or lag network has been developed for the Bode diagram [4, 5]. For a single-stage compensator,

$$G_c(s) = \frac{1 + \alpha\tau s}{1 + \tau s},$$ (10.75)

where $\alpha < 1$ yields a lag compensator and $\alpha > 1$ yields a lead compensator. The phase contribution of the compensator at the desired crossover frequency ω_c (see Equation 10.9) is given by

$$p = \tan\phi = \frac{\alpha\omega_c\tau - \omega_c\tau}{1 + (\omega_c\tau)^2\alpha}.$$ (10.76)

The magnitude M (in dB) of the compensator at ω_c is given by

$$c = 10^{M/10} = \frac{1 + (\omega_c\alpha\tau)^2}{1 + (\omega_c\tau)^2}.$$ (10.77)

Eliminating $\omega_c\tau$ from Equations (10.76) and (10.77), we obtain the nontrivial solution equation for α as

$$(p^2 - c + 1)\alpha^2 + 2p^2c\alpha + p^2c^2 + c^2 - c = 0.$$ (10.78)

For a single-stage compensator, it is necessary that $c > p^2 + 1$. If we solve for α from Equation (10.78), we can obtain τ from

$$\tau = \frac{1}{\omega_c}\sqrt{\frac{1 - c}{c - \alpha^2}}.$$ (10.79)

The design steps for a lead compensator are:

1. Select the desired ω_c.
2. Determine the phase margin desired and therefore the required phase ϕ for Equation (10.76).
3. Verify that the phase lead is applicable: $\phi > 0$ and $M > 0$.
4. Determine whether a single stage will be sufficient by testing $c > p^2 + 1$.
5. Determine α from Equation (10.78).
6. Determine τ from Equation (10.79).

If we need to design a single-lag compensator, then $\phi < 0$ and $M < 0$ (step 3). Step 4 will require $c < 1/(1 + p^2)$. Otherwise the method is the same.

EXAMPLE 10.10   Design using an analytical technique

Let us consider again the system of Example 10.1 and design a lead network by the analytical technique. Examine the uncompensated curves in Figure 10.9. We select $\omega_c = 5$. Then, as before, we desire a phase margin of 45°. The compensator must yield this phase, so

$$p = \tan 45° = 1.$$ (10.80)

The required magnitude contribution is 8 dB, or $M = 8$, so that

$$c = 10^{8/10} = 6.31.$$ (10.81)

Using c and p, we obtain

$$-4.31\alpha^2 + 12.62\alpha + 73.32 = 0. \tag{10.82}$$

Solving for α, we obtain $\alpha = 5.84$. Solving Equation (10.79), we obtain $\tau = 0.087$. Therefore, the compensator is

$$G_c(s) = \frac{1 + 0.515s}{1 + 0.087s}. \tag{10.83}$$

The pole is equal to 11.5, and the zero is 1.94. This design is similar to that obtained by the graphical technique of Section 10.4. ∎

10.10 SYSTEMS WITH A PREFILTER

In the earlier sections of this chapter, we utilized compensators of the form

$$G_c(s) = \frac{s + z}{s + p}$$

that alter the roots of the characteristic equation of the closed-loop system. However, the closed-loop transfer function $T(s)$ will contain the zero of $G_c(s)$ as a zero of $T(s)$. This zero will significantly affect the response of the system $T(s)$.

Let us consider the system shown in Figure 10.22, where

$$G(s) = \frac{1}{s}.$$

We will introduce a PI compensator, so that

$$G_c(s) = K_P + \frac{K_I}{s} = \frac{K_P s + K_I}{s}.$$

The closed-loop transfer function of the system with a prefilter (Figure 10.22) is

$$T(s) = \frac{(K_P s + K_I)G_p(s)}{s^2 + K_P s + K_I}. \tag{10.84}$$

For illustrative purposes, the specifications require a settling time (with a 2% criterion) of 0.5 second and an overshoot of approximately 4%. We use $\zeta = 1/\sqrt{2}$ and note that

$$T_s = \frac{4}{\zeta\omega_n}.$$

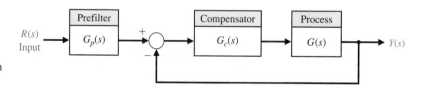

FIGURE 10.22
Control system with a prefilter $G_p(s)$.

Thus, we require that $\zeta\omega_n = 8$ or $\omega_n = 8\sqrt{2}$. We now obtain

$$K_P = 2\zeta\omega_n = 16 \quad \text{and} \quad K_I = \omega_n^2 = 128.$$

The closed-loop transfer function when $G_p(s) = 1$ is then

$$T(s) = \frac{16(s + 8)}{s^2 + 16s + 128}.$$

The effect of the zero on the step response is significant. Using Figure 5.13(a), we have $a/(\zeta\omega_n) = 1$ and $\zeta = 1/\sqrt{2}$, and the overshoot to a step as predicted from Figure 5.13(a) is 21%.

We use a prefilter $G_p(s)$ to eliminate the zero from $T(s)$ while maintaining the DC gain of 1, thus requiring that

$$G_p(s) = \frac{8}{s + 8}.$$

Then we have

$$T(s) = \frac{128}{s^2 + 16s + 128},$$

and the overshoot of this system is 4.5%, as expected.

Reviewing Figure 5.13(a), we note that the zero at $s = -a$ has a significant effect when $a/\zeta\omega_n < 5$, where $-a$ is the zero and $-\zeta\omega_n$ is the real part of the dominant roots of the characteristic equation of $T(s)$.

Let us now consider again Example 10.3, which includes the design of a lead compensator. The resulting closed-loop transfer function can be determined to be (using Figure 10.22)

$$T(s) = \frac{8.1(s + 1)G_p(s)}{(s + 1 + j2)(s + 1 - j2)(s + 1.62)}.$$

If $G_p(s) = 1$ (no prefilter), then we obtain a response with an overshoot of 46.6% and a settling time of 3.8 seconds. If we use a prefilter,

$$G_p(s) = \frac{1}{s + 1},$$

we obtain an overshoot of 6.7% and a settling time of 3.8 seconds. The real root at $s = -1.62$ helps to damp the step response. The prefilter is very useful in permitting the designer to introduce a compensator with a zero to adjust the root locations (poles) of the closed-loop transfer function while eliminating the effect of the zero incorporated in $T(s)$.

In general, we will add a prefilter for systems with lead networks or PI compensators. We will not use a prefilter for a system with a lag network, since we expect the effect of the zero to be insignificant. To check this assertion, let us consider again the design obtained in Example 10.6. The system with a phase-lag controller is

$$G(s)G_c(s) = \frac{5(s + 0.1)}{s(s + 2)(s + 0.0125)}.$$

The closed-loop transfer function is then

$$T(s) = \frac{5(s + 0.1)}{(s^2 + 1.98s + 5.1)(s + 0.095)} \approx \frac{5}{s^2 + 1.98s + 5.1},$$

since the zero at $s = -0.1$ and the pole at $s = -0.095$ approximately cancel. We expect an overshoot of 20% and a settling time (with a 2% criterion) of 4.0 seconds for the design parameters $\zeta = 0.45$ and $\zeta\omega_n = 1$. However, the actual response has an overshoot of 26% and a longer settling time of 5.8 seconds due to the effect of the real pole of $T(s)$ at $s = -0.095$. Thus, we usually do not use a prefilter with systems that utilize lag compensators.

EXAMPLE 10.11   **Design of a third-order system**

Consider a system of the form shown in Figure 10.22 with

$$G(s) = \frac{1}{s(s + 1)(s + 5)}.$$

Let us design a system that will yield a step response with an overshoot less than 2% and a settling time less than 3 seconds by using both $G_c(s)$ and $G_p(s)$ to achieve the desired response.

We use a lead compensation network

$$G_c(s) = \frac{K(s + 1.2)}{s + 10}$$

and select K to find the complex roots with $\zeta = 1/\sqrt{2}$. Then, with $K = 78.7$, the closed-loop transfer function is

$$
\begin{aligned}
T(s) &= \frac{78.7(s + 1.2)G_p(s)}{(s + 1.71 + j1.71)(s + 1.71 - j1.71)(s + 1.45)(s + 11.1)} \\
&\approx \frac{7.1(s + 1.2)G_p(s)}{(s^2 + 3.42s + 5.85)(s + 1.45)}.
\end{aligned}
$$

If we choose

$$G_p(s) = \frac{p}{s + p} \tag{10.85}$$

the closed-loop transfer function is

$$T(s) \approx \frac{7.1p(s + 1.2)}{(s^2 + 3.42s + 5.85)(s + 1.45)(s + p)}.$$

If $p = 1.2$, we cancel the effect of the zero. The response of the system with a prefilter is summarized in Table 10.1. We choose the appropriate value for p to achieve the response desired. Note that $p = 2.40$ will provide a response that may be desirable, since it effects a faster rise time than $p = 1.20$. The prefilter provides an additional parameter to select for design purposes. ∎

Table 10.1   Effect of a Prefilter on the Step Response

$G_p(s)$	$p = 1$	$p = 1.20$	$p = 2.4$
Percent overshoot	9.9%	0%	4.8%
90% rise time (seconds)	1.05	2.30	1.60
Settling time (seconds)	2.9	3.0	3.2

10.11 DESIGN FOR DEADBEAT RESPONSE

Often, the goal for a control system is to achieve a fast response to a step command with minimal overshoot. We define a **deadbeat response** as a response that proceeds rapidly to the desired level and holds at that level with minimal overshoot. We use the ±2% band at the desired level as the acceptable range of variation from the desired response. Then, if the response enters the band at time T_s, it has satisfied the settling time T_s upon entry to the band, as illustrated in Figure 10.23. A deadbeat response has the following characteristics:

1. Steady-state error = 0
2. Fast response → minimum rise time and settling time
3. $0.1\% \leq$ percent overshoot <2%
4. Percent undershoot <2%

Characteristics (3) and (4) require that the response remain within the ±2% band so that the entry to the band occurs at the settling time.

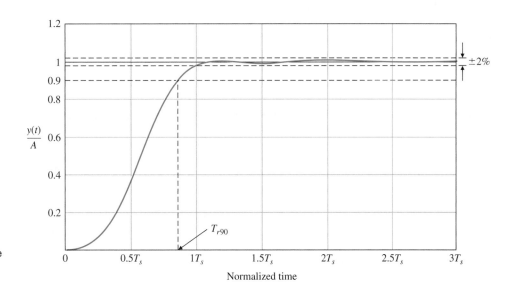

FIGURE 10.23
The deadbeat response. A is the magnitude of the step input.

We consider the transfer function $T(s)$ of a closed-loop system. To determine the coefficients that yield the optimal deadbeat response, the standard transfer function is first normalized. An example of this for a third-order system is

$$T(s) = \frac{\omega_n^3}{s^3 + \alpha\omega_n s^2 + \beta\omega_n^2 s + \omega_n^3}. \tag{10.86}$$

Dividing the numerator and denominator by ω_n^3 yields

$$T(s) = \frac{1}{\dfrac{s^3}{\omega_n^3} + \alpha\dfrac{s^2}{\omega_n^2} + \beta\dfrac{s}{\omega_n} + 1}. \tag{10.87}$$

Let $\bar{s} = s/\omega_n$ to obtain

$$T(s) = \frac{1}{\bar{s}^3 + \alpha\bar{s}^2 + \beta\bar{s} + 1}. \tag{10.88}$$

Equation (10.88) is the normalized, third-order, closed-loop transfer function. For a higher-order system, the same method is used to derive the normalized equation. The coefficients of the equation—α, β, γ, and so on—are then assigned the values necessary to meet the requirement of deadbeat response. The coefficients recorded in Table 10.2 were selected to achieve deadbeat response and minimize settling time and rise time T_r to 100% of the desired command. The form of Equation (10.88) is normalized since $\bar{s} = s/\omega_n$. Thus, we choose ω_n based on the desired settling time or rise time. Therefore, if we have a third-order system with a required settling time of 1.2 seconds, we note from Table 10.2 that the normalized settling time is

$$\omega_n T_s = 4.04.$$

Therefore, we require that

$$\omega_n = \frac{4.04}{T_s} = \frac{4.04}{1.2} = 3.37.$$

Once ω_n is chosen, the complete closed-loop transfer function is known, having the form of Equation (10.86). When designing a system to obtain a deadbeat response,

Table 10.2   Coefficients and Response Measures of a Deadbeat System

System Order	Coefficients					Percent Overshoot P.O.	Percent Undershoot P.U.	90% Rise Time T_{r90}	100% Rise Time T_r	Settling Time T_s
	α	β	γ	δ	ϵ					
2nd	1.82					0.10%	0.00%	3.47	6.58	4.82
3rd	1.90	2.20				1.65%	1.36%	3.48	4.32	4.04
4th	2.20	3.50	2.80			0.89%	0.95%	4.16	5.29	4.81
5th	2.70	4.90	5.40	3.40		1.29%	0.37%	4.84	5.73	5.43
6th	3.15	6.50	8.70	7.55	4.05	1.63%	0.94%	5.49	6.31	6.04

Note: All times are normalized.

the compensator is chosen, and the closed-loop transfer function is found. This compensated transfer function is then set equal to Equation (10.86), and the required compensator can be determined.

EXAMPLE 10.12   **Design of a system with a deadbeat response**

Let us consider a unity feedback system with a compensator $G_c(s)$ and a prefilter $G_p(s)$, as shown in Figure 10.22. The process is

$$G(s) = \frac{K}{s(s + 1)},$$

and the compensator is

$$G_c(s) = \frac{s + z}{s + p}.$$

Using the necessary prefilter yields

$$G_p(s) = \frac{z}{s + z}.$$

The closed-loop transfer function is

$$T(s) = \frac{Kz}{s^3 + (1 + p)s^2 + (K + p)s + Kz}.$$

We use Table 10.2 to determine the required coefficients, $\alpha = 1.90$ and $\beta = 2.20$. If we select a settling time (with a 2% criterion) of 2 seconds, then $\omega_n T_s = 4.04$, and thus $\omega_n = 2.02$. The required closed-loop system has the characteristic equation

$$q(s) = s^3 + \alpha\omega_n s^2 + \beta\omega_n^2 s + \omega_n^3 = s^3 + 3.84s^2 + 8.98s + 8.24.$$

Then, we determine that $p = 2.84$, $z = 1.34$, and $K = 6.14$. The response of this system will have $T_s = 2$ s, $T_r = 2.14$ s, and $T_{r90} = 1.72$ s. ∎

10.12 DESIGN EXAMPLES

In this section we present three illustrative examples. The first example is a rotor winder control system where both a lead and lag compensator are designed using root locus methods. The second example is an *x-y* plotter. In this example, three different controllers are designed, including a proportional controller, a lead compensator, and a proportional-derivative (PD) controller. In the third example, precise control of a milling machine used in manufacturing is employed to illustrate the design process. A lag compensator is designed using root locus methods to meet steady-state tracking error and percent overshoot specifications.

EXAMPLE 10.13   **Rotor winder control system**

Our goal is to replace a manual operation using a machine to wind copper wire onto the rotors of small motors. Each motor has three separate windings of several hundred turns of wire. It is important that the windings be consistent and that the

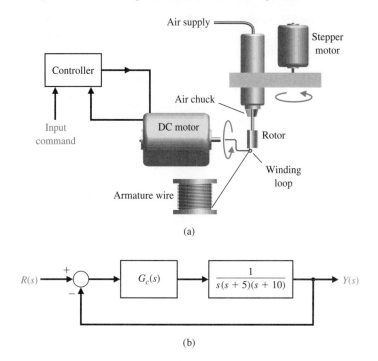

(a)

$$R(s) \xrightarrow{\;+\;} \bigcirc \xrightarrow{\quad} \boxed{G_c(s)} \xrightarrow{\quad} \boxed{\dfrac{1}{s(s + 5)(s + 10)}} \xrightarrow{\quad} Y(s)$$

FIGURE 10.24
(a) Rotor winder
control system.
(b) Block diagram.

(b)

throughout of the process be high. The operator simply inserts an unwound rotor, pushes a start button, and then removes the completely wound rotor. The DC motor is used to achieve accurate rapid windings. Thus, the goal is to achieve high steady-state accuracy for both position and velocity. The control system is shown in Figure 10.24(a) and the block diagram in Figure 10.24(b). This system has zero steady-state error for a step input, and the steady-state error for a ramp input is

$$e_{ss} = A/K_v,$$

where

$$K_v = \lim_{s \to 0} \frac{G_c(s)}{50}.$$

When $G_c(s) = K$, we have $K_v = K/50$. If we select $K = 500$, we will have $K_v = 10$, but the overshoot to a step is 70%, and the settling time is 8 seconds.

We first try a lead compensator so that

$$G_c(s) = \frac{K(s + z_1)}{s + p_1}. \tag{10.89}$$

Selecting $z_1 = 4$ and the pole p_1 so that the complex roots have a ζ of 0.6, we have (see Figure 10.25)

$$G_c(s) = \frac{191.2(s + 4)}{s + 7.3}. \tag{10.90}$$

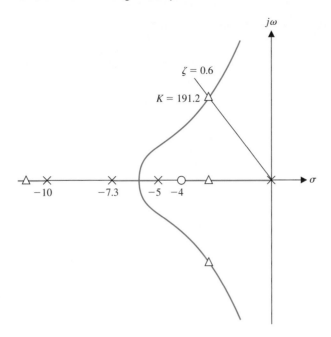

FIGURE 10.25
Root locus for lead compensator.

We find the response to a step input has a 3% overshoot and a settling time of 1.5 seconds. However, the velocity constant is

$$K_v = \frac{191.2(4)}{7.3(50)} = 2.1,$$

which is inadequate.

If we use a phase-lag design, we select

$$G_c(s) = \frac{K(s + z_2)}{s + p_2}$$

in order to achieve $K_v = 38$. Thus, the velocity constant of the lag-compensated system is

$$K_v = \frac{K z_2}{50 p_2}.$$

Using a root locus, we select $K = 105$ in order to achieve a reasonable uncompensated step response with an overshoot of less than or equal to 10%. We select $\alpha = z/p$ to achieve the desired K_v. We then have

$$\alpha = \frac{50 K_v}{K} = \frac{50(38)}{105} = 18.1.$$

Selecting $z_2 = 0.1$ to avoid affecting the uncompensated root locus, we have $p_2 = 0.0055$. We then obtain a step response with a 12% overshoot and a settling time of 2.5 seconds.

Table 10.3   Design Example Results

Controller	Gain, K	Lead Network	Lag Network	Lead-Lag Network
Step overshoot	70%	3%	12%	5%
Settling time (seconds)	8	1.5	2.5	2.0
Steady-state error for ramp	10%	48%	2.6%	4.8%
K_v	10	2.1	38	21

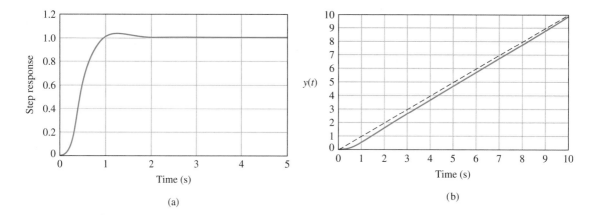

(a)

(b)

FIGURE 10.26   (a) Step response and (b) ramp response for rotor winder system.

The results for the simple gain, the lead network, and the lag network are summarized in Table 10.3.

Let us return to the lead-network system and add a cascade lag network, so that the compensator is

$$G_c(s) = \frac{K(s + z_1)(s + z_2)}{(s + p_1)(s + p_2)}. \tag{10.91}$$

The lead compensator of Equation (10.90) requires $K = 191.2$, $z_1 = 4$, and $p_1 = 7.3$. The root locus for the system is shown in Figure 10.25. We recall that this lead network resulted in $K_v = 2.1$ (see Table 10.3). To obtain $K_v = 21$, we use $\alpha = 10$ and select $z_2 = 0.1$ and $p_2 = 0.01$. Then the compensated loop transfer function is

$$G(s)G_c(s) = \frac{191.2(s + 4)(s + 0.1)}{s(s + 5)(s + 10)(s + 7.28)(s + 0.01)}. \tag{10.92}$$

The step response and ramp response of this system are shown in Figure 10.26 in parts (a) and (b), respectively, and are summarized in Table 10.3. Clearly, the lead-lag design is suitable for satisfaction of the design goals. ∎

EXAMPLE 10.14   **X-Y plotter**

Many physical phenomena are characterized by parameters that are transient or slowly varying. If recorded, these changes can be examined at leisure and stored for future reference or comparison. To accomplish such a recording, a number of electro-mechanical instruments have been developed, among them the *x-y* recorder. In this instrument, the displacement along the *x*-axis represents a variable of interest or time and the displacement along the *y*-axis varies as a function of another variable [6].

Such recorders can be found in many laboratories recording experimental data, such as changes in temperature, variations in transducer output levels, and stress versus applied strain, to name just a few. The *x-y* plotter produces graphs with ink pens by drawing lines from a graphics file or directly from input data. These output devices offer a resolution superior to a printer since the lines are actually drawn rather than being composed of tiny dots.

The purpose of a plotter is to accurately follow the input signal as it varies. We will consider the design of the movement of one axis, since the movement dynamics of both axes are identical. Thus, we will strive to control the position and the movement of the pen very accurately as it follows the input signal.

To achieve accurate results, our goal is to achieve (1) a step response with an overshoot of less than 5% and a settling time (with a 2% criterion) less than 0.5 second, and (2) a percentage steady-state error for a step equal to zero. If we achieve these specifications, we will have a fast and accurate response.

To move the pen, we select a DC motor as the actuator. The feedback sensor will be a 500-line optical encoder. By detecting all state changes of the two-channel quadrature output of the encoder, 2000 encoder counts per revolution of the motor shaft can be detected. This yields an encoder resolution of 0.001 inch at the pen tip. The encoder is mounted on the shaft of the motor. Since the encoder provides digital data, it is compared with the input signal by using a microprocessor. Next, we propose using the difference signal calculated by the microprocessor as the error signal and then using the microprocessor to calculate the necessary algorithm to obtain the designed compensator. The output of the compensator is then converted to an analog signal that will drive the motor.

The model of the feedback position control system is shown in Figure 10.27. Since the microprocessor calculation speed is very fast compared to the rate of change of the encoder and input signals, we assume that the continuous signal model is very accurate.

The model for the motor and pen carriage is

$$G(s) = \frac{1}{s(s + 10)(s + 1000)}, \tag{10.93}$$

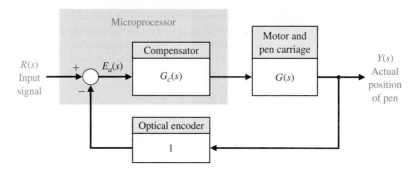

FIGURE 10.27
Model of the pen-plotter control system.

and our initial attempt at a specification of a compensator is to use a simple gain so that

$$G_c(s) = K.$$

In this case, we have only one parameter to adjust: K. To achieve a fast response, we have to adjust K so that it will provide two dominant s-plane roots with a damping ratio of 0.707, which will result in a step response overshoot of about 4.5%. A sketch of the root locus (note the break in the real axis) is shown in Figure 10.28.

Adjusting the gain to $K = 47,200$, we obtain a system with an overshoot of 3.6% to a step input and a settling time of 0.8 second. Since the transfer function has a pole at the origin, we have a steady-state error of zero for a step input.

This system does not meet our specifications, so we select a compensator that will reduce the settling time. Let us select a lead compensator so that

$$G_c(s) = \frac{K\alpha(s + z)}{(s + p)}, \tag{10.94}$$

where $p = \alpha z$. Let us use the method of Section 10.5, which selects the phase-lead compensator on the s-plane. Hence, we place the zero at $s = -20$ and determine the location of the pole, p, that will place the dominant roots on the line that has the damping ratio of $1/\sqrt{2}$. Thus, we find that $p = 60$ and $\alpha = 3$, so that

$$G(s)G_c(s) = \frac{142,600(s + 20)}{s(s + 10)(s + 60)(s + 1000)}. \tag{10.95}$$

Obtaining the actual step response, we determine that the percent overshoot is 2% and that the settling time is 0.35 second, which meet the specifications. The third design approach is to recognize that the encoder can be used to generate a velocity signal by counting the rate at which encoder lines pass by a fixed point, using the microprocessor. Since the position signal and the velocity signal are available, we can describe the compensator as

$$G_c(s) = K_P + K_D s, \tag{10.96}$$

where K_P is the gain for the error signal and K_D is the gain for the velocity signal.

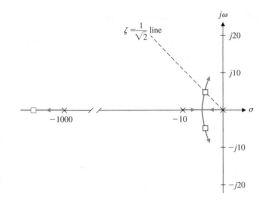

FIGURE 10.28
Root locus for the pen plotter, showing the roots with a damping ratio of $1/\sqrt{2}$. The dominant roots are $s = -4.9 \pm j4.9$.

Then we can write

$$G(s)G_c(s) = \frac{K_D(s + K_P/K_D)}{s(s + 10)(s + 1000)}.$$

If we set $K_P/K_D = 10$, we cancel the pole at $s = -10$ and obtain

$$G(s)G_c(s) = \frac{K_D}{s(s + 1000)}.$$

The characteristic equation for this system is

$$s^2 + 1000s + K_D = 0, \tag{10.97}$$

and we want $\zeta = 1/\sqrt{2}$. Noting that $2\zeta\omega_n = 1000$, we have $\omega_n = 707$ and $K_D = \omega_n^2$. Therefore, we obtain $K_D = 5 \times 10^5$, and the compensated system is

$$G_c(s)G(s) = \frac{5 \times 10^5}{s(s + 1000)}. \tag{10.98}$$

The response of this system will provide an overshoot of 4.3% and a settling time of 8 milliseconds.

The results for the three approaches to system design are compared in Table 10.4. The best design uses the velocity feedback. ∎

Table 10.4   Results for Three Designs

	Step Response	
System	Percent Overshoot	Settling Time (milliseconds)
Gain adjustment	3.6	800
Gain and lead compensator	2.0	350
Gain adjustment plus velocity signal multiplied by gain K_D	4.3	8

EXAMPLE 10.15   **Milling machine control system**

Smaller, lighter, less costly sensors are being developed by engineers for machining and other manufacturing processes. A milling machine table is depicted in Figure 10.29. This particular machine table has a new sensor that obtains information about the cutting process (that is, the depth-of-cut) from the acoustic emission (AE) signals. Acoustic emissions are low-amplitude, high-frequency stress waves that originate from the rapid release of strain energy in a continuous medium. The AE sensors are commonly piezoelectric amplitude sensitive in the 100 kHz to 1 MHz range; they are cost effective and can be mounted on most machine tools.

There is a relationship between the sensitivity of the AE power signal and small depth-of-cut changes ([16], [20], [21]). This relationship can be exploited to obtain a feedback signal or measurement of the depth-of-cut. A simplified block diagram of the feedback system is shown in Figure 10.30. The elements of the design process emphasized in this example are highlighted in Figure 10.31.

Since the acoustic emissions are sensitive to material, tool geometry, tool wear, and cutting parameters such as cutter rotational speed, the measurement of the depth-of-cut is modeled as being corrupted by noise, denoted by $N(s)$ in Figure 10.30. Also disturbances to the process, denoted by $T_d(s)$, are modeled. These could represent external disturbances resulting in unwanted motion of the cutter, fluctuations in the cutter rotation speed, and so forth.

The process model $G(s)$ is given by

$$G(s) = \frac{2}{s(s + 1)(s + 5)},$$ (10.99)

and represents the model of the cutter apparatus and the AE sensor dynamics. The input to $G(s)$ is a control signal to actuate an electromechanical device, which then applies downward pressure on the cutter.

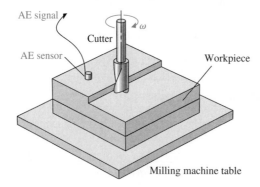

FIGURE 10.29
A depiction of the milling machine.

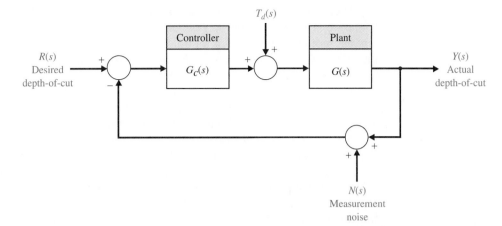

FIGURE 10.30
A simplified block
diagram of the
milling machine
feedback system.

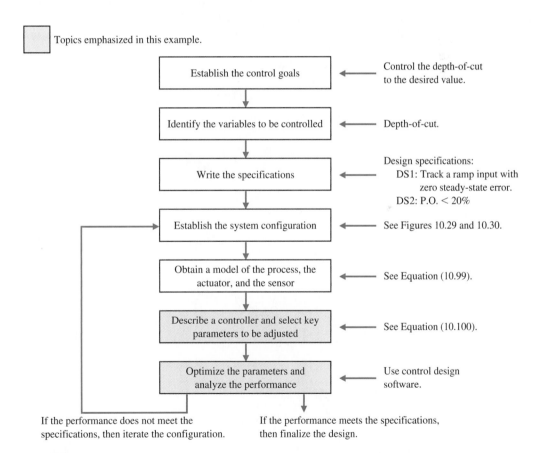

FIGURE 10.31   Elements of the control system design process emphasized in this milling machine
control system design example.

There are a variety of methods available to obtain the model represented by Equation (10.99). One approach would be to use basic principles to obtain a mathematical model in the form of a nonlinear differential equation, which can then be linearized about an operating point leading to a linear model (or equivalently, a transfer function). The basic principles include Newton's laws, the various conservation laws, and Kirchhoff's laws. Another approach would be to assume a form of the model (such as a second-order system) with unknown parameters (such as ω_n and ζ), and then experimentally obtain good values of the unknown parameters.

A third approach is to conduct a laboratory experiment to obtain the step or impulse response of the system. In other words we can apply an input (in this case, a voltage) to the system and measure the output—the depth-of-cut into the desired workpiece. Suppose, for example, we have the impulse response data shown in Figure 10.32 (the small circles on the graph represent the data). If we had access to the function $C_{imp}(t)$—the impulse response function of the milling machine—we could take the Laplace transform to obtain the transfer function model. There are many methods available for curve fitting the data to obtain the function $C_{imp}(t)$. We will not cover curve fitting here, but we can say a few words regarding the basic structure of the function.

From Figure 10.32 we see that the response approaches a steady-state value:

$$C_{imp}(t) \rightarrow C_{imp,ss} \approx \frac{2}{5} \text{ as } t \rightarrow \infty.$$

So we expect that

$$C_{imp}(t) = \frac{2}{5} + \Delta C_{imp}(t),$$

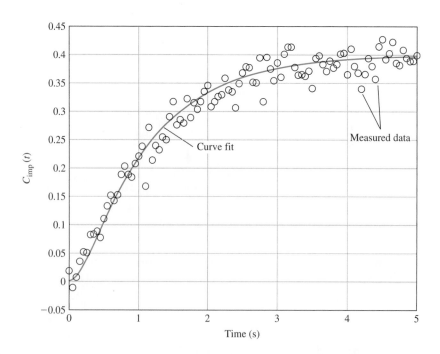

FIGURE 10.32
Hypothetical impulse response of the milling machine.

where $\Delta C_{imp}(t)$ is a function that goes to zero as t gets large. This leads us to consider $\Delta C_{imp}(t)$ as a sum of stable exponentials. Since the response does not oscillate, we might expect that the exponentials are, in fact, real exponentials,

$$\Delta C_{imp}(t) = \sum_i k_i e^{-\tau_i t},$$

where τ_i are positive real numbers. The data in Figure 10.32 can be fitted by the function

$$C_{imp}(t) = \frac{2}{5} + \frac{1}{10}e^{-5t} - \frac{1}{2}e^{-t},$$

for which the Laplace transform is

$$G(s) = \mathcal{L}\{C_{imp}(t)\} = \frac{2}{5}\frac{1}{s} + \frac{1}{10}\frac{1}{s+5} - \frac{1}{2}\frac{1}{s+1} = \frac{2}{s(s+1)(s+5)}.$$

Thus we can obtain the transfer function model of the milling machine.

The control goal is to develop a feedback system to track a desired step input. In this case the reference input is the desired depth-of-cut. The control goal is stated as

Control Goal
Control the depth-of-cut to the desired value.

The variable to be controlled is the depth-of-cut, or

Variable to Be Controlled
Depth-of-cut $y(t)$.

Since we are focusing on lead and lag controllers in this chapter, the key tuning parameters are the parameters associated with the compensator given in Equation. (10.100).

Select Key Tuning Parameters
Compensator variables: p, z, and K.

The control design specifications are

Control Design Specifications

DS1   Track a ramp input, $R(s) = a/s^2$, with a steady-state tracking error less than $a/8$, where a is the ramp velocity.

DS2   Percent overshoot to a step input less than 20%.

The lag compensator is given by

$$G_c(s) = \frac{K}{\alpha}\frac{s+z}{s+p}, \quad |p| < |z|, \tag{10.100}$$

where $\alpha = z/p$. The tracking error is

$$E(s) = R(s) - Y(s) = (1 - T(s))R(s),$$

where

$$T(s) = \frac{G_c(s)G(s)}{1 + G_c(s)G(s)}.$$

Therefore,

$$E(s) = \frac{1}{1 + G_c(s)G(s)}R(s).$$

With $R(s) = a/s^2$ and using the final value theorem, we find that

$$e_{ss} = \lim_{t\to\infty} e(t) = \lim_{s\to 0} sE(s) = \lim_{s\to 0} s\frac{1}{1 + G_c(s)G(s)}\frac{a}{s^2},$$

or equivalently,

$$\lim_{s\to 0} sE(s) = \frac{a}{\lim_{s\to 0} sG_c(s)G(s)}.$$

According to DS1, we require that

$$\frac{a}{\lim_{s\to 0} sG_c(s)G(s)} < \frac{a}{8},$$

or

$$\lim_{s\to 0} sG_c(s)G(s) > 8.$$

Substituting for $G(s)$ and $G_c(s)$ from Equations (10.99) and (10.100), respectively, we obtain the compensated velocity constant

$$K_{vcomp} = \frac{2}{5}\frac{K}{\alpha}\frac{z}{p} = \frac{2}{5}\hat{K}\frac{z}{p} > 8,$$

where $\hat{K} = K/\alpha$. The compensated velocity constant is the velocity constant of the system when the lag compensator is in the loop.

The loop transfer function is

$$L(s) = G_c(s)G(s) = \frac{s + z}{s + p}\frac{2\hat{K}}{s(s + 1)(s + 5)}.$$

We separate the lag compensator from the process and obtain the uncompensated root locus by considering the feedback loop with the gain $\hat{K}$, but not the lag compensator zero and pole factors. The uncompensated root locus for the characteristic equation

$$1 + \hat{K}\frac{2}{s(s + 1)(s + 5)} = 0$$

is shown in Figure 10.33.

From DS2 we determine that the target damping ratio of the dominant roots is $\zeta > 0.45$. We find that $\hat{K} \le 2.48$ at $\zeta \ge 0.45$. Then with $\hat{K} = 2.0$ the uncompensated velocity constant is

$$K_{v,unc} = \lim_{s\to 0} s\frac{2\hat{K}}{s(s + 1)(s + 5)} = \frac{2\hat{K}}{5} = 0.8.$$

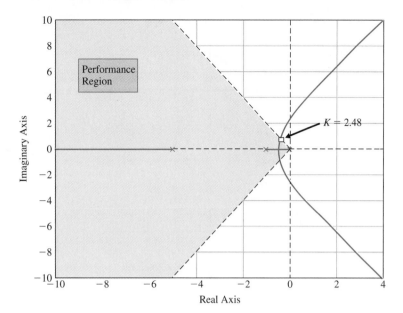

FIGURE 10.33
Root locus for the
uncompensated
system.

The compensated velocity constant is

$$K_{v,\text{comp}} = \lim_{s \to 0} s \frac{s + z}{s + p} \frac{2\hat{K}}{s(s + 1)(s + 5)} = \frac{z}{p} K_{v,\text{unc}}.$$

Therefore with $\alpha = z/p$, we obtain the relationship

$$\alpha = \frac{K_{v,\text{comp}}}{K_{v,\text{unc}}}.$$

We require $K_{v,\text{comp}} > 8$. A possible choice is $K_{v,\text{comp}} = 10$ as the desired velocity constant. Then

$$\alpha = \frac{K_{v,\text{comp}}}{K_{v,\text{unc}}} = \frac{10}{0.8} = 12.5$$

But $\alpha = z/p$, thus our lag compensator should have $p = 0.08z$. If we select $z = 0.01$ then $p \approx 0.0008$.

The compensated loop transfer function is given by

$$G_c(s)G(s) = \hat{K} \frac{s + z}{s + p} \frac{2}{s(s + 1)(s + 5)}.$$

The lag compensator with z and p as above is determined to be

$$G_c(s) = 2.0 \frac{s + 0.01}{s + 0.0008}. \qquad (10.101)$$

The step response is shown in Figure 10.34. The percent overshoot is approximately 20%. The velocity error constant is approximately 10, which satisfies DS1. ∎

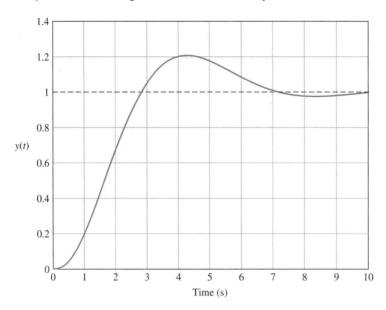

FIGURE 10.34
Step response for the compensated system.

10.13 SYSTEM DESIGN USING CONTROL DESIGN SOFTWARE

We want to use computers, when appropriate, to assist the designer in the selection of the parameters of a compensator. The development of algorithms for computer-aided design is an important alternative approach to the trial-and-error methods considered in earlier sections. Computer programs have been developed for the selection of suitable parameter values for compensators based on satisfaction of frequency response criteria such as the phase margin [3, 4].

In this section, the compensation of control systems is illustrated using frequency response and s-plane methods. We will consider again the rotor winder design example of Section 10.12 to illustrate the use of m-file scripts in designing and developing control systems with good performance characteristics. We examine both the lead and lag compensators for this design example and obtain the system response using computer-based analysis tools.

EXAMPLE 10.16   **Rotor winder control system**

Let us consider again the rotor winder control system shown in Figure 10.24. The design objective is to achieve high steady-state accuracy to a ramp input. The steady-state error to a unit ramp input $R(s) = 1/s^2$ is

$$e_{ss} = \frac{1}{K_v},$$

where

$$K_v = \lim_{s \to 0} \frac{G_c(s)}{50}.$$

The performance specification of overshoot and settling time must be considered, as must the steady-state tracking error. In all likelihood, a simple gain will not be satisfactory, so we will also consider compensation utilizing lead and lag compensators, using both Bode diagram and root locus plot design methods. Our approach is to develop a series of m-file scripts to aid in the compensator designs.

Consider a simple gain controller

$$G_c(s) = K.$$

Then the steady-state error is

$$e_{ss} = \frac{50}{K}.$$

The larger we make K, the smaller is the steady-state error e_{ss}. However, we must consider the effect that increasing K has on the transient response, as shown in Figure 10.35. When $K = 500$, our steady-state error for a ramp is 10%, but the overshoot is 70%, and the settling time is approximately 8 seconds for a step input. We consider this to be unacceptable performance and thus turn to compensation. The two important compensator types that we consider are lead and lag compensators.

First, we try a lead compensator

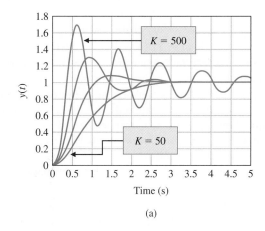

(a)

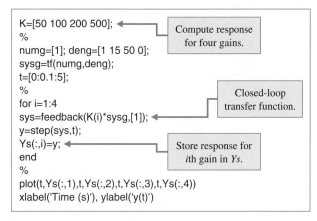

(b)

FIGURE 10.35
(a) Transient response for simple gain controller.
(b) m-file script.

$$G_c(s) = \frac{K(s + z)}{s + p},$$

where $|z| < |p|$. The lead compensator will give us the capability to improve the transient response. We will use a frequency-domain approach to design the lead compensator.

We want a steady-state error of less than 10% to a ramp input and $K_v = 10$. In addition to the steady-state specifications, we want to meet certain performance specifications: (1) settling time (with a 2% criterion) $T_s \leq 3$ s, and (2) percent overshoot for a step input $\leq 10\%$. Solving for ζ and ω_n using

$$P.O. = 100 \exp^{-\zeta\pi/\sqrt{1-\zeta^2}} = 10 \quad \text{and} \quad T_s = \frac{4}{\zeta\omega_n} = 3$$

yields $\zeta = 0.59$ and $\omega_n = 2.26$. We thus obtain the phase margin requirement:

$$\phi_{pm} = \frac{\zeta}{0.01} \approx 60°.$$

The steps leading to the final design are as follows:

1. Obtain the uncompensated system Bode diagram with $K = 500$, and compute the phase margin.
2. Determine the amount of necessary phase lead ϕ_m.
3. Evaluate α from $\sin \phi_m = (\alpha - 1)/(\alpha + 1)$.
4. Compute $10 \log \alpha$ and find the frequency ω_m on the uncompensated Bode diagram where the magnitude curve is equal to $-10 \log \alpha$.
5. In the neighborhood of ω_m on the uncompensated Bode diagram, draw a line through the 0-dB point at ω_m with slope equal to the current slope plus 20 dB/decade. Locate the intersection of the line with the uncompensated Bode diagram to determine the lead compensation zero location. Then calculate the lead compensator pole location as $p = \alpha z$.
6. Obtain the compensated Bode diagram and check the phase margin. Repeat any steps if necessary.
7. Raise the gain to account for the attenuation $1/\alpha$.
8. Verify the final design with simulation using step function inputs, and repeat any design steps if necessary.

We use three scripts in the design. The design scripts are shown in Figure 10.36–10.38. The script in Figure 10.36 is for the Bode diagram of the uncompensated system. The script in Figure 10.37 is for the detailed Bode diagram of the compensated system. The script in Figure 10.38 is for the step response analysis. The final lead compensator design is

$$G_c(s) = \frac{1800(s + 3.5)}{s + 25},$$

where $K = 1800$ was selected after iteratively using the m-file script.

The settling time and overshoot specifications are satisfied, but $K_v = 5$, resulting in a 20% steady-state error to a ramp input. It is possible to continue the design iteration and refine the compensator somewhat, although it should be clear that the lead compensator has added phase margin and improved the transient response as anticipated.

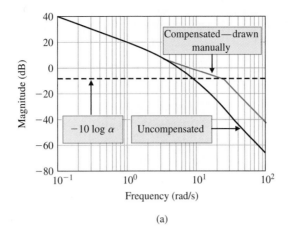

FIGURE 10.36
(a) Bode diagram.
(b) m-file script.

(a)

```
K=500;
numg=[1]; deng=[1 15 50 0]; sysg=tf(numg,deng);
sys=K*sysg;
%
[Gm,Pm,Wcg,Wcp]=margin(sys);        ◄——— Compute
%                                          phase margin.
Phi=(60-Pm)*pi/180;     ◄——————— Additional phase lead.
alpha=(1+sin(Phi))/(1-sin(Phi))
[mag,phase,w]=bode(sys);
mag_save(1,:)=mag(:,1,:);           ◄——— Compute α.
%
M=-10*log10(alpha)*ones(length(w),1);  ◄——— Plot −10 log(α) line to
%                                              aid in locating ωₘ.
semilogx(w,20*log10(mag_save),w,M), grid
xlabel('Frequency (rad/s)'), ylabel('Magnitude (dB)')
```

(b)

To reduce the steady-state error, we can consider the lag compensator, which has the form

$$G_c(s) = \frac{K(s + z)}{s + p},$$

where $|p| < |z|$. We will use a root locus approach to design the lag compensator, although it can be done using a Bode diagram as well. The desired root location region of the dominant roots is specified by

$$\zeta = 0.59 \quad \text{and} \quad \omega_n = 2.26.$$

The steps in the design are as follows:

1. Obtain the root locus of the uncompensated system.
2. Locate suitable root locations on the uncompensated system that lie in the region defined by $\zeta = 0.59$ and $\omega_n = 2.26$.
3. Calculate the loop gain at the desired root location and the system error constant, $K_{v,\text{unc}}$.
4. Compute $\alpha = K_{v,\text{comp}}/K_{v,\text{unc}}$, where $K_{v,\text{comp}} = 10$.

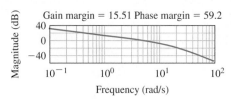

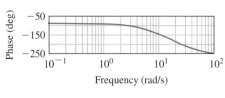

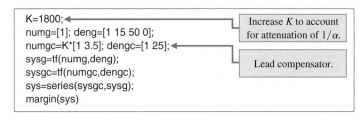

FIGURE 10.37
Lead compensator:
(a) compensated
Bode diagram,
(b) m-file script.

(b)

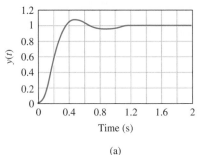

```
K=1800;
%
numg=[1]; deng=[1 15 50 0]; sysg=tf(numg,deng);
numgc=K*[1 3.5]; dengc=[1 25]; sysgc=tf(numgc,dengc);
%
syso=series(sysgc,sysg);
sys=feedback(syso,[1]);
%
t=[0:0.01:2];
step(sys,t)
ylabel ('y(t)')
```

FIGURE 10.38
Lead compensator:
(a) step response,
(b) m-file script.

(b)

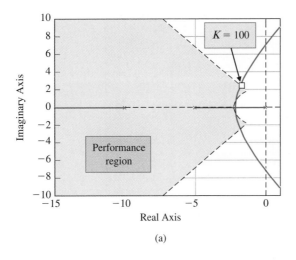

FIGURE 10.39
Lag compensator:
(a) uncompensated
root locus, (b) m-file
script.

5. With α known, determine suitable locations of the compensator pole and zero so that the compensated root locus still passes through the desired location.

6. Verify with simulation and repeat any steps if necessary.

The design methodology is illustrated in Figures 10.39–10.41. Using the rlocfind function, we can compute the gain K associated with the roots of our choice on the uncompensated root locus that lie in the performance region. We then compute α to ensure that we achieve the desired K_v. We place the lag compensator pole and zero to avoid affecting the uncompensated root locus. In Figure 10.40, the lag compensator pole and zero are very near the origin, at $z = -0.1$ and $p = -0.01$.

The settling time and overshoot specifications are not satisfied, but $K_v = 10$, as desired. It is possible to continue the design iteration and refine the compensator somewhat, although it should be clear that the lag compensator has improved the steady-state errors to a ramp input relative to the lead compensator design. The final lag compensator design is

$$G_c(s) = \frac{100(s + 0.1)}{s + 0.01}.$$

The resulting performance is summarized in Table 10.5. ∎

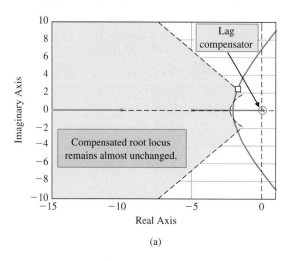

```
numg=[1]; deng=[1 15 50 0]; sysg=tf(numg,deng);
numgc=[1 0.1]; dengc=[1 0.01]; sysgc=tf(numgc,dengc);
sys=series(sysgc,sysg);
clf; rlocus(sys); hold on
%
zeta=0.5912; wn=2.2555;
x=[-10:0.1:-zeta*wn]; y=-(sqrt(1-zeta^2)/zeta)*x;
xc=[-10:0.1:-zeta*wn];c=sqrt(wn^2-xc.^2);
plot(x,y,':',x,-y,':',xc,c,':',xc,-c,':')
axis([-15,1,-10,10]);
```

FIGURE 10.40
Lag compensator:
(a) compensated
root locus, (b) m-file
script.

(b)

Table 10.5   Compensator Design Results

Controller	Gain, $K = 500$	Lead	Lag
Step overshoot	70%	8%	13%
Settling time (seconds)	8	1	9
Steady-state error for ramp	10%	20%	10%
K_v	10	5	10

10.14 SEQUENTIAL DESIGN EXAMPLE: DISK DRIVE READ SYSTEM

In this chapter, we design a proportional-derivative controller (PD) to achieve the specified response to a unit step input. The specifications are given in Table 10.6. The closed-loop system is shown in Figure 10.42. A prefilter is used to eliminate any undesired effects of the term $s + z$ introduced in the closed-loop transfer function. We will

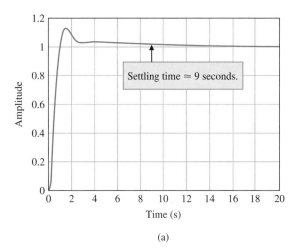

Settling time ≈ 9 seconds.

(a)

```
K=100;
%
numg=[1]; deng=[1 15 50 0]; sysg=tf(numg,deng);
numgc=K*[1 0.1]; dengc=[1 0.01]; sysgc=tf(numgc,dengc);
%
syso=series(sysgc,sysg);
sys=feedback(syso,[1]);
%
step(sys)
```

FIGURE 10.41
Lag compensator:
(a) step response,
(b) m-file response.

(b)

Table 10.6   Disk Drive Control System Specifications and Actual Performance

Performance Measure	Desired Value	Actual Response
Percent overshoot	Less than 5%	0.1%
Settling time	Less than 250 ms	40 ms
Maximum response to a unit disturbance	Less than 5×10^{-3}	6.9×10^{-5}

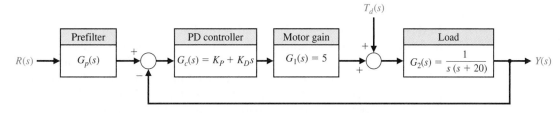

FIGURE 10.42   Disk drive control system with PD controller (second-order model).

use the deadbeat system of Section 10.11, where the desired closed-loop transfer function (Equation 10.86) is

$$T(s) = \frac{\omega_n^2}{s^2 + \alpha\omega_n s + \omega_n^2}. \tag{10.102}$$

For the second-order model shown in Figure 10.42, we require $\alpha = 1.82$ (see Table 10.2). Then the settling time is

$$\omega_n T_s = 4.82.$$

Since we want a settling time less than 50 ms, we will use $\omega_n = 120$. Then we expect $T_s = 40$ ms. Therefore, the denominator of Equation (10.102) is

$$s^2 + 218.4s + 14400. \tag{10.103}$$

The characteristic equation of the closed-loop system of Figure 10.42 is

$$s^2 + (20 + 5K_D)s + 5K_P = 0. \tag{10.104}$$

Equating Equations (10.103) and (10.104), we have

$$218.4 = 20 + 5K_D$$

and

$$14400 = 5K_P.$$

Therefore, $K_P = 2880$ and $K_D = 39.68$. Then we note that

$$G_c(s) = 39.68(s + 72.58).$$

The prefilter will then be

$$G_p(s) = \frac{72.58}{s + 72.58}.$$

The model neglected the motor field. Nevertheless, this design will be very accurate. The actual response is given in Table 10.6. All the specifications are satisfied.

10.15 SUMMARY

In this chapter, we have considered several alternative approaches to the design of feedback control systems. In the first two sections, we discussed the concepts of design and compensation and noted the several design cases that we completed in the preceding chapters. Then we examined the possibility of introducing cascade compensation networks within the feedback loops of control systems. The cascade compensation networks are useful for altering the shape of the root locus or frequency response of a system. The phase-lead network and the phase-lag network were considered in detail as candidates for system compensators. Then system compensation was studied by using a phase-lead s-plane network on the Bode diagram and the root locus s-plane. We noted that the phase-lead compensator increases the phase margin of the system and thus provides additional stability. When the design specifications include an error constant, the design of a phase-lead network is more readily accomplished on the Bode diagram. Alternatively, when an error constant is not specified but the settling time and overshoot for a step input are specified, the design of a phase-lead network is more readily carried out on the s-plane. When large error constants are specified for a feedback system, it is usually easier to compensate the system by using integration (phase-lag) networks. We also noted that the phase-lead compensation increases the system bandwidth, whereas the phase-lag compensation decreases the system bandwidth. The bandwidth may often be an important factor when noise is present at the input and generated within the system. Also, we noted that a satisfactory system is obtained when the asymptotic course for magnitude of the compensated system crosses the 0-dB line with a slope of -20 dB/decade. The characteristics of the phase-lead and phase-lag compensation networks are summarized in Table 10.7. Operational amplifier circuits for phase-lead and phase-lag and for PI and PD compensators are summarized in Table 10.8 [1]. The use of these controllers has been widely demonstrated in this and earlier chapters. These operational amplifier circuits are widely used in industrial practice to provide the compensator $G_c(s)$.

Table 10.7   A Summary of the Characteristics of Phase-Lead and Phase-Lag Compensation Networks

	Compensation	
	Phase-Lead	Phase-Lag
Approach	Addition of phase-lead angle near crossover frequency on Bode diagram. Add lead network to yield desired dominant roots in s-plane.	Addition of phase-lag to yield an increased error constant while maintaining desired dominant roots in s-plane or phase margin on Bode diagram
Results	1. Increases system bandwidth 2. Increases gain at higher frequencies	1. Decreases system bandwidth
Advantages	1. Yields desired response 2. Improves dynamic response	1. Suppresses high-frequency noise 2. Reduces steady-state error
Disadvantages	1. Requires additional amplifier gain 2. Increases bandwidth and thus susceptibility to noise 3. May require large values of components for RC network	1. Slows down transient response 2. May require large values of components for RC network
Applications	1. When fast transient response is desired	1. When error constants are specified
Situations not applicable	1. When phase decreases rapidly near crossover frequency	1. When no low-frequency range exists where phase is equal to desired phase margin

Table 10.8   Operational Amplifier Circuits for Compensators

Type of Controller	$G_c(s) = \dfrac{V_0(s)}{V_1(s)}$	
PD	$G_c = \dfrac{R_4 R_2}{R_3 R_1}(R_1 C_1 s + 1)$	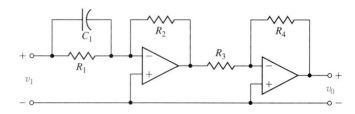
PI	$G_c = \dfrac{R_4 R_2 (R_2 C_2 s + 1)}{R_3 R_1 (R_2 C_2 s)}$	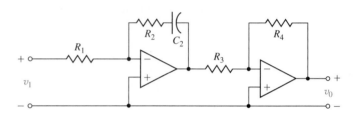
Lead or lag Lead if $R_1 C_1 > R_2 C_2$ Lag if $R_1 C_1 < R_2 C_2$	$G_c = \dfrac{R_4 R_2 (R_1 C_1 s + 1)}{R_3 R_1 (R_2 C_2 s + 1)}$	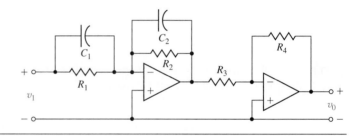

EXERCISES

E10.1   A negative feedback control system has a transfer function

$$G(s) = \frac{K}{s + 2}.$$

We select a compensator

$$G_c(s) = \frac{s + a}{s},$$

in order to achieve zero steady-state error for a step input. Select a and K so that the overshoot to a step is approximately 5% and the settling time (with a 2% criterion) is approximately 1 second.

Answer: $K = 6$, $a = 5.6$

E10.2   A control system with negative unity feedback has a process

$$G(s) = \frac{400}{s(s + 40)},$$

and we wish to use proportional plus integral compensation, where

$$G_c(s) = K_P + \frac{K_I}{s}.$$

Note that the steady-state error of this system for a ramp input is zero. (a) Set $K_I = 1$ and find a suitable value of K_P so the step response will have an overshoot of approximately 20%. (b) What is the expected settling time (with a 2% criterion) of the compensated system?

Answer : $K_P = 0.5$

E10.3   A unity negative feedback control system in a manufacturing system has a process transfer function

$$G(s) = \frac{e^{-s}}{s + 1},$$

and it is proposed that we use a compensator to achieve a 5% overshoot to a step input. The compensator is [4]

$$G_c(s) = K\left(1 + \frac{1}{\tau s}\right),$$

which provides proportional plus integral control. Show that one solution is $K = 0.5$ and $\tau = 1$.

E10.4   Consider a unity negative feedback system with

$$G(s) = \frac{K}{s(s + 5)(s + 10)},$$

where K is set equal to 100 in order to achieve a specified $K_v = 2$. We wish to add a lead-lag compensator

$$G_c(s) = \frac{(s + 0.15)(s + 0.7)}{(s + 0.015)(s + 7)}.$$

Show that the gain margin of the compensated system is 28.6 dB and that the phase margin is 75.4°.

E10.5   Consider a unity feedback system with the transfer function

$$G(s) = \frac{K}{s(s + 2)(s + 4)}.$$

We desire to obtain the dominant roots with $\omega_n = 3$ and $\zeta = 0.5$. We wish to obtain a $K_v = 2.7$. Show that we require a compensator

$$G_c(s) = \frac{7.53(s + 2.2)}{(s + 16.4)}.$$

Determine the value of K that should be selected.

Answer : $K = 22$

E10.6   Consider again the wind tunnel control system of Problem P7.31. When $K = 326$, find $T(s)$ and estimate the expected overshoot and settling time (with a 2% criterion). Compare your estimates with the actual overshoot of 60% and a settling time of 4 seconds. Explain the discrepancy in your estimates.

E10.7   NASA astronauts retrieved a satellite and brought it into the cargo bay of the space shuttle, as shown in Figure E10.7(a). A model of the feedback control system is shown in Figure. E10.7(b). Determine the value of K that will result in a phase margin of 45° when $T = 0.5$ s.

Answer : $K = 20.88$

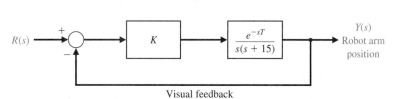

(a)

(b)

FIGURE E10.7
Retrieval of a
satellite.

E10.8   A negative unity feedback system has a plant

$$G(s) = \frac{2257}{s(\tau s + 1)},$$

where $\tau = 2.8$ ms. Select a compensator

$$G_c(s) = K_P + K_I/s,$$

so that the dominant roots of the characteristic equation have ζ equal to $1/\sqrt{2}$. Plot $y(t)$ for a step input.

E10.9   A control system with a controller is shown in Figure E10.9. Select K_P and K_I so that the overshoot to a step input is equal to 5% and the velocity constant K_v is equal to 5. Verify the results of your design.

E10.10   A control system with a controller is shown in Figure E10.10. We will select $K_I = 4$ in order to provide a reasonable steady-state error to a step [8]. Find K_P to obtain a phase margin of 60°. Find the peak time and percent overshoot of this system.

E10.11   A unity feedback system has

$$G(s) = \frac{1350}{s(s + 2)(s + 30)}.$$

A lead network is selected so that

$$G_c(s) = \frac{1 + 0.25s}{1 + 0.025s}.$$

Determine the peak magnitude and the bandwidth of the closed-loop frequency response using (a) the Nichols chart, and (b) a plot of the closed-loop frequency response.

Answer: $M_{pw} = 2.3$ dB, $\omega_B = 22$

E10.12   The control of an automobile ignition system has unity negative feedback and a loop transfer function $G_c(s)G(s)$, where

$$G(s) = \frac{K}{s(s + 5)} \quad \text{and} \quad G_c(s) = K_P + K_I/s.$$

A designer selects $K_I/K_P = 0.5$ and asks you to determine $K K_P$ so that the complex roots have a ζ of $1/\sqrt{2}$.

E10.13   The design of Example 10.3 determined a lead network in order to obtain desirable dominant root locations using a cascade compensator $G_c(s)$ in the system configuration shown in Figure 10.1(a). The same lead network would be obtained if we used the feedback compensation configuration of Figure 10.1(b). Determine the closed-loop transfer function $T(s) = Y(s)/R(s)$ of both the cascade and feedback configurations, and show how the transfer function of each configuration differs. Explain how the response to a step $R(s)$ will be different for each system.

E10.14   A robot will be operated by NASA to build a permanent lunar station. The position control system for the gripper tool is shown in Figure 10.1(a), where $H(s) = 1$, and

$$G(s) = \frac{5}{s(s + 1)(0.25s + 1)}.$$

Determine a compensator lag network $G_c(s)$ that will provide a phase margin of 45°.

Answer: $G_c(s) = \dfrac{1 + 7.5s}{1 + 110s}$

E10.15   A unity feedback control system has a plant transfer function

$$G(s) = \frac{40}{s(s + 2)}.$$

We desire to attain a steady-state error to a ramp $r(t) = At$ of less than $0.05A$ and a phase margin of 30°. We desire to have a crossover frequency ω_c of 10 rad/s.

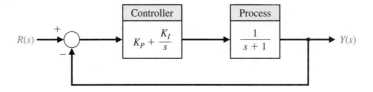

FIGURE E10.9
Design of a controller.

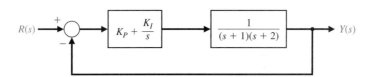

FIGURE E10.10
Design of a PI controller.

Use the methods of Section 10.9 to determine whether a lead or a lag compensator is required.

E10.16 Consider again the system and specifications of Exercise E10.15 when the required crossover frequency is 2 rad/s.

E10.17 Consider again the system of Exercise 10.9. Select K_P and K_I so that the step response is deadbeat and the settling time (with a 2% criterion) is less than 2 seconds.

E10.18 The nonunity feedback control system shown in Figure E10.18 has the transfer functions

$$G(s) = \frac{1}{s - 20} \quad \text{and} \quad H(s) = 10.$$

Design a compensator $G_c(s)$ and prefilter $G_p(s)$ so that the closed-loop system is stable and meets the following specifications: (i) a percent overshoot to a unit step input of less than 10%, (ii) a settling time of less than 2 seconds, and (iii) zero steady-state tracking error to a unit step.

E10.19 A unity feedback control system has the plant transfer function

$$G(s) = \frac{1}{s(s - 5)}.$$

Design a PID controller of the form

$$G_c(s) = K_p + K_D s + \frac{K_I}{s}$$

so that the closed-loop system has a settling time less than 1 second to a unit step input.

E10.20 Consider the system shown in Figure E10.20. Design the proportional-derivative controller $G_c(s) = K_P + K_D s$ such that the system has a phase margin of $P.M. \approx 45°$.

E10.21 Consider the unity feedback system shown in Figure E10.21. Design the controller gain, K, such that the maximum value of the output $y(t)$ in response to a unit step disturbance $T_d(s) = 1/s$ is less than 0.1.

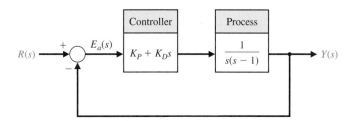

FIGURE E10.18
Nonunity feedback system with a prefilter.

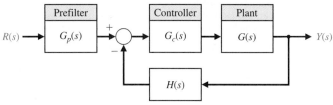

FIGURE E10.20
Unity feedback system with PD controller.

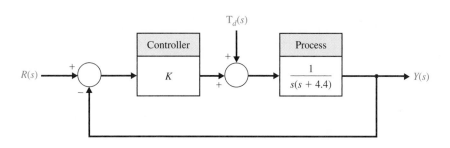

FIGURE E10.21
Closed-loop feedback system with a disturbance input.

PROBLEMS

P10.1   The design of a lunar excursion module (LEM) is an interesting control problem. The attitude control system for the lunar vehicle is shown in Figure P10.1. The vehicle damping is negligible, and the attitude is controlled by gas jets. The torque, as a first approximation, will be considered to be proportional to the signal $V(s)$ so that $T(s) = K_2V(s)$. The loop gain may be selected by the designer in order to provide a suitable damping. A damping ratio of $\zeta = 0.6$ with a settling time (with a 2% criterion) of less than 2.5 seconds is required. Using a lead network compensation, select the necessary compensator $G_c(s)$ by using (a) frequency response techniques and (b) root locus methods.

P10.2   A magnetic tape recorder transport for modern computers requires a high-accuracy, rapid-response control system. The requirements for a specific transport are as follows: (1) The tape must stop or start in 10 ms, and (2) it must be possible to read 45,000 characters per second. This system was discussed in Problem P7.11. We desire to set $J = 5 \times 10^{-3}$, and K_1 is set on the basis of the maximum error allowable for a velocity input. In this case, we desire to maintain a steady-state speed error of less than 5%. We will use a tachometer in this case and set $K_a = 50{,}000$ and $K_2 = 1$. To provide

a suitable performance, a compensator $G_c(s)$ is inserted immediately following the photocell transducer. Select a compensator $G_c(s)$ so that the overshoot of the system for a step input is less than 25%. We will assume that $\tau_1 = \tau_a = 0$.

P10.3   A simplified version of the attitude rate control for the F-94 or X-15 type aircraft is shown in Figure P10.3. When the vehicle is flying at four times the speed of sound (Mach 4) at an altitude of 100,000 ft, the parameters are [30]

$$\frac{1}{\tau_a} = 1.0, \qquad K_1 = 1.0,$$

$$\zeta\omega_a = 1.0, \qquad \text{and} \qquad \omega_a = 4$$

Design a compensator $G_c(s)$ so that the response to a step input has an overshoot of less than 5% and a settling time (with a 2% criterion) of less than 5 seconds.

P10.4   Magnetic particle clutches are useful actuator devices for high power requirements because they can typically provide a 200-W mechanical power output. The particle clutches provide a high torque-to-inertia ratio and fast time-constant response. A particle clutch positioning system for nuclear reactor rods is shown in Figure P10.4. The motor drives two counter-rotating clutch housings.

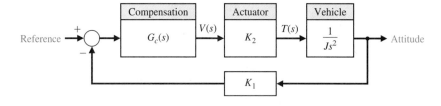

FIGURE P10.1
Attitude control system for a lunar excursion module.

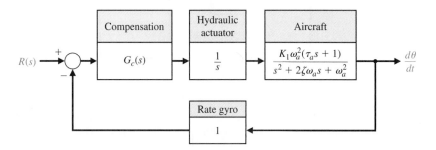

FIGURE P10.3
Aircraft attitude control.

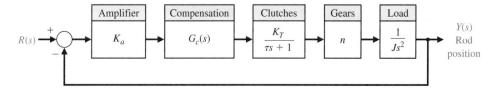

FIGURE P10.4
Nuclear reactor rod control.

The clutch housings are geared through parallel gear trains, and the direction of the servo output is dependent on the clutch that is energized. The time constant of a 200-W clutch is $\tau = 1/40$ s. The constants are such that $K_T n/J = 1$. We want the maximum overshoot for a step input to be in the range of 10% to 20%. Design a compensating network so that the system is adequately stabilized. The settling time (with a 2% criterion) of the system should be less than or equal to 2 seconds.

P10.5 A stabilized precision rate table uses a precision tachometer and a DC direct-drive torque motor, as shown in Figure P10.5. We want to maintain a high steady-state accuracy for the speed control. To obtain a zero steady-state error for a step command design, select a proportional plus integral compensator. Select the appropriate gain constants so that the system has an overshoot of approximately 10% and a settling time (with a 2% criterion) less than 1.5 seconds.

P10.6 Repeat Problem 10.5 by using a lead network compensator, and compare the results.

P10.7 The primary control loop of a nuclear power plant includes a time delay due to the need to transport the fluid from the reactor to the measurement point (see Figure P10.7). The transfer function of the controller is

$$G_c(s) = K_P + \frac{K_I}{s}.$$

The transfer function of the reactor and time delay is

$$G(s) = \frac{e^{-sT}}{\tau s + 1},$$

where $T = 0.4$ s and $\tau = 0.2$ s. Using frequency response methods, design the controller so that the overshoot of the system is less than 10%. Estimate the settling time (with a 2% criterion) of the system designed. Determine the actual overshoot and settling time.

P10.8 A chemical reactor process whose production rate is a function of catalyst addition is shown in block diagram form in Figure P10.8 [11]. The time delay is

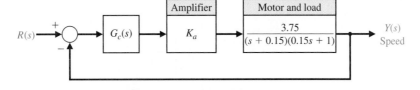

FIGURE P10.5
Stabilized rate table.

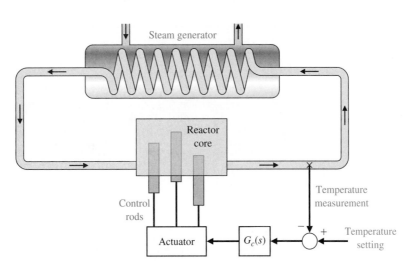

FIGURE P10.7
Nuclear reactor control.

FIGURE P10.8
Chemical reactor control.

$T = 50$ s, and the time constant τ is approximately 40 s. The gain of the process is $K = 1$. Design a compensation by using Bode diagram methods in order to provide a suitable system response. We want to have a steady-state error less than $0.10A$ for a step input $R(s) = A/s$. For the system with the compensation added, estimate the settling time of the system.

P10.9   A numerical path-controlled machine turret lathe is an interesting problem in attaining sufficient accuracy [2, 26]. A block diagram of a turret lathe control system is shown in Figure P10.9. The gear ratio is $n = 0.1$, $J = 10^{-3}$, and $b = 10^{-2}$. It is necessary to attain an accuracy of 5×10^{-4} in., and therefore a steady-state position accuracy of 2.5% is specified for a ramp input. Design a cascade compensator to be inserted before the silicon-controlled rectifiers in order to provide a response to a step command with an overshoot of less than 5%. A suitable damping ratio for this system is 0.7. The gain of the silicon-controlled rectifiers is $K_R = 5$. Design a suitable lag compensator by using the (a) Bode diagram method and (b) s-plane method.

P10.10   The Avemar ferry, shown in Figure P10.10(a), is a large 670-ton ferry hydrofoil built for Mediterranean ferry service. It is capable of 45 knots (52 mph) [33]. The boat's appearance, like its performance, derives from the innovative design of the narrow "wavepiercing" hulls which move through the water like racing shells. Between the hulls is a third quasihull which gives additional buoyancy in rough seas. Loaded with 900 passengers and crew, and a mix of cars, buses, and freight cars trucks, one of the boats can carry almost its own weight. The Avemar is capable of operating in seas with waves up to 8 ft in amplitude at a speed of 40 knots as a result of an automatic stabilization control system. Stabilization is achieved by means of flaps on the main foils and the adjustment of the aft foil. The stabilization control system maintains a level flight through rough seas. Thus, a system that minimizes deviations from a constant lift force or, equivalently, that minimizes the pitch angle θ has been designed. A block diagram of the lift control system is shown in Figure P10.10(b). The desired response of the system to wave disturbance is a constant-level travel of the

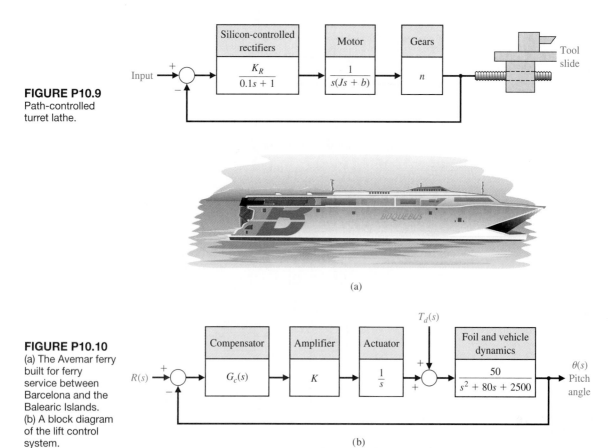

FIGURE P10.9
Path-controlled turret lathe.

(a)

FIGURE P10.10
(a) The Avemar ferry built for ferry service between Barcelona and the Balearic Islands.
(b) A block diagram of the lift control system.

(b)

craft. Establish a set of reasonable specifications and design a compensator $G_c(s)$ so that the performance of the system is suitable. Assume that the disturbance is due to waves with a frequency $\omega = 6$ rad/s.

P10.11 A unity feedback system of the form shown in Figure 10.1(a) has a plant

$$G(s) = \frac{5}{s(s^2 + 6s + 10)}.$$

(a) Determine the step response when $G_c(s) = 1$, and calculate the settling time and steady state for a ramp input $r(t) = t, t > 0$. (b) Design a lag network using the root locus method so that the velocity constant is increased to 10. Determine the settling time (with a 2% criterion) of the compensated system.

P10.12 A unity feedback control system of the form shown in Figure 10.1(a) has a plant

$$G(s) = \frac{160}{s^2}.$$

Select a lead-lag compensator so that the percent overshoot for a step input is less than 5% and the settling time (with a 2% criterion) is less than 1 second. It also is desired that the acceleration constant K_a be greater than 7500 (see Table 5.5).

P10.13 A unity feedback system has a plant

$$G(s) = \frac{20}{s(1 + 0.1s)(1 + 0.05s)}.$$

Select a compensator $G_c(s)$ so that the phase margin is at least 75°. Use a two-stage lead compensator

$$G_c(s) = \frac{K(1 + s/\omega_1)(1 + s/\omega_3)}{(1 + s/\omega_2)(1 + s/\omega_4)}.$$

It is required that the error for a ramp input be 0.5% of the magnitude of the ramp input ($K_v = 200$).

P10.14 Materials testing requires the design of control systems that can faithfully reproduce normal specimen operating environments over a range of specimen parameters [26]. From the control system design viewpoint, a materials-testing machine system can be considered a servomechanism in which we want to have the load waveform track the reference signal. The system is shown in Figure P10.14.

(a) Determine the phase margin of the system with $G_c(s) = K$, choosing K so that a phase margin of 50° is achieved. Determine the system bandwidth for this design.

(b) The additional requirement introduced is that the velocity constant K_v be equal to 2.0. Design a lag network so that the phase margin is 50° and $K_v = 2$.

P10.15 For the system described in Problem 10.14, the goal is to achieve a phase margin of 50° with the additional requirement that the time to settle (to within 2% of the final value) be less than 4 seconds. Design a lead network to meet the specifications. As before, we require $K_v = 2$.

P10.16 A robot with an extended arm has a heavy load, whose effect is a disturbance, as shown in Figure P10.16 [25]. Let $R(s) = 0$ and design $G_c(s)$ so that the effect of the disturbance is less than 20% of the open-loop system effect.

P10.17 A driver and car may be represented by the simplified model shown in Figure P10.17 [18]. The goal is to have the speed adjust to a step input with less than 10% overshoot and a settling time (with a 2% criterion) of 1 second. Select a proportional plus integral (PI) controller to yield these specifications. For the selected controller, determine the actual response (a) for $G_p(s) = 1$ and (b) with a prefilter $G_p(s)$ that removes the zero from the closed-loop transfer function $T(s)$.

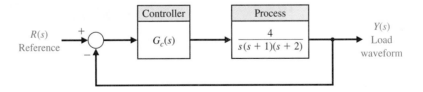

FIGURE P10.14

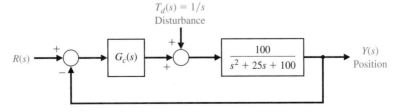

FIGURE P10.16
Robot control.

$R(s) \longrightarrow \boxed{G_p(s)} \xrightarrow{+} \bigcirc \xrightarrow{-} \boxed{G_c(s)} \longrightarrow \boxed{\dfrac{1}{s}} \longrightarrow \begin{array}{l} Y(s) \\ \text{Speed} \end{array}$

FIGURE P10.17
Speed control of an
automobile.

P10.18 A unity feedback control system for a robot submarine has a plant with a third-order transfer function [22]:

$$G(s) = \frac{K}{s(s + 10)(s + 50)}.$$

We want the overshoot to be approximately 7.5% for
a step input and the settling time (with a 2% criterion)
of the system be 400 ms. Find a suitable phase-lead
compensator by using root locus methods. Let the
zero of the compensator be located at $s = -15$, and
determine the compensator pole. Determine the
resulting system K_v.

P10.19 NASA is developing remote manipulators that
can be used to extend the hand and the power of
humankind through space by means of radio. A concept
of a remote manipulator is shown in Figure P10.19(a)
[12, 25]. The closed-loop control is shown schematically

in Figure P10.19(b). Assuming an average distance of
238,855 miles from Earth to the moon, the time delay T in
transmission of a communication signal is 1.28 seconds.
The operator uses a control stick to control remotely
the manipulator placed on the moon to assist in geological
experiments, and the TV display to access the response of
the manipulator. The time constant of the manipulator is
$\frac{1}{4}$ second.

(a) Set the gain K_1 so that the system has a phase margin of approximately 30°. Evaluate the percentage
steady-state error for this system for a step input.
(b) To reduce the steady-state error for a position
command input to 5%, add a lag compensation network in cascade with K_1. Plot the step response.

P10.20 There have been significant developments in the
application of robotics technology to nuclear power
plant maintenance problems. Thus far, robotics

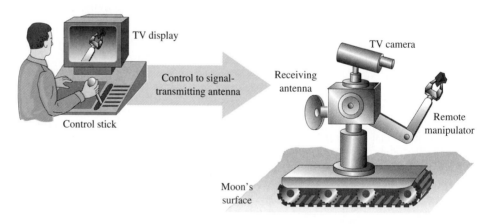

(a)

FIGURE P10.19
(a) Conceptual
diagram of a
remote manipulator
on the moon
controlled by a
person on the
Earth. (b) Feedback
diagram of the
remote manipulator
control system with
τ = transmission
time delay of the
video signal.

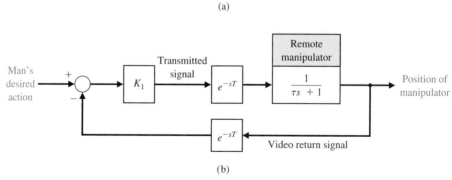

(b)

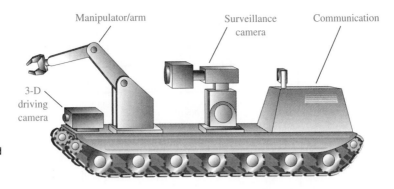

Manipulator/arm   Surveillance   Communication
camera

3-D
driving
camera

FIGURE P10.20
Remotely controlled
robot for nuclear
plants.

technology in the nuclear industry has been used primarily on spent-fuel reprocessing and waste management. Today, the industry is beginning to apply the technology to such areas as primary containment inspection, reactor maintenance, facility decontamination, and accident recovery activities. These developments suggest that the application of remotely operated devices can significantly reduce radiation exposure to personnel and improve maintenance-program performance.

Currently, an operational robotic system is under development to address particular operational problems within a nuclear power plant. This device, IRIS (Industrial Remote Inspection System), is a general-purpose surveillance system that conducts particular inspection and handling tasks with the goal of significantly reducing personnel exposure to high radiation fields [13]. The device is shown in Figure P10.20. The open-loop transfer function is

$$G(s) = \frac{Ke^{-sT}}{(s + 1)(s + 3)}.$$

(a) Determine a suitable gain K for the system when $T = 0.5$ s, so that the overshoot to a step input is less than 30%. Determine the steady-state error. (b) Design a compensator

$$G_c(s) = \frac{s + 2}{s + b}$$

to improve the step response for the system in part (a) so that the steady-state error is less than 12%. Assume the closed-loop system of Figure 10.1(a).

P10.21 An uncompensated control system with unity feedback has a plant transfer function

$$G(s) = \frac{K}{s(s/2 + 1)(s/6 + 1)}.$$

We want to have a velocity error constant of $K_v = 20$. We also want to have a phase margin of approximately 45° and a closed-loop bandwidth greater than $\omega = 4$ rad/s. Use two identical cascaded phase-lead networks to compensate the system.

P10.22 For the system of Problem 10.21, design a phase-lag network to yield the desired specifications, with the exception that a bandwidth equal to or greater than 2 rad/s will be acceptable.

P10.23 For the system of Problem 10.21, we wish to achieve the same phase margin and K_v, but in addition, we wish to limit the bandwidth to less than 10 rad/s but greater than 2 rad/s. Use a lead-lag compensation network to compensate the system. The lead-lag network could be of the form

$$G_c(s) = \frac{(1 + s/10a)(1 + s/b)}{(1 + s/a)(1 + s/10b)},$$

where a is to be selected for the lag portion of the compensator, and b is to be selected for the lead portion of the compensator. The ratio α is chosen to be 10 for both the lead and lag portions.

P10.24 A system of the form of Figure 10.1(a) with unity feedback has

$$G(s) = \frac{K}{(s + 4)^2}.$$

We desire the steady-state error to a step input to be approximately 5% and the phase margin of the system to be approximately 45°. Design a lag network to meet these specifications.

P10.25 The stability and performance of the rotation of a robot (similar to waist rotation) presents a challenging control problem. The system requires high gains in order to achieve high resolution; yet a large overshoot of the transient response cannot be tolerated. The block diagram of an electrohydraulic system for rotation control is shown in Figure P10.25 [16]. The arm-rotating dynamics are represented by

$$G(s) = \frac{100}{s(s^2/6400 + s/50 + 1)}.$$

We want to have $K_v = 20$ for the compensated system. Design a compensator that results in an overshoot to a step input of less than 10%.

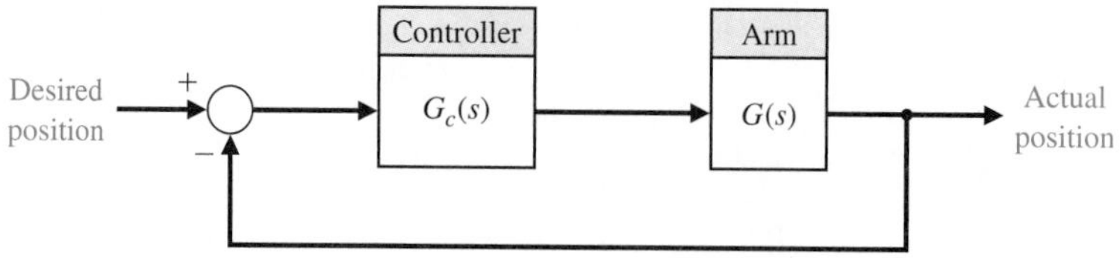

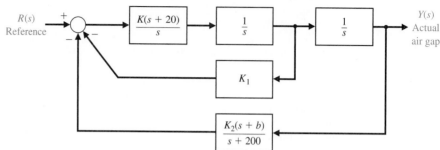

FIGURE P10.25
Robot position
control.

FIGURE P10.26
Airgap control of
train.

P10.26 The possibility of overcoming wheel friction, wear, and vibration by contactless suspension for passenger-carrying mass-transit vehicles is being investigated throughout the world. One design uses a magnetic suspension with an attraction force between the vehicle and the guideway with an accurately controlled airgap. A system is shown in Figure P10.26, which incorporates feedback compensation. Using root locus methods, select a suitable value for K_1 and b so the system has a damping ratio for the underdamped roots of $\zeta = 0.50$. Assume, if appropriate, that the pole of the air gap feedback loop ($s = -200$) can be neglected.

P10.27 A computer uses a printer as a fast output device. We desire to maintain accurate position control while moving the paper rapidly through the printer. Consider a system with unity feedback and a transfer function for the motor and amplifier of

$$G(s) = \frac{0.15}{s(s + 1)(5s + 1)}.$$

Design a lead network compensator so that the system bandwidth is 0.75 rad/s and the phase margin is 30°. Use a lead network with $\alpha = 10$.

P10.28 An engineering design team is attempting to control a process shown in Figure P10.28. The system has a controller $G_c(s)$, but the design team is unable to select $G_c(s)$ appropriately. It is agreed that a system with a phase margin of 50° is acceptable, but $G_c(s)$ is unknown. Determine $G_c(s)$.

First, let $G_c(s) = K$ and find (a) a value of K that yields a phase margin of 50° and the system's step response for this value of K. (b) Determine the settling time, percent overshoot, and the peak time. (c) Obtain the system's closed-loop frequency response, and determine $M_{p\omega}$ and the bandwidth.

The team has decided to let

$$G_c(s) = \frac{K(s + 12)}{(s + 20)}$$

and to repeat parts (a), (b), and (c). Determine the gain K that results in a phase margin of 50° and then proceed to evaluate the time response and the closed-loop frequency response. Prepare a table contrasting the results of the two selected controllers for $G_c(s)$ by comparing settling time (with a 2% criterion), percent overshoot, peak time, $M_{p\omega}$, and bandwidth.

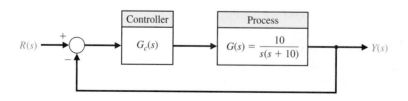

FIGURE P10.28
Controller design.

P10.29 An adaptive suspension vehicle uses a legged locomotion principle. The control of the leg can be represented by a unity feedback system with [13]

$$G(s) = \frac{K}{s(s + 10)(s + 14)}.$$

We desire to achieve a steady-state error for a ramp input of 10% and a damping ratio of the dominant roots of 0.707. Determine a suitable lag compensator, and determine the actual overshoot and the time to settle (to within 2% of the final value).

P10.30 A liquid-level control system (see Figure 9.32) has a loop transfer function

$$L(s) = G_c(s)G(s)H(s),$$

where $H(s) = 1$, $G_c(s)$ is a compensator, and the plant is

$$G(s) = \frac{10e^{-sT}}{s^2(s + 10)},$$

where $T = 50$ ms. Design a compensator so that $M_{p\omega}$ does not exceed 3.5 dB and ω_r is approximately 1.4 rad/s. Predict the overshoot and settling time (with a 2% criterion) of the compensated system when the input is a step. Plot the actual response.

P10.31 An automated guided vehicle (AGV) can be considered as an automated mobile conveyor designed to transport materials. Most AGVs require some type of guide path. The steering stability of the guidance control system has not been fully solved. The slight "snaking" of the AGV about the track generally has been acceptable, although it indicates instability of the steering guidance control system [9].

Most AGVs have a specification of maximum speed of about 1 m/s, although in practice they are usually operated at half that speed. In a fully automated manufacturing environment, there should be few personnel in the production area; therefore, the AGV should be able to be run at full speed. As the speed of the AGV increases, so does the difficulty in designing stable and smooth tracking controls.

A steering system for an AGV is shown in Figure P10.31, where $\tau_1 = 40$ ms and $\tau_2 = 1$ ms. We require

that the velocity constant K_v be 100 so that the steady-state error for a ramp input will be 1% of the slope of the ramp. Neglect τ_2 and design a lead compensator so that the phase margin is

$$45° \leq P.M. \leq 65°.$$

Attempt to obtain the two limiting cases for phase margin, and compare your results for the two designs by determining the actual percent overshoot and settling time for a step input.

P10.32 For the system of Problem 10.31, use a phase-lag compensator and attempt to achieve a phase margin of approximately 50°. Determine the actual overshoot and peak time for the compensated system.

P10.33 When a motor drives a flexible structure, the structure's natural frequencies, as compared to the bandwidth of the servodrive, determine the contribution of the structural flexibility to the errors of the resulting motion. In current industrial robots, the drives are often relatively slow, and the structures are relatively rigid, so that overshoots and other errors are caused mainly by the servodrive. However, depending on the accuracy required, the structural deflections of the driven members can become significant. Structural flexibility must be considered the major source of motion errors in space structures and manipulators. Because of weight restrictions in space, large arm lengths result in flexible structures. Furthermore, future industrial robots should require lighter and more flexible manipulators.

To investigate the effects of structural flexibility and how different control schemes can reduce unwanted oscillations, an experimental apparatus was constructed consisting of a DC motor driving a slender aluminum beam. The purpose of the experiments was to identify simple and effective control strategies to deal with the motion errors that occur when a servomotor is driving a very flexible structure [14].

The experimental apparatus is shown in Figure P10.33(a), and the control system is shown in Figure P10.33(b). The goal is that the system will have a K_v of 100. (a) When $G_c(s) = K$, determine K and plot the Bode diagram. Find the phase margin and gain margin. (b) Using the Nichols chart, find ω_r, $M_{p\omega}$, and ω_B.

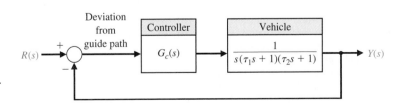

FIGURE P10.31
Steering control for vehicle.

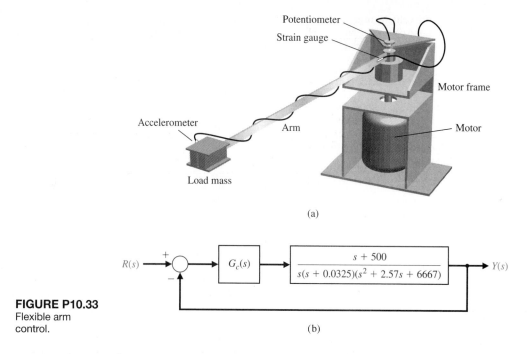

(a)

$$R(s) \longrightarrow \underset{-}{\overset{+}{\bigcirc}} \longrightarrow \boxed{G_c(s)} \longrightarrow \boxed{\dfrac{s + 500}{s(s + 0.0325)(s^2 + 2.57s + 6667)}} \longrightarrow Y(s)$$

FIGURE P10.33
Flexible arm
control.

(b)

(c) Select a compensator so that the phase margin is greater than $35°$ and find ω_r, $M_{p\omega}$, and ω_B for the compensated system.

P10.34 A human's ability to perform physical tasks is limited not by intellect but by physical strength. If, in an appropriate environment, a machine's mechanical power is closely integrated with a human arm's mechanical strength under the control of the human intellect, the resulting system will be superior to a loosely integrated combination of a human and a fully automated robot.

Extenders are defined as a class of robot manipulators that extend the strength of the human arm while maintaining human control of the task [15]. The defining characteristic of an extender is the transmission of both power and information signals. The extender is worn by the human; the physical contact between the extender and the human allows the direct transfer of mechanical power and information signals. Because of this unique interface, control of the extender trajectory can be accomplished without any type of joystick, keyboard, or master–slave system. The human provides a control system for the extender, while the extender actuators provide most of the strength necessary for the task. The human becomes a part of the extender and "feels" a scaled-down version of the load that the extender is carrying. The extender is distinguished from a conventional master–slave

system; in that type of system, the human operator is either at a remote location or close to the slave manipulator, but is not in direct physical contact with the slave in the sense of transfer of power. An extender is shown in Figure P10.34(a) [15]. The block diagram of the system is shown in Figure P10.34(b). The goal is that the compensated system will have a velocity constant K_v equal to 80, so that the settling time (with a 2% criterion) will be 1.6 seconds, and that the overshoot will be 16%, so that the dominant roots have a ζ of 0.5. Determine a lead-lag compensator using root locus methods.

P10.35 A magnetically levitated train is operating in Berlin, Germany. The M-Bahn 1600-m line represents the current state of worldwide systems. Fully automated trains can run at short intervals and operate with excellent energy efficiency. The control system for the levitation of the car is shown in Figure P10.35. Select a compensator so that the phase margin of the system is $45° \le P.M. \le 55°$. Predict the response of the system to a step command, and determine the actual step response for comparison.

P10.36 A unity feedback system has the loop transfer function

$$L(s) = G_c(s)G(s) = \frac{Ks + 0.54}{s(s + 1.76)}e^{-Ts},$$

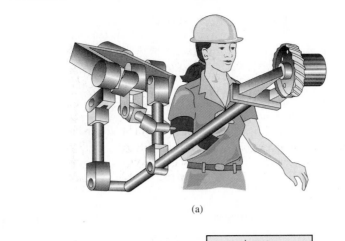

(a)

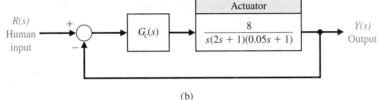

FIGURE P10.34
Extender robot
control.

(b)

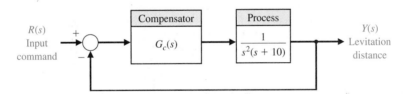

FIGURE P10.35
Magnetically
levitated train
control.

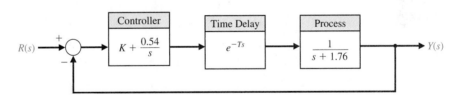

FIGURE P10.36
Unity feedback
system with a time
delay and PI
controller.

where T is a time delay and K is the controller proportional gain. The block diagram is illustrated in Figure P10.36. The nominal value of $K = 2$. Plot the phase margin of the system for $0 \le T \le 2$ s when $K = 2$. What happens to the phase margin as the time delay increases? What is the maximum time delay allowed before the system becomes unstable?

P10.37 A system's open-loop transfer function is a pure time delay of 0.5 s, so that $G(s) = e^{-s/2}$. Select a compensator $G_c(s)$ so that the steady-state error for a step input is less than 2% of the magnitude of the step

and the phase margin is greater than 30°. Determine the bandwidth of the compensated system and plot the step response.

P10.38 A unity negative feedback system has

$$G(s) = \frac{1}{s(s + 15)(s + 25)}.$$

The objective is that the dominant roots have a ζ equal to 0.707. Select a proportional plus integral (PI) controller so that the requirements are met. Determine the resulting peak time and settling time (with a 2% criterion) of the system.

P10.39 A unity feedback system of the form shown in Figure 10.1(a) has

$$G(s) = \frac{1}{(s + 1)(s + 10)}.$$

Design a compensator $G_c(s)$ so that the overshoot for a step input $R(s)$ is less than 10% and the steady-state error is less than 5%. Determine the bandwidth of the system.

P10.40 A unity feedback system has a plant

$$G(s) = \frac{40}{s(s + 2)}.$$

We desire to have a phase margin of 30° and a relatively large bandwidth. Select the crossover frequency $\omega_c = 10$ rad/s, and design a lead compensator using the analytical method of Section 10.9. Verify the results by plotting the compensated Bode diagram.

P10.41 A unity feedback system has a plant

$$G(s) = \frac{40}{s(s + 2)}.$$

We desire that the phase margin be equal to 30°. For a ramp input $r(t) = t$, we want the steady-state error to be equal to 0.05. Design a lag compensator to satisfy the requirements using the methods of Section 10.9. Verify the results by plotting the Bode diagram.

P10.42 For the system and requirements of Problem 10.41, determine the required compensator when the steady-state error for the ramp input must be equal to 0.02.

P10.43 Repeat Example 10.12 when we want the 100% rise time T_r be 1 second.

P10.44 Consider again the design for Example 10.4. Using a system as shown in Figure 10.22 and the compensator determined in Equation (10.46), select an appropriate prefilter. Compare the response of the system with and without the prefilter.

P10.45 Consider the system shown in Figure P10.45 and let $R(s) = 0$ and $T_d(s) = 0$. Design the controller $G_c(s) = K$ such that, in the steady-state, the response of the system $y(t)$ is less than -40 dB when the noise $N(s)$ is a sinusoidal input at a frequency of $\omega \geq 100$ rad/s.

P10.46 A unity feedback system has a loop transfer function

$$L(s) = G_c(s)G(s) = \frac{K(s + 10)}{s(s + 4)(s^2 + 2s + 1)}.$$

Plot the percent overshoot of the closed-loop system response to a unit step input for K in the range $0 < K \leq 100$. Explain the behavior of the system response for K in the range $0.7 < K \leq 57.3$.

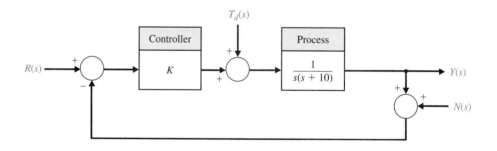

FIGURE 10.45
Unity feedback system with proportional controller and measurement noise.

ADVANCED PROBLEMS

AP10.1 A three-axis pick-and-place application requires the precise movement of a robotic arm in three-dimensional space, as shown in Figure AP10.1 for joint 2. The arm has specific linear paths it must follow to avoid other pieces of machinery. The overshoot for a step input should be less than 13%.
(a) Let $G_c(s) = K$, and determine the gain K that satisfies the requirement. Determine the resulting settling time (with a 2% criterion). (b) Use a lead network and reduce the settling time to less than 3 seconds.

AP10.2 The system of Advanced Problem AP10.1 is to have a percent overshoot less than 13%. In addition, we desire that the steady-state error for a unit ramp input will be less than 0.125 ($K_v = 8$) [28]. Design a lag network to meet the specifications. Check the resulting percent overshoot and settling time (with a 2% criterion) for the design.

AP10.3 The system of Advanced Problem AP 10.1 is required to have a percent overshoot less than 13% with a steady-state error for a unit ramp input less than 0.125 ($K_v = 8$).

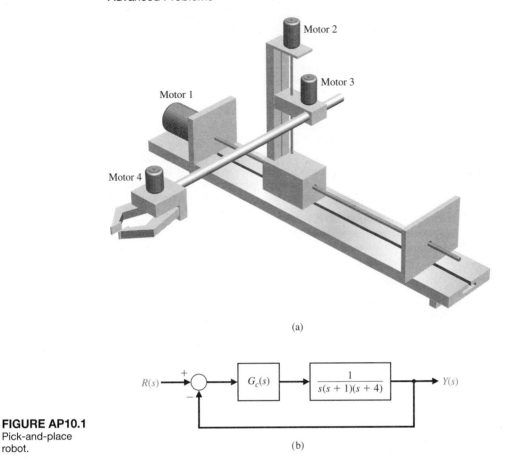

(a)

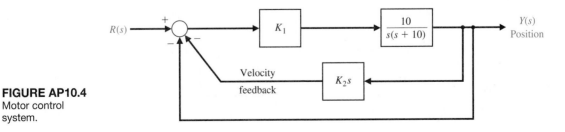

FIGURE AP10.1
Pick-and-place
robot.

(b)

Design a proportional plus integral (PI) controller to meet the specifications.

AP10.4 A DC motor control system with unity feedback has the form shown in Figure AP10.4. Select K_1 and K_2 so that the system response has a settling time (with a 2% criterion) less than 1 second and an overshoot less than 5% for a step input.

AP10.5 A unity feedback system is shown in Figure AP10.5. We want the step response of the system to

have an overshoot of about 16%, a fast response, and a settling time (with a 2% criterion) of about 1.8 seconds. (a) Design a lead compensator $G_c(s)$ to achieve the dominant roots desired. (b) Determine the step response of the system when $G_p(s) = 1$. (c) Select a prefilter $G_p(s)$, and determine the step response of the system with the prefilter.

AP10.6 Consider again Example 10.12 when we wish to minimize the settling time of the system while requiring

FIGURE AP10.4
Motor control
system.

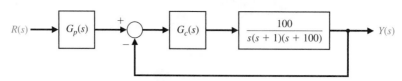

FIGURE AP10.5
Unity feedback with
a prefilter.

that $K < 52$. Determine the appropriate compensator that will minimize the settling time. Plot the system response.

AP10.7 A system has the form shown in Figure 10.22, with

$$G(s) = \frac{1}{s(s + 2)(s + 8)}.$$

A lead compensator is used, with

$$G_c(s) = \frac{K(s + 3)}{s + 28}.$$

Determine K so that the complex roots have $\zeta = 1/\sqrt{2}$. The prefilter is

$$G_p(s) = \frac{p}{s + p}.$$

(a) Determine the overshoot and rise time for $G_p(s) = 1$ and for $p = 3$. (b) Select an appropriate

value for p that will give an overshoot of 1% and compare the results.

AP10.8 The Manutec robot has large inertia and arm length resulting in a challenging control problem, as shown in Figure AP10.8(a). The block diagram model of the system is shown in Figure AP10.8(b). The plant dynamics are represented by

$$G(s) = \frac{250}{s(s + 2)(s + 40)(s + 45)}.$$

The percentage overshoot for a step input should be less than 20% with a rise time less than $\frac{1}{2}$ second and a settling time (with a 2% criterion) less than 1.2 seconds. Also, we desire that for a ramp input $K_v \geq 10$. Determine a suitable lead compensator.

AP10.9 The plant dynamics of a chemical process are represented by

$$G(s) = \frac{100}{s(s + 5)(s + 10)}.$$

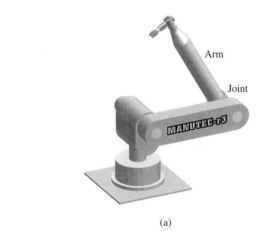

(a)

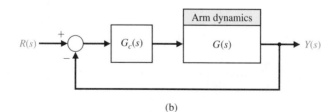

FIGURE AP10.8
(a) Manutec robot.
(b) Block diagram.

(b)

We desire that the system have a small steady-state error for a ramp input so that $K_v = 100$. For stability purposes, we desire a gain margin of 10 dB or greater and a phase margin of 40° or greater. Determine a lead-lag compensator that meets these specifications. Assume the system is of the form shown in Figure 10.1(a) with $H(s) = 1$.

AP10.10 An op-amp lead circuit is shown in Figure AP10.10.
(a) Determine the transfer function of this circuit.
(b) Sketch the frequency response of the circuit when $R_1 = 10\ k\Omega$, $R_2 = 10\ \Omega$, $C_1 = 0.1\ \mu F$, and $C_2 = 1\ mF$.

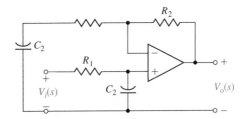

FIGURE AP10.10 Op-amp lead circuit.

DESIGN PROBLEMS

CDP10.1 The capstan-slide system of Figure CDP4.1 uses a PD controller. Determine the necessary values of the gain constants of the PD controller so that the deadbeat response is achieved. Also, we want the settling time (with a 2% criterion) to be less than 250 ms. Verify the results.

DP10.1 In Figure DP10.1, two robots are shown cooperating with each other to manipulate a long shaft to insert it into the hole in the block resting on the table. Long part insertion is a good example of a task that can benefit from cooperative control. The unity feedback control system of one robot joint has the process transfer function

$$G(s) = \frac{4}{s(s + 0.5)}.$$

The specifications require a steady-state error for a unit ramp input of 0.0125, and the step response has an

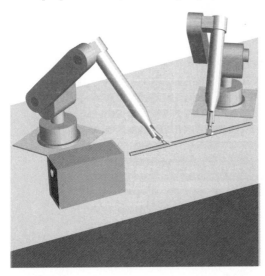

FIGURE DP10.1 Two robots cooperate to insert a shaft.

overshoot of less than 25% with a settling time (with a 2% criterion) of less than 2 seconds. Determine a lead-lag compensator that will meet the specifications, and plot the compensated and uncompensated responses for the ramp and step inputs.

DP10.2 The heading control of the traditional bi-wing aircraft, shown in Figure DP10.2(a), is represented by the block diagram of Figure DP10.2(b).

(a) Determine the minimum value of the gain K when $G_c(s) = K$, so that the steady-state effect of a unit step disturbance $T_d(s) = 1/s$ is less than or equal to 5% of the unit step ($y(\infty) = 0.05$).
(b) Determine whether the system using the gain of part (a) is stable.
(c) Design a compensator using one stage of lead compensation, so that the phase margin is 30°.
(d) Design a two-stage lead compensator so that the phase margin is 55°.
(e) Compare the bandwidth of the systems of parts (c) and (d).
(f) Plot the step response $y(t)$ for the systems of parts (c) and (d) and compare percent overshoot settling time (with a 2% criterion), and peak time.

DP10.3 NASA has identified the need for large deployable space structures, which will be constructed of lightweight materials and will contain large numbers of joints or structural connections. This need is evident for programs such as the space station. These deployable space structures may have precision shape requirements and a need for vibration suppression during in-orbit operations [17].

One such structure is the mast flight system, which is shown in Figure DP10.3(a). The intent of the system is to provide an experimental test bed for controls and dynamics. The basic element in the mast flight system is a 60.7-m-long truss beam structure, which is attached to the shuttle orbiter. Included at the tip of the truss structure are the primary actuators and collocated sensors. A deployment/retraction subsystem,

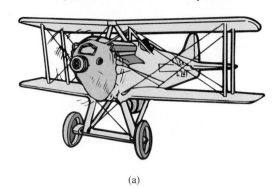

(a)

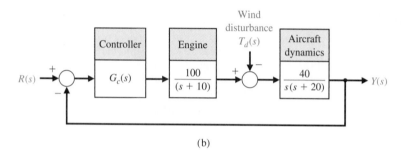

FIGURE DP10.2
(a) Bi-wing aircraft.
(*Source: The illustrated London News*, October 9, 1920.) (b) Control system.

(b)

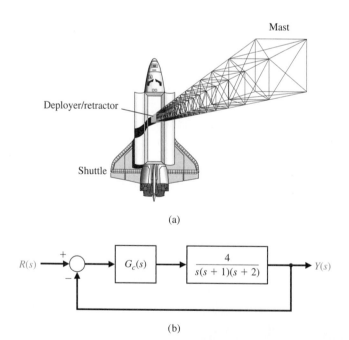

Mast

Deployer/retractor

Shuttle

(a)

$R(s) \longrightarrow$ + $\longrightarrow$ $G_c(s)$ $\longrightarrow$ $\dfrac{4}{s(s+1)(s+2)}$ $\longrightarrow Y(s)$

FIGURE DP10.3
Mast flight system.

(b)

which also secures the stowed beam package during launch and landing, is provided.

The system uses a large motor to move the structure and has the block diagram shown in Figure DP10.3(b). The goal is an overshoot to a step response of less than or equal to 16%; thus, we estimate the system ζ as 0.5 and the required phase margin as 50°. Design for $0.1 < K < 1$ and record overshoot, rise time, and phase margin for selected gains.

DP10.4 A mobile robot using a vision system as the measurement device is shown in Figure DP10.4 [22]. The control system is of the form shown in Figure 10.14, where

$$G(s) = \frac{1}{(s + 1)(0.5s + 1)},$$

and $G_c(s)$ is selected as a PI controller so that the steady-state error for a step input is equal to zero. We then have

$$G_c(s) = K_P + \frac{K_I}{s} = \frac{K_P s + K_I}{s}.$$

Determine a suitable G_c so that (a) the percent overshoot for a step input is 5% or less; (b) the settling time (with a 2% criterion) is less than 6 seconds; (c) the system K_v is greater than 0.9; and (d) the peak time for a step input is minimized.

DP10.5 A high-speed train is under development in Texas [23] with a design based on the French *Train à Grande Vitesse* (TGV). Train speeds of 186 miles per hour are foreseen. To achieve these speeds on tight curves, the train may use independent axles combined with the ability to tilt the train. Hydraulic cylinders connecting the passenger compartments to their wheeled bogies allow the train to lean into curves like a motorcycle. A pendulumlike device on the leading bogie of each car senses when it is entering a curve and feeds this information to the hydraulic system. Tilting does not make the train safer, but it does make passengers more comfortable.

Consider the tilt control shown in Figure DP10.5. Design a compensator $G_c(s)$ for a step-input command so that the overshoot is less than 5% and the settling time (with a 2% criterion) less than 0.6 second. We also desire that the steady-state error for a velocity (ramp) input be less than $0.15A$, where $r(t) = At, t > 0$. Verify the results for the design.

DP10.6 A large antenna, as shown in Figure DP10.6(a), is used to receive satellite signals and must accurately track the satellite as it moves across the sky. The control system uses an armature-controlled motor and a controller to be selected, as shown in Figure DP10.6(b). The system specifications require a steady-state error for a ramp input $r(t) = Bt$ be less than or equal to $0.01B$. We also seek an overshoot to a step

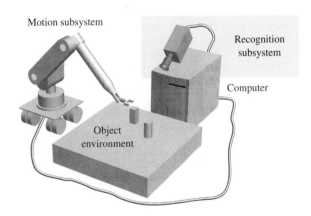

FIGURE DP10.4
A robot and vision system.

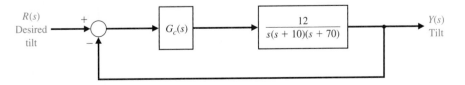

FIGURE DP10.5
High-speed train feedback control system.

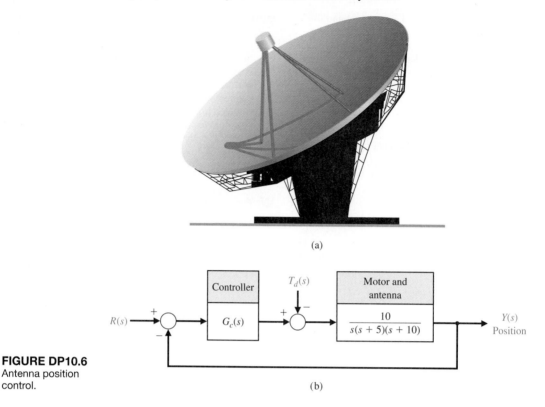

(a)

(b)

FIGURE DP10.6
Antenna position
control.

input less than 5% with a settling time (with a 2% criterion) less than 2 seconds. (a) Design a controller $G_c(s)$ and plot the resulting time response. (b) Determine the effect of the disturbance $D(s) = Q/s$ on the output $Y(s)$. (For simplicity, let $R(s) = 0$.)

DP10.7 High-performance tape transport systems are designed with a small capstan to pull the tape past the read/write heads and with take-up reels turned by DC motors. The tape is to be controlled at speeds up to 200 inches per second, with start-up as fast as possible, while preventing permanent distortion of the tape. Since we wish to control the speed and the tension of the tape, we will use a DC tachometer for the speed sensor and a potentiometer for the position sensor. We will use a DC motor for the actuator. Then the linear model for the system is a unity feedback system with

$$\frac{Y(s)}{E(s)} = G(s) =$$

$$\frac{K(s + 4000)}{s(s + 1000)(s + 3000)(s + p_1)(s + \hat{p}_1)},$$

where $p_1 = +2000 + j2000$, and $Y(s)$ is position.

The specifications for the system are (1) settling time of less than 12 ms, (2) an overshoot to a step

position command of less than 10%, and (3) a steady-state velocity error of less than .5%. Determine a compensator scheme to achieve these stringent specifications.

DP10.8 The past several years have witnessed a significant engine model-building activity in the automotive industry in a category referred to as "control-oriented" or "control design" models. These models contain representations of the throttle body, engine pumping phenomena, induction process dynamics, fuel system, engine torque generation, and rotating inertia.

The control of the fuel-to-air ratio in an automobile carburetor became of prime importance in the 1980s as automakers worked to reduce exhaust-pollution emissions. Thus, auto engine designers turned to the feedback control of the fuel-to-air ratio. Operation of an engine at or near a particular air-to-fuel ratio requires management of both air and fuel flow into the manifold system. The fuel command is considered the input and the engine speed is considered the output [9, 11].

The block diagram of the system is shown in Figure DP10.8, where $T = 0.066$ second. A compensator is required to yield zero steady-state error for a

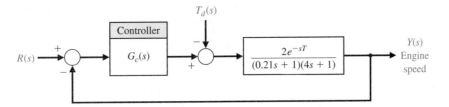

FIGURE DP10.8
Engine control
system.

step input and an overshoot of less than 10%. We also
desire that the settling time (with a 2% criterion) not
exceed 10 seconds.

DP10.9 A high-performance jet airplane is shown in
Figure DP10.9(a), and the roll-angle control system is
shown in Figure DP10.9(b). Design a controller $G_c(s)$
so that the step response is well behaved and the
steady-state error is zero.

DP10.10 A simple closed-loop control system has been
proposed to demonstrate proportional-integral (PI)
control of a windmill radiometer [31]. The windmill
radiometer is shown in Figure DP10.10(a) and the
control system is shown in Figure DP10.10(b). The
variable to be controlled is the angular velocity ω of
the windmill radiometer whose vanes turn when exposed to infrared radiation. An experimental setup

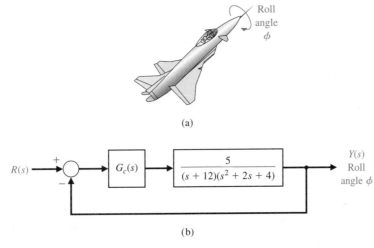

(a)

FIGURE DP10.9
Roll-angle control
of a jet airplane.

(b)

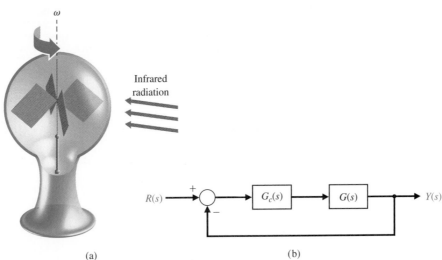

FIGURE DP10.10
(a) Radiometric
windmill. (b) Control
system.

(a)   (b)

using a reflexive photoelectric sensor and basic electronic circuitry makes possible the design and implementation of a highperformance control system.

The transfer function of the light source and radiometer is

$$G(s) = \frac{\tau}{\tau s + 1},$$

where $\tau = 20$ s. Design a PI controller so that the system achieves a deadbeat response with a settling time less than 25 s.

DP10.11   The feedback control system shown in Figure DP10.11 has the transfer function

$$G(s) = \frac{60}{(s^2 + 4s + 6)(s + 10)}.$$

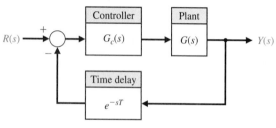

FIGURE DP10.11   Feedback control system with a time-delay.

Design a PID compensator $G_{c1}(s)$ and a lead-lag compensator $G_{c2}(s)$ such that, in each case, the closed-loop system is stable in the presence of a time-delay $T = 0.1$ s. Discuss the capability of each compensator to insure stability in the presence of an increase in the time-delay uncertainty of up to 0.2 second.

DP10.12   A unity feedback system has the process transfer function

$$G(s) = \frac{s + 1.59}{s(s + 3.7)(s^2 + 2.4s + 0.43)}.$$

Design the controller $G_c(s)$ such that the Bode magnitude plot of the loop transfer function $L(s) = G_c(s)G(s)$ is greater than 20 dB for $\omega \leq 0.01$ rad/s and less than -20 dB for $\omega \leq 10$ rad/s. The desired shape of the loop transfer function Bode plot magnitude is illustrated in Figure DP10.12. Explain why we would want the gain to be high at low-frequency and the gain to be low at high-frequency.

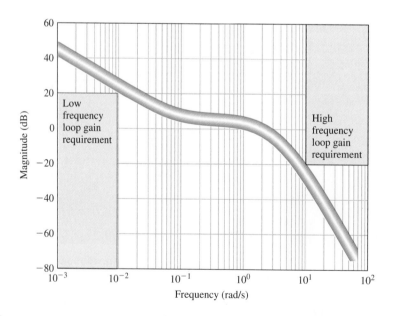

FIGURE DP10.12
Bode plot loop shaping requirements.

COMPUTER PROBLEMS

CP10.1 Consider the control system in Figure CP10.1, where

$$G(s) = \frac{1}{s + 1} \quad \text{and} \quad G_c(s) = \frac{5.5}{s}.$$

Develop an m-file to show that the phase margin is approximately 24° and that the percent overshoot to a unit step input is 50%.

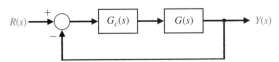

FIGURE CP10.1 A feedback control system with compensation.

CP10.2 A negative feedback control system is shown in Figure CP10.2 Design the proportional controller $G_c(s) = K$ so that the system has a 45° phase margin. Develop an m-file to obtain a Bode plot and verify that the design specification is satisfied.

CP10.3 Consider the system in Figure CP10.1, where

$$G(s) = \frac{1}{s(s + 2)}.$$

Design a compensator $G_c(s)$ so that the steady-state tracking error to a ramp input is zero and the settling time (with a 2% criterion) is less than 5 seconds. Obtain the response of the closed-loop system to the input $R(s) = 1/s^2$ and verify that the settling time

requirement has been satisfied and that the steady-state error is zero.

CP10.4 A fighter aircraft has the transfer function

$$\frac{\dot{\theta}}{\delta} = \frac{-10(s + 1)(s + 0.01)}{(s^2 + 2s + 2)(s^2 + 0.02s + 0.0101)},$$

where $\dot{\theta}$ is the pitch rate (rad/s) and δ is the elevator deflection (rad). The four poles represent the phugoid and short-period modes. The phugoid mode has a natural frequency of 0.1 rad/s, and the short period mode is 1.4 rad/s. The block diagram is shown in Figure CP10.4.
(a) Let the lead compensator be

$$G_c = K\frac{s + z}{s + p},$$

where $|z| < |p|$. Using Bode plot methods, design the lead compensator to meet the following specifications: (1) settling time (with a 2% criterion) to a unit step less than 2 seconds, and (2) percent overshoot less than 10%. (b) Simulate the closed-loop system with a step input of 10°/second, and show the time history of $\dot{\theta}$.

CP10.5 The pitch attitude motion of a rigid spacecraft is described by

$$J\ddot{\theta} = u,$$

where J is the principal moment of inertia, and u is the input torque on the vehicle [7]. Consider the PD controller

$$G_c(s) = K_P + K_D s.$$

(a) Obtain a block diagram of the control system. Design a control system to meet the following specifications:

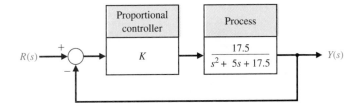

FIGURE CP10.2
Single-loop feedback system with proportional controller.

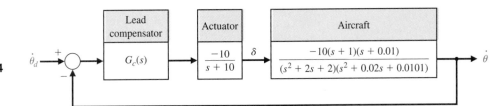

FIGURE CP10.4
An aircraft pitch rate feedback control system.

(1) closed-loop system bandwidth about 10 rad/s, and (2) percent overshoot less than 20% to a 10° step input. Complete the design by developing and using an interactive m-file script. (b) Verify the design by simulating the response to a 10° step input. (c) Include a closed-loop transfer function Bode plot to verify that the bandwidth requirement is satisfied.

CP10.6 Consider the control system shown in Figure CP10.6. Design a lag compensator using root locus methods to meet the following specifications: (1) steady-state error less than 10% for a step input, (2) phase margin greater than 45°, and (3) settling time (with a 2% criterion) less than 5 seconds for a unit step input.

(a) Design a lag compensator utilizing root locus methods to meet the design specifications. Develop a set of m-file scripts to assist in the design process. (b) Test the controller developed in part (a) by simulating the closed-loop system response to unit step input. Provide the time histories of the output $y(t)$. (c) Compute the phase margin using the margin function.

CP10.7 A lateral beam guidance system has an inner loop as shown in Figure CP10.7, where the transfer function for the coordinated aircraft is [30]

$$G(s) = \frac{23}{s + 23}.$$

Consider the PI controller

$$G_c(s) = K_P + \frac{K_I}{s}.$$

(a) Design a control system to meet the following specifications: (1) settling time (with a 2% criterion) to a unit step input of less than 1 second, and (2) steady-state tracking error for a unit ramp input of less than 0.1. (b) Verify the design by simulation.

CP10.8 Consider again the system and the lead compensator designed in Example 10.3. The actual overshoot of the compensated system will be 46%. We want to reduce the overshoot to 32%. Using a m-file script, determine an appropriate value for the zero of $G_c(s)$.

CP10.9 Plot the frequency response of the circuit of AP10.10.

CP10.10 The feedback control system shown in Figure CP10.10 has the transfer function

$$G(s) = \frac{K(s + 0.2)}{s^3 + 6s^2}.$$

The time delay is $T = 0.2$ s. Plot the phase margin for the system versus the gain in the range $0.1 \le K \le 10$. Determine the gain K that maximizes the phase margin.

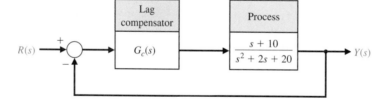

FIGURE CP10.6
A unity feedback control system.

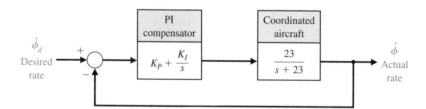

FIGURE CP10.7
A lateral beam guidance system inner loop.

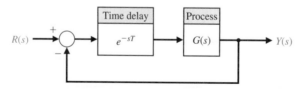

FIGURE CP10.10
Feedback control system with a time delay.

TERMS AND CONCEPTS

Cascade compensation network   A compensator network placed in cascade or series with the system process.

Compensation   The alteration or adjustment of a control system in order to provide a suitable performance.

Compensator   An additional component or circuit that is inserted into the system to compensate for a performance deficiency.

Deadbeat response   A system with a rapid response, minimal overshoot, and zero steady-state error for a step input.

Design of a control system   The arrangement or the plan of the system structure and the selection of suitable components and parameters.

Integration network   A network that acts, in part, like an integrator.

Lag network   *See* Phase-lag network.

Lead-lag network   A network with the characteristics of both a lead network and a lag network.

Lead network   *See* Phase-lead network.

Phase lag compensation   A widely-used compensator that possesses one zero and one pole with the pole closer to the origin of the s-plane. This compensator reduces the steady-state tracking errors.

Phase lead compensation   A widely-used compensator that possesses one zero and one pole with the zero closer to the origin of the s-plane. This compensator increases the system bandwidth and improves the dynamic response.

Phase-lag network   A network that provides a negative phase angle and a significant attenuation over the frequency range of interest.

Phase-lead network   A network that provides a positive phase angle over the frequency range of interest. Thus, phase lead can be used to cause a system to have an adequate phase margin.

PD controller   Controller with a proportional term and a derivative term (Proportional-Derivation).

PI controller   Controller with a proportional term and an integral term (Proportional-Integral).

Prefilter   A transfer function $G_p(s)$ that filters the input signal $R(s)$ prior to calculating the error signal.

CHAPTER 11

The Design of State Variable Feedback Systems

PREVIEW

The design of controllers utilizing state feedback is the subject of this chapter. We first present a system test for controllability and observability. Using the powerful notion of state variable feedback, we introduce the pole placement design technique. Ackermann's formula can be used to determine the state variable feedback gain matrix to place the system poles at the desired locations. The closed-loop system pole locations can be arbitrarily placed if and only if the system is controllable. When the full state is not available for feedback, we introduce an observer. The observer design process is described and the applicability of Ackermann's formula is established. The state variable compensator is obtained by connecting the full-state feedback law to the observer. We consider optimal control system design and then describe the use of internal model design to achieve prescribed steady-state response to selected input commands. The chapter concludes by revisiting the Sequential Design Example: Disk Drive Read System.

DESIRED OUTCOMES

Upon completion of Chapter 11, students should:

❑ Be familiar with the concepts of controllability and observability.
❑ Be able to design full-state feedback controllers and observers.
❑ Appreciate pole-placement methods and the application of Ackermann's formula.
❑ Understand the separation principle and how to construct state variable compensators.
❑ Have a working knowledge of reference inputs, optimal control, and internal model design.

11.1 INTRODUCTION

The time-domain method, expressed in terms of state variables, can also be used to design a suitable compensation scheme for a control system. Typically, we are interested in controlling the system with a control signal $\mathbf{u}(t)$ that is a function of several measurable state variables. Then we develop a state variable controller that operates on the information available in measured form. This type of system compensation is quite useful for system optimization and will be considered in this chapter.

State variable design typically comprises *three* steps. In the first step, we assume that all the state variables are measurable and utilize them in a **full-state feedback control law**. Full-state feedback is usually not practical because it is not possible (in general) to measure all the states. In practice, only certain states (or linear combinations thereof) are measured and provided as system outputs. The second step in state variable design is to construct an **observer** to estimate the states that are not directly sensed and available as outputs. Observers can either be full-state observers or reduced-order observers. Reduced-order observers account for the fact that certain states are already available as system outputs; hence they do not need to be estimated [29]. In this chapter, we consider only full-state observers. The final step in the design process is to appropriately connect the observer to the full-state feedback control law. It is common to refer to the state-variable controller (full-state control law plus the observer) as a **compensator**. The state variable design yields a compensator of the form depicted in Figure 11.1. Additionally, it is possible to consider reference inputs to the state variable compensator to complete the design. All three steps in the design process are discussed in the subsequent sections, as well as how to incorporate the reference inputs.

11.2 CONTROLLABILITY AND OBSERVABILITY

A key question that arises in the design of state variable compensators is whether or not all the poles of the closed-loop system can be arbitrarily placed in the complex plane. Recall that the poles of the closed-loop system are equivalent to the eigenvalues of the system matrix in state variable format. As we shall see, if the system is

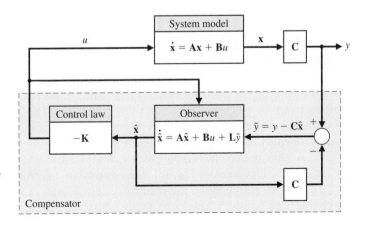

FIGURE 11.1
State variable compensator employing full-state feedback in series with a full-state observer.

controllable and **observable,** then we can accomplish the design objective of placing the poles precisely at the desired locations to meet the performance specifications. Full-state feedback design commonly relies on **pole-placement** techniques [2, 30]. Pole placement is discussed more fully in Section 11.3. It is important to note that a system must be completely controllable and completely observable to allow the flexibility to place *all* the closed-loop system poles arbitrarily. The concepts of controllability and observability (discussed in this section) were introduced by Kalman in the 1960s [31–33]. Rudolph Kalman was a central figure in the development of mathematical systems theory upon which much of the subject of state variable methods rests. Kalman is well known for his role in the development of the so-called Kalman filter, which was instrumental in the successful Apollo moon landings [34, 35].

> **A system is completely controllable if there exists an unconstrained control u(*t*) that can transfer any initial state x(*t*₀) to any other desired location x(*t*) in a finite time, $t_0 \leq t \leq T$.**

For the system

$$\dot{\mathbf{x}} = \mathbf{A}\mathbf{x} + \mathbf{B}u,$$

we can determine whether the system is controllable by examining the algebraic condition

$$\text{rank}[\mathbf{B} \quad \mathbf{AB} \quad \mathbf{A}^2\mathbf{B} \ \ldots \ \mathbf{A}^{n-1}\mathbf{B}] = n. \tag{11.1}$$

The matrix $\mathbf{A}$ is an $n \times n$ matrix and $\mathbf{B}$ is an $n \times 1$ matrix. For multi-input systems, $\mathbf{B}$ can be $n \times m$, where m is the number of inputs.

For a single-input, single-output system, the **controllability matrix $\mathbf{P}_c$** is described in terms of $\mathbf{A}$ and $\mathbf{B}$ as

$$\boxed{\mathbf{P}_c = [\mathbf{B} \quad \mathbf{AB} \quad \mathbf{A}^2\mathbf{B} \ \ldots \ \mathbf{A}^{n-1}\mathbf{B}],} \tag{11.2}$$

which is an $n \times n$ matrix. Therefore, if the determinant of $\mathbf{P}_c$ is nonzero, the system is controllable [12].

Advanced state variable design techniques can handle situations wherein the system is not completely controllable, but where the states (or linear combinations thereof) that cannot be controlled are inherently stable. These systems are classified as **stabilizable**. If a system is completely controllable, it is also stabilizable. The **Kalman state-space decomposition** provides a mechanism for partitioning the state-space so that it becomes apparent which states (or state combinations) are controllable and which are not [14, 20]. The controllable subspace is thus exposed, and if the system is stabilizable, the control system design can, in theory, proceed. In this chapter, we consider only completely controllable systems.

EXAMPLE 11.1   **Controllability of a system**

Let us consider the system

$$
\dot{\mathbf{x}} = \begin{bmatrix} 0 & 1 & 0 \\ 0 & 0 & 1 \\ -a_0 & -a_1 & -a_2 \end{bmatrix} \mathbf{x} + \begin{bmatrix} 0 \\ 0 \\ 1 \end{bmatrix} u.
$$

$$
y = [1 \quad 0 \quad 0]\mathbf{x} + [0]u
$$

The signal-flow graph and block diagram model are illustrated in Figure 11.2. Then we have

$$
\mathbf{A} = \begin{bmatrix} 0 & 1 & 0 \\ 0 & 0 & 1 \\ -a_0 & -a_1 & -a_2 \end{bmatrix}, \quad \mathbf{B} = \begin{bmatrix} 0 \\ 0 \\ 1 \end{bmatrix}, \quad \mathbf{AB} = \begin{bmatrix} 0 \\ 1 \\ -a_2 \end{bmatrix}, \quad \text{and} \quad \mathbf{A}^2\mathbf{B} = \begin{bmatrix} 1 \\ -a_2 \\ a_2^2 - a_1 \end{bmatrix}.
$$

Therefore, we obtain

$$
\mathbf{P}_c = [\mathbf{B} \quad \mathbf{AB} \quad \mathbf{A}^2\mathbf{B}] = \begin{bmatrix} 0 & 0 & 1 \\ 0 & 1 & -a_2 \\ 1 & -a_2 & a_2^2 - a_1 \end{bmatrix}.
$$

The determinant of $\mathbf{P}_c = -1 \neq 0$, hence this system is controllable.

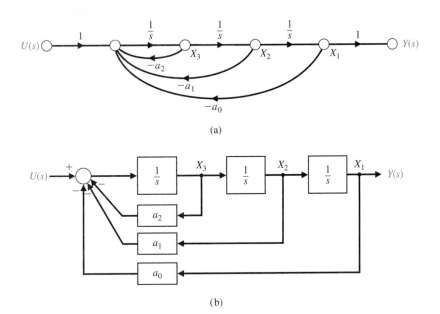

FIGURE 11.2
Third-order system.
(a) Signal-flow
graph model. (b)
Block diagram
model.

EXAMPLE 11.2   Controllability of a two-state system

Let us consider a system represented by the two state equations

$$\dot{x}_1 = -2x_1 + u, \quad \text{and} \quad \dot{x}_2 = -3x_2 + dx_1$$

and determine the condition for controllability. Also, we have $y = x_2$, as shown in Figure 11.3. The system state variable model is

$$\dot{\mathbf{x}} = \begin{bmatrix} -2 & 0 \\ d & -3 \end{bmatrix} \mathbf{x} + \begin{bmatrix} 1 \\ 0 \end{bmatrix} u,$$

$$y = \begin{bmatrix} 0 & 1 \end{bmatrix} \mathbf{x} + \begin{bmatrix} 0 \end{bmatrix} u.$$

We can determine the requirement on the parameter d by generating the matrix $\mathbf{P}_c$. So, with

$$\mathbf{B} = \begin{bmatrix} 1 \\ 0 \end{bmatrix} \quad \text{and} \quad \mathbf{AB} = \begin{bmatrix} -2 & 0 \\ d & -3 \end{bmatrix}\begin{bmatrix} 1 \\ 0 \end{bmatrix} = \begin{bmatrix} -2 \\ d \end{bmatrix},$$

we have

$$\mathbf{P}_c = \begin{bmatrix} 1 & -2 \\ 0 & d \end{bmatrix}.$$

The determinant of $\mathbf{P}_c$ is equal to d, which is nonzero whenever d is nonzero. ■

All the poles of the closed-loop system can be placed arbitrarily in the complex plane if and only if the system is observable and controllable. Observability refers to the ability to estimate a state variable.

> **A system is completely observable if and only if there exists a finite time T such that the initial state $x(0)$ can be determined from the observation history $y(t)$ given the control $u(t), 0 \leq t \leq T.$**

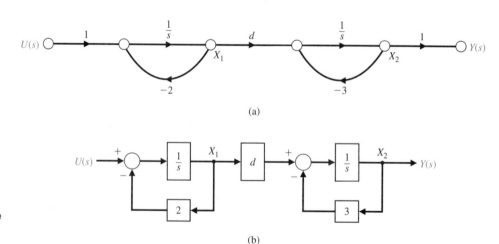

FIGURE 11.3
(a) Flow graph model for Example 11.2. (b) Block diagram model.

Consider the single-input, single-output system

$$\dot{\mathbf{x}} = \mathbf{A}\mathbf{x} + \mathbf{B}u \quad \text{and} \quad y = \mathbf{C}\mathbf{x},$$

where $\mathbf{C}$ is a $1 \times n$ row vector, and $\mathbf{x}$ is an $n \times 1$ column vector. This system is completely observable when the determinant of the **observability matrix** $\mathbf{P}_o$ is nonzero, where

$$\mathbf{P}_o = \begin{bmatrix} \mathbf{C} \\ \mathbf{CA} \\ \vdots \\ \mathbf{CA}^{n-1} \end{bmatrix}, \tag{11.3}$$

which is an $n \times n$ matrix.

As discussed in this section, advanced state variable design techniques can handle situations wherein the system is not completely controllable, as long as the system is stabilizable. These same techniques can handle cases wherein the system is not completely observable, but where the states (or linear combinations thereof) that cannot be observed are inherently stable. These systems are classified as **detectable**. If a system is completely observable, it is also detectable. The Kalman state-space decomposition provides a mechanism for partitioning the state-space so that it becomes apparent which states (or state combinations) are observable and which are not [14, 20]. The unobservable subspace is thus exposed, and if the system is detectable, the control system design can, in theory, proceed. In this chapter, we consider only completely observable systems. The approach to state-variable design involves first verifying that the system under consideration is completely controllable and completely observable. If so, the pole placement design technique considered here can provide acceptable closed-loop system performance.

EXAMPLE 11.3   **Observability of a system**

Consider again the system of Example 11.1. The model is shown in Figure 11.2. To construct $\mathbf{P}_o$, we use

$$\mathbf{A} = \begin{bmatrix} 0 & 1 & 0 \\ 0 & 0 & 1 \\ -a_0 & -a_1 & -a_2 \end{bmatrix} \quad \text{and} \quad \mathbf{C} = \begin{bmatrix} 1 & 0 & 0 \end{bmatrix}.$$

Therefore,

$$\mathbf{CA} = \begin{bmatrix} 0 & 1 & 0 \end{bmatrix} \quad \text{and} \quad \mathbf{CA}^2 = \begin{bmatrix} 0 & 0 & 1 \end{bmatrix}.$$

Thus, we obtain

$$\mathbf{P}_o = \begin{bmatrix} 1 & 0 & 0 \\ 0 & 1 & 0 \\ 0 & 0 & 1 \end{bmatrix}.$$

The det $\mathbf{P}_o = 1$, and the system is completely observable. ∎

EXAMPLE 11.4   **Observability of a two-state system**

Consider the system given by

$$\dot{\mathbf{x}} = \begin{bmatrix} 2 & 0 \\ -1 & 1 \end{bmatrix} \mathbf{x} + \begin{bmatrix} 1 \\ -1 \end{bmatrix} u \quad \text{and} \quad y = [1 \quad 1]\mathbf{x}.$$

The system is illustrated in Figure 11.4. We can check the system controllability and observability using the $\mathbf{P}_c$ and $\mathbf{P}_o$ matrices.

From the system definition, we obtain

$$\mathbf{B} = \begin{bmatrix} 1 \\ -1 \end{bmatrix} \quad \text{and} \quad \mathbf{AB} = \begin{bmatrix} 2 \\ -2 \end{bmatrix}.$$

Therefore, the controllability matrix is determined to be

$$\mathbf{P}_c = [\mathbf{B} \quad \mathbf{AB}] = \begin{bmatrix} 1 & 2 \\ -1 & -2 \end{bmatrix},$$

and det $\mathbf{P}_c = 0$. Thus, the system is not controllable.

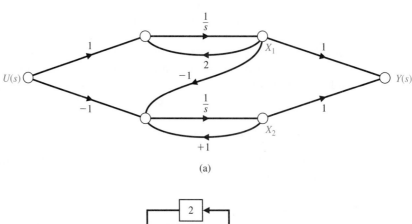

(a)

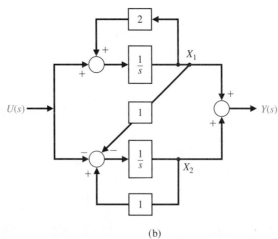

FIGURE 11.4
Two state system
model for Example
11.4. (a) Signal-flow
graph model. (b)
Block diagram
model.

(b)

From the system definition, we have

$$\mathbf{C} = [1 \quad 1] \quad \text{and} \quad \mathbf{CA} = [1 \quad 1].$$

Therefore, computing the observability matrix yields

$$\mathbf{P}_o = \begin{bmatrix} \mathbf{C} \\ \mathbf{CA} \end{bmatrix} = \begin{bmatrix} 1 & 1 \\ 1 & 1 \end{bmatrix},$$

and det $\mathbf{P}_o = 0$. Hence, the system is not observable.

If we look again at the state model, we note that

$$y = x_1 + x_2.$$

However,

$$\dot{x}_1 + \dot{x}_2 = 2x_1 + (x_2 - x_1) + u - u = x_1 + x_2.$$

Thus, the system state variables do not depend on u, and the system is not controllable. Similarly, the output $x_1 + x_2$ depends on $x_1(0)$ plus $x_2(0)$ and does not allow us to determine $x_1(0)$ and $x_2(0)$ independently. Consequently, the system is not observable.   ∎

11.3 FULL-STATE FEEDBACK CONTROL DESIGN

In this section, we consider full-state variable feedback to achieve the desired pole locations of the closed-loop system.

The first step in the state variable design process requires us to assume that all the states are available for feedback—that is, we have access to the complete state $\mathbf{x}(t)$ for all t. The system input $u(t)$ is given by

$$u = -\mathbf{Kx}. \tag{11.4}$$

Determining the gain matrix $\mathbf{K}$ is the objective of the full-state feedback design procedure. The beauty of the state variable design process is that the problem naturally separates into a full-state feedback component and an observer design component. These two design procedures can occur independently, and in fact, the **separation principle** provides the proof that this approach is optimal. We will show later that the stability of the closed-loop system is guaranteed if the full-state feedback control law stabilizes the system (under the assumption of access to the complete state) and the observer is stable (the tracking error is asymptotically stable). Observers are discussed in Section 11.4. The full-state feedback block diagram is illustrated in Figure 11.5. With the system defined by the state variable model

$$\dot{\mathbf{x}} = \mathbf{Ax} + \mathbf{B}u$$

and the control feedback given by

$$u = -\mathbf{Kx},$$

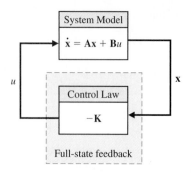

FIGURE 11.5
Full-state feedback
block diagram (with
no reference input).

we find the closed-loop system to be

$$\dot{\mathbf{x}} = \mathbf{A}\mathbf{x} + \mathbf{B}u = \mathbf{A}\mathbf{x} - \mathbf{B}\mathbf{K}\mathbf{x} = (\mathbf{A} - \mathbf{B}\mathbf{K})\mathbf{x}. \qquad (11.5)$$

As discussed in Section 6.4, the characteristic equation associated with Equation (11.5) is

$$\det(\lambda\mathbf{I} - (\mathbf{A} - \mathbf{B}\mathbf{K})) = 0.$$

If all the roots of the characteristic equation lie in the left half-plane, then the closed-loop system is stable. In other words, for any initial condition $\mathbf{x}(t_0)$, it follows that

$$\mathbf{x}(t) = e^{(\mathbf{A}-\mathbf{B}\mathbf{K})t}\mathbf{x}(t_0) \rightarrow 0 \qquad \text{as } t \rightarrow \infty.$$

Given the pair $(\mathbf{A}, \mathbf{B})$, we can always determine $\mathbf{K}$ to place *all* the system closed-loop poles in the left half-plane if and only if the system is completely controllable—that is, if and only if the controllability matrix $\mathbf{P}_c$ is full rank (for a single-input, single-output system, full rank implies that $\mathbf{P}_c$ is invertible).

The addition of a reference input can be written as

$$u(t) = -\mathbf{K}\mathbf{x}(t) + Nr(t),$$

where $r(t)$ is the reference input. The question of reference inputs is addressed in Section 11.6. When $r(t) = 0$ for all $t > t_0$, the control design problem is known as the **regulator problem**. That is, we want to compute $\mathbf{K}$ so that all initial conditions are driven to zero in a specified fashion (as determined by the design specifications).

When using this state variable feedback, the roots of the characteristic equation are placed where the transient performance meets the desired response.

EXAMPLE 11.5   **Design of a third-order system**

Let us consider the third-order system with the differential equation

$$\frac{d^3y}{dt^3} + 5\frac{d^2y}{dt^2} + 3\frac{dy}{dt} + 2y = u.$$

We can select the state variables as the phase variables (see Section 3.4) so that $x_1 = y$, $x_2 = dy/dt$, $x_3 = d^2y/dt^2$, and then

$$\dot{\mathbf{x}} = \begin{bmatrix} 0 & 1 & 0 \\ 0 & 0 & 1 \\ -2 & -3 & -5 \end{bmatrix} \mathbf{x} + \begin{bmatrix} 0 \\ 0 \\ 1 \end{bmatrix} u = \mathbf{Ax} + \mathbf{B}u$$

and

$$y = \begin{bmatrix} 1 & 0 & 0 \end{bmatrix} \mathbf{x}.$$

If the state variable feedback matrix is

$$\mathbf{K} = \begin{bmatrix} k_1 & k_2 & k_3 \end{bmatrix}$$

and

$$u = -\mathbf{Kx},$$

then the closed-loop system is

$$\dot{\mathbf{x}} = \mathbf{Ax} - \mathbf{BKx} = (\mathbf{A} - \mathbf{BK})\mathbf{x}.$$

The state feedback matrix is

$$[\mathbf{A} - \mathbf{BK}] = \begin{bmatrix} 0 & 1 & 0 \\ 0 & 0 & 1 \\ -2-k_1 & -3-k_2 & -5-k_3 \end{bmatrix},$$

and the characteristic equation is

$$\Delta(\lambda) = \det(\lambda\mathbf{I} - (\mathbf{A} - \mathbf{BK})) = \lambda^3 + (5+k_3)\lambda^2 + (3+k_2)\lambda + (2+k_1) = 0.$$
$$(11.6)$$

If we seek a rapid response with a low overshoot, we choose a desired characteristic equation such as (see Equation 5.18 and Table 5.2)

$$\Delta(\lambda) = (\lambda^2 + 2\zeta\omega_n\lambda + \omega_n^2)(\lambda + \zeta\omega_n).$$

We choose $\zeta = 0.8$ for minimal overshoot and ω_n to meet the settling time requirement. If we want a settling time (with a 2% criterion) equal to 1 second, then

$$T_s = \frac{4}{\zeta\omega_n} = \frac{4}{(0.8)\omega_n} \approx 1.$$

If we choose $\omega_n = 6$, the desired characteristic equation is

$$(\lambda^2 + 9.6\lambda + 36)(\lambda + 4.8) = \lambda^3 + 14.4\lambda^2 + 82.1\lambda + 172.8. \qquad (11.7)$$

Comparing Equations (11.6) and (11.7) yields the three equations

$$5 + k_3 = 14.4$$
$$3 + k_2 = 82.1$$
$$2 + k_1 = 172.8.$$

Therefore, we require that $k_3 = 9.4$, $k_2 = 79.1$, and $k_1 = 170.8$. The step response has no overshoot and a settling time of 1 second, as desired. ∎

EXAMPLE 11.6   Inverted pendulum control

Consider the control of the cart and the unstable inverted pendulum shown in Figure 3.22. We measure and utilize the state variables of the system in order to control the cart (see Example 3.4). Thus, if we want to measure the state variable $x_3 = \theta$, we could use a potentiometer connected to the shaft of the pendulum hinge. Similarly, we could measure the rate of change of the angle $x_4 = \dot{\theta}$ by using a tachometer generator. The state variables x_1 and x_2, which are the position and velocity of the cart, can also be measured by suitable sensors. If the state variables are all measured, then they can be used in a feedback controller so that $u = -\mathbf{Kx}$, where $\mathbf{K}$ is the feedback matrix. The state vector $\mathbf{x}$ represents the state of the system; therefore, knowledge of $\mathbf{x}(t)$ and the equations describing the system dynamics provide sufficient information for control and stabilization of a system [4, 5, 7].

To illustrate the use of state variable feedback, let us consider again the unstable portion of the inverted pendulum system and design a suitable state variable feedback control system. We begin by considering a reduced system. If we assume that the control signal is an acceleration signal and that the mass of the cart is negligible, we can focus on the unstable dynamics of the pendulum. When $u(t)$ is an acceleration signal, Equation (3.69) becomes

$$gx_3 - l\dot{x}_4 = \dot{x}_2 = \ddot{y} = u(t).$$

For the reduced system, where the control signal is an acceleration signal, the position and velocity of the cart are integral functions of $u(t)$. The portion of the state vector under consideration is $[x_3, x_4] = [\theta, \dot{\theta}]$. Thus, the state vector differential equation reduces to

$$\frac{d}{dt}\begin{bmatrix} x_3 \\ x_4 \end{bmatrix} = \begin{bmatrix} 0 & 1 \\ g/l & 0 \end{bmatrix}\begin{bmatrix} x_3 \\ x_4 \end{bmatrix} + \begin{bmatrix} 0 \\ -1/l \end{bmatrix}u(t). \tag{11.8}$$

The $\mathbf{A}$ matrix of Equation (11.8) is simply the lower right-hand portion of the $\mathbf{A}$ matrix of Equation (3.73), and the system has the characteristic equation $\lambda^2 - g/l = 0$ with one root in the right-hand s-plane. To stabilize the system, we generate a control signal that is a function of the two state variables, x_3 and x_4. Then we have

$$u(t) = -\mathbf{Kx} = -[k_1 \quad k_2]\begin{bmatrix} x_3 \\ x_4 \end{bmatrix} = -k_1 x_3 - k_2 x_4.$$

Substituting this control signal relationship into Equation (11.8), we have

$$\begin{bmatrix} \dot{x}_3 \\ \dot{x}_4 \end{bmatrix} = \begin{bmatrix} 0 & 1 \\ g/l & 0 \end{bmatrix} \begin{bmatrix} x_3 \\ x_4 \end{bmatrix} + \begin{bmatrix} 0 \\ (1/l)(k_1 x_3 + k_2 x_4) \end{bmatrix}.$$

Combining the two additive terms on the right side of the equation, we find that

$$\begin{bmatrix} \dot{x}_3 \\ \dot{x}_4 \end{bmatrix} = \begin{bmatrix} 0 & 1 \\ (g + k_1)/l & k_2/l \end{bmatrix} \begin{bmatrix} x_3 \\ x_4 \end{bmatrix}.$$

Obtaining the characteristic equation, we have

$$\begin{bmatrix} \lambda & -1 \\ -(g + k_1)/l & \lambda - k_2/l \end{bmatrix} = \lambda \left(\lambda - \frac{k_2}{l} \right) - \frac{g + k_1}{l}$$

$$= \lambda^2 - \left(\frac{k_2}{l} \right) \lambda + \frac{g + k_1}{l}. \qquad (11.9)$$

Thus, for the system to be stable, we require that $k_2/l < 0$ and $k_1 > -g$. Hence, we have stabilized an unstable system by measuring the state variables x_3 and x_4 and using the control function $u = -\mathbf{K}\mathbf{x}$ to obtain a stable system. If we wish to achieve a rapid response with modest overshoot, we select $\omega_n = 10$ and $\zeta = 0.8$. Then we require

$$\frac{k_2}{l} = -16 \quad \text{and} \quad \frac{k_1 + g}{l} = 100.$$

The step response would have an overshoot of 1.5% and a settling time of 0.5 second. ∎

Thus far, we have established an approach for the design of a feedback control system by using the state variables as the feedback variables in order to increase the stability of the system and obtain the desired system response. Now we face the task of computing the gain matrix $\mathbf{K}$ to place the poles at desired locations. For a single-input, single-output system, Ackermann's formula is useful for determining the state variable feedback matrix

$$\mathbf{K} = [k_1 \quad k_2 \ \dots \ k_n],$$

where

$$u = -\mathbf{K}\mathbf{x}.$$

Given the desired characteristic equation

$$q(\lambda) = \lambda^n + \alpha_{n-1}\lambda^{n-1} + \cdots + \alpha_o,$$

the state feedback gain matrix is

$$\boxed{\mathbf{K} = [0 \quad 0 \ \dots \ 0 \quad 1]\mathbf{P}_c^{-1} q(\mathbf{A}),} \qquad (11.10)$$

where

$$q(\mathbf{A}) = \mathbf{A}^n + \alpha_{n-1}\mathbf{A}^{n-1} + \cdots \alpha_1\mathbf{A} + \alpha_0\mathbf{I},$$

and $\mathbf{P}_c$ is the controllability matrix of Equation (11.2).

EXAMPLE 11.7   **Second-order system**

Consider the system

$$\frac{Y(s)}{U(s)} = G(s) = \frac{1}{s^2}$$

and determine the feedback gain to place the closed-loop poles at $s = -1 \pm j$. Therefore, we require that

$$q(\lambda) = \lambda^2 + 2\lambda + 2,$$

and $\alpha_1 = \alpha_2 = 2$. With $x_1 = y$ and $x_2 = \dot{y}$, the matrix equation for the system $G(s)$ is

$$\dot{\mathbf{x}} = \begin{bmatrix} 0 & 1 \\ 0 & 0 \end{bmatrix}\mathbf{x} + \begin{bmatrix} 0 \\ 1 \end{bmatrix}u.$$

The controllability matrix is

$$\mathbf{P}_c = [\mathbf{B} \quad \mathbf{AB}] = \begin{bmatrix} 0 & 1 \\ 1 & 0 \end{bmatrix}.$$

Thus, we obtain

$$\mathbf{K} = [0 \quad 1]\mathbf{P}_c^{-1}q(\mathbf{A}),$$

where

$$\mathbf{P}_c^{-1} = \frac{1}{-1}\begin{bmatrix} 0 & -1 \\ -1 & 0 \end{bmatrix} = \begin{bmatrix} 0 & 1 \\ 1 & 0 \end{bmatrix}$$

and

$$q(\mathbf{A}) = \begin{bmatrix} 0 & 1 \\ 0 & 0 \end{bmatrix}^2 + 2\begin{bmatrix} 0 & 1 \\ 0 & 0 \end{bmatrix} + 2\begin{bmatrix} 1 & 0 \\ 0 & 1 \end{bmatrix} = \begin{bmatrix} 2 & 2 \\ 0 & 2 \end{bmatrix}.$$

Then we have

$$\mathbf{K} = [0 \quad 1]\begin{bmatrix} 0 & 1 \\ 1 & 0 \end{bmatrix}\begin{bmatrix} 2 & 2 \\ 0 & 2 \end{bmatrix} = [0 \quad 1]\begin{bmatrix} 0 & 2 \\ 2 & 2 \end{bmatrix} = [2 \quad 2]. \quad \blacksquare$$

Note that computing the gain matrix $\mathbf{K}$ using Ackermann's formula requires the use of $\mathbf{P}_c^{-1}$. We see that complete controllability is essential because only then can we guarantee that the controllability matrix $\mathbf{P}_c$ has full rank and hence that $\mathbf{P}_c^{-1}$ exists.

11.4 OBSERVER DESIGN

In the full-state feedback design procedure discussed in Section 11.3, it was assumed that all the states were available for feedback at all times. This is a good assumption for the control law design process. However, generally speaking, only a subset of the states are readily measurable and available for feedback. Having all the states available for feedback implies that these states are measured with a sensor or sensor combinations. The cost and complexity of the control system increase as the number of required sensors increases. So, even in situations where extra sensors are available, it may not be cost effective to employ these extra sensors, if indeed, the control system design goals can be accomplished without them. Fortunately, if the system is completely observable with a given set of outputs, then it is possible to determine (or to estimate) the states that are not directly measured (or observed).

According to Luenberger [29], the full-state observer for the system

$$\dot{\mathbf{x}} = \mathbf{A}\mathbf{x} + \mathbf{B}u$$
$$y = \mathbf{C}\mathbf{x}$$

is given by

$$\dot{\hat{\mathbf{x}}} = \mathbf{A}\hat{\mathbf{x}} + \mathbf{B}u + \mathbf{L}(y - \mathbf{C}\hat{\mathbf{x}}) \qquad (11.11)$$

where $\hat{\mathbf{x}}$ denotes the estimate of the state $\mathbf{x}$. The matrix $\mathbf{L}$ is the observer gain matrix and is to be determined as part of the observer design procedure. The observer is depicted in Figure 11.6. The observer has two inputs, u and y, and one output, $\hat{\mathbf{x}}$.

The goal of the observer is to provide an estimate $\hat{\mathbf{x}}$ so that $\hat{\mathbf{x}} \to \mathbf{x}$ as $t \to \infty$. Remember that we do not know $\mathbf{x}(t_0)$ precisely; therefore we must provide an initial estimate $\hat{\mathbf{x}}(t_0)$ to the observer. Define the observer **estimation error** as

$$\mathbf{e}(t) = \mathbf{x}(t) - \hat{\mathbf{x}}(t). \qquad (11.12)$$

The observer design should produce an observer with the property that $\mathbf{e}(t) \to 0$ as $t \to \infty$. One of the main results of systems theory is that if the system is completely observable, we can always find $\mathbf{L}$ so that the tracking error is asymptotically stable, as desired.

Taking the time-derivative of the estimation error in Equation (11.12) yields

$$\dot{\mathbf{e}} = \dot{\mathbf{x}} - \dot{\hat{\mathbf{x}}}$$

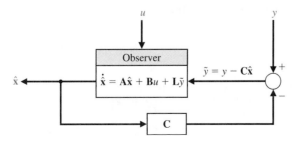

FIGURE 11.6
The full-state
observer.

and using the system model and the observer in Equation (11.11), we obtain

$$\dot{e} = Ax + Bu - A\hat{x} - Bu - L(y - C\hat{x})$$

or

$$\dot{e}(t) = (A - LC)e(t). \tag{11.13}$$

We can guarantee that $e(t) \rightarrow 0$ as $t \rightarrow \infty$ for any initial tracking error $e(t_0)$ if the characteristic equation

$$\det(\lambda I - (A - LC)) = 0 \tag{11.14}$$

has all its roots in the left half-plane. Therefore, the observer design process reduces to finding the matrix L such that the roots of the characteristic equation in Equation (11.14) lie in the left half-plane. This can always be accomplished if the system is completely observable; that is, if the observability matrix P_o has full rank (for a single-input, single-output system, full rank implies that P_o is invertible).

EXAMPLE 11.8   **Second-order system observer design**

Consider the second-order system

$$\dot{x} = \begin{bmatrix} 2 & 3 \\ -1 & 4 \end{bmatrix} x + \begin{bmatrix} 0 \\ 1 \end{bmatrix} u$$

$$y = [1 \quad 0]x.$$

In this example, we can only directly observe the state $y = x_1$. The observer will provide estimates of the second state x_2.

In this book, we only consider full-state observers, which implies that the observer will provide estimates of all the states. We might be inclined to suppose that since some states are directly measured, it may be possible to design an observer that provides just the estimates of the states not directly measured. This is, in fact, possible, and the resulting observers are known as reduced-order observers [14, 20]. However, since sensors are not noise free, even states that are directly measured are generally estimated in an effort to reduce the effect of sensor noise on the state estimate. The Kalman filter (which is a time-varying optimal observer) solves the observer problem in the presence of measurement noise (and process noise as well) [36, 37].

The observer design begins by checking the system observability to verify that an observer can be constructed to guarantee the stability of the estimation error. From the system model, we find that

$$A = \begin{bmatrix} 2 & 3 \\ -1 & 4 \end{bmatrix} \quad \text{and} \quad C = [1 \quad 0].$$

The corresponding observability matrix is

$$P_o = \begin{bmatrix} C \\ CA \end{bmatrix} = \begin{bmatrix} 1 & 0 \\ 2 & 3 \end{bmatrix}.$$

Since det $\mathbf{P}_o = 3 \neq 0$, the system is completely observable. Suppose that the desired characteristic equation is given by

$$\Delta_d(\lambda) = \lambda^2 + 2\zeta\omega_n\lambda + \omega_n^2. \tag{11.15}$$

We can select $\zeta = 0.8$ and $\omega_n = 10$, resulting in an expected settling time of less than 0.5 second. Computing the actual characteristic equation yields

$$\det(\lambda\mathbf{I} - (\mathbf{A} - \mathbf{LC})) = \lambda^2 + (L_1 - 6)\lambda - 4(L_1 - 2) + 3(L_2 + 1), \tag{11.16}$$

where $\mathbf{L} = [L_1 \quad L_2]^T$. Equating the coefficients in Equation (11.15) to those in Equation (11.16) yields the two equations

$$L_1 - 6 = 16$$
$$-4(L_1 - 2) + 3(L_2 + 1) = 100$$

which, when solved, produces

$$\mathbf{L} = \begin{bmatrix} L_1 \\ L_2 \end{bmatrix} = \begin{bmatrix} 22 \\ 59 \end{bmatrix}.$$

The observer is thus given by

$$\dot{\hat{\mathbf{x}}} = \begin{bmatrix} 2 & 3 \\ -1 & 4 \end{bmatrix}\hat{\mathbf{x}} + \begin{bmatrix} 0 \\ 1 \end{bmatrix}u + \begin{bmatrix} 22 \\ 59 \end{bmatrix}(y - \hat{x}_1).$$

The response of the estimation error to an initial error of

$$\mathbf{e} = \begin{bmatrix} 1 \\ -2 \end{bmatrix}$$

is shown in Figure 11.7. ■

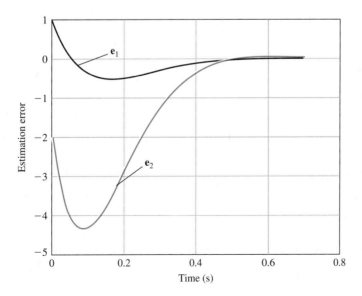

FIGURE 11.7
Second-order
observer response
to initial estimation
errors.

Ackermann's formula can also be employed to place the roots of the observer characteristic equation at the desired locations. Consider the observer gain matrix

$$\mathbf{L} = [L_1 \quad L_2 \quad \cdots \quad L_n]^T$$

and the desired observer characteristic equation

$$p(\lambda) = \lambda^n + \beta_{n-1}\lambda^{n-1} + \cdots + \beta_1\lambda + \beta_0.$$

The β's are selected to meet given performance specifications for the observer. The observer gain matrix is then computed via

$$\boxed{\mathbf{L} = p(\mathbf{A})\mathbf{P}_o^{-1}[0 \cdots 0 \quad 1]^T,} \tag{11.17}$$

where $\mathbf{P}_o$ is the observability matrix given in Equation (11.3) and

$$p(\mathbf{A}) = \mathbf{A}^n + \beta_{n-1}\mathbf{A}^{n-1} + \cdots + \beta_1\mathbf{A} + \beta_0\mathbf{I}.$$

EXAMPLE 11.9   **Second-order system observer design using Ackermann's formula**

Consider the second-order system in Example 11.8. The desired characteristic equation was given as

$$p(\lambda) = \lambda^2 + 2\zeta\omega_n\lambda + \omega_n^2,$$

where $\zeta = 0.8$ and $\omega_n = 10$; hence, $\beta_1 = 16$ and $\beta_2 = 100$. Computing $p(\mathbf{A})$ yields

$$p(\mathbf{A}) = \begin{bmatrix} 2 & 3 \\ -1 & 4 \end{bmatrix}^2 + 16\begin{bmatrix} 2 & 3 \\ -1 & 4 \end{bmatrix} + 100\begin{bmatrix} 1 & 0 \\ 0 & 1 \end{bmatrix} = \begin{bmatrix} 133 & 66 \\ -22 & 177 \end{bmatrix},$$

and from Example 11.8, we have the observability matrix

$$\mathbf{P}_o = \begin{bmatrix} 1 & 0 \\ 2 & 3 \end{bmatrix},$$

which implies that

$$\mathbf{P}_o^{-1} = \begin{bmatrix} 1 & 0 \\ -2/3 & 1/3 \end{bmatrix}.$$

Using Ackermann's formula in Equation (11.17) yields the observer gain matrix

$$\mathbf{L} = p(\mathbf{A})\mathbf{P}_o^{-1}[0 \quad \cdots \quad 0 \quad 1]^T = \begin{bmatrix} 133 & 66 \\ -22 & 177 \end{bmatrix}\begin{bmatrix} 1 & 0 \\ -2/3 & 1/3 \end{bmatrix}\begin{bmatrix} 0 \\ 1 \end{bmatrix} = \begin{bmatrix} 22 \\ 59 \end{bmatrix}.$$

This is the identical result obtained in Example 11.8 using other methods. ∎

11.5 INTEGRATED FULL-STATE FEEDBACK AND OBSERVER

The state variable compensator is constructed by appropriately connecting the full-state feedback control law (see Section 11.3) to the observer (see Section 11.4). The compensator is shown in Figure 11.1 (as discussed in Section 11.1). Our strategy was to design the state feedback control law as $u(t) = -\mathbf{K}\mathbf{x}(t)$, where we assumed that we had access to the complete state $\mathbf{x}(t)$. Then we designed an observer to provide an estimate of the state $\hat{\mathbf{x}}(t)$. It seems reasonable that we can employ the state estimate in the feedback control law in place of $\mathbf{x}(t)$. In other words, we can consider the feedback law

$$u(t) = -\mathbf{K}\hat{\mathbf{x}}(t). \tag{11.18}$$

But is this a good strategy? The feedback gain matrix $\mathbf{K}$ was designed to guarantee stability of the closed-loop system; that is, the roots of the characteristic equation

$$\det(\lambda\mathbf{I} - (\mathbf{A} - \mathbf{B}\mathbf{K})) = 0$$

are in the left half-plane. Under the assumption that the complete state $\mathbf{x}(t)$ is available for feedback, the feedback control law (with properly designed gain matrix $\mathbf{K}$) leads to the desired result that $\mathbf{x}(t) \rightarrow 0$ as $t \rightarrow \infty$ for any initial condition $\mathbf{x}(t_0)$. We need to verify that, when using the feedback control law in Equation (11.18), we retain the stability of the closed-loop system.

Consider the observer (from Section 11.4)

$$\dot{\hat{\mathbf{x}}} = \mathbf{A}\hat{\mathbf{x}} + \mathbf{B}u + \mathbf{L}(y - \mathbf{C}\hat{\mathbf{x}}).$$

Substituting the feedback law in Equation (11.18) and rearranging terms in the observer yields the compensator system

$$\dot{\hat{\mathbf{x}}} = (\mathbf{A} - \mathbf{B}\mathbf{K} - \mathbf{L}\mathbf{C})\hat{\mathbf{x}} + \mathbf{L}y$$
$$u = -\mathbf{K} . \tag{11.19}$$

Notice that the system in Equation (11.19) has the form of a state variable model with input y and output u, as illustrated in Figure 11.8.

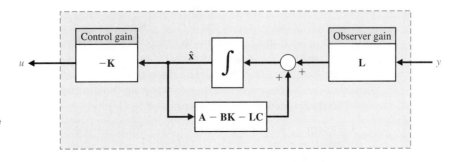

FIGURE 11.8
State variable compensator with integrated full-state feedback and observer.

Computing the estimation error using the compensator in Equation (11.19) yields

$$\dot{\mathbf{e}} = \dot{\mathbf{x}} - \dot{\hat{\mathbf{x}}} = \mathbf{A}\mathbf{x} + \mathbf{B}u - \mathbf{A}\hat{\mathbf{x}} - \mathbf{B}u - \mathbf{L}y + \mathbf{L}\mathbf{C}\hat{\mathbf{x}},$$

or

$$\dot{\mathbf{e}} = (\mathbf{A} - \mathbf{L}\mathbf{C})\mathbf{e}. \tag{11.20}$$

This is the same result as we obtained for the estimation error in Section 11.4. The estimation error does not depend on the input as seen in Equation (11.20), where the input terms cancel. Recall that the underlying system model is given by

$$\dot{\mathbf{x}} = \mathbf{A}\mathbf{x} + \mathbf{B}u$$
$$y = \mathbf{C}\mathbf{x}.$$

Substituting the feedback law $u(t) = -\mathbf{K}\hat{\mathbf{x}}(t)$ into the system model yields

$$\dot{\mathbf{x}} = \mathbf{A}\mathbf{x} + \mathbf{B}u = \mathbf{A}\mathbf{x} - \mathbf{B}\mathbf{K}\hat{\mathbf{x}},$$

and with $\hat{\mathbf{x}} = \mathbf{x} - \mathbf{e}$, we obtain

$$\dot{\mathbf{x}} = (\mathbf{A} - \mathbf{B}\mathbf{K})\mathbf{x} + \mathbf{B}\mathbf{K}\mathbf{e}. \tag{11.21}$$

Writing Equations (11.20) and (11.21) in matrix form, we have

$$\begin{pmatrix} \dot{\mathbf{x}} \\ \dot{\mathbf{e}} \end{pmatrix} = \begin{bmatrix} \mathbf{A} - \mathbf{B}\mathbf{K} & \mathbf{B}\mathbf{K} \\ \mathbf{0} & \mathbf{A} - \mathbf{L}\mathbf{C} \end{bmatrix} \begin{pmatrix} \mathbf{x} \\ \mathbf{e} \end{pmatrix}. \tag{11.22}$$

Recall that our goal is to verify that, with $u(t) = -\mathbf{K}\hat{\mathbf{x}}(t)$, we retain stability of the closed-loop system and the observer. The characteristic equation associated with Equation (11.22) is

$$\Delta(\lambda) = \det(\lambda\mathbf{I} - (\mathbf{A} - \mathbf{B}\mathbf{K})) \det(\lambda\mathbf{I} - (\mathbf{A} - \mathbf{L}\mathbf{C})).$$

So if the roots of $\det(\lambda\mathbf{I} - (\mathbf{A} - \mathbf{B}\mathbf{K})) = 0$ lie in the left half-plane (which they do by design of the full-state feedback law), and if the roots of $\det(\lambda\mathbf{I} - (\mathbf{A} - \mathbf{L}\mathbf{C})) = 0$ lie in the left half-plane (which they do by design of the observer), then the overall system is stable. Therefore, employing the strategy of using the state estimates for the feedback is in fact a good strategy.

In other words, when we use $u(t) = -\mathbf{K}\hat{\mathbf{x}}(t)$ where $\mathbf{K}$ is designed using the methods proposed in Section 11.3 and $\hat{\mathbf{x}}$ is derived from the observer discussed in Section 11.4, then $\mathbf{x}(t) \rightarrow 0$ as $t \rightarrow \infty$ for any initial condition $\mathbf{x}(t_0)$ and $\mathbf{e}(t) \rightarrow 0$ as $t \rightarrow \infty$ for any initial estimation error $\mathbf{e}(t_0)$. The fact that the full-state feedback law and the observer can be designed independently is an illustration of the **separation principle**.

The design procedure is summarized as follows:

1. Determine $\mathbf{K}$ such that $\det(\lambda\mathbf{I} - (\mathbf{A} - \mathbf{B}\mathbf{K})) = 0$ has roots in the left half-plane and place the poles appropriately to meet the control system design specifications. The ability to place the poles arbitrarily in the complex plane is guaranteed if the system is completely controllable.

2. Determine **L** such that $\det(\lambda\mathbf{I} - (\mathbf{A} - \mathbf{LC})) = 0$ has roots in the left half-plane and place the poles to achieve acceptable observer performance. The ability to place the observer poles arbitrarily in the complex plane is guaranteed if the system is completely observable.

3. Connect the observer to the full-state feedback law using

$$u(t) = -\mathbf{K}\hat{\mathbf{x}}(t).$$

Compensator Transfer Function. The compensator given in Equation (11.19) can be given equivalently in transfer function form with input $Y(s)$ and output $U(s)$. Taking the Laplace transform (with zero initial conditions) of the compensator yields

$$s\hat{\mathbf{X}}(s) = (\mathbf{A} - \mathbf{BK} - \mathbf{LC})\hat{\mathbf{X}}(s) + \mathbf{L}Y(s)$$
$$U(s) = -\mathbf{K}\hat{\mathbf{X}}(s),$$

and rearranging and solving for $U(s)$, we obtain the transfer function

$$\boxed{U(s) = [-\mathbf{K}(s\mathbf{I} - (\mathbf{A} - \mathbf{BK} - \mathbf{LC}))^{-1}\mathbf{L}]Y(s).} \qquad (11.23)$$

Note that the compensator transfer function itself (when viewed as a system) may or may not be stable. Even though $\mathbf{A} - \mathbf{BK}$ is stable and $\mathbf{A} - \mathbf{LC}$ is stable, it does not necessarily follow that $\mathbf{A} - \mathbf{BK} - \mathbf{LC}$ is stable. However, the overall closed-loop system is stable (as we proved in the previous discussions). The controller in Equation (11.23) is commonly referred to as a **stabilizing controller**.

EXAMPLE 11.10   **Compensator design for the inverted pendulum**

Consider the inverted pendulum of Example 3.4. The state variable model representing the inverted pendulum atop a moving cart is

$$\dot{\mathbf{x}} = \begin{bmatrix} 0 & 1 & 0 & 0 \\ 0 & 0 & \dfrac{-mg}{M} & 0 \\ 0 & 0 & 0 & 1 \\ 0 & 0 & \dfrac{g}{l} & 0 \end{bmatrix}\mathbf{x} + \begin{bmatrix} 0 \\ \dfrac{1}{M} \\ 0 \\ \dfrac{-1}{Ml} \end{bmatrix}u,$$

where $\mathbf{x} = (x_1, x_2, x_3, x_4)^T$, x_1 is the cart position, x_2 is the cart velocity, x_3 is the pendulum angular position (measured from the vertical), x_4 is the pendulum angular rate, and u is the input applied to the cart. As discussed in Example 11.6, we can measure the state variable $x_3 = \theta$ using a potentiometer attached to the shaft, or measure $x_4 = \dot{\theta}$ using a tachometer generator. However, suppose that we have a sensor available to measure the position of the cart. Is it possible to hold the angular position of the pendulum at the desired value ($\theta = 0°$) when only the output $y = x_1$ (the cart position) is available? In this case, we have the output equation

$$y = [1 \quad 0 \quad 0 \quad 0]\mathbf{x}.$$

Let the system parameters be

$$l = 0.098 \text{ m}$$
$$g = 9.8 \text{ m/s}^2$$
$$m = 0.825 \text{ kg}$$
$$M = 8.085 \text{ kg}.$$

Therefore, using the parameter values, the system state and input matrices are

$$\mathbf{A} = \begin{bmatrix} 0 & 1 & 0 & 0 \\ 0 & 0 & -1 & 0 \\ 0 & 0 & 0 & 1 \\ 0 & 0 & 100 & 0 \end{bmatrix} \quad \text{and} \quad \mathbf{B} = \begin{bmatrix} 0 \\ 0.1237 \\ 0 \\ -1.2621 \end{bmatrix}.$$

Checking controllability yields the controllability matrix

$$\mathbf{P}_c = \begin{bmatrix} 0 & 0.1237 & 0 & 1.2621 \\ 0.1237 & 0 & 1.2621 & 0 \\ 0 & -1.2621 & 0 & -126.21 \\ -1.2621 & 0 & -126.21 & 0 \end{bmatrix}.$$

Computing det $\mathbf{P}_c = 196.49 \neq 0$; hence, the system is completely controllable. Likewise, computing the observability matrix

$$\mathbf{P}_o = \begin{bmatrix} 1 & 0 & 0 & 0 \\ 0 & 1 & 0 & 0 \\ 0 & 0 & -1 & 0 \\ 0 & 0 & 0 & -1 \end{bmatrix}$$

and det $\mathbf{P}_o = 1 \neq 0$; hence, the system is completely observable. We can now proceed with the three-step design procedure knowing that we can determine a control gain matrix $\mathbf{K}$ and observer gain matrix $\mathbf{L}$ to place all the closed-loop system poles at desired locations.

STEP 1: Design the Full-State Feedback Control Law.
The open-loop system poles are located at $\lambda = 0, 0, -10$, and 10. It is evident that the open-loop system is unstable (there is a pole in the right half-plane). Suppose that the desired closed-loop system characteristic equation is given by

$$q(\lambda) = (\lambda^2 + 2\zeta\omega_n\lambda + \omega_n^2)(\lambda^2 + a\lambda + b),$$

where we choose (1) the pair (ζ, ω_n) so that these poles are the dominant poles and (2) the pair (a, b) farther in the left half-plane so as not to dominate the response. To obtain a settling time less than 10 seconds with low overshoot, we can select $(\zeta, \omega_n) = (0.8, 0.5)$. Then, we choose a separation factor of 20 between the dominant poles and the remaining poles, from which it follows that $(a, b) = (16, 100)$. Figure 11.9 shows the pole zero map for the system design. The separation factor

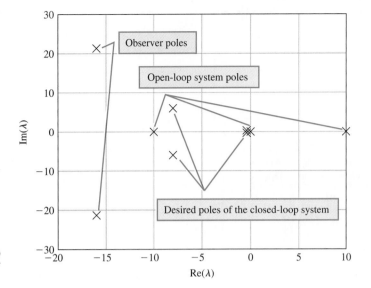

FIGURE 11.9
System pole map: open-loop poles, desired closed-loop poles, and observer poles.

between the dominant and nondominant poles is a parameter that can be varied as part of the design process. The larger the separation selected, the further left in the left half-plane the nondominant poles will be placed, and hence the larger the required control law gains. The desired roots are then specified to be

$$\det(\lambda \mathbf{I} - (\mathbf{A} - \mathbf{B}\mathbf{K})) = (\lambda + 8 \pm j6)(\lambda + 0.4 \pm j0.3)$$

The poles at $\lambda = -0.4 \pm 0.3j$ are the dominant poles. Using Ackermann's formula yields the feedback gain matrix

$$\mathbf{K} = [-2.2509 \quad -7.5631 \quad -169.0265 \quad -14.0523].$$

STEP 2: Observer Design
The observer needs to provide an estimate of the states that cannot be directly observed. The goal is to achieve an accurate estimate as fast as possible without resulting in too large a gain matrix $\mathbf{L}$. How large is too large depends on the problem under consideration. In particular, if there are significant levels of measurement noise (this is sensor dependent), then the magnitude of the observer matrix should be kept correspondingly low to avoid amplifying the measurement noise. The trade-off between the time required to obtain accurate observer performance and the amount of noise amplification is a primary design issue. For design purposes, we will attempt to insure a separation of the desired closed-loop system poles and the observer poles on the order of 2 to 10 (as illustrated in Figure 11.9). The desired observer characteristic equation is selected to be of the form

$$p(\lambda) = (\lambda^2 + c_1\lambda + c_2)^2,$$

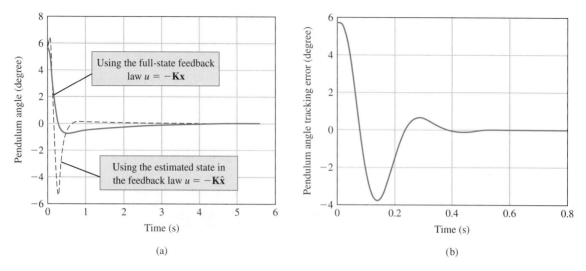

FIGURE 11.10   Pendulum performance under full-state feedback control with the observer in the loop.

where the constants c_1 and c_2 are appropriately chosen. As a first attempt, we select $c_1 = 32$ and $c_2 = 711.11$. These values should produce a response to an initial state estimation error that settles in less than 0.5 second with minimal overshoot. Using Ackermann's formula from Section 11.3, we determine that the observer gain that achieves the desired observer pole locations $\det(\lambda \mathbf{I} - (\mathbf{A} - \mathbf{LC})) = ((\lambda + 16 + j21.3)(\lambda + 16 - j21.3))^2$ is

$$\mathbf{L} = \begin{bmatrix} 64.0 \\ 2546.22 \\ -5.1911\text{E}04 \\ -7.6030\text{E}05 \end{bmatrix}.$$

STEP 3: Compensator Design

The final step in the design is to connect the observer to the full-state feedback control law via $u = -\mathbf{K}\hat{\mathbf{x}}$. As proved earlier, the closed-loop system will remain stable; however, we should not expect the closed-loop performance to be as good when using the state estimate from the observer. This makes sense, since it takes a finite amount of time for the observer to provide accurate state estimates. The response of the inverted pendulum design is shown in Figure 11.10. The pendulum is initially stationary at $\theta_0 = 5.72°$, and the cart is initially not moving. The initial state estimate in the observer is set to zero.

In Figure 11.10(a), we see that, indeed, the pendulum is balanced to the vertical in under 4 seconds. The response of the compensator (with the observer) is more oscillatory than without the observer in the loop—but this difference in performance is expected, since it takes about 0.4 second for the observer to converge to a minimal state tracking error, as seen in Figure 11.10(b). ∎

11.6 REFERENCE INPUTS

The feedback strategies discussed in the previous sections (and illustrated in Figure 11.1) were constructed without consideration of reference inputs. We referred to the design of state variable feedback compensators without reference inputs (i.e., $r(t) = 0$) as regulators. Since **command following** is also an important aspect of feedback design, it is important to consider how we can introduce a reference signal into the state variable feedback compensator. There are, in fact, many different techniques that can be employed to permit the tracking of a reference input. Two of the more common methods are discussed in this section.

The general form of the state variable feedback compensator is

$$\dot{\hat{\mathbf{x}}} = \mathbf{A}\hat{\mathbf{x}} + \mathbf{B}\widetilde{u} + \mathbf{L}\widetilde{y} + \mathbf{M}r$$
$$u = \widetilde{u} + Nr = -\mathbf{K}\hat{\mathbf{x}} + Nr, \qquad (11.24)$$

where $\widetilde{y} = y - \mathbf{C}\hat{\mathbf{x}}$ and $\widetilde{u} = -\mathbf{K}\hat{\mathbf{x}}$. The state variable compensator with the reference input is illustrated in Figure 11.11. Notice that when $\mathbf{M} = \mathbf{0}$ and $N = 0$, the compensator in Equation (11.24) reduces to the regulator described in Section 11.5 and illustrated in Figure 11.1.

The compensator key design parameters required to implement the command tracking of the reference input are $\mathbf{M}$ and N. When the reference input is a scalar signal (i.e., a single input), the parameter $\mathbf{M}$ is a column vector of length n, where n is the length of the state vector $\mathbf{x}$, and N is a scalar. Here, we consider two possibilities for selecting $\mathbf{M}$ and N. In the first case, we select $\mathbf{M}$ and N so that the estimation error $\mathbf{e}(t)$ is independent of the reference input $r(t)$. In the second case, we select $\mathbf{M}$ and N so that the tracking error $y(t) - r(t)$ is used as an input to the compensator. These two cases will result in implementations wherein the compensator is in the feedback loop in the first case and in the forward loop in the second case.

Employing the generalized compensator in Equation (11.24), the estimation error is found to be described by the differential equation

$$\dot{\mathbf{e}} = \dot{\mathbf{x}} - \dot{\hat{\mathbf{x}}} = \mathbf{A}\mathbf{x} + \mathbf{B}u - \mathbf{A}\hat{\mathbf{x}} - \mathbf{B}\widetilde{u} - \mathbf{L}\widetilde{y} - \mathbf{M}r,$$

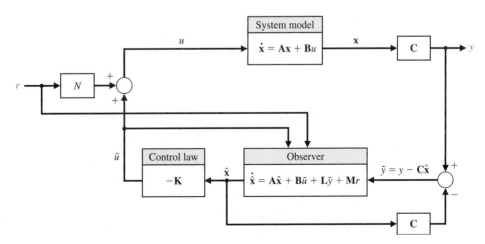

FIGURE 11.11
State variable compensator with a reference input.

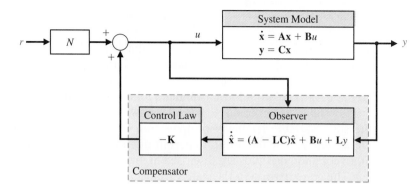

FIGURE 11.12
State variable
compensator with
reference input and
$\mathbf{M} = \mathbf{B}N$.

or

$$\dot{\mathbf{e}} = (\mathbf{A} - \mathbf{LC})\mathbf{e} + (\mathbf{B}N - \mathbf{M})r.$$

Suppose that we select

$$\mathbf{M} = \mathbf{B}N. \tag{11.25}$$

Then the corresponding estimation error is given by

$$\dot{\mathbf{e}} = (\mathbf{A} - \mathbf{LC})\mathbf{e}.$$

In this case, the estimation error is independent of the reference input $r(t)$. This is the identical result found in Section 11.4, where we considered the observer design assuming no reference inputs. The remaining task is to determine a suitable value of N, since the value of $\mathbf{M}$ follows from Equation (11.25). For example, we might choose N to obtain a zero steady-state tracking error to a step input $r(t)$.

With $\mathbf{M} = \mathbf{B}N$, we find that the compensator is given by

$$\dot{\hat{\mathbf{x}}} = \mathbf{A}\hat{\mathbf{x}} + \mathbf{B}u + \mathbf{L}\tilde{y}$$
$$u = -\mathbf{K}\hat{\mathbf{x}} + Nr.$$

This implementation of the state variable compensator is illustrated in Figure 11.12.

As an alternative approach, suppose that we select $N = 0$ and $\mathbf{M} = -\mathbf{L}$. Then, the compensator in Equation (11.24) is given by

$$\dot{\hat{\mathbf{x}}} = \mathbf{A}\hat{\mathbf{x}} + \mathbf{B}u + \mathbf{L}\tilde{y} - \mathbf{L}r$$
$$u = -\mathbf{K}\hat{\mathbf{x}},$$

which can be rewritten as

$$\dot{\hat{\mathbf{x}}} = (\mathbf{A} - \mathbf{BK} - \mathbf{LC})\hat{\mathbf{x}} + \mathbf{L}(y - r)$$
$$u = -\mathbf{K}\hat{\mathbf{x}}.$$

In this formulation, the observer is driven by the tracking error $y - r$. The reference input tracking implementation is illustrated in Figure 11.13.

Notice that in the first implementation (with $\mathbf{M} = \mathbf{B}N$) the compensator is in the feedback loop, whereas in the second implementation ($N = 0$ and $\mathbf{M} = -\mathbf{L}$) the

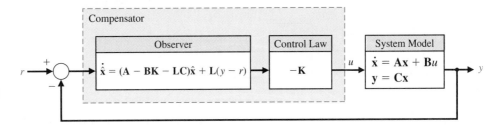

FIGURE 11.13
State variable
compensator with
reference input and
$N = 0$ and
$\mathbf{M} = -\mathbf{L}$.

compensator is in the forward path. These two implementations are representative of
the possibilities open to control system designers when considering reference inputs.

Depending on the choice of N and $\mathbf{M}$, other implementations are possible. For
example, Section 11.8 presents a method of tracking reference inputs with guaranteed steady-state tracking errors using **internal model design** techniques.

11.7 OPTIMAL CONTROL SYSTEMS

The design of optimal control systems is an important function of control engineering. The purpose of design is to realize a system with practical components that will
provide the desired operating performance. The desired performance can be readily
stated in terms of time-domain performance indices. For example, the maximum
overshoot and rise time for a step input are valuable time-domain indices. In the case
of steady-state and transient performance, the performance indices are normally
specified in the time domain; therefore, it is natural that we wish to develop design
procedures in the time domain.

The performance of a control system can be represented by integral performance
measures, as we found in Section 5.7. Therefore, the design of a system must be based
on minimizing a performance index, such as the integral of the squared error (ISE), as
in Section 5.7. Systems that are adjusted to provide a minimum performance index are
often called **optimal control systems**. In this section, we will consider the design of an
optimal control system that is described by a state variable formulation. We will consider the measurement of the state variables and their use in developing a control signal $u(t)$ so that the performance of the system is optimized.

The performance of a control system, written in terms of the state variables of a
system, can be expressed in general as

$$J = \int_0^{t_f} g(\mathbf{x}, \mathbf{u}, t)\, dt, \tag{11.26}$$

where $\mathbf{x}$ equals the state vector, $\mathbf{u}$ equals the control vector, and t_f equals the final time.[1]

We are interested in minimizing the error of the system; therefore, when the de-
sired state vector is represented as $\mathbf{x}_d = \mathbf{0}$, we are able to consider the error as identically equal to the value of the state vector. That is, we intend the system to be at

---

[1]Note that to denote the performance index, J is used instead of I, as in Chapter 5. This will enable the
reader to distinguish readily the performance index from the identity matrix, which is represented by the
boldfaced capital $\mathbf{I}$.

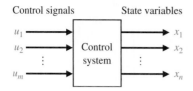

FIGURE 11.14
A control system in terms of x and u.

equilibrium, $\mathbf{x} = \mathbf{x}_d = \mathbf{0}$, and any deviation from equilibrium is considered an error. Therefore, in this section, we will consider the design of optimal control systems using state variable feedback and **error-squared performance indices** [1–3].

The control system we will consider is shown in Figure 11.14 and can be represented by the vector differential equation

$$\dot{\mathbf{x}} = \mathbf{Ax} + \mathbf{Bu}. \tag{11.27}$$

We will select a feedback controller so that $\mathbf{u}$ is some function of the measured state variables $\mathbf{x}$ and therefore

$$\mathbf{u} = -\mathbf{k}(\mathbf{x}).$$

For example, we might use

$$u_1 = -k_1 x_1, \qquad u_2 = -k_2 x_2, \qquad \ldots, \qquad u_m = -k_m x_m. \tag{11.28}$$

Alternatively, we might choose the control vector as

$$u_1 = -k_1(x_1 + x_2), \qquad u_2 = -k_2(x_2 + x_3), \qquad \ldots. \tag{11.29}$$

The choice of the control signals is somewhat arbitrary and depends partially on the actual desired performance and the complexity of the feedback structure allowable. Often, we are limited in the number of state variables available for feedback, since we are only able to use measurable state variables.

In our case, we limit the feedback function to a linear function so that $\mathbf{u} = -\mathbf{Kx}$, where $\mathbf{K}$ is an $m \times n$ matrix, as in Section 11.3. Therefore, in expanded form, we have

$$\begin{bmatrix} u_1 \\ u_2 \\ \vdots \\ u_m \end{bmatrix} = - \begin{bmatrix} k_{11} & \cdots & k_{1n} \\ \vdots & & \vdots \\ k_{m1} & \cdots & k_{mn} \end{bmatrix} \begin{bmatrix} x_1 \\ x_2 \\ \vdots \\ x_n \end{bmatrix}. \tag{11.30}$$

Substituting Equation (11.30) into Equation (11.27), we obtain

$$\dot{\mathbf{x}} = \mathbf{Ax} - \mathbf{BKx} = \mathbf{Hx}, \tag{11.31}$$

where $\mathbf{H}$ is the $n \times n$ matrix resulting from the addition of the elements of $\mathbf{A}$ and $-\mathbf{BK}$.

Now, returning to the error-squared performance index, we recall from Section 5.7 that the index for a single state variable, x_1, is written as

$$J = \int_0^{t_f} [x_1(t)]^2 \, dt. \tag{11.32}$$

A performance index written in terms of two state variables would then be

$$J = \int_0^{t_f} (x_1^2 + x_2^2)\, dt. \tag{11.33}$$

Since we wish to define the performance index in terms of an integral of the sum of the state variables squared, we will use the matrix operation

$$\mathbf{x}^T\mathbf{x} = [x_1, x_2, x_3, \ldots, x_n] \begin{bmatrix} x_1 \\ x_2 \\ \vdots \\ x_n \end{bmatrix} = x_1^2 + x_2^2 + x_3^2 + \cdots + x_n^2, \tag{11.34}$$

where $\mathbf{x}^T$ indicates the transpose of the $\mathbf{x}$ matrix.[†] Then the specific form of the performance index, in terms of the state vector, is

$$J = \int_0^{t_f} \mathbf{x}^T\mathbf{x}\, dt. \tag{11.35}$$

The general form of the performance index (Equation 11.26) incorporates a term with $\mathbf{u}$ that we have not included at this point, but we will do so later in this section.

Again considering Equation (11.35), we will let the final time of interest be $t_f = \infty$. To obtain the minimum value of J, we postulate the existence of an exact differential so that

$$\frac{d}{dt}(\mathbf{x}^T\mathbf{P}\mathbf{x}) = -\mathbf{x}^T\mathbf{x}, \tag{11.36}$$

where $\mathbf{P}$ is to be determined. A symmetric $\mathbf{P}$ matrix will be used to simplify the algebra without any loss of generality. Then, for a symmetric $\mathbf{P}$ matrix, $p_{ij} = p_{ji}$. Completing the differentiation indicated on the left-hand side of Equation (11.36), we have

$$\frac{d}{dt}(\mathbf{x}^T\mathbf{P}\mathbf{x}) = \dot{\mathbf{x}}^T\mathbf{P}\mathbf{x} + \mathbf{x}^T\mathbf{P}\dot{\mathbf{x}}.$$

Substituting Equation (11.31), we obtain

$$\frac{d}{dt}(\mathbf{x}^T\mathbf{P}\mathbf{x}) = (\mathbf{H}\mathbf{x})^T\mathbf{P}\mathbf{x} + \mathbf{x}^T\mathbf{P}(\mathbf{H}\mathbf{x})$$

$$= \mathbf{x}^T\mathbf{H}^T\mathbf{P}\mathbf{x} + \mathbf{x}^T\mathbf{P}\mathbf{H}\mathbf{x}$$

$$= \mathbf{x}^T(\mathbf{H}^T\mathbf{P} + \mathbf{P}\mathbf{H})\mathbf{x}, \tag{11.37}$$

where $(\mathbf{H}\mathbf{x})^T = \mathbf{x}^T\mathbf{H}^T$ by the definition of the transpose of a product. If we let $\mathbf{H}^T\mathbf{P} + \mathbf{P}\mathbf{H} = -\mathbf{I}$, then Equation (11.37) becomes

$$\frac{d}{dt}(\mathbf{x}^T\mathbf{P}\mathbf{x}) = -\mathbf{x}^T\mathbf{x}, \tag{11.38}$$

[†]The matrix operation $\mathbf{x}^T\mathbf{x}$ is discussed on the MCS website.

which is the exact differential we are seeking. Substituting Equation (11.38) into Equation (11.35), we obtain

$$J = \int_0^\infty -\frac{d}{dt}(\mathbf{x}^T \mathbf{P} \mathbf{x}) \, dt = -\mathbf{x}^T \mathbf{P} \mathbf{x} \Big|_0^\infty = \mathbf{x}^T(0)\mathbf{P}\mathbf{x}(0). \qquad (11.39)$$

In the evaluation of the limit at $t = \infty$, we have assumed that the system is stable, and hence $\mathbf{x}(\infty) = 0$, as desired. Therefore, to minimize the performance index J, we consider the two equations

$$J = \int_0^\infty \mathbf{x}^T \mathbf{x} \, dt = \mathbf{x}^T(0)\mathbf{P}\mathbf{x}(0) \qquad (11.40)$$

and

$$\mathbf{H}^T \mathbf{P} + \mathbf{P} \mathbf{H} = -\mathbf{I}. \qquad (11.41)$$

The design steps are then as follows:

1. Determine the matrix $\mathbf{P}$ that satisfies Equation (11.41), where $\mathbf{H}$ is known.
2. Minimize J by determining the minimum of Equation (11.40) by adjusting one or more unspecified system parameters.

EXAMPLE 11.11   State variable feedback

Consider the open-loop control system shown in Figure 11.15. The state variables are identified as x_1 and x_2. The performance of this system is quite unsatisfactory because an undamped response results for a step input. The vector differential equation of this system is

$$\frac{d}{dt}\begin{bmatrix} x_1 \\ x_2 \end{bmatrix} = \begin{bmatrix} 0 & 1 \\ 0 & 0 \end{bmatrix}\begin{bmatrix} x_1 \\ x_2 \end{bmatrix} + \begin{bmatrix} 0 \\ 1 \end{bmatrix}u(t), \qquad (11.42)$$

where

$$\mathbf{A} = \begin{bmatrix} 0 & 1 \\ 0 & 0 \end{bmatrix} \quad \text{and} \quad \mathbf{B} = \begin{bmatrix} 0 \\ 1 \end{bmatrix}.$$

We will choose a feedback control system so that

$$u(t) = -k_1 x_1 - k_2 x_2, \qquad (11.43)$$

FIGURE 11.15
Open-loop control system of Example 11.11.

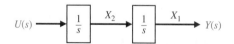

and therefore the control signal is a linear function of the two state variables. Then Equation (11.42) becomes

$$\dot{x}_1 = x_2,$$
$$\dot{x}_2 = -k_1 x_1 - k_2 x_2; \tag{11.44}$$

in matrix form, we have

$$\dot{\mathbf{X}} = \mathbf{Hx}$$
$$= \begin{bmatrix} 0 & 1 \\ -k_1 & -k_2 \end{bmatrix} \mathbf{x}. \tag{11.45}$$

We note that x_1 would represent the position of a position control system, and the transfer function of the system would be $G(s) = 1/(Ms^2)$, where $M = 1$ and the friction is negligible. We will let $k_1 = 1$ and determine a suitable value for k_2 so that the performance index is minimized. Writing Equation (11.41), we have

$$\mathbf{H}^T\mathbf{P} + \mathbf{PH} = -\mathbf{I},$$

and in expanded form

$$\begin{bmatrix} 0 & -1 \\ 1 & -k_2 \end{bmatrix}\begin{bmatrix} p_{11} & p_{12} \\ p_{12} & p_{22} \end{bmatrix} + \begin{bmatrix} p_{11} & p_{12} \\ p_{12} & p_{22} \end{bmatrix}\begin{bmatrix} 0 & 1 \\ -1 & -k_2 \end{bmatrix} = \begin{bmatrix} -1 & 0 \\ 0 & -1 \end{bmatrix}. \tag{11.46}$$

Completing the matrix multiplication and addition, we have

$$-p_{12} - p_{12} = -1,$$
$$p_{11} - k_2 p_{12} - p_{22} = 0,$$
$$p_{12} - k_2 p_{22} + p_{12} - k_2 p_{22} = -1. \tag{11.47}$$

Solving these simultaneous equations, we obtain

$$p_{12} = \frac{1}{2}, \qquad p_{22} = \frac{1}{k_2}, \qquad p_{11} = \frac{k_2^2 + 2}{2k_2}.$$

The integral performance index is then

$$J = \mathbf{x}^T(0)\mathbf{Px}(0), \tag{11.48}$$

and we will consider the case where each state is initially displaced one unit from equilibrium so that $\mathbf{x}^T(0) = [1, 1]$. Therefore Equation (11.48) becomes

$$J = \begin{bmatrix} 1 & 1 \end{bmatrix}\begin{bmatrix} p_{11} & p_{12} \\ p_{12} & p_{22} \end{bmatrix}\begin{bmatrix} 1 \\ 1 \end{bmatrix}$$

$$= \begin{bmatrix} 1 & 1 \end{bmatrix}\begin{bmatrix} p_{11} + p_{12} \\ p_{12} + p_{22} \end{bmatrix}$$

$$= (p_{11} + p_{12}) + (p_{12} + p_{22}) = p_{11} + 2p_{12} + p_{22}. \tag{11.49}$$

Substituting the values of the elements of $\mathbf{P}$, we have

$$J = \frac{k_2^2 + 2}{2k_2} + 1 + \frac{1}{k_2} = \frac{k_2^2 + 2k_2 + 4}{2k_2}. \tag{11.50}$$

To minimize as a function of k_2, we take the derivative with respect to k_2 and set it equal to zero:

$$\frac{dJ}{dk_2} = \frac{2k_2(2k_2 + 2) - 2(k_2^2 + 2k_2 + 4)}{(2k_2)^2} = 0. \tag{11.51}$$

Therefore, $k_2^2 = 4$, and $k_2 = 2$ when J is a minimum. The minimum value of J is obtained by substituting $k_2 = 2$ into Equation (11.50). Thus, we obtain

$$J_{min} = 3.$$

The system matrix $\mathbf{H}$, obtained for the compensated system, is then

$$\mathbf{H} = \begin{bmatrix} 0 & 1 \\ -1 & -2 \end{bmatrix}. \tag{11.52}$$

The characteristic equation of the compensated system is therefore

$$\det[\lambda \mathbf{I} - \mathbf{H}] = \det \begin{bmatrix} \lambda & -1 \\ 1 & \lambda + 2 \end{bmatrix} = \lambda^2 + 2\lambda + 1. \tag{11.53}$$

Because this is a second-order system, we note that the characteristic equation is of the form $s^2 + 2\zeta\omega_n s + \omega_n^2 = 0$, and therefore the damping ratio of the compensated system is $\zeta = 1.0$. This compensated system is considered to be an optimal system in that the compensated system results in a minimum value for the performance index when $k_1 = 1$ is fixed. Of course, we recognize that this system is optimal only for the specific set of initial conditions that were assumed. The compensated system is shown in Figure 11.16. A curve of the performance index as a function of k_2 is shown in Figure 11.17. It is clear that this system is not very sensitive to changes in k_2 and will maintain a near-minimum performance index if the k_2 is altered by some percentage. We define the sensitivity of an optimal system as

$$S_k^{opt} = \frac{\Delta J/J}{\Delta k/k}, \tag{11.54}$$

where k is the design parameter. Then, for this example, we have $k = k_2$, and considering $k_2 = 2.5$, for which $J = 3.05$, we obtain

$$S_{k_2}^{opt} \approx \frac{0.05/3}{0.5/2} = 0.07. \tag{11.55} \quad \blacksquare$$

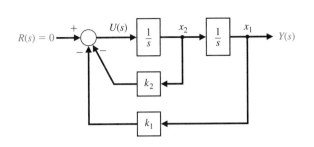

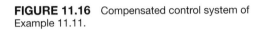

FIGURE 11.17
Performance index versus the
parameter k_2.

FIGURE 11.16   Compensated control system of
Example 11.11.

EXAMPLE 11.12   **Determination of an optimal system**

Now let us consider again the system of Example 11.11, where both the feedback gains, k_1 and k_2, are unspecified. To simplify the algebra without any loss in insight into the problem, let us set $k_1 = k_2 = k$. We can prove that if k_1 and k_2 are unspecified, then $k_1 = k_2$ when the minimum of the performance index (Equation 11.40) is obtained. Then, for the system of Example 11.11, Equation (11.45) becomes

$$\dot{\mathbf{x}} = \mathbf{Hx} = \begin{bmatrix} 0 & 1 \\ -k & -k \end{bmatrix}\mathbf{x}. \tag{11.56}$$

To determine the **P** matrix, we use Equation (11.41), which is

$$\mathbf{H}^T\mathbf{P} + \mathbf{PH} = -\mathbf{I}. \tag{11.57}$$

Solving the set of simultaneous equations resulting from Equation (11.57), we find that

$$p_{12} = \frac{1}{2k}, \quad p_{22} = \frac{k+1}{2k^2}, \quad \text{and} \quad p_{11} = \frac{1+2k}{2k}.$$

Let us consider the case where the system is initially displaced one unit from equilibrium so that $\mathbf{x}^T(0) = [1 \quad 0]$. Then the performance index (Equation 11.40) becomes

$$J = \int_0^\infty \mathbf{x}^T\mathbf{x}\, dt = \mathbf{x}^T(0)\mathbf{Px}(0) = p_{11}. \tag{11.58}$$

Thus, the performance index to be minimized is

$$J = p_{11} = \frac{1+2k}{2k} = 1 + \frac{1}{2k}. \tag{11.59}$$

The minimum value of J is obtained when k approaches infinity; the result is $J_{min} = 1$. A plot of J versus k, shown in Figure 11.18, illustrates that the performance index approaches a minimum asymptotically as k approaches an infinite

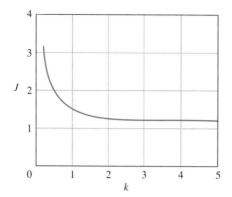

FIGURE 11.18
Performance index versus the feedback gain k for Example 11.12.

value. Now, we recognize that, in providing a very large gain k, we can cause the feedback signal

$$u(t) = -k[x_1(t) + x_2(t)]$$

to be very large. However, we are restricted to realizable magnitudes of the control signal $u(t)$. Therefore, we must introduce a constraint on $u(t)$ so that the gain k is not made too large. Then, for example, if we establish a constraint on $u(t)$ so that

$$|u(t)| \leq 50, \tag{11.60}$$

we require that the maximum acceptable value of k in this case be

$$k_{max} = \frac{|u|_{max}}{x_1(0)} = 50. \tag{11.61}$$

Then the minimum value of J is

$$J_{min} = 1 + \frac{1}{2k_{max}} = 1.01, \tag{11.62}$$

which is sufficiently close to the absolute minimum of J to satisfy our requirements.

Upon examining the performance index (Equation 11.35), we recognize that the reason the magnitude of the control signal is not accounted for in the original calculations is that $u(t)$ is not included within the expression for the performance index. However, in many cases, we are concerned with the expenditure of the control signal energy. For example, in an electric vehicle control system, $[u(t)]^2$ represents the expenditure of battery energy and must be restricted to conserve the energy for long periods of travel. To account for the expenditure of the energy of the control signal, we will use the performance index

$$\boxed{J = \int_0^\infty (\mathbf{x}^T \mathbf{I} \mathbf{x} + \lambda \mathbf{u}^T \mathbf{u})\, dt,} \tag{11.63}$$

computer may also provide a suitable approach for evaluating the minimum value of J for one or more parameters. However, the solution of Equation (11.68) may be difficult, especially when the system order is quite high ($n > 3$). An alternative method suitable for computer calculation is stated without proof in the following paragraph.

Consider the uncompensated single-input, single-output system with

$$\dot{\mathbf{x}} = \mathbf{A}\mathbf{x} + \mathbf{B}u$$

and feedback

$$u = -\mathbf{K}\mathbf{x} = -[k_1 \quad k_2 \dots k_n]\mathbf{x}.$$

The performance index is

$$J = \int_0^{\infty} (\mathbf{x}^T \mathbf{Q}\mathbf{x} + Ru^2)\, dt,$$

where R is the scalar weighting factor. This index is minimized when

$$\mathbf{K} = R^{-1}\mathbf{B}^T\mathbf{P}.$$

The $n \times n$ matrix $\mathbf{P}$ is determined from the solution of the equation

$$\mathbf{A}^T\mathbf{P} + \mathbf{P}\mathbf{A} - \mathbf{P}\mathbf{B}R^{-1}\mathbf{B}^T\mathbf{P} + \mathbf{Q} = \mathbf{0}. \tag{11.75}$$

Equation (11.75) can be easily programmed and solved using numerical methods. Equation (11.75) is often called the Riccati equation. This optimal control is called the **linear quadratic regulator** (LQR) [14, 22].

11.8   INTERNAL MODEL DESIGN

In this section, we consider the problem of designing a compensator that provides asymptotic tracking of a reference input with zero steady-state error. The reference inputs considered can include steps, ramps, and other persistent signals, such as sinusoids. For a step input, we know that zero steady-state tracking errors can be achieved with a type-one system. This idea is formalized here by introducing an **internal model** of the reference input in the compensator [5, 20].

Let us consider a state variable model of the plant given by

$$\dot{\mathbf{x}} = \mathbf{A}\mathbf{x} + \mathbf{B}u, \qquad y = \mathbf{C}\mathbf{x}, \tag{11.76}$$

where $\mathbf{x}$ is the state vector, u is the input, and y is the output. We will consider a reference input to be generated by a linear system of the form

$$\dot{\mathbf{x}}_r = \mathbf{A}_r\mathbf{x}_r, \qquad r = \mathbf{d}_r\mathbf{x}_r, \tag{11.77}$$

with unknown initial conditions. An equivalent model of the reference input $r(t)$ is

$$r^{(n)} = \alpha_{n-1}r^{(n-1)} + \alpha_{n-2}r^{(n-2)} + \cdots + \alpha_1\dot{r} + \alpha_0 r, \tag{11.78}$$

where $r^{(n)}$ is the nth derivative of $r(t)$.

We begin by considering a familiar design problem, namely, the design of a controller to enable the tracking of a step reference input with zero steady-state error. In this case, the reference input is generated by

$$\dot{x}_r = 0, \qquad r = x_r, \tag{11.79}$$

or equivalently

$$\dot{r} = 0, \tag{11.80}$$

and the tracking error e is defined as

$$e = y - r.$$

Taking the time derivative yields

$$\dot{e} = \dot{y} = \mathbf{C}\dot{\mathbf{x}},$$

where we have used the reference input model of Equation (11.80) and the process model of Equation (11.76). If we define the two intermediate variables

$$\mathbf{z} = \dot{\mathbf{x}} \quad \text{and} \quad w = \dot{u},$$

we have

$$\begin{pmatrix} \dot{e} \\ \dot{\mathbf{z}} \end{pmatrix} = \begin{bmatrix} 0 & \mathbf{C} \\ 0 & \mathbf{A} \end{bmatrix} \begin{pmatrix} e \\ \mathbf{z} \end{pmatrix} + \begin{bmatrix} 0 \\ \mathbf{B} \end{bmatrix} w. \tag{11.81}$$

If the system in Equation (11.81) is controllable, we can find a feedback of the form

$$w = -K_1 e - \mathbf{K}_2 \mathbf{z} \tag{11.82}$$

such that Equation (11.81) is stable. This implies that the tracking error e is stable; thus, we will have achieved the objective of asymptotic tracking with zero steady-state error. The control input, found by integrating Equation (11.82), is

$$u(t) = -K_1 \int_0^t e(\tau)\, d\tau - \mathbf{K}_2 \mathbf{x}(t).$$

The corresponding block diagram is shown in Figure 11.20. We see that the compensator includes an **internal model** (that is, an integrator) of the reference step input.

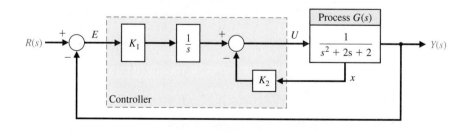

FIGURE 11.20
Internal model design for a step input.

EXAMPLE 11.14   **Internal model design for a unit step input**

Let us consider a process given by

$$\dot{\mathbf{x}} = \begin{bmatrix} 0 & 1 \\ -2 & -2 \end{bmatrix}\mathbf{x} + \begin{bmatrix} 0 \\ 1 \end{bmatrix}u, \qquad y = [1 \quad 0]\mathbf{x}. \qquad (11.83)$$

We want to design a controller for this system to track a reference step input with zero steady-state error. From Equation (11.81), we have

$$\begin{pmatrix} \dot{e} \\ \dot{\mathbf{z}} \end{pmatrix} = \begin{bmatrix} 0 & 1 & 0 \\ 0 & 0 & 1 \\ 0 & -2 & -2 \end{bmatrix}\begin{pmatrix} e \\ \mathbf{z} \end{pmatrix} + \begin{bmatrix} 0 \\ 0 \\ 1 \end{bmatrix}w. \qquad (11.85)$$

A check of controllability shows that the system described by Equation (11.85) is completely controllable. We use

$$K_1 = 20, \qquad \mathbf{K}_2 = [20 \quad 10],$$

in order to locate the roots of the characteristic equation of Equation (11.85) at $s = -1 \pm j, -10$. With w given in Equation (11.82), we have the system of Equation (11.85) as asymptotically stable. So for any initial tracking error $e(0)$ we are guaranteed that $e(t) \to 0$ as $t \to \infty$. The asymptotic stability of the tracking error is illustrated in Figure 11.21 for a step input. ∎

Consider the block diagram model of Figure 11.20 where the process is represented by $G(s)$ and the cascade controller is $G_c(s) = K_1/s$. The **internal model principle** states that if $G(s)G_c(s)$ contains $R(s)$, then $y(t)$ will track $r(t)$ asymptotically. In this case $R(s) = 1/s$, which is contained in $G(s)G_c(s)$, as we expect.

Consider the problem of designing a controller to provide asymptotic tracking of a ramp input with zero steady-state error $r(t) = Mt, t \geq 0$, where M is the ramp magnitude. In this case, the reference input model is

$$\dot{\mathbf{x}}_r = \mathbf{A}_r\mathbf{x}_r = \begin{bmatrix} 0 & 1 \\ 0 & 0 \end{bmatrix}\mathbf{x}_r$$

$$r = \mathbf{d}_r\mathbf{x}_r = [1 \quad 0]\mathbf{x}_r. \qquad (11.86)$$

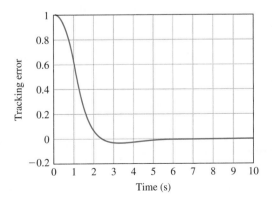

FIGURE 11.21
Internal model design response to an initial tracking error for a unit step input.

In input–output form, the reference model in Equation (11.86) is given by

$$\ddot{r} = 0.$$

Proceeding as before, we define the tracking error as

$$e = y - r,$$

and taking the time-derivative twice yields

$$\ddot{e} = \ddot{y} = \mathbf{C}\ddot{x}.$$

With the definitions

$$\mathbf{z} = \ddot{\mathbf{x}}, \qquad w = \dddot{u},$$

we have

$$\begin{pmatrix} \dot{e} \\ \ddot{e} \\ \dot{\mathbf{z}} \end{pmatrix} = \begin{bmatrix} 0 & 1 & 0 \\ 0 & 0 & \mathbf{C} \\ 0 & 0 & \mathbf{A} \end{bmatrix} \begin{pmatrix} e \\ \dot{e} \\ \mathbf{z} \end{pmatrix} + \begin{bmatrix} 0 \\ 0 \\ \mathbf{B} \end{bmatrix} w. \tag{11.87}$$

So if the system of Equation (11.87) is controllable, then we can compute $K_i, i = 1, 2, 3$, such that with

$$w = -[K_1 \quad K_2 \quad \mathbf{K}_3] \begin{bmatrix} e \\ \dot{e} \\ \mathbf{z} \end{bmatrix}, \tag{11.88}$$

the system represented by Equation (11.87) is asymptotically stable; hence, the tracking error $e(t) \to 0$ as $t \to \infty$, as desired. The control, u, is found by integrating Equation (11.88) twice. In Figure 11.22, we see that the resulting controller has a double integrator, which is the internal model of the reference ramp input.

The internal model approach can be extended to other reference inputs by following the same general procedure outlined for the step and ramp inputs. In addition, the internal model design can be used to reject persistent disturbances by including models of the disturbances in the compensator.

FIGURE 11.22
Internal model
design for a ramp
input. Note that
$G(s)G_c(s)$ contains
$1/s^2$, the reference
input $R(s)$.

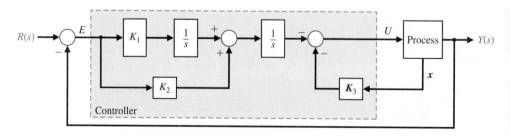

11.9 DESIGN EXAMPLES

In this section we present two illustrative examples. In the first example, a fourth-order state variable model of an automatic test system controller is used to illustrate the full-state feedback controller design to meet time-domain performance specifications. In the second example, a control system is designed to manage the speed of the electric motor shaft of a diesel electric locomotive. The design process focuses on the design of a full-state feedback control system using pole-placement methods.

EXAMPLE 11.15   **Automatic test system**

An automatic test and inspection system uses a DC motor to move a set of test probes, as shown in Figure 11.23. Low throughput and a high degree of error can occur from manually testing various panels of switches, relay, and indicator lights. Automating the test from a controller requires placing a plug across the leads of a part and testing for continuity, resistance, or functionality [19]. The system uses a DC motor with an encoded disk to measure position and velocity, as shown in Figure 11.24. The parameters of the system are shown in Figure 11.25 with K representing the required power amplifier.

We select the state variables as $x_1 = \theta$, $x_2 = d\theta/dt$, and $x_3 = i_f$, as shown in Figure 11.25. State variable feedback is available, and we let

$$u = [-K_1 \quad -K_2 \quad -K_3]\mathbf{x} + r,$$

or

$$u = -K_1 x_1 - K_2 x_2 - K_3 x_3 + r, \tag{11.89}$$

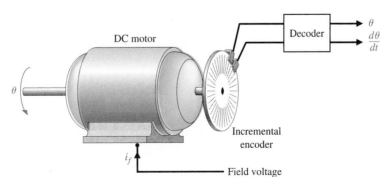

FIGURE 11.23
Automatic test system.

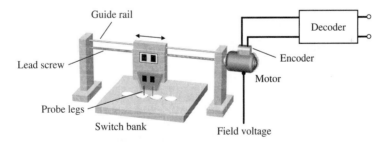

FIGURE 11.24
A DC motor with mounted encoder wheel.

FIGURE 11.25
Open-loop block diagram of the DC motor with mounted encoder wheel.

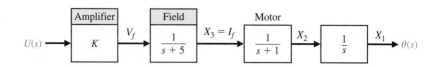

FIGURE 11.26
Closed-loop block diagram of the DC motor.

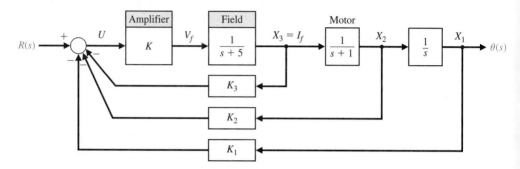

as shown in Figure 11.26. The goal is to select the gains so that the response to a step command has a settling time (with a 2% criterion) of less than 2 seconds and an overshoot of less than 4.0%.

To achieve an accurate output position, we let $K_1 = 1$ and determine K, K_2, and K_3. The characteristic equation of the system may be obtained in several ways. The state variable model associated with Figure 11.25 is given by

$$\dot{\mathbf{x}} = \mathbf{Ax} + \mathbf{B}u = \begin{bmatrix} 0 & 1 & 0 \\ 0 & -1 & 1 \\ 0 & 0 & -5 \end{bmatrix} \mathbf{x} + \begin{bmatrix} 0 \\ 0 \\ K \end{bmatrix} u.$$

$$y = \begin{bmatrix} 1 & 0 & 0 \end{bmatrix} \mathbf{x} \tag{11.90}$$

Substituting for u, as defined by Equation (11.89), we have

$$\dot{\mathbf{x}} = \begin{bmatrix} 0 & 1 & 0 \\ 0 & -1 & 1 \\ -K & -KK_2 & -(5 + K_3K) \end{bmatrix} \mathbf{x} + \begin{bmatrix} 0 \\ 0 \\ K \end{bmatrix} r \tag{11.91}$$

when $K_1 = 1$. The characteristic equation can be obtained from Equation (11.91) as

$$\det \begin{bmatrix} s & -1 & 0 \\ 0 & s+1 & -1 \\ K & KK_2 & s + (5 + K_3K) \end{bmatrix} = 0$$

yielding

$$s^3 + 6s^2 + 5s + K_3Ks^2 + K_3Ks + KK_2s + K = 0.$$

As will be shown in Section 11.10, we can plot a root locus for K_3K as

$$1 + \frac{KK_3(s^2 + as + b)}{s(s+1)(s+5)} = 0, \tag{11.92}$$

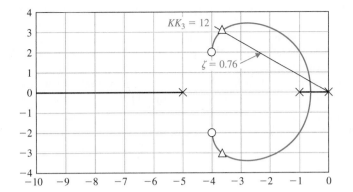

FIGURE 11.27
Root locus for the
automatic test
system.

where a and b are

$$a = (K_2 + K_3)/K_3$$

and

$$b = 1/K_3.$$

Setting $a = 8$ and $b = 20$, we place the zeros at $s = -4 \pm j2$ in order to pull the locus to the left in the s-plane. Then

$$\frac{K_2 + K_3}{K_3} = 8 \quad \text{and} \quad \frac{1}{K_3} = 20.$$

Therefore, $K_1 = 1$, $K_2 = 0.35$, and $K_3 = 0.05$. A plot of the root locus is shown in Figure 11.27. When $KK_3 = 12$, the roots lie on the $\zeta = 0.76$ line, as shown in Figure 11.27. Since $K_3 = 0.05$, we have $K = 240$. The roots at $K = 240$ are

$$s = -10.62, \quad \text{and} \quad s = -3.69 \pm j3.00.$$

The step response of this system is shown in Figure 11.28. The overshoot is 3%, and the settling time is 1.8 seconds. Thus the design is quite acceptable. ■

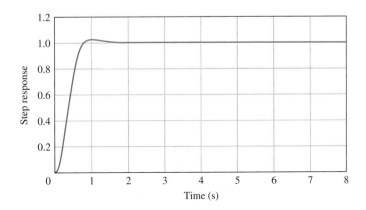

FIGURE 11.28
Step response of
the automatic test
system.

EXAMPLE 11.16   **Diesel electric locomotive control**

The diesel electric locomotive is depicted in Figure 11.29. The efficiency of the diesel engine is very sensitive to the speed of rotation of the motors. We want to design a control system that drives the electric motors of a diesel electric locomotive for use on railroad trains. The locomotive is driven by DC motors located on each of the axles. the throttle position (see Figure 11.29) is set by moving the input potentiometers. The elements of the design process emphasized in this example are highlighted in Figure 11.30.

The control objective is to regulate the shaft rotation speed ω_o to the desired value ω_r.

Control Goal

Regulate the shaft rotation speed to the desired value in the presence of external load torque disturbances.

The corresponding variable to be controlled is the shaft rotation speed ω_o.

Variable to Be Controlled

Shaft rotation speed ω_o.

The controlled speed ω_o is sensed by a tachometer, which supplies a feedback voltage v_o. The electronic amplifier amplifies the error signal, $v_r - v_o$, between the reference and feedback voltage signals and provides a voltage v_f that is supplied to the field winding of a DC generator.

The generator is run at a constant speed ω_d by the diesel engine and generates a voltage v_g that is supplied to the armature of a DC motor. The motor is armature

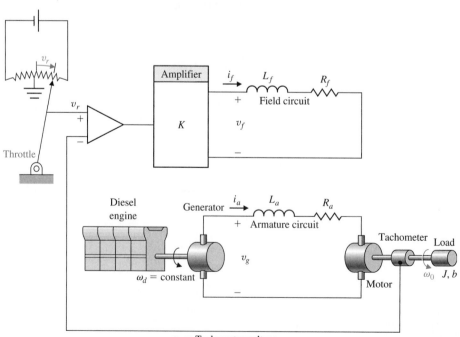

FIGURE 11.29
Diesel electric
locomotive system.

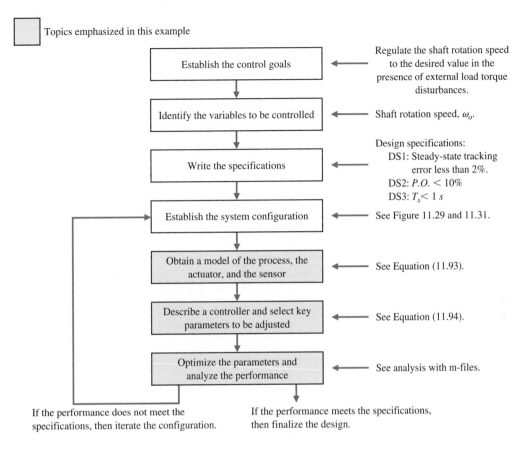

Topics emphasized in this example

FIGURE 11.30 Elements of the control system design process emphasized in this diesel electric locomotive example.

controlled, with a fixed current supplied to its field. As a result, the motor produces a torque T and drives the load connected to its shaft so that the controlled speed ω_o tends to equal the command speed ω_r.

A block diagram and signal flow graph of the system are shown in Figure 11.31. In Figure 11.31 we use L_t and R_t, which are defined as

$$L_t = L_a + L_g,$$

$$R_t = R_a + R_g.$$

Values for the parameters of the diesel electric locomotive are given in Table 11.1. Notice that the system has a feedback loop; we use the tachometer voltage v_o as a feedback signal to form an error signal $v_r - v_o$. Without additional state feedback, the only tuning parameter is the amplifier gain K. As a first step, we can investigate the system performance with tachometer voltage feedback only.

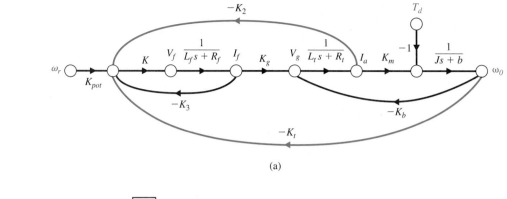

(a)

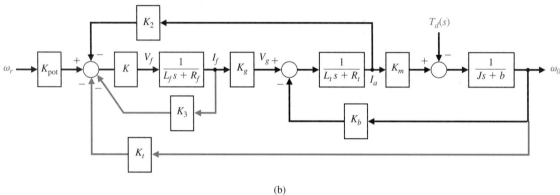

(b)

FIGURE 11.31   Signal flow graph of the diesel electric locomotive. (a) Signal flow graph. (b) Block diagram controller feedback loops are shown in light.

Table 11.1   Parameter Values for the Diesel Electric Locomotive

K_m	K_g	K_b	J	b	L_a	R_a	R_f	L_f	K_t	K_{pot}	L_g	R_g
10	100	0.62	1	1	0.2	1	1	0.1	1	1	0.1	1

The key tuning parameters are given by

Select Key Tuning Parameters
    K and $\mathbf{K}$

The matrix $\mathbf{K}$ is the state feedback gain matrix. The design specifications are

Design Specifications
    DS1   Steady-state tracking error less than 2% to a unit step input.
    DS2   Percent overshoot of $\omega_o(t)$ less than 10% to a unit step input $\omega_r(s) = 1/s$.
    DS3   Settling time less than 1 second to a unit step input.

The first step in the development of the vector differential equation that accurately describes the system is to choose a set of state variables. In practice the selection of state variables can be a difficult process, especially for complex systems. The state variables must be sufficient in number to determine the future behavior of the system when the present state and all future inputs are known. The selection of state variables is intimately related to the issue of complexity.

The diesel electric locomotive system has three major components: two electrical circuits and one mechanical system. It seems logical that the state vector will include state variables from both electrical circuits and from the mechanical system. One reasonable choice of state variables is $x_1 = \omega_o$, $x_2 = i_a$, and $x_3 = i_f$. This state variable selection is not unique. With the state variables defined above, the state variable model is

$$\dot{x}_1 = -\frac{b}{J}x_1 + \frac{K_m}{J}x_2 - \frac{1}{J}T_d,$$

$$\dot{x}_2 = -\frac{K_b}{L_t}x_1 - \frac{R_t}{L_t}x_2 + \frac{K_g}{L_t}x_3,$$

$$\dot{x}_3 = -\frac{R_f}{L_f}x_3 + \frac{1}{L_f}u,$$

where

$$u = KK_{\text{pot}}\omega_r.$$

In matrix form (with $T_d(s) = 0$), we have

$$\dot{\mathbf{x}} = \mathbf{A}\mathbf{x} + \mathbf{B}u,$$

$$y = \mathbf{C}\mathbf{x} + \mathbf{D}u, \tag{11.93}$$

where

$$\mathbf{A} = \begin{bmatrix} -\dfrac{b}{J} & \dfrac{K_m}{J} & 0 \\[2mm] -\dfrac{K_b}{L_t} & -\dfrac{R_t}{L_t} & \dfrac{K_g}{L_t} \\[2mm] 0 & 0 & -\dfrac{R_f}{L_f} \end{bmatrix}, \quad \mathbf{B} = \begin{bmatrix} 0 \\[1mm] 0 \\[1mm] \dfrac{1}{L_f} \end{bmatrix}, \text{ and}$$

$$\mathbf{C} = \begin{bmatrix} 1 & 0 & 0 \end{bmatrix}, \quad \mathbf{D} = \begin{bmatrix} 0 \end{bmatrix}.$$

The corresponding transfer function is

$$G(s) = \mathbf{C}(s\mathbf{I} - \mathbf{A})^{-1}\mathbf{B} = \frac{K_g K_m}{(R_f + L_f s)[(R_t + L_t s)(Js + b) + K_m K_b]}.$$

Begin by assuming the tachometer feedback in available, that is, that K_t is in the loop. If we take advantage of the fact that

$$K_{\text{pot}} = K_t = 1,$$

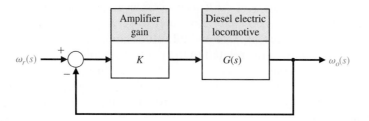

FIGURE 11.32
Block diagram
representation of
the diesel electric
locomotive.

then (from an input–output perspective) the system has the simple feedback config-uration shown in Figure 11.32.

Using the parameter values given in Table 11.1 and computing the steady-state tracking error for a unit step input yields

$$e_{ss} = \frac{1}{1 + KG(0)} = \frac{1}{1 + 121.95K}.$$

Using the Routh–Hurwith method, we also find that the closed-loop system is stable for

$$-0.008 < K < 0.0468.$$

The smallest steady-state tracking error is achieved for the largest value of K. At best we can obtain a 15% tracking error, which does not meet the design specifica-tion DS1. Also, as K gets larger, the response becomes unacceptably oscillatory.

We now consider a full state feedback controller design. The feedback loops are shown in Figure 11.31, which shows that ω_0, i_a, and i_f are available for feedback. With-out any loss of generality, we set $K = 1$. Any value of $K > 0$ would work as well.

The control input is

$$u = K_{pot}\omega_r - K_t x_1 - K_2 x_2 - K_3 x_3.$$

The feedback gains to be determined are K_t, K_2, and K_3. The tachometer gain, K_t, is now a key parameter of the design process. Also K_{pot} is a key variable for tuning. By adjusting the parameter K_{pot}, we have the freedom to scale the input ω_r. When we define

$$\mathbf{K} = [K_t \quad K_2 \quad K_3],$$

then

$$u = -\mathbf{K}\mathbf{x} + K_{pot}\omega_r. \tag{11.94}$$

The closed-loop system with state feedback is

$$\dot{\mathbf{x}} = (\mathbf{A} - \mathbf{B}\mathbf{K})\mathbf{x} + \mathbf{B}v,$$
$$y = \mathbf{C}\mathbf{x},$$

where

$$v = K_{pot}\omega_r.$$

We will use pole-placement methods to determine $\mathbf{K}$ such that the eigenvalues of $\mathbf{A} - \mathbf{BK}$ are in the desired locations. First we make sure the system is controllable. When $n = 3$ the controllability matrix is

$$\mathbf{P}_c = [\mathbf{B} \quad \mathbf{AB} \quad \mathbf{A}^2\mathbf{B}].$$

Computing the determinant of $\mathbf{P}_c$ yields

$$\det \mathbf{P}_c = -\frac{K_g^2 K_m}{J L_f^3 L_t^2}.$$

Since $K_g \neq 0$ and $K_m \neq 0$ and $J L_f^3 L_t^2$ is nonzero, we determine that

$$\det \mathbf{P}_c \neq 0.$$

Thus the system is controllable. We can place all the poles of the system appropriately to satisfy DS2 and DS3.

The desired region to place the eigenvalues of $\mathbf{A} - \mathbf{BK}$ is illustrated in Figure 11.33. The specific pole locations are selected to be

$$p_1 = -50,$$
$$p_2 = -4 + 3j,$$
$$p_3 = -4 - 3j.$$

Selecting $p_1 = -50$ allows for a good second-order response that is governed by p_2 and p_3.

The gain matrix $\mathbf{K}$ that achieves the desired closed-loop poles is

$$\mathbf{K} = [-0.0041 \quad 0.0035 \quad 4.0333].$$

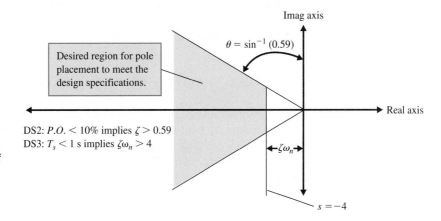

FIGURE 11.33
Desired location of the closed-loop poles (that is, the eigenvalues of $\mathbf{A} - \mathbf{BK}$).

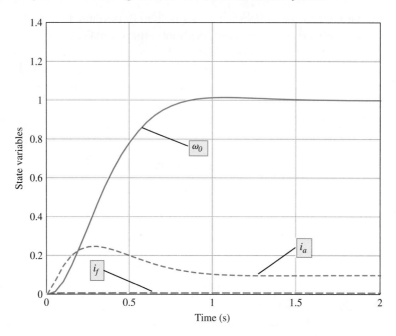

FIGURE 11.34
Closed-loop step response of the diesel electric locomotive.

To select the gain K_{pot}, we first compute the DC gain of the closed-loop transfer function. With the state feedback in place, the closed-loop transfer function is

$$T(s) = \mathbf{C}(s\mathbf{I} - \mathbf{A} + \mathbf{BK})^{-1}\mathbf{B}.$$

Then

$$K_{\text{pot}} = \frac{1}{T(0)}.$$

Using the gain K_{pot} in this manner effectively scales the closed-loop transfer function so that the DC gain is equal to 1. We then expect that a unit step input representing a 1°/s step command results in a 1°/s steady-state output at ω_o.

The step response of the system is shown in Figure 11.34. We can see that all the design specifications are satisfied. ∎

11.10 STATE VARIABLE DESIGN USING CONTROL DESIGN SOFTWARE

Controllability and observability of a system in state variable feedback form can be checked using the functions ctrb and obsv, respectively. The inputs to the ctrb function, shown in Figure 11.35, are the system matrix $\mathbf{A}$ and the input matrix $\mathbf{B}$; the output of ctrb is the controllability matrix $\mathbf{P}_c$. Similarly, the input to the obsv function, shown in Figure 11.35, is the system matrix $\mathbf{A}$ and the output matrix $\mathbf{C}$; the output of obsv is the observability matrix $\mathbf{P}_o$.

Notice that the controllability matrix $\mathbf{P}_c$ is a function only of $\mathbf{A}$ and $\mathbf{B}$, while the observability matrix $\mathbf{P}_o$ is a function only of $\mathbf{A}$ and $\mathbf{C}$.

EXAMPLE 11.17   **Satellite trajectory control**

Let us consider a satellite in a circular, equatorial orbit at an altitude of 250 nautical miles above the Earth, as illustrated in Figure 11.36 [16, 27]. The satellite motion (in the orbit plane) is described by the normalized state variable model

$$\dot{\mathbf{x}} = \begin{bmatrix} 0 & 1 & 0 & 0 \\ 3\omega^2 & 0 & 0 & 2\omega \\ 0 & 0 & 0 & 1 \\ 0 & -2\omega & 0 & 0 \end{bmatrix} \mathbf{x} + \begin{bmatrix} 0 \\ 1 \\ 0 \\ 0 \end{bmatrix} u_r + \begin{bmatrix} 0 \\ 0 \\ 0 \\ 1 \end{bmatrix} u_t, \tag{11.95}$$

where the state vector $\mathbf{x}$ represents normalized perturbations from the circular, equatorial orbit; u_r is the input from a radial thruster; u_t is the input from a tangential thruster; and $\omega = 0.0011$ rad/s (approximately one orbit of 90 minutes) is the orbital rate for the satellite at the specific altitude. In the absence of perturbations, the satellite will remain in the nominal circular equatorial orbit. However, disturbances such as aerodynamic drag can cause the satellite to deviate from its nominal path. The problem is to design a controller that commands the satellite thrusters in such a manner that the actual orbit remains near the desired circular orbit. Before commencing with the design, we check controllability. In this case, we investigate controllability using the radial and tangential thrusters independently.

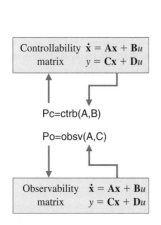

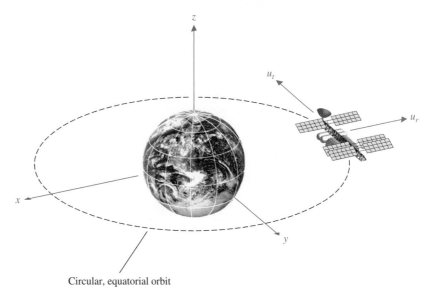

FIGURE 11.35   The **ctrb** and **obsv** functions.

FIGURE 11.36   The satellite in an equatorial circular orbit.

radial.m

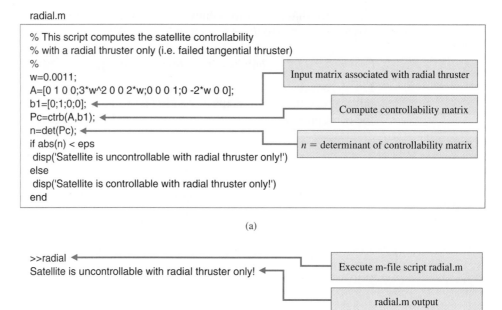

```
% This script computes the satellite controllability
% with a radial thruster only (i.e. failed tangential thruster)
%
w=0.0011;
A=[0 1 0 0;3*w^2 0 0 2*w;0 0 0 1;0 -2*w 0 0];
b1=[0;1;0;0];                          ← Input matrix associated with radial thruster
Pc=ctrb(A,b1);                         ← Compute controllability matrix
n=det(Pc);                             ← n = determinant of controllability matrix
if abs(n) < eps
 disp('Satellite is uncontrollable with radial thruster only!')
else
 disp('Satellite is controllable with radial thruster only!')
end
```

(a)

```
>>radial                                  ← Execute m-file script radial.m
Satellite is uncontrollable with radial thruster only!   ← radial.m output
```

(b)

FIGURE 11.37
Controllability with radial thrusters only: (a) m-file script, (b) output.

Suppose the tangential thruster fails (i.e., $u_t = 0$), and only the radial thruster is operational. Is the satellite controllable from u_r only? We answer this question by using an m-file script to determine the controllability. Using the script shown in Figure 11.37, we find that the determinant $\mathbf{P}_c$ is zero; thus, the satellite is not completely controllable when the tangential thruster fails.

Suppose now that the radial thruster fails (i.e., $u_r = 0$) and that the tangential thruster is functioning properly. Is the satellite controllable from u_t only? Using the script in Figure 11.38, we find that the satellite is completely controllable using the tangential thruster only. ∎

We conclude this section with a controller design for an automatic test system using state variable models. The design approach utilizes root locus methods and incorporates m-file scripts to assist in the procedure.

EXAMPLE 11.18   **Automatic test system**

The state-space representation for the automatic test system of Example 11.15 is

$$\dot{\mathbf{x}} = \mathbf{A}\mathbf{x} + \mathbf{B}u, \tag{11.96}$$

where

$$\mathbf{A} = \begin{bmatrix} 0 & 1 & 0 \\ 0 & -1 & 1 \\ 0 & 0 & -5 \end{bmatrix} \quad \text{and} \quad \mathbf{B} = \begin{bmatrix} 0 \\ 0 \\ K \end{bmatrix}.$$

Our design specifications are a step response with (1) a settling time (with a 2% criterion) less than 2 seconds and (2) an overshoot less than 4%. We assume that the

tangent.m

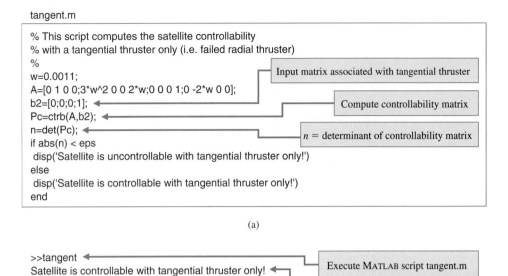

```
% This script computes the satellite controllability
% with a tangential thruster only (i.e. failed radial thruster)
%
w=0.0011;
A=[0 1 0 0;3*w^2 0 0 2*w;0 0 0 1;0 -2*w 0 0];
b2=[0;0;0;1];
Pc=ctrb(A,b2);
n=det(Pc);
if abs(n) < eps
 disp('Satellite is uncontrollable with tangential thruster only!')
else
 disp('Satellite is controllable with tangential thruster only!')
end
```

Input matrix associated with tangential thruster

Compute controllability matrix

n = determinant of controllability matrix

(a)

>>tangent
Satellite is controllable with tangential thruster only!

Execute MATLAB script tangent.m

Tangent.m output

FIGURE 11.38
Controllability with
tangential thrusters
only: (a) m-file
script, (b) output.

(b)

state variables are available for feedback, so that the control is given by

$$u = -[K_1 \quad K_2 \quad K_3]\mathbf{x} + r = -\mathbf{Kx} + r. \qquad (11.97)$$

We must select the gains K, K_1, K_2, and K_3 to meet the performance specifications. Using the design approximations

$$T_s = \frac{4}{\zeta\omega_n} < 2 \quad \text{and} \quad P.O. = 100e^{-\zeta\pi/\sqrt{1-\zeta^2}} < 4,$$

we find that

$$\zeta > 0.72 \quad \text{and} \quad \omega_n > 2.8.$$

This defines a region in the complex plane in which our dominant roots must lie, so that we expect to meet the design specifications, as shown in Figure 11.39. Substituting Equation (11.97) into Equation (11.96) yields

$$\dot{\mathbf{x}} = \begin{bmatrix} 0 & 1 & 0 \\ 0 & -1 & 1 \\ -KK_1 & -KK_2 & -(5 + KK_3) \end{bmatrix}\mathbf{x} + \begin{bmatrix} 0 \\ 0 \\ K \end{bmatrix}r = \mathbf{Hx} + \mathbf{B}r, \quad (11.98)$$

where $\mathbf{H} = \mathbf{A} - \mathbf{BK}$. The characteristic equation associated with Equation (11.98) can be obtained by evaluating $\det(s\mathbf{I} - \mathbf{H}) = 0$, resulting in

$$s(s + 1)(s + 5) + KK_3\left(s^2 + \frac{K_3 + K_2}{K_3}s + \frac{K_1}{K_3}\right) = 0. \qquad (11.99)$$

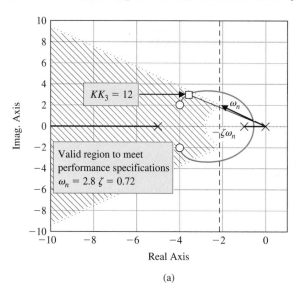

FIGURE 11.39
(a) Root locus for the automatic test system. (b) m-file script.

(a)

```
% Root locus script for the Automatic Test System
% including performance specs regions
num=[1 8 20]; den=[1 6 5 0]; sys=tf(num,den);
clf; rlocus(sys); hold on          ◄——— Hold plot to add
%                                         stability regions
zeta=0.72; wn=2.8;
x=[-10:0.1:-zeta*wn]; y=-(sqrt(1-zeta^2)/zeta)*x;
xc=[-10:0.1:-zeta*wn];c=sqrt(wn^2-xc.^2);
plot(x,y,':',x,-y,':',xc,c,':',xc,-c,':')
```

(b)

If we view KK_3 as a parameter and let $K_1 = 1$, then we can write Equation (11.99) as

$$1 + KK_3 \frac{s^2 + \dfrac{K_3 + K_2}{K_3}s + \dfrac{1}{K_3}}{s(s + 1)(s + 5)} = 0.$$

We place the zeros at $s = -4 \pm 2j$ in order to pull the locus to the left in the s-plane. Thus, our desired numerator polynomial is $s^2 + 8s + 20$. Comparing corresponding coefficients leads to

$$\frac{K_3 + K_2}{K_3} = 8 \quad \text{and} \quad \frac{1}{K_3} = 20.$$

Therefore, $K_2 = 0.35$ and $K_3 = 0.05$. We can now plot a root locus with KK_3 as the parameter, as shown in Figure 11.39.

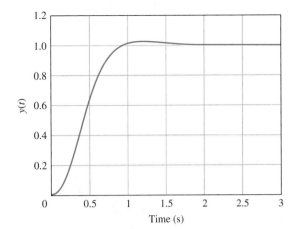

FIGURE 11.40
Step response for
the automatic test
system.

The characteristic equation, Equation (11.99), is

$$1 + KK_3 \frac{s^2 + 8s + 20}{s(s + 1)(s + 5)} = 0.$$

The roots for the selected gain, $KK_3 = 12$, lie in the performance region, as shown in Figure 11.39. The rlocfind function is used to determine the value of KK_3 at the selected point. The final gains are as follows:

$$K = 240.00,$$
$$K_1 = 1.00,$$
$$K_2 = 0.35,$$
$$K_3 = 0.05.$$

The controller design results in a settling time of about 1.8 seconds and an overshoot of 3%, as shown in Figure 11.40. ∎

In Section 11.4, we discussed Ackermann's formula to place the poles of the system at desired locations. The function acker calculates the gain matrix **K** to place the closed-loop poles at the desired locations. The acker function is illustrated in Figure 11.41.

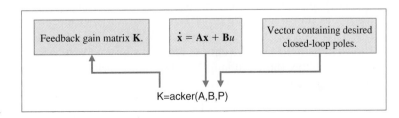

FIGURE 11.41
The **acker** function.

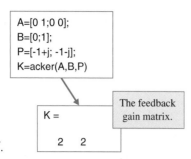

FIGURE 11.42
Using **acker** to
compute **K** to place
the poles at
$\mathbf{P} = [-1 + j \ -1 - j]^T$.

EXAMPLE 11.19   **Second-order system design using the acker function**

Consider again the second-order system in Example 11.7. The system model is

$$\dot{\mathbf{x}} = \begin{bmatrix} 0 & 1 \\ 0 & 0 \end{bmatrix} \mathbf{x} + \begin{bmatrix} 0 \\ 1 \end{bmatrix} u.$$

The desired closed-loop pole locations are $s_{1,2} = -1 \pm j$. To apply Ackermann's formula using the **acker** function, form the vector

$$\mathbf{P} = \begin{bmatrix} -1 + j \\ -1 - j \end{bmatrix}.$$

Then, with

$$\mathbf{A} = \begin{bmatrix} 0 & 1 \\ 0 & 0 \end{bmatrix} \quad \text{and} \quad \mathbf{B} = \begin{bmatrix} 0 \\ 1 \end{bmatrix},$$

the **acker** formula, illustrated in Figure 11.42, determines that the gain matrix that achieves the desired pole locations is

$$\mathbf{K} = \begin{bmatrix} 2 & 2 \end{bmatrix}.$$

This confirms the result in Example 11.7. ∎

11.11 SEQUENTIAL DESIGN EXAMPLE: DISK DRIVE READ SYSTEM

In this chapter, we will design a state variable feedback system that will achieve the desired system response. The specifications for the system are given in Table 11.2. The second-order open-loop model is shown in Figure 11.43. We will design the system for this second-order model and then test the system response for both the second-order and third-order models.

First, we select the two state variables as $x_1(t) = y(t)$ and $x_2(t) = dy/dt = dx_1/dt$, as shown in Figure 11.44. It is practical to measure these variables as the position and velocity of the reader head. We then add the state variable feedback, as shown in Figure 11.44. We choose $K_1 = 1$, since our goal is for $y(t)$ to closely and accurately follow the command $r(t)$. The state variable differential equation for the open-loop system is

$$\dot{\mathbf{x}} = \begin{bmatrix} 0 & 1 \\ 0 & -20 \end{bmatrix} \mathbf{x} + \begin{bmatrix} 0 \\ 5K_a \end{bmatrix} r(t).$$

Table 11.2   Disk Drive Control System Specifications and Actual Performance

Performance Measure	Desired Value	Response for Second-Order Model	Response for Third-Order Model
Percent overshoot	<5%	<1%	0%
Settling time	<50 ms	34.3 ms	34.2 ms
Maximum response for a unit step disturbance	$<5 \times 10^{-3}$	5.2×10^{-5}	5.2×10^{-5}

FIGURE 11.43
Open-loop model of head control system.

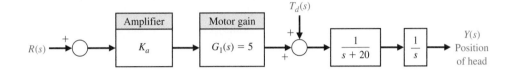

FIGURE 11.44
Closed-loop system with feedback of the two state variables.

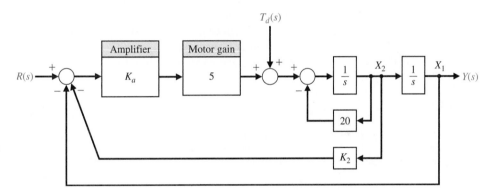

The closed-loop state variable differential equation obtained from Figure 11.44 is

$$\dot{\mathbf{x}} = \begin{bmatrix} 0 & 1 \\ -5K_1K_a & -(20 + 5K_2K_a) \end{bmatrix} \mathbf{x} + \begin{bmatrix} 0 \\ 5K_a \end{bmatrix} r(t).$$

The characteristic equation of the closed-loop system is

$$s^2 + (20 + 5K_2K_a)s + 5K_a = 0,$$

since $K_1 = 1$. In order to achieve the specifications, we select $\zeta = 0.90$ and $\zeta\omega_n = 125$. Then the desired closed-loop characteristic equation is

$$s^2 + 2\zeta\omega_n s + \omega_n^2 = s^2 + 250s + 19290 = 0.$$

Therefore, we require that $5K_a = 19290$ or $K_a = 3858$. Furthermore, we require that

$$20 + 5K_2K_a = 250,$$

or $K_2 = 0.012$.

The system with the second-order model has the desired response and meets all the specifications, as shown in Table 11.2. If we add the field inductance $L = 1$ mH, we have a third-order model with

$$G_1(s) = \frac{5000}{s + 1000}.$$

Using this model, which incorporates the field inductance, we test the response of the system with the feedback gains selected for the second-order model. The results are provided in Table 11.2, illustrating that the second-order model is a very good model of the system. The actual results of the third-order system meet the specifications.

11.12 SUMMARY

In this chapter, the design of control systems in the time domain was examined. The three-step design procedure for constructing state variable compensators was presented. The optimal design of a system using state variable feedback and an integral performance index was considered. Also, the s-plane design of systems utilizing state variable feedback was examined. Finally, internal model design was discussed.

EXERCISES

E11.1   The ability to balance actively is a key ingredient in the mobility of a device that hops and runs on one springy leg, as shown in Figure E11.1 [9]. The control of the attitude of the device uses a gyroscope and a feedback such that $u = \mathbf{K}\mathbf{x}$, where

$$\mathbf{K} = \begin{bmatrix} -1 & 0 \\ 0 & -k \end{bmatrix},$$

and

$$\dot{\mathbf{x}}(t) = \mathbf{A}\mathbf{x}(t) + \mathbf{B}u(t)$$

where

$$\mathbf{A} = \begin{bmatrix} 0 & 1 \\ 1 & 0 \end{bmatrix}, \quad \text{and} \quad \mathbf{B} = \mathbf{I}.$$

Determine a value for k so that the response of each hop is critically damped.

E11.2   A magnetically suspended steel ball can be described by the linear equation

$$\dot{\mathbf{x}} = \begin{bmatrix} 0 & 1 \\ 3 & 0 \end{bmatrix}\mathbf{x} + \begin{bmatrix} 0 \\ 1 \end{bmatrix}u.$$

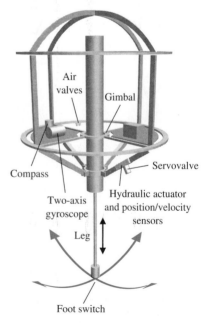

FIGURE E11.1   Single-leg control.

The state variables are x_1 = position and x_2 = velocity, and both are measurable. Select a feedback so that the system is critically damped and the settling time (with a 2% criterion) is 2 seconds. Choose the feedback in the form

$$u = -k_1 x_1 - k_2 x_2 + r$$

where r is the reference input and the gains k_1 and k_2 are to be determined.

E11.3 A system is described by the matrix equations

$$\dot{\mathbf{x}} = \begin{bmatrix} 0 & 1 \\ 0 & -3 \end{bmatrix}\mathbf{x} + \begin{bmatrix} 0 \\ 1 \end{bmatrix}u$$
$$y = [0 \quad 2]\mathbf{x}.$$

Determine whether the system is controllable and observable.

Answer: controllable, not observable

E11.4 A system is described by the matrix equations

$$\dot{\mathbf{x}} = \begin{bmatrix} -4 & 0 \\ 0 & -1 \end{bmatrix}\mathbf{x} + \begin{bmatrix} 0 \\ 1 \end{bmatrix}u$$
$$y = [1 \quad 0]\mathbf{x}.$$

Determine whether the system is controllable and observable.

E11.5 A system is described by the matrix equations

$$\dot{\mathbf{x}} = \begin{bmatrix} 0 & 1 \\ -1 & -2 \end{bmatrix}\mathbf{x} + \begin{bmatrix} 1 \\ -1 \end{bmatrix}u$$
$$y = [1 \quad 0]\mathbf{x}.$$

Determine whether the system is controllable and observable.

E11.6 A system is described by the matrix equations

$$\dot{\mathbf{x}} = \begin{bmatrix} 0 & 1 \\ -1 & -2 \end{bmatrix}\mathbf{x} + \begin{bmatrix} 0 \\ 1 \end{bmatrix}u$$
$$y = [1 \quad 0]\mathbf{x}.$$

Determine whether the system is controllable and observable.

Answer: controllable and observable

E11.7 Consider the system represented in state variable form

$$\dot{\mathbf{x}} = \mathbf{A}\mathbf{x} + \mathbf{B}u$$
$$y = \mathbf{C}\mathbf{x} + \mathbf{D}u,$$

where

$$\mathbf{A} = \begin{bmatrix} 0 & 1 \\ -2 & -7 \end{bmatrix}, \quad \mathbf{B} = \begin{bmatrix} 0 \\ 10 \end{bmatrix},$$
$$\mathbf{C} = [1 \quad -2], \quad \text{and} \quad \mathbf{D} = [0].$$

Sketch a block diagram model of the system.

E11.8 Consider the third-order system

$$\dot{\mathbf{x}} = \begin{bmatrix} 0 & 1 & 0 \\ 0 & 0 & 1 \\ -9 & -3 & -1 \end{bmatrix}\mathbf{x} + \begin{bmatrix} 0 \\ -1 \\ 4 \end{bmatrix}u$$
$$y = [2 \quad 8 \quad 10]\mathbf{x} + [1]u.$$

Sketch a block diagram model of the system.

E11.9 Consider the second-order system

$$\dot{\mathbf{x}} = \begin{bmatrix} 1 & -1 \\ -1 & 1 \end{bmatrix}\mathbf{x} + \begin{bmatrix} k_1 \\ k_2 \end{bmatrix}u$$
$$y = [1 \quad 0]\mathbf{x} + [0]u.$$

For what values of k_1 and k_2 is the system completely controllable?

E11.10 Consider the block diagram model in Figure E11.10. Write the corresponding state variable model in the form

$$\dot{\mathbf{x}} = \mathbf{A}\mathbf{x} + \mathbf{B}u$$
$$y = \mathbf{C}\mathbf{x} + \mathbf{D}u.$$

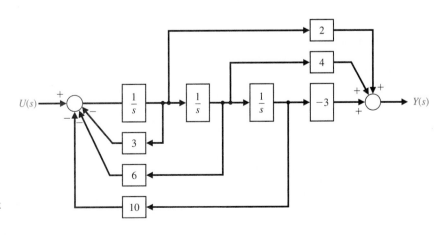

FIGURE E11.10
State variable block diagram.

E11.11   Consider the system shown in block diagram form in Figure E11.11. Obtain a state variable representation of the system. Determine if the system is controllable and observable.

E11.12   Consider the single-input, single-output system is described by

$$\dot{x}(t) = \mathbf{A}x(t) + \mathbf{B}u(t)$$

$$y(t) = \mathbf{C}x(t)$$

where

$$\mathbf{A} = \begin{bmatrix} 0 & 1 \\ -2 & -3 \end{bmatrix}, \mathbf{B} = \begin{bmatrix} 0 \\ 2 \end{bmatrix}, \mathbf{C} = [1 \quad 0].$$

Compute the corresponding transfer function representation of the system. If the initial conditions are zero (i.e., $x_1(0) = 0$ and $x_2(0) = 0$), determine the response when $u(t)$ is a unit step input for $t \geq 0$.

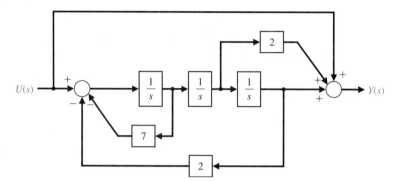

FIGURE E11.11
State variable block diagram with a feedforward term.

PROBLEMS

P11.1   A first-order system is represented by the time-domain differential equation

$$\dot{x} = x + u.$$

A feedback controller is to be designed such that

$$u(t) = -kx,$$

and the desired equilibrium condition is $x(t) = 0$ as $t \to \infty$. The performance integral is defined as

$$J = \int_0^\infty x^2 \, dt,$$

and the initial value of the state variable is $x(0) = \sqrt{2}$. Obtain the value of k in order to make J a minimum. Is this k physically realizable? Select a practical value for the gain k and evaluate the performance index with that gain. Is the system stable without the feedback due to $u(t)$?

P11.2   To account for the expenditure of energy and resources, the control signal is often included in the performance integral. Then the operation will not involve an unlimited control signal $u(t)$. One suitable performance index, which includes the effect of the magnitude of the control signal, is

$$J = \int_0^\infty (x^2(t) + \lambda u^2(t)) \, dt.$$

(a) Repeat Problem P11.1 for the performance index.
(b) If $\lambda = 2$, obtain the value of k that minimizes the performance index. Calculate the resulting minimum value of J.

P11.3   An unstable robot system is described by the vector differential equation [10]

$$\frac{d}{dt} \begin{bmatrix} x_1 \\ x_2 \end{bmatrix} = \begin{bmatrix} 1 & 0 \\ -1 & 2 \end{bmatrix} \begin{bmatrix} x_1 \\ x_2 \end{bmatrix} + \begin{bmatrix} 1 \\ 1 \end{bmatrix} u(t).$$

Both state variables are measurable, and so the control signal is set as $u(t) = -k(x_1 + x_2)$. Following the method of Section 11.7, design gain k so that the performance index is minimized. Evaluate the minimum value of the performance index. Determine the sensitivity of the performance to a change in k. Assume that the initial conditions are

$$\mathbf{x}(0) = \begin{bmatrix} 1 \\ 1 \end{bmatrix}.$$

Is the system stable without the feedback signals due to $u(t)$?

P11.4 Determine the feedback gain k of Example 11.12 that minimizes the performance index

$$J = \int_0^\infty \mathbf{x}^T \mathbf{x} \, dt$$

when $\mathbf{x}^T(0) = [1 \quad 1]$. Plot the performance index J versus the gain k.

P11.5 Determine the feedback gain k of Example 11.13 that minimizes the performance index

$$J = \int_0^\infty (\mathbf{x}^T \mathbf{x} + \mathbf{u}^T \mathbf{u}) \, dt$$

when $\mathbf{x}^T(0) = [1 \quad 1]$. Plot the performance index J versus the gain k.

P11.6 For the solutions of Problems P11.3, P11.4, and P11.5, determine the roots of the closed-loop optimal control system. Note that the resulting closed-loop roots depend on the performance index selected.

P11.7 A system has the vector differential equation as given in Equation (11.42). We want both state variables to be

used in the feedback so that $u(t) = -k_1 x_1 - k_2 x_2$. Also, we desire to have a natural frequency ω_n for this system equal to 2. Find a set of gains k_1 and k_2 in order to achieve an optimal system when J is given by Equation (11.63). Assume $\mathbf{x}^T(0) = [1 \quad 0]$.

P11.8 For the system of Example 11.11 determine the optimum value for k_2 when $k_1 = 1$ and $\mathbf{x}^T(0) = [1 \quad 0]$.

P11.9 An interesting mechanical system with a challenging control problem is the ball and beam, shown in Figure P11.9(a) [11]. It consists of a rigid beam that is free to rotate in the plane of the paper around a center pivot, with a solid ball rolling along a groove in the top of the beam. The control problem is to position the ball at a desired point on the beam using a torque applied to the beam as a control input at the pivot.

A linear model of the system with a measured value of the angle ϕ and its angular velocity $d\phi/dt = \omega$ is available. Select a feedback scheme so that the response of the closed-loop system has an overshoot of 4% and a settling time (with a 2% criterion) of 1 second for a step input.

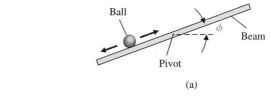

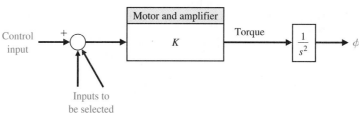

FIGURE P11.9
(a) Ball and beam.
(b) Model of the ball and beam.

P11.10 The dynamics of a rocket are represented by

$$\dot{\mathbf{x}} = \begin{bmatrix} 0 & 0 \\ 1 & 0 \end{bmatrix} \mathbf{x} + \begin{bmatrix} 1 \\ 0 \end{bmatrix} u$$

$$y = [0 \quad 1]\mathbf{x}$$

and state variable feedback is used, where $u = -8x_1 - 16x_2$. Determine the roots of the characteristic equation of this system and the response of the system when the initial conditions are $x_1(0) = 1$ and $x_2(0) = 0$.

P11.11 The state variable model of a plant to be controlled is

$$\dot{\mathbf{x}} = \begin{bmatrix} -5 & -2 \\ 2 & 0 \end{bmatrix} \mathbf{x} + \begin{bmatrix} 0.5 \\ 0 \end{bmatrix} u$$

$$y = [0 \quad 1]\mathbf{x} + [0]u.$$

Use state variable feedback and incorporate a command input $u = -\mathbf{Kx} + \alpha r$. Select the gains $\mathbf{K}$ and α so that the system has a rapid response with an overshoot of approximately 1%, a settling time (with a 2%

criterion) less than 1 second, and a zero steady-state error to a unit step input.

P11.12 A DC motor has the state variable model

$$\dot{x} = \begin{bmatrix} -3 & -2 & -0.75 & 0 & 0 \\ -3 & 0 & 0 & 0 & 0 \\ 0 & 2 & 0 & 0 & 0 \\ 0 & 0 & 1 & 0 & 0 \\ 0 & 0 & 0 & 2 & 0 \end{bmatrix} x + \begin{bmatrix} 1 \\ 0 \\ 0 \\ 0 \\ 0 \end{bmatrix} u$$

$$y = [0 \ 0 \ 0 \ 0 \ 2.75]x.$$

Determine whether this system is controllable and observable.

P11.13 A feedback system has a plant transfer function

$$\frac{Y(s)}{R(s)} = G(s) = \frac{K}{s(s + 70)}.$$

We want the velocity error constant K_v to be 35 and the overshoot to a step to be approximately 4% so that ζ is $1/\sqrt{2}$. The settling time (with a 2% criterion) desired is 0.11 second. Design an appropriate state variable feedback system for $r(t) = -k_1 x_1 - k_2 x_2$.

P11.14 A process has the transfer function

$$\dot{x} = \begin{bmatrix} -10 & 0 \\ 1 & 0 \end{bmatrix} x + \begin{bmatrix} 1 \\ 0 \end{bmatrix} u$$

$$y = [0 \ 1]x + [0]u.$$

Determine the state variable feedback gains to achieve a settling time (with a 2% criterion) of 1 second and an overshoot of about 10%. Also sketch the block diagram of the resulting system. Assume the complete state vector is available for feedback.

P11.15 A telerobot system has the matrix equations [18]

$$\dot{x} = \begin{bmatrix} -1 & 0 & 0 \\ 0 & -2 & 0 \\ 0 & 0 & -3 \end{bmatrix} x + \begin{bmatrix} 1 \\ 1 \\ 0 \end{bmatrix} u$$

and

$$y = [1 \ 0 \ 2]x.$$

(a) Determine the transfer function, $G(s) = Y(s)/U(s)$. (b) Draw the block diagram indicating the state variables. (c) Determine whether the system is controllable. (d) Determine whether the system is observable.

P11.16 Hydraulic power actuators were used to drive the dinosaurs of the movie *Jurassic Park* [23]. The motions of the large monsters required high-power actuators requiring 1200 watts.

One specific limb motion has dynamics represented by

$$\dot{x} = \begin{bmatrix} -2 & 0 \\ 1 & 0 \end{bmatrix} x + \begin{bmatrix} 1 \\ 0 \end{bmatrix} u$$

$$y = [0 \ 1]x + [0]u.$$

We want to place the closed-loop poles at $s = -2 \pm j2$. Determine the required state variable feedback using Ackermann's formula. Assume that the complete state vector is available for feedback.

P11.17 A system has a transfer function

$$\frac{Y(s)}{R(s)} = \frac{s + a}{s^4 + 9s^3 + 28s^2 + 38s + 24}.$$

Determine a real value of a so that the system is either uncontrollable or unobservable.

P11.18 A system has a plant

$$\frac{Y(s)}{U(s)} = G(s) = \frac{1}{(s + 1)^2}.$$

(a) Find the matrix differential equation to represent this system. Identify the state variables on a block diagram model. (b) Select a state variable feedback structure using $u(t)$, and select the feedback gains so that the response $y(t)$ of the unforced system is critically damped when the initial condition is $x_1(0) = 1$ and $x_2(0) = 0$, where $x_1 = y(t)$. The repeated roots are at $s = -\sqrt{2}$.

P11.19 The block diagram of a system is shown in Figure P11.19. Determine whether the system is controllable and observable.

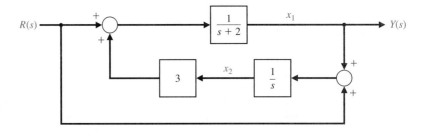

FIGURE P11.19
Multiloop feedback control system.

P11.20 Consider the automatic ship-steering system discussed in Problems P8.11 and P9.15. The state variable form of the system differential equation is

$$\dot{\mathbf{x}}(t) = \begin{bmatrix} -0.05 & -6 & 0 & 0 \\ -10^{-3} & -0.15 & 0 & 0 \\ 1 & 0 & 0 & 13 \\ 0 & 1 & 0 & 0 \end{bmatrix} \mathbf{x}(t) + \begin{bmatrix} -0.2 \\ 0.03 \\ 0 \\ 0 \end{bmatrix} \delta(t),$$

where $\mathbf{x}^T(t) = [v \quad \omega_s \quad y \quad \theta]$. The state variables are $x_1 = v =$ the transverse velocity; $x_2 = \omega_s =$ angular rate of ship's coordinate frame relative to response frame; $x_3 = y =$ deviation distance on an axis perpendicular to the track; $x_4 = \theta =$ deviation angle. (a) Determine whether the system is stable. (b) Feedback can be added so that

$$\delta(t) = -k_1 x_1 - k_3 x_3.$$

Determine whether this system is stable for suitable values of k_1 and k_3.

P11.21 An RL circuit is shown in Figure P11.21. (a) Select the two stable variables and obtain the vector differential equation where the output is $v_0(t)$. (b) Determine whether the state variables are observable when $R_1/L_1 = R_2/L_2$. (c) Find the conditions when the system has two equal roots.

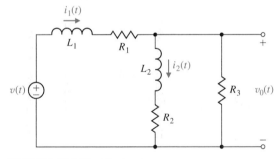

FIGURE P11.21 *RL* circuit.

P11.22 A manipulator control system has a loop transfer function of

$$G(s) = \frac{1}{s(s + 0.4)}$$

and negative unity feedback [17]. Represent this system by a state variable signal-flow graph or block diagram and a vector differential equation. (a) Plot the response of the closed-loop system to a step input. (b) Use state variable feedback so that the overshoot is 5% and the settling time (with a 2% criterion) is 1.35 seconds. (c) Plot the response of the state variable feedback system to a step input.

P11.23 Consider again the system of Example 11.7 when we desire that the steady-state error for a step input be zero and the desired roots of the characteristic equation be $s = -2 \pm j1$ and $s = -10$.

P11.24 Consider again the system of Example 11.7 when we desire that the steady-state error for a ramp input be zero and the roots of the characteristic equation be $s = -2 \pm j2$ and $s = -20$.

P11.25 Consider the system represented in state variable form

$$\dot{\mathbf{x}} = \mathbf{A}\mathbf{x} + \mathbf{B}u$$
$$y = \mathbf{C}\mathbf{x} + \mathbf{D}u,$$

where

$$\mathbf{A} = \begin{bmatrix} 1 & 4 \\ -5 & 10 \end{bmatrix}, \quad \mathbf{B} = \begin{bmatrix} 0 \\ 1 \end{bmatrix},$$
$$\mathbf{C} = [1 \quad -4], \quad \text{and} \quad \mathbf{D} = [0].$$

Verify that the system is observable. Then design a full-state observer by placing the observer poles at $s_{1,2} = -1$. Plot the response of the estimation error $\mathbf{e} = \mathbf{x} - \hat{\mathbf{x}}$ with an initial estimation error of $\mathbf{e}(0) = [1 \quad 1]^T$.

P11.26 Consider the third-order system

$$\dot{\mathbf{x}} = \begin{bmatrix} 0 & 1 & 0 \\ 0 & 0 & 1 \\ -1 & -2 & -3 \end{bmatrix} \mathbf{x} + \begin{bmatrix} 0 \\ 0 \\ 4 \end{bmatrix} u$$
$$y = [2 \quad -4 \quad 0]\mathbf{x} + [0]u.$$

Verify that the system is observable. If so, determine the observer gain matrix required to place the observer poles at $s_{1,2} = -1 \pm j2$ and $s_3 = -10$.

P11.27 Consider the second-order system

$$\dot{\mathbf{x}} = \begin{bmatrix} 1 & 0 \\ -3 & -2 \end{bmatrix} \mathbf{x} + \begin{bmatrix} 10 \\ 0 \end{bmatrix} u$$
$$y = [1 \quad 0]\mathbf{x} + [0]u.$$

Determine the observer gain matrix required to place the observer poles at $s_{1,2} = -1 \pm j$.

P11.28 Consider the single-input, single-output system is described by

$$\dot{\mathbf{x}}(t) = \mathbf{A}\mathbf{x}(t) + \mathbf{B}u(t)$$
$$y(t) = \mathbf{C}\mathbf{x}(t)$$

where

$$\mathbf{A} = \begin{bmatrix} 0 & 1 \\ -16 & -8 \end{bmatrix}, \mathbf{B} = \begin{bmatrix} 0 \\ K \end{bmatrix}, \mathbf{C} = [1 \quad 0].$$

(a) Determine the value of K resulting in a zero steady-state tracking error when $u(t)$ is a unit step input for $t \geq 0$. The tracking error is defined here as $e(t) = u(t) - y(t)$.

(b) Plot the response to a unit step input and verify that the tracking error is zero for the gain K determined in part (a).

P11.29   The block diagram shown in Figure P11.29 is an example of an interacting system. Determine a state variable representation of the system in the form

$$\dot{\mathbf{x}}(t) = \mathbf{A}\mathbf{x}(t) + \mathbf{B}u(t)$$

$$y(t) = \mathbf{C}\mathbf{x}(t) + \mathbf{D}u(t)$$

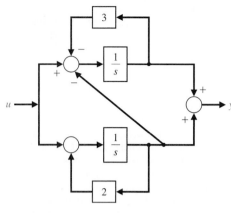

FIGURE P11.29   Interacting feedback system.

ADVANCED PROBLEMS

AP11.1   A DC motor control system has the form shown in Figure AP11.1 [6]. The three state variables are available for measurement; the output position is $x_1(t)$. Select the feedback gains so that the system has a steady-state error equal to zero for a step input and a response with a percent overshoot less than 3%.

AP11.2   A system has the model

$$\dot{\mathbf{x}} = \begin{bmatrix} -3 & -1.75 & -1.25 \\ 4 & 0 & 0 \\ 0 & 1 & 0 \end{bmatrix} \mathbf{x} + \begin{bmatrix} 2 \\ 0 \\ 0 \end{bmatrix} u.$$

Add state variable feedback so that the closed-loop poles are $s = -4, -4$, and -5.

AP11.3   A system has a matrix differential equation

$$\dot{\mathbf{x}} = \begin{bmatrix} 0 & 1 \\ 2 & 0 \end{bmatrix} \mathbf{x} + \begin{bmatrix} b_1 \\ b_2 \end{bmatrix} u.$$

What values for b_1 and b_2 are required so that the system is controllable?

AP11.4   The vector differential equation describing the inverted pendulum of Example 3.3 is

$$\frac{d\mathbf{x}}{dt} = \begin{bmatrix} 0 & 1 & 0 & 0 \\ 0 & 0 & -1 & 0 \\ 0 & 0 & 0 & 1 \\ 0 & 0 & 9.8 & 0 \end{bmatrix} \mathbf{x} + \begin{bmatrix} 0 \\ 1 \\ 0 \\ -1 \end{bmatrix} u.$$

Assume that all state variables are available for measurement and use state variable feedback. Place the system characteristic roots at $s = -2 \pm j, -5$, and -5.

AP11.5   An automobile suspension system has three physical state variables, as shown in Figure AP11.5 [15]. The state variable feedback structure is shown in the figure, with $K_1 = 1$. Select K_2 and K_3 so that the roots of the characteristic equation are three real roots lying between $s = -3$ and $s = -6$. Also, select K_p so that the steady-state error for a step input is equal to zero.

AP11.6   A system is represented by the differential equation

$$\frac{d^2 y}{dt^2} + 2\frac{dy}{dt} + y = \frac{du}{dt} + u,$$

where $y =$ output and $u =$ input.

(a) Develop a state variable representation and show that it is a controllable system. (b) Define the state variables as $x_1 = y$ and $x_2 = dy/dt - u$, and determine whether the system is controllable. Note that the controllability of a system depends on the definition of the state variables.

AP11.7   The new *Radisson Diamond* uses pontoons and stabilizers to damp out the effect of waves hitting the ship,

FIGURE AP11.1
Field-controlled DC motor.

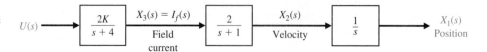

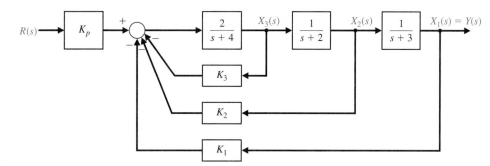

FIGURE AP11.5
Automobile
suspension system.

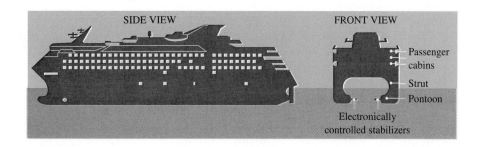

FIGURE AP11.7
(a) *Radisson
Diamond* (courtesy
of *Conde-Nast
Traveler*, July 1993,
23). (b) Control
system to reduce
the effect of the
disturbance.

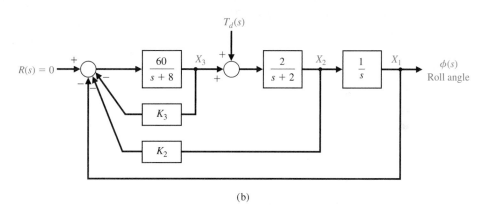

as shown in Figure AP11.7(a). The block diagram of the ship's roll control system is shown in Figure AP11.7(b). Determine the feedback gains K_2 and K_3 so that the characteristic roots are $s = -15$ and $s = -2 \pm j2$. Plot the roll output $\phi(t)$ for a unit step disturbance.

AP11.8 Consider again the liquid-level control system described in Problem P3.36.

(a) Design a state variable controller using only $h(t)$ as the feedback variable, so that the step response has an overshoot less than 10% and a settling time (with a 2% criterion) less than or equal to 5 seconds. (b) Design a state variable controller feedback using two state variables, level $h(t)$ and shaft position $\theta(t)$,

to satisfy the specifications of part (a). (c) Compare the results of parts (a) and (b).

AP11.9 The motion control of a lightweight hospital transport vehicle can be represented by a system of two masses, as shown in Figure AP11.9, where $m_1 = m_2 = 1$ and $k_1 = k_2 = 1$ [24]. (a) Determine the state vector differential equation. (b) Find the roots of the characteristic equation. (c) We wish to stabilize the system by letting $u = -kx_i$, where u is the force on the lower mass, and x_i is one of the state variables. Select an appropriate state variable x_i. (d) Choose a value for the gain k and sketch the root locus as k varies.

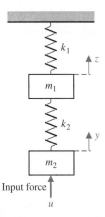

FIGURE AP11.9   Model of hospital vehicle.

AP11.10   Consider the inverted pendulum mounted to a motor, as shown in Figure AP11.10. The motor and load are assumed to have no friction damping. The pendulum to be balanced is attached to the horizontal shaft of a servomotor. The servomotor carries a tachogenerator, so that a velocity signal is available, but there is no position signal. When the motor is unpowered, the pendulum will hang vertically

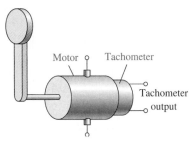

FIGURE AP11.10   Motor and inverted pendulum.

downward and, if slightly disturbed, will perform oscillations. If lifted to the top of its arc, the pendulum is unstable in that position. Devise a feedback compensator $G_c(s)$ using only the velocity signal from the tachometer.

AP11.11   Determine an internal model controller $G_c(s)$ for the system shown in Figure AP11.11. We want the steady-state error to a step input to be zero. We also want the settling time (with a 2% criterion) to be less than 5 seconds.

AP11.12   Repeat Advanced Problem AP11.11 when we want the steady-state error to a ramp input to be zero and the settling time (with a 2% criterion) of the ramp response to be less than 6 seconds.

AP11.13   Consider the system represented in state variable form

$$\dot{\mathbf{x}} = \mathbf{A}\mathbf{x} + \mathbf{B}u$$
$$y = \mathbf{C}\mathbf{x} + \mathbf{D}u,$$

where

$$\mathbf{A} = \begin{bmatrix} 1 & 2 \\ -5 & -10 \end{bmatrix}, \quad \mathbf{B} = \begin{bmatrix} -4 \\ 1 \end{bmatrix},$$
$$\mathbf{C} = [6 \quad -4], \quad \text{and} \quad \mathbf{D} = [0].$$

Verify that the system is observable and controllable. If so, design a full-state feedback law and an observer by placing the closed-loop system poles at $s_{1,2} = -1 \pm j$ and the observer poles at $s_{1,2} = -10$.

AP11.14   Consider the third-order system

$$\dot{\mathbf{x}} = \begin{bmatrix} 0 & 1 & 0 \\ 0 & 0 & 1 \\ -4 & -5 & -6 \end{bmatrix}\mathbf{x} + \begin{bmatrix} 0 \\ 0 \\ 4 \end{bmatrix}u$$
$$y = [2 \quad -9 \quad 2]\mathbf{x} + [0]u.$$

Verify that the system is observable and controllable. Then, design a full-state feedback law and an

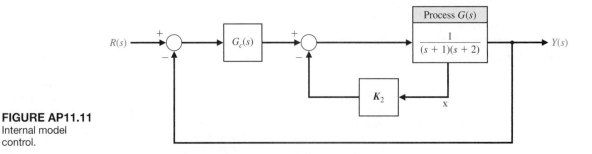

FIGURE AP11.11
Internal model
control.

observer by placing the closed-loop system poles at $s_{1,2} = -1 \pm j$, $s_3 = -2$ and the observer poles at $s_{1,2} = -10 \pm j2$, $s_3 = -20$.

AP11.15 Consider the system depicted in Figure AP11.15. Design a full-state observer for the system. Determine the observer gain matrix **L** to place the observer poles at $s_{1,2} = -10 \pm j10$.

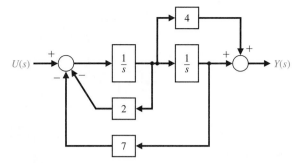

FIGURE AP11.15 A second-order system block diagram.

DESIGN PROBLEMS

CDP11.1 We wish to obtain a state variable feedback system for the capstan-slide system. Use the state variable model developed in CDP3.1 and determine the feedback system. The step response should have an overshoot less than 2% and a settling time less than 250 ms.

DP11.1 Consider the device for the magnetic levitation of a steel ball, as shown in Figures DP11.1(a) and (b). Obtain a design that will provide a stable response where the ball will remain within 10% of its desired position. Assume that y and dy/dt are measurable.

DP11.2 The control of the fuel-to-air ratio in an automobile carburetor became of prime importance in the 1980s as automakers worked to reduce exhaust-pollution emissions. Thus, auto engine designers turned to the feedback control of the fuel-to-air ratio. A sensor was placed in the exhaust stream and used as an input to a controller. The controller actually adjusts the orifice that controls the flow of fuel into the engine [3].

Select the devices and develop a linear model for the entire system. Assume that the sensor measures the actual fuel-to-air ratio with a negligible delay. With this model, determine the optimum controller when we desire a system with a zero steady-state error to a step input and an overshoot for a step command of less than 10%.

DP11.3 Consider the feedback system depicted in Figure DP11.3. The system model is given by

$$\dot{\mathbf{x}}(t) = \mathbf{A}\mathbf{x}(t) + \mathbf{B}u(t)$$
$$y(t) = \mathbf{C}\mathbf{x}(t)$$

where

$$\mathbf{A} = \begin{bmatrix} 0 & 1 \\ -10.5 & -11.3 \end{bmatrix}, \mathbf{B} = \begin{bmatrix} 0 \\ 0.55 \end{bmatrix}, \mathbf{C} = \begin{bmatrix} 1 & 0 \end{bmatrix}.$$

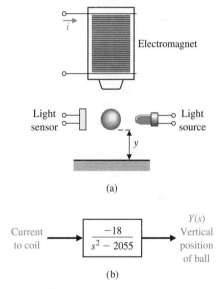

FIGURE DP11.1 (a) The levitation of a ball using an electromagnet. (b) The model of the electromagnet and the ball.

Design the compensator to meet the following specifications:

1. The steady-state error to a unit step input is zero.
2. The settling time $T_s < 1$ s and the percent overshoot is $P.O. < 5\%$.
3. Select initial conditions for **x** and different initial conditions for $\hat{\mathbf{x}}$ and simulate the response of the closed-loop system to a unit step input.

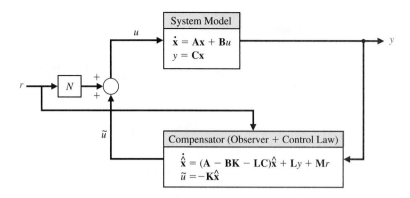

FIGURE DP11.3
Feedback system
constructed to
track a desired
input $r(t)$.

DP11.4 A high-performance helicopter has a model shown in Figure DP11.4. The goal is to control the pitch angle θ of the helicopter by adjusting the rotor thrust angle δ.

The equations of motion of the helicopter are

$$\frac{d^2\theta}{dt^2} = -\sigma_1\frac{d\theta}{dt} - \alpha_1\frac{dx}{dt} + n\delta$$

$$\frac{d^2x}{dt^2} = g\theta - \alpha_2\frac{d\theta}{dt} - \sigma_2\frac{dx}{dt} + g\delta,$$

where x is the translation in the horizontal direction. For a military high-performance helicopter, we find that

$$\sigma_1 = 0.415 \qquad \alpha_2 = 1.43$$
$$\sigma_2 = 0.0198 \qquad n = 6.27$$
$$\alpha_1 = 0.0111 \qquad g = 9.8$$

all in appropriate SI units.

Find (a) a state variable representation of this system and (b) the transfer function representation for $\theta(s)/\delta(s)$. (c) Use state variable feedback to achieve adequate performances for the controlled system.

Desired specifications include (1) a steady-state for an input step command for $\theta_d(s)$, the desired pitch angle, less than 20% of the input step magnitude; (2) an overshoot for a step input command less than 20%; and (3) a settling (with a 2% criterion) time for a step command of less than 1.5 seconds.

DP11.5 The headbox process is used in the manufacture of paper to transform the pulp slurry flow into a jet of 2 cm and then spread it onto a mesh belt [25]. To achieve desirable paper quality, the pulp slurry must be distributed as evenly as possible on the belt, and the relationship between the velocity of the jet and that of the belt, called the jet/belt ratio, must be maintained. One of the main control variables is the pressure in the headbox, which in turn controls the velocity of the slurry at the jet. The total pressure in the headbox is the sum of the liquid-level pressure and the air pressure that is pumped into the headbox. Because the pressurized headbox is a highly dynamic and coupled system, manual control would be difficult to maintain and could result in degradation in the sheet properties.

The state-space model of a typical headbox, linearized about a particular stationary point, is given by

$$\dot{\mathbf{x}} = \begin{bmatrix} -0.8 & +0.02 \\ -0.02 & 0 \end{bmatrix}\mathbf{x} + \begin{bmatrix} 0.05 \\ 0.001 \end{bmatrix}u$$

and $y = [1 \quad 0]\mathbf{x}$.

The state variables are $x_1 =$ liquid level and $x_2 =$ pressure. The control variable is $u_1 =$ pump current. (a) Design a state variable feedback system that has a characteristic equation with real roots with a magnitude greater than five. (b) Design an observer with observer poles located at least ten times farther in the left half-plane than the state variable feedback system. (c) Connect the observer and full-state feedback system and sketch the block diagram of the integrated system.

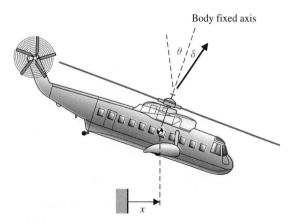

Body fixed axis

θ / δ

x

FIGURE DP11.4   Helicopter pitch angle, θ, control.

DP11.6 A coupled-drive apparatus is shown in Figure DP11.6. The coupled drives consist of two pulleys connected via an elastic belt, which is tensioned by a third pulley mounted on springs providing an under-damped dynamic mode. One of the main pulleys, pulley A, is driven by an electric DC motor. Both pulleys A and B are fitted with tachometers that generate measurable voltages proportional to the rate of rotation of the pulley. When a voltage is applied to the DC motor, pulley A will accelerate at a rate governed by the total inertia experienced by the system. Pulley B, at the other end of the elastic belt, will also accelerate owing to the applied voltage or torque, but with a lagging effect caused by the elasticity of the belt. Integration of the velocity signals measured at each pulley will provide an angular position estimate for the pulley [26].

The second-order model of a coupled-drive is

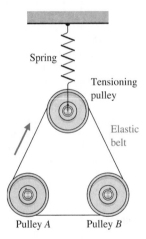

FIGURE DP11.6

$$\dot{\mathbf{x}} = \begin{bmatrix} 0 & 1 \\ -36 & -12 \end{bmatrix} \mathbf{x} + \begin{bmatrix} 0 \\ 1 \end{bmatrix} u$$

and $y = x_1$.

(a) Design a state variable feedback controller that will yield a step response with deadbeat response and a settling time (with a 2% criterion) less than 0.5 second. (b) Design an observer for the system by placing the observer poles appropriately in the left half-plane. (c) Draw the block diagram of the system including the compensator with the observer and state feedback. (d) Simulate the response to an initial state at $\mathbf{x}(0) = \begin{bmatrix} 1 & 0 \end{bmatrix}^T$ and $\hat{\mathbf{x}}(0) = \begin{bmatrix} 0 & 0 \end{bmatrix}^T$.

DP11.7 A closed-loop feedback system is to be designed to track a reference input. The desired feedback block diagram is shown in Figure DP11.3. The system model is given by

$$\dot{\mathbf{x}}(t) = \mathbf{A}\mathbf{x}(t) + \mathbf{B}u(t)$$

$$y(t) = \mathbf{C}\mathbf{x}(t)$$

where

$$\mathbf{A} = \begin{bmatrix} 0 & 1 & 0 \\ 0 & 0 & 1 \\ -2 & -5 & -10 \end{bmatrix}, \mathbf{B} = \begin{bmatrix} 0 \\ 0 \\ 1 \end{bmatrix}, \mathbf{C} = \begin{bmatrix} 1 & 0 & 0 \end{bmatrix}.$$

Design the observer and the control law to meet the following specifications:

1. The steady-state error of the closed-loop system to a unit step input is zero.
2. The gain margin $G.M. \geq 6$ dB.
3. The bandwidth of the closed-loop system $\omega_B \geq 10$ rad/s.
4. Select initial conditions for $\mathbf{x}$ and different initial conditions for $\hat{\mathbf{x}}$ and simulate the response of the closed-loop system to a unit step input. Verify that the tracking error is zero in the steady-state.

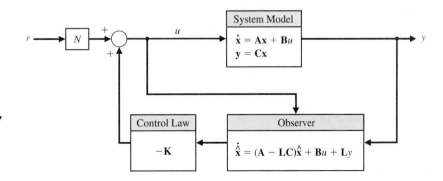

FIGURE DP11.7
Feedback system constructed to track a desired input $r(t)$.

COMPUTER PROBLEMS

CP11.1   Consider the system

$$\dot{x} = \begin{bmatrix} -4 & 2 & 0 \\ 4 & 0 & -6 \\ -10 & 1 & 11 \end{bmatrix} x + \begin{bmatrix} -3 \\ 0 \\ 1 \end{bmatrix} u,$$

$$y = [1 \quad 0 \quad 1]x.$$

Using the ctrb and obsv functions, show that the system is controllable and observable.

CP11.2   Consider the system

$$\dot{x} = \begin{bmatrix} 0 & 1 \\ -2 & -6.5 \end{bmatrix} x + \begin{bmatrix} 0 \\ 2 \end{bmatrix} u,$$

$$y = [1 \quad 0]x.$$

Determine if the system is controllable and observable. Compute the transfer function from u to y.

CP11.3   Find a gain matrix K so that the closed-loop poles of the system

$$\dot{x} = \begin{bmatrix} 0 & 1 \\ -1 & 2 \end{bmatrix} x + \begin{bmatrix} 1 \\ 1 \end{bmatrix} u,$$

$$y = [1 \quad -1]x$$

are $s_1 = -1$ and $s_2 = -2$. Use state feedback $u = -Kx$.

CP11.4   The following model has been proposed to describe the motion of a constant-velocity guided missile:

$$\dot{x} = \begin{bmatrix} 0 & 1 & 0 & 0 & 0 \\ -0.1 & -0.5 & 0 & 0 & 0 \\ 0.5 & 0 & 0 & 0 & 0 \\ 0 & 0 & 10 & 0 & 0 \\ 0.5 & 1 & 0 & 0 & 0 \end{bmatrix} x + \begin{bmatrix} 0 \\ 1 \\ 0 \\ 0 \\ 0 \end{bmatrix} u,$$

$$y = [0 \quad 0 \quad 0 \quad 1 \quad 0]x.$$

(a) Verify that the system is not controllable by analyzing the controllability matrix using the ctrb function.
(b) Develop a controllable state variable model by first computing the transfer function from u to y, then cancel any common factors in the numerator and denominator polynomials of the transfer function. With the modified transfer function just obtained, use the ss function to determine a modified state variable model for the system.
(c) Verify that the modified state variable model in part (b) is controllable.
(d) Is the constant velocity guided missile stable?
(e) Comment on the relationship between the controllability and the complexity of the state variable model (where complexity is measured by the number of state variables).

CP11.5   A linearized model of a vertical takeoff and landing (VTOL) aircraft is [27]

$$\dot{x} = Ax + B_1 u_1 + B_2 u_2,$$

where

$$A = \begin{bmatrix} -0.0389 & 0.0271 & 0.0188 & -0.4555 \\ 0.0482 & -1.0100 & 0.0019 & -4.0208 \\ 0.1024 & 0.3681 & -0.7070 & 1.4200 \\ 0 & 0 & 1 & 0 \end{bmatrix}$$

and

$$B_1 = \begin{bmatrix} 0.4422 \\ 3.5446 \\ -6.0214 \\ 0 \end{bmatrix}, \quad B_2 = \begin{bmatrix} 0.1291 \\ -7.5922 \\ 4.4900 \\ 0 \end{bmatrix}.$$

The state vector components are (i) x_1 is the horizontal velocity (knots), (ii) x_2 is the vertical velocity (knots), (iii) x_3 is the pitch rate (degrees/second), and (iv) x_4 is the pitch angle (degrees). The input u_1 is used mainly to control the vertical motion, and u_2 is used for the horizontal motion.

(a) Compute the eigenvalues of the system matrix A. Is the system stable? (b) Determine the characteristic polynomial associated with A using the poly function. Compute the roots of the characteristic equation, and compare them with the eigenvalues in part (a). (c) Is the system controllable from u_1 alone? What about from u_2 alone? Comment on the results.

CP11.6   In an effort to open up the far side of the moon to exploration, studies have been conducted to determine the feasibility of operating a communication satellite around the translunar equilibrium point in the Earth–Sun–Moon system. The desired satellite orbit, known as a halo orbit, is shown in Figure CP11.6. The objective of the controller is to keep the satellite on a halo orbit trajectory that can be seen from the earth so that the lines of communication are accessible at all times. The communication link is from the earth to the satellite and then to the far side of the moon.

The linearized (and normalized) equations of motion of the satellite around the translunar equilibrium point are [28]

$$\dot{x} = \begin{bmatrix} 0 & 0 & 0 & 1 & 0 & 0 \\ 0 & 0 & 0 & 0 & 1 & 0 \\ 0 & 0 & 0 & 0 & 0 & 1 \\ 7.3809 & 0 & 0 & 0 & 2 & 0 \\ 0 & -2.1904 & 0 & -2 & 0 & 0 \\ 0 & 0 & -3.1904 & 0 & 0 & 0 \end{bmatrix} x$$

$$+ \begin{bmatrix} 0 \\ 0 \\ 0 \\ 1 \\ 0 \\ 0 \end{bmatrix} u_1 + \begin{bmatrix} 0 \\ 0 \\ 0 \\ 0 \\ 1 \\ 0 \end{bmatrix} u_2 + \begin{bmatrix} 0 \\ 0 \\ 0 \\ 0 \\ 0 \\ 1 \end{bmatrix} u_3.$$

The state vector $\mathbf{x}$ is the satellite position and velocity, and the inputs u_i, $i = 1, 2, 3$, are the engine thrust accelerations in the ξ, η, and ζ directions, respectively.
(a) Is the translunar equilibrium point a stable location? (b) Is the system controllable from u_1 alone? (c) Repeat part (b) for u_2. (d) Repeat part (b) for u_3. (e) Suppose that we can observe the position in the η direction. Determine the transfer function from u_2 to η. (*Hint:* Let $y = [0 \quad 1 \quad 0 \quad 0 \quad 0 \quad 0]\mathbf{x}$.)
(f) Compute a state-space representation of the transfer function in part (e) using the ss function. Verify that the system is controllable. (g) Using state feedback

$$u_2 = -\mathbf{Kx},$$

design a controller (i.e., find $\mathbf{K}$) for the system in part (f) such that the closed-loop system poles are at $s_{1,2} = -1 \pm j$ and $s_{3,4} = -10$.

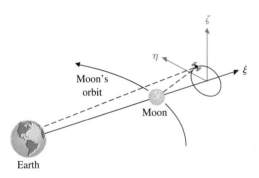

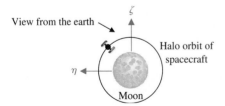

FIGURE CP11.6   The translunar satellite halo orbit.

CP11.7   Consider the system

$$\dot{\mathbf{x}}(t) = \begin{bmatrix} 0 & 1 & 0 \\ 0 & 0 & 1 \\ -2 & -4 & -6 \end{bmatrix} \mathbf{x}(t),$$

$$y(t) = [1 \quad 0 \quad 0]\mathbf{x}(t). \tag{CP11.1}$$

Suppose that we are given three observations $y(t_i)$, $i = 1, 2, 3$, as follows:

$$y(t_1) = 1 \quad \text{at} \quad t_1 = 0$$
$$y(t_2) = -0.0256 \quad \text{at} \quad t_2 = 2$$
$$y(t_3) = -0.2522 \quad \text{at} \quad t_3 = 4.$$

(a) Using the three observations, develop a method to determine the initial value of the state vector $\mathbf{x}(t_0)$ for the system in Equation (1) that will reproduce the three observations when simulated using the lsim function. (b) With the observations given, compute $\mathbf{x}(t_0)$ and discuss the condition under which this problem can be solved in general. (c) Verify the result by simulating the system response to the computed initial condition. (*Hint:* Recall that $\mathbf{x}(t) = e^{\mathbf{A}(t-t_0)}\mathbf{x}(t_0)$ for the system in Equation CP11.1.)

CP11.8   A system is described by a single-input state equation with

$$\mathbf{A} = \begin{bmatrix} 0 & 0 \\ -1 & 0 \end{bmatrix} \quad \text{and} \quad \mathbf{B} = \begin{bmatrix} 0 \\ 1 \end{bmatrix}.$$

Using the method of Section 11.7 (Equation 11.40) and a negative unity feedback, determine the optimal system when $\mathbf{x}^T(0) = [1 \quad 0]$.

CP11.9   A first-order system is given by

$$\dot{x} = -x + u$$

with the initial condition $x(0) = x_0$. We want to design a feedback controller

$$u = -kx$$

such that the performance index

$$J = \int_0^\infty (x^2(t) + \lambda u^2(t)) \, dt$$

is minimized.
(a) Let $\lambda = 1$. Develop a formula for J in terms of k, valid for any x_0, and use an m-file to plot J/x_0^2 versus k. From the plot, determine the approximate value of $k = k_{min}$ that minimizes J/x_0^2. (b) Verify the result in part (a) analytically. (c) Using the procedure developed in part (a), obtain a plot of k_{min} versus λ, where k_{min} is the gain that minimizes the performance index.

CP11.10   Consider the system represented in state variable form

$$\dot{\mathbf{x}} = \mathbf{Ax} + \mathbf{B}u$$
$$y = \mathbf{Cx} + \mathbf{D}u,$$

where

$$\mathbf{A} = \begin{bmatrix} 0 & 1 \\ -19.04 & -11.42 \end{bmatrix}, \quad \mathbf{B} = \begin{bmatrix} 12.8 \\ 24.6 \end{bmatrix},$$

$$\mathbf{C} = \begin{bmatrix} 1 & 0 \end{bmatrix} \quad \text{and} \quad \mathbf{D} = \begin{bmatrix} 0 \end{bmatrix}.$$

Using the acker function, determine a full-state feedback gain matrix and an observer gain matrix to place the closed-loop system poles at $s_{1,2} = -1$ and the observer poles at $s_{1,2} = -12 \pm j4$.

CP11.11   Consider the third-order system

$$\dot{\mathbf{x}} = \begin{bmatrix} 0 & 1 & 0 \\ 0 & 0 & 1 \\ -4.3 & -1.7 & -6.7 \end{bmatrix} \mathbf{x} + \begin{bmatrix} 0 \\ 0 \\ 0.35 \end{bmatrix} u$$

$$y = \begin{bmatrix} 0 & 1 & 0 \end{bmatrix} \mathbf{x} + \begin{bmatrix} 0 \end{bmatrix} u.$$

(a) Using the acker function, determine a full-state feedback gain matrix and an observer gain matrix to place the closed-loop system poles at $s_{1,2} = -1.4 \pm j1.4$, $s_3 = -2$ and the observer poles at $s_{1,2} = -18 \pm j5$, $s_3 = -20$. (b) Construct the state variable compensator using Figure 11.1 as a guide. (c) Simulate the closed-loop system with the state initial conditions $\mathbf{x}(0) = (1 \quad 0 \quad 0)^T$ and initial state estimate of $\hat{\mathbf{x}}(0) = (0.5 \quad 0.1 \quad 0.1)^T$.

CP11.12   Implement the system shown in Figure CP11.12 in an m-file. Obtain the step response of the system.

CP11.13   Consider the system in state variable form

$$\dot{\mathbf{x}} = \begin{bmatrix} 0 & 1 & 0 & 0 \\ 0 & 0 & 1 & 0 \\ 0 & 0 & 0 & 1 \\ -2 & -5 & -1 & -13 \end{bmatrix} \mathbf{x} + \begin{bmatrix} 0 \\ 0 \\ 0 \\ 1 \end{bmatrix} u$$

$$y = \begin{bmatrix} 1 & 0 & 0 & 0 \end{bmatrix} \mathbf{x} + \begin{bmatrix} 0 \end{bmatrix} u.$$

Design a full-state feedback gain matrix and an observer gain matrix to place the closed-loop system poles at $s_{1,2} = -1.4 \pm j1.4$, $s_{3,4} = -2 \pm j$ and the observer poles $s_{1,2} = -18 \pm j5$, $s_{3,4} = -20$. Construct the state variable compensator using Figure 11.1 as a guide and simulate the closed-loop system using Simulink. Select several values of initial states and initial state estimates in the observer and display the tracking results on an x-y-graph.

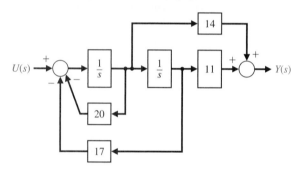

FIGURE CP11.12   Control system for Simulink implementation.

TERMS AND CONCEPTS

Command following   An important aspect of control system design wherein a nonzero reference input is tracked.

Controllability matrix   A linear system is (completely) controllable if and only if the controllability matrix $\mathbf{P}_c = [\mathbf{B} \quad \mathbf{AB} \quad \mathbf{A}^2\mathbf{B} \dots \mathbf{A}^{n-1}\mathbf{B}]$ has full rank, where $\mathbf{A}$ is an $n \times n$ matrix. For single-input, single-output linear systems, the system is controllable if and only if the determinant of the $n \times n$ controllability matrix $\mathbf{P}_c$ is nonzero.

Controllable system   A system is controllable on the interval $[t_0, t_f]$ if there exists a continuous input $u(t)$ such that any initial state $\mathbf{x}(t_0)$ can be driven to any arbitrary trial state $\mathbf{x}(t_f)$ in a finite time interval $t_f - t_0 > 0$.

Detectable   A system in which the states that are unobservable are naturally stable.

Estimation error   The difference between the actual state and the estimated state $\mathbf{e}(t) = \mathbf{x}(t) - \hat{\mathbf{x}}(t)$.

Full-state feedback control law   A control law of the form $u = -\mathbf{Kx}$ where $\mathbf{x}$ is the state of the system assumed known at all times.

Internal model design   A method of tracking reference inputs with guaranteed steady-state tracking errors.

Kalman state-space decomposition   A partition of the state space that illuminates the states that are controllable and unobservable, uncontrollable and unobservable, controllable and observable, and uncontrollable and observable.

Linear quadratic regulator   An optimal controller designed to minimize the quadratic performance index

$$J = \int_0^\infty (\mathbf{x}^T\mathbf{Qx} + \mathbf{u}^T\mathbf{Ru}) \, dt, \quad \text{where } \mathbf{Q} \text{ and } \mathbf{R} \text{ are}$$

design parameters.

Observability matrix   A linear system is (completely) observable if and only if the observability matrix $\mathbf{P}_o = [\mathbf{C}^T \quad (\mathbf{CA})^T \quad (\mathbf{CA}^2)^T \ldots (\mathbf{CA}^{n-1})^T]^T$ has full rank, where $\mathbf{A}$ is an $n \times n$ matrix. For single-input, single-output linear systems, the system is observable if and only if the determinant of the $n \times n$ observability matrix $\mathbf{P}_o$ is nonzero.

Observable system   A system is observable on the interval $[t_0, t_f]$ if any initial state $\mathbf{x}(t_0)$ is uniquely determined by observing the output $y(t)$ on the interval $[t_0, t_f]$.

Observer   A dynamic system used to estimate the state of another dynamic system given knowledge of the system inputs and measurements of the system outputs.

Optimal control system   A system whose parameters are adjusted so that the performance index reaches an extremum value.

Pole placement   A design methodology wherein the objective is to place the eigenvalues of the closed-loop system in desired regions of the complex plane.

Separation principle   The principle that states that the full-state feedback law and the observer can be designed independently and when connected will function as an integrated control system in the desired manner (i.e., stable).

Stabilizable   A system in which the states that are not controllable are naturally stable.

Stabilizing controller   A controller that stabilizes the closed-loop system.

State variable feedback   Occurs when the control signal u for the process is a direct function of all the state variables.

12

Robust Control Systems

P R E V I E W

Physical systems and the external environment in which they operate cannot be modeled precisely, may change in an unpredictable manner, and may be subject to significant disturbances. The design of control systems in the presence of significant uncertainty requires the designer to seek a robust system. Recent advances in robust control design methodologies can address stability robustness and performance robustness in the presence of uncertainty. In this chapter, we describe five methods for robust design, including root locus, frequency response, ITAE methods for a robust PID systems, internal model control, and pseudo-quantitative feedback methods. However, we should also realize that classical design techniques may also produce robust control systems. Control engineers who are aware of these issues can design robust PID controllers, robust lead-lag controllers, and so forth. The chapter concludes with a PID controller design for the Sequential Design Example: Disk Drive Read System.

DESIRED OUTCOMES

Upon completion of Chapter 12, students should:

❑   Appreciate the role of robustness in control system design.
❑   Be familiar with uncertainty models, including additive uncertainty, multiplicative uncertainty, and parameter uncertainty.
❑   Understand the various methods of tackling the robust control design problem using root locus, frequency response, ITAE methods for PID control, internal model, and pseudo-quantitative feedback methods.

12.1 INTRODUCTION

A control system designed using the methods and concepts of the preceding chapters assumes knowledge of the model of the process and controller and constant parameters. The process model will always be an inaccurate representation of the actual physical system because of

❑ parameter changes
❑ unmodeled dynamics
❑ unmodeled time delays
❑ changes in equilibrium point (operating point)
❑ sensor noise
❑ unpredicted disturbance inputs.

The goal of robust systems design is to retain assurance of system performance in spite of model inaccuracies and changes. A system is **robust** when the system has acceptable changes in performance due to model changes or inaccuracies.

> **A robust control system exhibits the desired performance despite the presence of significant process uncertainty.**

A system structure that incorporates potential system uncertainties is shown in Figure 12.1. This model includes the sensor noise $N(s)$, the unpredicted disturbance

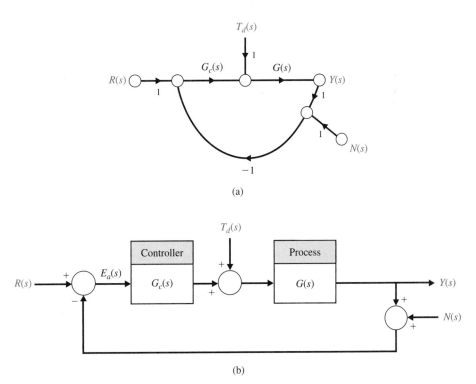

(a)

(b)

FIGURE 12.1
Closed-loop control system. (a) Signal flow graph; (b) Block diagram.

input $T_d(s)$, and a process $G(s)$ with potentially unmodeled dynamics or parameter changes. The unmodeled dynamics and parameter changes may be significant or very large, and for these systems the challenge is to create a design that retains the desired performance.

12.2 ROBUST CONTROL SYSTEMS AND SYSTEM SENSITIVITY

Designing highly accurate systems in the presence of significant plant uncertainty is a classical feedback design problem. The theoretical bases for the solution of this problem date back to the works of H. S. Black and H. W. Bode in the early 1930s, when this problem was referred to as the sensitivities design problem. A significant amount of literature has been published since then regarding the design of systems subject to large process uncertainty. The designer seeks to obtain a system that performs adequately over a large range of uncertain parameters. A system is said to be robust when it is durable, hardy, and resilient.

A control system is robust when (1) it has low sensitivities, (2) it is stable over the range of parameter variations, and (3) the performance continues to meet the specifications in the presence of a set of changes in the system parameters [3, 4]. Robustness is the low sensitivity to effects that are not considered in the analysis and design phase—for example, disturbances, measurement noise, and unmodeled dynamics. The system should be able to withstand these neglected effects when performing the tasks for which it was designed.

For small-parameter perturbations, we may use, as a measure of robustness, the differential sensitivities discussed in Sections 4.3 (system sensitivity) and Section 7.5 (root sensitivity) [6]. The **system sensitivity** is defined as

$$S_\alpha^T = \frac{\partial T/T}{\partial \alpha/\alpha}, \tag{12.1}$$

where α is the parameter and T the transfer function of the system. The **root sensitivity** is defined as

$$S_\alpha^{r_i} = \frac{\partial r_i}{\partial \alpha/\alpha}. \tag{12.2}$$

When the zeros of $T(s)$ are independent of the parameter α, we showed that

$$S_\alpha^T = -\sum_{i=1}^n S_\alpha^{r_i} \cdot \frac{1}{s + r_i}, \tag{12.3}$$

for an nth-order system. For example, if we have a closed-loop system, as shown in Figure 12.2, where the variable parameter is α, then $T(s) = 1/[s + (\alpha + 1)]$, and

$$S_\alpha^T = \frac{-\alpha}{s + \alpha + 1}. \tag{12.4}$$

This follows because $r_1 = +(\alpha + 1)$, and

$$-S_\alpha^{r_i} = -\alpha. \tag{12.5}$$

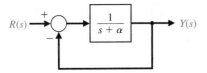

FIGURE 12.2
A first-order
system.

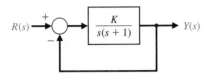

FIGURE 12.3
A second-order
system.

Therefore,

$$S_\alpha^T = -S_\alpha^{r_i}\frac{1}{s + \alpha + 1} = \frac{-\alpha}{s + \alpha + 1}. \qquad (12.6)$$

Let us examine the sensitivity of the second-order system shown in Figure 12.3. The transfer function of the closed-loop system is

$$T(s) = \frac{K}{s^2 + s + K}. \qquad (12.7)$$

The system sensitivity for K is

$$S(s) = S_K^T = \frac{s(s + 1)}{s^2 + s + K}. \qquad (12.8)$$

A Bode plot of the asymptotes of $20 \log|T(j\omega)|$ and $20 \log|S(j\omega)|$ is shown in Figure 12.4 for $K = 1/4$ (critical damping). Note that the sensitivity is small for lower frequencies, while the transfer function primarily passes low frequencies.

Of course, the sensitivity $S(s)$ only represents robustness for small changes in gain. If K changes from 1/4 within the range $K = 1/16$ to $K = 1$, the resulting range of step response is shown in Figure 12.5. This system, with an expected wide range of

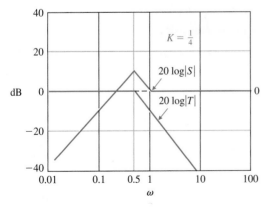

FIGURE 12.4   Sensitivity and $20 \log |T(j\omega)|$ for the second-order system in Figure 12.3. The asymptotic approximations are shown for $K = \frac{1}{4}$.

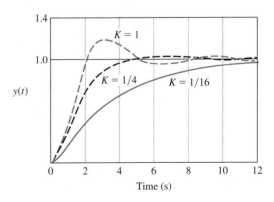

FIGURE 12.5   The step response for selected gain K.

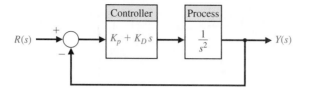

FIGURE 12.6
A system with a PD controller.

K, may not be considered adequately robust. A robust system would be expected to yield essentially the same (within an agreed-upon variation) response to a selected input.

EXAMPLE 12.1   Sensitivity of a controlled system

Consider the system shown in Figure 12.6, where $G(s) = 1/s^2$ and a PD controller $G_c(s) = K_p + K_D s$. Then the sensitivity with respect to changes in $G(s)$ is

$$S_G^T = \frac{1}{1 + G_c(s)G(s)} = \frac{s^2}{s^2 + K_D s + K_p}, \tag{12.9}$$

and

$$T(s) = \frac{K_D s + K_p}{s^2 + K_D s + K_p}. \tag{12.10}$$

Consider the normal condition $\zeta = 1$ and $\omega_n = \sqrt{K_p}$. Then, $K_D = 2\omega_n$ to achieve $\zeta = 1$. Therefore, we may plot $20 \log|S|$ and $20 \log|T|$ on a Bode diagram, as shown in Figure 12.7. Note that the frequency ω_n is an indicator on the boundary between the frequency region in which the sensitivity is the important design criterion and the region in which the stability margin is important. Thus, if we specify ω_n properly to take into consideration the extent of modeling error and the frequency of external disturbance, we can expect the system to have an acceptable amount of robustness. ■

EXAMPLE 12.2   System with a right-hand-plane zero

Consider the system shown in Figure 12.8, where the plant has a zero in the right-hand plane. The closed-loop transfer function is

$$T(s) = \frac{K(s - 1)}{s^2 + (2 + K)s + (1 - K)}. \tag{12.11}$$

The system is stable for a gain $-2 < K < 1$. The steady-state error due to a negative unit step input $R(s) = -1/s$ is

$$e_{ss} = \frac{1 - 2K}{1 - K}, \tag{12.12}$$

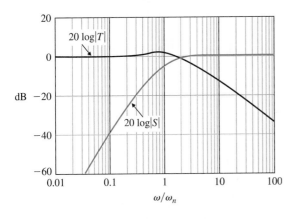

FIGURE 12.7   Sensitivity and $T(s)$ for the second-order system in Figure 12.6.

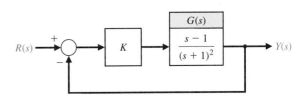

FIGURE 12.8   A second-order system.

FIGURE 12.9
Step response of the system in Figure 12.8 with $K = \frac{1}{2}$.

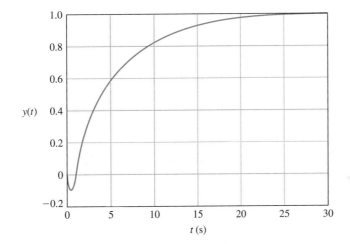

and $e_{ss} = 0$ when $K = 1/2$. The response is shown in Figure 12.9. Note the initial undershoot at $t = 1$ s. This system is sensitive to changes in K, as recorded in Table 12.1. The performance of this system might be considered barely acceptable for a change of gain of only $\pm 10\%$. Thus, this system would not be considered robust. The steady-state error of this system changes greatly as K changes. ∎

Table 12.1   Results for Example 12.2

K	0.25	0.45	0.50	0.55	0.75
$|e_{ss}|$	0.67	0.18	0	0.22	1.0
Undershoot	5%	9%	10%	11%	15%
Settling time (seconds)	15	24	27	30	45

12.3 ANALYSIS OF ROBUSTNESS

Consider the closed-loop system shown in Figure 12.1. System goals include maintaining a small tracking error $[E(s) = R(s) - Y(s)]$ for an input $R(s)$ and keeping the output $Y(s)$ small for a disturbance $T_d(s)$.

Following the discussion in Section 4.1, the **sensitivity function** is

$$S(s) = [1 + G_c(s)G(s)]^{-1},$$

and the **complementary sensitivity function** is

$$C(s) = \frac{G_c(s)G(s)}{1 + G_c(s)G(s)}.$$

We also have the relationship

$$S(s) + C(s) = 1. \tag{12.13}$$

For physically realizable systems, the loop gain $L(s) = G_c(s)G(s)$ must be small for high frequencies. This means that $S(j\omega)$ approaches 1 at high frequencies.

An **additive perturbation** characterizes the set of possible processes as follows (here we assume that $G_c(s) = 1$):

$$G_a(s) = G(s) + A(s),$$

where $G(s)$ is the nominal process, and $A(s)$ is the perturbation that is bounded in magnitude. We assume that $G_a(s)$ and $G(s)$ have the same number of poles in the right-hand s-plane (if any) [36]. Then the system stability will not change if

$$|A(j\omega)| < |1 + G(j\omega)| \qquad \text{for all } \omega. \tag{12.14}$$

This assures stability but not dynamic performance.

A **multiplicative perturbation** is modeled as

$$G_m(s) = G(s)[1 + M(s)].$$

The perturbation is bounded in magnitude, and it is again assumed that $G_m(s)$ and $G(s)$ have the same number of poles in the right-hand s-plane. Then the system stability will not change if

$$|M(j\omega)| < \left|1 + \frac{1}{G(j\omega)}\right| \qquad \text{for all } \omega. \tag{12.15}$$

Equation (12.15) is called the **robust stability criterion**. This is a test for robustness with respect to a multiplicative perturbation. This form of perturbation is often used because it satisfies the intuitive properties of (1) being small at low frequencies, where the nominal process model is usually well known, and (2) being large at high frequencies, where the nominal model is always inexact.

EXAMPLE 12.3   System with multiplicative perturbation

Consider the system of Figure 12.1 with $G_c = K$, and

$$G(s) = \frac{170,000\,(s + 0.1)}{s(s + 3)(s^2 + 10s + 10,000)}.$$

The system is unstable with $K = 1$, but a reduction in gain to $K = 0.5$ will stabilize it. Now, consider the effect of an unmodeled pole at 50 rad/s. In this case, the multiplicative perturbation is determined from

$$1 + M(s) = \frac{50}{s + 50},$$

or $M(s) = -s/(s + 50)$. The magnitude bound is then

$$|M(j\omega)| = \left| \frac{-j\omega}{j\omega + 50} \right|.$$

$|M(j\omega)|$ and $|1 + 1/(KG(j\omega))|$ are plotted in Figure 12.10(a), where it is seen that the criterion of Equation (12.15) is not satisfied. Thus, the system may not be stable.

If we use a lag compensator

$$G_c(s) = \frac{0.15(s + 25)}{s + 2.5},$$

the loop transfer function is $1 + G_c(s)G(s)$, and we reshape the function $G_c(j\omega)G(j\omega)$ in the frequency range $2 < \omega < 25$. Then we have the altered magnitude

$$\left| 1 + \frac{1}{G_c(j\omega)G(j\omega)} \right|,$$

as plotted in Figure 12.10(b). Here the robustness inequality is satisfied, and the system remains stable. ■

The control objective is to design a compensator $G_c(s)$ so that the transient, steady-state, and frequency-domain specifications are achieved and the cost of feedback measured by the bandwidth of the compensator $G_c(j\omega)$ is sufficiently small. This bandwidth constraint is needed mainly because of noise that is inevitable in measuring the system output. A large noise amplification can saturate either the latter stages of $G_c(s)$ or the early process stages. In subsequent sections, we can add a pre-filter in a two-degree-of-freedom configuration to help achieve the design goals.

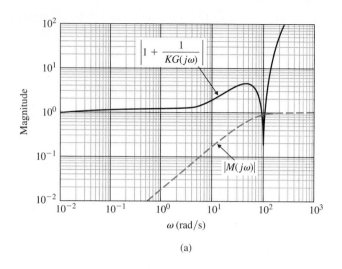

(a)

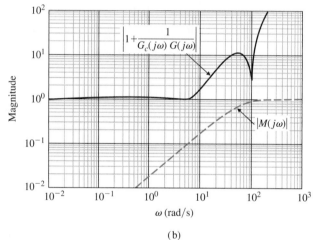

FIGURE 12.10
The robust stability
criterion for
Example 12.3.

(b)

12.4 SYSTEMS WITH UNCERTAIN PARAMETERS

Many systems have several parameters that are constants but uncertain within a range. For example, consider a system with a characteristic equation

$$s^n + a_{n-1}s^{n-1} + a_{n-2}s^{n-2} + \cdots + a_0 = 0 \tag{12.16}$$

with known coefficients within bounds

$$\alpha_i \le a_i \le \beta_i \text{ and } i = 0, \ldots, n,$$

where $a_n = 1$.

To ascertain the stability of the system, we might have to investigate all possible combinations of parameters. Fortunately, it is possible to investigate a limited number of worst-case polynomials [22]. The analysis of only four polynomials is sufficient, and they are readily defined for a third-order system with a characteristic equation

$$s^3 + a_2 s^2 + a_1 s + a_0 = 0. \tag{12.17}$$

Then the four polynomials are

$$q_1(s) = s^3 + \alpha_2 s^2 + \beta_1 s + \beta_0,$$
$$q_2(s) = s^3 + \beta_2 s^2 + \alpha_1 s + \alpha_0,$$
$$q_3(s) = s^3 + \beta_2 s^2 + \beta_1 s + \alpha_0,$$
$$q_4(s) = s^3 + \alpha_2 s^2 + \alpha_1 s + \beta_0.$$

One of the four polynomials represents the **worst case** and may indicate either unstable performance or at least the worst performance for the system in that case.

EXAMPLE 12.4   Third-order system with uncertain coefficients

Consider a third-order system with uncertain coefficients such that

$$8 \le a_0 \le 60 \Rightarrow \alpha_0 = 8, \beta_0 = 60;$$
$$12 \le a_1 \le 100 \Rightarrow \alpha_1 = 12, \beta_1 = 100;$$
$$7 \le a_2 \le 25 \Rightarrow \alpha_2 = 7, \beta_2 = 25.$$

The four polynomials are

$$q_1(s) = s^3 + 7s^2 + 100s + 60,$$
$$q_2(s) = s^3 + 25s^2 + 12s + 8,$$
$$q_3(s) = s^3 + 25s^2 + 100s + 8,$$
$$q_4(s) = s^3 + 7s^2 + 12s + 60.$$

We then proceed to check these four polynomials by means of the Routh–Hurwitz criterion, and hence we determine that the system is stable for all the range of uncertain parameters. ■

EXAMPLE 12.5   Stability of uncertain system

Consider a unity feedback system with a process transfer function (under nominal conditions)

$$G(s) = \frac{4.5}{s(s + 1)(s + 2)}.$$

The nominal characteristic equation is then

$$q(s) = s^3 + 3s^2 + 2s + 4.5 = 0.$$

Using the Routh–Hurwitz criterion, we find that this system is nominally stable. However, if the system has uncertain coefficients such that

$$4 \le a_0 \le 5 \Rightarrow \alpha_0 = 4, \beta_0 = 5;$$
$$1 \le a_1 \le 3 \Rightarrow \alpha_1 = 1, \beta_1 = 3; \text{ and}$$
$$2 \le a_2 \le 4 \Rightarrow \alpha_2 = 2, \beta_2 = 4,$$

then we must examine the four polynomials:

$$q_1(s) = s^3 + 2s^2 + 3s + 5,$$
$$q_2(s) = s^3 + 4s^2 + 1s + 4,$$
$$q_3(s) = s^3 + 4s^2 + 3s + 4,$$
$$q_4(s) = s^3 + 2s^2 + 1s + 5.$$

Using the Routh–Hurwitz criterion, $q_1(s)$ and $q_3(s)$ are stable and $q_2(s)$ is marginally stable. For $q_4(s)$, we have

$$
\begin{array}{c|cc}
s^3 & 1 & 1 \\
s^2 & 2 & 5 \\
s^1 & -3/2 & \\
s^0 & 5 &
\end{array}
$$

Therefore, the system is unstable for the worst case, where $\alpha_2 = $ minimum, $\alpha_1 = $ minimum, and $\beta_0 = $ maximum. This occurs when the process has changed to

$$G(s) = \frac{5}{s(s+1)(s+1)}.$$

Note that the third pole has moved toward the $j\omega$-axis to its limit at $s = -1$ and that the gain has increased to its limit at $K = 5$. Often, we are able to examine the transfer function $G(s)$ and predict the worst-case conditions. ∎

12.5 THE DESIGN OF ROBUST CONTROL SYSTEMS

The design of robust control systems is based on two tasks: determining the structure of the controller and adjusting the controller's parameters to give an "optimal" system performance. This design process is normally done with "assumed complete knowledge" of the process. Furthermore, the process is normally described by a linear time-invariant continuous model. The structure of the controller is chosen such that the system's response can meet certain performance criteria.

One possible objective in the design of a control system is that the controlled system's output should exactly and instantaneously reproduce its input. That is, the system transfer function should be unity:

$$T(s) = \frac{Y(s)}{R(s)} = 1. \tag{12.18}$$

In other words, the system should be presentable on a Bode gain versus frequency diagram with a 0-dB gain of infinite bandwidth and zero phase shift. In practice, this is not possible, since every system will contain inductive- and capacitive-type components that store energy in some form. These elements and their interconnections with energy-dissipative components produce the system's dynamic response characteristics. Such systems reproduce some inputs almost exactly, while other inputs are not reproduced at all, signifying that the system bandwidth is less than infinite.

Once we recognize that the system dynamics cannot be ignored, we need a new design objective. One possible design objective is to maintain the magnitude response curve as flat and as close to unity for as large a bandwidth as possible for a given plant and controller combination [22].

Another important goal of a control system design is that the effect on the output of the system due to disturbances is minimized. Thus, we wish to minimize $Y(s)/T_d(s)$ over a range of frequency.

Consider the control system shown in Figure 12.11, where $G(s) = G_1(s)G_2(s)$ is the plant and $T_d(s)$ is the disturbance. We then have

$$T(s) = \frac{Y(s)}{R(s)} = \frac{G_c(s)G_1(s)G_2(s)}{1 + G_c(s)G_1(s)G_2(s)}, \quad (12.19)$$

and

$$\frac{Y(s)}{T_d(s)} = \frac{G_2(s)}{1 + G_c(s)G_1(s)G_2(s)}. \quad (12.20)$$

Note that both the reference and disturbance transfer functions have the same denominator; in other words, they have the same characteristic equation—namely,

$$1 + G_c(s)G_1(s)G_2(s) = 1 + L(s) = 0. \quad (12.21)$$

Recall that the sensitivity of $T(s)$ with respect to $G(s)$ is

$$S_G^T = \frac{1}{1 + G_c(s)G_1(s)G_2(s)}, \quad (12.22)$$

and the characteristic equation is the influencing factor on the sensitivity. Equation (12.22) shows that for low sensitivity S, we require a high value of loop gain $L(j\omega)$, but it is known that a high gain could cause instability or poor responsiveness of $T(s)$. Thus, we seek the following:

1. $T(s)$ with wide bandwidth and faithful reproduction of $R(s)$.
2. Large loop gain $L(s)$ in order to minimize sensitivity S.
3. Large loop gain $L(s)$ attained primarily by $G_c(s)G_1(s)$, since $Y(s)/T_d(s) \simeq 1/G_c(s)G_1(s)$.

Setting the design of robust systems in frequency-domain terms, we must find a proper compensator $G_c(s)$ such that the closed-loop sensitivity is less than some tolerance value. But sensitivity minimization involves finding a proper compensator such that the closed-loop sensitivity equals or is arbitrarily close to the minimal attainable

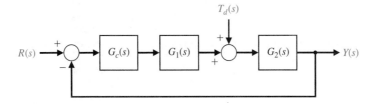

FIGURE 12.11
A system with a disturbance.

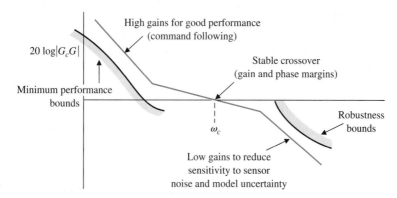

FIGURE 12.12
Bode diagram for
$20 \log |G_c(j\omega)G(j\omega)|$.

sensitivity. Similarly, the gain margin problem is to find a proper compensator to achieve some prescribed gain margin. But gain margin maximization involves finding a proper compensator to achieve the maximal attainable gain margin. For the frequency-domain specifications, we require the following conditions for the Bode diagram of $G_c(j\omega)G(j\omega)$, shown in Figure 12.12:

1. For relative stability, $G_c(j\omega)G(j\omega)$ must have, for an adequate range of ω, not more than a -20-dB/decade slope at or near the crossover frequency ω_c.
2. Steady-state accuracy achieved by the low frequency gain.
3. Accuracy over a bandwidth ω_B, by maintaining $|G_c(j\omega)G(j\omega)|$ above a prescribed level.
4. Disturbance rejection by a high gain for $G_c(j\omega)$ over the system bandwidth.

Using the root sensitivity concept, we can state that S_a^r must be minimized while attaining $T(s)$ with dominant roots that will provide the appropriate response and minimize the effect of $T_d(s)$. Again, we see that the goal is to have the gain of the loop primarily attained by $G_c(s)$. As an example, let $G_c(s) = K$, $G_1(s) = 1$, and $G_2(s) = 1/(s(s + 1))$ for the system in Figure 12.11. This system has two roots, and we select a gain K so that $Y(s)/T_d(s)$ is minimized, S_K^r is minimized, and $T(s)$ has desirable dominant roots. The sensitivity is

$$S_K^r = \frac{dr}{dK} \cdot \frac{K}{r} = \frac{ds}{dK}\bigg|_{s=r} \cdot \frac{K}{r}, \tag{12.23}$$

and the characteristic equation is

$$s(s + 1) + K = 0. \tag{12.24}$$

Therefore, $dK/ds = -(2s + 1)$, since $K = -s(s + 1)$. We then obtain

$$S_K^r = \frac{-1}{2s + 1} \frac{-s(s + 1)}{s}\bigg|_{s=r}. \tag{12.25}$$

When $\zeta < 1$, the roots are complex and $r = -0.5 + j\omega$. Then,

$$|S_K^r| = \left(\frac{0.25 + \omega^2}{4\omega^2}\right)^{1/2}. \tag{12.26}$$

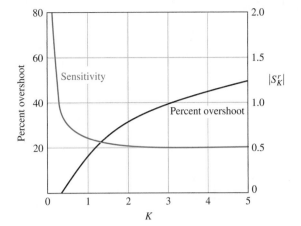

FIGURE 12.13
Sensitivity and percent overshoot for a second-order system.

The magnitude of the sensitivity is plotted in Figure 12.13 for $K = 0.2$ to $K = 5$. The percent overshoot to a step is also shown. It is best to reduce the sensitivity while limiting K to 1.5 or less. We then attain the majority of the attainable reduction in sensitivity while maintaining good performance for the step response. In general, we can use the design procedure as follows:

1. Sketch the root locus of the compensated system with $G_c(s)$ chosen to attain the desired location for the dominant roots.
2. Maximize the gain of $G_c(s)$ to reduce the effect of the disturbance.
3. Determine S_a^r and attain the minimum value of the sensitivity consistent with the transient response required, as described in Step 1.

EXAMPLE 12.6   Sensitivity and compensation

Let us consider again the system in Example 10.1 when $G(s) = 1/s^2$, $H(s) = 1$, and $G_c(s)$ is to be selected by frequency response methods. Therefore, the compensator is to be selected to achieve an appropriate gain and phase margin while minimizing sensitivity and the effect of the disturbance. Thus, we choose

$$G_c(s) = \frac{K(s/z + 1)}{s/p + 1}. \tag{12.27}$$

As in Example 10.1, we choose $K = 10$ to reduce the effect of the disturbance. To attain a phase margin of $45°$, we select $z = 2.0$ and $p = 12.0$. We then attain the compensated diagram shown in Figure 10.9 and repeated in Figure 12.14. Recall that the closed-loop bandwidth is $\omega_B = 1.6\omega_c$. Thus, we will increase the bandwidth by using the compensator and improve the fidelity of reproduction of the input signals.

The sensitivity at ω_c may be ascertained as

$$|S_G^T(j\omega_c)| = \left| \frac{1}{1 + G_c(j\omega)G(j\omega)} \right|_{\omega_c}. \tag{12.28}$$

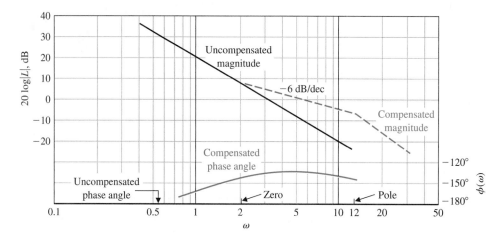

FIGURE 12.14
Bode diagram for
Example 12.6.

To estimate $|S_G^T|$, we recall that the Nichols chart enables us to obtain

$$|T(j\omega)| = \left| \frac{G_c(j\omega)G(j\omega)}{1 + G_c(j\omega)G(j\omega)} \right|. \tag{12.29}$$

Thus, we can plot a few points of $G_c(j\omega)G(j\omega)$ on the Nichols chart and then read $T(\omega)$ from the chart. Then

$$|S_G^T(j\omega_1)| = \frac{|T(j\omega_1)|}{|G_c(j\omega_1)G(j\omega_1)|}, \tag{12.30}$$

where ω_1 is chosen arbitrarily as $\omega_c/2.5$. In general, we choose a frequency below ω_c to determine the value of $|S(\omega_1)|$. Of course, we desire a low value of sensitivity. The Nichols chart for the compensated system is shown in Figure 12.15. For $\omega_1 = \omega_c/2.5 = 2$, we have $20 \log |T(j\omega_1)| = 2.5$ dB and $20 \log |G_c(j\omega_1)G(j\omega_1)| = 9$ dB. Therefore,

$$|S(j\omega_1)| = \frac{|T(j\omega_1)|}{|G_c(j\omega_1)G(j\omega_1)|} = \frac{1.33}{2.8} = 0.47. \ \blacksquare$$

EXAMPLE 12.7   **Sensitivity with a lead compensator**

Let us again consider the system in Example 12.6, using the root locus design obtained in Example 10.3. The compensator was chosen as

$$G_c(s) = \frac{8.1(s + 1)}{s + 3.6}, \tag{12.31}$$

for the system of Figure 12.16. The dominant roots are thus $s = -1 \pm j2$. Because the gain is 8.1, the effect of the disturbance is reduced, and the time response meets the specifications. The sensitivity at a root r may be obtained by assuming that the

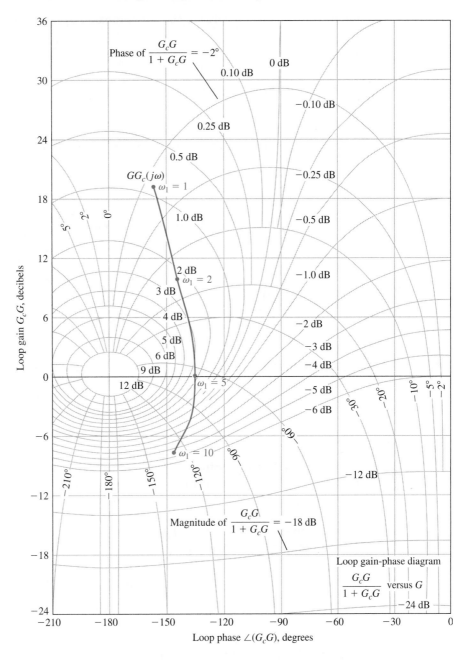

FIGURE 12.15
Nichols chart for
Example 12.7.

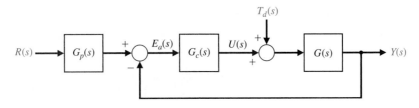

FIGURE 12.16
Feedback control
system with a
desired input $R(s)$
and an undesired
input $T_d(s)$.

system, with dominant roots, may be approximated by the second-order system

$$T(s) = \frac{K}{s^2 + 2\zeta\omega_n s + K} = \frac{K}{s^2 + 2s + K},$$

since $\zeta\omega_n = 1$. The characteristic equation is thus

$$s^2 + 2s + K = 0.$$

Then $dK/ds = -(2s + 2)$, since $K = -(s^2 + 2s)$. Therefore,

$$S_K^r = \frac{-1}{2s + 2} \cdot \frac{-(s^2 + 2s)}{1}\bigg|_{s=r} = \frac{s(s + 2)}{(2s + 2)}\bigg|_{s=r}, \tag{12.32}$$

where $r = -1 + j2$. Then, substituting $s = r$, we obtain

$$|S_K^r| = 1.25.$$

If we raise the gain to in Equation (12.31) from 8.1 to 10, we expect $r \simeq -1.1 \pm j2.4$. Then the sensitivity is

$$|S_K^r| = 1.4. \quad \blacksquare$$

12.6 THE DESIGN OF ROBUST PID-CONTROLLED SYSTEMS

The PID controller has the transfer function

$$G_c(s) = K_P + \frac{K_I}{s} + K_D s.$$

The popularity of PID controllers can be attributed partly to their robust performance in a wide range of operating conditions and partly to their functional simplicity, which allows engineers to operate them in a simple straightforward manner. To implement such a controller, three parameters must be determined for the given process: proportional gain, integral gain, and derivative gain [35].

Consider the PID controller

$$G_c(s) = K_P + \frac{K_I}{s} + K_D s = \frac{K_D s^2 + K_P s + K_I}{s}$$

$$= \frac{K_D(s^2 + as + b)}{s} = \frac{K_D(s + z_1)(s + z_2)}{s}, \tag{12.33}$$

where $a = K_P/K_D$ and $b = K_I/K_D$. Therefore, a PID controller introduces a transfer function with one pole at the origin and two zeros that can be located anywhere in the left-hand s-plane.

Recall that a root locus begins at the poles and ends at the zeros. If we have a system as shown in Figure 12.16 with

$$G(s) = \frac{1}{(s + 2)(s + 5)},$$

and we use a PID controller with complex zeros, we can plot the root locus as shown in Figure 12.17. As the gain K_D of the controller is increased, the complex roots approach the zeros. The closed-loop transfer function is

$$T(s) = \frac{G(s)G_c(s)G_p(s)}{1 + G(s)G_c(s)}$$

$$= \frac{K_D(s + z_1)(s + \hat{z}_1)}{(s + r_2)(s + r_1)(s + \hat{r}_1)}G_p(s)$$

$$\simeq \frac{K_D G_p(s)}{s + r_2}, \tag{12.34}$$

because the zeros and the complex roots are approximately equal ($r_1 \approx z_1$). Setting $G_p(s) = 1$, we have

$$T(s) = \frac{K_D}{s + r_2} \approx \frac{K_D}{s + K_D} \tag{12.35}$$

when $K_D \gg 1$. The only limiting factor is the allowable magnitude of $U(s)$ (Figure 12.16) when K_D is large. If K_D is 100, the system has a fast response and zero steady-state error. Furthermore, the effect of the disturbance is reduced significantly.

In general, we note that PID controllers are particularly useful for reducing steady-state error and improving the transient response when $G(s)$ has one or two poles (or may be approximated by a second-order process).

The selection of the three coefficients of PID controllers is basically a search problem in a three-dimensional space. Points in the search space correspond to different selections of a PID controller parameters. By choosing different points of the parameter space, we can produce, for example, different step responses for a step input. A PID controller can be determined by moving in this search space on a trial-and-error basis.

The main problem in the selection of the three coefficients is that these coefficients do not readily translate into the desired performance and robustness characteristics that the control system designer has in mind. Several rules and methods

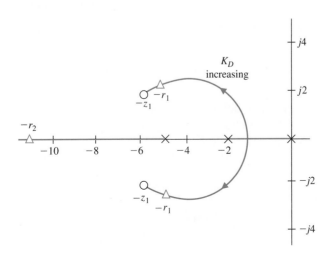

FIGURE 12.17
Root locus with
$-z_1 = -6 + j2$.

have been proposed to solve this problem. In this section, we consider several design methods using root locus and performance indices.

The first design method uses the ITAE performance index of Section 5.7 and the optimum coefficients of Table 5.6 for a step input or Table 5.7 for a ramp input. Hence, we select the three PID coefficients to minimize the ITAE performance index, which produces an excellent transient response to a step (Figure 5.30c) or a ramp. The design procedure consists of three steps:

1. Select the ω_n of the closed-loop system by specifying the settling time.

2. Determine the three coefficients using the appropriate optimum equation (Table 5.6) and the ω_n of step 1 to obtain $G_c(s)$.

3. Determine a prefilter $G_p(s)$ so that the closed-loop system transfer function, $T(s)$, does not have any zeros, as required by Equation (5.47).

EXAMPLE 12.8   **Robust control of temperature**

Consider a temperature controller with a control system as shown in Figure 12.16 and a process

$$G(s) = \frac{1}{(s + 1)^2}. \tag{12.36}$$

If $G_c(s) = 1$, the steady-state error is 50%, and the settling time (with a 2% criterion) is 3.2 seconds for a step input. We want to obtain an optimum ITAE performance for a step input and a settling time of less than 0.5 second. Using a PID controller, we have

$$G_c(s) = \frac{K_D s^2 + K_P s + K_I}{s}. \tag{12.37}$$

Therefore, the closed-loop transfer function without prefiltering $[G_p(s) = 1]$ is

$$T_1(s) = \frac{Y(s)}{R(s)} = \frac{G_c(s)G(s)}{1 + G_c(s)G(s)}$$

$$= \frac{K_D s^2 + K_P s + K_I}{s^3 + (2 + K_D)s^2 + (1 + K_P)s + K_I}. \tag{12.38}$$

The optimum coefficients of the characteristic equation for ITAE are obtained from Table 5.6 as

$$s^3 + 1.75\omega_n s^2 + 2.15\omega_n^2 s + \omega_n^3. \tag{12.39}$$

We need to select ω_n in order to meet the settling time requirement. Since $T_s = 4/(\zeta\omega_n)$ and ζ is unknown but near 0.8, we set $\omega_n = 10$. Then, equating the denominator of Equation (12.38) to Equation (12.39), we obtain the three coefficients as $K_P = 214$, $K_D = 15.5$, and $K_I = 1000$.

Then Equation (12.38) becomes

$$T_1(s) = \frac{15.5s^2 + 214s + 1000}{s^3 + 17.5s^2 + 215s + 1000}$$

$$= \frac{15.5(s + 6.9 + j4.1)(s + 6.9 - j4.1)}{s^3 + 17.5s^2 + 215s + 1000}. \tag{12.40}$$

The response of this system to a step input has an overshoot of 32%, as recorded in Table 12.2.

We select a prefilter $G_p(s)$ so that we achieve the desired ITAE response with

$$T(s) = \frac{G_c(s)G(s)G_p(s)}{1 + G_c(s)G(s)} = \frac{1000}{s^3 + 17.5s^2 + 215s + 1000}. \tag{12.41}$$

Therefore, we require that

$$G_p(s) = \frac{64.5}{s^2 + 13.8s + 64.5} \tag{12.42}$$

in order to eliminate the zeros in Equation (12.40) and bring the overall numerator to 1000. The response of the system $T(s)$ to a step input is indicated in Table 12.2. The system has a small overshoot, a settling time of less than $\frac{1}{2}$ second, and zero steady-state error. Furthermore, for a disturbance $T_d(s) = 1/s$, the maximum value of $y(t)$ due to the disturbance is 0.4% of the magnitude of the disturbance. This is a very favorable design. ∎

EXAMPLE 12.9   **Robust system design**

Let us consider again the system in Example 12.8 when the plant varies significantly, so that

$$G(s) = \frac{K}{(\tau s + 1)^2}, \tag{12.43}$$

where $0.5 \le \tau \le 1$ and $1 \le K \le 2$. We want to achieve robust behavior using an ITAE optimum system with a prefilter while attaining an overshoot of less than 4% and a settling time (with a 2% criterion) of less than 2 seconds, while $G(s)$ can attain any value in the range indicated. We select $\omega_n = 8$ in order to attain the settling time and determine the ITAE coefficients for $K = 1$ and $\tau = 1$. Completing the calculation, we obtain the system without a prefilter $[G_p(s) = 1]$ as

$$T_1(s) = \frac{12(s^2 + 11.38s + 42.67)}{s^3 + 14s^2 + 137.6s + 512}, \tag{12.44}$$

Table 12.2   Results for Example 12.8

Controller	$G_c(s) = 1$	PID and $G_p(s) = 1$	PID with $G_p(s)$ Prefilter
Percent overshoot	0	31.7%	1.9%
Settling time (seconds)	3.2	0.20	0.45
Steady-state error	50.1%	0.0%	0.0%
Disturbance error	52%	0.4%	0.4%

and

$$G_c(s) = \frac{12(s^2 + 11.38s + 42.67)}{s}. \tag{12.45}$$

We select a prefilter

$$G_p(s) = \frac{42.67}{s^2 + 11.38s + 42.67} \tag{12.46}$$

to obtain the optimum ITAE transfer function

$$T(s) = \frac{512}{s^3 + 14s^2 + 137.6s + 512}. \tag{12.47}$$

We then obtain the step response for the four conditions: $\tau = 1$, $K = 1$; $\tau = 0.5$, $K = 1$; $\tau = 1$, $K = 2$; and $\tau = 0.5$, $K = 2$. The results are summarized in Table 12.3. This is a very robust system. ∎

The value of ω_n that can be chosen will be limited by considering the maximum allowable $u(t)$, where $u(t)$ is the output of the controller, as shown in Figure 12.16. If the maximum value of $e_a(t)$ is 1, then $u(t)$ would normally be limited to 100 or less. As an example, consider the system in Figure 12.16 with a PID controller, $G(s) = 1/(s(s + 1))$, and the necessary prefilter $G_p(s)$ to achieve ITAE performance. If we select $\omega_n = 10, 20$, and 40, the maximum value of $u(t)$ is as recorded in Table 12.4. If we wish to limit $u(t)$ to a maximum equal to 100, we need to limit ω_n to 16. Thus, we are limited in the settling time we can achieve.

Let us consider the design of a PID compensator using frequency response techniques for a system with a time delay so that

$$G(s) = \frac{Ke^{-Ts}}{\tau s + 1}. \tag{12.48}$$

This type of system represents many industrial processes that incorporate a time delay. We use a PID compensator to introduce two equal zeros so that

$$G_c(s) = \frac{K_I(\tau_1 s + 1)^2}{s}. \tag{12.49}$$

Table 12.3   Results for Example 12.9 with $\omega_n = 8$

Plant Conditions	$\tau = 1$ $K = 1$	$\tau = 0.5$, $K = 1$	$\tau = 1$, $K = 2$	$\tau = 0.5$, $K = 2$
Percent overshoot	2%	0%	0%	1%
Settling time (seconds)	1.25	0.8	0.8	0.9

Table 12.4   Maximum Value of Plant Input

ω_n	10	20	40
$u(t)$ maximum for $R(s) = 1/s$	35	135	550
Settling time (seconds)	0.9	0.5	0.3

The design method is as follows:

1. Plot the uncompensated Bode diagram for $K_I G(s)/s$ with a gain K_I that satisfies the steady-state error requirement.

2. Place the two equal zeros at or near the crossover frequency ω_c.

3. Test the results and adjust K_I or the zero locations, if necessary.

EXAMPLE 12.10   PID control of a system with a delay

Consider the system of Figure 12.16 when

$$G(s) = \frac{Ke^{-0.1s}}{0.1s + 1}, \tag{12.50}$$

where $K = 20$ is selected to achieve a small steady-state error for a step input, and where $G_p(s) = 1$. We want an overshoot to a step input of less than 5%.

Plotting the Bode diagram for $G(j\omega)$, we find that the uncompensated system has a negative phase margin and that the system is unstable.

We will use a PID controller of the form of Equation (12.49) to attain a desirable phase margin of 70°. Then the loop transfer function is

$$G_c(s)G(s) = \frac{20e^{-0.1s}(\tau_1 s + 1)^2}{s(0.1s + 1)}, \tag{12.51}$$

where $K_I K = 20$. We plot the Bode diagram without the two zeros, as shown in Figure 12.18. The phase margin is $-32°$, and the system is unstable prior to the introduction of the zeros.

Because we have introduced a pole at the origin due to the integration term in the PID compensator, we may reduce the gain $K_I K$ because e_{ss} is now zero. We

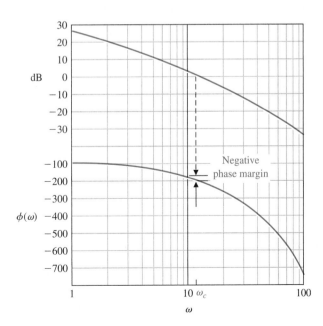

FIGURE 12.18
Bode diagram for
$G(s)/s$ for Example
12.10.

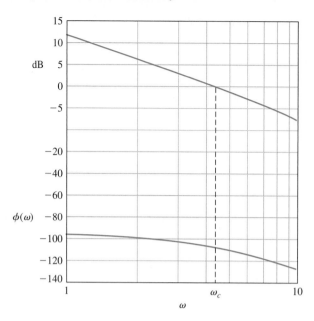

FIGURE 12.19
Bode diagram for
$G_c(s)G(s)$ for
Example 12.10.

place the two zeros at or near the crossover $\omega_c = 11$. We choose to set $\tau_1 = 0.06$ so that the two zeros are set at $\omega = 16.7$. Also, we reduce the gain to $K_I K = 4.5$. Then we obtain the frequency response shown in Fig. 12.19, where

$$G_c(s)G(s) = \frac{4.5(0.06s + 1)^2 e^{-0.1s}}{s(0.1s + 1)}. \tag{12.52}$$

The new crossover frequency is $\omega_c = 4.5$, and the phase margin is $70°$. The step response of this system has no overshoot and has a settling time (with a 2% criterion) of 0.80 second. This response satisfies the requirements. However, if we wanted to adjust the system further, we could raise $K_I K$ to 10 and achieve a somewhat faster response with an overshoot of less than 5%. ∎

As a final consideration of the design of robust control systems using a PID controller, we turn to an s-plane root locus method. This design approach may be simply stated as follows:

1. Place the poles and zeros of $G(s)/s$ on the s-plane.

2. Select a location for the zeros of $G_c(s)$ that will result in an acceptable root locus and suitable dominant roots.

3. Test the transient response of the compensated system and iterate Step 2, if necessary.

12.7 THE ROBUST INTERNAL MODEL CONTROL SYSTEM

The internal model control system is shown in Figure 12.20 and was previously considered in Section 11.8. We now consider again the use of the internal model design with special attention to robust system performance. The **internal model principle**

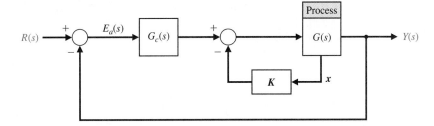

FIGURE 12.20
The internal model
control system.

states that if $G_c(s)G(s)$ contains $R(s)$ then $y(t)$ will track $r(t)$ asymptotically (in the steady state), and the tracking is robust.

Examining the system of Figure 12.20, we note that for lower-order processes, state variable feedback will not be required, and a suitable $G_c(s)$ can be obtained. However, with higher-order systems, the feedback of all state variables may be required.

Consider a simple system with $G(s) = 1/s$, for which we seek a ramp response with a steady-state error of zero. A PI controller is sufficient, and we let $\mathbf{K} = \mathbf{0}$ (no state variable feedback). Then we have

$$G_c(s)G(s) = \left(K_p + \frac{K_I}{s}\right)\frac{1}{s} = \frac{K_ps + K_I}{s^2}. \tag{12.53}$$

Note that for a ramp, $R(s) = 1/s^2$, which is contained as a factor of Equation (12.53), and the closed-loop transfer function is

$$T(s) = \frac{K_ps + K_I}{s^2 + K_ps + K_I}. \tag{12.54}$$

Using the ITAE specifications for a ramp response (Table 5.7), we require that

$$T(s) = \frac{3.2\omega_n s + \omega_n^2}{s^2 + 3.2\omega_n s + \omega_n^2}. \tag{12.55}$$

We select ω_n to satisfy a specification for the settling time. For a settling time (with a 2% criterion) of 1 second, we select $\omega_n = 5$. Then we require $K_p = 16$ and $K_I = 25$. The response of this system settles in 1 second and then tracks the ramp with zero steady-state error. If this system (designed for a ramp input) receives a step input, the response has an overshoot of 5% and a settling time of 1.5 seconds. This system is very robust to changes in the plant. For example, if $G(s) = K/s$ changes gain so that K shifts from $K = 1$ by $\pm50\%$, the change in the ramp response is insignificant.

EXAMPLE 12.11   **Design of an internal model control system**

Consider the system of Figure 12.21 with state variable feedback and a compensator $G_c(s)$. We wish to track a step input with zero steady-state error. Here, we select a PID controller for $G_c(s)$. We then have

$$G_c(s) = \frac{K_Ds^2 + K_Ps + K_I}{s},$$

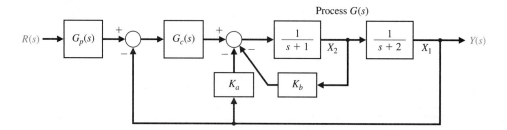

FIGURE 12.21
An internal model control with state variable feedback and $G_c(s)$.

and $G(s)G_c(s)$ will contain $R(s) = 1/s$, the input command. Note that we feed back both state variables and add these additional signals after $G_c(s)$ in order to retain the integrator in $G_c(s)$.

The goal is to achieve a settling time (to within 2% of the final value) in less than 1 second and a deadbeat response (see Section 10.11) while retaining a robust response. Here, we assume that the two poles of $G(s)$ can change by $\pm50\%$. Then the worst-case condition is

$$\hat{G}(s) = \frac{1}{(s + 0.5)(s + 1)}.$$

One design approach is to design the control for this worst-case condition. Another approach, which we use here, is to design for the nominal $G(s)$ and one-half the desired settling time. Then we expect to meet the settling time requirement and attain a very fast, highly robust system. Note that the prefilter $G_P(s)$ is used to attain the desired form for $T(s)$.

The response desired is deadbeat (see Table 10.2), so we use a third-order transfer function as

$$T(s) = \frac{\omega_n{}^3}{s^3 + 1.9\omega_n s^2 + 2.20\omega_n{}^2 s + \omega_n{}^3}, \tag{12.56}$$

and the settling time (with a 2% criterion) is $T_s = 4.04/\omega_n$. For a settling time of $\frac{1}{2}$ second, we use $\omega_n = 8.08$.

The closed-loop transfer function of the system of Figure 12.21 with the appropriate $G_P(s)$ is

$$T(s) = \frac{K_I}{s^3 + (3 + K_D + K_b)s^2 + (2 + K_P + K_a + 2K_b)s + K_I}. \tag{12.57}$$

We let $K_a = 10$, $K_b = 2$, $K_P = 127.6$, $K_I = 527.5$, and $K_D = 10.35$. Note that $T(s)$ could be achieved with other gains, including $K_b = 0$.

The step response of this system has a deadbeat response with an overshoot of 1.65% and a settling time of 0.5 second. When the poles of $G(s)$ change by $\pm50\%$, the overshoot changes to 1.86%, and the settling time becomes 0.95 second. This is an outstanding design of a very robust, deadbeat response system. ■

12.8 DESIGN EXAMPLES

In this section we present five illustrative examples. In the first example, an aircraft autopilot is analyzed using root locus methods. In the second example, a PI controller and a PID controller are designed for a space telescope control system in the presence of time delays. The third example is the design of a robust bobbin drive using robust PID controller design approach with ITAE optimal performance objectives. The fourth example illustrates the design of two degree-of-freedom controllers (that is, two separate controllers) for an ultra-precision diamond turning machine. In the fifth and final design example, we consider the practical problem of designing a controller in the presence of an uncertain time delay. The specific problem under investigation is a PID controller for a digital audio tape drive. The design process is highlighted with an emphasis on robustness.

EXAMPLE 12.12   Aircraft autopilot

A typical aircraft autopilot control system consists of electrical, mechanical, and hydraulic devices that move the flaps, elevators, fuel-flow controllers, and other components that cause the aircraft to vary its flight. Sensors provide information on velocity, heading, rate of rotation, and other flight data. This information is combined with the desired flight characteristics (commands) available electronically to the autopilot. The autopilot should be able to fly the aircraft on a heading and under conditions set by the pilot. The command often consists of a predetermined heading. Design often focuses on a forward-moving aircraft that moves somewhat up or down without moving right or left and without rolling (tipping the wings). Such a study is called pitch axis design. The aircraft is represented by a process [26]

$$G(s) = \frac{K}{s(s + 1/\tau)(s^2 + 2\zeta_1\omega_1 s + \omega_1^2)}, \tag{12.58}$$

where τ is the time constant of the actuator. Let $\tau = \frac{1}{4}$, $\omega_1 = 2$, and $\zeta = \frac{1}{2}$. Then the s-plane plot has two complex poles, a pole at the origin, and a pole at $s = -4$, as shown in Figure 12.22. The complex poles, representing the aircraft dynamics, can vary within the dashed-line box shown in the figure. We then choose the zeros of the controller as $s = -1.3 \pm j2$, as shown. We select the gain K so that the roots r_2 and $\hat{r}_2$ are complex with a ζ of $1/\sqrt{2}$. The other roots, r_1 and $\hat{r}_1$, lie very near the zeros. Therefore, the closed-loop transfer function is approximately

$$T(s) \approx \frac{\omega_n^2}{s^2 + 2\zeta\omega_n s + \omega_n^2} = \frac{5}{s^2 + 3.16s + 5}, \tag{12.59}$$

with $\omega_n = \sqrt{5}$ and $\zeta = 1/\sqrt{2}$. The resulting response to a step input has an overshoot of 4.5% and a settling time (with a 2% criterion) of 2.5 seconds, as expected. ∎

EXAMPLE 12.13   Space telescope control system

Scientists have proposed the operation of a space vehicle as a space-based research laboratory and test bed for equipment to be used on a manned space station. The industrial space facility (ISF) would remain in space, and the astronauts would be able

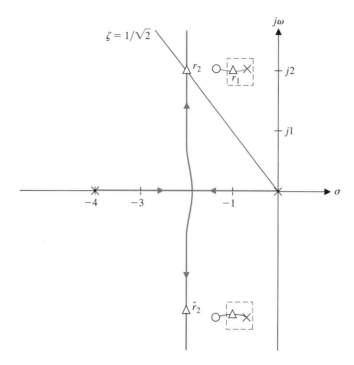

FIGURE 12.22
Root locus for aircraft autopilot. The complex poles can vary within the dashed-line box.

to use it only when the shuttle is attached [16, 21]. The ISF will be the first permanent, human-operated commercial space facility designed for R&D, testing, and, eventually, processing in the space environment.

We will consider an experiment operated in space but controlled from Earth. The goal is to manipulate and position a small telescope to accurately point at a planet. We want to have a steady-state error equal to zero, while maintaining a fast response to a step with an overshoot of less than 5%. The actuator chosen is a low-power actuator, and the model of the combined actuator and telescope is shown in Figure 12.23. The command signal is received from an Earth station with a delay of $\pi/16$ seconds. A sensor will measure the pointing direction of the telescope accurately. However, this measurement is relayed back to Earth with a delay of $\pi/16$ seconds. Thus, the total transfer function of the telescope, actuator, sensor, and round-trip delay (Figure 12.24) is

$$G(s) = \frac{e^{-s\pi/8}}{(s + 1)^2}.$$ (12.60)

We propose a PID controller where

$$G_c(s) = K_P + \frac{K_I}{s} + K_D s = \frac{K_P s + K_I + K_D s^2}{s}.$$ (12.61)

FIGURE 12.23
Model of a low-power actuator and telescope.

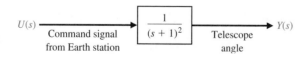

$U(s) \longrightarrow$ | Command signal from Earth station | $\dfrac{1}{(s + 1)^2}$ Telescope angle | $\longrightarrow Y(s)$

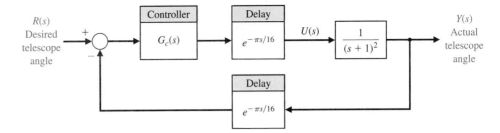

FIGURE 12.24
Feedback control system for the telescope experiment.

The use of only the proportional term will not be acceptable since we require a steady-state error of zero for a step input. Thus, we must use a nonzero value of K_I, and hence we may elect to use either a proportional plus integral control (PI) or a proportional plus integral plus derivative control (PID).

We will first try PI control, so that

$$G_c(s) = K_P + \frac{K_I}{s} = \frac{K_P s + K_I}{s}. \tag{12.62}$$

Since we have a pure delay e^{-sT}, we use the frequency response methods for the design process. Thus, we will translate the overshoot specification to the frequency domain. If we have two dominant characteristic roots, the overshoot to a step is 5% when $\zeta = 0.7$, or the phase margin requirement is about 70°.

If we choose $K_P = 0.022$ and $K_I = 0.22$, we have

$$G_c(s)G(s) = \frac{0.22(0.1s + 1)e^{-s\pi/8}}{s(s + 1)^2}, \tag{12.63}$$

and the Bode diagram is shown in Figure 12.25. The location of the zero at $s = -10$ was chosen to add a phase lead angle in order to attain the desired phase margin. An iterative procedure yields a series of trials for K_1 and K_2 until the desired phase margin is achieved. Note that we have achieved a phase margin of about 63°. The actual step response was plotted, and we determine that the overshoot was 4.7% with a settling time (with a 2% criterion) of 16 seconds, as recorded in Table 12.5.

The proportional plus integral plus derivative controller is

$$G_c(s) = \frac{K_P s + K_I + K_D s^2}{s}. \tag{12.64}$$

We now have three parameters to vary to achieve the desired phase margin. If we select, after some iteration, $K_P = 0.8$, $K_I = 0.5$, and $K_D = 10^{-3}$, we obtain a phase margin of 64°. The percentage overshoot is 3.7%, and the settling time (with a 2% criterion) is 5.8 seconds. Perhaps the easiest way to select the gain constants is to let K_D be a small, but nonzero, number initially and $K_P = K_I = 0$, then plot the frequency response. In this case, we choose $K_D = 10^{-3}$ and obtain a Bode plot. We then use $K_P \approx K_I$ and iterate to obtain the appropriate values of these unspecified gains.

The performance of the PI- and the PID-compensated systems is recorded in Table 12.5. The PID controller is the most desirable, since it provides a shorter settling time. ∎

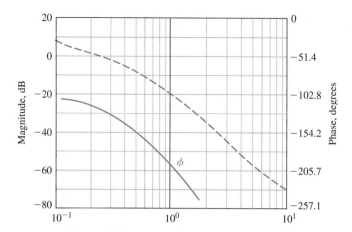

FIGURE 12.25
Bode diagram for the system with the PI controller.

Table 12.5   Step Response of the Space Telescope for Two Controllers

	Steady-State Error	Percent Overshoot	Settling Time (seconds)
PI controller	0	4.7	16.0
PID controller	0	3.7	5.8

EXAMPLE 12.14   **Robust bobbin drive**

Monofilament nylon is produced by an extrusion process that outputs filament at a constant rate. The product is wound onto a bobbin that rotates at a maximum speed of 2000 rpm. The tension in the filament must be held between 0.2 and 0.6 pound to ensure that it is not stretched. The winding diameter varies between 2 to 4 inches.

The filament is laid onto the bobbin by a ballscrew-driven arm that oscillates back and forth at constant speed, as shown in Figure 12.26(a). The arm must reverse rapidly at the end of the move. The required ballscrew speed is 60 rpm. The prime requirement of the bobbin drive is to provide a controlled tension. Since the winding diameter varies by 2 to 1, the tension will fall by 50% from start to finish.

The control system will have a system structure as shown in Figure 12.26(b), for which we select a PID controller. The parameter variations are $1.5 \leq K_m \leq 2.5$ and $3 \leq p \leq 5$ with the nominal conditions $K_m = 2$ and $p = 4$. Furthermore, a third pole at $s = -50$ has been omitted from the model. The requirements are an overshoot less than 2.5% and a settling time (with a 2% criterion) less than 0.4 second. The magnitude of $u(t)$ must be less than 100.

Using a PID controller, the ITAE design, and the nominal parameters, we determine ω_n from the settling time requirement. Since we expect that $\zeta \approx 0.8$, we use

$$T_s = \frac{4}{0.8\omega_n} < 0.4.$$

We select $\omega_n = 23$ as the maximum allowable for $|u| < 100$. Then, for

$$G_c(s) = K_P + \frac{K_I}{s} + K_D s,$$

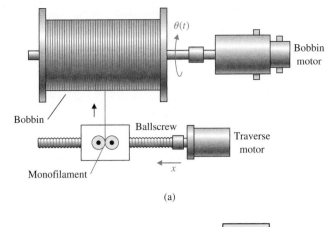

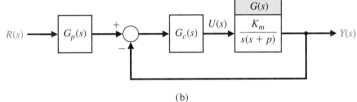

FIGURE 12.26
A monofilament
bobbin winder.

we obtain $K_P = 568.68$, $K_I = 6083.5$, and $K_D = 18.13$. Using the appropriate pre-filter, we obtain the response recorded in Table 12.6. The system does not offer robust performance since the overshoot requirement is not satisfied when the worst-case parameters are considered.

We also examine the performance of the system with the nominal parameters but with the unmodeled pole added, so that the actual process is

$$G(s) = \frac{2(50)}{s(s + 4)(s + 50)}. \tag{12.65}$$

Table 12.6   Response of the Bobbin Drive System for a Unit Step Input (original design)

	Parameters	Percent Overshoot	Settling Time	$\left\lvert\dfrac{u(t)}{r(t)}\right\rvert_{\text{maximum}}$
Nominal parameters	$K_m = 2$, $p = 4$	1.96%	0.318	98
Worst-case parameters	$K_m = 1.5$, $p = 3$	7.48%	0.375	95
Nominal parameters and added third pole at $s = -50$		9.82%	0.732	90

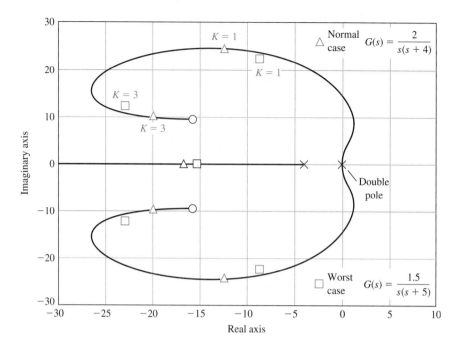

FIGURE 12.27
Root locus for the normal case and the worst case for $K = 1$ and $K = 3$.

(Note that the DC gain, $\lim_{s \to 0} sG(s)$, remains 0.5.) The response of the PID controller with the added pole is recorded in Table 12.6. Again, the system fails the requirement of robust performance.

We need to adjust the system so that the performance with the worst-case parameters is acceptable. Examine the root locus for the nominal parameters shown in Figure 12.27. Insert a cascade gain K prior to $G_c(s)$ so that we have $KG_c(s)G(s)$. Then the roots for $K = 1$ and $K = 3$ are shown on the locus. Since the worst-case response occurs when the motor constant K_m drops to 1.5, we use the cascade gain $K = 3$ to move the roots to the left on the s-plane. Then, when the gain K_m drops to 1.5, the roots still are in the desired region. The response of the system with $K = 3$ is recorded in Table 12.7 for the nominal and worst-case conditions, as well as with the added pole. This system meets all the specifications. This approach uses a cascade gain that, when adjusted correctly, will drive the dominant roots near the complex zeros of the PID controller. Then, when the worst parameter change occurs, the system will still maintain the required performance. ■

EXAMPLE 12.15   Ultra-precision diamond turning machine

The design of an ultra-precision diamond turning machine has been studied at Lawrence Livermore National Laboratory. This machine shapes optical devices such as mirrors with ultra-high precision using a diamond tool as the cutting device. In this discussion, we will consider only the z-axis control. Using frequency response identification with sinusoidal input to the actuator we determined that

$$G(s) = \frac{4500}{s + 60}. \tag{12.66}$$

Table 12.7   Response of the Bobbin Drive System for a Unit Step with Additional Cascade Gain $K = 3$

	Percent Overshoot	Settling Time (seconds)
Nominal parameters	0.12%	0.218
Worst-case parameters	0.47%	0.214
Nominal parameters and third pole	0.50%	0.242

The system can accommodate high gains, such as 4500, since the input command, $r(t)$, is a series of step commands of very small magnitude (a fraction of a micron). The system has an outer loop for position feedback using a laser interferometer with an accuracy of 0.1 micron (10^{-7} m). An inner feedback loop is also used for velocity feedback, as shown in Figure 12.28.

We want to select the controllers, $G_1(s)$ and $G_2(s)$, to obtain an overdamped, highly robust, high-bandwidth system. The robust system must accommodate changes in $G(s)$ due to varying loads, materials, and cutting requirements. Thus, we seek a large phase margin and gain margin for the inner and outer loops, and low root sensitivity. The specifications are summarized in Table 12.8.

Since we want zero steady-state error for the velocity loop, we use a velocity loop controller $G_2(s) = G_3(s)G_4(s)$, where $G_3(s)$ is a PI controller and $G_4(s)$ is a lead controller. We use

$$G_2(s) = G_3(s)G_4(s) = \left(K_p + \frac{K_I}{s} \right) \cdot \frac{1 + K_4 s}{\alpha \left(1 + \dfrac{K_4}{\alpha} s \right)}$$

and choose $K_P/K_I = 0.00532$, $K_4 = 0.00272$, and $\alpha = 2.95$. We now have

$$G_2(s) = K_P \frac{s + 188}{s} \cdot \frac{s + 368}{s + 1085}.$$

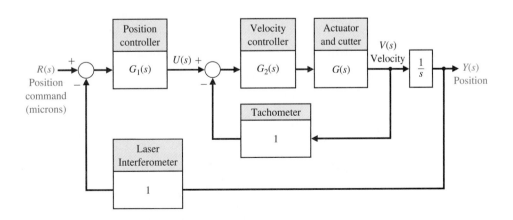

FIGURE 12.28
Turning machine control system.

Table 12.8   Specifications for Turning Machine Control System

| | Transfer Function | |
Specification	Velocity, $V(s)/U(s)$	Position $Y(s)/R(s)$
Minimum bandwidth	950 rad/s	95 rad/s
Steady-state error to a step	0	0
Minimum damping ratio ζ	0.8	0.9
Maximum root sensitivity $|S_K^r|$	1.0	1.5
Minimum phase margin	90°	75°
Minimum gain margin	40 dB	60 dB

The root locus for $G_2(s)G(s)$ is shown in Figure 12.29. When $K_P = 2$, we have, for the velocity closed-loop transfer function,

$$T_2(s) = \frac{V(s)}{U(s)} = \frac{9000(s + 188)(s + 368)}{(s + 205)(s + 305)(s + 10^4)} \approx \frac{10^4}{(s + 10^4)}, \qquad (12.67)$$

which is a large-bandwidth system. The actual bandwidth and root sensitivity are summarized in Table 12.9. Note that we have exceeded the specifications for the velocity transfer function.

We will use a lead network for the position loop of the form

$$G_1(s) = K_1 \frac{1 + K_5 s}{\alpha\left(1 + \dfrac{K_5}{\alpha} s\right)},$$

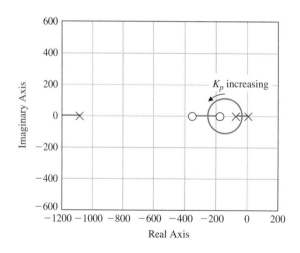

FIGURE 12.29
Root locus for velocity loop as K_p varies.

Table 12.9   Design Results for Turning Machine Control System

Achieved Result	Velocity Transfer Function	Position Transfer Function $Y(s)/R(s)$
Closed-loop bandwidth	4000 rad/s	1000 rad/s
Steady-state error	0	0
Damping ratio, ζ	1.0	1.0
Root sensitivity, $|S_K^r|$	0.92	1.2
Phase margin	93°	85°
Gain margin	Infinite	76 dB

and we choose $\alpha = 2.0$ and $K_5 = 0.0185$ so that

$$G_1(s) = \frac{K_1(s + 54)}{s + 108}.$$

We then plot the root locus for

$$G_1(s) \cdot T_2(s) \cdot \frac{1}{s}.$$

If we use the approximate $T_2(s)$ of Equation (12.67), we have the root locus of Figure 12.30(a). Using the actual $T_2(s)$, we get the close-up of the root locus shown in Figure 12.30(b). We select $K_P = 1000$ and achieve the actual results for the total system transfer function as recorded in Table 12.9. The total system has a high phase margin, has a low sensitivity, and is overdamped with a large bandwidth. This system is very robust. ∎

EXAMPLE 12.16   **Digital audio tape controller**

Consider the feedback control system shown in Figure 12.31, where

$$G_d(s) = e^{-Ts}.$$

The exact value of the time delay is uncertain, but it is known to lie in the interval $T_1 \le T \le T_2$. For example, if a robot on Mars is being remotely controlled from earth, the time it takes the signals to reach the planetary robot is not precisely known since transient time depends on the distance between the transmitter and the planetary robot, the atmospheric medium through which the signals travel, interplanetary space effects, and so forth—all of which are time varying and cannot be precisely modeled.
    Define

$$G_m(s) = e^{-Ts}G(s).$$

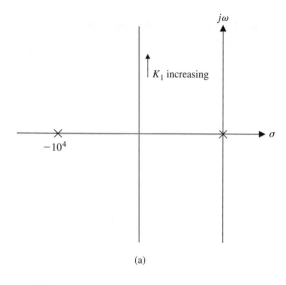

(a)

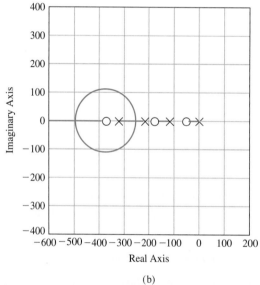

FIGURE 12.30
The root locus for $K_1 > 0$ for (a) overview and (b) close-up near origin of the s-plane.

(b)

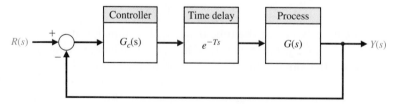

FIGURE 12.31
A feedback system with a time delay in the loop.

Then

$$G_m(s) - G(s) = e^{-Ts}G(s) - G(s) = (e^{-Ts} - 1)G(s),$$

or

$$\frac{G_m(s)}{G(s)} - 1 = e^{-Ts} - 1.$$

If we define

$$M(s) = e^{-Ts} - 1,$$

then we have

$$G_m(s) = (1 + M(s))G(s). \tag{12.68}$$

In the development of a robust stability controller, we would like to represent the time-delay uncertainty in the form shown in Figure 12.32 where we need to determine a function $M(s)$ that approximately models the time delay. This will lead to the establishment of a straightforward method of testing the system for stability robustness in the presence of the uncertain time-delay. The uncertainty model is known as a multiplicative uncertainty representation, as discussed in Section 12.3.

Since we are concerned with stability, we can consider $R(s) = 0$. Then we can manipulate the block diagram in Figure 12.32 to obtain the form shown in Figure 12.33. Using the so-called small gain theorem, we have the condition that the closed-loop system is stable if

$$|M(j\omega)|\left|\frac{G_c(j\omega)G(j\omega)}{1 + G_c(j\omega)G(j\omega)}\right| < 1,$$

or equivalently (see Equation (12.15))

$$|M(j\omega)| < \left|1 + \frac{1}{G_c(j\omega)G(j\omega)}\right| \quad \text{for all } \omega.$$

The problem is that the time delay T is not known exactly. One approach to solving the problem is to find a weighting function, denoted by $W(s)$, such that

$$|e^{-j\omega T} - 1| < |W(j\omega)| \quad \text{for all } \omega \text{ and } T_1 \le T \le T_2.$$

If $W(s)$ satisfies the above inequality, it follows that

$$|M(j\omega)| < |W(j\omega)|.$$

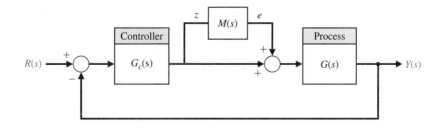

FIGURE 12.32
Multiplicative
uncertainty
representation.

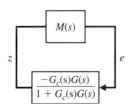

FIGURE 12.33
Equivalent block
diagram depiction
of the multiplicative
uncertainty.

Therefore, the robust stability condition can be satisfied by

$$\left| W(j\omega) \right| < \left| 1 + \frac{1}{G_c(j\omega)G(j\omega)} \right| \quad \text{for all } \omega. \qquad (12.69)$$

This is a conservative bound. If the condition in Eq. (12.69) is satisfied, then stability is guaranteed in the presence of any time delay in the range $T_1 \le T \le T_2$ [5], [36]. If the condition is not satisfied, the system may or may not be stable.

Suppose we have an uncertain time delay that is known to lie in the range $0.1 \le T \le 1$. We can determine a suitable weighting function $W(s)$ by plotting the magnitude of $e^{-j\omega T} - 1$, as shown in Figure 12.34 for various values of T in the range $T_1 \le T \le T_2$. A reasonable weighting function obtained by trial and error is

$$W(s) = \frac{2.5s}{1.2s + 1}.$$

This function satisfies the condition

$$\left| e^{-j\omega T} - 1 \right| < \left| W(j\omega) \right|.$$

Keep in mind that the selection of the weighting function is not unique.

A digital audio tape (DAT) stores 1.3 gigabytes of data in a package the size of a credit card—roughly nine times more than a half-inch-wide reel-to-reel tape or quarter-inch-wide cartridge tape. A DAT sells for about the same amount as a floppy disk, even though it can store 1000 times more data. A DAT can record for two

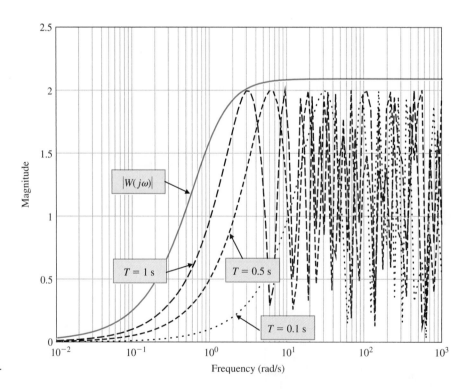

FIGURE 12.34
Magnitude plot of $\left| e^{-j\omega T} - 1 \right|$ for $T = 0.1, 0.5,$ and 1.

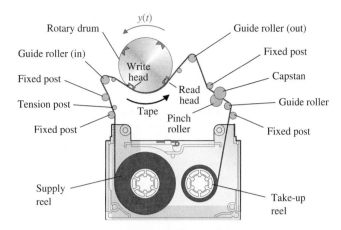

FIGURE 12.35
Digital audio tape
driver mechanism.

hours (longer than either reel-to-reel or cartridge tape), which means that it can run longer unattended and requires fewer changes and hence fewer interruptions of data transfer. DAT gives access to a given data file within 20 seconds, on the average, compared with several minutes for either cartridge or reel-to-reel tape [2].

The tape drive electronically controls the relative speeds of the drum and tape so that the heads follow the tracks on the tape, as shown in Figure 12.35. The control system is much more complex than that for a CD-ROM because more motors have to be accurately controlled: capstan, take-up and supply reels, drum, and tension control. The elements of the design process emphasized in this example are highlighted in Figure 12.36.

Consider the speed control system shown in Figure 12.37. The motor and load transfer function varies because the tape moves from one reel to the other. The transfer function is

$$G(s) = \frac{K_m}{(s + p_1)(s + p_2)}, \tag{12.70}$$

where nominal values are $K_m = 4$, $p_1 = 1$, and $p_2 = 4$.

However, the range of variation is $3 \le K_m \le 5, 0.5 \le p_1 \le 1.5$, and $3.5 \le p_2 \le 4.5$. Thus, the process belongs to a family of processes, where each member corresponds to different values of K_m, p_1, and p_2. The design goal is

Design Goal
Control the DAT speed to the desired value in the presence of significant process uncertainties.

Associated with the design goal we have the variable to be controlled defined as the tape speed:

Variable to Be Controlled
DAT speed $Y(s)$.

The design specifications are

Design Specifications

DS1   Percent overshoot less than 13% and settling time less than 2 seconds for a unit step input.

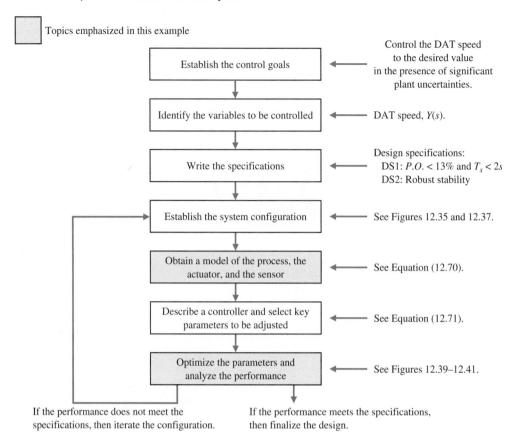

Topics emphasized in this example

FIGURE 12.36   Elements of the control system design process emphasized in this digital audio tape speed control design.

DS2   Robust stability in the presence of a time delay at the plant input. The time delay value is uncertain but known to be in the range $0 \leq T \leq 0.1$.

Design specification DS1 must be satisfied for all process in the family. Design specification DS2 must be satisfied by the nominal process ($K_m = 4$, $p_1 = 1$, $p_2 = 4$).
    The following constraints on the design are given:

❏ Fast peak time requires that an overdamped condition is not acceptable.

❏ Use a PID controller:

$$G_c(s) = K_P + \frac{K_I}{s} + K_D s. \tag{12.71}$$

❏ $K_m K_D \leq 20$ when $K_m = 4$.

The key tuning parameters are the PID gains:

Select Key Tuning Parameters
    K_P, K_I, and K_D.

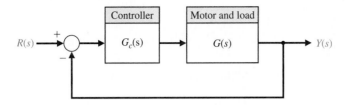

FIGURE 12.37
Block diagram of
the digital audio
tape speed control
system.

Since we are constrained to have $K_m K_D \le 20$ when $K_m = 4$, we must select $K_D \le 5$. We will design the PID controller using nominal values for K_m, p_1, and p_2. We will analyze the performance of the controlled system for the various values of the process parameters, using a simulation to check that DS1 is satisfied. The nominal process is given by

$$G(s) = \frac{4}{(s + 1)(s + 4)}.$$

The closed-loop transfer function is

$$T(s) = \frac{4K_D s^2 + 4K_P s + 4K_I}{s^3 + (5 + 4K_D)s^2 + (4 + 4K_P)s + 4K_I}.$$

If we choose $K_D = 5$, then we write the characteristic equation as

$$s^3 + 25s^2 + 4s + 4(K_P s + K_I) = 0,$$

or

$$1 + \frac{4K_P(s + K_I/K_P)}{s(s^2 + 25s + 4)} = 0.$$

Per specifications, we try to place the dominant poles in the region defined by $\zeta \omega_n > 2$ and $\zeta > 0.55$. We need to select a value of $\tau = K_I/K_P$, and then we can plot the root locus with the gain $4K_P$ as the varying parameter. After several iterations, we choose a reasonable value of $\tau = 3$. The root locus is shown in Figure 12.38. We determine that $4K_P \approx 120$ represents a valid selection since the roots lie inside the desired performance region. This value of $4K_P$ has been rounded off from the exact value on the root locus plot of $4K_P = 121.7683$. We obtain $K_P = 30$, and $K_I = \tau K_P = 90$. The PID controller is then given by

$$G_c(s) = 30 + \frac{90}{s} + 5s. \qquad (12.72)$$

The step response (for the process with nominal parameter values) is shown in Figure 12.39. A family of responses is shown in Figure 12.40 for various values of

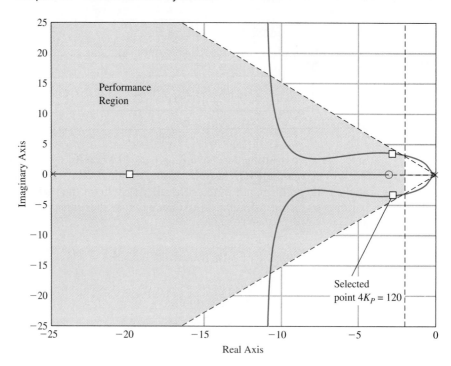

FIGURE 12.38
Root locus for the
DAT system with
$K_D = 5$ and
$\tau = K_I/K_P = 3$.

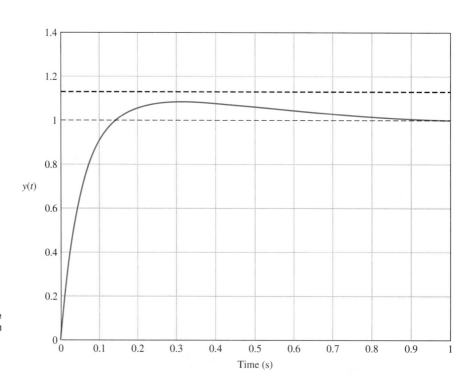

FIGURE 12.39
Unit step response
for the DAT system
with $K_P = 30$,
$K_D = 5$, and
$K_I = 90$.

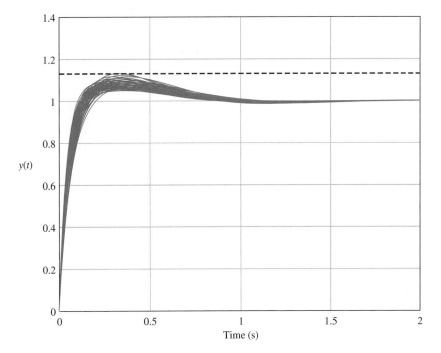

FIGURE 12.40
A family of step responses for the DAT system for various values of the process parameters K_m, p_1, and p_2.

K_m, p_1, and p_2. None of the responses suggests a percent overshoot over the specified value of 13%, and the settling times are all under the 2 second specification as well. As we can see in Figure 12.40, all of the tested processes in the family are adequately controlled by the single PID controller in Equation (12.72). Therefore DS1 is satisfied for all processes in the family.

Suppose the system has a time delay at the input to the process. The actual time delay is uncertain but known to be in the range $0 \leq T \leq 0.1$ s. Following the method discussed previously, we determine that a reasonable function $W(s)$ which bounds the plots of $|e^{-j\omega T} - 1|$ for various values of T is

$$W(s) = \frac{0.29s}{0.28s + 1}.$$

To check the stability robustness property, we need to verify that

$$|W(j\omega)| < \left| 1 + \frac{1}{G_c(j\omega)G(j\omega)} \right| \quad \text{for all } \omega. \qquad (12.73)$$

The plot of both $|W(j\omega)|$ and $\left| 1 + \dfrac{1}{G_c(j\omega)G(j\omega)} \right|$ is shown in Figure 12.41. It can be seen that the condition in Equation (12.73) is indeed satisfied. Therefore, we expect that the nominal system will remain stable in the presence of time-delays up to 0.1 seconds. ∎

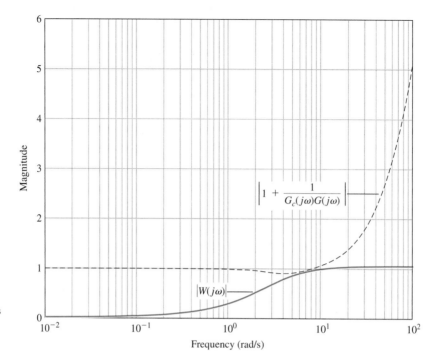

$$\left| 1 + \frac{1}{G_c(j\omega)G(j\omega)} \right|$$

$|W(j\omega)|$

Magnitude

Frequency (rad/s)

FIGURE 12.41
Stability robustness
to a time delay of
uncertain
magnitude.

12.9 THE PSEUDO-QUANTITATIVE FEEDBACK SYSTEM

Quantitative feedback theory (QFT) uses a controller, as shown in Figure 12.42, to achieve robust performance. The goal is to achieve a wide bandwidth for the closed-loop transfer function with a high loop gain K. Typical QFT design methods use graphical and numerical methods in conjunction with the Nichols chart. Generally, QFT design seeks a high loop gain and large phase margin so that robust performance is achieved [27–29, 32].

In this section, we pursue a simple method of achieving the goals of QFT with an s-plane, root locus approach to the selection of the gain K and the compensator $G_c(s)$. This approach, dubbed pseudo-QFT, follows these steps:

1. Place the n poles and m zeros of $G(s)$ on the s-plane for the nth order $G(s)$. Also, add any poles of $G_c(s)$.

2. Starting near the origin, place the zeros of $G_c(s)$ immediately to the left of each of the $(n - 1)$ poles on the left-hand s-plane. This leaves one pole far to the left of the left-hand side of the s-plane.

3. Increase the gain K so that the roots of the characteristic equation (poles of the closed-loop transfer function) are close to the zeros of $G_c(s)G(s)$.

This method introduces zeros so that all but one of the root loci end on finite zeros. If the gain K is sufficiently large, then the poles of $T(s)$ are almost equal to the zeros of $G_c(s)G(s)$. This leaves one pole of $T(s)$ with a significant partial fraction residue and the system with a phase margin of approximately 90° (actually about 85°).

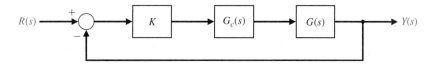

FIGURE 12.42
Feedback system.

EXAMPLE 12.17 **Design using the pseudo-QFT method**

Consider the system of Figure 12.42 with

$$G(s) = \frac{1}{(s + p_1)(s + p_2)},$$

where the nominal case is $p_1 = 1$ and $p_2 = 2$, with $\pm 50\%$ variation. The worst case is with $p_1 = 0.5$ and $p_2 = 1$. We wish to design the system for zero steady-state error for a step input, so we use the PID controller

$$G_c(s) = \frac{(s + z_1)(s + z_2)}{s}.$$

We then invoke the internal model principle, with $R(s) = 1/s$ incorporated within $G_c(s)G(s)$. Using Step 1, we place the poles of $G_c(s)G(s)$ on the s-plane, as shown in Figure 12.43. There are three poles (at $s = 0$, -1, and -2), as shown. Step 2 calls for placing a zero to the left of the pole at the origin and at the pole at $s = -1$, as shown in Figure 12.43.

The compensator is thus

$$G_c(s) = \frac{(s + 0.8)(s + 1.8)}{s}. \tag{12.74}$$

We select $K = 100$, so that the roots of the characteristic equation are close to the zeros. The closed-loop transfer function is

$$T(s) = \frac{100(s + 0.80)(s + 1.80)}{(s + 0.798)(s + 1.797)(s + 100.4)} \approx \frac{100}{s + 100}. \tag{12.75}$$

This closed-loop system provides a fast response and possesses a phase margin of approximately $85°$. The performance is summarized in Table 12.10.

When the worst-case conditions are realized ($p_1 = 0.5$ and $p_2 = 1$), the performance remains essentially unchanged, as shown in Table 12.10. Pseudo-QFT design results in very robust systems. ∎

12.10 ROBUST CONTROL SYSTEMS USING CONTROL DESIGN SOFTWARE

In this section, we will investigate robust control systems using control design software. In particular, we will consider the commonly used PID controller in the feedback control system shown in Figure 12.16. Note that the system has a prefilter $G_p(s)$. The contribution of the prefilter to optimum performance is discussed in Section 10.10.

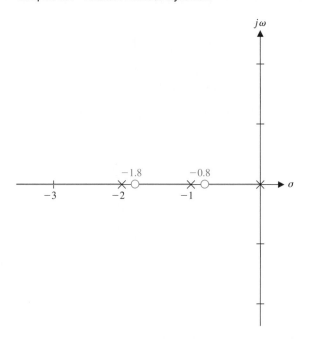

FIGURE 12.43
Root locus for
$KG_c(s)G(s)$.

Table 12.10   Performance of Pseudo-QFT Design

	Percent Overshoot	Settling Time
Nominal $G(s)$	0.01%	40 ms
Worst-case $G(s)$	0.97%	40 ms

The PID controller has the form

$$G_c(s) = \frac{K_D s^2 + K_P s + K_I}{s}.$$

Note that the PID controller is not a proper rational function (i.e., the degree of the numerator polynomial is greater than the degree of the denominator polynomial). The objective is to choose the parameters K_P, K_I, and K_D to meet the performance specifications and have desirable robustness properties. Unfortunately, it is not immediately clear how to choose the parameters in the PID controller to obtain certain robustness characteristics. An illustrative example will show that it is possible to choose the parameters iteratively and verify the robustness by simulation. Using the computer helps in this process, because the entire design and simulation can be automated using scripts and can easily be executed repeatedly.

EXAMPLE 12.18   **Robust control of temperature**

Consider the feedback control system in Figure 12.16, where

$$G(s) = \frac{1}{(s + c_0)^2},$$

and the nominal value is $c_0 = 1$, and $G_p(s) = 1$. We will design a compensator based on $c_0 = 1$ and check robustness by simulation. Our design specifications include

1. A settling time (with a 2% criterion) $T_s \leq 0.5$ s, and

2. An optimum ITAE performance for a step input.

For this design, we will not use a prefilter to meet specification (2), but will instead show that acceptable performance (i.e., low overshoot) can be obtained by increasing a cascade gain.

The closed-loop transfer function is

$$T(s) = \frac{K_D s^2 + K_P s + K_I}{s^3 + (2 + K_D)s^2 + (1 + K_P)s + K_I}. \tag{12.76}$$

The associated root locus equation is

$$1 + \hat{K}\left(\frac{s^2 + as + b}{s^3}\right) = 0,$$

where

$$\hat{K} = K_D + 2, \quad a = \frac{1 + K_P}{2 + K_D}, \quad \text{and} \quad b = \frac{K_I}{2 + K_D}.$$

The settling time requirement $T_s < 0.5$ s leads us to choose the roots of $s^2 + as + b$ to the left of the $s = -\zeta\omega_n = -8$ line in the s-plane, as shown in Figure 12.44, to ensure that the locus travels into the required s-plane region. We have chosen $a = 16$ and $b = 70$ to ensure the locus travels past the $s = -8$ line. We select a point on the root locus in the performance region, and using the rlocfind function, we find the associated gain $\hat{K}$ and the associated value of ω_n. For

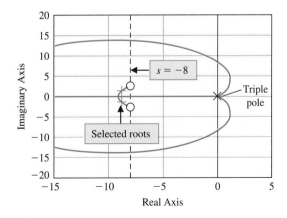

FIGURE 12.44
Root locus for the PID-compensated temperature controller as $\hat{K}$ varies.

```
>>a=16; b=70;  num=[1 a b]; den=[1 0 0 0]; sys=tf(num,den);
>>rlocus(sys)
>>rlocfind(sys)
```

our chosen point, we find that

$$\hat{K} = 118.$$

Then, with $\hat{K}$, a, and b, we can solve for the PID coefficients as follows:

$$K_D = \hat{K} - 2 = 116,$$
$$K_P = a(2 + K_D) - 1 = 1887,$$
$$K_I = b(2 + K_D) = 8260.$$

To meet the overshoot performance requirements for a step input, we will use a cascade gain K that will be chosen by iterative methods using the step function, as illustrated in Figure 12.45. The step response corresponding to $K = 5$ has an acceptable overshoot of 2%. With the addition of the gain $K = 5$, the final PID controller is

$$G_c(s) = K\frac{K_D s^2 + K_P s + K_I}{s} = 5\frac{116s^2 + 1887s + 8260}{s}. \tag{12.77}$$

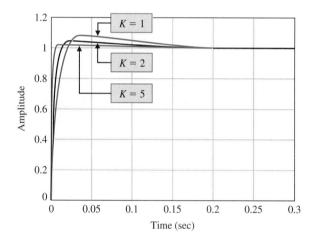

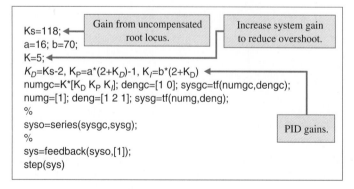

FIGURE 12.45
Step response of the PID temperature controller.

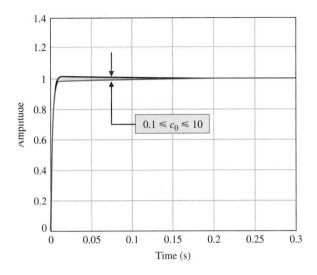

FIGURE 12.46
Robust PID
controller analysis
with variations in c_0.

The figure shows a step response plot with Amplitude on the y-axis (0 to 1.4) and Time (s) on the x-axis (0 to 0.3), with annotation $0.1 \leq c_0 \leq 10$.

```
c0=10                           Specify process parameter.
numg=[1]; deng=[1 2*c0 c0^2];
numgc=5*[116 1887 8260]; dengc=[1 0];
sysg=tf(numg,deng);
sysgc=tf(numgc,dengc);
%
syso=series(sysgc,sysg);
%
sys=feedback(syso,[1]);
%
step(sys)
```

We do not use the prefilter, as in Example 12.8. Instead, we increase the cascade gain K to obtain satisfactory transient response. Now we can consider the question of robustness to changes in the plant parameter c_0.

The investigation into the robustness of the design consists of a step response analysis using the PID controller given in Equation (12.77) for a range of plant parameter variations of $0.1 \leq c_0 \leq 10$. The results are displayed in Figure 12.46. The script is written to compute the step response for a given c_0. It can be convenient to place the input of c_0 at the command prompt level to make the script more interactive.

The simulation results indicate that the PID design is robust with respect to changes in c_0. The differences in the step responses for $0.1 \leq c_0 \leq 10$ are barely discernible on the plot. If the results showed otherwise, it would be possible to iterate on the design until an acceptable performance was achieved. The interactive capability of the m-file allows us to check the robustness by simulation. ∎

12.11 SEQUENTIAL DESIGN EXAMPLE: DISK DRIVE READ SYSTEM

In this section, we will design a PID controller to achieve the desired system response. Many actual disk drive head control systems use a PID controller and use a command signal $r(t)$ that utilizes an ideal velocity profile at the maximum allowable velocity until the head arrives near the desired track, when $r(t)$ is switched to a step-type input. Thus, we want zero steady-state error for a ramp (velocity) signal and a step signal. Examining the system shown in Figure 12.47, we note that the forward path possesses two pure integrations, and we expect zero steady-state error for a velocity input $r(t) = At, t > 0$.

The PID controller is

$$G_c(s) = K_P + \frac{K_I}{s} + K_D s = \frac{K_D(s + z_1)(s + \hat{z}_1)}{s}.$$

The motor field transfer function is

$$G_1(s) = \frac{5000}{(s + 1000)} \approx 5.$$

The second-order model uses $G_1(s) = 5$, and the design is determined for this model.

We use the second-order model and the PID controller for the s-plane design technique illustrated in Section 12.6. The poles and zeros of the system are shown in the s-plane in Figure 12.48 for the second-order model and $G_1(s) = 5$. Then we have

$$G_c(s)G_1(s)G_2(s) = \frac{5K_D(s + z_1)(s + \hat{z}_1)}{s^2(s + 20)}.$$

We select $-z_1 = -120 + j40$ and determine $5K_D$ so that the roots are to the left of the line $s = -100$. If we achieve that requirement, then

$$T_s < \frac{4}{100},$$

and the overshoot to a step input is (ideally) less than 2% since ζ of the complex roots is approximately 0.8. Of course, this sketch is only a first step. As a second

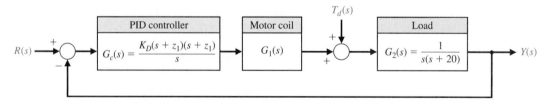

FIGURE 12.47   Disk drive feedback system with a PID controller.

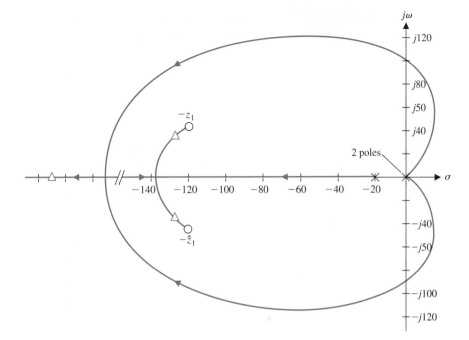

FIGURE 12.48
A sketch of a root locus at K_3 increases for estimated root locations with a desirable system response.

step, we recompute determine K_D. We then obtain the actual root locus as shown in Figure 12.49 with $K_D = 800$. The system response is recorded in Table 12.11. The system meets all the specifications.

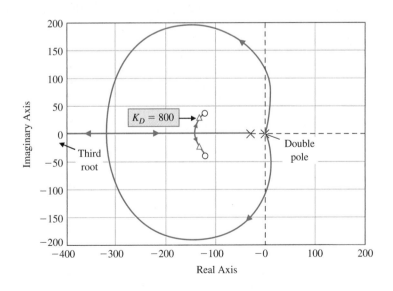

FIGURE 12.49
Actual root locus for the second-order model.

Table 12.11   Disk Drive Control System Specifications and Actual Performance

Performance Measure	Desired Value	Response for Second-Order Model
Percent overshoot	$<5\%$	4.5%
Settling time for step input	<50 ms	6 ms
Maximum response for a unit step disturbance	$<5 \times 10^{-3}$	7.7×10^{-7}

12.12 SUMMARY

The design of highly accurate control systems in the presence of significant plant uncertainty requires the designer to seek a robust control system. A robust control system exhibits low sensitivities to parameter change and is stable over a wide range of parameter variations.

The three-mode, or PID, controller was considered as a compensator to aid in the design of robust control systems. The design issue for a PID controller is the selection of the gain and two zeros of the controller transfer function. We used three design methods for the selection of the controller: the root locus method, the frequency response method, and the ITAE performance index method. An operational amplifier circuit used for a PID controller is shown in Figure 12.50. In general, the use of a PID controller will enable the designer to attain a robust control system.

The internal model control system with state variable feedback and a controller $G_c(s)$ was used to obtain a robust control system. Finally, the robust nature of a pseudo-QFT control system was demonstrated.

$$G_c(s) = \frac{V_0(s)}{V_1(s)} = \frac{R_4 R_2 (R_1 C_1 s + 1)(R_2 C_2 s + 1)}{R_3 R_1 (R_2 C_2 s)}$$

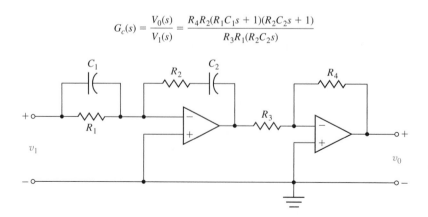

FIGURE 12.50
Operational amplifier circuit used for PID controller.

> **A robust control system provides stable, consistent performance as specified by the designer in spite of the wide variation of plant parameters and disturbances. It also provides a highly robust response to command inputs and a steady-state tracking error equal to zero.**

For systems with uncertain parameters, the need for robust systems will require the incorporation of advanced machine intelligence, as shown in Figure 12.51.

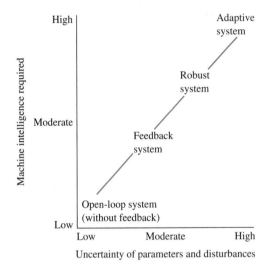

FIGURE 12.51
Intelligence required versus uncertainty for modern control systems.

EXERCISES

E12.1 Consider a system of the form shown in Figure 12.1, where

$$G(s) = \frac{3}{(s + 3)}.$$

Using the ITAE performance method for a step input, determine the required $G_c(s)$. Assume $\omega_n = 30$ for Table 5.6. Determine the step response with and without a prefilter $G_p(s)$.

E12.2 For the ITAE design obtained in Exercise E12.1, determine the response due to a disturbance $T_d(s) = 1/s$.

E12.3 A closed-loop unity feedback system has the loop transfer function

$$G_c(s)G(s) = \frac{18}{s(s + b)},$$

where b is normally equal to 5. Determine S_b^T and plot $|T(j\omega)|$ and $|S(j\omega)|$ on a Bode plot.

Answer: $S_b^T = \dfrac{-bs}{s^2 + bs + 18}$

E12.4 A PID controller is used in the system in Figure 12.1, where

$$G(s) = \frac{1}{(s + 2)(s + 8)}.$$

The gain K_D of the controller (Equation (12.33)) is limited to 180. Select a set of compensator zeros so that the pair of closed-loop roots is approximately equal to the zeros. Find the step response for the approximation in Equation (12.35) and the actual response, and compare them.

E12.5 A system has a process function

$$G(s) = \frac{1000}{s\left(\dfrac{s}{10} + 1\right)\left(\dfrac{s}{20} + 1\right)}$$

and negative unity feedback with a PD compensator

$$G_c(s) = K_p + K_D s.$$

The objective is to design $G_c(s)$ so that the overshoot to a step is less than 5% and the settling time (with a 2% criterion) is less than 50 ms. Find a suitable $G_c(s)$.

E12.6 Consider the control system shown in Figure E12.6 when $G(s) = 1/(s + 5)^2$, and select a PID controller so that the settling time (with a 2% criterion) is less than 1.5 second for an ITAE step response. Plot $y(t)$ for a step input $r(t)$ with and without a prefilter. Determine and plot $y(t)$ for a step disturbance. Discuss the effectiveness of the system.

Answer: One possible controller is

$$G_c(s) = \frac{0.5s^2 + 52.4s + 216}{s}.$$

E12.7 For the control system of Figure E12.6 with $G(s) = 1/(s + 4)^2$, select a PID controller to achieve a settling time (with a 2% criterion) of less than 1.0 second for an ITAE step response. Plot $y(t)$ for a step input $r(t)$ with and without a prefilter. Determine and plot $y(t)$ for a step disturbance. Discuss the effectiveness of the system.

E12.8 Repeat Exercise 12.6, striving to achieve a minimum settling time while adding the constraint that $|u(t)| \leq 80$ for $t > 0$ for a unit step input, $r(t) = 1, t \geq 0$.

Answer: $G_c(s) = \dfrac{3600 + 80s}{s}.$

E12.9 A system has the form shown in Figure E12.6 with

$$G(s) = \frac{K}{s(s + 1)(s + 4)},$$

where $K = 1$. Design a PD controller to place the dominant closed-loop poles at $s = -1.5 \pm j2$. Determine

the step response of the system. Predict the effect of a change in K of $\pm 50\%$. Estimate the step response of the worst-case system.

E12.10 A system has the form shown in Figure E12.6 with

$$G(s) = \frac{K}{s(s + 1)(s + 4)},$$

where $K = 1$. Design a PI controller so that the dominant roots are at $s = -0.365 \pm j0.514$. Determine the step response of the system. Predict the effect of a change in K of $\pm 50\%$. Estimate the step response of the worst-case system.

E12.11 Consider the closed-loop system represented in state variable form

$$\dot{x} = Ax + Br$$
$$y = Cx + Dr,$$

where

$$A = \begin{bmatrix} 0 & 1 \\ -3 & -k \end{bmatrix},$$

$$B = \begin{bmatrix} 0 \\ 1 \end{bmatrix}, C = [3 \ \ 0], \text{ and } D = [0].$$

The nominal value of $k = 2$. However, the value of k can vary in the range $0.1 \leq k \leq 6$. Plot the percent overshoot to a unit step input as k varies from 0.1 to 6.

E12.12 Consider the second-order system

$$\dot{x} = \begin{bmatrix} 0 & 1 \\ -a & -b \end{bmatrix} x + \begin{bmatrix} c_1 \\ c_2 \end{bmatrix} u$$

$$y = [1 \ \ 0]x + [0]u.$$

The parameters $a, b, c_1,$ and c_2 are unknown *a priori*. Under what conditions is the system completely controllable? Select valid values of $a, b, c_1,$ and c_2 to ensure controllability and plot the step response.

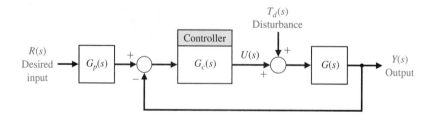

FIGURE E12.6
System with controller.

PROBLEMS

P12.1 Interest in unmanned underwater vehicles (UUVs) has been increasing recently, with a large number of possible applications being considered. These include intelligence-gathering, mine detection, and surveillance applications. Regardless of the intended mission, a strong need exists for reliable and robust control of the vehicle. The proposed vehicle is shown in Figure P12.1(a) [13].

We want to control the vehicle through a range of operating conditions. The vehicle is 30 feet long with a vertical sail near the front. The control inputs are stern-plane, rudder, and shaft speed commands. In this case, we wish to control the vehicle roll by using the stern planes. The control system is shown in Figure P12.1(b), where $R(s) = 0$, the desired roll angle, and $T_d(s) = 1/s$. We select $G_c(s) = K(s + 2)$, where $K = 4$. (a) Plot $20 \log|T|$ and $20 \log|S_K^T|$ on a Bode diagram. (b) Evaluate $|S_K^T|$ at $\omega_B, \omega_{B/2}$, and $\omega_{B/4}$ ($T(s) = Y(s)/R(s)$).

P12.2 A new suspended, mobile, remote-controlled video-camera system to bring three-dimensional mobility to professional NFL football is shown in Figure P12.2(a) [24]. The camera can be moved over the field, as well as up and down. The motor control on each pulley is represented by the system in Figure P12.2(b), where $\tau_1 = 20$ ms and $\tau_2 = 2$ ms.

(a) Select K so that $M_{p\omega} = 1.84$. (b) Plot $20 \log|T|$ and $20 \log|S_K^T|$ on one Bode diagram. (c) Evaluate $|S_K^T|$ at $\omega_B, \omega_{B/2}$, and $\omega_{B/4}$. (d) Let $R(s) = 0$ and determine the effect of $T_d(s) = 1/s$ for the gain K of part (a) by plotting $y(t)$.

P12.3 Magnetic levitation (maglev) trains may replace airplanes on routes shorter than 200 miles. The maglev train developed by a German firm uses electromagnetic attraction to propel and levitate heavy vehicles, carrying up to 400 passengers at 300-mph speeds. But the $\frac{1}{4}$-inch gap between car and track is difficult to maintain [7, 12, 17].

The air-gap control system is shown in Figure P12.3(a). The block diagram of the air-gap control system is shown in Figure P12.3(b). The compensator is

$$G_c(s) = \frac{K(s + 5)}{(s + 10)}.$$

(a) Find the range of K for a stable system. (b) Select a gain so that the steady-state error of the system is zero for a step input command. (c) Find $y(t)$ for the gain of part (b). (d) Find $y(t)$ when K varies $\pm 15\%$ from the gain of part (b).

P12.4 Computer control of a robot to spray-paint an automobile is accomplished by the system shown in

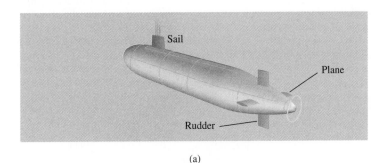

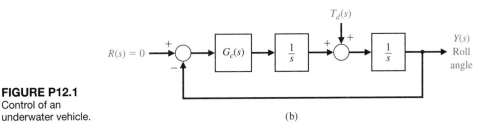

FIGURE P12.1
Control of an underwater vehicle.

(a)

(b)

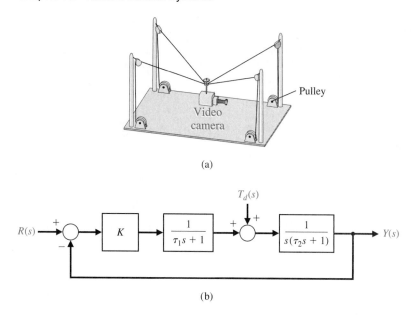

(a)

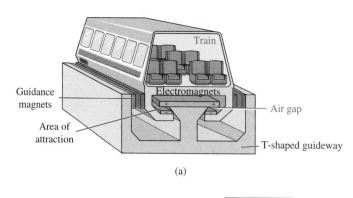

(b)

FIGURE P12.2
Remote-controlled
TV camera.

Figure P12.4(a) [1]. We wish to investigate the system when $K = 1, 10,$ and 20. (a) For the three values of K, determine ζ, ω_n, percent overshoot, settling time (with a 2% criterion), and steady-state error for a step input. Record your results in a table. (b) Determine the sensitivity $|S_K^r|$ for the three values of K. (c) Select the best of the three values of K. (d) For the value selected in part (c), determine $y(t)$ for a disturbance $T_d(s) = 1/s$ when $R(s) = 0$.

P12.5 An automatically guided vehicle is shown in Figure P12.5(a) and its control system is shown in Figure P12.5(b). The goal is to track the guide wire accurately, to be insensitive to changes in the gain K_1, and to reduce the effect of the disturbance [15, 25]. The gain K_1 is normally equal to 1 and $\tau_1 = 1/25$ second.
(a) Select a compensator $G_c(s)$ so that the percent overshoot to a step input is less than or equal to 10%, the settling time (with a 2% criterion) is less

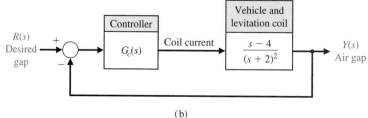

(a)

FIGURE P12.3
Maglev train
control.

(b)

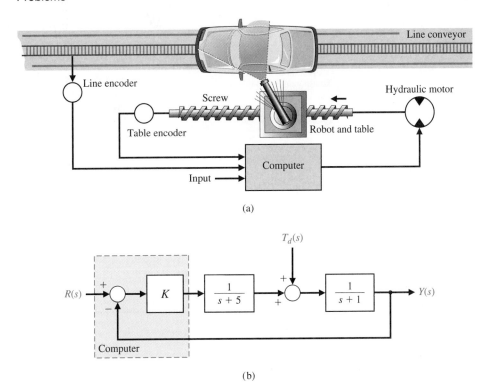

FIGURE P12.4
Spray-paint robot.

(a)

(b)

than 100 ms, and the velocity constant K_v for a ramp input is 100.

(b) For the compensator selected in part (a), determine the sensitivity of the system to small changes in K_1 by determining $S^r_{K_1}$ or $S^T_{K_1}$.

(c) If K_1 changes to 2 while $G_c(s)$ of part (a) remains unchanged, find the step response of the system and compare selected performance figures with those obtained in part (a).

(d) Determine the effect of $T_d(s) = 1/s$ by plotting $y(t)$ when $R(s) = 0$.

P12.6 A roll-wrapping machine (RWM) receives, wraps, and labels large paper rolls produced in a paper mill [9, 16]. The RWM consists of several major stations: positioning station, waiting station, wrapping station, and so forth. We will focus on the positioning station shown in Figure P12.6(a). The positioning station is the first station that sees a paper roll. This station is responsible for receiving and weighing the roll, measuring its diameter and width, determining the desired wrap for the roll, positioning it for downstream processing, and finally ejecting it from the station.

Functionally, the RWM can be categorized as a complex operation because each functional step (e.g., measuring the width) involves a large number of field device actions and relies upon a number of accompanying sensors.

The control system for accurately positioning the width-measuring arm is shown in Figure P12.6(b). The negative pole p of the positioning arm is normally equal to 2, but it is subject to change because of loading and misalignment of the machine. (a) For $p = 2$, design a compensator so that the complex roots are $s = -2 \pm j2\sqrt{3}$. (b) Plot $y(t)$ for a step input $R(s) = 1/s$. (c) Plot $y(t)$ for a disturbance $T_d(s) = 1/s$, with $R(s) = 0$. (d) Repeat parts (b) and (c) when p changes to 1 and $G_c(s)$ remains as designed in part (a). Compare the results for the two values of the negative pole p.

P12.7 The function of a steel plate mill is to roll reheated slabs into plates of scheduled thickness and dimension [5, 10]. The final products are of rectangular plane view shapes having a width of up to 3300 mm and a thickness of 180 mm.

A schematic layout of the mill is shown in Figure P12.7(a). The mill has two major rolling stands, denoted No. 1 and No. 2. These are equipped with large rolls (up to 508 mm in diameter), which are driven by high-power electric motors (up to 4470 kW). Roll gaps and forces are maintained by large hydraulic cylinders.

Typical operation of the mill can be described as follows. Slabs coming from the reheating furnace initially go through the No. 1 stand, whose function is to reduce the slabs to the required width. The slabs proceed

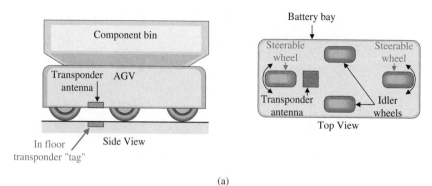

(a)

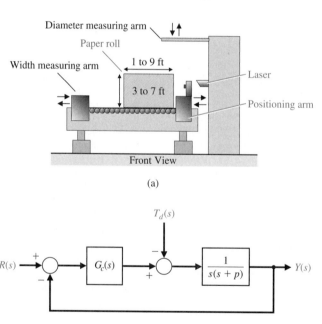

FIGURE P12.5
Automatically
guided vehicle.

(b)

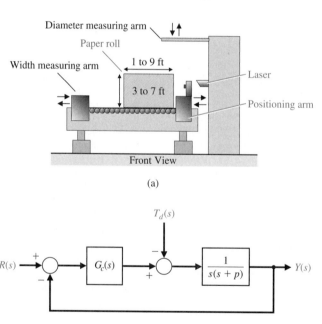

(a)

FIGURE P12.6
Roll-wrapping
machine control.

(b)

through the No. 2 stand, where finishing passes are carried out to produce the required slab thickness. Finally, they go through the hot plate leveller, which gives each plate a smooth finish.

One of the key systems controls the thickness of the plates by adjusting the rolls. The block diagram of this control system is shown in Figure P12.7(b). The plant is represented by

$$G(s) = \frac{1}{s(s^2 + 4s + 5)}.$$

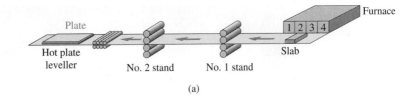

(a)

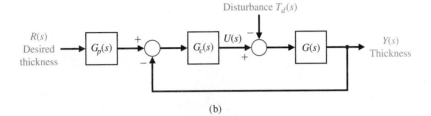

(b)

FIGURE P12.7
Steel-rolling mill
control.

The controller is a PID with two equal real zeros. (a) Select the PID zeros and the gains so that the closed-loop system has two pairs of equal roots. (b) For the design of part (a), obtain the step response without a prefilter $(G_p(s) = 1)$. (c) Repeat part (b) for an appropriate prefilter. (d) For the system, determine the effect of a unit step disturbance by evaluating $y(t)$ with $r(t) = 0$.

P12.8 A motor and load with negligible friction and a voltage-to-current amplifier K_a is used in the feedback control system, shown in Figure P12.8. A designer selects a PID controller

$$G_c(s) = K_P + \frac{K_I}{s} + K_D s,$$

where $K_P = 5$, $K_I = 500$, and $K_D = 0.0475$.
(a) Determine the appropriate value of K_a so that the phase margin of the system is 30°. (b) For the gain K_a, plot the root locus of the system and determine the roots of the system for the K_a of part (a). (c) Determine the maximum value of $y(t)$ when $T_d(s) = 1/s$ and $R(s) = 0$ for the K_a of part (a). (d) Determine the response to a step input $r(t)$, with and without a prefilter.

P12.9 A unity feedback system has a nominal characteristic equation

$$q(s) = s^3 + 3s^2 + 3s + 6 = 0.$$

The coefficients vary as follows:

$$2 \le a_2 \le 4, \quad 1 \le a_1 \le 4,$$
$$4 \le a_0 \le 5.$$

Determine whether the system is stable for these uncertain coefficients.

P12.10 Future astronauts may drive on the moon in a pressurized vehicle, shown in Figure P12.10(a), that would have a range of 620 miles and could be used for missions of up to six months. Boeing Company engineers first analyzed the Apollo-era Lunar Roving Vehicle, then designed the new vehicle, incorporating improvements in radiation and thermal protection, shock and vibration control, and lubrication and sealants.

The steering control of the moon buggy is shown in Figure P12.10(b). The objective of the control design is to achieve a step response to a steering command with zero steady-state error, an overshoot less than 20%, and a peak time less than 0.3 second with a $|u(t)| \le 50$. It is also necessary to determine the effect of a step disturbance $T_d(s) = 1/s$ when $R(s) = 0$, in order to ensure the reduction of moon surface effects. Using (a) a PI controller and (b) a PID controller, design an acceptable controller. Record the results for each design in a table. Compare the performance of each design. Use a prefilter $G_p(s)$ if necessary.

FIGURE P12.8
PID controller for
the motor and load
system.

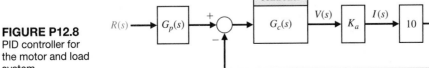

(a)

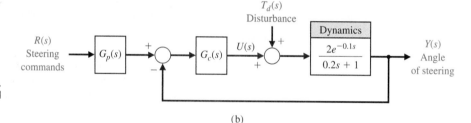

(b)

FIGURE P12.10
(a) A moon vehicle.
(b) Steering control
for the moon
vehicle.

P12.11 A plant has a transfer function

$$G(s) = \frac{25}{s^2}.$$

We want to use a negative unity feedback with a PID controller and a prefilter. The goal is to achieve a peak time of 1 second with ITAE-type performance. Predict the system overshoot and settling time (with a 2% criterion) for a step input.

P12.12 A three-dimensional cam for generating a function of two variables is shown in Figure P12.12(a). Both x and θ may be controlled using a position control system [19]. The control of x may be achieved with a DC motor and position feedback of the form shown in Figure P12.12(b), with the DC motor and load represented by

$$G(s) = \frac{K}{s(s + p)(s + 4)},$$

where $1 \le K \le 3$ and $1 \le p \le 3$. Normally $K = 2.5$ and $p = 2$. Design an ITAE system with a PID controller so that the peak time response to a step input is less than 3 seconds for the worst-case performance.

P12.13 Consider the closed-loop second-order system

$$\dot{\mathbf{x}} = \begin{bmatrix} 0 & 3 \\ -5 & -K \end{bmatrix}\mathbf{x} + \begin{bmatrix} 0 \\ 1 \end{bmatrix}r$$

$$y = [2 \quad 0]\mathbf{x} + [0]u.$$

Compute the sensitivity of the closed-loop system to variations in the parameter K.

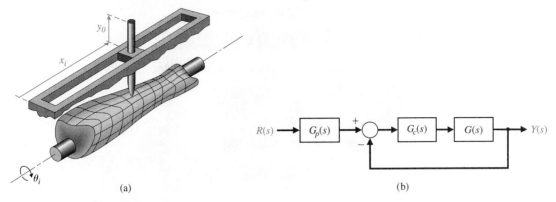

(a)   (b)

FIGURE P12.12   (a) Three-dimensional cam. (b) *x*-axis control system.

ADVANCED PROBLEMS

AP12.1   To minimize vibrational effects, a telescope is magnetically levitated. This method also eliminates friction in the azimuth magnetic drive system. The photodetectors for the sensing system require electrical connections. The system block diagram is shown in Figure AP12.1. Design a PID controller so that the velocity error constant is $K_v = 100$ and the maximum overshoot for a step input is less than 10%.

AP12.2   One promising solution to traffic gridlock is a magnetic levitation (maglev) system. Vehicles are suspended on a guideway above the highway and guided by magnetic forces instead of relying on wheels or aerodynamic forces. Magnets provide the propulsion for the vehicles [7, 12, 17]. Ideally, maglev can offer the environmental and safety advantages of a high-speed train, the speed and low friction of an airplane, and the convenience of an automobile. All these shared attributes notwithstanding, the maglev system is truly a new mode of travel and will enhance the other modes of travel by relieving congestion and providing connections among them. Maglev travel would be fast, operating at 150 to 300 miles per hour.

The tilt control of a maglev vehicle is illustrated in Figures AP12.2(a) and (b). The dynamics of the plant $G(s)$ are subject to variation so that the poles will lie within the boxes shown in Figure AP12.2(c), and $1 \le K \le 2$.

The objective is to achieve a robust system with a step response possessing an overshoot less than 10%, as well as a settling time (with a 2% criterion) less than 2 seconds when $|u(t)| \le 100$. Obtain a design with a PI, PD, and PID controller and compare the results. Use a prefilter $G_p(s)$ if necessary.

AP12.3   Antiskid braking systems present a challenging control problem, since brake/automotive system parameter variations can vary significantly (e.g., due to the brake-pad coefficient of friction changes or road slope variations) and environmental influences (e.g., due to adverse road conditions). The objective of the antiskid system is to regulate wheel slip to maximize the coefficient of friction between the tire and road for any given road surface [8]. As we expect, the braking coefficient of friction is greatest for dry asphalt, slightly reduced for wet asphalt, and greatly reduced for ice.

One simplified model of the braking system is represented by a plant transfer function $G(s)$ with a system as shown in Figure 12.16 with

$$G(s) = \frac{Y(s)}{U(s)} = \frac{1}{(s + a)(s + b)},$$

where normally $a = 1$ and $b = 4$.

FIGURE AP12.1
Magnetically levitated telescope position control system.

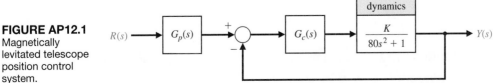

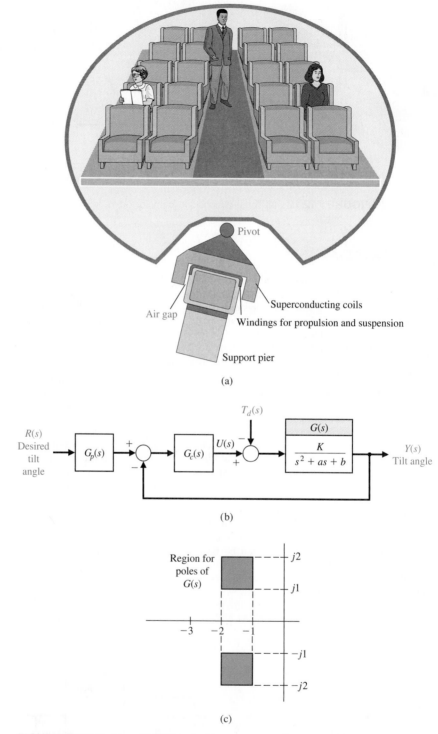

FIGURE AP12.2   (a) and (b) Tilt control for a maglev vehicle. (c) Plant dynamics.

(a) Using a PID controller, design a very robust system where, for a step input, the overshoot is less than 4% and the settling time (with a 2% criterion) is 1 second or less. The steady-state error must be less than 1% for a step. We expect a and b to vary by $\pm 50\%$.

(b) Design a system to yield the specifications of part (a) using an ITAE performance index. Predict the overshoot and settling time for this design.

AP12.4 A robot has been designed to aid in hip-replacement surgery. The device, called RoBoDoc, is used to precisely orient and mill the femoral cavity for acceptance of the prosthetic hip implant. Clearly, we want a very robust surgical tool control, because there is no opportunity to redrill a bone [23, 30]. The control system will be as shown in Figure 12.1 with

$$G(s) = \frac{b}{s^2 + as + b},$$

where $1 \le a \le 2$, and $4 \le b \le 12$.

Select a PID controller so that the system is very robust. Use the s-plane root locus method. Select the appropriate $G_p(s)$ and plot the response to a step input.

AP12.5 A spacecraft with a camera is shown in Figure AP12.5. The camera slews about 16° in a canted plane relative to the base. Reaction jets stabilize the base against the reaction torques from the slewing motors. The rotational speed control for the camera slewing has a plant transfer function

$$G(s) = \frac{1}{(s + 1)(s + 2)(s + 4)}.$$

A PID controller is used in a system as shown in Figure 12.1, where

$$G_c(s) = \frac{K(s + \sigma + j1)(s + \sigma - j1)}{s}.$$

Determine K and the resulting roots so that all the roots have ζ less than or equal to $1/\sqrt{2}$ when $1 < \sigma < 2$ (vary σ in increments of 0.25). Determine the root sensitivity to changes in K.

AP12.6 Consider the system of Figure 12.1 with

$$G(s) = \frac{K_1}{s(s + 10)},$$

where $K_1 = 1$ under normal conditions. Design a PID controller to achieve a phase margin of 50°. The controller is

$$G_c(s) = \frac{K(s^2 + 20s + b)}{s}$$

with complex zeros. Select an appropriate prefilter. Determine the effect of a change of $\pm 25\%$ in K_1 by developing a tabular record of the system performance.

AP12.7 Consider the system of Figure 12.1 with

$$G(s) = \frac{K_1}{s(\tau s + 1)},$$

where $K_1 = 1.5$ and $\tau \approx 0.001$ second, which may be neglected. (Check this later in the design process.) Select a PID controller so that the settling time (with a 2% criterion) for a step input is less than 1 second and the overshoot is less than 10%. Also, the effect of a

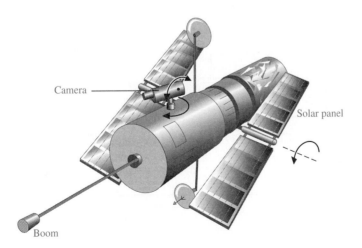

FIGURE AP12.5
Spacecraft with a camera.

Camera

Solar panel

Boom

disturbance at the output must be reduced to less than 5% of the magnitude of the disturbance. Select ω_n, and use the ITAE design method.

AP12.8   Consider the system of Figure 12.1 with

$$G(s) = \frac{1}{s}.$$

The goal is to select a PI controller using the ITAE design criterion while constraining the control signal as $|u(t)| \leq 1$ for a unit step input. Determine the appropriate PI controller and the settling time (with a 2% criterion) for a step input. Use a prefilter.

AP12.9   Consider the system of Figure 12.1 with

$$G(s) = \frac{3}{s(s^2 + 4s + 5)}.$$

Design a PID controller to achieve (a) an acceleration constant $K_a = 2$, (b) a phase margin equal to 45°, and (c) a bandwidth greater than 2.8 rad/s. Select an appropriate prefilter and plot the response to a step input.

AP12.10   A machine tool control system is shown in Figure AP12.10. The transfer function of the power amplifier, prime mover, moving carriage, and tool bit is

$$G(s) = \frac{50}{s(s + 1)(s + 4)(s + 5)}.$$

The goal is to have an overshoot less than 25% for a step input while achieving a peak time less than 3 seconds. Determine a suitable controller using (a) PD control, (b) PI control, and (c) PID control. (d) Then select the best controller.

AP12.11   Consider a system with the structure shown in Figure 12.1 with

$$G(s) = \frac{K}{s^2 + 2as + a^2},$$

where $1 \leq a \leq 3$ and $2 \leq K \leq 4$.

Use a PID controller and design the controller for the worst-case condition. We desire that the settling time (with a 2% criterion) be less than 0.8 second with an ITAE performance.

AP12.12   A system of the form shown in Figure 12.1 has

$$G(s) = \frac{s + r}{(s + p)(s + q)},$$

where $3 \leq p \leq 5, 0 \leq q \leq 1$, and $1 \leq r \leq 2$. We will use a compensator

$$G_c(s) = \frac{K(s + z_1)(s + z_2)}{(s + p_1)(s + p_2)},$$

with all real poles and zeros. Select an appropriate compensator to achieve robust performance.

AP12.13   A system of the form shown in Figure 12.44 has a plant

$$G(s) = \frac{1}{(s + 2)(s + 4)(s + 6)}.$$

We want to attain a steady-state error for a step input. Select a compensator $G_c(s)$ and gain K, using the pseudo-QFT method, and determine the performance of the system when all the poles of $G(s)$ change by -50%. Describe the robust nature of the system.

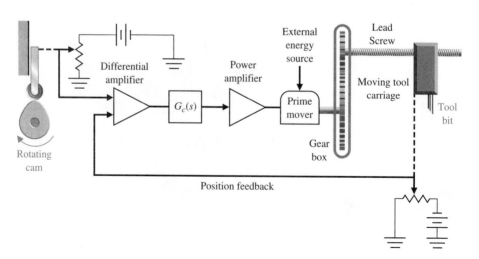

FIGURE AP12.10
A machine tool control system.

DESIGN PROBLEMS

CDP12.1 Design a PID controller for the capstan-slide system of Figure CDP4.1. The percent overshoot should be less than 3% and the settling time should be (with a 2% criterion) less than 250 ms for a step input $r(t)$. Determine the response to a disturbance for the designed system.

DP12.1 A position control system for a large turntable is shown in Figure DP12.1(a), and the block diagram of the system is shown in Figure DP12.1(b) [11, 14]. This system uses a large torque motor with $K_m = 15$. The objective is to reduce the steady-state effect of a step change in the load disturbance to 5% of the magnitude of the step disturbance while maintaining a fast response to a step input command $R(s)$, with less than 5% overshoot. Select K_1 and the compensator when (a) $G_c(s) = K$ and (b) $G_c(s) = K_P + K_D s$ (a PD compensator). Plot the step response for the disturbance and the input for both compensators.

Determine whether a prefilter is required to meet the overshoot requirement.

DP12.2 Consider the closed-loop system depicted in Figure DP12.2. The process has a parameter K that is nominally $K = 1$. Design a controller that results in a percent overshoot $P.O. \leq 15\%$ for a unit step input for all K in the range $0.1 \leq K \leq 2$.

DP12.3 Many university and government laboratories have constructed robot hands capable of grasping and manipulating objects. But teaching the artificial devices to perform even simple tasks required formidable computer programming. Now, however, the Dexterous Hand Master (DHM) can be worn over a human hand to record the side-to-side and bending motions of finger joints. Each joint is fitted with a sensor that changes its signal depending on position. The signals from all the sensors are translated into computer data and used to operate robot hands [1].

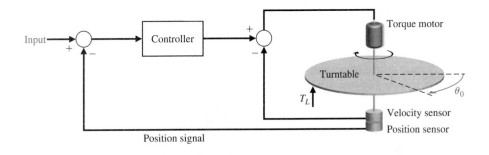

(a)

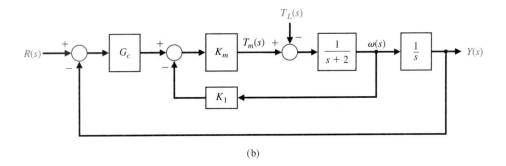

FIGURE DP12.1
Turntable control.

(b)

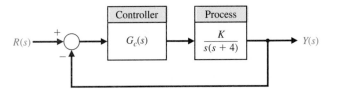

FIGURE DP12.2
A unity feedback system with a process with varying parameter K.

The DHM is shown in parts (a) and (b) of Figure DP12.3. The joint angle control system is shown in part (c). The normal value of K_m is 1.0. The goal is to design a PID controller so that the steady-state error for a ramp input is zero. Also, the settling time (with a 2% criterion) must be less than 3 seconds for the ramp input. We want the controller to be

$$G_c(s) = \frac{K_D(s^2 + 6s + 18)}{s}.$$

(a) Select K_D and obtain the ramp response. Plot the root locus as K_D varies. (b) If K_m changes to one-half of its normal value and $G_c(s)$ remains as designed in part (a), obtain the ramp response of the system. Compare the results of parts (a) and (b) and discuss the robustness of the system.

DP12.4   Objects smaller than the wavelengths of visible light are a staple of contemporary science and technology. Biologists study single molecules of protein or DNA; materials scientists examine atomic-scale flaws in crystals; microelectronics engineers lay out circuit patterns only a few tenths of atoms thick. Until recently, this minute world could be seen only

by cumbersome, often destructive methods, such as electron microscopy and X-ray diffraction. It lay beyond the reach of any instrument as simple and direct as the familiar light microscope. New microscopes, typified by the scanning tunneling microscope (STM), are now available [3].

The precision of position control required is in the order of nanometers. The STM relies on piezo-electric sensors that change size when an electric voltage across the material is changed. The "aperture" in the STM is a tiny tungsten probe, its tip ground so fine that it may consist of only a single atom and measure just 0.2 nanometer in width. Piezoelectric controls maneuver the tip to within a nanometer or two of the surface of a conducting specimen—so close that the electron clouds of the atom at the probe tip and of the nearest atom of the specimen overlap. A feedback mechanism senses the variations in tunneling current and varies the voltage applied to a third, z-axis, control. The z-axis piezo-electric moves the probe vertically to stabilize the current and to maintain a constant gap between the microscope's tip and the surface. The control system is shown in Figure DP12.4(a), and the block diagram

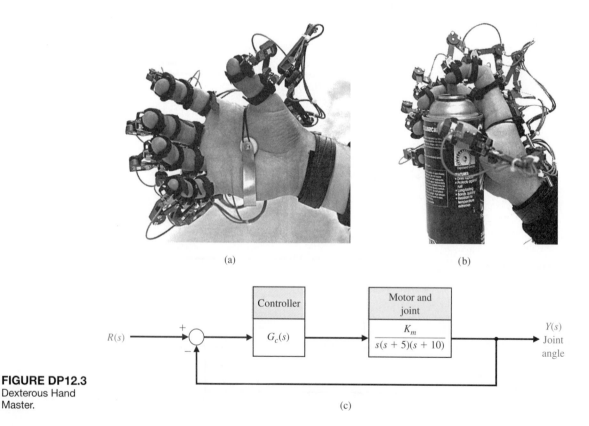

(a)   (b)

FIGURE DP12.3
Dexterous Hand
Master.

(c)

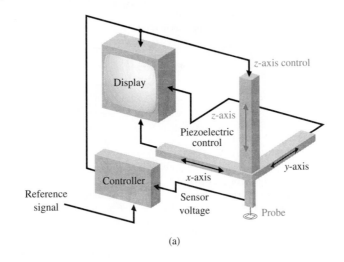

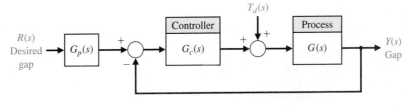

FIGURE DP12.4
Microscope control.

(b)

is shown in Figure DP12.4(b). The process is

$$G(s) = \frac{17,640}{s(s^2 + 59.4s + 1764)},$$

and the controller is chosen to have two real, unequal zeros so that we have

$$G_c(s) = \frac{K_I(\tau_1 s + 1)(\tau_2 s + 1)}{s}.$$

(a) Use the ITAE design method to determine $G_c(s)$. (b) Determine the step response of the system with and without a prefilter $G_p(s)$. (c) Determine the response of the system to a disturbance when $T_d(s) = 1/s$. (d) Using the prefilter and $G_c(s)$ of parts (a) and (b), determine the actual response when the process changes to

$$G(s) = \frac{16,000}{s(s^2 + 40s + 1600)}.$$

DP12.5 The system described in DP12.4 is to be designed using the frequency response techniques described in Section 12.6 with

$$G_c(s) = \frac{K_I(\tau_1 s + 1)(\tau_2 s + 1)}{s}.$$

Select the coefficients of $G_c(s)$ so that the phase margin is approximately 45°. Obtain the step response of the system with and without a prefilter $G_p(s)$.

DP12.6 The use of control theory to provide insight into neurophysiology has a long history. As early as the beginning of the last century, many investigators described a muscle control phenomenon caused by the feedback action of muscle spindles and by sensors based on a combination of muscle length and rate of change of muscle length.

This analysis of muscle regulation has been based on the theory of single-input, single-output control systems. An example is a proposal that the stretch reflex is an experimental observation of a motor control strategy, namely, control of individual muscle length by the spindles. Others later proposed the regulation of individual muscle stiffness (by sensors of both length and force) as the motor control strategy [34].

One model of the human standing-balance mechanism is shown in Figure DP12.6. Consider the case of a paraplegic who has lost control of his standing mechanism. We propose to add an artificial controller to enable the person to stand and move his legs. (a) Design a controller when the normal values of the parameters are $K = 10$, $a = 12$, and $b = 100$,

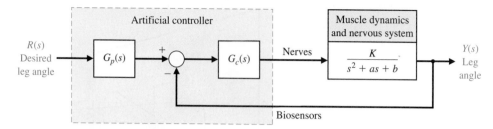

FIGURE DP12.6
Artificial control of standing and leg articulation.

in order to achieve a step response with percent overshoot less than 10%, steady-state error less than 5%, and a settling time (with a 2% criterion) less than 2 seconds. Try a controller with proportional gain, PI, PD, and PID. (b) When the person is fatigued, the parameters may change to $K = 15$, $a = 8$, and $b = 144$. Examine the performance of this system with the controllers of part (a). Prepare a table contrasting the results of parts (a) and (b).

DP12.7 The goal is to design an elevator control system so that the elevator will move from floor to floor rapidly and stop accurately at the selected floor (Figure DP12.7). The elevator will contain from one to three occupants. However, the weight of the elevator should be greater than the weight of the occupants; you may assume that the elevator weighs 1000 pounds and each occupant weighs 150 pounds. Design a system to accurately control the elevator to within one centimeter. Assume that the large DC motor is field-controlled. Also, assume that the time constant of the motor and load is one second, the

time constant of the power amplifier driving the motor is one-half second, and the time constant of the field is negligible. We seek an overshoot less than 6% and a settling time (with a 2% criterion) less than 4 seconds.

DP12.8 Patients with a cardiological illness and less than normal heart muscle strength can benefit from an assistance device. An electric ventricular assist device (EVAD) converts electric power into blood flow by moving a pusher plate against a flexible blood sac. The pusher plate reciprocates to eject blood in systole and to allow the sac to fill in diastole. The EVAD will be implanted in tandem or in parallel with the intact natural heart as shown in Figure DP12.8(a). The EVAD is driven by rechargeable batteries, and the electric power is transmitted inductively across the skin through a transmission system. The batteries and the transmission system limit the electric energy storage and the transmitted peak power. Consequently, we desire to drive the EVAD in a fashion that minimizes its electric power consumption [29].

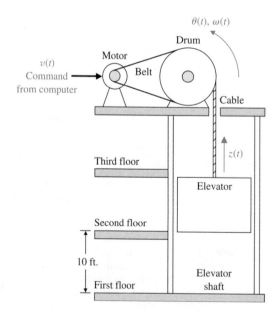

FIGURE DP12.7
Elevator position control.

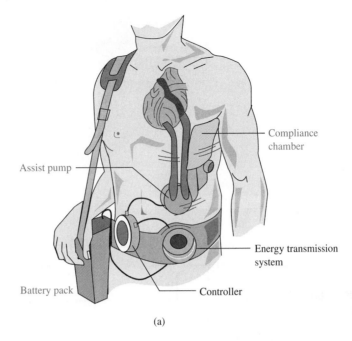

(a)

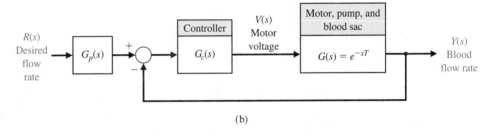

FIGURE DP12.8
(a) An electric
ventricular assist
device for
cardiology patients.
(b) Feedback
control system.

(b)

The EVAD has a single input, the applied motor voltage, and a single output, the blood flow rate. The control system of the EVAD performs two main tasks: It adjusts the motor voltage to drive the pusher plate through its desired stroke, and it varies the EVAD's blood flow to meet the body's cardiac output demand. The blood flow controller adjusts the blood flow rate by varying the EVAD's beat rate.

A model of the feedback control system is shown in Figure DP12.8(b). The motor, pump, and blood sac can be modeled by a time delay with $T = 1$ s. The goal is to achieve a step response with less than 5% steady-state error and less than 10% overshoot. Furthermore, to prolong the batteries, the voltage is limited to 30 V. Design a controller using (a) $G_c(s) = K/s$, (b) a PI controller, and (c) a PID controller. Compare the results for the three controllers by recording in a table the percent overshoot, peak time, settling time (with 2% criterion) and the maximum value of $v(t)$.

DP12.9 One arm of a space robot is shown in Figure DP12.9(a). The block diagram for the control of the arm is shown in Figure DP12.9(b). The transfer function of the motor and arm is

$$G(s) = \frac{1}{s(s + 10)}.$$

(a) If $G_c(s) = K$, determine the gain necessary for an overshoot of 4.5%, and plot the step response. (b) Design a proportional plus derivative (PD) controller using the ITAE method and $\omega_n = 10$. Determine the required prefilter $G_p(s)$. (c) Design a PI controller and a prefilter using the ITAE method. (d) Design a PID controller and a prefilter using the ITAE method with $\omega_n = 10$. (e) Determine the effect of a unit step disturbance for each design. Record the maximum value of $y(t)$ and the final value of $y(t)$ for the disturbance input. (f) Determine the overshoot, peak time, and settling time (with a 2% criterion) step $R(s)$ for

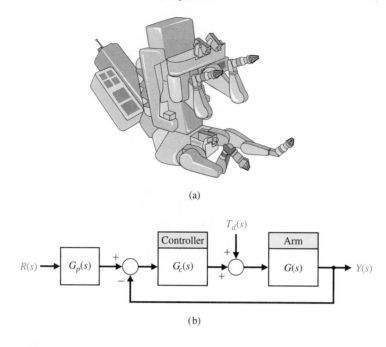

(a)

FIGURE DP12.9
Space robot
control.

(b)

each design above. (g) The process is subject to variation due to load changes. Find the magnitude of the
sensitivity at $\omega = 5$, $|S_G^T(j5)|$, where

$$T = \frac{G_c(s)G(s)}{1 + G_c(s)G(s)}.$$

(h) Based on the results of parts (e), (f), and (g), select
the best controller.

DP12.10   A photovoltaic system is mounted on a space
station in order to develop the power for the station.
The photovoltaic panels should follow the sun with
good accuracy in order to maximize the energy from
the panels. The system uses a DC motor, so that the
transfer function of the panel mount and the motor is

$$G(s) = \frac{1}{s(s + 19)}.$$

We will select a controller $G_c(s)$ assuming that an
optical sensor is available to accurately track the sun's
position, and thus $H(s) = 1$.

The goal is to design $G_c(s)$ so that (1) the percent
overshoot to a step is less than 7% and (2) the steady-
state error to a ramp input is less than or equal to 1%.
Determine the best phase-lead controller. Examine
the robustness of the system to a 10% variation in the
motor time constant.

DP12.11   Electromagnetic suspension systems for air-
cushioned trains are known as magnetic levitation
(maglev) trains. One maglev train uses a superconducting magnet system [17]. It uses superconducting
coils, and the levitation distance $x(t)$ is inherently

unstable. The model of the levitation is

$$G(s) = \frac{X(s)}{V(s)} = \frac{K}{(s\tau_1 + 1)(s^2 - \omega_1^2)},$$

where $V(s)$ is the coil voltage; τ_1 is the magnet time
constant; and ω_1 is the natural frequency. The system
uses a position sensor with a negligible time constant.
A train traveling at 250 km/hr would have $\tau_1 = 0.75$ s
and $\omega_1 = 75$ rad/s. Determine a controller that can
maintain steady, accurate levitation when disturbances occur along the railway. Use the system model
of Figure 12.1.

DP12.12   Consider again the Mars rover problem described in DP6.2. The system uses a PID controller, and
a robust system is desired. The specifications are (1)
maximum overshoot equal to 18%, (2) settling time
(with a 2% criterion) less than 2 seconds, (3) rise time
equal to or greater than 0.20 to limit the power requirements, (4) phase margin greater than 65°, (5) gain
margin greater than 8 dB, (6) maximum root sensitivity
(magnitude of real and imaginary parts) less than 1.
Select the best value of the gain K.

DP12.13   A benchmark problem consists of the mass–
spring system shown in Figure DP12.13, which represents a flexible structure. Let $m_1 = m_2 = 1$ and
$0.5 \le k \le 2.0$ [33]. It is possible to measure x_3 and
x_2 and use a controller prior to $u(t)$. Obtain the system
description, choose a control structure, and design a
robust system. Determine the response of the system
to a unit step disturbance. Assume that the output
$x_2(t)$ is the variable to be controlled.

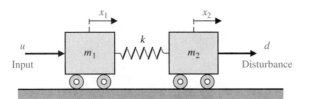

FIGURE DP12.13
Two-mass cart
system.

 COMPUTER PROBLEMS

CP12.1 A closed-loop feedback system is shown in Figure CP12.1. Use an m-file to obtain a plot of $|S_K^T|$ versus ω. Plot $|T(s)|$ versus ω, where $T(s)$ is the closed-loop transfer function.

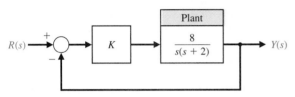

FIGURE CP12.1 Closed-loop feedback system with gain K.

CP12.2 An aircraft aileron can be modeled as a first-order system

$$G(s) = \frac{p}{s + p},$$

where p depends on the aircraft. Obtain a family of step responses for the aileron system in the feedback configuration shown in Figure CP12.2.

The nominal value of $p = 8$. Compute a reasonable value of K so that the step response (with $p = 8$) has $P.O. < 5\%$ and $T_s < 0.1$s. Then, use an m-file to

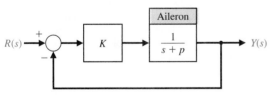

FIGURE CP12.2 Closed-loop control system for the aircraft aileron.

obtain the step responses for $0.1 < p < 20$, with the gain K as determined above.

CP12.3 Consider the control system in Figure CP12.3, where

$$G(s) = \frac{1}{Js^2}.$$

The value of J is known to change slowly with time, although, for design purposes, the nominal value is chosen to be $J = 10$.

(a) Design a PID compensator (denoted by $G_c(s)$) to achieve a phase margin greater than 45° and a bandwidth less than 5 rad/s. (b) Using the PID controller designed in part (a), develop an m-file script to generate a plot of the phase margin as J varies from 1 to 30.

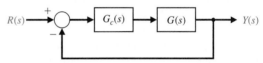

FIGURE CP12.3 A feedback control system with compensation.

CP12.4 Consider the feedback control system in Figure CP12.4. The exact value of parameter b is unknown; however, for design purposes, the nominal value is taken to be $b = 4$. The value of $a = 8$ is known very precisely.

(a) Design the proportional controller K so that the closed-loop system response to a unit step input has a settling time (with a 2% criterion) less than 5 seconds and an overshoot of less than 10%. Use the nominal value of b in the design.

(b) Investigate the effects of variations in the parameter b on the closed-loop system unit step response.

FIGURE CP12.4
A feedback control
system with
uncertain
parameter b.

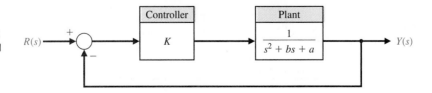

Let $b = 0, 1, 4,$ and $40,$ and co-plot the step response associated with each value of b. In all cases, use the proportional controller from part (a). Discuss the results.

CP12.5 A model of a flexible structure is given by

$$G(s) = \frac{(1 + k\omega_n^2)s^2 + 2\zeta\omega_n s + \omega_n^2}{s^2(s^2 + 2\zeta\omega_n s + \omega_n^2)},$$

where ω_n is the natural frequency of the flexible mode, and ζ is the corresponding damping ratio. In general, it is difficult to know the structural damping precisely, while the natural frequency can be predicted more accurately using well-established modeling techniques. Assume the nominal values of $\omega_n = 2$ rad/s, $\zeta = 0.005,$ and $k = 0.1.$

(a) Design a lead compensator to meet the following specifications: (1) a closed-loop system response to a unit step input with a settling time (with a 2% criterion) less than 200 seconds and (2) an overshoot of less than 50%.

(b) With the controller from part (a), investigate the closed-loop system unit step response with $\zeta = 0, 0.005, 0.1,$ and 1. Co-plot the various unit step responses and discuss the results.

(c) From a control system point of view, is it preferable to have the actual flexible structure damping less than or greater than the design value? Explain.

CP12.6 The industrial process shown in Figure CP12.6 is known to have a time delay in the loop. In practice, it is often the case that the magnitude of system time delays cannot be precisely determined. The magnitude of the time delay may change in an unpredictable manner depending on the process environment. A robust control system should be able to operate satisfactorily in the presence of the system time delays.

(a) Develop an m-file script to compute and plot the phase margin for the industrial process in Figure CP12.6 when the time delay, T, varies between 0 and 5 seconds. Use the pade function with a second-order approximation to approximate the time delay. Plot the phase margin as a function of the time delay.

(b) Determine the maximum time delay allowable for system stability. Use the plot generated in part (a) to compute the maximum time delay approximately.

CP12.7 A unity negative feedback loop has the loop transfer function

$$G_c(s)G(s) = \frac{a(s - 0.5)}{s^2 + 2s + 1}.$$

We know from the underlying physics of the problem that the parameter a can vary only between $0 < a < 1.$ Develop an m-file script to generate the following plots:

(a) The steady-state tracking error due to a negative unit step input (i.e., $R(s) = -1/s$) versus the parameter a.

(b) The maximum percent initial undershoot (or overshoot) versus parameter a.

(c) The gain margin versus the parameter a.

(d) Based on the results in parts (a)–(c), comment on the robustness of the system to changes in parameter a in terms of steady-state errors, stability, and transient time response.

CP12.8 The Gamma-Ray Imaging Device (GRID) is a NASA experiment to be flown on a long-duration, high-altitude balloon during the coming solar maximum. The GRID on a balloon is an instrument that will qualitatively improve hard X-ray imaging and carry out the first gamma-ray imaging for the study of solar high-energy phenomena in the next phase of peak solar activity. From its long-duration balloon platform, GRID will observe numerous hard X-ray bursts, coronal hard X-ray sources, "superhot" thermal events, and microflares [2]. Figure CP12.8(a) depicts the GRID payload attached to the balloon. The major components of the GRID experiment consist of a 5.2-meter canister and mounting gondola, a high-altitude balloon, and a cable connecting the gondola and balloon. The instrument–sun pointing requirements of the experiment are 0.1 degree pointing accuracy and 0.2 arcsecond per 4 ms pointing stability.

An optical sun sensor provides a measure of the sun–instrument angle and is modeled as a first-order system with a DC gain and a pole at $s = -500.$ A torque motor actuates the canister/gondola assembly. The azimuth angle control system is shown in Figure CP12.8(b). The PID controller is selected by the design team so that

$$G_c(s) = \frac{K_D(s^2 + as + b)}{s},$$

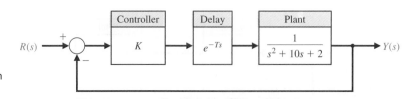

FIGURE CP12.6
An industrial controlled process with a time delay in the loop.

$R(s)$ —→ ⊕ —→ | **Controller** K | —→ | **Delay** e^{-Ts} | —→ | **Plant** $\dfrac{1}{s^2 + 10s + 2}$ | —→ $Y(s)$

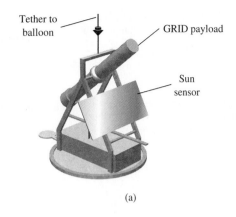

Tether to balloon

GRID payload

Sun sensor

(a)

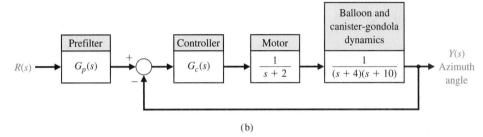

Prefilter		Controller		Motor		Balloon and canister-gondola dynamics

$R(s) \rightarrow$ $G_p(s)$ $\rightarrow$ + $\ominus$ $\rightarrow$ $G_c(s)$ $\rightarrow$ $\dfrac{1}{s+2}$ $\rightarrow$ $\dfrac{1}{(s+4)(s+10)}$ $\rightarrow$ $Y(s)$ Azimuth angle

FIGURE CP12.8
The GRID device.

(b)

where a and b are to be selected. A prefilter is used as shown in Figure CP12.8(b). Determine the value of K_3, a, and b so that the dominant roots have a ζ of 0.8 and the overshoot to a step input is less than 3%.

Develop a simulation to study the control system performance. Use a step response to confirm the percent overshoot meets the specification.

TERMS AND CONCEPTS

Additive perturbation A system perturbation model expressed in the additive form $G_a(s) = G(s) + A(s)$, where $G(s)$ is the nominal process function, $A(s)$ is the perturbation that is bounded in magnitude, and $G_a(s)$ is the family of perturbed process functions.

Complementary sensitivity function The function
$$T(s) = \frac{G_c(s)G(s)}{1 + G_c(s)G(s)}$$ that satisfies the relationship $S(s) + T(s) = 1$, where $S(s)$ is the sensitivity function. The function $T(s)$ is the closed-loop transfer function.

Internal model principle The principle that states that if $G_c(s)G(s)$ contains the input $R(s)$, then the output $y(t)$ will track $R(s)$ asymptotically (in the steady-state) and the tracking is robust.

Multiplicative perturbation A system perturbation model expressed in the multiplicative form $G_m(s) = G(s)(1 + M(s))$, where $G(s)$ is the nominal process function, $M(s)$ is the perturbation that is bounded in

magnitude, and $G_m(s)$ is the family of perturbed process functions.

PID controller A controller with three terms in which the output is the sum of a proportional term, an integrating term, and a differentiating term, with an adjustable gain for each term.

Prefilter A transfer function $G_p(s)$ that filters the input signal $R(s)$ prior to the calculation of the error signal.

Process controller *See* PID controller.

Robust control system A system that exhibits the desired performance in the presence of significant plant uncertainty.

Robust stability criterion A test for robustness with respect to multiplicative perturbations in which stability is guaranteed if $|M(j\omega)| < \left|1 + \dfrac{1}{G(j\omega)}\right|$, for all ω, where $M(s)$ is the multiplicative perturbation.

Root sensitivity   A measure of the sensitivity of the roots (i.e., the poles and zeros) of the system to changes in a parameter defined by $S_\alpha^{r_i} = \dfrac{\partial r_i}{\partial \alpha / \alpha}$, where α is the parameter and r_i is the root.

Sensitivity function   The function $S(s) = [1 + G_c(s)G(s)]^{-1}$ that satisfies the relationship $S(s) + T(s) = 1$, where $T(s)$ is the complementary sensitivity function.

System sensitivity   A measure of the system sensitivity to changes in a parameter defined by $S_\alpha^T = \dfrac{\partial T / T}{\partial \alpha / \alpha}$, where α is the parameter and T is the system transfer function.

Three-mode controller   *See* PID controller.

Three-term controller   *See* PID controller.

Digital Control Systems

PREVIEW

A digital computer often hosts the controller algorithm in a feedback control system. Since the computer receives data only at specific intervals, it is necessary to develop a method for describing and analyzing the performance of computer control systems. In this chapter, we provide an introduction to the topic of digital control systems. The notion of a sampled-data system is presented followed by a discussion of the z-transform. We may use the z-transform of a transfer function to analyze the stability and transient response of a system. The basics of closed-loop stability with a digital controller in the loop are covered with a short presentation on the role of root locus in the design process. This chapter concludes with the design of a digital controller for the Sequential Design Example: Disk Drive Read System.

DESIRED OUTCOMES

Upon completion of Chapter 13, students should:

❏ Understand the role of digital computers in control system design and application.

❏ Be familiar with the z-transform and sampled-data systems.

❏ Be able to design digital controllers using root locus methods.

❏ Appreciate the issues associated with the implementation of digital controllers.

13.1 INTRODUCTION

The use of a digital computer as a compensator (controller) device has grown during the past three decades as the price and reliability of digital computers have improved dramatically [1, 2]. A block diagram of a single-loop digital control system is shown in Figure 13.1. The digital computer in this system configuration receives the error in digital form and performs calculations in order to provide an output in digital form. The computer may be programmed to provide an output so that the performance of the process is near or equal to the desired performance. Many computers are able to receive and manipulate several inputs, so a digital computer control system can often be a multivariable system.

A digital computer receives and operates on signals in digital (numerical) form, as contrasted to continuous signals [3]. A **digital control system** uses digital signals and a digital computer to control a process. The measurement data are converted from analog form to digital form by means of the analog-to-digital converter shown in Figure 13.1. After processing the inputs, the digital computer provides an output in digital form. This output is then converted to analog form by the digital-to-analog converter shown in Figure 13.1.

13.2 DIGITAL COMPUTER CONTROL SYSTEM APPLICATIONS

The total number of computer control systems installed in industry has grown over the past three decades [2]. Currently, there are approximately 100 million control systems using computers, although the computer size and power may vary significantly. If we consider only computer control systems of a relatively complex nature, such as chemical process control or aircraft control, the number of computer control systems is approximately 20 million.

A digital computer consists of a central processing unit (CPU), input–output units, and a memory unit. The size and power of a computer will vary according to the size, speed, and power of the CPU, as well as the size, speed, and organization of the memory unit. Small computers, called **minicomputers**, have become increasingly common since 1980. Powerful but inexpensive computers, called **microcomputers**, which use a 16-bit word or 32-bit word, have become readily available. These systems use a microprocessor as a CPU. Therefore, the nature of the control task, the extent of the data required in memory, and the speed of calculation required will dictate the selection of the computer within the range of available computers.

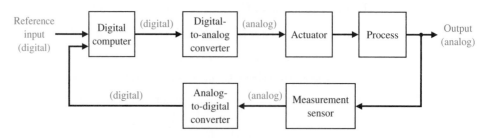

FIGURE 13.1
A block diagram of a computer control system, including the signal converters. The signal is indicated as digital or analog.

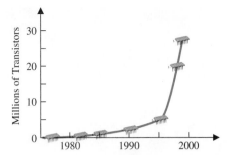

FIGURE 13.2
The development of INTEL microprocessors measured in millions of transistors. (*Source:* INTEL.)

The size of computers and the cost for the active logic devices used to construct them have both declined exponentially. The active components per cubic centimeter have increased so that the actual computer can be reduced in size to the point where relatively inexpensive, powerful laptop computers are providing mobile high-performance computational capability to students and professionals alike, and are, in many instances, replacing traditional desktop microcomputers. The speed of computers has also increased exponentially. The transistor density (a measure of computational performance) on INTEL microprocessor integrated circuits has increased exponentially over the last 30 years, as illustrated in Figure 13.2. In fact, according to "Moore's law," the transistor density doubles every year, and will probably continue to do so for the next twenty years. A simple calculation shows that by 2012, microprocessors will contain over a billion transistors with operating speeds approaching 10 GHz! In 1976, the popular 8086 central processing units containing only 29,000 transistors and operating at 10 MHz were introduced. Since then, significant progress in computation capability has been and will continue to be made. Clearly, improvements in computational capability have revolutionized the application of control theory and design in the modern era. With the availability of fast, low-priced, and small-sized microprocessors, much of the control of industrial and commercial processes is moving toward the use of computers within the control system.

Digital control systems are used in many applications: for machine tools, metal-working processes, chemical processes, aircraft control, and automobile traffic control, and others [4–8]. An example of a computer control system used in the aircraft industry is shown in Figure 13.3. Automatic computer-controlled systems are used for purposes as diverse as measuring the objective refraction of the human eye and controlling the engine spark timing or air–fuel ratio of automobile engines. The latter innovations are necessary to reduce automobile emissions and increase fuel economy.

The advantages of using digital control include: improved measurement sensitivity; the use of digitally coded signals, digital sensors and transducers, and microprocessors; reduced sensitivity to signal noise; and the capability to easily reconfigure the control algorithm in software. Improved sensitivity results from the low-energy signals required by digital sensors and devices. The use of digitally coded signals permits the wide application of digital devices and communications. Digital sensors and transducers can effectively measure, transmit, and couple signals and devices. In addition, many systems are inherently digital because they send out pulse signals. Examples of such a digital system are a radar tracking system and a space satellite.

FIGURE 13.3   The flight deck of the Boeing 757 and 767 features digital control electronics, including an engine indicating system and a crew alerting system. All systems controls are within reach of either pilot. The system includes an inertial reference system making use of laser gyroscopes and an electronic attitude director indicator. A flight-management computer system integrates navigation, guidance, and performance data functions. When coupled with the automatic flight control system (automatic pilot), the flight-management system provides accurate engine thrust settings and flight-path guidance during all phases of flight from immediately after takeoff to final approach and landing. The system can predict the speeds and altitudes that will result in the best fuel economy and command the airplane to follow the most fuel-efficient or the "least time" flight paths. (Courtesy of Boeing Airplane Co.)

13.3 SAMPLED-DATA SYSTEMS

Computers used in control systems are interconnected to the actuator and the process by means of signal converters. The output of the computer is processed by a digital-to-analog converter. We will assume that all the numbers that enter or leave the computer do so at the same fixed period T, called the **sampling period**. Thus, for example, the reference input shown in Figure 13.4 is a sequence of sample values $r(kT)$. The variables $r(kT)$, $m(kT)$, and $u(kT)$ are discrete signals in contrast to $m(t)$ and $y(t)$, which are continuous functions of time.

> **Sampled data (or a discrete signal) are data obtained for the system variables only at discrete intervals and are denoted as $x(kT)$.**

A sampler is basically a switch that closes every T seconds for one instant of time. Consider an ideal sampler, as shown in Figure 13.5. The input is $r(t)$, and the

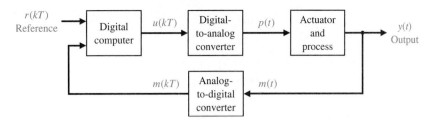

FIGURE 13.4
A digital control
system.

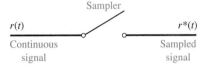

FIGURE 13.5
An ideal sampler
with an input $r(t)$.

output is $r^*(t)$, where nT is the current sample time, and the current value of $r^*(t)$ is $r(nT)$. We then have $r^*(t) = r(nT)\delta(t - nT)$, where δ is the impulse function.

Let us assume that we sample a signal $r(t)$, as shown in Figure 13.5, and obtain $r^*(t)$. Then, we portray the series for $r^*(t)$ as a string of impulses starting at $t = 0$, spaced at T seconds, and of amplitude $r(kT)$. For example, consider the input signal $r(t)$ shown in Figure 13.6(a). The sampled signal is shown in Figure 13.6(b) with an impulse represented by a vertical arrow of magnitude $r(kT)$.

A digital-to-analog converter serves as a device that converts the sampled signal $r^*(t)$ to a continuous signal $p(t)$. The digital-to-analog converter can usually be represented by a zero-order hold circuit, as shown in Figure 13.7. The zero-order hold takes the value $r(kT)$ and holds it constant for $kT \le t < (k + 1)T$, as shown in Figure 13.8 for $k = 0$. Thus, we use $r(kT)$ during the sampling period.

A sampler and zero-order hold can accurately follow the input signal if T is small compared to the transient changes in the signal. The response of a sampler

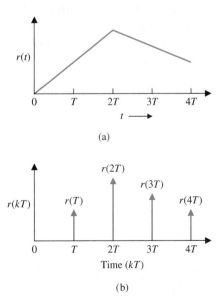

FIGURE 13.6
(a) An input signal
$r(t)$. (b) The sampled
signal $r^*(t) =$
$\sum_{k=0}^{X} r(kT)\delta(t - kT)$.
The vertical arrow
represents an
impulse.

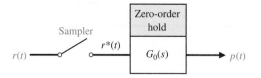

FIGURE 13.7
A sampler and
zero-order hold
circuit.

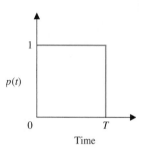

FIGURE 13.8
The response of a
zero-order hold to
an impulse input
$r(kT)$, which equals
unity when $k = 0$
and equals zero
when $k \neq 0$, so that
$r*(t) = r(0)\delta(t)$.

and zero-order hold for a ramp input is shown in Figure 13.9. Finally, the response of a sampler and zero-order hold for an exponentially decaying signal is shown in Figure 13.10 for two values of the sampling period. Clearly, the output $p(t)$ will approach the input $r(t)$ as T approaches zero, meaning that we sample frequently.

The impulse response of a zero-order hold is shown in Figure 13.8. The transfer function of the **zero-order hold** is

$$G_0(s) = \frac{1}{s} - \frac{1}{s}e^{-sT} = \frac{1 - e^{-sT}}{s}. \tag{13.1}$$

The precision of the digital computer and the associated signal converters is limited (see Figure 13.4). **Precision** is the degree of exactness or discrimination with which a quantity is stated. The precision of the computer is limited by a finite word length. The precision of the analog-to-digital converter is limited by an ability to store its output only in digital logic composed of a finite number of binary digits. The converted signal $m(kT)$ is then said to include an **amplitude quantization error**. When the quantization error and the error due to a computer's finite word size are

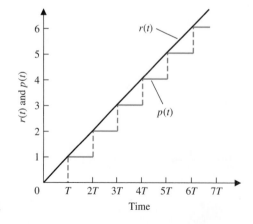

FIGURE 13.9
The response of a
sampler and zero-
order hold for a
ramp input $r(t) = t$.

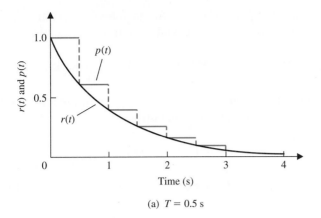

(a) $T = 0.5$ s

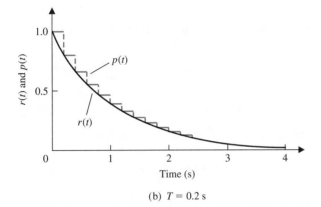

FIGURE 13.10
The response of a
sampler and zero-
order hold to an
input $r(t) = e^{-t}$ for
two values of
sampling period T.

(b) $T = 0.2$ s

small relative to the amplitude of the signal [13, 19], the system is sufficiently precise, and the precision limitations can be neglected.

13.4 THE z-TRANSFORM

Because the output of the ideal sampler, $r^*(t)$, is a series of impulses with values
$r(kT)$, we have

$$r^*(t) = \sum_{k=0}^{\infty} r(kT)\delta(t - kT), \tag{13.2}$$

for a signal for $t > 0$. Using the Laplace transform, we have

$$\mathcal{L}\{r^*(t)\} = \sum_{k=0}^{\infty} r(kT)e^{-ksT}. \tag{13.3}$$

We now have an infinite series that involves multiples of e^{sT} and its powers. We
define

$$\boxed{z = e^{sT},} \tag{13.4}$$

where this relationship involves a conformal mapping from the s-plane to the z-plane. We then define a new transform, called the z-transform, so that

$$Z\{r(t)\} = Z\{r^*(t)\} = \sum_{k=0}^{\infty} r(kT)z^{-k}. \tag{13.5}$$

As an example, let us determine the z-transform of the unit step function $u(t)$ (not to be confused with the control signal $u(t)$). We obtain

$$Z\{u(t)\} = \sum_{k=0}^{\infty} u(kT)z^{-k} = \sum_{k=0}^{\infty} z^{-k}, \tag{13.6}$$

since $u(kT) = 1$ for $k \geq 0$. This series can be written in closed form as[1]

$$U(z) = \frac{1}{1 - z^{-1}} = \frac{z}{z - 1}. \tag{13.7}$$

In general, we will define the **z-transform** of a function $f(t)$ as

$$\boxed{Z\{f(t)\} = F(z) = \sum_{k=0}^{\infty} f(kT)z^{-k}.} \tag{13.8}$$

EXAMPLE 13.1   **Transform of an exponential**

Let us determine the z-transform of $f(t) = e^{-at}$ for $t \geq 0$. Then

$$Z\{e^{-at}\} = F(z) = \sum_{k=0}^{\infty} e^{-akT}z^{-k} = \sum_{k=0}^{\infty} (ze^{+aT})^{-k}. \tag{13.9}$$

Again, this series can be written in closed form as

$$F(z) = \frac{1}{1 - (ze^{aT})^{-1}} = \frac{z}{z - e^{-aT}}. \tag{13.10}$$

In general, we may show that

$$Z\{e^{-at}f(t)\} = F(e^{aT}z). \quad \blacksquare$$

EXAMPLE 13.2   **Transform of a sinusoid**

Let us determine the z-transform of $f(t) = \sin(\omega t)$ for $t \geq 0$. We can write $\sin(\omega t)$ as

$$\sin(\omega t) = \frac{e^{j\omega T} - e^{-j\omega T}}{2j}.$$

Therefore,

$$\sin(\omega t) = \frac{e^{j\omega T}}{2j} - \frac{e^{-j\omega T}}{2j}. \tag{13.11}$$

---

[1]Recall that the infinite geometric series may be written $(1 - bx)^{-1} = 1 + bx + (bx)^2 + (bx)^3 + \ldots$, if $|bx| < 1$.

Table 13.1   z-Transforms

$x(t)$	$X(s)$	$X(z)$
$\delta(t) = \begin{cases} \dfrac{1}{\epsilon}, & t < \epsilon, \epsilon \to 0 \\ 0 & \text{otherwise} \end{cases}$	1	—
$\delta(t - a) = \begin{cases} \dfrac{1}{\epsilon}, & a < t < a + \epsilon, \epsilon \to 0 \\ 0 & \text{otherwise} \end{cases}$	e^{-as}	—
$\delta_o(t) = \begin{cases} 1 & t = 0, \\ 0 & t = kT, k \neq 0 \end{cases}$	—	1
$\delta_o(t - kT) = \begin{cases} 1 & t = kT, \\ 0 & t \neq kT \end{cases}$	—	z^{-k}
$u(t)$, unit step	$1/s$	$\dfrac{z}{z - 1}$
t	$1/s^2$	$\dfrac{Tz}{(z - 1)^2}$
e^{-at}	$\dfrac{1}{s + a}$	$\dfrac{z}{z - e^{-aT}}$
$1 - e^{-at}$	$\dfrac{1}{s(s + a)}$	$\dfrac{(1 - e^{-aT})z}{(z - 1)(z - e^{-aT})}$
$\sin(\omega t)$	$\dfrac{\omega}{s^2 + \omega^2}$	$\dfrac{z \sin(\omega T)}{z^2 - 2z \cos(\omega T) + 1}$
$\cos(\omega t)$	$\dfrac{s}{s^2 + \omega^2}$	$\dfrac{z(z - \cos(\omega T))}{z^2 - 2z \cos(\omega T) + 1}$
$e^{-at} \sin(\omega t)$	$\dfrac{\omega}{(s + a)^2 + \omega^2}$	$\dfrac{(ze^{-aT} \sin(\omega T))}{z^2 - 2ze^{-aT} \cos(\omega T) + e^{-2aT}}$
$e^{-at} \cos(\omega t)$	$\dfrac{s + a}{(s + a)^2 + \omega^2}$	$\dfrac{z^2 - ze^{-aT} \cos(\omega T)}{z^2 - 2ze^{-aT} \cos(\omega T) + e^{-2aT}}$

Then

$$
\begin{aligned}
F(z) &= \frac{1}{2j}\left(\frac{z}{z - e^{j\omega T}} - \frac{z}{z - e^{-j\omega T}} \right) \\
&= \frac{1}{2j}\left(\frac{z(e^{j\omega T} - e^{-j\omega T})}{z^2 - z(e^{j\omega T} + e^{-j\omega T}) + 1} \right) \qquad (13.12) \\
&= \frac{z \sin(\omega T)}{z^2 - 2z \cos(\omega T) + 1}. \blacksquare
\end{aligned}
$$

A table of z-transforms is given in Table 13.1 and at the MCS website. Note that we use the same letter to denote both the Laplace and z-transforms, distinguishing

Table 13.2　Properties of the *z*-Transform

	$x(t)$	$X(z)$
1.	$kx(t)$	$kX(z)$
2.	$x_1(t) + x_2(t)$	$X_1(z) + X_2(z)$
3.	$x(t + T)$	$zX(z) - zx(0)$
4.	$tx(t)$	$-Tz\dfrac{dX(z)}{dz}$
5.	$e^{-at}x(t)$	$X(ze^{aT})$
6.	$x(0)$, initial value	$\lim\limits_{z\to\infty} X(z)$ if the limit exists
7.	$x(\infty)$, final value	$\lim\limits_{z\to 1}(z - 1)X(z)$ if the limit exists and the system is stable; that is, if all poles of $(z - 1)X(z)$ are inside the unit circle $\|z\| = 1$ on z-plane.

them by the argument s or z. A table of properties of the z-transform is given in Table 13.2. As in the case of Laplace transforms, we are ultimately interested in the output $y(t)$ of the system. Therefore, we must use an inverse transform to obtain $y(t)$ from $Y(z)$. We may obtain the output by (1) expanding $Y(z)$ in a power series, (2) expanding $Y(z)$ into partial fractions and using Table 13.1 to obtain the inverse of each term, or (3) obtaining the inverse z-transform by an inversion integral. We will limit our methods to (1) and (2) in this limited discussion.

EXAMPLE 13.3　Transfer function of an open-loop system

Let us consider the system shown in Figure 13.11 for $T = 1$. The transfer function of the zero-order hold (Equation 13.1) is

$$G_0(s) = \frac{1 - e^{-sT}}{s}.$$

Therefore, the transfer function $Y(s)/R^*(s)$ is

$$\frac{Y(s)}{R^*(s)} = G_0(s)G_p(s) = G(s) = \frac{1 - e^{-sT}}{s^2(s + 1)}. \qquad (13.13)$$

Expanding into partial fractions, we have

$$G(s) = (1 - e^{-sT})\left(\frac{1}{s^2} - \frac{1}{s} + \frac{1}{s + 1}\right) \qquad (13.14)$$

$$G(z) = Z\{G(s)\} = (1 - z^{-1})Z\left(\frac{1}{s^2} - \frac{1}{s} + \frac{1}{s + 1}\right). \qquad (13.15)$$

FIGURE 13.11
An open-loop, sampled-data system (without feedback).

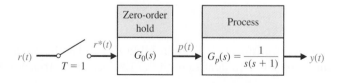

Using the entries of Table 13.1 to convert from the Laplace transform to the corresponding z-transform of each term, we have

$$G(z) = (1 - z^{-1})\left[\frac{Tz}{(z-1)^2} - \frac{z}{z-1} + \frac{z}{z-e^{-T}}\right]$$

$$= \frac{(ze^{-T} - z + Tz) + (1 - e^{-T} - Te^{-T})}{(z-1)(z-e^{-T})}.$$

When $T = 1$, we obtain

$$G(z) = \frac{ze^{-1} + 1 - 2e^{-1}}{(z-1)(z-e^{-1})}$$

$$= \frac{0.3678z + 0.2644}{(z-1)(z-0.3678)} = \frac{0.3678z + 0.2644}{z^2 - 1.3678z + 0.3678}. \qquad (13.16)$$

The response of this system to a unit impulse is obtained for $R(z) = 1$ so that $Y(z) = G(z) \cdot 1$. We may obtain $Y(z)$ by dividing the denominator into the numerator:

$$
\begin{array}{r}
0.3678z^{-1} + 0.7675z^{-2} + 0.9145z^{-3} + \ldots = Y(z) \\
z^2 - 1.3678z + 0.3678 \overline{)0.3678z \quad + 0.2644} \\
\underline{0.3678z \quad - 0.5031 \quad + 0.1353z^{-1}} \\
+ 0.7675 \quad - 0.1353z^{-1} \\
\underline{+ 0.7675 \quad - 1.0497z^{-1} + 0.2823z^{-2}} \\
0.9145z^{-1} - 0.2823z^{-2}
\end{array}
\qquad (13.17)
$$

This calculation yields the response at the sampling instants and can be carried as far as is needed for $Y(z)$. From Equation (13.5), we have

$$Y(z) = \sum_{k=0}^{\infty} y(kT)z^{-k}.$$

In this case, we have obtained $y(kT)$ as follows: $y(0) = 0$, $y(T) = 0.3678$, $y(2T) = 0.7675$, and $y(3T) = 0.9145$. Note that $y(kT)$ provides the values of $y(t)$ at $t = kT$. ∎

We have determined $Y(z)$, the z-transform of the output sampled signal. The z-transform of the input sampled signal is $R(z)$. The transfer function in the z-domain is

$$\frac{Y(z)}{R(z)} = G(z). \qquad (13.18)$$

Since we determined the sampled output, we can use an output sampler to depict this condition, as shown in Figure 13.12; this represents the system of Figure 13.11

FIGURE 13.12
System with
sampled output.

$r(t)$ ———o $R(z)$ $G(z)$ ——o Y(z)

FIGURE 13.13
The z-transform
transfer function in
block diagram form.

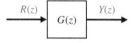

$R(z)$ → $G(z)$ → $Y(z)$

with the sampled input passing to the process. We assume that both samplers have the same sampling period and operate synchronously. Then

$$Y(z) = G(z)R(z), \tag{13.19}$$

as required. We may represent Equation (13.19), which is a z-transform equation, by the block diagram of Figure 13.13.

13.5 CLOSED-LOOP FEEDBACK SAMPLED-DATA SYSTEMS

In this section, we consider closed-loop, sampled-data control systems. Consider the system shown in Figure 13.14(a). The sampled-data z-transform model of this figure with a sampled-output signal $Y(z)$ is shown in Figure 13.14(b). The closed-loop transfer function (using block diagram reduction) is

$$\frac{Y(z)}{R(z)} = T(z) = \frac{G(z)}{1 + G(z)}. \tag{13.20}$$

Here, we assume that the $G(z)$ is the z-transform of $G(s) = G_0(s)G_p(s)$, where $G_0(s)$ is the zero-order hold, and $G_p(s)$ is the process transfer function.

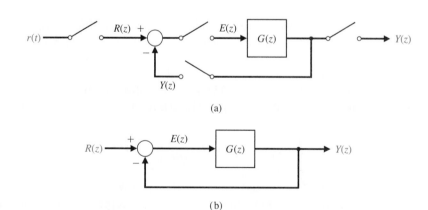

FIGURE 13.14
Feedback control
system with unity
feedback. G(z) is
the z-transform
corresponding to
G(s), which
represents the
process and the
zero-order hold.

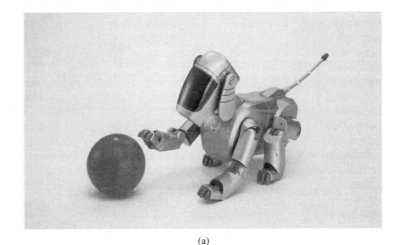

FIGURE 13.15
(a) Aibo is a sophisticated entertainment robot. Aibo looks like a Chihuahua and wags its tail, does tricks, and goes for walks. Aibo depends on a wide range of sensors, including touch, color CCD camera, range finder, and velocity sensors. A 64-bit RISC micro-processor and 16MB of memory are built in. It has 18 joints powered by 18 motors. Photo courtesy of Sony Electronics Inc. (b) Feedback control system with a digital controller. (c) Block diagram model. Note that $G(z) = Z\{G_0(s)G_p(s)\}$.

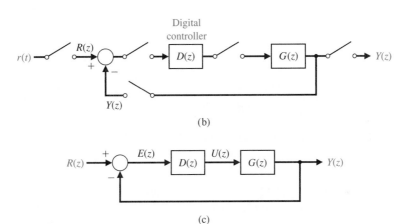

An example of a digital control system is the robotic dog Aibo, shown in Figure 13.15(a). The feedback control system of one joint with a digital controller is shown in Figure 13.15(b). The z-transform block diagram model is shown in Figure 13.15(c). The closed-loop transfer function is

$$\frac{Y(z)}{R(z)} = T(z) = \frac{G(z)D(z)}{1 + G(z)D(z)}. \qquad (13.21)$$

EXAMPLE 13.4 Response of a closed-loop system

Now, let us consider the closed-loop system, as shown in Figure 13.16. We have obtained the z-transform model of this system, as shown in Figure 13.14. Therefore, we have

$$\frac{Y(z)}{R(z)} = \frac{G(z)}{1 + G(z)}. \qquad (13.22)$$

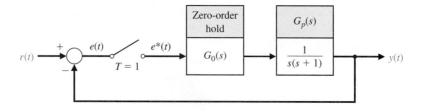

FIGURE 13.16
A closed-loop,
sampled-data
system.

In Example 13.3, we obtained $G(z)$ as Equation (13.16) when $T = 1$ s. Substituting $G(z)$ into Equation (13.22), we obtain

$$\frac{Y(z)}{R(z)} = \frac{0.3678z + 0.2644}{z^2 - z + 0.6322}. \tag{13.23}$$

Since the input is a unit step,

$$R(z) = \frac{z}{z - 1}, \tag{13.24}$$

it follows that

$$Y(z) = \frac{z(0.3678z + 0.2644)}{(z - 1)(z^2 - z + 0.6322)} = \frac{0.3678z^2 + 0.2644z}{z^3 - 2z^2 + 1.6322z - 0.6322}.$$

Completing the division, we have

$$Y(z) = 0.3678z^{-1} + z^{-2} + 1.4z^{-3} + 1.4z^{-4} + 1.147z^{-5}.... \tag{13.25}$$

The values of $y(kT)$ are shown in Figure 13.17, using the symbol □. The complete response of the sampled-data, closed-loop system is shown and contrasted to the response of a continuous system (when $T = 0$). The overshoot of the sampled system is 45%, in contrast to 17% for the continuous system. Furthermore, the settling time of the sampled system is twice as long as that of the continuous system. ■

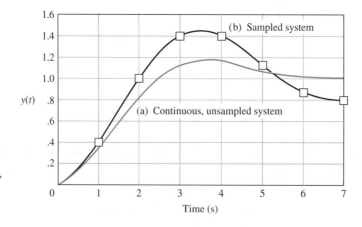

FIGURE 13.17
The response of a
second-order
system: (a)
continuous ($T = 0$),
not sampled; (b)
sampled system,
$T = 1$ s.

A linear continuous feedback control system is stable if all poles of the closed-loop transfer function $T(s)$ lie in the left half of the s-plane. The z-plane is related to the s-plane by the transformation

$$z = e^{sT} = e^{(\sigma + j\omega)T}. \tag{13.26}$$

We may also write this relationship as

$$|z| = e^{\sigma T}$$

and

$$\angle z = \omega T. \tag{13.27}$$

In the left-hand s-plane, $\sigma < 0$; therefore, the related magnitude of z varies between 0 and 1. Thus, the imaginary axis of the s-plane corresponds to the unit circle in the z-plane, and the inside of the unit circle corresponds to the left half of the s-plane [15].

Therefore, we can state that **a sampled system is stable if all the poles of the closed-loop transfer function $T(z)$ lie within the unit circle of the z-plane**.

EXAMPLE 13.5   **Stability of a closed-loop system**

Let us consider the system shown in Figure 13.18 when $T = 1$ and

$$G_p(s) = \frac{K}{s(s + 1)}. \tag{13.28}$$

Recalling Equation (13.16), we note that

$$G(z) = \frac{K(0.3678z + 0.2644)}{z^2 - 1.3678z + 0.3678} = \frac{K(az + b)}{z^2 - (1 + a)z + a}, \tag{13.29}$$

where $a = 0.3678$ and $b = 0.2644$.

The poles of the closed-loop transfer function $T(z)$ are the roots of the equation $1 + G(z) = 0$. We call $q(z) = 1 + G(z) = 0$ the characteristic equation. Therefore, we obtain

$$q(z) = 1 + G(z) = z^2 - (1 + a)z + a + Kaz + Kb = 0. \tag{13.30}$$

When $K = 1$, we have

$$\begin{aligned} q(z) &= z^2 - z + 0.6322 \\ &= (z - 0.50 + j0.6182)(z - 0.50 - j0.6182) = 0. \end{aligned} \tag{13.31}$$

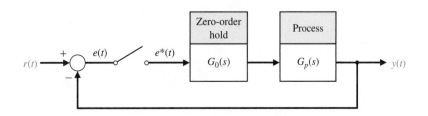

FIGURE 13.18
A closed-loop
sampled system.

Therefore, the system is stable because the roots lie within the unit circle. When $K = 10$, we have

$$q(z) = z^2 + 2.310z + 3.012$$

$$= (z + 1.155 + j1.295)(z + 1.155 - j1.295), \qquad (13.32)$$

and the system is unstable because both roots lie outside the unit circle. This system is stable for $0 < K < 2.39$. The locus of the roots as K varies is discussed in Section 13.8.

We notice that a second-order sampled system can be unstable with increasing gain where a second-order continuous system is stable for all values of gain (assuming both the poles of the open-loop system lie in the left half s-plane). ∎

13.6 PERFORMANCE OF A SAMPLED-DATA, SECOND-ORDER SYSTEM

Let us consider the performance of a sampled second-order system with a zero-order hold, as shown in Figure 13.18, when the process is

$$G_p(s) = \frac{K}{s(\tau s + 1)}. \qquad (13.33)$$

We then obtain $G(z)$ for the arbitrary sampling period T as

$$G(z) = \frac{K\{(z - E)[T - \tau(z - 1)] + \tau(z - 1)^2\}}{(z - 1)(z - E)}, \qquad (13.34)$$

where $E = e^{-T/\tau}$. The stability of the system is analyzed by considering the characteristic equation

$$q(z) = z^2 + z\{K[T - \tau(1 - E)] - (1 + E)\} + K[\tau(1 - E) - TE] + E = 0. \qquad (13.35)$$

Because the polynomial $q(z)$ is a quadratic and has real coefficients, the necessary and sufficient conditions for $q(z)$ to have all its roots within the unit circle are

$$|q(0)| < 1, \qquad q(1) > 0, \quad \text{and} \quad q(-1) > 0.$$

These stability conditions for a second-order system can be established by mapping the z-plane characteristic equation into the s-plane and checking for positive coefficients of $q(s)$. Using these conditions, we establish the necessary conditions from Equation (13.35) as

$$K\tau < \frac{1 - E}{1 - E - (T/\tau)E}, \qquad (13.36)$$

$$K\tau < \frac{2(1 + E)}{(T/\tau)(1 + E) - 2(1 - E)}, \qquad (13.37)$$

and $K > 0, T > 0$. For this system, we can calculate the maximum gain permissible for a stable system. The maximum gain allowable is given in Table 13.3 for several

Table 13.3   Maximum Gain for a Second-Order Sampled System

T/τ	0	0.1	0.5	1	2
Maximum $K\tau$	∞	20.4	4.0	2.32	1.45

values of T/τ. If the computer system has sufficient speed of computation and data handling, it is possible to set $T/\tau = 0.1$ and obtain system characteristics approaching those of a continuous (nonsampled) system.

The maximum overshoot of the second-order system for a unit step input is shown in Figure 13.19.

The performance criterion, integral squared error, can be written as

$$I = \frac{1}{\tau} \int_0^\infty e^2(t)\, dt. \tag{13.38}$$

The loci of this criterion are given in Figure 13.20 for constant values of I. For a given value of T/τ, we can determine the minimum value of I and the required

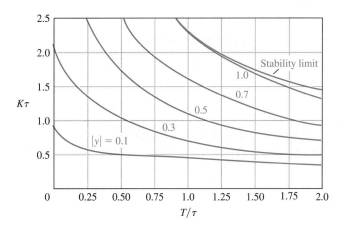

FIGURE 13.19
The maximum overshoot $|y|$ for a second-order sampled system for a unit step input.

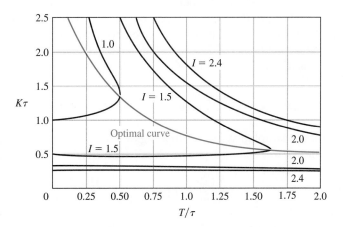

FIGURE 13.20
The loci of integral squared error for a second-order sampled system for constant values of I.

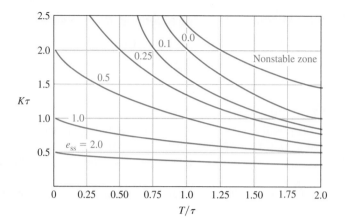

FIGURE 13.21
The steady-state error of a second-order sampled system for a unit ramp input $r(t) = t, t > 0$.

value of $K\tau$. The optimal curve shown in Figure 13.20 indicates the required $K\tau$ for a specified T/τ that minimizes I. For example, when $T/\tau = 0.75$, we require $K\tau = 1$ in order to minimize the performance criterion I.

The steady-state error for a unit ramp input $r(t) = t$ is shown in Figure 13.21. For a given T/τ, we can reduce the steady-state error, but then the system yields a greater overshoot and settling time for a step input.

EXAMPLE 13.6   Design of a sampled system

Let us consider a closed-loop sampled system as shown in Figure 13.18 when

$$G_p(s) = \frac{K}{s(0.1s + 1)(0.005s + 1)} \tag{13.39}$$

and we need to select T and K for suitable performance. As an approximation, we neglect the effects of the time constant $\tau_2 = 0.005$ s, because it is only 5% of the primary time constant $\tau_1 = 0.1$. Then we can use Figures 13.19, 13.20, and 13.21 to select K and T. Limiting the overshoot to 30% for the step input, we select $T/\tau = 0.25$, yielding $K\tau = 1.4$. For these values, the steady-state error for a unit ramp input is approximately 0.6 (see Figure 13.21).

Because $\tau = 0.1$, we then set $T = 0.025$ s and $K = 14$. The sampling rate is then required to be 40 samples per second.

The overshoot to the step input and the steady-state error for a ramp input may be reduced if we set T/τ to 0.1. The overshoot to a step input will be 25% for $K\tau = 1.6$. Using Figure 13.21, we estimate that the steady-state error for a unit ramp input is 0.55 for $K\tau = 1.6$. ■

13.7 CLOSED-LOOP SYSTEMS WITH DIGITAL COMPUTER COMPENSATION

A closed-loop, sampled system with a digital computer used to improve the performance is shown in Figure 13.15. The closed-loop transfer function is

$$\frac{Y(z)}{R(z)} = T(z) = \frac{G(z)D(z)}{1 + G(z)D(z)}. \tag{13.40}$$

The transfer function of the computer is represented by

$$\frac{U(z)}{E(z)} = D(z). \tag{13.41}$$

In our prior calculations, $D(z)$ was represented simply by a gain K. As an illustration of the power of the computer as a compensator, we will consider again the second-order system with a zero-order hold and process

$$G_p(s) = \frac{1}{s(s+1)} \text{ when } T = 1.$$

Then (see Equation 13.16)

$$G(z) = \frac{0.3678(z+0.7189)}{(z-1)(z-0.3678)}. \tag{13.42}$$

If we select

$$D(z) = \frac{K(z-0.3678)}{z+r}, \tag{13.43}$$

we cancel the pole of $G(z)$ at $z = 0.3678$ and have to set two parameters, r and K. If we select

$$D(z) = \frac{1.359(z-0.3678)}{z+0.240}, \tag{13.44}$$

we have

$$G(z)D(z) = \frac{0.50(z+0.7189)}{(z-1)(z+0.240)}. \tag{13.45}$$

If we calculate the response of the system to a unit step, we find that the output is equal to the input at the fourth sampling instant and thereafter. The responses for both the uncompensated and the compensated system are shown in Figure 13.22.

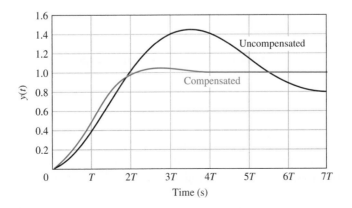

FIGURE 13.22
The response of a
sampled-data
second-order
system to a unit
step input.

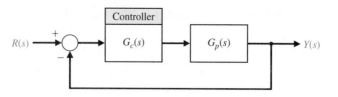

FIGURE 13.23
The continuous system model of a sampled system.

The overshoot of the compensated system is 4%, whereas the overshoot of the uncompensated system is 45%. It is beyond the objective of this book to discuss all the extensive methods for the analytical selection of the parameters of $D(z)$; other texts [2–4] can provide further information. However, we will consider two methods of compensator design: (1) the $G_c(s)$-to-$D(z)$ conversion method (in the following paragraphs) and (2) the root locus z-plane method (in Section 13.8).

One method for determining $D(z)$ first determines a controller $G_c(s)$ for a given process $G_p(s)$ for the system shown in Figure 13.23. Then, the controller is converted to $D(z)$ for the given sampling period T. The methods described in Chapter 10 are used to determine $G_c(s)$. This design method is called the $G_c(s)$-to-$D(z)$ conversion method. It converts the $G_c(s)$ of Figure 13.23 to $D(z)$ of Figure 13.15 [7].

We consider a first-order compensator

$$G_c(s) = K\frac{s + a}{s + b} \tag{13.46}$$

and a **digital controller**

$$D(z) = C\frac{z - A}{z - B}. \tag{13.47}$$

We determine the z-transform corresponding to $G_c(s)$ and set it equal to $D(z)$ as

$$Z\{G_c(s)\} = D(z). \tag{13.48}$$

Then the relationship between the two transfer functions is $A = e^{-aT}, B = e^{-bT}$, and when $s = 0$, we require that

$$C\frac{1 - A}{1 - B} = K\frac{a}{b}. \tag{13.49}$$

EXAMPLE 13.7   **Design to meet a phase margin specification**

Consider a system with a process

$$G_p(s) = \frac{1740}{s(0.25s + 1)}. \tag{13.50}$$

We will attempt to design $G_c(s)$ so that we achieve a phase margin of 45° with a crossover frequency $\omega_c = 125$ rad/s. Using the Bode diagram of $G_p(s)$, we find that the phase margin is 2°. Using the method of Section 10.4, we find that the required pole–zero ratio is $\alpha = 6.25$. It is specified that $\omega_c = 125$, so we note that $\omega_c = (ab)^{1/2}$. Therefore, $a = 50$ and $b = 312$. The lead compensator is then

$$G_c(s) = \frac{K(s + 50)}{s + 312}. \tag{13.51}$$

We select K in order to yield $|GG_c(j\omega)| = 1$ when $\omega = \omega_c = 125$ rad/s. Then we find that $K = 5.6$. The compensator $G_c(s)$ is to be realized by $D(z)$, so we solve the relationships with a selected sampling period. Setting $T = 0.001$ s, we have

$$A = e^{-0.05} = 0.95, \quad B = e^{-0.312} = 0.73, \quad \text{and} \quad C = 4.85.$$

Then we have

$$D(z) = \frac{4.85(z - 0.95)}{z - 0.73}. \tag{13.52}$$

Of course, if we select another value for the sampling period, then the coefficients of $D(z)$ would differ. ∎

In general, we select a small sampling period so that the design based on the continuous system will accurately carry over to the z-plane. However, we should not select too small a T, or the computation requirements may be more than necessary. In general, we use a sampling period $T \approx 1/(10f_B)$, where $f_B = \omega_B/(2\pi)$, and ω_B is the bandwidth of the closed-loop continuous system.

The bandwidth of the system designed in Example 13.7 is $\omega_B = 180$ rad/s or $f_B = 28.6$ Hz. Thus we select a period $T = 0.003$ s. Note that $T = 0.001$ s was used in Example 13.7.

13.8 THE ROOT LOCUS OF DIGITAL CONTROL SYSTEMS

Let us consider the transfer function of the system shown in Figure 13.24. Recall that $G(s) = G_0(s)G_p(s)$. The closed-loop transfer function is

$$\frac{Y(z)}{R(z)} = \frac{KG(z)D(z)}{1 + KG(z)D(z)}. \tag{13.53}$$

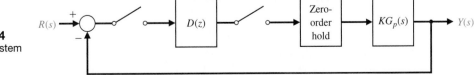

FIGURE 13.24
Closed-loop system with a digital controller.

Table 13.4   Root Locus in the z-Plane

1. The root locus starts at the poles and progresses to the zeros.
2. The root locus lies on a section of the real axis to the left of an odd number of poles and zeros.
3. The root locus is symmetrical with respect to the horizontal real axis.
4. The root locus may break away from the real axis and may reenter the real axis. The breakaway and entry points are determined from the equation

$$K = -\frac{N(z)}{D(z)} = F(z),$$

with $z = \sigma$. Then obtain the solution of $\dfrac{dF(\sigma)}{d\sigma} = 0.$

5. Plot the locus of roots that satisfy

$$1 + KG(z)D(z) = 0,$$

or

$$|KG(z)D(z)| = 1$$

and

$$\underline{/G(z)D(z)} = 180° \pm k360°, \qquad k = 0, 1, 2, \ldots$$

The **characteristic equation** is

$$\boxed{1 + KG(z)D(z) = 0,}$$

which is analogous to the characteristic equation for the s-plane analysis of $KG(s)$. Thus, we can plot the root locus for the characteristic equation of the sampled system as K varies. The rules for obtaining the root locus are summarized in Table 13.4.

EXAMPLE 13.8   **Root locus of a second-order system**

Consider the system shown in Figure 13.24 with $D(z) = 1$ and $G_p(s) = 1/s^2$. Then we obtain

$$KG(z) = \frac{T^2}{2} \frac{K(z + 1)}{(z - 1)^2}.$$

Let $T = \sqrt{2}$ and plot the root locus. We now have

$$KG(z) = \frac{K(z + 1)}{(z - 1)^2},$$

and the poles and zeros are shown on the z-plane in Figure 13.25. The characteristic equation is

$$1 + KG(z) = 1 + \frac{K(z + 1)}{(z - 1)^2} = 0.$$

Let $z = \sigma$ and solve for K to obtain

$$K = -\frac{(\sigma - 1)^2}{\sigma + 1} = F(\sigma).$$

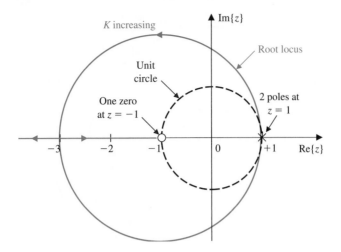

FIGURE 13.25
Root locus for
Example 13.8.

Then obtain the derivative $dF(\sigma)/d\sigma = 0$ and calculate the roots as $\sigma_1 = -3$ and $\sigma_2 = 1$. The locus leaves the two poles at $\sigma_2 = 1$ and reenters at $\sigma_1 = -3$, as shown in Figure 13.25. The unit circle is also shown in Figure 13.25. The system always has two roots outside the unit circle and is always unstable for all $K > 0$. ∎

We now turn to the design of a digital controller $D(z)$ to achieve a specified response utilizing a root locus method. We will select a controller

$$D(z) = \frac{z - a}{z - b}.$$

We then use $z - a$ to cancel one pole at $G(z)$ that lies on the positive real axis of the z-plane. Then we select $z - b$ so that the locus of the compensated system will give a set of complex roots at a desired point within the unit circle on the z-plane.

EXAMPLE 13.9   **Design of a digital compensator**

Let us design a compensator $D(z)$ that will result in a stable system when $G_p(s)$ is as described in Example 13.8. With $D(z) = 1$, we have an unstable system. Select

$$D(z) = \frac{z - a}{z - b}$$

so that

$$KG(z)D(z) = \frac{K(z + 1)(z - a)}{(z - 1)^2(z - b)}.$$

If we set $a = 1$ and $b = 0.2$, we have

$$KG(z)D(z) = \frac{K(z + 1)}{(z - 1)(z - 0.2)}.$$

Using the equation for $F(\sigma)$, we obtain the entry point as $z = -2.56$, as shown in Figure 13.26. The root locus is on the unit circle at $K = 0.8$. Thus, the system is stable

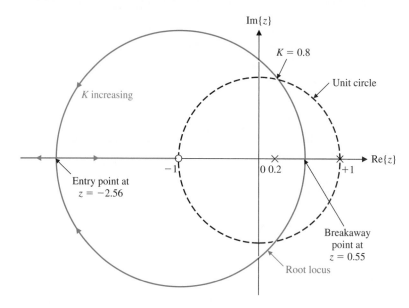

FIGURE 13.26
Root locus for
Example 13.9.

for $K < 0.8$. If we select $K = 0.25$, we find that the step response has an overshoot of 20% and a settling time (with a 2% criterion) equal to 8.5 seconds.

If the system performance were inadequate, we would improve the root locus by selecting $a = 1$ and $b = -0.98$ so that

$$KG(z)D(z) = \frac{K(z + 1)}{(z - 1)(z + 0.98)} \approx \frac{K}{z - 1}.$$

Then the root locus would lie on the real axis of the z-plane. When $K = 1$, the root of the characteristic equation is at the origin, and $T(z) = 1/z = z^{-1}$. Then the response of the sampled system (at the sampling instants) is the input step delayed by one sampling period. ∎

We can draw lines of constant ζ on the z-plane. The mapping between the s-plane and the z-plane is obtained by the relation $z = e^{sT}$. The lines of constant ζ on the s-plane are radial lines with

$$\frac{\sigma}{\omega} = -\tan \theta = -\tan(\sin^{-1} \zeta) = -\frac{\zeta}{\sqrt{1 - \zeta^2}}.$$

Since $s = \sigma + j\omega$, we have

$$z = e^{\sigma T} e^{j\omega T},$$

where

$$\sigma = -\frac{\zeta}{\sqrt{1 - \zeta^2}} \omega.$$

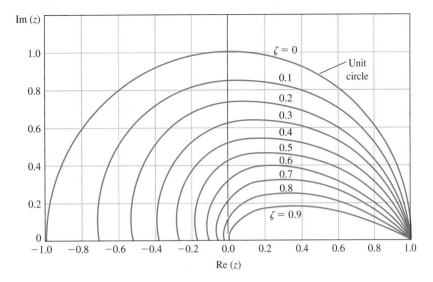

FIGURE 13.27
Curves of constant ζ on the z-plane.

The plot of these lines for constant ζ is shown in Figure 13.27 for a range of T. A common value of ζ for many design specifications is $\zeta = 1/\sqrt{2}$. Then we have $\sigma = -\omega$ and

$$z = e^{-\omega T} e^{j\omega T} = e^{-\omega T} \underline{/\theta},$$

where $\theta = \omega T$.

13.9   IMPLEMENTATION OF DIGITAL CONTROLLERS

We will consider the PID controller with an s-domain transfer function

$$\frac{U(s)}{X(s)} = G_c(s) = K_P + \frac{K_I}{s} + K_D s. \tag{13.54}$$

We can determine a digital implementation of this controller by using a discrete approximation for the derivative and integration. For the time derivative, we use the **backward difference rule**

$$u(kT) = \left.\frac{dx}{dt}\right|_{t=kT} = \frac{1}{T}(x(kT) - x[(k-1)T]). \tag{13.55}$$

The z-transform of Equation (13.55) is then

$$U(z) = \frac{1 - z^{-1}}{T} X(z) = \frac{z-1}{Tz} X(z).$$

The integration of $x(t)$ can be represented by the **forward-rectangular integration** at $t = kT$ as

$$u(kT) = u[(k-1)T] + Tx(kT), \tag{13.56}$$

where $u(kT)$ is the output of the integrator at $t = kT$. The z-transform of Equation (13.56) is

$$U(z) = z^{-1}U(z) + TX(z),$$

and the transfer function is then

$$\frac{U(z)}{X(z)} = \frac{Tz}{z - 1}.$$

Hence, the z-domain transfer function of the **PID controller** is

$$G_c(z) = K_P + \frac{K_I Tz}{z - 1} + K_D \frac{z - 1}{Tz}. \qquad (13.57)$$

The complete difference equation algorithm that provides the PID controller is obtained by adding the three terms to obtain [we use $x(kT) = x(k)$]

$$
\begin{aligned}
u(k) &= K_P x(k) + K_I[u(k - 1) + Tx(k)] + (K_D/T)[x(k) - x(k - 1)] \\
&= [K_P + K_I T + (K_D/T)]x(k) - K_D Tx(k - 1) + K_I u(k - 1). \qquad (13.58)
\end{aligned}
$$

Equation (13.58) can be implemented using a digital computer or microprocessor. Of course, we can obtain a PI or PD controller by setting an appropriate gain equal to zero.

13.10 DESIGN EXAMPLES

In this section we present two illustrative examples. In the first example, two controllers are designed to control the motor and lead screw of a movable worktable. Using a zero-order hold formulation, a proportional controller and a lead compensator are obtained and their performance compared. In the second example, a control system is designed control an aircraft control surface as part of a fly-by-wire system. Using root locus methods, the design process focuses on the design of a digital controller to meet settling time and percent overshoot performance specifications.

EXAMPLE 13.10   **Worktable motion control system**

An important positioning system in manufacturing systems is a worktable motion control system. The system controls the motion of a worktable at a certain location [21]. We assume that the table is activated in each axis by a motor and lead screw, as shown in Figure 13.28(a). We consider the x-axis and examine the motion control for a feedback system, as shown in Figure 13.28(b). The goal is to obtain a fast response with a rapid rise time and settling time to a step command while not exceeding an overshoot of 5%.

The specifications are then (1) a percent overshoot equal to 5% and (2) a minimum settling time (with a 2% criterion) and rise time. Rise time is defined as the time to reach the magnitude of the command and is illustrated in Figure 5.7 by T_R.

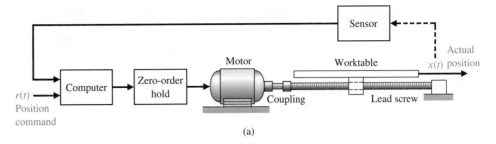

FIGURE 13.28
A table motion control system: (a) actuator and table; (b) block diagram.

To configure the system, we choose a power amplifier and motor so that the system is described by Figure 13.29. Obtaining the transfer function of the motor and power amplifier, we have

$$G_p(s) = \frac{1}{s(s + 10)(s + 20)}. \tag{13.59}$$

We will initially use a continuous system and design $G_c(s)$ as described in Section 13.8. We then obtain $D(z)$ from $G_c(s)$. First, we select the controller as a simple gain K in order to determine the response that can be achieved without a compensator. Plotting the root locus, we find that when $K = 700$, the dominant complex roots have a damping ratio of 0.707, and we expect a 5% overshoot. Then, using a simulation, we find that the overshoot is 5%, the rise time is 0.48 second, and the settling time (with a 2% criterion) is 1.12 seconds. These values are recorded as item 1 in Table 13.5.

The next step is to introduce a lead compensator, so that

$$G_c(s) = \frac{K(s + a)}{s + b}. \tag{13.60}$$

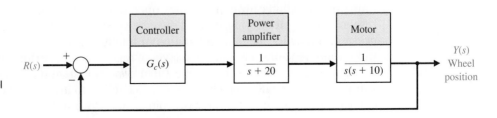

FIGURE 13.29
Model of the wheel control for a work table.

Table 13.5   Performance for Two Controllers

Compensator $G_c(s)$	K	Percent Overshoot	Settling Time (seconds)	Rise Time (seconds)
1. K	700	5.0	1.12	0.40
2. $K(s + 11)/(s + 62)$	8000	5.0	0.60	0.25

We will select the zero at $s = -11$ so that the complex roots near the origin dominate. Using the method of Section 10.5, we find that we require the pole at $s = -62$. Evaluating the gain at the roots, we find that $K = 8000$. Then the step response has a rise time of 0.25 second and a settling time (with a 2% criterion) of 0.60 second. This is an improved response, and we finalize this system as acceptable.

It now remains to select the sampling period and then use the method of Section 13.7 to obtain $D(z)$. The rise time of the compensated continuous system is 0.25 second. Then we require $T \ll T_R$ in order to obtain a system response predicted by the design of the continuous system. Let us select $T = 0.01$ s. We have

$$G_c(s) = \frac{8000(s + 11)}{s + 62}.$$

Then

$$D(z) = C\frac{z - A}{z - B},$$

where

$$A = e^{-11T} = 0.8958 \quad \text{and} \quad B = e^{-62T} = 0.5379.$$

We now have

$$C = K\frac{a(1 - B)}{b(1 - A)} = \frac{8000(11)(0.462)}{62(0.1042)} = 6293.$$

Using this $D(z)$, we expect a response very similar to that obtained for the continuous system model. ∎

EXAMPLE 13.11   **Fly-by-wire aircraft control surface**

Increasing constraints on weight, performance, fuel consumption, and reliability created a need for a new type of flight control system known as fly-by-wire. This approach implies that particular system components are interconnected electrically rather than mechanically and that they operate under the supervision of a computer responsible for monitoring, controlling, and coordinating the tasks. The fly-by-wire principle allows for the implementation of totally digital and highly redundant control systems reaching a remarkable level of reliability and performance [24].

Operational characteristics of a flight control system depend on the dynamic stiffness of an actuator, which represents its ability to maintain the position of the control surface in spite of the disturbing effects of random external forces. One

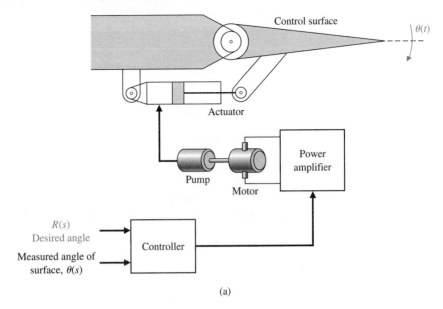

FIGURE 13.30
(a) Fly-by-wire
aircraft control
surface system and
(b) block diagram.
The sampling
period is 0.1
second.

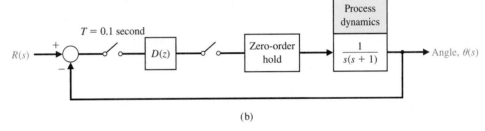

flight actuator system consists of a special type of DC motor, driven by a power amplifier, which drives a hydraulic pump that is connected to either side of a hydraulic cylinder. The piston of the hydraulic cylinder is directly connected to a control surface of an aircraft through some appropriate mechanical linkage, as shown in Figure 13.30. The elements of the design process emphasized in this example are highlighted in Figure 13.31.

The process model is given by

$$G_p(s) = \frac{1}{s(s+1)}. \tag{13.61}$$

The zero-order hold is modeled by

$$G_o(s) = \frac{1 - e^{-sT}}{s}. \tag{13.62}$$

Combining the process and the zero-order hold in series yields

$$G(s) = G_o(s)G_p(s) = \frac{1 - e^{-sT}}{s^2(s+1)}. \tag{13.63}$$

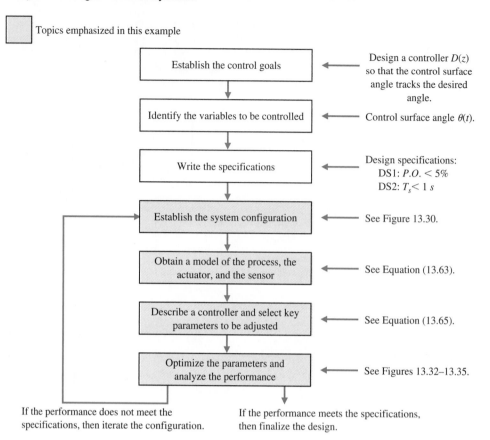

☐ Topics emphasized in this example

Establish the control goals ← Design a controller $D(z)$ so that the control surface angle tracks the desired angle.

Identify the variables to be controlled ← Control surface angle $\theta(t)$.

Write the specifications ← Design specifications:
DS1: *P.O.* < 5%
DS2: T_s < 1 *s*

Establish the system configuration ← See Figure 13.30.

Obtain a model of the process, the actuator, and the sensor ← See Equation (13.63).

Describe a controller and select key parameters to be adjusted ← See Equation (13.65).

Optimize the parameters and analyze the performance ← See Figures 13.32–13.35.

If the performance does not meet the specifications, then iterate the configuration.

If the performance meets the specifications, then finalize the design.

FIGURE 13.31
Elements of the control system design process emphasized in this fly-by-wire aircraft control surface example.

The control goal is to design a compensator, $D(z)$, so that the control surface angle $Y(s) = \theta(s)$ tracks the desired angle, denoted by $R(s)$. We state the control goal as

Control Goal

Design a controller $D(z)$ so that the control surface angle tracks the desired angle.

The variable to be controlled is the control surface angle $\theta(t)$:

Variable to Be Controlled

Control surface angle $\theta(t)$.

The design specifications are as follows:

Design Specifications

DS1   Percent overshoot less than 5% to a unit step input.

DS2   Settling time less than 1 second to a unit step input.

We begin the design process by determining $G(z)$ from $G(s)$. Expanding $G(s)$ in Equation (13.63) in partial fractions yields

$$G(s) = (1 - e^{-sT})\left(\frac{1}{s^2} - \frac{1}{s} + \frac{1}{s+1}\right),$$

and

$$G(z) = Z\{G(s)\} = \frac{ze^{-T} - z + Tz + 1 - e^{-T} - Te^{-T}}{(z-1)(z-e^{-T})},$$

where $Z\{\cdot\}$ represents the z-transform. Choosing $T = 0.1$, we have

$$G(z) = \frac{0.004837z + 0.004679}{(z-1)(z-0.9048)}. \tag{13.64}$$

For a simple compensator, $D(z) = K$, the root locus is shown in Figure 13.32. For stability we require $K < 21$. Note that the stability region for discrete-time systems is inside the unit circle in the complex plane. Recall that for continuous-time systems, the stability region is the left half-plane.

Using an iterative approach we discover that as $K \to 21$, the step response is very oscillatory, and the percent overshoot is too large; conversely, as K gets smaller, the settling time gets too long, although the percent overshoot decreases. In any case the design specifications cannot be satisfied with a simple proportional controller, $D(z) = K$. We need to utilize a more sophisticated controller.

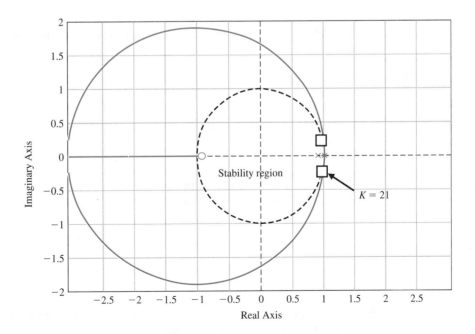

FIGURE 13.32
Root locus for
$D(z) = K$.

We have the freedom to select the controller type. As with control design for continuous-time systems, the choice of compensator is always a challenge and problem-dependent. Here we choose a compensator with the general structure

$$D(z) = K\frac{z - a}{z - b}. \tag{13.65}$$

Therefore, the key tuning parameters are the compensation parameters:

Select Key Tuning Parameters
    K, a, and b.

For continuous systems we know that a design rule-of-thumb formula for the settling time is

$$T_s = \frac{4}{\zeta\omega_n},$$

where we use a 2% bound to define settling. This design rule-of-thumb is valid for second-order systems with no zeros. So to meet the T_s requirement, we want

$$-\text{Re}(s_i) = \zeta\omega_n > \frac{4}{T_s}, \tag{13.66}$$

where s_i, $i = 1, 2$ are the dominant complex-conjugate poles. In the definition of the desired region of the z-plane for placing the dominant poles, we use the transform

$$z = e^{s_iT} = e^{\left(-\zeta\omega_n \pm j\omega_n\sqrt{(1-\zeta^2)}\right)T} = e^{-\zeta\omega_nT}e^{\pm j\omega_nT\sqrt{(1-\zeta^2)}}.$$

Computing the magnitude of z yields

$$r_o = |z| = e^{-\zeta\omega_nT}.$$

To meet the settling time specification, we need the z-plane poles to be inside the circle defined by

$$r_o = e^{-4T/T_s}, \tag{13.67}$$

where we have used the result in Equation (13.66).

Consider the settling time requirement $T_s < 1$ s. In our case $T = 0.1$ s. From Equation (13.67) we determine that the dominant z-plane poles should lie inside the circle defined by

$$r_o = e^{-0.4/1} = 0.67.$$

As shown previously we can draw lines of constant ζ on the z-plane. The lines of constant ζ on the s-plane are radial lines with

$$\sigma = -\omega \tan(\sin^{-1}\zeta) = -\frac{\zeta}{\sqrt{1 - \zeta^2}}\omega.$$

Then, with $s = \sigma + j\omega$ and using the transform $z = e^{sT}$, we have

$$z = e^{-\sigma\omega T}e^{j\omega T}. \tag{13.68}$$

For a given ζ, we can plot $\text{Re}(z)$ vs $\text{Im}(z)$ for z given in Equation (13.68).

If we were working with a second-order transfer function in the s-domain, we would need to have the damping ratio associated with the dominant roots be greater

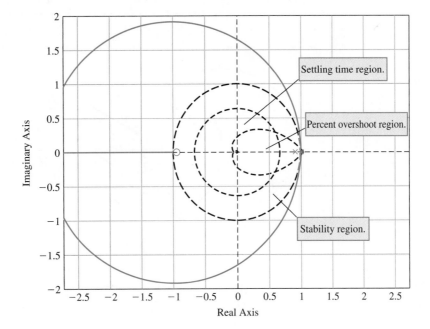

FIGURE 13.33
Root locus for
$D(z) = K$ with the
stability and
performance
regions shown.

than $\zeta \geq 0.69$. When $\zeta \geq 0.69$, the percent overshoot for a second-order system (with no zeros) will be less than 5%. The curves of constant ζ on the z-plane will define the region in the z-plane where we need to place the dominant z-plane poles to meet the percent overshoot specification.

The root locus in Figure 13.32 is repeated in Figure 13.33 with the stability and desired performance regions included. We can see that the root locus does not lie in the intersection of the stability and performance regions. The question is how to select the controller parameters K, a, and b so that the root locus lies in the desired regions.

One approach to the design is to choose a such that the pole of $G(z)$ at $z = 0.9048$ is cancelled. Then we must select b so that the root locus lies in the desired region. For example, when $a = -0.9048$ and $b = 0.25$, the compensated root locus appears as shown in Figure 13.34. The root locus lies inside the perfor-mance region, as desired.

A valid value of K is $K = 70$. Thus the compensator is

$$D(z) = 70 \frac{s - 0.9048}{s + 0.25}.$$

The closed-loop step response is shown in Figure 13.35. Notice that the per-cent overshoot specification ($P.O. \leq 5\%$) is satisfied, and the system response settles in less than 10 samples (10 samples = 1 second because the sampling time is 0.1 second). ∎

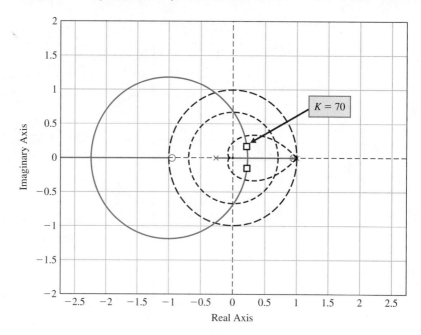

FIGURE 13.34
Compensated root locus.

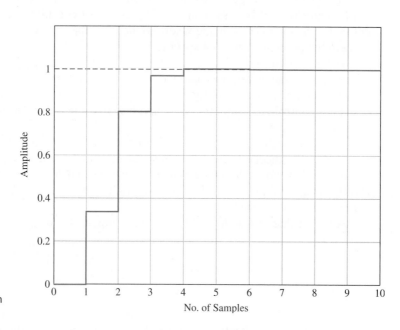

FIGURE 13.35
Closed-loop system step response.

13.11 DIGITAL CONTROL SYSTEMS USING CONTROL DESIGN SOFTWARE

The process of designing and analyzing sampled-data systems is enhanced with the use of interactive computer tools. Many of the control design functions for continuous-time control design have equivalent counterparts for sampled-data systems. Discrete-time transfer function model objects are obtained with the tf function, similar to continuous time models discussed in Chapter 2. Figure 13.36 illustrates the use of tf. Model conversion can be accomplished with the functions c2d and d2c, shown in Figure 13.36. The function c2d converts continuous-time systems to discrete-time systems; the function d2c converts discrete-time systems to continuous-time systems. For example, consider the process transfer function

$$G_p(s) = \frac{1}{s(s + 1)},$$

as shown in Figure 13.16. For a sampling period of $T = 1$ s, we know from Equation (13.16) that

$$G(z) = \frac{0.3678(z + 0.7189)}{(z - 1)(z - 0.3680)} = \frac{0.3679z + 0.2644}{z^2 - 1.368z + 0.3680}. \qquad (13.69)$$

We can use an m-file script to obtain the $G(z)$, as shown in Figure 13.37.

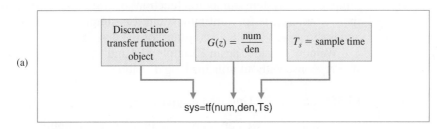

(a)

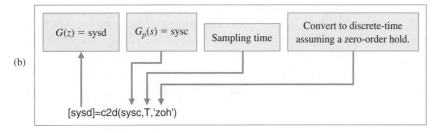

(b)

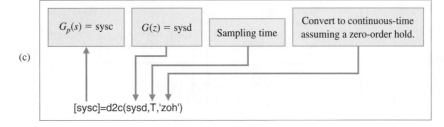

(c)

FIGURE 13.36
(a) The **tf** function.
(b) The **c2d** function. (c) The **d2c** function.

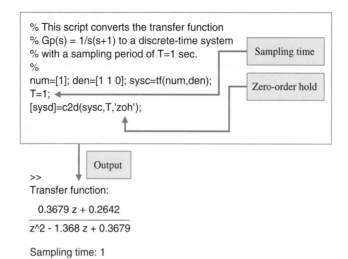

FIGURE 13.37
Using the **c2d** function to convert $G(s) = G_0(s)G_p(s)$ to $G(z)$.

The functions **step, impulse,** and **lsim** are used for simulation of sampled-data systems. The unit step response is generated by **step**. The **step** function format is shown in Figure 13.38. The unit impulse response is generated by the function **impulse,** and the response to an arbitrary input is obtained by the **lsim** function. The **impulse** and **lsim** functions are shown in Figure 13.39 and 13.40, respectively. These sampled-data system simulation functions operate in essentially the same manner as their counterparts for continuous-time (unsampled) systems. The output is $y(kT)$ and is shown as $y(kT)$ held constant for the period T.

We now consider again Example 13.4 and approach the problem of obtaining a step response without utilizing long division.

EXAMPLE 13.12   **Unit step response**

In Example 13.4, we considered the problem of computing the step response of a closed-loop sampled-data system. In that example, the response, $y(kT)$, was computed using long division. We can compute the response $y(kT)$ using the **step** function,

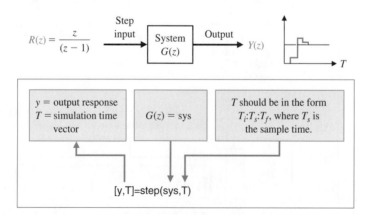

FIGURE 13.38
The **step** function generates the output $y(kT)$ for a step input.

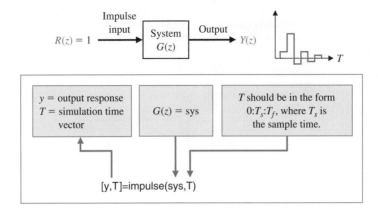

FIGURE 13.39
The **impulse** function generates the output $y(kT)$ for an impulse input.

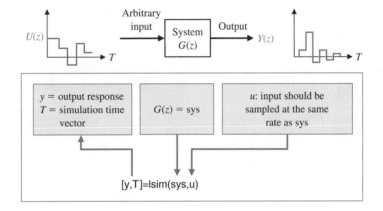

FIGURE 13.40
The **lsim** function generates the output $y(kT)$ for an arbitrary input.

shown in Figure 13.38. With the closed-loop transfer function given by

$$\frac{Y(z)}{R(z)} = \frac{0.3678z + 0.2644}{z^2 - z + 0.6322},$$

the associated closed-loop step response is shown in Figure 13.41. The discrete step response shown in this figure is also shown in Figure 13.17. To determine the actual continuous response $y(t)$, we use the m-file script as shown in Figure 13.42. The zero-order hold is modeled by the transfer function

$$G_0(s) = \frac{1 - e^{-sT}}{s}.$$

In the m-file script in Figure 13.42, we approximate the e^{-sT} term using the **pade** function with a second-order approximation and a sampling time of 1 second. We then compute an approximation for $G_0(s)$ based on the Padé approximation of e^{-sT}. ∎

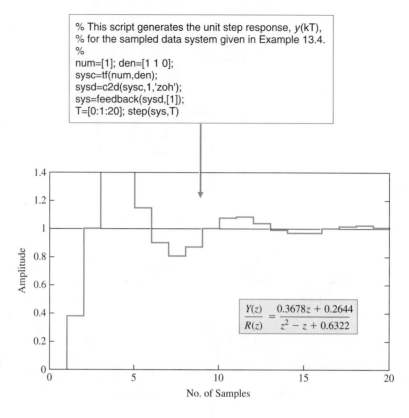

```
% This script generates the unit step response, y(kT),
% for the sampled data system given in Example 13.4.
%
num=[1]; den=[1 1 0];
sysc=tf(num,den);
sysd=c2d(sysc,1,'zoh');
sys=feedback(sysd,[1]);
T=[0:1:20]; step(sys,T)
```

$$\frac{Y(z)}{R(z)} = \frac{0.3678z + 0.2644}{z^2 - z + 0.6322}$$

No. of Samples

FIGURE 13.41
The discrete
response, $y(kT)$, of a
sampled second-
order system to a
unit step.

The subject of digital computer compensation was discussed in Section 13.7. In the next example, we consider again the subject utilizing control design software.

EXAMPLE 13.13   **Root locus of a digital control system**

Recall from Equation (13.16) that the process was given by

$$G(z) = \frac{0.3678(z + 0.7189)}{(z - 1)(z - 0.3680)}.$$

The compensator is selected to be

$$D(z) = \frac{K(z - 0.3678)}{z + 0.2400},$$

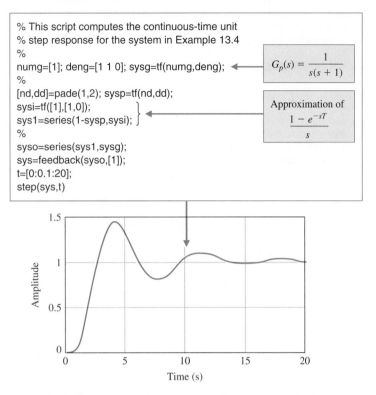

```
% This script computes the continuous-time unit
% step response for the system in Example 13.4
%
numg=[1]; deng=[1 1 0]; sysg=tf(numg,deng);   ←————  $G_p(s) = \dfrac{1}{s(s+1)}$
%
[nd,dd]=pade(1,2); sysp=tf(nd,dd);
sysi=tf([1],[1,0]);           ⎤
sys1=series(1-sysp,sysi);     ⎦  ←——————————  Approximation of $\dfrac{1 - e^{-sT}}{s}$
%
syso=series(sys1,sysg);
sys=feedback(syso,[1]);
t=[0:0.1:20];
step(sys,t)
```

FIGURE 13.42   The continuous response $y(t)$ to a unit step for the system of Figure 13.16.

with the parameter K as a variable yet to be determined. When

$$G(z)D(z) = K\frac{0.3678(z + 0.7189)}{(z - 1)(z + 0.2400)},\tag{13.70}$$

we have the problem in a form for which the root locus method is directly applicable. The rlocus function works for discrete-time systems in the same way as for continuous-time systems. Using a m-file script, the root locus associated with Equation (13.70) is easily generated, as shown in Figure 13.43. Remember that the stability region is defined by the unit circle in the complex plane. The function rlocfind can be used with the discrete-time system root locus in exactly the same way as for continuous-time systems to determine the value of the system gain associated with any point on the locus. Using rlocfind, we determine that $K = 4.639$ places the roots on the unit circle. ■

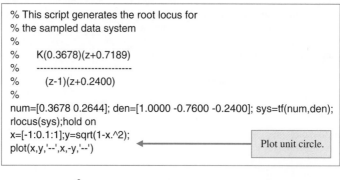

```
% This script generates the root locus for
% the sampled data system
%
%      K(0.3678)(z+0.7189)
%      ----------------------------
%         (z-1)(z+0.2400)
%
num=[0.3678 0.2644]; den=[1.0000 -0.7600 -0.2400]; sys=tf(num,den);
rlocus(sys);hold on
x=[-1:0.1:1];y=sqrt(1-x.^2);
plot(x,y,'--',x,-y,'--')
```

Plot unit circle.

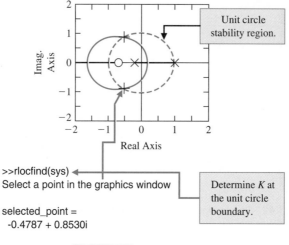

Unit circle stability region.

```
>>rlocfind(sys)
Select a point in the graphics window
```

Determine K at the unit circle boundary.

```
selected_point =
  -0.4787 + 0.8530i
```

FIGURE 13.43
The **rlocus** function for sampled data systems.

```
ans =
   4.6390
```

$K = 4.639$

13.12 SEQUENTIAL DESIGN EXAMPLE: DISK DRIVE READ SYSTEM

In this chapter, we will design a digital controller for the disk drive system. As the disk rotates, the sensor head reads the patterns used to provide the reference error information. This error information pattern is read intermittently as the head reads the stored data, and then the pattern in turn. Because the disk is rotating at a constant speed, the time T between position-error readings is a constant. This sampling period is typically 100 μs to 1 ms [25]. Thus, we have sampled error information. We may also use a digital controller, as shown in Figure 13.44, to achieve a satisfactory system response. In this section, we will design $D(z)$.

First, we determine

$$G(z) = Z[G_0(s)G_p(s)].$$

Since

$$G_p(s) = \frac{5}{s(s + 20)}, \tag{13.71}$$

FIGURE 13.44
Feedback control
system with a
digital controller.
Note that $G(z) = Z[G_0(s)G_p(s)]$.

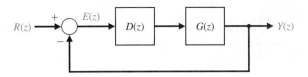

we have

$$G_0(s)G_p(s) = \frac{1 - e^{-sT}}{s}\frac{5}{s(s + 20)}.$$

We note that for $s = 20$ and $T = 1$ ms, e^{-sT} is equal to 0.98. Then we see that the pole at $s = -20$ in Equation (13.71) has an insignificant effect. Therefore, we could approximate

$$G_p(s) \approx \frac{0.25}{s}.$$

Then we need

$$G(z) = Z\left[\frac{1 - e^{-sT}}{s}\frac{0.25}{s}\right]$$

$$= (1 - z^{-1})(0.25)Z\left[\frac{1}{s^2}\right]$$

$$= (1 - z^{-1})(0.25)\frac{Tz}{(z - 1)^2}$$

$$= \frac{0.25T}{z - 1} = \frac{0.25 \times 10^{-3}}{z - 1}.$$

We need to select the digital controller $D(z)$ so that the desired response is achieved for a step input. If we set $D(z) = K$, then we have

$$D(z)G(z) = \frac{K(0.25 \times 10^{-3})}{z - 1}.$$

The root locus for this system is shown in Figure 13.45. When $K = 4000$,

$$D(z)G(z) = \frac{1}{z - 1}.$$

Therefore, the closed-loop transfer function is

$$T(z) = \frac{D(z)G(z)}{1 + D(z)G(z)} = \frac{1}{z}.$$

We expect a rapid response for the system. The percent overshoot to a step input is 0%, and the settling time is 2 ms.

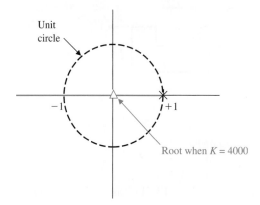

FIGURE 13.45
Root locus.

13.13 SUMMARY

The use of a digital computer as the compensation device for a closed-loop control system has grown during the past two decades as the price and reliability of computers have improved dramatically. A computer can be used to complete many calculations during the sampling interval T and to provide an output signal that is used to drive an actuator of a process. Computer control is used today for chemical processes, aircraft control, machine tools, and many common processes.

The z-transform can be used to analyze the stability and response of a sampled system and to design appropriate systems incorporating a computer. Computer control systems have become increasingly common as low-cost computers have become readily available.

EXERCISES

E13.1 State whether the following signals are discrete or continuous:

(a) Elevation contours on a map.
(b) Temperature in a room.
(c) Digital clock display.
(d) The score of a basketball game.
(e) The output of a loudspeaker.

E13.2 (a) Find the values $y(kT)$ when

$$Y(z) = \frac{z}{z^2 - 3z + 2}$$

for $k = 0$ to 4.

(b) Obtain a closed form of solution for $y(kT)$ as a function of k.

Answer : $y(0) = 0, y(T) = 1, y(2T) = 3, y(3T) = 7,$
$y(4T) = 15$

E13.3 A system has a response $y(kT) = kT$ for $k \geq 0$. Find $Y(z)$ for this response.

$$\textbf{\textit{Answer :}}\ Y(z) = \frac{Tz}{(z - 1)^2}$$

E13.4 We have a function

$$Y(s) = \frac{5}{s(s + 2)(s + 10)}.$$

Using a partial fraction expansion of $Y(s)$ and Table 13.1, find $Y(z)$ when $T = 0.1$ s.

E13.5 The space shuttle, with its robotic arm, is shown in Figure E13.5(a). An astronaut controls the robotic arm and gripper by using a window and the TV cameras [9]. Discuss the use of digital control for this system and sketch a block diagram for the system, including a computer for display generation and control.

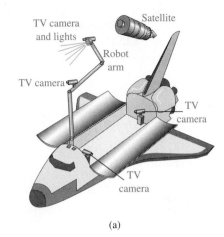

(a)

FIGURE E13.5
(a) Space shuttle
and robotic arm.
(b) Astronaut
control of the arm.

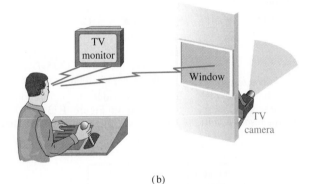

(b)

E13.6 Computer control of a robot to spraypaint an automobile is shown by the system in Figure E13.6 [1]. The system is of the type shown in Figure 13.24, where

$$KG_p(s) = \frac{20}{s(s/2 + 1)},$$

and we want a phase margin of 45°. A compensator for this system was obtained in Section 10.8. Obtain the $D(z)$ required when $T = 0.001$ s.

E13.7 Find the response for the first four sampling instants for

$$Y(z) = \frac{z^3 + 2z^2 + 1}{z^3 - 1.5z^2 + 0.5z}.$$

Then find $y(0), y(1), y(2),$ and $y(3)$.

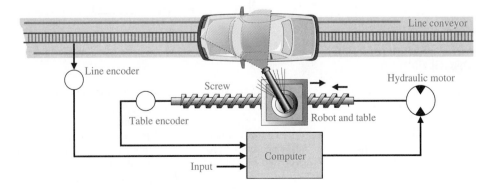

FIGURE E13.6
Automobile
spraypaint system.

E13.8 Determine whether the closed-loop system with $T(z)$ is stable when

$$T(z) = \frac{z}{z^2 + 0.2z - 0.4}.$$

Answer: stable

E13.9 (a) Determine $y(kT)$ for $k = 0$ to 3 when

$$Y(z) = \frac{z+1}{z^2-1}.$$

(b) Determine the closed form solution for $y(kT)$ as a function of k.

E13.10 A system has $G(z)$ as described by Equation (13.34) with $T = 0.01$ s and $\tau = 0.008$ s. (a) Find K so that the overshoot is less than 40%. (b) Determine the steady-state error in response to a unit ramp input. (c) Determine K to minimize the integral squared error.

E13.11 A system has a process transfer function

$$G_p(s) = \frac{100}{s^2 + 100}.$$

(a) Determine $G(z)$ for $G_p(s)$ preceded by a zero-order hold with $T = 0.05$ s. (b) Determine whether the digital system is stable. (c) Plot the impulse response of $G(z)$ for the first 15 samples. (d) Plot the response for a sine wave input with the same frequency as the natural frequency of the system.

E13.12 Find the z-transform of

$$X(s) = \frac{s+1}{s^2 + 5s + 6}$$

when the sampling period is 1 second.

E13.13 The characteristic equation of a sampled system is

$$z^2 + (K - 2)z + 0.75 = 0.$$

Find the range of K so that the system is stable.

Answer: $0.25 < K < 3.75$

E13.14 A unity feedback system, as shown in Figure 13.18, has a plant

$$G_p(s) = \frac{K}{s(s+3)},$$

with $T = 0.5$. Determine whether the system is stable when $K = 5$. Determine the maximum value of K for stability.

E13.15 Consider the open-loop sampled-data system shown in Figure E13.15. Determine the transfer function $G(z)$ when the sampling time is $T = 1$ s.

E13.16 Consider the open-loop sampled-data system shown in Figure E13.16. Determine the transfer function $G(z)$ and when the sampling time $T = 0.5$ s.

FIGURE E13.15
An open-loop sampled-data system with sampling time $T = 1$ s.

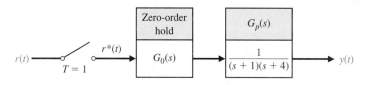

FIGURE E13.16
An open-loop sampled-data system with sampling time $T = 0.5$ s.

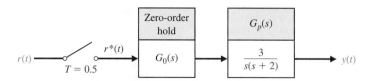

PROBLEMS

P13.1 The input to a sampler is $r(t) = \sin(\omega t)$, where $\omega = 1/\pi$. Plot the input to the sampler and the output $r^*(t)$ for the first 2 seconds when $T = 0.25$ s.

P13.2 The input to a sampler is $r(t) = \sin(\omega t)$, where $\omega = 1/\pi$. The output of the sampler enters a zero-order hold, as shown in Figure 13.7. Plot the output

of the hold circuit $p(t)$ for the first 2 seconds when $T = 0.25$ s.

P13.3 A unit ramp $r(t) = t, t > 0$, is used as an input to a process where $G(s) = 1/(s + 1)$, as shown in Figure P13.3. Determine the output $y(kT)$ for the first four sampling instants.

$r(t)$ — $r^*(t)$ — $G(s)$ — $y(t)$

FIGURE P13.3 Sampling system.

P13.4 A closed-loop system has a hold circuit and process as shown in Figure 13.18. Determine $G(z)$ when $T = 1$ and

$$G_p(s) = \frac{2}{s + 2}.$$

P13.5 For the system in Problem P13.4, let $r(t)$ be a unit step input and calculate the response of the system by synthetic division.

P13.6 For the output of the system in Problem P13.4, find the initial and final values of the output directly from $Y(z)$.

P13.7 A closed-loop system is shown in Figure 13.18. This system represents the pitch control of an aircraft. The process transfer function is $G_p(s) = K/[s(0.5s + 1)]$. Select a gain K and sampling period T so that the overshoot is limited to 0.3 for a unit step input and the steady-state error for a unit ramp input is less than 1.0.

P13.8 Consider the computer-compensated system shown in Figure 13.24 when $T = 1$ and

$$KG_p(s) = \frac{K}{s(s + 10)}.$$

Select the parameters K and r of $D(z)$ when

$$D(z) = \frac{z - 0.3678}{z + r}.$$

Select within the range $1 < K < 2$ and $0 < r < 1$. Determine the response of the compensated system and compare it with the uncompensated system.

P13.9 A suspended, mobile, remote-controlled system to bring three-dimensional mobility to professional NFL football is shown in Figure P13.9. The camera can be moved over the field as well as up and down. The motor control on each pulley is represented by Figure 13.18 with

$$G_p(s) = \frac{10}{s(s + 1)(s/10 + 1)}.$$

We wish to achieve a phase margin of $45°$ using $G_c(s)$. Select a suitable crossover frequency and sampling period to obtain $D(z)$. Use the $G_c(s)$-to-$D(z)$ conversion method.

P13.10 Consider a system as shown in Figure 13.15 with a zero-order hold, a process

$$G_p(s) = \frac{1}{s(s + 10)},$$

and $T = 0.1$ s.

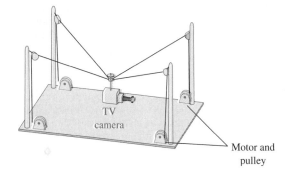

FIGURE P13.9 Mobile camera for football field.

(a) Let $D(z) = K$ and determine the transfer function $G(z)D(z)$. (b) Determine the characteristic equation of the closed-loop system. (c) Calculate the maximum value of K for a stable system. (d) Determine K such that the overshoot is less than 30%. (e) Calculate the closed-loop transfer function $T(z)$ for K of part (d) and plot the step response. (f) Determine the location of the closed-loop roots and the overshoot if K is one-half of the value determined in part (c). (g) Plot the step response for the K of part (f).

P13.11 (a) For the system described in Problem 13.10, design a lag compensator $G_c(s)$ using the methods of Chapter 10 to achieve an overshoot less than 30% and a steady-state error less than 0.01 for a ramp input. Assume a continuous nonsampled system with $G_p(s)$. (b) Determine a suitable $D(z)$ to satisfy the requirements of part (a) with a sampling period $T = 0.1$ s. Assume a zero-order hold and sampler, and use the $G_c(s)$-to-$D(z)$ conversion method. (c) Plot the step response of the system with the continuous-time compensator $G_c(s)$ of part (a) and of the digital system with the $D(z)$ of part (b). Compare the results. (d) Repeat part (b) for $T = 0.01$ s and then repeat part (c). (e) Plot the ramp response for $D(z)$ with $T = 0.1$ s and compare it with the continuous-system response.

P13.12 The transfer function of a plant and a zero-order hold (Figure 13.18) is

$$G(z) = \frac{K(z + 0.2)}{z(z - 1)}.$$

(a) Plot the root locus. (b) Determine the range of gain K for a stable system.

P13.13   The space station orientation controller described in Exercise E7.6 is implemented with a sampler and hold and has the transfer function (Figure 13.18)

$$G(z) = \frac{K(z^2 + 1.1206z - 0.0364)}{z^3 - 1.7358z^2 + 0.8711z - 0.1353}.$$

(a) Plot the root locus. (b) Determine the value of K so that two of the roots of the characteristic equation are equal. (c) Determine all the roots of the characteristic equation for the gain of part (b).

P13.14   A sampled-data system with a sampling period $T = 0.05$s (Figure 13.18) is

$$G(z) = \frac{K(z^3 + 10.3614z^2 + 9.758z + 0.8353)}{z^4 - 3.7123z^3 + 5.1644z^2 - 3.195z + 0.7408}.$$

(a) Plot the root locus. (b) Determine K when the two real poles break away from the real axis. (c) Calculate the maximum K for stability.

P13.15   A closed-loop system with a sampler and hold, as shown in Figure 13.18, has a process transfer function

$$G_p(s) = \frac{20}{s - 5}.$$

Calculate and plot $y(kT)$ for $0 \le T \le 0.6$ when $T = 0.1$ s. The input signal is a unit step.

P13.16   A closed-loop system as shown in Figure 13.18 has

$$G_p(s) = \frac{1}{s(s + 3)}.$$

Calculate and plot $y(kT)$ for $0 \le k \le 8$ when $T = 1$ s and the input is a unit step.

P13.17   A closed-loop system, as shown in Figure 13.18, has

$$G_p(s) = \frac{K}{s(s + 0.5)}$$

and $T = 1$ s. Plot the root locus for $K \ge 0$, and determine the gain K that results in the two roots of the characteristic equation on the z-circle (at the stability limit).

P13.18   A unity feedback system, as shown in Figure 13.18, has

$$G_p(s) = \frac{K}{s(s + 1)}.$$

If the system is continuous ($T = 0$), then $K = 1$ yields a step response with an overshoot of 16% and a settling time (with a 2% criterion) of 8 seconds. Plot the response for $0 \le T \le 1.2$, varying T by increments of 0.2 when $K = 1$. Complete a table recording overshoot and settling time versus T.

ADVANCED PROBLEMS

AP13.1   A closed-loop system, as shown in Figure 13.18, has a process

$$G_p(s) = \frac{K(1 + as)}{s^2},$$

where a is adjustable to achieve a suitable response. Plot the root locus when $a = 10$. Determine the range of K for stability when $T = 1$ s.

AP13.2   A manufacturer uses an adhesive to form a seam along the edge of the material, as shown in Figure AP13.2. It is critical that the glue be applied evenly to avoid flaws; however, the speed at which the material passes beneath the dispensing head is not constant. The glue needs to be dispensed at a rate proportional to the varying speed of the material. The controller adjusts the valve that dispenses the glue [12].

The system can be represented by the block diagram shown in Figure 13.15, where $G_p(s) = 2/(0.03s + 1)$ with a zero-order hold $G_0(s)$. Use a controller

$$D(z) = \frac{KT}{1 - z^{-1}} = \frac{KTz}{z - 1}.$$

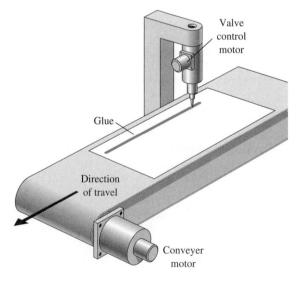

FIGURE AP13.2   A glue control system.

that represents an integral controller. Determine $G(z)D(z)$ for $T = 30$ ms, and plot the root locus. Select an appropriate gain K and plot the step response.

AP13.3   A system of the form shown in Figure 13.15 has $D(z) = K$ and

$$G_p(s) = \frac{8}{s(s + 8)}.$$

When $T = 0.04$, find a suitable K for a rapid step response with an overshoot less than 10%.

AP13.4   A system of the form shown in Figure 13.18 has

$$G_p(s) = \frac{10}{s + 1}.$$

Determine the range of sampling period T for which the system is stable. Select a sampling period T so that the system is stable and provides a rapid response.

AP13.5   Consider the closed-loop sampled-data system shown in Figure AP13.5. Determine the acceptable range of the parameter K for closed-loop stability.

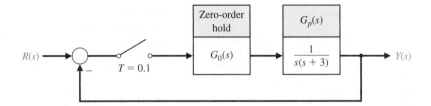

FIGURE AP13.5
A closed-loop sampled-data system with sampling time $T = 0.1$ s.

DESIGN PROBLEMS

CDP13.1   Design a digital controller for the system using the second-order model of the motor-capstan-slide as described in CDP2.1 and CDP4.1. Use a sampling period of $T = 1$ ms and select a suitable $D(z)$ for the system shown in Figure 13.15. Determine the response of the designed system to a step input $r(t)$.

DP13.1   A temperature system, as shown in Figure 13.15, has a process transfer function

$$G_p(s) = \frac{0.8}{3s + 1}$$

and a sampling period T of 0.5 second.
(a) Using $D(z) = K$, select a gain K so that the system is stable. (b) The system may be slow and overdamped, and thus we seek to design a lead network using the method of Section 10.5. Determine a suitable controller $G_c(s)$ and then calculate $D(z)$. (c) Verify the design obtained in part (b) by plotting the step response of the system for the selected $D(z)$.

DP13.2   A disk drive read-write head-positioning system has a system as shown in Figure 13.15 [11]. The process transfer function is

$$G_p(s) = \frac{9}{s^2 + 0.85s + 788}.$$

Accurate control using a digital compensator is required. Let $T = 10$ ms and design a compensator, $D(z)$, using (a) the $G_c(s)$-to-$D(z)$ conversion method and (b) the root locus method.

DP13.3   Vehicle traction control, which includes antiskid braking and antispin acceleration, can enhance vehicle performance and handling. The objective of this control is to maximize tire traction by preventing the wheels from locking during braking and from spinning during acceleration.

Wheel slip, the difference between the vehicle speed and the wheel speed (normalized by the vehicle speed for braking and the wheel speed for acceleration), is chosen as the controlled variable for most of the traction-control algorithm because of its strong influence on the tractive force between the tire and the road [20].

A model for one wheel is shown in Figure DP13.3 when y is the wheel slip. The goal is to minimize the slip when a disturbance occurs due to road conditions. Design a controller $D(z)$ so that the ζ of the system is $1/\sqrt{2}$, and determine the resulting K. Assume $T = 0.1$ s. Plot the resulting step response, and find the overshoot and settling time (with a 2% criterion).

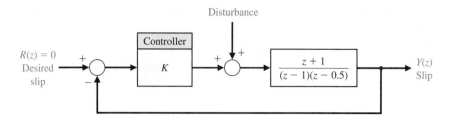

FIGURE DP13.3
Vehicle fraction
control system.

DP13.4 A machine-tool system has the form shown in Figure 13.28 with [10]

$$G_p(s) = \frac{0.1}{s(s + 0.1)}.$$

The sampling rate is chosen as $T = 1$ s. We desire the step response to have an overshoot of 16% or less and a settling time (with a 2% criterion) of 12 seconds or less. Also, the error to a unit ramp input, $r(t) = t$, must be less than or equal to 1. Design a $D(z)$ to achieve these specifications.

DP13.5 Plastic extrusion is a well-established method widely used in the polymer processing industry [12]. Such extruders typically consist of a large barrel divided into several temperature zones, with a hopper at one end and a die at the other. Polymer is fed into the barrel in raw and solid form from the hopper and is pushed forward by a powerful screw. Simultaneously, it is gradually heated while passing through the various temperature zones set in gradually increasing

temperatures. The heat produced by the heaters in the barrel, together with the heat released from the friction between the raw polymer and the surfaces of the barrel and the screw, eventually causes the melting of the polymer, which is then pushed by the screw out from the die, to be processed further for various purposes.

The output variables are the outflow from the die and the polymer temperature. The main controlling variable is the screw speed, since the response of the process to it is rapid.

The control system for the output polymer temperature is shown in Figure DP13.5. Select a gain K and a sampling period T to obtain a step overshoot of 10% while reducing the steady-state error for a ramp input.

DP13.6 A sampled-data system closed-loop block diagram is shown in Figure DP13.6. Design $D(z)$ to such that the closed-loop system response to a unit step response has a percent overshoot $P.O. \le 15\%$ and a settling time $T_s \le 20$ s.

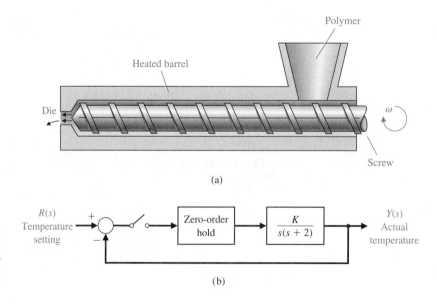

FIGURE DP13.5
Control system for
an extruder.

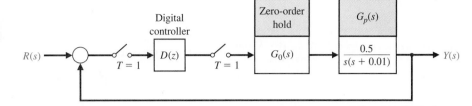

FIGURE DP13.6
A closed-loop
sampled-data
system with
sampling time
$T = 1$ s.

COMPUTER PROBLEMS

CP13.1 Develop an m-file to plot the unit step response of the system

$$G(z) = \frac{0.2145z + 0.1609}{z^2 - 0.75z + 0.125}.$$

Verify graphically that the steady-state value of the output is 1.

CP13.2 Convert the following continuous-time transfer functions to sampled-data systems using the c2d function. Assume a sample period of 1 second and a zero-order hold $G_0(s)$.

(a) $G_p(s) = \dfrac{1}{s}$

(b) $G_p(s) = \dfrac{s}{s^2 + 2}$

(c) $G_p(s) = \dfrac{s + 4}{s + 3}$

(d) $G_p(s) = \dfrac{1}{s(s + 8)}$

CP13.3 The closed-loop transfer function of a sampled-data system is given by

$$T(z) = \frac{Y(z)}{R(z)} = \frac{1.7(z + 0.46)}{z^2 + z + 0.5}.$$

(a) Compute the unit step response of the system using the **step** function. (b) Determine the continuous-time transfer function equivalent of $T(z)$ using the d2c function and assume a sampling period of $T = 0.1$ s. (c) Compute the unit step response of the continuous (nonsampled) system using the **step** function, and compare the plot with part (a).

CP13.4 Plot the root locus for the system

$$G(z)D(z) = K\frac{z}{z^2 - z + 0.1}.$$

Find the range of K for stability.

CP13.5 Consider the feedback system in Figure CP13.5. Obtain the root locus and determine the range of K for stability.

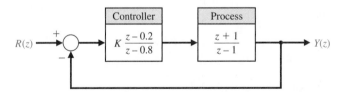

FIGURE CP13.5
Control system with
a digital controller.

CP13.6 Consider the sampled data system with the loop transfer function

$$G(z)D(z) = K\frac{z^2 + 3z + 3.75}{z^2 - 0.2z - 1.9}.$$

(a) Plot the root locus using the rlocus function.
(b) From the root locus, determine the range of K for stability. Use the rlocfind function.

CP13.7 An industrial grinding process is given by the transfer function [17]

$$G_p(s) = \frac{10}{s(s + 5)}.$$

The objective is to use a digital computer to improve the performance, where the transfer function of the computer is represented by $D(z)$. The design specifications are (1) phase margin greater than $45°$, and (2) settling time (with a 2% criterion) less than 1 second.

(a) Design a controller

$$G_c(s) = K \frac{s + a}{s + b}$$

to meet the design specifications. (b) Assuming a sampling time of $T = 0.02$ s, convert $G_c(s)$ to $D(z)$.

(c) Simulate the continuous-time, closed-loop system with a unit step input. (d) Simulate the sampled-data, closed-loop system with a unit step input. (e) Compare the results in parts (c) and (d) and comment.

TERMS AND CONCEPTS

Amplitude quantization error   The sampled signal available only with a limited precision. The error between the actual signal and the sampled signal.

Backward difference rule   A computational method of approximating the time derivative of a function given by $\dot{x}(kT) \approx \dfrac{x(kT) - x((k-1)T)}{T}$, where $t = kT$, T is the sample time, and $k = 1, 2, \ldots$.

Digital computer compensator   A system that uses a digital computer as the compensator element.

Digital control system   A control system using digital signals and a digital computer to control a process.

Forward rectangular integration   A computational method of approximating the integration of a function given by $x(kT) \approx x((k-1)T) + T\dot{x}((k-1)T)$, where $t = kT$, T is the sample time, and $k = 1, 2, \ldots$.

Microcomputer   A small personal computer (PC) based on a microprocessor.

Minicomputer   A stand-alone computer with size and performance between a microcomputer and a large mainframe. The term is not commonly used today, and computers in this class are now often known as mid-range servers.

PID controller   A controller with three terms in which the output is the sum of a proportional term, an integrating term, and a differentiating

term, with an adjustable gain for each term, given by

$$G_c(z) = K_1 + \frac{K_2 T s}{z - 1} + K_3 \frac{z - 1}{T z}.$$

Precision   The degree of exactness or discrimination with which a quantity is stated.

Sampled data   Data obtained for the system variables only at discrete intervals. Data obtained once every sampling period.

Sampled-data system   A system where part of the system acts on sampled data (sampled variables).

Sampling period   The period when all the numbers leave or enter the computer. The period for which the sampled variable is held constant.

Stability of a sampled-data system   The stable condition exists when all the poles of the closed-loop transfer function $T(z)$ are within the unit circle on the z-plane.

z-plane   The plane with the vertical axis equal to the imaginary part of z and the horizontal axis equal to the real part of z.

z-transform   A conformal mapping from the s-plane to the z-plane by the relation $z = e^{sT}$. A transform from the s-domain to the z-domain.

Zero-order hold   A mathematical model of a sample and data hold operation whose input–output transfer function is represented by $G_o(s) = \dfrac{1 - e^{-sT}}{s}$.

Appendixes

MATLAB Basics

A.1 INTRODUCTION

MATLAB is an interactive program for scientific and engineering calculations. The MATLAB family of programs includes the base program plus a variety of **toolboxes**, a collection of special files called *m-files* that extend the functionality of the base program [1–8]. Together, the base program plus the *Control System Toolbox* provide the capability to use MATLAB for control system design and analysis. Whenever MATLAB is referenced in this book, it means the base program plus the *Control System Toolbox*.

Most of the statements, functions, and commands are computer-platform-independent. Regardless of what particular computer system you use, your interaction with MATLAB is basically the same. This appendix concentrates on this computer platform–independent interaction. A typical session will utilize a variety of objects that allow you to interact with the program: (1) statements and variables, (2) matrices, (3) graphics, and (4) scripts. MATLAB interprets and acts on input in the form of one or more of these objects. The goal in this appendix is to introduce each of the four objects in preparation for our ultimate goal of using MATLAB for control system design and analysis.

The manner in which MATLAB interacts with a specific computer system is computer-platform-dependent. Examples of computer-dependent functions include installation, the file structure, hard-copy generation of the graphics, the invoking and exiting of a session, and memory allocation. Questions related to platform-dependent issues are not addressed here. This does not mean that they are not important, but rather that there are better sources of information such as the MATLAB *User's Guide* or the local resident expert.

The remainder of this appendix consists of four sections corresponding to the four objects already listed. In the first section, we present the basics of **statements** and **variables**. Following that is the subject of **matrices**. The third section presents an introduction to **graphics**, and the fourth section is a discussion on the important topic of **scripts** and **m-files**. All the figures in this appendix can be constructed using the m-files found at the MCS website.

A.2 STATEMENTS AND VARIABLES

Statements have the form shown in Figure A.1. MATLAB uses the assignment so that equals ("=") implies the assignment of the expression to the variable. The command

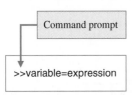

Command prompt

>>variable=expression

FIGURE A.1   MATLAB
statement form.

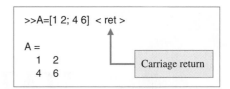

>>A=[1 2; 4 6] < ret >

A =
    1   2
    4   6

Carriage return

FIGURE A.2   Entering and displaying a
matrix **A**.

prompt is two right arrows, " $\gg$." A typical statement is shown in Figure A.2, where we are entering a 2×2 matrix to which we attach the variable name **A**. The statement is executed after the carriage return (or enter key) is pressed. The carriage return is not explicitly denoted in the remaining examples in this appendix.

The matrix **A** is automatically displayed after the statement is executed following the carriage return. If the statement is followed by a semicolon (;), the output matrix **A** is suppressed, as seen in Figure A.3. The assignment of the variable **A** has been carried out even though the output is suppressed by the semicolon. It is often the case that your MATLAB sessions will include intermediate calculations for which the output is of little interest. Use the semicolon whenever you have a need to reduce the amount of output. Output management has the added benefit of increasing the execution speed of the calculations since displaying screen output takes time.

The usual mathematical operators can be used in expressions. The common operators are shown in Table A.1. The order of the arithmetic operations can be altered by using parentheses.

The example in Figure A.4 illustrates that MATLAB can be used in a "calculator" mode. When the variable name and "=" are omitted from an expression, the result is assigned to the generic variable *ans*. MATLAB has available most of the trigonometric and elementary math functions of a common scientific calculator. Type help elfun at the command prompt to view a complete list of available trigonometric and elementary math functions; the more common ones are summarized in Table A.2.

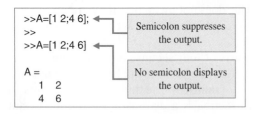

>>A=[1 2;4 6];
>>
>>A=[1 2;4 6]

A =
    1   2
    4   6

Semicolon suppresses
the output.

No semicolon displays
the output.

FIGURE A.3   Using semicolons to suppress the
output.

Table A.1   Mathematical Operators

+	Addition
−	Subtraction
*	Multiplication
/	Division
^	Power

FIGURE A.4
Using the calculator
mode.

>>12.4/6.9

ans =
    1.7971

Table A.2   Common Mathematical Functions

sin(x)	Sine	acoth(x)	Inverse hyperbolic cotangent
sinh(x)	Hyperbolic sine	exp(x)	Exponential
asin(x)	Inverse sine	log(x)	Natural logarithm
asinh(x)	Inverse hyperbolic sine	log10(x)	Common (base 10) logarithm
cos(x)	Cosine	log2(x)	Base 2 logarithm and dissect floating point number
cosh(x)	Hyperbolic cosine	pow2(x)	Base 2 power and scale floating point number
acos(x)	Inverse cosine	sqrt(x)	Square root
acosh(x)	Inverse hyperbolic cosine	nextpow2(x)	Next higher power of 2
tan(x)	Tangent	abs(x)	Absolute value
tanh(x)	Hyperbolic tangent	angle(x)	Phase angle
atan(x)	Inverse tangent	complex(x,y)	Construct complex data from real and imaginary parts
atan2(y,x)	Four quadrant inverse tangent	conj(x)	Complex conjugate
atanh(x)	Inverse hyperbolic tangent	imag(x)	Complex imaginary part
sec(x)	Secant	real(x)	Complex real part
sech(x)	Hyperbolic secant	unwrap(x)	Unwrap phase angle
asec(x)	Inverse secant	isreal(x)	True for real array
asech(x)	Inverse hyperbolic secant	cplxpair(x)	Sort numbers into complex conjugate pairs
csc(x)	Cosecant	fix(x)	Round towards zero
csch(x)	Hyperbolic cosecant	floor(x)	Round towards minus infinity
acsc(x)	Inverse cosecant	ceil(x)	Round towards plus infinity
acsch(x)	Inverse hyperbolic cosecant	round(x)	Round towards nearest integer
cot(x)	Cotangent	mod(x,y)	Modulus (signed remainder after division)
coth(x)	Hyperbolic cotangent	rem(x,y)	Remainder after division
acot(x)	Inverse cotangent		

Variable names begin with a letter and are followed by any number of letters and numbers (including underscores). Keep the name length to N characters, since MATLAB remembers only the first N characters, where N = namelengthmax. It is a good practice to use variable names that describe the quantity they represent. For example, we might use the variable name *vel* to represent the quantity *aircraft velocity*. Generally, we do not use extremely long variable names even though they may be legal MATLAB names.

Since MATLAB is **case sensitive**, the variables *M* and *m* are not the same. By **case**, we mean upper- and lowercase, as illustrated in Figure A.5. The variables *M* and *m* are recognized as different quantities.

MATLAB has several predefined variables, including *pi*, *Inf*, *NaN*, *i*, and *j*. Three examples are shown in Figure A.6. *NaN* stands for *Not-a-Number* and results from undefined operations. *Inf* represents $+\infty$, and *pi* represents π. The variable $i = \sqrt{-1}$ is used to represent complex numbers. The variable $j = \sqrt{-1}$ can be used for complex arithmetic by those who prefer it over *i*. These predefined variables can be inadvertently overwritten. Of course, they can also be purposely overwritten in order to free the variable name for other uses. For instance, you might want to use *i* as an integer and reserve *j* for complex arithmetic. Be safe and leave these predefined variables alone, as

FIGURE A.5
Variables are case sensitive.

```
>>M=[1 2];
>>m=[3 5 7];
```

```
>>z=3+4*i
z =
   3.0000 + 4.0000i

>>Inf
ans =
   Inf

>>0/0
Warning:  Divide by zero
ans =
   NaN
```

FIGURE A.6
Three predefined variables *i*, *Inf*, and *NaN*.

```
>>who
Your variables are:
A    M    ans    m    z
```

FIGURE A.7   Using the **who** function to display variables.

there are plenty of alternative names that can be used. Predefined variables can be reset to their default values by using clear *name* (e.g., clear *pi*).

The matrix **A** and the variable *ans*, in Figures A.3 and A.4, respectively, are stored in the **workspace**. Variables in the workspace are automatically saved for later use in your session. The who function gives a list of the variables in the workspace, as shown in Fig. A.7. MATLAB has a host of built-in functions. Refer to the MATLAB *User's Guide* for a complete list or use the MATLAB help browser. Each function will be described as the need arises.

The whos function lists the variables in the workspace and gives additional information regarding variable dimension, type, and memory allocation. Figure A.8 gives an example of the whos function. The memory allocation information given by the whos function can be interpreted as follows: Each element of the 2×2 matrix **A** requires 8 bytes of memory for a total of 32 bytes, the 1×1 variable *ans* requires 8 bytes, and so forth. All the variables in the workspace use a total of 96 bytes.

Variables can be removed from the workspace with the clear function. Using the function clear, by itself, removes all items (variables and functions) from the workspace; clear variables removes all variables from the workspace; clear *name1 name2 . . .* removes the variables *name1*, *name2*, and so forth. The procedure for removing the matrix **A** from the workspace is shown in Figure A.9.

```
>>whos
         Name    Size    Bytes    Class     Attributes

         A       2x2     32       double
         M       1x2     16       double
         ans     1x1     8        double
         m       1x3     24       double
         z       1x1     16       double    complex
```

FIGURE A.8
Using the **whos** function to display variables.

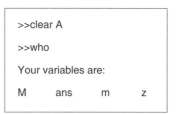

FIGURE A.9
Removing the matrix **A** from the workspace.

```
>>clear A

>>who

Your variables are:

M      ans      m      z
```

Computations in MATLAB are performed in **double precision**. However, the screen output can be displayed in several formats. The default output format contains four digits past the decimal point for nonintegers. This can be changed by using the format function shown in Figure A.10. Once a particular format has been specified, it remains in effect until altered by a different format input. The output format does not affect internal MATLAB computations. On the other hand, the number of digits displayed does not necessarily reflect the number of significant digits of the number. This is problem-dependent, and only the user can know the true accuracy of the numbers input and displayed by MATLAB. Other display formats (not shown in Figure A.10) include format long g (best of fixed or floating point format with 15 digits after the decimal point), format short g (same as format long g but with 4 digits after the decimal point), format hex (hexidecimal format), format bank (fixed format for dollars and cents), format rat (ratio of small integers) and format (same as format short).

Since MATLAB is case sensitive, the functions who and WHO are not the same functions. The first function, who, is a built-in function, and typing who lists the variables in the workspace. On the other hand, typing the uppercase WHO results in the error message shown in Figure A.11. Case sensitivity applies to all functions.

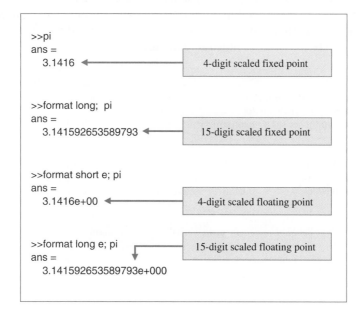

FIGURE A.10
Output format control illustrates the four forms of output.

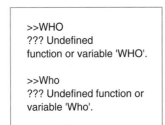

FIGURE A.11
Function names are
case sensitive.

A.3 MATRICES

MATLAB is short for **matrix laboratory**. Although we will not emphasize the matrix routines underlying our calculations, we will learn how to use the interactive capability to assist us in the control system design and analysis. We begin by introducing the basic concepts associated with manipulating matrices and vectors.

The basic computational unit is the matrix. Vectors and scalars can be viewed as special cases of matrices. A typical matrix expression is enclosed in square brackets, $[\,\cdot\,]$. The column elements are separated by blanks or commas, and the rows are separated by semicolons or carriage returns. Suppose we want to input the matrix

$$\mathbf{A} = \begin{bmatrix} 1 & -4j & \sqrt{2} \\ \log(-1) & \sin(\pi/2) & \cos(\pi/3) \\ \operatorname{asin}(0.8) & \operatorname{acos}(0.8) & \exp(0.8) \end{bmatrix}.$$

One way to input $\mathbf{A}$ is shown in Figure A.12. The input style in Figure A.12 is not unique.

Matrices can be input across multiple lines by using a carriage return following the semicolon or in place of the semicolon. This practice is useful for entering large matrices. Different combinations of spaces and commas can be used to separate the columns, and different combinations of semicolons and carriage returns can be used to separate the rows, as illustrated in Figure A.12.

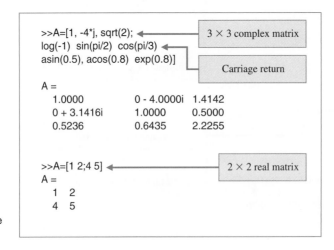

FIGURE A.12
Complex and real
matrix input with
automatic
dimension and type
adjustment.

No dimension statements or type statements are necessary when using matrices; memory is allocated automatically. Notice in the example in Figure A.12 that the size of the matrix **A** is automatically adjusted when the input matrix is redefined. Also notice that the matrix elements can contain trigonometric and elementary math functions, as well as complex numbers.

The important basic matrix operations are addition and subtraction, multiplication, transpose, powers, and the so-called array operations, which are element-to-element operations. The mathematical operators given in Table A.1 apply to matrices. We will not discuss matrix division, but be aware that MATLAB has a left- and right-matrix division capability.

Matrix operations require that the matrix dimensions be compatible. For matrix addition and subtraction, this means that the matrices must have the same dimensions. If **A** is an $n \times m$ matrix and **B** is a $p \times r$ matrix, then $\mathbf{A} \pm \mathbf{B}$ is permitted only if $n = p$ and $m = r$. Matrix multiplication, given by $\mathbf{A} * \mathbf{B}$, is permitted only if $m = p$. Matrix-vector multiplication is a special case of matrix multiplication. Suppose **b** is a vector of length p. Multiplication of the vector **b** by the matrix **A**, where **A** is an $n \times m$ matrix, is allowed if $m = p$. Thus, $\mathbf{y} = \mathbf{A} * \mathbf{b}$ is the $n \times 1$ vector solution of $\mathbf{A} * \mathbf{b}$. Examples of three basic matrix-vector operations are given in Figure A.13.

The matrix transpose is formed with the apostrophe ('). We can use the matrix transpose and multiplication operation to create a vector **inner product** in the following manner. Suppose **w** and **v** are $m \times 1$ vectors. Then the inner product (also known as the dot product) is given by $\mathbf{w}' * \mathbf{v}$. The inner product of two vectors is a scalar. The **outer product** of two vectors can similarly be computed as $\mathbf{w} * \mathbf{v}'$. The outer product of two $m \times 1$ vectors is an $m \times m$ matrix of rank 1. Examples of inner and outer products are given in Figure A.14.

The basic matrix operations can be modified for element-by-element operations by preceding the operator with a period. The modified matrix operations are known

FIGURE A.13
Three basic matrix operations: addition, multiplication, and transpose.

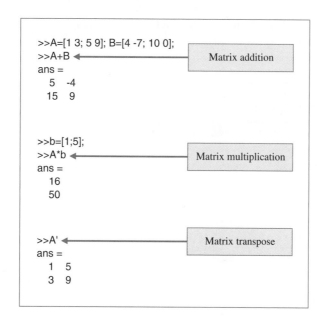

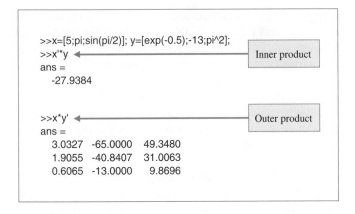

FIGURE A.14   Inner and outer products of two vectors.

Table A.3   Mathematical Array Operators

+	Addition
−	Subtraction
.*	Multiplication
./	Division
.^	Power

as **array operations**. The commonly used array operators are given in Table A.3. Matrix addition and subtraction are already element-by-element operations and do not require the additional period preceding the operator. However, array multiplication, division, and power do require the preceding dot, as shown in Table A.3.

Consider **A** and **B** as 2×2 matrices given by

$$\mathbf{A} = \begin{bmatrix} a_{11} & a_{12} \\ a_{21} & a_{22} \end{bmatrix}, \quad \mathbf{B} = \begin{bmatrix} b_{11} & b_{12} \\ b_{21} & b_{22} \end{bmatrix}.$$

Then, using the array multiplication operator, we have

$$\mathbf{A}.* \mathbf{B} = \begin{bmatrix} a_{11}b_{11} & a_{12}b_{12} \\ a_{21}b_{21} & a_{22}b_{22} \end{bmatrix}.$$

The elements of $\mathbf{A}.* \mathbf{B}$ are the products of the corresponding elements of **A** and **B**. A numerical example of two array operations is given in Figure A.15.

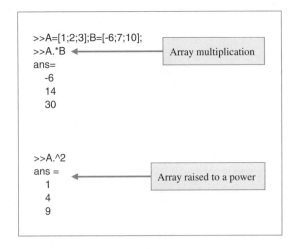

FIGURE A.15
Array operations.

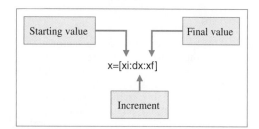

FIGURE A.16
The colon notation.

Before proceeding to the important topic of graphics, we need to introduce the notion of **subscripting using colon notation**. The colon notation, shown in Figure A.16, allows us to generate a row vector containing the numbers from a given starting value, x_i, to a final value, x_f, with a specified increment, dx.

We can easily generate vectors using the colon notation, and as we shall soon see, this is quite useful for developing *x-y* **plots**. Suppose our objective is to generate a plot of $y = x \sin(x)$ versus x for $x = 0, 0.1, 0.2, \ldots, 1.0$. Our first step is to generate a table of *x-y* data. We can generate a vector containing the values of x at which the values of $y(x)$ are desired using the colon notation, as illustrated in Figure A.17. Given the desired x vector, the vector $y(x)$ is computed using the multiplication array operation. Creating a plot of $y = x \sin(x)$ versus x is a simple step once the table of *x-y* data is generated.

A.4 GRAPHICS

Graphics plays an important role in both the design and analysis of control systems. An important component of an **interactive** control system design and analysis tool is an effective graphical capability. A complete solution to the control system design and analysis will eventually require a detailed look at a multitude of data types in many formats. The objective of this section is to acquaint the reader with the basic

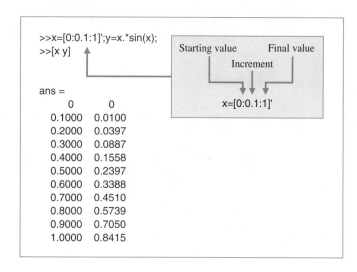

FIGURE A.17
Generating vectors using the colon notation.

Table A.4   Plot Formats

plot(x,y)	Plots the vector **x** versus the vector **y**.
semilogx(x,y)	Plots the vector **x** versus the vector **y**. The x-axis is $\log_{10}$; the y-axis is linear.
semilogy(x,y)	Plots the vector **x** versus the vector **y**. The x-axis is linear; the y-axis is $\log_{10}$.
loglog(x,y)	Plots the vector **x** versus the vector **y**. Creates a plot with $\log_{10}$ scales on both axes.

x-y plotting capability of MATLAB. More advanced graphics topics are addressed in the chapter sections on MATLAB.

MATLAB uses a **graph display** to present plots. The graph display is activated automatically when a plot is generated using any function that generates a plot (e.g., the plot function). The plot function opens a graph display, called a **FIGURE** window. You can also create a new figure window with the figure function. Multiple figure windows can exist in a single MATLAB session; the function figure (n) makes n the current figure. The plot in the graph display is cleared by the clf function at the command prompt. The shg function brings the current figure window forward.

There are two basic groups of graphics functions. The first group, shown in Table A.4, specifies the type of plot. The list of available plot types includes the x-y plot, semilog plots, and log plots. The second group of functions, shown in Table A.5, allows us to customize the plots by adding titles, axis labels, and text to the plots and to change the scales and display multiple plots in subwindows.

The standard x-y plot is created using the plot function. The x-y data in Figure A.17 are plotted using the plot function, as shown in Figure A.18. The axis scales and line types are automatically chosen. The axes are labeled with the xlabel and ylabel functions; the title is applied with the title function. The legend function puts a legend on the current figure. A grid can be placed on the plot by using the grid on function. A basic x-y plot is generated with the combination of functions plot, legend, xlabel, ylabel, title, and grid on.

Multiple lines can be placed on the graph by using the plot function with multiple arguments, as shown in Figure A.19. The default line types can also be altered. The available line types are shown in Table A.6. The line types will be automatically

Table A.5   Functions for Customized Plots

title('text')	Puts 'text' at the top of the plot
legend (string1, string2, ...)	Puts a legend on current plot using specified strings as labels
xlabel('text')	Labels the x-axis with 'text'
ylabel('text')	Labels the y-axis with 'text'
text(p1,p2, 'text')	Adds 'text' to location (p1,p2), where (p1,p2) is in units from the current plot
subplot	Subdivides the graphics window
grid on	Adds grid lines to the current figure
grid off	Removes grid lines from the current figure
grid	Toggles the grid state

Table A.6   Commands for Line Types for Customized Plots

-	Solid line
- -	Dashed line
:	Dotted line
-.	Dashdot line

chosen unless specified by the user. The use of the text function and the changing of line types are illustrated in Figure A.19.

The other graphics functions—loglog, semilogx, and semilogy—are used in a fashion similar to that of plot. To obtain an *x-y* plot where the *x*-axis is a linear scale and the *y*-axis is a $\log_{10}$ scale, you would use the semilogy function in place of the plot function. The customizing features listed in Table A.5 can also be utilized with the loglog, semilogx, and semilogy functions.

The graph display can be subdivided into smaller subwindows. The function subplot(m,n,p) subdivides the graph display into an $m \times n$ grid of smaller subwindows. The integer *p* specifies the window, numbered left to right, top to bottom, as illustrated in Figure A.20, where the graphics window is subdivided into four subwindows.

```
>>x=[0:0.1:1]';
>>y=x.*sin(x);
>>plot(x,y)
>>title('Plot of x sin(x) vs x ')
>>xlabel('x')
>>ylabel('y')
>>grid on
```

(a)

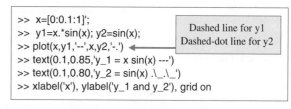

(a)

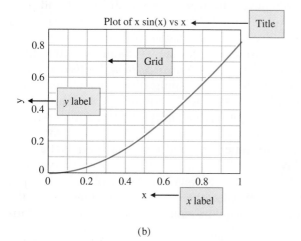

(b)

FIGURE A.18 (a) MATLAB commands. (b) A basic *x-y* plot of *x* sin(x) versus *x*.

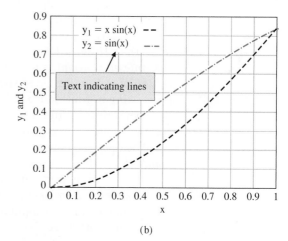

(b)

FIGURE A.19 (a) MATLAB commands. (b) A basic *x-y* plot with multiple lines.

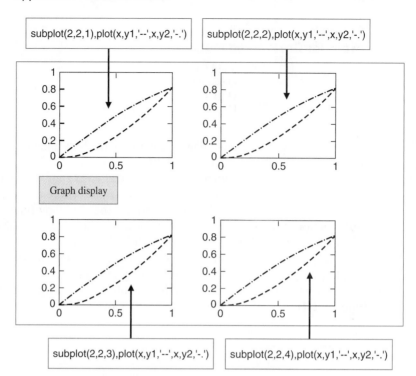

subplot(2,2,1),plot(x,y1,'--',x,y2,'-.')   subplot(2,2,2),plot(x,y1,'--',x,y2,'-.')

Graph display

subplot(2,2,3),plot(x,y1,'--',x,y2,'-.')   subplot(2,2,4),plot(x,y1,'--',x,y2,'-.')

FIGURE A.20
Using **subplot** to create a 2 × 2 partition of the graph display.

A.5 SCRIPTS

Up to this point, all of our interaction with MATLAB has been at the command prompt. We entered statements and functions at the command prompt, and MATLAB interpreted our input and took the appropriate action. This is the preferable mode of operation whenever the work sessions are short and non-repetitive. However, the real power of MATLAB for control system design and analysis derives from its ability to execute a long sequence of commands stored in a file. These files are called **m-files**, since the filename has the form *filename.m*. A **script** is one type of m-file. The *Control System Toolbox* is a collection of m-files designed specifically for control applications. In addition to the preexisting m-files delivered with MATLAB and the toolboxes, we can develop our own scripts for our applications. Scripts are ordinary ASCII text files and are created using a text editor.

A script is a sequence of ordinary statements and functions used at the command prompt level. A script is invoked at the command prompt level by typing in the filename or by using the pull-down menu. Scripts can also invoke other scripts. When the script is invoked, MATLAB executes the statements and functions in the file without waiting for input at the command prompt. The script operates on variables in the workspace.

Suppose we want to plot the function $y(t) = \sin \alpha t$, where α is a variable that we want to vary. Using a text editor, we write a script that we call plotdata.m, as

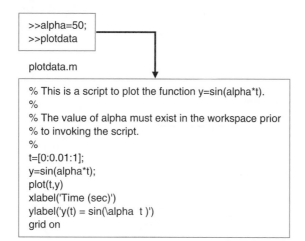

FIGURE A.21
A simple script to plot the function $y(t) = \sin \alpha t$.

shown in Figure A.21, then input a value of α at the command prompt, placing α in the workspace. Then we execute the script by typing in plotdata at the command prompt; the script plotdata.m will use the most recent value of α in the workspace. After executing the script, we can enter another value of α at the command prompt and execute the script again.

Your scripts should be well documented with **comments**, which begin with a %. Put a **header** in the script; make sure the header includes several descriptive comments regarding the function of the script, and then use the help function to display the header comments and describe the script to the user, as illustrated in Figure A.22.

Use plotdata.m to develop an interactive capability with α as a variable, as shown in Figure A.23. At the command prompt, input a value of $\alpha = 10$ followed by the script filename, which in this case is plotdata. The graph of $y(t) = \sin \alpha t$ is automatically generated. You can now go back to the command prompt, enter a value of $\alpha = 50$, and run the script again to obtain the updated plot.

A limited subset of TeX[1] characters are available to allow you to annotate plots with symbols and mathematical characters. Table A.7 shows the available symbols. Figure A.21 illustrates the use of '\alpha' to generate the α character in the y-axis label. The '\' character preceeds all TeX sequences. Also, you can modify the characters with the following modifiers:

❑ \bf—bold font

❑ \it—italics font

❑ \rm—normal font

❑ \fontname—specify the name of the font family to use

❑ \fontsize—specify the font size

[1]TeX is a trademark of the American Mathematical Society.

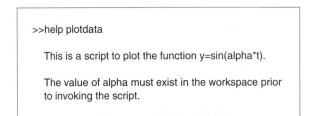

FIGURE A.22
Using the **help**
function.

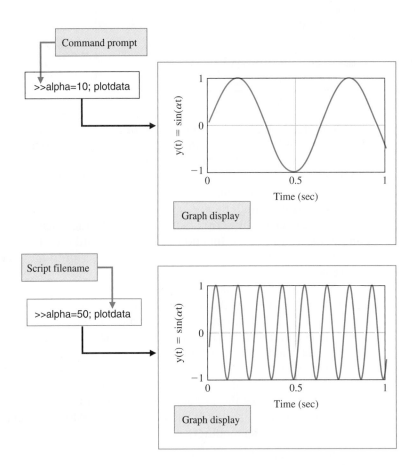

FIGURE A.23
An interactive
session using a
script to plot
the function
$y(t) = \sin \alpha t$.

Subscripts and superscripts are obtained with "_" and "^", respectively. For exam-ple, ylabel('y_1 and y_2') generates the y-axis label shown in Figure A.19.

The graphics capability of MATLAB extends beyond the introductory material presented here. A table of MATLAB functions used in this book is provided in Table A.8.

Table A.7   TeX Symbols and Mathematics Characters

Character Sequence	Symbol	Character Sequence	Symbol	Character Sequence	Symbol
\alpha	α	\upsilon	υ	\sim	∼
\beta	β	\phi	φ	\leq	≤
\gamma	γ	\chi	χ	\infty	∞
\delta	δ	\psi	ψ	\clubsuit	♣
\epsilon	ε	\omega	ω	\diamondsuit	♦
\zeta	ζ	\Gamma	Γ	\heartsuit	♥
\eta	η	\Delta	Δ	\spadesuit	♠
\theta	θ	\Theta	Θ	\leftrightarrow	↔
\vartheta	ϑ	\Lambda	Λ	\leftarrow	←
\iota	ι	\Xi	Θ	\uparrow	↑
\kappa	κ	\Pi	Π	\rightarrow	→
\lambda	λ	\Sigma	Σ	\downarrow	↓
\mu	μ	\Upsilon	Υ	\circ	∘
\nu	ν	\Phi	Φ	\pm	±
\xi	ξ	\Psi	Ψ	\geq	≥
\pi	π	\Omega	Ω	\propto	∝
\rho	ρ	\forall	∀	\partial	∂
\sigma	σ	\exist	∃	\bullet	·
\varsigma	ζ	\ni	∋	\div	÷
\tau	τ	\cong	≅	\neq	≠
\equiv	≡	\approx	≈	\aleph	ℵ
\Im	ℑ	\Re	ℜ	\wp	℘
\otimes	⊗	\oplus	⊕	\oslash	∅
\cap	∩	\cup	∪	\supseteq	⊇
\supset	⊃	\subseteq	⊆	\subset	⊂
\int	∫	\in	∋	\o	ο

Table A.8   MATLAB Functions

Function Name	Function Description
abs	Computes the absolute value
acos	Computes the arccosine
ans	Variable created for expressions
asin	Computes the arcsine
atan	Computes the arctangent (2 quadrant)
atan2	Computes the arctangent (4 quadrant)
axis	Specifies the manual axis scaling on plots
bode	Generates Bode frequency response plots
c2d	Converts a continuous-time state variable system representation to a discrete-time system representation
clear	Clears the workspace
clf	Clears the graph window
conj	Computes the complex conjugate
conv	Multiplies two polynomials (convolution)
cos	Computes the cosine
ctrb	Computes the controllability matrix
diary	Saves the session in a disk file
d2c	Converts a discrete-time state variable system representation to a continuous-time system representation
eig	Computes the eigenvalues and eigenvectors
end	Terminates control structures
exp	Computes the exponential with base e
expm	Computes the matrix exponential with base e
eye	Generates an identity matrix
feedback	Computes the feedback interconnection of two systems
for	Generates a loop
format	Sets the output display format
grid on	Adds a grid to the current graph
help	Prints a list of HELP topics
hold on	Holds the current graph on the screen
i	$\sqrt{-1}$
imag	Computes the imaginary part of a complex number
impulse	Computes the unit impulse response of a system
inf	Represents infinity
j	$\sqrt{-1}$
legend	Puts a legend on the current plot
linspace	Generates linearly spaced vectors
load	Loads variables saved in a file
log	Computes the natural logarithm
log10	Computes the logarithm base 10
loglog	Generates log-log plots
logspace	Generates logarithmically spaced vectors
lsim	Computes the time response of a system to an arbitrary input and initial conditions
margin	Computes the gain margin, phase margin, and associated crossover frequencies from frequency response data
max	Determines the maximum value
mesh	Creates three-dimensional mesh surfaces
meshgrid	Generates arrays for use with the mesh function
min	Determines the minimum value
minreal	Transfer function pole–zero cancellation

Table A.8 continued

Table A.8 *Continued*

Function Name	Function Description
NaN	Representation for Not-a-Number
ngrid	Draws grid lines on a Nichols chart
nichols	Computes a Nichols frequency response plot
num2str	Converts numbers to strings
nyquist	Calculates the Nyquist frequency response
obsv	Computes the observability matrix
ones	Generates a matrix of integers where all the integers are 1
pade	Computes an nth-order Padé approximation to a time delay
parallel	Computes a parallel system connection
plot	Generates a linear plot
pole	Computes the poles of a system
poly	Computes a polynomial from roots
polyval	Evaluates a polynomial
printsys	Prints state variable and transfer function representations of linear systems in a pretty form
pzmap	Plots the pole–zero map of a linear system
rank	Calculates the rank of a matrix
real	Computes the real part of a complex number
residue	Computes a partial fraction expansion
rlocfind	Finds the gain associated with a given set of roots on a root locus plot
rlocus	Computes the root locus
roots	Determines the roots of a polynomial
semilogx	Generates an x-y plot using semilog scales with the x-axis $\log_{10}$ and the y-axis linear
semilogy	Generates an x-y plot using semilog scales with the y-axis $\log_{10}$ and the x-axis linear
series	Computes a series system connection
shg	Shows graph window
sin	Computes the sine
sqrt	Computes the square root
ss	Creates a state-space model object
step	Calculates the unit step response of a system
subplot	Splits the graph window into subwindows
tan	Computes the tangent
text	Adds text to the current graph
title	Adds a title to the current graph
tf	Creates a transfer function model object
who	Lists the variables currently in memory
whos	Lists the current variables and sizes
xlabel	Adds a label to the x-axis of the current graph
ylabel	Adds a label to the y-axis of the current graph
zero	Computes the zeros of a system
zeros	Generates a matrix of zeros

MATLAB BASICS: PROBLEMS

A.1   Consider the two matrices

$$A = \begin{bmatrix} 4 & 2\pi \\ 6j & 10 + \sqrt{2}j \end{bmatrix}$$

$$B = \begin{bmatrix} 6j & -13\pi \\ \pi & 16 \end{bmatrix}.$$

Using MATLAB, compute the following:

(a) $A + B$ (b) AB
(c) A^2 (d) A'
(e) B^{-1} (f) $B'A'$
(g) $A^2 + B^2 - AB$

A.2   Consider the following set of linear algebraic equations:

$$5x + 6y + 10z = 4,$$
$$-3x + 14z = 10,$$
$$-7y + 21z = 0.$$

Determine the values of x, y, and z so that the set of algebraic equations is satisfied. (*Hint:* Write the equations in matrix vector form.)

A.3   Generate a plot of

$$y(x) = e^{-0.5x} \sin \omega x,$$

where $\omega = 10$ rad/s, and $0 \le x \le 10$. Utilize the colon notation to generate the x vector in increments of 0.1.

A.4   Develop a MATLAB script to plot the function

$$y(x) = \frac{4}{\pi} \cos \omega x + \frac{4}{9\pi} \cos 3\omega x.$$

where ω is a variable input at the command prompt. Label the x-axis with *time(sec)* and the y-axis with $y(x) = (4/\pi) * \cos(\omega x) + (4/9\pi) * \cos(3\omega x)$. Include a descriptive header in the script, and verify that the help function will display the header. Choose $\omega = 1, 3, 10$ rad/s and test the script.

A.5   Consider the function

$$y(x) = 10 + 5e^{-x} \cos(\omega x + 0.5).$$

Develop a script to co-plot $y(x)$ for the three values of $\omega = 1, 3, 10$ rad/s with $0 \le x \le 5$ seconds. The final plot should have the following attributes:

Title	$y(x) = 10 + 5 \exp(-x) * \cos(\omega x + 0.5)$
x-axis label	Time (sec)
y-axis label	$y(x)$
Line type	$\omega = 1$: solid line
	$\omega = 3$: dashed line
	$\omega = 10$: dotted line
Grid	grid on

MathScript Basics

B.1 INTRODUCTION

LabVIEW is short for Laboratory Virtual Instrument Engineering Workbench. It is a flexible graphical development environment from National Instruments, Inc. Engineers and scientists in research, development, production, test, and service industries as diverse as automotive, semiconductor, aerospace, electronics, chemical, telecommunications, and pharmaceutical use LabVIEW, especially in the area of testing and measurements, industrial automation, and data analysis. Users of Lab-VIEW are familiar with the use of the graphical programming language to create programs relying on graphic symbols to describe programming actions. An important new development introduced in LabVIEW 8.0 or higher is the MathScript environment. LabVIEW MathScript is a text-based command line environment using m-files and command line prompts. It is assumed here that the reader has LabVIEW 8 installed and knows how to access the LabVIEW Getting Started window. This appendix only provides an introduction to MathScript. Readers should refer to *Learning with LabVIEW*[1] for a more complete introduction to LabVIEW and MathScript.

In this appendix, we discuss the MathScript Interactive Window. The essentials of creating user-defined functions and scripts, of saving and loading data files, and of using the MathScript Node are presented. With the MathScript Interactive Window, students will be able to interact with LabVIEW through a command prompt.

B.2 WHAT IS MATHSCRIPT?

MathScript is a high-level, text-based programming language with an easily accessible syntax and ample functionality to address programming tasks related to signal processing, analysis, and mathematics. MathScript includes more than 500 built-in functions. There are linear algebra functions, curve fitting function, digital filters, functions for solving differential equations, and probability and statistics functions. And since MathScript employs a commonly used syntax, it follows that you can work with many of your previously developed mathematical computation scripts, or any of those openly available in engineering textbooks or on the Internet.

---

[1] Bishop, R. H., *Learning with LabVIEW*, Prentice Hall Publishing, 2007.

The fundamental math-oriented data types in MathScript are matrices with built-in operators for generating data and accessing elements. You can extend MathScript by defining your own custom functions. You can find more information on MathScript at http://ni.com/mathscript including lists of built-in MathScript functions and links to online examples.

B.3 ACCESSING THE MATHSCRIPT INTERACTIVE WINDOW

The interactive interface is provided by the MathScript Interactive Window. You can access the interactive window from the **Getting Started** window or any VI by selecting **Tools»MathScript Window**…, as illustrated in Figure B.1. The MathScript Interactive Window is a user interface comprised of a **Command Window** (the user command inputs), an **Output Window** echoing the inputs and showing the resulting outputs, a **Script Editor** window (for loading, saving, compiling, and running scripts), a **Variables** window (showing variables, dimensions, and type), and a **Command History** window providing a historical account of commands and interactions with MathScript. A new MathScript Interactive Window is shown in Figure B.2 with the various components highlighted.

As you work, the Output Window updates to show your inputs and the subsequent results. The Command History window tracks your commands. The history view is very useful because there you can access and reuse your previously executed commands by double clicking a previous command to execute it again. You can also navigate up and down through the previous commands (which will appear in the Command Window) by using the "↑" and "↓" keys. In the Script Editor window, you can enter and execute groups of commands and then save the commands in a file (called a script) for use in a later LabVIEW session.

Select MathScript Window

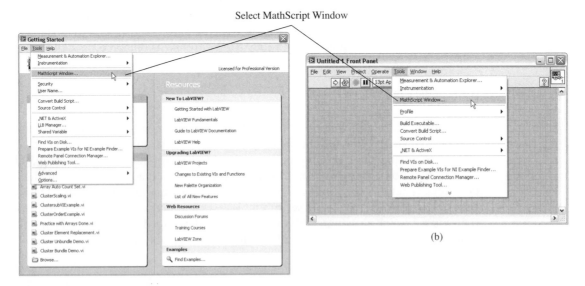

(b)

FIGURE B.1 Accessing the MathScript Interactive Window from (a) the Getting Started window, or (b) the **Tools** pull-down menu on the front panel or block diagram.

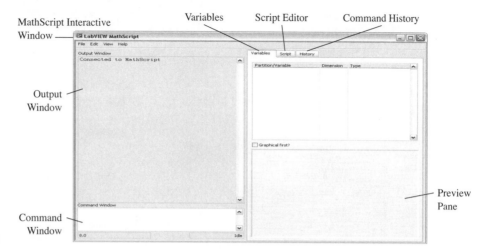

FIGURE B.2
The basic components of the MathScript Interactive Window.

Clearing the Command History Window

The commands entered in previous sessions using the MathScript Interactive Window will reappear in subsequent sessions. In the Command History window you will find a header that shows the day and time that you entered the commands. This feature allows you to easily discern when the commands were entered. If the Command History window gets too full and you want to clear it out, you can right-click the Command History window and select **Clear History** from the shortcut menu. This process is illustrated in Figure B.3.

Clearing the Output Window

In a manner similar to the clearing the Command History window, you can clear the Output Window. To accomplish this task, right-click the Output Window and select **Clear** from the shortcut menu, as shown in Figure B.3. You also can use the MathScript function clc to clear the Output Window by typing in clc in the Command Window.

Copying Output Window Data

You can copy data from the Output Window and paste it in the Script Editor window or a text editor. Right-click the Output Window and select **Copy Data** from the shortcut menu to copy the contents of the Output Window to the clipboard. You also can highlight text in the Output Window and select **Edit»Copy** or press the <Ctrl-C> keys to copy the selected text to the clipboard.

Viewing Data in a Variety of Formats

In the MathScript Interactive Window you can view the variables in a variety of formats, as shown in Figure B.4. Depending on the variable type, the available formats

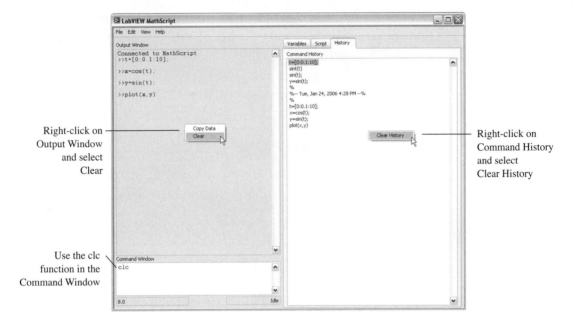

FIGURE B.3   Clearing the Command History and the Output Window.

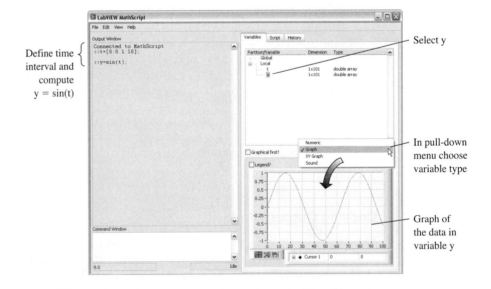

FIGURE B.4
Showing the data
type in various
formats.

include: numeric, string, graph, XY graph, sound, surface, and picture. You can edit
a variable in the **Preview Pane** when the display type is **Numeric** or **String.** Selecting
Sound plays the data as a sound, but works for one-dimensional variables only. The
remaining display types show the data as graphs of one sort or another: **Graph** displays the data on a waveform graph, **XY Graph** displays the data on an XY graph,
Surface displays the data on a 3D surface graph, and **Picture** displays the data on an
intensity graph.

Open a new MathScript Interactive Window from the Getting Started window by selecting **Tools»MathScript Window...**, as illustrated in Figure B.1. In the Command Window input the time from $t = 0$ seconds to $t = 10$ seconds in increments of 0.1 seconds, as follows:

$$t = [0:0.1:10];$$

Then, compute the $y = \cos(t)$ as follows:

$$y = \cos(t);$$

Notice that in the Variables window the two variables t and y appear, as illustrated in Figure B.5a. Select the variable y and note that in the Preview Pane the variable appears in the numeric format. Now, in the pull-down menu above the Preview Pane, select **Graph**. The data will now be shown in graphical form, as illustrated in Figure B.5b. The graph can be undocked from the Preview Pane for re-sizing and customization. To undock the graph, right-click on the graph and select **Undock Window**. The window can now be re-sized and the plot can be customized interactively and printed.

As an alternative to using the Preview Pane, you can also obtain a plot of the y versus t programmatically using the **plot** command:

$$plot(t,y)$$

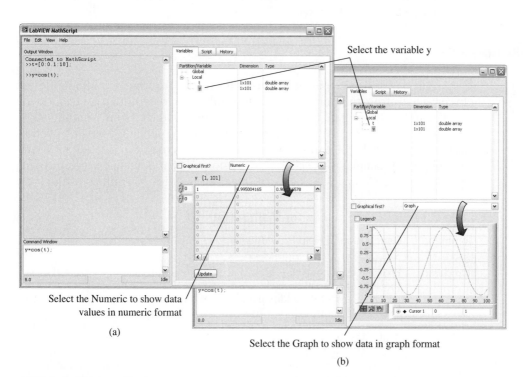

(a)

(b)

FIGURE B.5 (a) Entering the time, computing $y = \cos(t)$ and viewing the variable y in numerical form. (b) Viewing the variable y in graphical form.

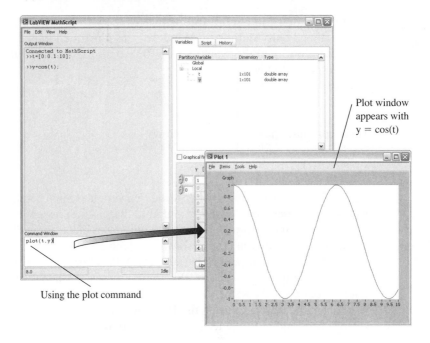

Plot window
appears with
y = cos(t)

FIGURE B.6
Obtaining a plot of
the cosine function
using the **plot**
command.

Using the plot command

The process is illustrated in Figure B.6. A new window appears that presents the
graph of *y* versus *t*. Following the same procedure, see if you can obtain a plot of
$y = \cos(\omega t)$ where $\omega = 4$ rad/sec.

B.4 MATHSCRIPT HELP

You can display several types of help content for MathScript by calling different
help commands from the Command Window. Table B.1 lists the help commands you
can call and the type of help these commands display in the Output Window.

As illustrated in Figure B.7, entering help classes in the Command Window
launches the LabVIEW Help showing all classes of functions and commands that Math-
Script supports. Examples of the classes of functions are basic and matrixops. Entering
help basic in the Command Window results in a list of the members of the basic

Table B.1   Help Commands for MathScript

Command	Description of Help Provided
help	Provides an overview of the MathScript window.
help classes	Provides a list of all classes of MathScript functions and topics as well as a short description of each class.
help cdt classes	Provides a list of the additional classes of functions that are installed with the LabVIEW Control Design Toolkit.
help *class*	Provides a list of the names and short descriptions of all functions in a particular MathScript class. Example: help basic
help *function*	Provides reference help for a particular MathScript function or topic, including its name, syntax, description, inputs and outputs, examples to type in the Command Window, and related functions or topics. Example: help abs

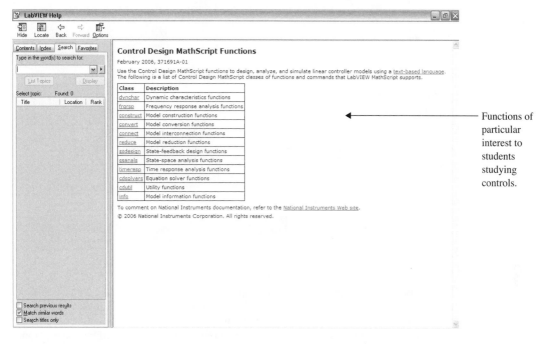

FIGURE B.7 Accessing the help for MathScript classes, members, and functions.

class, including abs (i.e., the absolute value), conj (i.e., the complex conjugate function), and exp (i.e., the exponential function). Then, entering help abs in the Command Window will result in an output that contains a description of the abs function, including examples of its usage and related topics.

B.5 SYNTAX

The syntax associated with MathScript is straightforward. Most students with some experience programming a text-based language will be comfortable with the programming constructs in MathScript. If you need help getting started with MathScript, you can access help by selecting Help » Search the LabVIEW Help from the MathScript Interactive Window and typing mathscript in the search window.

Eleven basic MathScript syntax guidelines are:

1. **Scalar operations:** MathScript is ideally suited for quick mathematical operations, such as addition, subtraction, multiplication, and division. For example, consider the addition of two scalar numbers, 16 and 3. This is a simple operation that you might perform on a calculator. This can be accomplished using the MathScript command:

```
>>16+3
ans=

        19
```

Output Window display

In MathScript, if you perform any calculation or function without assigning the result to a variable, the default variable ans is used. If you want to assign the value of the addition of two scalars to the variable x, enter the following command:

```
>>x=16+3
X=
        19
```

In the same manner, you can add two scalar variables y and z by entering the following commands:

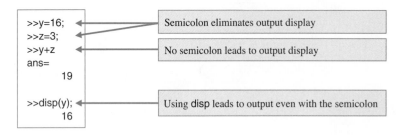

```
>>y=16;     ◄────── Semicolon eliminates output display
>>z=3;      ◄──────
>>y+z       ◄────── No semicolon leads to output display
ans=
        19

>>disp(y);  ◄────── Using disp leads to output even with the semicolon
        16
```

Notice that in the previous example a semicolon was used for the first two lines, and no output was displayed. In MathScript, if you end a command line with a semicolon, the MathScript Interactive Window does not display the output for that command. Some functions display output even if you end the command line with a semicolon. For example, the disp function displays an output even if followed by a semicolon.

You use the symbol '-' for subtraction, the symbol '/' for division, and the symbol '*' for multiplication, as illustrated below:

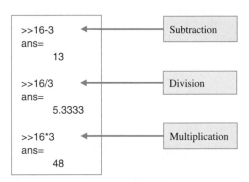

```
>>16-3      ◄────── Subtraction
ans=
        13

>>16/3      ◄────── Division
ans=
        5.3333

>>16*3      ◄────── Multiplication
ans=
        48
```

2. **Creating matrices and vectors:** To create row or column vectors and matrices, use white space or commas to separate elements, and use semicolons to separate rows. Consider for example, the matrix **A** (a column vector),

$$\mathbf{A} = \begin{bmatrix} 1 \\ 2 \\ 3 \end{bmatrix}.$$

In MathScript syntax, you would form the matrix as

$$A = [1; 2; 3]$$

Consider for example, the matrix **B** (a row vector)

$$\mathbf{B} = \begin{bmatrix} 1 & -2 & 7 \end{bmatrix}.$$

In MathScript syntax, you would form the matrix as

$$B = [1, -2, 7] \text{ or } B = [1 \ -2 \ 7]$$

As a final example, consider the matrix **C** (a 3 × 3 matrix)

$$\mathbf{C} = \begin{bmatrix} -1 & 2 & 0 \\ 4 & 10 & -2 \\ 1 & 0 & 6 \end{bmatrix}.$$

In MathScript syntax, you would form the matrix as

$$C = [-1 \ 2 \ 0; 4 \ 10 \ -2; 1 \ 0 \ 6] \text{ or } C = [-1, 2, 0; 4, 10, -2; 1, 0, 6]$$

3. **Creating vectors using the colon operator:** There are several ways to create a one-dimensional array of equally spaced elements. For example, you will often need to create a vector of elements representing time. To create a one-dimensional array equally spaced and incremented by 1, use the MathScript syntax

```
≫ t = 1:10
t =
        1    2    3    4    5    6    7    8    9    10
```

To create a one-dimensional array equally spaced and incremented by 0.5, use the MathScript syntax

```
≫ t = 1:0.5:10
t =
        1    1.5    2    2.5    3    3.5    4    4.5    5    5.5
    6    6.5    7    7.5    8    8.5    9    9.5    10
```

4. **Accessing individual elements of a vector or matrix:** You may want to access specific elements or subsets of a vector or matrix. Consider the 3 × 3 matrix **C**:

$$\mathbf{C} = \begin{bmatrix} -1 & 2 & 0 \\ 4 & 10 & -2 \\ 1 & 0 & 6 \end{bmatrix}.$$

In MathScript syntax, you can access the element in the second row and third column of the matrix **C**, as follows:

```
>>C = [-1 2 0;4 10 -2;1 0 6]
C=
    -1    2    0
     4   10   -2
     1    0    6

>>C(2,3)
ans=

    -2
```

C(2,3) denotes the second row of the third column

You can assign this value to a new variable by entering the following command:

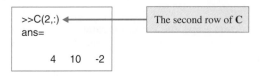

```
>>F=C(2,3)
F=
        -2
```

You can also access an entire row or an entire column of a matrix using the colon operator. In MathScript syntax, if you want to access the entire second row of matrix **C**, enter the following command:

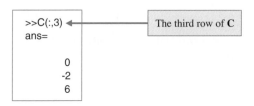

```
>>C(2,:)  ←——————  The second row of C
ans=

       4    10   -2
```

In the same way, if you wish to access the entire third column of matrix **C**, enter the following command:

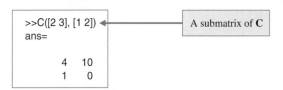

```
>>C(:,3)  ←——————  The third row of C
ans=

        0
       -2
        6
```

Suppose you want to extract the 2 × 2 submatrix from **C** consisting of rows 2 and 3 and columns 1 and 2. You use brackets to specify groups of rows and columns to access a subset of data as follows:

```
>>C([2 3], [1 2])  ←——————  A submatrix of C
ans=

       4    10
       1     0
```

5. **Calling functions in MathScript:** You can call MathScript functions from the Command Window. Consider the creation of a vector of a certain number of elements that are equally distributed in a given interval. To accomplish this in MathScript syntax, you can use the built-in function linspace. Using the command help linspace you find that this function uses the syntax

$$linspace(a, b, n)$$

where a specifies the start of the interval, b specifies the end of the interval, and n identifies the number of elements. Thus, to create a vector of $n = 13$ numbers equally distributed between $a = 1$ and $b = 10$, use the following command:

```
>>G = linspace (1, 10, 13)
G=
    1      1.75     2.5     3.25     4     4.75     5.5     6.25     7

   7.75    8.5     9.25     10
```

If you do not specify a value for n, the linspace command will automatically return a vector of 100 elements. To select a subset of **G** that consists of all elements after a specified index location, you can use the syntax described in guideline 4 and the end function to specify the end of the vector. For example, the following command will return all elements of **G** from the fifth element to the final element:

```
>>H=G(5: end(G))
H =
    4      4.75     5.5     6.25     7     7.75     8.5     9.25     10
```

The function linspace is an example of a built-in MathScript function. Calling user-defined functions are discussed further in Section B.4.1.

6. **Assigning data types to variables:** MathScript variables adapt to data types. For example, if

```
a = sin(3 * pi/2)
```

then a is a double-precision floating-point number. If

```
a = 'temperature'
```

then a is a string.

7. **Using complex numbers:** You can use either i or j to represent the imaginary unit equal to the square root of -1. If you assign values to either i or j in your scripts, then those variable names are no longer complex numbers. For example, if you let $y = 4 + j$, then y is a complex number with real part equal to 4 and imaginary part equal to $+j$. If however, you assign $j = 3$, and then compute $y = 4 + j$, the result is $y = 7$, a real number.

8. **Matrix operations:** Many of the same mathematical functions used on scalars can also be applied to matrices and vectors. Consider adding two matrices **K** and **L**, where

$$\mathbf{K} = \begin{bmatrix} -1 & 2 & 0 \\ 4 & 10 & -2 \\ 1 & 0 & 6 \end{bmatrix} \text{ and } \mathbf{L} = \begin{bmatrix} 1 & 0 & 0 \\ 0 & 1 & 0 \\ 0 & 0 & 1 \end{bmatrix}.$$

To add the two matrices **K** and **L**, element by element, enter the following MathScript command:

```
>>K+L
ans =
    0     2     0
    4    11    -2
    1     0     7
```

In a similar fashion, you can also multiply two matrices **K** and **L**, as follows:

```
>>K*L
ans =
    -1     2     0
     4    10    -2
     1     0     6
```

Consider the 3×1 matrix **M** (column vector) and the 1×3 matrix **N** (row vector)

$$\mathbf{M} = \begin{bmatrix} 1 \\ 2 \\ 3 \end{bmatrix} \text{ and } \mathbf{N} = [0 \quad 1 \quad 2].$$

Then, the product **M** *N** is the 3×3 matrix

```
>>M*N
ans =
     0     1     2
     0     2     4
     0     3     6
```

and the product **N** *M** is a scalar

```
>>N*M
ans =
     8
```

To multiply two matrices, they must be of compatible dimensions. For example, suppose a matrix **M** is of dimension $m \times n$, and a second matrix **N** is of dimension $n \times p$. Then you can multiply $\mathbf{M} \times \mathbf{N}$ resulting in an $m \times p$ matrix. In the example above, the 3×1 matrix **M** (column vector) was multiplied with the 1×3 matrix **N** (row vector) resulting in a 3×3 matrix. You cannot multiply $\mathbf{N} \times \mathbf{M}$ unless $m = p$. In the example above, the 1×3 matrix **N** (row vector) was multiplied with the 3×1 matrix **M** (column vector) resulting in a 1×1 matrix (a scalar), so in this case, $m = p = 1$.

When working with vectors and matrices in MathScript, it is often useful to perform mathematical operations element-wise. For example, consider the two vectors

$$\mathbf{M} = \begin{bmatrix} -1 \\ 4 \\ 0 \end{bmatrix} \text{ and } \mathbf{N} = \begin{bmatrix} 2 \\ -2 \\ 1 \end{bmatrix}.$$

In light of our previous discussion, it is not possible to compute **M** *N** since the dimensions are not compatible. However, you can multiply the vectors element-wise using the syntax '.*' for the multiplication operator, as follows:

$$\mathbf{M}.*\mathbf{N} = \begin{bmatrix} -1*2 \\ 4*(-2) \\ 0*1 \end{bmatrix} = \begin{bmatrix} -2 \\ -8 \\ 0 \end{bmatrix}.$$

By definition, matrix addition and subtraction occurs element-wise. However, it is also possible to perform division element-wise. For **M** and **N** as above, we find that element-wise division yields

$$\mathbf{M./N} = \begin{bmatrix} -1/2 \\ 4/(-2) \\ 0/1 \end{bmatrix} = \begin{bmatrix} -0.5 \\ -2 \\ 0 \end{bmatrix}.$$

Element-wise operations are useful in plotting functions. For example, suppose that you wanted to plot $y = t \sin(t)$ for $t = [0:0.1:10]$. This would be achieved via the commands

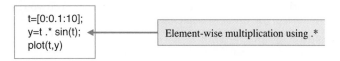

9. **Logical expressions:** MathScript can evaluate logical expressions such as EQUAL, NOT EQUAL, AND, and OR. To perform an equality comparison, use the statement

$$a == b$$

If a is equal to b, MathScript will return a 1 (indicating True); if a and b are not equal, MathScript will return a 0 (indicating False). To perform an inequality comparison, use the statement

$$a \sim= b$$

If a is not equal to b, MathScript will return a 1 (indicating True); if a and b are equal, MathScript will return a 0 (indicating False).

In other scenarios, you may want to use MathScript to evaluate compound logical expressions, such as when at least one expression of many is True (OR), or when all of your expressions are True (AND). The compound logical expression AND is executed using the '&' command. The compound logical expression OR is executed using the '|' command.

10. **Control flow constructs:** Table B.2 provides the MathScript syntax for commonly used programming constructs.

11. **Adding comments:** To add comments to your scripts, precede each line of documentation with a % character. For example, consider a script that has two inputs, x and y, and computes the addition of x and y as the output variable z.

```
% In this script, the inputs are x and y
% and the output is z.              } Comments
% z is the addition of x and y
z = x + y;
```

The script shown above has three comments, all preceded by the % character. In the next section, we will discuss more details on how to use comments to provide help documentation.

Some considerations that have a bearing on your usage of MathScript follow:

1. You cannot define variables that begin with an underscore, white space, or digit. For example, you can name a variable *time*, but you cannot name a variable *4time* or *_time*.

2. MathScript variables are case sensitive. The variables X and x are not the same variables.

Table B.2   MathScript syntax for commonly used constructs

Construct	Grammar	Example

Case-Switch Statement	**switch**
	expression
	case expression
	statement-list
	[**case**
	expression
	statement-list]
	. . .
	[**otherwise**
	statement-list]
	end
		switch mode
		case 'start'
		a = 0;
		case 'end'
		a = −1;
		otherwise
		a = a + 1;
		end
		When a case in a case-switch statement executes, LabVIEW does not select the next case automatically. Therefore, you do not need to use break statements as in C.

For Loop	**for** expression statement-list **end**	for k = 1:10 a = sin(2*pi*k/10) end
If-Else Statement	**if** expression statement-list [**elseif** expression statement-list] . . . [**else** statement-list] **end**	if b == 1 c = 3 else c = 4 end
Range	start**:**[step**:**]end	t = 0:0.1:10 or t = [0:0.1:10] t returns an array of numbers $0 \le t \le 10$ with a step size of 0.1 If you do not specify a step size, LabVIEW uses a step size of 1.
While Loop	**while** expression statement-list **end**	while k < 10 a = cos(2*pi*k/10) k = k + 1; end

Key MathScript Functions

MathScript offers more than 500 textual functions for math, signal processing, and analysis. These are in addition to the more than 600 graphical functions for signal processing, analysis, and math that are available as VIs within LabVIEW. Table B.3 lists many of the key areas with supporting MathScript functions. For a comprehensive function list, visit the National Instruments website at  http://www.ni.com/mathscript or see the online help.

B.6 DEFINING FUNCTIONS AND CREATING SCRIPTS

You can define functions and create scripts to use in the MathScript Interactive Window. Functions and scripts can be created in the Script Editor window on the MathScript Interactive Window (see Figure B.2). You can also use your favorite text editor to create functions and scripts. Once your function or script is complete, you should save it for use later. The filename for a function must be the same as the name of the function and must have a lowercase .m extension. For example, the filename for a

Table B.3   MathScript Function Classes

Function Classes	Brief Description
Control Design and Analysis	Classical and state-space control design and analysis functions. Dynamic characteristics, root locus, frequency response, Bode, Nyquist, Nichols, model contstruction, connection, reduction and more.
Plots (2D and 3D)	Standard x-y plot; mesh plot; 3D plot; surface plot; subplots; stairstep plot; logarithmic plots; stem plot and more.
Digital Signal Processing (DSP)	Signal synthesis; Butterworth, Chebyshev, Parks-McClellan, windowed FIR, elliptic (Cauer), lattice and other filter designs; FFT (1D/2D); inverse FFT (1D/2D); Hilbert transform; Hamming, Hanning, Kaiser-Bessel and other windows; pole/zero plotting and others.
Approximation (Curve Fitting & Interpolation)	Cubic spline, cubic Hermite and linear interpolation; exponential, linear and power fit; rational approximation and others.
Ordinary Differential Equation (ODE) Solvers	Adams-Moulton, Runge-Kutta, Rosenbrock and other continuous ordinary differential equation (ODE) solvers.
Polynomial Operations	Convolution; deconvolution; polynomial fit; piecewise polynomial; partial fraction expansion and others.
Linear Algebra	LU, QR, QZ, Cholesky, Schur decomposition; SVD; determinant; inverse; transpose; orthogonalization; solutions to special matrices; Taylor series; real and complex eigenvalues and eigenvectors; polynomial eigenvalue and more.
Matrix Operations	Hankel, Hilbert, Rosser, Vandermonde special matrices; inverse; multiplication; division; unary operations and others.
Vector Operations	Cross product; curl and angular velocity; gradient; Kronecker tensor product and more.
Probability and Statistics	Mean; median; Poisson, Rayleigh, chi-squared, Weibull, T, gamma distributions; covariance; variance; standard deviation; cross correlation; histogram; numerous types of white noise distributions and other functions.
Optimization	Quasi-Newton, quadratic, Simplex methods and more.
Advanced Functions	Bessel, spherical Bessel, Psi, Airy, Legendre, Jacobi functions; trapezoidal, elliptic exponential integral functions and more.
Basic	Absolute value; Cartesian to polar and spherical and other coordinate conversions; least common multiple; modulo; exponentials; logarithmic functions; complex conjugates and more.
Trigonometric	Standard cosine, sine and tangent; inverse hyperbolic cosine, cotangent, cosecant, secant, sine and tangent; hyperbolic cosine cotangent, cosecant, secant, sine and tangent; exponential; natural logarithm and more.
Boolean and Bit Operations	AND, OR, NOT and other logic operations; bitwise shift, bitwise OR and other bitwise operations.
Data Acquisition/Generation	Perform analog and digital I/O using National Instruments devices.
Other	Programming primitives such as if, for and while loops; unsigned and signed datatype conversions; file I/O; benchmarking and other timing functions; various set and string operations and more.

user-defined starlight[2] function must be starlight.m. Use unique names for all functions and scripts and save them in a directory that you specified in the **Path** section of the **File»MathScript Preferences** dialog box.

---

[2]The name starlight does not represent a real function. It is used here for illustrative purposes only.

User-Defined Functions

MathScript offers more than 500 textual functions for math, signal processing, and analysis. But what if you have a special purpose function that you want to add to your personal library? This function may be particular to your area of study or research, and is one that you need to call as part of a larger program. With MathScript it is simple to create a function once you understand the basic syntax.

A MathScript function definition must use the following syntax:

> function *outputs* $=$ *function_name*(*inputs*)
> % *documentation*
> *script*

An example of a user-defined function definition utilizing the proper syntax is

> function ave $=$ compute_average(x, y)
> % compute_average determines the average of the two inputs x and y.
> ave $=$ (x $+$ y)/2;

Begin each function definition with the term **function**. The *outputs* lists the output variables of the function. If the function has more than one output variable, enclose the variables in square brackets and separate the variables with white space or commas. The *function_name* is the name of the function you want to define and is the name that you use when calling the function. The *inputs* lists the input variables to the function. Use commas to separate the input variables. The *documentation* is the set of comments that you want MathScript to return for the function when you execute the help command. Comments are preceded with a % character. You can place comments anywhere in the function; however, LabVIEW returns only the first comment block in the Output Window to provide the help to the user. All other comment blocks are for internal documentation. The *script* defines the executable body of the function.

Checking the help on the function compute_average.m and then executing the function with $x = 2$ and $y = 4$ as inputs yields

> »help compute_average
> compute_average determines the average of the two inputs x and y.
> »x $=$ 2; y $=$ 4; compute_average(x, y)
> ans $=$
>   3

Note that there is a MathScript function named mean that can also be used to compute the average of two inputs, as follows:

> »mean([2 4])
> ans $=$
>   3

Functions can be edited in the Script Editor window and saved for later use. In Figure B.8, the buttons **Load Script, Save Script As, Save & Compile Script As** and **Run**

Script are shown. You can compile a function when you save it to decrease the run-time compilation time. Selecting **Load Script** will open a window to browse for the desired function (or script) to load into MathScript. Similarly, selecting **Save Script As** will open a browser to navigate to the desired folder to save the function.

In Figure B.8, the function compute_average is used to compute the average of two arrays. Notice that the function computes the average element-wise. If compute_average had inadvertently been named mean, then LabVIEW would execute the user-defined function instead of the built-in function. Generally it is not a good idea to redefine LabVIEW functions, and students should avoid doing so. If you define a function with the same name as a built-in MathScript function, Lab-VIEW executes the function you defined instead of the original MathScript function. When you execute the help command, LabVIEW returns help content for the function you defined and not the help content for the original MathScript function.

Other examples of valid function syntax for the starlight function include:

function starlight	% No inputs and no outputs
function a = starlight	% No inputs and one output
function [a b] = starlight	% No inputs and two outputs
function starlight (g)	% One input and no outputs
function a = starlight (g)	% One input and one output
function [a b] = starlight (g)	% One input and two outputs
function starlight (g, h)	% No inputs and two outputs
function a = starlight (g, h)	% One input and two outputs
function [a b] = starlight (g, h)	% Two inputs and two outputs

There are several restrictions on the use of functions. First, if you define multiple functions in one MathScript file, all functions following the first are **subfunctions** and are accessible only to the main function. A function can call only those functions that you define below it. Second, you cannot call functions recursively. For example, the function starlight cannot call starlight. And third, LabVIEW also does not allow circular recursive function calls. For example, the function starlight cannot call the function bar if bar calls starlight.

Scripts

A script is a sequence of MathScript commands that you want to perform to accomplish a task. For convenience and reusability, once you have created a script, you can save it and load it into another session of LabVIEW at a later time. Also, often you can use a script designed for a different task as a starting point for the development of a new script. Since the scripts themselves are saved as common ascii text and editable with any text editor (including the one found in the MathScript Interactive Window), it is easy to do this. The MathScript functions as well as the user-defined functions can be employed in scripts.

Continuing the example above, suppose that we used a script to compute the average of two numbers. The compute_average function could be used within the script. Once saved, the script can subsequently be loaded into MathScript for use in another session. A script using the compute_average function is shown in Figure B.9.

Select Load to browse for a
function to load in MathScript

Select Save to browse for a
directory to save the function

Running the function
for x = [1 4] and y = [2 6]
yields the average [1.5 5]

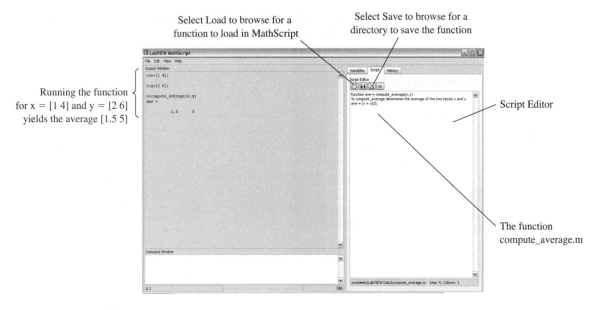

Script Editor

The function
compute_average.m

FIGURE B.8   Loading and saving functions.

B.7 SAVING AND LOADING DATA FILES AND SCRIPTS

In MathScript you can save and load data files in the MathScript Interactive Window. A data file contains numerical values for variables. Being able to save and load djata gives you the flexibility to save important data output from a MathScript session for use in external programs. There are two ways to save data files. The first

Select Run

Results of
executing script
are displayed

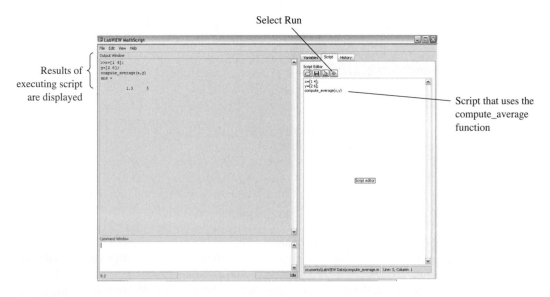

Script that uses the
compute_average
function

FIGURE B.9   Editing, saving, and running a script to compute the average of two arrays element-wise.

method saves the data for *all* the variables in the workspace, and the second method allows you to select the variables to save to a file.

To save all the variables in the workspace, select **File»Save Data** in the Math-Script Interactive Window. You also can right-click the **Variable List** on the Variables window and select **Save Data** from the shortcut menu. In the file dialog box, navigate to the directory in which you want to save the data file. Enter a name for the data file in the **File name** field and click the **OK** button to save the data file.

The second method allows you to select the variables to save. In this case, in the Command Window, enter the command **save filename var1, var2, ..., varn**, where **filename** is the name of the file to store the data and **var1, var2, ... varn** are the variables that you want to save. In this case, the data will be saved in filename in the LabVIEW Data directory in the path specified in **File»MathScript Preferences....** In Figure B.10 the process of saving data is illustrated. In Figure B.10a, all the variables are saved in the file save_all.mlv after navigating to the folder LabVIEW Data. In Figure B.10b, the variable **x** is saved in the file save_x.mlv.

You also can load existing data files into your MathScript session. In the Math-Script Interactive Window, select **File»Load Data** or right-click the **Variable List** on

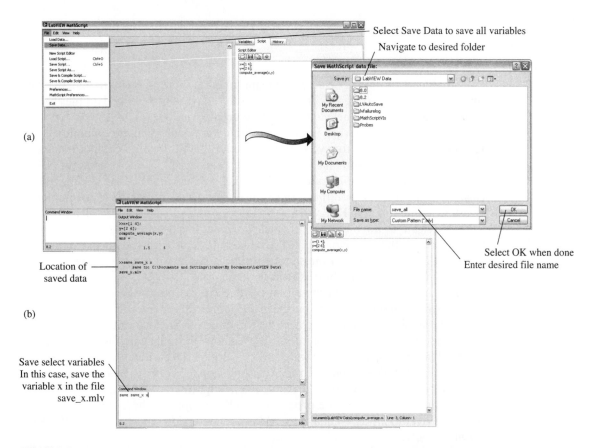

FIGURE B.10   Saving data files. (a) Saving all the variables in the workspace. (b) Saving select variables.

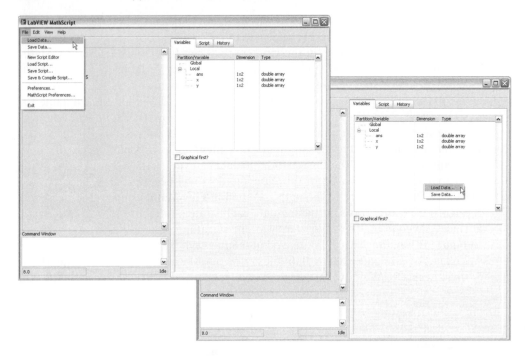

FIGURE B.11   Loading data from a previous MathScript session.

the Variables window and select **Load Data** from the shortcut menu to load the data file you want, as illustrated in Figure B.11. Note that you must save data files before you can load them into the MathScript Interactive Window.

Being able to save scripts is an important feature giving you the capability to develop a library of scripts that you can readily access in future MathScript sessions. To save a script that you have created in the Script Editor window, select **File»Save Script As**, as illustrated in Figure B.12a. You can also save your script by clicking the **Save** button on the Script Editor window of the MathScript Interactive Window, as illustrated in Figure B.12b. In both cases, a file dialog box will appear for you to navigate to the directory in which you want to save the script. Enter a name for the script in the **File name** field. The name must have a lowercase .m extension if you want LabVIEW to run the script (in this example, we use the name average_example.m). Click the **OK** button to save the script.

You can compile a script by selecting **File»Save & Compile Script** or by clicking on the **Save & Compile Script As** button in the MathScript Interactive Window. This will save and compile the script to decrease the run-time compilation time. You can load existing scripts into the MathScript Interactive Window. This will be useful upon returning to a MathScript session or if you want to use a script in the current session that was developed in a previous session. To load an existing script, select **File»Load Script** or click the **Load** button on the **Script** page on the MathScript Interactive Window.

Figure B.13 illustrates the process of loading scripts. In the example, the script compute_average.m is loaded into a MathScript session, and then using the **Run Script As** button, the script is executed.

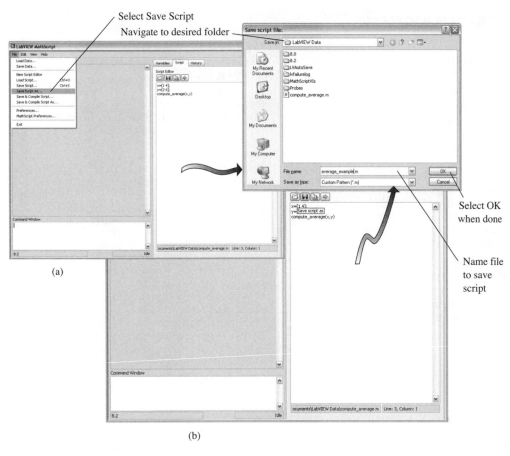

FIGURE B.12 (a) Saving a script using the **File»Save Script As** pull-down menu. (b) Saving a script using the **Save Script As** button on the MathScript Interactive Window.

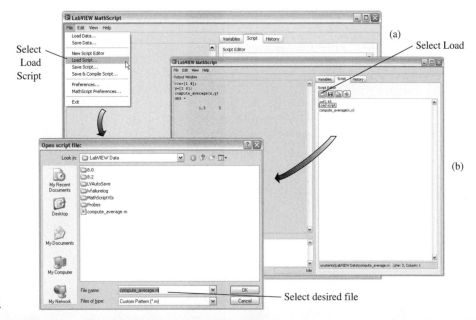

FIGURE B.13
(a) Loading a script using the
File»Load Script
pull-down menu.
(b) Loading a script
using the **Load
Script** button on
the MathScript
Interactive Window.

MATHSCRIPT BASICS: PROBLEMS

B.1   Write a script to generate a 3 × 2 matrix **M** of random numbers using the rand function. Use the help command for syntax help on the rand function. Verify that each time you run the script the matrix **M** changes.

B.2   In the MathScript Interactive Window, create a script that generates a time vector over the interval 0 to 10 with a step size of 0.5 and creates a second vector, y, according to the equation

$$y = e^{-t}(0.5 \sin 0.1t - 0.25 \cos 0.2t).$$

Add the plot function to generate a graph of y versus t. Type the script in the Script Editor window and, when done, use the **Save** button in the Script Editor to save the script. Clear the script from the Script Editor window, and then **Load** the script back into the Script Editor and select **Run**.

B.3   Create a plot of the cosine function, $y = \cos(t)$, where t varies from 0 to π, with an increment of $\pi/20$.

B.4   Open the MathScript Interactive Window. In the Command Window, create the matrices **A** and **B**:

$$\mathbf{A} = \begin{bmatrix} 1 & -2 \\ 0 & 3 \\ -1 & 5 \end{bmatrix} \text{ and } \mathbf{B} = \begin{bmatrix} 1 & -1 & 7 \\ 2 & 0 & -2 \end{bmatrix}.$$

Is it possible to perform the following math operations on the matrices? If so, what is the result?
(a) **A** * **B**
(b) **B** * **A**
(c) **A** + **B**
(d) **A** + **B**' (where **B**' is the transpose of **B**)
(e) **A**./**B**'

B.5   Using MathScript, generate a plot of a sine wave of frequency $\omega = 10$ rad/sec. Use the linspace function to generate the time vector starting at $t = 0$ and ending at $t = 10$. Label the x-axis as **Time (sec)**. Label the y-axis as **sin(w * t).** Add the following title to the plot: **Sine wave with frequency w = 10 rad/sec.**

B.6   The rand function generates uniformly distributed random numbers between 0 and 1. This means that the average of all of the random values generated by the rand function should approach 0.5 as the number of random numbers increases. Using the rand function, generate random vectors of length 5, 100, 500, and 1000. Confirm that as the number of elements increases, the average of the random numbers approaches 0.5. Generate a plot of the average of the random numbers as a function of the number of random numbers. Use both rand and mean functions in your script.

References

Chapter 1

1. O. Mayr, *The Origins of Feedback Control*, MIT Press, Cambridge, Mass., 1970.
2. O. Mayr, "The Origins of Feedback Control," *Scientific American*, 223, 4, October 1970, pp. 110–118.
3. O. Mayr, *Feedback Mechanisms in the Historical Collections of the National Museum of History and Technology*, Smithsonian Institution Press, Washington, D. C., 1971.
4. E. P. Popov, *The Dynamics of Automatic Control Systems*, Gostekhizdat, Moscow, 1956; Addison-Wesley, Reading, Mass., 1962.
5. J. C. Maxwell, "On Governors," *Proc. of the Royal Society of London*, 16, 1868; in *Selected Papers on Mathematical Trends in Control Theory*. Dover, New York, 1964, pp. 270–283.
6. I. A. Vyshnegradskii, "On Controllers of Direct Action," *Izv. SPB Tekhnolog. Inst.*, 1877.
7. H. W. Bode, "Feedback—The History of an Idea," in *Selected Papers on Mathematical Trends in Control Theory*. Dover, New York, 1964, pp. 106–123.
8. H. S. Black, "Inventing the Negative Feedback Amplifier," *IEEE Spectrum*, December 1977, pp. 55–60.
9. J. E. Brittain, *Turning Points in American Electrical History*, IEEE Press, New York, 1977, Sect. II-E.
10. W. S. Levine, *The Control Handbook*, CRC Press, Boca Raton, Fla., 1996.
11. G. Newton, L. Gould, and J. Kaiser, *Analytical Design of Linear Feedback Controls*, John Wiley & Sons, New York, 1957.
12. M. D. Fagen, *A History of Engineering and Science on the Bell Systems*, Bell Telephone Laboratories, 1978, Chapter 3.
13. G. Zorpette, "Parkinson's Gun Director," *IEEE Spectrum*, April 1989, p. 43.
14. R. C. Dorf and A. Kusiak, *Handbook of Automation and Manufacturing*, John Wiley & Sons, New York, 1994.
15. W. D. Rasmussen, "The Mechanization of Agriculture," *Scientific American*, September 1982, pp. 26–37.
16. M. M. Gupta, *Intelligent Control*, IEEE Press, Piscataway, N. J., 1995.
17. A. G. Ulsoy, "Control of Machining Processes," *Journal of Dynamic Systems*, ASME, June 1993, pp. 301–307.
18. M. P. Groover, *Fundamentals of Modern Manufacturing*, Prentice Hall, Englewood Cliffs, N. J., 1996.
19. R. T. O'Brien, "Vertical Lateral Control for Automated Highway Systems," *IEEE Transactions on Control Systems Technology*, May 1996, pp. 266–273.
20. J. G. Kassakian, "Automotive Electrical Systems circa 2005," *IEEE Spectrum*, August 1996, pp. 22–27.
21. P. M. Moretti and L. V. Divone, "Modern Windmills," *Scientific American*, June 1986, pp. 110–118.
22. B. Preising and T. C. Hsia, "Robots in Medicine," *IEEE Engineering in Medicine and Biology*, June 1991, pp. 13–22.
23. R. C. Dorf and J. Unmack, "A Time-Domain Model of the Heart Rate Control System," *Proceedings of the San Diego Symposium for Biomedical Engineering*, 1965, pp. 43–47.
24. R. C. Dorf, *Electrical Engineering Handbook*, 2nd ed., CRC Press, Boca Raton, Fla., 1998.
25. R. C. Dorf, *Introduction to Computers and Computer Science*, Boyd and Fraser, San Francisco, 3rd ed., 1982, Chapters 13, 14.
26. K. Sutton, "Productivity," in *Encyclopedia of Engineering*, McGraw-Hill, New York, pp. 947–948.
27. R. C. Dorf, *The Encyclopedia of Robotics*, John Wiley & Sons, New York, 1988.
28. R. C. Dorf, *Robotics and Automated Manufacturing*, Reston Publishing, Reston, Va., 1983.
29. S. S. Hacisalihzade, "Control Engineering and Therapeutic Drug Delivery," *IEEE Control Systems*, June 1989, pp. 44–46.
30. E. R. Carson and T. Deutsch, "A Spectrum of Approaches for Controlling Diabetes," *IEEE Control Systems*, December 1992, pp. 25–30.
31. J. R. Sankey and H. Kaufman, "Robust Considerations of a Drug Infusion System," *Proceedings of the American Control Conference*, San Francisco, Calif., June 1993, pp. 1689–1695.
32. W. S. Levine, *The Control Handbook*, CRC Press, Boca Raton, Fla., 1996.

33. D. Auslander and C. J. Kempf, *Mechatronics*, Prentice Hall, Englewood Cliffs, N. J., 1996.

34. "Things that Go Bump in Your Flight," *The Economist*, July 3, 1999, pp. 69–70.

35. P. J. Brancazio, "Science and the Game of Baseball," *Science Digest*, July 1984, pp. 66–70.

36. D. Dooling, "Smoother Sailing," *IEEE Spectrum*, August 1996, pp. 52–59.

37. C. Klomp, et al., "Development of an Autonomous Cow-Milking Robot Control System," *IEEE Control Systems*, October 1990, pp. 11–19.

38. M. B. Tischler et al., "Flight Control of a Helicopter," *IEEE Control Systems*, August 1999, pp. 22–32.

39. M. Bodson, "Emerging Technologies in Control Engineering," *IEEE Control Systems*, December 1995, pp. 6–9.

40. C. O'Malley, "Rapid Rails," *Popular Science*, June 1992, pp. 74–79.

41. G. B. Gordon and J. C. Roark, "ORCA: An Optimized Robot for Chemical Analysis," *Hewlett-Packard Journal*, June 1993, pp. 6–19.

42. K. Hollenback, "Destabilizing Effects of Muscular Contraction in Human-Machine Interaction," *Proceedings of the American Control Conference*, San Francisco, Calif., 1993, pp. 736–740.

43. L. Scivicco and B. Siciliano, *Modeling and Control of Robot Manipulators*, McGraw-Hill, New York, 1996.

44. O. Mayr, "Adam Smith and the Concept of the Feedback System," *Technology and Culture*, 12, 1, January 1971, pp. 1–22.

45. A. Goldsmith, "Autofocus Cameras," *Popular Science*, March 1988, pp. 70–72.

46. R. Johansson, *System Modeling and Identification*, Prentice Hall, Englewood Cliffs, N. J., 1993.

47. M. DiChristina, "Telescope Tune-Up," *Popular Science*, September 1999, pp. 66–68.

48. K. Capek, *Rossum's Universal Robots*, English version by P. Selver and N. Playfair, Doubleday, Page, New York, 1923.

49. D. Hancock, "Prototyping the Hubble Fix," *IEEE Spectrum*, October 1993, pp. 34–39.

50. A. K. Naj, "Engineers Want New Buildings to Behave Like Human Beings," *Wall Street Journal*, January 20, 1994, p. B1.

51. E. H. Maslen, et al., "Feedback Control Applications in Artifical Hearts," *IEEE Control Systems*, December 1998, pp. 26–30.

52. M. DiChristina, "What's Next for Hubble?" *Popular Science*, March 1998, pp. 56–59.

53. L. L. Bucciarelli, *Designing Engineers*, MIT Press, Cambridge, Mass., 1996.

54. K. G. Ashar, *Magnetic Disk Drive Technology*, IEEE Press, Piscataway, N. J., 1997.

55. K. D. Fisher and W. L. Abbott, "PRML Detection Boosts Hard Disk Drive Technology," *IEEE Spectrum*, November 1996, pp. 70–76.

56. R. Stone, "Putting a Human Face on a New Breed of Robot," *Science*, October 11, 1996, p. 182.

57. P. I. Ro, "Nanometric Motion Control of a Traction Drive," *ASME Dynamic Systems and Control*, vol. 55.2, 1994, pp. 879–883.

58. M. Allen, "May the Cornering Force Be with You," *Popular Mechanics*, December 1995, pp. 74–77.

59. D. Dooling, "Smoother Sailing," *IEEE Spectrum*, August 1996, pp. 52–59.

60. K. C. Cheok, "A Smart Automatic Windshield Wiper," *IEEE Control Systems Magazine*, December 1996, pp. 28–34.

61. D. Dooling, "Transportation," *IEEE Spectrum*, January 1996, pp. 82–86.

62. G. Norris, "Boeing's Seventh Wonder," *IEEE Spectrum*, October 1995, pp. 20–23.

63. D. Hughes, "Fly-by-Wire 777 Keeps Traditional Cockpit," *Aviation Week & Space Technology*, McGraw-Hill Publication, May 1, 1995, pp. 42–48.

64. R. Reck, "Design Engineering," *Aerospace America*, American Institute of Aeronautics and Astronautics, December 1994, p. 74.

65. C. Rist, "Angling for Momentum," *Discover*, September 1999, p. 37.

66. S. J. Elliott, "Down With Noise," *IEEE Spectrum*, June 1999, pp. 54–62.

67. W. Ailor, "Controlling Space Traffic," *AIAA Aerospace America*, November 1999, pp. 34–38.

68. J. W. Toigo, "Avoiding a Data Crunch," *Scientific American*, May 2000, pp. 58–74.

69. G. F. Hughes, "Wise Drives," *IEEE Spectrum*, August 2002, pp. 37–41.

70. R. H. Bishop, *The Mechatronics Handbook*, CRC Press, Inc., Boca Raton, Fla., 2002.

71. N. Kyura and H. Oho, "Mechatronics—An Industrial Perspective," *IEEE/ASME Transactions on Mechatronics*, Vol. 1, No. 1, 1996, pp. 10–15.

72. T. Mori, "Mecha-tronics," *Yasakawa Internal Trademark Application Memo 21.131.01*, July 12, 1969.

73. F. Harshama, M. Tomizuka, and T. Fukuda, "Mechatronics—What is it, Why, and How?—An Editorial," *IEEE/ASME Transactions on Mechatronics*, Vol. 1, No. 1, 1996, pp. 1–4.

74. D. M. Auslander and C. J. Kempf, *Mechatronics: Mechanical System Interfacing*, Prentice Hall, Upper Saddle River, N. J., 1996.

75. D. Shetty and R. A. Kolk, *Mechatronic System Design*, PWS Publishing Company, Boston, Mass., 1997.

76. W. Bolton, *Mechatronics: Electrical Control Systems in Mechanical and Electrical Engineering*, 2nd ed., Addison Wesley Longman, Harlow, England, 1999.

77. D. Tomkinson and J. Horne, *Mechatronics Engineering*, McGraw-Hill, New York, 1996.

78. B. Jorgensen, "Shifting Gears," Auto Electronics, *Electronic Business*, February 2001.

79. M. B. Barron and W. F. Powers, "The Role of Electronic Controls for Future Automotive Mechatronic Systems," *IEEE/ASME Transactions on Mechatronics*, Vol. 1, No. 1, 1996, pp. 80–88.

80. G. Kobe, "Electronics: What's Driving The Growth," *Automotive Industries*, August 2000.

81. H. Kobayashi, Guest Editorial, *IEEE/ASME Transactions on Mechatronics*, Vol. 2, No. 4, 1997, pp. 217.

82. D. S. Bernstein, "What Makes Some Control Problems Hard?" *IEEE Control Systems Magazine*, August 2002, pp. 8–19.

83. S. Douglas, "No Pilot Required," *Popular Science*, June 2001, pp. 40–44.

84. O. Zerbinati, "A Direct Methanol Fuel Cell," *Journal of Chemical Education*, Vol. 79. No. 7, July 2002, p. 829.

85. D. Basmadjian, *Mathematical Modeling of Physical Systems: An Introduction*, Oxford University Press, New York, NY, 2003.

86. D. W. Boyd, *Systems Analysis and Modeling: A Macro-to-Micro Approach with Multidisciplinary Applications*, Academic Press, San Diego, CA, 2001.

87. F. Bullo and A. D. Lewis, *Geometric Control of Mechanical Systems: Modeling, Analysis, and Design for Simple Mechanical Control Systems*, Springer Verlag, New York, NY, 2004.

88. P. D. Cha, J. J. Rosenberg, and C. L. Dym, *Fundamentals of Modeling and Analyzing Engineering Systems*, Cambridge University Press, Cambridge, United Kingdom, 2000.

89. P. H. Zipfel, *Modeling and Simulation of Aerospace Vehicle Dynamics*, AIAA Education Series, American Institute of Aeronautics & Astronautics, Inc., Reston, Virginia, 2001.

90. D. Hristu-Varsakelis and W. S. Levin (Eds.), *Handbook of Networked and Embedded Control Systems*, Series: Control Engineering Series, Birkhäuser, Boston, MA, 2005.

Chapter 2

1. R. C. Dorf, *Electric Circuits*, 4th ed., John Wiley & Sons, New York, 1999.

2. I. Cochin, *Analysis and Design of Dynamic Systems*, Addison-Wesley Publishing Co., Reading, Mass., 1997.

3. J. W. Nilsson, *Electric Circuits*, 5th ed., Addison-Wesley, Reading, Mass., 1996.

4. E. W. Kamen and B. S. Heck, *Fundamentals of Signals and Systems Using MATLAB*, Prentice Hall, Upper Saddle River, N. J., 1997.

5. F. Raven, *Automatic Control Engineering*, 3rd ed., McGraw-Hill, New York, 1994.

6. S. Y. Nof, *Handbook of Industrial Robotics*, John Wiley & Sons, New York, 1999.

7. R. R. Kadiyala, "A Toolbox for Approximate Linearization of Nonlinear Systems," *IEEE Control Systems*, April 1993, pp. 47–56.

8. R. Smith and R. Dorf, *Circuits, Devices and Systems*, 5th ed., John Wiley & Sons, New York, 1992.

9. Y. M. Pulyer, *Electromagnetic Devices for Motion Control*, Springer-Verlag, New York, 1992.

10. B. C. Kuo, *Automatic Control Systems*, 5th ed., Prentice Hall, Englewood Cliffs, N. J., 1996.

11. F. E. Udwadia, *Analytical Dynamics*, Cambridge Univ. Press, New York, 1996.

12. R. C. Dorf, *Electrical Engineering Handbook*, 2nd ed., CRC Press, Boca Raton, Fla., 1998.

13. H. Saadat, *Computational Aids in Control Systems Using MATLAB*, McGraw-Hill, New York, 1993.

14. S. M. Ross, *Simulation*, 2nd ed., Academic Press, Orlando, Fla., 1996.

15. G. B. Gordon, "ORCA: Optimized Robot for Chemical Analysis," *Hewlett-Packard Journal*, June 1993, pp. 6–19.

16. P. E. Sarachik, *Principles of Linear Systems*, Cambridge Univ. Press, New York, 1997.

17. S. Bennett, "Nicholas Minorsky and the Automatic Steering of Ships," *IEEE Control Systems*, November 1984, pp. 10–15.

18. P. Gawthorp, *Metamodeling: Bond Graphs and Dynamic Systems*, Prentice Hall, Englewood Cliffs, N. J., 1996.

19. C. M. Close and D. K. Frederick, *Modeling and Analysis of Dynamic Systems*, 2nd ed., Houghton Mifflin, Boston, 1995.

20. H. S. Black, "Stabilized Feed-Back Amplifiers," *Electrical Engineering*, 53, January 1934, pp. 114–120. Also in *Turning Points in American History*, J. E. Brittain, ed., IEEE Press, New York, 1977, pp. 359–361.

21. P. L. Corke, *Visual Control of Robots*, John Wiley & Sons, New York, 1997.

22. W. J. Rugh, *Linear System Theory*, 2nd ed., Prentice Hall, Englewood Cliffs, N. J., 1997.

23. S. Pannu and H. Kazerooni, "Control for a Walking Robot," *IEEE Control Systems*, February 1996, pp. 20–25.

24. K. Ogata, *Modern Control Engineering*, 3rd ed., Prentice Hall, Upper Saddle River, N. J., 1997.

25. S. P. Parker, *Encyclopedia of Engineering*, 2nd ed., McGraw-Hill, New York, 1993.

26. M. Vidyasagar, *Nonlinear System Analysis*, Prentice Hall, Englewood Cliffs, N. J., 1993.

27. G. T. Pope, "Living-Room Levitation," *Discover*, June 1993, p. 24.

28. D. J. Bak, "Fast Trains Face Off," *Design News*, September 20, 1993, pp. 78–85.

29. B. B. Muvdi, *Dynamics for Engineers*, Springer-Verlag, New York, 1997.

30. G. Rowell and D. Wormley, *System Dynamics*, Prentice Hall, Upper Saddle River, N. J., 1997.

31. Y. Y. Tzou, "AC Induction Servo Drive for Motion Control," *IEEE Transactions on Control Systems Technology*, November 1996, pp. 614–25.

32. R. E. Ziemer, *Signals and Systems*, 4th ed., Prentice Hall, Upper Saddle River, N. J., 1998.

33. R. H. Bishop, *The Mechatronics Handbook*, CRC Press, Inc., Boca Raton, Fla., 2002.

34. C. N. Dorny, *Understanding Dynamic Systems: Approaches to Modeling, Analysis, and Design*, Prentice-Hall, Englewood Cliffs, New Jersey, 1993.

35. T. D. Burton, *Introduction to Dynamic Systems Analysis*, McGraw-Hill, Inc., New York, 1994.

36. K. Ogata, *System Dynamics*, 4th Edition, Prentice-Hall, Englewood Cliffs, New Jersey, 2003.

37. J. D. Anderson, *Fundamentals of Aerodynamics*, 4th Edition, McGraw-Hill, Inc., New York, 2005.

38. G. Emanuel, *Gasdynamics Theory and Applications*, AIAA Education Series, New York, 1986.

39. A. M. Kuethe and C-Y. Chow, *Foundations of Aerodynamics: Bases of Aerodynamic Design*, 5th Edition, John Wiley & Sons, New York, 1997.

Chapter 3

1. R. C. Dorf, *Electric Circuits*, 3rd ed., John Wiley & Sons, New York, 1997.

2. W. J. Rugh, *Linear System Theory*, 2nd ed., Prentice Hall, Englewood Cliffs, N. J., 1996.

3. H. Kajiwara, et al., "LPV Techniques for Control of an Inverted Pendulum," *IEEE Control Systems*, February 1999, pp. 44–47.

4. R. C. Dorf, *Encyclopedia of Robotics*, John Wiley & Sons, New York, 1988.

5. A. V. Oppenheim et al., *Signals and Systems*, Prentice Hall, Englewood Cliffs, N. J., 1996.

6. J. L. Stein, "Modeling and State Estimator Design Issues for Model Based Monitoring Systems," *Journal of Dynamic Systems*, ASME, June 1993, pp. 318–326.

7. I. Cochin, *Analysis and Design of Dynamic Systems*, Addison-Wesley, Reading, Mass., 1997.

8. R. C. Dorf, *Electrical Engineering Handbook*, CRC Press, Boca Raton, Fla., 1993.

9. Y. M. Pulyer, *Electromagnetic Devices for Motion Control*, Springer-Verlag, New York, 1992.

10. C. M. Close and D. K. Frederick, *Modeling and Analysis of Dynamic Systems*, 2nd ed., Houghton Mifflin, Boston, 1995.

11. R. C. Durbeck, "Computer Output Printer Technologies," in *Electrical Engineering Handbook*, R. C. Dorf, Ed., CRC Press, Boca Raton, Fla., 1998, pp. 1958–1975.

12. B. Wie et al., "New Approach to Attitude/Momentum Control for the Space Station," *AIAA Journal of Guidance, Control, and Dynamics* 12, No. 5, 1989, pp. 714–722.

13. B. Etkin and L. D. Reid, *Dynamics of Flight*, John Wiley & Sons, New York, 1996.

14. J. J. Bertin, *Aerodynamics for Engineers*, 3rd ed., Prentice Hall, Upper Saddle River, N. J., 1998.

15. H. Ramirez, "Feedback Controlled Landing Maneuvers," *IEEE Transactions on Automatic Control*, April 1992, pp. 518–523.

16. C. A. Canudas De Wit, *Theory of Robot Control*, Springer-Verlag, New York, 1996.

17. R. R. Kadiyala, "A Toolbox for Approximate Linearization of Nonlinear Systems," *IEEE Control Systems*, April 1993, pp. 47–56.

18. B. C. Crandall, *Nanotechnology*, MIT Press, Cambridge, Mass., 1996.

19. E. Colle, "Localization of an Autonomous Vehicle with Beacons," *International Journal of Robotics and Automation* 8, No. 1, 1993, pp. 30–38.

20. W. Leventon, "Mountain Bike Suspension Allows Easy Adjustment," *Design News*, July 19, 1993, pp. 75–77.

21. A. Cavallo et al., *Using* MATLAB, SIMULINK, *and Control System Toolbox*, Prentice Hall, Englewood Cliffs, N. J., 1996.

22. M. Jamshidi, *Computer Aided Analysis and Design of Linear Control Systems*, Prentice Hall, Englewood Cliffs, N. J., 1992.

23. G. E. Carlson, *Signal and Linear System Analysis*, John Wiley & Sons, New York, 1998.

24. M. Fujita, "Synthesis of an Electromagnetic Suspension System," *Proceedings of the Conference on Decision and Control*, IEEE Press, New York, 1992.

25. D. Cho, "Magnetic Levitation Systems," *IEEE Control Systems*, February 1993, pp. 42–48.

26. W. J. Palm, *Modeling, Analysis, Control of Dynamic Systems*, 2nd ed., John Wiley & Sons, New York, 2000.

27. H. Kazerooni, "Human Extenders," *Journal of Dynamic Systems*, ASME, June 1993, pp. 281–290.

28. C. N. Dorny, *Understanding Dynamic Systems*, Prentice Hall, Englewood Cliffs, N. J., 1993.

29. Chen, Chi-Tsing, *Linear System Theory and Design*, 3rd ed., Oxford Univ. Press, New York, 1998.

30. M. Kaplan, *Modern Spacecraft Dynamics and Control*, John Wiley and Sons, New York, 1976.

31. J. Wertz, Ed., *Spacecraft Attitude Determination and Control*, Kluwer Academic Publishers, Dordrecht, The Netherlands, 1978 (reprinted in 1990).

32. W. E. Wiesel, *Spaceflight Dynamics*, McGraw-Hill, New York, 1989.

33. B. Wie, K. W. Byun, V. W. Warren, D. Geller, D. Long and J. Sunkel, "New Approach to Attitude/Momentum Control for the Space Station," *AIAA Journal Guidance, Control, and Dynamics*, Vol. 12, No. 5, 1989, pp. 714–722.

34. L. R. Bishop, R. H. Bishop, and K. L. Lindsay, "Proposed CMG Momentum Management Scheme for Space Station," *AIAA Guidance Navigation and Controls Conference Proceedings*, Vol. 2, No. 87-2528, 1987, pp. 1229–1236.

35. H. H. Woo, H. D. Morgan, and E. T. Falangas, "Momentum Management and Attitude Control Design for a Space Station," *AIAA Journal of Guidance, Control, and Dynamics*, Vol. 11, No. 1, 1988, pp. 19–25.

36. J. W. Sunkel and L. S. Shieh, "An Optimal Momentum Management Controller for the Space Station," *AIAA Journal of Guidance, Control, and Dynamics*, Vol. 13, No. 4, 1990, pp. 659–668.

37. V. W. Warren, B. Wie, and D. Geller, "Periodic-Disturbance Accommodating Control of the Space Station," *AIAA Journal of Guidance, Control, and Dynamics*, Vol. 13, No. 6, 1990, pp. 984–992.

38. B. Wie, A. Hu, and R. Singh, "Multi-Body Interaction Effects on Space Station Attitude Control and Momentum Management," *AIAA Journal of Guidance, Control, and Dynamics*, Vol. 13, No. 6, 1990, pp. 993–999.

39. J. W. Sunkel and L. S. Shieh, "Multistage Design of an Optimal Momentum Management Controller for the Space Station," *AIAA Journal of Guidance, Control, and Dynamics*, Vol. 14, No. 3, 1991, pp. 492–502.

40. K. W. Byun, B. Wie, D. Geller, and J. Sunkel, "Robust H_∞ Control Design for the Space Station with Structured Parameter Uncertainty," *AIAA Journal of Guidance, Control, and Dynamics*, Vol. 14, No. 6, 1991, pp. 1115–1122.

41. E. Elgersma, G. Stein, M. Jackson, and J. Yeichner, "Robust Controllers for Space Station Momentum Management," *IEEE Control Systems Magazine*, Vol. 12, No. 2, October 1992, pp. 14–22.

42. G. J. Balas, A. K. Packard, and J. T. Harduvel, "Application of μ-Synthesis Technique to Momentum Management and Attitude Control of the Space Station," *Proceedings of 1991 AIAA Guidance, Navigation, and Control Conference*, New Orleans, Louisiana, pp. 565–575.

43. Rhee and J. L. Speyer, "Robust Momentum Management and Attitude Control System for the Space Station," *AIAA Journal of Guidance, Control, and Dynamics*, Vol. 15, No. 2, 1992, pp. 342–351.

44. T. F. Burns and H. Flashner, "Adaptive Control Applied to Momentum Unloading Using the Low Earth Orbital Environment," *AIAA Journal of Guidance, Control, and Dynamics*, Vol. 15, No. 2, 1992, pp. 325–333.

45. X. M. Zhao, L. S. Shieh, J. W. Sunkel, and Z. Z. Yuan, "Self-Tuning Control of Attitude and Momentum Management for the Space Station," *AIAA Journal of Guidance, Control, and Dynamics*, Vol. 15, No. 1, 1992, pp. 17–27.

46. G. Parlos and J. W. Sunkel, "Adaptive Attitude Control and Momentum Management for Large-Angle Spacecraft Maneuvers," *AIAA Journal of Guidance, Control, and Dynamics*, Vol. 15, No. 4, 1992, pp. 1018–1028.

47. R. H. Bishop, S. J. Paynter, and J. W. Sunkel, "Adaptive Control of Space Station with Control Moment Gyros," *IEEE Control Systems Magazine*, Vol. 12, No. 2, October 1992, pp. 23–28.

48. S. R. Vadali and H. S. Oh, "Space Station Attitude Control and Momentum Management: A Nonlinear Look," *AIAA Journal of Guidance, Control, and Dynamics*, Vol. 15, No. 3, 1992, pp. 577–586.

49. S. N. Singh and T. C. Bossart, "Feedback Linearization and Nonlinear Ultimate Boundedness Control of the Space Station Using CMG," *AIAA Guidance Navigation and Controls Conference Proceedings*, Vol. 1, No. 90-3354-CP, 1990, pp. 369–376.

50. S. N. Singh and T. C. Bossart, "Invertibility of Map, Zero Dynamics and Nonlinear Control of Space Station," *AIAA Guidance Navigation and Controls Conference Proceedings*, Vol. 1, No. 91-2663-CP, 1991, pp. 576–584.

51. S. N. Singh and A. Iyer, "Nonlinear Regulation of Space Station: A Geometric Approach," *AIAA Journal of Guidance, Control, and Dynamics*, Vol. 17, No. 2, 1994, pp. 242–249.

52. J. J. Sheen and R. H. Bishop, "Spacecraft Nonlinear Control," *The Journal of Astronautical Sciences*, Vol. 42, No. 3, 1994, pp. 361–377.

53. J. Dzielski, E. Bergmann, J. Paradiso, D. Rowell, and D. Wormley, "Approach to Control Moment Gyroscope Steering Using Feedback Linearization," *AIAA Journal of Guidance, Control, and Dynamics*, Vol. 14, No. 1, 1991, pp. 96–106.

54. J. J. Sheen and R. H. Bishop, "Adaptive Nonlinear Control of Spacecraft," *The Journal of Astronautical Sciences*, Vol. 42, No. 4, 1994, pp. 451–472.

55. S. N. Singh and T. C. Bossart, "Exact Feedback Linearization and Control of Space Station Using CMG," *IEEE Transactions on Automatic Control*, Vol. Ac-38, No. 1, 1993, pp. 184–187.

Chapter 4

1. R. C. Dorf, *Electrical Engineering Handbook*, 2nd ed., CRC Press, Boca Raton, Fla., 1998.

2. R. C. Dorf, *Electric Circuits*, 3rd ed., John Wiley & Sons, New York, 1996.

3. C. E. Rohrs, J. L. Melsa, and D. Schultz, *Linear Control Systems*, McGraw-Hill, New York, 1993.

4. P. E. Sarachik, *Principles of Linear Systems*, Cambridge Univ. Press, New York, 1997.

5. B. K. Bose, *Power Electronics and Variable Frequency Drives*, IEEE Press, Piscataway, N. J., 1997.

6. J. C. Nelson, *Operational Amplifier Circuits*, Butterworth, New York, 1995.

7. *Motomatic Speed Control*, Electro-Craft Corp., Hopkins, Minn., 1999.

8. M. W. Spong et al., *Robot Control Dynamics, Motion Planning and Analysis*, IEEE Press, New York, 1993.

9. R. C. Dorf, *Encyclopedia of Robotics*, John Wiley & Sons, New York, 1988.

10. D. J. Bak, "Dancer Arm Feedback Regulates Tension Control," *Design News*, April 6, 1987, pp. 132–133.

11. "The Smart Projector Demystified," *Science Digest*, May 1985, p. 76.

12. J. M. Maciejowski, *Multivariable Feedback Design*, Addison-Wesley, Wokingham, England, 1989.

13. L. Fortuna and G. Muscato, "A Roll Stabilization System for a Monohull Ship," *IEEE Transactions on Control Systems Technology*, January 1996, pp. 18–28.

14. C. N. Dorny, *Understanding Dynamic Systems*, Prentice Hall, Englewood Cliffs, N. J., 1993.

15. D. W. Clarke, "Sensor, Actuator, and Loop Validation," *IEEE Control Systems*, August 1995, pp. 39–45.

16. S. P. Parker, *Encyclopedia of Engineering*, 2nd ed., McGraw-Hill, New York, 1993.

17. M. S. Markow, "An Automated Laser System for Eye Surgery," *IEEE Engineering in Medicine and Biology*, December 1989, pp. 24–29.

18. M. Eslami, *Theory of Sensitivity in Dynamic Systems*, Springer-Verlag, New York, 1994.

19. Y. M. Pulyer, *Electromagnetic Devices for Motion Control*, Springer-Verlag, New York, 1992.

20. J. R. Layne, "Control for Cargo Ship Steering," *IEEE Control Systems*, December 1993, pp. 23–33.

21. S. Begley, "Greetings From Mars," *Newsweek*, July 14, 1997, pp. 23–29.

22. M. Carroll, "Assault on the Red Planet," *Popular Science*, January 1997, pp. 44–49.

23. C. S. Powell, "Lilliput in Space," *Discover*, March 1999, p. 38.

24. The American Medical Association, *Home Medical Encyclopedia*, Volume 1, Random House, New York, 1989, pp. 104–106.

25. J. B. Slate, L. C. Sheppard, V. C. Rideout, and E. H. Blackstone, "Closed-loop Nitroprusside Infusion: Modeling and Control Theory for Clinical Applications," *Proceedings IEEE International Symposium on Circuits and Systems*, 1980, pp. 482–488.

26. B. C. McInnis and L. Z. Deng, "Automatic Control of Blood Pressures with Multiple Drug Inputs," *Annals of Biomedical Engineering*, Vol. 13, 1985, pp. 217–225.

27. R. Meier, J. Nieuwland, A. M. Zbinden, and S. S. Hacisalihzade, "Fuzzy Logic Control of Blood Pressure During Anesthesia," *IEEE Control Systems*, December 1992, pp. 12–17.

28. L. C. Sheppard, "Computer Control of the Infusion of Vasoactive Drugs," *Proceedings IEEE International Symposium on Circuits and Systems*, 1980, pp. 469–473.

29. R. Vishnoi and R. J. Roy, "Adaptive Control of Closed-Circuit Anesthesia," *IEEE Transactions on Biomedical Engineering*, Vol. 38, No. 1, 1991, pp. 39–47.

Chapter 5

1. C. M. Close and D. K. Frederick, *Modeling and Analysis of Dynamic Systems*, 2nd ed., Houghton Mifflin, Boston, 1993.
2. R. C. Dorf, *Electric Circuits*, 3rd ed., John Wiley & Sons, New York, 1996.
3. B. K. Bose, *Power Electronics and Variable Frequency Drives*, IEEE Press, Piscataway, N. J., 1997.
4. P. R. Clement, "A Note on Third-Order Linear Systems," *IRE Transactions on Automatic Control*, June 1960, p. 151.
5. R. N. Clark, *Introduction to Automatic Control Systems*, John Wiley & Sons, New York, 1962, pp. 115–124.
6. D. Graham and R. C. Lathrop, "The Synthesis of Optimum Response: Criteria and Standard Forms, Part 2," *Trans. of the AIEE* 72, November 1953, pp. 273–288.
7. T. L. Floyd and D. Buchla, *Basic Operational Amplifiers*, Prentice Hall, Upper Saddle River, N. J., 1999.
8. R. C. Dorf, *Encyclopedia of Robotics*, John Wiley & Sons, New York, 1988.
9. L. E. Ryan, "Control of an Impact Printer Hammer," *ASME Journal of Dynamic Systems*, March 1990, pp. 69–75.
10. T. C. Hsia, "On the Simplification of Linear Systems," *IEEE Transactions on Automatic Control*, June 1972, pp. 372–374.
11. E. J. Davison, "A Method for Simplifying Linear Dynamic Systems," *IEEE Transactions on Automatic Control*, January 1966, pp. 93–101.
12. R. C. Dorf, *Electrical Engineering Handbook*, CRC Press, Boca Raton, Fla., 1998.
13. A. G. Ulsoy, "Control of Machining Processes," *ASME Journal of Dynamic Systems*, June 1993, pp. 301–310.
14. I. Cochin, *Analysis and Design of Dynamic Systems*, Addison-Wesley, Reading, Mass., 1997.
15. W. J. Rugh, *Linear System Theory*, 2nd ed., Prentice Hall, Englewood Cliffs, N.J., 1997.
16. W. J. Book, "Controlled Motion in an Elastic World," *Journal of Dynamic Systems*, June 1993, pp. 252–260.
17. C. E. Rohrs, J. L. Melsa, D. Schultz, *Linear Control Systems*, McGraw-Hill, New York, 1993.
18. S. Lee, "Intelligent Sensing and Control for Advanced Teleoperation," *IEEE Control Systems*, June 1993, pp. 19–28.
19. R. Rosen and L. J. Williams, "The Rebirth of the Supersonic Transport," *Technology Review*, March 1993, pp. 22–29.
20. T. J. Lueck, "Amtrack Unveils Its Bullet to Boston," *New York Times*, March 10, 1999, p. B2.
21. M. DiChristina, "Telescope Tune-Up," *Popular Science*, September 1999, pp. 66–68.
22. M. Hutton and M. Rabins, "Simplification of Higher-Order Mechanical Systems Using the Routh Approximation," *Journal of Dynamic Systems*, ASME, December 1975, pp. 383–392.
23. E. W. Kamen and B. S. Heck, *Fundamentals of Signals and Systems Using MATLAB*, Prentice Hall, Upper Saddle River, N. J., 1997.
24. M. DiChristina, "What's Next for Hubble?" *Popular Science*, March 1998, pp. 56–59.
25. Aaron Edsinger-Gonzales and Jeff Weber. "Domo: A Force Sensing Humanoid Robot for Manipulation Research," *Proceedings of the IEEE/RSJ International Conference on Humanoid Robotics*, 2004.
26. Aaron Edsinger-Gonzales, "Design of a Compliant and Force Sensing Hand for a Humanoid Robot," *Proceedings of the International Conference on Intelligent Manipulation and Grasping*, 2004.
27. B. L. Stevens and F. L. Lewis, *Aircraft Control and Simulation, 2nd Edition*, John Wiley & Sons, New York, 2003.
28. B. Etkin and L. D. Reid, *Dynamics of Flight, 3rd Edition*, John Wiley & Sons, New York, 1996.
29. G. E. Cooper and R. P. Harper, Jr., "The Use of Pilot Rating in the Evaluation of Aircraft Handling Qualities," NASA TN D-5153, 1969 (see also http://flighttest.navair.navy.mil/unrestricted/ch.pdf).
30. USAF, "Flying Qualities of Piloted Vehicles," USAF Spec., MIL-F-8785C, 1980.

Chapter 6

1. R. C. Dorf, *Electrical Engineering Handbook*, 2nd ed., CRC Press, Boca Raton, Fla., 1998.
2. R. C. Dorf, *Electric Circuits*, 3rd ed., John Wiley & Sons, New York, 1996.
3. W. J. Palm, *Modeling, Analysis and Control*, 2nd ed., John Wiley & Sons, New York, 2000.
4. W. J. Rugh, *Linear System Theory*, 2nd ed., Prentice Hall, Englewood Cliffs, N. J., 1997.
5. F. B. Farquharson, "Aerodynamic Stability of Suspension Bridges, with Special Reference to the Tacoma Narrows Bridge," *Bulletin 116, Part I*, The Engineering Experiment Station, University of Washington, 1950.
6. A. Hurwitz, "On the Conditions under which an Equation Has Only Roots with Negative Real Parts," *Mathematische Annalen* 46, 1895, pp. 273–284. Also

in *Selected Papers on Mathematical Trends in Control Theory*, Dover, New York, 1964, pp. 70–82.

7. E. J. Routh, *Dynamics of a System of Rigid Bodies*, Macmillan, New York, 1892.

8. G. G. Wang, "Design of Turning Control for a Tracked Vehicle," *IEEE Control Systems*, April 1990, pp. 122–125.

9. N. Mohan, *Power Electronics*, John Wiley & Sons, New York, 1995.

10. *World Robotics 2003–Statistics, Market Analysis, Forecasts, Case Studies, and Profitability of Robot Investment*, U. N. Economic Commission, ISBN No. 92-1-101059-4, 2003.

11. R. C. Dorf and A. Kusiak, *Handbook of Manufacturing and Automation*, John Wiley & Sons, New York, 1994.

12. A. N. Michel, "Stability: The Common Thread in the Evolution of Control," *IEEE Control Systems*, June 1996, pp. 50–60.

13. S. P. Parker, *Encyclopedia of Engineering*, 2nd ed., McGraw-Hill, New York, 1933.

14. J. Levine et al., "Control of Magnetic Bearings," *IEEE Transactions on Control Systems Technology*, September 1996, pp. 524–544.

15. F. S. Ho, "Traffic Flow Modeling and Control," *IEEE Control Systems*, October 1996, pp. 16–24.

16. G. E. Young and K. N. Reid, "Control of Moving Webs," *Journal of Dynamics*, ASME, June 1993, pp. 309–316.

17. D. W. Freeman, "Jump-Jet Airliner," *Popular Mechanics*, June 1993, pp. 38–40.

18. B. Sweetman, "Venture Star–21st-Century Space Shuttle," *Popular Science*, October 1996, pp. 43–47.

19. S. Lee, "Intelligent Sensing and Control for Advanced Teleoperation," *IEEE Control Systems*, June 1993, pp. 19–28.

20. "Uplifting," *The Economist*, July 10, 1993, p. 79.

21. R. N. Clark, "The Routh-Hurwitz Stability Criterion, Revisited," *IEEE Control Systems*, June 1992, pp. 119–120.

22. Gregory Mone, "5 Paths to the Walking, Talking, Pie-Baking Humanoid Robot," *Popular Science*, September 2006.

23. P. Haase, "Breakthrough in Stability Assessment," *EPRI Journal*, August 1996, pp. 25–30.

24. E. I. Jury, "Remembering Force Stability Theory Pioneers of the Nineteenth Century," *IEEE Transactions on Automatic Control*, September 1996, pp. 1242–43.

25. L. Hatvani, "Adaptive Control: Stabilization," *Applied Control*, edited by Spyros G. Tzafestas, Marcel Decker, New York, 1993, pp. 273–287.

Chapter 7

1. W. R. Evans, "Graphical Analysis of Control Systems," *Transactions of the AIEE*, 67, 1948, pp. 547–551. Also in G. J. Thaler, ed., *Automatic Control*, Dowden, Hutchinson, and Ross, Stroudsburg, Pa., 1974, pp. 417–421.

2. W. R. Evans, "Control System Synthesis by Root Locus Method," *Transactions of the AIEE*, 69, 1950, pp. 1–4. Also in *Automatic Control*, G. J. Thaler, ed., Dowden, Hutchinson, and Ross, Stroudsburg, Pa., 1974, pp. 423–425.

3. W. R. Evans, *Control System Dynamics*, McGraw-Hill, New York, 1954.

4. R. C. Dorf, *Electrical Engineering Handbook*, 2nd ed., CRC Press, Boca Raton, Fla., 1998.

5. J. G. Goldberg, *Automatic Controls*, Allyn and Bacon, Boston, 1965.

6. R. C. Dorf, *The Encyclopedia of Robotics*, John Wiley & Sons, New York, 1988.

7. H. Ur, "Root Locus Properties and Sensitivity Relations in Control Systems," *I.R.E. Trans. on Automatic Control*, January 1960, pp. 57–65.

8. T. R. Kurfess and M. L. Nagurka, "Understanding the Root Locus Using Gain Plots," *IEEE Control Systems*, August 1991, pp. 37–40.

9. T. R. Kurfess and M. L. Nagurka, "Foundations of Classical Control Theory," *The Franklin Institute*, 330, No. 2, 1993, pp. 213–227.

10. R. C. Dorf and A. Kusiak, *Handbook of Manufacturing and Automation*, John Wiley & Sons, New York, 1994.

11. D. K. Lindner, *Introduction to Signals and Systems*, McGraw Hill, New York, 1999.

12. S. Ashley, "Putting a Suspension through Its Paces," *Mechanical Engineering*, April 1993, pp. 56–57.

13. B. K. Bose, *Modern Power Electronics*, IEEE Press, New York, 1992.

14. P. Varaiya, "Smart Cars on Smart Roads," *IEEE Transactions on Automatic Control*, February 1993, pp. 195–207.

15. J. L. Jones and A. M. Flynn, *Mobile Robots*, A. K. Peters Publishing, New York, 1993.

16. B. Sweetman, "21st Century SST," *Popular Science*, April 1998, pp. 56–60.

17. L. V. Merritt, "Tape Transport Head Positioning Servo Using Positive Feedback," *Motion*, April 1993, pp. 19–22.

18. G. E. Young and K. N. Reid, "Control of Moving Webs," *Journal of Dynamic Systems* ASME, June 1993, pp. 309–316.

19. S. P. Parker, *Encyclopedia of Engineering*, 2nd ed., McGraw-Hill, New York, 1993.

20. A. J. Calise and R. T. Rysdyk, "Nonlinear Adaptive Flight Control Using Neural Networks," *IEEE Control Systems*, December 1998, pp. 14–23.

21. T. B. Sheridan, *Telerobotics, Automation and Control*, MIT Press, Cambridge, Mass., 1992.

22. L. W. Couch, *Digital and Analog Communication Systems*, 5th ed., Macmillan, New York, 1997.

23. D. Hrovat, "Applications of Optimal Control to Automotive Suspension Design," *Journal of Dynamic Systems*, ASME, June 1993, pp. 328–342.

24. T. J. Lueck, "Amtrak Unveils Its Bullet to Boston," *New York Times*, March 10, 1999.

25. M. van de Panne, "A Controller for the Dynamic Walk of a Biped," *Proceedings of the Conference on Decision and Control*, IEEE, December 1992, pp. 2668–2673.

26. R. C. Dorf, *Electric Circuits*, 3rd ed., John Wiley & Sons, New York, 1996.

27. S. Begley, "Mission to Mars," *Newsweek*, September 23, 1996, pp. 52–58.

28. G. Padfield, *Helicopter Flight Dynamics*, AIAA Press, New York, 1996.

29. H. Tennekes, *The Simple Science of Flight*, MIT Press, Cambridge, Mass., 1996.

30. W. J. Cook, "The International Space Station Takes Shape," *US News and World Report*, December 7, 1998, pp. 56–59.

31. "Batwings and Dragonfies," *The Economist*, July 2002, pp. 66–67.

32. Reed Electronics Research, *Automotive Electronics—A Profile of International Markets and Suppliers to 2010*, #IN0603375RE, See the press release at http://www.instat.com/press.asp?ID=1752&sku=IN 0603375RE for more information, September 2006.

33. J. G. Kassakian, H.-C. Wolf, J. M. Miller, and C. J. Hurton, "Automotive Electrical Systems Circa 2005," Vol. 33, No. 8, *IEEE Spectrum*, August, 1996, pp. 22–27.

34. M. B. Barron and W. F. Powers, "The Role of Electronic Controls for Future Automotive Mechatronic Systems," *IEEE/ASME Transactions on Mechatronics*, Vol. 1, No. 1, 1996, pp. 80–88.

Chapter 8

1. R. C. Dorf, *Electrical Engineering Handbook*, 2nd ed., CRC Press, Boca Raton, Fla., 1998.

2. I. Cochin and H. J. Plass, *Analysis and Design of Dynamic Systems*, John Wiley & Sons, New York, 1997.

3. R. C. Dorf, *Electric Circuits*, 3rd ed., John Wiley & Sons, New York, 1996.

4. H. W. Bode, "Relations Between Attenuation and Phase in Feedback Amplifier Design," *Bell System Tech. J.*, July 1940, pp. 421–454. Also in *Automatic Control: Classical Linear Theory*, G. J. Thaler, ed., Dowden, Hutchinson, and Ross, Stroudsburg, Pa., 1974, pp. 145–178.

5. M. D. Fagen, *A History of Engineering and Science in the Bell System*, Bell Telephone Laboratories, Murray Hill, N.J., 1978, Chapter 3.

6. D. K. Lindner, *Introduction to Signals and Systems*, McGraw Hill, New York., 1999.

7. R. C. Dorf and A. Kusiak, *Handbook of Manufacturing and Automation*, John Wiley & Sons, New York, 1994.

8. R. C. Dorf, *The Encyclopedia of Robotics*, John Wiley & Sons, New York, 1988.

9. T. B. Sheridan, *Telerobotics, Automation and Control*, MIT Press, Cambridge, Mass., 1992.

10. J. L. Jones and A. M. Flynn, *Mobile Robots*, A. K. Peters Publishing, New York, 1993.

11. D. McLean, *Automatic Flight Control Systems*, Prentice Hall, Englewood Cliffs, N. J., 1990.

12. G. Leitman, "Aircraft Control Under Conditions of Windshear," *Proceedings of IEEE Conference on Decision and Control*, December 1990, pp. 747–749.

13. S. Lee, "Intelligent Sensing and Control for Advanced Teleoperation," *IEEE Control Systems*, June 1993, pp. 19–28.

14. R. A. Hess, "A Control Theoretic Model of Driver Steering Behavior," *IEEE Control Systems*, August 1990, pp. 3–8.

15. J. Winters, "Personal Trains," *Discover*, July 1999, pp. 32–33.

16. J. Ackermann and W. Sienel, "Robust Yaw Damping of Cars with Front and Rear Wheel Steering," *IEEE Transactions on Control Systems Technology*, March 1993, pp. 15–20.

17. L. V. Merritt, "Differential Drive Film Transport," *Motion*, June 1993, pp. 12–21.

18. S. Ashley, "Putting a Suspension through Its Paces," *Mechanical Engineering*, April 1993, pp. 56–57.

19. D. A. Linkens, "Anaesthesia Simulators," *Computing and Control Engineering Journal*, IEEE, April 1993, pp. 55–62.

20. J. R. Layne, "Control for Cargo Ship Steering," *IEEE Control Systems*, December 1993, pp. 58–64.

21. A. Titli, "Three Control Approaches for the Design of Car Semi-active Suspension," *IEEE Proceedings of Conference on Decision and Control*, December 1993, pp. 2962–2963.

22. H. H. Ottesen, "Future Servo Technologies for Hard Disk Drives," *Journal of the Magnetics Society of Japan*, Vol. 18, 1994, pp. 31–36.

23. D. Leonard, "Ambler Ramblin," Vol. 2, No. 7, *Ad Astra*, July–August 1990, pp. 7–9.

Chapter 9

1. H. Nyquist, "Regeneration Theory," *Bell Systems Tech. J.*, January 1932, pp. 126–147. Also in *Automatic Control: Classical Linear Theory*, G. J. Thaler, ed., Dowden, Hutchinson, and Ross, Stroudsburg, Pa., 1932, pp. 105–126.

2. M. D. Fagen, *A History of Engineering and Science in the Bell System*, Bell Telephone Laboratories, Inc., Murray Hill, N. J., 1978, Chapter 5.

3. H. M. James, N. B. Nichols, and R. S. Phillips, *Theory of Servomechanisms*, McGraw-Hill, New York, 1947.

4. W. J. Rugh, *Linear System Theory*, 2nd ed., Prentice Hall, Englewood Cliffs, N. J., 1996.

5. D. A. Linkens, *CAD for Control Systems*, Marcel Dekker, New York, 1993.

6. A. Cavallo, *Using MATLAB, SIMULINK, and Control System Toolbox*, Prentice Hall, Englewood Cliffs, N. J., 1996.

7. R. C. Dorf, *Electrical Engineering Handbook*, 2nd ed., CRC Press, Boca Raton, Fla., 1998.

8. D. Sbarbaro-Hofer, "Control of a Steel Rolling Mill," *IEEE Control Systems*, June 1993, pp. 69–75.

9. R. C. Dorf and A. Kusiak, *Handbook of Manufacturing and Automation*, John Wiley & Sons, New York, 1994.

10. J. J. Gribble, "Systems with Time Delay," *IEEE Control Systems*, February 1993, pp. 54–55.

11. C. N. Dorny, *Understanding Dynamic Systems*, Prentice Hall, Englewood Cliffs, N. J., 1993.

12. R. C. Dorf, *Electric Circuits*, 3rd ed., John Wiley & Sons, New York, 1996.

13. J. Yan and S. E. Salcudean, "Teleoperation Controller Design," *IEEE Transactions on Control Systems Technology*, May 1996, pp. 244–247.

14. K. K. Chew, "Control of Errors in Disk Drive Systems," *IEEE Control Systems*, January 1990, pp. 16–19.

15. R. C. Dorf, *The Encyclopedia of Robotics*, John Wiley & Sons, New York, 1988.

16. D. W. Freeman, "Jump-Jet Airliner," *Popular Mechanics*, June 1993, pp. 38–40.

17. F. D. Norvelle, *Electrohydraulic Control Systems*, Prentice Hall, Upper Saddle River, N. J., 2000.

18. B. K. Bose, *Power Electronics and Variable Frequency Drives*, IEEE Press, Piscataway, N. J., 1997.

19. C. S. Bonaventura and K. W. Lilly, "A Constrained Motion Algorithm for the Shuttle Remote Manipulator System," *IEEE Control Systems*, October 1995, pp. 6–16.

20. A. T. Bahill and L. Stark, "The Trajectories of Saccadic Eye Movements," *Scientific American*, January 1979, pp. 108–117.

21. T. L. Floyd and D. Buchla, *Basic Operational Amplifiers*, Prentice Hall, Upper Saddle River, N. J., 1999.

22. A. G. Ulsoy, "Control of Machining Processes," ASME, *Journal of Dynamic Systems*, June 1993, pp. 301–310.

23. C. E. Rohrs, J. L. Melsa, and D. Schultz, *Linear Control Systems*, McGraw-Hill, New York, 1993.

24. J. L. Jones and A. M. Flynn, *Mobile Robots*, A. K. Peters Publishing, New York, 1993.

25. D. A. Linkens, "Adaptive and Intelligent Control in Anesthesia," *IEEE Control Systems*, December 1992, pp. 6–10.

26. R. H. Bishop, "Adaptive Control of Space Station with Control Moment Gyros," *IEEE Control Systems*, October 1992, pp. 23–27.

27. J. B. Song, "Application of Adaptive Control to Arc Welding Processes," *Proceedings of the American Control Conference*, IEEE, June 1993, pp. 1751–1755.

28. X. G. Wang, "Estimation in Paper Machine Control," *IEEE Control Systems*, August 1993, pp. 34–43.

29. R. Patton, "Mag Lift," *Scientific American*, October 1993, pp. 108–109.

30. P. Ferreira, "Concerning the Nyquist Plots of Rational Functions of Nonzero Type," *IEEE Transaction on Education*, Vol. 42, No. 3, 1999, pp. 228–229.

31. J. Pretolve, "Stereo Vision," *Industrial Robot*, Vol. 21, No. 2, 1994, pp. 24–31.

32. M. W. Spong and M. Vidyasagar, *Robot Dynamics and Control*, John Wiley & Sons, New York, 1989.

Chapter 10

1. R. C. Dorf, *Electrical Engineering Handbook*, 2nd ed., CRC Press, Boca Raton, Fla., 1998.

2. Z. Gajic and M. Lelic, *Modern Control System Engineering*, Prentice Hall, Englewood Cliffs, N. J., 1996.

3. K. S. Yeung, et al., "A Non-trial and Error Method for Lag-Lead Compensator Design," *IEEE Transactions on Education*, February 1998, pp. 76–80.

4. W. R. Wakeland, "Bode Compensator Design," *IEEE Transactions on Automatic Control*, October 1976, pp. 771–773.

5. J. R. Mitchell, "Comments on Bode Compensator Design," *IEEE Transactions on Automatic Control*, October 1977, pp. 869–870.

6. S. T. Van Voorhis, "Digital Control of Measurement Graphics," *Hewlett-Packard Journal*, January 1986, pp. 24–26.

7. R. H. Bishop, "Adaptive Control of Space Station with Control Moment Gyros," *IEEE Control Systems*, October 1992, pp. 23–27.

8. C. L. Phillips, "Analytical Bode Design of Controllers," *IEEE Transactions on Education*, February 1985, pp. 43–44.

9. R. C. Garcia and B. S. Heck, "Enhancing Classical Controls Education via Interactive Design," *IEEE Control Systems*, June 1999, pp. 77–82.

10. B. Sridhar, "Kalman Filters for Helicopter Flight," *IEEE Control Systems*, August 1993, pp. 26–33.

11. J. D. Powell, N. P. Fekete, and C-F. Chang, "Observer-Based Air-Fuel Ratio Control," *IEEE Control Systems*, October 1998, p. 72.

12. T. B. Sheridan, *Telerobotics, Automation and Control*, MIT Press, Cambridge, Mass., 1992.

13. R. C. Dorf, *The Encyclopedia of Robotics*, John Wiley & Sons, New York, 1988.

14. R. L. Wells, "Control of a Flexible Robot Arm," *IEEE Control Systems*, January 1990, pp. 9–15.

15. H. Kazerooni, "Human Extenders," *Journal of Dynamic Systems*, ASME, June 1993, pp. 281–290.

16. R. C. Dorf and A. Kusiak, *Handbook of Manufacturing and Automation*, John Wiley & Sons, New York, 1994.

17. F. M. Ham, S. Greeley, and B. Henniges, "Active Vibration Suppression for the Mast Flight System," *IEEE Control System Magazine*, Vol. 9, No. 1, 1989, pp- 85–90.

18. K. Pfeiffer and R. Isermann, "Driver Simulation in Dynamical Engine Test Stands," *Proceedings of the American Control Conference*, IEEE, 1993, pp. 721–725.

19. H. Raza, "Control Design for Automated Highway Systems," *IEEE Control Systems*, December 1996, pp. 43–60.

20. A. G. Ulsoy, "Control of Machining Processes," ASME, *Journal of Dynamic Systems*, June 1993, pp. 301–310.

21. B. K. Bose, *Modern Power Electronics*, IEEE Press, New York, 1992.

22. F. G. Martin, *The Art of Robotics*, Prentice Hall, Upper Saddle River, N. J., 1999.

23. J. M. Weiss, "The TGV Comes to Texas," *Europe*, March 1993, pp. 18–20.

24. P. Schmidt, "A Parameter Optimization Approach to Controller Partioning for Flight Control," *IEEE Transactions on Control Systems Technology*, March 1993, pp. 21–36.

25. H. Kazerooni, "A Controller Design Framework for Telerobotic Systems," *IEEE Transactions on Control Systems Technology*, March 1993, pp. 50–62.

26. W. H. Zhu, "Industrial Manipulators," *IEEE Control Systems*, April 1999, pp. 24–28.

27. J. Ackermann, "Robust Yaw Damping of Cars with Front and Rear Wheel Steering," *IEEE Transactions on Control Systems Technology*, March 1993, pp. 15–20.

28. E. W. Kamen and B. S. Heck, *Fundamentals of Signals and Systems Using MATLAB*, Prentice Hall, Upper Saddle River, N. J., 1997.

29. C. T. Chen, *Analog and Digital Control Systems Design*, Oxford Univ. Press, New York, 1996.

30. M. J. Sidi, *Spacecraft Dynamics and Control*, Cambridge Univ. Press, New York, 1997.

31. A. Arenas, et al., "Angular Velocity Control for a Windmill Radiometer," *IEEE Transactions on Education*, May 1999, pp. 147–152.

32. M. Berenguel, et al., "Temperature Control of a Solar Furnace," *IEEE Control Systems*, February 1999, pp. 8–19.

33. A. H. Moore, "The Shipping News: Fast Ferries," *Fortune*, December 6, 1999, pp. 240–249.

Chapter 11

1. R. C. Dorf, *Electrical Engineering Handbook*, 2nd ed., CRC Press, Boca Raton, Fla., 1998.

2. G. Goodwin, S. Graebe, and M. Salgado, *Control System Design*, Prentice Hall, Saddle River, N.J., 2001.

3. A. E. Bryson, "Optimal Control," *IEEE Control Systems*, June 1996, pp. 26–33.

4. J. Farrell, "Using Learning Techniques to Accommodate Unanticipated Faults," *IEEE Control Systems*, June 1993, pp. 40–48.

5. M. Jamshidi, *Design of Intelligent Manufacturing Systems*, Prentice Hall, Upper Saddle River, N. J., 1998.

6. M. Bodson, "High Performance Control of a Permanent Magnet Stepper Motor," *IEEE Transactions on Control Systems Technology*, March 1993, pp. 5–14.

7. G. W. Van der Linden, "Control of an Inverted Pendulum," *IEEE Control Systems*, August 1993, pp. 44–50.

8. Y. Ishii, "Joint Connection Mechanism for Reconfigurable Manipulator," *IEEE Control Systems*, August 1993, pp. 73–78.

9. W. J. Book, "Controlled Motion in an Elastic World," *Journal of Dynamic Systems*, June 1993, pp. 252–260.

10. E. W. Kamen, *Introduction to Industrial Control*, Academic Press, San Diego, 1999.

11. M. Jamshidi, *Large-Scale Systems*, Prentice Hall, Upper Saddle River, N. J., 1997.

12. W. J. Rugh, *Linear System Theory*, 2nd ed., Prentice Hall, Englewood Cliffs, N. J., 1996.

13. S. B. Niku, *Introduction to Robotics: Analysis, Systems, Applications*, Prentice Hall, Upper Saddle River, N. J., 2002.

14. J. B. Burl, *Linear Optimal Control*, Prentice Hall, Upper Saddle River, N. J., 1999.

15. D. Hrovat, "Applications of Optimal Control to Automotive Suspension Design," *Journal of Dynamic Systems*, ASME, June 1993, pp. 328–342.

16. R. H. Bishop, "Adaptive Control of Space Station with Control Moment Gyros," *IEEE Control Systems*, October 1992, pp. 23–27.

17. R. C. Dorf, *Encyclopedia of Robotics*, John Wiley & Sons, New York, 1988.

18. T. B. Sheridan, *Telerobotics, Automation and Control*, MIT Press, Cambridge, Mass., 1992.

19. R. C. Dorf and A. Kusiak, *Handbook of Manufacturing and Automation*, John Wiley & Sons, New York, 1994.

20. C. T. Chen, *Linear System Theory and Design*, 3rd ed., Oxford University Press, New York, 1999.

21. K. E. Drexler, *Nanosystems*, John Wiley & Sons, New York, 1993.

22. F. L. Chernousko, *State Estimation for Dynamic Systems*, CRC Press, Boca Raton, Fla., 1993.

23. M. A. Gottschalk, "Dino-Adventure Duels Jurassic Park," *Design News*, August 16, 1993, pp. 52–58.

24. Y. Z. Tsypkin, "Robust Internal Model Control," *Journal of Dynamic Systems*, ASME, June 1993, pp. 419–425.

25. J. D. Irwin, *The Industrial Electronics Handbook*, CRC Press, Boca Raton, Fla., 1997.

26. J. K. Pieper, "Control of a Coupled-Drive Apparatus," *IEE Proceedings*, March 1993, pp. 70–79.

27. Rama K. Yedavalli, "Robust Control Design for Aerospace Applications," *IEEE Transactions of Aerospace and Electronic Systems*, Vol. 25, No. 3, 1989, pp. 314–324.

28. Bryan L. Jones and Robert H. Bishop, "H_2 Optimal Halo Orbit Guidance," *Journal of Guidance, Control, and Dynamics, AIAA*, 16, No. 6, 1993, pp. 1118–1124.

29. D. G. Luenberger, "Observing the State of a Linear System," *IEEE Transactions on Military Electronics*, 1964, pp. 74–80.

30. G. F. Franklin, J. D. Powell, and A. Emami-Naeini, *Feedback Control of Dynamic Systems*, 4th ed., Prentice Hall, Upper Saddle River, N. J., 2002.

31. R. E. Kalman, "Mathematical description of linear dynamical systems," *SIAM J. Control*, Vol. 1, 1963, pp. 152–192.

32. R. E. Kalman, "A New Approach to Linear Filtering and Prediction Problems," *Journal of Basic Engineering*, 1960, pp. 35–45.

33. R. E. Kalman and R. S. Bucy, "New Results in Linear Filtering and Prediction Theory," Transactions of the American Society of Mechanical Engineering, Series D, *Journal of Basic Engineering*, 1961, pp. 95–108.

34. B. Cipra, "Engineers Look to Kalman Filtering for Guidance," *SIAM News*, Vol. 26, No. 5, August 1993.

35. R. H. Battin, "Theodore von Karman Lecture: Some Funny Things Happened on the Way to the Moon," 27th Aerospace Sciences Meeting, Reno, Nevada, AIAA-89-0861, 1989.

36. R. G. Brown and P. Y. C. Hwang, *Introduction to Random Signal Analysis and Kalman Filtering with Matlab Exercises and Solutions*, John Wiley and Sons, Inc., 1996.

37. M. S. Grewal, and A. P. Andrews, *Kalman Filtering: Theory and Practice Using MATLAB, 2nd ed.*, Wiley-Interscience, 2001.

Chapter 12

1. R. C. Dorf, *The Encyclopedia of Robotics*, John Wiley & Sons, New York, 1988.

2. R. C. Dorf, *Electrical Engineering Handbook*, 2nd ed., CRC Press, Boca Raton, Fla., 1998.

3. R. S. Sanchez-Pena and M. Sznaier, *Robust Systems Theory and Applications*, John Wiley & Sons, New York, 1998.

4. G. Zames, "Input-Output Feedback Stability and Robustness," *IEEE Control Systems*, June 1996, pp. 61–66.

5. K. Zhou and J. C. Doyle, *Essentials of Robust Control*, Prentice Hall, Upper Saddle River, N. J., 1998.

6. C. M. Close and D. K. Frederick, *Modeling and Analysis of Dynamic Systems*, 2nd ed., Houghton Mifflin, Boston, 1993.

7. A. Charara, "Nonlinear Control of a Magnetic Levitation System," *IEEE Transactions on Control System Technology*, September 1996, pp. 513–523.

8. J. Yen, *Fuzzy Logic: Intelligence and Control*, Prentice Hall, Upper Saddle River, N. J., 1998.

9. X. G. Wang, "Estimation in Paper Machine Control," *IEEE Control Systems*, August 1993, pp. 34–43.

10. D. Sbarbaro-Hofer, "Control of a Steel Rolling Mill," *IEEE Control Systems*, June 1993, pp. 69–75.

11. N. Mohan, *Power Electronics*, John Wiley & Sons, New York, 1995.

12. J. M. Weiss, "The TGV Comes to Texas," *Europe*, March 1993, pp. 18–20.

13. S. Lee, "Intelligent Sensing and Control for Advanced Teleoperation," *IEEE Control Systems*, June 1993, pp. 19–28.

14. J. V. Wait and L. P. Huelsman, *Operational Amplifier Theory*, 2nd ed., McGraw-Hill, New York, 1992.

15. F. G. Martin, *The Art of Robotics*, Prentice Hall, Upper Saddle River, N. J., 1999.

16. R. Shoureshi, "Intelligent Control Systems," *Journal of Dynamic Systems*, June 1993, pp. 392–400.

17. A. Butar and R. Sales, "Control for MagLev Vehicles," *IEEE Control Systems*, August 1998, pp. 18–25.

18. L. W. Couch, *Digital and Analog Communication Systems*, 5th ed., Macmillan, New York, 1997.

19. H. Paraci and M. Jamshidi, *Design and Implementation of Intelligent Manufacturing Systems*, Prentice Hall, Upper Saddle River, N. J., 1997.

20. C. T. Chen, *Analog and Digital Control System Design*, Harcourt Brace Jovanovich, Orlando, Fla., 1993.

21. B. Johnstone, "Japan's Friendly Robots," *Technology Review*, June 1999, pp. 66–69.

22. W. J. Grantham and T. L. Vincent, *Modern Control Systems Analysis and Design*, John Wiley & Sons, New York, 1993.

23. K. Capek, *Rossum's Universal Robots*, English edition by P. Selver and N. Playfair, Doubleday, Page, New York, 1923.

24. L. L. Cone, "Skycam: An Aerial Robotic Camera System," *Byte*, October 1985, pp. 122–128.

25. H. Kazerooni, "Human Extenders," *Journal of Dynamic Systems*, ASME, June 1993, pp. 281–290.

26. C. Lapiska, "Flight Simulation," *Aerospace America*, August 1993, pp. 14–17.

27. D. E. Bossert, "A Root-Locus Analysis of Quantitative Feedback Theory," *Proceedings of the American Control Conference*, June 1993, pp. 1698–1705.

28. J. A. Gutierrez and M. Rabins, "A Computer Loop-shaping Algorithm for Controllers," *Proceedings of the American Control Conference*, June 1993, pp. 1711–1715.

29. J. W. Song, "Synthesis of Compensators in Linear Uncertain Plants," *Proceedings of the Conference on Decision and Control*, December 1992, pp. 2882–2883.

30. M. Gottschalk, "Part Surgeon–Part Robot," *Design News*, June 7, 1993, pp. 68–75.

31. D. E. Whitney, "From Robots to Design," *Journal of Dynamic Systems*, ASME, June 1993, pp. 262–270.

32. S. Jayasuriya, "Frequency Domain Design for Robust Performance Under Uncertainties," *Journal of Dynamic Systems*, June 1993, pp. 439–450.

33. L. S. Shieh, "Control of Uncertain Systems," *IEE Proceedings*, March 1993, pp. 99–110.

34. M. van de Panne, "A Controller for the Dynamic Walk of a Biped," *Proceedings of the Conference on Decision and Control*, IEEE, December 1992, pp. 2668–2673.

35. S. Bennett, "The Development of the PID Controller," *IEEE Control Systems*, December 1993, pp. 58–64.

36. J. C. Doyle, A. B. Francis, and A. R. Tannenbaum, *Feedback Control Theory*, Macmillan, New York, 1992.

Chapter 13

1. R. C. Dorf, *The Encyclopedia of Robotics*, John Wiley & Sons, New York, 1988.

2. C. L. Phillips and H. T. Nagle, *Digital Control Systems*, Prentice Hall, Englewood Cliffs, N. J., 1995.

3. G. F. Franklin, et al., *Digital Control of Dynamic Systems*, 2nd ed., Prentice Hall, Upper Saddle River, N.J., 1998.

4. S. H. Zak, "Ripple-Free Deadbeat Control," *IEEE Control Systems*, August 1993, pp. 51–56.

5. C. Lapiska, "Flight Simulation," *Aerospace America*, August 1993, pp. 14–17.

6. F. G. Martin, *The Art of Robotics*, Prentice Hall, Upper Saddle River, N. J., 1999.

7. D. Raviv and E. W. Djaja, "Discretized Controllers," *IEEE Control Systems*, June 1999, pp. 52–58.

8. R. C. Dorf, *Electrical Engineering Handbook*, 2nd ed., CRC Press, Boca Raton, Fla., 1998.

9. T. M. Foley, "Engineering the Space Station," *Aerospace America*, October 1996, pp. 26–32.

10. A. G. Ulsoy, "Control of Machining Processes," ASME, *Journal of Dynamic Systems*, June 1993, pp. 301–310.

11. K. J. Astrom, *Computer-Controlled Systems*, Prentice Hall, Upper Saddle River, N. J., 1997.

12. R. C. Dorf and A. Kusiak, *Handbook of Manufacturing and Automation*, John Wiley & Sons, New York, 1994.

13. L. W. Couch, *Digital and Analog Communication Systems*, 5th ed., Macmillan, New York, 1995.

14. M. van de Panne, "A Controller for the Dynamic Walk of a Biped," *Proceedings of the Conference on*

Decision and Control, IEEE, December 1992, pp. 2668–2673.

15. K. S. Yeung and H. M. Lai, "A Reformation of the Nyquist Criterion for Discrete Systems," *IEEE Transactions on Education*, February 1988, pp. 32–34.

16. M. Eslami, *Theory of Sensitivity in Dynamic Systems*, Springer-Verlag, New York, 1994.

17. T. R. Kurfess, "Predictive Control of a Robotic Grinding System," *Journal of Engineering for Industry*, ASME, November 1992, pp. 412–420.

18. T. Studt, "Why You Really Need a DSP," *R&D Magazine*, June 1993, pp. 32–34.

19. D. M. Auslander, *Mechatronics*, Prentice Hall, Englewood Cliffs, N. J., 1996.

20. R. Shoureshi, "Intelligent Control Systems," *Journal of Dynamic Systems*, June 1993, pp. 392–400.

21. D. J. Leo, "Control of a Flexible Frame in Slewing," *Proceedings of American Control Conference*, 1992, pp. 2535–2540.

22. M. Alpert, "A Bullet Train for the U.S.," *Fortune*, May 1993, p. 32.

23. J. L. Jones and A. M. Flynn, *Mobile Robots*, A. K. Peters Publishing, New York, 1993.

24. V. Skormin, "On-Line Diagnostics of a Self-Contained Flight Actuator," *IEEE Transactions on Aerospace and Electronic Systems*, January 1994, pp. 130–141.

25. H. H. Ottesen, "Future Servo Technologies for Hard Disk Drives," *J. of the Magnetics Society of Japan*, vol. 18, 1994, pp. 31–36.

Appendix A

1. A. Gilat, MATLAB: *An Introduction with Applications*, 3rd ed., Wiley and Sons, NJ, 2008.

2. D. Hanselman and B. Littlefield, *Mastering MATLAB 7*, The MATLAB Curriculum Series, Prentice Hall, Upper Saddle River, N. J., 2004.

3. R. Pratap, *Getting Started with MATLAB 7*, Oxford University Press, New York, 2005.

4. *The Student Edition of MATLAB 7 User's Guide*, The Mathworks, Inc., Prentice Hall, Upper Saddle River, N. J., 2007.

5. D. J. Higham and N. J. Higham, *MATLAB Guide*, SIAM, Society for Industrial and Applied Mathematics, 2005.

6. W. J. Palm III, *Introduction to MATLAB 7 for Engineers*, McGraw-Hill, Boston, 2003.

7. T. A. Davis and K. Sigmon, *MATLAB Primer 7e*, Chapman and Hall/CRC, Boca Raton, FL, 2004.

8. D. M. Etter, *Engineering Problem Solving with MATLAB*, 3rd ed., Prentice Hall, Upper Saddle River, N. J., 2006.

Index

Design Process

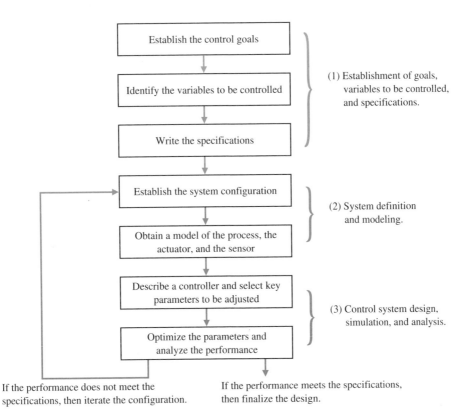

EXAMPLES

Selected Tables and Formulas for Design

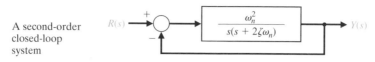

A second-order closed-loop system

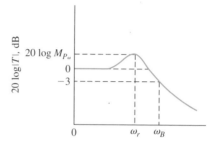

UNIT STEP RESPONSE

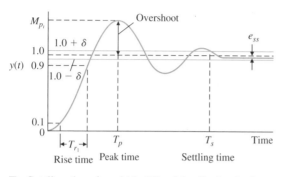

Rise time   Peak time   Settling time

CLOSED-LOOP MAGNITUDE PLOT

❑ Settling time (to within 2% of the final value)

$$T_s = \frac{4}{\zeta \omega_n}$$

❑ Maximum magnitude ($\zeta \leq 0.7$)

$$M_{p_\omega} = \frac{1}{2\zeta\sqrt{1 - \zeta^2}}$$

❑ Percent overshoot

$$M_{p_t} = 1 + e^{-\zeta\pi/\sqrt{1-\zeta^2}}$$ and $$P.O. = 100e^{-\pi\zeta/\sqrt{1-\zeta^2}}$$

❑ Time-to-peak

$$T_p = \frac{\pi}{\omega_n\sqrt{1 - \zeta^2}}$$

❑ Resonant frequency ($\zeta \leq 0.7$)

$$\omega_r = \omega_n\sqrt{1 - 2\zeta^2}$$

❑ Rise time (time to rise from 10% to 90% of final value)

$$T_{r_1} = \frac{2.16\zeta + 0.60}{\omega_n}$$ $(0.3 \leq \zeta \leq 0.8)$

❑ Bandwidth ($0.3 \leq \zeta \leq 0.8$)

$$\omega_B = (-1.196\zeta + 1.85)\omega_n$$

PID Controller: $$G_c(s) = K_P + K_D s + \frac{K_1}{s} = \frac{(s + z_1)(s + z_2)}{s}$$